Values of fundamental constants

Quantity	Symbol	Value
Speed of light in vacuum	c_0	3.00×10^8 m/s
Gravitational constant	G	6.6738×10^{-11} N · m²/kg²
Avogadro's number	N_A	6.0221413×10^{23} mol^{-1}
Boltzmann's constant	k_B	1.380×10^{-23} J/K
Charge on electron	e	1.60×10^{-19} C
Permittivity constant	ϵ_0	$8.85418782 \times 10^{-12}$ C^2/(N · m^2)
Permeability constant	μ_0	$4\pi \times 10^{-7}$ T · m/A
Planck's constant	h	6.626×10^{-34} J · s
Electron mass	m_e	9.11×10^{-31} kg
Proton mass	m_p	1.6726×10^{-27} kg
Neutron mass	m_n	1.6749×10^{-27} kg
Atomic mass unit	amu	1.6605×10^{-27} kg

Other useful numbers

Number or quantity	Value
π	3.1415927
e	2.7182818
1 radian	57.2957795°
Absolute zero ($T = 0$)	−273.15 °C
Average acceleration g due to gravity near Earth's surface	9.8 m/s^2
Speed of sound in air at 20 °C	343 m/s
Density of dry air at atmospheric pressure and 20 °C	1.29 kg/m^3
Earth's mass	5.97×10^{24} kg
Earth's radius (mean)	6.38×10^6 m
Earth–Moon distance (mean)	3.84×10^8 m

PRINCIPLES & PRACTICE OF PHYSICS

VOLUME 2

Eric Mazur
Harvard University

With contributions from

Catherine H. Crouch
Swarthmore College

Peter A. Dourmashkin
Massachusetts Institute of Technology

PEARSON

Boston Columbus Indianapolis New York San Francisco Hoboken
Amsterdam Cape Town Dubai London Madrid Milan Munich Paris Montréal Toronto
Delhi Mexico City São Paulo Sydney Hong Kong Seoul Singapore Taipei Tokyo

Executive Editor: Becky Ruden
Publisher: Jim Smith
Project Managers: Beth Collins and Martha Steele
Program Manager: Katie Conley
Managing Development Editor: Cathy Murphy
Senior Development Editor: Margot Otway
Development Editor: Irene Nunes
Editorial Assistant: Sarah Kaubisch
Text Permissions Project Manager: Liz Kincaid
Text Permissions Specialist: Paul Sarkis
Associate Content Producer: Megan Power
Production Management: Rose Kernan
Copyeditor: Carol Reitz
Compositor: Cenveo® Publisher Services
Design Manager: Mark Ong
Interior Designer: Hespenheide Design
Cover Designer: Tandem Creative, Inc.
Illustrators: Rolin Graphics
Photo Permissions Management: Maya Melenchuk
Photo Researcher: Eric Schrader
Manufacturing Buyer: Jeff Sargent
Vice-President of Marketing: Christy Lesko
Marketing Manager: Will Moore
Senior Marketing Development Manager: Michelle Cadden

Cover Photo Credit: Franklin Kappa

Credits and acknowledgments borrowed from other sources and reproduced, with permission, in this textbook appear on the appropriate page within the text or on p. C-1.

Library of Congress Cataloging-in-Publication Data on file.

ISBN 10: 0-321-95842-X; ISBN 13: 978-0-321-95842-6

1 2 3 4 5 6 7 8 9 10—DOW—18 17 16 15 14

www.pearsonhighered.com

Brief Contents

Volume 1 of *Principles of Physics* includes Chapters 1–21. Volume 2 of *Principles of Physics* includes Chapters 22–34.

About the Author

Eric Mazur is the Balkanski Professor of Physics and Applied Physics at Harvard University and Area Dean of Applied Physics. Dr. Mazur is a renowned scientist and researcher in optical physics and in education research, and a sought-after author and speaker.

Dr. Mazur joined the faculty at Harvard shortly after obtaining his Ph.D. at the University of Leiden in the Netherlands. In 2012 he was awarded an Honorary Doctorate from the École Polytechnique and the University of Montreal. He is a Member of the Royal Academy of Sciences of the Netherlands and holds honorary professorships at the Institute of Semiconductor Physics of the Chinese Academy of Sciences in Beijing, the Institute of Laser Engineering at the Beijing University of Technology, and the Beijing Normal University.

Dr. Mazur has held appointments as Visiting Professor or Distinguished Lecturer at Carnegie Mellon University, the Ohio State University, the Pennsylvania State University, Princeton University, Vanderbilt University, Hong Kong University, the University of Leuven in Belgium, and National Taiwan University in Taiwan, among others.

In addition to his work in optical physics, Dr. Mazur is interested in education, science policy, outreach, and the public perception of science. In 1990 he began developing peer instruction, a method for teaching large lecture classes interactively. This teaching method has developed a large following, both nationally and internationally, and has been adopted across many science disciplines.

Dr. Mazur is author or co-author of over 250 scientific publications and holds two dozen patents. He has also written on education and is the author of *Peer Instruction: A User's Manual* (Pearson, 1997), a book that explains how to teach large lecture classes interactively. In 2006 he helped produce the award-winning DVD *Interactive Teaching*. He is the co-founder of Learning Catalytics, a platform for promoting interactive problem solving in the classroom, which is available in MasteringPhysics®.

To the Student

Let me tell you a bit about myself.

I always knew exactly what I wanted to do. It just never worked out that way.

When I was seven years old, my grandfather gave me a book about astronomy. Growing up in the Netherlands I became fascinated by the structure of the solar system, the Milky Way, the universe. I remember struggling with the concept of infinite space and asking endless questions without getting satisfactory answers. I developed an early passion for space and space exploration. I knew I was going to be an astronomer. In high school I was good at physics, but when I entered university and had to choose a major, I chose astronomy.

It took only a few months for my romance with the heavens to unravel. Instead of teaching me about the mysteries and structure of the universe, astronomy had been reduced to a mind-numbing web of facts, from declinations and right ascensions to semi-major axes and eccentricities. Disillusioned about astronomy, I switched majors to physics. Physics initially turned out to be no better than astronomy, and I struggled to remain engaged. I managed to make it through my courses, often by rote memorization, but the beauty of science eluded me.

It wasn't until doing research in graduate school that I rediscovered the beauty of science. I knew one thing for sure, though: I was never going to be an academic. I was going to do something useful in my life. Just before obtaining my doctorate, I lined up my dream job working on the development of the compact disc, but I decided to spend one year doing postdoctoral research first.

It was a long year. After my postdoc, I accepted a junior faculty position and started teaching. That's when I discovered that the combination of doing research—uncovering the mysteries of the universe—and teaching—helping others to see the beauty of the universe—is a wonderful combination.

When I started teaching, I did what all teachers did at the time: lecture. It took almost a decade to discover that my award-winning lecturing did for my students exactly what the courses I took in college had done for me: It turned the subject that I was teaching into a collection of facts that my students memorized by rote. Instead of transmitting the beauty of my field, I was essentially regurgitating facts to my students.

When I discovered that my students were not mastering even the most basic principles, I decided to completely change my approach to teaching. Instead of lecturing, I asked students to read my lecture notes at home, and then, in class, I taught by questioning—by asking my students to reflect on concepts, discuss in pairs, and experience their own "aha!" moments.

Over the course of more than twenty years, the lecture notes have evolved into this book. Consider this book to be my best possible "lecturing" to you. But instead of listening to me without having the opportunity to reflect and think, this book will permit you to pause and think; to hopefully experience many "aha!" moments on your own.

I hope this book will help you develop the thinking skills that will make you successful in your career. And remember: your future may be—and likely will be—very different from what you imagine.

I welcome any feedback you have. Feel free to send me email or tweets.

I wrote this book for you.

Eric Mazur
@eric_mazur
mazur@harvard.edu
Cambridge, MA

To the Instructor

They say that the person who teaches is the one who learns the most in the classroom. Indeed, teaching led me to many unexpected insights. So, also, with the writing of this book, which has been a formidably exciting intellectual journey.

Why write a new physics text?

In May 1993 I was driving to Troy, NY, to speak at a meeting held in honor of Robert Resnick's retirement. In the car with me was a dear friend and colleague, Albert Altman, professor at the University of Massachusetts, Lowell. He asked me if I was familiar with the approach to physics taken by Ernst Mach in his popular lectures. I wasn't. Mach treats conservation of momentum before discussing the laws of motion, and his formulation of mechanics had a profound influence on Einstein.

The idea of using conservation principles derived from experimental observations as the basis for a text—rather than Newton's laws and the concept of force—appealed to me immediately. After all, most physicists never use the concept of force because it relates only to mechanics. It has no role in quantum physics, for example. The conservation principles, however, hold throughout all of physics. In that sense they are much more fundamental than Newton's laws. Furthermore, conservation principles involve only algebra, whereas Newton's second law is a differential equation.

It occurred to me, however, that Mach's approach could be taken further. Wouldn't it be nice to start with conservation of both momentum *and* energy, and only *later* bring in the concept of force? After all, physics education research has shown that the concept of force is fraught with pitfalls. What's more, after tediously deriving many results using kinematics and dynamics, most physics textbooks show that you can derive the same results from conservation principles in just one or two lines. Why not do it the easy way first?

It took me many years to reorganize introductory physics around the conservation principles, but the resulting approach is one that is much more unified and modern—the conservation principles are the theme that runs throughout this entire book.

Additional motives for writing this text came from my own teaching. Most textbooks focus on the acquisition of information and on the development of procedural knowledge. This focus comes at the expense of conceptual understanding or the ability to transfer knowledge to a new context. As explained below, I have structured this text to redress that balance. I also have drawn deeply on the results of physics education research, including that of my own research group.

I have written this text to be accessible and easy for students to understand. My hope is that it can take on the burden of basic teaching, freeing class time for synthesis, discussion, and problem solving.

Setting a new standard

The tenacity of the standard approach in textbooks can be attributed to a combination of inertia and familiarity. Teaching large introductory courses is a major chore, and once a course is developed, changing it is not easy. Furthermore, the standard texts worked for *us*, so it's natural to feel that they should work for our students, too.

The fallacy in the latter line of reasoning is now well-known thanks to education research. Very few of our students are like us at all. Most take physics because they are required to do so; many will take no physics beyond the introductory course. Physics education research makes it clear that the standard approach fails these students.

Because of pressure on physics departments to deliver better education to non-majors, changes are occurring in the way physics is taught. These changes, in turn, create a need for a textbook that embodies a new educational philosophy in both format and presentation.

Organization of this book

As I considered the best way to convey the conceptual framework of mechanics, it became clear that the standard curriculum truly deserved to be rethought. For example, standard texts are forced to redefine certain concepts more than once—a strategy that we know befuddles students. (Examples are *work,* the standard definition of which is incompatible with the first law of thermodynamics, and *energy,* which is redefined when modern physics is discussed.)

Another point that has always bothered me is the arbitrary division between "modern" and "classical" physics. In most texts, the first thirty-odd chapters present physics essentially as it was known at the end of the 19th century; "modern physics" gets tacked on at the end. There's no need for this separation. Our goal should be to explain physics in the way that works best for students, using our full contemporary understanding. *All* physics is modern!

That is why my table of contents departs from the "standard organization" in the following specific ways.

Emphasis on conservation laws. As mentioned earlier, this book introduces the conservation laws early and treats them the way they should be: as the backbone of physics. The advantages of this shift are many. First, it avoids many of the standard pitfalls related to the concept of force, and it leads naturally to the two-body character of forces and the laws of motion. Second, the conservation laws enable students to solve a wide variety of problems without any calculus. Indeed, for complex systems, the conservation laws are often the natural (or only) way to solve problems. Third, the book deduces the conservation laws from observations, helping to make clear their connection with the world around us.

Table 1 Scheduling matrix

Topic	Chapters	Can be inserted after chapter...	Chapters that can be omitted without affecting continuity
Mechanics	1–14		6, 13–14
Waves	15–17	12	16–17
Fluids	18	9	
Thermal Physics	19–21	10	21
Electricity & Magnetism	22–30	12 (but 17 is needed for 29–30)	29–30
Circuits	31–32	26 (but 30 is needed for 32)	32
Optics	33–34	17	34

I and several other instructors have tested this approach extensively in our classes and found markedly improved performance on problems involving momentum and energy, with large gains on assessment instruments like the Force Concept Inventory.

Early emphasis on the concept of system. Fundamental to most physical models is the separation of a system from its environment. This separation is so basic that physicists tend to carry it out unconsciously, and traditional texts largely gloss over it. This text introduces the concept in the context of conservation principles and uses it consistently.

Postponement of vectors. Most introductory physics concerns phenomena that take place along one dimension. Problems that involve more than one dimension can be broken down into one-dimensional problems using vectorial notation. So a solid understanding of physics in one dimension is of fundamental importance. However, by introducing vectors in more than one dimension from the start, standard texts distract the student from the basic concepts of kinematics.

In this book, I develop the complete framework of mechanics for motions and interactions in one dimension. I introduce the second dimension when it is needed, starting with rotational motion. Hence, students are free to concentrate on the actual physics.

Just-in-time introduction of concepts. Wherever possible, I introduce concepts only when they are necessary. This approach allows students to put ideas into immediate practice, leading to better assimilation.

Integration of modern physics. A survey of syllabi shows that less than half the calculus-based courses in the United States cover modern physics. I have therefore integrated selected "modern" topics throughout the text. For example, special relativity is covered in Chapter 14, at the end of mechanics. Chapter 32, Electronics, includes sections on semiconductors and semiconductor devices. Chapter 34, Wave and Particle Optics, contains sections on quantization and photons.

Modularity. I have written the book in a modular fashion so it can accommodate a variety of curricula (See Table 1, "Scheduling matrix").

The book contains two major parts, Mechanics and Electricity and Magnetism, plus five shorter parts. The two major parts by themselves can support an in-depth two-semester or three-quarter course that presents a complete picture of physics embodying the fundamental ideas of modern physics. Additional parts can be added for a longer or faster-paced course. The five shorter parts are more or less self-contained, although they do build on previous material, so their placement is flexible. Within each part or chapter, more advanced or difficult material is placed at the end.

Pedagogy

This text draws on many models and techniques derived from my own teaching and from physics education research. The following are major themes that I have incorporated throughout.

Separation of conceptual and mathematical frameworks. Each chapter is divided into two parts: Concepts and Quantitative Tools. The first part, Concepts, develops the full conceptual framework of the topic and addresses many of the common questions students have. It concentrates on the underlying ideas and paints the big picture, whenever possible without equations. The second part of the chapter, Quantitative Tools, then develops the mathematical framework.

Deductive approach; focus on ideas before names and equations. To the extent possible, this text develops arguments deductively, starting from observations, rather than stating principles and then "deriving" them. This approach makes the material easier to assimilate for students. In the same vein, this text introduces and explains each idea before giving it a formal name or mathematical definition.

Stronger connection to experiment and experience. Physics stems from observations, and this text is structured so that it can do the same. As much as possible, I develop the material from experimental observations (and preferably those that students can make) rather than assertions. Most chapters use actual data in developing ideas, and new notions are always introduced by going from the specific to the general—whenever possible by interpreting everyday examples.

By contrast, standard texts often introduce laws in their most general form and then show that these laws are consistent with specific (and often highly idealized) cases. Consequently the world of physics and the "real" world remain two different things in the minds of students.

Addressing physical complications. I also strongly oppose presenting unnatural situations; real life complications must always be confronted head-on. For example, the use of unphysical words like *frictionless* or *massless* sends a message to the students that physics is unrealistic or, worse, that the world of physics and the *real* world are unrelated entities. This can easily be avoided by pointing out that friction or mass may be neglected under certain circumstances and pointing out *why* this may be done.

Engaging the student. Education is more than just transfer of information. Engaging the student's mind so the information can be assimilated is essential. To this end, the text is written as a dialog between author and reader (often invoking the reader—*you*—in examples) and is punctuated by Checkpoints—questions that require the reader to stop and think. The text following a Checkpoint often refers directly to its conclusions. Students will find complete solutions to all the Checkpoints at the back of the book; these solutions are written to emphasize physical reasoning and discovery.

Visualization. Visual representations are central to physics, so I developed each chapter by designing the figures before writing the text. Many figures use multiple representations to help students make connections (for example, a sketch may be combined with a graph and a bar diagram). Also, in accordance with research, the illustration style is spare and simple, putting the emphasis on the ideas and relationships rather than on irrelevant details. The figures do not use perspective unless it is needed, for instance.

Structure of this text

Division into *Principles* and *Practice* books

I've divided this text into a *Principles* book, which teaches the physics, and a *Practice* book, which puts the physics into practice and develops problem-solving skills. This division helps address two separate intellectually demanding tasks: understanding the physics and learning to solve problems. When these two tasks are mixed together, as they are in standard texts, students are easily overwhelmed. Consequently many students focus disproportionately on worked examples and procedural knowledge, at the expense of the physics.

Structure of *Principles* chapters

As pointed out earlier, each *Principles* chapter is divided into two parts. The first part (Concepts) develops the conceptual framework in an accessible way, relying primarily on qualitative descriptions and illustrations. In addition to including Checkpoints, each Concepts section ends with a one-page Self-quiz consisting of qualitative questions.

The second part of each chapter (Quantitative Tools) formalizes the ideas developed in the first part in mathematical terms. While concise, it is relatively traditional in nature—teachers should be able to continue to use material developed for earlier courses. To avoid creating the impression that equations are more important than the concepts behind them, no equations are highlighted or boxed.

Both parts of the *Principles* chapters contain worked examples to help students develop problem-solving skills.

Structure of the *Practice* chapters

This book contains material to put into practice the concepts and principles developed in the corresponding chapters in the *Principles* book. Each chapter contains the following sections:

1. *Chapter Summary.* This section provides a brief tabular summary of the material presented in the corresponding *Principles* chapter.
2. *Review Questions.* The goal of this section is to allow students to quickly review the corresponding *Principles* chapter. The questions are straightforward one-liners starting with "what" and "how" (rather than "why" or "what if").
3. *Developing a Feel.* The goals of this section are to develop a quantitative feel for the quantities introduced in the chapter; to connect the subject of the chapter to the real world; to train students in making estimates and assumptions; to bolster students' confidence in dealing with unfamiliar material. It can be used for self-study or for a homework or recitation assignment. This section, which has no equivalent in existing books, combines a number of ideas (specifically, Fermi problems and tutoring in the style of the *Princeton Learning Guide*). The idea is to start with simple estimation problems and then build up to Fermi problems (in early chapters Fermi problems are hard to compose because few concepts have been introduced). Because students initially find these questions hard, the section provides many hints, which take the form of questions. A key then provides answers to these "hints."
4. *Worked and Guided Problems.* This section contains complex worked examples whose primary goal is to teach problem solving. The Worked Problems are fully solved; the Guided Problems have a list of questions and suggestions to help the student think about how to solve the problem. Typically, each Worked Problem is followed by a related Guided Problem.
5. *Questions and Problems.* This is the chapter's problem set. The problems 1) offer a range of levels, 2) include problems relating to client disciplines (life sciences, engineering, chemistry, astronomy, etc.), 3) use the second person as much as possible to draw in the student, and 4) do not spoon-feed the students with information and unnecessary diagrams. The problems are classified into three levels as follows: (•) application of single concept; numerical plug-and-chug; (••) nonobvious application of single concept or application of multiple concepts from current chapter; straightforward numerical or algebraic computation; (•••) application of multiple concepts, possibly spanning multiple chapters. Context-rich problems are designated CR.

As I was developing and class-testing this book, my students provided extensive feedback. I have endeavored to incorporate all of their feedback to make the book as useful as possible for future generations of students. In addition, the book was class-tested at a large number of institutions, and many of these institutions have reported significant increases in learning gains after switching to this manuscript. I am confident the book will help increase the learning gains in your class as well. It will help you, as the instructor, coach your students to be the best they can be.

Instructor supplements

The **Instructor Resource DVD** (ISBN 978-0-321-56175-6/0-321-56175-9) includes an Image Library, the Procedure and special topic boxes from *Principles*, and a library of presentation applets from **ActivPhysics**, PhET simulations, and PhET Clicker Questions. **Lecture Outlines** with embedded **Clicker Questions in PowerPoint®** are provided, as well as the ***Instructor's Guide*** and ***Instructor's Solutions Manual***.

The ***Instructor's Guide*** (ISBN 978-0-321-94993-6/0-321-94993-5) provides chapter-by-chapter ideas for lesson planning using *Principles & Practice of Physics* in class, including strategies for addressing common student difficulties.

The ***Instructor's Solutions Manual*** (ISBN 978-0-321-95053-6/0-321-95053-4) is a comprehensive solutions manual containing complete answers and solutions to all Developing a Feel questions, Guided Problems, and Questions and Problems from the *Practice* book. The solutions to the Guided Problems use the book's four-step problem-solving strategy (Getting Started, Devise Plan, Execute Plan, Evaluate Result).

MasteringPhysics® is the leading online homework, tutorial, and assessment product designed to improve results by helping students quickly master concepts. Students benefit from self-paced tutorials that feature specific wrong-answer feedback, hints, and a wide variety of educationally effective content to keep them engaged and on track. Robust diagnostics and unrivalled gradebook reporting allow instructors to pinpoint the weaknesses and misconceptions of a student or class to provide timely intervention.

MasteringPhysics enables instructors to:

- Easily assign **tutorials** that provide individualized coaching.
- Mastering's hallmark **Hints** and **Feedback** offer scaffolded instruction similar to what students would experience in an office hour.
- **Hints** (declarative and Socratic) can provide problem-solving strategies or break the main problem into simpler exercises.
- **Feedback** lets the student know precisely what misconception or misunderstanding is evident from their answer and offers ideas to consider when attempting the problem again.

Learning Catalytics™ is a "bring your own device" student engagement, assessment, and classroom intelligence system available within MasteringPhysics. With Learning Catalytics you can:

- Assess students in real time, using open-ended tasks to probe student understanding.
- Understand immediately where students are and adjust your lecture accordingly.
- Improve your students' critical-thinking skills.
- Access rich analytics to understand student performance.
- Add your own questions to make Learning Catalytics fit your course exactly.
- Manage student interactions with intelligent grouping and timing.

The **Test Bank** (ISBN 978-0-130-64688-0/0-130-64688-1) contains more than 2000 high-quality problems, with a range of multiple-choice, true-false, short-answer, and conceptual questions correlated to *Principles & Practice of Physics* chapters. Test files are provided in both TestGen® and Microsoft® Word for Mac and PC.

Instructor supplements are available on the Instructor Resource DVD, the Instructor Resource Center at www.pearsonhighered.com/irc, and in the Instructor Resource area at www.masteringphysics.com.

Student supplements

MasteringPhysics (www.masteringphysics.com) is designed to provide students with customized coaching and individualized feedback to help improve problem-solving skills. Students complete homework efficiently and effectively with tutorials that provide targeted help.

Interactive eText allows you to highlight text, add your own study notes, and review your instructor's personalized notes, 24/7. The eText is available through MasteringPhysics, www.masteringphysics.com.

Acknowledgments

This book would not exist without the contributions from many people. It was Tim Bozik, currently President, Higher Education at Pearson plc, who first approached me about writing a physics textbook. If it wasn't for his persuasion and his belief in me, I don't think I would have ever undertaken the writing of a textbook. Tim's suggestion to develop the art electronically also had a major impact on my approach to the development of the visual part of this book.

Albert Altman pointed out Ernst Mach's approach to developing mechanics starting with the law of conservation of momentum. Al encouraged me throughout the years as I struggled to reorganize the material around the conservation principles.

I am thankful to Irene Nunes, who served as Development Editor through several iterations of the manuscript. Irene forced me to continuously rethink what I had written and her insights in physics kept surprising me. Her incessant questioning taught me that one doesn't need to be a science major to obtain a deep understanding of how the world around us works and that it is possible to explain physics in a way that makes sense for non-physics majors.

Catherine Crouch helped write the final chapters of electricity and magnetism and the chapters on circuits and optics, permitting me to focus on the overall approach and the art program. Peter Dourmashkin helped me write the chapters on special relativity and thermodynamics. Without his help, I would not have been able to rethink how to introduce the ideas of modern physics in a consistent way.

Many people provided feedback during the development of the manuscript. I am particularly indebted to the late Ronald Newburgh and to Edward Ginsberg, who meticulously checked many of the chapters. I am also grateful to Edwin Taylor for his critical feedback on the special relativity chapter and to my colleague Gary Feldman for his suggestions for improving that chapter.

Lisa Morris provided material for many of the Self-quizzes and my graduate students James Carey, Mark Winkler, and Ben Franta helped with data analysis and the appendices. I would also like to thank my uncle, Erich Lessing, for letting me use some of his beautiful pictures as chapter openers.

Many people helped put together the *Practice* book. Without Daryl Pedigo's hard work authoring and editing content, as well as coordinating the contributions to that book, the manuscript would never have taken shape. Along with Daryl, the following people provided the material for the *Practice* book: Wayne Anderson, Bill Ashmanskas, Linda Barton, Ronald Bieniek, Michael Boss, Anthony Buffa, Catherine Crouch, Peter Dourmashkin, Paul Draper, Andrew Duffy, Edward Ginsberg, William Hogan, Gerd Kortemeyer, Rafael Lopez-Mobilia, Christopher Porter, David Rosengrant, Gay Stewart, Christopher Watts, Lawrence Weinstein, Fred Wietfeldt, and Michael Wofsey.

I would also like to thank the editorial and production staff at Pearson. Margot Otway helped realize my vision for the art program. Martha Steele and Beth Collins made sure the production stayed on track. In addition, I would like to thank Frank Chmely for his meticulous accuracy checking of the manuscript. I am indebted to Jim Smith and Becky Ruden for supporting me through the final stages of this process and to Carol Trueheart, Alison Reeves, and Christian Botting of Prentice Hall for keeping me on track during the early stages of the writing of this book. Finally, I am grateful to Will Moore for his enthusiasm in developing the marketing program for this book.

I am also grateful to the participants of the NSF Faculty Development Conference "Teaching Physics Conservation Laws First" held in Cambridge, MA, in 1997. This conference helped validate and cement the approach in this book.

Finally, I am indebted to the hundreds of students in Physics 1, Physics 11, and Applied Physics 50 who used early versions of this text in their course and provided the feedback that ended up turning my manuscript into a text that works not just for instructors but, more importantly, for students.

Reviewers of *Principles & Practice of Physics*

Over the years many people reviewed and class-tested the manuscript. The author and publisher are grateful for all of the feedback the reviewers provided, and we apologize if there are any names on this list that have been inadvertently omitted.

Edward Adelson, *Ohio State University*
Albert Altman, *University of Massachusetts, Lowell*
Susan Amador Kane, *Haverford College*
James Andrews, *Youngstown State University*
Arnold Arons, *University of Washington*
Robert Beichner, *North Carolina State University*
Bruce Birkett, *University of California, Berkeley*
David Branning, *Trinity College*
Bernard Chasan, *Boston University*
Stéphane Coutu, *Pennsylvania State University*
Corbin Covault, *Case Western Reserve University*
Catherine Crouch, *Swarthmore College*
Paul D'Alessandris, *Monroe Community College*
Paul Debevec, *University of Illinois at Urbana-Champaign*
N. John DiNardo, *Drexel University*
Margaret Dobrowolska-Furdyna, *Notre Dame University*
Paul Draper, *University of Texas, Arlington*
David Elmore, *Purdue University*
Robert Endorf, *University of Cincinnati*
Thomas Furtak, *Colorado School of Mines*
Ian Gatland, *Georgia Institute of Technology*
J. David Gavenda, *University of Texas, Austin*
Edward Ginsberg, *University of Massachusetts, Boston*
Gary Gladding, *University of Illinois*
Christopher Gould, *University of Southern California*
Victoria Greene, *Vanderbilt University*
Benjamin Grinstein, *University of California, San Diego*
Kenneth Hardy, *Florida International University*
Gregory Hassold, *Kettering University*
Peter Heller, *Brandeis University*
Laurent Hodges, *Iowa State University*
Mark Holtz, *Texas Tech University*
Zafar Ismail, *Daemen College*
Ramanathan Jambunathan, *University of Wisconsin Oshkosh*
Brad Johnson, *Western Washington University*
Dorina Kosztin, *University of Missouri Columbia*
Arthur Kovacs, *Rochester Institute of Technology* (deceased)
Dale Long, *Virginia Polytechnic Institute* (deceased)
John Lyon, *Dartmouth College*
Trecia Markes, *University of Nebraska, Kearney*
Peter Markowitz, *Florida International University*
Bruce Mason, *University of Oklahoma*
John McCullen, *University of Arizona*
James McGuire, *Tulane University*
Timothy McKay, *University of Michigan*
Carl Michal, *University of British Columbia*
Kimball Milton, *University of Oklahoma*
Charles Misner, *University of Maryland, College Park*
Sudipa Mitra-Kirtley, *Rose-Hulman Institute of Technology*
Delo Mook, *Dartmouth College*
Lisa Morris, *Washington State University*
Edmund Myers, *Florida State University*
Alan Nathan, *University of Illinois*
K.W. Nicholson, *Central Alabama Community College*
Fredrick Olness, *Southern Methodist University*
Dugan O'Neil, *Simon Fraser University*
Patrick Papin, *San Diego State University*
George Parker, *North Carolina State University*
Claude Penchina, *University of Massachusetts, Amherst*
William Pollard, *Valdosta State University*
Amy Pope, *Clemson University*
Joseph Priest, *Miami University* (deceased)
Joel Primack, *University of California, Santa Cruz*
Rex Ramsier, *University of Akron*
Steven Rauseo, *University of Pittsburgh*
Lawrence Rees, *Brigham Young University*
Carl Rotter, *West Virginia University*
Leonard Scarfone, *University of Vermont*
Michael Schatz, *Georgia Institute of Technology*
Cindy Schwarz, *Vassar College*
Hugh Scott, *Illinois Institute of Technology*
Janet Segar, *Creighton University*
Shahid Shaheen, *Florida State University*
David Sokoloff, *University of Oregon*
Gay Stewart, *University of Arkansas*
Roger Stockbauer, *Louisiana State University*
William Sturrus, *Youngstown State University*
Carl Tomizuka, *University of Illinois*
Mani Tripathi, *University of California–Davis*
Rebecca Trousil, *Skidmore College*
Christopher Watts, *Auburn University*
Robert Weidman, *Michigan Technological University*
Ranjith Wijesinghe, *Ball State University*
Augden Windelborn, *Northern Illinois University*

Detailed Contents

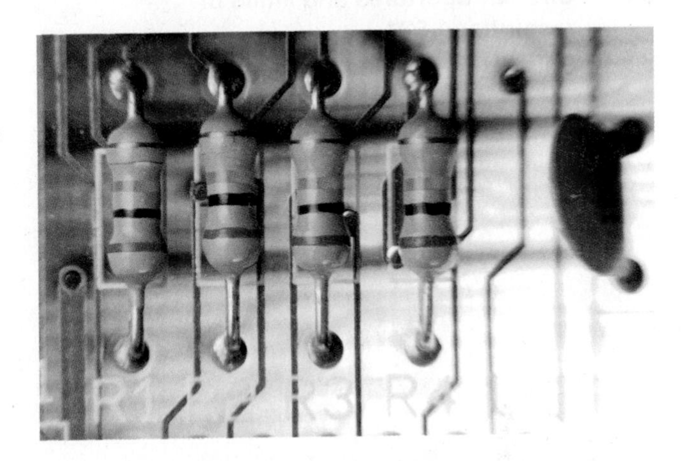
R1
R3
R4

22 Electric Interactions

Electricity is a familiar term—outlets, batteries, light bulbs, computers all involve electricity. It is no understatement to say that modern life depends on electricity, but what exactly *is* electricity? We all know what electricity does, but it's not that easy to explain what electricity *is*.

Electricity manifests itself in many ways: from the sparks that fly when you scuff your feet across a carpet on a dry winter day to the electricity we use in our homes to the transmission of radio and television programs. Even the attraction between magnets has to do with electricity. In this chapter, we begin our treatment of electricity with a discussion of static electricity.

22.1 Static electricity

When you tear off some plastic wrap from its roll, the wrap is attracted to anything that gets close: your hand, the countertop, a dish. This interaction between the plastic wrap and other objects doesn't have to involve any physical contact. For example, you can feel the presence of a piece of freshly torn-off plastic wrap with your cheek or the back of your hand even when your face or hand is held some distance away from the piece. You may have experienced many similar interactions: Styrofoam peanuts are attracted to your arms when you unpack a box full of them (**Figure 22.1**). Running a comb through your hair on a dry day causes the comb to attract your hair. After rubbing a balloon against a woolen sweater, you can hold the balloon close to a wall and *see* the attraction as the balloon moves toward the wall. In all these instances, the mass of the objects is too small for the interactions to be gravitational. What, then, is this interaction?

You may never have thought of these interactions as being particularly strong, but consider this: If you rub a comb through your hair and then pass the comb over some small bits of paper, the bits of paper jump up to your comb and stick to it. In other words, the bits of paper accelerate upward, which means the force exerted by your comb on them must be *greater* than the gravitational force exerted on them by Earth!

Now try this: Quickly pull a 20-cm strip of transparent tape* out of a dispenser and suspend it from the edge of a table (just be sure the table is not metal). Notice how the tape is attracted to anything brought nearby. It might even take some practice to prevent the tape from curling up and sticking to the underside of the table or to your hand. Bring a few objects near the suspended tape and notice the attractive interaction between them.† Go ahead—experiment!

Figure 22.1 Styrofoam peanuts cling to the cat's fur because of static electricity.

22.1 Suspend a freshly pulled piece of transparent tape from the edge of your desk. (*a*) What happens when you hold a battery near the tape? Does it matter whether you point the + side or the − side of the battery toward the tape? Does a spent battery yield a different result? Does a wooden object yield a different result? (*b*) What happens when you hold a strip of freshly pulled tape near the power cord of a lamp? Does it make any difference if the lamp is on or off?

All these interactions involving static electricity are examples of **electric interactions.** The experiment you just did tells you there is no obvious connection between electric interactions and the electricity we think of as "flowing" in electric circuits and batteries. In Chapter 31 we shall see, however, that the two are connected.

Objects that participate in electric interactions exert an **electric force** on each other. The electric force is a field force (see Section 8.3): Objects exerting electric forces on each other need not be physically touching. As you may have noticed from the interaction between the strips of tape and various nearby objects, the magnitude of the electric force depends on distance: It decreases as you increase the separation.

22.2 Suspend a freshly pulled strip of transparent tape from the edge of your desk. (*a*) Pull a second strip of tape out of the dispenser and hold it near the first strip. What do you notice? (*b*) Does it matter which sides of the strips you orient toward each other?

As Checkpoint 22.2 makes clear, not all electric interactions are attractive. Even if you increase the mass of the strips by suspending paper clips from them, the repulsion between the strips is great enough to keep the paper clips apart (**Figure 22.2**). Now place your hand between two repelling strips and notice how both strips fly toward your hand! Then run each strip of tape several times between your fingers and notice how the electric interaction diminishes or even disappears.

22.3 Suspend two freshly pulled 20-cm strips of transparent tape from the edge of your desk. Cut two 20-cm strips of paper, making each strip the same width as the tape, and investigate the interactions between the paper strips and the tape by bringing them near each other. Which of the following combinations display an electric interaction: paper-paper, tape-paper, tape-tape?

*For best results, use the type called "magic" tape.

†If you find something that *repels* the tape, wipe the entire surface of the object with your hand and see if it still repels—it shouldn't. Mystified? Hang on! We'll soon be able to resolve your questions.

Figure 22.2 Strips of tape just pulled out of a dispenser repel each other. The repulsive force is great enough to keep the strips apart even when they are weighted down by paper clips.

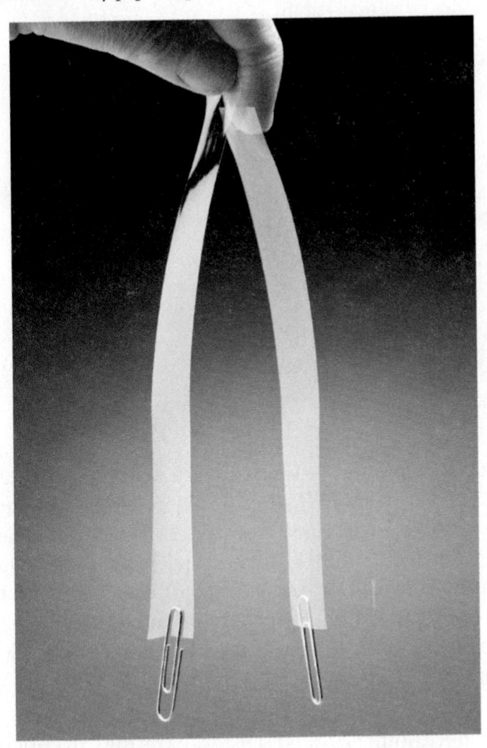

Figure 22.4

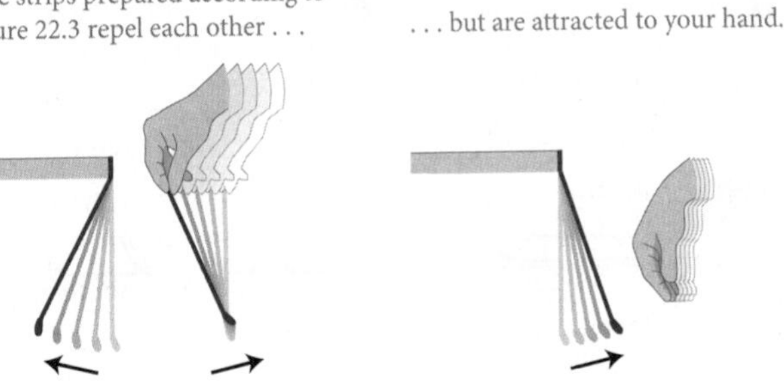

22.2 Electrical charge

As we saw in the preceding section, electric interactions are sometimes attractive and sometimes repulsive. In addition, the experiment you performed in Checkpoint 22.3 demonstrates that paper strips, which do not interact electrically with each other, do interact electrically with transparent tape. What causes these interactions? To answer this question, we need to carry out a systematic sequence of experiments.

Figure 22.3 illustrates a simple procedure for reproducibly creating strips of tape that interact electrically. A suspended strip created according to this procedure interacts in the following ways: It repels another strip created in the same manner, and it attracts any other object that does not itself interact electrically with other objects (Figure 22.4).

Let us call the attribute responsible for the electric interaction **electrical charge,** or simply **charge.** Saying that something carries an electrical charge is just another way of saying that that object interacts electrically with other objects that carry electrical charge. Freshly pulled strips of tape carry electrical charge, and two such strips interact because each possesses an electrical charge, just as your body and Earth interact because each possesses mass. The general term for any microscopic object that carries an electrical charge, such as an electron or ion, is **charge carrier.**

It is not immediately clear what attributes to assign to objects that do not interact electrically with each other but do interact with a charged tape strip—a strip of paper, your hand, an eraser, you name it. All we know for now is that the interaction between these objects and a charged tape is attractive rather than repulsive.

The electric charge on an object is not a permanent property; if you let a charged strip of tape hang for a while, it loses its ability to interact electrically. In other words, the strip is no longer charged—it is *discharged.* Depending on the humidity of the air, the discharging can take minutes or hours, but you can speed up the discharging by rubbing your fingers a few times over the entire length of a suspended charged strip of tape.* (The rubbing allows the charge to "leak away" from the tape by distributing itself over your body.)

*If rubbing your fingers along the tape doesn't do the job, try licking them before rubbing them over the tape.

Figure 22.3 Procedure for making strips of transparent tape that interact electrically. The purpose of the foundation strip is simply to provide a standard surface.

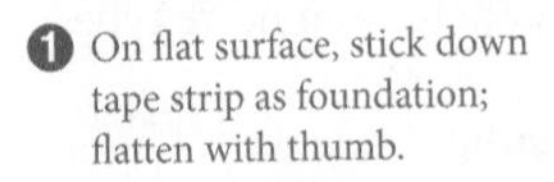

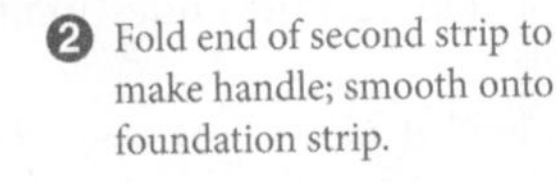

1 On flat surface, stick down tape strip as foundation; flatten with thumb.

2 Fold end of second strip to make handle; smooth onto foundation strip.

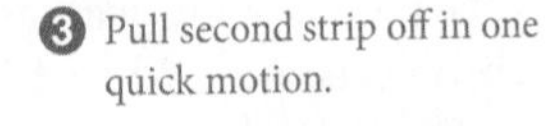
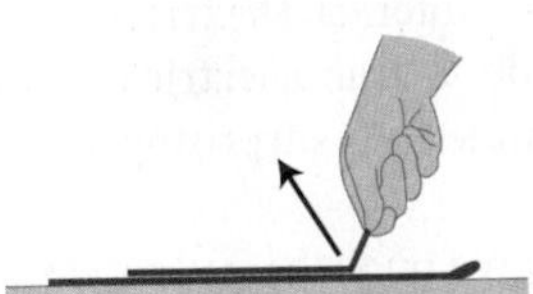

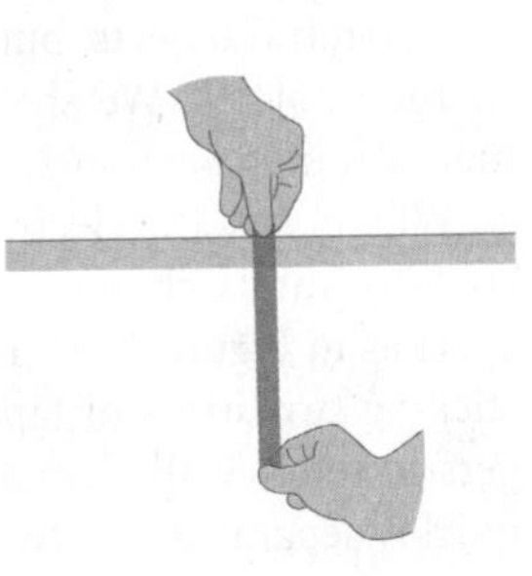

3 Pull second strip off in one quick motion.

4 Holding both ends of strip to prevent curling, hang strip on table edge.

Figure 22.5 Procedure for making strips of transparent tape that carry opposite charges.

1. Layer two tape strips onto foundation strip; smooth down & label as indicated.
2. Pull combined T and B strips *very slowly* off foundation strip.
3. Hang combined strips on table edge. Rub with thumb to discharge, then pull T strip quickly off B strip.
4. Hang strips 0.5 m apart on table edge.

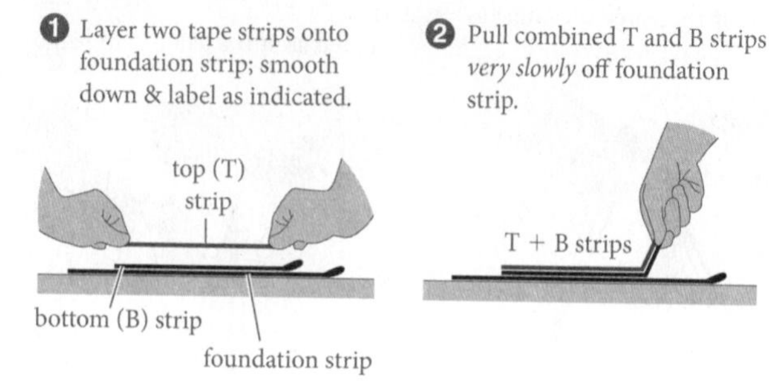

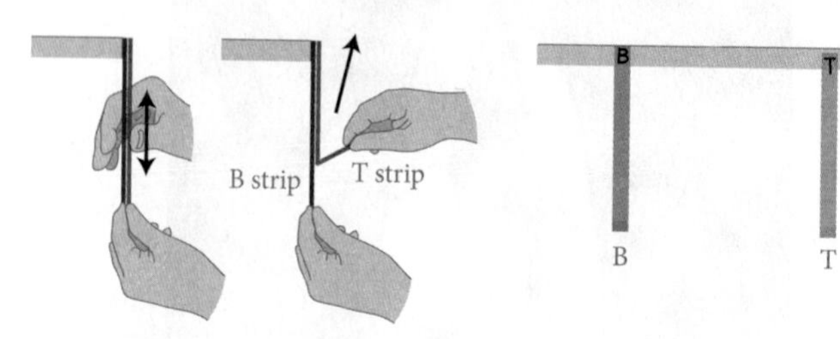

22.4 (*a*) Prepare a charged strip of transparent tape as described in Figure 22.3 and then suspend the strip from the edge of your desk. Verify that the tape interacts as you would expect with your hand, with a strip of paper, and with another charged strip of tape. (*b*) Rub your fingers along the hanging strip to remove all the charge from it, and then verify that it no longer interacts with your hand. If it does interact, rub again until it no longer interacts. (*c*) Predict and then verify experimentally how the uncharged suspended strip interacts with a strip of paper and with a charged strip of tape.

To restore the charge on a discharged strip, stick the strip on top of the foundation strip from which you pulled it off (step 1 in Figure 22.3), smooth it out, and then quickly pull it off again. You can recharge a strip quite a few times before it loses its adhesive properties. Once the tape does lose its adhesiveness, however, recharging it becomes impossible. It is generally a good idea to rub your finger over the foundation strip before you reuse it to make sure that it, too, is uncharged.

22.5 Recharge the discharged strip from Checkpoint 22.4 and verify that it interacts as before with your hand, with a strip of paper, and with another charged strip of tape.

A discharged tape strip interacts in the same way as objects that carry no charge. Such objects are said to be electrically **neutral.** They do not interact electrically with other neutral objects, but they do interact electrically with charged objects. We shall examine this surprising fact in more detail in Section 22.4.

Where does the electrical charge on a charged tape strip come from? Is charge *created* when two strips are pulled apart as in Figure 22.3? This is something we can check by sticking two strips of tape together, rubbing with our fingers to remove all charge from the combination, and then quickly separating the two strips (**Figure 22.5**).

22.6 Follow the procedure illustrated in Figure 22.5 to separate a pair of charged strips. (*a*) How does strip B interact with a neutral object? How does strip T interact with a neutral object? (*b*) Create a third charged strip and see how it interacts with strip B and with strip T. (*c*) Is strip T charged? (*d*) Is strip B charged? (*e*) Check what happens to the interactions with B and T strips when you discharge a B or a T strip by rubbing your fingers along its length.

As Checkpoint 22.6 shows, separating an uncharged pair of strips produces two charged ones, but the behavior of strip B is different from that of the other strips we have encountered so far!

22.7 Make two charged pairs of strips (B and T) following the procedure illustrated in Figure 22.5. Investigate the interaction of B with T, T with T, and B with B.

The interactions between the B and T strips are illustrated in **Figure 22.6**: Strips of the same type repel each other, while strips of different types attract each other. This series of experiments leads us to conclude that there are two types of charge on the tapes, one type on B strips and another type on T strips. Strips that carry the same type of charge, called *like charges,* exert repulsive forces on each other; strips that carry different types of charge, called *opposite charges,* exert attractive forces on each other.

Having determined that two types of electrical charge exist, a logical next question is: Are there even more types?

22.8 (*a*) Prepare one charged strip of tape according to Figure 22.3 and hang it from the edge of your desk. Hang a narrow strip of paper from the desk edge also, about 0.5 m away from the tape strip. Pass a *plastic* comb six times quickly through your hair and then show that the comb is charged. Be sure to use a plastic comb; combs made from other materials do not acquire a charge when passed through hair. The cheapest type of comb usually works best. (*b*) Make a pair of oppositely charged B and T strips (Figure 22.5) and investigate how they interact with a charged comb. (*c*) Does your comb behave like a B strip, a T strip, or neither?

Figure 22.6 Interactions of B and T charged strips.

Strips of same type repel each other.

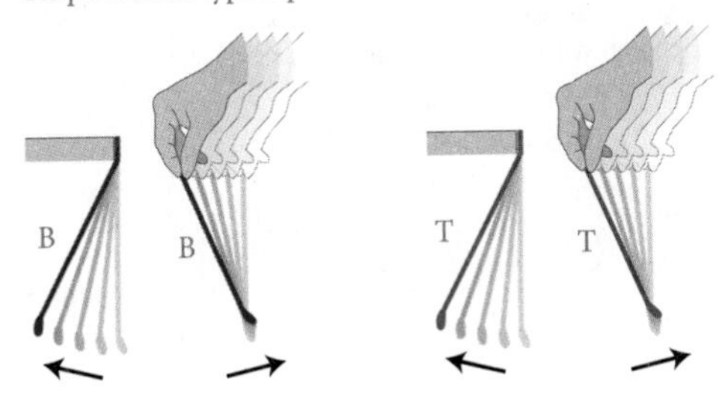

Strips of different types attract each other.

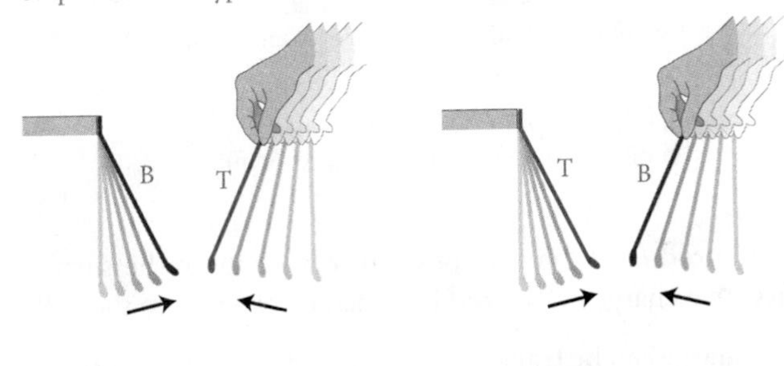

Experiments show that *any* charged object—obtained by rubbing objects together or otherwise—always attracts either a B strip or a T strip and repels the other. No one has ever found a charged object that repels or attracts *both* types of strips. In other words:

There are two and only two types of charge. Objects that carry like charges repel each other; objects that carry opposite charges attract each other.

The two types of charge never appear independently of each other: Whenever two neutral objects are either rubbed together and then separated or, if an adhesive surface is involved, stuck together and then separated and one of them acquires a charge of one type, the other object always acquires a charge of the other type. The generation of opposite charges is obvious when you separate a neutral pair of tape strips. When you pass a comb through your hair, the comb acquires a charge of one type and your hair acquires a charge of the other type. On a dry day, you may have noticed that some hair strands stand up away from your head. Each charged strand is being repelled by the other charged strands, and so they are all getting as far away from one another as possible.

It can be shown that when two tape strips are separated, the forces exerted by the B strip and the T strip on a third charged strip are equal in magnitude, although one is attractive and the other repulsive. Furthermore, when the B and T strips are recombined, the combination is neutral again. These observations suggest that after you rub and then separate a pair of objects, the objects carry equal amounts of opposite charge. Combining these equal amounts of opposite charge produces zero charge. These observations indicate that all neutral matter contains equal amounts of positive and negative charge. The two types of charge are called **positive** and **negative charges.** The definition of negative charge is as follows:*

Negative charge is the type of charge acquired by a plastic comb that has been passed through hair a few times.

22.9 Does the B strip you created in Checkpoint 22.8 carry a positive charge or a negative charge?

When two neutral objects touch, some charge can be transferred from one object to the other, with the result that one object ends up with a surplus of one type of charge and the other object ends up with an equal surplus of the other type of charge. For example, when a neutral piece of styrofoam is rubbed with a neutral piece of plastic wrap, the styrofoam acquires a positive charge (meaning it contains more positive than negative charge) and the plastic wrap acquires a negative charge (it has a surplus of negative charge). Without further information, however, we cannot tell whether positive charge has been transferred from the wrap to the styrofoam, or negative charge has been transferred from the styrofoam to the plastic wrap, or a combination of these two. (See **Figure 22.7** on the next page.) Summarizing:

All neutral matter contains equal amounts of positive and negative charge; charged objects contain unequal amounts of positive and negative charge.

In illustrations, surplus charge is represented by plus or minus signs. Keep in mind, however, that these signs never represent the only type of charge in an object. The plus signs on the positively charged styrofoam in Figure 22.7, for example, mean only that the styrofoam contains more positive than negative charge, either because some of its negative charge has been removed or because some positive charge has been added. In addition to the 12 positive charge carriers shown in Figure 22.7, the styrofoam contains millions and millions of positive charge carriers paired with millions and millions of negative charge carriers. A drawing such as Figure 22.7, shows only *unpaired* charge carriers (usually referred to as *surplus charge*).

As our observations in Figure 22.6 show, oppositely charged B and T strips attract each other. The interaction between positive and negative charge tends to bring positive and negative charge carriers as close together as possible. Because combining equal amounts of positive and negative charge results in zero charge, we can say that charge carriers always tend to arrange themselves in such a way as to produce uncharged objects—indeed, all matter around us tends to be neutral.

*Historically, negative charge was (arbitrarily) defined by Benjamin Franklin (1706–1790) as the charge acquired by a rubber rod rubbed with cat fur. Because plastic combs and hair are more easily accessible than rubber rods and cat fur, the definition of negative charge given here is more convenient.

Figure 22.7 Rubbing neutral styrofoam with neutral plastic wrap leaves the two objects with equal charges of opposite types.

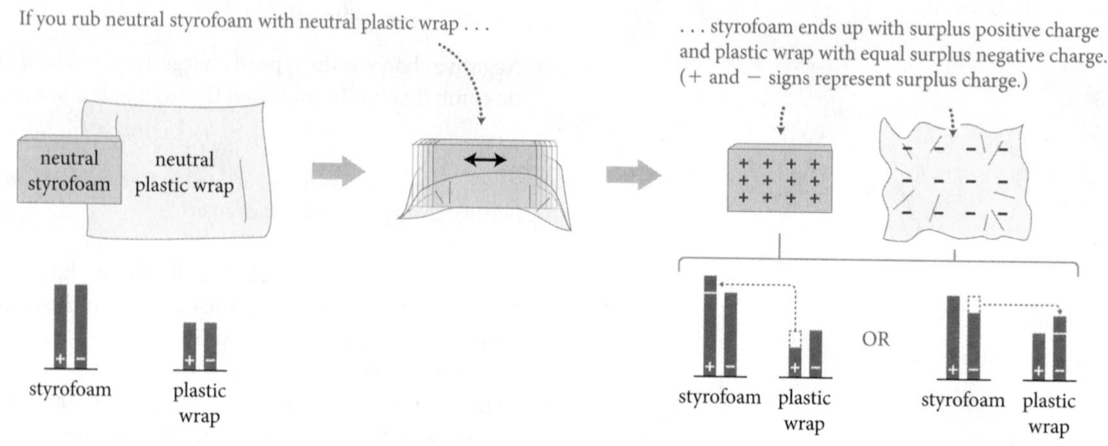
If you rub neutral styrofoam with neutral plastic wrap . . .
. . . styrofoam ends up with surplus positive charge and plastic wrap with equal surplus negative charge. (+ and − signs represent surplus charge.)
neutral styrofoam
neutral plastic wrap
styrofoam
plastic wrap
OR
styrofoam
plastic wrap
styrofoam
plastic wrap
Bars signify content of + and − charge. Neutral objects contain large but equal amounts of + and − charge.
Charge surplus can result from transfer of positive charge or negative charge (or both).

22.10 Imagine having a collection of charged marbles that retain their charge even when they touch other objects. Red marbles are positively charged, and blue marbles are negatively charged. (*a*) What happens if you place a bunch of red marbles close together on a flat horizontal surface? (*b*) What happens if you do the same with a bunch of blue marbles? (*c*) What happens if you do the same with an equal mixture of red and blue marbles? (*d*) What happens in part *c* if you have a few more red marbles than blue ones? (*e*) As a whole, is the collection of marbles in part *d* positively charged, negatively charged, or neither?

22.3 Mobility of charge carriers

To gain a better understanding of electrical charge, many additional experiments are required, most of which require items not easily found at home. A rubber rod rubbed with a piece of cat fur acquires—by Benjamin Franklin's original definition—a negative charge (and the fur acquires a positive charge). A glass rod rubbed with silk acquires a positive charge (and the silk a negative charge). Other materials also acquire a charge upon contact or rubbing, but these two combinations of rubber/fur and glass/silk provide the most convenient means of generating relatively large amounts of charge.

Interesting things happen when a charged rubber rod is brought into contact with an uncharged pith ball.* As the rod is brought near the ball, the ball moves toward the rod because of the attraction between the charged rod and the neutral ball (**Figure 22.8*a***). As the ball touches the rod, however (Figure 22.8*b*), the crackling sounds of tiny sparks may be heard. The ball suddenly jumps away from the rod (Figure 22.8*c*), indicating that the interaction between rod and ball has become repulsive. This repulsive interaction indicates that the ball has acquired the same type of charge as the rod (negative). In other words, some of the surplus negative charge on the rod has been transferred to the ball.

Charge can be transferred from one object to another by bringing the two into contact.

We can use this phenomenon to investigate the electrical behavior of different kinds of materials. For example, if we transfer some charge to one end of an uncharged rubber rod and then extend the charged end toward an uncharged pith ball, the two interact electrically, as shown in **Figure 22.9*a***. If we hold the *uncharged* end near the pith ball, as in Figure 22.9*b*, however, no interaction occurs. This tells us that the charge does not flow from one end of the rubber rod to the other; instead, it remains near the spot where it has been deposited. Any material in which charge doesn't flow (or moves only with great difficulty) is called an **electrical insulator.**

Electrical insulators are materials through which charge carriers cannot flow easily. Any charge transferred to an insulator remains near the spot at which it was deposited.

Figure 22.8 A charged rubber rod can transfer charge to a neutral pith ball.

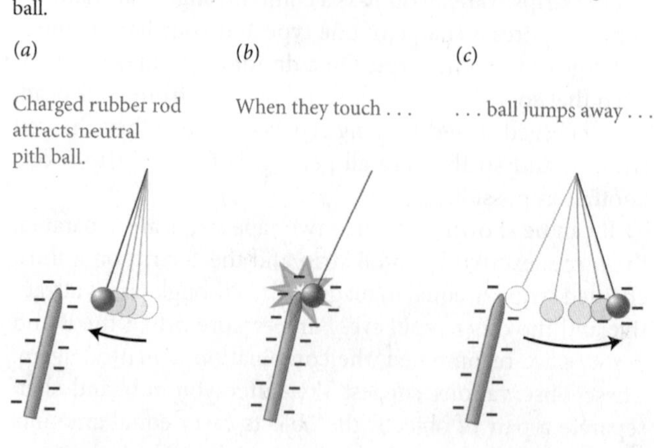
(*a*)
(*b*)
(*c*)
Charged rubber rod attracts neutral pith ball.
When they touch . . .
. . . ball jumps away . . .
. . . which tells us that rod & ball have same type of charge.

*Pith is the soft, lightweight, spongelike material that makes up the interior of the stems of flowering plants.

CONCEPTS

Figure 22.9 A rubber rod is an example of an electrical insulator.

(*a*) Charged end of rubber rod attracts pith ball . . .

(*b*) . . . uncharged end doesn't.

So, rubber is electrical insulator: Charge does not flow through it.

Figure 22.11 A metal rod is an electrical conductor.

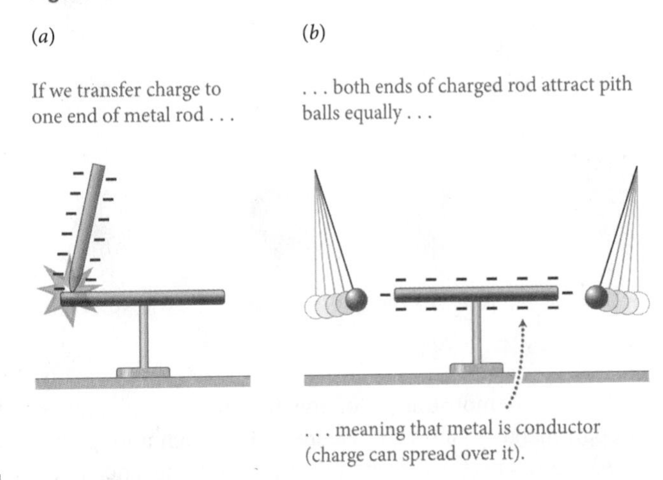

(a)
(b)
If we transfer charge to one end of metal rod . . .
. . . both ends of charged rod attract pith balls equally . . .
. . . meaning that metal is conductor (charge can spread over it).

Glass, rubber, wood, and plastic are examples of electrical insulators. Air, particularly dry air, is also an insulator, although the presence of large amounts of charge can cause charge carriers to "jump" from one object to another, causing sparks.

Exercise 22.1 Electric forces

(*a*) Draw a free-body diagram for the pith ball in Figure 22.9*a*. (*b*) Two identical neutral pith balls A and B are suspended side by side from two vertical strings. After some charge is transferred from a charged rod to A, A and B interact. (B remains neutral because the two balls never come into contact with each other.) Sketch the orientation of A and B after the charge has been transferred to A. (*c*) Draw a free-body diagram for each ball.

SOLUTION (*a*) The ball is subject to three forces: a gravitational force, a contact force exerted by the string, and an attractive electric force exerted by the charged particles in the rod. This last force is directed horizontally toward the rod. Because the ball is at rest, I know that the vector sum of these three forces is zero. So the horizontal component of the force exerted by the string on the ball must be equal in magnitude to the electric force exerted by the rod on the ball, and the vertical component of the force exerted by the string on the ball must be equal in magnitude to the gravitational force exerted by Earth on the ball (**Figure 22.10*a***).

Figure 22.10

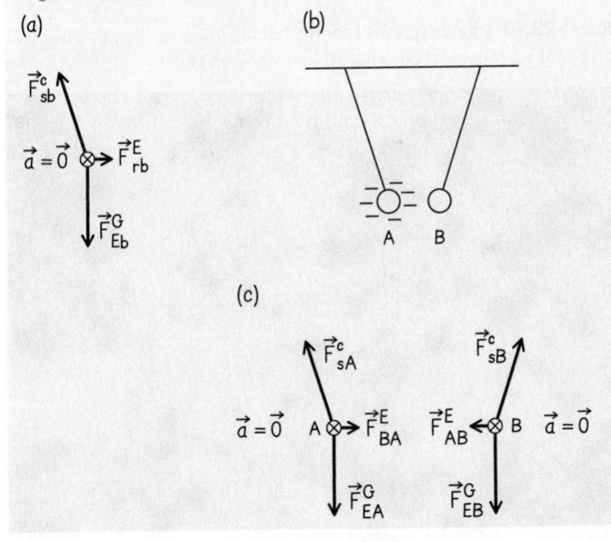

(*b*) As we saw in Section 22.2, a neutral object interacts electrically with a charged object. The electric force exerted by A on B and that exerted by B on A form an interaction pair and so their magnitudes are equal. Because the masses of the pith balls are the same, each is pulled in by the same distance. Thus my sketch is as shown in Figure 22.10*b*. (*c*) See Figure 22.10*c*.

In **Figure 22.11**, a charged rod is brought into contact with an uncharged metal rod supported on an electrically insulating stand. Once the charged rod has touched the metal rod, *all* points on the surface of the metal rod interact electrically with other objects, indicating that the charge spreads out over the metal rod. The tendency of charge to spread out over metal objects can be demonstrated with an *electroscope* (**Figure 22.12*a***). Two strips made of metal foil are suspended from a small metal rod in an electrically insulating enclosure; the rod is connected to a metal ball on top of the enclosure. When the metal ball is charged by an exterior source, the strips move away from each other. The explanation for this movement is that the added charge quickly moves from the metal ball through the metal rod and onto the two metal strips. Once the strips carry the same type of charge, they repel each other (Figure 22.12*b*).

Figure 22.12 An electroscope depends on electrical conduction.

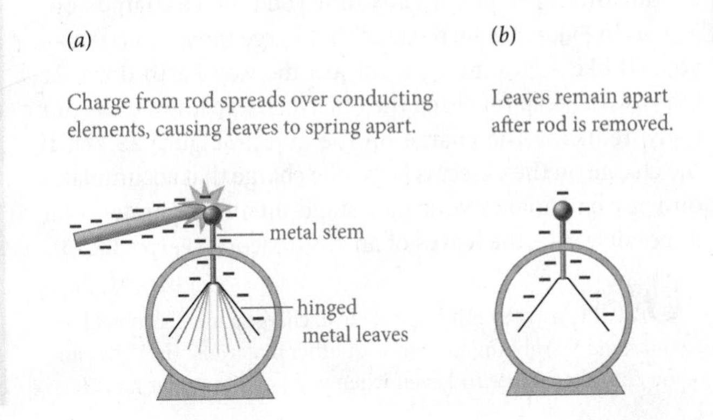

Charge from rod spreads over conducting elements, causing leaves to spring apart.
Leaves remain apart after rod is removed.
metal stem
hinged metal leaves

Figure 22.13 A conducting wire distributes charge between two conducting spheres.

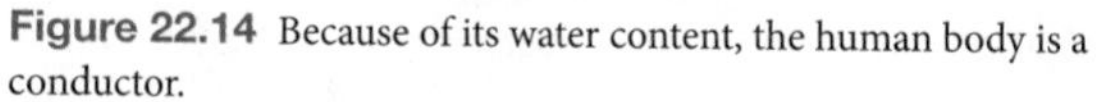
(a)
Use charged rod . . .
(b)
. . . to charge one metal sphere.
(c)
Connect spheres with wire.
(d)
Charge distributes equally.

Another demonstration of the free motion of charges through metals is shown in **Figure 22.13**: When a long wire is used to connect a charged metal sphere to an uncharged metal sphere, charged particles flow from the charged sphere to the uncharged one. Because wires are made of metal, this experiment shows that, in contrast to what happens with electrical insulators, charge moves easily through a metal and across a metal-to-metal contact. Materials through which charge carriers can flow are called electrical **conductors,** and the flow of charge through conductors is called **conduction.**

Electrical conductors are materials through which charge carriers can flow easily. Any charge transferred to a conductor spreads out over the conductor and over any other conductors in contact with it.

Metals are the only solid materials that are conductors at room temperature. (As noted earlier, glass, plastic, and most other solids are electrical insulators.) Although charge does not flow easily through pure water, minute amounts of impurities turn water into a fairly good conductor. Because most water contains some impurities, water is therefore usually considered a conductor.

Except for the outer layer of soil, Earth is also a good conductor. Consequently, when a charged, conducting object is connected to Earth by a wire, a process called **grounding,** charge carriers can flow between Earth ("ground") and the object. Because Earth is so large, it can supply or absorb a nearly unlimited number of charge carriers. In the absence of other nearby electrical influences, the grounded object is left with no surplus of either type of charge.

Because of its high water content, the human body is a conductor. Consequently, any time you touch a charged object, as in **Figure 22.14**, some of the charge moves into you—you act like a grounding agent just the way Earth does. As long as you keep touching the object, charge flows into your body, reducing the charge on the object (Figure 22.14). If the charge on the object is large, the charge that accumulates on your hair makes your hair stand up and separate as far as possible, like the leaves of an electroscope (**Figure 22.15**).

22.11 (*a*) Why is it impossible to charge a metal rod held in your hand by rubbing the rod with other materials? (*b*) Why can you charge a rubber rod even when you hold it in your hand?

Figure 22.14 Because of its water content, the human body is a conductor.

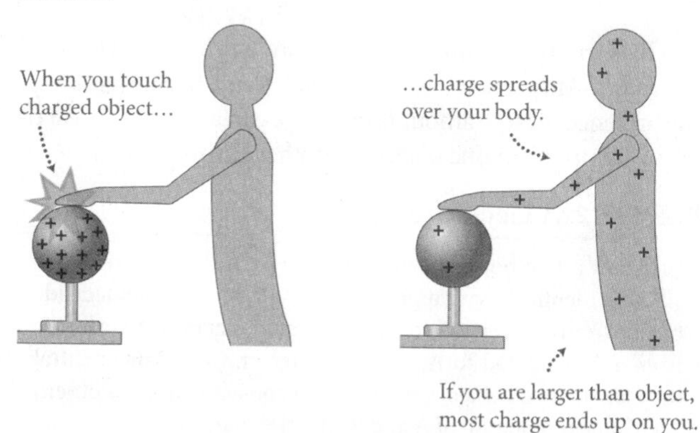
When you touch charged object...
...charge spreads over your body.
If you are larger than object, most charge ends up on you.

Is charge some sort of fluid that flows from one object to another, or is it composed of small particles that can be peeled off or stuck onto objects? To answer this question, we must look at the atomic structure of matter. All matter consists of atoms (see Section 1.3), the structure of which is schematically illustrated in **Figure 22.16**. Nearly all the atom's mass is concentrated in the extremely small nucleus at the center. The nucleus is composed of protons and neutrons. The region surrounding the nucleus, representing most of the atom's volume, is a cloud of electrons.

Figure 22.15 Charge spreads over the human body, so a large charge will cause your hairs to repel one another and stand on end.

Danger
Keep Off
Danger

CONCEPTS

Figure 22.16 Structure of the atom (not to scale).

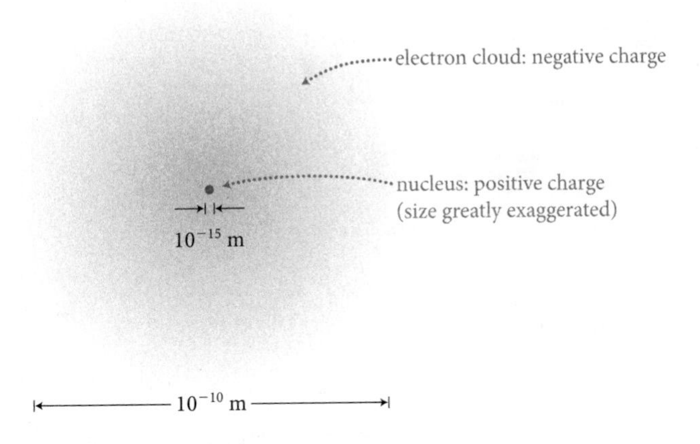

Experiments show that electrons have a negative electrical charge—they are repelled by a charged comb* and by other electrons. Protons carry a positive charge, and neutrons carry no charge. The protons and neutrons in the nucleus are held together tightly by the strong interaction (see Chapter 7), which is great enough to overcome the electrical repulsion between the positively charged protons. The electrical attraction between the positively charged nucleus and the negatively charged electrons is responsible for keeping the electrons bound to the nucleus. The electron cloud does not collapse on the nucleus because of additional constraints imposed on the electrons by the laws of quantum mechanics.

Charge is an inherent property of the electron, which means it is impossible to remove the charge from an electron—there is no such thing as a discharged electron. Experiments show that

All electrical charge comes in whole-number multiples of the electrical charge on the electron.

For this reason, the magnitude of the charge on the electron, designated by the letter *e*, is called the **elementary charge.**

Every atom contains equal numbers of electrons and protons. Because atoms are neutral, the fact that they contain equal numbers of electrons and protons tells us that the magnitude of the positive charge on the proton is also *e*. The charge on an electron is $-e$, and that on a proton $+e$. As with electrons, the charge cannot be removed from a proton.

Given that macroscopic objects contain an immense number of atoms and that each atom can contain dozens of electrons and protons, we see that ordinary objects contain an immense number of positively charged protons, exactly balanced by an equal number of negatively charged electrons. A surplus of just a minute fraction of these numbers is sufficient to give rise to a noticeable macroscopic charge. For example, when you pull apart two strips of transparent tape, the separation causes a surplus of less than one in a trillion (10^{12}) electrons. (Because there are about 10^{22} electrons in the strip, that fraction represents some 10^{10}, or ten billion electrons.)

When two atoms are brought close together, they may form a chemical bond by transferring one or more electrons from one atom to the other. Once such an electron transfer takes place, both atoms contain unequal numbers of electrons and protons and are now called **ions** instead of atoms. One of the two ions has gained one or more electrons, meaning it contains more electrons than protons and therefore carries a negative charge. The other ion, the one that lost electrons, contains more protons than electrons and so carries a positive charge.

Ions in solids are always immobile, but ions in liquids can move freely. For instance, in table salt, a compound made of pairs of sodium (Na^+) and chloride (Cl^-) ions, the charged ions hardly move at all, meaning that solid table salt is an electrical insulator. Dissolve table salt in water, however, and the solution contains large quantities of positively charged sodium ions and negatively charged chloride ions. Because these ions can move freely, the solution is an electrical conductor.

Some solids are made not of paired ions the way sodium chloride is but rather of individual atoms. In atomic solids that are electrical insulators, the electrons in the atoms are unable to move because each electron is bound to a specific atom. Diamond (made of the element carbon) and glass are two familiar examples. Metals are also atomic solids rather than ionic solids, but in metals, each atom gives up one or more electrons to a shared "gas" of electrons that spreads throughout the volume of the metal. The metal as a whole is still neutral: The negative charge of the electron gas is exactly balanced by the positive charge of the ions. The electrons in the gas are called *free electrons* because they can move freely inside the metal; these electrons are responsible for the easy flow of charge through a metal.

Nearly all electrical phenomena are due to the transfer of electrons—and therefore charge—from one atom to another. For example, when the sticky side of one strip of transparent tape is applied on top of the nonsticky side of a second strip, atoms in the adhesive from the top strip form chemical bonds with atoms in the nonsticky surface of the bottom strip by transferring electrons, as shown in Figure 22.17 on the next page. These bonds are responsible for the adhesion of one strip to the other. When the strips are pulled apart quickly, the bonds are broken, but not all electrons manage to get back to the top strip. The bottom strip thus ends up with a surplus of electrons, making it negatively charged, and the top one with a deficit of electrons, making it positively charged.*

*Recall from our discussion of positive and negative charge carriers in Section 22.2 that a plastic comb carries a negative charge.

*Depending on the type of adhesive and the material of the backing, the transfer of electrons can also be in the other direction.

Figure 22.17 How strips of tape can acquire opposite charges when pulled apart.

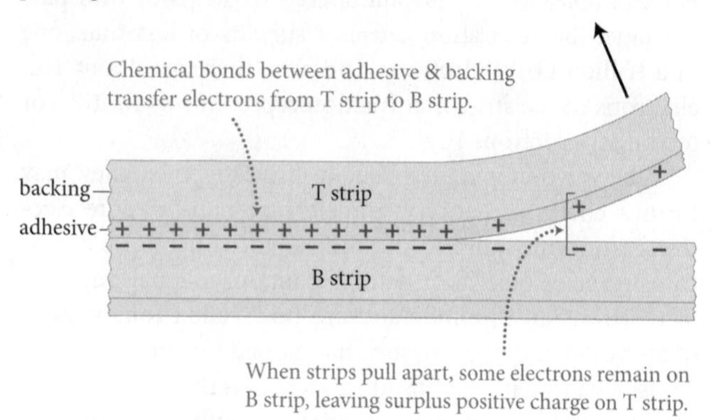

As we saw in Section 10.4, friction between two surfaces also involves the breaking of chemical bonds. As with the separation of the tape strips, this bond breaking sometimes leaves surplus charge on the surfaces. When you touch a piece of plastic food wrap to a piece of styrofoam, for instance, chemical bonds form between atoms on the two surfaces. In these bonds, electrons from the styrofoam move to the wrap. If you then rub the surfaces against each other, these bonds are broken, and some of the electrons originally on the styrofoam stay on the wrap. If the breaking of the chemical bonds occurs slowly, the electrons migrate back and no surplus charge builds up. For that reason it is necessary to rub vigorously or to separate strips of tape quickly. The key point is:

Any two dissimilar materials become charged when brought into contact with each other. When they are separated rapidly, small amounts of opposite charge may be left behind on each material.

Because charging by breaking of chemical bonds is due to a transfer of charge, we now see that for every surplus of negative charge that appears in one place, an equal surplus of positive charge appears somewhere else. After the two strips in Figure 22.17 are separated, the sum of the positive charge on the T strip and the negative charge on the B strip is still zero. No creation or destruction of charge is involved, suggesting that electrical charge—like momentum, energy, and angular momentum—is a conserved quantity. The principle of **conservation of charge** states:

Electrical charge can be created or destroyed only in identical positive-negative pairs such that the charge of a closed system always remains constant.

No process has ever been found to violate this principle. Even when charged subatomic particles, such as electrons and protons, are created or destroyed—a process that can be observed in high-energy particle accelerators—charge is conserved. For example, when an electron (charge $-e$) collides with a subatomic particle called the *positron* (charge $+e$), both particles are destroyed, leaving nothing but a flash of highly energetic radiation (Figure 22.18a). The charge of the electron-positron system before the collision is $(-e) + (+e) = 0$, and it is still zero after the collision. Likewise, in a process called beta decay, when a free neutron (carrying zero charge and made up of two down quarks and one up quark) decays into a proton (charge $+e$, one down and two up quarks), an electron (charge $-e$), and a neutral subatomic particle called the antineutrino (zero charge), the charge of the system comprising the neutron and the particles into which it decays remains zero (Figure 22.18*b* and Section 7.6).

Figure 22.18 Because charge is conserved, the charge of a closed system does not change even when particles are created or destroyed.

(*a*) Electron-positron annihilation

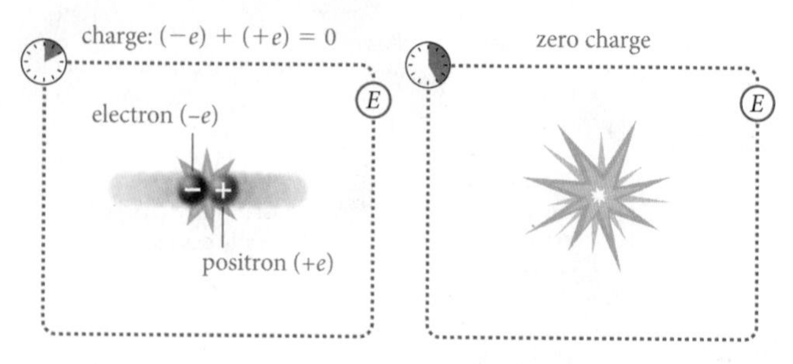

(*b*) Decay of a free neutron

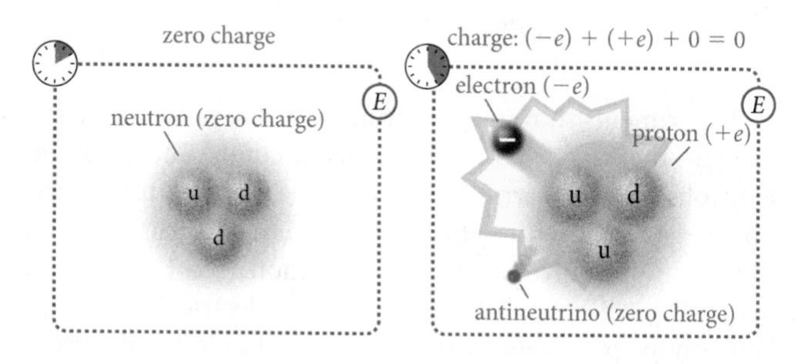

22.12 When two objects made of the same material are rubbed together, friction occurs but neither material acquires surplus charge. Why?

22.4 Charge polarization

Let us now reexamine the interaction between a charged object and a neutral one. Figure 22.19 shows the interaction between a charged rubber rod and an uncharged electroscope. With the rod far away (Figure 22.19*a*), the leaves of the electroscope hang straight down. When the rod is brought near the ball of the electroscope (Figure 22.19*b*), the leaves separate even without any contact between the rod and the electroscope. As the distance between the rod and the electroscope is increased again, the leaves drop down, showing that no charge has been transferred from the rod to the electroscope.

Figure 22.19 In (*a*) and (*b*), a charged rod induces polarization in an electroscope. (*c*) A schematic atom-level view.

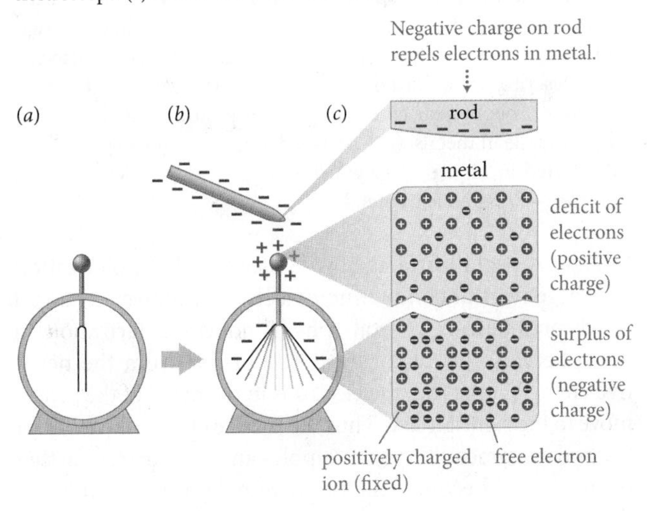

Why do the leaves separate even though the electroscope remains neutral? They separate because the negative charge on the rod repels the free electrons in the metallic parts of the electroscope: The free electrons are pushed as far away as possible from the rod (Figure 22.19*c*) and pile up in the leaves. This redistribution of charge is nearly instantaneous. The top of the electroscope thus ends up with a deficit of electrons—a positive charge—and the leaves end up with a surplus of electrons—a negative charge. The negative charge on the leaves is responsible for the repulsion between them. When the rod is removed, the electrons, being repelled by one another and attracted to the positive charge on the electroscope ball, immediately flow back to their normal positions, evening out the distribution of positive and negative charge.

22.13 (*a*) In Figure 22.19*b*, is the electroscope as a whole positively charged, negatively charged, or neutral? (*b*) How does the magnitude of the positive charge on the electroscope ball compare with the magnitude of the negative charge on the leaves? (*c*) Is the force exerted by the rod on the electroscope ball attractive or repulsive? Is the force exerted by the rod on the leaves attractive or repulsive? (*d*) How do you expect the magnitude of the force the rod exerts on the ball to compare with the magnitude of the force the rod exerts on the leaves?

Any separation of charge carriers in an object is called **charge polarization,** or simply **polarization,** and an object in which charge polarization occurs is said to be *polarized*. The electroscope of Figure 22.19*b*, for instance, is polarized by the nearby charged rod. In any object in which charge is polarized, there are two charged *poles*, one positive and the other negative. In the electroscope of Figure 22.19*b*, the positive pole is at the ball and the negative pole is in the foil strips.

In metals, the polarization induced by the presence of a nearby charged object is very great because the free electrons in the metal move easily in response to the presence of the charged object. Even in electrical insulators, however, where there are no free electrons moving about, a nearby charged object induces some polarization. The basic reason for the polarization of insulators is illustrated in **Figure 22.20**: In the presence of an external charge, the center of the electron cloud and the nucleus of an atom shift away from each other, causing the atom to become polarized. So, when a negatively charged comb is brought near a small piece of paper, each atom in the paper becomes polarized—the electron clouds are pushed away from the comb, and the nuclei are pulled toward the comb. If we consider the paper as consisting of two overlapping parts that have the same shape but carry opposite charges, the positively charged part is pulled a bit toward the comb and the negatively charged part is pushed away, as shown in **Figure 22.21*a*** on the next page. This leaves the central part of the paper neutral but creates a sliver of surplus positive charge on the side facing the comb and an equal amount of surplus negative charge on the opposite side, and so the paper is polarized.

22.14 In an atom, what limits the separation between the electron cloud and the nucleus in the presence of an external charge? Why, for example, isn't the electron cloud in Figure 22.20*b* pulled all the way to the location of the external positive charge?

The polarization of atoms is responsible for the attraction between charged and neutral objects. In Figure 22.21, for example, the positively charged side of the paper is closer to the comb than the negatively charged side. Because the electric force decreases with increasing distance, the magnitude of the attractive force exerted on the positive side is greater than the magnitude of the repulsive force exerted on the negatively charged side (Figure 22.21*b*). Consequently, the vector sum of the electric forces exerted by the comb on the neutral piece of paper points toward the comb and the paper is pulled toward the comb.

Figure 22.20 Polarization of a neutral atom.

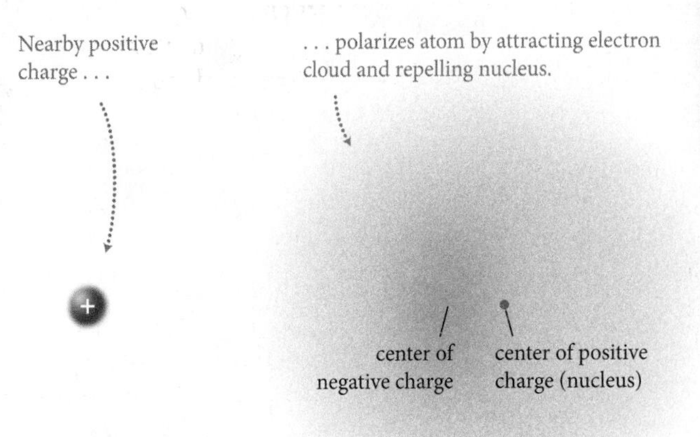

CONCEPTS

Figure 22.21 Polarization of a neutral insulator (bits of paper) by a charged comb. In (*a*), a single bit of paper is modeled as two offset sheets with opposite charges.

Charged comb picks up neutral paper

(*a*) Schematic model of interaction between comb and paper

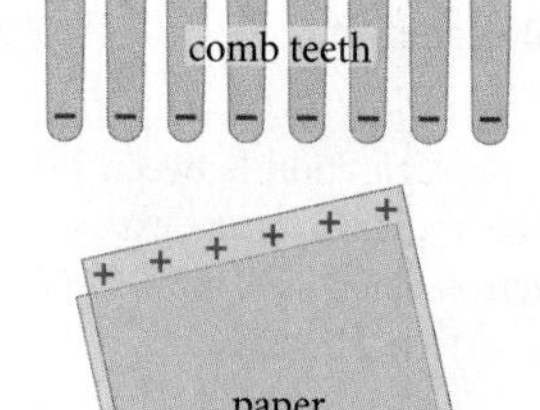

(*b*) Free-body diagram for paper

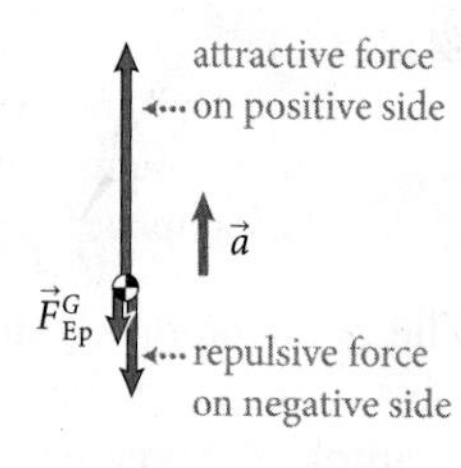

22.15 (*a*) When a positively charged object is brought near a neutral piece of paper, is the vector sum of the forces exerted by the charged object on the paper attractive or repulsive? (*b*) Describe what would happen when a negatively charged comb is brought near an electroscope if protons, not electrons, were mobile in metals. (*c*) Can you deduce from the experiment illustrated in Figure 22.19 which charge carriers—electrons or protons—are mobile in metals?

Figure 22.22 illustrates how you can exploit polarization to charge neutral conducting objects. A negatively charged rod brought near a metal sphere induces polarization in the sphere. To get as far away as possible from the negative charge on the rod, the electrons in the metal of the sphere move to the right surface. Thus the surface of the sphere nearer the rod becomes the positive pole, and the surface farther from the rod becomes the negative pole. When you touch the metal, the electrons can move into your body, thereby getting even farther away from the negative charge on the rod. In essence, you have become the negative pole, while the sphere is the positive pole. If you remove your hand from the sphere, the sphere is left with a deficit of electrons (they stayed inside you!) and so carries a positive charge—it has become charged without ever touching a charged object. (Likewise, you now have a surplus of negative charge and carry a negative charge, which is spread so thin that it is hardly noticeable.) This process is called **charging by induction.**

Figure 22.22 Polarization can be exploited to charge neutral conducting objects.

1. Charged rod induces polarization in metal sphere.

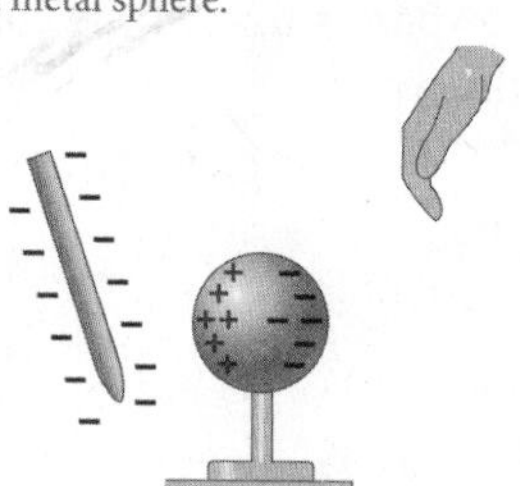

2. When you touch sphere, negative charge gets farther from rod by spreading onto you.

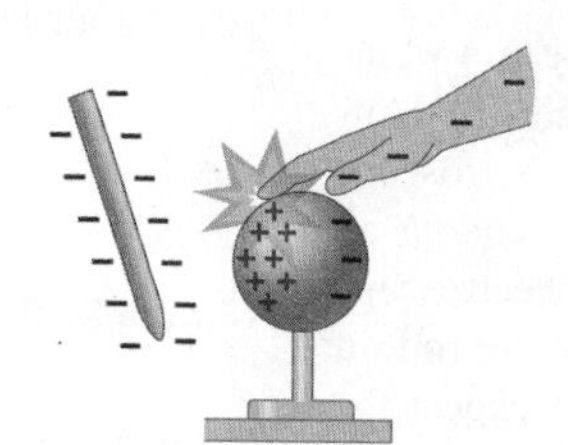

3. When you let go, you retain surplus of one type of charge and sphere retains surplus of opposite type of charge.

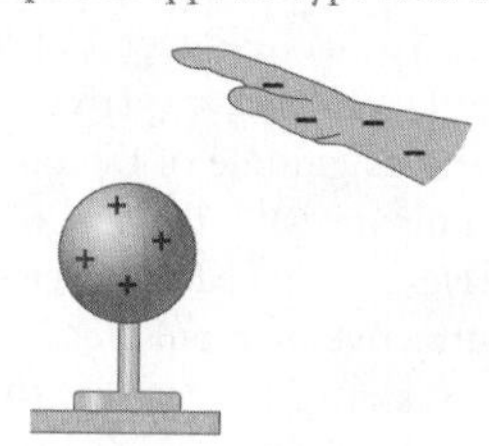

Self-quiz

1. You can use a positively charged object to charge a neutral object (*i*) by conduction or (*ii*) by induction. For each process, which type of charge (positive, negative, or neither) does the neutral object acquire?
2. Because we observe two types of electric interactions, attractive and repulsive, we postulate two types of charge. Do you think there are also two types of mass? Why do you think this? Do you think there are two types of magnetic pole? Why do you think this?
3. Is the statement *A plastic comb that has been passed through the hair a few times carries a negative charge* a physical law or a definition? What are some of the differences between a law and a definition?
4. A balloon rubbed in your hair or on your clothes sticks to a wall. If you place the rubbed side of the balloon against the wall, it sticks to the wall immediately. Try this, however: After rubbing the balloon in your hair, place the side of the balloon you did not rub against the wall and notice how the balloon turns until the rubbed side is touching the wall (*a*) Draw a free-body diagram for the balloon sticking to the wall. (*b*) Given that the balloon rotates so that the rubbed area is against the wall, do you think the balloon is an electrical insulator or conductor? (*c*) Was charge created in either the balloon or the wall in order for the sticking to occur? (*d*) Is any charge transferred from the balloon to the wall? Why or why not?
5. Air can act as both an insulator and a conductor. Consider reaching for a metal doorknob after scuffing your feet over a carpet. As your hand approaches the knob, a spark jumps between your hand and the knob. Explain how air acts as both insulator and conductor in this situation.

Answers

1. (*i*) Positive. When the two objects are brought into contact with each other (a necessary condition for conduction), surplus positive charge moves from the charged object to the neutral one. Thus the neutral object acquires the same type of charge as the charged object. (*ii*) Negative. Because the objects don't touch during charging by induction, charge carriers of the same type as the charged object escape from the neutral object during grounding.
2. Mass is the quantity responsible for gravitational interactions. Because all gravitational interactions are attractive, we can assume there is only one type of mass. Magnetic poles are responsible for magnetic interactions, which can be attractive or repulsive. Therefore there must be two types of magnetic pole (called north and south).
3. Definition. A law arises from observable phenomena and is found to be true in all cases that have been tested or observed (see Section 1.1). A definition cannot be tested. There is no test that allows us to observe that the charge on a comb passed through hair is negative. All we can do is show that the charged comb behaves in a fashion similar to other objects whose charge we call negative.
4. (*a*) See **Figure 22.23**. (*b*) Insulator. Because the only portion of the balloon that is attracted to the wall is the rubbed area, we know that the charge created by the rubbing does not spread out over the balloon surface. (*c*) No. Electrical charge can never be created. The surplus charge on the balloon was transferred from your hair or clothing. (*d*) No. If change were transferred, the balloon and the wall would repel each other.

Figure 22.23

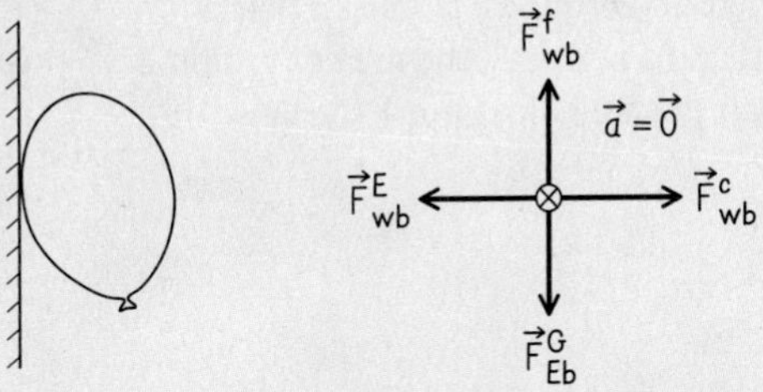

$\vec{F}^{f}_{wb}$, $\vec{a} = \vec{0}$, $\vec{F}^{E}_{wb}$, $\vec{F}^{c}_{wb}$, $\vec{F}^{G}_{Eb}$

5. Scuffing your feet on the carpet transfers charge from the carpet to you. Before you get near the knob, the air insulates your charged body from the knob. As you move nearer and nearer to the knob, the magnitude of the electric force between the charge carriers in your hand and those in the knob increases until the forces are so great that the air molecules are ionized, thereby producing a conducting pathway between your hand and the knob. Now the ionized air acts as a conductor for the jumping charge.

22.5 Coulomb's law

Quantitative experiments with electrical charge are difficult to carry out because objects lose their charge and because charge carriers on objects tend to rearrange themselves in the presence of other charged objects. In the 18th century, however, the English clergyman and scientist Joseph Priestley carried out a remarkable experiment: He charged a hollow sphere and showed that no electric force was exerted on a small piece of charged cork placed inside the sphere. Remembering that Newton had proven that no gravitational force exists inside a hollow sphere (see the box "Zero gravitational force inside a spherical shell" and **Figure 22.24**, below) because the gravitational force decreases with the square of distance, Priestley proposed that the electric force, too, decreases as $1/r^2$.

In 1785, Charles Coulomb, a French physicist, provided direct evidence for an inverse-square law by measuring how the electric force between two charged spheres changes as the distance between the spheres changes. The basic apparatus for Coulomb's experiment is shown in **Figure 22.25**. A small dumbbell is suspended from a long fiber. When spheres A and B are charged, the electric

Zero gravitational force inside a spherical shell

A consequence of the inverse-square dependence on distance of the gravitational force is that a uniform spherical shell exerts *no force at all* on a mass placed anywhere inside it. This result is very important in electrostatics because it provides a strong test of the law describing the electric force between charged objects.

Consider the uniform spherical shell in Figure 22.24*a*. A particle of mass m is placed off-center inside the shell. To determine the force exerted by the shell on the particle, consider first the force exerted by region 1, defined as a very small region of the shell surface. Let the mass of this region be m_1 and its distance from the particle be d_1, which means the magnitude of the gravitational force exerted by this region on the particle is Gm_1m/d_1^2. Extending the cone defined by the particle and region 1 to the opposite side of the shell gives us a small region 2, of mass m_2 and at distance d_2 from the particle. The magnitude of the gravitational force exerted by this region on the particle is Gm_2m/d_2^2. If the particle is closer to region 1 than to region 2, the area of region 2 must be greater than the area of region 1 (because of how we defined region 2 as being an extension of the cone formed by region 1 and the particle), which means region 2 contains more mass: $m_2 > m_1$.

Because the mass of the shell is distributed uniformly over the shell, the mass of each of our two regions is proportional to its area: $m_1/m_2 = a_1/a_2$. Because they are marked out by similar cones, the areas of the two regions are proportional to the squares of the distances to the particle: $a_1/a_2 = d_1^2/d_2^2$, which means that

$$\frac{m_1}{m_2} = \frac{d_1^2}{d_2^2}.$$

Rearranging terms, we get $m_1/d_1^2 = m_2/d_2^2$ *regardless* of the position of the particle, and so we see that the forces cancel: The two regions exert forces of equal magnitude in opposite directions on the particle (Figure 22.24*b*). We can now apply the same arguments to other pairs of small regions on either side of the particle, each yielding equal and opposite forces on the particle (Figure 22.24*c*). The vector sum of all the forces exerted on the particle by all the small regions making up the spherical shell is thus zero.

Figure 22.24

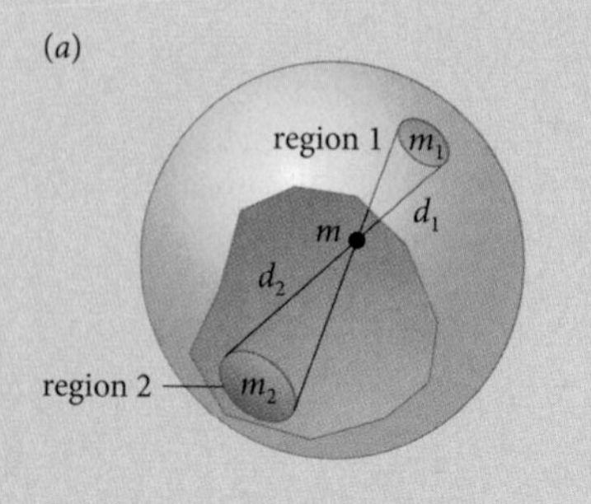

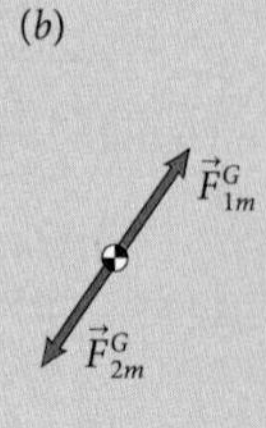

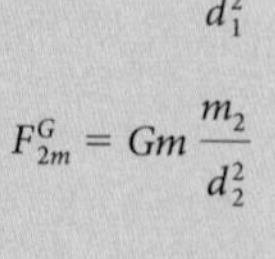

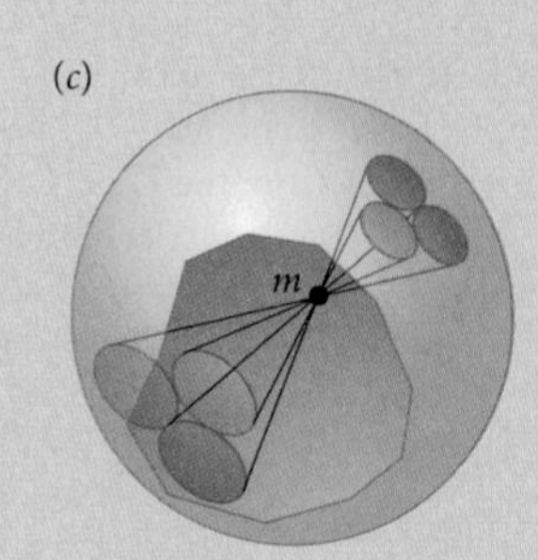

force between them twists the fiber. The amount of twist is a measure of the magnitude of the force between the two spheres (see also Section 15.7). A similar arrangement was used a few years later by Cavendish to study gravitational interactions (see Section 13.5).

Coulomb also devised a method for systematically varying the "quantity of charge" q on a metal sphere. He found that when a charged metal sphere is brought into contact with an identical uncharged metal sphere, the final charge is the same on each sphere—both exert a force of equal magnitude on a third charged object. In other words, each sphere gets half the original charge. By sharing charge among several identical metal spheres, Coulomb could produce spheres whose charge was one-half, one-quarter, one-eighth, and so on of the original charge (**Figure 22.26**).

By thus varying the charges on spheres A and B of his apparatus, Coulomb found that the **electric force** is proportional to the charge on each sphere. We can summarize these findings in one equation, called **Coulomb's law,** which gives the magnitude of the electric force exerted by two charged particles separated by a distance r_{12} and carrying charges q_1 and q_2:

$$F_{12}^{E} = k\frac{|q_1|\,|q_2|}{r_{12}^{2}}. \tag{22.1}$$

As we shall see in Chapter 27, the interaction between charged particles becomes more complicated when the particles are not at rest. For this reason the force in Coulomb's law is sometimes called the *electrostatic* force and the branch of physics that deals with stationary distributions of charge is called *electrostatics*.

If the positions of the two charged particles are given by the vectors $\vec{r}_1$ and $\vec{r}_2$, respectively, then the distance between them is $r_{12} = |\vec{r}_2 - \vec{r}_1|$. The value of the constant of proportionality k depends on the units used for charge, force, and length. The absolute-value signs around the charges in Eq. 22.1 are necessary because q_1 and q_2 can be negative but the *magnitude* F_{12}^{E} of the electric force must always be positive.

Coulomb's law bears a striking resemblance to Newton's law of gravity (Eq. 13.1):

$$F_{12}^{G} = G\frac{m_1 m_2}{r_{12}^{2}}. \tag{22.2}$$

Why these two laws have the same mathematical form remains a mystery. The main differences between the two are that mass is always positive but electrical charge can be positive or negative, which means that the gravitational force is always attractive but the electric force can be attractive or repulsive.

The derived SI unit of charge, called the **coulomb** (C), is defined as the quantity of electrical charge transported in 1 s by a current of 1 ampere, a quantity and unit we shall define in Chapter 27. One coulomb is equal to the magnitude of the charge on about 6.24×10^{18} electrons. Conversely, the magnitude of the charge of the electron is

$$e = 1/6.24 \times 10^{18}\text{ C} = 1.60 \times 10^{-19}\text{ C}. \tag{22.3}$$

The charge on any object comes in only whole-number multiples of this elementary charge:

$$q = ne, \quad n = 0, \pm 1, \pm 2, \pm 3, \ldots. \tag{22.6}$$

This means that an object can have charge $q = 0$, $q = +7e$, $q = -4e$, and so forth, but not, for instance, $+1.2e$. Because the elementary charge is very small,

Figure 22.25 Schematic diagram of Coulomb's apparatus for measuring the electric force between two charged spheres.

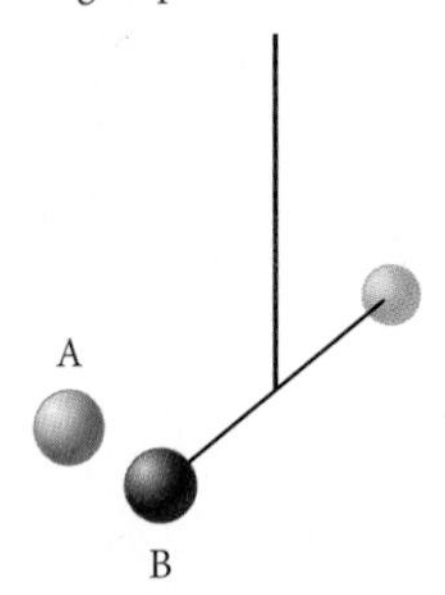

Figure 22.26 By successively allowing a charged sphere to touch an initially uncharged neighbor, we can distribute an amount of charge in ever-lessening amounts over a number of spheres.

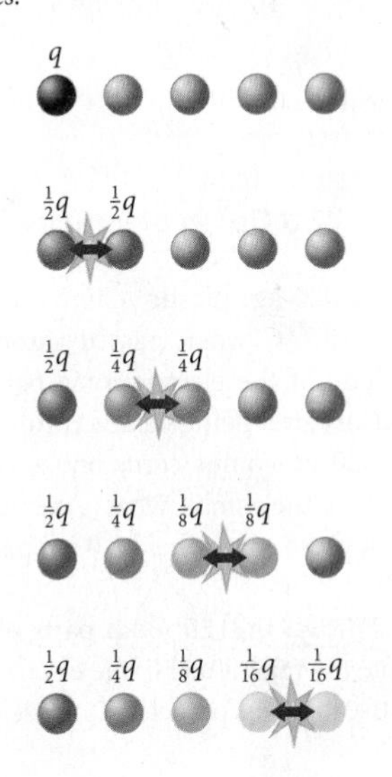

the fact that charge exists only as whole-number multiples of the elementary charge isn't noticeable under ordinary circumstances. For example, running a comb through your hair easily gives the comb a surplus of about 10^{12} electrons, and the quantity of electrons flowing through a 100-W light bulb each second is about 10^{19}. These are such large numbers that the fact that charge comes in only whole-number multiples of the elementary charge normally remains unnoticed.

Using the coulomb as the unit of charge, we can determine the value of k in Eq. 22.1 experimentally by measuring the force between two known charged particles separated by a known distance:

$$k = 9.0 \times 10^9 \text{ N} \cdot \text{m}^2/\text{C}^2. \tag{22.5}$$

The value of this constant shows how large a unit the coulomb is: Two particles, each carrying a charge of 1 C, separated by 1 m exert on each other a force of 9 billion newtons—equal to the gravitational force exerted by Earth on several dozen loaded supertankers! It is very difficult to build up a charge of this magnitude on all but very large objects because things get ripped apart by the enormous forces. The largest accumulations of charge we know of occur in the atmosphere: Large clouds that accumulate a charge of about 50 C discharge through the air to Earth, causing lightning.

Example 22.2 Gravity versus electricity

Compare the magnitudes of the gravitational and electric forces exerted by the nucleus of a hydrogen atom—a single proton ($m_p = 1.7 \times 10^{-27}$ kg)—on an electron ($m_e = 9.1 \times 10^{-31}$ kg) when the two are 0.50×10^{-10} m apart.

❶ **GETTING STARTED** For simplicity, I assume I can treat the proton and electron as particles. I also assume they are at rest so that I can use the principles of electrostatics.

❷ **DEVISE PLAN** I can use Eq. 22.2 to calculate the magnitude of the gravitational force and Eq. 22.1 to calculate the magnitude of the electric force.

❸ **EXECUTE PLAN**

$$F^G_{pe} = G\frac{m_p m_e}{r^2_{pe}}$$

$$= (6.7 \times 10^{-11} \text{ N} \cdot \text{m}^2/\text{kg}^2)\frac{(9.1 \times 10^{-31} \text{ kg})(1.7 \times 10^{-27} \text{ kg})}{(0.50 \times 10^{-10} \text{ m})^2}$$

$$= 4.1 \times 10^{-47} \text{ N}$$

and

$$F^E_{pe} = k\frac{|q_p|\,|q_e|}{r^2_{pe}}$$

$$= (9.0 \times 10^9 \text{ N} \cdot \text{m}^2/\text{C}^2)\frac{(1.6 \times 10^{-19} \text{ C})(1.6 \times 10^{-19} \text{ C})}{(0.50 \times 10^{-10} \text{ m})^2}$$

$$= 9.2 \times 10^{-8} \text{ N}.$$

The electric force exerted by the proton on the electron is $(9.2 \times 10^{-8} \text{ N})/(4.1 \times 10^{-47} \text{ N}) \approx 10^{39}$ times greater than the gravitational force exerted by the proton on the electron. ✔

❹ **EVALUATE RESULT** The difference in magnitudes is in agreement with the information given in Table 7.1.

Example 22.3 Comb electricity

(*a*) A 0.020-kg plastic comb acquires a charge of about -1.0×10^{-8} C when passed through your hair. What is the magnitude of the electric force between two such combs held 1.0 m apart after being passed through your hair? (*b*) If two identical 0.020-kg combs carry one surplus electron for every 10^{11} electrons in the combs, what is the magnitude of the electric force between these combs held 1.0 m apart?

❶ **GETTING STARTED** Both parts of the problem require me to calculate the magnitude of the electric force between the combs. If I treat the combs as particles, I can use Eq. 22.1 to calculate this force.

❷ **DEVISE PLAN** To calculate the magnitude of the electric force between two charged objects, I need to know the charge on each object and their separation distance. I know these data for part *a*: $q1 = q2 = -1.0 \times 10^{-8}$ C and $r_{12} = 1.0$ m, where the subscripts 1 and 2 denote the two combs. For part *b* I am given only the separation distance, and so I need to determine the charge on each comb.

I am given the fraction of electrons added, and I know the charge on one electron. So, to determine the charge on each comb, I need to determine how many electrons each comb contains. The number of electrons in each comb is equal to the

QUANTITATIVE TOOLS

number of protons in the comb: $N_e = N_p$. I am given the mass of the comb and I know that the mass is determined by the protons and neutrons in all the atoms making up the comb (the electrons contribute very little). Given that the protons and neutrons have almost identical mass ($m_p = m_n = 1.7 \times 10^{-27}$ kg), I can determine the number N of protons and neutrons by dividing the mass of the comb by m_p. Given that most atoms contain roughly equal numbers of protons and neutrons, I can say that the number of protons is $N_p \approx N/2$.

❸ **EXECUTE PLAN** (*a*) Substituting the values given into Eq. 22.1, I get

$$F^E_{12} = k\frac{|q_1|\,|q_2|}{r^2_{12}}$$

$$= (9.0 \times 10^9\ \text{N}\cdot\text{m}^2/\ \text{C}^2)\frac{(1.0 \times 10^{-8}\ \text{C})(1.0 \times 10^{-8}\ \text{C})}{(1.0\ \text{m})^2}$$

$$= 9.0 \times 10^{-7}\ \text{N.} ✔$$

(*b*) The number of protons plus neutrons in the comb is

$$N = \frac{0.020\ \text{kg}}{1.7 \times 10^{-27}\ \text{kg}} = 1.2 \times 10^{25},$$

and so $N_p \approx N/2 = 6 \times 10^{24}$. The number of electrons is equal to the number of protons, and so there are 6×10^{24} electrons in each comb to begin with. Adding one surplus electron for every 10^{11} electrons means adding $(6 \times 10^{24})/(1 \times 10^{11}) = 6 \times 10^{13}$ electrons to each comb; these electrons carry a combined charge of $(6 \times 10^{13})(-1.6 \times 10^{-19}\ \text{C}) = -9.6 \times 10^{-6}$ C. The magnitude of the repulsive electric force between the combs is then

$$F^E_{12} = k\frac{|q_1|\,|q_2|}{r^2_{12}}$$

$$= (9.0 \times 10^9\ \text{N}\cdot\text{m}^2/\text{C}^2)\frac{(9.6 \times 10^{-6}\ \text{C})(9.6 \times 10^{-6}\ \text{C})}{(1.0\ \text{m})^2}$$

$$\approx 1\ \text{N.} ✔$$

❹ **EVALUATE RESULT** My answer to part *a* is a force too small to be felt, which is what I expect based on experience (two combs passed through hair don't exert an appreciable force on each other). In contrast, my answer to part *b* is phenomenally large for an electric force. The magnitude of the initial acceleration acquired by the combs would be $a_1 = F^E_{21}/m_1 = (1\ \text{N})/(0.020\ \text{kg}) = 50\ \text{m/s}^2$, or about five times the acceleration due to gravity! Even though the fraction of electrons removed—one in 100 billion—is very small, the factor k in Eq. 22.1 is so great that the resulting force is also great. Indeed, I learned in Table 7.1 that the electromagnetic interaction is 36 orders of magnitude stronger than the gravitational interaction, so my answer is not unreasonable.

22.16 Two identical conducting spheres, one carrying charge $+q$ and the other carrying charge $+3q$, are initially held a distance d apart. The spheres are allowed to touch briefly and then returned to separation distance d. Is the magnitude of the force they exert on each other after the touching greater than, smaller than, or the same as the magnitude of the force they exerted on each other before the touching?

Like the gravitational force, the electric force is *central;* that is, its line of action is along the line connecting the two interacting charged particles. Consider, for example, the two particles carrying charges q_1 and q_2 shown in **Figure 22.27a**. The vector $\vec{r}_{12} \equiv \vec{r}_2 - \vec{r}_1$ gives the position of particle 2 relative to particle 1; this vector points from particle 1 to particle 2. We can define a unit vector pointing in this direction by dividing the vector $\vec{r}_{12}$ by its magnitude:

$$\hat{r}_{12} \equiv \frac{\vec{r}_2 - \vec{r}_1}{r_{12}}. \qquad (22.6)$$

Depending on the algebraic sign of the charges, the electric force can be attractive or repulsive. For like charges ($q_1q_2 > 0$), the force is repulsive. In this case, the force $\vec{F}^E_{12}$ exerted by particle 1 on particle 2 points in the same direction as the unit vector $\hat{r}_{12}$ (Figure 22.27*b*). For opposite charges ($q_1q_2 < 0$), the force is attractive, and so $\vec{F}^E_{12}$ points in the direction opposite the direction of $\hat{r}_{12}$ (Figure 22.27*c*). In either case, $\vec{F}^E_{12}$ can be written in the form

$$\vec{F}^E_{12} = k\frac{q_1q_2}{r^2_{12}}\hat{r}_{12}. \qquad (22.7)$$

Because $r_{12} = r_{21}$ and because $\hat{r}_{21}$ points in the direction opposite the direction of $\hat{r}_{12}$, the force $\vec{F}^E_{21}$ exerted *by* particle 2 *on* particle 1, which points in the direction

Figure 22.27 (*a*) Position vectors for two charged particles. (*b*) Repulsive forces exerted on each other by two particles carrying like charges. (*c*) Attractive forces exerted on each other by two particles carrying opposite charges.

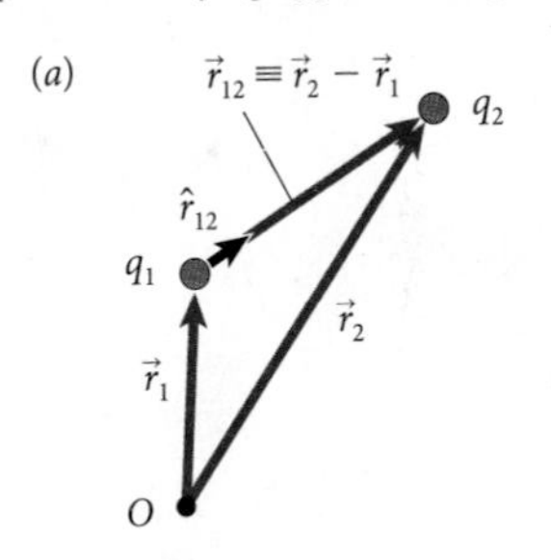

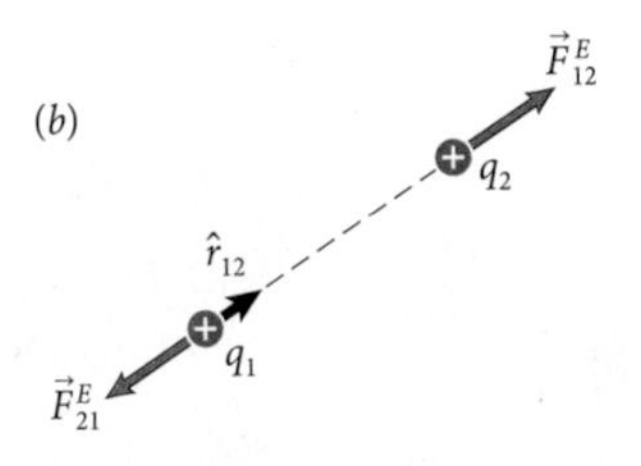

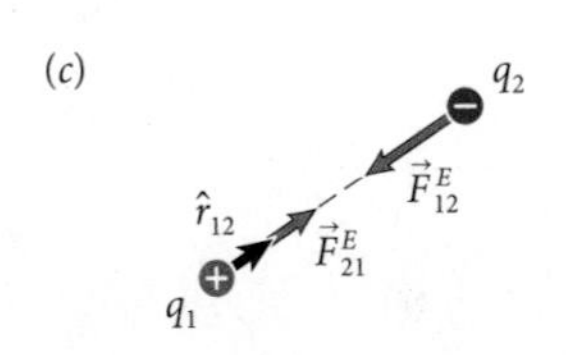

QUANTITATIVE TOOLS

Figure 22.28 The reason Coulomb's law does not apply in a strict sense to macroscopic charged objects. The law is approximately correct if the objects are far apart relative to their radii.

(*a*) Charged spheres separated by a distance large compared to the sphere radii

Distance between centers of charge distributions same as distance between sphere centers

(*b*) Charged spheres separated by a distance small compared to the sphere radii

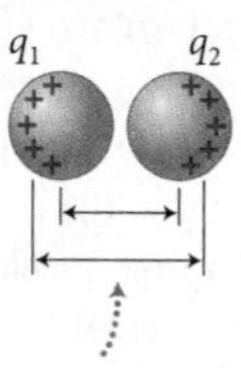

Charge repulsion causes distance between centers of charge distributions to differ from distance between sphere centers.

opposite the direction of $\vec{F}^E_{12}$, is obtained by simply switching the indices 1 and 2 in Eq. 22.7:

$$\vec{F}^E_{21} = k\frac{q_2q_1}{r^2_{21}}\hat{r}_{21} = k\frac{q_1q_2}{r^2_{12}}(-\hat{r}_{12}) = -\vec{F}^E_{12}, \quad (22.8)$$

as we would expect for an interaction pair (see Eq. 8.15). Equation 22.7 is the vectorial form of Eq. 22.1.

22.17 Using your knowledge about work and potential energy, determine whether the potential energy of a closed system of two charged particles carrying like charge increases, decreases, or stays the same when the distance between the two is increased. Repeat for two particles carrying opposite charge.

Before going on, I should mention a limitation to Coulomb's law. Strictly speaking, it is applicable only to charged particles. This is so because the distance r_{12} is well defined only when the size of the charged objects is negligibly small compared with their separation distance. When the charged objects are not particles, the distance r_{12} is not equal to the center-to-center distance. You can see why with the help of Figure 22.28. In **Figure 22.28a**, the charge is distributed uniformly over the surface of each of two widely separated metal spheres. The way in which a collection of charge carriers is spread out over a macroscopic object is called a **charge distribution.** Because the charge distributions over the metal spheres in Figure 22.28*a* are uniform, the center of each charge distribution coincides with the center of the sphere, and so r_{12} is well defined. When we bring the spheres close together, as in Figure 22.28*b*, the like charge carriers repel one another and move to the far side of each sphere. Now the centers of the charge distributions no longer coincide with the spheres' centers, so r_{12} (the center-to-center distance of the two charge distributions) is not simply the distance separating the centers of the two conductors.

22.18 (*a*) Is the magnitude of the electric force between the two conducting spheres in Figure 22.28*b* greater or smaller than that obtained from Coulomb's law, which assumes the charge is concentrated at the center of each sphere? (*b*) Is the answer to part *a* the same if the charge on one of the conductors is negative instead of positive?

22.6 Forces exerted by distributions of charge carriers

Coulomb's law deals only with *pairs* of charged objects. To calculate the force exerted by an assembly of objects carrying charges $q_2, q_3, q_4, \ldots$ on an object 1 carrying a charge q_1, we take the vector sum of all the forces exerted on object 1 by each of the other charged objects independently:

$$\Sigma\vec{F}^E_1 = \vec{F}^E_{21} + \vec{F}^E_{31} + \vec{F}^E_{41} + \cdots, \quad (22.9)$$

where each term is given by Coulomb's law:

$$\Sigma\vec{F}^E_1 = k\frac{q_2q_1}{r^2_{21}}\hat{r}_{21} + k\frac{q_3q_1}{r^2_{31}}\hat{r}_{31} + k\frac{q_4q_1}{r^2_{41}}\hat{r}_{41} + \cdots. \quad (22.10)$$

In other words, we calculate the force exerted by object 2 on object 1, then calculate the force exerted by object 3 on object 1, and so forth, and then add the forces. This means that if we know the details of some distribution of charged objects, we can calculate the force exerted by this distribution of charged objects on a single charged particle. For distributions that contain large numbers of charged objects, the summation can be accomplished via an integration. We will limit our discussion here to simple cases involving only a few charged objects.

22.19 Figure 22.29 shows how a charged particle 1 interacts with two other charged particles 2 and 3. Determine the direction of the vector sum of the electric forces exerted on particle 2.

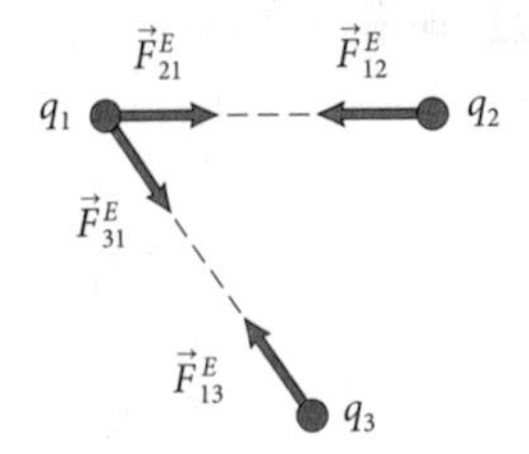

Figure 22.29 Forces exerted by two charged particles 2 and 3 on charged particle 1.

The basic limitation of Coulomb's law continues to apply when we are analyzing collections of charged objects: Eq. 22.10 is valid only for charged *particles*, not for charged extended bodies. Suppose, for example, that we replace each charged particle in Figure 22.29 by a conducting sphere carrying the same charge as each particle. Let us first consider just the interaction between spheres 1 and 2. When these oppositely charged spheres are placed near each other, the charge carriers on the two spheres rearrange themselves to be as close as possible to each other (**Figure 22.30*a***). A similar type of rearrangement takes place when just spheres 1 and 3 are placed near each other (Figure 22.30*b*). When all three spheres are placed near one another, the positive charge on sphere 3 pushes the positive charge on sphere 2 up and pulls the negative charge on sphere 1 down (Figure 22.30*c*). Consequently, the forces that the spheres exert on one another are not the same as the forces exerted by the individual pairs (compare the forces in Figures 22.29 and 22.30).

Figure 22.30

(*a*) Sphere 1 interacts with just sphere 2

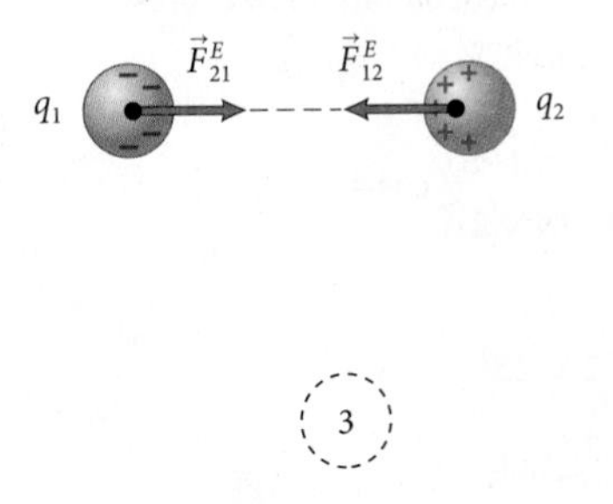

(*b*) Sphere 1 interacts with just sphere 3

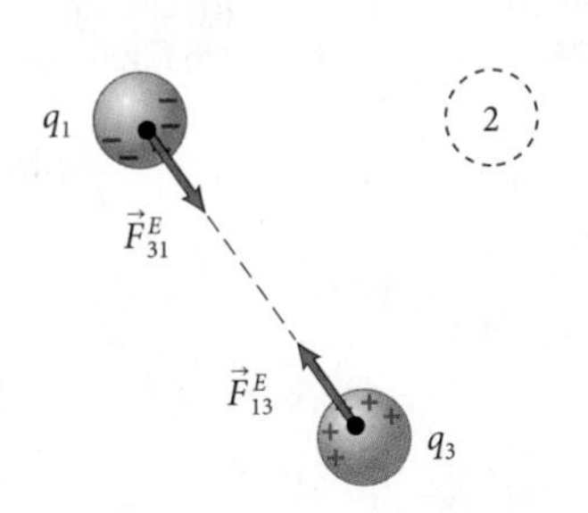

(*c*) Sphere 1 interacts with both spheres

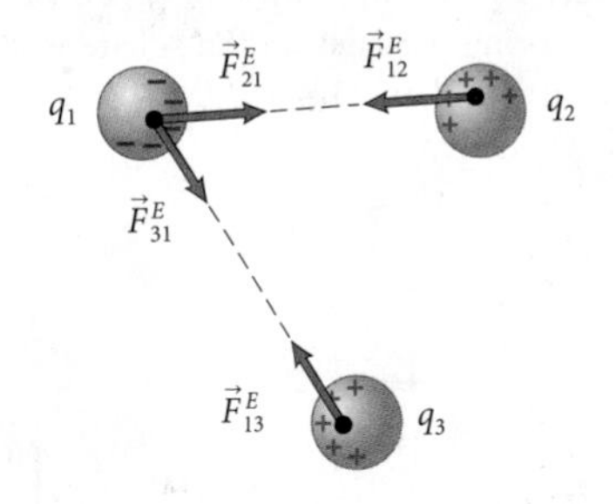

Example 22.4 Electric tug of war

You are given three charged particles. Particles 1 and 2 carry charge $+q$ and particle 3 carries charge $-4q$. (*a*) Determine the relative values of the separation distances r_{12} and r_{13} when the three particles are arranged along a straight line in such a way that the vector sum of the forces exerted on particle 1 is zero. (*b*) With the particles arranged this way, are the vector sums of the forces exerted on particles 2 and 3 also zero?

❶ GETTING STARTED Particle 2 exerts a repulsive force on particle 1; particle 3 exerts an attractive force on particle 1. In order for the two forces exerted on particle 1 to cancel, particles 2 and 3 must be on the same side of particle 1. Because the magnitude of the charge on particle 2 is smaller than that of the charge on particle 3, the distance r_{12} must be shorter than the distance r_{13}.

❷ DEVISE PLAN I choose my x axis vertically up along the line defined by the particles (**Figure 22.31*a***). To determine the relative values of the separation distances r_{12} and r_{13} when the vector sum of the forces exerted on particle 1 is zero, I draw a free-body diagram for particle 1 (Figure 22.31*b*). The magnitude of each force exerted on this particle is given by Eq. 22.1, and so by setting the sum of the x components of the forces equal to zero, I have an expression containing r_{12} and r_{13} and I can manipulate the expression to get the relative values of r_{12} and r_{13}.

Figure 22.31

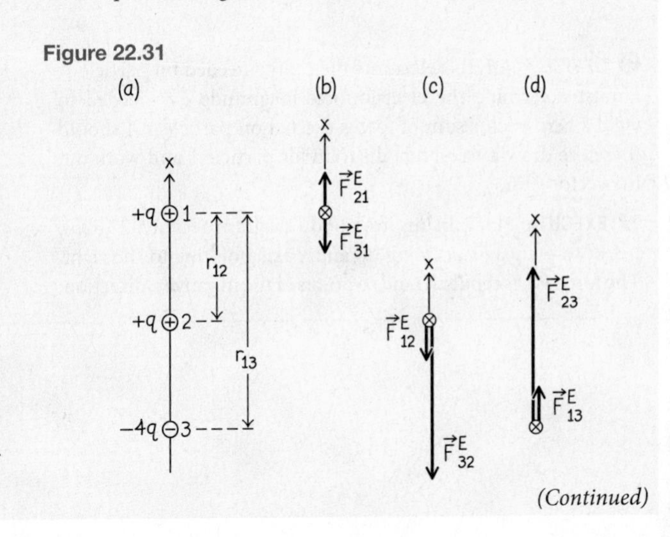

(Continued)

❸ **EXECUTE PLAN** (*a*) The x components of the two forces in Figure 22.31*b* must add to zero, so

$$\Sigma F_{1x} = F^E_{21x} + F^E_{31x}$$

$$= +k\frac{|q_1||q_2|}{r^2_{12}} - k\frac{|q_1||q_3|}{r^2_{13}} = k\frac{qq}{r^2_{12}} - k\frac{(q)(4q)}{r^2_{13}}$$

$$= +k\frac{q^2}{r^2_{12}} - k\frac{4q^2}{r^2_{13}} = 0.$$

I therefore must have $\dfrac{q^2}{r^2_{12}} = \dfrac{4q^2}{r^2_{13}}$,

or $r^2_{13} = 4r^2_{12}$, and so $r_{13} = 2r_{12}$. ✔

(*b*) The forces exerted by particles 1 and 3 on particle 2 both point in the negative x direction (Figure 22.31*c*) and so cannot sum to zero. The forces exerted by particles 1 and 2 on particle 3 both point in the positive x direction (Figure 22.31*d*) and so cannot sum to zero. ✔

❹ **EVALUATE RESULT** My answer for part *a* makes sense because the force exerted by each particle is inversely proportional to the square of the separation distance, and this force varies directly with the quantity of charge. Because the charge on particle 3 is four times greater than that on particle 2, the square of the separation distance r_{13} must be four times r_{12}, and so $r_{13} = 2r_{12}$, as I found.

Example 22.5 Electrostatic equilibrium

Consider four charged particles placed at the corners of a square whose sides have length d. Particles 1, 2, and 4 carry identical positive charges. In order for the vector sum of the forces exerted on particle 1 to be zero, what charge must be given to particle 3, which is in the corner diametrically opposite particle 1?

❶ **GETTING STARTED** I begin by making a sketch of the situation (**Figure 22.32*a***). Because particles 1, 2, and 4 all carry identical positive charge, I write $q_1 = q_2 = q_4 = +q$. The separation between neighboring particles on the square is d. The separation between particles 1 and 3 is $\sqrt{2}d$.

Figure 22.32

(a)

$q_1 = +q$ $q_2 = +q$ $\sqrt{2}d$ d $q_4 = +q$ $q_3 = ?$

(b)

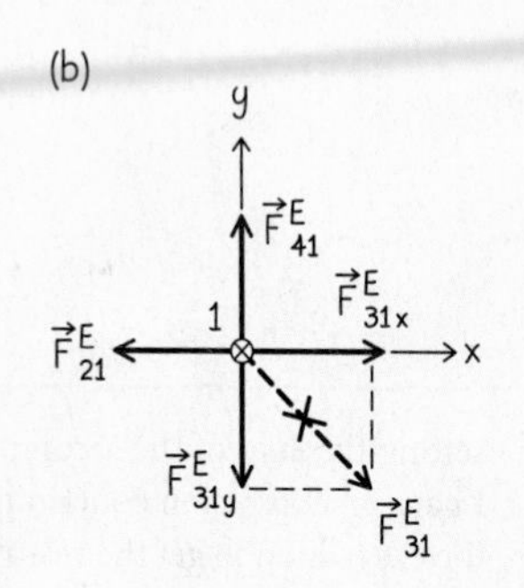

❷ **DEVISE PLAN** To determine the charge needed on particle 3, I must determine the electric force magnitude F^E_{31} needed to yield a zero vector sum of forces exerted on particle 1. I should therefore draw a free-body diagram for particle 1 and work out the vector sum.

❸ **EXECUTE PLAN** In my free-body diagram (Figure 22.32*b*), I choose my y axis pointing up and x axis pointing to the right. The force $\vec{F}^E_{21}$ is repulsive and so points in the negative x direction; the force $\vec{F}^E_{41}$ is repulsive and points in the positive y direction. To make the vector sum of the forces exerted on 1 zero, $\vec{F}^E_{31}$ must be such that its components cancel $\vec{F}^E_{21}$ and $\vec{F}^E_{41}$. Along the x axis, I therefore have

$$\Sigma F_{1x} = F^E_{21x} + F^E_{31x} + F^E_{41x} = -F^E_{21} + F^E_{31}\cos 45° + 0 = 0,$$

so $F^E_{21} = F^E_{31}\cos 45°$. Substituting the Coulomb's law expressions for these two force magnitudes, I have

$$k\frac{|q_2||q_1|}{d^2} = k\frac{|q_3||q_1|}{(d\sqrt{2})^2}\cos 45°$$

$$\frac{q^2}{d^2} = \frac{|q_3|q}{2\sqrt{2}d^2}$$

$$|q_3| = 2\sqrt{2}\,q.$$

From Figure 22.32*b*, I also know that $\vec{F}^E_{31}$ must be an attractive force, so

$$q_3 = -2\sqrt{2}q. ✔$$

❹ **EVALUATE RESULT** My result indicates that the charge on particle 3 is greater than q. This makes sense because the attractive force exerted by this particle must balance the forces exerted by particles 2 and 4, which are both closer to particle 1. To obtain my answer I solved only for the x component of $\vec{F}^E_{31}$ and did not consider the y component. However, my free-body diagram shows that the magnitudes of the components along the y axis are the same as those along the x axis, so analyzing the y components would have given me the same result.

Example 22.6 Electric trajectory

Consider the arrangement of charged particles shown in **Figure 22.33**. The charge magnitudes are the same in all three cases, but q_1 and q_2 are positive and q_3 is negative. Sketch the trajectory of particle 1 if it is released while particles 2 and 3 are held fixed. Ignore any gravitational force exerted by Earth on the particle.

Figure 22.33 Example 22.6

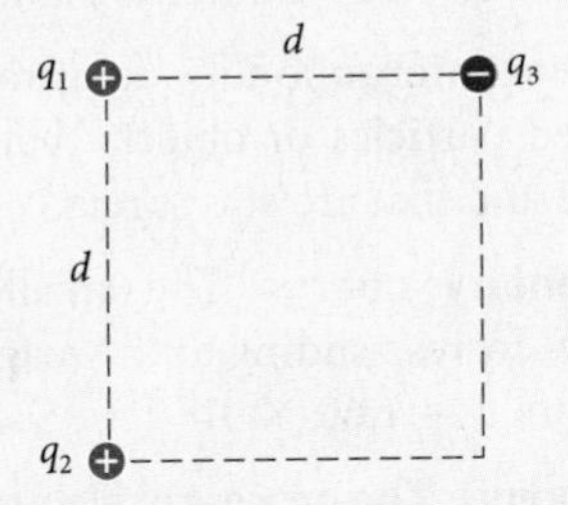

❶ **GETTING STARTED** I begin by drawing a free-body diagram for particle 1, choosing the x axis to the right and the y axis up (**Figure 22.34*a***). Because the charges on the two particles have the same magnitude and because the separation distances are the same, the magnitudes F^E_{21} and F^E_{31} are the same.

Figure 22.34

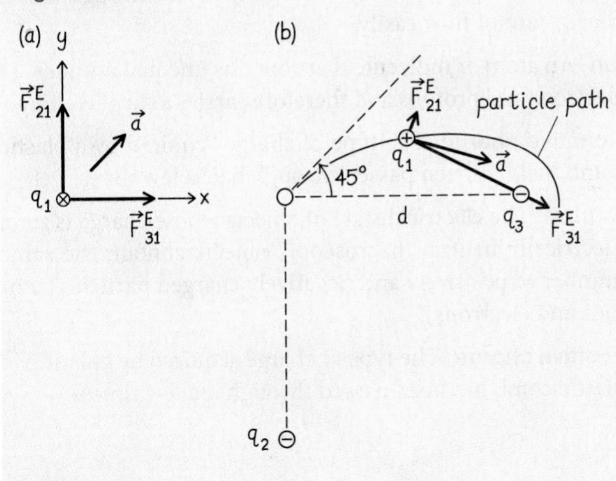

❷ **DEVISE PLAN** The direction in which particle 1 accelerates is determined by the vector sum of the forces exerted on it, and I can determine this direction from my free-body diagram. As the particle moves along its trajectory, however, the forces exerted on it change in both direction and magnitude, and so I must consider these changes in formulating my answer.

❸ **EXECUTE PLAN** Because the magnitudes F^E_{21} and F^E_{31} are the same, the vector sum of the two forces exerted on particle 1 bisects the angle between the two forces, and so the initial acceleration of 1 (which points in the same direction as the vector sum of forces) points up and to the right at an angle of 45°, as illustrated in Figure 22.34*a*. As particle 1 moves in the direction indicated in Figure 22.34*a*, $\vec{F}^E_{21}$ and $\vec{F}^E_{31}$ change in both direction and magnitude. The magnitude of $\vec{F}^E_{21}$ decreases because the distance between 1 and 2 increases, and the magnitude of $\vec{F}^E_{31}$ increases because the distance between 1 and 3 decreases. The direction of the acceleration of particle 1 is the same as the direction of the vector sum of these two forces. The resultant motion is qualitatively illustrated in Figure 22.34*b*. ✔

❹ **EVALUATE RESULT** My sketch makes sense: Particle 1 first moves up and to the right because it is repelled by particle 2 and attracted by particle 3. As it moves away from 2 and approaches 3, the effect of the attraction increases and so the trajectory curves and heads toward 3.

22.20 Seven small metal spheres are arranged in a hexagonal pattern as illustrated in **Figure 22.35**. Spheres 1 and 7 carry equal amounts of positive charge; the other spheres are uncharged. (*a*) To give sphere 7 an acceleration $\vec{a}$ that points to the right, what (single) other sphere must be charged? There may be more than one possibility. (*b*) What are the sign and magnitude of that charge?

Figure 22.35 Checkpoint 22.20.

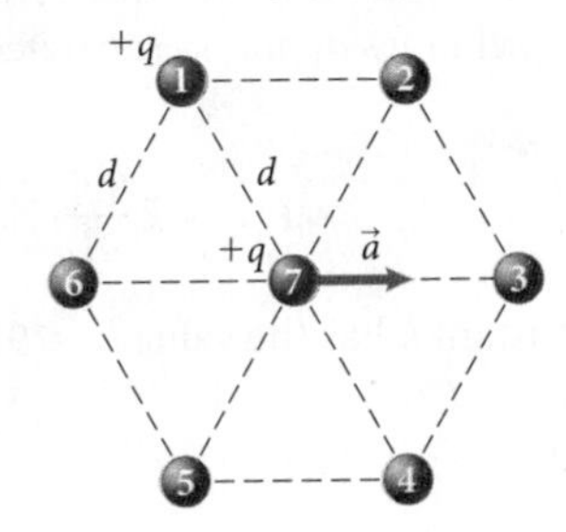

QUANTITATIVE TOOLS

Chapter Glossary

SI units of physical quantities are given in parentheses.

Charge (electrical) q (C) A scalar that represents the attribute responsible for electromagnetic interactions, including electric interactions. There are two types of charge: *positive* ($q > 0$) and *negative* ($q < 0$). Two objects that carry the same type of charge exert repulsive forces on each other; objects that carry different types of charge exert attractive forces on each other.

Charge carrier Any microscopic object that carries an electrical charge.

Charge distribution The way in which a collection of charge carriers is distributed in space.

Charge polarization A spatial separation of the positive and negative charge carriers in an object. The polarization of neutral objects induced by the presence of external charged objects is responsible for the electric interaction between charged and neutral objects.

Charging by induction A method of charging a neutral object using a charged object, with no physical contact between them.

Conduction The flow of charge carriers through a material.

Conductor (electrical) Any material or object through which charge carriers can flow easily.

Conservation of charge The principle that the charge of a closed system cannot change. Thus charge can be transferred from one object to another and can be created or destroyed only in identical positive-negative pairs.

Coulomb (C) The derived SI unit of charge equal to the magnitude of the charge on about 6.24×10^{18} electrons. (The coulomb is defined as the quantity of electrical charge transported in 1 s by a current of 1 ampere, a unit we shall define in Chapter 27).

Coulomb's law The force law that gives the direction and magnitude of the electric force between two particles at rest carrying charges q_1 and q_2 separated by a distance r_{12}:

$$\vec{F}^E_{12} = k\frac{q_1 q_2}{r^2_{12}}\hat{r}_{12}. \tag{22.7}$$

The constant k has the value $k = 9.0 \times 10^9\ \text{N} \cdot \text{m}^2/\text{C}^2$.

Electric force $\vec{F}^E$ (N) The force that charge carriers (and macroscopic objects that carry a surplus electrical charge) exert on each other. The magnitude and direction of this force are given by Coulomb's law.

Electric interaction A long-range interaction between charged particles or objects that carry a surplus electrical charge and that are at rest relative to the observer.

Elementary charge The smallest observed quantity of charge, corresponding to the magnitude of the charge of the electron: $e = 1.60 \times 10^{-19}$ C. See also *Coulomb*.

Grounding The process of electrically connecting an object to Earth ("ground"). Grounding permits the exchange of charge carriers with Earth, a huge reservoir of charge carriers. A charged, conducting object that is grounded will retain no surplus of either type of charge, assuming no other nearby electrical influences.

Insulator (electrical) Any material or object through which charge cannot flow easily.

Ion An atom or molecule that contains unequal numbers of electrons and protons and therefore carries a surplus charge.

Negative charge The type of charge acquired by a plastic comb that has been passed through hair a few times.

Neutral The electrical state of objects whose charge is zero. Electrically neutral macroscopic objects contain the same number of positively and negatively charged particles (protons and electrons).

Positive charge The type of charge acquired by hair after a plastic comb has been passed through it a few times.

23 The Electric Field

In this chapter we revisit an issue we discussed briefly in Chapter 7: the long-range nature of electric and gravitational interactions. How does one charged object "reach out" and affect another charged object? What are the invisible "springs" that pull us—and everything around us—toward Earth's surface? You can describe these long-range interactions by saying that every charged object and every object that has mass has a "sphere of influence" surrounding it. The modern word for this sphere of influence is *field*. Fields are not imaginary. That sensation you felt when, while reading the beginning of Chapter 22, you held a piece of plastic food wrap near your face was the sensation of a field created by the charged particles in the wrap. The closer the wrap is to your skin, the stronger the sensation.

The concept of field is important for two reasons. First, it is impossible to describe the interaction between moving charged particles without it. Second, as you will soon see, it is often easier to deal with fields than with distributions of charge because frequently more is known about fields than about the way charge is distributed.

23.1 The field model

Newton's law of gravity and Coulomb's law describing the electric force between charged particles successfully account for the magnitudes of the gravitational and electric forces between stationary objects. However, they do not address the fundamental puzzle of how objects separated in space can interact without any mediator of the interaction (such an interaction is called *action at a distance*). Worse, they share a fundamental flaw: Both imply that the action of one object on another is instantaneous everywhere throughout space. Consider, for example, the two metal rods in **Figure 23.1**. Even if both are electrically neutral, their electrons interact. Suppose you quickly drive the electrons in rod A down to the bottom, as in Figure 23.1*b*. According to Coulomb's law, doing this will instantly change the force exerted by the electrons in A on those in B, regardless of how far apart the rods are. This means that it would be possible—in principle—to be standing at one position in space and instantly detect a change that occurs at some far distant position.

Figure 23.1 Newton's and Coulomb's laws imply that forces are exerted instantaneously across a distance—but experiments show that they do not.

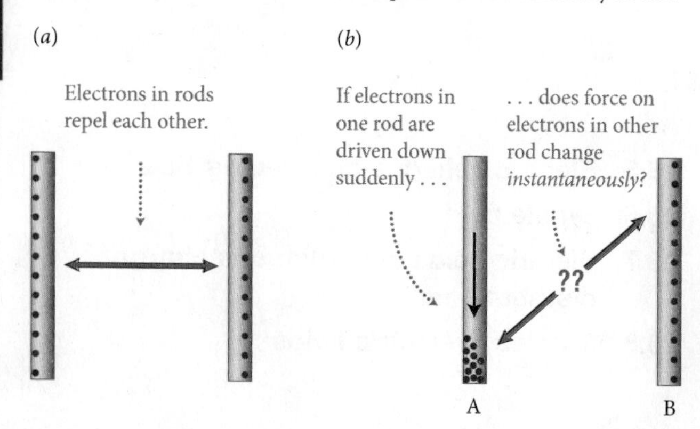
(a)
Electrons in rods repel each other.
(b)
If electrons in one rod are driven down suddenly . . .
. . . does force on electrons in other rod change instantaneously?
??
A
B

The idea that an object can directly and instantly influence another object regardless of their separation was troubling in the 19th century but became untenable in the early 20th century when it was demonstrated experimentally that the interaction between charged objects is not instantaneous. The principle illustrated in Figure 23.1, for example, is what makes possible the transmission of radio signals from a transmitting antenna (rod A) to a receiving antenna (rod B). We know from many experiments that such transmission is not instantaneous. As just one example, a radio signal takes about 0.1 s to travel from Earth's surface to an orbiting communications satellite. In other words, the picture conveyed by Newton's law of gravity and Coulomb's law—that one object directly and instantly affects other objects regardless of the distance between them—cannot be correct.

Instead we must adopt another model of long-range interactions, a model in which interactions take place through the intermediary of an **interaction field** (or simply a **field**). In the field model, an interacting object fills the space around itself with a field. When an object A is placed in the field of an object B, A can "feel" the presence of B's field. Instead of the two objects interacting directly as illustrated in **Figure 23.2*a***, it is the field created by each object that acts on the other object (Figure 23.2*b*). The stronger the field, the greater the magnitude of the force resulting from the interaction.

In Figure 23.1, the electrons in rod B feel the field set up by the electrons in rod A. When the electrons in A accelerate, their motion causes a disturbance in A's field, and this disturbance propagates outward through space like the ripples on the surface of a pond. Only when these ripples in the field reach rod B can the motion of A's electrons be detected by those in rod B. In Chapter 30 we shall study the propagation of disturbances in fields due to accelerating charge carriers. For now, we shall concentrate on fields created by stationary objects.

The field model applies equally well to gravitational and electric interactions, with each interaction having its own type of field. The space around any object that has mass is filled with a *gravitational field*, and the space around any electrically charged object is filled with an *electric field*. Gravitational fields exert forces on objects that have mass, and electric fields exert forces on objects that either carry a charge or can be polarized. Let's begin by developing the concept of gravitational field.

Before we attempt to obtain a physical quantity we can use to describe any gravitational field, we should note a number of things. First, for any object A located in a gravitational field created by an object S (S is called the *source* of the field), the magnitude of the field felt by A depends only on the properties of S and on the position of A relative to S; the field magnitude does not depend in any way on the properties of A. Second, the field of an object is always there, even when the object is not interacting with anything else. A field therefore must be represented by a set of numerical

Figure 23.2 The field model for interaction at a distance.

(*a*) Model of direct interaction at a distance

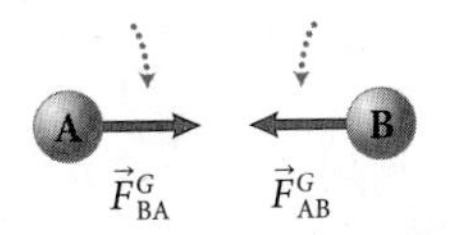

(*b*) Field model of interaction at a distance

field of A

force due to field of A at location of B

A

B

$\vec{F}^G_{AB}$

AND

force due to field of B at location of A

field of B

A

B

$\vec{F}^G_{BA}$

Fields of A and B shown separately for clarity; both are present at same time.

values that cover all the space outside the field source, with every point of this space having a different numerical value. One field representation you are already familiar with is that of a "temperature field," where the temperature across the surface of a region has a specific value at each location (**Figure 23.3**). Third, for stationary objects the field model must give the same forces as Newton's law of gravity and Coulomb's law. In particular, the field model must still yield forces between the two objects that are equal in magnitude and opposite in direction. It is not immediately obvious that the field model preserves this symmetry in the forces because a field is not something shared by two interacting objects—each object has its own field.

For example, in the gravitational interaction between a ball and Earth, the gravitational force exerted by Earth on the ball is due to Earth's gravitational field and the gravitational force exerted by the ball on Earth is due to the ball's field. These two gravitational fields are very different from each other: Earth's field pulls strongly on, say, a paper clip, while the effect of the ball's field on that paper clip is unmeasurably small. As you will see shortly, however, the symmetry of the interaction is preserved despite the asymmetry in the fields (see Checkpoint 23.3).

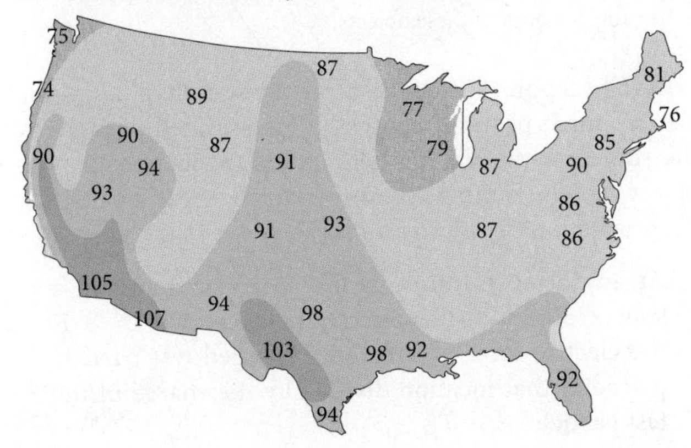

Figure 23.3 The temperature across a region is specified by a set of values, with a specific temperature value for every position in that region. Such a set of values is called a *field*.

What physical quantity can we use to describe the gravitational field of an object? How about the gravitational force exerted by the object? Let's examine this possibility using Earth as our object. As we saw in Section 8.8, the magnitude of the gravitational force exerted by Earth on an object of mass m near Earth's surface is $F^G_{Eo} = mg$. This force is not a good quantity for describing Earth's field because the force depends not only on the source—Earth (which determines g)—but also on the mass m of the object placed in the field. As illustrated in **Figure 23.4**, two objects that have different masses m_1 and m_2 but are placed at the same height above Earth's surface are subject to different gravitational forces m_1g and m_2g. The quantity $g = F^G_E/m$, however—the gravitational force per unit of mass—is the same for any object.* This quantity is determined solely by the properties of Earth and is independent of those of any object that experiences a gravitational force exerted by Earth.

23.1 Two objects 1 and 2, of mass m_1 and m_2, are released from rest far from Earth, at a location where the magnitude of the acceleration due to gravity is much less than $g = 9.8 \text{ m/s}^2$. (*a*) What is the ratio F^G_{E1}/F^G_{E2}? (*b*) For these two objects, is the magnitude of the gravitational force exerted by Earth per unit mass independent of the properties of the objects?

Figure 23.4 Comparison between gravitational force and gravitational acceleration on objects of different mass at the same distance from Earth.

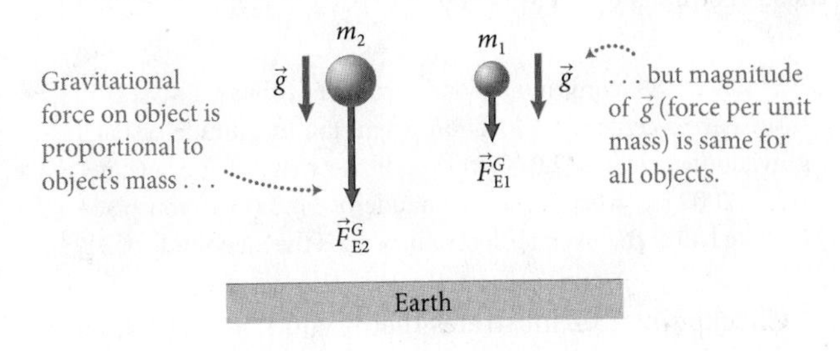

*This is so because mass and inertia are equivalent (see Section 13.1), and so the force per unit of mass is equal to force divided by inertia, which is acceleration.

CONCEPTS

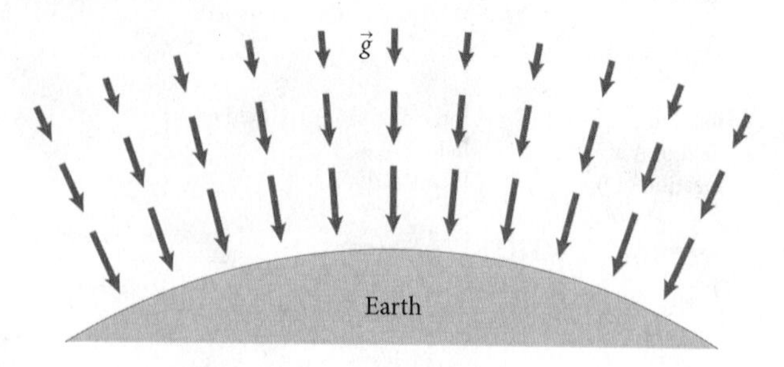

Figure 23.5 Vector field diagram for the gravitational field in a region near Earth.

We can use the gravitational force per unit mass exerted by an object as a measure of the magnitude of the object's gravitational field. For example, near Earth's surface, because $g = 9.8\ \text{m/s}^2$, the magnitude of Earth's gravitational field is 9.8 N/kg; near the surface of the Moon, where $g_{\text{moon}} = 1.6\ \text{m/s}^2$, the magnitude of the Moon's gravitational field is 1.6 N/kg. (Remember that $1\ \text{N/kg} = 1\ (\text{kg} \cdot \text{m/s}^2)/\text{kg} = 1\ \text{m/s}^2$.)

At any given location in the space surrounding a source object S, the magnitude of the gravitational field created by S is the magnitude of the gravitational force exerted on an object B placed at that location divided by the mass of B.

Unlike the temperature field in Figure 23.3, which is a *scalar field*, the gravitational field is a *vector field:* At every position, it has both a magnitude and a direction. Figure 23.5, for example, shows a **vector field diagram** representing the gravitational field near Earth. You can determine the magnitude and direction of this field in the space surrounding Earth using a **test particle** (an idealized particle whose mass is small enough that its presence does not perturb the object whose gravitational field we are measuring). Measure, at each location, the gravitational force exerted by Earth on the test particle, and then divide that force by the mass of the test particle to obtain the direction and the magnitude of the gravitational field at that location. As you can see from Figure 23.5, Earth's gravitational field, which can be represented at each position by a vector $\vec{g}$, always points toward the center of Earth, and its magnitude decreases with increasing distance away from Earth. Near Earth's surface, the magnitude of these vectors is $g = 9.8\ \text{N/kg}$.

23.2 A communications satellite orbits 1.4×10^7 m from Earth's center, at a location where the magnitude of Earth's gravitational field is 2.0 N/kg. (*a*) If the mass of the satellite is $m_s = 2000$ kg, what is the magnitude of $\vec{F}^G_{\text{Es}}$? (*b*) If you place a 0.20-kg ball at the satellite's location, what is the magnitude of $\vec{F}^G_{\text{Eb}}$?

Checkpoint 23.2 illustrates that if you know the gravitational field at a certain position, you can easily calculate the gravitational force exerted by the source of that field on any object at that position by taking the product of the magnitude of the gravitational field at the location of the object and the mass of the object.

Figure 23.6 Electric force exerted on two objects of different inertia m and charge q by the electric fields created by two identical charged particles.

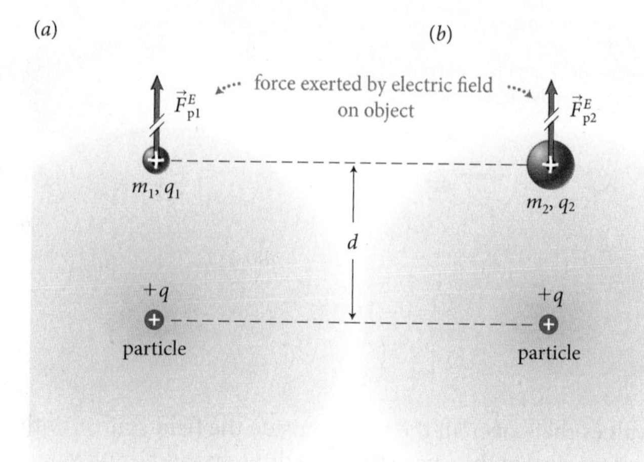

23.3 (*a*) Is the magnitude of the gravitational force exerted by Earth on a ball greater than, equal to, or smaller than the magnitude of the gravitational force exerted by the ball on Earth? (*b*) Is the magnitude of Earth's gravitational field at the position of the ball greater than, equal to, or smaller than the magnitude of the gravitational field of the ball at a distance equal to Earth's radius? (*c*) Explain how the answers to parts *a* and *b* can both be correct.

23.2 Electric field diagrams

Let us now apply the same ideas to electric interactions. Figure 23.6*a* shows object 1 of mass m_1 and charge q_1 a distance d from a particle that carries a charge $+q$. What is the electric field $\vec{E}$ created by the particle at the position of object 1? Before answering this question, answer the next checkpoint, which concerns the interactions of the particle with object 1 and with an object 2 of mass m_2 and charge q_2 (Figure 23.6*b*; $m_2 \neq m_1$ and $q_2 \neq q_1$).

23.4 (*a*) Are the electric forces $\vec{F}^E_{\text{p1}}$ and $\vec{F}^E_{\text{p2}}$ in Figure 23.6 equal? (*b*) What does the quantity $\vec{F}^E_{\text{p}i}/m_i$ represent? (*c*) Is this quantity the same for objects 1 and 2? If not, what quantity is the same for both of these objects?

As Checkpoint 23.4 makes clear, the quantity $\vec{F}^E_{\text{p}i}/q_i$—the electric force per unit charge—is determined entirely by the source of the electric field and is independent of the object on which the field exerts a force. So, in analogy to the gravitational field, we can say:

At any given location in the space surrounding a source object S, the electric field created by S is the electric force exerted on a charged test particle placed at that location divided by the charge of the test particle: $\vec{E}_{\text{S}} \equiv \vec{F}^E_{\text{St}}/q_{\text{t}}$.

Like gravitational fields, electric fields are vector fields. There is one difference between the two types of fields, however. Electric interactions can be either repulsive or attractive, and so the direction of the electric force—and hence the direction of the electric field—depends on the sign of the charge. Our rule is that:

The direction of the electric field at a given location is the same as the direction of the electric force exerted on a positively charged object at that location.

23.5 (*a*) If the particle in Figure 23.6 carries a negative charge $q < 0$ and q_1 and q_2 are positive, what are the directions of $\vec{F}^E_{p1}$ and $\vec{F}^E_{p2}$? (*b*) Does the electric field created by the particle point toward or away from the particle? (*c*) If q and q_2 are negative, what are the direction of $\vec{F}^E_{p2}$ and the direction of the electric field created by the particle at the location of object 2? (*d*) If q is positive, does the electric field created by the particle point toward or away from the particle? (*e*) How does the magnitude of the electric field created by a particle that carries a charge $+q$ ($q > 0$) compare with the magnitude of the electric field created by a particle that carries a charge $-q$ of identical magnitude at a distance d from each particle?

Figure 23.7 shows the vector field diagrams for the electric fields of particles that carry positive and negative charges. Because it is impossible to draw electric field vectors at *all* locations, the diagrams show vectors at only certain positions; from these representative vectors you can get an idea of how the electric field looks as a whole. In addition, the drawing is limited to two dimensions, but you should visualize the electric field as spreading out in all three dimensions.

Figure 23.7 Vector field diagrams for positively and negatively charged particles. The lengths of the vectors show that the electric field magnitude decreases with increasing distance from the source.

(*a*) Electric field of positively charged particle

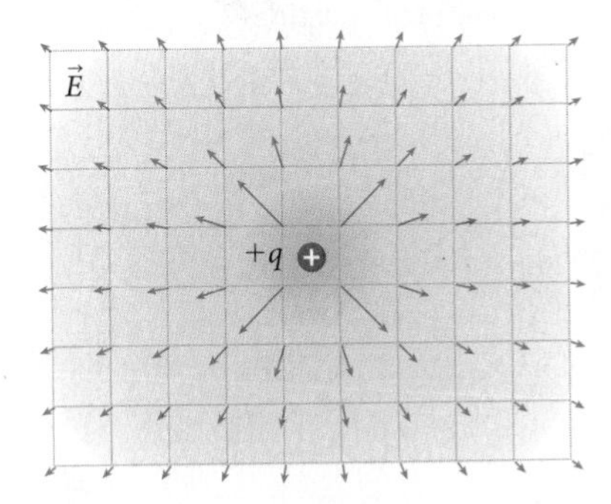
$\vec{E}$
$+q$
Electric field is directed away from positive source . . .

(*b*) Electric field of negatively charged particle

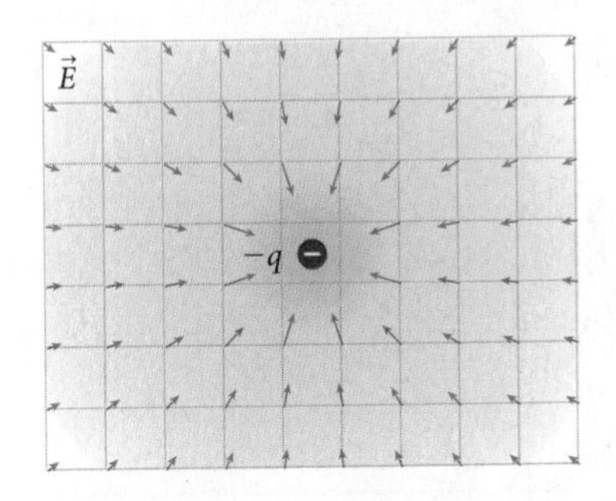
$\vec{E}$
$-q$
. . . and toward negative source.

Figure 23.8 Electric field pattern created by a small charged object in a solution that contains plastic fibers. The fibers align with the direction of the electric field created by the charged object.

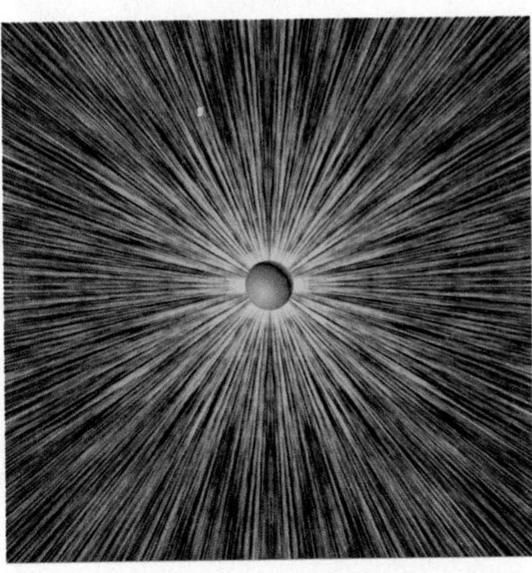

Electric fields can be made visible by putting charged objects in a (nonconducting) liquid that contains small uncharged plastic fibers or grass seed. Each fiber aligns itself in the direction of the electric field at the fiber's location (**Figure 23.8**).

23.6 If you know the electric field $\vec{E}$ at some location, how can you determine the magnitude and direction of the electric force exerted by that field on an object carrying a charge q and placed at that location?

23.3 Superposition of electric fields

The concept of electric field becomes especially useful when we consider the combined electric field that results from more than one charged object. Suppose we are interested in the electric field created by two particles that carry charges of equal magnitude but opposite sign. To determine the electric field created by the particles at a point P, we place a test particle* carrying a positive charge q_t at P and measure the vector sum of the forces exerted on it by the two charged source particles (**Figure 23.9** on the next page). The electric field at P is then equal to this vector sum divided by q_t.

Figure 23.9 illustrates the **superposition of electric fields:**

The combined electric field created by a collection of charged objects is equal to the vector sum of the electric fields created by the individual objects.

*When measuring electric fields, we assume that the charge q_t on the test particle is so small that the particle does not perturb the particles or objects that generate the electric field we are measuring.

Figure 23.9 The electric field due to multiple charged objects (here, a pair of charged particles) is the vector sum of the fields created by the individual objects.

(*a*)

To find electric field at P . . .

P

q_1

q_2

(*b*)

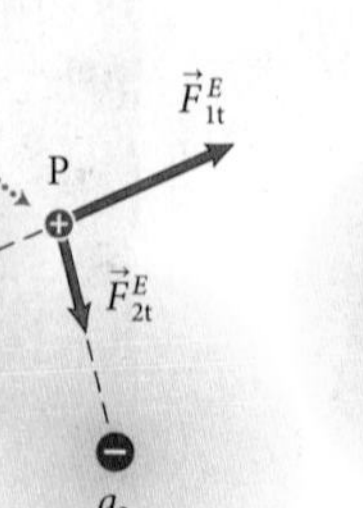

(*c*)

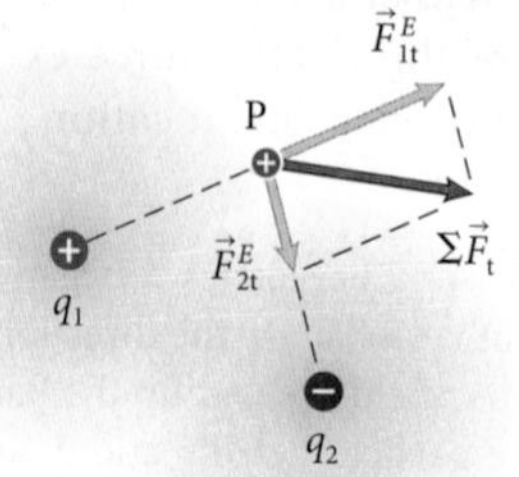

(*d*)

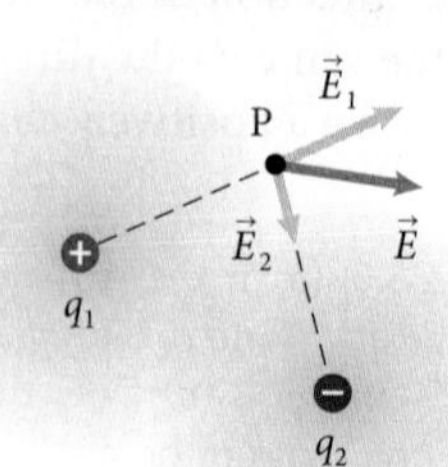

The superposition principle holds regardless of the number of sources. Because of the vectorial nature of the electric interaction, electric forces add vectorially (Eq. 22.9). Consequently the electric field at P is equal to $(\vec{F}^E_{1t} + \vec{F}^E_{2t})/q_t = \vec{F}^E_{1t}/q_t + \vec{F}^E_{2t}/q_t$, which is the vector sum of the electric fields created by the two sources individually.

The only caveat is the one I pointed out in Figure 22.30. When we deal with conductors, the distribution of charge on the individual conductors in isolation might be different from what it is when the conductors are placed close together.

Exercise 23.1 Electric field of two positively charged particles

Consider two identical particles 1 and 2 carrying charges $q_1 = q_2 > 0$ (**Figure 23.10**). What is the direction of the combined electric field at points P_1 through P_4?

Figure 23.10 Exercise 23.1.

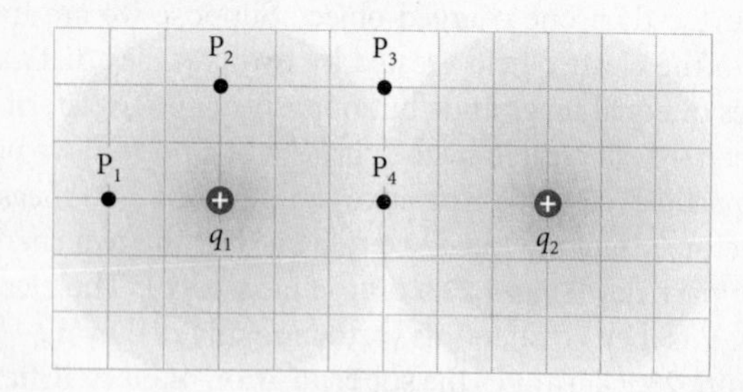

SOLUTION I place a positively charged test particle at each location and determine the vector sum of the (repulsive) forces exerted by 1 and 2 on each test particle. Because $\vec{E} = \Sigma\vec{F}/q_{test}$, the direction of $\vec{E}$ is the same as the direction of $\Sigma\vec{F}$ (**Figure 23.11**).

Figure 23.11

(a) Electric forces on test particles

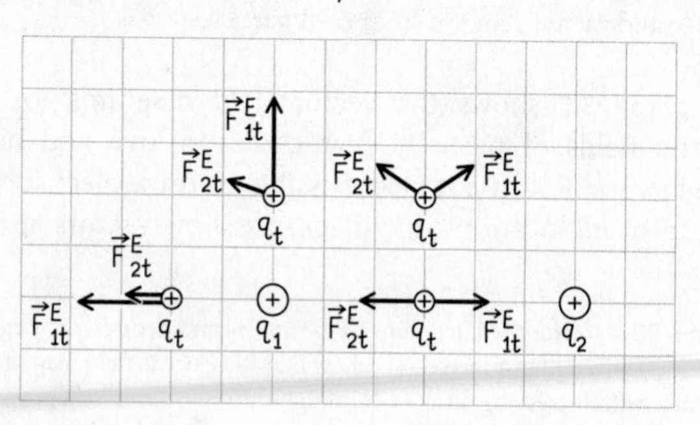

(b) Vector sum of forces on each test particle

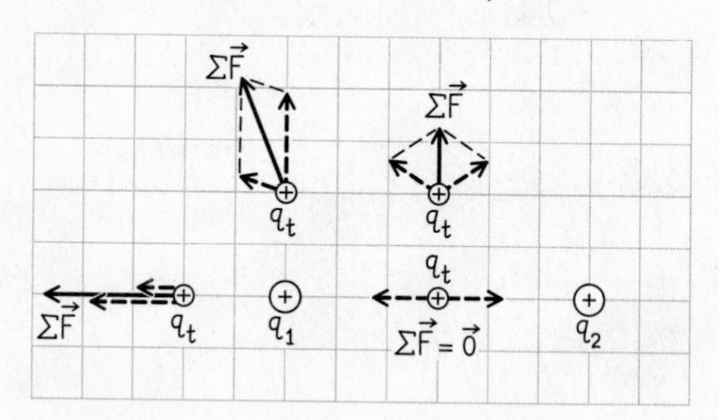

(c) Electric field at each tested point

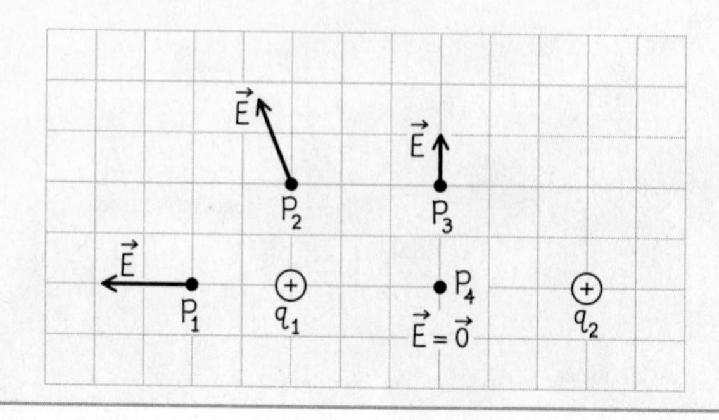

Figure 23.12 Vector field diagrams showing the superposition of the electric fields of the two charged particles of Figure 23.10.

(*a*) Electric field of particle 1

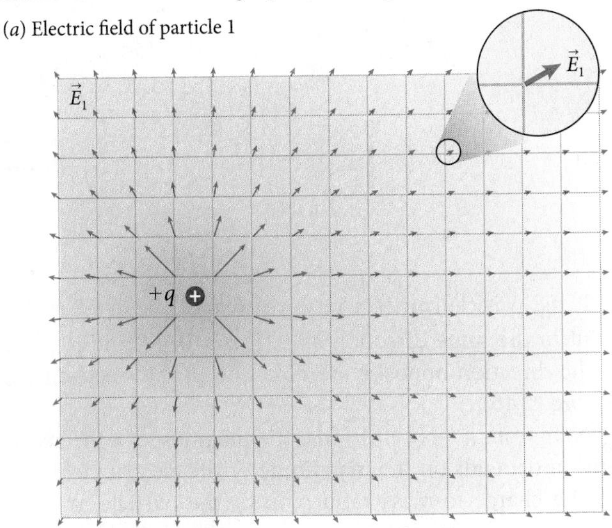

(*b*) Electric field of particle 2

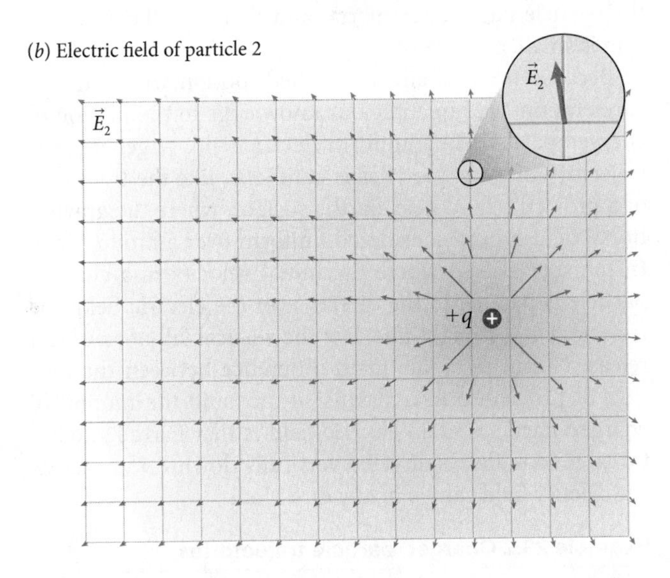

(*c*) Electric field of both particles

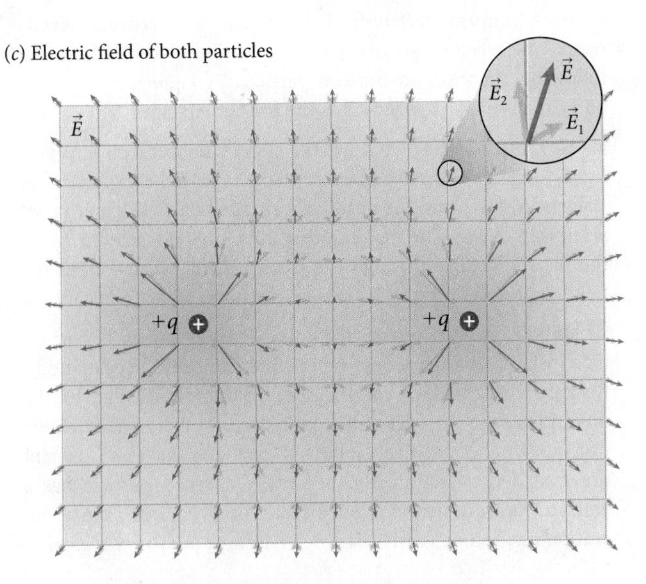

Figure 23.13 Pattern created by two identical charged particles in a liquid containing plastic fibers. Compare with the vector field diagram in Figure 23.12*c*.

23.7 (*a*) If the charge on particle 2 in Exercise 23.1 is doubled so that $q_2 = 2_{q1}$, what happens to the direction of the electric field at points P_1 through P_4? (*b*) If the charge on particle 2 is negative so that $q_2 = -q_1$, what is the direction of the electric field at points P_1 through P_4?

Figure 23.11 provides a limited view of the electric field created by the two particles. A more complete view is given in **Figure 23.12**. This diagram is obtained by vectorially adding, for each grid point, the electric field vectors for the individual particles. Note how the pattern of vectors resembles the pattern created by two identically charged particles in a solution of plastic fibers (**Figure 23.13**).

Using the superposition principle, we can determine the electric field produced by any system of charged particles. **Figure 23.14**, for example, shows a vector diagram for the electric field generated by three charged particles. Because every charged object is made up of charged particles—electrons and protons—we can determine the electric field of any object at any position in space. For a real object, the calculation might be very tedious or even intractable because of the large number of charged particles, but the basic principle is as given above.

Figure 23.14 Vector field diagram of the electric field created by three charged objects.

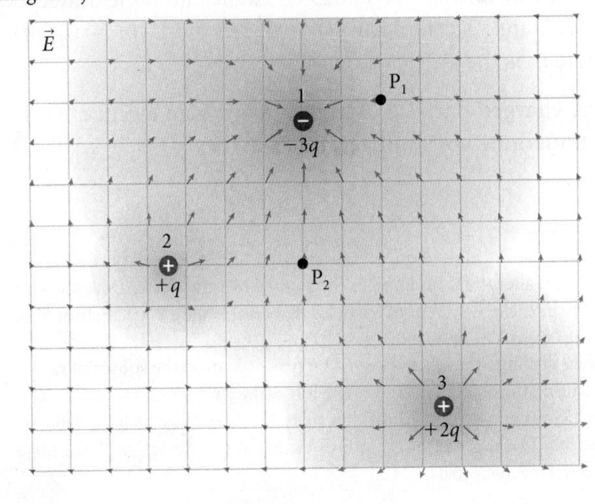

23.8 (*a*) In Figure 23.14, what is the direction of the force $\Sigma\vec{F}^E$ exerted on a particle carrying a charge $+q$ and placed at P_1? (*b*) How does $\Sigma\vec{F}^E$ change when the particle at P_1 carries a charge $+3q$? (*c*) Do the magnitude and direction of $\Sigma\vec{F}^E$ on a charged particle at P_1 change if the $+2q$ charge on object 3 is halved? (*d*) What is the direction of $\Sigma\vec{F}^E$ exerted on a particle carrying a charge $-q$ and placed at point P_2?

23.4 Electric fields and forces

Before developing additional techniques for determining the electric fields created by systems of charged particles, let us consider this question: What are the forces exerted by an electric field on charged or polarized objects? One of the advantages of working with electric fields is that, for any system of charged particles, once we know the electric field the system creates at some point P in space, we can determine the force exerted by the system on any other charged particle placed at P without worrying about any of the individual source particles in the system anymore.* In Chapter 22 we used the action-at-a-distance model to discuss the forces exerted by charged objects either on other charged objects or on polarized objects. Now we can use the field model to do the same thing. For stationary charged particles, both methods must yield the same result.

When we study the forces exerted by electric fields, it is useful to distinguish between uniform and nonuniform. In a *uniform electric field,* the direction and magnitude of the electric field are the same everywhere. No electric field is ever uniform throughout all space, but as we shall see in Section 23.7 it is possible to create regions of space where the electric field is uniform. In a *nonuniform electric field,* the direction and magnitude of the electric field vary from position to position. (All the electric fields we have considered so far are nonuniform.)

Let us first consider what happens to a charged particle placed in a uniform electric field. Because the electric field is defined as the electric force per unit of charge, the force $\vec{F}^E_p$ exerted by an electric field $\vec{E}$ on a particle carrying a charge q is $\vec{F}^E_p = q\vec{E}$.† Because $\vec{E}$ is the same everywhere, the force $\vec{F}^E_p$ exerted on the particle is constant and so it undergoes a constant acceleration $\vec{a} = \vec{F}^E_p/m = q\vec{E}/m = (q/m)\vec{E}$, where m is the particle's mass.

A charged particle placed in a uniform electric field undergoes constant acceleration.

*This procedure is valid only if the presence of the charged particle at P does not alter the way charge is distributed over the system. We shall refer to the particle or system of particles that creates the electric field at P as being "fixed."

†When dealing with forces exerted by fields, we drop the subscript representing the object that exerts the force (the "by" subscript) because the field is due to *all* objects other than the object on which the force is exerted. The superscript E reminds us that we are dealing with the force exerted by an electric field.

Figure 23.15 Forces exerted by a uniform electric field on a positively and a negatively charged particle.

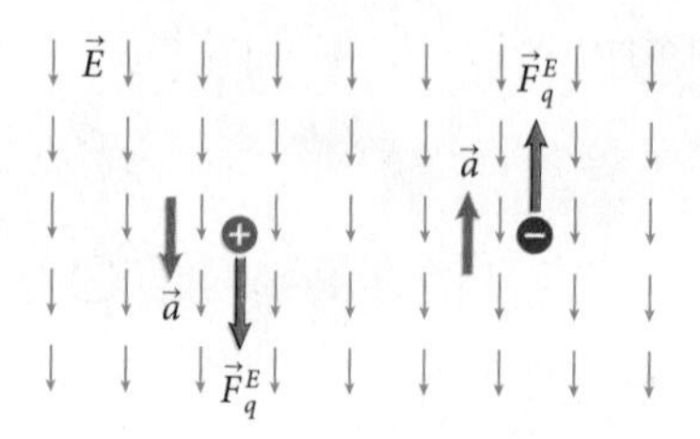

If the particle carries a positive charge, $q > 0$, $\vec{F}^E_p$ and $\vec{a}$ point in the same direction as $\vec{E}$. If $q < 0$, $\vec{F}^E_p$ and $\vec{a}$ point in the direction opposite the direction of the electric field (**Figure 23.15**).

Note from $\vec{a} = (q/m)\vec{E}$ that the magnitude of the acceleration depends on the magnitude of the electric field and on the charge-to-mass ratio q/m of the particle. A large charge q causes a greater force to be exerted on the particle and therefore a greater acceleration; a larger mass m means the particle has greater inertia and therefore the acceleration is smaller.

Because we have already studied motion with constant acceleration, we can apply our knowledge to the motion of charged particles in a uniform electric field. In general, the trajectory of these particles is parabolic, like the trajectory of a projectile fired near Earth's surface, where the gravitational field can be considered uniform over a limited area. In the special case where the initial velocity of a charged particle is parallel to the direction of the electric field, the trajectory is a straight line, like the vertical fall of an object released from rest. The main difference between the motion of projectiles near Earth's surface and the motion of charged particles in an electric field is that Earth's gravitational field is always directed vertically downward, whereas the electric field can be in any direction.

Example 23.2 Charged particle trajectories

Four charged particles are fired with a horizontal initial velocity $\vec{v}$ into a uniform electric field that is directed vertically downward. The effect of gravity is negligible. The particles have the following charges and masses: particle 1 $(+q, m)$; 2 $(+q, 2m)$; 3 $(+2q, 2m)$; 4 $(-q, m)$. Sketch the four trajectories.

❶ **GETTING STARTED** Because the electric field direction is vertically down, the three positively charged particles experience a downward force and the negatively charged particle experiences an upward force. The magnitude of this force doesn't change as the particles move through the electric field because both the field magnitude and the charges on the particles are constant.

❷ **DEVISE PLAN** Because the force exerted on each particle is constant, the particles experience constant accelerations. Because the direction of the force is perpendicular to the direction of the particles' initial motion, they all have a parabolic trajectory. The positively charged particles have a constant downward acceleration; the negatively charged particle has a constant upward acceleration. Because $\vec{a} = (q/m)\vec{E}$, the acceleration magnitude is greatest when q is large and/or m is small.

CONCEPTS

❸ **EXECUTE PLAN** I draw trajectories that curve down for 1, 2, and 3 and up for 4 (**Figure 23.16**). The magnitude of the electric force exerted on particle 2 is the same as that exerted on particle 1, but the acceleration of particle 2 is smaller because this particle has the greater mass. I indicate this difference in acceleration by making trajectory 1 more curved than trajectory 2. The magnitude of the electric force exerted on particle 3 is twice as great as that exerted on particle 1, but 3's mass is also twice as great, and so the two particles have the same charge-to-mass ratio and therefore the same acceleration and trajectory. The magnitude of the electric force exerted on particle 4 is the same as that exerted on particle 1 but points in the opposite direction, and so trajectories 1 and 4 are identical in shape but curve in opposite directions. ✔

Figure 23.16

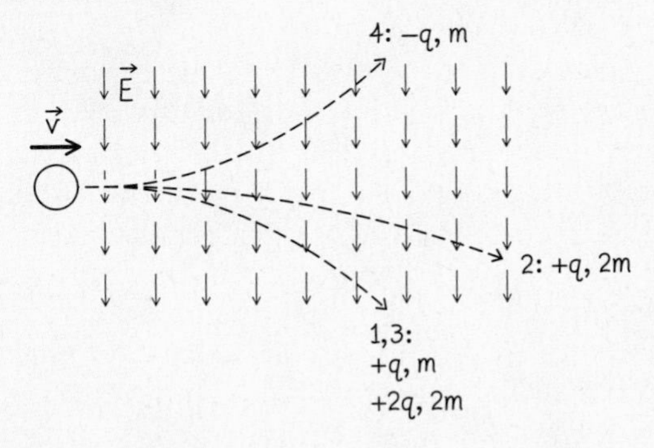

❹ **EVALUATE RESULT** My sketch indicates that particles with increasingly positive charge-to-mass ratios curve increasingly downward. Conversely, particles with increasingly negative charge-to-mass ratios curve increasingly upward. This is what I expect because a particle's deflection is a function of both its charge, which determines the magnitude of the force exerted by the electric field on it (greater charge, greater deflection), and its mass, which relates the particle's acceleration to the force exerted on it (greater mass, smaller deflection).

23.9 A water droplet carrying a positive charge is released from rest in a uniform horizontal electric field near Earth's surface. The horizontal electric force is comparable in magnitude to the gravitational force exerted by Earth. Describe the droplet's trajectory.

In a nonuniform electric field, the force exerted on a charged particle varies from one position to another, so we cannot easily specify the particle's trajectory without knowing more about the electric field. As in a uniform electric field, however:

> **A positively charged particle placed in a nonuniform electric field has an acceleration in the same direction as the electric field; a negatively charged particle placed in such a field has an acceleration in the opposite direction.**

Figure 23.17 Extended free-body diagram for a permanent dipole placed in a uniform electric field.

dipole in electric field — extended free-body diagram

$\vec{E}$ $\vec{F}^E_-$ $\vec{F}^E_+$ ϑ $\Sigma\vec{F}_d = \vec{0}$ $\vec{a}_{cm} = \vec{0}$

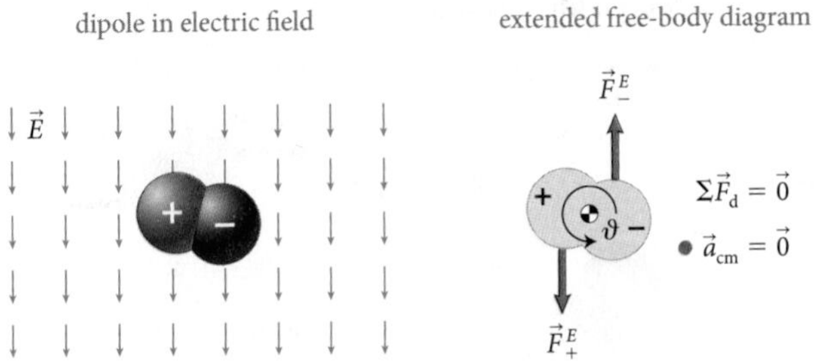

In Chapter 22 we found that charged objects can polarize electrically neutral objects by separating the centers of positive and negative charge in the latter. The resulting configuration of charge—equal amounts of positive and negative charge separated by a small distance—is called an **electric dipole** or simply **dipole.** Many molecules, such as water molecules, are *permanent dipoles;* that is to say, the centers of positive and negative charge are kept separated by some internal mechanism. **Figure 23.17** illustrates the forces exerted on a permanent electric dipole in a uniform electric field. Because the electric field is uniform and the magnitude of the charge on the positive end of the dipole is equal to the magnitude of the charge on the negative end, the forces exerted on the two ends are equal in magnitude but opposite in direction, making their vector sum zero. However, the forces exerted on the two ends cause a torque (see Chapter 12).

23.10 (*a*) What effect does the torque caused by the electric field have on the electric dipole in Figure 23.17? (*b*) Is the torque the same for every orientation of the molecule?

The orientation of an electric dipole can be characterized by a vector, the **dipole moment,** that, by definition, points from the center of negative charge to the center of positive charge, as shown in **Figure 23.18** on the next page. As Checkpoint 23.10 illustrates, the electric forces create a torque on the dipole that tends to align the dipole moment with the electric field.

In a nonuniform electric field, the situation is more complicated because the two ends of the dipole are now subject to forces that have different magnitudes as well as different directions. Consider, for example, the nonuniform electric field in Figure 23.18*a*, which is due to a positively charged particle to the left side of the figure. The magnitude of $\vec{F}^E_-$ is greater than the magnitude of $\vec{F}^E_+$ because the negative end of the dipole is closer to the positively charged particle. Thus the vector sum of the forces exerted on the

Figure 23.18 Extended free-body diagrams for permanent dipoles in nonuniform electric fields. The electric field shown is due to a positively charged particle to the left of the figure.

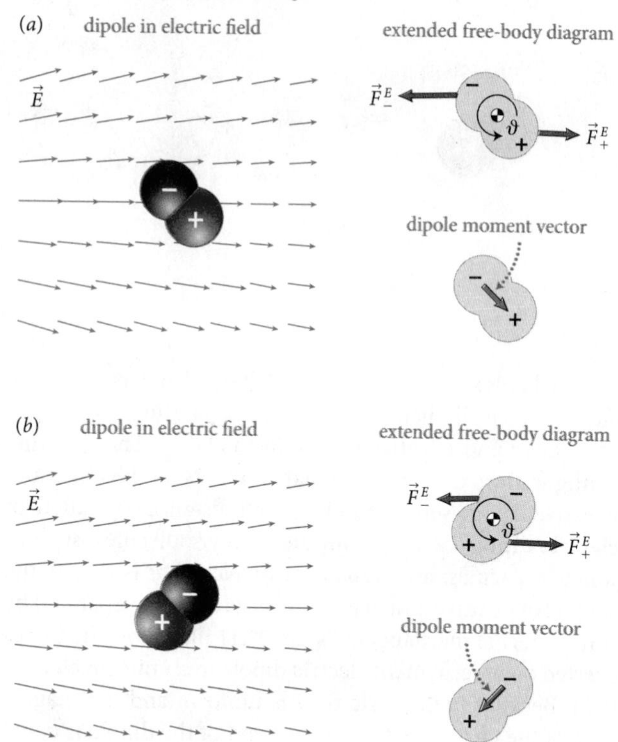

two ends is nonzero, and so the dipole experiences an acceleration whose magnitude and direction depend on its orientation with respect to the electric field. In addition, the forces create a torque about the dipole's center of mass. As in a uniform electric field:

> **A permanent electric dipole placed in an electric field is subject to a torque that tends to align the dipole moment with the direction of the electric field. If the field is uniform, the dipole has zero acceleration; if the electric field is nonuniform, the dipole has a nonzero acceleration.**

23.11 (*a*) Draw a free-body diagram for the dipole in Figure 23.18*a* and determine the direction of the dipole's center-of-mass acceleration. (*b*) Draw a free-body diagram for the dipole in Figure 23.18*b* and qualitatively describe the dipole's motion.

Self-quiz

1. Suppose someone discovers that blue and yellow objects attract each other, that two blue objects repel each other, and that two yellow objects repel each other. The strength of this "chromatic interaction" is found to depend on color depth: The deeper the color, the greater the magnitude of the interaction. How would you define the magnitude and direction of the "chromatic field" of an object?
2. (*a*) Does an electrically neutral particle that has mass interact with an electric field? (*b*) Does a charged particle interact with a gravitational field?
3. The two particles in **Figure 23.19** have the same mass, carry charges of the same magnitude ($q_1 = -q_2 > 0$), and are equidistant from point P. (*a*) What is the electric field direction at P? (*b*) At P, what is the direction of the gravitational field due to the two particles? Ignore Earth's gravitational field.

Figure 23.19

1 ⊕ q_1

P

2 ⊖ q_2

4. Can electric and gravitational fields exist in the same place at the same time?
5. What are the directions of the acceleration of each particle in **Figure 23.20**? Describe the resulting motions.

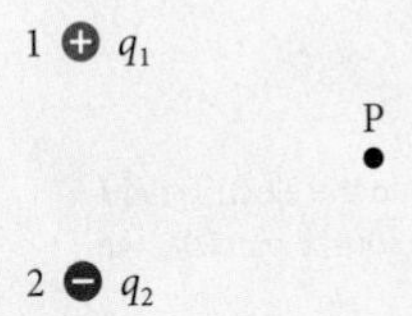

Figure 23.20

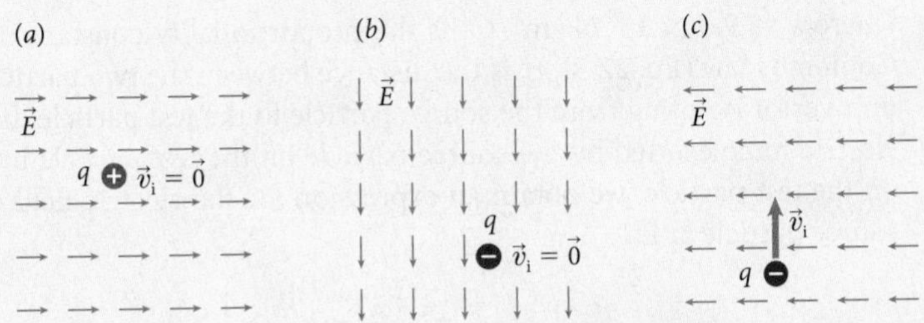

Answers

1. The gravitational field is defined as the gravitational force per unit of mass, with the field direction the same as the direction of the force. The electric field is defined as the electric force per unit of charge, with the field direction parallel to that of the force exerted on a positively charged particle. Therefore, the chromatic field can be defined as the chromatic force per unit of color, with the field direction parallel to that of the force exerted on a particle carrying some chosen color.
2. (*a*) No. Uncharged particles don't interact with electric fields. (Remember that a particle has no extent and therefore cannot be polarized.) (*b*) Yes, because any particle or object, charged or uncharged, interacts with a gravitational field.
3. (*a*) The electric field of particle 1 points away from the particle, which means that at P it points to the right and down. The electric field of particle 2 points toward the particle, meaning to the left and down at P. The vector sum of the electric fields at P therefore points straight down (**Figure 23.21*a***). (*b*) The gravitational fields of the two particles point toward them from P, and so their vector sum points to the left (Figure 23.21*b*).

Figure 23.21

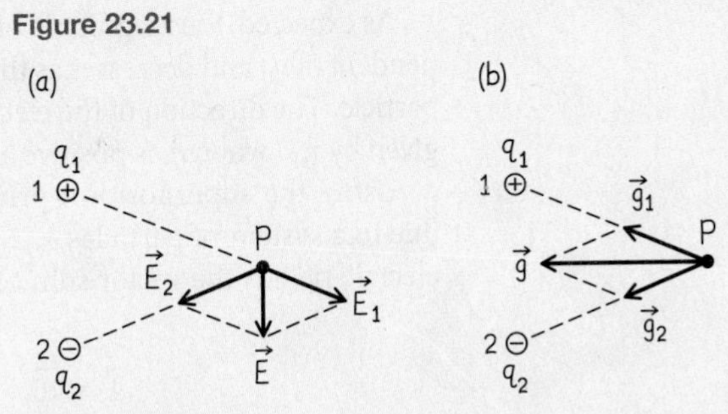

4. Yes. Consider a charged object near Earth's surface. This object is surrounded by an electric field, but it is also surrounded by Earth's (and to a lesser extent its own) gravitational field. These electric and gravitational fields exist in the same place at the same time.
5. (*a*) Recall from the discussion following Checkpoint 23.4 that the direction of the electric field at a given location is the same as the direction of the electric force exerted on a positively charged particle at that location. In Figure 23.20*a*, therefore, the positively charged particle experiences a force directed to the right. Because its initial velocity is zero, the particle moves in a straight line in the direction of the electric field. (*b*) The negatively charged particle in Figure 23.20*b* experiences an acceleration up the page, opposite the direction of the electric field. This particle moves in a straight line up the page. (*c*) The negatively charged particle in Figure 23.20*c* experiences an acceleration to the right, in the direction opposite the direction of the electric field. Because its initial velocity is perpendicular to the direction of the electric field, the particle travels in a parabolic trajectory up the page and curving to the right.

23.5 Electric field of a charged particle

In Section 23.2 we defined the **electric field** at a certain point P in space as the electric force experienced at P by a test particle carrying a charge q_t divided by the charge of the test particle:

$$\vec{E} \equiv \frac{\vec{F}^E_t}{q_t}. \tag{23.1}$$

The SI unit of electric field is the newton per coulomb (N/C).

Equation 23.1 requires no knowledge of the charge distribution that causes the electric field: It gives a prescription for determining the electric field at a given position in space. We can use Coulomb's law, however, to derive an expression for the electric field created at some point P due to a source particle carrying a charge q_s at position $\vec{r}_s$ (**Figure 23.22**). If we place a test particle carrying a charge q_t at P, Coulomb's law (Eq. 22.7) tells us that the force exerted on the test particle is

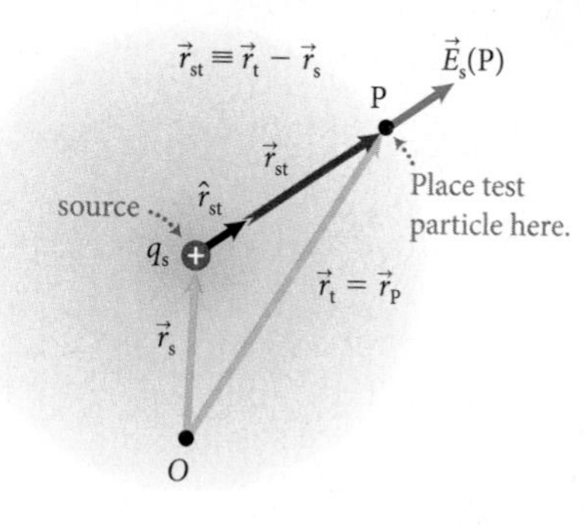

Figure 23.22 To determine the electric field at P generated by a charged source particle, we place a test particle at P.

$$\vec{F}^E_{st} = k\frac{q_s q_t}{r^2_{st}}\hat{r}_{st}, \tag{23.2}$$

where $k = 9.0 \times 10^9 \text{ N} \cdot \text{m}^2/\text{C}^2$ is the proportionality constant that appears in Coulomb's law (Eq. 22.5), r_{st} is the distance between the two particles, and $\hat{r}_{st}$ is a unit vector pointing from the source particle to the test particle. If we divide the electric force exerted by the source particle on the test particle by the charge q_t on the test particle, we obtain an expression for the electric field created by the source particle at P:

$$\vec{E}_s = \frac{\vec{F}^E_{st}}{q_t} = k\frac{q_s}{r^2_{st}}\hat{r}_{st}. \tag{23.3}$$

Because the test particle has nothing to do with this electric field, we can omit any reference to it by writing $\vec{r}_{st} = \vec{r}_{sP}$ and referring only to the position of point P:

$$\vec{E}_s(\text{P}) = k\frac{q_s}{r^2_{sP}}\hat{r}_{sP}. \tag{23.4}$$

This expression represents the electric field at P due to a source particle carrying a charge q_s at position $\vec{r}_s$.

As expected, the magnitude of the electric field at P is proportional to q_s, is independent of q_t, and decreases as the inverse square of the distance r_{sP} from the source particle. The direction of the electric field is outward (that is to say, in the direction given by $\hat{r}_{sP}$) when q_s is positive and inward (antiparallel to $\hat{r}_{sP}$) when q_s is negative.

Using the superposition principle, we now can determine the electric field due to a system of particles 1, 2, . . . carrying charges $q_1, q_2, \ldots$. The combined electric field is the vector sum of the individual electric fields:

$$\vec{E} = \vec{E}_1 + \vec{E}_2 + \cdots = \sum k\frac{q_i\hat{r}_{iP}}{r^2_{iP}}. \tag{23.5}$$

Once the electric field at a certain position is known, the force exerted on any particle carrying charge q placed at that position can be found from

$$\vec{F}^E_p = q\vec{E}. \tag{23.6}$$

(Remember that we omit the "by" subscript on the force when the force is exerted by a field.) If q is positive, the force exerted on the particle is in the same direction as the electric field; if q is negative, the force exerted on the particle is in the direction opposite the direction of the electric field.

Example 23.3 Electric field due to two charged particles

A point P is located at $x_P = 2.0$ m, $y_P = 3.0$ m. What are the magnitude and direction of the electric field at P due to a particle 1 carrying charge $q_1 = +10\ \mu$C and located at $x_1 = 1.0$ m, $y_1 = 0$ and a particle 2 carrying charge $q_2 = +20\ \mu$C and located at $x_2 = -1.0$ m, $y_2 = 0$?

❶ GETTING STARTED I begin by making a sketch of the situation (**Figure 23.23**). Each particle carries a positive charge, so the electric field due to each particle points away from the particle.

Figure 23.23

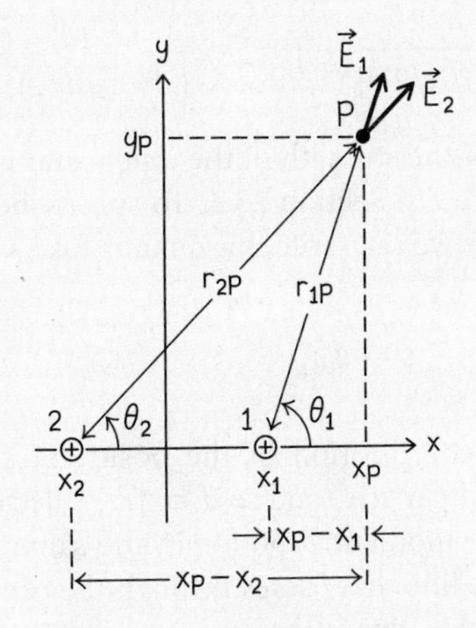

❷ DEVISE PLAN To determine the electric field $\vec{E}_P$ at P, I must take the vector sum of $\vec{E}_1$ and $\vec{E}_2$ at P. I can use Eq. 23.4 to calculate the magnitudes E_1 and E_2. To obtain the vector sum of the two fields, I add their x and y components.

❸ EXECUTE PLAN The distances from the particles to P are $r_{1P} = \sqrt{(x_P - x_1)^2 + y_P^2} = \sqrt{10\ \text{m}^2} = 3.2$ m and $r_{2P} = \sqrt{(x_P - x_2)^2 + y_P^2} = \sqrt{18\ \text{m}^2} = 4.2$ m. The magnitudes of the electric fields created by the particles at P are thus

$$E_1 = k\frac{|q_1|}{r_{1P}^2} = (9.0 \times 10^9\ \text{N}\cdot\text{m}^2/\text{C}^2)\frac{(1.0 \times 10^{-5}\ \text{C})}{10\ \text{m}^2}$$
$$= 0.90 \times 10^4\ \text{N/C}$$

$$E_2 = k\frac{|q_2|}{r_{2P}^2} = (9.0 \times 10^9\ \text{N}\cdot\text{m}^2/\text{C}^2)\frac{(2.0 \times 10^{-5}\ \text{C})}{18\ \text{m}^2}$$
$$= 1.0 \times 10^4\ \text{N/C}.$$

To calculate $\vec{E}_P$, I take the vector sum of $\vec{E}_1$ and $\vec{E}_2$ at P. In component form, I have

$$E_{Px} = E_{1x} + E_{2x} = E_1 \cos\theta_1 + E_2 \cos\theta_2$$
$$= E_1\frac{(x_P - x_1)}{r_{1P}} + E_2\frac{(x_P - x_2)}{r_{2P}}$$

$$E_{Py} = E_{1y} + E_{2y} = E_1 \sin\theta_1 + E_2 \sin\theta_2$$
$$= E_1\frac{y_P}{r_{1P}} + E_2\frac{y_P}{r_{2P}}.$$

Substituting the values given, I have

$$E_{Px} = (0.9 \times 10^4\ \text{N/C})\frac{1.0\ \text{m}}{3.2\ \text{m}} + (1.0 \times 10^4\ \text{N/C})\frac{3.0\ \text{m}}{4.2\ \text{m}}$$
$$= +1.0 \times 10^4\ \text{N/C}$$

$$E_{Py} = (0.9 \times 10^4\ \text{N/C})\frac{3.0\ \text{m}}{3.2\ \text{m}} + (1.0 \times 10^4\ \text{N/C})\frac{3.0\ \text{m}}{4.2\ \text{m}}$$
$$= +1.6 \times 10^4\ \text{N/C}.$$

Finally, I write this in vector form as

$$\vec{E}_P = (+1.0 \times 10^4\ \text{N/C})\hat{\imath} + (+1.6 \times 10^4\ \text{N/C})\hat{\jmath}.\ ✔$$

❹ EVALUATE RESULT Both E_{Px} and E_{Py} are positive, as I expect based on my sketch. The magnitudes of $\vec{E}_1$ and $\vec{E}_2$ are comparable, which is what I would expect: Particle 2 carries twice the charge of particle 1, but the square of its distance to P is greater by a factor of 1.8.

23.12 What is the magnitude of the electric force exerted by the electric field on an electron placed at point P in Figure 23.23? What is the initial acceleration of the electron if it is released from rest from that point? [$e = 1.6 \times 10^{-19}$ C; $m_e = 9.1 \times 10^{-31}$ kg]

23.6 Dipole field

Next we examine the electric field due to a permanent electric dipole. **Figure 23.24** on the next page shows a dipole that consists of a particle carrying a charge $+q_p$ at $x = 0, y = +\frac{1}{2}d$, and another particle carrying a charge $-q_p$ at $x = 0, y = -\frac{1}{2}d$, where d is the distance between the two particles. The charge q_p of the positively charged pole is called the *dipole charge*, and the distance d is called the *dipole separation*. Each particle creates an electric field at all positions in space, so the two fields overlap everywhere. We can determine the combined electric field at any position by adding the two fields vectorially. Let us do this for two general locations: anywhere along the x axis and anywhere along the y axis.

Figure 23.24 Calculating the electric field due to a dipole.

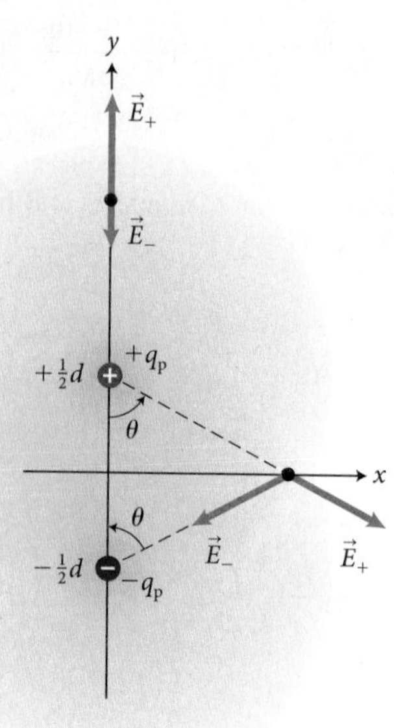

Along the x axis, which bisects the dipole, the magnitudes of the electric fields due to the two ends of the dipole are equal:

$$E_+ = E_- = k\frac{q_p}{x^2 + (d/2)^2}. \tag{23.7}$$

The x components of these two electric fields point in opposite directions and so add to zero. The magnitude of the combined electric field is thus equal to the sum of the y components:

$$E_y = E_{+y} + E_{-y} = -(E_+ + E_-)\cos\theta$$

$$= -\left(2k\frac{q_p}{x^2 + (d/2)^2}\right)\left(\frac{d/2}{[x^2 + (d/2)^2]^{1/2}}\right) = -k\frac{q_p d}{[x^2 + (d/2)^2]^{3/2}}. \tag{23.8}$$

The product $q_p d$ is a measure of the strength of the dipole and is the magnitude of the dipole moment introduced in Section 23.4. To specify both the strength and the orientation of the dipole we can write this quantity as a vector, called the **dipole moment:**

$$\vec{p} \equiv q_p \vec{r}_p, \tag{23.9}$$

where $\vec{r}_p \equiv \vec{r}_{-+} = \vec{r}_+ - \vec{r}_-$ is the position of the positively charged particle relative to the negatively charged particle (and so $d = |\vec{r}_p|$). Because q_p is always taken to be positive, the dipole moment $\vec{p}$ points in the same direction as $\vec{r}_p$: along the axis of the dipole (the line that passes through the center of each particle), in the direction from the negative to the positive pole (**Figure 23.25**). Large permanent dipole moments can be caused either by a large dipole separation d or by a large dipole charge q_p. Conceptually you can think of the magnitude of the dipole moment as a measure of how strongly the dipole wants to align itself in the direction of an electric field. The SI unit of dipole moment is the C · m.

Figure 23.25 The dipole moment $\vec{p}$ points along the axis of the dipole from the negative to the positive pole.

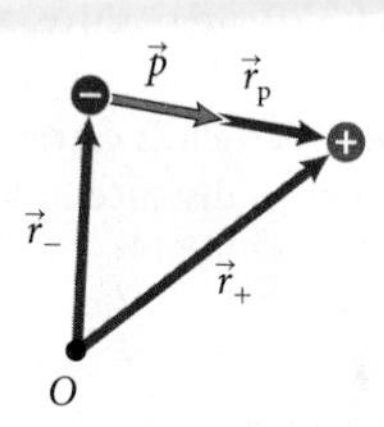

For distances far from the dipole ($x \gg d/2$), we may ignore $d/2$, and so $[x^2 + (d/2)^2]^{3/2} \rightarrow x^3$. Equation 23.8 thus becomes

$$E_y \approx -k\frac{p}{|x^3|} \quad \text{(far from dipole along the positive } x \text{ axis).} \tag{23.10}$$

The right side of this equation is negative for both positive and negative x, and so anywhere along the x axis the dipole's electric field $\vec{E}$ points in the negative y direction, opposite the direction of the dipole moment. Equation 23.10 also shows that the magnitude of the electric field is inversely proportional to x^3, in contrast to the electric field of a charged particle, which is inversely proportional to x^2 (Eq. 23.4). The reason the electric field of a dipole approaches zero faster as x increases is that the angle between $\vec{E}_+$ and $\vec{E}_-$ in Figure 23.24 approaches 180° as x increases, and so the electric fields of the two poles tend to cancel each other more and more.

Along the y axis, the electric field created by either end of the dipole is directed along the y axis. Thus to determine the y component of the electric field of the dipole at any position along the y axis, we must add the y components of the fields from each particle. For $y > +d/2$:

$$E_y = E_{+y} + E_{-y} = k\frac{q_p}{[y - (d/2)]^2} - k\frac{q_p}{[y + (d/2)]^2}. \tag{23.11}$$

After some algebra, this can be rewritten in the form

$$E_y = k\frac{q_p}{y^2}\left[\left(1 - \frac{d}{2y}\right)^{-2} - \left(1 + \frac{d}{2y}\right)^{-2}\right] \quad (y > +d/2). \tag{23.12}$$

QUANTITATIVE TOOLS

For distances far from the dipole, $y \gg +d/2$, so we can use the binomial series expansion, which states that for $x \ll 1$, $(1 + x)^n \approx 1 + nx$ (see Appendix B). Applying this expansion to the two terms inside the square brackets in Eq. 23.12, we get

$$E_y \approx k\frac{q_p}{y^2}\left[\left(1 + 2\frac{d}{2y}\right) - \left(1 - 2\frac{d}{2y}\right)\right]$$

$$= k\frac{q_p}{y^2}\left[\frac{2d}{y}\right] = 2k\frac{q_p d}{y^3} = 2k\frac{p}{y^3} \quad (y \gg d/2). \tag{23.13}$$

The right side of this equation has the same algebraic sign as y, so the dipole's electric field $\vec{E}$ points in the positive y direction. (Carrying out the same calculation for $y < -d/2$, you can show that the electric field still points in the positive y direction. In between the two charged particles, the electric field points in the negative y direction.) The magnitude of the electric field is inversely proportional to y^3—just as along the x axis, the electric fields of each of the two poles tend to cancel each other more and more as the distance from the dipole increases. One can show that the electric field of the dipole depends on $1/r^3$ for all positions far from the dipole (where r is the distance between the point under consideration and the center of the dipole). The reason is that the electric fields of the positive and negative ends of the dipole partially cancel each other, and this cancellation becomes more complete far from the dipole: The farther you are from the dipole, the smaller the separation between the charged particles appears to be.

23.13 The magnitude of the electric field created by dipole A at a certain point P is E_A. If the dipole is replaced with another dipole B that has its dipole moment oriented in the same direction, the magnitude of the electric field at point P is found to be greater: $E_B > E_A$. Which dipole has the greater dipole moment? For which of these two dipoles is the dipole charge q_p greater?

23.7 Electric fields of continuous charge distributions

So far we have dealt with only charged particles because Coulomb's law applies only to charged particles. However, most charged objects of interest—from charged combs to electrical components—are not particles. Instead, they are extended bodies. Although every macroscopic object consists of very large numbers of charged particles—protons and electrons—it is not practical to calculate the individual field of each of these particles and then add them vectorially. Instead, we shall treat any macroscopic charged object as having a continuous charge distribution and calculate the electric field created by the object by dividing the charge distribution on the object into infinitesimally small segments that may be considered charged source particles carrying a charge dq_s. For the charged macroscopic object shown in **Figure 23.26**, for example, we can use Coulomb's law to obtain the infinitesimal portion of the electric field at point P contributed by a segment:

$$d\vec{E}_s(\text{P}) = k\frac{dq_s}{r_{sP}^2}\hat{r}_{sP}. \tag{23.14}$$

Using the principle of superposition, we can then sum the contributions of all the segments that make up the object. Because the segments are infinitesimally small, this sum corresponds to an integral:

$$\vec{E} = \int d\vec{E}_s = k\int\frac{dq_s}{r_{sP}^2}\hat{r}_{sP}. \tag{23.15}$$

Figure 23.26 To calculate the electric field created at P by a continuous charge distribution, we divide the distribution into infinitesimally small segments that can be treated as charged source particles carrying charge dq_s.

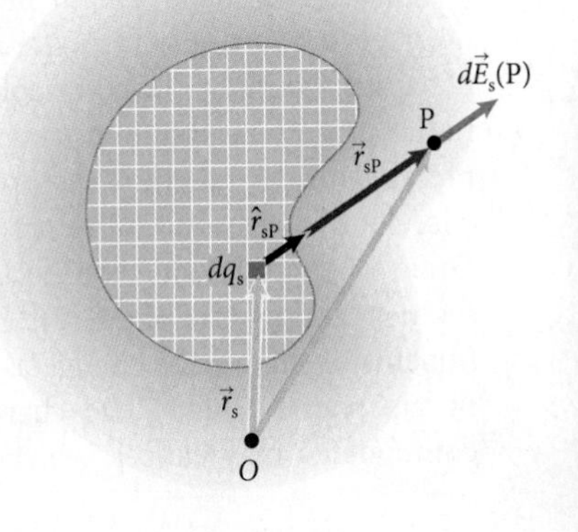

QUANTITATIVE TOOLS

In order to evaluate this integral, we must express dq_s, $1/r_{sP}^2$, and $\hat{r}_{sP}$ in terms of the same coordinate(s). To do so, it is necessary to express the charge on the object in terms of a **charge density**—the amount of charge per unit of length, per unit of surface area, or per unit of volume. For a one-dimensional object, such as a thin charged wire of length ℓ carrying a charge q uniformly distributed along the wire, the *linear charge density*—the amount of charge per unit of length (in coulombs per meter)—is given by

$$\lambda \equiv \frac{q}{\ell} \quad \text{(uniform charge distribution).} \tag{23.16}$$

For uniformly charged two-dimensional objects, we use the *surface charge density*—the amount of charge per unit of area (in coulombs per square meter). For example, the surface charge density of a flat plate of area A carrying a uniformly distributed charge q is

$$\sigma \equiv \frac{q}{A} \quad \text{(uniform charge distribution).} \tag{23.17}$$

For a uniformly charged three-dimensional object, we use the *volume charge density:*

$$\rho \equiv \frac{q}{V} \quad \text{(uniform charge distribution),} \tag{23.18}$$

which gives the amount of charge per cubic meter.

The procedure on this page provides some helpful steps for carrying out the integral in Eq. 23.15, and the next four examples show how to put the procedure into practice.

Procedure: Calculating the electric field of continuous charge distributions by integration

To calculate the electric field of a continuous charge distribution, you need to evaluate the integral in Eq. 23.15. The following steps will help you evaluate the integral.

1. Begin by making a sketch of the charge distribution. Mentally divide the distribution into small segments. Indicate one such segment that carries a charge dq_s in your drawing.
2. Choose a coordinate system that allows you to express the position of the segment in terms of a minimum number of coordinates (x, y, z, r, or θ). These coordinates are the integration variables. For example, use a radial coordinate system for a charge distribution with radial symmetry. Unless the problem specifies otherwise, let the origin be at the center of the object.
3. Draw a vector showing the electric field caused by the segment at the point of interest. Examine how the components of this vector change as you vary the position of the segment along the charge distribution. Some components may cancel, which greatly simplifies the calculation. If you can determine the direction of the resulting electric field, you may need to calculate only one component. Otherwise express $\hat{r}_{sP}$ in terms of your integration variable(s) and evaluate the integrals for each component of the field separately.
4. Determine whether the charge distribution is one-dimensional (a straight or curved wire), two-dimensional (a flat or curved surface), or three-dimensional (any bulk object). Express dq_s in terms of the corresponding charge density of the object and the integration variable(s).
5. Express the factor $1/r_{sP}^2$, where r_{sP} is the distance between dq_s and the point of interest, in terms of the integration variable(s).

At this point you can substitute your expressions for dq_s and $1/r_{sP}^2$ into Eq. 23.15 and carry out the integral (or component integrals), using what you determined about the direction of the electric field (or substituting your expression for $\hat{r}_{sP}$).

QUANTITATIVE TOOLS

Example 23.4 Electric field created by a uniformly charged thin rod

A thin rod of length ℓ carries a uniformly distributed charge q. What is the electric field at a point P along a line that is perpendicular to the long axis of the rod and passes through the rod's midpoint?

❶ **GETTING STARTED** I begin by making a sketch of the situation. After drawing a set of axes, I place the rod along the y axis, with the origin at the rod center and point P on the positive x axis (**Figure 23.27**).

Figure 23.27

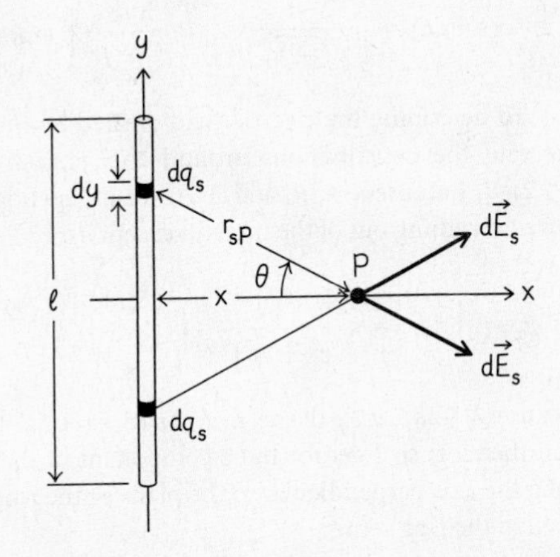

❷ **DEVISE PLAN** The word *thin* implies that I can treat the rod as a one-dimensional object. Because the rod is uniformly charged, I can thus use Eq. 23.16 to determine the linear charge density along the rod. To determine the electric field at P, I divide the rod lengthwise into a large number of infinitesimally small segments, each of length dy. Each segment contributes to the electric field at P an amount given by Eq. 23.14. For each segment above the x axis there is a corresponding segment below the axis at the same distance from P. The y components of the electric fields $d\vec{E}_s$ due to these two segments add up to zero, so I need to calculate only the x component dE_{sx}. To get the electric field created by the entire rod, I use Eq. 23.15 to integrate my result over the length of the rod.

❸ **EXECUTE PLAN** The charge dq_s on each segment dy is $dq_s = \lambda\, dy = (q/\ell)\, dy$. The x component of the electric field created by each segment at P is thus

$$dE_{sx} = k\frac{dq_s}{r_{sP}^2}\cos\theta = k\frac{q}{\ell r_{sP}^2}\cos\theta\, dy, \tag{1}$$

where θ is the angle between the x axis and the line that connects the segment dy with P. Both θ and r_{sP} depend on the position y of the segment, so I must choose one integration variable and express the others in terms of that variable. I choose θ as the integration variable, which means I must express the factor dy/r_{sP}^2 in Eq. 1 in terms of θ. Using trigonometry, I have

$$\cos\theta = \frac{x}{r_{sP}} \tag{2}$$

and

$$\tan\theta = \frac{y}{x}, \tag{3}$$

where x is the x coordinate of point P. Differentiating Eq. 3 yields

$$dy = x\, d(\tan\theta) = \frac{x}{\cos^2\theta}d\theta. \tag{4}$$

Next I divide Eq. 4 by r_{sP}^2 to obtain the factor dy/r_{sP}^2 I need. I use r_{sP}^2 on the left, but on the right I use Eq. 2 to write r_{sP}^2 in the form $x^2/\cos^2\theta$, yielding

$$\frac{dy}{r_{sP}^2} = \left(\frac{\cos^2\theta}{x^2}\right)\left(\frac{x}{\cos^2\theta}d\theta\right) = \frac{1}{x}d\theta.$$

Substituting this result into Eq. 1 and integrating over the entire rod yield

$$E_x = k\frac{q}{\ell}\int_{-\theta_{max}}^{+\theta_{max}} \frac{\cos\theta}{x}d\theta = \frac{kq}{\ell x}\int_{-\theta_{max}}^{+\theta_{max}} \cos\theta\, d\theta$$

$$= \frac{kq}{\ell x}\sin\theta\Big|_{-\theta_{max}}^{+\theta_{max}} = \frac{2kq}{\ell x}\sin\theta_{max},$$

where θ_{max}, the maximum value of θ, is the angle between the x axis and the line that connects the top end of the rod with P. Substituting $\sin\theta_{max} = y/r_{sP} = \frac{1}{2}\ell/\sqrt{(\ell/2)^2 + x^2}$ finally yields

$$E_x = \frac{kq}{x\sqrt{\ell^2/4 + x^2}};\ E_y = 0;\ E_z = 0. ✔$$

❹ **EVALUATE RESULT** Very far from the rod along the positive x axis, $x \ll \ell$, so the rod looks like a particle. In this case, I can ignore the ℓ^2 term in the denominator and my result becomes identical to Eq. 23.4, the equation for a particle that carries a charge q ($E = kq/r^2$, or using the symbols from this problem, $E_x = kq/x^2$).

When P is very close to the rod, $x \ll \ell$, I can ignore the x^2 term in the denominator and write

$$E_x = \frac{2k(q/\ell)}{x} = \frac{2k\lambda}{x}. \tag{5}$$

In this $x \ll \ell$ case, the distance from P to either end of the rod is much greater than the distance from P to the closest point on the rod (which is the rod midpoint, located at the origin), and thus the rod essentially looks "infinitely long" to an observer at P. Indeed, Eq. 5 shows that my result no longer depends on ℓ.

I also note that the rod's electric field is now inversely proportional to x rather than x^2. I saw in Chapter 17 that the amplitudes of waves that spread out in three dimensions are inversely proportional to x^2, whereas the amplitudes of waves that spread out in two dimensions are inversely proportional to x. My result therefore make sense because the electric field that emanates from a charged particle "spreads out" in three dimensions, but the field that emanates from an infinitely long charged rod spreads out in just two dimensions.

Example 23.5 Electric field created by a uniformly charged thin ring

A thin ring of radius R carries a uniformly distributed charge q. What is the electric field at point P along an axis that is perpendicular to the plane of the ring and passes through its center?

❶ GETTING STARTED I begin by making a sketch of the situation. I let the ring be in the xy plane, with the origin at the center of the ring, and I place P on the positive z axis (**Figure 23.28**).

Figure 23.28

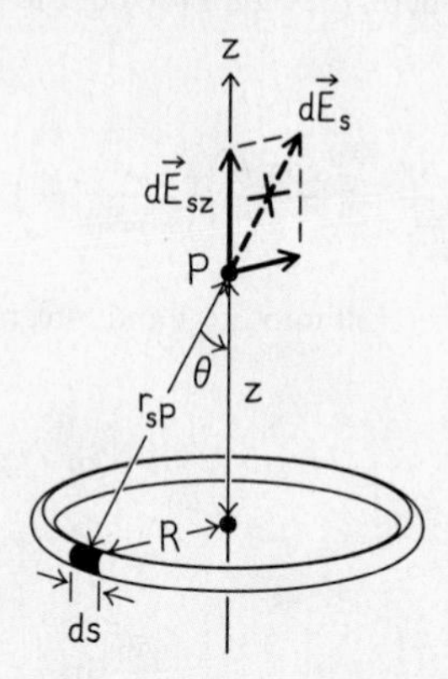

❷ DEVISE PLAN Because the ring is thin, I can use Eq. 23.16 to determine its linear charge density. To determine the electric field at P, I divide the ring into a large number of infinitesimally small segments, each of arc length ds. Each segment contributes to the electric field an amount given by Eq. 23.14. Because all segments are at the same distance from P, all contribute an electric field $d\vec{E}_s$ of the same magnitude. As shown in my sketch, the contribution $d\vec{E}_s$ makes an angle θ with the z axis, and so each segment ds produces a component parallel to the z axis and a component perpendicular to it. For each pair of segments on opposite sides of the ring, the components of $d\vec{E}_s$ perpendicular to the z axis add up to zero, and so I am concerned only with the z components.

To get the electric field created by the ring, I can use Eq. 23.15 to integrate my $d\vec{E}_s$ result over the circumference of the ring.

❸ EXECUTE PLAN Each segment carries a charge $dq_s = \lambda\, ds$, where $\lambda = q/2\pi R$ is the linear charge density along the ring. The magnitude of each segment's contribution to the electric field is, from Eq. 23.14,

$$dE_s = k\frac{dq_s}{r_{sP}^2} = k\frac{\lambda\, ds}{r_{sP}^2} = k\frac{\lambda\, ds}{z^2 + R^2}.$$

For the z component of $d\vec{E}_s$, I see from Figure 23.28 that the angle between the vectors $d\vec{E}_{sz}$ and $d\vec{E}_s$ is also θ, and so $\cos\theta = dE_{sz}/dE_s$. Then, combining this expression with the relationship

$$\cos\theta = \frac{z}{r_{sP}} = \frac{z}{\sqrt{z^2 + R^2}},$$

I have

$$dE_{sz} = \cos\theta\, dE_s = \left[\frac{z}{\sqrt{z^2 + R^2}}\right]\left[k\frac{\lambda\, ds}{z^2 + R^2}\right] = k\frac{z\lambda}{[z^2 + R^2]^{3/2}}ds.$$

To determine the electric field created by the ring, I must integrate the contributions around the ring, from $s = 0$ to $s = 2\pi R$. Because k, z, R, and λ are all independent of s, I can move everything out of the integral except ds:

$$E_z = \int dE_{sz} = k\frac{z\lambda}{[z^2 + R^2]^{3/2}}\int_0^{2\pi R} ds = k\frac{z\lambda(2\pi R)}{[z^2 + R^2]^{3/2}}.$$

Because $\lambda = q/2\pi R$, the term $\lambda(2\pi R)$ is equal to the charge q on the ring, so I get for the z component of the electric field along the axis perpendicular to the plane of the ring and passing through the ring center

$$E_x = 0;\ E_y = 0;\ E_z = k\frac{qz}{[z^2 + R^2]^{3/2}}. ✔$$

❹ EVALUATE RESULT Very far from the ring my expression for E_z should become the same as that for a charged particle. Indeed, when $z \gg R$, I can ignore the R^2 term in my result, and so

$$E_z \approx k\frac{qz}{[z^2]^{3/2}} = k\frac{qz}{z^3} = k\frac{q}{z^2},$$

as I expect.

At the center of the ring, $z = 0$ and so my expression yields $E_z = 0$. This result is reasonable because at the center of the ring the electric forces exerted by segments on opposite sides of the ring on a charged test particle add to zero. When the vector sum of these forces is zero, the electric field must be zero also.

Example 23.6 Electric field created by a uniformly charged disk

A thin disk of radius R carries a uniformly distributed charge. The surface charge density on the disk is σ. What is the electric field at a point P along the perpendicular axis through the disk center?

❶ GETTING STARTED I begin with a sketch, placing the disk in the xy plane, with the disk center at the origin. I let point P lie on the positive z axis (**Figure 23.29**).

Figure 23.29

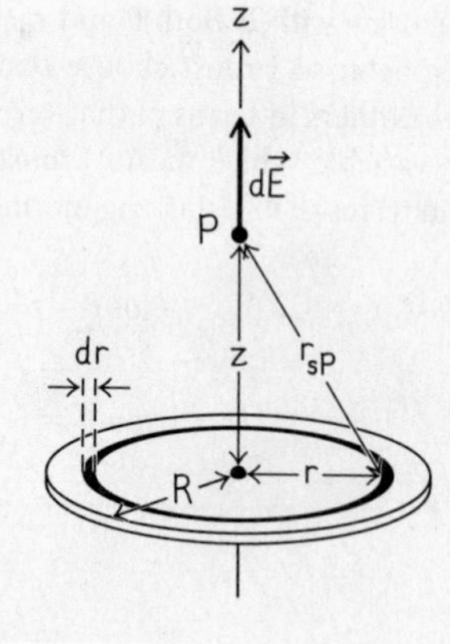

❷ DEVISE PLAN Because of the circular symmetry of the disk, I divide it into a large number of ring-shaped segments, each of radius r and width dr. The charge on each ring is the product of the ring surface area (circumference times width) and the surface charge density: $dq_s = (2\pi r)dr\,\sigma$. The contribution of each

QUANTITATIVE TOOLS

ring to the electric field at P is given by the expression for E_z I obtained for a uniformly charged thin ring in Example 23.5 with $2\pi r\,\sigma\,dr$ substituted for q and r substituted for R. So all I need to do is integrate over the entire disk.

❸ EXECUTE PLAN Substituting $q = 2\pi r\,\sigma\,dr$ and $R = r$ into the Example 23.5 expression for E_z and integrating the result over the disk from $r = 0$ to $r = R$, I have

$$E_z = \int dE_z = k\int_0^R \frac{2\pi r\,\sigma z}{(z^2 + r^2)^{3/2}}dr.$$

Because σ and z are independent of r, I can move them out of the integral:

$$E_x = 0;\ E_y = 0;\ E_z = k\pi\sigma z\int_0^R \frac{2r\,dr}{(z^2 + r^2)^{3/2}}$$

$$= k\pi\sigma z\int_0^R \frac{d(r^2)}{(z^2 + r^2)^{3/2}} = k\pi\sigma z\frac{-2}{(z^2 + r^2)^{1/2}}\Bigg|_0^R$$

$$= 2k\pi\sigma z\left[\frac{1}{(z^2)^{1/2}} - \frac{1}{(z^2 + R^2)^{1/2}}\right]. ✔ \quad (1)$$

❹ EVALUATE RESULT Let me evaluate my result for $z \gg R$, where, to an observer at P, the disk looks like a particle. For positive z, I can write the E_z in Eq. 1 as

$$E_z = 2k\pi\sigma\left[1 - \frac{z}{(z^2 + R^2)^{1/2}}\right]. \quad (2)$$

From the binomial series expansion (see Appendix B), I get in the case that $z \gg R$,

$$\frac{z}{(z^2 + R^2)^{1/2}} = \left(1 + \frac{R^2}{z^2}\right)^{-1/2} \approx 1 - \frac{1}{2}\frac{R^2}{z^2}.$$

Substituting this result into Eq. 2 and writing $q = \sigma(\pi R^2)$ for the charge on the disk, I get

$$E_z = 2k\pi\sigma\left(\frac{1}{2}\frac{R^2}{z^2}\right) = k\frac{\sigma\pi R^2}{z^2} = k\frac{q}{z^2},$$

which is the result for a charged particle, as I expect.

I can also evaluate my result for $z \approx 0$, where, to an observer at that location, the disk looks like it has an infinite radius (that is to say, it looks like an infinite flat sheet). In that case, the second term inside the brackets in Eq. 2 vanishes and $E_z = 2k\pi\sigma$. This tells me that the electric field of an infinite flat charged sheet is independent of z and constant throughout space, as I have sketched in **Figure 23.30**. (In other words, it is uniform.) While this lack of dependence on z is somewhat counterintuitive, it agrees with what I concluded earlier: The electric field that emanates from a charged particle "spreads out" in three dimensions and its amplitude is inversely proportional to the square of the distance from the particle, whereas the electric field that emanates from an infinitely long charged rod spreads out in just two dimensions and its amplitude is inversely proportional to the distance from the rod. As my sketch shows, the electric field that emanates from an infinite plane can't spread out at all (if the plane is truly infinite), and therefore its amplitude is independent of distance.

Figure 23.30

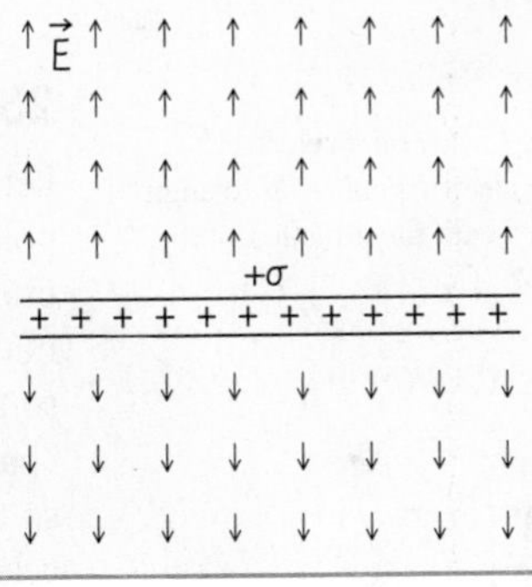

23.14 (a) Describe the electric field between two infinitely large parallel charged sheets if the charge density of one sheet is $+\sigma$ and that of the other is $-\sigma$. (b) Describe the electric field outside the sheets.

Example 23.7 Electric field created by a uniformly charged sphere

A solid sphere of radius R carries a fixed, uniformly distributed charge q. Exploiting the analogy between Newton's law of gravity and Coulomb's law, use the result obtained in Section 13.8 to obtain an expression for the magnitude of the electric field created by the sphere at a point P outside the sphere.

❶ GETTING STARTED To determine the electric field magnitude at point P, I need to determine the magnitude of the electric force exerted by the sphere on a test particle carrying a charge q_t at P and then divide that force magnitude by q_t.

❷ DEVISE PLAN I can follow the same procedure as in Section 13.8 to calculate the gravitational force of a spherical object: I first divide the sphere into a series of thin concentric shells that resemble the layers in an onion and then divide each shell into a series of vertical rings (**Figure 23.31**). I then calculate the contribution of each ring to the electric field at P and integrate first over each shell and then over the sphere. The expression I get for F^E_{sphere} must be of the same form as that for the gravitational sphere, F^G_{sphere} (Eq. 13.37), because the gravitational force and the electric force are both inversely proportional to the square of the distance between the interacting particles. So all I need to do is replace G in Eq. 13.37 by k, M_{sphere} by q, and m by q_t to obtain the magnitude of the electric force exerted by the sphere on the test particle. To obtain an expression for the electric field, I then divide the result by q_t.

(Continued)

Figure 23.31

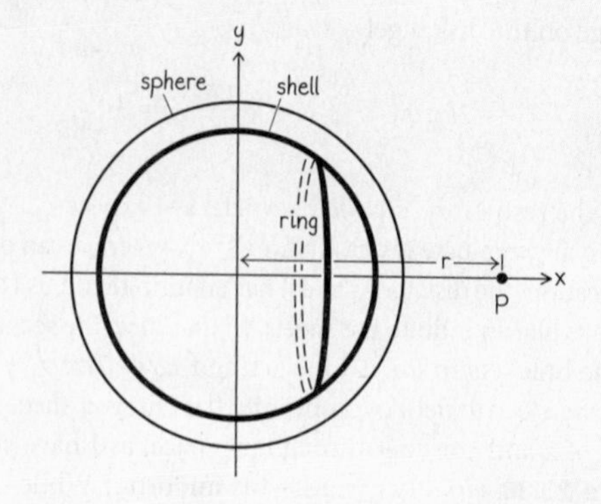

❸ **EXECUTE PLAN** In analogy to Eq. 13.37, I write

$$F^G_{\text{sp}} = G\frac{mM_{\text{sphere}}}{r^2} \rightarrow F^E_{\text{st}} = k\frac{q_t q}{r^2},$$

where r is the distance from the center of the sphere to P. To obtain the magnitude of the electric field, I divide the electric force by q_t:

$$E_{\text{sphere}} = k\frac{q}{r^2}. ✔$$

❹ **EVALUATE RESULT** Comparing my result with Eq. 23.3 (in scalar form), $E_s = kq_s/r^2_{\text{st}}$, shows that outside a uniformly charged solid sphere the magnitude of the electric field is the same as that surrounding a particle carrying the same charge and located at the center of the sphere. The result is independent of the radius R of the sphere and similar to the result obtained in Section 13.8: A solid sphere exerts a gravitational force as if the entire mass of the sphere were concentrated at the center. It makes sense that I obtain a similar result for the charged sphere because the gravitational force and the electric force are both proportional to $1/r^2$.

23.15 How does the electric field inside a uniformly charged sphere vary with distance from the sphere center? [Hint: What is the electric field inside a hollow uniformly charged sphere?]

23.8 Dipoles in electric fields

Figure 23.32 The torque on an electric dipole caused by an electric field tends to align the dipole moment $\vec{p}$ with the direction of the electric field.

(a) Electric dipole in electric field

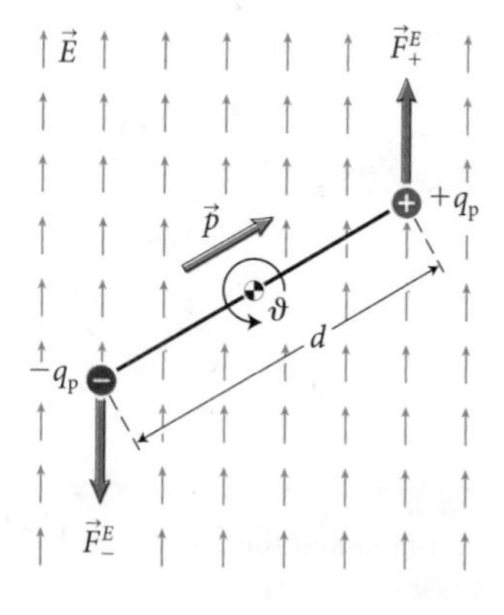

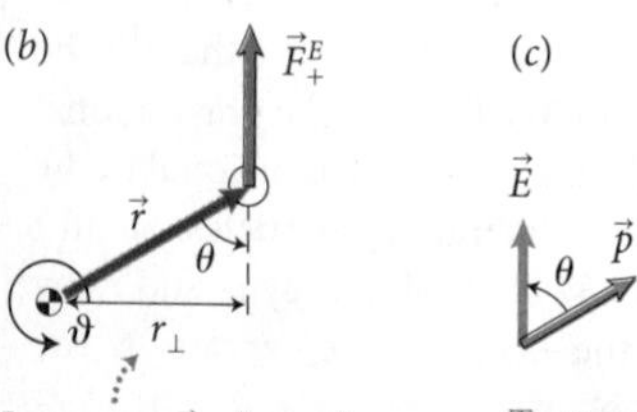

Lever arm $\vec{r}_\perp$ depends on angle θ of dipole with respect to electric field.

Torque on dipole is vector product $\Sigma\vec{\tau} = \vec{p} \times \vec{E}$.

Let us end this chapter by considering the forces exerted by electric fields on dipoles. **Figure 23.32a** shows a dipole consisting of two particles that carry charges of equal magnitude but opposite sign connected by a rod of length d; the dipole makes an angle θ with a uniform electric field $\vec{E}$ created by some unseen distant source. As we saw in Section 23.4, the forces exerted by the electric field on the charged ends of the dipole are equal in magnitude but opposite in direction, and so the vector sum of the forces exerted on the dipole is zero. Consequently the acceleration of the center of mass of the dipole is zero. Because the forces are exerted on opposite ends of the dipole, however, they create torques that cause the dipole to rotate counterclockwise about its center of mass. Figure 23.32*b* shows that the force exerted on the positive end causes a counterclockwise torque of magnitude

$$\tau_+ = r_\perp F^E_+ = (\tfrac{1}{2}d \sin\theta)(q_p E), \tag{23.19}$$

where $r_\perp$ is the lever arm of the force. The force exerted on the negative end causes an identical torque because the lever arm and the magnitude of the force are the same. The electric field thus causes a torque on the dipole equal to

$$\Sigma\tau_\vartheta = 2(\tfrac{1}{2}d \sin\theta)(q_p E) = (q_p d)E \sin\theta \equiv pE \sin\theta. \tag{23.20}$$

This can be written in vectorial form as

$$\Sigma\vec{\tau} = \vec{p} \times \vec{E}, \tag{23.21}$$

where $\vec{p}$ is the dipole moment, which by definition points from the negative end of the dipole to the positive end and whose magnitude is given by Eq. 23.9. According to the right-hand rule (see Section 12.4), the vector product $\vec{p} \times \vec{E}$ in Eq. 23.21 gives a torque that points out of the plane of the drawing in

Figure 23.33 A dipole interacts with a charged particle.

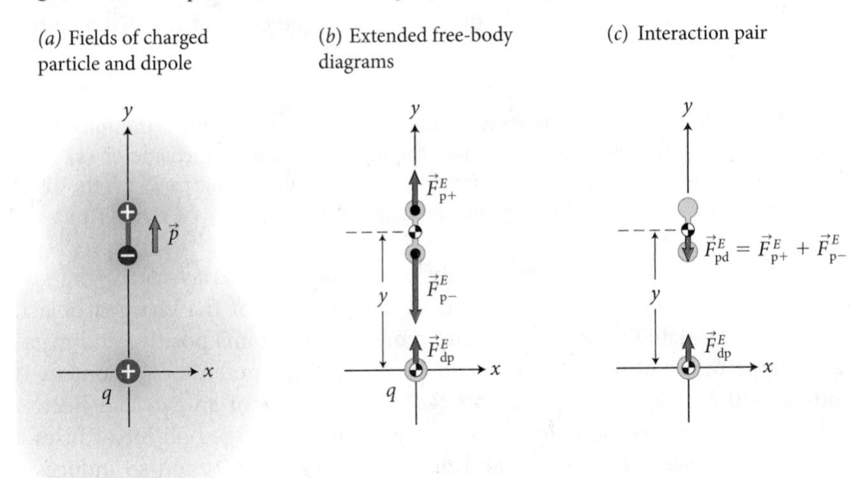

Figure 23.32*c*. As we saw in Section 12.8, such a torque indeed causes a counterclockwise rotation. The torque on the dipole is maximum when the dipole moment is perpendicular to the electric field and zero when it is parallel or antiparallel to the electric field.

23.16 Is Eq. 23.20 valid if the center of mass is not in the middle of the dipole?

As we saw in Section 23.4, the vector sum of the forces exerted on dipoles in nonuniform electric fields is not zero. Consider, for example, the situation illustrated in **Figure 23.33a**. A dipole with its dipole moment $\vec{p}$ aligned along the y axis is placed in the nonuniform electric field generated by a particle carrying a charge q and located at the origin. Because the distance between the negative end of the dipole and the particle is smaller than the distance between the positive end and the particle, the magnitude of the attractive force $\vec{F}^E_{p-}$ on the negative end is greater than the repulsive force $\vec{F}^E_{p+}$ on the positive end. Consequently the vector sum of the forces exerted by the nonuniform field on the dipole is nonzero, and the dipole is attracted to the particle. How does this attraction vary with the position y of the dipole?

To answer this question, we can write an expression for the vector sum of the forces $\sum \vec{F}^E_d$ exerted on the two ends of the dipole and examine how this sum varies with y. Alternatively, we can calculate the force $\vec{F}^E_{dp}$ exerted by the dipole on the particle, using our results from Section 23.6. This force and the vector sum of the forces exerted by the particle on the dipole form an interaction pair, and so their magnitudes are the same. Equation 23.13 tells us that, along the dipole axis, the magnitude of the electric field created by the dipole is $2k(p/y^3)$, and so the magnitude of the force exerted by the dipole on the particle is

$$F^E_{dp} = qE_d = 2k\frac{pq}{y^3}. \qquad (23.22)$$

The magnitude of the force exerted by the particle on the dipole, being equal in magnitude to F^E_{dp}, is thus

$$F^E_{pd} = F^E_{p-} - F^E_{p+} = 2k\frac{pq}{y^3}. \qquad (23.23)$$

Like the electric field of a dipole, the forces between a charged object carrying charge q and a dipole is inversely proportional to the cube of the distance between them.

23.17 How does doubling each of the following quantities affect the force between a dipole and a particle placed near the dipole and carrying charge q? (*a*) the charge q, (*b*) the dipole separation d of the dipole, (*c*) the dipole charge q_p, (*d*) the distance between the dipole and the charged particle

As we saw in Chapter 22, electrically neutral objects interact with a charged object because they become polarized in the presence of the charged object. Consider an isolated neutral atom. The centers of the atom's positive and negative charge distributions coincide, and therefore the atom's dipole moment is zero: $d = 0$ and so $\vec{p} = \vec{0}$ (**Figure 23.34a**).The presence of an external electric field—that is, an electric field created by some other charged object—causes a separation between the positive and negative charge centers and so induces a dipole moment (Figure 23.34*b*). To understand the interaction between charged objects and neutral ones, we must therefore study the interaction between a charged particle and what is called an **induced dipole**. The first question to ask is: How does the magnitude of the induced dipole moment depend on the presence of a charged particle?

When a neutral atom is placed in an electric field $\vec{E}$, it is found that, as long as the electric forces exerted by that field on the charged particles in the atom are not too large, the induced dipole separation d_{ind} in the atom obeys Hooke's law. In other words, the induced dipole separation is proportional to the magnitude of the applied electric force, $F^E_d = cd_{ind}$, with c being the "spring constant" of the atom. We can rewrite this as $d_{ind} = (1/c)F^E_d$, and because d_{ind} is proportional to the magnitude of the induced dipole moment p_{ind} and the magnitude of the force exerted on the dipole is proportional to the magnitude E of the electric field at the position of the dipole, the **induced dipole moment** is proportional to the field at the position of the dipole:

$$\vec{p}_{ind} = \alpha\vec{E} \quad (\vec{E} \text{ not too large}), \tag{23.24}$$

where α, the **polarizability** of the atom, is a constant that expresses how easily the charge distributions in the atom are displaced from each other. The SI unit of polarizability is $C^2 \cdot m/N$.

Figure 23.34 A charged particle induces a dipole in an electrically neutral atom.

(*a*) Neutral atom

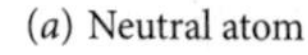

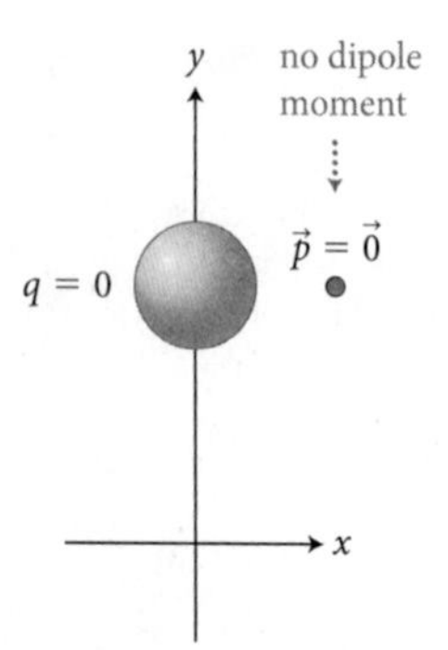

(b) Charged particle induces dipole in atom

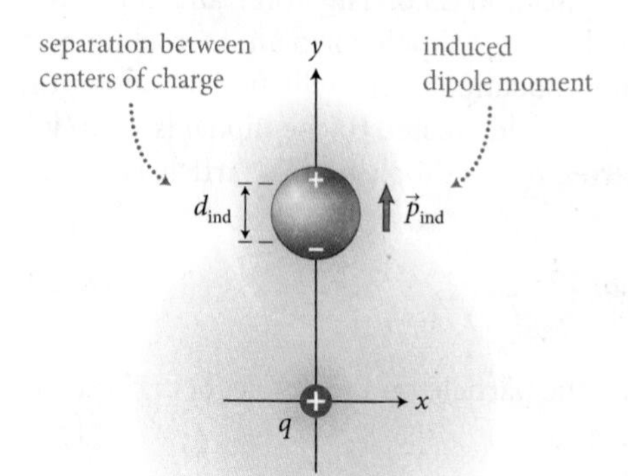

(*c*) Charged particle and dipole interact

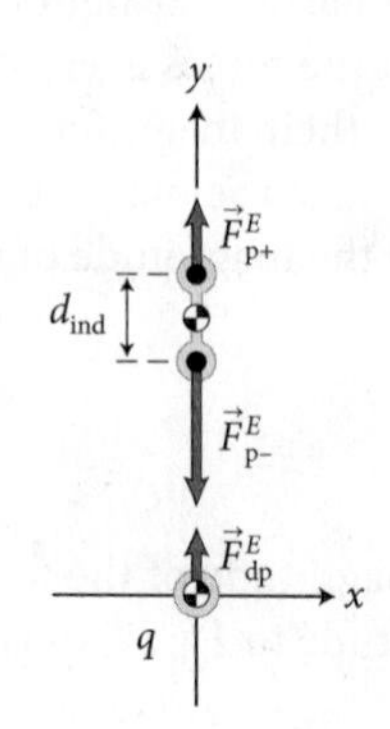

23.18 Given that the induced dipole moment $\vec{p}_{\text{ind}}$ points from the negative to the positive end of an induced dipole and the electric field $\vec{E}$ displaces the positive charge center in the direction of the electric field and the negative charge center in the opposite direction, do you expect the polarizability α to be positive or negative?

The electric field of a charged particle is given by Eq. 23.4, so the magnitude of the induced dipole moment is proportional to the inverse square of the distance between the particle and the dipole:

$$p_{\text{ind}} = \alpha E = \alpha\, k\frac{q}{y^2}. \tag{23.25}$$

In contrast, the dipole moment of a permanent dipole is constant.

We can now substitute the induced-dipole result of Eq. 23.25 into Eq. 23.23 to determine the force exerted by a charged particle on an induced dipole:

$$F^E_{\text{pd}} = 2k\frac{p_{\text{ind}}q}{y^3} = \alpha\frac{2k^2q^2}{y^5}. \tag{23.26}$$

This result shows that the interaction between a charged particle and a polarized object depends much more strongly on the distance between them ($1/y^5$) than does the interaction between two charged objects ($1/y^2$). You may have noticed this in Chapter 22 when comparing the attraction between two charged strips of tape with the attraction between a charged strip and a neutral object.* As the neutral object approaches the charged strip, the force varies so fast with distance that it is often difficult to prevent the tape from sticking to the neutral object.

23.19 (*a*) How does doubling the charge q_A carried by an object A affect the force exerted by A on another charged particle? (*b*) How does doubling q_A affect the force exerted by A on an induced dipole? (*c*) Explain why your answers to parts *a* and *b* are the same or different. (*d*) Can the force exerted by a charged particle cause a torque on an induced dipole?

*Try it! Pull two strips of transparent tape out of a dispenser, suspend one from the edge of a table and then move the other slowly toward it. Notice how the interaction between the strips varies relatively smoothly as a function of separation. Next, move your hand slowly toward the suspended strip and note how the force increases rapidly.

Chapter Glossary

SI units of physical quantities are given in parentheses.

Charge density, linear λ (C/m), surface σ (C/m^2), or volume ρ (C/m^3): A scalar that is a measure of the amount of charge per unit of length, area, or volume on a one-, two-, or three-dimensional object, respectively.

Dipole (electric) A neutral charge configuration in which the center of positive charge is separated from the center of negative charge by a small distance. Dipoles can be *permanent*, or they can be *induced* by an external electric field.

Dipole moment (electric) $\vec{p}$ (C · m) A vector defined as the product of the *dipole charge* q_p (the positive charge of the dipole) and the vector $\vec{r}_p$ that points from the center of negative charge to the center of positive charge:

$$\vec{p} \equiv q_p \vec{r}_p. \tag{23.9}$$

Electric field $\vec{E}$ (N/C) A vector equal to the electric force exerted on a charged test particle divided by the charge on the test particle:

$$\vec{E} \equiv \frac{\vec{F}_t^E}{q_t}. \tag{23.1}$$

Induced dipole A separation of the positive and negative charge centers in an electrically neutral object caused by an external electric field.

Induced dipole moment $\vec{p}_{ind}$ (C · m) A dipole moment induced by an external electric field in an electrically neutral object. For small electric fields, the induced dipole moment in an atom is proportional to the applied electric field:

$$\vec{p}_{ind} = \alpha \vec{E}, \tag{23.24}$$

where α is the *polarizability* of the atom.

Interaction field or **field** A physical quantity surrounding objects that mediates an interaction. Objects that have mass are surrounded by a *gravitational field*; those that carry an electrical charge are surrounded by an *electric field*. Both are *vector fields* specified by a direction and a magnitude at each position in space.

Polarizability α (C^2 · m/N) A scalar measure of the amount of charge separation that occurs in an atom or molecule in the presence of an externally applied electric field.

Superposition of electric fields The electric field of a collection of charged particles is equal to the vector sum of the electric fields created by the individual charged particles:

$$\vec{E} = \vec{E}_1 + \vec{E}_2 + \cdots. \tag{23.5}$$

Test particle An idealized particle whose physical properties (mass or charge) are so small that the particle does not perturb the particles or objects generating the field we are measuring.

Vector field diagram A diagram that represents a vector field, obtained by plotting field vectors at a series of locations.

24 Gauss's Law

In principle, Coulomb's law allows us to calculate the electric field produced by any discrete or continuous distribution of charged objects. In practice, however, the calculation is often so complicated that the sums or integrals that arise might require numerical evaluation on a computer. For this reason, it pays to search for additional methods to determine the electric field produced by a charge distribution. In this chapter we develop a relationship between an electric field and its source, known as *Gauss's law,* that can be used to determine the electric fields due to charge distributions that exhibit certain simple symmetries. These symmetries appear in many common applications, which makes Gauss's law an important tool in calculating electric fields. As we shall see in Chapter 30, Gauss's law is one of the fundamental equations of *electromagnetism*—the theory that describes electromagnetic interactions and electromagnetic waves.

24.1 Electric field lines

In Chapter 23 we used vector field diagrams to visualize electric fields. Another way to visualize electric fields, which will help us reach some new insights, is to draw **electric field lines.** These lines are drawn so that at any location the electric field $\vec{E}$ is tangent to them. Because the electric field is a vector, we assign to field lines a direction that corresponds with the direction of the electric field.

To draw an electric field line, imagine placing a test particle carrying a positive charge q_t somewhere near a charge distribution. Then move the test particle a small distance in the direction of the electric force exerted on it. (Remember from Chapter 23 that the electric field points in the same direction as the electric force exerted on a positively charged test particle.) Repeat the procedure to trace out a line (**Figure 24.1**). We label the field lines with the symbol E to remind us that they represent an electric field.

Figure 24.1 Using a positively charged test particle to trace out an electric field line.

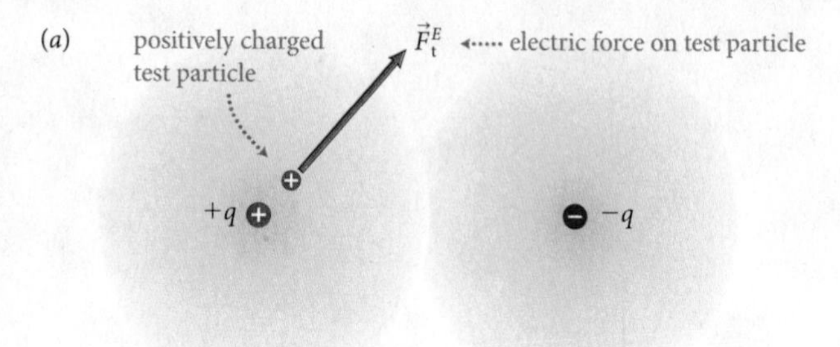
(a)
positively charged test particle
$\vec{F}^E_t$
electric force on test particle
$+q$
$-q$

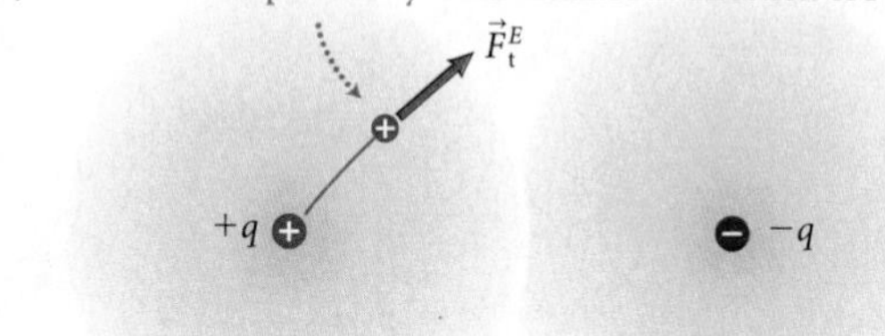
(b)
Move test particle by small distance in direction of $\vec{F}^E_t$.
$\vec{F}^E_t$
$+q$
$-q$

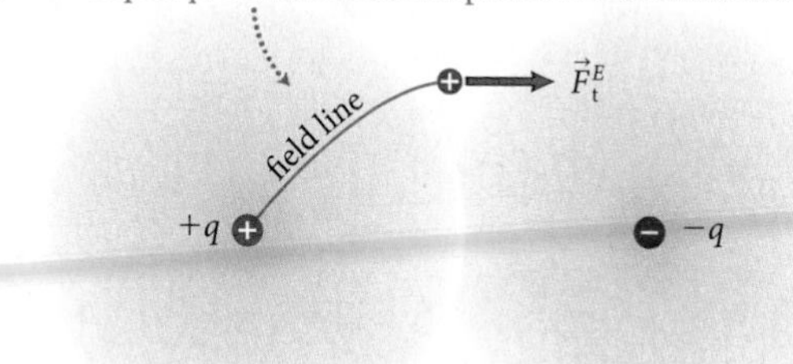
(c)
Repeat procedure. Particle's path traces out field line.
field line
$\vec{F}^E_t$
$+q$
$-q$

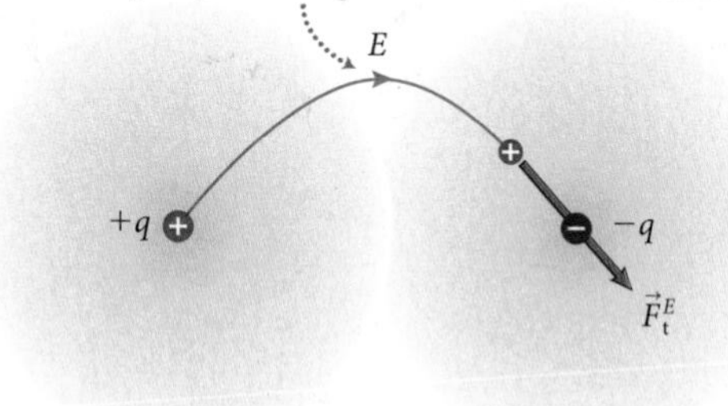
(d)
At each point, field line points in direction of electric force.
E
$+q$
$-q$
$\vec{F}^E_t$

24.1 Draw several field lines representing the electric field of an isolated positively charged particle. Repeat for a negatively charged particle.

As Checkpoint 24.1 illustrates, the field line diagrams for a positive and for a negative isolated charged particle are similar, even though the electric fields point in opposite directions. They point radially outward from a positively charged particle and point radially inward toward a negatively charged particle. This direction means that electric field lines always start from a positively charged object and always end on a negatively charged object, never the other way around.

Because an electric field is present everywhere around a charged object, a field line passes through every location in space. In practice, we draw only a finite number of field lines to represent the entire field. **Figure 24.2**, for example, shows the pattern of field lines created by a pair of oppositely charged particles (that is, a dipole). Sixteen field lines emanate from the positively charged particle on the left, and 16 field lines terminate on the negatively charged particle on the right. Notice the correspondence between this pattern and the corresponding vector field diagram in **Figure 24.3a** and the pattern created by the fibers in Figure 24.3*b*.

The number of field lines that emanate from a positively charged object is arbitrary; we could have chosen some number other than 16 for Figure 24.2. However, in a given field line diagram, the number of field lines is always proportional to the magnitude of the charge carried by the

CONCEPTS

Figure 24.2 Electric field line diagram for an electric dipole.

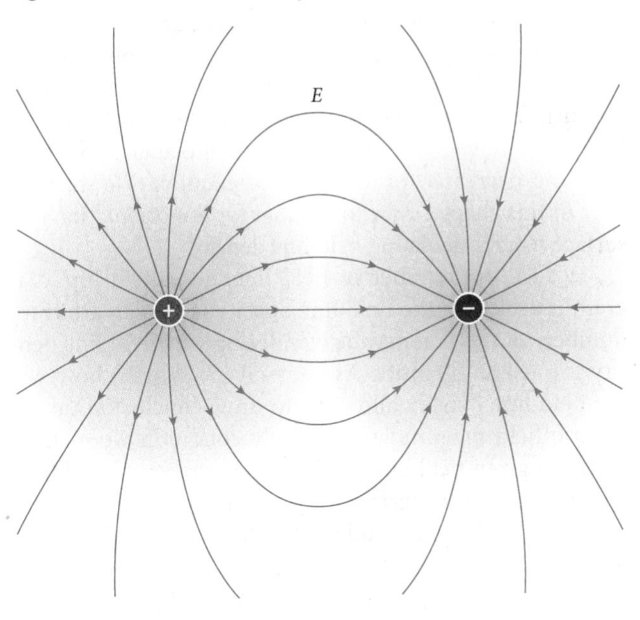

object. If, for example, 16 field lines emanate from an object that carries a charge $+q$, then 32 lines emanate from an object that carries a charge $+2q$ and eight lines terminate on an object that carries a charge $-q/2$.

The number of field lines that emanate from a positively charged object or terminate on a negatively charged object is proportional to the charge carried by the object.

Exercise 24.1 Field lines of infinite charged plate

Draw a field line diagram for an infinite plate that carries a uniform positive charge distribution.

SOLUTION I know from Chapter 23 that the electric field produced by a charged plate of infinite area is always perpendicular to the plate (see Figure 23.30). Thus, the field lines must be straight lines perpendicular to the plate. Because the plate is positively charged, I draw the field lines perpendicular to and away from the plate on either side (**Figure 24.4**). ✔

Figure 24.4

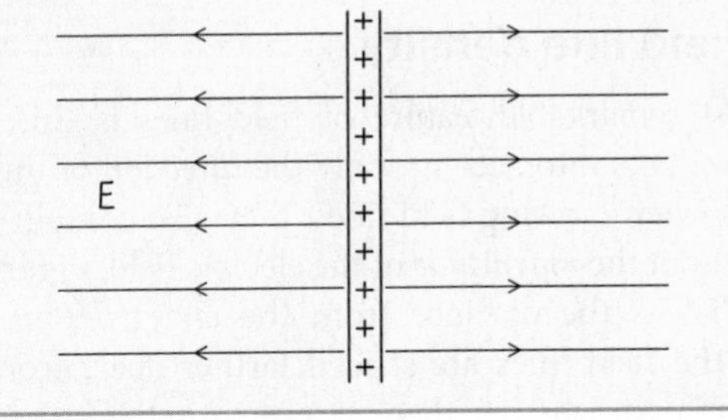

Like vector field diagrams, field line diagrams provide an incomplete view of the electric field and are awkward to draw for all but the simplest charge distributions. Both types of diagram are limited by the two-dimensional nature of illustrations. In particular, you should keep in mind that field lines emanate in three dimensions (**Figure 24.5**), not just in the plane of the drawing.

Figure 24.5 Although we generally use two-dimensional representations of field line diagrams, field lines emanate in three dimensions.

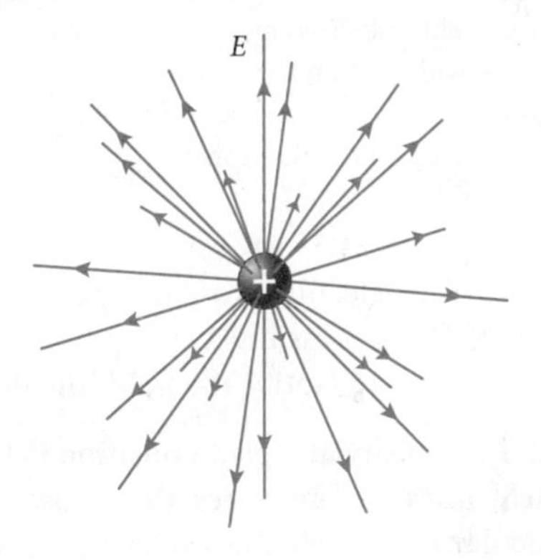

Figure 24.3 Two representations of the electric field of an electric dipole.

(*a*) Vector field diagram of an electric dipole

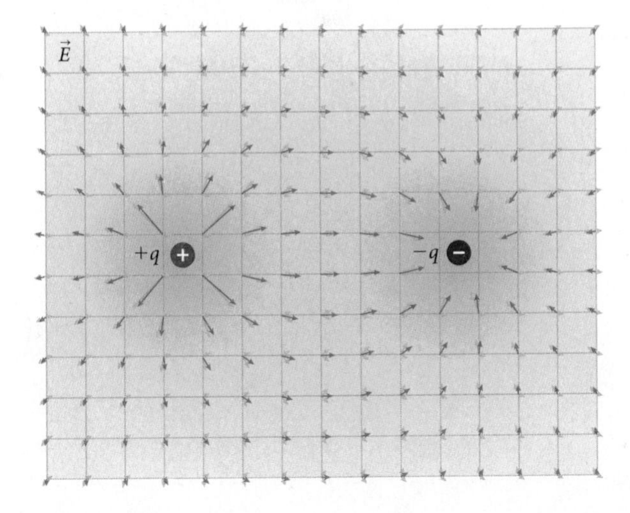

(*b*) Pattern created by electric dipole in a suspension of plastic fibers that align with the electric field

24.2 (*a*) Is it possible for two electric field lines to cross? (Hint: What is the direction of the electric field at the point of intersection?) (*b*) Can two electric field lines touch?

24.2 Field line density

The most remarkable feature of field lines is this: Even though we take into account only the *direction* of the electric field when drawing field lines, they also contain information about the *magnitude* of the electric field. Figure 24.5 shows that as the distance from the charged object increases, the field lines are spaced farther apart from one another. To see whether there is a quantitative correspondence between field line spacing and electric field magnitude, complete the next checkpoint.

24.3 Imagine a hollow sphere enclosing the charged object in Figure 24.5, centered on the object. (*a*) Given that 26 field lines emanate from the charged object, how many field lines cross the surface of the hollow sphere? (*b*) If the radius of the hollow sphere is *R*, what is the number of field line crossings per unit surface area? (*c*) Now consider a second sphere with radius 2*R*, also centered on the charged object. How many field lines cross the surface of this second sphere? (*d*) How does the number of field line crossings per unit area on the second sphere compare with that on the first sphere? (*e*) How does the electric field at a location on the second sphere compare with the field at a location on the first sphere?

From Checkpoint 24.3, we see that the electric field and the number of field line crossings per unit area both decrease as $1/r^2$. To express this correspondence quantitatively, we define a new quantity, the **field line density:**

The field line density at a given position is the number of field lines per unit area that cross a surface perpendicular to the field lines at that position.

Figure 24.6 illustrates why the surface through which the field lines pass must be perpendicular to the field lines. The field represented by the field lines in the figure is uniform—its magnitude and direction are the same everywhere. As you can see in the figure, the number of field lines that cross the surface depends on the orientation of the surface. The number of field lines that cross the surface is maximum when the surface is perpendicular to the field lines and decreases for any other orientation. We shall see later in this chapter how to account for the orientation of a surface when calculating field line density.

Because the number of field lines in a field line diagram is arbitrary, the field line density is also an arbitrary number, and so you may be wondering why field line density is a useful quantity. As you will see shortly, however, the field line density allows us to draw conclusions about electric field magnitudes. The only condition we make is that, in a given field line diagram, the number of field lines emanating from or terminating on charged objects is proportional to the magnitude of the charge carried by these objects.

24.4 (*a*) In Figure 24.6, for what orientation is the number of field lines that cross the surface a minimum? (*b*) How many field lines cross a plane surface of area 0.5 m^2 placed perpendicular to the field lines in Figure 24.6? (*c*) Using your answer to part *b*, what is the number of field line crossings *per unit area* through the 0.5-m^2 surface? (*d*) How does this compare to the number of field line crossings per unit area for the 1-m^2 surface in Figure 24.6*a*?

For the spherical surfaces of Checkpoint 24.3, the field lines are all perpendicular to the surface because the field lines are radial. The number of field line crossings you calculated per unit area *is* the field line density. These results lead us to conclude:

At every position in a field line diagram, the magnitude of the electric field is proportional to the field line density at that position.

The box "Properties of electric field lines" on page 643 summarizes the properties of electric field lines.

Figure 24.6 The number of field lines that cross a given surface depends on the orientation of the surface relative to the field lines.

(*a*)

Plane perpendicular to field lines intersects maximum number of field lines.

area = 1 m^2

E

(*b*)

Same plane at any other orientation intersects fewer field lines.

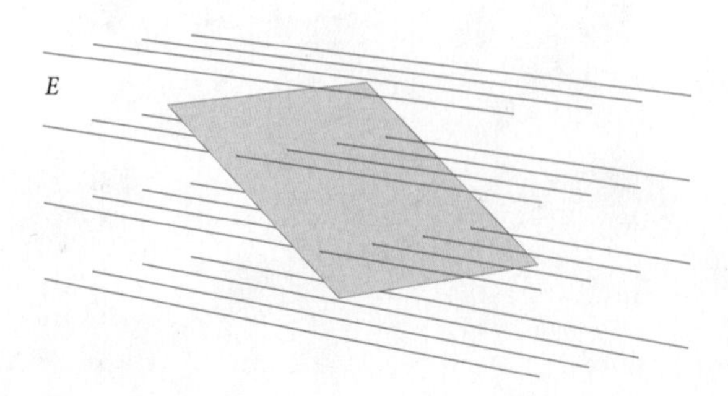

Properties of electric field lines

When working with electric field lines, keep the following points in mind:

1. Field lines emanate from positively charged objects and terminate on negatively charged objects.
2. At every position, the direction of the electric field is given by the direction of the tangent to the electric field line through that position.
3. Field lines never intersect or touch.
4. The number of field lines emanating from or terminating on a charged object is proportional to the magnitude of the charge on the object.
5. At every position, the magnitude of the electric field is proportional to the field line density.

Exercise 24.2 Field strength from field lines

Consider the field line diagram shown in **Figure 24.7**. (*a*) What are the signs of the charges on the two small spherical objects? (*b*) What are the relative magnitudes of these charges? (*c*) What is the ratio of the magnitudes of the electric fields at points P and R? (*d*) Is the electric field zero anywhere in the region shown?

Figure 24.7 Exercise 24.2.

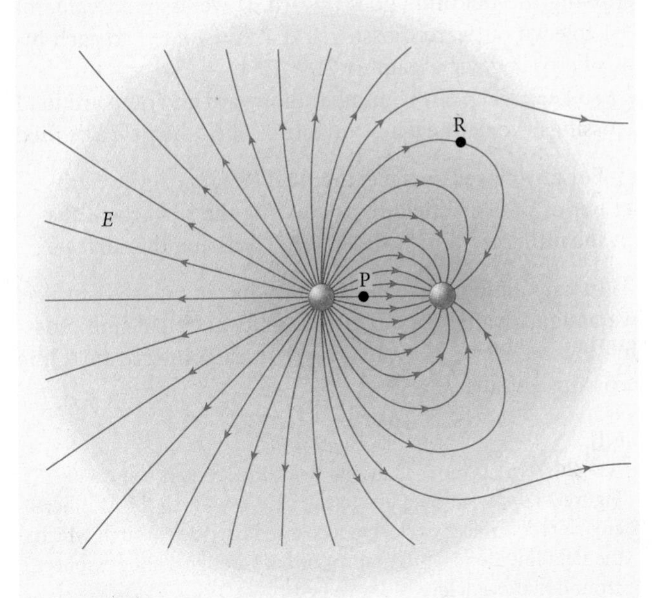

SOLUTION (*a*) Because field lines leave the left object and terminate on the right one, the left object carries a positive charge and the right object carries a negative charge. ✔

(*b*) From Figure 24.7 I see that about twice as many field lines leave the left object as end on the right one. Thus, the charge on the left object is about twice that on the right object.* ✔

(*c*) The magnitude of the electric field at each position is proportional to the field line density at that position. The field line density is equal to the number of field lines per unit length, which is proportional to the inverse of the distance between adjacent field lines. Measuring with a ruler, I see that the distances between adjacent field lines at points P and R are $d_P \approx 1$ mm and $d_R \approx 6$ mm, so

$$\frac{E_P}{E_R} = \frac{d_R}{d_P} \approx \frac{6\text{ mm}}{1\text{ mm}} = 6. ✔$$

(*d*) The absence of field lines on the right suggests that the electric field is small (or even zero). Indeed, if a test particle carrying a positive charge is placed in that region, it is subject to a repulsive force exerted by the positively charged object on the left and an attractive force exerted by the negatively charged object on the right. If the test particle is to the right of the particles and $\sqrt{2}$ as far from the positively charged particle as it is from the negatively charged particle, the vector sum of the two forces is zero and so the electric field at that position is zero. ✔

*If you answered that the magnitude of the charge on the left is four times that on the right because in three dimensions there must be four times as many field lines radiating outward from the left object, don't worry—we've hit on one of the shortcomings of field line representations. In general we shall go by the number of dimensions represented in the drawing (in this case, two).

Note that in Exercise 24.2, half of the field lines leave the area of interest. These field lines either eventually terminate on a negatively charged object (not shown) or continue out to "infinity."

24.5 Imagine moving the hollow sphere of radius R of Checkpoint 24.3*a* sideways so that the charged object is no longer at the center of the sphere (but still within it). (*a*) How does the number of field line crossings through the surface of the sphere change as it is moved? (*b*) How does the average number of field line crossings per unit surface area of the sphere change? (*c*) Does the electric field at a fixed position on the surface of the sphere change or remain the same as the sphere is moved? (*d*) Are your answers to parts *b* and *c* in contradiction, given the relationship between the electric field magnitude and field line crossings per unit area?

24.3 Closed surfaces

Checkpoint 24.5 leads us to another result that will be important in deriving Gauss's law: Whenever a charged particle is placed inside a hollow spherical surface, the number of field lines that pierce the surface is the same *regardless of where inside the surface the particle is placed.* This is true simply because so long as the charged particle is inside the surface, all the field lines emanating from the particle must go through the spherical surface. In fact, we don't even need to use a spherical surface—a

Figure 24.8 Any surface that encloses a positively charged particle is pierced by all the field lines that emanate from that particle, regardless of the shape of the surface and the position of the particle within the surface.

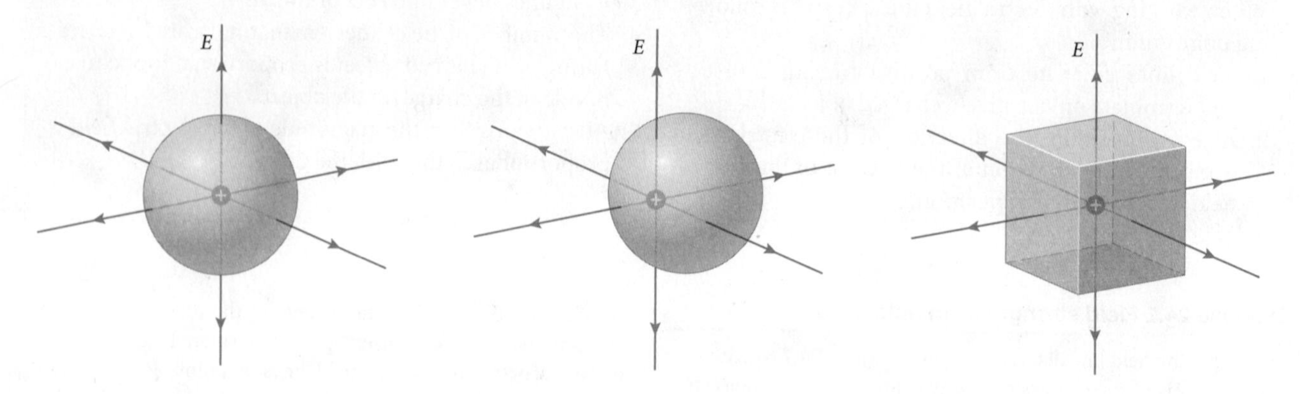

cube-shaped surface or any other surface enclosing the charged particle will do (**Figure 24.8**). In each case, the number of field lines that pierce the surface is equal to the number of field lines that emanate from or terminate on the charged particle enclosed by the surface.

> **24.6** Suppose eight field lines emanate from an object carrying a charge $+q$. How many field lines pierce the surface of a hollow sphere if the sphere contains (*a*) a single object carrying a charge $+2q$ and (*b*) two separate objects, each carrying a charge $+q$? (*c*) If the sphere is pierced by 20 field lines, what can you deduce about the combined charge on objects inside the sphere?

A surface that completely encloses a volume is called a **closed surface**. Checkpoint 24.6 suggests that a direct relationship exists between the number of field lines that cross a closed surface and the **enclosed charge**—the sum of all charge enclosed by that surface. However, what happens if a field line reenters the closed surface, as illustrated in **Figure 24.9a**? Field line 4 now crosses the closed surface *three* times, so the number of field line crossings is not six but eight. If you look closely at the figure, however, you will discover that not all the crossings are the same. For seven of the crossings the field line goes outward (from the inside of the closed surface to the outside), while for the eighth crossing the field line goes inward. If we assign a value of $+1$ to each outward crossing and a value of -1 to each inward crossing, we obtain $(+7) + (-1) = 6$.

To keep track of the number of inward and outward field crossings, we define a new quantity called the *field line flux*:

For any closed surface, the *field line flux* is the number of outward field lines crossing the surface minus the number of inward field lines crossing the surface.

In calculating the field line flux for any closed surface, we assign a value of $+1$ to each outward field line crossing the surface and a value of -1 to each inward field line crossing the surface

> **24.7** (*a*) If more than one field line reenters the donut in Figure 24.9a, what happens to the field line flux? (*b*) Are there any closed surfaces enclosing a charged particle through which the field line flux is different from that through a simple sphere around that particle?

Figure 24.9 The number of field lines exiting a closed surface minus the number entering it is always equal to the number of field lines generated inside the surface.

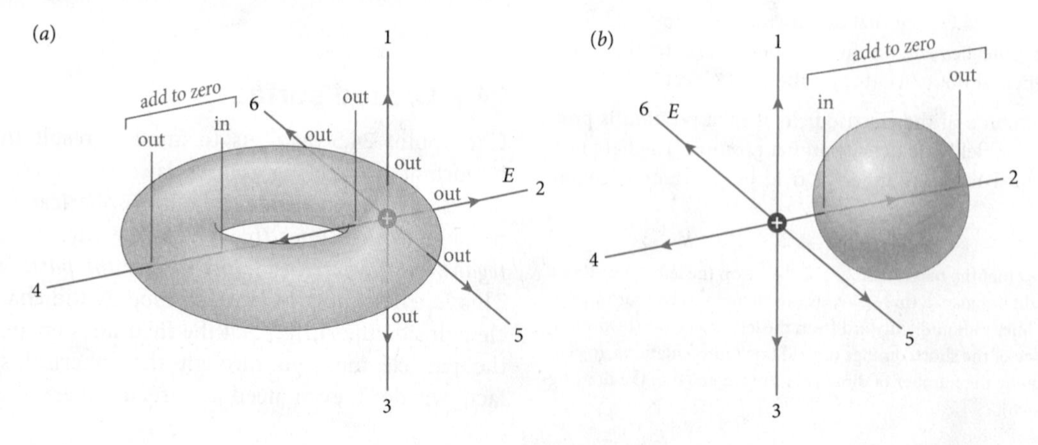

The field line flux through any closed surface is always equal to the number of field lines that originate from within that surface minus the number of field lines that terminate on charged objects within that surface. So far, we have drawn an arbitrary number of field lines, but we can make our statement more precise:

The *field line flux* through a closed surface is equal to the charge enclosed by the surface multiplied by the number of field lines per unit charge.

What about charged objects outside the closed surface? To see what effect such charged objects have, complete the following checkpoint.

24.8 (*a*) What is the field line flux through the closed spherical surface in Figure 24.9*b* due to a charged particle outside the sphere? (*b*) Does your answer to part *a* change if we move the particle around (but keep it outside the volume enclosed by the surface)?

Checkpoint 24.8 demonstrates a very important point:

The field line flux through a closed surface due to charged objects outside the volume enclosed by that surface is always zero.

This means that if we know the field line flux through a closed surface, then we can determine the charge enclosed by that surface, regardless of the distribution of charge outside the surface. This statement is a form of Gauss's law, which we shall describe mathematically in Section 24.7.

Example 24.3 Flux of an electric dipole

Consider the three-dimensional dipole field line diagram shown in **Figure 24.10**. Six field lines emanate from the positively charged end, and six terminate on the negatively charged end. (*a*) What is the field line flux through the surface of the cube that encloses the positively charged end shown in the figure? (*b*) What is the field line flux through the surface of a similar cube that encloses the negatively charged end?

Figure 24.10 Example 24.3.

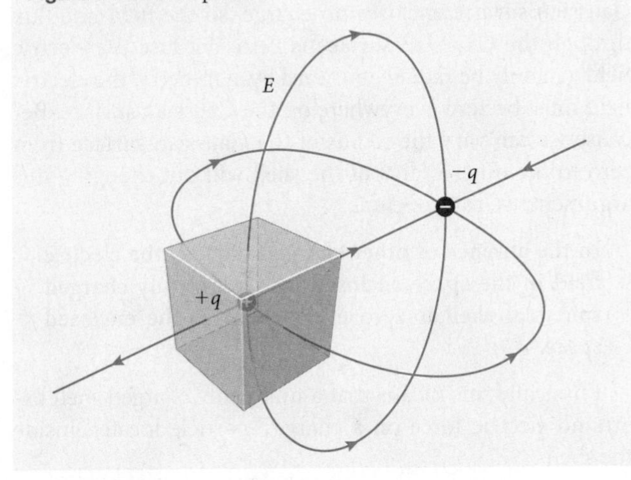

SOLUTION (*a*) Six field lines emanate from the positively charged particle. Each line crosses the surface of the cube in the outward direction and thus contributes a value of +1 to the field line flux. The field line flux is +6. ✔

(*b*) Six field lines terminate on the negatively charged particle, so there are again six field line crossings. However, these field lines are directed inward and so the field line flux is −6. ✔

24.9 What is the field line flux through the surface of a rectangular box that encloses *both* ends of an electric dipole?

The relationship between the field line flux through a closed surface and the enclosed charge is important because it can help us determine one from a knowledge of the other. For example, in the next section we shall use this relationship to derive two important theorems about isolated conducting objects.

24.10 Consider the two-dimensional field line diagram in **Figure 24.11**, part of which is hidden from view. (*a*) If the object in the top left carries a charge of +1 C, what is the charge enclosed in the region that is hidden? (*b*) What is the field line flux through a surface that encloses the entire area represented by the diagram?

Figure 24.11 Checkpoint 24.10: What is the charge inside the dashed region?

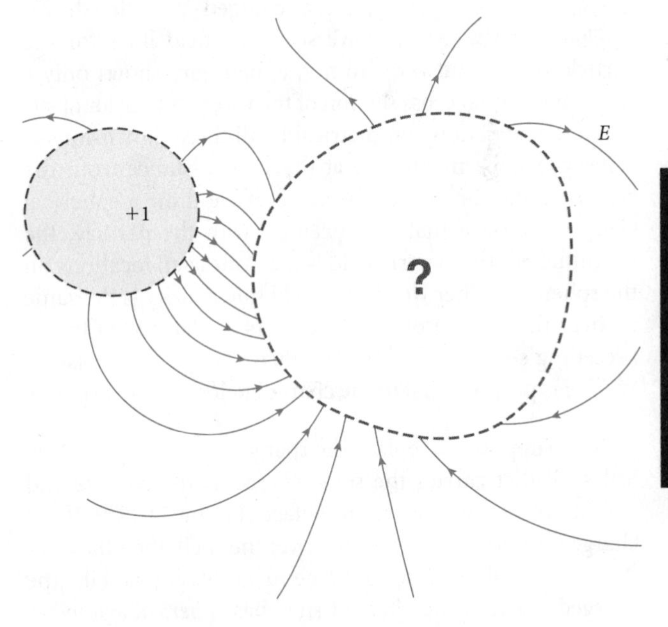

24.4 Symmetry and Gaussian surfaces

The relationship between field line flux and enclosed charge allows us to reach several important conclusions about charged objects and their electric fields without having to do

Figure 24.12 Using spherical Gaussian surfaces to examine the electric fields of a charged particle and a uniformly charged spherical shell. The electric fields, Gaussian surfaces, and charged shell are spherical and are shown here in cross section.

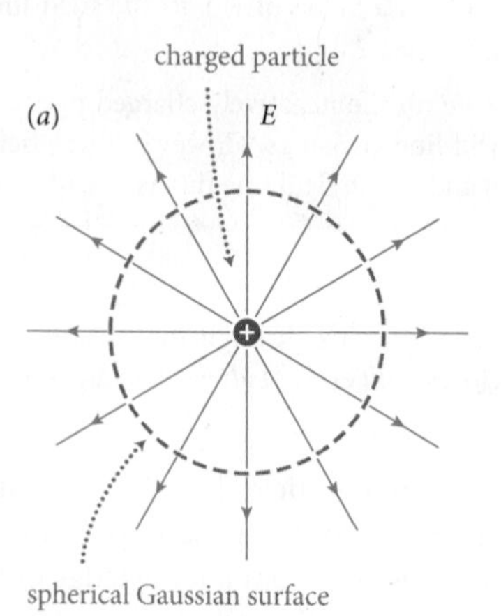
charged particle
(a)
E
spherical Gaussian surface

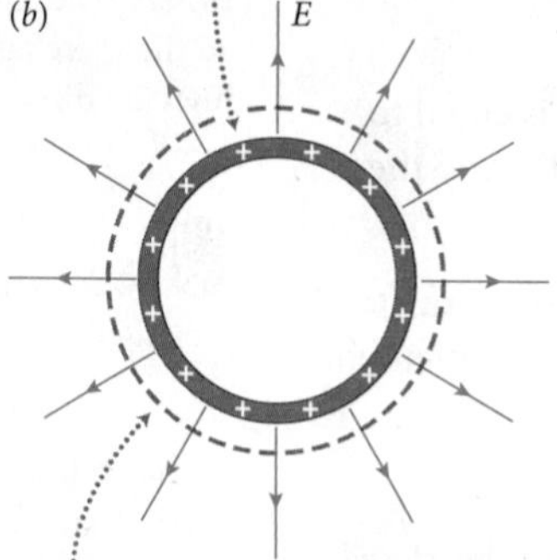
hollow shell with same charge as particle
(b)
E
Gaussian surface outside shell: electric field is same as for particle.

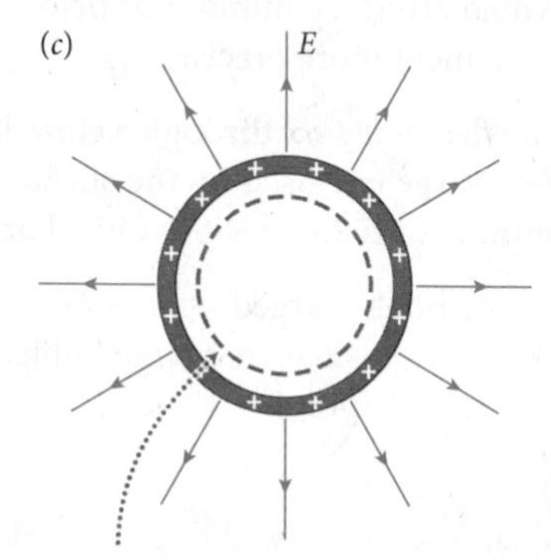
(c)
E
Gaussian surface inside shell: encloses no charge, so electric field must be zero.

any calculations. To apply this relationship in a given situation, we first need to select a closed surface. This surface need not correspond to a real object—any surface, real or imagined, will do. We'll refer to these closed surfaces as **Gaussian surfaces.** The choice of surface is dictated by the symmetry of the situation at hand. As a rule of thumb, we choose a surface such that the electric field is the same (and possibly zero) everywhere along as many regions of the surface as possible, because such a choice makes it easy to determine the field line flux through the surface.

Consider, for example, the charged particle shown in Figure 24.12*a*. As we have seen, the field lines for the particle radiate outward from it. (The figure shows only a two-dimensional cross section of the three-dimensional situation.) The field is symmetrical in all three dimensions—it has the same magnitude at the same distance from the center in any direction. Therefore, if we draw a spherical Gaussian surface that is concentric with the particle, the magnitude of the electric field is the same at all locations on the sphere. In other words, the field line density is the same all over the surface of the sphere. As we have seen in the preceding section, the field line flux through the Gaussian surface is proportional to the charge enclosed by the sphere.

Now suppose we replace the charged particle by a spherical shell that carries the same charge as the particle and still fits within our Gaussian surface (Figure 24.12*b*). If the charge is uniformly distributed over the shell, then the electric field should still be the same in all directions. Like the charged particle, the charged shell has *spherical symmetry* (see the box "Symmetry and Gauss's law" on page 647): Reorienting the spherical shell by rotating it over an angle about any axis does not change the charge configuration and so should not change the electric field at a given location. This means that the field lines should again be straight lines radiating uniformly outward. Also, the field line flux through the Gaussian surface should still be the same because the surface encloses the same amount of charge. The only way that the field line fluxes through the Gaussian surfaces in Figures 24.12*a* and *b* can be uniform *and* equal in magnitude is if the electric fields are the same at every position on the spherical Gaussian surface. Because this argument holds for a spherical Gaussian surface of any radius, we can conclude:

The electric field outside a uniformly charged spherical shell is the same as the electric field due to a particle that carries an equal charge located at the center of the shell.

This means that a uniformly charged shell exerts a electric force on a charged particle outside the shell as if all the shell's charge were concentrated at the center of the shell. Because a sphere may be viewed as a collection of shells, we can extend this statement to uniformly charged spheres.

Let us now turn our attention to the electric field in the space enclosed by the shell. We draw a spherical Gaussian surface that fits within the shell (Figure 24.12*c*). This Gaussian surface encloses no charge, so the field line flux through the Gaussian surface is zero. Because the electric field can only be radially outward by symmetry, the electric field must be zero everywhere on the Gaussian surface. Because we can vary the radius of the Gaussian surface from zero to the inner radius of the shell without changing this argument, we can conclude:

In the absence of other charged objects, the electric field in the space enclosed by a uniformly charged spherical shell is zero everywhere in the enclosed space.

Physically this means that a uniformly charged shell exerts no electric force on a charged particle located inside the shell.

Symmetry and Gauss's law

The symmetry of an object is determined by its *symmetry operations*—manipulations that leave its appearance unchanged (see Section 1.2). A sphere, for example, looks the same if we reorient it by rotating it about any axis (Figure 24.13*a*). This type of symmetry is called **spherical symmetry.** An infinitely long, cylindrical rod does not look any different if we rotate it, reverse it, or translate it about its long axis (Figure 24.13*b*). The rod is said to have **cylindrical symmetry.** An infinite flat sheet has **planar symmetry:** It remains unchanged if it is rotated about an axis perpendicular to the sheet or translated along either of the two axes perpendicular to this axis (Figure 24.13*c*).

Many other types of symmetry may occur, but these three types play an important role in electrostatics. For charge configurations that exhibit any of these three symmetries, we can calculate the electric field due to the charge distribution directly using Gauss's law.

Because objects are never infinite, they cannot exhibit true cylindrical or planar symmetry. However, for a long straight wire or a large flat sheet we can often obtain good results by assuming they have cylindrical or planar symmetry. When we work problems, the words *long* and *large* imply that you may assume the object has infinite dimensions compared to other length scales of interest.

Figure 24.13 Three symmetries important for applications of Gauss's law.

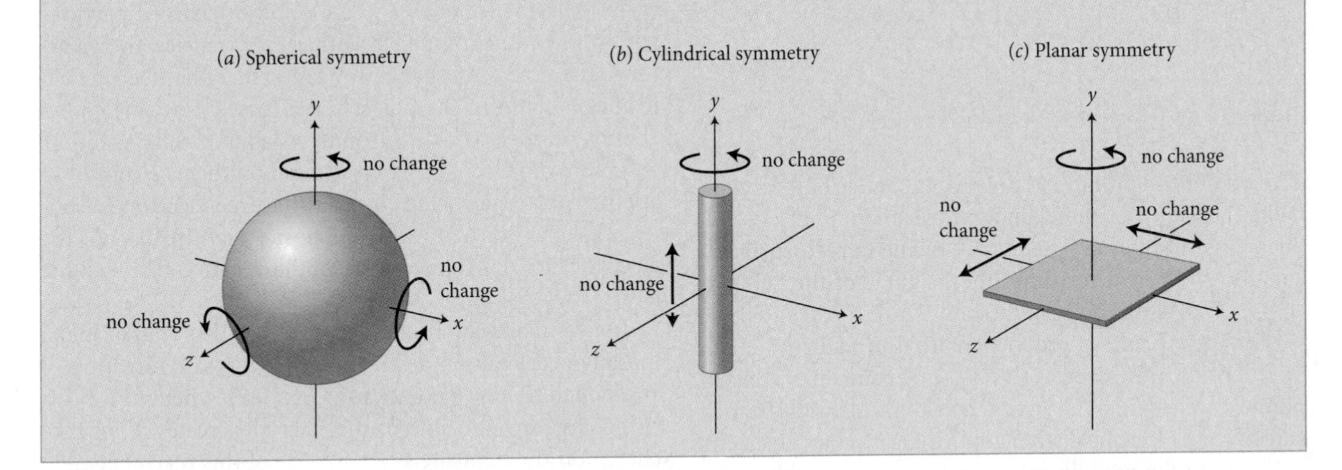

24.11 There are two reasons the field line flux through a closed surface may be zero: because the field is zero everywhere or because the outward flux is balanced by an equal inward flux. Why can't the latter situation be true for the Gaussian surface in Figure 24.12*c*?

Particles, shells, and spheres are the only objects that exhibit spherical symmetry. **Figure 24.14** illustrates a different type of symmetry: the *cylindrical symmetry* of an infinitely long, uniformly charged straight wire.* Because of this symmetry, rotating the wire about its axis or moving it along the axis should not have any effect on the electric field at any position in space. For this to be the case, the field lines must be arranged radially along planes that are perpendicular to the wire (Figure 24.14). We can take advantage of this symmetry by drawing a Gaussian surface in the shape of a cylinder that is concentric with the wire, as shown in Figure 24.14.

*For the wire to exhibit cylindrical symmetry, it has to be infinitely long. If the wire has finite length, you can tell when it is moved along its axis.

Figure 24.14 The electric field of an infinite uniformly charged wire exhibits cylindrical symmetry. We can examine this field by surrounding the wire with a concentric cylindrical Gaussian surface.

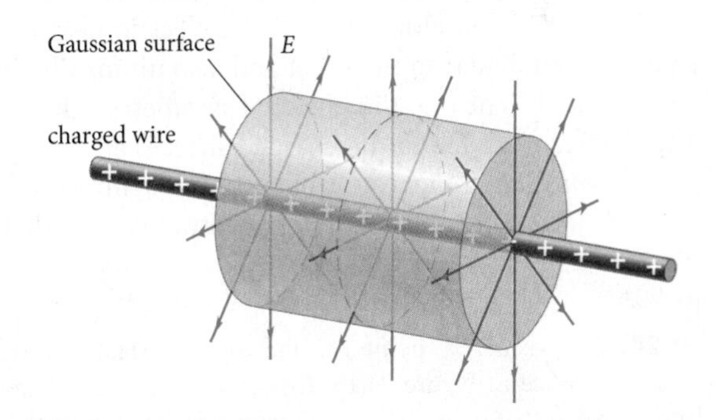

24.12 Consider a point on the curved part of the Gaussian surface in Figure 24.14. Does the magnitude of the electric field at that point increase, decrease, or stay the same if you (*a*) change the location of the point on the curved surface or (*b*) increase the radius of the Gaussian surface? (*c*) What is the field line flux through the left and right surfaces of the Gaussian surface?

We can use the cylindrical Gaussian surface to determine how the electric field due to the charged wire decreases with

Figure 24.15 The electric field of a uniformly charged sheet exhibits planar symmetry. We examine its field by drawing a cylindrical Gaussian surface that straddles the sheet.

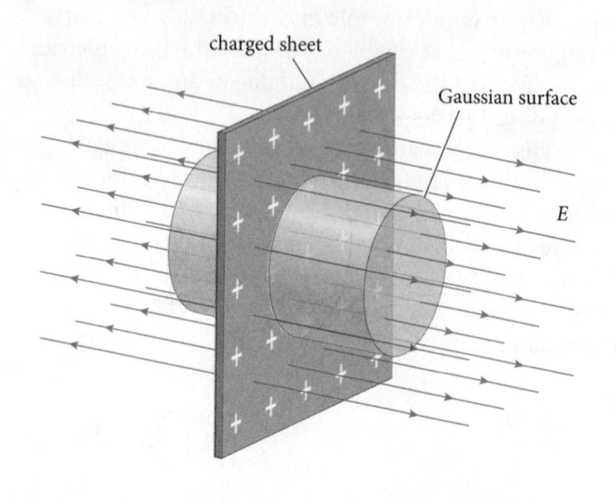

distance from the wire. As you can see from the figure, the field line flux through the curved surface of the cylinder is independent of its radius r. No matter how large we make the radius of the cylinder, the same number of field lines pass through it. The area A of the curved surface is equal to the perimeter times the height h of the cylinder: $A = 2\pi rh$. As we increase the radius r of the cylinder, the surface area increases proportionally to r, and so the field line density must decrease as $1/r$ to maintain a constant number of field lines. Because the field line density is a measure of the electric field strength, this means that the electric field due to the wire decreases as $1/r$, as we established in Example 23.4. A quantitative expression for the electric field due to a charged rod is given in Section 24.8.

Figure 24.15 shows a situation with a different symmetry: a charged sheet. If the sheet is very large and the charge is uniformly distributed along it, then the electric field lines must be perpendicular to the sheet and also uniformly distributed along it. The one-dimensional symmetry exhibited by the electric field due to the charged sheet is an example of *planar symmetry*. To take advantage of this symmetry, we draw a cylindrical Gaussian surface that straddles the sheet, as illustrated in Figure 24.15.

24.13 Consider a point on the right surface of the Gaussian surface in Figure 24.15. Does the magnitude of the electric field at that point increase, decrease, or stay the same if you (*a*) change the location of the point on the right surface, or (*b*) increase the height h of the Gaussian surface? (*c*) Is the field line flux through the right surface of the Gaussian surface positive, negative, or zero? (*d*) How does the field line flux through the right surface compare to that through the left surface? (*e*) Is the field line flux through the curved surface of the Gaussian surface positive, negative, or zero?

Because the area of the right surface and the field line flux through that surface don't change as we change the height of the cylinder, we conclude that the electric field line density doesn't change with distance to the plane. Hence the magnitude of the electric field due to the charged sheet is the same everywhere (see also Example 23.6).

24.5 Charged conducting objects

Let as now apply the relationship between field line flux and enclosed charge to charged conducting objects. As we saw in Chapter 22, conducting materials permit the free flow of charge carriers within the bulk of the material. Conducting objects typically contain many charge carriers that are free to move, such as electrons (in a metal) or ions (in a liquid conductor). The material as a whole can still be electrically neutral; a neutral piece of metal, for example, contains as many positively charged protons as negatively charged electrons.

A consequence of this free motion of charged particles within a conducting object is that the particles always arrange themselves in such a way as to make the electric field inside the bulk of the object zero. To see how this comes about, consider a free electron in a slab of metal. If no field is present, no electric force is exerted on the electron. If we apply an external field, however, the free electron is subject to a force in a direction opposite the direction of the electric field (opposite because of the negative charge of the electron).

In a similar way, all the free electrons in a slab of metal initially accelerate in a direction opposite the direction of an applied field (**Figure 24.16**). This leaves behind a positive charge on one side of the slab and creates a negative charge on the opposite side. Because of this rearrangement of charge, an induced electric field builds up in a direction opposite the direction of the external field. As a result, the electric field inside the slab, which is the sum of the external electric field and the induced electric field, decreases. As this field decreases, so does the force exerted on the free electrons in the slab. When enough charge carriers have accumulated on each side of the slab to make the electric field inside the slab zero, the electric force exerted on the free electrons in the metal becomes zero and the material reaches **electrostatic equilibrium**—the condition in which the distribution of charge in a system does not change. The time interval it takes for a metal to reach electrostatic equilibrium is very short (about 10^{-16} s), so the rearrangement of charge carriers is virtually instantaneous. The important point to remember is:

The electric field inside a conducting object that is in electrostatic equilibrium is zero.

Keep in mind that this statement holds *only* in electrostatic equilibrium. When charge carriers are made to flow through a conducting object—as in any electric or electronic device, like your stereo or a refrigerator—the electric field is *not* zero inside the object!

Suppose now we add charge to a conducting object. It makes sense to assume that the charged particles will

Figure 24.16 Why the electric field inside the bulk of a conducting object is zero when the object is in electrostatic equilibrium.

(*a*) No electric field

neutral metal

(*b*) Electric field just switched on

External field accelerates free electrons . . .

$E_{external}$

(*c*) Electrostatic equilibrium established

. . . creating charge separation . . .

. . . which induces internal electric field opposing external field.

$E_{external}$

$E_{induced}$

$E_{external}$

Once field within metal is zero, $\vec{E}_{inside} = \vec{E}_{external} + \vec{E}_{induced} = \vec{0}$, equilibrium is reached.

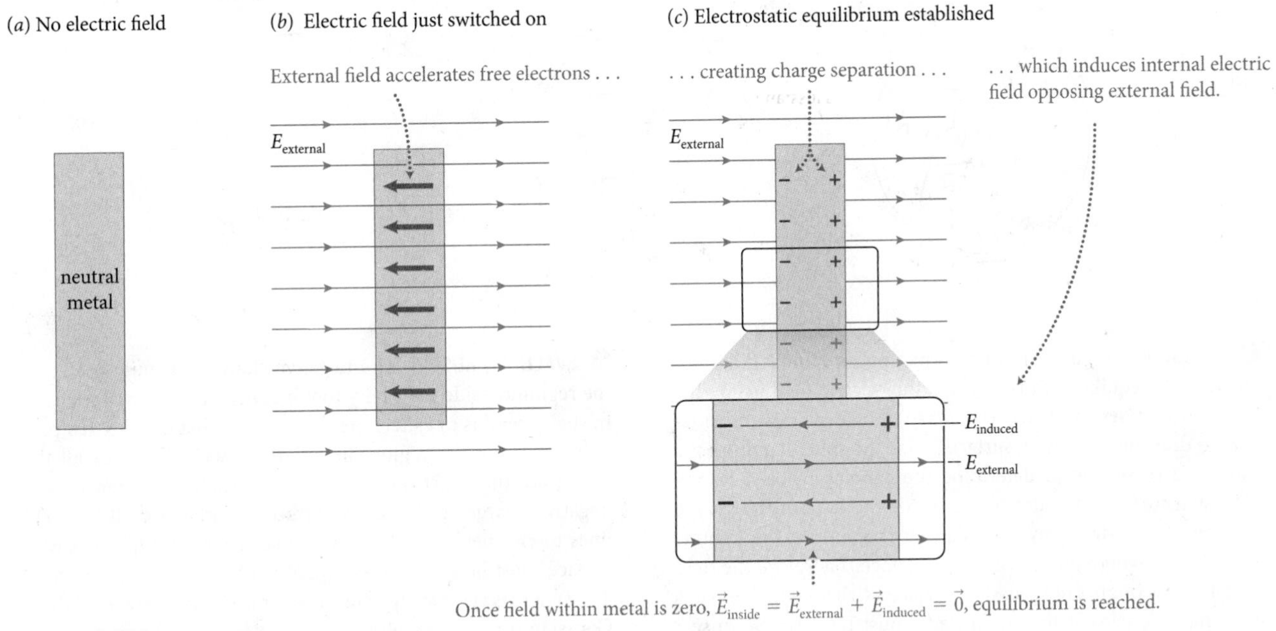

arrange themselves over the object in such a way as to spread out as far as possible from one another, given that particles carrying like charges repel. We can use a Gaussian surface to obtain a better understanding of where the charged particles go.

24.14 Consider a spherical Gaussian surface inside a positively charged conducting object that has reached electrostatic equilibrium. (*a*) Is the field line flux through the Gaussian surface positive, negative, or zero? (*b*) What can you conclude from your answer to part *a* about the charge enclosed by the Gaussian surface?

We can extend the result of Checkpoint 24.14 to conducting objects of any shape. Consider, for example, the irregularly shaped, charged conducting object shown in **Figure 24.17**. Draw a Gaussian surface of the same shape as the object, just below its surface. Given that the field is zero everywhere inside the conducting object, the field line flux through the Gaussian surface is zero and the charge enclosed by the Gaussian surface is also zero. Because we can choose the Gaussian surface arbitrarily close to the surface of the object, we conclude:

Any surplus charge placed on an isolated conducting object arranges itself at the surface of the object. No surplus charge remains in the body of the conducting object once it has reached electrostatic equilibrium.

Figure 24.17 Because the electric field inside a conducting object in electrostatic equilibrium is zero, we conclude that there cannot be any surplus charge inside the object.

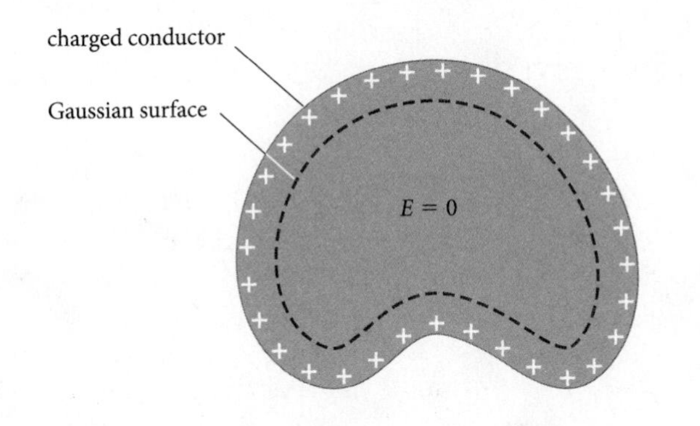

24.15 Suppose the charged conducting object in Figure 24.17 contains an empty cavity. Does any surplus charge reside on the inner surface of the cavity?

Example 24.4 Charged particle in a cavity

An electrically neutral, conducting sphere contains an irregularly shaped cavity. Inside the cavity is a particle carrying a positive charge $+q$. What are the sign and magnitude of the charge on the sphere's outer surface?

❶ GETTING STARTED I am told that a sphere made of material that is an electrical conductor has a cavity in its interior and that a particle in the cavity carries charge $+q$. My task is to determine the sign and magnitude of any charge residing on the sphere's outer surface. I begin by sketching a vertical cross section through the sphere showing the cavity and the charged particle inside it (**Figure 24.18*a*** on the next page). The problem states that the sphere is electrically neutral, but the question posed implies that some charge resides on its outer surface. Because $\vec{E} = \vec{0}$ inside the conducting material, I know that an equal quantity of the opposite charge must accumulate somewhere else on the sphere.

Figure 24.18

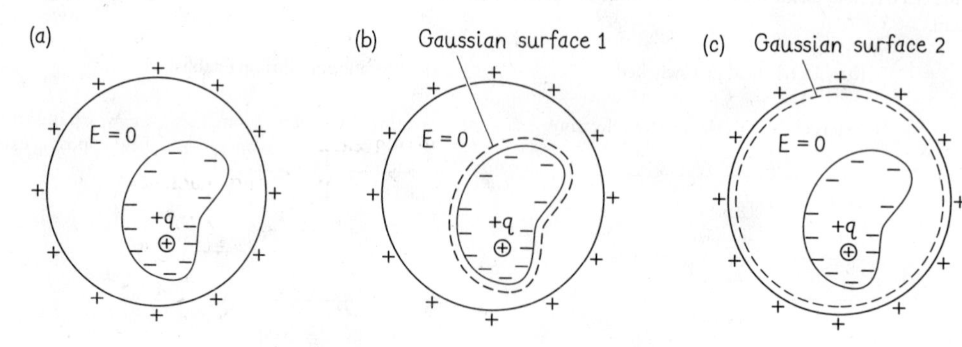

❷ **DEVISE PLAN** The sphere is conducting, so I know that once electrostatic equilibrium is reached, the electric field inside the bulk of the sphere must be zero: $\vec{E} = \vec{0}$. I can use this information to draw any Gaussian surface inside the sphere and use the following reasoning to determine the charge enclosed by my Gaussian surface: Because $\vec{E} = \vec{0}$ inside the bulk of the sphere and because I draw my Gaussian surface inside the sphere, $\vec{E} = \vec{0}$ everywhere on the Gaussian surface. Therefore the field line flux through the Gaussian surface is zero, which means the charge enclosed by this surface must be zero. Because I can draw my Gaussian surface anywhere inside the bulk of the conductor, I can use this information to determine the distribution of charge on the sphere.

❸ **EXECUTE PLAN** I begin by drawing a Gaussian surface 1 enclosing the cavity (Figure 24.18*b*). Because the field line flux through this surface is zero, the charge enclosed by the surface must be zero. There is charge $+q$ inside the surface (in the charged particle), however, and so in order for the charge enclosed by the surface to be zero, a quantity of charge $-q$ must have migrated from someplace in the region surrounding the cavity and accumulated on the inner cavity surface.

Because the sphere is electrically neutral, the charge $-q$ that migrated to the inner cavity surface must leave a charge $+q$ behind somewhere else on the sphere. If I now draw Gaussian surface 2 just inside the sphere's outer surface (Figure 24.18*c*), I see that, because the field line flux through this Gaussian surface has to be zero, the charge enclosed by this surface is also zero. The positive charge $+q$ that results from the migration of charge $-q$ to the cavity surface must therefore reside outside Gaussian surface 2. Because I can draw surface 2 arbitrarily close to the sphere's outer surface, I conclude that the sphere's outer surface carries a charge $+q$. ✔

❹ **EVALUATE RESULT** The negative charge that migrates from the region outside the cavity to the cavity surface arranges itself in such a way as to cancel the electric field that the charged particle creates in the region outside the cavity. Therefore all the field lines that start on the charged particle must end on the negative charge at the cavity surface. In order for all the field lines to end here, the quantity of negative charge on the cavity surface must be equal to the quantity of charge on the particle. The sphere is electrically neutral, and my choice for where I draw Gaussian surface 2 requires that all the positive charge resulting from the migration of negative charge to the cavity surface must accumulate outside Gaussian surface 2, which means right at the sphere's outer surface. Thus my answer makes sense.

That the electric field inside any conductor is zero in electrostatic equilibrium allows us to draw one additional important conclusion. Because the electric fields must be zero everywhere, including at the surface of a conducting object, there cannot be any component of the electric field parallel to the surface of the object, and therefore we can conclude:

In electrostatic equilibrium, the electric field at the surface of a conducting object is perpendicular to that surface.

If there were a component of the electric field parallel to the surface, that component would cause any free charge carrier to move along the surface, which means the conductor is not in electrostatic equilibrium.

24.16 In Example 24.4, is the electric field inside the cavity zero?

CONCEPTS

Self-quiz

(For this self-quiz assume all situations are two-dimensional.)

1. In Figure 24.19, which of the two charged spheres carries a charge of greater magnitude?
2. Consider Gaussian surfaces 1–3 in Figure 24.19. Determine the field line flux through each surface.
3. In Figure 24.19, is the field line density greater at point A or point B? At which of these locations is the magnitude of the electric field greater? Is the field line density at point C zero or nonzero?
4. The electric field lines in Figure 24.20 tell you there must be one or more charged particles inside the Gaussian surface defined by the dashed line. Could the electric field shown be due to a single particle inside the Gaussian surface? What must the signs and relative magnitudes of the charged particle(s) be in order to create the electric field lines shown?
5. Figure 24.21 shows a small ball that carries a charge of $+q$ inside a conducting metal shell that carries a charge of $+2q$. (*a*) What are the sign and magnitude of the charge on the inner surface of the shell? (*b*) What are the sign and magnitude of the charge on the outer surface of the shell?

Figure 24.19

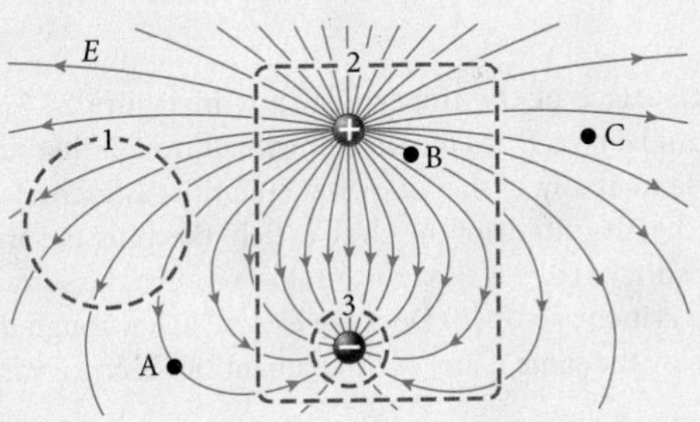

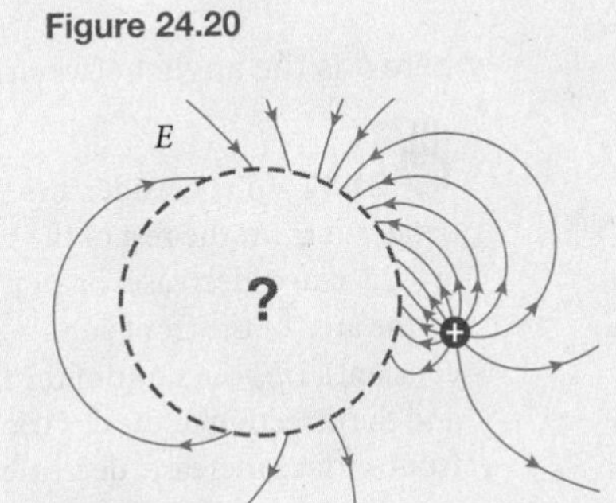

Figure 24.20

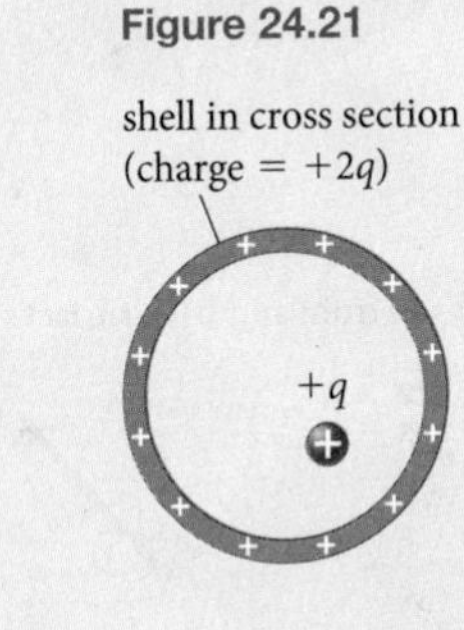

Figure 24.21

Answers

1. The number of field lines is proportional to the charge on the object. Because more lines emanate from the charged object on the top, that object must carry a greater charge.
2. For surface 1, all lines that enter the surface also exit the surface, so the field line flux is zero. For surface 2, 25 lines exit the surface and 6 lines enter the surface. The field line flux through surface 2 is thus $25 - 6 = 19$. Fifteen field lines cross surface 3, with all lines entering from the outside. The field line flux for surface 3 is -15.
3. Point B has the greater field line density because the lines are closer together at B than they are at A. The magnitude of the electric field is greater at point B because electric field strength is proportional to the field line density. Even though C is not on a field line, the field line density, which is represented by the spacing of the field lines *around* point C, is nonzero.
4. Because electric field lines converge on one point near the top of the area enclosed by the surface and diverge from a point near the bottom, there must be objects that carry both negative and positive charges inside the surface. Because more field lines enter the surface than exit the surface, the negatively charged object(s) must carry a charge of greater magnitude than the positively charged object(s).
5. (*a*) Because the electric field in the conducting shell is zero, the field line flux through a Gaussian surface drawn inside the material of the conducting shell must be zero. According to Gauss's law, the charge enclosed in the surface must also be zero. The charge on the inner surface of the shell must therefore be $-q$, which added to $+q$ gives zero. (*b*) For a neutral shell, a charge of $-q$ on the inside surface of the shell would leave a surplus of $+q$ on the outside surface of the shell. The surplus charge of $+2q$ that was placed on the shell also resides on the outside surface. The charge residing on the outside surface of the shell is thus $+3q$.

24.6 Electric flux

We introduced two important concepts in the first part of this chapter: the field line density, which is proportional to the strength of the electric field, and the field line flux, which represents the number of field lines going outward through a closed surface minus the number of field lines going inward. In this section we'll turn these concepts into quantities we can calculate.

Consider, for example, a trapezoidal box in a uniform electric field $\vec{E}$ (**Figure 24.22a**). The field line flux through the closed surface of the trapezoidal box is zero: Twenty field lines go into the back surface and 20 come out through the front surface. Another way of putting this is to say that the field line flux into the back surface is equal in magnitude to the field line flux out of the front surface. Instead of using field lines, however, whose number is chosen arbitrarily, we'll work with a quantity called the **electric flux,** represented by the symbol Φ_E (Φ is the Greek capital phi). The magnitude of the electric flux through a surface with area A in a uniform electric field of magnitude E is defined as

$$\Phi_E \equiv EA\cos\theta \quad \text{(uniform electric field)}, \tag{24.1}$$

where θ is the angle between the electric field and the normal to the surface.

Figure 24.22 Determining the electric flux through a flat surface.

(*a*) Trapezoidal box in uniform electric field

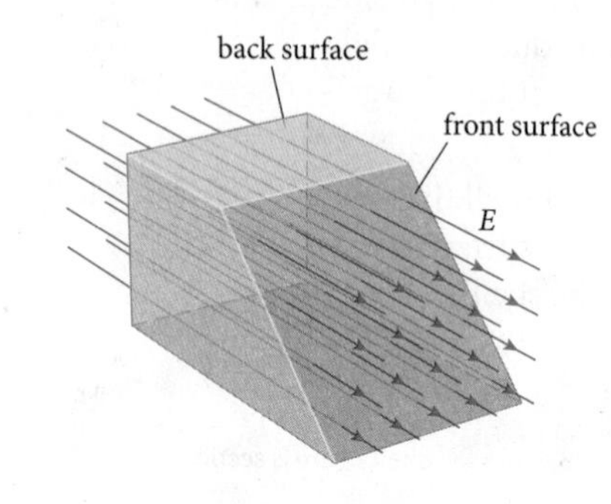

(*b*) Vector areas of front and back surfaces

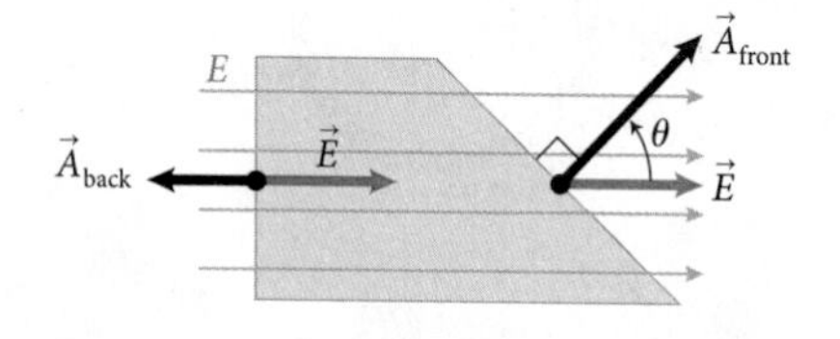

24.17 (*a*) Consider the front surface of the trapezoidal box in Figure 24.22*a*, detached from the rest of the trapezoidal box. Does the field line flux through that surface increase, decrease, or stay the same if any of the following quantities is increased: (*i*) the area of the front surface, (*ii*) the magnitude of the electric field (keeping the area constant), (*iii*) the slope of the front surface (that is to say, the angle between the surface and the direction of the electric field is increased)? (*b*) Does the *electric flux* through the front surface increase, decrease, or stay the same if any of these quantities is changed?

As Checkpoint 24.17 shows, electric flux, as we've defined it in Eq. 24.1, behaves like the field line flux. To make the correspondence more precise, we define an *area vector* $\vec{A}$ for a flat surface area as a vector whose magnitude is equal to the surface area A and whose direction is normal to the plane of the area. On closed surfaces we choose $\vec{A}$ to point outward (we'll deal with open surfaces later). Figure 24.22*b* shows the area vectors associated with the front and back surfaces of the (closed) trapezoidal box. With this definition, Eq. 24.1 can be written as a scalar product (Eq. 10.33):

$$\Phi_E \equiv EA\cos\theta = \vec{E}\cdot\vec{A} \quad \text{(uniform electric field)}, \tag{24.2}$$

where θ is the angle between $\vec{E}$ and $\vec{A}$. Electric flux is a scalar, and the SI unit of electric flux is $\text{N}\cdot\text{m}^2/\text{C}$.

24.18 Let the area of the back surface of the trapezoidal box in Figure 24.22 be $1.0\ \text{m}^2$, the magnitude of the electric field be $1.0\ \text{N/C}$, and $\theta = 30°$ for the front surface. (*a*) What are the magnitudes of the area vectors for the front and back surfaces of the trapezoidal box? (*b*) What are the electric fluxes through the front and back surfaces?

Figure 24.23 To obtain the electric flux through an irregularly shaped, nonplanar surface and/or for a nonuniform electric field, we divide the surface into small segments. For very small segments, each segment is essentially flat and the field through each segment is essentially uniform. The flux through the entire surface is then given by the sum of all of the contributions through each segment.

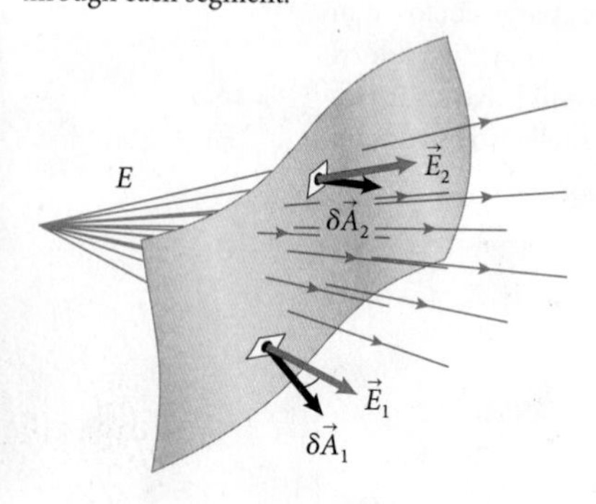

The above definition of electric flux applies only for uniform electric fields and flat surfaces. Let us therefore consider the more general case of an irregular surface in a nonuniform field (**Figure 24.23**). To calculate the electric flux through that surface, we divide the entire surface into small segments of surface area δA_i, with each segment being small enough that we can consider it to be essentially flat, and we can define an area vector $\delta\vec{A}_i$ whose magnitude is equal to the surface area δA_i of the segment and whose direction is normal to the segment. This allows us to apply Eq. 24.2 to each individual segment. The electric flux through a single segment is then $\Phi_{Ei} = \vec{E}_i \cdot \delta\vec{A}_i$, where $\vec{E}_i$ is the electric field vector at the location of the segment. To calculate the electric flux through the entire surface we must sum the electric flux through all the surface segments:

$$\Phi_E = \Sigma\vec{E}_i \cdot \delta\vec{A}_i. \tag{24.3}$$

If we let the area of each segment approach zero, then the number of segments approaches infinity and the sum is replaced by an integral:

$$\Phi_E = \lim_{\delta A_i \to 0} \Sigma\vec{E}_i \cdot \delta\vec{A}_i = \int \vec{E} \cdot d\vec{A}. \tag{24.4}$$

The integral in Eq. 24.4 is called a *surface integral* (see Appendix B); $d\vec{A}$ is the area vector of an infinitesimally small surface segment. If the surface is closed, this surface integral is written as

$$\Phi_E = \oint \vec{E} \cdot d\vec{A}, \tag{24.5}$$

where the circle through the integral sign indicates that the integration is to be taken over the entire closed surface and $d\vec{A}$ is chosen to point outward. Because evaluating a surface integral is mathematically more complicated than single-variable integration, it is important to exploit any symmetry to simplify the calculation.

Example 24.5 Cylindrical Gaussian surface in a uniform electric field

Consider a cylindrical Gaussian surface of radius r and length ℓ in a uniform electric field $\vec{E}$, with the length axis of the cylinder parallel to the electric field (**Figure 24.24**). What is the electric flux Φ_E through this Gaussian surface?

Figure 24.24 Example 24.5.

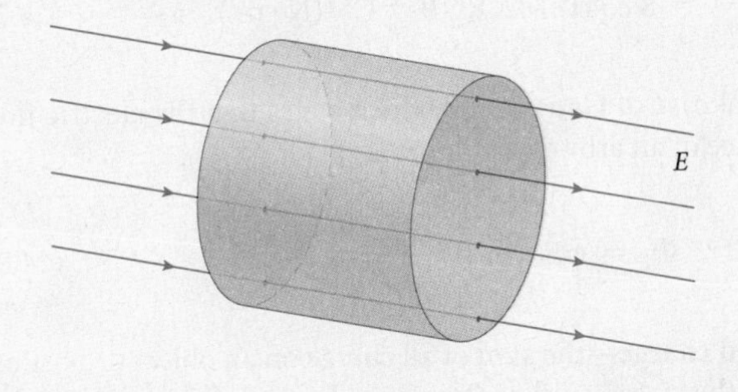

❶ **GETTING STARTED** From Figure 24.24 I see that a cylindrical Gaussian surface consists of three regions: front and back flat surfaces and a curved surface joining them. The electric field is perpendicular to the front and back flat surfaces and parallel to the curved surface.

❷ **DEVISE PLAN** The electric flux is given by Eq. 24.5, so I can calculate the electric flux through the Gaussian surface by applying Eq. 24.5 to each of the three regions and then summing the three contributions to the electric flux. In order to evaluate the scalar product $\vec{E} \cdot d\vec{A}$ for each region, I sketch a side view of the Gaussian surface showing the vectors $\vec{E}$ and $d\vec{A}$ for each of the three regions (**Figure 24.25**).

Figure 24.25

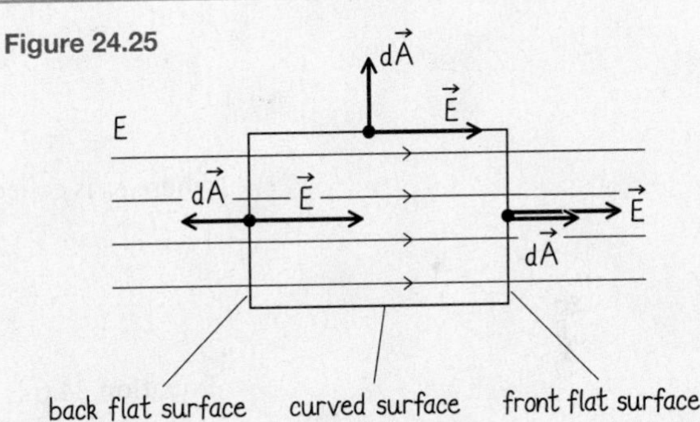

❸ **EXECUTE PLAN** Applying Eq. 24.5 to the three regions, I can write the flux through the Gaussian surface as the sum of three surface integrals: one over the back surface, one over the curved surface, and one over the front surface:

$$\Phi_E = \oint \vec{E} \cdot d\vec{A} = \int_{\text{back flat surface}} \vec{E} \cdot d\vec{A} + \int_{\text{curved surface}} \vec{E} \cdot d\vec{A} + \int_{\text{front flat surface}} \vec{E} \cdot d\vec{A}. \tag{1}$$

From my sketch, I see that on the back flat surface the angle between $\vec{E}$ and $d\vec{A}$ is 180°, so $\vec{E} \cdot d\vec{A} = E(\cos 180°)\,dA = -E\,dA$. Because the magnitude of the electric field is the same everywhere, I can pull E out of the integral:

$$\int_{\text{back flat surface}} \vec{E} \cdot d\vec{A} = \int (-E)\,dA = -E\int dA = -E(\pi r^2),$$

where πr^2 is the area of the back flat surface.

(Continued)

The integral over the curved region of the Gaussian surface yields a value of zero because the angle between between $\vec{E}$ and $d\vec{A}$ is 90° everywhere on the curved region, so $\vec{E} \cdot d\vec{A} = E(\cos 90°)\, dA = 0$.

Finally, for the front flat surface I have $\vec{E} \cdot d\vec{A} = E(\cos 0°)\, dA = E\, dA$, so

$$\int_{\text{front flat surface}} \vec{E} \cdot d\vec{A} = \int E\, dA = E \int dA = E(\pi r^2).$$

Adding up the three terms in Eq. 1 yields

$$\Phi_E = -E(\pi r^2) + 0 + E(\pi r^2) = 0. ✔$$

❹ **EVALUATE RESULT** Because there is no charge enclosed by the Gaussian surface, I know that the field line flux through the surface must be zero, a fact I confirm by looking at Figure 24.24: The four field lines shown contribute a flux of -4 on the back flat surface, $+4$ on the front flat surface, and 0 along the curved surface, for a total of $-4 + 4 + 0 = 0$. It therefore makes sense that the electric flux through this Gaussian surface also is zero.

24.19 Consider a spherical Gaussian surface of radius r with a particle that carries a charge $+q$ at its center. (*a*) What is the magnitude of the electric field due to the particle at the Gaussian surface? (*b*) What is the electric flux through the sphere due to the charged particle? (*c*) Combining your answers to parts *a* and *b*, what is the relationship between the electric flux through the sphere and the enclosed charge q_{enc}? (*d*) Would this relationship change if you doubled the radius r of the sphere?

24.7 Deriving Gauss's Law

Checkpoint 24.19 shows that the electric flux through a spherical Gaussian surface is equal to the charge q enclosed by the sphere times $4\pi k$, where $k = 9.0 \times 10^9\ \text{N} \cdot \text{m}^2/\text{C}^2$ is the proportionality constant that appears in Coulomb's law (see Eqs. 22.1 and 22.5). This relationship is usually written in the form

$$\Phi_E = 4\pi k q = \frac{q}{\epsilon_0}, \tag{24.6}$$

where ϵ_0 is called the **electric constant:**

$$\epsilon_0 \equiv \frac{1}{4\pi k} = 8.85418782 \times 10^{-12}\ \text{C}^2/(\text{N} \cdot \text{m}^2). \tag{24.7}$$

Equation 24.6 is a special case of **Gauss's law,** which states that the electric flux through the closed surface of an arbitrary volume is

$$\Phi_E = \oint \vec{E} \cdot d\vec{A} = \frac{q_{\text{enc}}}{\epsilon_0}, \tag{24.8}$$

where q_{enc} is the **enclosed charge**—the sum of all charge on an object or portion of an object enclosed by the closed surface. The formal proof of Gauss's law is an extension of the calculation you performed in Checkpoint 24.19 and is shown in the box "Electric flux though an arbitrary closed surface".

Gauss's law is a direct consequence of Coulomb's law with its $1/r^2$ dependence and the superposition of electric fields. In that respect, it contains nothing new. However, as the next section shows, Gauss's law greatly simplifies the calculation of electric fields due to charge distributions that exhibit one of the three symmetries we discussed in Section 24.4.

24.20 Suppose Coulomb's law showed a $1/r^{2.00001}$ dependence instead of a $1/r^2$ dependence. (*a*) Calculate the electric flux through a spherical Gaussian surface of radius R centered on a particle carrying a charge $+q$. (*b*) Substitute your result in Eq. 24.8. What do you notice?

QUANTITATIVE TOOLS

Electric flux through an arbitrary closed surface

Consider a particle that carries a positive charge $+q$, surrounded by the irregularly shaped closed surface shown in **Figure 24.26a**.

1. To determine the electric flux through the irregular surface, we divide the volume enclosed by the surface into small square wedges that taper to a point at the charged particle, one of which is shown. We calculate the electric flux through the surface segment dA cut out by each wedge and then sum the contributions from all the wedges.
2. To determine the electric flux through dA in Figure 24.26*a*, we draw two spherical Gaussian surfaces around q: one with a radius r_1 equal to the distance between dA and q, the other with an arbitrary radius r_2. Our wedge from step 1 now defines two other small surface segments dA_1 and dA_2 on these two spheres.
3. If the segment dA is made very small, then the field lines in the wedge are nearly parallel to one another. Therefore, according to what we found in Section 24.6, the electric flux through dA is equal to that through dA_1 (Figure 24.26*b*). In addition, as you showed in Checkpoint 24.19, this flux is also equal to that through surface segment dA_2. In fact, the electric flux is the same through *any* surface that cuts through the wedge. Put differently, any surface that cuts through the wedge intercepts the same number of field lines.
4. We can repeat this procedure for each wedge. Each time, we see that the electric flux through a segment dA on the irregular surface is equal to that through a corresponding segment dA_2. As we add the contributions of all the wedges, we conclude that the electric flux through the closed irregular surface is equal to that through a sphere of arbitrary radius r centered on q.
5. If there is more than one charged particle inside the irregularly shaped surface, we can use the above arguments for each particle individually and then use the superposition of electric fields (Section 23.3):

$$\vec{E} = \Sigma \vec{E}_i,$$

where $\vec{E}_i$ is the electric field due to particle i alone. The electric flux due to all charged particles is then the sum of the electric fluxes due to the individual electric fields:

$$\Phi_E = \Sigma \Phi_{Ei} = \Sigma \frac{q_i}{\epsilon_0} = \frac{q_{\text{enc}}}{\epsilon_0}.$$

24.21 Suppose q is *outside* the irregularly shaped surface in Figure 24.26. Show that the electric flux due to q through the closed surface is zero. (Hint: Draw a small wedge from q through the surface and determine the electric flux through the two intersections between the wedge and the surface.)

Figure 24.26 Formal proof of Gauss's law.

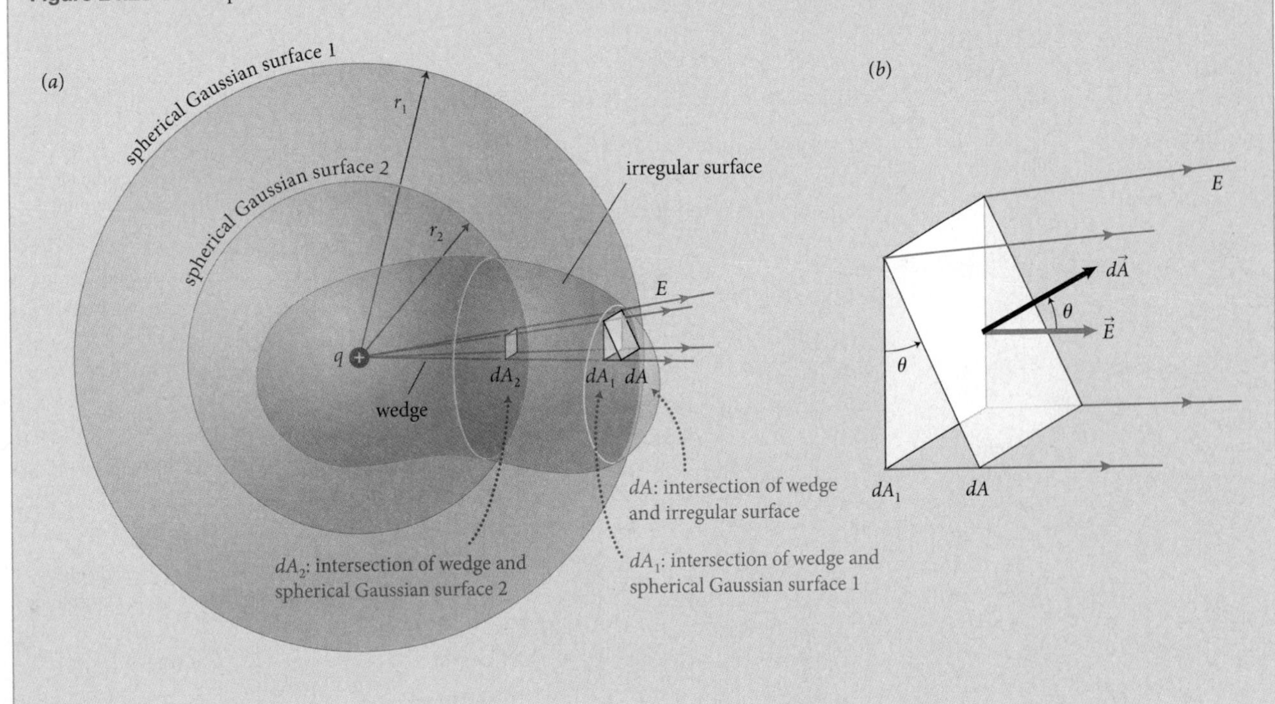

24.8 Applying Gauss's Law

Gauss's law relates the electric flux through a closed surface to the charge enclosed by it. In Section 24.5 we encountered one important application of Gauss's law: If we can choose a Gaussian surface such that the electric field is zero everywhere on that surface (inside a conducting material, for example), then we know that the charge enclosed by that surface must be zero. In this section we show how Gauss's law can be used to avoid having to carry out any integrations to calculate the electric field. In principle one can calculate the electric flux through any surface, but the calculation is not trivial in general. For the surfaces and charged objects shown in Section 24.4 and summarized in **Figure 24.27**, however, the electric flux is easy to calculate because the field lines are either parallel to the surface (in which case the electric flux is zero) or perpendicular to it and the magnitude of the electric field is constant (in which case the electric flux is simply the product of the magnitude of the electric field E and the surface area A).

To see the benefit of Gauss's law, consider a charged spherical shell of radius R that carries a uniformly distributed positive charge q. The electric field due to this charged shell can be calculated using the procedure outlined in Section 23.7: Divide the shell into infinitesimally small segments that carry a charge dq, apply Coulomb's law to each small segment, and integrate over the entire shell (Eq. 23.15):

$$\vec{E}_{sP} = k\int \frac{dq_s}{r_{sP}^2}\hat{r}_{sP}. \tag{24.9}$$

Figure 24.27 Applying Gauss's law to determine the electric fields of symmetrical charge distributions.

Symmetry of charge distribution	Electric field geometry	Gaussian surface	To find electric flux
Spherical (charged sphere)	$\vec{E}$ radiates uniformly outward in three dimensions.	Concentric sphere 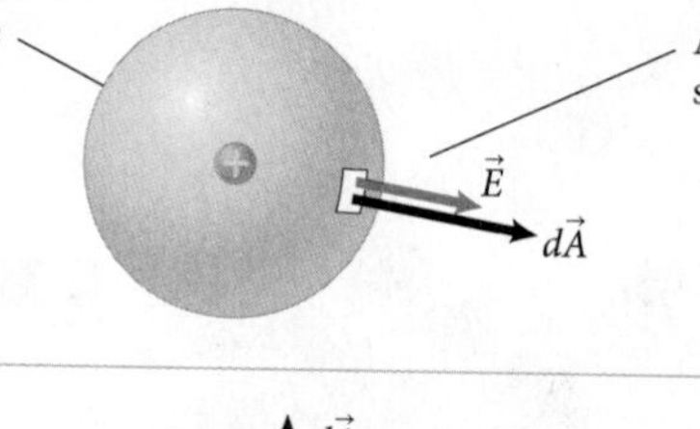	At all points, $\vec{E}$ is perpendicular to surface and has same magnitude.
Cylindrical (infinite charged rod)	$\vec{E}$ radiates uniformly outward perpendicular to axis.	Coaxial cylinder 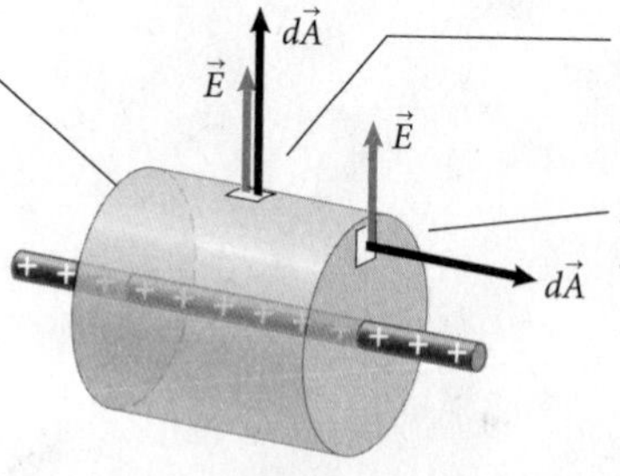	*Cylindrical surface:* At all points, $\vec{E}$ is perpendicular to surface and has same magnitude. *End faces:* $\vec{E}$ is parallel to face, so flux is zero.
Planar (infinite charged sheet)	$\vec{E}$ is uniform and perpendicular to plane.	Cylinder or box perpendicular to plane 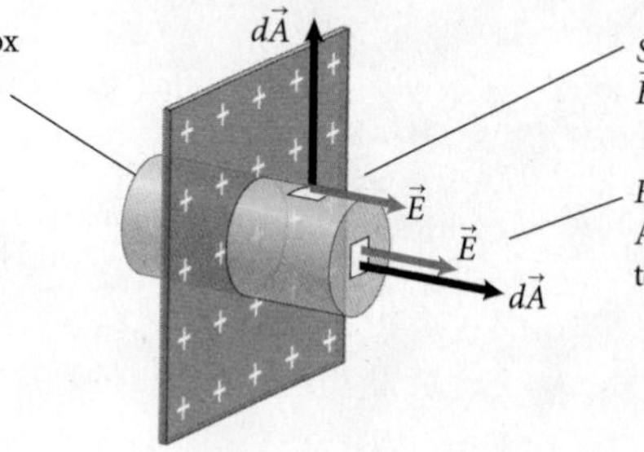	*Surface perpendicular to plane:* $\vec{E}$ is parallel to face, so flux is zero. *Faces parallel to plane:* At all points, $\vec{E}$ is perpendicular to surface and has same magnitude.

As you may imagine, this so-called *direct integration* is no simple matter (see Section 13.8 for a similar integral), even though the result, which we derived qualitatively in Section 24.4, is surprisingly simple. Taking advantage of the symmetry of the problem, however, we can use Gauss's law to calculate the answer in just two steps.

We begin by drawing a concentric spherical Gaussian surface of radius $r > R$ around the shell (Figure 24.28a). According to Gauss's law, the flux through the Gaussian surface is equal to the enclosed charge divided by ϵ_0:

$$\Phi_E = \frac{q_{\text{enc}}}{\epsilon_0} = \frac{q}{\epsilon_0}. \tag{24.10}$$

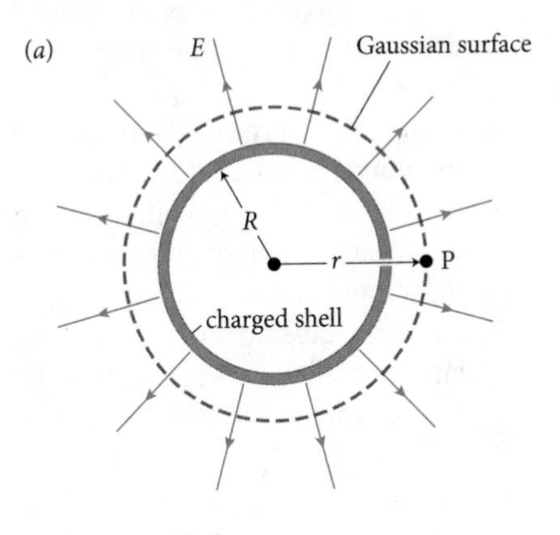

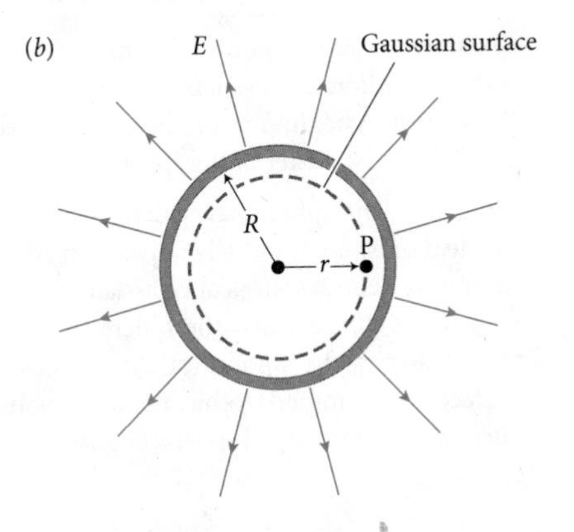

Figure 24.28 Applying Gauss's law to a charged spherical shell.

In addition, we know that because of the spherical symmetry the electric field has the same magnitude E at each position on the Gaussian surface and the field is perpendicular to the surface. Because the electric field is perpendicular, we have $\vec{E} \cdot d\vec{A} = E\,dA$, and because E has the same value everywhere on the Gaussian surface, we can pull the electric field out of the integral in Eq. 24.5:

$$\Phi_E = \oint \vec{E} \cdot d\vec{A} = \oint E\,dA = E \oint dA = EA, \tag{24.11}$$

where A is the area of the spherical Gaussian surface:

$$A = 4\pi r^2. \tag{24.12}$$

Substituting Eq. 24.12 into Eq. 24.11, we obtain

$$\Phi_E = 4\pi r^2 E. \tag{24.13}$$

Combining Eqs. 24.10 and 24.13, we obtain

$$4\pi r^2 E = \frac{q}{\epsilon_0} \tag{24.14}$$

or

$$E = \frac{1}{4\pi\epsilon_0}\frac{q}{r^2} = k\frac{q}{r^2}. \tag{24.15}$$

This is exactly the magnitude of the electric field due to a particle carrying a charge q located at the center of the shell, as we concluded in Example 23.7 for a solid sphere.

We can also use Gauss's law to determine the electric field inside the shell by drawing a concentric spherical Gaussian surface with radius $r < R$ (Figure 24.28*b*). For this surface the enclosed charge is zero, $q_{\text{enc}} = 0$, so the right side of Eq. 24.14 becomes zero. Consequently, the electric field inside the uniformly charged spherical shell must be zero.

Note that our calculation did not involve working out any integrals, even though Eq. 24.5 does contain an integral—the symmetry of the problem allows us to bypass the integration. The procedure box on page 658 shows how to calculate the electric field using Gauss's law for charge distributions that exhibit one of the symmetries listed in Figure 24.27. In the next three exercises we apply this procedure to calculate the electric field of a number of different charge distributions.

Procedure: Calculating the electric field using Gauss's Law

Gauss's law allows you to calculate the electric field for charge distributions that exhibit spherical, cylindrical, or planar symmetry without having to carry out any integrations.

1. Identify the symmetry of the charge distribution. This symmetry determines the general pattern of the electric field and the type of Gaussian surface you should use (see Figure 24.27).
2. Sketch the charge distribution and the electric field by drawing a number of field lines, remembering that the field lines start on positively charged objects and end on negatively charged ones. A two-dimensional drawing should suffice.
3. Draw a Gaussian surface such that the electric field is either parallel or perpendicular (and constant) to each face of the surface. If the charge distribution divides space into distinct regions, draw a Gaussian surface in each region where you wish to calculate the electric field.
4. For each Gaussian surface determine the charge q_{enc} enclosed by the surface.
5. For each Gaussian surface calculate the electric flux Φ_E through the surface. Express the electric flux in terms of the unknown electric field E.
6. Use Gauss's law (Eq. 24.8) to relate q_{enc} and Φ_E and solve for E.

You can use the same general approach to determine the charge carried by a charge distribution given the electric field of a charge distribution exhibiting one of the three symmetries in Figure 24.27. Follow the same procedure, but in steps 4–6, express q_{enc} in terms of the unknown charge q and solve for q.

Exercise 24.6 Electric field inside uniformly charged sphere

Consider a charged sphere of radius R carrying a positive charge q that is uniformly distributed over the volume of the sphere. What is the magnitude of the electric field a radial distance $r < R$ from the center of the sphere?

SOLUTION The sphere has spherical symmetry, so I know that the field must point radially outward in all directions. I therefore draw a concentric spherical Gaussian surface with a radius $r < R$ (**Figure 24.29**). Because the sphere carries a uniformly distributed charge q, the amount of charge enclosed by the Gaussian surface is determined by the ratio of the volumes of the Gaussian surface and the charged sphere:

$$q_{enc} = \frac{\frac{4}{3}\pi r^3}{\frac{4}{3}\pi R^3} q = \frac{r^3}{R^3} q. \qquad (1)$$

Figure 24.29

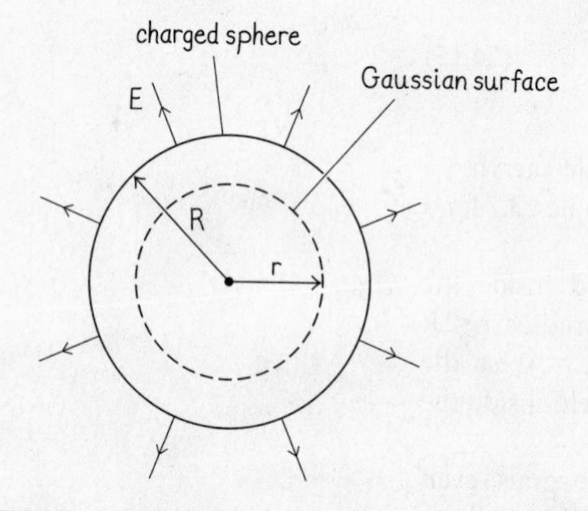

The electric flux is given by the product of the magnitude of the electric field $E(r)$ and the surface area A of the Gaussian surface (Eq. 24.13):

$$\Phi_E = 4\pi r^2 E. \qquad (2)$$

Substituting Eqs. 1 and 2 into Eq. 24.8, I obtain

$$4\pi r^2 E = \frac{r^3}{R^3}\frac{q}{\epsilon_0}$$

or

$$E = \frac{1}{4\pi\epsilon_0}\frac{q}{R^3} r = k\frac{q}{R^3} r, ✔$$

the same result I obtained in Checkpoint 23.15.

24.22 What is the electric field outside a solid sphere carrying a charge $+q$ uniformly distributed throughout its volume?

QUANTITATIVE TOOLS

Exercise 24.7 Electric field of an infinitely long charged thin rod

What is the electric field magnitude a radial distance r from the central length axis of an infinitely long thin rod carrying a positive charge per unit length λ?

SOLUTION An infinitely long rod has cylindrical symmetry. I assume the rod's diameter is vanishingly small. From the symmetry, I know that the electric field points radially outward (see Section 24.4). I therefore make a sketch showing the rod and a few representative field lines. I then draw a cylindrical Gaussian surface of radius r and height h around the rod (**Figure 24.30**). The cylinder encloses a length h of the rod, so the enclosed charge is

$$q_{\text{enc}} = \lambda h. \tag{1}$$

Figure 24.30

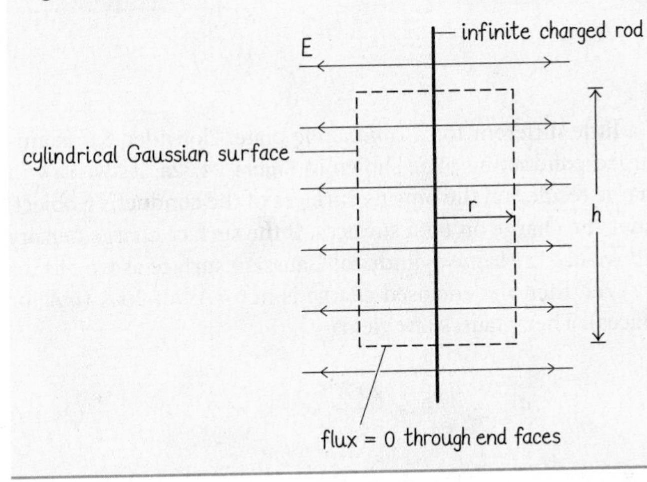

The electric flux through the top and bottom faces of the Gaussian surface is zero because the electric field is parallel to those faces. I also know that symmetry requires the electric field to have the same magnitude E at each position on the cylindrical part of the surface. I can therefore pull the electric field out of the integral. The electric flux through that part of the surface is

$$\Phi_E = \int_{\text{cyl surface}} \vec{E} \cdot d\vec{A} = \int E\,dA = E\int dA. \tag{2}$$

The area of the cylindrical surface is equal to the circumference of the cylinder, $2\pi r$, times its height h, so Eq. 2 becomes

$$\Phi_E = E(2\pi rh). \tag{3}$$

Substituting Eqs. 1 and 3 into Eq. 24.8, I obtain

$$E(2\pi rh) = \frac{\lambda h}{\epsilon_0}$$

or

$$E = \frac{\lambda}{2\pi\epsilon_0 r} = \frac{2k\lambda}{r}. ✔$$

This is the same result I obtained by direct integration in Example 23.4 for a finite charged rod in the limit that I am close to the rod (which makes the rod appear infinitely long). The direct integration, however, took almost two pages of work!

24.23 The direct integration procedure also yields an expression for a rod of *finite* length (see Example 23.4). Can you use Gauss's law to derive this expression as well? Why or why not?

Exercise 24.8 Electric field of an infinite charged sheet

What is the electric field a distance d from a thin, infinite nonconducting sheet with a uniform positive surface charge density σ?

SOLUTION An infinite sheet has planar symmetry. From the symmetry I know that the electric field points away from the sheet and that the magnitude of the electric field is the same everywhere (see Section 24.4). I make a sketch of the sheet and the electric field and then draw a Gaussian surface in the form of a cylinder that straddles the sheet (**Figure 24.31**). If the cross section of the cylinder has area A, then the cylinder encloses a piece of the sheet of area A and the enclosed charge is

$$q_{\text{enc}} = \sigma A. \tag{1}$$

Figure 24.31

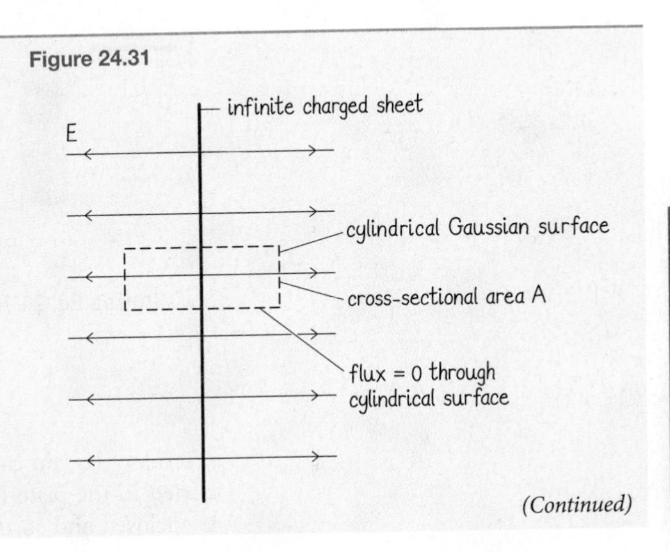

(Continued)

The electric flux through the cylindrical part of the Gaussian surface is zero because the field lines are parallel to that surface. The field is perpendicular to the two ends and points outward, however, so the electric flux through those ends is the product of the area of each end, A, and the magnitude of the electric field:

$$\Phi_E = 2EA. \tag{2}$$

Substituting Eqs. 1 and 2 into Eq. 24.8, I get

$$2EA = \frac{\sigma A}{\epsilon_0}$$

or

$$E = \frac{\sigma}{2\epsilon_0}. ✔$$

Because $1/(2\epsilon_0) = 2\pi k$, I can also write this as $E = 2\pi k\sigma$, which is the same result I obtained by direct integration in Example 23.6 for a uniformly charged disk in the limit that I am very close to the disk (which makes the disk appear like an infinite sheet).

The situation is a little different for a *conducting* plate. Consider, for example, the infinite charged conducting plate shown in **Figure 24.32*a***. As we saw in Section 24.5, any charge resides on the outside surfaces of the conducting object, so we have to consider the charge on *both* surfaces. If the surface charge density on the plate is σ and we use the same cylindrical Gaussian surface as we did for the nonconducting sheet, then the enclosed charge is not σA but $2\sigma A$ (σA for each of the two surfaces). Then Gauss's law yields

$$\Phi_E = \frac{q_{enc}}{\epsilon_0} = \frac{2\sigma A}{\epsilon_0}. \tag{24.16}$$

Figure 24.32 Applying Gauss's law to an infinite charged conducting plate.

(*a*)

infinite conducting plate

σ σ Gaussian surface

E $\vec{A}$ $\vec{E}$ $\vec{A}$ $\vec{E}$

(*b*)

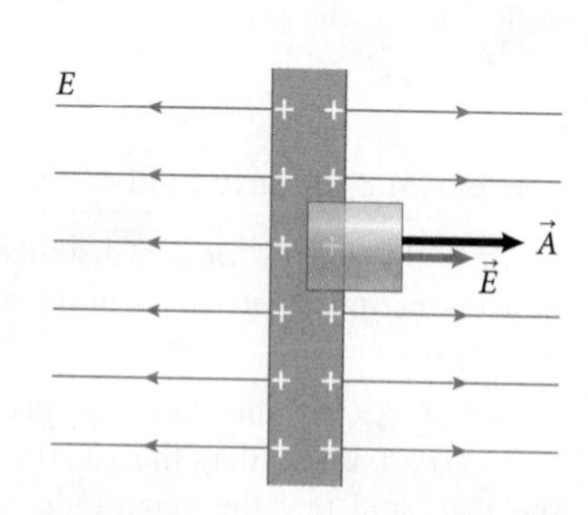

Substituting Eq. 24.16 into Eq. 2 of Exercise 24.8 yields

$$E = \frac{\sigma}{\epsilon_0} \quad \text{(infinite conducting plate).} \tag{24.17}$$

Alternatively, you can choose a cylindrical Gaussian surface that has one end buried in the plate (Figure 24.32*b*). In that case only one of the two surfaces is enclosed and so the enclosed charge is σA. However, now the electric flux through the left end is zero because $E = 0$ inside the bulk of the plate, so the electric flux through the Gaussian surface is not $2EA$ but EA. Substituting these values into Gauss's law yields again Eq. 24.17.

24.24 (*a*) A very large metal plate of surface area A carries a positive charge q. What is the surface charge density of the plate? What is the magnitude of the field created by the plate? (*b*) A very large, thin nonconducting sheet of surface area A carries a fixed, uniformly distributed positive charge q. What is the surface charge density of the sheet? What is the magnitude of the field created by the sheet?

Chapter Glossary

SI units of physical quantities are given in parentheses.

Closed surface Any surface that completely encloses a volume.

Cylindrical symmetry A configuration that remains unchanged if rotated or translated about one axis exhibits cylindrical symmetry.

Electric constant ϵ_0 ($C^2/(N \cdot m^2)$) A constant that relates the electric flux to the enclosed charge in Gauss's law:

$$\epsilon_0 \equiv \frac{1}{4\pi k} = 8.85418782 \times 10^{-12}\ C^2/(N \cdot m^2). \quad (24.7)$$

Electric field lines A representation of electric fields using lines of which the tangent to the line at every position gives the direction of the electric field at that position.

Electric flux Φ_E ($N \cdot m^2/C$) A scalar that provides a quantitative measure of the number of electric field lines that pass through an area. The electric flux through a surface is given by the surface integral

$$\Phi_E \equiv \int \vec{E} \cdot d\vec{A}. \quad (24.4)$$

Electrostatic equilibrium The condition in which the distribution of charge in a system does not change.

Enclosed charge The sum of all the charge within a given closed surface.

Field line density The number of field lines per unit area that cross a surface perpendicular to the field lines at that position. The field line density at a given position in a field line diagram is proportional to the magnitude of the electric field.

Field line flux The number of outward field line crossings through a closed surface minus the number of inward field line crossings. The field line flux through a closed surface is equal to the charge enclosed by the surface multiplied by the number of field lines per unit charge.

Gauss's law The relationship between the electric flux through a closed surface and the charge enclosed by that surface:

$$\Phi_E = \oint \vec{E} \cdot d\vec{A} = \frac{q_{enc}}{\epsilon_0}. \quad (24.8)$$

Gaussian surface Any closed surface used to apply Gauss's law.

Planar symmetry A configuration that remains unchanged if rotated about one axis or translated about any axis perpendicular to the axis of rotation exhibits planar symmetry.

Spherical symmetry A configuration that remains unchanged when rotated about any axis through its center exhibits spherical symmetry.

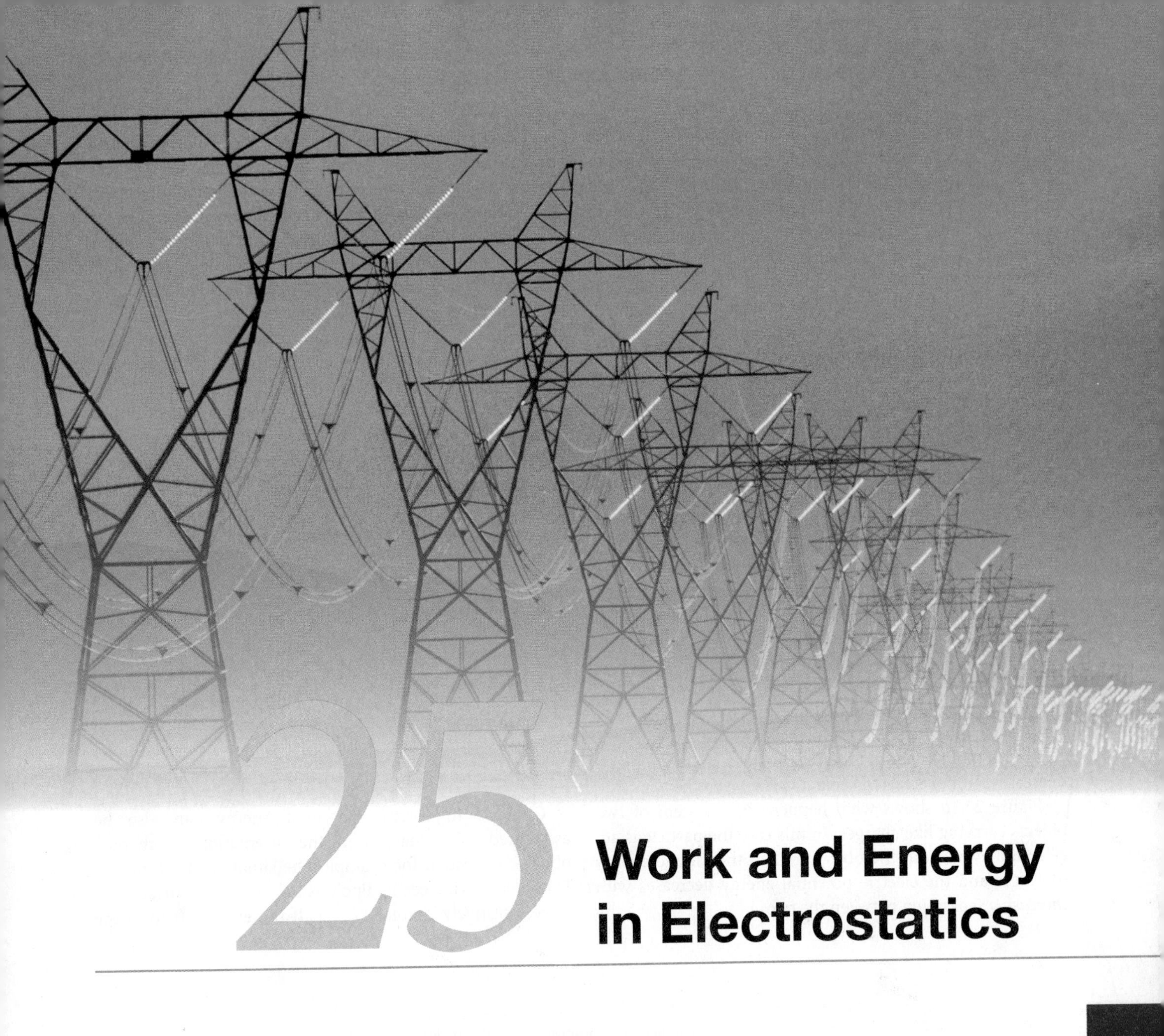

25 Work and Energy in Electrostatics

As we saw in our study of mechanics, it is often easier to solve problems by using the concepts of energy and work than by using forces. In this chapter we study how to apply energy considerations to electric interactions. Because there are two types of charge—positive and negative—the energy changes associated with changes in charge configurations are a bit more complicated than those associated with changes in gravitational configurations. We first analyze the potential energy associated with a stationary charge distribution and then introduce a new quantity, *potential difference,* that is related to potential energy and that plays an important role in electronics because, unlike potential energy, it can be measured directly.

25.1 Electric potential energy

Figure 25.1 shows the energy changes that occur in closed systems of two charged objects. In Figure 25.1*a*, a positively charged particle is released from rest in the constant electric field of a large stationary object carrying a negative charge. The attractive electric interaction between the particle and the object accelerates the particle toward the object, and so the kinetic energy of the system increases. This increase in kinetic energy must be due to a decrease in **electric potential energy,** the potential energy associated with the relative positions of charged objects. As we can see, the electric potential energy of two oppositely charged objects decreases with decreasing separation between the two.

Figure 25.1*b* shows what happens in a system of two objects carrying like charges. In this case the particle is accelerated away from the object, and so the kinetic energy increases and the electric potential energy decreases with *increasing* separation between the two.

The situation depicted in Figure 25.1*a* is the electric equivalent of free fall. While all objects in free fall near Earth's surface experience the same acceleration, objects in electric fields experience *different* accelerations. Consider, for example, two particles with different masses in free fall near Earth's surface (**Figure 25.2a**). Even though the particle with the greater mass is subject to a greater gravitational force, its acceleration is the same as that of the particle with the smaller mass. The reason is that the particle's inertia (its resistance to acceleration) is equal to its mass (see Chapter 13). Now consider the situation illustrated in Figure 25.2*b*. Two particles carrying the same charge $+q$, but of different mass, are released near the surface of a large negatively charged object. The electric forces exerted by the negatively charged object on the two particles are equal in magnitude, but because the masses of the two particles are different, the particles' accelerations are different as well.

25.1 Suppose both particles in Figure 25.2*b* are released from rest. Let $m_2 > m_1$ and consider only electric interactions. (*a*) How do their kinetic energies compare after they have both undergone the same displacement? (*b*) How do their momenta compare? (*c*) How do their kinetic energies and momenta compare at some fixed instant after they have been released? (*d*) How would you need to adjust the charges on the particles in order for the two particles to have the same acceleration upon release?

Changes in electric potential energy can also be associated with changes in the orientation of charged objects. Consider, for example, the situation illustrated in **Figure 25.3**. An electric dipole is held near the surface of a large, positively charged object. If the electric field of the

Figure 25.1 Energy diagrams for closed systems in which a positively charged particle is released from rest near a large stationary object that carries (*a*) a negative or (*b*) positive charge.

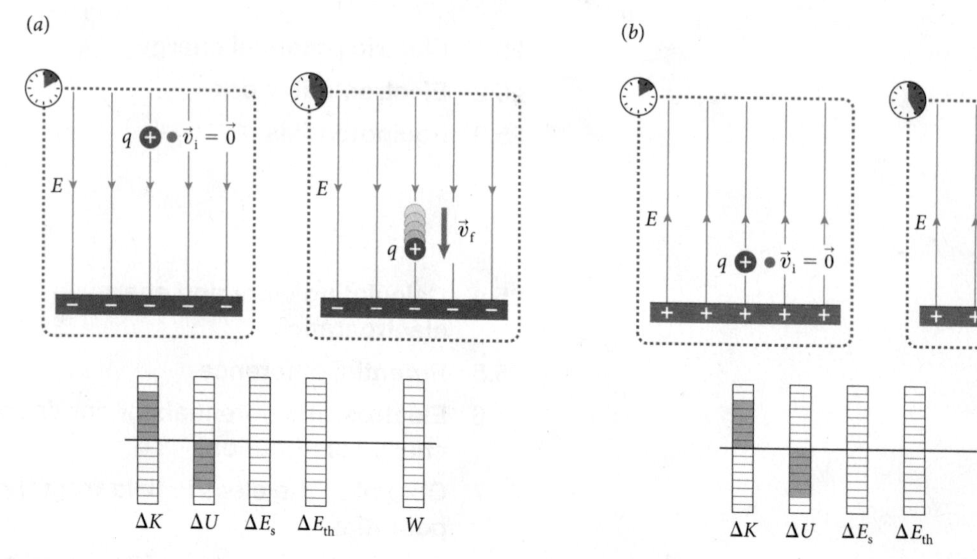

CONCEPTS

Figure 25.2 The free motion of charged particles in an electric field is different from the free fall of objects in a gravitational field.

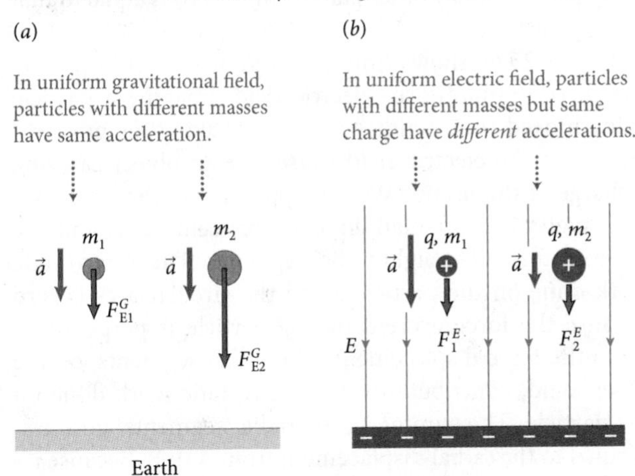

large charged object is uniform, then the dipole begins rotating as shown (see Sections 23.4 and 23.8). As it begins rotating, the dipole gains rotational kinetic energy. This means that the electric potential energy of the system must be changing as the orientation of the dipole changes. This change occurs because the positive side of the dipole gets farther away from the positively charged object, while the negative side gets closer to it; as we've seen, both of these motions correspond to a decrease of electric potential energy.

25.2 As the dipole in Figure 25.3 continues to rotate, it reaches the point where its axis is aligned with the electric field of the large object. (*a*) What happens to the electric potential energy as the dipole moves beyond that point? (*b*) Describe the motion of the dipole beyond that point. (*c*) How would the motion of the dipole change if it were released with a different orientation from the one shown in Figure 25.3?

Figure 25.3 Energy diagram for a system in which a dipole is released from rest near a positively charged stationary object.

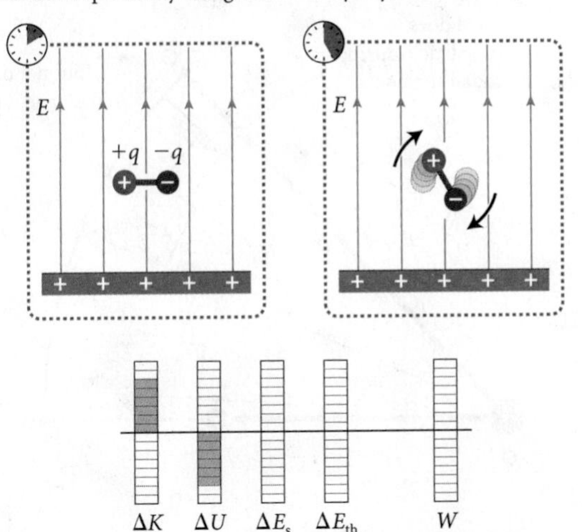

Figure 25.4 Energy diagram for a positively charged particle in the uniform electric field of a stationary, negatively charged object, that is not part of the system.

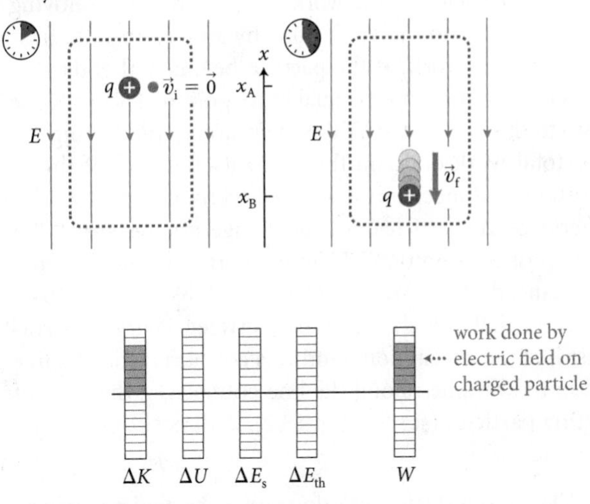

25.2 Electrostatic work

In general, we shall be considering the motion of a charged object through the constant electric field created by other stationary charged objects (such a field is sometimes called an *electrostatic field*). Therefore, our system—the charged object whose motion we are considering—is not closed. The energy of the system is not constant, and we must take into account the work done by the electric field on the system. For example, considering just the particle in Figure 25.1*a* as our system, we obtain the energy diagram shown in **Figure 25.4**. The particle can have only kinetic energy, so its increase in kinetic energy is now due to work done by the electric field on it.

25.3 (*a*) Suppose the particle in Figure 25.4 moves along the x axis from point A at $x = x_A$ to point B at $x = x_B$. How much work is done by the electric field $\vec{E}$ on it? (*b*) Suppose now that an external agent moves the particle back from B to A, starting and ending at rest as shown in **Figure 25.5**. How much work does the electric field do on the particle as it is moved? (*c*) How much work does the agent do on the particle while it is moved? (*d*) What is the combined work done by the agent and by the electric field on the particle as it is moved? (*e*) Draw an energy diagram to illustrate the energy changes of the particle as the agent moves it from B to A.

Figure 25.5 Checkpoint 25.3.

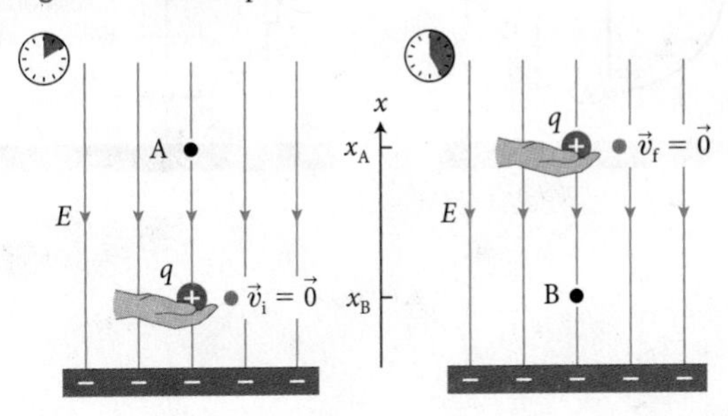

Checkpoint 25.3 illustrates the importance of distinguishing between the work done by the electric field on a charged particle and the work done by the agent moving it. We shall refer to the work done by an electrostatic field as **electrostatic work.** If the particle begins and ends at rest, the electrostatic work is equal in magnitude and opposite in sign to the work done by the agent doing the moving. Then the total work done on the particle—the sum of the electrostatic and mechanical work—is zero. Indeed, the kinetic energy of the particle does not change, and because the particle possesses no other form of energy, its energy remains unchanged: $\Delta E = W = 0$.

Note that the electric force between charged particles, just like the gravitational force, is a central force: Its line of action always lies along the line connecting the two interacting particles (see Section 13.2). A direct consequence of this fact is:

The electrostatic work done on a charged particle as it moves from one point to another is independent of the path taken by the particle and depends on only the positions of the endpoints of the path.

The proof of this statement parallels the one for gravitational forces in Section 13.6. Imagine, for example, lifting a particle from A to B along curved path 2 in **Figure 25.6a** instead of along straight path 1. As shown in Figure 25.6*b*, the path can be approximated by small straight horizontal and vertical segments. Along the horizontal segments, the electric force is perpendicular to the force displacement, so the electrostatic work on the particle along these segments is zero. Along the vertical segments, the electrostatic work on the particle is nonzero, but note that each vertical segment corresponds to an equivalent vertical segment on path 1. Thus, the displacements along all the vertical segments of path 2 add up to precisely the displacement along path 1. In other words, the electrostatic work on the particle along path 2 (or any other path from A to B) is equal to that along path 1.

Figure 25.6*c* shows how this argument can be generalized to a nonuniform electric field. Imagine a particle being moved from point A to point C along the gray trajectory in the electric field caused by an object carrying a charge at the origin. We can approximate the trajectory by a succession of small circular arcs centered about the origin and small straight radial segments. The electrostatic work done on the particle along the circular arcs is zero because the force exerted on the particle is perpendicular to the force displacement. The radial segments, on the other hand, contribute to the electrostatic work done on the particle. The sum of all the radial segments, however, is equal to the radial displacement from A to B. Because no electrostatic work is done on the particle along the circular path from B to C, we thus see that the electrostatic work done on the particle along the gray trajectory from A to C is equal to the electrostatic work done along the path from A to B. The electrostatic work done on the particle along *any* path from A to C is thus the same as the electrostatic work done from A to B.

25.4 Suppose the electrostatic work done on a charged particle as it moves along the gray path from A to C in Figure 25.6*c* is W. What is the electrostatic work done on the particle (*a*) along the path from C to B to A and (*b*) along the closed path from A to C to B and back to A?

Checkpoint 25.4 shows that the electrostatic work done on a charged particle that moves around a closed path—*any* closed path—in an electrostatic field is zero. We obtained a similar result for the work done by the gravitational force (Eq. 13.17). The physical reason for this result is that the

CONCEPTS

Figure 25.6 The electrostatic work done on a charged particle as the particle moves from point A to point B is independent of the path taken; it depends only on the positions of the endpoints of the path.

(*a*) Two paths by which particle can move from A to B

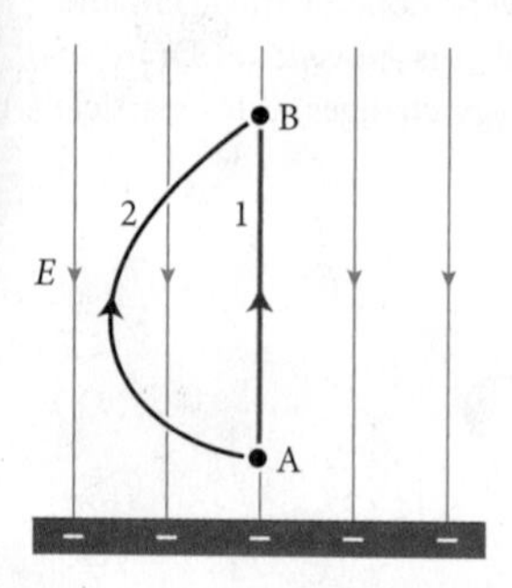
B
2
1
E
A

(*b*) Path 2 approximated by vertical & horizontal segments

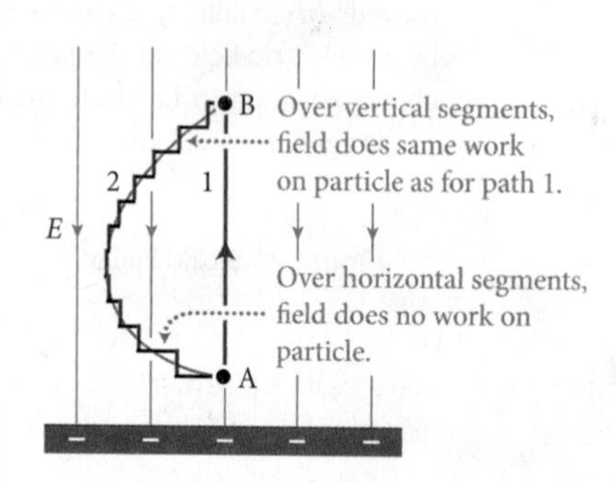
B
Over vertical segments, field does same work on particle as for path 1.
2
1
E
Over horizontal segments, field does no work on particle.
A

(*c*) Same argument applied to nonuniform electric field

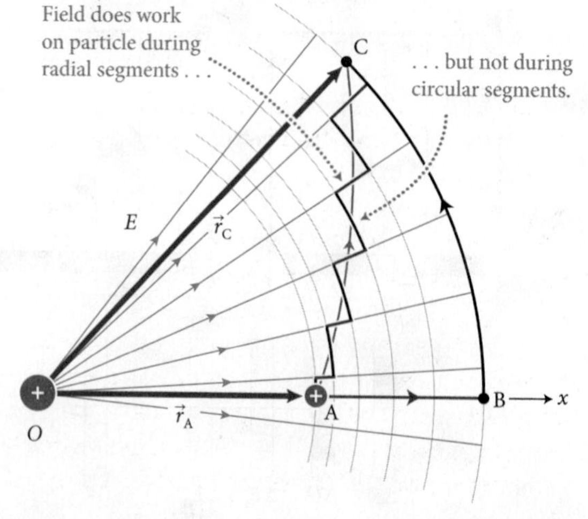
Field does work on particle during radial segments . . .
C
. . . but not during circular segments.
E
$\vec{r}_C$
$\vec{r}_A$
A
B
x
O

electric force between two charged particles, like the gravitational force between two objects that have mass, is nondissipative: When charged particles are moved around in electrostatic fields, no energy is irreversibly converted to other forms of energy. This is so because the electric interaction is nondissipative.

25.5 (*a*) If the electrostatic work done on a charged particle as it is moves around a closed path is zero, does this also mean that the electrostatic work done on the particle is zero as it moves along a piece of the closed path? (*b*) In Checkpoint 25.3*a* we found that the electrostatic work done on the particle is the product of the *x* components of the electric force and the force displacement. Is the same true for the electrostatic work done on a particle that is moved along path AB in Figure 25.6*c*? (*c*) What happens to this electrostatic work if (*i*) the charge on the particle and (*ii*) the mass of the particle is doubled?

The third part of Checkpoint 25.5 demonstrates a very important point: The electrostatic work done on a charged particle is proportional to the charge carried by that particle. This means that once we have calculated the electrostatic work done on a particle carrying a charge q along some path in an electrostatic field, we don't need to carry out the whole calculation again if we are interested in the electrostatic work done on another particle carrying a charge $2q$. We know that the electrostatic work done on the second particle is twice the electrostatic work done on the first. Thus, if we know the electrostatic work done on a particle carrying a unit positive charge along some path, then we know the electrostatic work done on a particle carrying *any* charge between the same two points! We therefore introduce a new quantity, called **electrostatic potential difference** (or simply **potential difference**), defined as:

The potential difference between point A and point B in an electrostatic field is equal to the negative of the electrostatic work per unit charge done on a charged particle as it moves from A to B.

The potential difference is a scalar, and because the electrostatic work done on a charged particle as it moves from one position to another can be positive or negative, the potential difference can also be positive or negative; the potential difference between any two points B and A is the negative of that between points A and B. It is important to keep in mind that potential difference *is not a form of energy*—it is electrostatic work done per unit charge and therefore has SI units of J/C.

You may be wondering why the potential difference is defined in terms of the *negative* of the electrostatic work done on a particle and not in terms of the energy required to move the particle, which has the opposite sign. The reason is that the energy required to move the particle depends on the change in the particle's kinetic energy. If the particle starts at rest and ends at a nonzero speed, the particle's kinetic energy increases and so the energy required is greater than when it starts and ends at rest. The electrostatic work done on a particle, on the other hand, is independent of any change in the particle's kinetic energy.

25.6 (*a*) Is the potential difference along any path from A to C in Figure 25.6*c* positive, negative, or zero? (*b*) Along any path from C to B? (*c*) Along the straight path from B to A? (*d*) In Figure 25.4, is the potential difference between the particle's initial and final positions positive, negative, or zero? (*e*) Express this potential difference in terms of the change in the particle's kinetic energy ΔK and its charge q.

In principle, only the potential *difference* between the endpoints of a path is meaningful. We can, however, assign a value to the **potential** at each of these endpoints by choosing a reference point. Specifically, if there is a positive potential difference between points A and B, then A is at a lower potential than B; by assigning a value to the potential at one of the two points, the value of the potential at the other point is fixed. Potential and potential difference, which are immensely useful in solving problems in electrostatics, are discussed in more detail in Section 25.5.

25.3 Equipotentials

As we have seen, the electrostatic work done on a charged particle along the horizontal segments in Figure 25.6*b* and the circular arcs in Figure 25.6*c* is zero. Consequently, the potential difference between any two points on such an arc or horizontal segment is zero. In other words, the potential has the same value at all points along these arcs or segments. Such paths are said to be **equipotential lines:**

An equipotential line is a line along which the value of the electrostatic potential does not change. The electrostatic work done on a charged particle as it moves along an equipotential line is zero.

The equipotential lines in Figure 25.6 are much like contour lines on a topographical map. For example, the contour lines in Figure 25.7 on the next page connect points of equal elevation. If you follow a contour line, your elevation remains the same, and so the gravitational potential energy associated with the separation between Earth and your body remains constant. Consequently, no gravitational work is done by Earth on you while you follow contour lines—these lines represent "gravitational equipotential lines."

Returning to the electrostatic case, note that Figure 25.6 is a two-dimensional representation of a three-dimensional situation. Thus, the equipotential line segments shown are really parts of **equipotential surfaces.** In Figure 25.6*a*, for example, the electrostatic work done on a charged particle is zero along *any* displacement parallel to the surface of the negatively charged object, into or out of the plane of the drawing. Consequently, as shown in Figure 25.8 on the next page, any surface parallel to the surface of the charged sheet causing the electric field is an equipotential surface. Often we'll use the term *equipotential* to denote an equipotential line or surface.

Figure 25.7 Contour lines on a map are analogous to equipotential lines. If you hike along a contour line, you neither gain nor lose gravitational potential energy.

(*a*)

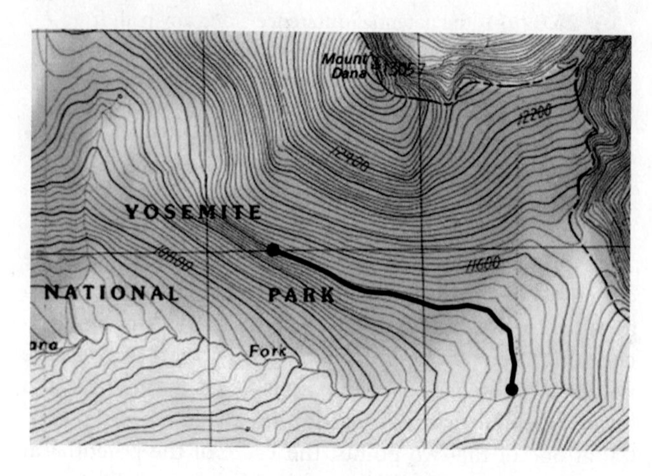

(*b*)

Just as with vector fields and field line diagrams, it is impossible to draw equipotential surfaces at all locations. In general, equipotentials are drawn with some fixed potential difference between them, just as contour lines represent a fixed difference in altitude. Keep in mind, however, that at any point between the equipotential surfaces shown in a figure we can, in principle, draw another equipotential surface.

25.7 Consider a single charged particle. Are there any equipotential lines or equipotential surfaces surrounding this particle?

The equipotential surfaces in Figure 25.8 and those in Checkpoint 25.7 are perpendicular to the electric field lines. This is true for *any* stationary charge distribution:

The equipotential surfaces of a stationary charge distribution are everywhere perpendicular to the corresponding electric field lines.

The proof of this statement is straightforward: If the electric field line were not perpendicular to the equipotential surface, then the electric field would have a nonzero component along the surface. This means there would be a nonzero component of electric force along the surface. By definition, however, the electrostatic work done on a charged particle is zero along an equipotential surface, and so there cannot be such a component.

Figure 25.9 shows a two-dimensional view of the equipotential surfaces of a more complicated stationary charge distribution. Note how, at every point in the diagram, the equipotentials are, indeed, perpendicular to the field lines.

Recall from Section 24.5 that in electrostatic equilibrium, the electric field inside the bulk of a conducting object is zero, regardless of the shape of the object or any charge carried by it. This means that no electrostatic work is done on a charged particle inside a charged or uncharged conducting object. Thus, the entire volume of the conducting object

Figure 25.8 Equipotential surfaces in a uniform electric field in (*a*) two dimensions; and (*b*) three dimensions.

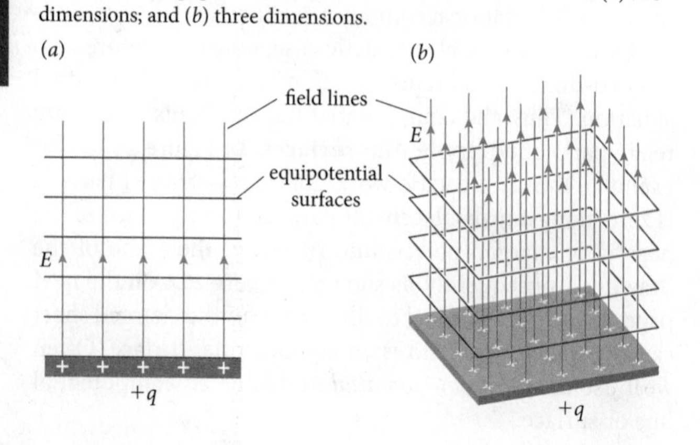

Figure 25.9 Field lines and equipotentials for three stationary charged particles.

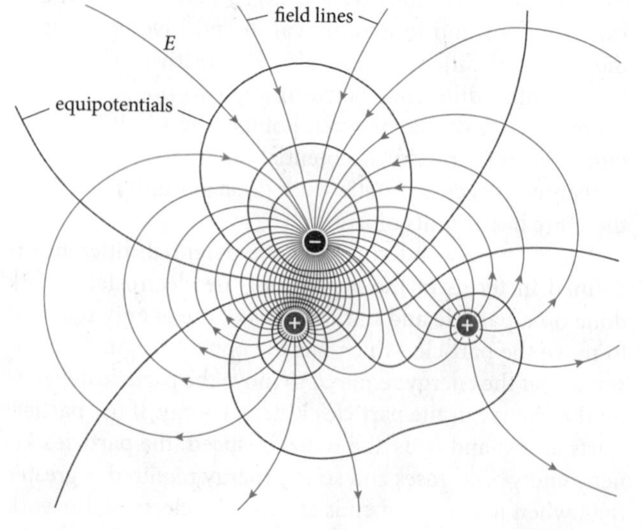

CONCEPTS

is an *equipotential volume*. In electrostatic equilibrium, all points within a conducting object are at the same electrostatic potential.

Example 25.1 Potential differences

Two metallic spheres A and B are placed on nonconducting stands. Sphere A carries a positive charge, and sphere B is electrically neutral. The two spheres are connected to each other via a wire, and the charge carriers reach a new electrostatic equilibrium. (*a*) Is the electric potential energy of the charge configuration after the spheres are connected greater than, smaller than, or equal to that of the original configuration? (*b*) Before the spheres are connected, is the potential difference between A and B positive, negative, or zero? Is it positive, negative, or zero after the spheres are connected?

❶ GETTING STARTED I need to evaluate the electric potential energy—the energy associated with the configuration of charge carriers—and the potential difference—the negative of the electrostatic work per unit charge done on a charge carrier. I begin with two sketches: one showing the charge distribution before the spheres are connected, and one showing the distribution after they are connected (**Figure 25.10*a*** and *b*). Once the spheres are connected, they form one conducting object, and the positive charge initially on A spreads out over both spheres.

Figure 25.10.

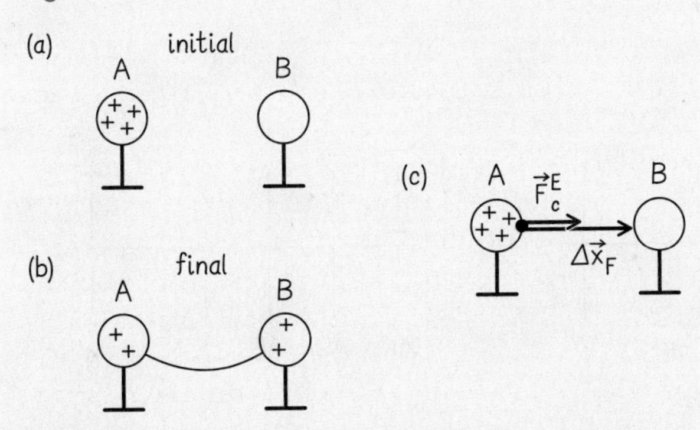

❷ DEVISE PLAN To evaluate the potential difference between A and B, I need to determine the negative of the electrostatic work done on a charged particle as it is moved from A to B and then divide the result by the charge on the particle.

❸ EXECUTE PLAN (*a*) After the two spheres are connected, the charge carriers spread out over both spheres, so they are farther apart than they were before the connection was made. Because I know that energy is required to push positively charged particles together, I conclude that the electric potential energy associated with the charge configuration in Figure 25.10*b* is smaller than that associated with the configuration in Figure 25.10*a*. ✔

(*b*) When a positive charge carrier is moved from A to B, the electrostatic work done on the carrier is *positive* because the electric force exerted on the carrier (directed away from sphere A) and the force displacement (from A to B) point in the same direction (Figure 25.10*c*). Consequently, the potential difference between A and B must be negative. ✔

Once the spheres are connected, they form one conducting object. Because the electric field is zero inside the entire object, no energy is required to move charge carriers around inside A and B, or from A to B, or vice versa. Thus, the potential difference between A and B after they are connected is zero. ✔

❹ EVALUATE RESULT I know that a closed system always tends to arrange itself so as to lower the system's potential energy (see Section 7.8). It therefore makes sense that the electric potential energy of the system is smaller after the spheres are connected. That the potential difference between A and B is negative before they are connected is a direct consequence of the definition of potential difference. After A and B are connected, they form one large equipotential volume and so it makes sense that the potential difference is zero.

Example 25.1 allows us to make another very useful observation:

An electrostatic field is directed from points of higher potential to points of lower potential.

Because the electric field gives the direction of the force exerted on a positively charged particle, a direct consequence of this fact is:

In an electrostatic field, positively charged particles tend to move toward regions of lower potential, whereas negatively charged particles tend to move toward regions of higher potential.

25.8 When you hold a positively charged rod above a metallic sphere without touching it, a surplus of negative charge carriers accumulates at the top of the sphere, leaving a surplus of positive charge carriers at the bottom. Is the potential difference between the top and the bottom of the sphere positive, negative, or zero?

Self-quiz

1. Consider the situation illustrated in **Figure 25.11**. A positively charged particle is lifted against the uniform electric field of a negatively charged plate. Ignoring any gravitational interactions, draw energy diagrams for the following choices of systems: (*a*) particle and plate, $v_f = 0$; (*b*) particle only, $v_f = 0$; (*c*) particle, plate, and person lifting, $v_f = 0$; (*d*) particle and plate, $v_f \neq 0$.

2. A positively charged particle is moved from point A to point B in the electric field of the large, stationary, positively charged object in **Figure 25.12**. (*a*) Is the electrostatic work done on the particle positive, negative or zero? (*b*) How is the electrostatic work done on the particle along the straight path from A to B different from the electrostatic work done on the particle along a path from A to B via C?

3. **Figure 25.13** shows both the electric field lines and the equipotentials associated with the given charge distribution. (*a*) Is the potential at point A higher than, lower than, or the same as the potential at point B? (*b*) Is the potential at point C higher than, lower than, or the same as the potential at point B? (*c*) Is the potential at point C higher than, lower than, or the same as the potential at point A?

Figure 25.11

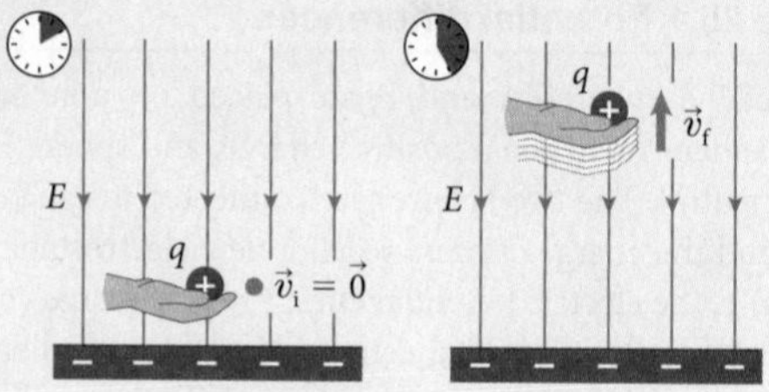

Figure 25.12

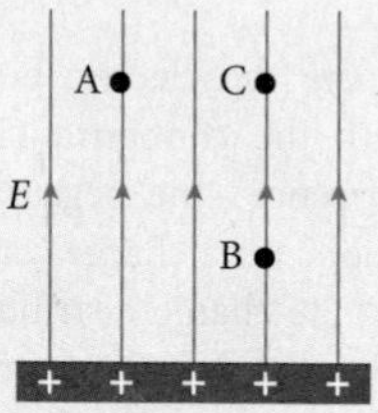

Figure 25.13

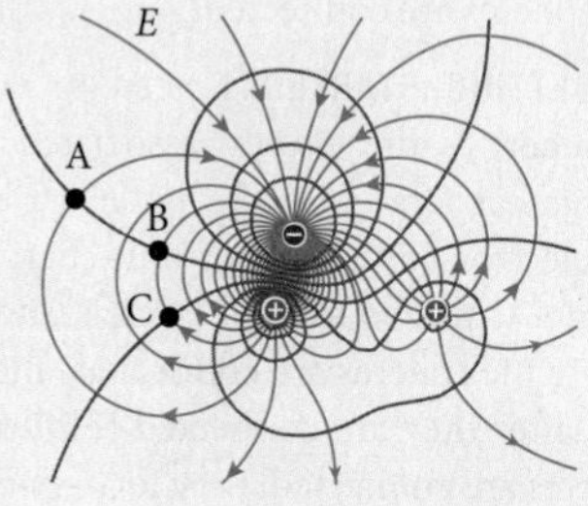

Figure 25.14

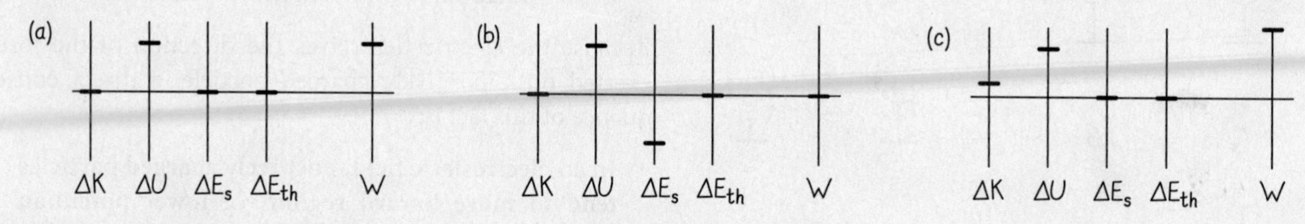

Answers

1. Only the person involves the conversion of source energy, so $\Delta E_s = 0$ in cases *a*, *b*, and *d*. When $v_f = 0$, $\Delta K = 0$; in *d*, $v_f \neq 0$ and so $\Delta K > 0$. For case *a*, the electric potential energy of the system increases as a result of positive work done by the agent (hand) on the particle, see **Figure 25.14*a***. For case *b*, the (kinetic) energy of the particle alone does not change because the positive work done by the agent on the particle and the negative electrostatic work done on it are equal in magnitude, so all bars are zero. For case *c*, a decrease in the source energy (provided by the agent) is responsible for the increase in electric potential energy of the system (Figure 25.14*b*). For case *d*, the positive work done by the agent increases both the kinetic energy of the particle and the electric potential energy of the system (Figure 25.14*c*).

2. (*a*) The electrostatic work done on the particle while it moves from point A to point B is negative because the angle between the force exerted on the particle and the force displacement is between 90° and 180° (see Eqs. 10.35 and 10.33). (*b*) The electrostatic work done along the two paths is the same because the electrostatic work done on a particle between two points is independent of the path taken.

3. (*a*) The same, because these points lie along an equipotential surface. (*b*) Higher. The potential increases in a direction opposite to the direction of the electric field, so point C is at a higher potential than point B. (*c*) Higher. Points A and B are at the same potential, and point C is at a higher potential than point B.

25.4 Calculating work and energy in electrostatics

To quantify the electrostatic work done on a charged particle, consider charged particles 1 and 2 in **Figure 25.15**. Particle 2 is moved from point A to point B through the nonuniform electric field of particle 1, which is held stationary. The electrostatic work done by particle 1 on particle 2 as it is moved along the solid path from A to B is given by (Eq. 10.44)

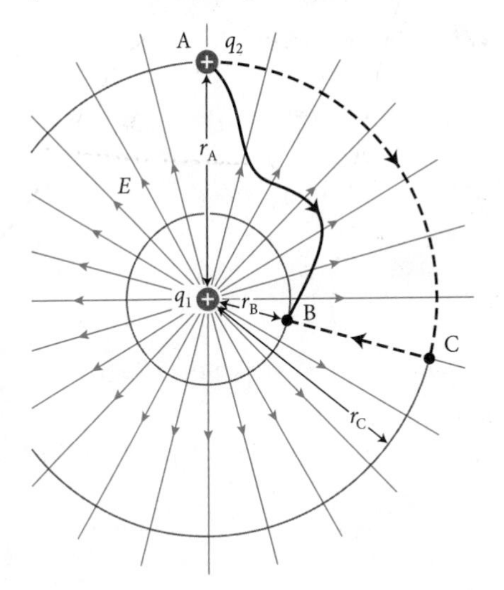

Figure 25.15 The electrostatic work done by particle 1 on particle 2 as the latter is moved from A to B is the same for the meandering solid path and for the dashed path ACB.

$$W_{12}(\mathrm{A} \to \mathrm{B}) = \int_{\mathrm{A}}^{\mathrm{B}} \vec{F}^{E}_{12} \cdot d\vec{\ell}, \tag{25.1}$$

where $\vec{F}^{E}_{12}$ is the electric force exerted by particle 1 on particle 2 and $d\vec{\ell}$ is an infinitesimal segment of the path. This line integral (see Appendix B) is generally not easy to calculate because the magnitude of the electric force and the angle between the force and the path segment $d\vec{\ell}$ vary along the path. However, as we saw in Section 25.2, the electrostatic work done on particle 2 along the solid path from A to B in Figure 25.15 is the same as that done along the dashed path from A to C to B. The circular path from A to C is along an equipotential (see Checkpoint 25.7), so the electrostatic work done on the particle along that path is zero. The electric force is given by Coulomb's law (Eq. 22.7). If we take particle 1 to be at the origin, we can write $r_{12} = r$ for the distance between the two particles and $\hat{r}_{12} = \hat{r}$ for the unit vector pointing from particle 1 to particle 2. Along the radial path from C to B, an infinitesimal segment of the path can be written as $d\vec{\ell} = dr\,\hat{r}$, and so with $k = 1/(4\pi\epsilon_0)$ (Eq. 24.7), the integrand in Eq. 25.1 becomes

$$\vec{F}^{E}_{12} \cdot d\vec{\ell} = \frac{1}{4\pi\epsilon_0}\frac{q_1 q_2}{r^2}\,\hat{r} \cdot (dr\,\hat{r}) = \frac{1}{4\pi\epsilon_0}\frac{q_1 q_2}{r^2}\,dr. \tag{25.2}$$

The line integral in Eq. 25.1 thus becomes

$$W_{12}(\mathrm{A} \to \mathrm{B}) = W_{12}(\mathrm{C} \to \mathrm{B}) = \frac{q_1 q_2}{4\pi\epsilon_0}\int_{r_\mathrm{C}}^{r_\mathrm{B}} \frac{1}{r^2}\,dr$$

$$= -\frac{q_1 q_2}{4\pi\epsilon_0}\left[\frac{1}{r}\right]_{r_\mathrm{C}}^{r_\mathrm{B}} = \frac{q_1 q_2}{4\pi\epsilon_0}\left[\frac{1}{r_\mathrm{C}} - \frac{1}{r_\mathrm{B}}\right]. \tag{25.3}$$

The distance from particle 1 to point A is the same as the distance from particle 1 to point C, so we have $r_\mathrm{A} = r_\mathrm{C}$ and

$$W_{12}(\mathrm{A} \to \mathrm{B}) = \frac{q_1 q_2}{4\pi\epsilon_0}\left[\frac{1}{r_\mathrm{A}} - \frac{1}{r_\mathrm{B}}\right]. \tag{25.4}$$

Generalizing this result to arbitrary initial and final points, we get

$$W_{12} = \frac{q_1 q_2}{4\pi\epsilon_0}\left[\frac{1}{r_{12,\mathrm{i}}} - \frac{1}{r_{12,\mathrm{f}}}\right], \tag{25.5}$$

where $r_{12,\mathrm{i}}$ and $r_{12,\mathrm{f}}$ are the initial and final values of the distance separating particles 1 and 2. Note that this expression is independent of the path taken: The electrostatic work done on particle 2 depends on only the distance between the two particles at the endpoints. Equation 25.5 also does not require particle 1 to be at the origin.

25.9 (*a*) Using Eq. 25.5, determine whether the electrostatic work done on particle 2 along path CB in Figure 25.15 is positive, negative, or zero. (*b*) Does moving the particle carrying charge q_2 along path CB involve positive, negative, or zero mechanical work done on the particle? Assume the particle begins and ends at rest; verify the consistency of your answer with part *a*. (*c*) By how much does the electric potential energy of the two charged particles change as particle 2 is moved from C to B? Is this change positive, negative, or zero?

As Checkpoint 25.9 demonstrates, the change in electric potential energy of the system that comprises the two particles in Figure 25.15 is the negative of the electrostatic work done by particle 1 on the system that comprises particle 2 only. (This relationship between electrostatic work and change in electric potential energy can also be seen in Figures 25.1 and Figure 25.4.) Taking the negative of Eq. 25.5, we thus obtain

$$\Delta U^E = \frac{q_1 q_2}{4\pi\epsilon_0}\left[\frac{1}{r_{12,\text{f}}} - \frac{1}{r_{12,\text{i}}}\right]. \quad (25.6)$$

Equation 25.6 gives only the *change* in electric potential energy, not *the* potential energy for a given configuration of charge. To obtain such an expression, we must first choose a zero of potential energy. It is customary to choose this zero to correspond to the configuration for which the force between the interacting particles is zero—that is to say, for infinite separation: $U^E = 0$ when $r_{12} = \infty$. Substituting this choice of reference point for the initial point in Eq. 25.6, we obtain

$$\Delta U^E \equiv U^E_\text{f} - U^E_\text{i} = U^E_\text{f} - 0 = \frac{q_1 q_2}{4\pi\epsilon_0}\left[\frac{1}{r_{12,\text{f}}} - 0\right]. \quad (25.7)$$

Thus, the **electric potential energy** for two particles carrying charges q_1 and q_2 and separated by distance r_{12} is

$$U^E = \frac{q_1 q_2}{4\pi\epsilon_0}\frac{1}{r_{12}} \quad (U^E \text{ zero at infinite separation}). \quad (25.8)$$

As this expression shows, the electric potential energy associated with two positively charged particles is positive: If they are brought together starting from infinite separation, the electric potential energy of the two particles increases. This is as it should be, because the two particles repel each other and so energy must be added to the system to bring it together ($W > 0$). The same idea applies to two negatively charged particles because the product $q_1 q_2$ is still positive. For a system of oppositely charged particles, on the other hand, $q_1 q_2 < 0$ and so $U^E < 0$. Indeed, the interaction is attractive and the potential energy decreases as the two are brought together.

25.10 Suppose we keep particle 2 stationary and move particle 1 in so that their final separation is the same as that in Figure 25.15. Is the electrostatic work done on particle 1 as it is moved in also given by the right-hand side of Eq. 25.4?

QUANTITATIVE TOOLS

Example 25.2 Putting two charged particles together

Two small pith balls, initially separated by a large distance, are each given a positive charge of 5.0 nC. By how much does the electric potential energy of the two-ball system change if the balls are brought together to a separation distance of 2.0 mm?

❶ GETTING STARTED I am asked to calculate a change in the electric potential energy of a system. I take *large* to mean a separation distance great enough that the balls don't interact. So the initial state is one in which the two pith balls have infinite separation. In the final state they are 2.0 mm apart.

❷ **DEVISE PLAN** To calculate the change in electric potential energy I can use Eq. 25.6.

❸ **EXECUTE PLAN** Substituting $q_1 = q_2 = 5.0 \times 10^{-9}$ C, $r_{12,i} \approx \infty$, $r_{12,f} = 2.0 \times 10^{-3}$ m, and $k = 1/(4\pi\epsilon_0) = 9.0 \times 10^9\ \text{N}\cdot\text{m}^2/\text{C}^2$ into Eq. 25.6, I get

$$\Delta U^E = (9.0 \times 10^9\ \text{N}\cdot\text{m}^2/\text{C}^2)(5.0 \times 10^{-9}\ \text{C})^2\left[\frac{1}{2.0 \times 10^{-3}\ \text{m}} - 0\right]$$
$$= 1.1 \times 10^{-4}\ \text{J}.$$

❹ **EVALUATE RESULT** My answer is positive, which makes sense because the balls repel each other and so work must be done on the system as they are brought together. This work increases the potential energy of the system. The magnitude of the potential energy change is small, but so is the magnitude of the force between the balls: Substituting q_1, q_2, and $r_{12,f}$ into Coulomb's law (Eq. 22.1), I obtain a force magnitude of 0.056 N. This is the maximum force between the two balls, so it makes sense that the energy associated with this interaction is small.

We can readily generalize the expressions for electrostatic work and electric potential energy for situations involving more than two charged particles. To determine the electric potential energy of a system of three charged particles in a certain configuration, for example, we calculate the electrostatic work done while assembling the system in its final configuration, starting from a situation where all three particles are far apart. Placing the first particle, carrying charge q_1, in its final position involves no electrostatic work because the other two particles are far away and therefore not interacting with particle 1. Next we bring in particle 2, carrying charge q_2, as illustrated in **Figure 25.16a**. The electrostatic work done while moving particle 2 can be found by substituting infinity for $r_{12,i}$ and r_{12} for $r_{12,f}$ in Eq. 25.5:

$$W_{12} = -\frac{q_1 q_2}{4\pi\epsilon_0}\frac{1}{r_{12}}. \tag{25.9}$$

Finally we bring in particle 3, carrying charge q_3, as shown in Figure 25.16*b*. The electrostatic work done while moving particle 3 is given by Eq. 25.1:

$$W_3 = \int_i^f (\Sigma\vec{F}_3^E)\cdot d\vec{\ell}, \tag{25.10}$$

where $\Sigma\vec{F}_3^E$ is the vector sum of the forces exerted on particle 3. Particle 3 is subject to two forces, one exerted by particle 1 and one by particle 2, so

$$\int_i^f (\Sigma\vec{F}_3^E)\cdot d\vec{\ell} = \int_i^f (\vec{F}_{13}^E + \vec{F}_{23}^E)\cdot d\vec{\ell}$$

$$= \int_i^f \vec{F}_{13}^E\cdot d\vec{\ell} + \int_i^f \vec{F}_{23}^E\cdot d\vec{\ell} = W_{13} + W_{23}. \tag{25.11}$$

In other words, the electrostatic work done as 3 is moved to its final position is the sum of the electrostatic work done when only 1 is present plus that done when only 2 is present. Now we can apply Eq. 25.5 to each term in Eq. 25.11:

$$W_{13} + W_{23} = -\frac{q_1 q_3}{4\pi\epsilon_0}\frac{1}{r_{13}} - \frac{q_2 q_3}{4\pi\epsilon_0}\frac{1}{r_{23}}. \tag{25.12}$$

The total electrostatic work done on the three-particle system while assembling the charge configuration is the sum of Eqs. 25.9 and Eq. 25.12:

$$W = -\frac{q_1 q_2}{4\pi\epsilon_0}\frac{1}{r_{12}} - \frac{q_1 q_3}{4\pi\epsilon_0}\frac{1}{r_{13}} - \frac{q_2 q_3}{4\pi\epsilon_0}\frac{1}{r_{23}}. \tag{25.13}$$

Figure 25.16 To obtain the electrostatic potential energy of a system of three charged particles, we assemble the system one particle at a time.

(*a*) We bring in second charged particle

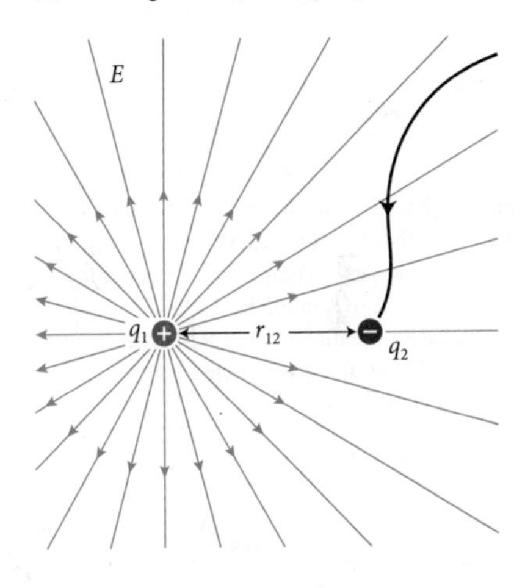

(*b*) We bring in third charged particle

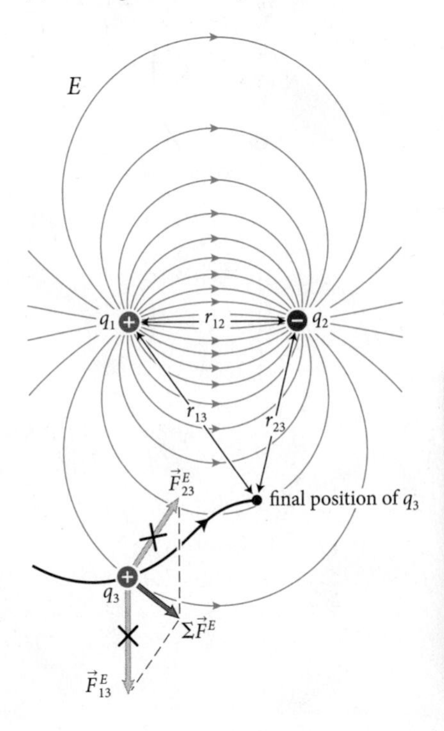

QUANTITATIVE TOOLS

Alternatively, with our choice of zero at infinity, the electric potential energy of the system is

$$U^E = \frac{q_1 q_2}{4\pi\epsilon_0}\frac{1}{r_{12}} + \frac{q_1 q_3}{4\pi\epsilon_0}\frac{1}{r_{13}} + \frac{q_2 q_3}{4\pi\epsilon_0}\frac{1}{r_{23}}$$

$$(U^E \text{ zero at infinite separation}). \quad (25.14)$$

In words, to determine the electric potential energy of a system of charged particles, we need to add the electric potential energies for each pair of particles.

25.11 Suppose the charged particles in Figure 25.16 are assembled in a different order—say, 3 first, then 1, and finally 2. Do you obtain the same result as in Eq. 25.13 and Eq. 25.14?

25.5 Potential difference

The negative of the electrostatic work per unit charge done on a particle that carries a positive charge q from one point to another is defined as the **potential difference** between those points:

$$V_{AB} \equiv V_B - V_A \equiv \frac{-W_q(A \to B)}{q}. \quad (25.15)$$

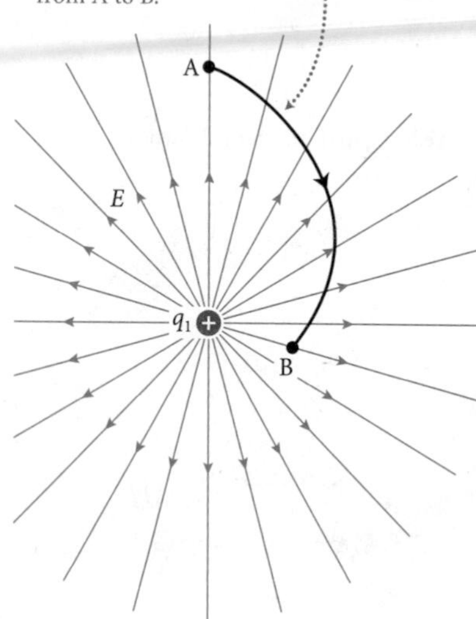
Potential difference between A and B, $V_{AB} = V_B - V_A$, is negative of electrostatic work per unit charge done along a path from A to B.
A
E
q_1
B

Figure 25.17 Once we know the electrostatic work done by a particle carrying a charge q_1 on a particle carrying a charge q_2 as the latter is moved from A to B in Figure 25.15, we can determine the potential difference between A and B in the electric field of particle 1.

Potential difference is a scalar, and the SI units of potential difference are joules per coulomb (J/C). In honor of Alessandro Volta (1745–1827), who developed the first battery, this derived unit is given the name **volt** (V):

$$1\text{ V} \equiv 1\text{ J/C}. \quad (25.16)$$

For example, if the electric field in **Figure 25.17** does -12 J of electrostatic work on a particle carrying charge $q_2 = +2.0$ C as it is moved from point A to point B (in other words, it requires $+12$ J of work by an external agent without the particle gaining any kinetic energy), then the potential difference V_{AB} between A and B is $-(-12\text{ J})/(2.0\text{ C}) = +6.0$ V. That is, the potential at B is 6.0 V higher than that at A.

Once we know the potential difference V_{AB} between A and B, we can obtain the electrostatic work done on *any* object carrying a charge q as it is moved along any path from A to B:

$$W_q(A \to B) = -qV_{AB}. \quad (25.17)$$

Keep in mind that the subscripts AB mean "from A to B." Because B is the final position, we write $V_{AB} \equiv V_B - V_A$. So when we refer to the "potential difference between A and B," we always mean the potential at B minus the potential at A.

Potential is important in practical applications because the potential difference between two points can be measured readily with a device called a *voltmeter* (we'll encounter these devices when we discuss electrical circuits in Chapters 31 and 32.) In the next chapter we shall discuss the operation of another familiar device, the *battery*, which allows one to maintain a constant potential difference between two points. A 9-V battery, for example, maintains a +9-V-potential difference between its negative and positive terminals. (The positive terminal is at the higher potential, and thus the potential difference is *positive* when going from − to +.) For example, when a particle carrying charge +1 C is moved from the negative terminal of a 9-V battery to the positive terminal, the particle undergoes a potential difference $V_{-+} = V_+ - V_- = +9$ V. Equation 25.17 tells us that the electrostatic work done on the particle is equal to

$$W_q(- \to +) = -q\,V_{-+} = -(+1\text{ C})(+9\text{ V}) = -9\text{ J}. \tag{25.18}$$

That this quantity is negative indicates that the agent moving the particle must do a positive amount of work on the particle. Likewise, when a particle carrying a charge of -2 C is moved from the $-$ terminal to the $+$ terminal of a 9-V battery, the electrostatic work done on the particle is $-(-2\text{ C})(+9\text{ V}) = +18$ J.

For simplicity, we shall denote the magnitude of the potential difference between the terminals of a battery by V_{batt}. Therefore

$$V_{\text{batt}} = V_+ - V_-. \tag{25.19}$$

Just as with potential energy, only potential *differences* are physically relevant. If we choose a reference point, however, we can determine the value for the potential at any other point. In the preceding section we chose infinity as the reference point for the electric potential energy for charged particles because they do not interact at infinite distance $U^E(\infty) = 0$. The same choice can be made for the potential, but when we deal with electrical circuits it is customary to assign zero potential to Earth (ground) because Earth is a good and very large conducting object through which the motion of charge carriers requires negligible energy.

Exercise 25.3 Potential and potential difference

The negative terminal of a 9-V battery is connected to ground via a wire. (*a*) What is the potential of the negative terminal? (*b*) What is the potential of the positive terminal? (*c*) What is the potential of the negative terminal if the positive terminal is connected to ground?

SOLUTION (*a*) I know from Section 25.3 that any conducting objects that are in electrical contact with each other form an equipotential. Once they are in contact, therefore, the ground (which is conducting), the wire, and the negative terminal are all at the same potential. If the potential of the ground is zero (an arbitrary but customary choice), then the potential of the negative terminal is also zero. ✔

(*b*) The potential difference between the negative and positive terminals is $+9$ V, meaning that the potential of the positive terminal is 9 V higher than that of the negative terminal: $V_{\text{batt}} = V_+ - V_- = +9$ V. If V_- is zero, then the potential of the positive terminal is $V_+ = +9\text{ V} + V_- = (+9\text{ V}) + (0\text{ V}) = +9$ V. ✔

(*c*) With the positive terminal connected to ground, that terminal's potential becomes zero. Because the battery maintains a potential difference of $+9$ V between the negative and positive terminals, I now have $V_{\text{batt}} = V_+ - V_- = 0 - V_- = +9$ V and so the negative terminal is at a potential of -9 V. ✔

To obtain an explicit expression for the potential difference between two points A and B in the electric field of particle 1 carrying charge q_1, we start with the expression for the electrostatic work done by particle 1 on a particle 2 carrying charge q_2 as it is moved from A to B (Eq. 25.4). All we need to do is add a minus sign and divide by q_2:

$$V_{\text{AB}} \equiv \frac{-W_{12}(\text{A} \to \text{B})}{q_2} = \frac{q_1}{4\pi\epsilon_0}\left[\frac{1}{r_{\text{B}}} - \frac{1}{r_{\text{A}}}\right]. \tag{25.20}$$

Figure 25.18 Equipotentials, field lines, and graph of potential for a charged particle.

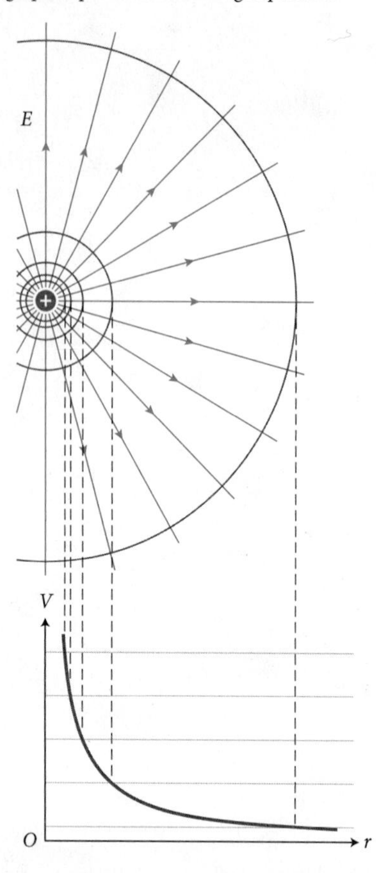

For charged particles, the choice of ground as a zero for the potential is not very meaningful. However, if we set the zero for potential at infinity and let r_A be at infinity, we obtain for the potential at a distance $r = r_B$ from a single charged particle

$$V(r) = \frac{1}{4\pi\epsilon_0}\frac{q_1}{r} \quad \text{(potential zero at infinity).} \tag{25.21}$$

Equation 25.21 confirms that the potential is constant on any spherical surface centered on the charged particle (that is, any surface for which r is constant), as we concluded in Checkpoint 25.7. **Figure 25.18** shows the potential for a charged particle in graphical form. The bottom of the figure shows the $1/r$ dependence of the potential; the top shows a two-dimensional view of the equipotentials around the charged particle. The $1/r$ dependence of the potential is reflected by the increasing spacing of the equipotentials (with a fixed potential difference between them) as r increases.

25.12 (*a*) Using Eq. 25.20, determine whether the potential difference between A and B in Figure 25.17 is positive, negative, or zero. (*b*) From the directions of the electric force and the force displacement, is the electrostatic work done on a positively charged particle as it is moved along a straight path from A to B positive, negative, or zero? Verify that your answer is consistent with the answer in part *a*.

Example 25.4 Atomic potential difference

A (simplistic) model of the hydrogen atom treats the electron as a particle carrying a charge $-e$ orbiting a proton (a particle carrying a charge $+e$) in a circle of radius $r_H = 0.53 \times 10^{-10}$ m. (*a*) How much energy is required to completely separate the electron from the proton? For simplicity, ignore the electron's kinetic energy. (*b*) Across what potential difference does the electron travel as it is separated from the proton?

❶ **GETTING STARTED** To completely separate the electron from the proton, I must increase the distance between them from r_H to infinity (at which point their interaction is reduced to zero). By increasing the separation, I increase the potential energy of the electron-proton system. The energy I must add to the system in order to separate the particles is equal to the increase in potential energy of the system.

❷ **DEVISE PLAN** I can calculate the potential energy increase from Eq. 25.6, setting $r_i = r_H$ and $r_f = \infty$. I can obtain the potential difference between the initial and final positions of the electron by taking the negative of the electrostatic work done by the electric field of the proton on the electron during the separation and dividing this value by the charge on the electron.

❸ **EXECUTE PLAN** (*a*) The change in the electric potential energy of the two-particle system is

$$\Delta U^E = \frac{q_1 q_2}{4\pi\epsilon_0}\left[\frac{1}{r_f} - \frac{1}{r_i}\right] = \frac{-e^2}{4\pi\epsilon_0}\left[0 - \frac{1}{r_H}\right]$$

$$= (9.0 \times 10^9\ \text{N}\cdot\text{m}^2/\text{C}^2)\frac{(1.6 \times 10^{-19}\,\text{C})^2}{0.53 \times 10^{-10}\,\text{m}} = 4.3 \times 10^{-18}\ \text{J}.$$

Because there are no changes in any other forms of energy in the system, the energy required to separate the electron and the proton in a hydrogen atom is 4.3×10^{-18} J. ✔

(*b*) The answer I obtained in part *a* is the mechanical work an external agent must do on the electron to separate it from the proton. I know that the energy of the electron does not change because the electron gains no kinetic energy. Considering just the electron as my system, I thus have $\Delta E = 0$ and so the work done on the electron is zero. This work has two parts: mechanical work done by the external agent and electrostatic work done by the electric field of the proton. Because the sum of these two parts is zero, I know that the electrostatic work done by the electric field of the proton on the electron must be the negative of the mechanical work done by the external agent, so $W_{pe}(r_H \rightarrow \infty) = -4.3 \times 10^{-18}$ J. The potential difference is thus (Eq. 25.15)

$$V_{H\infty} = \frac{-W_{pe}(r_H \rightarrow \infty)}{q} = \frac{-(-4.3 \times 10^{-18}\ \text{J})}{-1.6 \times 10^{-19}\ \text{C}} = -27\ \text{V.} ✔$$

❹ **EVALUATE RESULT** The positive sign on ΔU^E in part *a* means that the potential energy of the proton-electron system increases when the two are moved apart. Therefore mechanical work must be done to pull them apart, as I expect because the electron and proton attract each other. The value I obtain is extremely small, but I know that 1 m^3 of matter contains about 10^{29} atoms (see Exercise 1.6), so the electric potential energy in a cubic meter of matter is on the order of $(10^{28})(4.3 \times 10^{-18}\text{ J}) \approx 10^{11}$ J. In the box "Coherent versus incoherent energy" on page 158 I learned that the amount of chemical energy in a pencil is on the order of 10^5 J. Given that chemical energy is derived from electric potential energy stored in chemical bonds and that the volume of a pencil is about 10^{-5} m^3, my answer for part *a* is not unreasonable.

For part *b*, I could have used Eq. 25.20 directly. Substituting the values given into that equation, I obtain the same answer, which gives me confidence in the answer I obtained.

To obtain a more general result for the potential difference between one point and another in an arbitrary electric field $\vec{E}$, consider the situation illustrated in **Figure 25.19**. A particle carrying a charge q is moved from point A to point B in an electric field due to some charge distribution (not visible in the illustration). The electrostatic work done on the particle is

$$W_q(\text{A} \rightarrow \text{B}) = \int_{\text{A}}^{\text{B}} \vec{F}_q^E \cdot d\vec{\ell}. \tag{25.22}$$

The vector sum of the forces exerted on the particle is equal to the product of the electric field and the charge q (Eq. 23.6):

$$\vec{F}_q^E = q\vec{E}, \tag{25.23}$$

so

$$W_q(\text{A} \rightarrow \text{B}) = q\int_{\text{A}}^{\text{B}} \vec{E} \cdot d\vec{\ell}. \tag{25.24}$$

The potential difference between point A and point B is therefore

$$V_{\text{AB}} \equiv \frac{-W_q(\text{A} \rightarrow \text{B})}{q} = -\int_{\text{A}}^{\text{B}} \vec{E} \cdot d\vec{\ell}. \tag{25.25}$$

In evaluating this line integral, we must keep in mind that the integral does not depend on the path taken but only on the endpoints. It therefore pays to choose a path that facilitates evaluating the integral. The Procedure box below and the next example will help you gain practice calculating potential differences between two points.

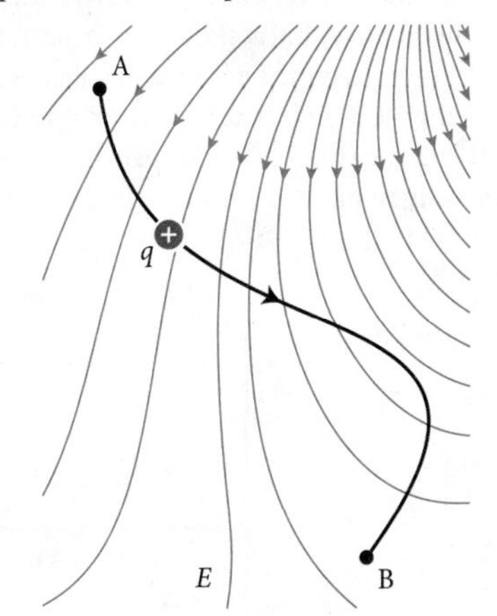

Figure 25.19 To determine the potential difference between two points in an electric field, we must evaluate the electrostatic work done on a charged particle as the particle is moved along a path between those points.

Procedure: Calculating the potential difference between two points in an electric field

The potential difference between two points in an electric field is given by Eq. 25.25. The following steps will help you evaluate the integral:

1. Begin by making a sketch of the electric field, indicating the points corresponding to the two points between which you wish to determine the potential difference.
2. To facilitate evaluating the scalar product $\vec{E} \cdot d\vec{\ell}$, choose a path between the two points so that $\vec{E}$ is either parallel or perpendicular to the path. If necessary, break the path into segments. If $\vec{E}$ has a constant value along the path (or a segment of the path), you can pull it out of the integral; the remaining integral is then equal to the length of the corresponding path (or the segment of the path).
3. Remember that to determine V_{AB} ("the potential difference between points A and B"), your path begins at *A* and ends at *B*. The vector $d\vec{\ell}$ therefore is tangent to the path, in the direction that leads from A to B (see also Appendix B).

At this point you can substitute the expression for the electric field and carry out the integral. Once you are done, you may want to verify the algebraic sign of the result you obtained: *negative* when a positively charged particle moves along the path in the direction of the electric field, and *positive* when it moves in the opposite direction.

Example 25.5 Electrostatic potential in a uniform field

Consider a uniform electric field of magnitude E between two parallel charged plates separated by a distance d. (*a*) What is the potential difference between the positive plate and the negative plate? (*b*) What is the value of the potential at a point P that lies between the plates and is a distance $a < d$ from the positive plate, if the potential of the negative plate is zero?

❶ **GETTING STARTED** I begin by making a sketch of the two parallel plates and the electric field (**Figure 25.20**). For part *a* I must calculate the potential difference between the positive and negative plates, so I choose a path that begins at the positive plate and ends at the negative plate. For part *b* I am asked to determine, relative to the potential at the negative plate, the potential at a point P located between the plates a distance a from the positive plate, so I choose a path that runs from P to the negative plate. I indicate the endpoints for both paths in my drawing and choose my x axis to be parallel with the electric field with the positive plate at $x = 0$.

Figure 25.20

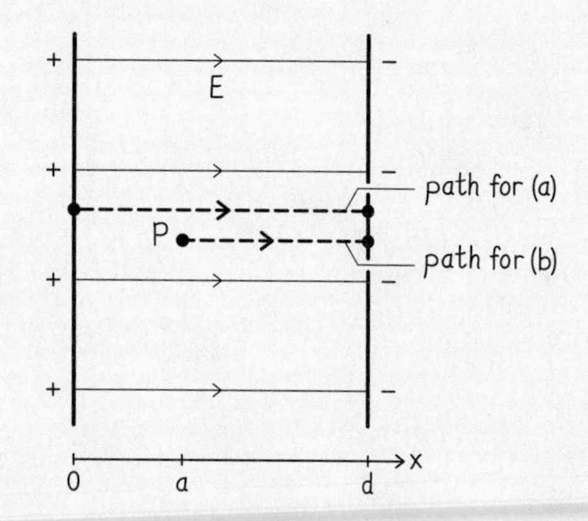

❷ **DEVISE PLAN** To obtain the potential difference between the endpoints of each path, I apply Eq. 25.25. With my choice of x axis the $d\vec{\ell}$ factor in Eq. 25.25 becomes $d\vec{\ell} = dx\,\hat{\imath}$. Because $\vec{E} = E\,\hat{\imath}$ I thus have $\vec{E}\cdot d\vec{\ell} = E\,dx$, and because the field is uniform, E is constant. For part *b*, once I have the potential difference between the endpoints of the path from P to the negative plate, I can determine the potential at P because I am told that the potential at the negative plate is zero.

❸ **EXECUTE PLAN** (*a*) Applying Eq. 25.25 to the path from $x_i = 0$ (positive plate) to $x_f = d$ (negative plate), I get

$$V_{0d} = -\int_0^d E\,dx = -E\int_0^d dx = -Ed, \checkmark \quad (1)$$

where I have pulled E out of the integral because it is constant.

(*b*) If I start the line integral in Eq. 1 at point P, I obtain

$$V(d) - V(a) = -E\int_a^d dx = -E(d - a)$$

or, because $V(d) = 0$,

$$V(a) = E(d - a). \checkmark$$

❹ **EVALUATE RESULT** The negative sign in Eq. 1 means that the potential at the end of the path (that is, at the negative plate) is *lower* than the potential at the beginning (the positive plate). This negative potential difference is in agreement with the sign of the electrostatic work done on a positively charged particle: Positive electrostatic work is done on the particle as it is moved from the positive plate to the negative plate, and so, according to Eq. 25.15, the potential difference should indeed be negative.

My result for part *b* indicates that the potential is positive at $x = a$ $(a < d)$, decreases linearly with the distance a to the positive plate, and goes to zero at $a = d$. That makes sense, because I know that the potential of the positive plate must be higher than that of the negative plate and if $a = d$, I get $V(d) = 0$, as expected.

With an appropriate choice of reference point for the potential, Eq. 25.25 allows us to assign values to the potential at every point surrounding a charge distribution. This "potential field" is related to the electric field: Each can be determined from the other. A drawing that shows a set of equipotentials for a charge distribution is equivalent to a drawing that shows a set of field lines for that charge distribution. In Section 25.7 we shall show how the electric field can be derived from the potential field. The potential field, however, has advantages over the electric field. First, it is a scalar field whereas the electric field is a vector field, and so calculations involving the potential are generally simpler. Second, while no devices exist to measure electric field strength directly, we can measure the potential difference between two points with a voltmeter.

25.13 Verify that Eq. 25.25 is consistent with Eq. 25.20 by substituting the expression for the electric field of a charged particle.

By following the same procedure as in Checkpoint 25.13, we can now obtain the potential for a group of charged particles. Recall that the electric field due to a group of charged particles is equal to the sum of the electric fields of the individual charged particles (Eq. 23.5),

$$\vec{E} = \sum_n \vec{E}_n. \tag{25.26}$$

This gives us

$$V_{AB} = -\int_A^B \vec{E} \cdot d\vec{\ell} = -\int_A^B \left(\sum_n \vec{E}_n\right) \cdot d\vec{\ell}. \tag{25.27}$$

Because the integral of a sum is equal to a sum of integrals, we have

$$-\int_A^B \left(\sum_n \vec{E}_n\right) \cdot d\vec{\ell} = -\sum_n \int_A^B \vec{E}_n \cdot d\vec{\ell}. \tag{25.28}$$

The line integral after the summation sign is the negative of the potential difference due to the field $\vec{E}_n$, so we see that the total potential difference is the sum of the potential differences caused by the individual particles:

$$V_{AB} = \sum_n V_{AB,n}, \tag{25.29}$$

where $V_{AB,n}$ is the potential difference caused by particle n. Substituting the potential of a single particle, Eq. 25.21, and again letting the potential at infinity be zero, we get

$$V_P = \frac{1}{4\pi\epsilon_0} \sum_n \frac{q_n}{r_{nP}} \quad \text{(potential zero at infinity)}, \tag{25.30}$$

where q_n is the charge carried by particle n and r_{nP} is the distance of particle n from the point P at which we are evaluating the potential (**Figure 25.21**).

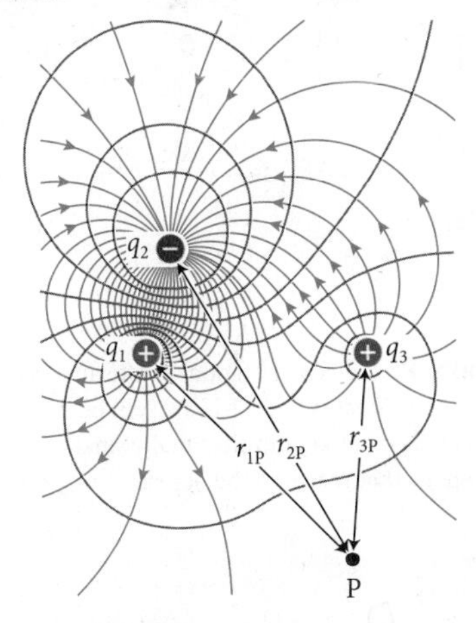

Figure 25.21 The potential at point P due to a group of charged particles is the algebraic sum of the potentials at P due to the individual particles.

The line integral on the right in Eq. 25.25 has an important significance. As we argued in Section 25.2, the electrostatic work done on a charged particle moving around a closed path (as in **Figure 25.22**) is zero. Therefore, Eq. 25.24 must yield zero for a closed path:

$$W_q(\text{closed path}) = q\oint \vec{E} \cdot d\vec{\ell} = 0, \tag{25.31}$$

where the circle through the integral sign indicates that the integration is to be taken around a closed path. Because Eq. 25.31 holds for any value of q, we have

$$\oint \vec{E} \cdot d\vec{\ell} = 0 \quad \text{(electrostatic field)}. \tag{25.32}$$

In other words, for any electrostatic field, the line integral of the electric field around a closed path is zero. This is equivalent to saying that you cannot extract energy from an electrostatic field by moving a charged particle around a closed path. In terms of potential, this means that if we start at some point P on the closed path and the potential has a value V_P at that point, then the potential can take on other values as we go around the closed path, but as we return to P, the value of the potential must once again be V_P. As we shall see later in Chapter 30, it *is* possible to get energy out of electric fields, however. Here we have shown only that it is *not* possible to extract energy by moving a charged particle around a closed path in *electrostatic* fields—that is, those due to stationary charge distributions.

Figure 25.22 The electrostatic work done on a charged particle as the particle is moved around a closed path starting and ending at some point P is zero.

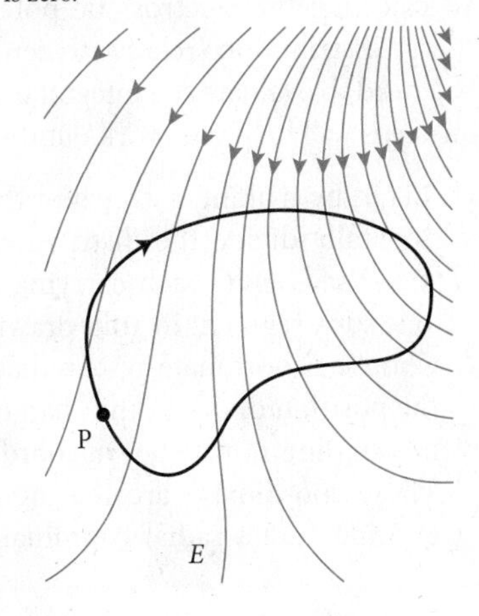

25.14 Describe how the potential varies as you go around the closed path in Figure 25.22 in the direction shown, starting from a potential V_P at P.

25.6 Electrostatic potentials of continuous charge distributions

For extended objects with continuous charge distributions, we cannot use Eq. 25.30 directly to calculate the potential. Instead we must divide the object into infinitesimally small segments, each carrying charge dq_s (which we can treat as a charged particle), and then integrate over the entire object.

Consider, for example the object shown in **Figure 25.23**. Let the zero of potential again be at infinity. Treating each segment as a charged particle, we calculate its contribution to the potential at P (Eq. 25.21):

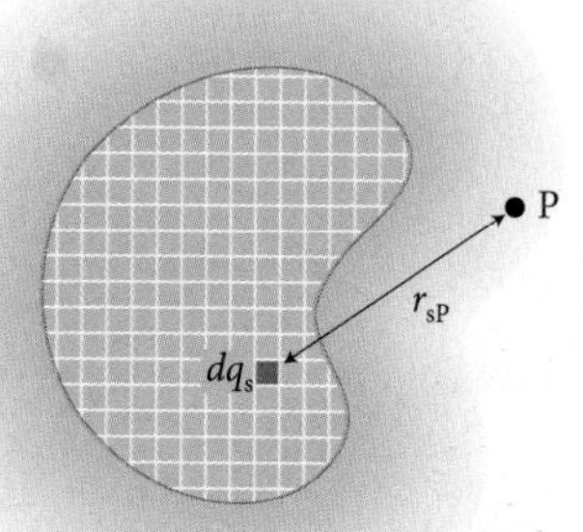

Figure 25.23 The potential due to an extended object is the algebraic sum of the potentials of all the infinitesimally small segments that make up the object.

$$dV_s = \frac{1}{4\pi\epsilon_0}\frac{dq_s}{r_{sP}}, \tag{25.33}$$

where r_{sP} is the distance between P and dq_s. The potential due to the entire object is then given by the sum over all the segments that make up the object. For infinitesimally small segments, this yields the integral

$$V_P = \int dV_s = \frac{1}{4\pi\epsilon_0}\int\frac{dq_s}{r_{sP}}, \quad \text{(potential zero at infinity)}, \tag{25.34}$$

where the integral is taken over the entire object. Note the parallel between this integral and the integral in Eq. 23.15 for calculating the electric field of a continuous charge distribution. Indeed, the procedure for calculating the potential of a continuous charge distribution is very similar to that for calculating the electric field of a continuous charge distribution (see Section 23.7). However, the potential in Eq. 25.34 is much easier to evaluate because it is a scalar and does not involve any unit vectors. The Procedure box below provides some helpful steps in evaluating Eq. 25.34, and the next two examples show how to put the procedure into practice.

PROCEDURE: Calculating the electrostatic potentials of continuous charge distributions

To calculate the electrostatic potential of a continuous charge distribution (relative to zero potential at infinity), you need to evaluate the integral in Eq. 25.34. The following steps will help you work out the integral:

1. Begin by making a sketch of the charge distribution. Mentally divide the distribution into infinitesimally small segments, each carrying a charge dq_s. Indicate one such segment in your drawing.
2. Choose a coordinate system that allows you to express the position of dq_s in the charge distribution in terms of a minimum number of coordinates (x, y, z, r, or θ). These coordinates are the integration variables. For example, use a radial coordinate system for a charge distribution with radial symmetry. Never place the representative segment dq_s at the origin.
3. Indicate the point at which you wish to determine the potential. Express the factor $1/r_{sP}$, where r_{sP} is the distance between dq_s and the point of interest, in terms of the integration variable(s).
4. Determine whether the charge distribution is one dimensional (a straight or curved wire), two dimensional (a flat or curved surface), or three dimensional (any bulk object). Express dq_s in terms of the corresponding charge density of the object and the integration variable(s).

At this point you can substitute your expressions for dq_s and $1/r_{sP}$ into Eq. 25.34 and work out the integral.

Example 25.6 Electrostatic potential of a uniformly charged thin rod

A thin rod of length ℓ carries a uniformly distributed charge q. What is the potential V_P at point P a distance d from the rod along a line that runs perpendicular to the long axis of the rod and passes through one end of the rod?

❶ GETTING STARTED I begin by making a sketch. I let the y axis be along the rod, with the origin at the bottom of the rod. Point P then lies on the x axis (**Figure 25.24**). What I must calculate is the electrostatic potential V_P at P.

Figure 25.24

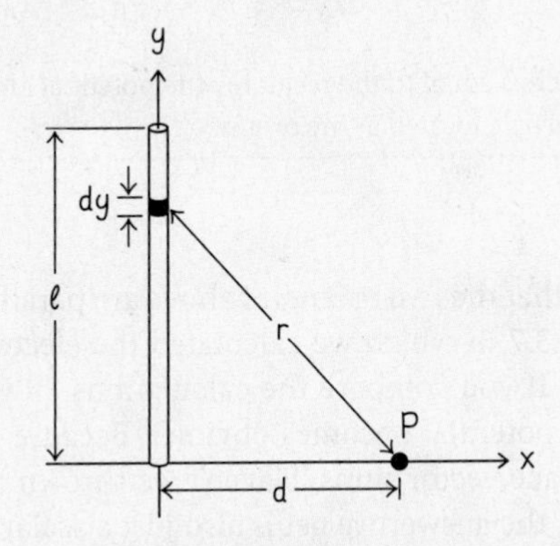

❷ DEVISE PLAN Because the rod is thin and uniformly charged, I can treat it as a one-dimensional object that has a linear charge density $\lambda = q/\ell$. To determine the electrostatic potential at point P, I divide the rod lengthwise into a large number of infinitesimally small segments, each of length dy. Each segment contributes to the potential at P an amount given by Eq. 25.33. To calculate the potential at P due to the entire rod, I can then use Eq. 25.34 to integrate my result over the entire length of the rod.

❸ EXECUTE PLAN Each length segment dy carries a charge $dq_s = \lambda dy = (q/\ell)dy$. To calculate the potential at P due to the whole rod, I substitute the distance between each segment dy and P, $r_{sP} = \sqrt{y^2+d^2}$, and my expression for dq_s into Eq. 25.34 and integrate from the bottom of the rod at $y = 0$ to the top at $y = \ell$:

$$V_P = \frac{1}{4\pi\epsilon_0}\frac{q}{\ell}\int_0^\ell \frac{dy}{\sqrt{y^2+d^2}}.$$

Looking up the solution of the integral, I obtain

$$V_P = \frac{1}{4\pi\epsilon_0}\frac{q}{\ell}\left[\ln(y+\sqrt{y^2+d^2})\right]_0^\ell$$

$$= \frac{1}{4\pi\epsilon_0}\frac{q}{\ell}\left[\ln(\ell+\sqrt{\ell^2+d^2}) - \ln d\right]$$

$$= \frac{1}{4\pi\epsilon_0}\frac{q}{\ell}\ln\left(\frac{\ell+\sqrt{\ell^2+d^2}}{d}\right). ✔$$

❹ EVALUATE RESULT: If I let ℓ go to zero, my answer should become the same as that for a particle (Eq. 25.21). When $\ell \ll d$, I can ignore the term ℓ^2 in my answer and so the argument of the logarithm becomes $(\ell + d)/d = 1 + \ell/d$. Because $\ln(1+\epsilon) \approx \epsilon$ when $\epsilon \ll 1$ (see Appendix B), my answer becomes

$$V_P \approx \frac{1}{4\pi\epsilon_0}\frac{q}{\ell}\ln\left(\frac{\ell+d}{d}\right) \approx \frac{1}{4\pi\epsilon_0}\frac{q}{\ell}\frac{\ell}{d} = \frac{1}{4\pi\epsilon_0}\frac{q}{d},$$

which is indeed equal to the electrostatic potential at a distance $r = d$ from a particle.

Example 25.7 Electrostatic potential of a uniformly charged disk

A thin disk of radius R carries a uniformly distributed charge. The surface charge density on the disk is σ. What is the electrostatic potential due to the disk at point P that lies a distance z from the plane of the disk along an axis that runs through the disk center and is perpendicular to the plane of the disk?

❶ GETTING STARTED I begin by making a sketch of the disk (**Figure 25.25**). I let the disk be in the xy plane, with the origin at the center of the disk and point P on the z axis.

Figure 25.25

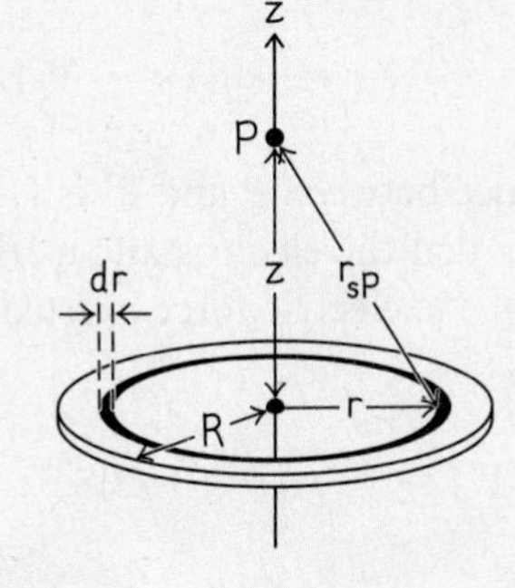

❷ DEVISE PLAN Because of the circular symmetry of the disk, I divide it into a large number of thin circular ring segments, each of radius r and thickness dr. All parts of a given ring are the same distance r_{sP} from point P, so each part makes the same contribution to the potential at P. I can therefore calculate the contribution of an entire ring segment to the potential at P using Eq. 25.33, substituting the charge dq_s on the segment and the distance from a point on the segment to P. The charge dq_s on the ring segment is given by the product of its area (circumference times thickness) and the surface charge density: $dq_s = (2\pi r)\,dr\,\sigma$. To calculate the potential at point P due to the entire disk, I can use Eq. 25.34 to integrate my result over the radius of the disk, using r as my integration variable.

❸ EXECUTE PLAN: The distance from a point on any ring segment to P is $r_{sP} = \sqrt{z^2+r^2}$, and so each segment's contribution to the potential is given by

$$dV_P = \frac{1}{4\pi\epsilon_0}\frac{(2\pi r)\sigma\, dr}{\sqrt{z^2+r^2}}.$$

(Continued)

To calculate the potential at P due to the whole disk, I integrate this expression from $r = 0$ to $r = R$:

$$V_P = \int dV_P = \frac{1}{4\pi\epsilon_0}\int_0^R \frac{2\pi r\sigma}{\sqrt{z^2 + r^2}}\,dr.$$

Looking up the solution of the integral, I get

$$V_P = \frac{2\pi\sigma}{4\pi\epsilon_0}\int_0^R \frac{r\,dr}{\sqrt{z^2 + r^2}} = \frac{\sigma}{2\epsilon_0}\left(\sqrt{z^2 + R^2} - |z|\right). ✔$$

❹ **EVALUATE RESULT** When z is very large relative to R, the disk should resemble a particle and my result should reduce to that for a particle (Eq. 25.21). For large $z > 0$ I can use the binomial expansion (see Appendix B) to write the factor in parentheses as

$$\sqrt{z^2(1 + R^2/z^2)} - z \approx z\left(1 + \tfrac{1}{2}\frac{R^2}{z^2}\right) - z = \tfrac{1}{2}\frac{R^2}{z}.$$

Because $\sigma\pi R^2$ is equal to the charge q on the disk, my expression for V becomes

$$V_P \approx \frac{\sigma}{2\epsilon_0}\left(\frac{\tfrac{1}{2}R^2}{z}\right) = \frac{\sigma\pi R^2}{4\pi\epsilon_0 z} = \frac{1}{4\pi\epsilon_0}\frac{q}{z},$$

which is equal to the result for the potential along a z axis due to a particle located at the origin.

You may have noticed that the two examples above are parallel to corresponding examples in Section 23.7 in which we calculated the electric fields due to a thin charged rod or disk. If you compare the calculations, however, the advantage of working with the potential becomes obvious: Because the potential is a scalar you don't need to take vector sums. Haven't we thrown away some information, though? After all, the answer we get is also just a scalar, not a vector like the electric field. Figure 25.9 gives us some idea of the answer to this question: Because field lines and equipotentials are always perpendicular to each other, you can draw equipotentials if you know the field line pattern or, conversely, draw field lines if you know the equipotentials. It turns out that even though the potential is a scalar and the electric field is a vector, it is possible to determine one from the other.

25.15 Verify that the potentials obtained in Examples 25.6 and 25.7 have the correct sign.

25.7 Obtaining the electric field from the potential

For the potential to be a useful quantity, we must be able to determine the electric field (and therefore the forces exerted by this field) from the potential. For example, let the equipotentials in **Figure 25.26** represent the potential of some charge distribution. How can we use the known potential of the charge distribution to determine the value of the electric field at any point P?

Figure 25.26 To determine the component of electric field along an axis, we calculate the electrostatic work done on a charged particle as the particle is moved over a short segment along that axis.

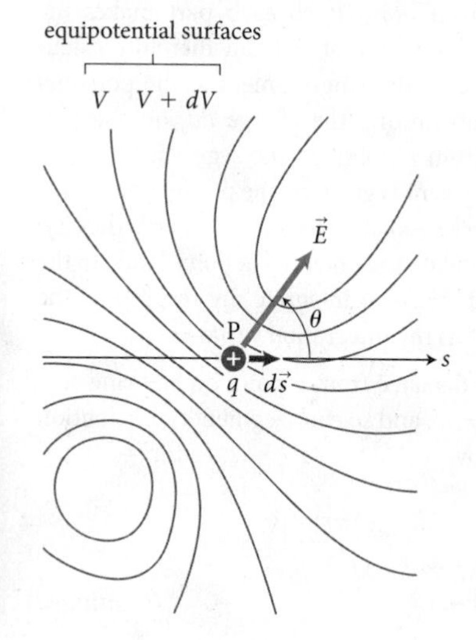

We know that the electric field is perpendicular to the equipotentials, and so the electric field at P must be along the direction indicated in the figure. To determine the magnitude of the electric field, imagine moving a particle carrying a charge q over an infinitesimally small displacement $d\vec{s}$ along some arbitrary axis s. Let the particle be displaced from P, where the potential is V, to a point P′ where the potential is $V + dV$. According to Eq. 25.17, the electrostatic work done on the particle is

$$W_q(\text{P} \rightarrow \text{P}') = -qV_{\text{PP}'} = -q(V_{\text{P}'} - V_\text{P}) = -q\,dV \quad (25.35)$$

because the potential difference between P and P′ is $(V + dV) - V = dV$. On the other hand, we also know that the electrostatic work done on the particle is equal to the scalar product of the electric force exerted on the particle and the force displacement $d\vec{r}_F = d\vec{s}$:

$$W_q(\text{P} \rightarrow \text{P}') = \vec{F}^E_q \cdot d\vec{s} = (q\vec{E}) \cdot d\vec{s}$$
$$= q(\vec{E} \cdot d\vec{s}) = qE\cos\theta\,ds, \quad (25.36)$$

where we have assumed that the force displacement is small enough that $\vec{E}$ can be considered constant between P and P′. Note that θ is the angle between $\vec{E}$ and the s axis, so $E \cos\theta$ is the component of the electric field along the s axis. We can write $E\cos\theta = E_s$, and equating the two expressions for the electrostatic work done on the particle, Eqs. 25.35 and 25. 36, we get

$$-q\,dV = qE_s\,ds \tag{25.37}$$

or

$$E_s = -\frac{dV}{ds}. \tag{25.38}$$

In other words, the component of the electric field along the s axis is given by the negative of the derivative of the potential with respect to s. The faster V varies (the more closely spaced the equipotentials), the greater the magnitude of the electric field.

Equation 25.38 gives only the component of the electric field along the (arbitrary) axis s. To determine all the components of the electric field, we must repeat this procedure for each of the three Cartesian coordinates:

$$E_x = -\frac{\partial V}{\partial x},\quad E_y = -\frac{\partial V}{\partial y},\quad E_z = -\frac{\partial V}{\partial z}. \tag{25.39}$$

Note the partial derivatives; these are necessary because the function $V(x, y, z)$ depends on all three Cartesian coordinates. When you take the partial derivative with respect to one coordinate, just remember to keep the other coordinates constant. (For example, if $V = x^2y$, then $\partial V/\partial x = 2xy$, $\partial V/\partial y = x^2$, and $\partial V/\partial z = 0$.)

Once the components of the electric field are determined, we can write the electric field in vectorial form:

$$\vec{E} = -\frac{\partial V}{\partial x}\hat{\imath} - \frac{\partial V}{\partial y}\hat{\jmath} - \frac{\partial V}{\partial z}\hat{k}. \tag{25.40}$$

Equation 25.40 then tells us how to obtain the electric field from the potential.

25.16 Apply Eq. 25.40 to the potential you obtained for the uniform field between two parallel charged plates in Example 25.5, and verify that you get the correct expression for the electric field between the plates.

Example 25.8 The electrostatic potential and electric field due to a dipole

A permanent dipole consists of a particle carrying a charge $+q_p$ at $x = 0, y = +\frac{1}{2}d$ and another particle carrying a charge $-q_p$ at $x = 0, y = -\frac{1}{2}d$. Use the electrostatic potential at a point P on the axis of the dipole to determine the electric field at that point.

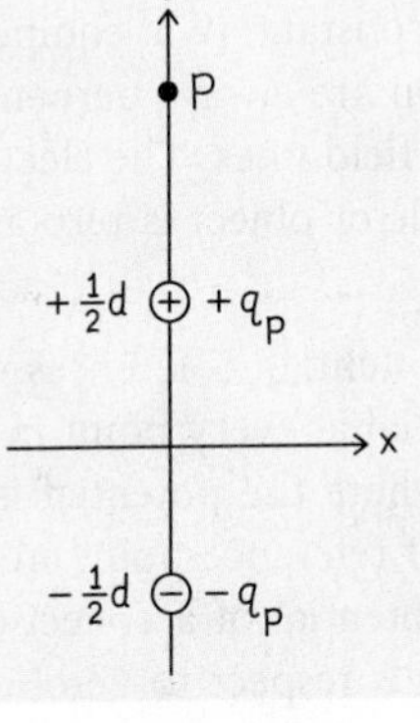

❶ GETTING STARTED I begin by making a sketch of the dipole (**Figure 25.27**). I place the dipole along my y axis, with the midpoint of the dipole length at the origin, so that the particles are at the coordinates given in the problem. I choose a point P on the positive y axis, and I let the y coordinate of P be y. Because the two charged particles lie on the y axis, the electric field they create at P must be directed along the y axis.

Figure 25.27

(Continued)

❷ DEVISE PLAN The potential at P is the sum of the potentials due to the individual charged particles. To calculate these potentials, I can use Eq. 25.30. Once I have calculated the potential at P, I can use Eq. 25.40 to determine the electric field.

❸ EXECUTE PLAN The potential at P is

$$V_P = \frac{1}{4\pi\epsilon_0}\Sigma\frac{q_n}{r_{nP}} = \frac{1}{4\pi\epsilon_0}\left[\frac{+q_p}{y - \frac{1}{2}d} + \frac{-q_p}{y + \frac{1}{2}d}\right]. \quad (1)$$

Now I use this result to determine the electric field. Because I am working with the y axis, I work with the y component of Eq. 25.40:

$$E_y = -\frac{\partial V}{\partial y} = \frac{1}{4\pi\epsilon_0}\left[\frac{q_p}{(y - \frac{1}{2}d)^2} - \frac{q_p}{(y + \frac{1}{2}d)^2}\right]. ✔$$

Because the electric field at P is directed along the y axis, the other components of the electric field are zero: $E_x = E_z = 0$. ✔

❹ EVALUATE RESULT: This is the same result as in Eq. 23.11. (Remember $k = 1/(4\pi\epsilon_0)$.)

In comparing the derivation in Example 25.8 with the derivation of the electric field in Section 23.6, note that the calculation of the electric field via the potential does not involve any vector addition. The scalar nature of the potential therefore greatly simplifies calculations.

25.17 Calculate the electric field at any point on the axis of a thin charged disk from the potential we obtained in Example 25.7. Compare your answer to the result we obtained by direct integration in Section 23.7.

Chapter Glossary

SI units of physical quantities are given in parentheses.

Electric potential energy U^E (J) The form of potential energy associated with the configuration of stationary objects that carry electrical charge. When the reference point for the electric potential energy is set at infinity, the potential energy for two particles carrying charges q_1 and q_2 and separated by a distance r_{12} is

$$U^E = \frac{q_1 q_2}{4\pi\epsilon_0}\frac{1}{r_{12}} \quad (U^E \text{ zero at infinite separation}). \quad (25.8)$$

Electrostatic work W_q (J) Work done by an electrostatic field on a charged particle or object moving through that field. The electrostatic work depends on only the endpoints of the path. For a particle of charge q that is moved from point A to point B in an electric field, the electrostatic work is

$$W_q(\mathrm{A} \to \mathrm{B}) = q\int_{\mathrm{A}}^{\mathrm{B}} \vec{E} \cdot d\vec{\ell}. \quad (25.24)$$

Equipotentials Lines or surfaces along which the value of the potential is constant. The equipotential surfaces of a charge distribution are always perpendicular to the corresponding electric field lines. The electrostatic work done on a charged particle or object is zero as it is moved along an equipotential.

Potential V_P (V) Potential differences can be turned into values of the potential at every point in space by choosing a reference point where the potential is taken to be zero. Common choices of reference point are Earth (or *ground*) and infinity. The potential of a collection of charged particles (measured with respect to zero at infinity) at some point P can be found by taking the algebraic sum of the potentials due to the individual particles at P:

$$V_P = \frac{1}{4\pi\epsilon_0}\Sigma\frac{q_n}{r_{nP}} \quad \text{(potential zero at infinity)}, \quad (25.30)$$

where q_n is the charge carried by particle n and r_{nP} is the distance from P to that particle. For continuous charge distributions, the sum can be replaced by an integral:

$$V_P = \frac{1}{4\pi\epsilon_0}\int\frac{dq_s}{r_{sP}} \quad \text{(potential zero at infinity)}. \quad (25.34)$$

The electric field can be obtained from the potential by taking the partial derivatives:

$$\vec{E} = -\frac{\partial V}{\partial x}\hat{\imath} - \frac{\partial V}{\partial y}\hat{\jmath} - \frac{\partial V}{\partial z}\hat{k}. \quad (25.40)$$

Potential difference V_{AB} (V) The potential difference between points A and B is equal to the negative of the electrostatic work per unit charge done on a charged particle as it is moved along a path from A to B:

$$V_{AB} \equiv \frac{-W_q(\mathrm{A} \to \mathrm{B})}{q} = -\int_{\mathrm{A}}^{\mathrm{B}} \vec{E} \cdot d\vec{\ell}. \quad (25.25)$$

For electrostatic fields, the potential difference around a closed path is zero:

$$\oint \vec{E} \cdot d\vec{\ell} = 0 \quad \text{(electrostatic field)}. \quad (25.32)$$

Volt (V) The derived SI unit of potential defined as $1\,\mathrm{V} \equiv 1\,\mathrm{J/C}$.

26 Charge Separation and Storage

This chapter deals with generating and storing electric potential energy. To produce charged objects, positive and negative charge carriers must be pulled apart and then kept separate. Work is required to pull apart charge carriers, just as work is required to stretch a spring. In each case, this work results in energy storage in the system. We now look at what kind of changes in energy are involved in the separation of positive and negative charge carriers and how charge carriers that have been separated can be stored in simple arrangements of conductors.

26.1 Charge separation

Whenever objects are "charged" (by separating strips of Scotch tape, rubbing objects against each other, using batteries, etc.), the basic phenomenon is the same: Some process (pulling, rubbing, chemical reactions) separates positive and negative charge carriers from one another. As a concrete example, consider a rubber rod and a piece of fur. If you rub the two together and then separate them, they become oppositely charged because the rod pulls electrons away from the fur: The rod ends up with a surplus of electrons and the fur with a deficit. Provided none of the electrons on the rod leak away (to the air, your hand, etc.), the magnitude of the negative charge on the rod is equal to that of the positive charge on the fur.

What is the change in energy associated with this charge separation? Consider the rubber-fur system in its initial and final states (**Figure 26.1*a***). To separate the positive and negative charge carriers, they must be pulled apart against an attractive electric force, just as the ends of a stretched spring are pulled apart against an elastic force. Because work must be done on the rod-fur system, the electric potential energy of the system is greater in the final state. This energy is supplied by you while you rub the two objects together and then increase their separation. Not all of the energy you put into the system goes into electric potential energy; the friction involved in the rubbing not only produces charge separation but also heats up the rod and the fur, so part of the work you do on the system increases the thermal energy. An energy diagram for the rod-fur system is shown in Figure 26.1*b*.

CONCEPTS

> **26.1** Suppose you repeat the charging (starting again with uncharged rod and fur), but this time you rub longer and twice as much charge accumulates at each point on the two objects. How do the following quantities compare to what they were after the first charging: (*i*) the direction and magnitude of the electric field at point P in Figure 26.1*a*; (*ii*) the potential difference between two fixed points on the rod and the fur; and (*iii*) the electric potential energy in the rod-fur system?

Checkpoint 26.1 highlights the essence of this chapter. Be sure not to confuse *potential difference* and *electric potential energy*:

- The system's electric potential energy depends on the configuration of the positive and negative charge carriers in the system.

Figure 26.1 When we charge a rubber rod and a piece of fur by rubbing them together, we do work to separate charge and hence increase the electric potential energy of the system comprising the rod and fur.

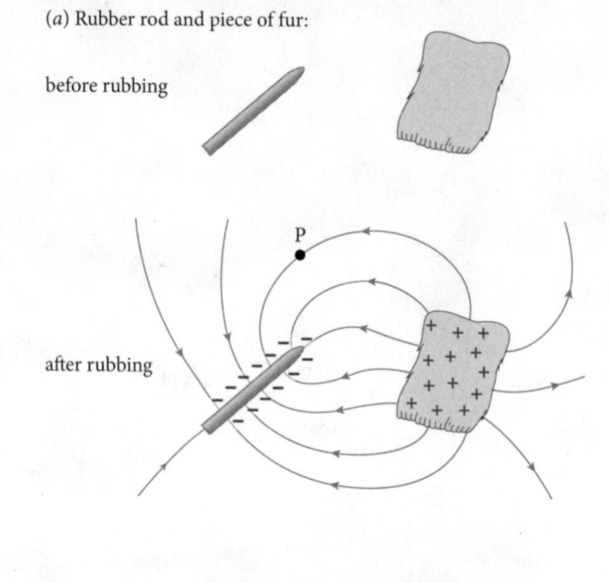

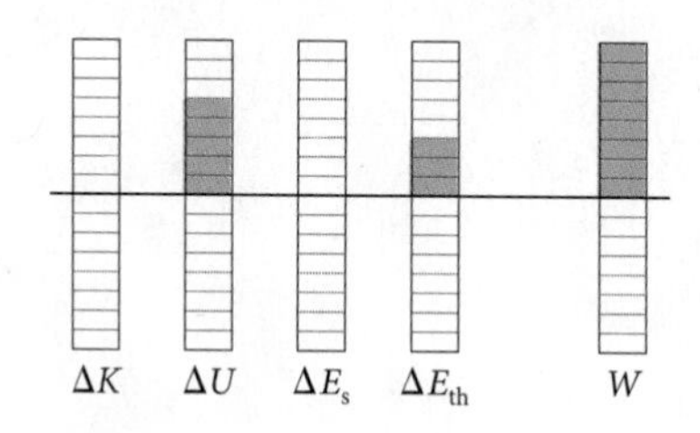

- The potential difference between points on the rod and the fur is a measure of the electrostatic work done on a particle carrying a unit of charge (not part of the system) while moving between those points.

In the next section we examine the proportionality between potential difference and charge separation in detail. In this section we concentrate on the relationship between charge separation and electric potential energy.

The crucial point to take away from Checkpoint 26.1 is:

Positive work must be done on a system to cause a charge separation of the positive and negative charge carriers in the system. This work increases the system's electric potential energy.

> **26.2** If you double the separation between the charged rod and fur in Figure 26.1*a*, does the electric potential energy of the rod-fur system increase, decrease, or stay the same?

The amount of stored electric potential energy depends on the amount of charge that is separated and the distance that separates the charge carriers. More charge or a greater separation means more electric potential energy is stored. These arguments apply to *all* devices that separate charge, such as Van de Graaff generators (see the box below) and batteries (see Section 26.4). Once electric potential energy has been generated by separating charge carriers, this energy can be used for other processes, such as lighting a lamp, operating a radio, and so on.

Every **charge-separating device** (or **charging device**) has some mechanism that moves charge carriers *against* an electric field—a process that requires work to be done on the system of charge carriers. For the rod-fur system, this work is mechanical and is supplied by the person doing the rubbing and separating the objects. In a battery, chemical reactions drive charge carriers through a region where the electric field opposes their motion.

26.3 If you include the person doing the rubbing in the system considered in Figure 26.1, what is the resulting energy diagram?

Where is the electric potential energy of the rod-fur system stored? As you may recall from Section 7.2, potential energy is stored in reversible changes in the configuration of interacting components of a system. Electric potential energy, therefore, is associated with the configuration of the charge carriers in a system. A look at the electric field pattern suggests an alternative view, however. **Figure 26.2** shows how the electric field line pattern changes as the distance between the charged rod and the fur increases. Note how more of the space around the system becomes filled with field lines (indicating that the magnitude of the electric field there increases), while the density of the field lines between

Figure 26.2 Change in the electric field pattern as the distance between the rod and fur increases.

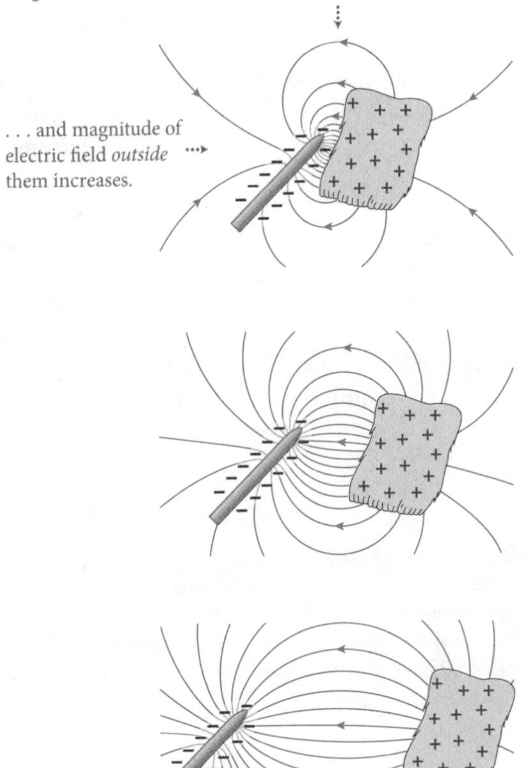
As rod and fur get farther apart:
Magnitude of electric field *between* them decreases . . .
. . . and magnitude of electric field *outside* them increases.

Van de Graaff generator

Figure 26.3 shows a schematic diagram of a Van de Graaff generator—a mechanical device invented in the 1930s by Robert J. Van de Graaff to separate large amounts of electrical charge. The basic principle is extremely simple: A nonconducting belt delivers charge carriers to a hollow conducting dome that rests on a nonconducting support. Machines of this type are used to generate the very large potential differences required in particle accelerators and for the generation of x rays.

Operation of the generator involves three important steps. The first step is a transfer of charge carriers to the belt at A. This transfer can be done by literally "spraying" charged particles onto the belt or simply by rubbing the rubber belt against some appropriate material.

The second step transports the charge carriers to the dome. This step is possible because the belt is nonconducting, so the charge carriers are not mobile—they are stuck to the belt, which is driven by a motor around a pulley inside the dome. The motor must do work on the charge carriers to move them against the electric field of the dome. (In the example shown in the diagram, positive charge carriers at B must be transported upward against the downward electric field of the positively charged dome.)

Figure 26.3 Schematic diagram of a Van de Graaff generator.

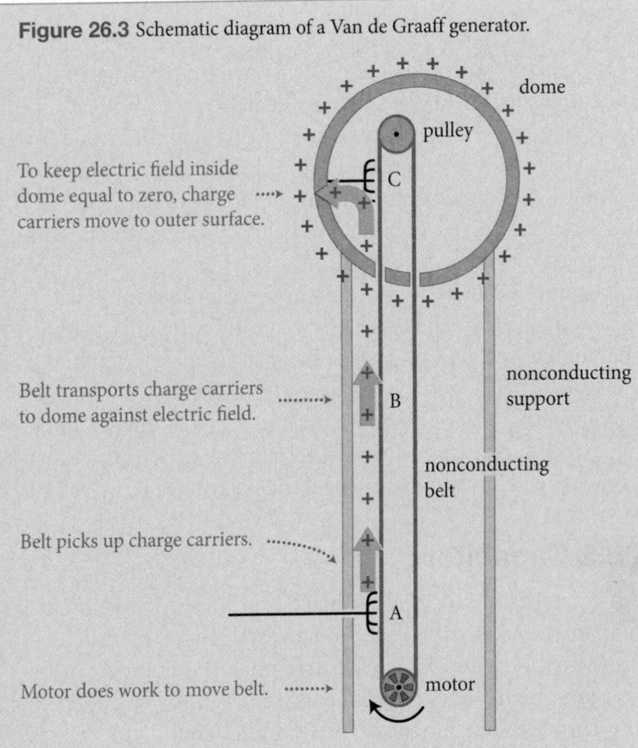
dome
pulley
C
To keep electric field inside dome equal to zero, charge carriers move to outer surface.
nonconducting support
Belt transports charge carriers to dome against electric field.
B
nonconducting belt
Belt picks up charge carriers.
A
Motor does work to move belt.
motor

CONCEPTS

The third step transfers the charge carriers from the belt onto the dome. As we saw in Section 24.5, the electric field inside a hollow conductor is always zero and any charge inside a conductor moves toward the outer surface. Therefore, once the charge carriers are inside the dome, they tend to move to the outer surface of the dome. For this purpose, a comb of conducting needles is placed close to the belt at C. If the charge carriers on the belt are electrons, the electrons hop onto the comb and move via the connecting wire toward the outside of the metal dome, causing the dome to acquire a negative charge. Alternatively, the charge carriers on the belt can be positively ionized air molecules, in which case electrons in the comb are attracted toward the ions. These electrons then jump from the comb onto the belt, neutralizing the ions on the belt while leaving a positive charge behind on the outside of the dome.

Construction of a huge double Van de Graaff generator for the MIT Physics Department in South Dartmouth, Massachusetts, in 1933. These generators, currently at the Boston Museum of Science, generated opposite charge and were able to produce potential differences of 10,000,000 V between the two 4.5-m domes.

the rod and the fur decreases. Thus, as the distance between the rod and the fur changes, the electric potential energy changes and the electric field changes.

As we shall see later in this chapter, we can relate a change in the electric potential energy of a system to a change in the system's electric field (integrated over all of space), which suggests that the electric potential energy of a system is stored in its electric field. In other words, the electric potential energy of the rod-fur system is spread throughout the space around it. As long as the two objects are held stationary, the field is stationary, so the exact "location" of the energy is not very important because there is no way we can determine it experimentally. If we shake the charged rod or fur, however, the shaking causes a wavelike disturbance in the electric field. This disturbance propagates away from the rod or fur and carries with it energy that we can detect. For now we don't need to concern ourselves with such waves. It suffices to know that electric fields store electric potential energy.

26.2 Capacitors

Any system of two charged objects, such as the rod-fur system in the preceding section, stores electric potential energy. To study how much electric potential energy can be stored in a system of two objects, let's begin by considering the simple arrangement of two parallel conducting plates shown in **Figure 26.4a**. A system for storing electric potential energy that consists of two conductors is called a **capacitor**; the arrangement in Figure 26.4*a* is called a *parallel-plate capacitor*

Figure 26.4*b* illustrates a simple method for charging such a capacitor. Each plate is connected by a wire to a terminal of a battery, which maintains a fixed potential difference between its terminals. **Figure 26.5** shows what happens when the connection is made between the battery and the capacitor. If the capacitor plates are far enough away from

Figure 26.4 A parallel-plate capacitor.

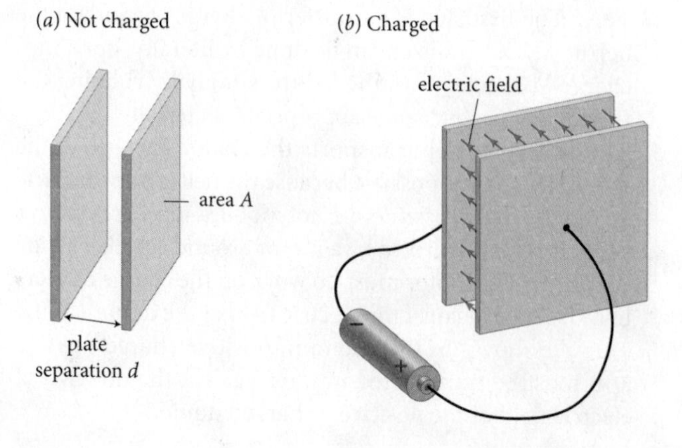

CONCEPTS

Figure 26.5 Charging a capacitor.

(*a*) Capacitor not connected to battery

Zero potential difference between uncharged plates.

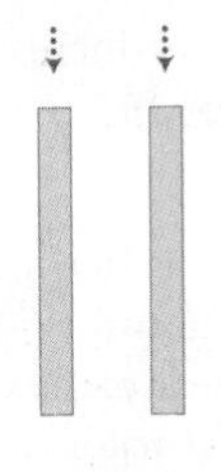

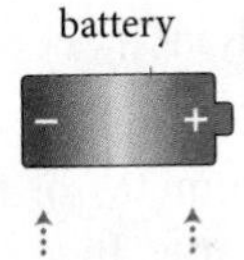

Battery maintains potential difference between terminals.

(*b*) Capacitor being charged

Electrons flow along wires in direction of higher potential.

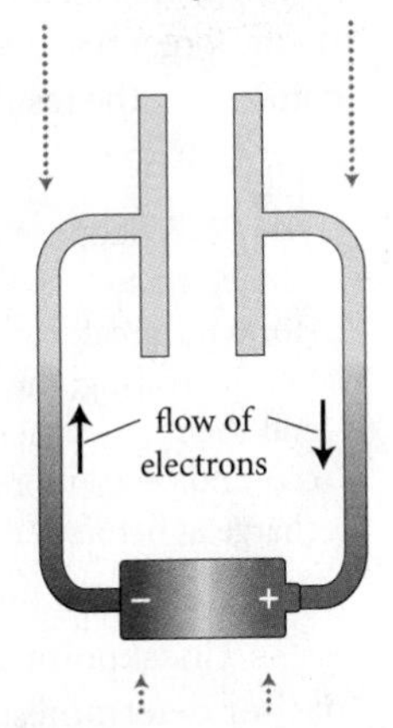

Chemical reactions in battery supply charge to terminals, keeping potential difference fixed.

(*c*) Capacitor fully charged

Potential of each plate now identical to that of corresponding battery terminal.

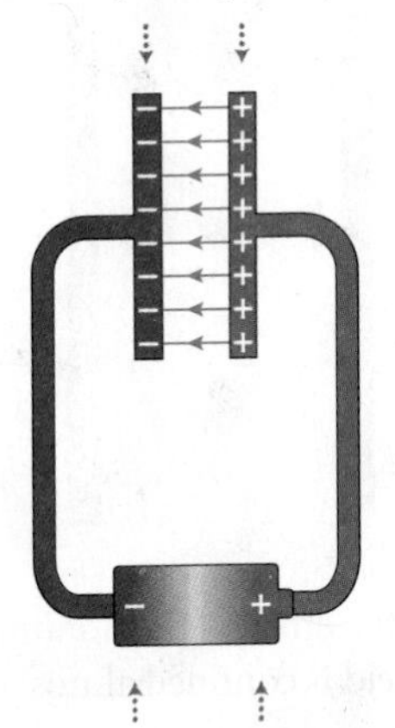

Potential difference between terminals stays the same.

the battery, the potential difference between the plates initially is zero (Figure 26.5*a*). Immediately after the wires are connected, there is a potential difference between the ends of each wire. This difference in potential causes electrons (which are mobile in metal) in the wires to flow as indicated by the arrows in Figure 26.5*b*. A positive charge builds up on the plate connected to the positive terminal, and a negative charge of equal magnitude builds up on the other plate. As electrons leave one plate and accumulate on the other, the potential difference between the plates changes. This process continues until the potential is the same at both ends of each wire—that is, when the potential difference between the plates is equal to that between the terminals of the battery (Figure 26.5*c*). Because there is no longer any potential difference from one end of the wire to the other, the flow of electrons stops and the capacitor is said to be *fully charged*. In the process of achieving this state, the battery has done work on the electrons; this work has now become electric potential energy stored in the capacitor.

The time interval it takes to fully charge a capacitor depends on the properties of the capacitor, the battery, and the way the capacitor is connected to the battery. Typically, only a fraction of a second is needed for charging, although the time interval it takes to charge very large capacitors can be minutes (more on this in Chapter 32).

26.4 (*a*) Suppose that we disconnect the wires from the plates after the capacitor is charged as shown in Figure 26.5*c*. How does the potential difference between the plates after the wires are disconnected compare to that just before they are disconnected? (*b*) If we replace the battery in Figure 26.5 by a battery that maintains a greater potential difference between its terminals, is the magnitude of the charge on the plates greater than, smaller than, or the same as when the first battery is connected?

When a capacitor is not connected to anything, as in Checkpoint 26.4*a*, it is said to be *isolated*. For an isolated capacitor, the *quantity of charge on each plate* is fixed because the charge carriers have nowhere to go. In contrast, for a capacitor that is connected to a battery, the *potential difference across the capacitor* is fixed—the charge carriers on the plates always adjust themselves in such a way as to ensure that the potential difference across the capacitor is equal to that across the battery.

Figure 26.6 shows the electric fields of two isolated charged parallel-plate capacitors. The field is nearly uniform in the region between the plates, but it is nonuniform at the edges. When the spacing between the plates is small compared to the

Figure 26.6 Effect of plate separation in relation to plate area on the field of a parallel-plate capacitor.

Plate separation small compared to plate area:

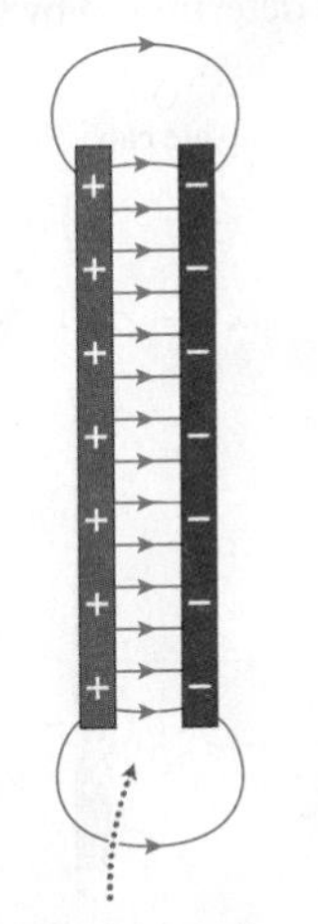

Electric field nearly uniform, localized mainly between plates.

As plate separation becomes greater compared to plate area . . .

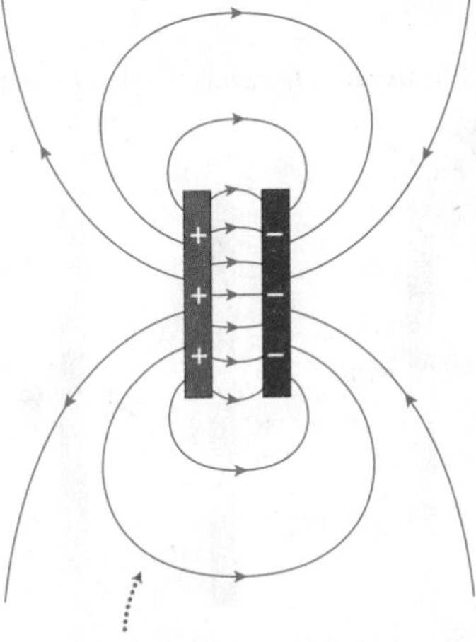

. . . more electric field "escapes" from between plates.

CONCEPTS

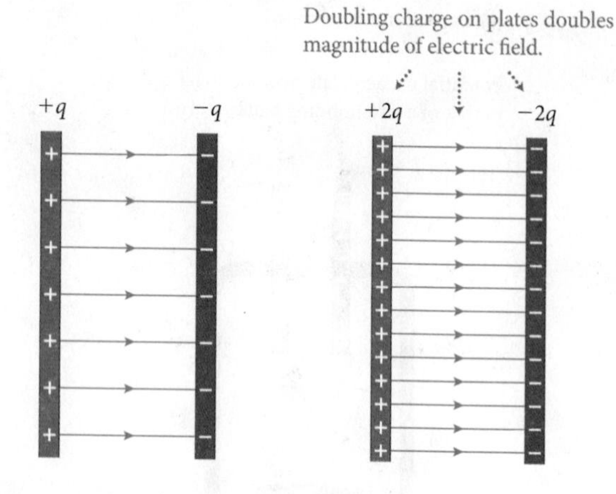
Doubling charge on plates doubles magnitude of electric field.
$+q$
$-q$
$+2q$
$-2q$

Figure 26.7 Doubling the charge on a parallel-plate capacitor.

area of the plates, the effect of the nonuniform field is negligible: The electric field is confined almost entirely to the region between the plates, and for all practical purposes we can consider this field to be uniform. When discussing parallel-plate capacitors, we shall ignore the nonuniform fields at the edges and assume the electric field is entirely uniform between the plates. This simplification is justified by the geometry of most capacitors used in electronic applications.

Let us now examine the relationship between the magnitude of the charge on the plates of a parallel-plate capacitor and the potential difference between them. **Figure 26.7** shows an isolated parallel-plate capacitor carrying a positive charge $+q$ on one plate and a negative charge $-q$ on the other. If we double the magnitude of the charge on each plate, then the electric field between the plates doubles, too (see Checkpoint 23.14). Consequently, the electric force exerted on a charged particle between the plates doubles, so the electrostatic work done in moving a charged particle from one plate to the other also doubles. According to Eq. 25.15, the potential difference between the plates doubles as well. In other words, the potential difference between the plates is proportional to the magnitude of the charge on the plates.

What happens if we increase the plate separation of an isolated parallel-plate capacitor, as illustrated in **Figure 26.8**? The electric field remains the same because it is determined by the surface charge density on the plates, which doesn't change (see Checkpoint 23.14). Because the distance between the plates increases, however, the electrostatic work done in moving a charged particle from one plate to the other increases—more work is required to move the particle over a greater distance—so the potential difference between the plates increases too, confirming the result we obtained in Example 25.5.

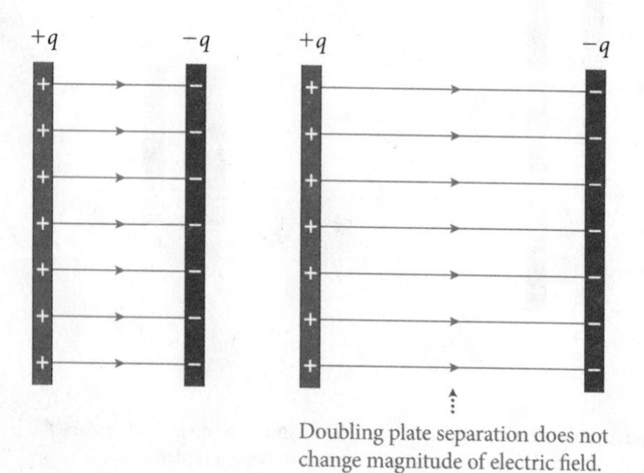
$+q$
$-q$
$+q$
$-q$
Doubling plate separation does not change magnitude of electric field.

Figure 26.8 Doubling the plate separation of a parallel-plate capacitor.

26.5 Suppose the two capacitors in Figure 26.8 are each connected to a 9-V battery. (*a*) Which of the two capacitors stores the greater amount of charge? (*b*) If, instead of the separation increasing, the area of the plates of the capacitor is halved and then the capacitor is connected to a 9-V battery, does the capacitor store more charge, less charge, or the same amount of charge as before the area of the plates was halved?

As Checkpoint 26.5 illustrates, the geometry of the capacitor determines its capacity to store charge. In general:

For a given potential difference between the plates of a parallel-plate capacitor, the amount of charge stored on its plates increases with increasing plate area and decreases with increasing plate separation.

Does this mean that we can increase the amount of charge stored on a parallel-plate capacitor indefinitely simply by making the plate separation infinitesimally small? The answer is *no*, because if the plate spacing is decreased while the potential difference between the capacitor plates is fixed, the charge on each plate increases and thus the magnitude of the electric field in the capacitor increases. When the electric field is about 3×10^6 V/m, the air molecules between the plates become *ionized* and the air becomes conducting, allowing a direct transfer of charge carriers between the plates. Once such a so-called **electrical breakdown** occurs, the capacitor loses all its stored energy in the form of a spark.

The opening page of this chapter shows an electrical breakdown of air between the charged dome of a very large Van de Graaff generator and a nearby metal object. The breakdown limits the maximum potential difference across a capacitor and thus the maximum amount of charge that can be stored on it. The electric field at which electrical breakdown occurs is called the *breakdown threshold.*

The breakdown threshold can be raised by inserting a nonconducting material between the capacitor plates. As we shall see in the next section, such a nonconducting material also greatly increases the amount of charge that can be stored by a capacitor. To understand why this is so, we begin by considering a simpler situation: the insertion of a conductor between the plates of a parallel-plate capacitor.

Figure 26.9*a* shows an isolated charged capacitor. Suppose we now insert a conducting slab between the plates of this capacitor (Figure 26.9*b*). As we saw in Section 24.5, the charge carriers in the conductor rearrange themselves in such a fashion as to eliminate the field inside the bulk of the conductor.

Figure 26.9 Inserting a conducting slab between the plates of a parallel-plate capacitor.

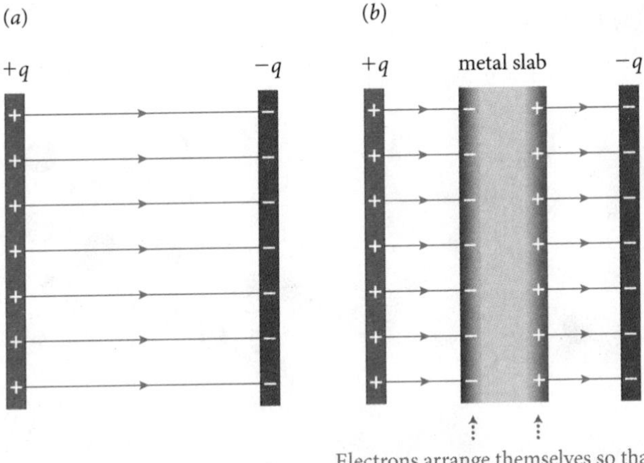

26.6 Suppose the capacitor in Figure 26.9*a* is charged and then disconnected from the battery. (*a*) As the conducting slab is inserted in the capacitor, as in Figure 26.9*b*, does the amount of charge on the capacitor plates increase, decrease, or stay the same? (*b*) How much charge accumulates on each side of the slab, once it is inserted? (*c*) What is the potential difference across the metal slab? (*d*) As the slab is inserted, does the magnitude of the potential difference between the capacitor plates increase, decrease, or stay the same?

As Checkpoint 26.6 demonstrates, the effect of the slab is to make the electric field in part of the space between the capacitor plates zero, thus reducing the magnitude of the potential difference between the plates for a given amount of charge on them. As the next example shows, the converse of this fact is that, for a given potential difference between its plates, the capacitor can store more energy with the slab inserted than it can without the slab. In other words, the slab increases the capacitor's capacity to store charge.

Example 26.1 Metal-slab capacitor

Suppose the capacitor in Figure 26.9 has a plate separation distance d and the plates carry charges $+q$ and $-q$ when the capacitor is connected to a battery that maintains a potential difference V_{batt} between its terminals. If a metal slab of thickness $d/2$ is inserted midway between the plates while the battery remains connected, what happens to (*a*) the magnitude of the electric field between the plates and (*b*) the quantity of charge on the plates?

❶ GETTING STARTED I am given the plate separation distance and plate charges for a capacitor connected to a battery, and I must determine how the electric field magnitude between the plates and the quantity of charge on each plate change when a metal slab is inserted. Because the capacitor remains connected to the battery, the potential difference across the capacitor is fixed. Because the slab is conducting, the electric field inside the slab is always zero.

❷ DEVISE PLAN The electric field magnitude determines the electrostatic work, which in turn determines the potential difference between the plates (which I know). To determine how the electric field magnitude changes when the metal slab is inserted, therefore, I must first determine how the potential between the plates changes when the slab is inserted. Once I know the electric field, I can determine how the slab affects the charge on the plates because I know that the magnitude of the electric field between the plates is proportional to the surface charge density σ (see Checkpoint 23.14, $E = 4k\pi\sigma$).

❸ EXECUTE PLAN To compare the potentials before and after the slab is inserted, I must plot V as a function of position between the plates (**Figure 26.10**). I choose my x axis to be parallel to the electric field, with the positive plate at $x = 0$ and the negative plate at $x = d$ as the zero of potential. In the absence of the metal slab, the field is uniform between the plates, and so the potential decreases linearly from $x = 0$ to $x = d$ (Figure 26.10*a*). Because the electric field inside the slab is zero, the potential does not vary across the slab (it is an equipotential; see Section 25.3). Because the battery keeps the potential at each plate constant, the potential-versus-distance curve must take on the zigzag form shown in Figure 26.10*b*.

Figure 26.10

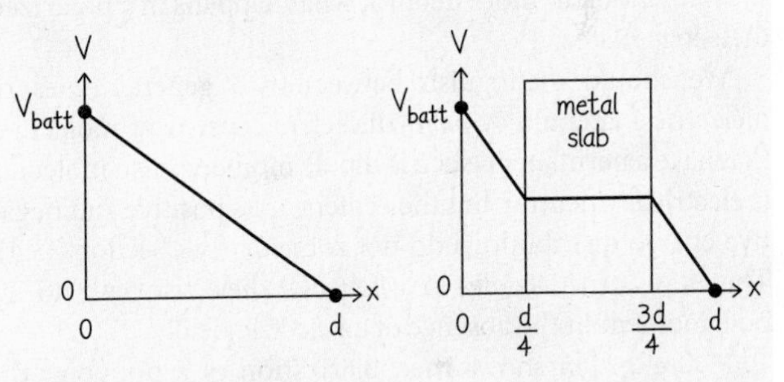

(*a*) Insertion of the metal slab affects the electric field between the plates because the slab is an electrical conductor and so the electric field inside it must always be zero. However, because V_{batt} is constant, I know that the electrostatic work, $W_q = F^E d = qV_{batt}$, done on a particle carrying a charge q to move the particle from one plate to the other must be the same whether or not the slab is in place there. Because no electrostatic work is done to move the particle through the slab where the electric field is zero, the field outside the slab must be greater to make up for the smaller distance over which the particle is moved. More precisely, because the distance over which the electric field is nonzero is reduced to $d/2$, the magnitude of the field must be twice what it was before the slab was inserted. ✔

(*b*) If the field doubles, then the charge per unit area must also double. Given that the area of the plates does not change, this means that the charge on the plates must double. ✔

❹ EVALUATE RESULT Inserting the metal slab with the battery connected is equivalent to halving the separation distance between the plates while keeping the potential difference constant. I know from Checkpoint 26.5 that, for a constant potential difference, the quantity of charge stored on a capacitor increases with decreasing plate separation, as I found.

26.7 (*a*) Does the position of the slab in Figure 26.9 affect the potential difference across the capacitor? Consider, in particular, the case in which the slab is moved all the way to one side and makes electrical contact with one of the plates. (*b*) Sketch the potential $V(x)$ as a function of x, with the slab off-center.

26.3 Dielectrics

As we just saw, decreasing the space inside an isolated charged capacitor where the electric field is nonzero increases its capacity to store electrical charge for a given potential difference across its plates. With a conducting slab inserted, however, the gap between either plate and the slab face nearest it is smaller than the plate-to-plate gap before the slab was inserted. Because a decreased gap with the potential difference held constant means E increases, we still have the problem of electrical breakdown.

Suppose, however, that we insert a nonconducting material—a **dielectric**—between the plates of a capacitor. As we discussed in Section 22.4, the electric field between the plates of the capacitor polarizes the dielectric. What effect does this polarization have? To answer this question we must first look in more detail at what happens in a polarized dielectric material.

We should distinguish between two general types of dielectric materials. A *polar* dielectric consists of molecules that have a permanent electric dipole moment; each molecule is electrically neutral, but the centers of its positive and negative charge distributions do not coincide (see Section 23.4). The atoms or molecules in a *nonpolar* dielectric have no dipole moment in the absence of an electric field.

Figure 26.11a shows the polarization of a nonpolar dielectric. In the presence of an electric field, the electrons in a nonpolar dielectric are displaced in the direction opposite to $\vec{E}$, inducing a dipole moment on each molecule.

26.8 Why are the electrons displaced in a direction *opposite* the electric field?

The polarization of a polar dielectric is shown in Figure 26.11*b*. In the absence of an electric field, the individual molecules' dipole moments are randomly aligned, so the material as a whole is not polarized. In the presence of an electric field, however, the molecular dipoles are subject to a torque (see Section 23.8) that tends to align the dipoles with the electric field, giving rise to a macroscopic polarization. In general, the polarization of polar dielectrics is much greater than that of nonpolar ones because the permanent dipole moments of the molecules in a polar dielectric are much greater than the induced dipole moments in a nonpolar dielectric.

Figure 26.12 illustrates the effect of the uniform polarization of the atoms or molecules in a polar or nonpolar dielectric. The charge enclosed by any volume that lies entirely inside the polarized dielectric is zero: The positive and negative charge carriers may not coincide exactly, but on average they occur in equal numbers. However, this is not true for a small volume at the surface of the dielectric. In the volume on the right in Figure 26.12, for example, a surplus of positive charge appears at the surface of the material. Thus, a polarized dielectric has a very thin sliver of surplus positive charge on one side and a sliver containing an equal amount of surplus negative charge on the other side (see also Figure 22.21). The surface charge on either side of the polarized dielectric is said to be **bound** because the charge carriers that cause it are not free to roam around in the material. In contrast, the charge on the capacitor plates is **free**

Figure 26.11 Polarization of nonpolar and polar molecules in an electric field.

(a) Polarization of nonpolar molecules

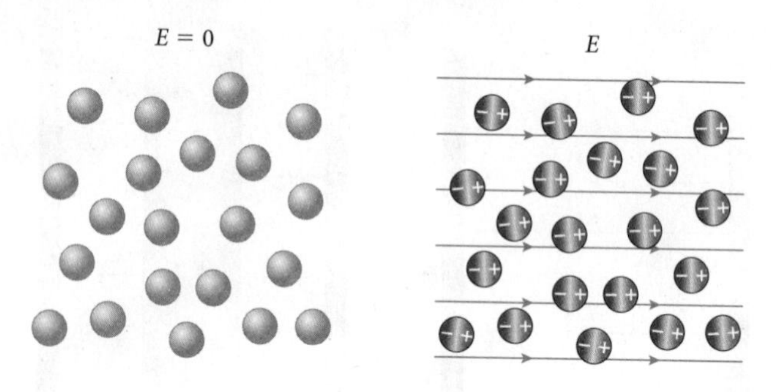
$E = 0$
E

(b) Polarization of polar molecules

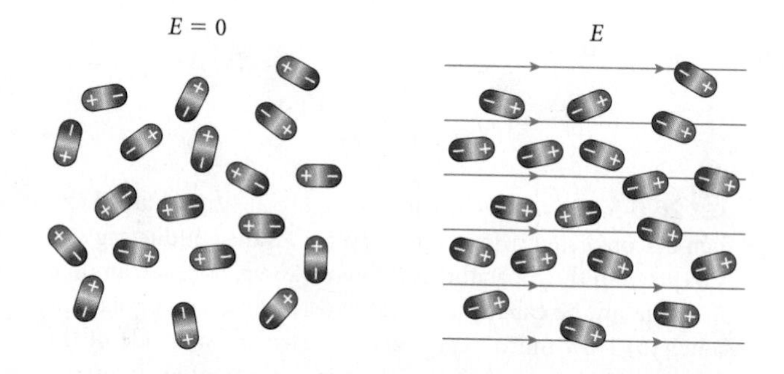
$E = 0$
E

Figure 26.12 The reason a polarized dielectric exhibits a macroscopic polarization.

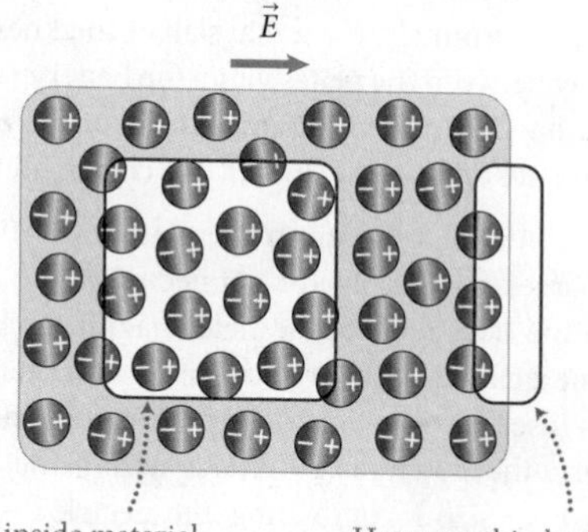
$\vec{E}$
For volume inside material, enclosed charge sums to zero.
However, thin layer at surface has surplus of positive charge.

CONCEPTS

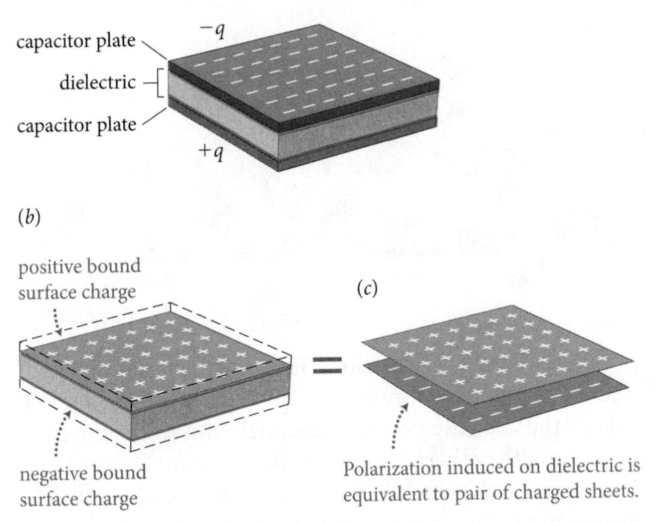
(a) Dielectric sandwiched between capacitor plates
capacitor plate
dielectric
capacitor plate
−q
+q
(b)
positive bound surface charge
negative bound surface charge
(c)
Polarization induced on dielectric is equivalent to pair of charged sheets.

Figure 26.13 The polarization induced on a dielectric in a parallel-plate capacitor is equivalent to two thin sheets carrying opposite charge.

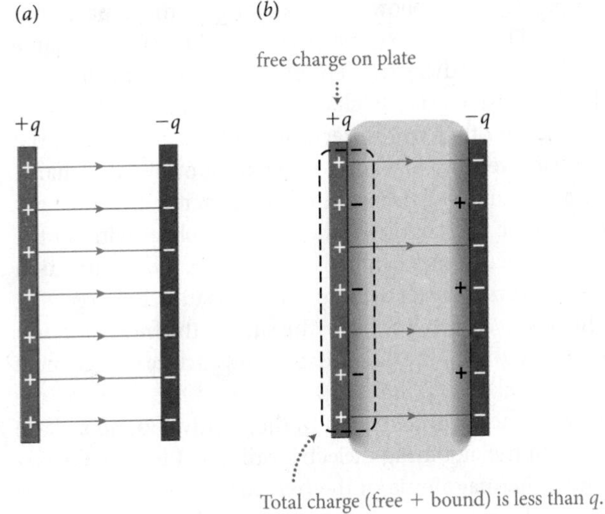
(a)
(b)
free charge on plate
+q
−q
Total charge (free + bound) is less than q.

Figure 26.14 The presence of a polarized dielectric reduces the strength of the electric field between the plates of a capacitor.

because the charge carriers that cause it can move around freely. From a macroscopic point of view, a uniformly polarized dielectric differs from an unpolarized dielectric only by the presence of this bound surface charge.

What is the effect of the bound surface charge on the electric field inside the capacitor? Consider a dielectric slab inside an isolated charged capacitor (**Figure 26.13a**). The dielectric is polarized by the electric field of the capacitor; that is, a positive bound surface charge appears on the surface near the negatively charged capacitor plate, and a negative bound surface charge appears on the other side (Figure 26.13*b*). Imagine now that we could "freeze in" the polarization and consider just the slab by itself. Except for two sheets of charge at the top and bottom, the bulk of the dielectric is neutral. Thus, for all practical purposes, the polarized dielectric is equivalent to two very thin sheets carrying opposite charge (Figure 26.13*c*).

26.9 (*a*) In which direction does the electric field due to the bound surface charge point at a location above the top surface in Figure 26.13*c*? (*b*) In which direction does it point at a location between the top and bottom surfaces?

We can now obtain the electric field of the capacitor with the dielectric by superposition: It is equal to the electric field of the capacitor without the dielectric, plus the electric field of the polarized dielectric by itself. As you found in Checkpoint 26.9, the direction of the electric field due to the polarized dielectric is opposite that of the capacitor, so the presence of the dielectric decreases the electric field strength in the capacitor. Alternatively, we can say that each of the bound surface charges compensates for part of the free charge on the adjoining capacitor plate, so, in effect, the total charge (free and bound) on each side of the capacitor is reduced (**Figure 26.14**). This reduction in charge, in turn, gives rise to a smaller electric field inside the capacitor. For some materials, the field inside can be reduced by a factor of several thousand.

26.10 (*a*) If the magnitude of the bound surface charge on the dielectric slab in Figure 26.14*b* were equal to the magnitude of the free charge on the capacitor plates, what would be the electric field inside the capacitor? (*b*) Could the magnitude of the bound surface charge ever be *greater* than the magnitude of the free charge on the plates?

Figure 26.15 shows what happens when a dielectric-filled capacitor is connected to a battery. The battery keeps the potential difference between the capacitor plates the same regardless of the presence of the dielectric. Because

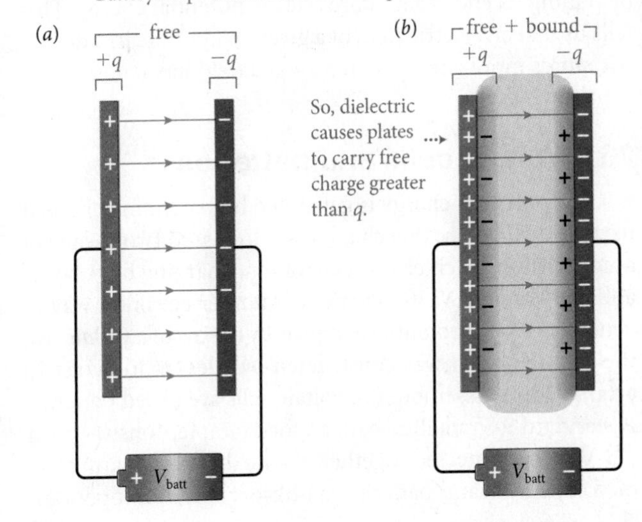
Battery keeps electric field between plates same in both cases:
(a)
free
+q
−q
(b)
free + bound
+q
−q
So, dielectric causes plates to carry free charge greater than q.
V_{batt}

Figure 26.15 The presence of a polarized dielectric increases the charge on the plates of a capacitor connected to a battery.

the electric field in the capacitor is equal to the potential difference divided by the distance between the plates (see Example 25.5), it follows that, as long as the capacitor is connected to the battery, the electric field must be the same regardless of the dielectric. The electric field can be the same only if the distribution of charge causing the electric field is the same. In other words, regardless of the presence of the dielectric, we must have the same amount of total charge (free and bound) on each side of the capacitor. Let the magnitude of the free charge on the capacitor plates without the dielectric be q. As shown in Figure 26.15*b*, the polarization of the dielectric causes a negative bound surface charge next to the positive capacitor plate; the sum of the free charge on the conductor and the adjoining bound surface charge must still be equal to $+q$. Similarly, the sum of the negative free charge on the opposite plate and the positive bound surface charge on the adjoining dielectric still must be $-q$. Consequently, the magnitude of the free charge on each plate by itself can be much greater than without the dielectric. This extra charge is supplied by the battery.

26.11 Given that the electric field is the same in both capacitors in Figure 26.15, which stores the greater amount of electric potential energy?

If the answer to Checkpoint 26.11 surprises you—after all, the electric fields are the same in the two capacitors—remember that the amount of *charge separation* is not the same. The charge on the capacitor plates polarizes the dielectric, and this polarization is the result of charge separation in the molecules of the dielectric. Thus, instead of empty space without charge separation between the capacitor plates, we now have (in addition to a greater charge on the plates) a lot of additional charge separation on the microscopic scale. Most of the energy stored in the capacitor is not due to the separation of charge on the plates, but to the separation of charge in the dielectric between the plates. The pulling apart of the positive and negative charge distributions in the dielectric increases the electric potential energy stored in the dielectric, much like stretching a spring by pulling its ends apart stores elastic potential energy. This tells us that an electric field of a given magnitude in a dielectric stores more energy than an equal field in vacuum.

26.4 Voltaic cells and batteries

Electric potential energy is generated by separating charged particles. Earlier in this chapter we discussed two means of accomplishing such charge separation: charging by rubbing and the Van de Graaff generator. Another common way to generate electric potential energy is by means of a *voltaic cell*, the first of which was constructed by Alessandro Volta in around 1800. Assemblies of voltaic cells are called *batteries*. A standard 9-V alkaline battery, for example, consists of six 1.5-V cells connected together. While there are many types of voltaic cells and batteries, all have a common operating principle: Chemical reactions turn chemical energy into electric potential energy by accumulating electrons on one side of the cell (the negative terminal) and removing electrons from the other side (the positive terminal).

Figure 26.16 General operating principle of a voltaic cell. Electrons flow when the cell is connected to an electronic device.

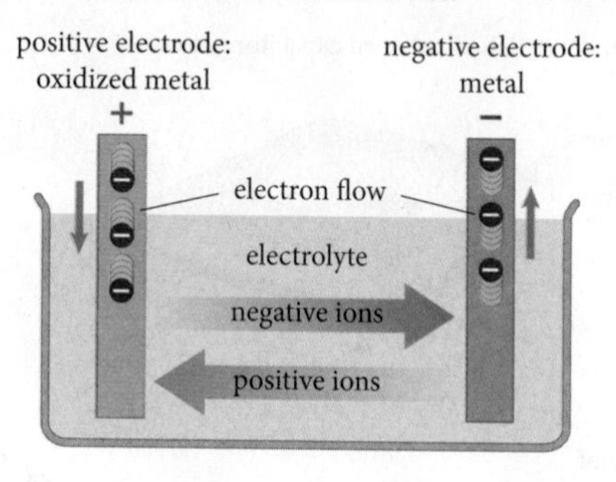

The general principle of a voltaic cell is illustrated in **Figure 26.16**. Two conducting terminals, or *electrodes*, are submerged in an *electrolyte*—a solvent that contains mobile ions. One electrode is usually made from an oxidized metal; the oxidized metal reacts by accepting positive ions from the electrolyte and electrons from the electrode. The other electrode is generally metallic; it oxidizes by taking in negative ions and giving up electrons. Because of these reactions, a surplus of electrons builds up on the metallic terminal and a deficit of electrons builds up on the oxidized-metal terminal, causing a potential difference between the two. The reactions stop when the potential difference between the electrodes reaches a certain value called the *cell potential difference*. This value is determined by the type of chemicals in the cell and is typically on the order of a few volts. As we saw in Section 25.5, this potential difference can be used to do electrostatic work on charge carriers when a battery is connected to some device—such as a light bulb, a motor, or a capacitor.

As long as the cell is not connected to anything and its chemicals do not deteriorate, the cell remains in the same state indefinitely. When the cell is connected to a capacitor or to some other device, however, the surplus of electrons is removed from the negative electrode and electrons are supplied to the positive electrode. Then the chemical reactions resume in order to maintain the cell potential difference between the terminals. As the reactions proceed, the electrolyte becomes more dilute and the compositions of the electrodes change. The cell is exhausted when all the ions in the electrolyte have been depleted. For some types of cells, the chemical reactions can be reversed by supplying a potential difference to the terminals of the cell; electric potential energy is then converted back to chemical energy. Such cells are used in rechargeable batteries, which can be reused repeatedly to store and recover electric potential energy.

As we noted in Section 26.1, any charge-separating device involves the motion of charge carriers against the direction of an electric field. This is where work is done on the charge

Figure 26.17 Schematic diagram of a lead-acid cell and of the reactions taking place at the positive and negative electrodes.

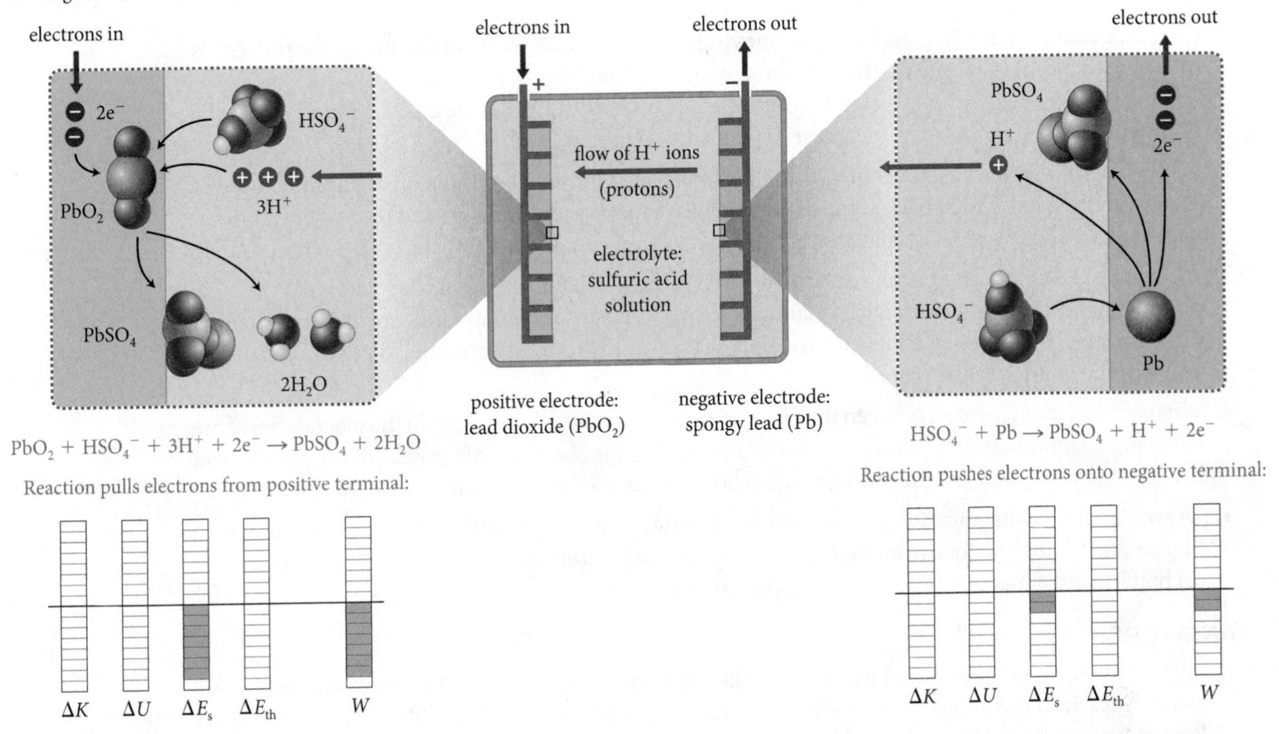

carriers and where some form of energy is converted to electric potential energy. Inside a voltaic cell, electrons must be pulled away from the positively charged terminal and deposited onto the negatively charged terminal. This process occurs at the surface of the electrodes where the chemical reactions take place. The chemical reactions move charge carriers against strong opposing electric forces. The chemical energy released in the reactions provides the energy necessary to move the charge carriers against the electric field. The work done per unit charge is called the **emf** (pronounced e-m-f)* of the device:

The emf of a charge-separating device is the work per unit charge done by nonelectrostatic interactions in separating positive and negative charge carriers inside the device.

All charge-separating devices—batteries, voltaic cells, generators, solar cells—have some *nonelectrostatic* means to separate charge carriers and thereby create a potential difference across the terminals of the device.

26.12 As electrons leave one terminal and are added to the other, ions in the electrolyte must flow in the direction indicated in Figure 26.16 to maintain an even distribution of charge. What must be the direction of the electric field in the bulk of the electrolyte to cause this flow?

Figure 26.17 illustrates the operation of a lead-acid cell used in automobile batteries. A 12-V automobile battery consists of six such cells, each producing a potential difference of 2.1 V. The negative electrode of a lead-acid cell is composed of spongy lead (Pb) packed on a metal grid; the positive electrode contains lead dioxide (PbO_2) packed on a metal grid. The electrodes are immersed in sulfuric acid and chemical reactions convert the lead, the lead dioxide, and the sulfuric acid into lead sulfate ($PbSO_4$) and water. For every molecule of lead sulfate that is produced in these reactions, one electron is removed from the positive terminal and one is added to the negative terminal. The left and right sides of Figure 26.17 show energy diagrams for the species undergoing chemical reactions at each of the electrodes. For each reaction, the chemical energy of the species involved in the reaction decreases; this energy is used to do work on the electrons in the electrodes.

26.13 Given that the cell does *positive* work on the electrons, why is it that the work in both energy diagrams in Figure 26.17 is *negative*?

**emf* stands for *electromotive force*, a misnomer because this quantity bears no relation to the concept of force. For this reason we shall always refer to this quantity by its abbreviation, rather than its original meaning.

Self-quiz

1. Consider again Figure 26.2 and imagine moving one more electron from the fur to the rod. (*a*) Is the work that must be done on the rod-fur system to accomplish this transfer positive, negative, or zero? (*b*) Is the electrostatic work positive, negative, or zero? (*c*) Does the electric potential energy of the rod-fur system increase, decrease, or remain the same?
2. You have probably seen pictures in which a person's hair stands out from his or her head because of "electrostatic charge." Look back at the discussion of Van de Graaff generators and discuss how this can happen when a person makes contact with the globe of the generator but is insulated from the ground.
3. A parallel-plate capacitor is connected to a battery. If the distance between the plates is decreased, do the magnitudes of the following quantities increase, decrease, or stay the same: (*i*) the potential difference between the negative plate and the positive plate, (*ii*) the electric field between the plates, and (*iii*) the charge on the plates?
4. When a dielectric is inserted between the plates of an isolated charged capacitor, do the magnitudes of the following quantities increase, decrease, or stay the same: (*i*) the charge on the plates, (*ii*) the electric field between the plates, and (*iii*) the potential difference between the negative plate and the positive plate?
5. Draw an energy diagram for the process of charging a capacitor with a dielectric as shown in Figure 26.15*b* for the following systems: (*a*) battery, capacitor, and dielectric; (*b*) dielectric only; (*c*) battery and capacitor. Ignore any dissipation of energy.

ANSWERS:

1. (*a*) To displace the electron toward the rod, you must apply a force directed toward the rod. Because the force and force displacement are in the same direction, you (an external agent) must do positive work. (*b*) The electric force exerted on the electron is directed toward the fur, opposite the direction of the force displacement, so the electrostatic work is negative. (*c*) The electric potential energy of the system increases because separating charge carriers increases a system's electric potential energy.
2. If a person is in contact with the globe of the generator but insulated from the ground, then the person acts as an extension of the globe. Electrical charge spreads out over the surface of the person, including the surface of each hair as well. Because each hair has a surplus of the same type of charge, the hairs repel each other and stand out, getting as far away from each other as possible.
3. (*i*) Stays the same. The battery keeps the potential difference across the capacitor constant. (*ii*) To keep a constant potential difference when the distance between the plates decreases, the magnitude of the electric field between the plates must increase because $Ed = V_{batt}$ (see Example 25.5). (*iii*) For the magnitude of the electric field to increase, the charge on the plates must increase.
4. (*i*) Because the capacitor is isolated, the charge on the plates must remain the same—there is no path for the charge to travel elsewhere. (*ii*) When the dielectric is inserted, the electric field due to the bound surface charge is in the opposite direction of the electric field due to the free charge on the plates and decreases the magnitude of the electric field between the plates. (*iii*) Because the magnitude of the electric field decreases and the separation between the plates is constant, the magnitude of the potential difference between the negative plate and the positive plate must also decrease.
5. See **Figure 26.18**. (*a*) During charging, a decrease in source energy (from the battery) increases the electric potential energy (more charge separation in the dielectric and on the capacitor plates). (*b*) The electric potential energy of the dielectric increases due to work done on it by the battery and the capacitor. The electric potential energy stored in the dielectric is smaller than that stored in part *a* because some electric potential energy is stored on the capacitor plates. (*c*) The decrease in source energy is the same as in part *a*. The electric potential energy stored on the capacitor is smaller than in part *a* because most of the converted source energy ends up in the dielectric, which is not part of the system considered. This energy leaves the system as negative work.

Figure 26.18

(a)

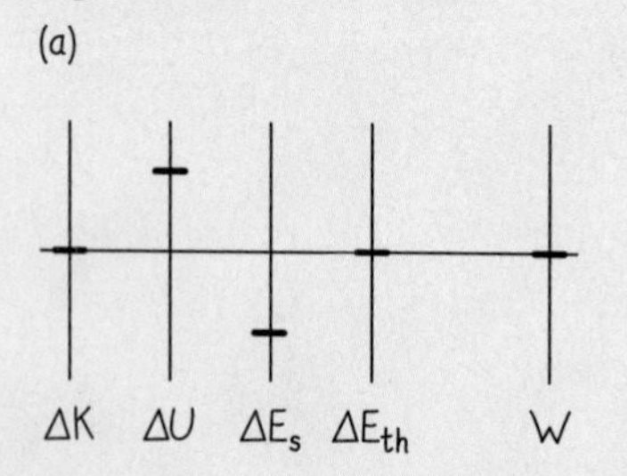

(b)

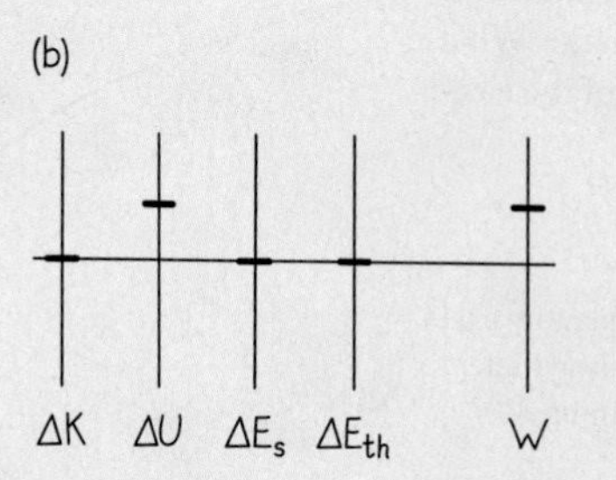

(c)

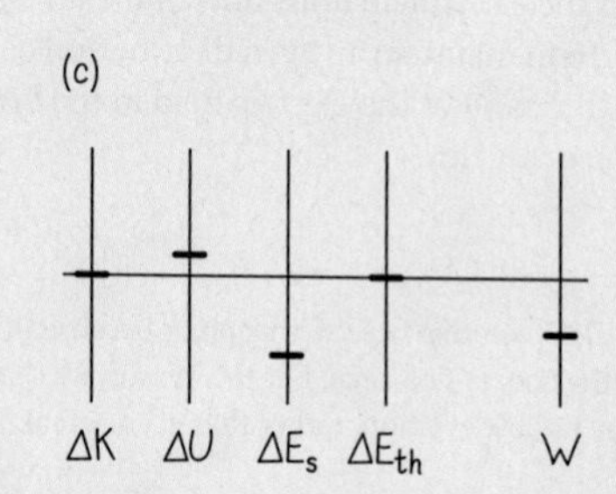

SELF-QUIZ

26.5 Capacitance

Figure 26.19 shows three capacitors, each one consisting of a pair of conducting objects carrying opposite charges of magnitude q. For each arrangement, the potential difference between the objects is proportional to q; that is, doubling q doubles the potential difference across the capacitor. The ratio of the magnitude of the charge on one of the objects to the magnitude of the potential difference across them is defined as the **capacitance** of the arrangement:

$$C \equiv \frac{q}{V_{\text{cap}}}. \tag{26.1}$$

In Eq. 26.1, q represents *the magnitude of the charge on each conducting object* and V_{cap} is *the magnitude of the potential difference between the conducting objects.* Because both these quantities are positive, C is always positive.

The value of C depends on the size, shape, and separation of the conductors. In Figure 26.19, for example, the values of V would typically be different for the three capacitors, even though q is the same for each. Below we'll examine how to determine C for a given set of conductors.

26.14 Two capacitors, A and B, are each connected to a 9-V battery. If $C_A > C_B$, which capacitor stores the greater amount of charge?

Figure 26.19 The electric fields and potential differences of three different capacitors.

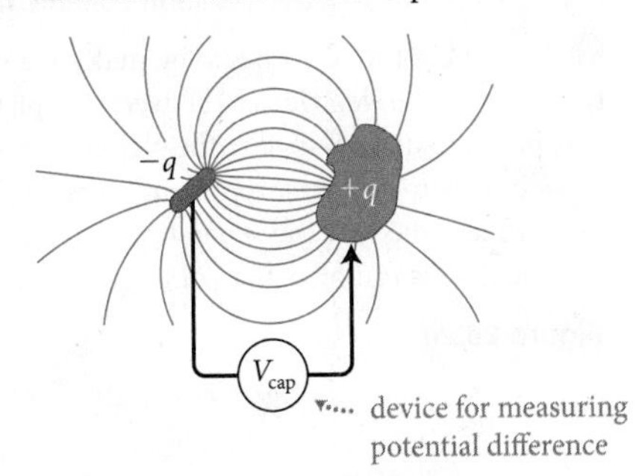

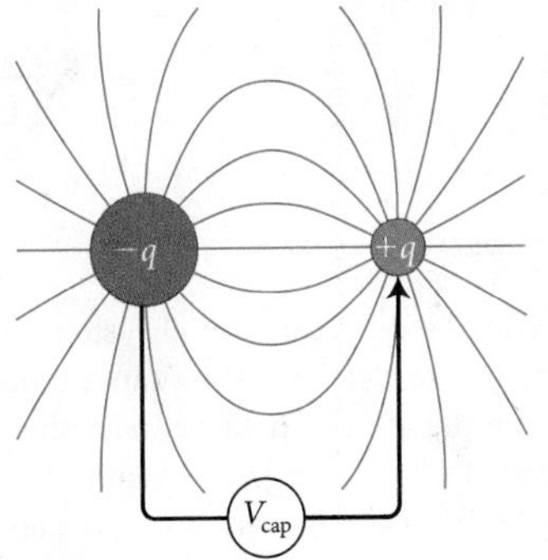

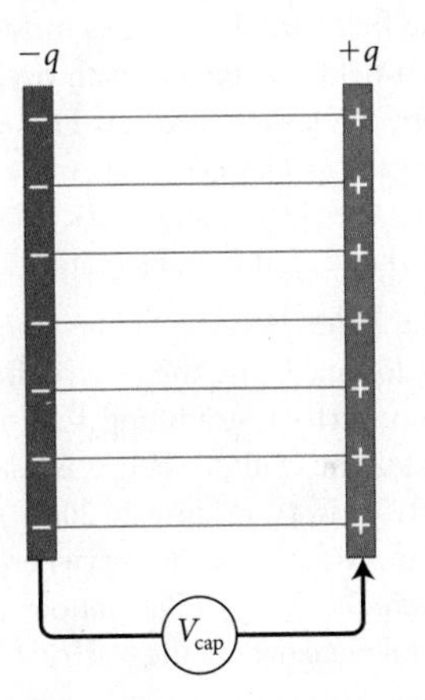

The answer to Checkpoint 26.14 suggests a simple interpretation of C. As its name suggests, C represents the capacitor's *capacity to store charge*: The greater C, the greater the amount of charge stored for a given value of V_{cap}.

As you can see from Eq. 26.1, capacitance has SI units of coulomb per volt. This derived unit is given the name **farad** (F), in honor of the English physicist Michael Faraday:

$$1 \text{ F} \equiv 1 \text{ C/V}.$$

As you will see in Checkpoint 26.15, a capacitance of 1 F is enormous. The capacitance of capacitors commonly found in electronic devices is expressed in microfarads ($1\ \mu\text{F} = 1 \times 10^{-6}$ F) and picofarads ($1 \text{ pF} = 1 \times 10^{-12}$ F).

Figure 26.19 suggests a simple procedure for determining the capacitance of a given set of conductors: Determine the potential difference V_{cap} between the two conductors when they carry some given charge q, and use Eq. 26.1 to calculate C. Note that because conductors are equipotentials, V_{cap} represents the potential difference between *any* two points on the conductors measured along *any* path. The Procedure box below gives one procedure for determining the capacitance of a given set of conductors. In the next examples we apply this procedure to some simple configurations of conductors.

Procedure: Calculating the capacitance of a pair of conductors

To calculate the capacitance of a pair of conductors:

1. Let the conductors carry opposite charges of magnitude q.
2. Use Gauss's law, Coulomb's law, or direct integration to determine the electric field along a path leading from the negatively charged conductor to the positively charged conductor.
3. Calculate the electrostatic work W done on a test particle carrying a charge q_t along this path (Eq. 25.24) and determine the potential difference across the capacitor from Eq. 25.15:

$$V_{\text{cap}} = -W_{q_t}(- \rightarrow +)/q_t.$$

4. Use Eq. 26.1, $C \equiv q/V_{\text{cap}}$, to determine C.

Example 26.2 Parallel-plate capacitor

What is the capacitance of a parallel-plate capacitor that has a plate area A and a plate separation distance d?

❶ GETTING STARTED I begin by making a sketch of the capacitor, showing the electric field between the plates (**Figure 26.20a**). The problem doesn't specify the plate shape, so I simply show the capacitor from the side, representing each plate by a horizontal line. If I assume that the separation distance d is small, then the electric field is uniform and confined between the plates.

Figure 26.20

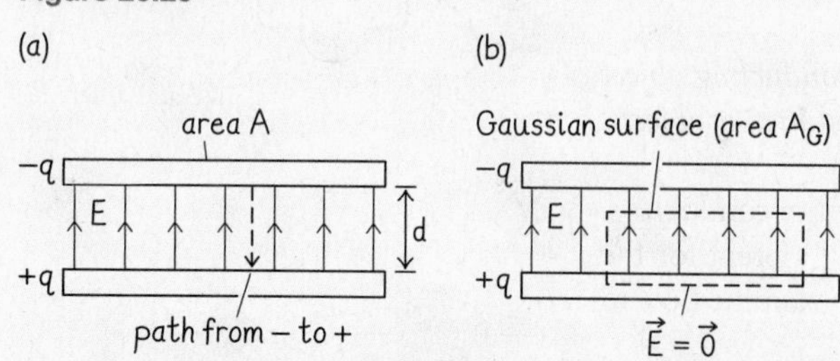

❷ DEVISE PLAN I can use the steps of the Procedure box on page 697 to determine the capacitance. The first step is to determine the electric field between the capacitor plates when they carry opposite charges of magnitude q. The second step is to obtain the electrostatic work done on a test particle moved from one plate to the other; I can use Eq. 25.24 to obtain this work. Because the field is uniform, it is most convenient to choose a path along a field line for the path over which the electrostatic work is done. As specified in the Procedure box, the path runs from the negatively charged plate to the positively charged plate. Once I know the electrostatic work, I know the potential difference across the capacitor and so can calculate the capacitance.

❸ EXECUTE PLAN Because of the planar symmetry, I can use Gauss's law to determine the electric field. I choose a cylindrical Gaussian surface straddling the surface of the positively charged plate. The cylinder height is less than d, and the area of the end surfaces is A_G (Figure 26.20b). The electric flux through this Gaussian surface is zero everywhere except through the top surface, where $\Phi = EA_G$. (The bottom surface is inside the conducting metal plate, where the electric field is zero.)

To apply Gauss's law, I also need to know the charge enclosed by the Gaussian surface. The positive plate carries a charge $+q$ distributed over a surface of area A, so the surface charge density is $\sigma = +q/A$ and the charge enclosed by the Gaussian surface is $q_{enc} = \sigma A_G = (q/A)A_G$. Applying Gauss's law, $\Phi = q_{enc}/\epsilon_0$ (Eq. 24.8), I get

$$EA_G = \frac{q}{\epsilon_0 A}A_G \quad \text{or} \quad E = \frac{q}{\epsilon_0 A} = \frac{\sigma}{\epsilon_0} \tag{1}$$

in agreement with Eq. 24.17.

Now that I know E, I can calculate the electrostatic work required to move a test particle carrying a charge $+q_t$ from the negatively charged to the positively charged plate. The electric force exerted on the test particle is upward in Figure 26.20, and the force displacement is downward because the particle moves from negative plate to positive plate. Because these two vectors point in opposite directions, the electrostatic work done on the test particle is negative, $W_{q_t} = -q_t Ed$, and the potential difference across the capacitor is, from Eq. 25.15,

$$V_{cap} \equiv \frac{-W_{q_t}(- \to +)}{q_t} = \frac{q_t Ed}{q_t} = Ed$$

or, substituting E from Eq. 1,

$$V_{cap} = \frac{qd}{\epsilon_0 A}.$$

Note that the potential difference is proportional to the magnitude of the charge on each plate, q. Using the definition of capacitance, I obtain

$$C \equiv \frac{q}{V_{cap}} = \frac{q}{qd/(\epsilon_0 A)} = \frac{\epsilon_0 A}{d}. \checkmark$$

❹ EVALUATE RESULT My result agrees with the conclusions we drew in Section 26.2: The capacitance (or quantity of charge stored for a given potential difference) increases with increasing plate area A and decreasing plate separation distance d. Also, I note that the electric field—and therefore the capacitance—do not depend on the plate: Circular or square plates give the same result.

26.15 The plate spacing in a typical parallel-plate capacitor is about 50 μm. (a) What is the plate area in a 1-μF capacitor? (b) Given that the electric field at which electrical breakdown occurs in air is about 3×10^6 V/m, what is the maximum charge that this capacitor can hold? (c) How many electrons does this charge correspond to? (The electron's charge is $e = 1.6 \times 10^{-19}$ C.)

Figure 26.21

(a) One way to design a compact capacitor with a large surface area

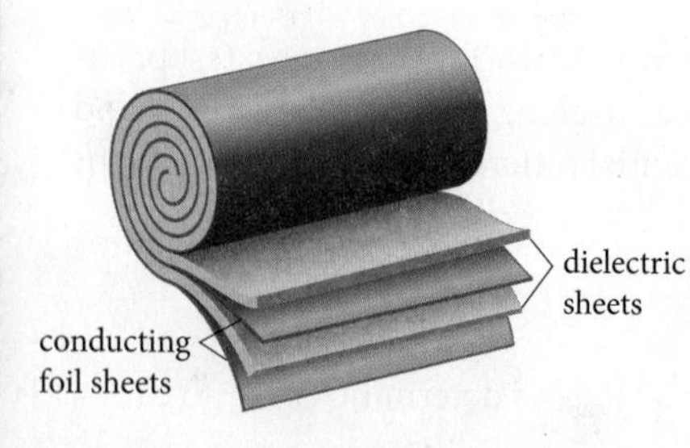

(b) Some capacitors used in electronic circuits

As Checkpoint 26.15 shows, even modest capacitances require very large plate areas. Various techniques are used to keep the overall size of capacitors small, one of which involves rolling up two thin conducting sheets that are separated by thin sheets of a dielectric material (**Figure 26.21a**). Figure 26.21b shows a number of different capacitors used in electronic circuits.

Example 26.3 Coaxial cylindrical capacitor

Figure 26.22 shows a *coaxial capacitor* consisting of two concentric metal cylinders 1 and 2, of radii R_1 and $R_2 > R_1$, and both of length $\ell \gg R_2$. Both cylinders are made of metal. What is the capacitance of this arrangement?

Figure 26.22 Example 26.3.

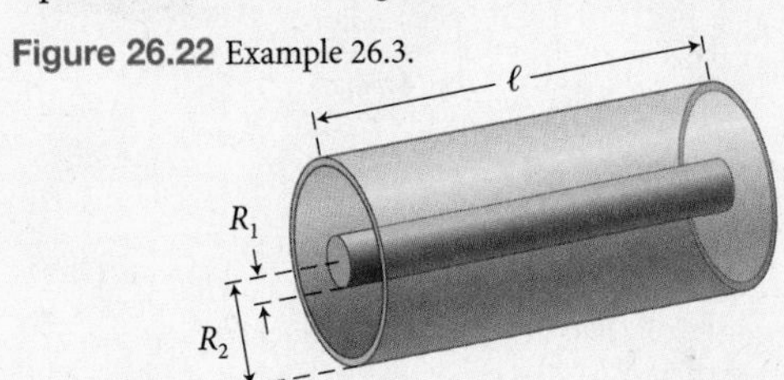

❶ **GETTING STARTED** To determine the capacitance, I must let the two cylinders carry opposite charges of magnitude q, which I assume to be uniformly distributed over each cylinder. If I let cylinder 1 carry a charge $+q$ and cylinder 2 carry a charge $-q$, the electric field points radially outward from cylinder 1 to cylinder 2 (**Figure 26.23**). Because the cylinders are very long relative to their separation distance $R_2 - R_1$, I assume that the electric field is confined to the volume between the cylinders.

Figure 26.23

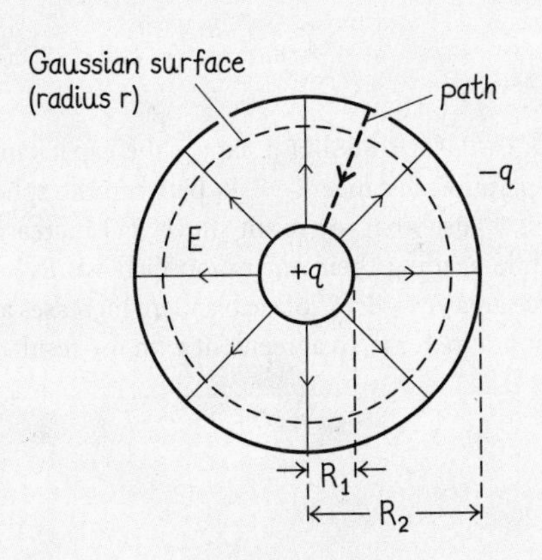

❷ **DEVISE PLAN** Again I refer to the Procedure box on page 697 to calculate the capacitance. For the path over which electrostatic work is done, I choose a straight path that goes radially from cylinder 2 to cylinder 1.

❸ **EXECUTE PLAN** Because of the cylindrical symmetry, I choose a cylindrical Gaussian surface (Figure 26.23). The length of the Gaussian surface is ℓ_G, and its radius is r ($R_2 > r > R_1$). The electric flux Φ through the curved portion of the Gaussian surface is equal to the product of the electric field strength E_r at a distance r from the common axis of cylinders 1 and 2 and the surface area of the Gaussian surface $A_G = (2\pi r)\ell_G$. Therefore $\Phi = 2\pi r \ell_G E_r$. Because the linear charge density on cylinder 1 is $+q/\ell$, the quantity of charge enclosed by the Gaussian surface is given by the product of the linear charge density and the length of the Gaussian surface: $q_{enc} = +(q/\ell)\ell_G$. Applying Gauss's law, I get

$$2\pi r \ell_G E_r = \frac{q\ell_G}{\epsilon_0 \ell},$$

or $E_r = q/(2\pi\epsilon_0 \ell r)$, in agreement with the result we obtained in Exercise 24.7, $E = 2k\lambda/r$, because $k = 1/(4\pi\epsilon_0)$ and $q/\ell = \lambda$.

Now that I know E_r, I can calculate the electrostatic work required to move a test particle carrying a charge q_t from cylinder 2 to cylinder 1. Integrating the electric force exerted on the test particle over the force displacement from cylinder 2 (negatively charged) to cylinder 1 (positively charged), I get

$$W_{q_t} = \int_{R_2}^{R_1} \frac{qq_t}{2\pi\epsilon_0 \ell r}\,dr.$$

Working out the integral, I obtain

$$W_{q_t} = \frac{qq_t}{2\pi\epsilon_0 \ell}\Big[\ln r\Big]_{R_2}^{R_1} = \frac{qq_t}{2\pi\epsilon_0 \ell}\ln\left(\frac{R_1}{R_2}\right).$$

The potential difference between the negative cylinder 2 and the positive cylinder 1 is thus

$$V_{cap} \equiv \frac{-W_{q_t}}{q_t} = -\frac{q}{2\pi\epsilon_0 \ell}\ln\left(\frac{R_1}{R_2}\right) = \frac{q}{2\pi\epsilon_0 \ell}\ln\left(\frac{R_2}{R_1}\right).$$

Because $R_2 > R_1$, the logarithm is positive and therefore V_{cap} is positive, as it should be because I am bringing a quantity q_t of positive charge from a location of low potential on negatively charged cylinder 2 to a location of high potential on positively charged cylinder 1.

According to Eq. 26.1, the capacitance of the coaxial capacitor is thus

$$C \equiv \frac{q}{V_{cap}} = \frac{2\pi\epsilon_0 \ell}{\ln(R_2/R_1)}. ✔$$

❹ **EVALUATE RESULT** My result shows that the capacitance is proportional to ℓ, which makes sense: The longer the coaxial cylinders, the greater the quantity of charge that can be stored on them. Decreasing R_1 or increasing R_2 is equivalent to increasing the plate separation distance d in a parallel-plate capacitor, which decreases the capacitance. Indeed, my result shows a decreasing capacitance for decreasing R_1 or increasing R_2. (The dependence on R_1 and R_2 is a bit more complicated than the dependence on d in a parallel-plate capacitor because the electric field in the coaxial capacitor is nonuniform and because changing the radii of the cylinders affects their surface areas.)

26.16 Coaxial cables used for cable television typically have a central metallic core of 0.20-mm radius, surrounded by a cylindrical metallic sheath of 2.0-mm radius. The two are separated by a plastic spacer. If the effect of the spacer can be ignored (that is, assuming the two conductors are separated by air), what is the capacitance of a 100-m-long cable?

Example 26.4 Spherical capacitor

What is the capacitance of a spherical capacitor consisting of two concentric conducting spherical shells of radii R_1 and $R_2 > R_1$?

❶ GETTING STARTED If I let the inner sphere carry a positive charge $+q$ and the outer one a negative charge $-q$, my sketch of the capacitor looks identical to the sketch of the coaxial capacitor of Example 26.3 (**Figure 26.24**). The calculation, however, will not be the same because now the electric field has spherical, not cylindrical, symmetry.

Figure 26.24

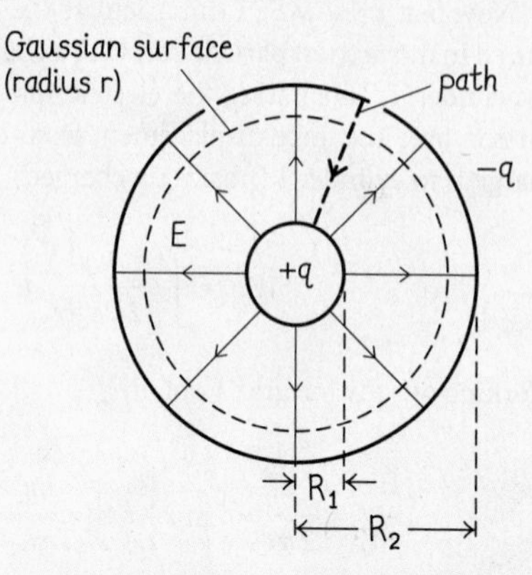

❷ DEVISE PLAN For the path over which electrostatic work is done, I choose a straight path that goes radially from the outer sphere to the inner sphere. The outer sphere does not contribute to the electric field between the spheres because the field inside a hollow conductor is always zero (see Section 24.5). The electric field created by the inner sphere is the same as that created by a charged particle (Eq. 24.15, $E = kq/r^2$), so I can use this expression to follow steps 3 and 4 in the Procedure box on page 697.

❸ EXECUTE PLAN The electrostatic work done in moving a test particle carrying a charge q_t from the outer sphere to the inner sphere is

$$W_{q_t}(- \to +) = \int_{R_2}^{R_1} \frac{qq_t}{4\pi\epsilon_0 r^2}\,dr.$$

Working out the integral, I get

$$W_{q_t}(- \to +) = -\frac{qq_t}{4\pi\epsilon_0}\left[\frac{1}{r}\right]_{R_2}^{R_1} = -\frac{qq_t}{4\pi\epsilon_0}\left[\frac{1}{R_1} - \frac{1}{R_2}\right].$$

The potential difference between the outer and inner spheres is thus

$$V_{\text{cap}} \equiv \frac{-W_{q_t}(- \to +)}{q_t} = \frac{q}{4\pi\epsilon_0}\left[\frac{1}{R_1} - \frac{1}{R_2}\right],$$

so the capacitance is

$$C \equiv \frac{q}{V_{\text{cap}}} = 4\pi\epsilon_0\left[\frac{1}{R_1} - \frac{1}{R_2}\right]^{-1} = 4\pi\epsilon_0\left[\frac{R_2 - R_1}{R_1R_2}\right]^{-1}$$

$$= 4\pi\epsilon_0\left[\frac{R_1R_2}{R_2 - R_1}\right]. \checkmark$$

❹ EVALUATE RESULT I expect the capacitance to go up as the separation distance $R_2 - R_1$ between the spheres decreases, and this is just what my result shows. If I increase the spheres' radii while keeping their separation distance $R_2 - R_1$ fixed, the surface area $A = 4\pi R^2$ of each sphere increases and the capacitance should increase, in agreement with my result.

26.17 (*a*) To calculate the "capacitance" of an isolated sphere, evaluate the result we obtained in Example 26.4 in the limit that R_2 goes to infinity. (*b*) What is the capacitance of the spherical metal dome of a Van de Graaff generator like the one shown in the chapter-opening photo, which has a radius of about 2.5 m. (*c*) Given that air breaks down in an electric field with a magnitude of about 3.0×10^6 V/m, what is the maximum amount of charge that can be stored on the dome before the air breaks down?

26.6 Electric field energy and emf

How much electric potential energy is stored in a charged capacitor? To answer this question, consider a simple capacitor consisting of two conducting objects. In order to charge the capacitor, some charge-separating device must transfer charge from one conductor to the other (**Figure 26.25**). During this transfer, the charge-separating device does work on the capacitor and this energy ends up as electric potential energy "stored in the capacitor."* One complication in the calculation of the work done by the charge-separating device is that as the magnitude of the charge on each conductor increases, the potential difference increases too, so the work required to transfer a unit of charge increases.

Figure 26.25 To determine the electric potential energy stored in a capacitor, we calculate the energy required to transfer charge from one conductor to the other.

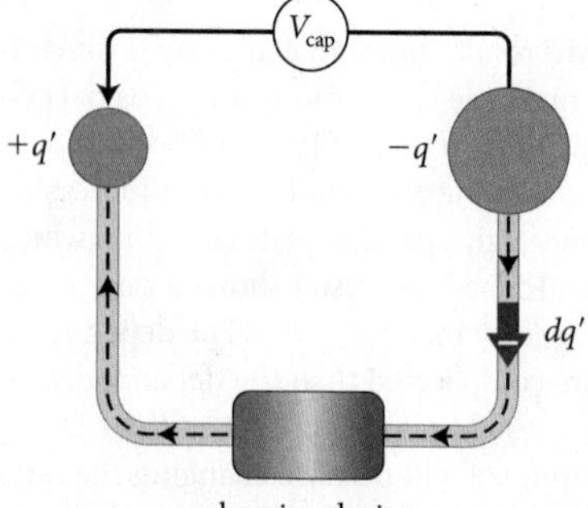

*We are assuming that there is no dissipation of energy, so that all the work done on the system ends up as electric potential energy. In practice this is a reasonable assumption.

Let us therefore break down the transfer of charge from one conductor to the other into small increments of charge dq', so that the potential difference is essentially constant during the transfer of a single increment. Consider some instant during the charging when the magnitude of the charge on each conductor is q'. The potential difference between the negative and positive conductors is then given by Eq. 26.1: $V_{\text{cap}} = q'/C$. As an additional increment of charge dq' is moved from the negative to the positive conductor, the electrostatic work done on it is $dW = -dq' V_{\text{cap}}$ (Eq. 25.17). Because the charge-separating device must do work on charge carriers against the electric force, the work done by the charge-separating device on the charge carriers is the negative of the electrostatic work, so the change in electric potential energy of the capacitor during the transfer is

$$dU^E = -dW = V_{\text{cap}}\,dq' = \frac{q'}{C}\,dq'. \tag{26.2}$$

When the magnitude of the charge on each conductor has increased from zero to its final value q, the electric potential energy stored in the capacitor is

$$U^E = \int dU^E = \int_0^q \frac{q'}{C}\,dq' = \frac{1}{C}\int_0^q q'\,dq' = \frac{1}{2}\frac{q^2}{C}. \tag{26.3}$$

Often it is more convenient to express the electric potential energy not in terms of the magnitude of the charge q on the capacitor, but in terms of the potential difference across it:

$$U^E = \frac{1}{2}\frac{q^2}{C} = \frac{1}{2}CV_{\text{cap}}^2 = \frac{1}{2}qV_{\text{cap}}. \tag{26.4}$$

Note that Eqs. 26.3 and 26.4 hold for any type of capacitor, regardless of the configuration of the conductors. All that enters into these expressions besides the charge or the potential difference is the capacitance, which depends on the size, shape, and the separation of the conductors.

26.18 A 1.0-μF parallel-plate capacitor with a plate spacing of 50 μm is charged up to the breakdown threshold. *(a)* If the electric field in the air between the capacitor plates is 3.0×10^6 V/m, how much energy is stored in the capacitor? Express your answer in joules. *(b)* How high must you raise this book ($m \approx 2$ kg) to increase the gravitational potential energy of the Earth-book system by the same amount?

As we discussed in Section 26.1, we can imagine electric potential energy to be stored either in the configuration of charge in the capacitor or in the electric field. We can use Eq. 26.4 and our knowledge about the electric field in a capacitor to relate electric potential energy to the electric field. From Example 25.5 we know that the magnitude of the potential difference between the plates of a parallel-plate capacitor is given by Ed, so, using the expression for C in Example 26.2 and Eq. 26.4, we can write for the electric potential energy stored in a parallel-plate capacitor

$$U^E = \frac{1}{2}CV_{\text{cap}}^2 = \frac{1}{2}\left(\frac{\epsilon_0 A}{d}\right)(Ed)^2 = \frac{1}{2}\epsilon_0 E^2(Ad). \tag{26.5}$$

The term in parentheses on the right side, Ad, is equal to the volume of the space between the capacitor plates—that is, the region to which the electric field is confined. Therefore, the energy per unit volume stored in the electric field—the **energy density** of the electric field—is

$$u_E \equiv \frac{U^E}{\text{volume}} = \tfrac{1}{2}\,\epsilon_0 E^2. \tag{26.6}$$

Although we derived this expression for the special case of a parallel-plate capacitor, it holds true for any electric field in vacuum. Any given region of space where a uniform electric field is present can be viewed as containing an amount of electric potential energy equal to $\frac{1}{2}\epsilon_0$ times the square of the magnitude of the electric field in that region times the volume. If the electric field is nonuniform, we must subdivide the volume of interest into small enough segments that E can be considered uniform within each segment, then apply Eq. 26.6 to each segment and take the sum of all the contributions. (This corresponds to integrating the energy density over the volume of the region that contains the electric field.)

26.19 A parallel-plate capacitor has plates of area A separated by a distance d. The magnitude of the charge on each plate is q. (*a*) Determine the magnitude of the force exerted by the positively charged plate on the negatively charged one. (*b*) Suppose you increase the separation between the plates by an amount Δx. How much work do you need to do on the capacitor to achieve this increase? (*c*) What is the change in the electric potential energy of the capacitor? (*d*) Moving the plate adds additional space with electric field between the plates. Show that the energy stored in the electric field in this additional space is equal to the work done on the capacitor.

The energy stored in a capacitor is supplied to it by the charging device—such as a generator, a battery, or a solar cell. Inside this device, nonelectrostatic interactions cause a separation of charge by doing work on charged particles. The work per unit charge done by the nonelectrostatic interactions on the charge carriers inside the device is called the emf and is denoted by $\mathscr{E}$:

$$\mathscr{E} \equiv \frac{W_{\text{nonelectrostatic}}}{q}. \tag{26.7}$$

The SI unit of emf is the same as that of potential: the volt. The rating of a battery—1.5 V or 9 V—gives its emf.*

If no energy is dissipated inside the charging device, *all* of the energy can be transferred to charge carriers outside the device. This transfer takes place through electric interactions. In Figure 26.25, for example, electric forces remove electrons from one object and push them onto the other, charging the capacitor. In the absence of any energy dissipation, the nonelectrostatic work done on charge carriers inside the device is equal to the electrostatic work done on charge carriers outside it. Because the electrostatic work per unit charge is the potential difference between the negative and positive terminals of the charging device, we have, for an ideal charging device,

$$V_{\text{device}} = \mathscr{E} \quad \text{(ideal device)}. \tag{26.8}$$

In practice, some energy is always dissipated inside the device, so not all of the nonelectrostatic work done on charge carriers inside the device can be turned into electrostatic work. Consequently, for most devices, $V_{\text{device}} < \mathscr{E}$.

*The term *voltage* is sometimes used to refer to a potential difference or to an emf (such as the rating of a battery). Potential difference, however, is related to *electrostatic* work done on charge carriers, whereas emf deals with *nonelectrostatic* work done on them. Thus, electrostatic work brings opposite charge carriers together, while nonelectrostatic work causes charge separation. To maintain this important distinction, we shall avoid the term *voltage*.

Example 26.5 Van de Graaff energy

The radius of the dome on the Van de Gr
on the opening page of this chapter is abou
down when the field magnitude is about
much electric potential energy is stored i
rounding the dome just before the air the

❶ **GETTING STARTED** I am given the ra
dome and asked to calculate how much
electric field surrounding the dome ju
the air to break down. If I approximate
charged spherical shell, then the elect
the same as that surrounding a particl
[Eq. 24.15, $E = q/(4\pi\epsilon_0 r^2)$]. The ma
the dome is greatest at the dome surf
as the maximum value just before th

❷ **DEVISE PLAN** Equation 26.6 giv
the electric field around the dome. I can su...
expression for the electric field of a sphere carrying a charge q into Eq. 26.6 to obtain an expression for the energy density of the electric field at an arbitrary distance r from the dome center. Because the electric field has the same magnitude at any location a distance r from the dome center, I can divide the space outside the dome into a series of thin spherical shells, each of thickness dr and all concentric with the dome (**Figure 26.26**), and then integrate over all shells from $r = R$ to $r = \infty$ to obtain the energy stored in the electric field surrounding the dome in terms of the charge q. I can then use Eq. 24.15 to eliminate q from my result and express the energy stored in terms of E_{surf}, which is given.

Figure 26.26

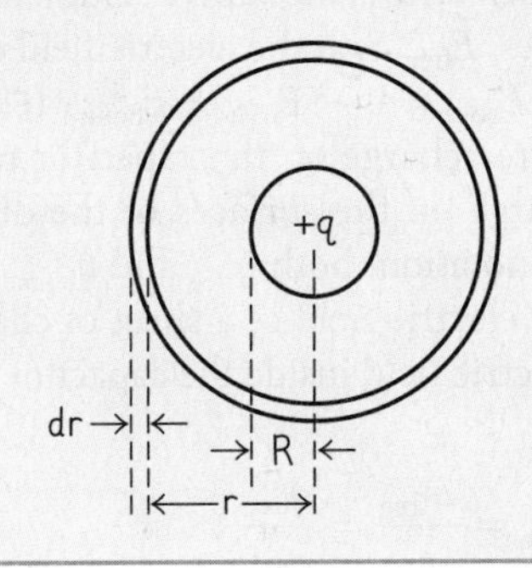

q. 24.15 into Eq. 26.6, I get

$$\tfrac{1}{2}\epsilon_0\left(\frac{q}{4\pi\epsilon_0 r^2}\right)^2.$$

shell of radius r and thickness dr
to the surface area of a sphere of
ness: $(4\pi r^2)dr$. The energy in that
$r^2)dr$, and so the electric potential
e dome is

$$r^2)\,dr = \tfrac{1}{2}\epsilon_0\int_R^\infty\left(\frac{q}{4\pi\epsilon_0 r^2}\right)^2(4\pi r^2)dr$$

$$\frac{q^2}{8\pi\epsilon_0}\left[\frac{-1}{r}\right]_R^\infty = \frac{q^2}{8\pi\epsilon_0 R}. \quad (1)$$

for U^E, the quantity I must determine, but I have no value fo... Given that the electric field at the dome surface is given by Eq. 24.15, $E_{\text{surf}} = q/(4\pi\epsilon_0 R^2)$, I can rearrange this expression to $q = E_{\text{surf}}(4\pi\epsilon_0 R^2)$ and rewrite Eq. 1 as $U^E = 2\pi\epsilon_0 E_{\text{surf}}^2 R^3$. Because E_{surf} is the maximum electric field magnitude around the dome, I know that this magnitude must be the breakdown value for air. Substituting the values given, I get

$$U^E = 2\pi(8.85 \times 10^{-12}\,\text{C}^2/(\text{N}\cdot\text{m}^2))(3.0 \times 10^6\,\text{V/m})^2(2.5\,\text{m})^3$$
$$= 7.8\,\text{kJ}. ✔$$

❹ **EVALUATE RESULT** As a check on my work, I can calculate the electric potential energy of the charge stored on the dome using Eq. 26.4. In Checkpoint 26.17, you found that the capacitance of an isolated sphere is $C_{\text{sphere}} = 4\pi\epsilon_0 R$ and that the potential of a sphere is related to the electric field at its surface by $V_{\text{cap}} = ER$. Therefore $U^E = \tfrac{1}{2}CV_{\text{cap}}^2 = 2\pi\epsilon_0 R^3 E_{\text{surf}}^2$, which is the same result I obtained.

Add Derivations for Parallel e Series capacitors

26.20 The flash unit on a typical camera uses a 100-μF capacitor to store electric potential energy. The capacitor is charged to a potential of 300 V. When the flash is fired, the energy in the capacitor is released to a bulb in a burst of about 1.0-ms duration. (*a*) How much energy is stored in the fully charged capacitor before it is fired? (*b*) What is the average power of the flash firing?

26.7 Dielectric constant

As we saw in Section 26.3, the capacitance of a capacitor can be increased by inserting a dielectric between the two conductors. For example, inserting a slab of mica between the plates of an isolated charged capacitor (**Figure 26.27**) decreases the potential difference across the capacitor by a factor of 5. This tells us that the mica reduces the electric field inside the isolated capacitor by a factor of 5. By definition, the magnitude of the potential difference V_0 across the isolated

Figure 26.27 The potential difference across an isolated parallel-plate capacitor is greater (*a*) without a dielectric between the plates than it is (*b*) with a dielectric.

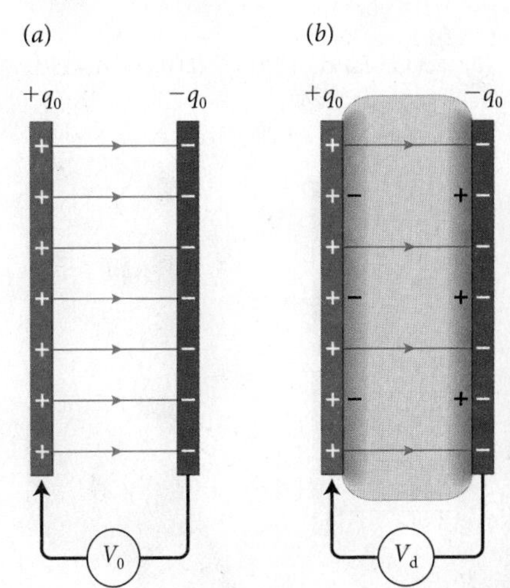

capacitor without a dielectric divided by the magnitude of the potential difference V_d with the dielectric is called the **dielectric constant** κ:

$$\kappa \equiv \frac{V_0}{V_d}. \tag{26.9}$$

Given that the magnitude q_0 of the charge on each plate of the isolated capacitor is not affected by the dielectric, we see from Eqs. 26.9 and 26.1 that

$$\kappa \equiv \frac{V_0}{V_d} = \frac{V_0/q_0}{V_d/q_0} = \frac{1/C_0}{1/C_d} = \frac{C_d}{C_0}, \tag{26.10}$$

where C_d is the capacitance of the capacitor with the dielectric and C_0 that without a dielectric. Therefore, the capacitance changes by the factor κ when a dielectric is inserted:

$$C_d = \kappa C_0. \tag{26.11}$$

The dielectric constant is always greater than 1 ($\kappa > 1$) because the presence of a dielectric decreases the electric field inside the capacitor. The greater the polarization of the dielectric material, the more reduced the electric field inside the dielectric and the greater the dielectric constant κ. **Table 26.1** gives the dielectric constants for several commonly used dielectric materials. The dielectric constant for vacuum—that is, no material between the plates—is unity by definition. Because air is very dilute, the dielectric constant of air is nearly unity as well. If the dielectric is composed of polar molecules that can align themselves (such as the water molecules in liquid water), then the overall polarization is much greater than in nonpolar dielectrics, so the dielectric constant is large. For some polar materials, the dielectric constant can be in the thousands.

The electric field $\vec{E}$ inside the dielectric is the superposition of the electric field due to the free charge on the plates, $\vec{E}_{free}$, and the electric field due to the bound surface charge on the dielectric, $\vec{E}_{bound}$: $\vec{E} = \vec{E}_{free} + \vec{E}_{bound}$ (**Figure 26.28a**). We designate the magnitude of the free charge on the capacitor plates by q_{free} and the magnitude of the bound charge on the surfaces of the dielectric by q_{bound} (Figure 26.28*b* and *c*). With this notation, both q_{free} and q_{bound} are always positive. Using the expression for the electric field of a sheet of charge, we can thus write for the magnitude of the electric field inside the capacitor in the absence of a dielectric

$$E_{free} = \frac{\sigma_{free}}{\epsilon_0} = \frac{q_{free}}{\epsilon_0 A}, \tag{26.12}$$

Figure 26.28 (*a*) The electric field inside a dielectric-filled capacitor is the vector sum of the electric field due to the charged plates and that due to the polarized dielectric. (*b*) and (*c*) Bound and free charge on a vacuum-filled and a dielectric-filled isolated parallel-plate capacitor.

(*a*)

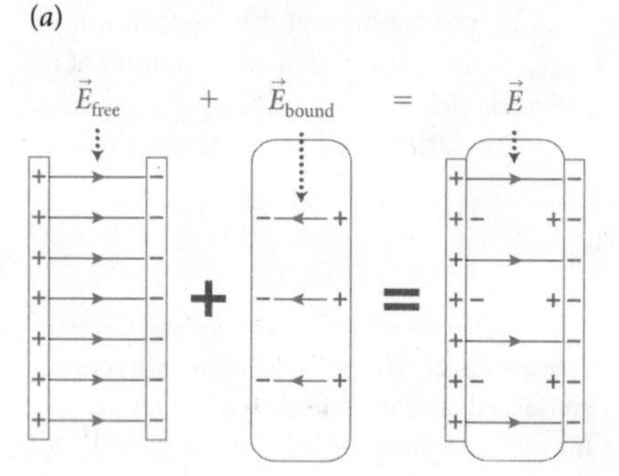

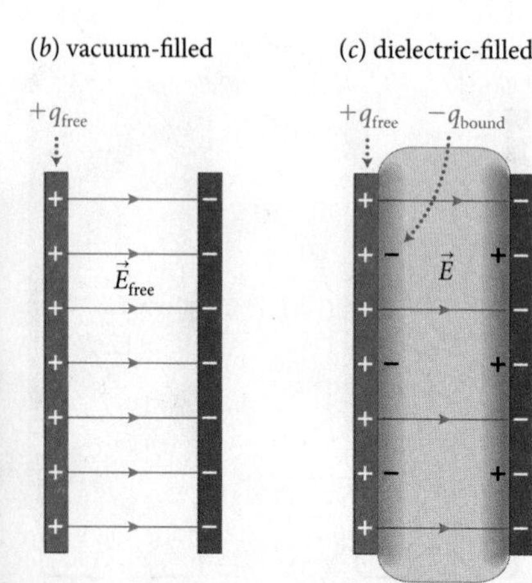

Table 26.1 Dielectric properties

Material	Dielectric constant κ	Breakdown threshold E_{max} (V/m)
Air (1 atm)	1.00059	3.0×10^6
Paper	1.5–3	4.0×10^7
Mylar (polyester)	3.3	4.3×10^8
Quartz	4.3	8×10^6
Mica	5	2×10^8
Oil	2.2–2.7	
Porcelain	6–8	
Water (distilled, 20 °C)	80.2	$6.5–7 \times 10^7$
Titania ceramic	126	8×10^6
Strontium titanate	322	
Barium titanate	1200	8×10^7

where $\sigma_{free} = q_{free}/A$ is the magnitude of the free surface charge density and A is the area of either capacitor plate. Likewise, the magnitude of the electric field due to the bound surface charge is

$$E_{bound} = \frac{\sigma_{bound}}{\epsilon_0} = \frac{q_{bound}}{\epsilon_0 A}, \tag{26.13}$$

where σ_{bound} is the magnitude of the bound surface charge density. Because $\vec{E}_{free}$ and $\vec{E}_{bound}$ point in opposite directions, the magnitude of the electric field $\vec{E}$ inside the dielectric is then

$$E = E_{free} - E_{bound} = \frac{\sigma_{free} - \sigma_{bound}}{\epsilon_0} = \frac{q_{free} - q_{bound}}{\epsilon_0 A}. \tag{26.14}$$

Let us determine the magnitude of the bound charge q_{bound}. If the plate separation is d, we can write Eq. 26.10 in the form

$$\kappa \equiv \frac{V_0}{V_d} = \frac{E_{free}\, d}{E\, d} = \frac{E_{free}}{E}, \tag{26.15}$$

where E_{free} is the magnitude of the electric field due to the free charge only, and E is the magnitude of the electric field inside the dielectric. In other words, the dielectric reduces the electric field by the factor κ: $E = E_{free}/\kappa$. Substituting Eqs. 26.12 and 26.14 into this expression, we get

$$\frac{q_{free}}{\kappa \epsilon_0 A} = \frac{q_{free} - q_{bound}}{\epsilon_0 A} \tag{26.16}$$

or

$$\frac{q_{free}}{\kappa} = q_{free} - q_{bound}. \tag{26.17}$$

Solving this expression for q_{bound}:

$$q_{bound} = \frac{\kappa - 1}{\kappa} q_{free}. \tag{26.18}$$

Because κ is always greater than 1, we see that the magnitude of the bound surface charge is always smaller than the magnitude of the free charge that causes it.

Next we consider the situation of a capacitor connected to a battery (**Figure 26.29**). In this situation, the potential difference across the capacitor is constant, but the charge on the plates changes when the dielectric is inserted. As we have seen in Section 26.3, the electric field inside the dielectric must be the same as before the dielectric was inserted. In other words, the sum of the free and bound charges must still be equal to q_0; that is, $q_0 = q_{free} - q_{bound}$ (Figure 26.29). Because the definition of capacitance involves only the charge on the capacitor plates, we can write

$$C_d \equiv \frac{q_{free}}{V_d} \tag{26.19}$$

and likewise

$$C_0 \equiv \frac{q_0}{V_0}. \tag{26.20}$$

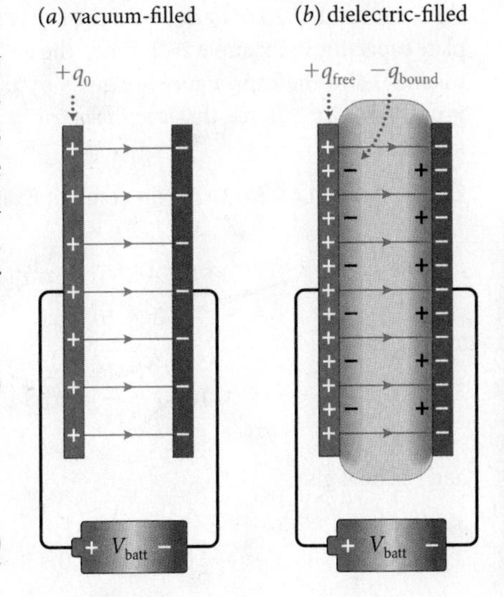

Figure 26.29 Bound and free charge on a vacuum-filled and a dielectric-filled parallel-plate capacitor connected to a battery.

Note the difference between q_{free} and q_0. Even though both represent free charge, they are not equal because when the dielectric is inserted, the battery increases the charge on the plate in order to maintain a constant potential difference across the capacitor, so $q_{free} > q_0$. Because $C_d = \kappa C_0$ (Eq. 26.11), we see from Eqs. 26.19 and 26.20 that the dielectric increases the charge on the capacitor plates by the factor κ:

$$q_{free} = \kappa q_0. \tag{26.21}$$

26.21 (*a*) In Figure 26.29, what is the magnitude of q_{bound}? Express your answer in terms of q_0 and the properties of the dielectric. (*b*) What is the bound surface charge density on the dielectric? Express your answer in terms of the electric field E.

Example 26.6 Capacitor with dielectric

A parallel-plate capacitor consists of two conducting plates with a surface area of 1.0 m² and a plate separation distance of 50 μm. (*a*) Determine the capacitance and the energy stored in the capacitor when it is charged by connecting it to a 9.0-V battery. (*b*) With the capacitor fully charged and disconnected from the battery, a 50-μm-thick sheet of Mylar is inserted between the plates. Determine the potential difference across the capacitor and the energy stored in it. (*c*) If the Mylar-filled capacitor is connected to the battery, how much work does the battery do to fully charge the capacitor?

❶ **GETTING STARTED** I am given information about a capacitor and a battery used to charge it. From this information, I must determine the capacitance and the energy stored in the capacitor with and without a sheet of Mylar between the plates connected to the battery, and determine what happens to the potential with and without a sheet of Mylar between the plates and with and without the battery connected to the capacitor. When connected to the capacitor, the battery maintains a constant potential across the capacitor. When the battery is not connected to the capacitor, the charge on the plates remains constant.

❷ **DEVISE PLAN** To calculate the energy stored in the capacitor I can use Eq. 26.4; I calculated the capacitance of a parallel-plate capacitor in Example 26.2. When the dielectric is added, the potential and the capacitance are given by Eqs. 26.9 and 26.11. From Table 26.1, I see that the dielectric constant of Mylar is $\kappa = 3.3$.

❸ **EXECUTE PLAN** (*a*) Using the result of Example 26.2, I get

$$C = \frac{[8.85 \times 10^{-12}\,\mathrm{C^2/(N\cdot m^2)}](1.0\,\mathrm{m^2})}{50 \times 10^{-6}\,\mathrm{m}}$$

$$= 0.18 \times 10^{-6}\frac{\mathrm{C^2}}{\mathrm{N\cdot m}} = 0.18\,\mu\mathrm{F}, ✔$$

and Eq. 26.4 gives

$$U^E = \tfrac{1}{2}C(V_0)^2 = \tfrac{1}{2}(0.18\,\mu\mathrm{F})(9.0\,\mathrm{V})^2 = 7.2\,\mu\mathrm{J}. ✔$$

(*b*) Because of the bound surface charge on the dielectric, the electric field between the capacitor plates decreases, and so the potential difference across the capacitor decreases, too. From the definition of the dielectric constant (Eq. 26.9), I have

$$V_d = \frac{V_0}{\kappa} = \frac{9.0\,\mathrm{V}}{3.3} = 2.7\,\mathrm{V}, ✔$$

where I obtained my value for κ from Table 26.1. To calculate the energy in the presence of the dielectric, I must first obtain an expression for the capacitance of the dielectric-filled capacitor. Substituting the expression for the capacitance of a parallel-plate capacitor (see Example 26.2) into Eq. 26.11 yields

$$C_d = \kappa C_0 = \frac{\kappa\epsilon_0 A}{d}$$

$$= \frac{(3.3)[8.85 \times 10^{-12}\,\mathrm{C^2/(N\cdot m^2)}](1.0\,\mathrm{m^2})}{50 \times 10^{-6}\,\mathrm{m}}$$

$$= 0.58\,\mu\mathrm{F}.$$

The stored energy is thus

$$U^E = \tfrac{1}{2}CV_d^2 = \tfrac{1}{2}(0.58\,\mu\mathrm{F})(2.9\,\mathrm{V})^2 = 2.2\,\mu\mathrm{J}. ✔$$

(*c*) The energy stored in the fully charged dielectric-filled capacitor is

$$U^E = \tfrac{1}{2}CV_{batt}^2 = \tfrac{1}{2}(0.58\,\mu\mathrm{F})(9.0\,\mathrm{V})^2 = 24\,\mu\mathrm{J}.$$

From part *b* I know that before it was connected to the battery, the capacitor stored 2.2 μJ, and so the work done by the battery in charging the capacitor must be 24 μJ − 2.2 μJ = 22 μJ. ✔

❹ **EVALUATE RESULT** My answers to parts *a* and *b* show that the amount of energy stored decreases when the dielectric is inserted. That makes sense because, as the dielectric is brought near the plates, the charged plates induce a polarization on the dielectric and consequently it is pulled into the space between the plates (**Figure 26.30**). Therefore, the capacitor does *positive* work on whoever is holding the dielectric, and the energy in the capacitor decreases as the dielectric enters the space between the plates. This work is equal to the difference in energy between parts *a* and *b*: $W = 7.2\ \mu\text{J} - 2.2\ \mu\text{J} = 5.0\ \mu\text{J}$. My answer to part *c* is about three times greater than the value I calculated for U^E in part *a*, which is what I expect given that the capacitance is increased by the factor $\kappa = 3.3$ once the dielectric is inserted.

Figure 26.30

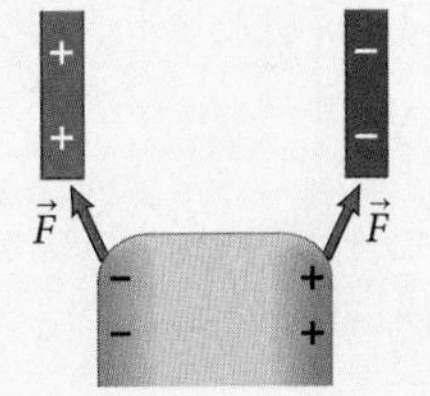

26.22 Verify that in the solution to part *a* of Example 26.6, (*a*) the ratio of units $\text{C}^2/(\text{N} \cdot \text{m})$ is equivalent to the unit F and (*b*) the product of units $\text{F} \cdot \text{V}^2$ is equivalent to the unit J.

26.8 Gauss's law in dielectrics

Can we apply Gauss's law to calculate the electric fields inside dielectric materials? The answer is *yes,* because Gauss's law is a fundamental law that follows directly from the $1/r^2$ dependence of Coulomb's law. Thus, the presence of a dielectric cannot affect its validity.

Consider the situation illustrated in **Figure 26.31**. To determine the magnitude of the electric field E inside the dielectric, we consider the cylindrical Gaussian surface with cross-sectional area A shown in the figure. The electric flux is zero except through the right flat surface of the cylinder, so

$$\oint \vec{E} \cdot d\vec{A} = EA. \qquad (26.22)$$

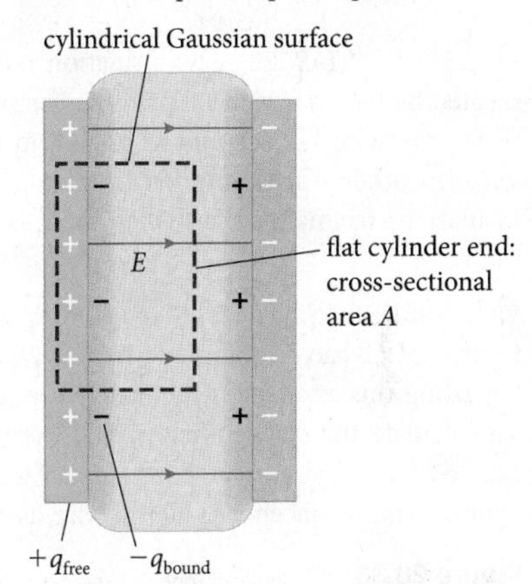

Figure 26.31 A cylindrical Gaussian surface used to calculate the electric field inside a dielectric-filled parallel-plate capacitor.

The charge enclosed by the Gaussian surface is not just the enclosed charge on the plate—we must also take into account the enclosed bound charge on the dielectric. The enclosed charge is thus $q_{\text{enc}} = q_{\text{free, enc}} - q_{\text{bound, enc}}$, and Gauss's law then gives

$$\oint \vec{E} \cdot d\vec{A} = EA = \frac{q_{\text{free, enc}} - q_{\text{bound, enc}}}{\epsilon_0}. \qquad (26.23)$$

In this form, Gauss's law is not very useful, because in order to extract E from Eq. 26.23, we need to know the magnitude of the bound surface charge. Generally, we don't know the contribution from the bound charge in a given situation.

Substituting the relationship between the free and bound charges (Eq. 26.18), however, we can rewrite Eq. 26.23 in the form

$$\oint \vec{E} \cdot d\vec{A} = \frac{q_{\text{free, enc}}}{\epsilon_0 \kappa}. \qquad (26.24)$$

This result—Gauss's law in dielectrics—is remarkable. The left side contains the electric flux of the electric field *inside the dielectric*. We can obtain this field, however, just by accounting for the enclosed *free* charge (and we already know how to deal with that charge). This relationship is valid because the effect of the bound charge is completely accounted for by the dielectric constant in the denominator. As Eq. 26.17 shows, dividing q_{free} by κ gives the difference of the free and bound charges.

Because the dielectric constant affects the value of the electric field, **Gauss's law in matter** is usually written in the form

$$\oint \kappa \vec{E} \cdot d\vec{A} = \frac{q_{\text{free, enc}}}{\epsilon_0}. \tag{26.25}$$

This form of Gauss's law is very general: Even though we derived it for the special case of a parallel-plate capacitor, it holds in any situation, even one without a dielectric. In the absence of matter (that is, in vacuum), $\kappa = 1$, and because there is no bound charge we have $q_{\text{free, enc}} = q_{\text{enc}}$. Then Eq. 26.25 becomes identical to the familiar form of Gauss's law (Eq. 24.8).

Example 26.7 Electric field surrounding a charged insulated wire

A thin, long, straight wire is surrounded by plastic insulation of radius R and dielectric constant κ (**Figure 26.32**). The wire carries a uniform distribution of charge with a positive linear charge density λ. If the wire has a diameter d, what is the potential difference between the outer surface of the wire and the outer surface of the insulation?

Figure 26.32 Example 26.7.

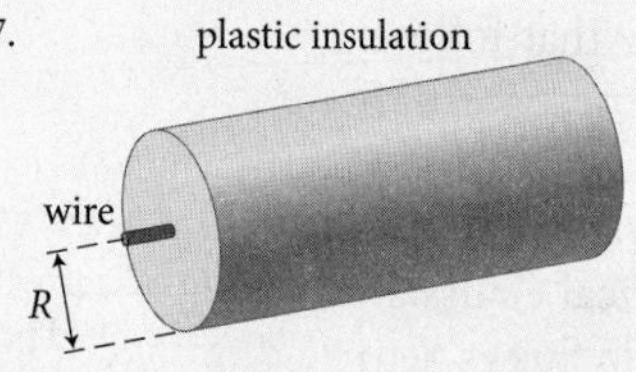

❶ **GETTING STARTED:** The insulation reduces the electric field created by the charge in the wire, so the potential difference between the wire surface and any location a distance R from the wire (in other words, any location on the outer surface of the insulation) is smaller than when there is no insulation around the wire.

❷ **DEVISE PLAN:** The potential difference between two locations A and B can be obtained from Eq. 25.25, $V_{AB} = -\int_A^B \vec{E} \cdot d\vec{\ell}$, but using this expression requires me to know the electric field. To calculate the electric field inside the insulation, I can apply Eq. 26.25 to a cylindrical Gaussian surface that has radius r and length L and is concentric with the wire, as shown in **Figure 26.33**.

Figure 26.33

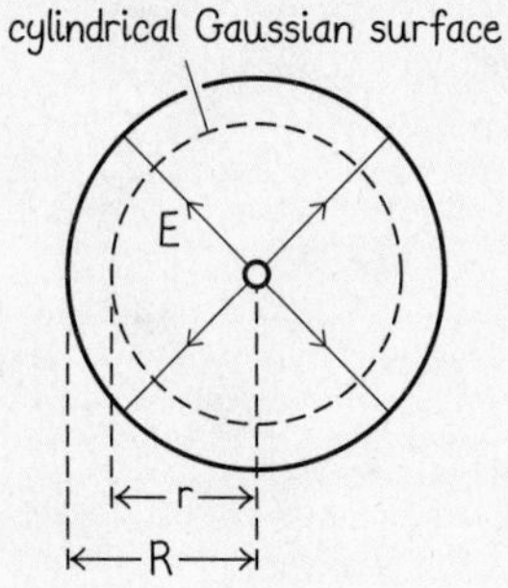

❸ **EXECUTE PLAN:** Because of the cylindrical symmetry of the wire, the electric field has the same magnitude E everywhere on the curved region of the Gaussian surface. The electric flux through that region of the Gaussian surface is equal to the product of the electric field at a distance r from the wire and the surface area: $\Phi = EA = E(2\pi rL)$ (Eq. 3 in Exercise 24.7). The free charge enclosed by the cylinder is λL, and with these substitutions Eq. 26.25 becomes

$$\kappa E(2\pi rL) = \frac{\lambda L}{\epsilon_0}$$

and

$$E = \frac{\lambda}{2\pi\kappa\epsilon_0 r}. \tag{1}$$

Substituting this expression for E into Eq. 25.25, I obtain for the potential difference between the outer surface of the wire and the outer surface of the insulation

$$V_{dR} = -\int_{d/2}^{R} \vec{E} \cdot d\vec{r} = -\frac{\lambda}{2\pi\kappa\epsilon_0}\int_{d/2}^{R} \frac{1}{r}\,dr$$

$$= -\frac{\lambda}{2\pi\kappa\epsilon_0}\ln\frac{2R}{d}. \checkmark$$

❹ **EVALUATE RESULT:** Because $\ln(2R/d)$ is positive, the potential difference is negative, as it should be because I am moving away from the positively charged wire. As an additional check, if I set $\kappa = 1$ in Eq. 1, my result for the electric field becomes identical to the result I obtained for a thin wire without insulation in Exercise 24.7.

26.23 Show that, if you account for the free and bound charges, Gauss's law in vacuum (Eq. 24.8) yields the same result for the electric field outside the insulation as Gauss's law in matter (Eq. 26.25) does.

Chapter Glossary

SI units of physical quantities are given in parentheses.

Bound charge A surplus of charge in polarized matter due to charge carriers that are bound to atoms and cannot move freely within the bulk of the material.

Capacitance C (F) The ratio of the magnitude of the charge q on a pair of oppositely charged conductors and the magnitude V_{cap} of the potential difference between them:

$$C \equiv \frac{q}{V_{cap}}. \quad (26.1)$$

The capacitance is a measure of a capacitor's capacity to store charge (or, equivalently, electric potential energy).

Capacitor A pair of conducting objects separated by a nonconducting material or vacuum. Any such pair of objects stores electric potential energy when charge has been transferred from one object to the other.

Charge-separating device A device that transfers charge from one object to another. To achieve this charge transfer, the device must move charge carriers against an electric field, requiring the device to do work on the charge carriers. This work can be supplied from a variety of sources, such as mechanical or chemical energy. Examples of charge-separating devices are voltaic cells, batteries, and Van de Graaff generators.

Dielectric A nonconducting material inserted between the plates of a capacitor. Often used more broadly to describe any nonconducting material. *Polar* dielectrics are made up of molecules that have a nonzero dipole moment, whereas *nonpolar* dielectrics consist of nonpolar molecules.

Dielectric constant κ (unitless) The factor by which the potential across an isolated capacitor is reduced by the insertion of a dielectric:

$$\kappa \equiv \frac{V_0}{V_d}. \quad (26.9)$$

Electrical breakdown When a dielectric material is subject to a very large electric field, the molecules in the material may ionize, temporarily turning the dielectric into a conductor. The electric field magnitude at which breakdown occurs is called the *breakdown threshold*.

Emf $\mathscr{E}$ (V) The emf of a charge-separating device is the work per unit charge done by nonelectrostatic interactions in separating positive and negative charge carriers inside the device:

$$\mathscr{E} \equiv \frac{W_{nonelectrostatic}}{q}. \quad (26.7)$$

Energy density of the electric field u_E (J/m^3) The energy per unit volume contained in an electric field. In vacuum:

$$u_E = \tfrac{1}{2}\epsilon_0 E^2. \quad (26.6)$$

Farad (F) The derived SI unit of capacitance:

$$1\ \text{F} \equiv 1\ \text{C/V}.$$

Free charge A surplus of charge due to charge carriers that can move freely within the bulk of a material.

Gauss's law in matter For electric fields inside matter, Gauss's law can be written in the form

$$\oint \kappa \vec{E} \cdot d\vec{A} = \frac{q_{free,\,enc}}{\epsilon_0}. \quad (26.25)$$

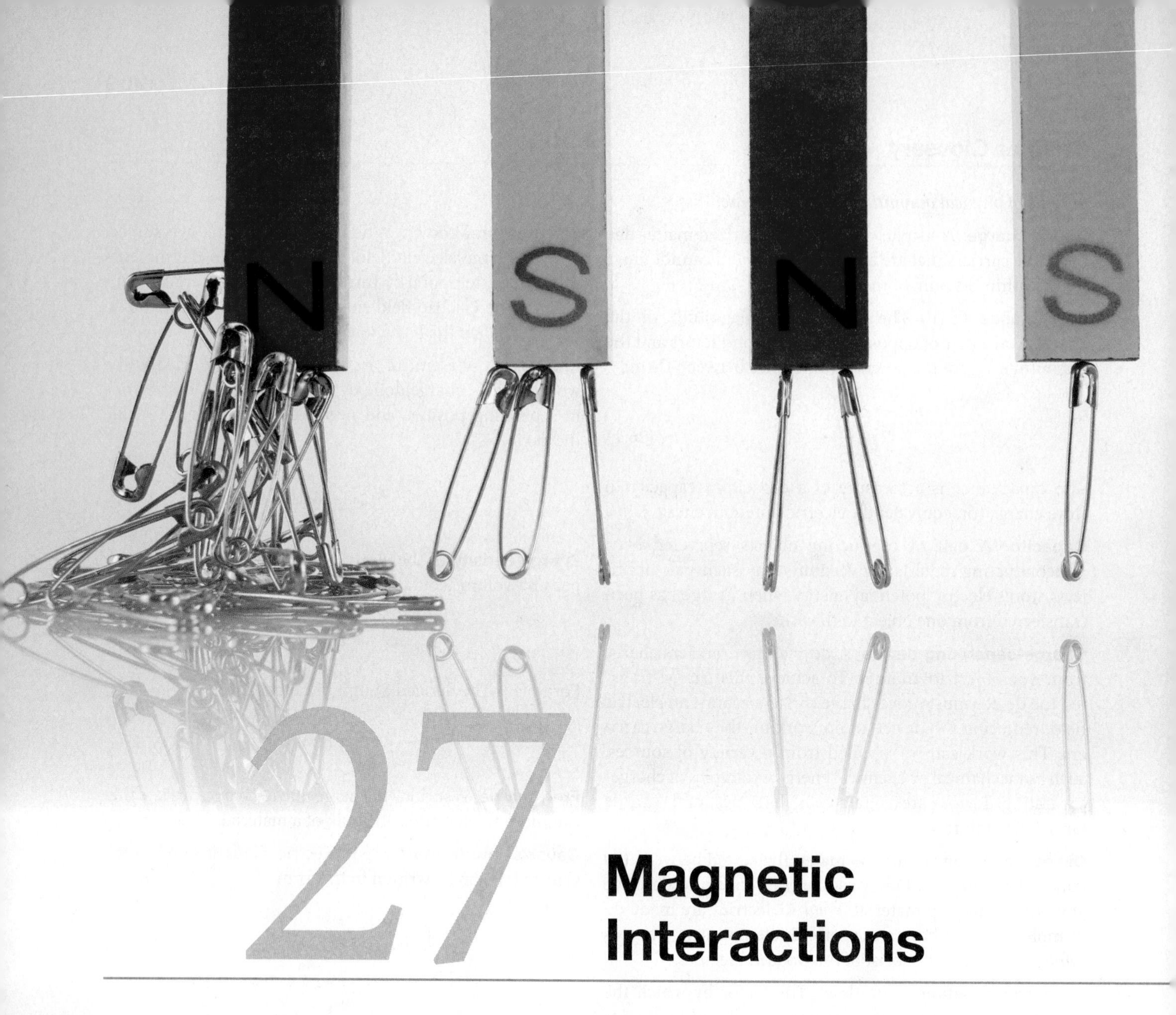

27 Magnetic Interactions

CONCEPTS

QUANTITATIVE TOOLS

Th[illegible]*gnetism* is derived from Magnesia, a province where the ancient Greeks mined *magnetite,* a mineral that attracts iron. The interactions between a magnet and a paper clip or a refrigerator door are familiar ones. These interactions may seem to have nothing to do with the subject of earlier chapters, but as we shall see, electricity and magnetism are closely related phenomena. They are two manifestations of one interaction, called the *electromagnetic interaction*. The discovery of the connection between electricity and magnetism in the 19th century opened the door to important technological breakthroughs, such as electric motors and generators, the transmission of radio signals, and the electronics and telecommunications industries.

In this chapter, we discuss interactions between magnets and introduce the concept of a magnetic field. We also begin to explore the connection between electricity and magnetism.

27.1 Magnetism

One simple definition of **magnet** is any object that attracts pieces of iron, such as iron filings or paper clips. Magnets come in many shapes and sizes, as **Figure 27.1** shows.

If you examine the surface of a magnet interacting with an object that contains iron, such as a paper clip, you will discover that certain parts of the magnets, called *magnetic poles,* interact more strongly with the paper clip than do other parts of the magnet. Disk-shaped magnets (Figure 27.1*b*) usually have poles on the two faces; paper clips stick to the flat faces but not to the curved surface. Bar magnets (Figure 27.1*c*) have poles at the ends, and most of the length of the bar doesn't interact very strongly with a paper clip. A horseshoe magnet (Figure 27.1*d*) is simply a bent bar magnet with poles at the ends of the bent bar.

Figure 27.1 Magnets come in many shapes and sizes.

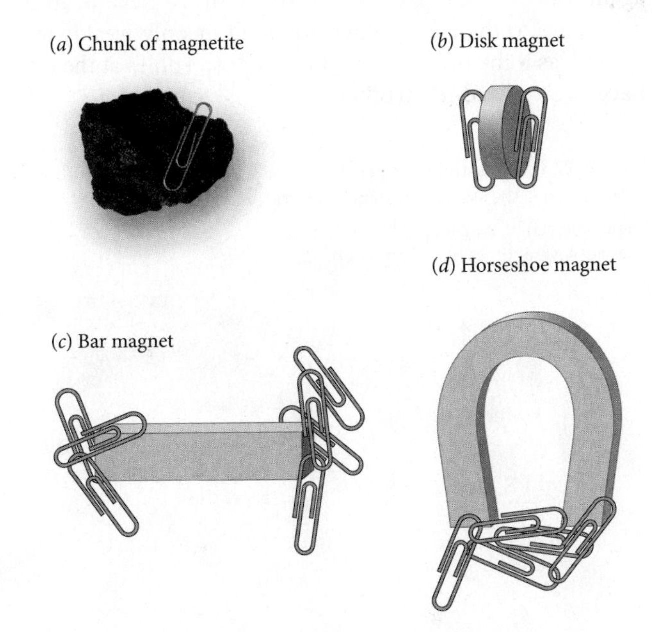

Figure 27.2 The needle of a compass is a small bar magnet that is free to rotate. When the compass is held horizontally, the needle points in the direction of Earth's North Pole. This end of the needle is defined as its north pole N.

Depending on which of their poles you hold near each another, two magnets can attract or repel without touching. This tells us that these interactions are long-range and that there are two types of magnetic poles. The interaction between magnets is one example of what are called **magnetic interactions.**

When a bar magnet is suspended and free to rotate, it aligns itself so that its poles lie on a line roughly north-south along Earth's surface. This north-south alignment of a freely rotating magnet is the basic operational principle of a compass needle, which is simply a freely rotating magnetic needle (**Figure 27.2**). This alignment provides a means of distinguishing between the two types of poles:

The pole of a freely suspended bar magnet that settles toward north is defined as being the north pole of the magnet (denoted N); the opposite pole is defined as being the south pole of the magnet (denoted S).

To understand what causes this alignment, we must first examine how one magnetic pole interacts with another. If you hold the poles of two magnets near each other, you discover that the magnetic interaction between the two types of poles follows a rule very similar to that between the two types of charge (**Figure 27.3** on the next page):

Like magnetic poles repel each other; opposite magnetic poles attract each other.

27.1 Because we cannot see any obvious difference between the ends of a bar magnet, could it be that *like* poles attract each other and *unlike* poles repel each other?

How do the poles of a magnet interact with objects that are not magnets? For most materials that are not magnets, the answer to this question is: not at all. Try picking up a penny, a wooden tooth pick, a piece of aluminum, or a piece

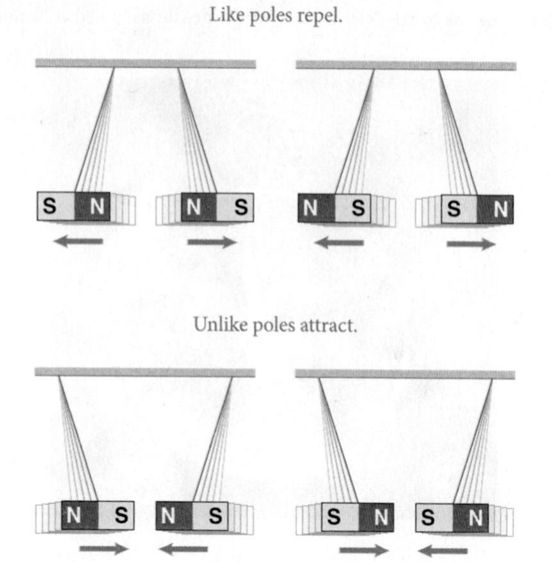

Figure 27.3 Interactions between magnetic poles.

of plastic with a magnet. It won't work. However, both the north pole and the south pole of a magnet attract an iron paper clip (**Figure 27.4**).

Materials attracted by both types of magnetic poles include iron, nickel, cobalt, and certain alloys, such as steel. These materials are called *magnetic materials.* Normally, two objects made from these materials do not interact with each other. For example, one iron paper clip does not attract another one. However, just as a charged object can induce charge polarization on a neutral object, the presence of a magnet induces a **magnetic polarization** in a paper clip or any other object made from a magnetic material (**Figure 27.5**). The poles of the magnet then interact with the induced poles of the magnetized paper clip.

27.2 (*a*) Is the interaction between a charged object and an electrically neutral object always attractive? Why or why not? (*b*) In Figure 27.4, which type of magnetic pole is induced at the top of each paper clip?

Figure 27.4 Both the north and south poles of a magnet attract an unmagnetized iron object.

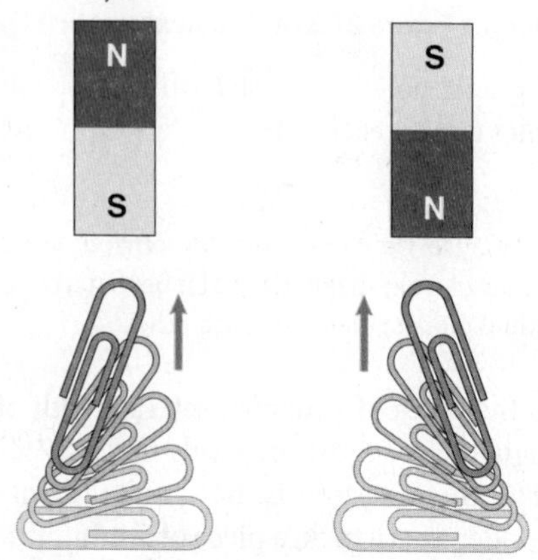

Figure 27.5 Comparing induced electric and m[illegible]ation. Magnetic polarization can be induced only in an object made from magnetic material.

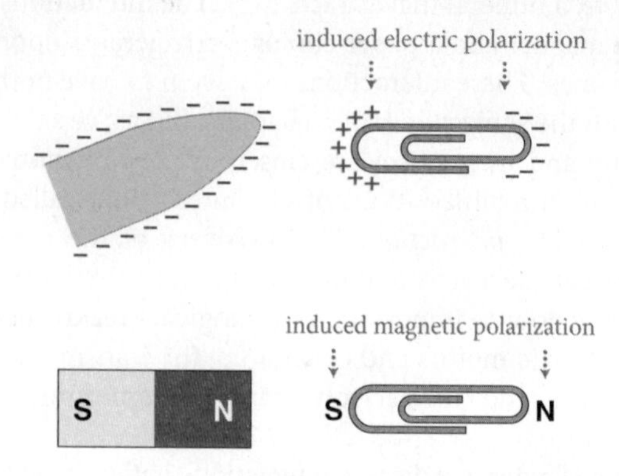

Unlike induced electric polarization, which vanishes as soon as the charged object is removed, some magnetic materials retain their induced magnetic polarization. For example, sewing needles left unmoved for long periods become magnetized by Earth's magnetic field. If you stroke a paper clip several times in the same direction with a magnet, the paper clip remains magnetically polarized even after you remove the magnet. In fact, you may even be able to pick up one paper clip with another (**Figure 27.6**). The clip can be *demagnetized* again—its magnetic polarization undone—by heating it, dropping it, or stroking it with a magnet in random directions.

In addition to the retention of magnetic polarization, there is another fundamental difference between magnetic and electric interactions: It is not possible to isolate one pole of a magnet the way we can separate a positively charged particle and a negatively charged particle from each other. For example, if we cut a bar magnet in two, we see that each of the resulting pieces has two opposite poles (**Figure 27.7**). The cutting has created an additional *pair* of opposite poles. Remarkably, if we carefully place the two pieces together again, the two new poles seem to vanish. A piece of iron held close to the cut is either not at all or only weakly attracted, as if the two newly formed opposite poles at the cut have "neutralized" each other.

Figure 27.6 If you stroke a paper clip several times in the same direction with a magnet, the clip retains some magnetic polarization.

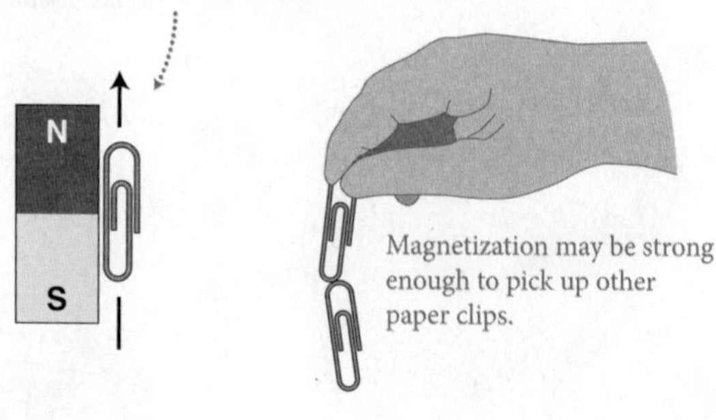

CONCEPTS

Figure 27.7 When a magnet is cut in two, each piece retains both an N and S pole.

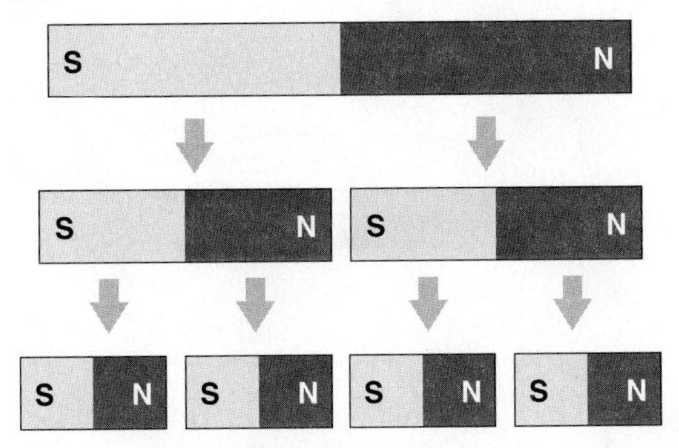

Each of the two halves can again be cut in half, but regardless of how small the pieces into which we divide the bar, each piece always has two poles. As we continue to divide the bar into smaller and smaller pieces, we come to a point where each part is a single atom. Experiments show that the iron, nickel, or cobalt atoms that make up magnets still behave like magnets. Even elementary particles behave like magnets—we shall refer to these as *elementary magnets*. Thus, a magnet consists of a large number of elementary magnets whose alignment is such that their combined effect is reinforced at the poles (Figure 27.8*a*). Because we cannot cut the elementary magnets, a pair of new poles appears every time we cut through the magnet (Figure 27.8*b*).

In spite of extensive searches, isolated magnetic poles, called *magnetic monopoles* (say, a north pole in the absence of a south pole), have never been found. Magnetism is therefore not due to magnetic monopoles, the way electricity is due to electrical charge. Although the pole of a magnet is not a monopole, we will occasionally rely on the picture of a bar magnet as a pair of opposite monopoles separated by a small distance to get an intuitive feel for magnetism. We call such an arrangement a **magnetic dipole.**

In an object made of magnetic material, the elementary magnets are randomly oriented relative to one another (Figure 27.9*a*). Some push or pull in one direction, while others push or pull in other directions. As a result, their effects cancel and the object as a whole does not act like a magnet. When a magnet is brought nearby, however, the elementary magnets in the object align themselves with the poles of the magnet, and the object becomes magnetized (Figure 27.9*b*). With some magnetic materials, the alignment is (partially) maintained even after the nearby magnet is removed, leaving the object magnetized.

You can use this model to visualize demagnetization. If a magnet is heated or handled roughly, the elementary magnets are jarred around and lose their alignment.

Figure 27.8 The concept of elementary magnets explains why cutting a magnet in two reveals new N and S poles.

(*a*)

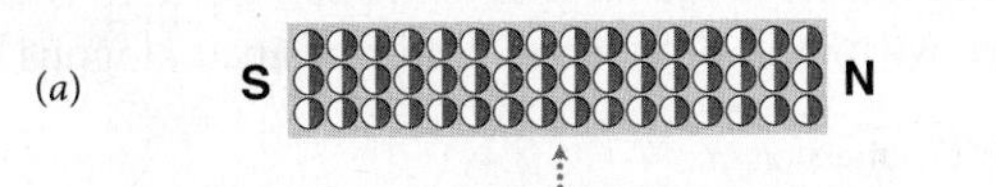

In magnetized material, elementary magnets (◑) are aligned . . .

(*b*)

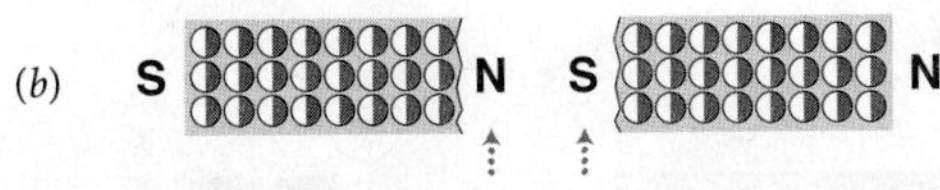

. . . so splitting magnet exposes two new poles.

Figure 27.9 (*a*) Unmagnetized and (*b*) magnetized pieces of magnetic material.

(*a*) Unmagnetized material: atoms oriented randomly

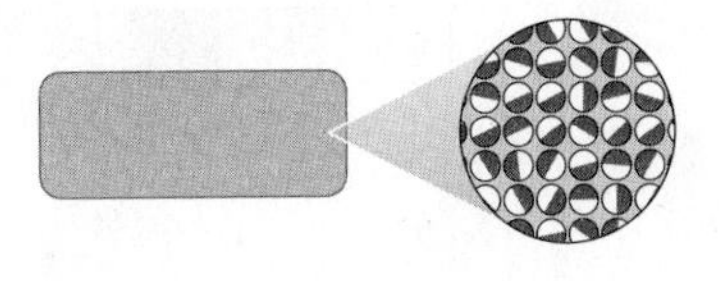

(*b*) Magnetized material: atoms aligned

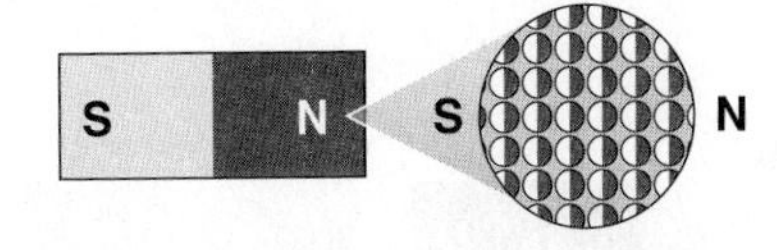

27.3 (*a*) Draw the elementary magnets inside a bar magnet and a horseshoe magnet, using the half-filled-circle format shown in Figures 27.8 and 27.9. (*b*) How many poles does the magnetized ring in Figure 27.10 have? (*c*) If someone gave you such a ring, how could you verify that it is indeed magnetized as illustrated?

Figure 27.10 A magnetized ring (Checkpoint 27.3).

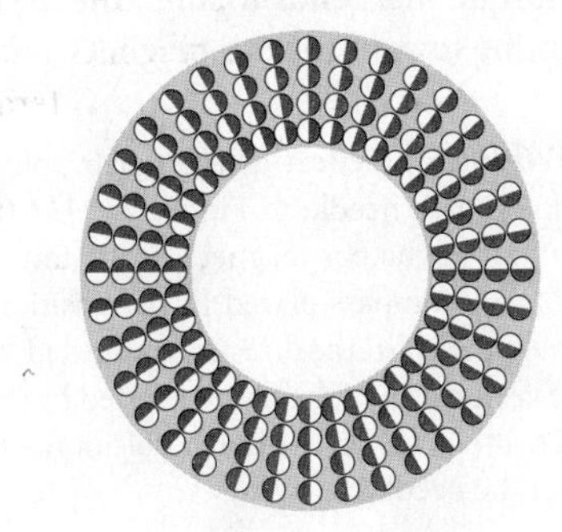

27.2 Magnetic fields

The long-range nature of magnetic interactions suggests that we can introduce the concept of a **magnetic field,** denoted by $\vec{B}$.* In analogy to the field model for electric interactions, a magnet is surrounded by a magnetic field. This magnetic field exerts a force on the poles of another magnet. Because there is no such thing as a "magnetic charge," it is not possible to map out a magnetic field using a "test magnetic

*The unintuitive symbol $\vec{B}$ was introduced early on in the description of magnetism. Whenever you see a B in an illustration or force superscript, remember that it stands for a magnetic field or force. The force exerted by a magnet on a paper clip, for example, is denoted by $\vec{F}^{B}_{\text{mc}}$.

Figure 27.11 Effect of a bar magnet on a compass needle.

(*a*) Forces exerted by magnet on poles of compass needle cause a torque.

(*b*) North pole of compass needle points toward south pole of magnet

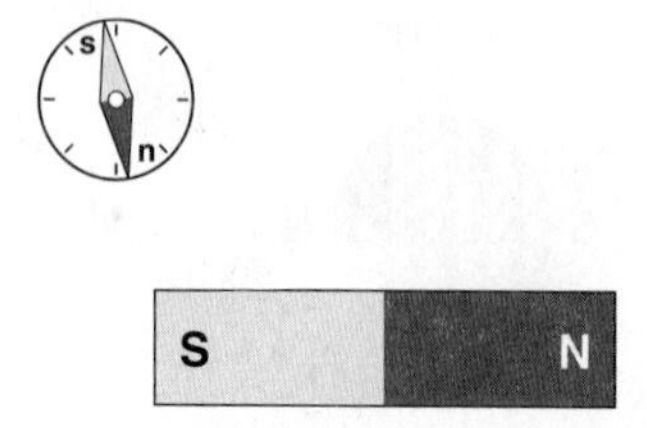

Figure 27.13 Magnetic field line pattern surrounding a bar magnet.

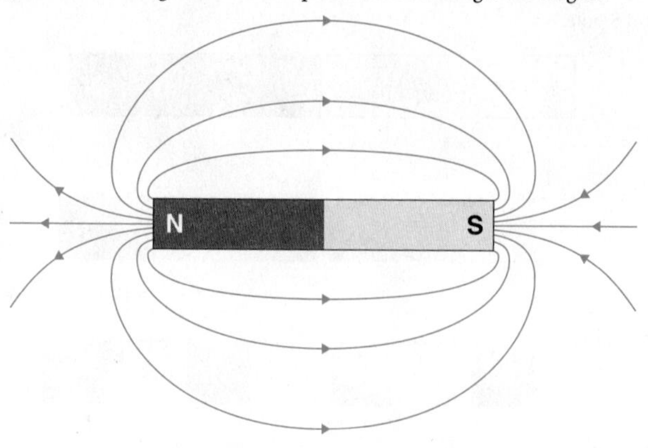

charge." We can, however, use a compass to map out the direction of the magnetic field.

27.4 (*a*) Which end of a compass needle is a north pole: the end that points toward Earth's North Pole or the other end? (*b*) If you place a compass near the north pole of a magnet, what happens to the compass needle? (*c*) Is Earth's geographic North Pole a magnetic north pole?

When a compass is placed near a bar magnet as illustrated in Figure 27.11*a*, the south pole of the bar magnet exerts an attractive force on the north pole of the compass needle and a repulsive force on the needle's south pole. These two forces cause a torque that tends to align the needle with its north pole pointing toward the bar magnet's south pole.

27.5 (*a*) What is the effect of the north pole of the bar magnet on the compass needle in Figure 27.11*a*? (*b*) What is the combined effect of the bar magnet's north and south poles on the needle? (*c*) A compass placed in the position shown in Figure 27.11*b* aligns itself in the direction indicated. What is the direction of the vector sum of the forces exerted by the bar magnet's north and south poles on the north pole of the needle? On the south pole of the needle?

We can now use the alignment of a compass needle to map out **magnetic field lines** by placing a compass somewhere near a bar magnet, moving the compass a small distance in the direction of the needle, and repeating this procedure to trace out a line, as illustrated in Figure 27.12. At any location, the direction of the magnetic field $\vec{B}$ is tangent to the magnetic field line passing through the location. By convention:

Near a magnet, magnetic field lines point away from north poles and toward south poles.

A more complete diagram of the magnetic field pattern of a bar magnet is shown in Figure 27.13. Notice the similarity with an electric dipole field (Figure 24.2).

The field line patterns around a magnet can be made visible by sprinkling iron filings on a piece of paper placed over the magnetic. The little pieces of iron become magnetized and act like compass needles that align themselves along field lines (Figure 27.14).

The similarity between electric and magnetic field patterns suggests that we may be able to carry over many of the concepts we developed for electric fields into our study of magnetism. In particular, we can associate a magnetic field magnitude with the density of magnetic field lines:

At every location in a magnetic field line diagram, the magnitude of the magnetic field is proportional to the field line density at that location.

In the field pattern shown in Figure 27.15*a*, for example, the magnitude of the magnetic field is greatest near the bottom and smallest near the top. Occasionally we shall work with magnetic fields that are perpendicular to the plane of the drawing, and so we need to be able to represent them. Figure 27.15*b* and *c* show conventions for representing field lines perpendicular to the page.

In electrostatics, the concept of field line flux led to Gauss's law, and so we may ask ourselves if there is an analogous law

Figure 27.12 A magnetic field line can be traced by moving a compass small distances in the direction in which its needle points.

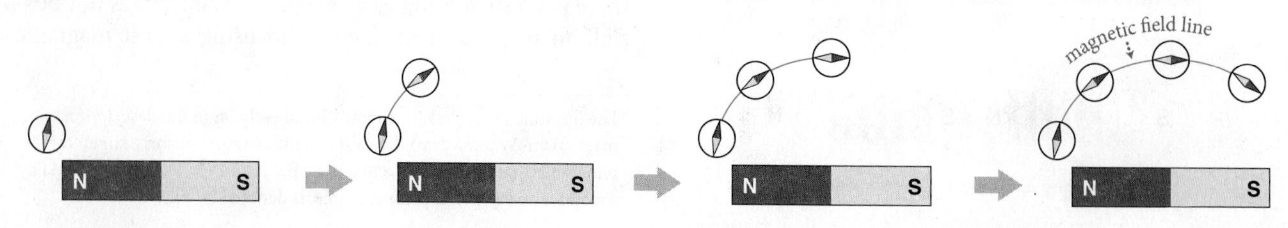

Figure 27.14 The magnetic field line pattern surrounding a bar magnet made visible by sprinkled iron filings around the magnet.

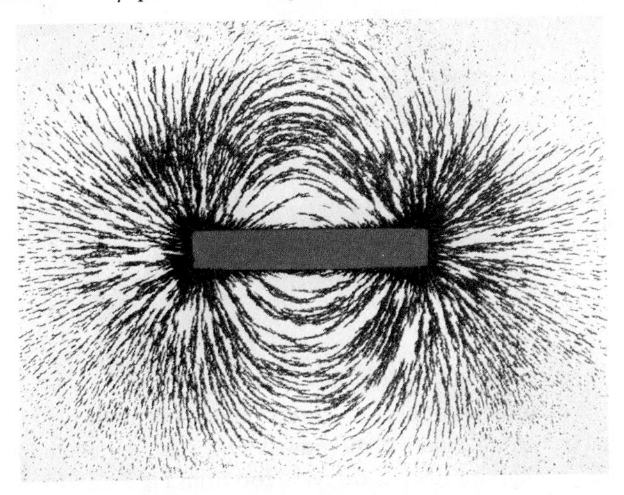

for magnetism. The answer is *yes,* but the result is somewhat different from that in electrostatics. To see why, complete this checkpoint.

> **27.6** (*a*) Consider a single elementary magnet inside a closed surface. Given that elementary magnets are particles without spatial extent, is the magnetic field line flux through the closed surface positive, negative, or zero? (*b*) Does adding a second elementary magnet inside the closed surface change your answer? (*c*) Consider a bar magnet inside a closed surface. Is the magnetic field line flux through the closed surface positive, negative, or zero? (*d*) Does your answer change if the closed surface cuts through the magnet?

The fundamental point of Checkpoint 27.6 is that each field line leaving a closed surface that encloses an elementary magnet reenters the surface somewhere else, so the magnetic field line flux due to an elementary magnet inside a closed surface is zero. Generalizing this statement to a collection of many elementary magnets, we say:

Figure 27.15 Conventions for representing a magnetic field.

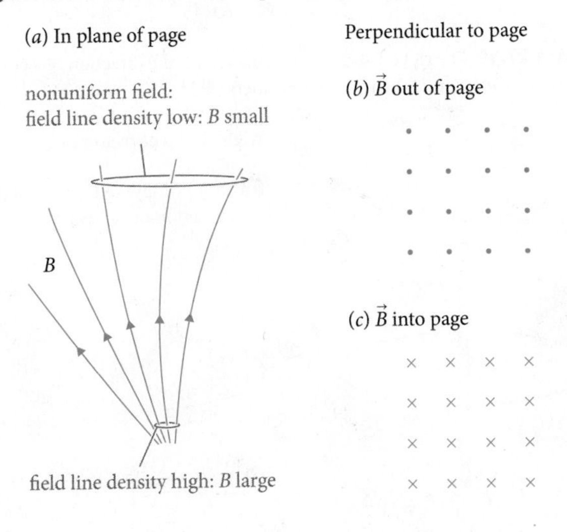

The magnetic field line flux through a closed surface is always zero.

In electrostatics, electric field lines originate or terminate on electrical charge. Because of the absence of magnetic monopoles, however, we have no "magnetic charge" on which magnetic field lines could originate or terminate. Magnetic field lines must therefore always form loops that close on themselves. The statement that the magnetic field line flux through a closed surface is always zero is therefore a direct consequence of the absence of magnetic monopoles.

> **27.7** What is the direction of the magnetic field lines *inside* the bar magnet of Figure 27.13?

27.3 Charge flow and magnetism

The first indication of a connection between electricity and magnetism came in 1820, when the Danish physicist Hans Christian Ørsted discovered that a flow of charge carriers deflects the needle of a compass. This effect is illustrated in **Figure 27.16**, which shows battery terminals connected to a conducting rod.

The top of the rod in Figure 27.16 is connected to the negative battery terminal, and the bottom is connected to the positive terminal. As we saw in Chapter 26, this means that the bottom of the rod is at a higher potential than the top and an upward-pointing electric field arises inside the rod.*

The upward electric field in the rod exerts a downward electric force on the negatively charged electrons in the rod, causing a flow of charge carriers, called **current,** through the rod. The battery maintains a constant potential difference across the rod by adding electrons at the top of the rod and removing them at the bottom. Consequently, the battery maintains a constant current through the rod.

*In Section 24.5 we concluded that the electric field inside a conductor in electrostatic equilibrium is zero. The rod in Figure 27.16, however, is not in electrostatic equilibrium because the battery maintains a potential difference across it, causing free charge carriers in the rod to move.

Figure 27.16 A flow of charge carriers through a conducting rod causes a circular alignment of compass needles.

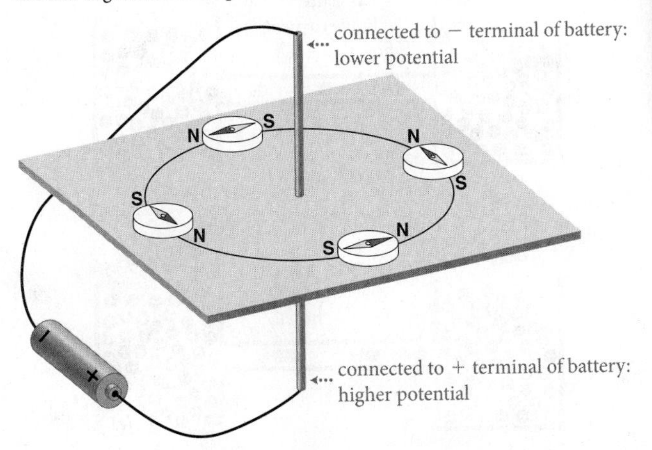

27.8 (*a*) Is the rod in Figure 27.16 electrically charged while connected to the battery? (*b*) Is there an electric field due to the rod at the positions of the compasses in Figure 27.16?

When the electrons flow through the rod in Figure 27.16, the compass needles align themselves in a circular pattern around the rod. This observation leads to a conclusion:

A flow of charged particles causes a magnetic field.

If the direction of the flow is reversed by reversing the battery, the compass needles turn around until they point in the opposite direction. If the flow of charge carriers is stopped, the needles point toward Earth's geographic North Pole, as usual.

27.9 Sketch the magnetic field line pattern in the horizontal plane around the rod in Figure 27.16.

It is important to note that a flow of positive charge carriers in one direction is equivalent to a flow of negative charge carriers in the opposite direction. In **Figure 27.17**, for example, the charge on the right increases and the charge on the left decreases regardless of which flow occurs. Experiments also show that the two types of flow are equivalent in terms of the magnetic field:

A flow of positive charge in one direction produces the same magnetic field as an equal flow of negative charge in the opposite direction.

Therefore, regardless of the actual movement of charge carriers through a current-carrying wire, we shall always denote the direction of decreasing potential (that is, the direction in which positive charge carriers would flow) with an arrow labeled with the symbol for current (I) next to the wire, as shown in **Figure 27.18**. We'll call this direction the *direction of current*.

Figure 27.17 A flow of positive charge carriers in one direction is equivalent to a flow of negative charge carriers in the other direction.

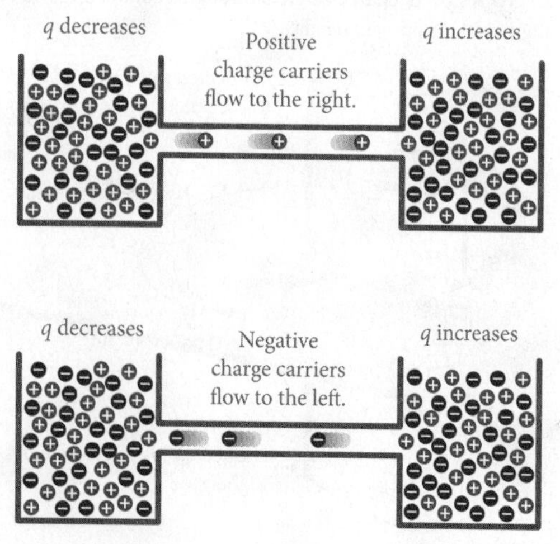

Figure 27.18 By definition, current has the direction in which positive charge carriers would flow, even if it is actually carried by negative charge carriers moving in the opposite direction.

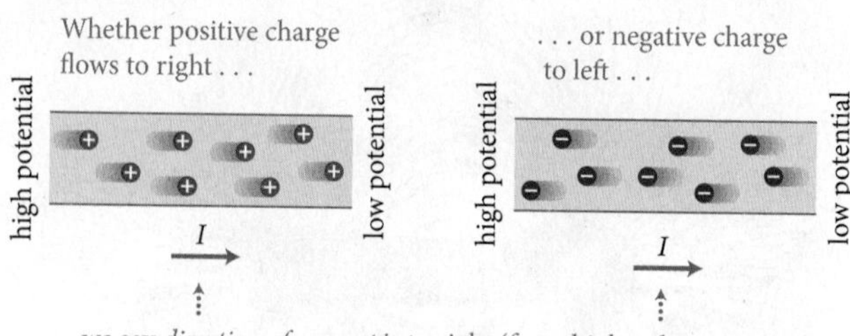

We can now connect the direction of a magnetic field to that of current. **Figure 27.19** shows a single circular magnetic field line for the current-carrying rod from Figure 27.16. As seen from the top, the magnetic field curls in a counterclockwise direction around the rod. The downward flow of negative electrons through the rod corresponds to an upward current, and so the direction of the magnetic field is connected to that of the current by the *right-hand current rule*:

If you point the thumb of your right hand in the direction of current, your fingers curl in the direction of the magnetic field produced by that current.

27.10 Can you replace the current-carrying rod of Figure 27.16 by a magnet and get the same magnetic field?

The relationship between the magnetic field produced by a magnet and that produced by a current-carrying wire is not immediately obvious. We shall study this relationship in more detail in the next chapter. Before doing so, however, we still need to answer the question: How does a bar magnet interact with a straight current-carrying wire? The answer, which follows from the experiment illustrated in **Figure 27.20**, describes an interaction that is very different

Figure 27.19 The right-hand current rule relates the direction of a current to the direction of the resulting magnetic field.

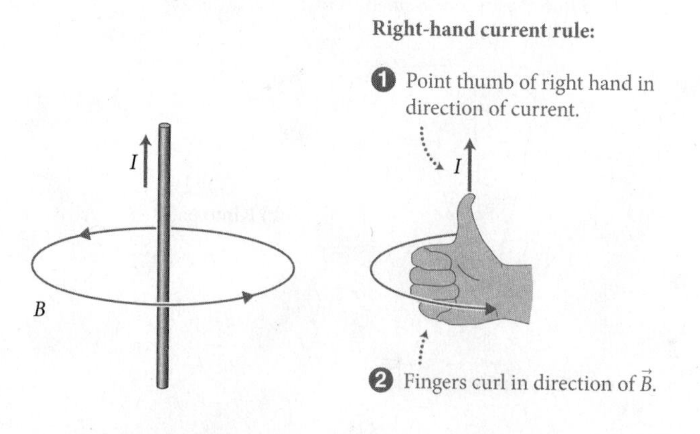

CONCEPTS

Figure 27.20 The magnetic force exerted by a bar magnet on a current-carrying wire depends on the magnet's orientation.

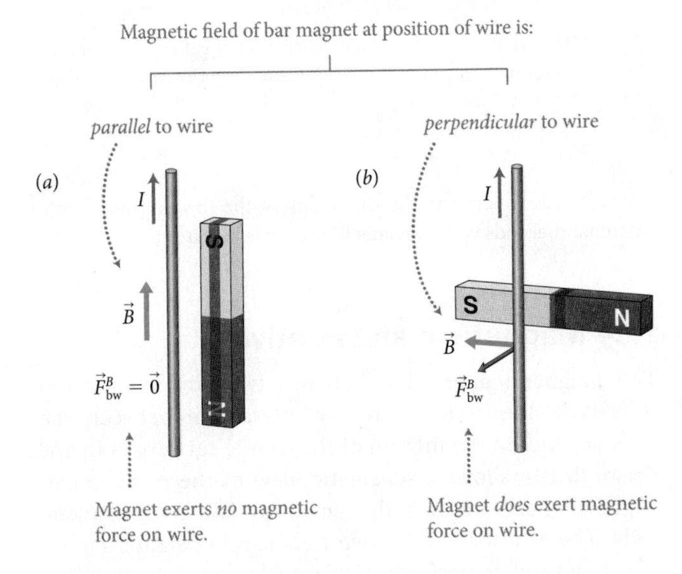
Magnetic field of bar magnet at position of wire is:
parallel to wire
perpendicular to wire
(a)
(b)
I
B
S
N
$\vec{F}^B_{bw} = \vec{0}$
$\vec{B}$
$\vec{F}^B_{bw}$
Magnet exerts *no* magnetic force on wire.
Magnet *does* exert magnetic force on wire.

Figure 27.22 When a magnet exerts a force on a current-carrying wire, the right-hand force rule relates the direction of the force to that of the current and the magnet's magnetic field.

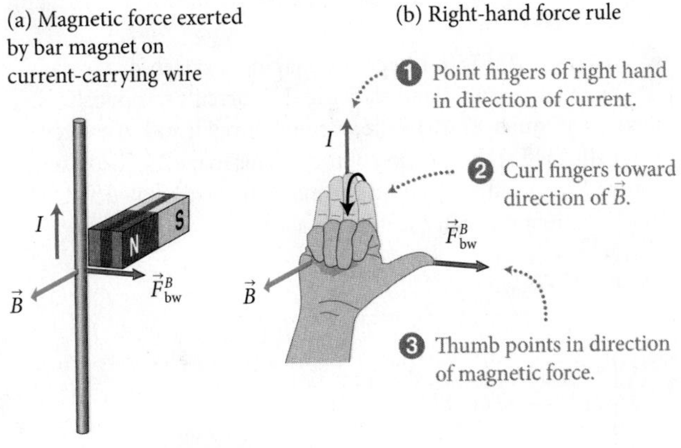
(a) Magnetic force exerted by bar magnet on current-carrying wire
(b) Right-hand force rule
1 Point fingers of right hand in direction of current.
2 Curl fingers toward direction of $\vec{B}$.
3 Thumb points in direction of magnetic force.
I
N
S
$\vec{B}$
$\vec{F}^B_{bw}$

from all the interactions we have encountered so far. When a bar magnet is held parallel to a current-carrying wire, the magnetic field of the magnet at the position of the wire is parallel to the wire (Figure 27.20*a*). In this position, the magnet exerts *no force* on the wire! When we rotate the magnet 90°, so that its long axis is perpendicular to the long axis of the wire, the magnet exerts a repulsive force on the wire (Figure 27.20*b*). The most striking aspect of this interaction is that the direction of the magnetic force is *perpendicular* to the magnetic field.

27.11 Use Figure 27.20 to determine the direction of the magnetic force exerted by the magnet on the wire when the magnet is in the orientations shown (*a*) in **Figure 27.21*a*** and (*b*) in Figure 27.21*b*. (Hint: First determine the direction of the magnetic field due to the magnet at the wire; then use the observations from Figure 27.20 to determine the direction of the magnetic force.)

The surprising *sideways* force exerted by the bar magnet on the wire in Figure 27.21*b* (see Checkpoint 27.11) is a result of the fact that the magnetic force exerted by a bar magnet on a current-carrying wire is always at right angles to both the magnetic field and the direction of current. This observation suggests that these three directions can be connected by a right-hand rule. **Figure 27.22** shows that if you orient the fingers of your right hand in the direction of current in such a way that you can curl them so that the fingertips point in the direction of the magnetic field, then your right thumb points in the direction of the magnetic force exerted on the wire. Take a minute to verify the direction of the force in Figure 27.20*b* using this procedure, commonly called the *right-hand force rule*:*

The direction of the magnetic force exerted by a magnetic field on a current-carrying wire is given by the direction of the right-hand thumb when the fingers of that hand are placed along the direction of current in such a way that they can be curled toward the magnetic field.

Table 27.1 summarizes the two right-hand rules we have encountered so far.

Figure 27.21 Checkpoint 27.11.

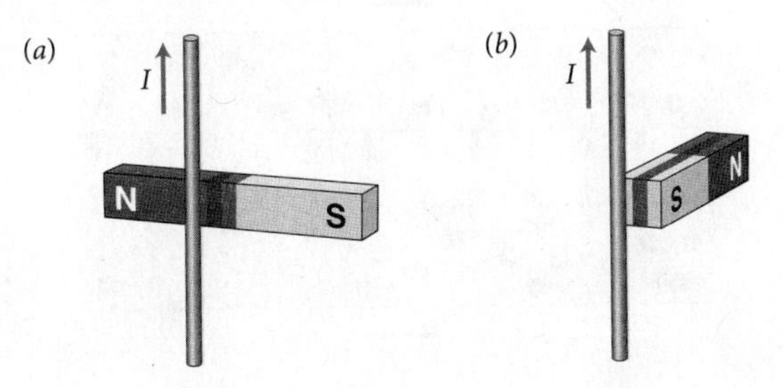
(a)
(b)
I
N
S
S
N

Table 27.1 Right-hand rules in magnetism

Right-hand rule	thumb points along	fingers curl
current rule	current	along *B*-field
force rule	magnetic force	from current to *B*-field

*There is nothing magical about right hands. You could use your left hand and change the rule to read, "The direction . . . given by the direction of the left-hand thumb when the fingers of that hand are placed along the direction of the magnetic field in such a way that they can be curled toward the direction of current." The same applies to determining the direction of a magnetic field surrounding a current-carrying rod. In Figure 27.19, for instance, orienting your left hand so that the thumb points in the direction *opposite* the direction of current automatically curls the fingers in the direction of the magnetic field. A consistent convention is what is important.

CONCEPTS

Example 27.1 Current-carrying rods

Two parallel rods carry currents in opposite directions. Determine the direction of the magnetic force exerted by each rod on the other rod.

❶ **GETTING STARTED** I begin by making a sketch of the rods, labeling them 1 and 2 and showing the currents in opposite directions (**Figure 27.23**). The current through rod 2 creates a magnetic field at the location of rod 1. This magnetic field exerts a force $\vec{F}^B_{21}$ on rod 1. Likewise, the magnetic field created at rod 2 by the current through rod 1 exerts a force $\vec{F}^B_{12}$ on rod 2.

Figure 27.23

(*a*)

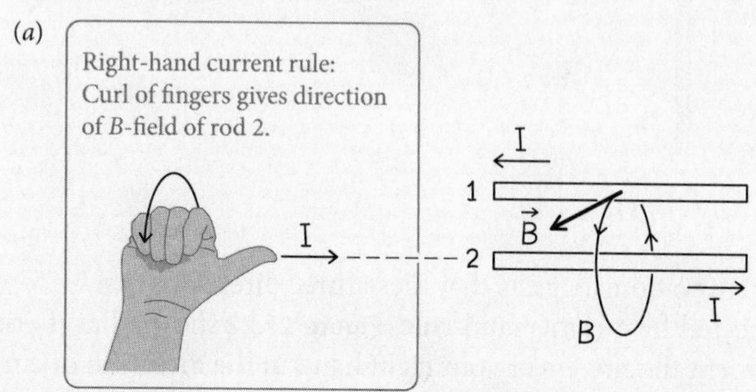
Right-hand current rule:
Curl of fingers gives direction
of B-field of rod 2.
I
1
2
B
I
B
I

(*b*)

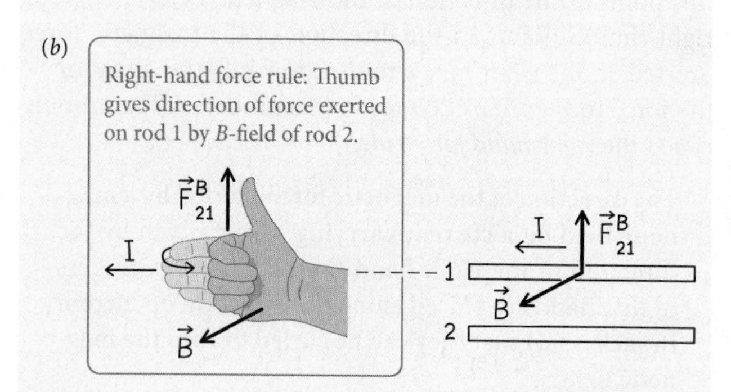
Right-hand force rule: Thumb
gives direction of force exerted
on rod 1 by B-field of rod 2.
F B 21
I
B
I
F B 21
1
B
2

❷ **DEVISE PLAN** To determine the direction of each force, I need to apply the right-hand force rule. To apply this rule, I need to know the direction of the magnetic field created by each rod at the location of the other. I can determine these directions using the right-hand current rule.

❸ **EXECUTE PLAN** I begin with the right-hand current rule to determine the direction of the magnetic field due to the current through rod 2 at the location of rod 1. When I align my right thumb with the direction of current through rod 2, my fingers curl out of the page (Figure 27.23*a*), telling me that this is the direction of the magnetic field due to rod 2. Next I apply the right-hand force rule to determine the direction of the magnetic force $\vec{F}^B_{21}$ exerted by rod 2 on rod 1. I align the fingers of my right hand with the current through rod 1 in such a way that they can curl toward $\vec{B}$ (Figure 27.23*b*). My right thumb now points up in the plane of the page, telling me that this is the direction of $\vec{F}^B_{21}$. ✔

To determine the direction of the magnetic force $\vec{F}^B_{12}$ exerted by rod 1 on rod 2, I again begin with the right-hand current rule, which tells me that, at the location of rod 2, the magnetic field due to rod 1 is directed out of the page. Applying the force rule, I place my fingers along the direction of I_2 and see that curling them in the direction of $\vec{B}$ makes my thumb point downward in the plane of the page, so this is the direction of $\vec{F}^B_{12}$. ✔

The force $\vec{F}^B_{21}$ points upward, the force $\vec{F}^B_{12}$ points downward, and therefore the rods repel each other.

❹ **EVALUATE RESULT** It makes sense that the two rods exert forces on each other that point in opposite directions because the forces $\vec{F}^B_{12}$ and $\vec{F}^B_{21}$ form an interaction pair.

27.12 Determine the directions of the forces exerted by two parallel rods with currents in the same direction.

27.4 Magnetism and relativity

The magnetic interaction between two current-carrying rods is baffling. How can any interaction between the rods depend on the motion of the charge carriers in them? **Figure 27.24a** shows a schematic view of the positive and negative charge carriers that make up two parallel metal rods. The electrons in the two rods repel one another, but this repulsion is perfectly balanced by the attraction between the electrons in one rod and the positively charged ions in the other rod. Once the electrons are set in motion, however, this balance appears to be disturbed, and, depending on the signs of the currents, the rods either attract (Figure 27.24*b*) or repel each other.

To analyze this interaction, consider the simpler situation of two pairs of electrons. Relative to Earth, pair A is at rest and pair B is moving at constant velocity $\vec{v}$ (**Figure 27.25a**). To observer E, who is also at rest relative to Earth, the interaction between the two electrons of pair A is completely described by Coulomb's law: As we saw in Chapter 22, the

Figure 27.24 Schematic view of the interaction between two current-carrying rods.

(*a*) Metal rods don't interact when they carry no current

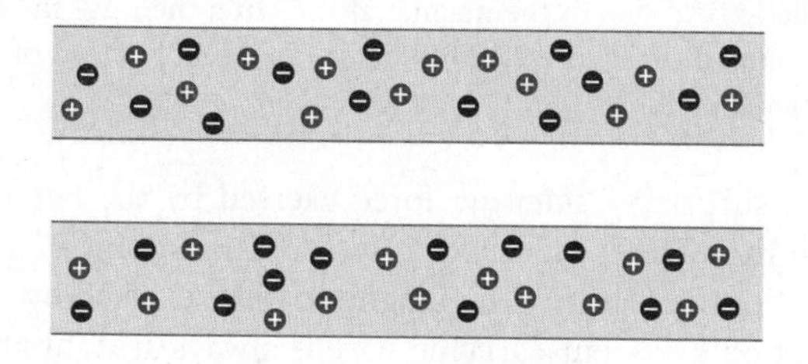

(*b*) When carrying a current, they do interact

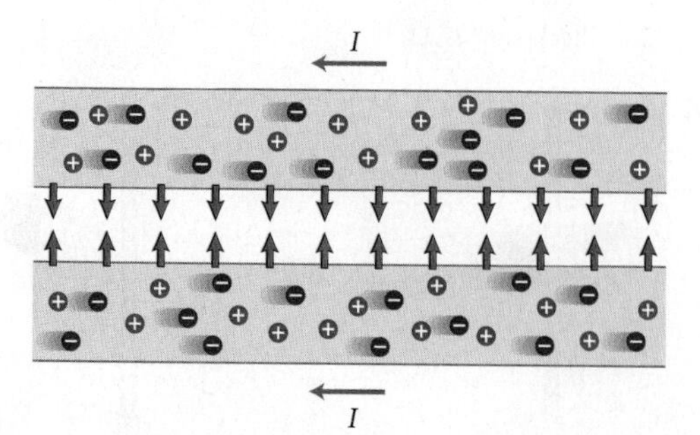
I
I

Figure 27.25 A relativistic view of the interaction between two charged particles.

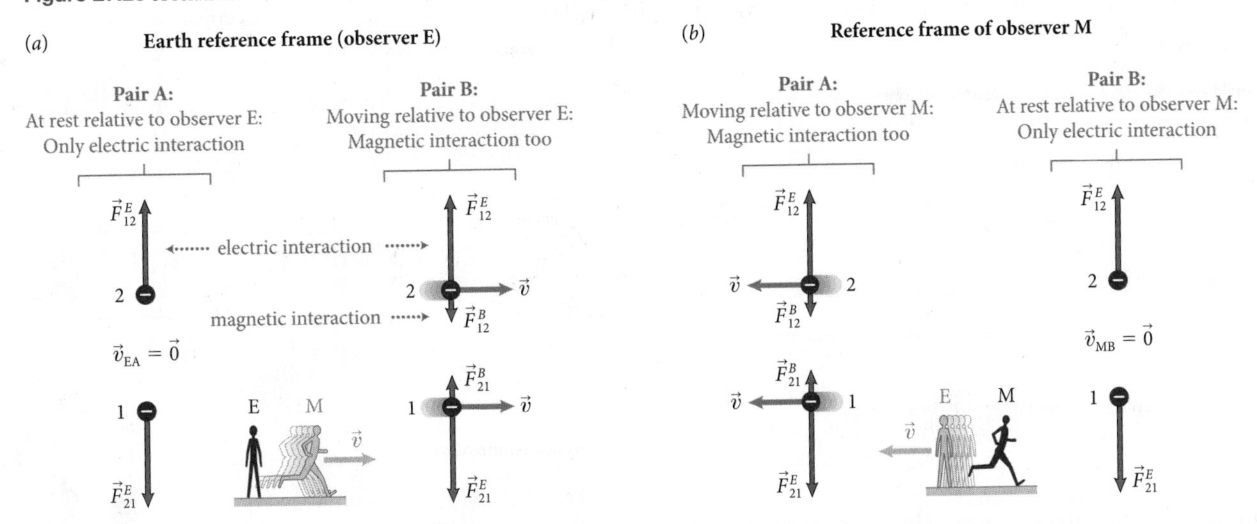

electrons repel each other due to an electric interaction. According to observer E, the interaction between the electrons of pair B is *different* from that between the pair A electrons. In addition to the Coulomb repulsion, observer E perceives a magnetic interaction between the electrons of pair B because the moving electrons constitute a current.

The conclusions drawn by an observer M moving along with pair B are startlingly different (Figure 27.25*b*). To this observer, pair B is at rest and pair A is moving to the left at constant velocity. Consequently, to M the interaction between the electrons of pair B is purely electric, while that between the electrons of pair A must have a magnetic component as well. This odd conclusion is a direct consequence of the fact that magnetism is motion-dependent and therefore, like velocity, must be relative:

The observed interaction between charge carriers depends on their motion relative to the observer: The interaction can be purely electric, purely magnetic, or a combination of the two.

At first sight, this statement appears to be a violation of the principle of relativity. If reference frame M moves at constant velocity relative to reference frame E, then both observers should agree on their observations (see Section 6.3). The resolution of this problem lies in the fact that the magnetic interaction between two moving charged carriers is a direct consequence of special relativity: *Magnetism is a relativistic effect.*

As explained in Chapter 14, relativistic effects tend to be so extraordinarily small that they cannot be observed under most ordinary conditions. So why is it that we can feel this relativistic effect every time we stick a note to a refrigerator with a magnet? To see why, let us return to the two current-carrying rods in Figure 27.24. Two 1-m rods, each 0.5 mm in diameter, contain an immense number of free electrons—more than 10^{22} of them. To appreciate how enormous this number is, imagine that these electrons were not accompanied by an equal number of positively charged ions. The force with which two such collections of unbalanced electrons separated by a distance of 5 mm would repel each other is a phenomenal 10^{19} N—enough to lift up thousands of average-sized mountains. The fact that two electrically neutral metal rods do not exert any force on each other shows how incredibly accurate is the balance between positive and negative charge carriers in matter. In contrast, if we run large currents through the rods—by applying and maintaining a potential difference of, say, several volts across the length of each rod—then the magnetic force exerted by each rod on the other is a mere 10^{-2} N. This small but measurable magnetic force is 10^{21} times smaller than the electric repulsion between the electrons in the rods! Thus, if magnetism is indeed a relativistic effect (see below and Section 27.8 for details), we can now understand why we can so readily observe it. First, the electric force is incredibly large, and so even a small relativistic correction becomes measurable. Second, because electrical charge is so well balanced in all matter, electric forces are balanced, leaving only the small "magnetic" correction. In a sense, therefore, magnetism provides the most direct observation of a relativistic effect!

Let us now try to understand qualitatively how special relativity requires a magnetic interaction between charge carriers in motion. The treatment that follows relies on very little knowledge of special relativity—only the concept of length contraction is needed (see Section 14.6): When an object moves relative to an observer, the observer measures the length of the object in the direction of its motion to be shorter than the proper length of the object.

A direct result of length contraction is that the charge density of an object depends on its motion relative to the observer. Consider, for example, a rod carrying a positive charge q. At rest in the Earth reference frame, the rod has length ℓ_{proper} and its "proper" charge density is $\lambda_{\text{proper}} = q/\ell_{\text{proper}}$. If the rod is set in motion along its

Figure 27.26 Because a rod appears shorter when it is moving relative to the observer, its charge density appears larger than when the rod is at rest.

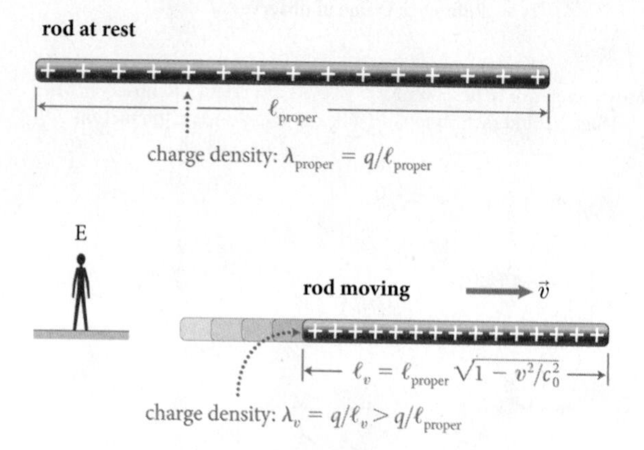

rod at rest
ℓ_{proper}
charge density: $\lambda_{proper} = q/\ell_{proper}$
E
rod moving
$\vec{v}$
$\ell_v = \ell_{proper}\sqrt{1 - v^2/c_0^2}$
charge density: $\lambda_v = q/\ell_v > q/\ell_{proper}$

lengthwise axis, its length measured by an observer at rest is smaller than ℓ_{proper} (**Figure 27.26**). Because the charge on the rod is still q, the observer at rest sees a greater charge density on the moving rod.*

27.13 Consider the two identical rods in Figure 27.26, one moving and the other at rest relative to observer E, at rest in the Earth reference frame. Suppose a second observer M moves along with the moving rod. (*a*) Which rod has the greater charge density according to observer M? (*b*) Suppose the charge on the rod moving in the Earth reference frame is adjusted so that its charge *density* as seen by observer E is the same as the charge density of the rod at rest in the Earth reference frame. Is the charge density on each rod as seen by observer M greater than, equal to, or smaller than $\lambda_{proper} = q/\ell_{proper}$?

Understanding the changes in charge density that occur when charge carriers move relative to the observer is the key point for the remainder of this section. Thus if you did not complete the preceding checkpoint, please go back and do it now before reading on.

Let us now apply these ideas to the interaction between a current-carrying wire and a charged particle. **Figure 27.27a** schematically shows a current-carrying wire consisting of fixed positively charged ions (red) and negatively charged electrons (blue) traveling to the right at velocity $\vec{v}$. (For clarity, the ions and electrons are drawn side by side rather than intermingled as they are in a real conductor.)

To an observer E at rest relative to the ions, the wire does not carry a surplus charge, and so any length of wire contains the same number of ions and electrons. That is, to E the linear charge density of the ions, λ_{proper}, is equal in magnitude to that of the electrons, $-\lambda_{proper}$. Consequently, there is no electric field outside the wire, and the wire cannot exert an electric force on a charged particle placed near the wire.

Suppose, however, that a positively charged particle moves alongside the wire—for simplicity, we let it move at the same velocity $\vec{v}$ relative to the wire as the electrons in the wire. First we consider the particle and the wire from

Figure 27.27 Observers E (at rest in the Earth reference frame) and M (moving in the Earth reference frame) observe a current-carrying wire.

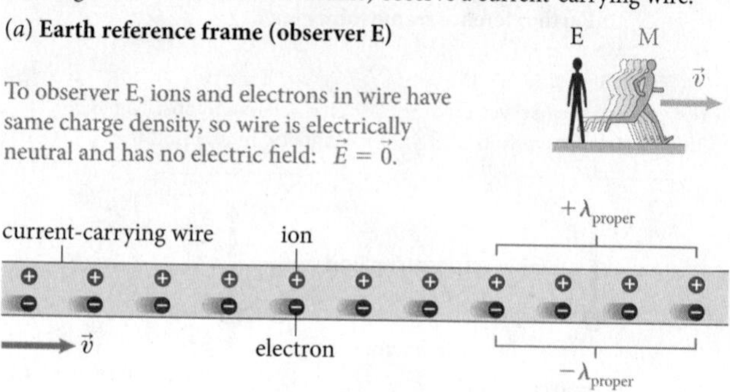

(*a*) Earth reference frame (observer E)
E M
$\vec{v}$
To observer E, ions and electrons in wire have same charge density, so wire is electrically neutral and has no electric field: $\vec{E} = \vec{0}$.
$+\lambda_{proper}$
current-carrying wire
ion
$\vec{v}$
electron
$-\lambda_{proper}$

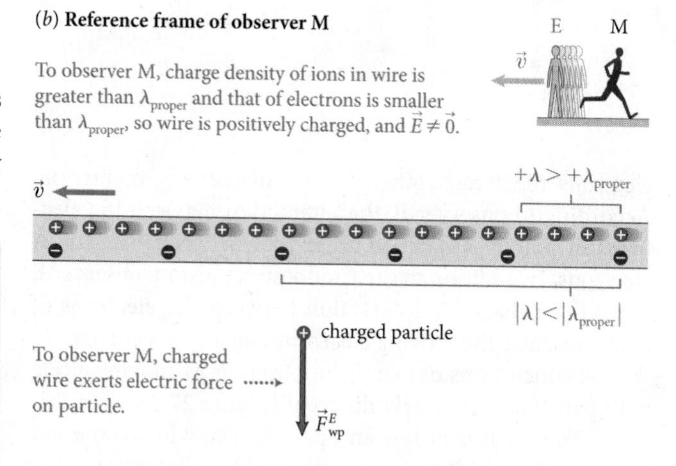

(*b*) Reference frame of observer M
E M
$\vec{v}$
To observer M, charge density of ions in wire is greater than λ_{proper} and that of electrons is smaller than λ_{proper}, so wire is positively charged, and $\vec{E} \neq \vec{0}$.
$+\lambda > +\lambda_{proper}$
$\vec{v}$
$|\lambda| < |\lambda_{proper}|$
charged particle
To observer M, charged wire exerts electric force on particle.
$\vec{F}^E_{wp}$

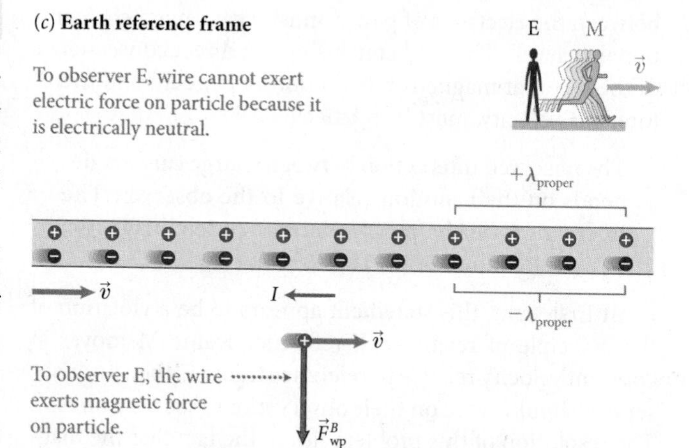

(*c*) Earth reference frame
E M
$\vec{v}$
To observer E, wire cannot exert electric force on particle because it is electrically neutral.
$+\lambda_{proper}$
$\vec{v}$
I
$-\lambda_{proper}$
$\vec{v}$
To observer E, the wire exerts magnetic force on particle.
$\vec{F}^B_{wp}$

the point of view of an observer M moving along with the particle. Because the particle is at rest relative to M, it experiences no magnetic force according to this observer. From M's point of view, the electrons in the wire are at rest and the ions move at velocity $\vec{v}$ to the left (Figure 27.27*b*). In addition, as you saw in Checkpoint 27.13, the charge densities

*It is not immediately obvious that the charge of a system is not affected by the motion of the charge carriers. Experiments, however, show that charge is, indeed, an invariant (that is, its value does not depend on the choice of reference frame).

CONCEPTS

seen by M are different from those seen by E. To M, the ions are closer together because they are moving, which means their charge density is greater than λ_{proper}. The electrons, on the other hand, are at rest relative to M and so farther apart. According to observer M the magnitude of their linear charge density therefore must be smaller than that observed by E relative to whom they are moving.

The different charge densities mean that although the wire appears electrically neutral to E, it cannot also appear neutral to M. According to M, the magnitude of the electron density is smaller than λ_{proper}, and the magnitude of the ion density is greater than λ_{proper}. Thus, the wire appears positively charged to observer M. Observer M therefore sees a downward electric field due to the wire at the location of the charged particle, with the wire and the particle repelling each other.

In the reference frame of the fixed ions, a very different picture emerges. Observer E, seeing no electric field (Figure 27.27*a*), cannot attribute the repulsion between the wire and the particle to an electric interaction. Instead, E attributes this repulsion to a magnetic interaction between two currents, one in the wire and the other caused by the moving charged particle (Figure 27.27*c*). As we shall see in Section 27.8, this magnetic interaction can be completely accounted for by the Coulomb interaction and special relativity, and in principle any other magnetic interaction can also be explained this way. However, because transforming back and forth from one moving reference frame to another is cumbersome, it is easier to develop a separate treatment for magnetism that does not require reference-frame transformations and that ignores any relativistic effects. It is important to keep in mind, however, that magnetic and electric interactions are two aspects of one *electromagnetic* interaction, with magnetism being a relativistic correction to the electric interaction.

27.14 What is the direction of current through the rod in (*a*) Figure 27.27*b* and (*b*) Figure 27.27*c*? (*c*) Is the direction of the force in Figure 27.27*c* in agreement with what we learned about the interaction of two parallel current-carrying wires in Section 27.3? (*d*) If the particle moving alongside the wire in Figure 27.27 carries a negative charge, is the force exerted by the wire on the particle attractive or repulsive? (*e*) Is this direction consistent with the direction of the forces exerted on each other by two parallel current-carrying wires?

CONCEPTS

Self-quiz

1. A compass sits on a table with its needle pointing to Earth's North Pole. A bar magnet with its long axis oriented along an east-west line is brought toward the compass from the right. If the needle turns clockwise, which pole of the bar magnet is nearer the compass?
2. Draw the magnetic field lines associated with the magnets in Figure 27.1*b–d* (both outside and inside the magnets). Assume that the pole on the left of each magnet is the north pole.
3. For each situation shown in **Figure 27.28**, apply the appropriate right-hand rule to determine the direction at position P of the magnetic field generated by the current-carrying wire.

Figure 27.28

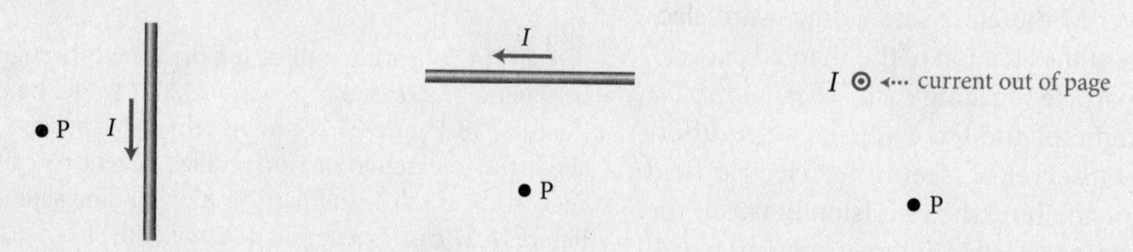

4. For each situation in Figures 27.20*a*, 27.20*b*, and 27.21*a*, reverse the polarity of the magnet and then apply the appropriate right-hand rule to determine the direction of the force exerted by the bar magnet on the current-carrying wire.

Answers

1. South pole. The clockwise rotation means the needle tip (which is a north pole by definition) moves toward the bar magnet. Because opposite poles attract, the bar magnet's south pole must be the closer one. (Remember that Earth's geographic North Pole is a magnetic south pole and so attracts the north pole of any compass needle.)
2. See **Figure 27.29**.

Figure 27.29

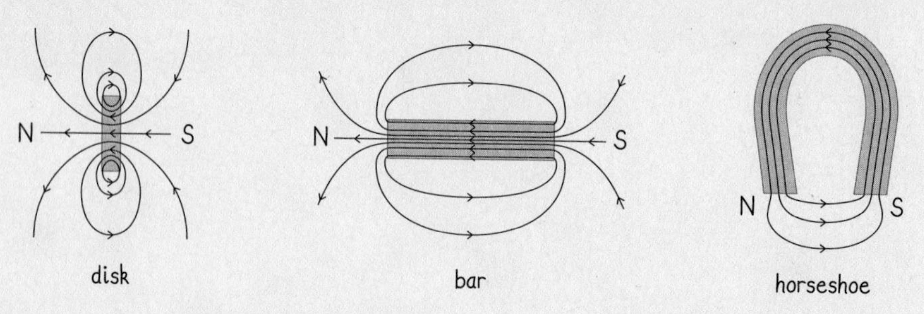

3. See **Figure 27.30**. (See Checkpoint 27.7 if you do not understand this answer.)

Figure 27.30

4. Reversing the magnet polarity reverses the direction of the force exerted on the wire, unless the force is zero, in which case it remains zero.

SELF-QUIZ

27.5 Current and magnetism

As we saw in the first part of this chapter, currents exert magnetic forces on one another. That is, a current creates a magnetic field, and magnetic fields exert forces on currents. Let us therefore begin this part of the chapter by introducing a quantitative definition of current. As mentioned in Section 27.3, **current** I is the rate at which charged particles cross a section of a conductor in a given direction. For a constant current, we have

$$I \equiv \frac{q}{\Delta t} \quad \text{(constant current)}, \tag{27.1}$$

where q is the quantity of charge passing a given position in a time interval Δt. If the current is not constant, we evaluate the flow of charged particles over infinitesimally small time intervals, yielding

$$I \equiv \frac{dq}{dt}, \tag{27.2}$$

where dq is the infinitesimal quantity of charge crossing a given section of a conductor in an infinitesimally small time interval dt. Note that I can be positive or negative; by definition it is positive in the direction from high to low potential because that is the direction in which positive charge carriers flow (which we defined earlier as the direction of current).

The SI unit of current is the **ampere** (A). This base unit is defined to be the current through two parallel straight thin wires of infinite length separated by 1 m in vacuum when the wires exert a force of 2×10^{-7} N per meter of length on each other. As we saw in Chapter 22, the coulomb is derived from the ampere: 1 C corresponds to the quantity of charge transported by a current of 1 A through a chosen section in a time interval of 1 s, or

$$1\ \text{C} \equiv 1\ \text{A} \cdot \text{s}. \tag{27.3}$$

Even though a charge of 1 C is extremely large by ordinary standards, currents of several amperes are quite common. Simple devices, which we shall discuss in Chapter 31, allow us to measure the current through a conductor quite readily. In the remainder of this chapter, we shall discuss constant currents—that is, currents whose magnitude does not change with time.

Experiments show that when a straight wire carrying a current I is placed in a uniform external magnetic field,* the magnetic force $\vec{F}^B_{\text{w}}$ exerted by the magnetic field on the wire is proportional to the length ℓ of wire in the magnetic field and to the magnitude of the current I. The force also depends on the angle θ between the direction of the current through the wire and the magnetic field (**Figure 27.31**). When the wire is parallel to the magnetic field ($\theta = 0$), the magnetic force exerted on the wire is zero. When the wire and the magnetic field are perpendicular to each other ($\theta = 90°$), the force is maximum:

$$F^B_{\text{w, max}} = |I|\ell B \quad \text{(straight wire perpendicular to uniform magnetic field)}. \tag{27.4}$$

Because I is a signed quantity, we need to put absolute-value symbols around it to indicate its magnitude.

Equation 27.4 defines the magnitude B of the magnetic field. If we measure the magnetic force exerted by a magnetic field on a wire of known length carrying a known current, we can determine B from Eq. 27.4:

$$B \equiv \frac{F^B_{\text{w, max}}}{|I|\ell} \quad \text{(straight wire perpendicular to uniform magnetic field)}. \tag{27.5}$$

*This magnetic field does not include the magnetic field of the current-carrying wire.

Figure 27.31 A current-carrying wire in an external magnetic field (that is, a magnetic field created by an object other than the wire).

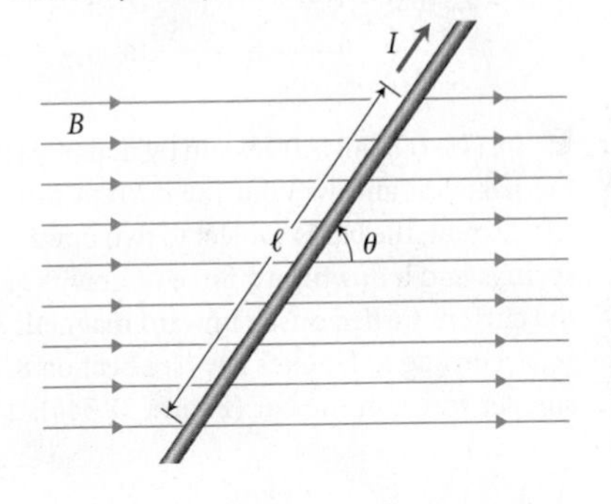

Table 27.2 Magnetic fields

Object	B (T)
Earth's surface	5×10^{-5}
small bar magnet	0.01
neodymium magnet	0.2
laboratory magnet	10
neutron star surface	10^8

It follows from this expression that the magnetic field has SI units of N/(A · m), a derived unit called the **tesla:**

$$1\text{ T} \equiv 1\text{ N/(A}\cdot\text{m)} = 1\text{ kg/(s}^2\cdot\text{A)}. \quad (27.6)$$

A magnetic field of 1 T is relatively large. Earth's magnetic field at Earth's surface varies between 3×10^{-5} T and 6×10^{-5} T. **Table 27.2** provides some examples of the magnitudes of various magnetic fields.

At intermediate angles θ, the magnitude of the magnetic force exerted on the wire is proportional to $\sin\theta$, and so for an arbitrary angle θ between 0 and 180°, the magnitude of the magnetic force exerted on the wire is

$$F^B_{\text{w}} = |I|\ell B\sin\theta \quad (0 < \theta < 180°). \quad (27.7)$$

As we saw in Figure 27.22, the direction of the magnetic force exerted by a magnet on a current-carrying wire is always perpendicular to both the direction of the current I and the magnetic field $\vec{B}$, and is given by the right-hand force rule. In Figure 27.31, for example, the force is directed into the plane of the page. (If you line up the fingers of your right hand with the direction of current and then curl them, over the smallest angle, toward the direction of the magnetic field, your thumb points into the page.) If we define a vector $I\vec{\ell}$, whose magnitude is given by the product of the magnitude of the current and the length ℓ of the wire, and whose direction is given by the direction of current through the wire, we can write the **magnetic force** in Eq 27.7 as the vector product of two vectors (see Section 12.8):

$$\vec{F}^B_{\text{w}} = I\vec{\ell} \times \vec{B} \quad \text{(straight wire in uniform magnetic field)}. \quad (27.8)$$

This vector product represents a force that has the magnitude given in Eq. 27.7. The direction of $\vec{F}^B_{\text{w}}$ is always perpendicular to both $I\vec{\ell}$ (the direction of current) and $\vec{B}$ and is obtained by curling the fingers of the right hand from the direction of current to $\vec{B}$ (**Figure 27.32**).

Figure 27.32 The right-hand rule for the direction of the vector product $I\vec{\ell} \times \vec{B}$.

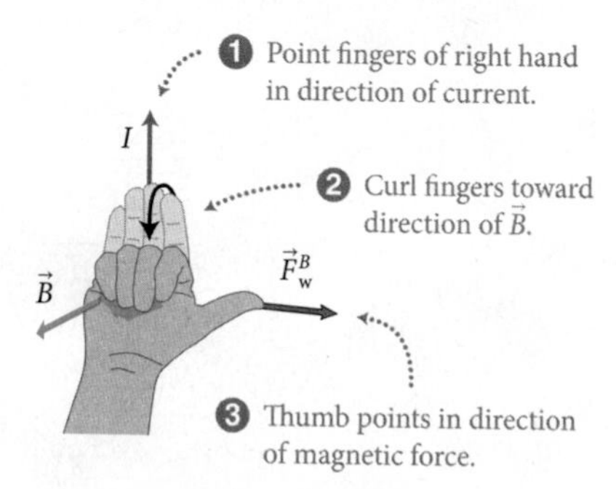

Example 27.2 Magnetic field meter

A metal bar 0.20 m long is suspended from two springs, each with spring constant $k = 0.10$ N/m, and the bar is in an external magnetic field directed perpendicular to the bar length (**Figure 27.33**). With a current of 0.45 A in the bar, the bar rises a distance $d = 1.5$ mm. (*a*) In what direction is the current? (*b*) What is the magnitude B of the external magnetic field?

Figure 27.33 Example 27.2.

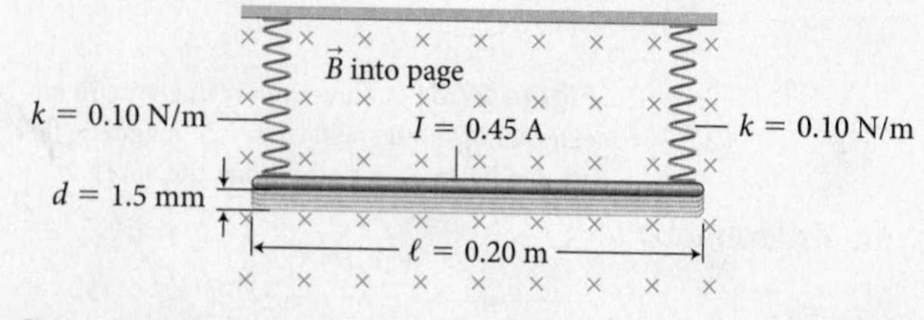

❶ GETTING STARTED I begin by making a free-body diagram for the bar with and without the current through the bar. Without the current, the bar is subject to two upward forces exerted by the springs and a downward force of gravity (**Figure 27.34*a***). With the current turned on, an upward magnetic force lifts the bar, and so, according to Hooke's law (see Section 8.9), the springs exert a smaller force on the bar (Figure 27.34*b*). The amount by which the bar rises determines by how much the force exerted by the springs is reduced, which in turn gives me the magnitude of the upward magnetic force exerted on the bar.

Figure 27.34

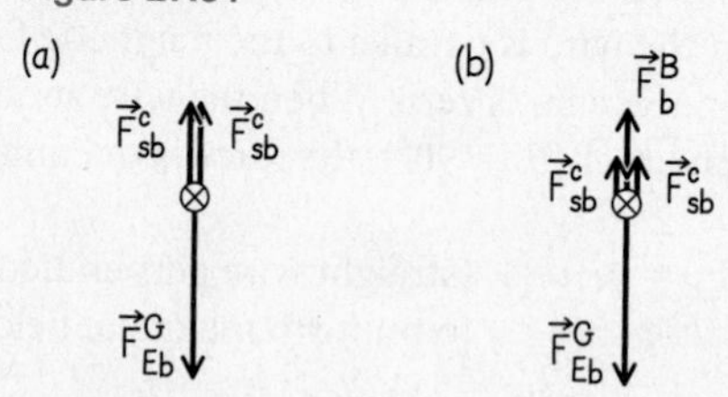

❷ DEVISE PLAN I know the directions of the external magnetic field (into the page) and of the force (upward) exerted by that field, so I can determine the direction of the current from the right-hand force rule. Because the bar is straight and perpendicular to the external magnetic field, I can use Eq. 27.5 to relate the magnitude of the magnetic field to the magnitude of the magnetic force. I know both the current and the length of the bar but not the magnitude of the force exerted on the bar. I know, however, that this force magnitude is equal to the change in the magnitude

QUANTITATIVE TOOLS

of the force exerted by the springs, which I can calculate using Hooke's law (Eq. 8.20).

❸ EXECUTE PLAN (*a*) The external magnetic field is into the plane of the page (Figure 27.33) and the magnetic force is directed upward. I therefore lay my right hand on top of Figure 27.33 with my thumb pointing up to represent the upward force. I can lay my hand on the page palm up or palm down. In the palm-up position, curling my fingers draws the tips away from the direction of $\vec{B}$. In the palm-down position, curling them draws the tips toward the direction of $\vec{B}$. Therefore that's the position I want, telling me the current is to the right, in the direction my fingers point before I curl them. ✔

(*b*) From Eq. 8.20, I know that raising the bar a distance d reduces the spring force by an amount kd, and so the magnitude of the magnetic force is $F^B_b = 2kd$. Substituting this result into Eq. 27.5 gives

$$B = \frac{2kd}{|I|\ell} = \frac{2(0.10\ \text{N/m})(1.5 \times 10^{-3}\ \text{m})}{(0.45\ \text{A})(0.20\ \text{m})} = 3.3 \times 10^{-3}\ \text{T}. ✔$$

❹ EVALUATE RESULT: The magnitude of the magnetic field I obtain is about 100 times greater than Earth's magnetic field and of the same order of magnitude as that of a bar magnet and therefore not unreasonable.

27.15 Suppose the charge carriers flowing through the horizontal bar in Example 27.2 are negatively charged. In which direction must they flow so that the magnetic force exerted on the wire is still directed upward?

27.6 Magnetic flux

From the field line picture for electrostatics, we obtained Gauss's law, which states that the electric flux Φ_E through any closed surface is proportional to the enclosed charge. Gauss's law allows us to determine the electric field of certain symmetrical charge distributions with great ease.

Because magnetic fields can also be described by field lines, we can follow a similar treatment for magnetism. Consider a surface of area A in a uniform magnetic field B (Figure 27.35). If a line normal to the surface makes an angle θ with the field, we can define a **magnetic flux** in analogy to the electric flux defined in Eq. 24.2:

$$\Phi_B \equiv BA\cos\theta = \vec{B}\cdot\vec{A} \quad \text{(uniform magnetic field)}, \qquad (27.9)$$

Figure 27.35 The magnetic flux through a surface of area A is given by the scalar product of the area vector $\vec{A}$ and the magnetic field $\vec{B}$.

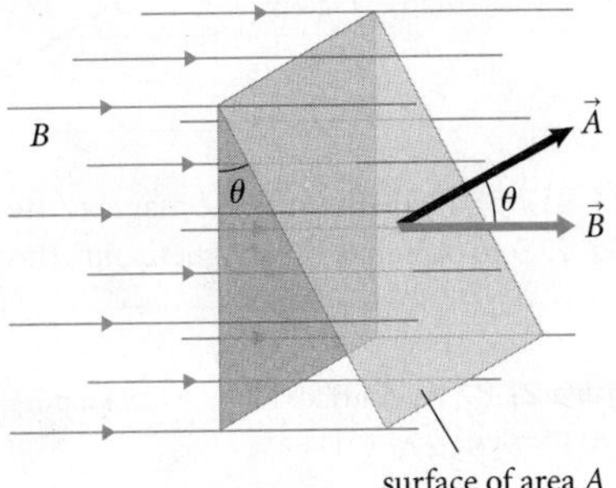

where $\vec{A}$ is an area vector (see Section 24.6) whose magnitude is equal to the area A of the surface and whose direction is normal to the surface. In the case of a closed surface, the direction of $\vec{A}$ is always chosen to be outward. Magnetic flux is a scalar, and as you can see from Eq. 27.9, magnetic flux has SI units of $\text{T}\cdot\text{m}^2$. This derived unit is given the name **weber** (Wb), in honor of the German physicist Wilhelm Weber:

$$1\ \text{Wb} \equiv 1\ \text{T}\cdot\text{m}^2 = 1\ \text{m}^2\cdot\text{kg}/(\text{s}^2\cdot\text{A}).$$

If the field is nonuniform or the surface is not flat, we follow the same procedure as for the electric flux and divide the surface into small surface elements, apply Eq. 27.9 to each surface element, and then sum the magnetic flux through all the elements. In the limit that the area of each element approaches zero, the sum is replaced by a surface integral (Eq. 24.4):

$$\Phi_B \equiv \int \vec{B}\cdot d\vec{A}, \qquad (27.10)$$

where $d\vec{A}$ is the area vector of an infinitesimally small segment of the surface. The meaning of magnetic flux Φ_B is similar to that of electric flux. It is a quantitative measure of the number of magnetic field lines crossing the surface specified in the integration.

As we saw in Section 27.2, magnetic field lines always form loops. Thus, the magnetic flux through any closed surface must always be zero because every field line that exits the surface must enter it somewhere else:

$$\Phi_B = \oint \vec{B} \cdot d\vec{A} = 0. \tag{27.11}$$

The zero reflects the fact that magnetic field lines always form loops or, equivalently, that there is no magnetic equivalent of an isolated charged particle. Equation 27.11 is called **Gauss's law for magnetism.** However, because of the zero result and because magnetic fields generally do not exhibit the same type of symmetry as electric fields, this expression does not allow us to determine magnetic fields in the same way that Gauss's law allows us to determine electric fields. As we shall see in Chapter 29, however, the magnetic flux is an important quantity when we consider changing magnetic fields.

Example 27.3 Magnetic flux through a loop

A square loop 0.20 m on each side is placed in a uniform magnetic field of magnitude 0.50 T. The plane of the loop makes a 30° angle with the magnetic field. What is the magnetic flux through the loop?

❶ **GETTING STARTED** I begin by drawing a side view to visualize the situation (**Figure 27.36**). I have to calculate the magnetic flux through the flat surface defined by the loop.

Figure 27.36

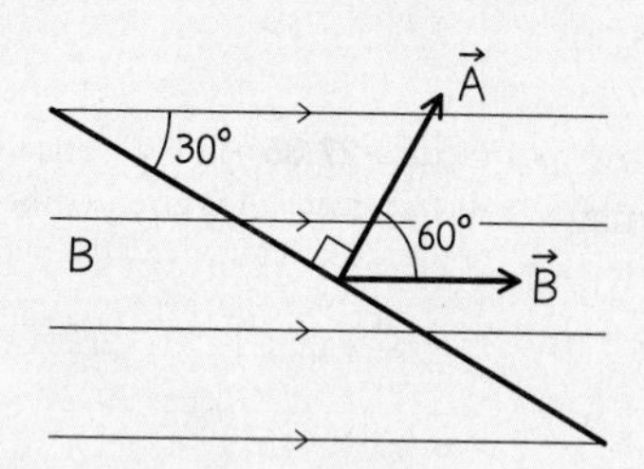

❷ **DEVISE PLAN** Because the magnetic field is uniform, I can use Eq. 27.9 to calculate the magnetic flux through the loop.

❸ **EXECUTE PLAN** The area of the loop is $A = (0.20\text{ m})^2 = 0.040\text{ m}^2$. Because the loop makes a 30° angle with the magnetic field, the angle between the magnetic field and a normal to the plane of the loop is 60°. Therefore

$$\Phi_B = AB\cos\theta = (0.040\text{ m}^2)(0.50\text{ T})\cos(60°)$$
$$= 1.0 \times 10^{-2}\text{ Wb.} ✔$$

❹ **EVALUATE RESULT** I arbitrarily chose the area vector $\vec{A}$ to point upward and to the right. Had I chosen to point it downward and to the left, the angle between $\vec{A}$ and $\vec{B}$ would have been 120° and, given that $\cos 120° = -\frac{1}{2}$, I would have obtained $\Phi_B = -0.010$ Wb. The flat surface defined by the loop doesn't constitute a closed surface, however, and so there is no unique direction for $\vec{A}$. The sign of the magnetic flux is therefore not determined by the information given in the question. I have no means of evaluating the magnitude of the flux, so I carefully check my calculations one more time.

Figure 27.37 Any surface bounded by loop L yields the same magnetic flux.

(*a*)

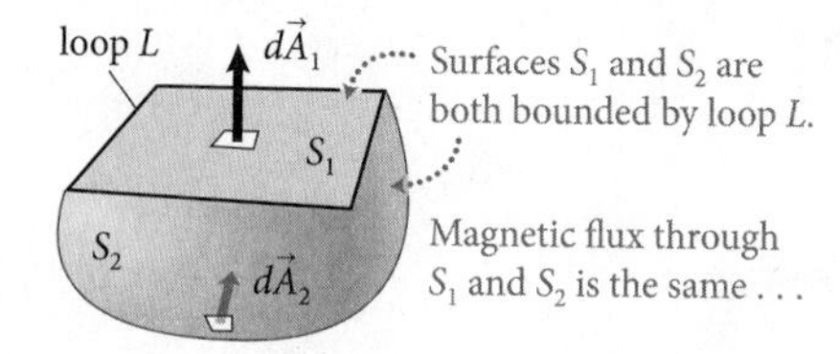

(*b*)

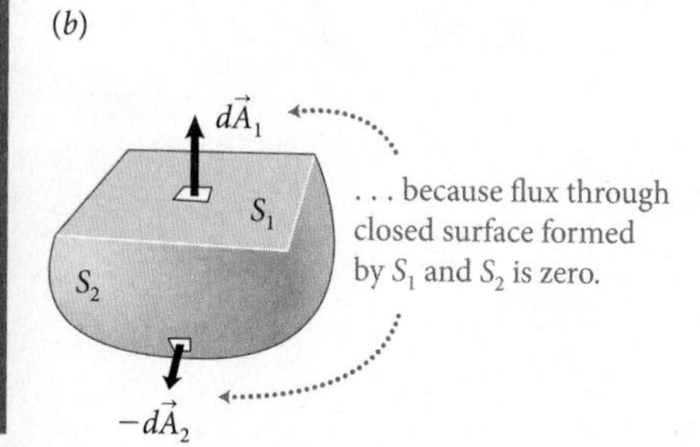

Because we can draw an infinite number of surfaces that are bounded by a loop, the expression "flux through the loop" in Example 27.3 appears open to interpretation. **Figure 27.37*a***, for example, shows two surfaces S_1 and S_2 bounded by the same loop L. Which surface should we use when computing the magnetic flux? The answer is: *any* surface bounded by L yields the same magnetic flux. To see why, let us examine the magnetic fluxes through S_1 and S_2. In both cases we use a surface normal that points up through loop L.

The magnetic flux through S_2 is

$$\Phi_B = \int_{S_2} \vec{B} \cdot d\vec{A}_2. \tag{27.12}$$

To see how this flux relates to that through S_1, consider the magnetic flux through the closed surface made up of S_1 and S_2 (Figure 27.37*b*). Because this surface is closed, we know the magnetic flux through it is zero (Eq. 27.11):

$$\oint_{S_1+S_2} \vec{B} \cdot d\vec{A} = \int_{S_1} \vec{B} \cdot d\vec{A} + \int_{S_2} \vec{B} \cdot d\vec{A} = 0, \tag{27.13}$$

where the area vector $d\vec{A}$ points *outward* from the closed surface. In terms of the surface normals $d\vec{A}_1$ and $d\vec{A}_2$ shown in Figure 27.37*a*, Eq. 27.13 becomes

$$\oint_{S_1+S_2} \vec{B} \cdot d\vec{A} = \int_{S_1} \vec{B} \cdot d\vec{A}_1 + \int_{S_2} \vec{B} \cdot (-d\vec{A}_2)$$

$$= \int_{S_1} \vec{B} \cdot d\vec{A}_1 - \int_{S_2} \vec{B} \cdot d\vec{A}_2. \quad (27.14)$$

The term $\int_{S_1} \vec{B} \cdot d\vec{A}_1$ represents the magnetic flux through the flat surface S_1; the term $\int_{S_2} \vec{B} \cdot d\vec{A}_2$ represents the magnetic flux through the curved surface S_2. Because the magnetic flux through the closed surface is zero (Eq. 27.11), the right side of Eq. 27.14 must also be zero, and so

$$\int_{S_1} \vec{B} \cdot d\vec{A}_1 = \int_{S_2} \vec{B} \cdot d\vec{A}_2. \quad (27.15)$$

27.16 A cube 1.0 m on each side is placed in a 1.0-T magnetic field with the field perpendicular to one surface of the cube. What are (*a*) the magnetic flux through the side through which the field enters the cube and (*b*) the magnetic flux through the entire surface of the cube?

27.7 Moving particles in electric and magnetic fields

The magnetic force exerted by a magnetic field on a straight current-carrying wire is really the sum of the magnetic forces exerted on many individual charge carriers moving through the wire. By examining how a current I through a conductor is related to the properties of the charge carriers causing that current, we can therefore deduce the magnetic force acting on a single charge carrier.

Consider a constant current caused by a flow of charge carriers, each carrying a quantity of charge q through a wire of cross-sectional area A (**Figure 27.38**). If the charge carriers flow at speed v, then in a time interval Δt each advances a distance $\ell = v\Delta t$. Thus all the charge carriers in the shaded volume $V = A\ell$ pass through a cross section of the wire in a time interval Δt. If the wire contains n charge carriers per unit volume, the shaded volume contains nV charge carriers and the charge flowing through the cross section is $Q = nVq = nA\ell q = nA(v\Delta t)q$. The current through the wire is thus

$$I \equiv \frac{Q}{\Delta t} = \frac{nA(v\Delta t)q}{\Delta t} = nAqv. \quad (27.16)$$

Figure 27.38 If the charge carriers in a straight current-carrying wire move at speed v, they advance a distance $\ell = v\Delta t$ in a time interval Δt. All the charge carriers in the shaded volume pass through the cross-sectional area A in that time interval.

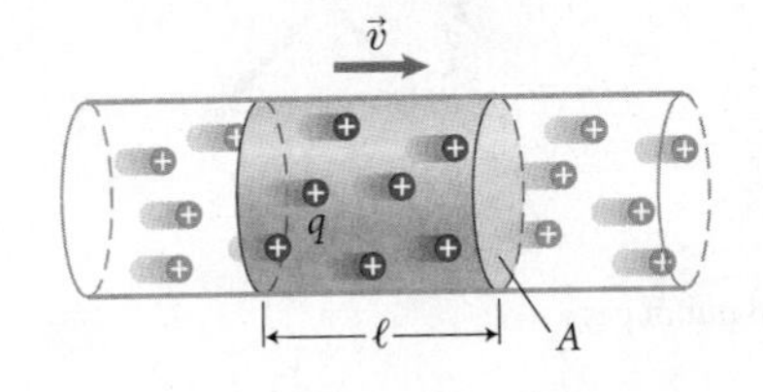

Substituting this result into Eq. 27.7, we can write for the magnitude of the magnetic force exerted on the current-carrying wire

$$F^B_w = |nA\ell qv|B \sin\theta = nA\ell|q|vB \sin\theta, \quad (27.17)$$

where θ is the angle between the velocity of the charge carriers and the magnetic field. Because n is the number of charge carriers per unit volume and $A\ell$ is the volume of a length ℓ of the wire, the quantity $N = nA\ell$ represents the number of charge carriers in a length ℓ of the wire. The magnetic force exerted on that length of the wire can thus be written $F^B_w = N|q|vB \sin\theta$. Because there are N charge carriers in the length ℓ, the magnitude of the **magnetic force** exerted on a single particle carrying a charge q moving at velocity $\vec{v}$ is

$$F^B_p = |q|vB \sin\theta \quad (27.18)$$

or, in vector form,

$$\vec{F}^B_p = q\vec{v} \times \vec{B}. \quad (27.19)$$

Note that the vector $q\vec{v}$ always points in the direction of current—that is, from high potential to low potential—regardless of the sign of q. As we saw in Section 25.3, a negatively charged particle moves from low to high potential, so $\vec{v}$ points opposite the direction of current, but because $q < 0$, the vector $q\vec{v}$ points in the direction of current.

This is the fundamental expression for the magnetic force acting on a moving charge carrier. Even though we derived Eq. 27.19 for a positively charged particle, it holds for any kind of charge. Because of its velocity dependence, this force is very different from any of the other forces we have encountered so far. In particular, if the charge carrier is at rest *relative to the reference frame in which the magnetic field is measured*, the magnetic force vanishes. (In contrast, the electric force exerted on a charged particle is independent of the motion of the charged particle relative to the reference frame in which the electric field is measured.)

In the presence of both electric and magnetic fields, the **electromagnetic force** exerted on each charge carrier is

$$\vec{F}^{EB}_{\mathrm{p}} = q\vec{E} + q\vec{v} \times \vec{B} = q(\vec{E} + \vec{v} \times \vec{B}). \tag{27.20}$$

Figure 27.39 A charged particle moving in a uniform magnetic field travels (*a*) in a straight line when its velocity is parallel to the field and (*b*) in a circle when the two are perpendicular.

(*a*) Particle's velocity is parallel to field

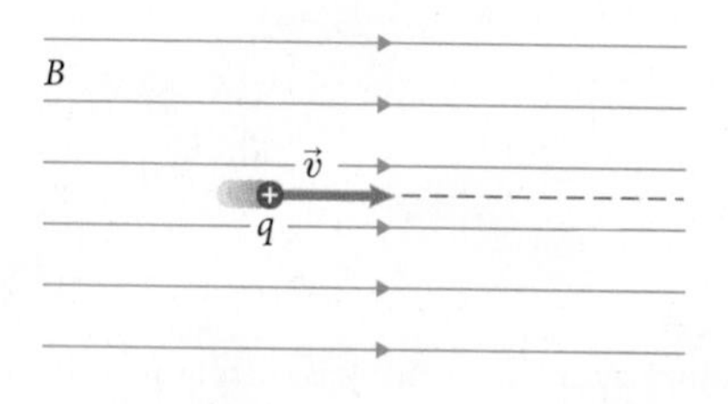

(*b*) Particle's velocity is perpendicular to field

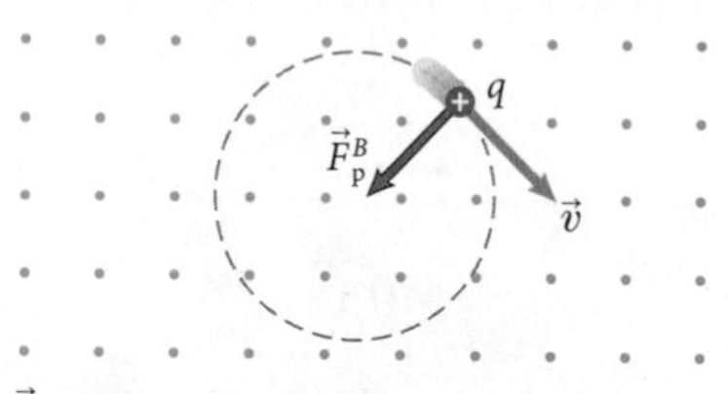

$\vec{B}$ out of page

Let us now examine what kind of trajectory a charged particle follows when it travels through a region of uniform magnetic and electric fields. We begin by examining the two special cases of a positive charge carrier traveling into a region of uniform magnetic field. When the carrier travels parallel to the direction of $\vec{B}$, as in Figure 27.39a, the magnetic force is zero because the vector product in Eq. 27.19 is zero when $\vec{v}$ is parallel to $\vec{B}$. (Alternatively, you can visualize the moving charge carrier as a current. As we have seen in Section 27.3, the magnetic force is zero when the current is parallel to the magnetic field.)

When the charge carrier moves perpendicular to a uniform magnetic field, as in Figure 27.39*b*, the magnetic force is nonzero. Because the magnetic force is always perpendicular to both $\vec{B}$ and $\vec{v}$, it lies in the plane of the drawing and is perpendicular to $\vec{v}$. Because the force acting on the charge carrier always remains perpendicular to the direction of motion, the speed of the charge carrier does not change and we have the condition for circular motion at constant speed. The magnetic force, always directed toward the center of the circular path in which the charge carrier moves, provides the centripetal acceleration.

The equation of motion for the charge carrier is

$$\Sigma\vec{F} = m\vec{a}. \tag{27.21}$$

If we ignore the force of gravity exerted on the charge carrier, the only force exerted on it is the magnetic force. Because $\vec{B}$ and $\vec{v}$ are perpendicular, the magnitude of this force is $F^B_{\mathrm{p}} = |q|vB$ (Eq. 27.18). The magnitude of the charge carrier's centripetal acceleration is given by Eq. 11.15, $a_{\mathrm{c}} = v^2/R$, where R is the radius of its circular trajectory. Therefore

$$|q|vB = \frac{mv^2}{R}. \tag{27.22}$$

Solving for R, we obtain for the radius of the trajectory

$$R = \frac{mv}{|q|B}. \tag{27.23}$$

Because the ratio $m/|q|$ is fixed for a given charge carrier, we see that the radius of its trajectory depends on the charge carrier's speed. In a given magnetic field, fast charge carriers move in larger circles than slow ones of the same type. Interestingly, the time interval needed for one full revolution is independent of the carrier's speed. This time interval is equal to the circumference of the trajectory divided by the carrier's speed:

$$T = \frac{2\pi R}{v} = \frac{2\pi}{v}\frac{mv}{|q|B} = \frac{2\pi m}{|q|B}. \quad (27.24)$$

The corresponding angular frequency is

$$\omega = 2\pi f = \frac{2\pi}{T} = \frac{|q|B}{m}. \quad (27.25)$$

This angular frequency is sometimes called the *cyclotron frequency* after a type of particle accelerator, called a *cyclotron*, in which particles are accelerated between successive semicircular trajectories.

27.17 A proton and an electron travel through a region of uniform magnetic field B. If their speeds are the same, what is the ratio R_p/R_e of the radii of their circular paths through the field?

Example 27.4 Mass spectrometer

Figure 27.40 shows a schematic of a device, called a *mass spectrometer*, for determining the mass of ions or other charged particles. The ions that enter the mass spectrometer are first accelerated by an electric field and then deflected by a magnetic field. The mass of the ions is obtained from the position at which they hit a detector after being deflected by the magnetic field.

Figure 27.40 Example 27.4.

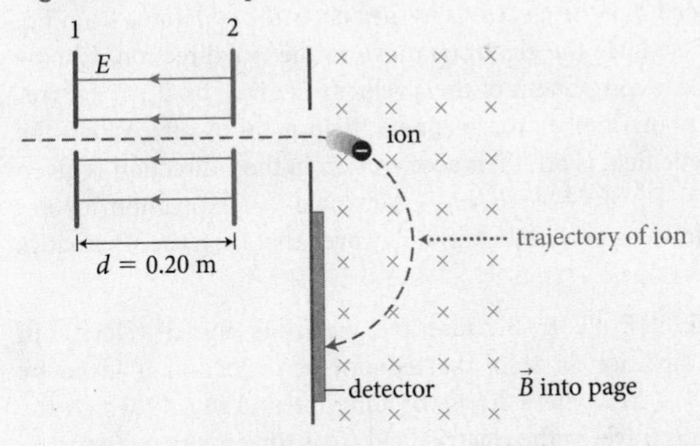

In a certain mass spectrometer, the electric field is caused by a potential difference of 10 kV across a distance $d = 0.20$ m between plates 1 and 2, and the magnitude of the magnetic field is $B = 0.20$ T. An oxygen ion with a charge q of $-2e$ ($1e = 1.6022 \times 10^{-19}$ C) and a mass m of 16 atomic mass units (1 atomic mass unit = 1 u = 1.6605×10^{-27} kg) enters the electric field with negligible initial velocity. At what distance from the point of entry into the magnetic field does the ion hit the detector?

❶ GETTING STARTED From the schematic I see that the ion's motion consists of two parts: a linear motion at constant acceleration in the uniform electric field, followed by circular motion at constant speed in the uniform magnetic field. Because the drawing shows the ion entering the magnetic field perpendicular to the plane of the detector, the ion traces out a half circle before hitting the detector. The distance we must determine is therefore equal to the diameter of the circular trajectory of the ion.

❷ DEVISE PLAN Equation 27.23 gives the radius of the circular trajectory. I am given B, m, and q, but I don't know the speed v at which the ion enters the magnetic field. This speed is equal to the final speed the ion acquires after accelerating in the electric field. I know from Eq. 25.17 how much electrostatic work is done on the ion as it traverses the potential difference between plates 1 and 2 that set up the electric field. According to Eq. 25.17, I have $W_{ion}(1 \rightarrow 2) = -qV_{12}$. This work increases the kinetic energy of the ion. Because the initial kinetic energy is zero, I have $W_{ion}(1 \rightarrow 2) = \Delta K = K_f$. I know q, V_{12}, and m, so I can now calculate v from $K_f = \frac{1}{2}mv^2$ and then use that value in Eq. 27.23 to obtain the answer to this question.

❸ EXECUTE PLAN The electrostatic work done on the ion is

$$W_{ion}(1 \rightarrow 2) = -qV_{12} = 2eV_{12}.$$

From the expression for kinetic energy, I obtain for the ion's speed as it enters the magnetic field

$$v = \sqrt{\frac{K_f}{\frac{1}{2}m}} = \sqrt{\frac{4eV_{12}}{m}}.$$

Substituting this speed and the magnitude of the charge q on the oxygen ion into Eq 27.23, I obtain for the diameter of the ion's circular trajectory

$$2R = \frac{2mv}{|q|B} = \frac{2m}{2eB}\sqrt{\frac{4eV_{12}}{m}} = \frac{2}{B}\sqrt{\frac{mV_{12}}{e}}$$

$$= \frac{2}{0.20\text{ T}}\sqrt{\frac{16(1.66 \times 10^{-27}\text{ kg})(1.0 \times 10^4\text{ V})}{1.60 \times 10^{-19}\text{ C}}}$$

$$= 0.41\text{ m.} ✔$$

❹ EVALUATE RESULT To be measurable in a laboratory, the distance from the point of entry at which the ions hit the detector must be neither too small nor too great. Given these constraints, the distance I obtained appears reasonable.

Figure 27.41 Charged particles whose speed satisfies Eq. 27.26 move in a straight line through magnetic and electric fields that are oriented perpendicular to each other.

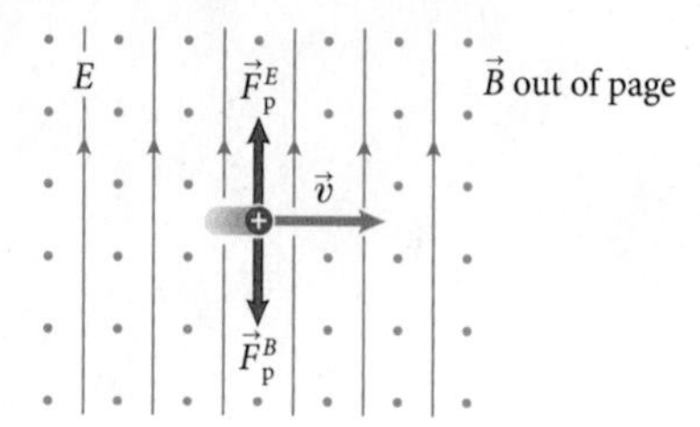

Let us next consider the trajectory of a charged particle in combined uniform electric and magnetic fields. If the charged particle has a velocity component perpendicular to the magnetic field, it is subject to a magnetic force in addition to an electric force. With an appropriate choice of fields, it is possible for the two forces to add up to zero, leaving the charged particle undisturbed. This arrangement is illustrated in **Figure 27.41**: A positive charged particle moves at right angles to both $\vec{E}$ and $\vec{B}$, which are perpendicular to each other. If the two forces are of equal magnitude, we have $F^B_p = F^E_p$, and so from Eqs. 27.18 and 23.6 $|q|vB = |q|E$ or

$$v = \frac{E}{B} \quad \text{(electric and magnetic force cancel).} \tag{27.26}$$

Thus, if we adjust the magnitudes of the electric and magnetic fields so that one charged particle goes through undeflected, then any other charged particle moving at the same velocity also passes through the fields undeflected, *regardless of its mass or charge*. Consequently, the setup shown in Figure 27.41 serves as a *velocity selector*: If charge carriers traveling at different speeds are injected from the left into a region where an electric field and a magnetic field are perpendicular to each other, only those whose speed satisfies Eq. 27.26 make it through without being deviated. Another important application of the cancellation of electric and magnetic forces is discussed in the box "The Hall effect" on page 731.

27.18 In Figure 27.41, do $\vec{F}^B_p$ and $\vec{F}^E_p$ still cancel if the charged particle (*a*) carries a negative charge, (*b*) travels in the opposite direction, (*c*) travels at a slight angle to the two fields?

Example 27.5 The mass of the electron

Figure 27.42 shows schematically part of the apparatus used in 1897 by J. J. Thomson to determine the charge-to-mass ratio of the electron. A beam of electrons, all moving at the same speed v, enters a region of electric and/or magnetic fields. When an electric field of magnitude 1.0 kV/m and a magnetic field of magnitude 1.2×10^{-4} T are turned on, the electrons go through the device undeflected. When the magnetic field is turned off, the electrons are deflected by 3.2 mm in the negative y direction after traveling the length $\ell = 0.050$ m of the apparatus. Given that the charge of the electron is $-e = -1.60 \times 10^{-19}$ C, what is the mass of each electron?

Figure 27.42 Example 27.5.

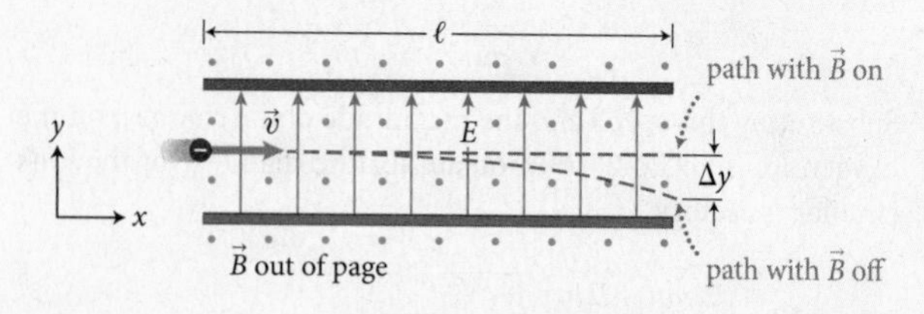

❶ **GETTING STARTED** When the magnetic field is turned on, the electrons travel in a straight line, which means that the forces exerted by the electric and magnetic fields on the electron cancel and the electrons travel at constant speed in the $+x$ direction. When the magnetic field is turned off, the electrons undergo a constant acceleration in the $-y$ direction due to the electric field while continuing to travel at constant speed in the $+x$ direction. I know from the expression $\vec{a} = (q/m)\vec{E}$ we derived in Section 23.4 that the acceleration of the electrons in the electric field depends on their charge-to-mass ratio. If I can determine this ratio from the electrons' trajectory, I can obtain their mass.

❷ **DEVISE PLAN** Because I know the magnitudes of the electric and magnetic fields that yield a straight trajectory, I can obtain the speed v of the electrons as they enter the apparatus from Eq. 27.26. Because the electrons move in the $+x$ direction, I know that the x component of their velocity is given by $v_x = +v$ (regardless of whether the magnetic field is on or off). When the magnetic field is off, their acceleration in the y direction is given by $a_y = F^E_y/m_e = -eE/m_e$. I can solve this expression for m_e, but I don't know a_y. I do know, however, that I can use kinematics to determine a_y.

❸ **EXECUTE PLAN** Because the electron's initial velocity in the y direction is zero, the amount of deflection is given by $\Delta y = \frac{1}{2}a_y\Delta t^2$, where Δt is the time interval during which the electrons travel in the electric field. This time interval is equal to $\Delta t = \ell/v_x = \ell/v$, so that

$$a_y = \frac{2\Delta y}{(\Delta t)^2} = \frac{2\Delta y}{(\ell/v)^2} = \frac{2\Delta y\, v^2}{\ell^2} = \frac{2\Delta y(E/B)^2}{\ell^2}, \tag{1}$$

where I have substituted Eq. 27.26 for the speed v of the electrons. Substituting $a_y = -eE/m_e$ into Eq. 1 and solving for m_e, I get

$$m_e = \frac{-eE}{a_y} = \frac{-e\ell^2B^2}{2\Delta yE}$$

$$= \frac{(-1.60 \times 10^{-19}\ \text{C})(0.050\ \text{m})^2(1.2 \times 10^{-4}\ \text{T})^2}{2(-3.2 \times 10^{-3}\ \text{m})(1.0 \times 10^3\ \text{V/m})}$$

$$= 9.0 \times 10^{-31}\ \text{kg.} ✔$$

❹ **EVALUATE RESULT** The value I obtain is close to the published value of the electron mass ($m_e = 9.10938291 \times 10^{-31}$ kg), giving me confidence in my calculation.

QUANTITATIVE TOOLS

The Hall effect

The canceling of electric and magnetic forces exerted on a charge carrier makes it possible to determine whether the mobile charge carriers in a conductor are positively or negatively charged. This determination is done using a phenomenon called the *Hall effect*.

Consider the rectangular conducting strip carrying an upward current illustrated in **Figure 27.43** (that is, the bottom of the strip is at a higher potential than the top). The strip is placed in a magnetic field directed into the page. If the current is caused by positive charge carriers (Figure 27.43*a*), the carriers move upward and so, according to the right-hand rule, the magnetic force acting on them is to the left. This force deflects the carriers toward the left side of the strip, where they pile up. This accumulation of positive charge carriers on the left causes an electric field to the right across the strip that exerts on the carriers an electric force that pulls them to the right. As more carriers accumulate on the left, the magnitude of this electric force increases until it becomes equal in magnitude to the magnetic force. Once the two forces are equal in magnitude, the carriers are no longer deflected and so no more carriers accumulate on the left. Any subsequent charge carriers travel in a straight line.

Figure 27.43

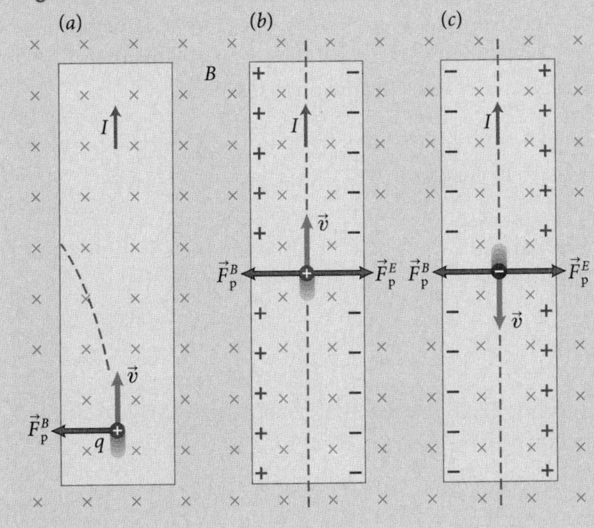

The accumulation of charge carriers can be determined by measuring the potential difference between the left (L) and right (R) sides of the strip. If, as illustrated in Figure 27.43*b*, positive carriers accumulate on the left, this potential difference is positive:

$$V_{RL} = V_L - V_R > 0.$$

As illustrated in Figure 27.43*c*, however, if the mobile charge carriers causing the current are negatively charged, V_{RL} is *negative*. This is so because an upward current means that negative charge carriers move downward and, as you can verify using the right-hand rule, the magnetic force exerted on these carriers is also to the left. Thus, regardless of the sign of their charge, the mobile charge carriers pile up on the left. If they are positively charged, V_{RL} is positive; if they are negatively charged, V_{RL} is negative. Experiments on strips of common metals always yield a negative V_{RL}, showing that the mobile charge carriers in these metals are negatively charged, as we stated earlier (see Section 22.3 and Checkpoint 22.15).

27.19 (*a*) Express the magnitude of the electric field inside the strip in Figure 27.43 in terms of the width w of the strip and the potential difference V_{RL}. (*b*) Given the magnitude B of the magnetic field, what is the speed at which the charge carriers travel? (*c*) Show how this information, together with Eq. 27.16, can be used to determine the number density n of the charge carriers.

27.8 Magnetism and electricity unified

In this section we quantitatively examine the situation discussed in the latter part of Section 27.4—the interaction between a current-carrying wire and a positively charged particle moving parallel to the wire. For simplicity, we again let the charged particle move at the same speed v as the electrons in the wire.* In the Earth reference frame in Figure 27.27 the wire is electrically neutral: The linear charge density of the fixed positively charged ions $\lambda_{proper} > 0$ is equal in magnitude to that of the electrons, $-\lambda_{proper}$. The interaction between the wire and the charged particle is thus purely magnetic.

As we saw in Section 27.4, the wire appears to be positively charged to observer M, who moves along with the electrons. According to M, the average distance between the positively charged ions has decreased by the factor

$$\gamma = \frac{1}{\sqrt{1 - v^2/c_0^2}} \quad (27.27)$$

*In the more general case where the charged particle does not move at the same speed as the electrons in the wire, the algebra becomes more complicated but the conclusions remain the same.

QUANTITATIVE TOOLS

Figure 27.44 A positively charged particle moving parallel to a current-carrying wire, as seen by two observers in motion relative to each other.

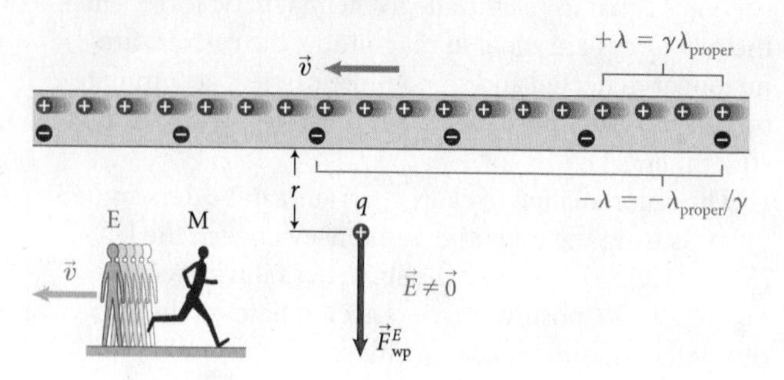

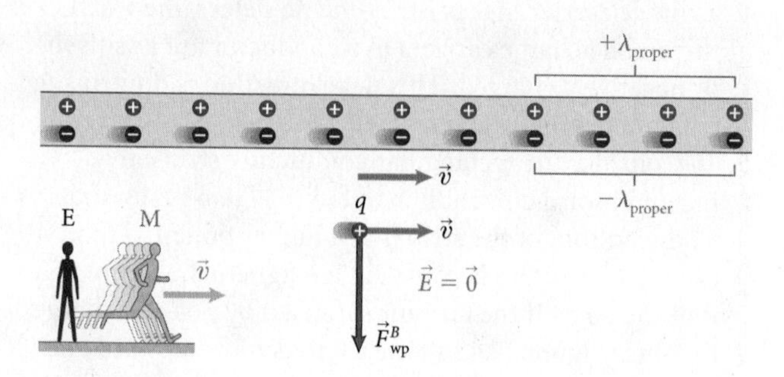

due to length contraction (Eqs. 14.28 and 14.6; $\gamma > 1$). Thus the charge density of the ions is increased and has a magnitude $\lambda_{\text{M ions}} = \gamma\lambda_{\text{proper}}$ (**Figure 27.44a**). The electrons, on the other hand, are at rest relative to M and moving relative to E. Therefore the average distance between the electrons according to E must be decreased by the factor γ in Eq. 27.27 relative to the distance observed by M. Because to observer E in the Earth reference frame the wire is electrically neutral, the charge density of the electrons in that reference frame must be the negative of that of the ions: $-\lambda_{\text{proper}}$. According to M, therefore, the charge density must be $\lambda_{\text{M electrons}} = -\lambda_{\text{proper}}/\gamma$. Consequently, the combined charge density of the ions and the electrons in the moving reference frame of observer M is

$$\lambda_{\text{M}} = \lambda_{\text{M ions}} + \lambda_{\text{M electrons}} = \gamma\lambda_{\text{proper}} - \frac{\lambda_{\text{proper}}}{\gamma} = \lambda_{\text{proper}}\left(\gamma - \frac{1}{\gamma}\right). \tag{27.28}$$

Because $\gamma > 1$, the term in parentheses is positive, making the combined charge density positive. Using Eq. 14.9 we can rewrite the term in parentheses as

$$\gamma - \frac{1}{\gamma} = \gamma\left(1 - \frac{1}{\gamma^2}\right) = \gamma\frac{v^2}{c_0^2}. \tag{27.29}$$

Substituting Eq. 27.29 into Eq. 27.28, we see that according to M the wire has a (nonzero) charge density equal to

$$\lambda_{\text{M}} = \lambda_{\text{proper}}\gamma\frac{v^2}{c_0^2}. \tag{27.30}$$

Using the expression $E = 2k\lambda/r$ for the magnitude of the electric field of a charged wire (see Exercise 24.7), we obtain for the magnitude of the electric field at a distance r from the wire according to M

$$E_{\text{M}} = \frac{2k\lambda_{\text{M}}}{r} = \frac{2k\lambda_{\text{proper}}\gamma v^2}{rc_0^2}. \tag{27.31}$$

(Because the distance to the wire is perpendicular to the direction of motion, both observers measure the same distance r and so we need no subscript on r.)

According to observer M, the electric force exerted by the wire on the charged particle therefore has a magnitude

$$F^E_{\text{Mwp}} = |q|E_{\text{M}} = |q|\frac{2k\lambda_{\text{proper}}\gamma v^2}{rc_0^2} \tag{27.32}$$

and points perpendicular to and away from the wire.

Now we return to the Earth reference frame, as illustrated in Figure 27.44*b*. In this reference frame, the combined charge density of the ions and electrons is

$$\lambda_{\text{ions}} + \lambda_{\text{electrons}} = \lambda_{\text{proper}} - \lambda_{\text{proper}} = 0, \tag{27.33}$$

and so the electric field outside the wire is zero. Consequently, the force between the wire and the charged particle cannot be due to an electric interaction. We know experimentally, however, that according to observer E the current-carrying wire exerts a *magnetic* force on the particle. The direction of this force is also perpendicular to and away from the wire. Is this magnetic force the same as the electric force seen by M?

To answer this question we first need to determine the relationship between the forces measured by two observers in motion relative to each other. To determine this relationship, we shall use the definition of force given in Eq. 8.4, $\sum \vec{F} \equiv d\vec{p}/dt$, where $\sum \vec{F} = \vec{F}_{wp}$. The vector $d\vec{p}$ points in the same direction as the force—perpendicular to the relative velocity of the two frames—and so must be the same for both observers: $d\vec{p} = d\vec{p}_M$. Because of time dilation, however, we have from Eq. 14.13, $dt = \gamma dt_M$, where dt is the infinitesimal time interval measured by observer E and dt_M is the corresponding time interval measured by observer M (because M moves along with the particle, this is a proper time interval). The magnitude of the force exerted by the wire on the particle measured by observer E is thus

$$F_{wp} = \frac{dp}{dt} = \frac{dp_M}{\gamma dt_M} = \frac{F_{Mwp}}{\gamma}, \tag{27.34}$$

where F_{Mwp} is the magnitude of the force exerted by the wire on the particle measured by observer M. To observer M this force is electric in nature and its magnitude is given by Eq. 27.32. Substituting Eq. 27.32 on the right in Eq. 27.34 thus yields

$$F_{wp} = |q| \frac{2k\lambda_{proper}v^2}{rc_0^2}. \tag{27.35}$$

The quantity $\lambda_{proper}v$ is just the current I because if the electrons advance by a distance d in a time interval Δt, a quantity of charge $|q| = \lambda_{proper}d$ flows through the wire in that time interval. So the rate at which the charge carriers flow is

$$I \equiv \left|\frac{q}{\Delta t}\right| = \lambda_{proper}\frac{d}{\Delta t} = \lambda_{proper}v. \tag{27.36}$$

This result means that we can write Eq. 27.35 as

$$F_{wp} = |q|v\frac{2kI}{rc_0^2}. \tag{27.37}$$

Equation 27.37 gives the magnitude of the force exerted by the current-carrying wire on the particle measured by observer E. The magnitude depends on the current I through the wire, on the speed v and charge q of the particle, and on the distance r between the wire and the particle. Observer E interprets this force as a magnetic force due to the magnetic field caused by the current through the wire. From Eq. 27.18 we see that the magnitude of the magnetic force exerted by the wire on the moving particle is $F^B_{wp} = |q|vB$ (because $\vec{v}$ and $\vec{B}$ are at right angles, $\sin\theta = 1$). Substituting this expression into Eq. 27.37 and dividing both sides by qv, we see that the magnitude of the magnetic field according to observer E is

$$B = \frac{2kI}{rc_0^2} = \frac{2k}{c_0^2}\frac{I}{r}. \tag{27.38}$$

In words, the magnitude B of the magnetic field is proportional to the current I and inversely proportional to the distance r to the wire, with the proportionality constant being $2k/c_0^2$. As we shall see in the next chapter, experiments confirm this dependence. Here we see that Coulomb's law, together with special relativity, *requires* a magnetic field of this form. In other words, electricity and magnetism are two aspects of the same interaction, not two different interactions.

27.20 Explain why the $1/r$ dependence expressed in Eq. 27.38 is consistent with the symmetry of the wire causing the magnetic field.

Chapter Glossary

SI units of physical quantities are given in parentheses.

Ampere (A) The SI base unit of current, defined as the constant current through two straight parallel thin wires of infinite length placed 1 meter apart that produces a force between the wires of magnitude 2×10^{-7} N for each meter length of the wires. The coulomb and the ampere are related by

$$1\ \text{C} \equiv 1\ \text{A} \cdot \text{s}. \qquad (27.3)$$

Current I (A) A scalar that gives the rate at which charge carriers cross a section of a conductor in a given direction:

$$I \equiv \frac{dq}{dt}. \qquad (27.2)$$

The *direction of current* through a conductor is the direction in which the potential decreases. In this direction I is positive.

Electromagnetic force $\vec{F}^{EB}$ (N) The force exerted on a moving charged particle in the presence of both an electric field and a magnetic field:

$$\vec{F}^{EB}_{\text{p}} = q(\vec{E} + \vec{v} \times \vec{B}), \qquad (27.20)$$

Gauss's law for magnetism The magnetic flux through a closed surface is always zero:

$$\Phi_B = \oint \vec{B} \cdot d\vec{A} = 0. \qquad (27.11)$$

Magnet Any object that attracts pieces of iron, such as iron filings or paper clips.

Magnetic dipole An object with a pair of opposite magnetic poles separated by a small distance.

Magnetic field $\vec{B}$ (T) A vector that provides a measure of the magnetic interaction of objects. The magnitude of a uniform magnetic field can be determined by measuring the force exerted by that magnetic field on a straight wire of length ℓ carrying a current I:

$$B \equiv \frac{F^B_{\text{w, max}}}{|I|\ell} \quad \text{(straight wire perpendicular to uniform magnetic field).} \qquad (27.5)$$

The direction of the magnetic field at a certain location is given by the direction in which a compass needle points at that location.

Magnetic field line A representation of magnetic fields using lines of which the tangent at every position gives the direction of the magnetic field at that position. Near a magnet, magnetic field lines point away from north poles and toward south poles.

Magnetic flux Φ_B (Wb) A scalar that provides a quantitative measure of the number of magnetic field lines passing through an area. The magnetic flux through a surface is given by the surface integral

$$\Phi_B \equiv \int \vec{B} \cdot d\vec{A}. \qquad (27.10)$$

Magnetic force $\vec{F}^B$ (N) The force exerted by magnets, current-carrying wires, and moving charged particles on each other. The magnetic force exerted on a straight wire of length ℓ carrying a current I and placed in a uniform magnetic field B is

$$\vec{F}^B_{\text{w}} = I\vec{\ell} \times \vec{B}, \qquad (27.8)$$

where $I\vec{\ell}$ is a vector whose magnitude is given by the product of the magnitude of the current and the length ℓ of the wire, and whose direction is given by the direction of current through the wire. The magnetic force exerted on a particle carrying a charge q and moving at a velocity $\vec{v}$ through a magnetic field $\vec{B}$ is

$$\vec{F}^B_{\text{p}} = q\vec{v} \times \vec{B}. \qquad (27.19)$$

Magnetic interaction The long-range interaction between magnets and/or current-carrying wires that are at rest relative to the observer.

Magnetic polarization The magnetic state induced in a piece of magnetic material because of the presence of a magnet.

Tesla (T) The SI derived unit of magnetic field:

$$1\ \text{T} \equiv 1\ \text{N}/(\text{A} \cdot \text{m}) = 1\ \text{kg}/(\text{s}^2 \cdot \text{A}). \qquad (27.6)$$

Weber (Wb) The SI derived unit of magnetic flux:

$$1\ \text{Wb} \equiv 1\ \text{T} \cdot \text{m}^2 = 1\ \text{m}^2 \cdot \text{kg}/(\text{s}^2 \cdot \text{A}).$$

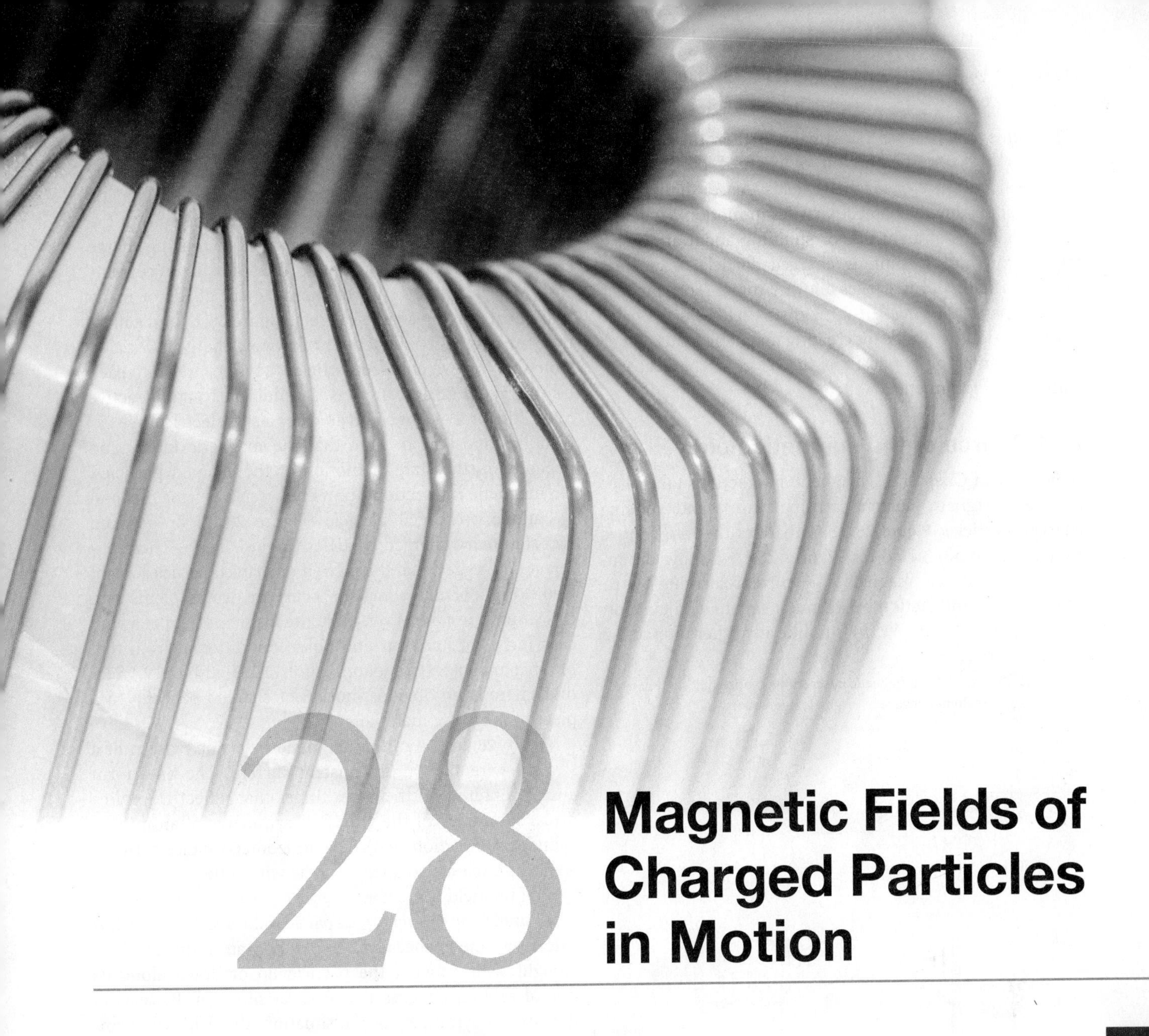

28 Magnetic Fields of Charged Particles in Motion

CONCEPTS

QUANTITATIVE TOOLS

In this chapter we investigate further the relationship between the motion of charged particles and the occurrence of magnetic fields. As we shall see, *all* magnetism is due to charged particles in motion—whether moving along a straight line or spinning about an axis. It takes a moving or spinning charged particle to create a magnetic field, and it a takes another moving or spinning charged particle to "feel" that magnetic field. We shall also discuss various methods for creating magnetic fields, which have wide-ranging applications in electromechanical machines and instruments.

28.1 Source of the magnetic field

As we saw in Chapter 27, magnetic interactions take place between magnets, current-carrying wires, and moving charged particles. **Figure 28.1** summarizes the interactions we have encountered so far. Figures 28.1*a–c* show the interactions between magnets and current-carrying wires. The sideways interaction between a magnet and a current-carrying wire (Figure 28.1*b*) is unlike any other interaction we have encountered. The forces between the wire and the magnet are not central; they do not point directly from one object to the other. As we saw in Section 27.7, the magnetic force exerted on a current-carrying wire is the sum of the magnetic forces exerted on many individual moving charge carriers. Similarly the magnetic field due to a current-carrying wire is the sum of the magnetic fields of many individual moving charge carriers. Figure 28.1*d* and *e* illustrate the magnetic interactions of moving charged particles. Note that for two charged particles moving parallel to each other (Figure 28.1*e*), there is, in addition to an attractive magnetic force, a (much greater) repulsive electric force.

It is important to note that the magnetic interaction depends on the state of motion of the charged particles. No magnetic interaction occurs between a bar magnet and a stationary charged particle (Figure 28.1*f*) or between two stationary charged particles (Figure 28.1*g*). These observations suggest that the motion of charged particles might be the origin of *all* magnetism. There are two problems with this assumption, however. First, the magnetic field of a wire carrying a constant current looks very different from that of a bar magnet. (Compare Figures 27.13 and 27.19.) Second, there is no obvious motion of charged particles in a piece of magnetic material.

Figure 28.2*a* shows the magnetic field line pattern of a straight wire carrying a constant current. The lines form circles centered on the wire, circles that reflect the cylindrical symmetry of the wire (the symmetry of an infinite cylinder, see Section 24.4). The horizontal distance between adjacent circles is smaller near the wire, where the magnitude of the field is greater.

A single moving charged particle does not have cylindrical symmetry because, unlike for an infinitely long straight wire, moving the particle up or down along its line of motion changes the physical situation. Because of the *circular* symmetry of the situation, the field still forms circles around the line of motion, but the magnitude of the field at a fixed distance from the particle's line of motion

Figure 28.1 Summary of magnetic interactions. Notice that stationary charged particles do not engage in magnetic interactions.

(*a*)

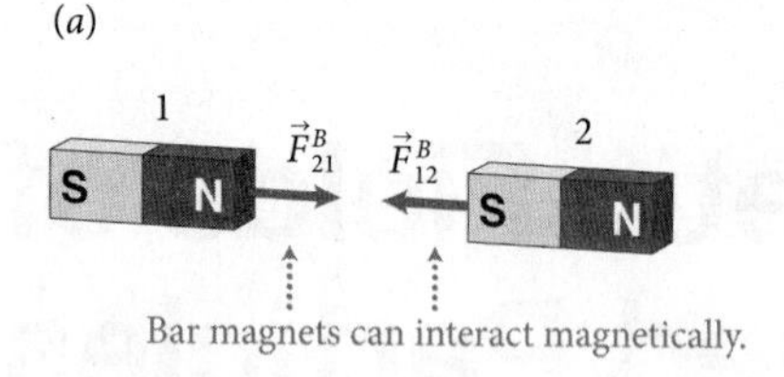

(*b*) (*c*)

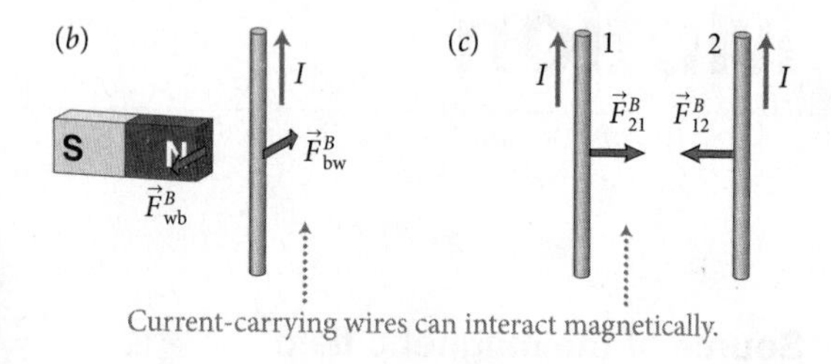

(*d*) (*e*)

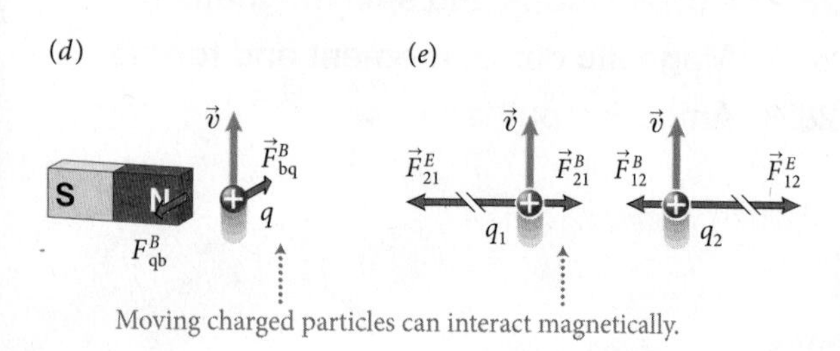

(*f*) (*g*)

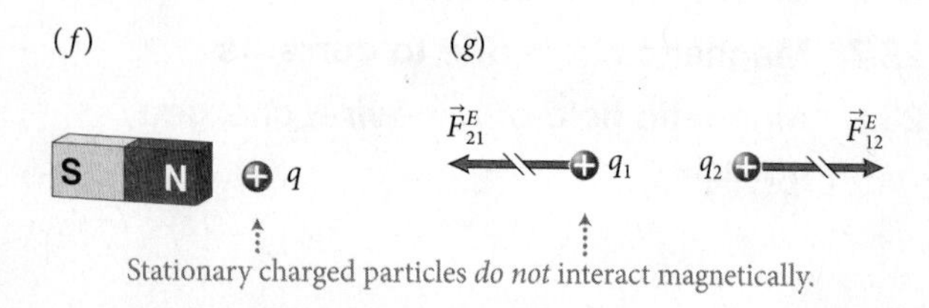

Figure 28.2 Comparing the magnetic fields of a current-carrying wire and a moving charged particle.

(*a*) Magnetic field of a wire carrying a constant current

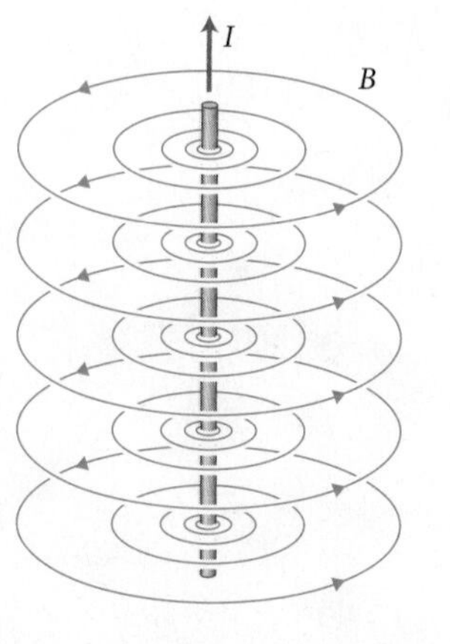

(*b*) Magnetic field of a moving charged particle

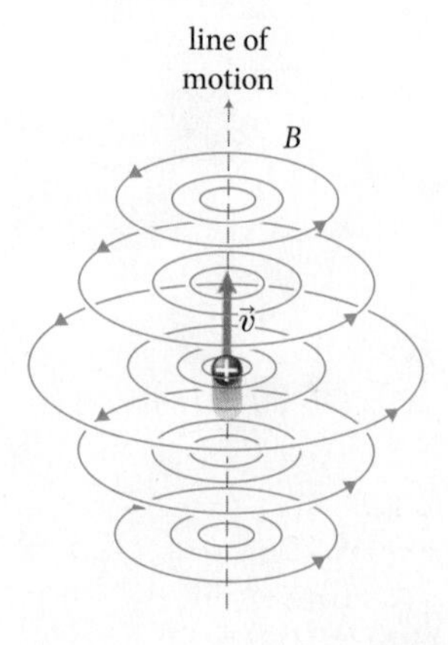

CONCEPTS

decreases if one moves away from the particle (Figure 28.2*b*). It is not at all obvious how the field pattern shown in Figure 28.2*b* could give rise to the magnetic field of a bar magnet; there are certainly no poles in the magnetic field of the moving charged particle.

28.1 Make a sketch showing the directions of the magnetic forces exerted on each other by (*a*) an electron moving in the same direction as the current through a wire, (*b*) a moving charged particle and a stationary charged particle, and (*c*) two current-carrying wires at right angle to each other as illustrated in **Figure 28.3**. (Hint: Determine the forces exerted at points P_1 through P_5).

Figure 28.3 Checkpoint 28.1*c*.

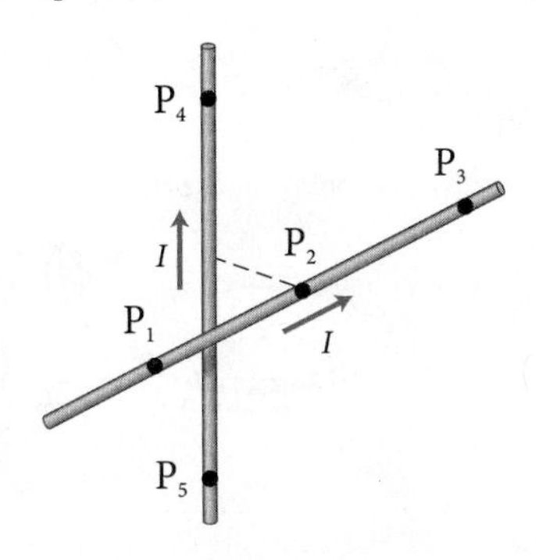

28.2 Current loops and spin magnetism

The circular pattern of magnetic field lines around a wire carrying a constant current suggests a method for generating a strong magnetic field: If a wire carrying a constant current is bent into a loop as shown in **Figure 28.4**, all the magnetic field lines inside the loop point in the same direction, reinforcing one another.*

Figure 28.4 The magnetic field of a wire loop carrying a constant current. The magnetic fields from all parts of the loop reinforce one another in the center of the loop.

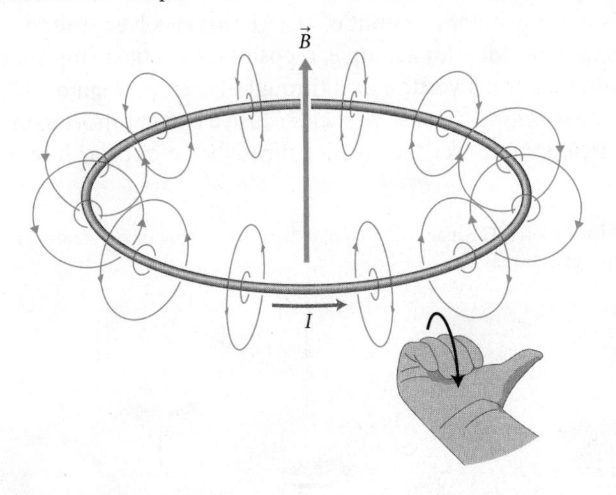

*For now, we'll ignore how to make charge carriers flow through such a loop. In Section 28.6 we'll discuss physical arrangements that accomplish the situation illustrated in Figure 28.4.

What does the magnetic field of such a current-carrying loop, called a **current loop** for short, look like? To answer this question, we treat the current loop as a collection of small segments of a current-carrying wire and determine the direction of the magnetic field at various points around the loop. As we did in Chapter 27, we shall assume all currents to be constant in the remainder of this chapter.

We begin by considering the magnetic field due to a small segment of the current loop at a point on the central axis that passes perpendicularly through the face of the loop (point A in **Figure 28.5** on the next page). Segment 1 carries a current that points into the page. The magnetic field lines of this segment are concentric circles centered on the segment. Using the right-hand current rule, we see that the magnetic field curls clockwise in the plane of the drawing, and so the magnetic field due to segment 1 at A points up and to the right. Figure 28.5*b* shows a magnetic field line through A due to segment 2. Because the current through segment 2 points out of the page, this field line curls counterclockwise, and so the magnetic field due to segment 2 at A points up and to the left.

Figure 28.5*c* shows the contributions from segments 1 and 2 together; because their horizontal components cancel, the vector sum of $\vec{B}_1$ and $\vec{B}_2$ points vertically up. The same arguments can be applied to any other pair of segments lying on opposite sides of the current loop. Therefore the magnetic field due to the entire current loop points vertically up at A.

Now consider point C at the center of the current loop. As illustrated in Figure 28.5*d*, the magnetic fields due to segments 1 and 2 point straight up there, too. The same is true for all other segments, and so the field at point C also points straight up. We can repeat the procedure for point D below the current loop, and as shown in Figure 28.5*e*, the magnetic field there also points up.

To determine the direction of the magnetic field at a point outside the current loop, consider point G in Figure 28.5*f*. Using the right-hand current rule, you can verify that the field due to segment 1 points vertically down and the field due to segment 2 points vertically up. Because G is closer to 1 than it is to 2, the magnetic field due to 1 is stronger, and so the sum of the two fields points down.

28.2 What is the direction of the magnetic field at a point vertically (*a*) above and (*b*) below segment 1 in Figure 28.5?

The complete magnetic field pattern of the current loop, obtained by determining the magnetic fields for many more points, is shown in **Figure 28.6*a*** on the next page. Close to the wire, the field lines are circular, but as you move farther away from the wire, the circles get squashed inside the loop and stretched outside due to the contributions of other parts of the loop to the magnetic field.

Figure 28.5 Mapping the magnetic field of a current loop. The magnetic field contributions from (*a*) segment 1 and (*b*) segment 2 at A (*c*) add up to a vertical field. Magnetic fields at (*d*) point C at the center of the ring, (*e*) point D below the ring, and (*f*) point G to the right of the ring. Note that in all cases the magnetic field of each segment is perpendicular to the line connecting that segment to the point at which we are determining the field.

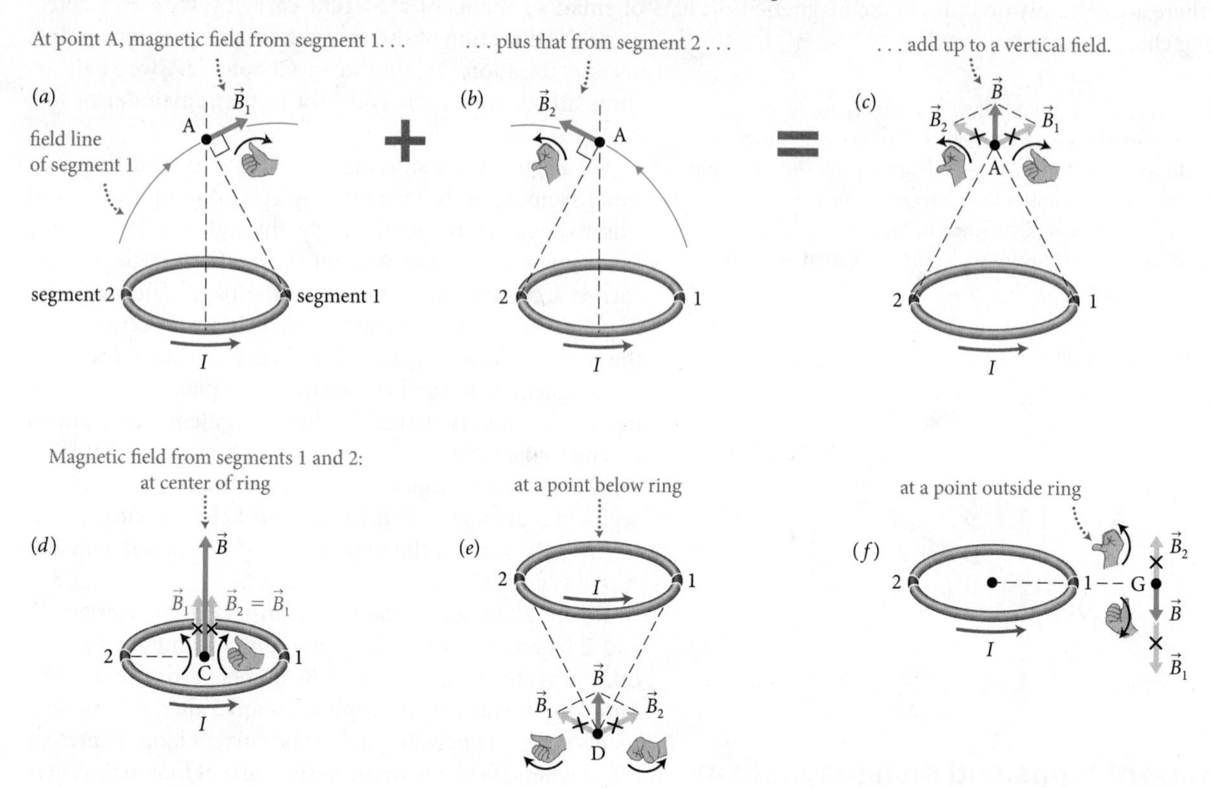

At point A, magnetic field from segment 1...
... plus that from segment 2...
... add up to a vertical field.
(a)
(b)
(c)
field line
of segment 1
segment 2
segment 1
A
I
Magnetic field from segments 1 and 2:
at center of ring
at a point below ring
at a point outside ring
(d)
(e)
(f)
C
D
G

As you may have noticed, the field line pattern of a current loop resembles that of a bar magnet (Figure 28.6*b*). Indeed, if you shrink the size of both the current loop and the bar magnet, their magnetic field patterns become identical (Figure 28.6*c*). The magnetic field pattern in Figure 28.6*c* is that of an infinitesimally small magnetic dipole.

Figure 28.6 The magnetic field of a current loop (*a*) resembles that of a dipole (*b*, *c*).

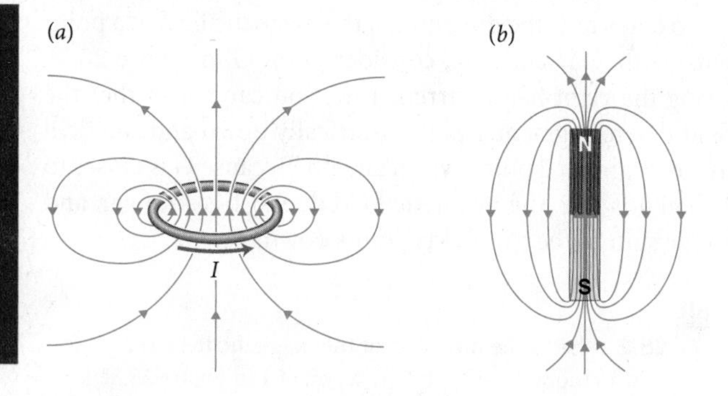

(a)
(b)
N
S
I
Magnetic field of current loop...
... resembles that of bar magnet.

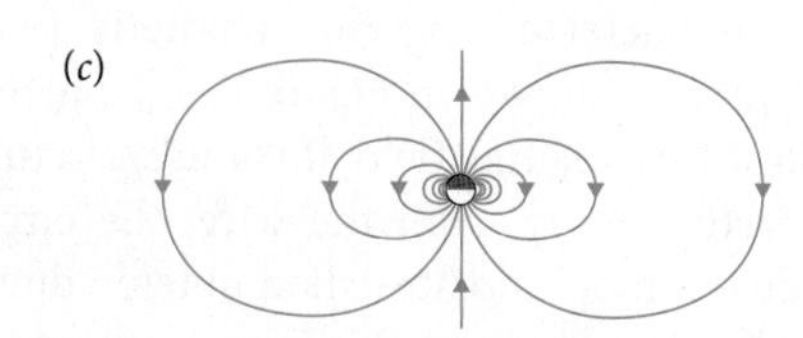

(c)
Shrinking either one yields field of infinitesimally small magnetic dipole.

Given that a current loop produces a magnetic field similar to that of a bar magnet, is the magnetic field of a bar magnet then perhaps due to tiny current loops inside the magnet? More precisely, are elementary magnets (see Section 27.1) simply tiny current loops?

The connection between current loops and elementary magnets becomes clearer once we realize that a current loop is nothing but an amount of charge that revolves around an axis. Consider, for example, a positively charged ring spinning around a vertical axis through its center (**Figure 28.7**). The spinning charged particles cause a current moving in a circle exactly like the current through the circular loop in

Figure 28.7 The magnetic field of a charged spinning ring is identical to that of a current loop.

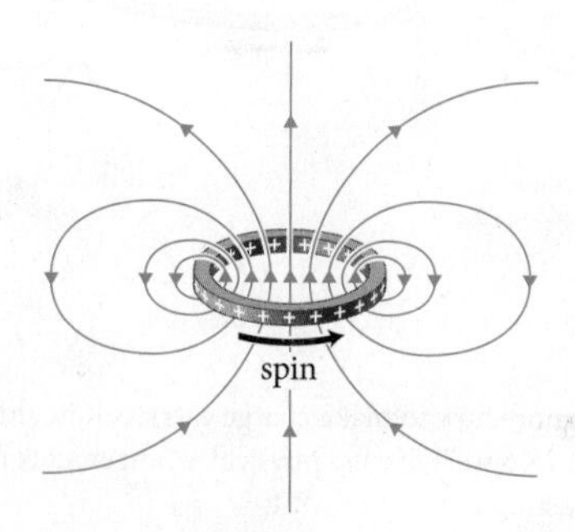

spin

Figure 28.5. If we let the radius of the ring approach zero, the ring becomes a spinning charged particle, and its magnetic field pattern approaches that of the magnetic dipole illustrated in Figure 28.6*c*. This surprising result tells us:

A spinning charged particle has a magnetic field identical to that of an infinitesimally small magnetic dipole.

Experiments show that most elementary particles, such as electrons and protons, possess an intrinsic angular momentum—as if they permanently spin around—and such spinning motion indeed would produce a magnetic field of the form shown in Figure 28.6*c*. Because of the combined intrinsic angular momentum of these elementary particles inside atoms, certain atoms have a magnetic field, causing them to be the elementary magnets we discussed earlier. The reason we cannot separate north and south magnetic poles is therefore a direct consequence of the fact that the magnetic field of a particle with intrinsic angular momentum is that of an infinitesimally small magnetic dipole.

28.3 Suppose a negatively charged ring is placed directly above the positively charged ring in Figure 28.7. If both rings spin in the same direction, is the magnetic interaction between them attractive or repulsive?

28.3 Magnetic dipole moment and torque

To specify the orientation of a magnetic dipole we introduce the **magnetic dipole moment.** This vector, represented by the Greek letter μ (mu), is defined to point, like a compass needle, along the direction of the magnetic field through the center of the dipole (**Figure 28.8**). For a bar magnet the magnetic dipole moment points from the south pole to the north pole. To determine the direction of $\vec{\mu}$ for a current loop, you can use a right-hand rule: When you curl the fingers of your right hand along the direction of the current, your thumb points in the direction of $\vec{\mu}$.

Figure 28.8 The magnetic dipole moment points in the direction of the magnetic field through the center of the dipole.

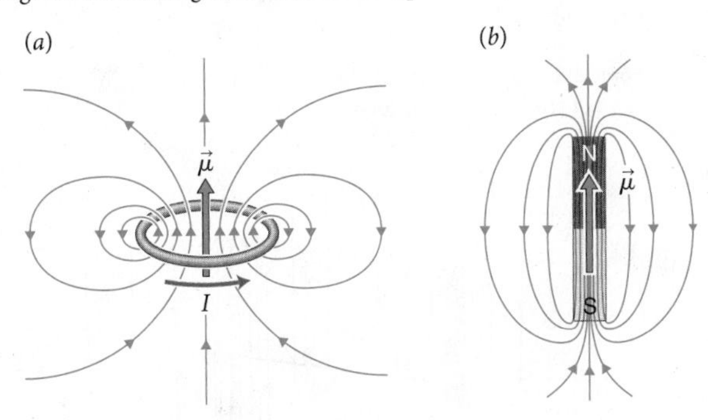
(a)
$\vec{\mu}$
I
(b)
N
$\vec{\mu}$
S

We now have three right-hand rules in magnetism, illustrated in **Figure 28.9** and summarized in **Table 28.1** together with the right-hand vector product rule. What helps distinguish these right-hand rules from one another is looking at which quantity curls: For a current-carrying wire, it is the magnetic field that curls, whereas for a current loop, it is the current that curls. The only additional thing you need to remember is that in the case of the magnetic force exerted on a current-carrying wire, the fingers are associated with the curl from the direction of the current to the direction of the magnetic field. Note, also, that the order of application

Table 28.1 Right-hand rules

Right-hand rule	thumb points along	fingers curl
vector product	$\vec{C} = \vec{A} \times \vec{B}$	from $\vec{A}$ to $\vec{B}$
current rule	current	along *B*-field
force rule	magnetic force	from current to *B*-field
dipole rule	$\vec{\mu}$ (parallel to $\vec{B}$)	along current loop

Figure 28.9 Right-hand rules in magnetism.

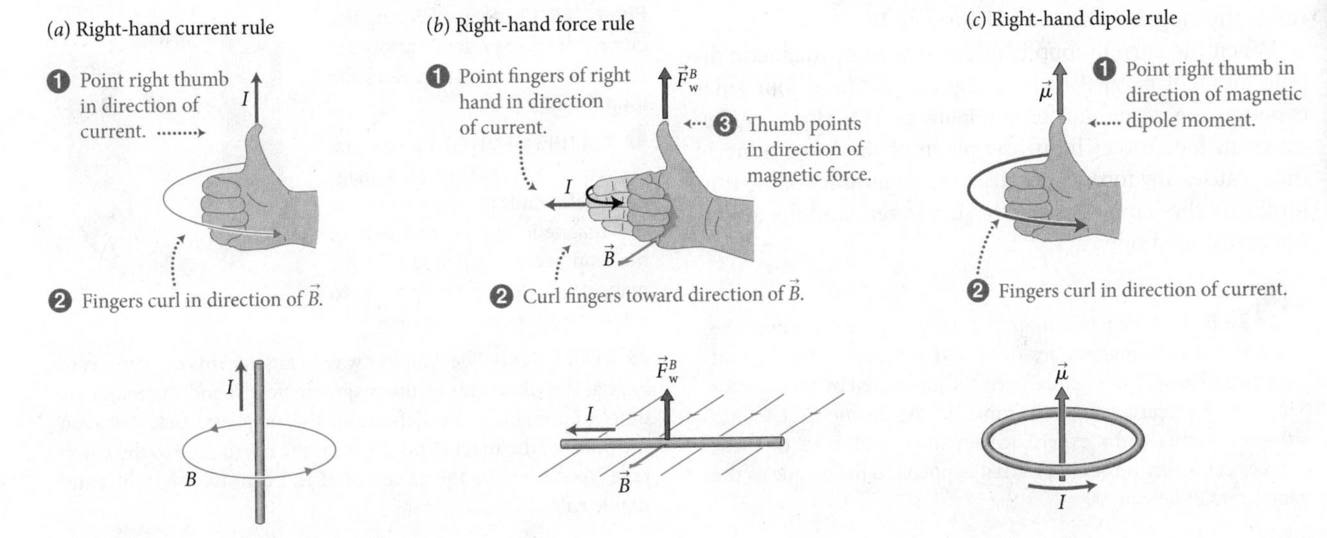
(a) Right-hand current rule
1 Point right thumb in direction of current.
I
2 Fingers curl in direction of $\vec{B}$.
I
B
(b) Right-hand force rule
1 Point fingers of right hand in direction of current.
$\vec{F}^B_w$
3 Thumb points in direction of magnetic force.
I
$\vec{B}$
2 Curl fingers toward direction of $\vec{B}$.
$\vec{F}^B_w$
I
$\vec{B}$
(c) Right-hand dipole rule
1 Point right thumb in direction of magnetic dipole moment.
$\vec{\mu}$
2 Fingers curl in direction of current.
$\vec{\mu}$
I

CONCEPTS

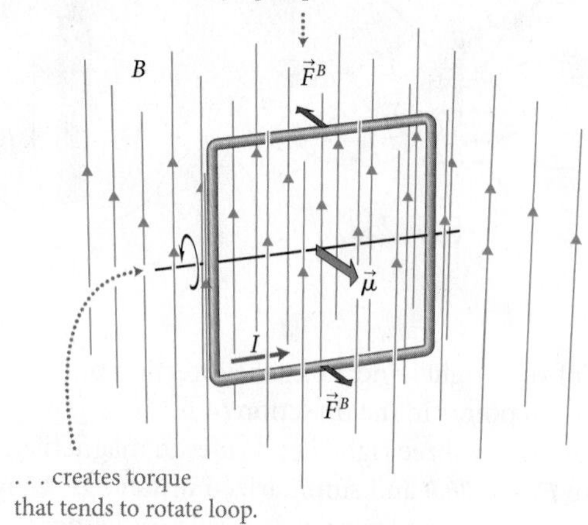
Force exerted by magnetic field on top and bottom sides of current-carrying loop . . .
B
$\vec{F}^B$
$\vec{\mu}$
I
$\vec{F}^B$
. . . creates torque that tends to rotate loop.

Figure 28.10 Magnetic forces exerted on a square current loop that is oriented so that its magnetic dipole moment is perpendicular to an external magnetic field.

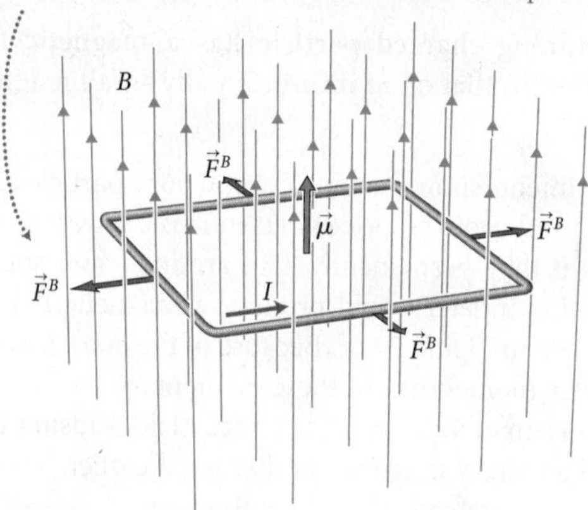
Magnetic field exerts force on all four sides of loop . . .
B
$\vec{F}^B$
$\vec{\mu}$
$\vec{F}^B$
$\vec{F}^B$
I
$\vec{F}^B$
. . . but forces all lie in plane of loop, so cause no torque.

Figure 28.11 Magnetic forces exerted on a square current loop that is oriented so that its magnetic dipole moment is parallel to an external magnetic field.

of each rule in Figure 28.9 can be reversed to solve for a different variable. If you know the direction in which the magnetic field lines curl, for example, you can use the right-hand current rule to determine the current direction.

28.4 Does the direction of the electric field along the axis inside an electric dipole coincide with the direction of the electric dipole moment?

What happens when a current loop is placed in a magnetic field? To find out, consider a square current loop of wire placed in a uniform magnetic field with its magnetic dipole moment perpendicular to the magnetic field (**Figure 28.10**). The current loop experiences magnetic forces on the top and bottom sides but not on the vertical sides because they are parallel to the direction of the magnetic field (see Section 27.3). Using the right-hand force rule of Figure 28.9*b*, we see that the magnetic forces exerted on the top and bottom sides cause a torque that tends to rotate the loop as indicated in Figure 28.10.

When the current loop is oriented with its magnetic dipole moment parallel to the magnetic field, all four sides experience a magnetic force (**Figure 28.11**). However, because all four forces lie in the plane of the loop, none of them causes any torque. Because the magnitudes of the four forces are the same, their vector sum is zero and the loop is not accelerated sideways.

28.5 As the current loop in Figure 28.10 rotates over the first 90°, do the magnitudes of the (*a*) magnetic force exerted on the horizontal sides and (*b*) the torque caused by these forces increase, decrease, or stay the same? (*c*) As the loop rotates, do the two vertical sides experience any force, and, if so, do these forces cause any torque? (*d*) What happens to the torque as the loop rotates beyond 90°?

Summarizing the results of Checkpoint 28.5:

A current loop placed in a magnetic field tends to rotate such that the magnetic dipole moment of the loop becomes aligned with the magnetic field.

This alignment is completely analogous to the alignment of the electric dipole moment in the direction of an external electric field, which we studied in Section 23.8 (see Figure 23.32).

28.6 Suppose the square current loop in Figure 28.10 is replaced by a circular loop with a diameter equal to the width of the square loop and with the same current. Does the circular loop experience a torque? If not, why not? If so, how does this torque compare with that on the square loop?

Example 28.1 Current loop torque

When placed between the poles of a horseshoe magnet as shown in **Figure 28.12**, does a rectangular current loop experience a torque? If so, in which direction does the loop rotate?

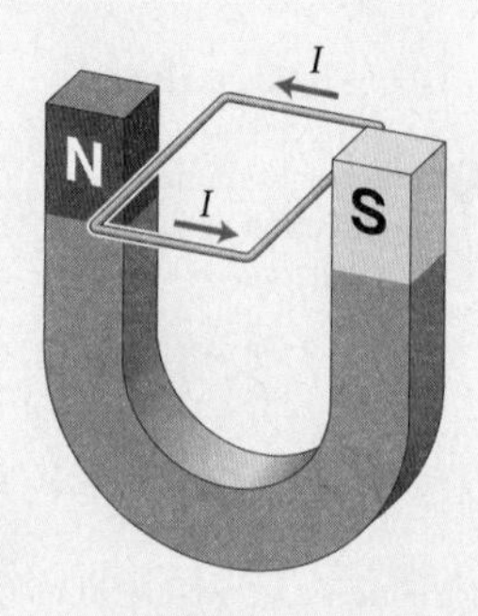
I
N
I
S

Figure 28.12 Example 28.1

❶ GETTING STARTED I know that a current loop placed in a magnetic field tends to rotate such that the magnetic dipole moment of the loop becomes aligned with the magnetic field. If the loop tends to rotate, it experiences a torque.

❷ DEVISE PLAN The simplest way to answer this question is to look at the directions of the magnetic field $\vec{B}$ and the magnetic dipole moment $\vec{\mu}$. By definition, the magnetic field between the poles of the magnet points from the north pole to the south pole. To determine the direction of $\vec{\mu}$, I can use the right-hand dipole rule.

CONCEPTS

❸ **EXECUTE PLAN** I begin by sketching the loop and indicating the direction of the magnetic field (**Figure 28.13*a***). To determine the direction of $\vec{\mu}$, I curl the fingers of my right hand along the direction of the current through the loop. My thumb shows that $\vec{\mu}$ points straight up. To align $\vec{\mu}$ with $\vec{B}$, therefore, the loop rotates in the direction shown by the curved arrow in Figure 28.13*a*. The loop must experience a torque in order for this rotation to occur. ✔

Figure 28.13

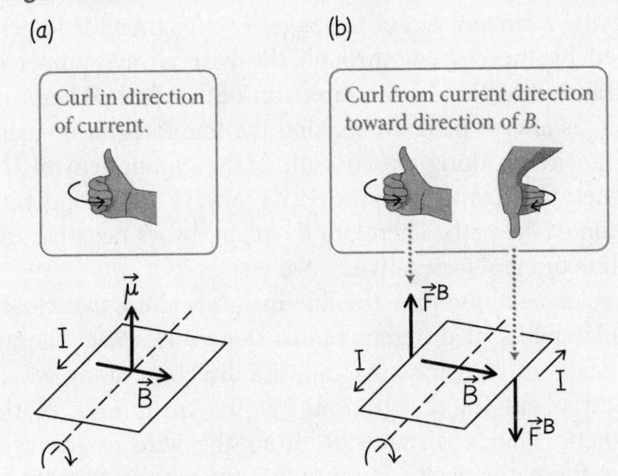

❹ **EVALUATE RESULT** I can verify my answer by determining the force exerted by the magnetic field on each side of the loop and seeing if these forces cause a torque (Figure 28.13*b*). The front and rear sides experience no force because the current through them is either parallel or antiparallel to $\vec{B}$. To determine the direction of the force exerted on the left side of the loop, I point my right-hand fingers along the direction of the current through that side and curl them toward the direction of the magnetic field. My upward-pointing thumb indicates that the force exerted on the left side is upward. Applying the right-hand force rule to the right side of the loop tells me that the magnetic force exerted on that side is directed downward. The magnetic forces exerted on the left and right sides thus cause a torque that makes the loop rotate in the same direction I determined earlier.

The alignment of the magnetic dipole moment of current loops in magnetic fields is responsible for the operation of any device involving an electric motor (see the box below titled Electric motors).

28.7 Describe the motion of the current loop in Figure 28.12 if the magnitude of the magnetic field between the poles of the magnet is greater on the left than it is on the right.

Electric motors

The torque caused by the forces exerted on a current loop in a magnetic field is the basic operating mechanism of an electric motor. A problem with the arrangement shown in Figure 28.12, however, is that once the magnetic dipole moment is aligned with the magnetic field, the torque disappears. Also if the current loop overshoots this equilibrium position, the torque reverses direction (see Checkpoint 28.5*d*). The most common way to overcome this problem is with a *commutator*—an arrangement of two curved plates that reverses the current through the current loop each half turn. The result is that the current loop keeps rotating.

The basic operation is illustrated in **Figure 28.14**. Each half of the commutator is connected to one terminal of a battery and is in contact with one end of the current loop. In the position illustrated in Figure 28.14*a*, the black end is in contact with the negative commutator (that is, the half of the commutator connected to the negative battery terminal). Consequently, the current direction is counterclockwise as seen from above through the current loop, and the magnetic dipole moment $\vec{\mu}$ points up and to the left. The vertical magnetic field therefore causes a torque that turns the loop clockwise to align $\vec{\mu}$ with $\vec{B}$.

Once the loop has rotated to the position shown in Figure 28.14*b*, $\vec{\mu}$ is aligned with $\vec{B}$ but the contact between the ends of the current loop and the commutator is broken, which means there is no current in the loop. The loop overshoots this equilibrium position, but as soon as it does so, the current direction reverses because now the black end of the loop is in contact with the positive commutator (Figure 28.14*c*). The current direction is now clockwise as seen from above, and so $\vec{\mu}$ is reversed. Consequently, the current loop continues to rotate clockwise, as illustrated in Figure 28.14*c*–*e*. After half a revolution, we reach the initial situation again and the sequence repeats.

Figure 28.14 Operating principle of an electric motor. At instant (b), the current direction through the current loop is reversed.

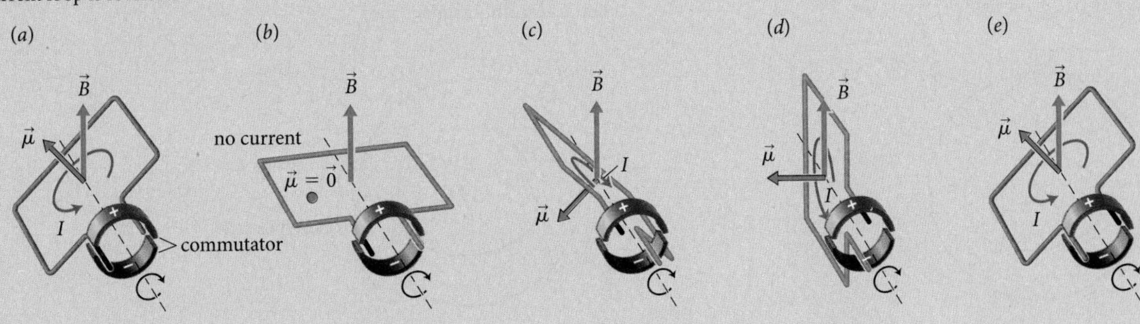

28.4 Ampèrian paths

In electrostatics, Gauss's law provides a powerful tool for determining the electric field of a charge distribution: The electric flux through a closed surface is determined by the amount of charge enclosed by that surface (see Section 24.3). The basic reason for Gauss's law is illustrated in **Figure 28.15a**: Electric field lines originate or terminate on charged particles, and the number of "field line piercings" through a closed surface is proportional to the amount of charge enclosed by that surface. As you saw in Checkpoint 27.6, however, Gauss's law for magnetism is not as helpful for determining magnetic fields. The reason is illustrated in Figure 28.15*b*: Magnetic field lines form loops, so if they exit a closed surface, they must reenter it at some other point. Consequently, the magnetic flux through a closed surface is always zero, regardless of whether or not the surface encloses any magnets or current-carrying wires. In mathematical language, the surface integral of the magnetic field over a closed surface is always zero.

Let us next consider line integrals of electric and magnetic fields. As we saw in Section 25.5, the line integral of the electrostatic field around a closed path is always zero. Consider, for example, the line integral of the electrostatic field around closed path 1 in Figure 28.15*c*. Because the electrostatic field generated by the charged particle at the center of path 1 is always perpendicular to the path, $\vec{E}$ is perpendicular to $d\vec{\ell}$. Therefore $\vec{E}\cdot d\vec{\ell} = 0$ and so the line integral is always zero. Any other path, such as closed path 2 in Figure 28.15*c*, can be broken down into small radial and circular segments. As we saw in Section 25.5, going once around a closed path, all the nonzero contributions along the radial segments add up to zero. Physically this means that as you move a charged particle around a closed path through an electrostatic field, the work done by the electrostatic field on the particle is zero.

Consider now, however, the line integral around closed path 1 in Figure 28.15*d*, which is concentric with a wire carrying a current out of the page. The magnetic field generated by the current through the wire always points in the same direction as the direction of this path. Therefore $\vec{B}\cdot d\vec{\ell}$ is always positive, making the line integral nonzero and positive. Along closed path 2, the component of the magnetic field tangent to the path is always opposite the direction of the path. Therefore $\vec{B}\cdot d\vec{\ell}$ is always negative and the line integral is negative.

Let us next compare the line integrals along two closed circular paths of different radius (**Figure 28.16a**). The arrowheads in the paths indicate the direction along which we carry out the integration. Let the magnitude of the magnetic field a distance R_1 from the wire at the center of the paths be B_1. Because this magnitude is constant along the entire circular path, the line integral along this path is the product of the field magnitude and the length of the path: $B_1(2\pi R_1) = 2\pi B_1 R_1$. Along closed path 2 we obtain $2\pi B_2 R_2$, where B_2 is the field magnitude a distance R_2 from the wire. As you may suspect from the cylindrical symmetry of the wire, the magnitude of the magnetic field

Figure 28.15 Surface and line integrals of electric and magnetic fields.

(*a*) Surface integral of electric field (Gauss's law)

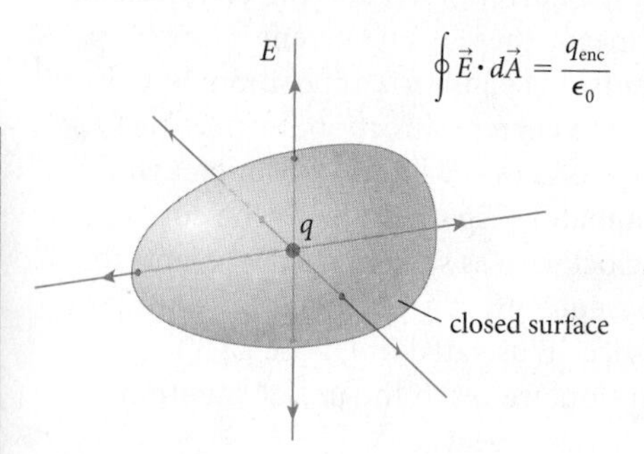

(*b*) Surface integral of magnetic field (Gauss's law for magnetism)

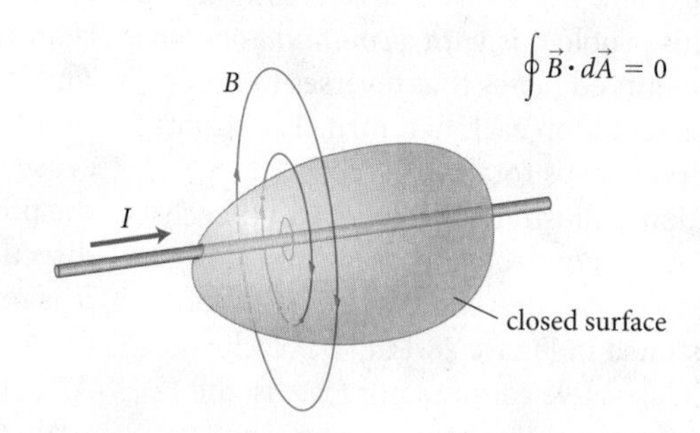

(*c*) Line integral of electric field

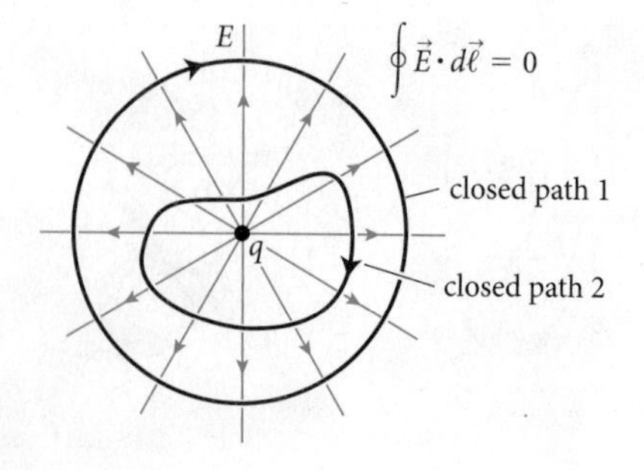

(*d*) Line integral of magnetic field

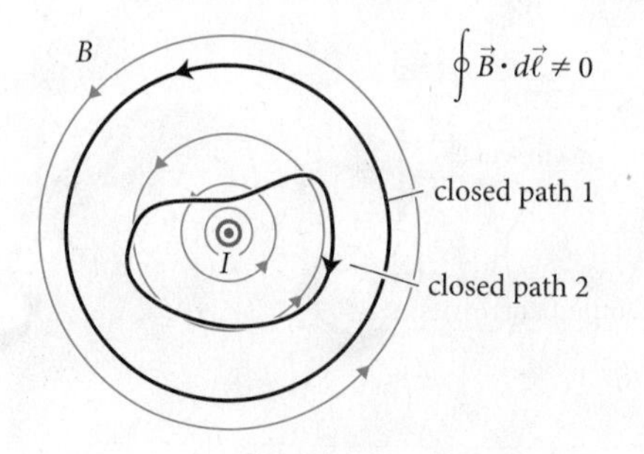

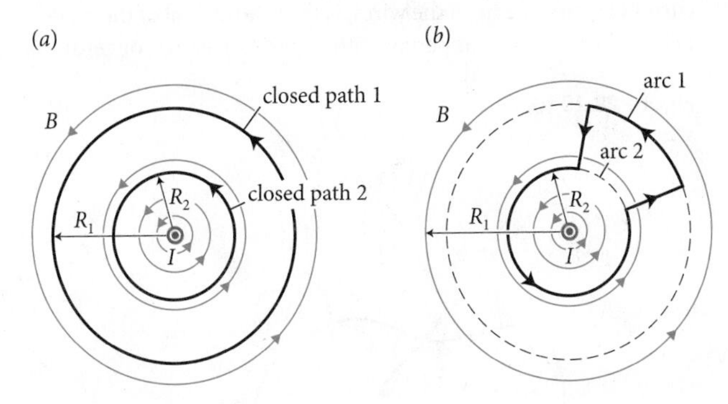

Figure 28.16 (*a*) Two closed circular paths concentric with a wire that carries a current directed out of the page. (*b*) A noncircular path encircling the current-carrying wire. The two arcs each represent one-eighth of a circle.

decreases as $1/r$ with distance r from the wire (we'll confirm this dependence in Example 28.6). This means that as we go from R_2 to R_1, the field decreases by a factor R_2/R_1. In other words, $B_1 = B_2(R_2/R_1)$ or $B_1R_1 = B_2R_2$. Thus we see that the line integrals along the two paths are equal: $2\pi B_1R_1 = 2\pi B_2R_2$. The same argument can be applied to any other closed circular path centered on the wire, from which we conclude that the line integral of the magnetic field of a straight current-carrying wire over any circular path centered on the wire has the same value.

28.8 If the magnitude of the current I through a wire is increased, do you expect the line integral of the magnetic field around a closed path around the wire to increase, decrease, or stay the same?

Now consider the noncircular path illustrated in Figure 28.16*b*. Most of the path lies along a circle of radius R_2, with the exception of one-eighth of a revolution, which is along two radial segments and an arc of radius R_1. Because the radial segments are perpendicular to the magnetic field, they do not contribute to the line integral. How do the line integrals along arcs 1 and 2 compare? As we just saw, the line integrals along the two closed circular paths in Figure 28.16*a* are equal, so the line integrals along one-eighth of each closed path in Figure 28.16*a* must also be equal. The same must be true in Figure 28.16*b*, which means the line integrals along arcs 1 and 2 are identical, and so the line integral along the noncircular path in Figure 28.16*b* is equal to that along any circular path centered on the wire.

We can make the deviations from a circular path progressively more complicated, but as illustrated in **Figure 28.17**, any path can always be broken down into small segments that are either radial or circular and concentric with the wire. The radial segments never contribute to the line integral because the magnetic field is always perpendicular to them, while the circular segments always add up to a single complete revolution. So, in conclusion:

Figure 28.17 (*a*) A noncircular closed path encircling a current-carrying wire. (*b*) We can approximate the path by using small arcs and radial segments.

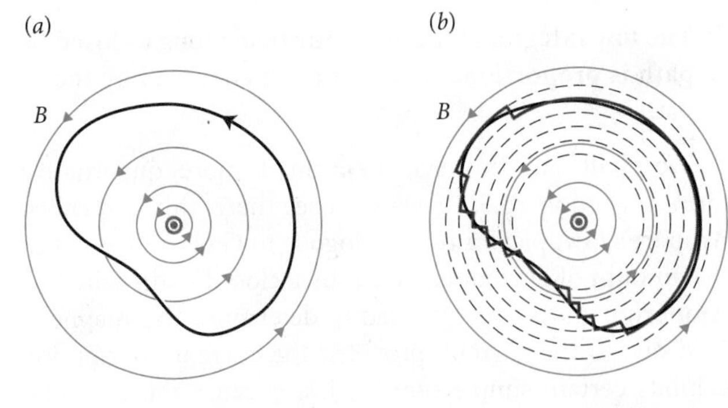

The value of the line integral of the magnetic field along a closed path encircling a current-carrying wire is independent of the shape of the path.

28.9 What happens to the value of the line integral along the closed path in Figure 28.17*a* when (*a*) the direction of the current through the wire is reversed; (*b*) a second wire carrying an identical current is added parallel to and to the right of the first one (but still inside the path); and (*c*) the current through the second wire is reversed?

Next let's examine the line integral along a closed path near a current-carrying wire lying outside the path. One such path is shown in **Figure 28.18** as two arcs joined by two radial segments. As before, the radial segments do not contribute to the line integral. The magnitudes of the line integrals along arcs 1 and 2 are equal, but because the direction of arc 2 is opposite the direction of the magnetic field, the line integral along that arc is negative. Consequently, the line integral along the entire path adds up to zero. We can again extend this statement to a path of a different form, but as long as the path does not encircle the current-carrying wire, the line integral is zero:

The line integral of the magnetic field along a closed path that does not encircle any current-carrying wire is zero.

Figure 28.18 A noncircular closed path not encircling a current-carrying wire.

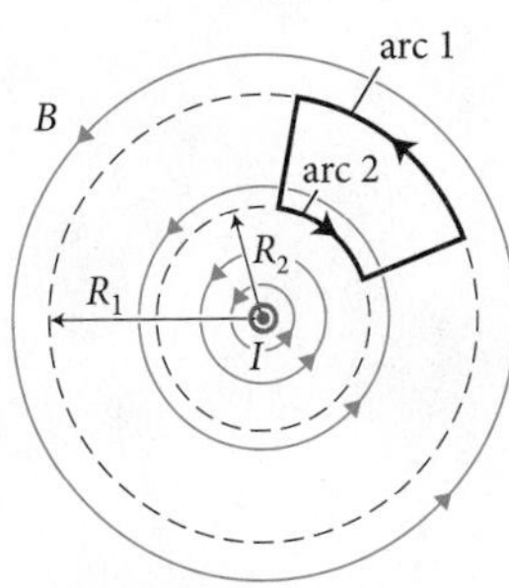

Putting the above results together, we see that the line integral of the magnetic field along a closed path tells us something about the amount of current encircled by the path:

The line integral of the magnetic field along a closed path is proportional to the current encircled by the path.

We shall put this statement in a more quantitative form in Section 28.5. As we shall see there, this law, called **Ampère's law,** plays a role analogous to Gauss's law: Given the amount of current encircled by a closed path, called an **Ampèrian path,** we can readily determine the magnetic field due to the current, provided the current distribution exhibits certain simple symmetries. Because the line integral along a closed path depends on the direction of integration, we must always choose a direction along the path when specifying an Ampèrian path. Exercise 28.2 illustrates the importance of the direction of the Ampèrian path.

28.10 Suppose the path in Figure 28.17 were tilted instead of being in a plane perpendicular to the current-carrying wire. Would this tilt change the value of the line integral of the magnetic field around the path?

Exercise 28.2 Crossed wires

Consider the Ampèrian path going through the collection of current-carrying wires in **Figure 28.19**. If the magnitude of the current is the same in all the wires, is the line integral of the magnetic field along the Ampèrian path positive, negative, or zero?

Figure 28.19 Exercise 28.2

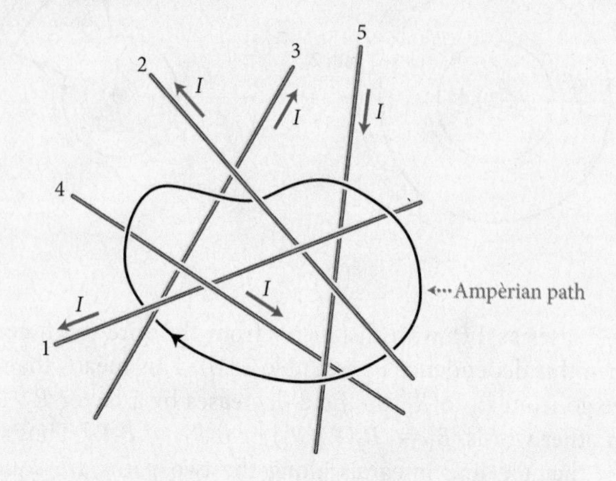

SOLUTION For each wire, I must determine whether or not the path encircles the wire and, if it does, whether the component of that wire's magnetic field tangent to the path points in the same direction as the path. I see that wires 1 and 3 go through the path but the other three wires lie either on top of the path or beneath it.

The direction of current through wire 1 is forward out of the plane of the page, so the magnetic field lines around this wire curl counterclockwise—opposite the direction of the Ampèrian path—giving a negative contribution to the line integral. The direction of current through wire 3 is down into the plane of the page, so it yields a positive contribution to the line integral. Because the two currents are equal in magnitude, the contributions to the line integral add up to zero. ✔

28.11 How do the following changes affect the answer to Exercise 28.2: (*a*) reversing the current through wire 1, (*b*) reversing the current through wire 2, (*c*) reversing the direction of the Ampèrian path?

CONCEPTS

Self-quiz

1. Determine the direction of the magnetic force exerted at the center of the wire or on the particles in **Figure 28.20**.

Figure 28.20

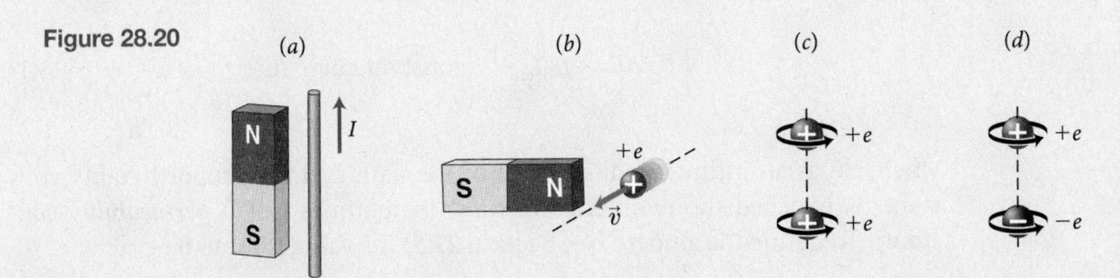

2. Determine the direction of the magnetic field at P due to (a) the current loop in **Figure 28.21a** and (b) segments A and C of the current loop in Figure 28.21b.

Figure 28.21

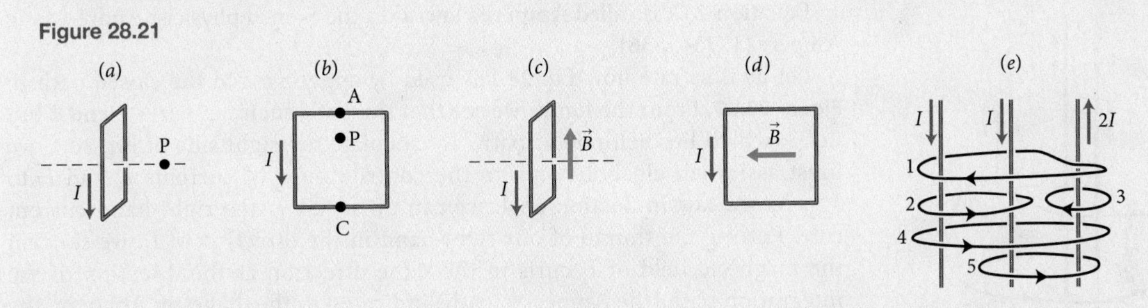

3. Determine in which direction the current loop rotates (a) in Figure 28.21c and (b) in Figure 28.21d.

4. (a) Determine the currents encircled by the five Ampèrian paths in Figure 28.21e. (b) Rank the paths according to the magnitudes of the line integral of the magnetic field along each path, greatest first.

Answers:

1. (a) No magnetic force is exerted by the magnet on the wire because the magnetic field at the location of the wire and the current are antiparallel. (b) The magnetic force acting on the particle is upward. To see this, consider the moving positively charged particle to be current in the direction of the velocity of the particle and then use the right-hand force rule, which makes your thumb point upward. (c) The magnetic dipole moments of the two spinning particles both point up and so the particles attract each other, just like two bar magnets oriented the way the spinning particles are (**Figure 28.22a**). (d) The magnetic dipole moment of the negative particle points down, that of the positive particle points up, and so the two particles repel each other. The comparable bar magnet orientation is shown in Figure 28.22b.

Figure 28.22

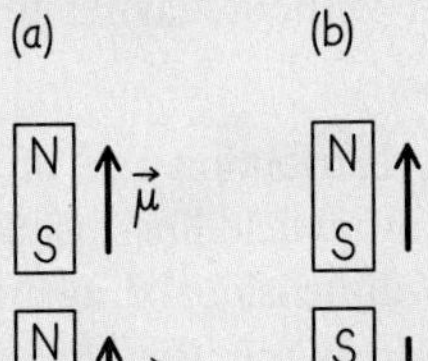

(a)	(b)
N / S $\uparrow \vec{\mu}$	N / S $\uparrow \vec{\mu}$
N / S $\uparrow \vec{\mu}$	S / N $\downarrow \vec{\mu}$

2. (a) The right-hand dipole rule tells you that the magnetic dipole of the loop and the magnetic field produced by the loop point to the right at P. (b) Segments A and C both contribute a magnetic field that points out of the page at P, so the magnetic field due to both segments also points out of the page at P.

3. (a) $\vec{\mu}$ for the current loop points to the right, so the current loop rotates counterclockwise about an axis perpendicular to the page and through the center of the loop. (b) $\vec{\mu}$ for the current loop points out of the page, so the loop rotates about an axis aligned with the vertical sides of the loop. The right side of the loop moves up out of the page, and the left side moves down into the page.

4. (a) Path 1 encircles two currents I in the same direction as the magnetic field: $+2I$. Path 2 encircles the same two currents in the opposite direction: $-2I$. Path 3 encircles $2I$ in the direction opposite the magnetic field direction: $-2I$. Path 4 encircles all three currents, which add up to zero. Path 5 encircles I in the direction opposite the magnetic field direction and $2I$ in the same direction as the magnetic field: $+I$. (b) Each line integral is proportional to the current encircled, making the ranking $1 = 2 = 3 > 5 > 4$.

SELF-QUIZ

28.5 Ampère's law

In the preceding section, we saw that the line integral of the magnetic field around a closed path, called an Ampèrian path, is proportional to the current encircled by the path, I_{enc}. This can be expressed mathematically as

$$\oint \vec{B} \cdot d\vec{\ell} = \mu_0 I_{enc} \quad \text{(constant currents)}, \tag{28.1}$$

where $d\vec{\ell}$ is an infinitesimal segment of the path and the proportionality constant μ_0 is called the **magnetic constant** (sometimes called *permeability constant*). To define the ampere (see Section 27.5), its value is set to be exactly

$$\mu_0 = 4\pi \times 10^{-7}\ \text{T} \cdot \text{m/A}.$$

Figure 28.23 A closed path encircling two of three straight current-carrying wires.

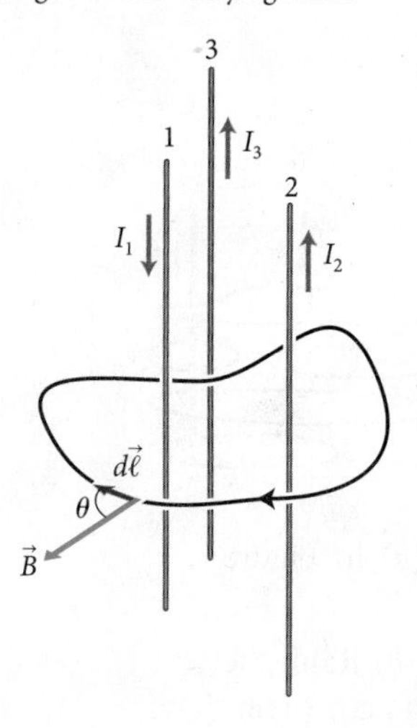

Equation 28.1 is called **Ampère's law,** after the French physicist André-Marie Ampère (1775–1836).

Let us illustrate how Eq. 28.1 is used by applying it to the closed path in Figure 28.23. From the figure we see that the path encircles wires 1 and 2 but not 3, which lies behind the path. To calculate the right side of Eq. 28.1, we must assign an algebraic sign to the contributions of currents I_1 and I_2 to I_{enc}. As we saw in Section 28.4, we can do so using the right-hand current rule. Putting the thumb of our right hand in the direction of I_1, we see that the magnetic field of I_1 curls in the same direction as the direction of the integration path (the Ampèrian path) indicated in the diagram. Applying the same rule to I_2 tells us that the magnetic field of this current curls in the opposite direction. Thus I_1 yields a positive contribution and I_2 yields a negative contribution to the line integral. The right side of Eq. 28.1 thus becomes $\mu_0(I_1 - I_2)$.

Procedure: Calculating the magnetic field using Ampère's law

For magnetic fields with straight or circular field lines, Ampère's law allows you to calculate the magnitude of the magnetic field without having to carry out any integrations.

1. Sketch the current distribution and the magnetic field by drawing one or more field lines using the right-hand current rule. A two-dimensional drawing should suffice.
2. If the field lines form circles, the Ampèrian path should be a circle. If the field lines are straight, the path should be rectangular.
3. Position the Ampèrian path in your drawing such that the magnetic field is either perpendicular or tangent to the path and constant in magnitude. Choose the direction of the Ampèrian path so that, where it runs parallel to the magnetic field lines, it points in the same direction as the field. If the current distribution divides space into distinct regions, draw an Ampèrian path in each region where you wish to calculate the magnetic field.
4. Use the right-hand current rule to determine the direction of the magnetic field of each current encircled by the path. If this magnetic field and the Ampèrian path have the same direction, the contribution of the current to I_{enc} is positive. If they have opposite directions, the contribution is negative.
5. For each Ampèrian path, calculate the line integral of the magnetic field along the path. Express your result in terms of the unknown magnitude of the magnetic field B along the Ampèrian path.
6. Use Ampère's law (Eq. 28.1) to relate I_{enc} and the line integral of the magnetic field and solve for B. (If your calculation yields a negative value for B, then the magnetic field points in the direction opposite the direction you assumed in step 1.)

You can use the same general approach to determine the current given the magnetic field of a current distribution. Follow the same procedure, but in steps 4–6, express I_{enc} in terms of the unknown current I and solve for I.

To calculate the line integral on the left side of Eq. 28.1, we first divide the closed path into infinitesimally small segments $d\vec{\ell}$, one of which is shown in Figure 28.23. The segments are directed tangentially along the path in the direction of integration. For each segment $d\vec{\ell}$, we take the scalar product of $d\vec{\ell}$ with the magnetic field $\vec{B}$ at the location of that segment, $\vec{B} \cdot d\vec{\ell}$, and then we add up the scalar products for all segments of the closed path. In the limit that the segment lengths approach zero, this summation becomes the line integral on the left in Eq. 28.1. In the situation illustrated in Figure 28.23, we cannot easily carry out this integration because of the irregular shape of the path. Just like Gauss's law in electrostatics, however, Ampère's law allows us to easily determine the magnetic field for highly symmetrical current configurations. The general procedure is outlined in the Procedure box on the previous page. The next two examples illustrate how this procedure can be applied to simplify calculating the line integral in Ampère's law.

Example 28.3 Magnetic field generated by a long straight current-carrying wire

A long straight wire carries a current of magnitude I, and this current creates a magnetic field $\vec{B}$. Derive an expression for the magnitude of the magnetic field a radial distance r from the wire.

❶ **GETTING STARTED** I begin by making a sketch of the wire, arbitrarily orienting it vertically (**Figure 28.24**). I know that the magnetic field is circular, so I draw one circular field line, centering it on the wire and giving it a radius r. Using the right-hand current rule, I determine the direction in which the magnetic field points along the circle and indicate that with an arrowhead in my drawing.

Figure 28.24

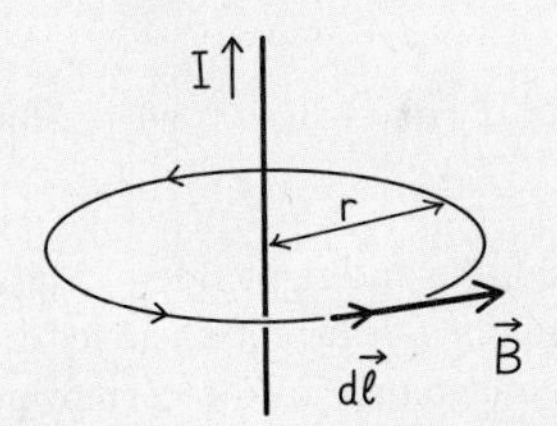

❷ **DEVISE PLAN** To begin my derivation I can use Ampère's law (Eq. 28.1). If I let the field line in my sketch be the Ampèrian path, the magnetic field is constant in magnitude and tangential all along the circular path, simplifying the integral on the left in Eq. 28.1. I let the direction of the Ampèrian path be the same as that of the magnetic field. That is to say, the infinitesimal path segment $d\vec{\ell}$ points in the same direction as $\vec{B}$ all along the Ampèrian path.

❸ **EXECUTE PLAN** With my choice of Ampèrian path, $d\vec{\ell}$ and $\vec{B}$ are always pointing in the same direction, and so $\vec{B} \cdot d\vec{\ell} = B\,dl$.

Because the magnitude B is the same all around the path, I can write for the left side of Eq. 28.1

$$\oint \vec{B} \cdot d\vec{\ell} = \oint B\,d\ell = B \oint d\ell.$$

The line integral on the right here is the sum of the lengths of all the segments $d\vec{\ell}$ around the Ampèrian path; that is to say, it is equal to the circumference of the circle. Therefore I have

$$\oint \vec{B} \cdot d\vec{\ell} = B \oint d\ell = B(2\pi r). \qquad (1)$$

Because the direction in which the Ampèrian path encircles the current I is the same as the direction of the magnetic field generated by the current, the right side of Eq. 28.1 yields

$$\mu_0 I_{enc} = +\mu_0 I. \qquad (2)$$

Substituting the right side of Eq. 2 and Eq. 1 into Eq. 28.1, I get $B(2\pi r) = \mu_0 I$ or

$$B = \frac{\mu_0 I}{2\pi r}. ✔$$

❹ **EVALUATE RESULT** My result shows that the magnitude of the magnetic field is proportional to I, as I expect (doubling the current should double the magnetic field), and inversely proportional to the radial distance r from the wire. I know from Chapter 24 that the electric field is also inversely proportional to r in cases that exhibit cylindrical symmetry, another indication that my result here makes sense.

28.12 Suppose the wire in Example 28.3 has a radius R and the current is uniformly distributed throughout the volume of the wire. Follow the procedure of Example 28.3 to calculate the magnitude of the magnetic field inside ($r < R$) and outside ($r > R$) the wire.

Example 28.4 Magnetic field generated by a large current-carrying sheet

A large flat metal sheet carries a current. The magnitude of the current per unit of sheet width is K. What is the magnitude of the magnetic field a distance d above the sheet?

❶ GETTING STARTED I begin by drawing the sheet and indicating a point P a distance d above it where I am to determine the magnetic field magnitude (**Figure 28.25a**). I draw a lengthwise arrow to show the current through the sheet.

Figure 28.25

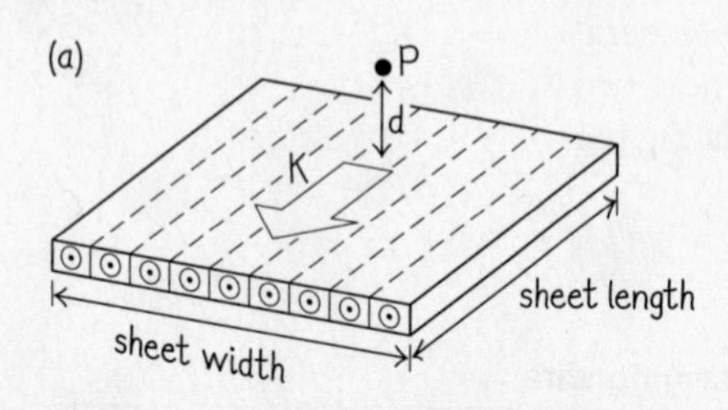

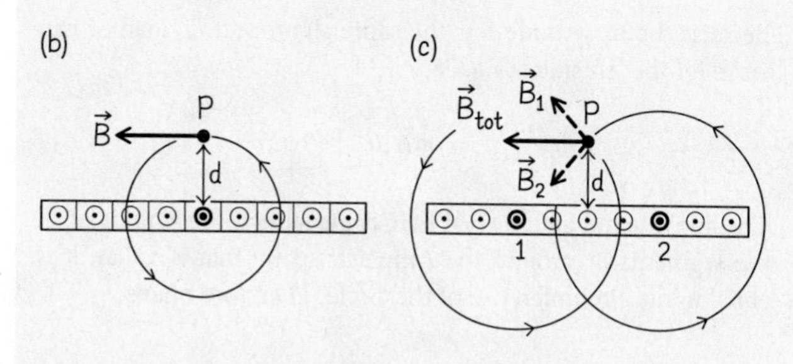

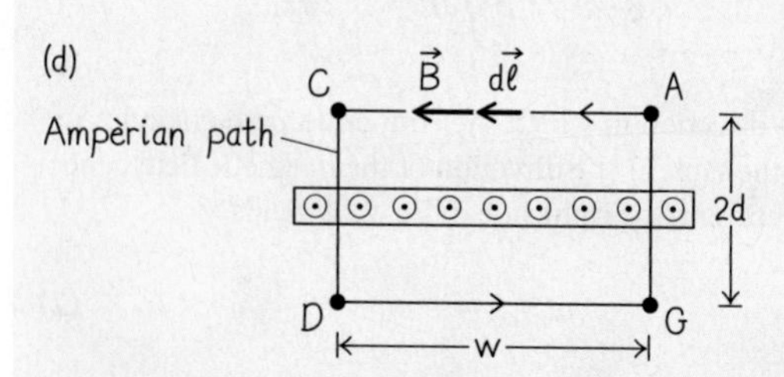

❷ DEVISE PLAN I'll solve this problem using Ampère's law (Eq. 28.1). I first need to determine the direction of the magnetic field on either side of the sheet. To this end, I divide the sheet into thin parallel strips, as indicated by the dashed lines in Figure 28.25*a*. I can then treat each strip as a current-carrying wire.

Figure 28.25*b* is a cross-sectional view of the sheet after I have divided it into strips. The perspective here is looking into the sheet in the direction opposite the direction of the current. Using the right-hand current rule, I determine that the strip right underneath point P contributes a magnetic field at P that points parallel to the sheet and to the left. Next I look at the contributions from the two strips labeled 1 and 2 in Figure 28.25*c*, equidistant on either side of P. The two large circles show the magnetic field lines from these two strips that go through P. Because the magnitudes of the contributions from the two strips are equal, the vector sum of their contributions to the magnetic field also points parallel to the sheet and to the left. The same argument can be applied to any other pair of strips that are equidistant on either side of P. Thus, at P the magnetic field due to the entire sheet must be parallel to the sheet and to the left. Similar reasoning shows that the magnetic field below the sheet is also parallel to the sheet and points to the right.

To exploit what I know about the magnetic field, I choose the rectangular path ACDG in Figure 28.25*d* as my Ampèrian path, making the direction of the path the same as that of the magnetic field. The width of this path is w, and its height is $2d$.

❸ EXECUTE PLAN I can write the line integral around the Ampèrian path as the sum of four line integrals, one over each side of the path:

$$\oint \vec{B}\cdot d\vec{\ell} = \int_A^C \vec{B}\cdot d\vec{\ell} + \int_C^D \vec{B}\cdot d\vec{\ell} + \int_D^G \vec{B}\cdot d\vec{\ell} + \int_G^A \vec{B}\cdot d\vec{\ell}.$$

Along sides CD and GA the magnetic field is perpendicular to the Ampèrian path and so $\vec{B}\cdot d\vec{\ell}$ is zero. Along each of the two horizontal sides, $d\vec{\ell}$ and $\vec{B}$ point in the same direction, and so $\vec{B}\cdot d\vec{\ell} = B\,d\ell$.

Symmetry requires that the magnitude of the magnetic field a distance d below the sheet be the same as the magnitude a distance d above it (because flipping the sheet upside down does not alter the physical situation). Therefore I can take the magnitude B out of the integral:

$$\oint \vec{B}\cdot d\vec{\ell} = \int_A^C B\,d\ell + \int_D^G B\,d\ell = B\int_A^C d\ell + B\int_D^G d\ell.$$

These two line integrals $\int d\ell$ yield the length of the two horizontal sides, which is the Ampèrian path width w, and so the left side of Eq. 28.1 becomes

$$\oint \vec{B}\cdot d\vec{\ell} = B(w + w) = 2Bw. \tag{1}$$

To get an expression for the right side of Eq. 28.1, I must first determine the amount of current encircled by the Ampèrian path. Because the magnitude of the current per unit of width through the sheet is K, the magnitude of the current through the Ampèrian path of width w is Kw. The right side of Eq. 28.1 thus becomes

$$\mu_0 I_{enc} = \mu_0 K w. \tag{2}$$

Substituting the right sides of Eqs. 1 and 2 into Ampère's law, I get $2Bw = \mu_0 K w$, or

$$B = \tfrac{1}{2}\mu_0 K. ✔$$

❹ EVALUATE RESULT Because the sheet has planar symmetry (see Section 24.4), I expect that, by analogy with the electric field around a flat charged sheet, the magnetic field on either side of my sheet here is uniform. That is, the field magnitude does not depend on the distance from the sheet, and this is just what my result shows. It also makes sense that my result shows that the magnitude of the magnetic field is proportional to the current per unit width K through the sheet.

28.13 (*a*) What are the direction and magnitude of the magnetic field between the parallel current-carrying sheets of **Figure 28.26*a***? What is the direction of $\vec{B}$ outside these sheets? (*b*) Repeat for the sheets of Figure 28.26*b*.

Figure 28.26 Checkpoint 28.13.

(*a*)

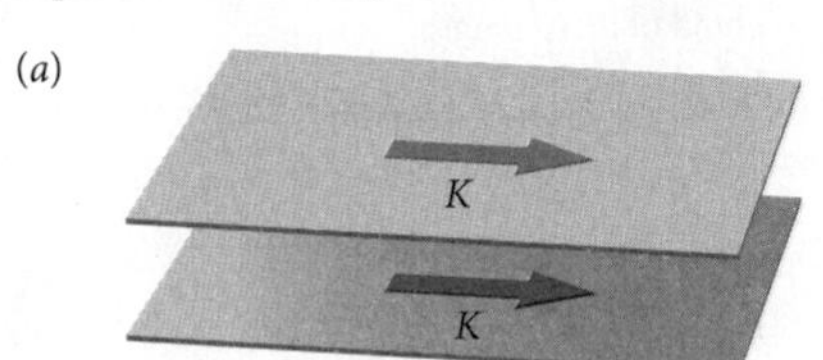

(*b*)

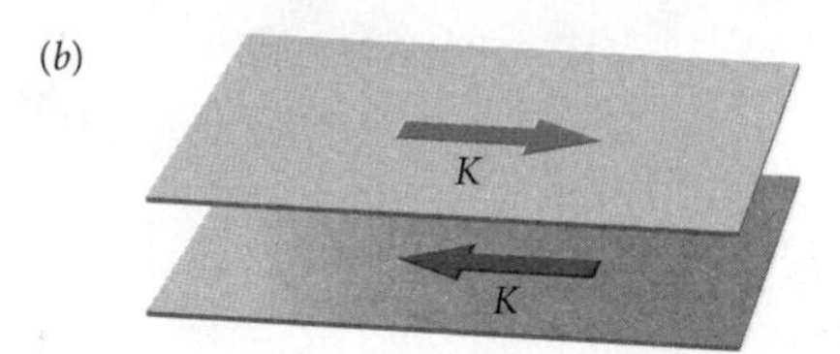

28.6 Solenoids and toroids

A **solenoid** is a long, tightly wound helical coil of wire (**Figure 28.27*a***). In general, the diameter of the coil is much smaller than the length of the coil. When a current enters a solenoid at one end and exits at the other end, the solenoid generates a strong magnetic field. If a magnetic core is placed in the solenoid, the solenoid exerts a strong magnetic force on the core, turning electrical energy into motion. Solenoids are therefore often used in electrical valves and actuators. Because solenoids are generally very tightly wound, we can treat the windings of a solenoid as a stack of closely spaced current loops (Figure 28.27*b*).

Figure 28.28*a* on the next page shows that the magnetic field inside a long solenoid must be directed along the axis of the solenoid. Consider, for example, point P inside the solenoid. The figure shows the field lines of two of the loops, one on either side of P and equidistant from that point. The magnetic field contributions of the two loops give rise to a magnetic field that points along the axis of the solenoid. The same argument can be applied to any other pair of loops and to any other point inside the solenoid. Therefore the magnetic field everywhere inside a long solenoid must be directed parallel to the solenoid axis.

Figure 28.27 A solenoid is a tightly wound helical coil of wire.

(*a*) A solenoid

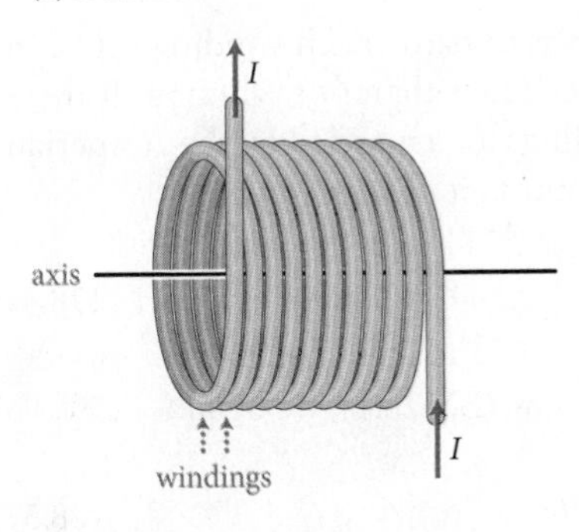

(*b*) A solenoid approximated as a stack of parallel current loops

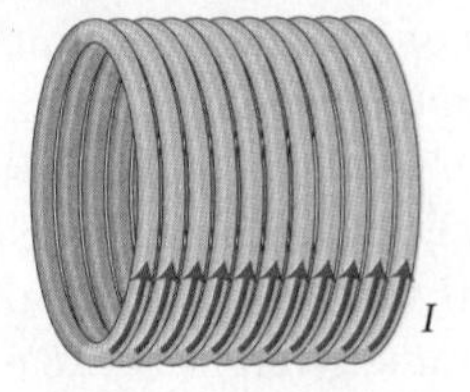

Figure 28.28 The magnetic field of a solenoid that carries a current *I*.

(*a*) Cross section of a solenoid showing the contribution of two loops to the magnetic field at a point inside the solenoid

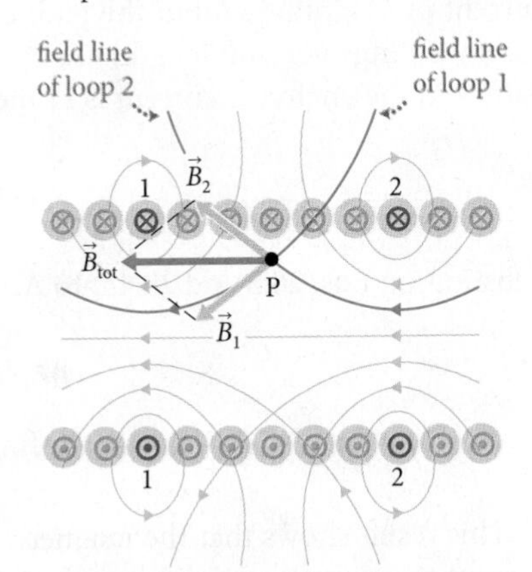

(*b*) Ampèrian path for calculating the magnetic field inside the solenoid

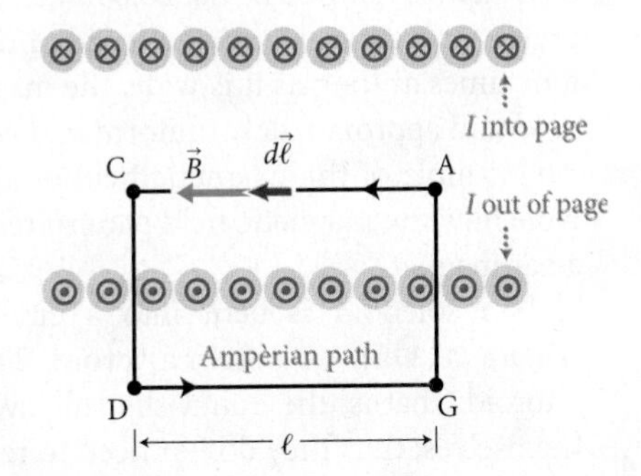

Figure 28.29 Magnetic field line pattern in a solenoid of finite length.

(*a*) Magnetic field of a solenoid, shown in cross section

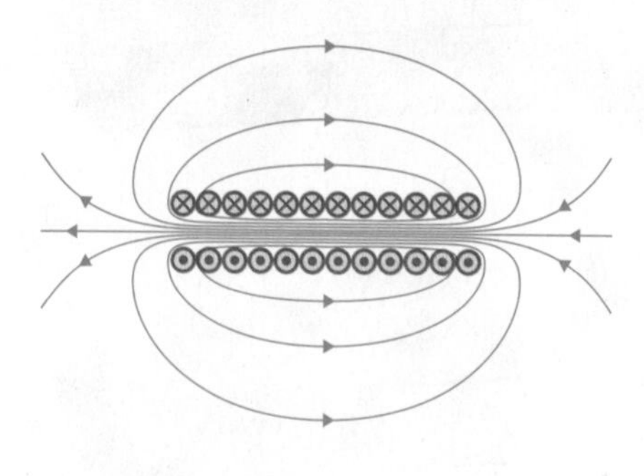

(*b*) The magnetic field of a solenoid

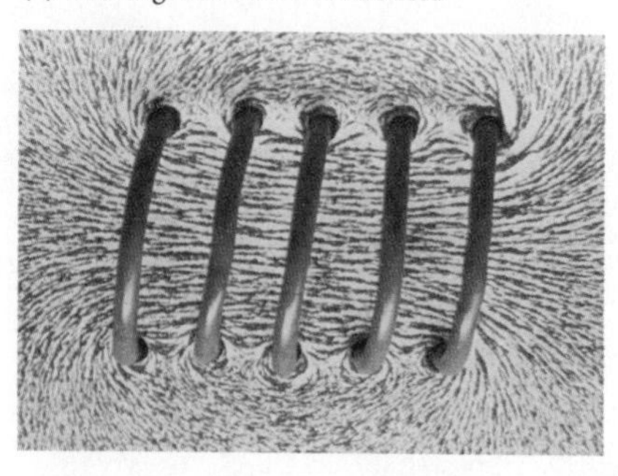

Figure 28.30 A toroid is a solenoid bent into a ring.

(*a*) Toroid

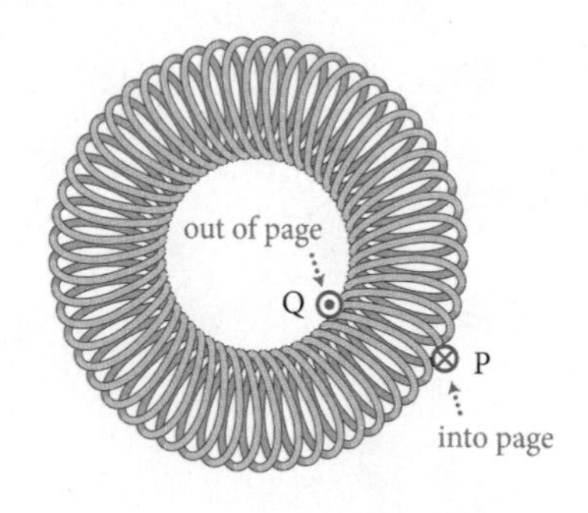

(*b*) Cross section showing magnetic field and a choice of Ampèrian path

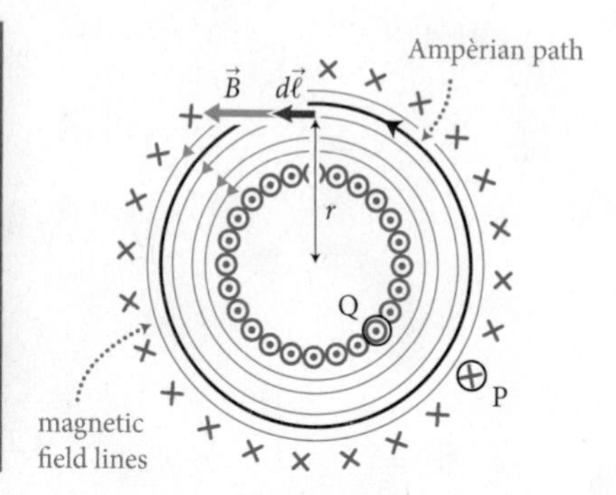

Because magnetic field lines form loops, all the lines that go through the solenoid must loop back from one end of the solenoid to the other. Because there is much more space available outside the solenoid, the density of the field lines as they loop back is much smaller than the field line density inside the solenoid. The longer the solenoid, the smaller the field line density in the immediate vicinity outside the solenoid. In the limit of an infinitely long solenoid, we can expect the magnetic field outside the solenoid to approach zero.

We can now use Ampère's law to determine the magnitude of the magnetic field when there is a current of magnitude I through the solenoid. Exploiting what we know about the direction of the magnetic field, we choose the rectangular path ACDG in Figure 28.28*b* as the Ampèrian path. Along AC, the magnetic field is parallel to the path, and so $\vec{B} \cdot d\vec{\ell} = B\,d\ell$. Along CD and GA, the magnetic field is perpendicular to the path, which means that these segments do not contribute: $\vec{B} \cdot d\vec{\ell} = 0$. Because the magnetic field is zero outside the solenoid, the segment DG also does not contribute. The line integral on the left side of Ampère's law (Eq. 28.1) thus becomes

$$\oint \vec{B} \cdot d\vec{\ell} = \int_A^C B\,d\ell. \tag{28.2}$$

The cylindrical symmetry of the solenoid requires the magnitude of the magnetic field to be constant along AC, so we can pull B out of the integral:

$$\oint \vec{B} \cdot d\vec{\ell} = \int_A^C B\,d\ell = B\int_A^C d\ell = B\ell, \tag{28.3}$$

where ℓ is the length of side AC.

What is the current encircled by the Ampèrian path? Each winding carries a current of magnitude I, but the path encircles more than one winding. If there are n windings per unit length, then $n\ell$ windings are encircled by the Ampèrian path, and the encircled current is I times the number of windings:

$$I_{enc} = n\ell I. \tag{28.4}$$

Substituting Eqs. 28.3 and 28.4 into Ampère's law (Eq. 28.1), we obtain

$$B\ell = \mu_0 n\ell I \tag{28.5}$$

$$B = \mu_0 nI \quad \text{(infinitely long solenoid)}. \tag{28.6}$$

This result shows that the magnetic field inside the solenoid depends on the current through the windings and on the number of windings per unit length. The field within the solenoid is uniform—it does not depend on position inside the solenoid. Although Eq. 28.6 holds for an infinitely long solenoid, the result is pretty accurate even for a solenoid of finite length. For a solenoid that is at least four times as long as it is wide, the magnetic field is very weak outside the solenoid and approximately uniform and equal to the value given in Eq. 28.6 inside. An example of the magnetic field of a finite solenoid is shown in **Figure 28.29**. Note how the magnetic field pattern resembles that of the magnetic field around a bar magnet.

If a solenoid is bent into a circle so that its two ends are connected (**Figure 28.30*a***), we obtain a **toroid**. The magnetic field lines in the interior of a toroid (that is, the donut-shaped cavity enclosed by the coiled wire) close on themselves; thus they do not need to reconnect outside the toroid, as in the case of a solenoid. The entire magnetic field is contained inside the cavity. Symmetry requires the field lines to form circles inside the cavity; the field lines run in the direction of your right thumb when you curl the fingers of your right hand along the direction of the current through the windings. Figure 28.30*b* shows a few representative magnetic field lines in a cross section of the toroid where the current goes into the page on the outside rim of the toroid (as at point P, for example) and out of the page on the inside rim (as at point Q).

To determine the magnitude of the magnetic field, we apply Ampère's law (Eq. 28.1) to a circular path of radius r that coincides with a magnetic field line (Figure 28.30*b*). Because the field is tangential to the integration path, we have $\vec{B}\cdot d\vec{\ell} = B\,d\ell$. Furthermore symmetry requires the magnitude of the magnetic field to be the same all along the field line, and so we can pull B out of the integration:

$$\oint \vec{B}\cdot d\vec{\ell} = \oint B\,d\ell = B\oint dl = B(2\pi r). \tag{28.7}$$

The Ampèrian path encircles one side of all of the windings, so if there are N windings, the encircled current is

$$I_{\text{enc}} = NI. \tag{28.8}$$

Substituting these last two equations into Ampère's law (Eq. 28.1), we obtain

$$B = \mu_0 \frac{NI}{2\pi r} \quad \text{(toroid)}. \tag{28.9}$$

This result tells us that in contrast to a solenoid, the magnitude of the magnetic field in a toroid is not constant—it depends on the distance r to the axis through the center of the toroid.

Example 28.5 Square toroid

The toroid in **Figure 28.31** has 1000 windings carrying a current of 1.5 mA. Each winding is a square of side length 10 mm, and the toroid's inner radius is 10 mm. What is the magnitude of the magnetic field at the center of the winding squares?

Figure 28.31 Example 28.5

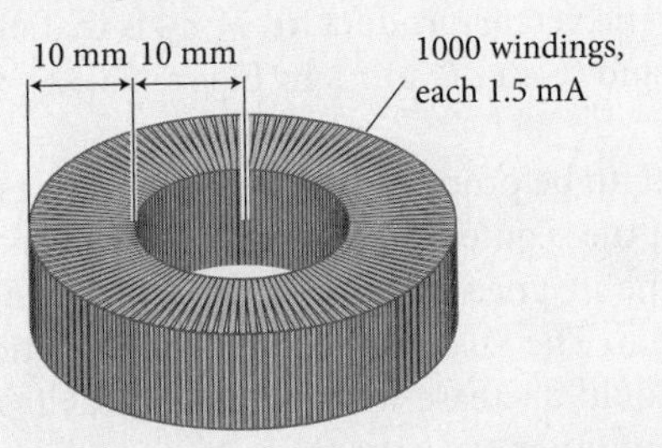

❶ **GETTING STARTED** The fact that the windings are square does not change the magnetic field pattern. The magnetic field is still circular as in Figure 28.30*b*, and the arguments given in the derivation of Eq. 28.9 still apply.

❷ **DEVISE PLAN** I can use Eq. 28.9 to determine the magnitude of the magnetic field.

❸ **EXECUTE PLAN** The distance from the center of the toroid to the center of each winding is 10 mm + 5 mm = 15 mm, so

$$B = (4\pi \times 10^{-7}\ \text{T}\cdot\text{m/A})\frac{(1000)(1.5 \times 10^{-3}\ \text{A})}{2\pi(0.015\ \text{m})}$$

$$= 2.0 \times 10^{-5}\ \text{T}. ✔$$

❹ **EVALUATE RESULT** The magnetic field magnitude I obtain is small—comparable to the magnitude of Earth's magnetic field at ground level—but the current through the toroid is very small, so my answer is not unreasonable.

28.14 Use Ampère's law to determine the magnetic field outside a toroid at a distance r from the center of the toroid (*a*) when r is greater than the toroid's outer radius and (*b*) when r is smaller than the toroid's inner radius.

Figure 28.32 The magnetic field at point P due to an arbitrarily shaped current-carrying wire cannot be obtained from Ampère's law. As shown in (*c*), the direction of $d\vec{B}_s$ can be found by taking the vector product of $d\vec{\ell}$, which has length $d\ell$ and direction given by the current through the wire, and $\hat{r}_{sP}$, which points from the segment to point P: $d\vec{\ell}_{sP} \times \hat{r}_{sP}$.

(*a*) Arbitrarily shaped current-carrying wire

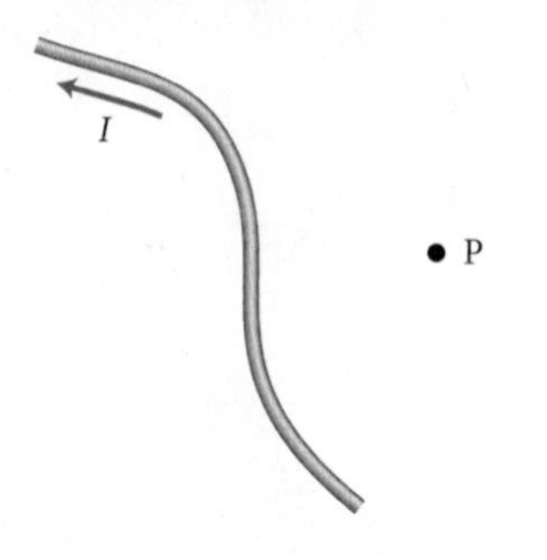

(*b*) Small segment $d\vec{\ell}$ contributes magnetic field $d\vec{B}_s$ at point P

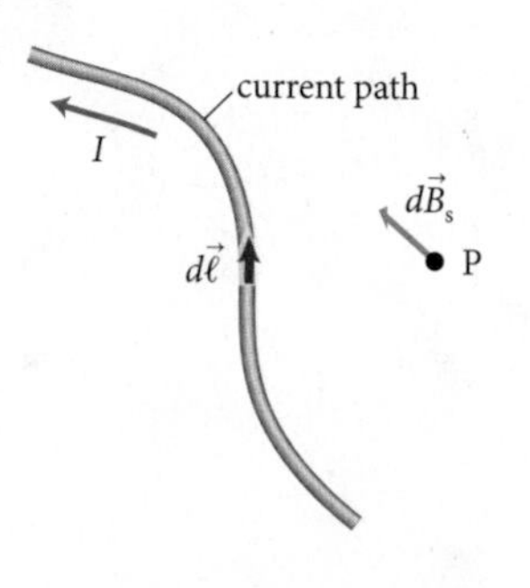

(*c*) Determining the direction of $d\vec{B}_s$

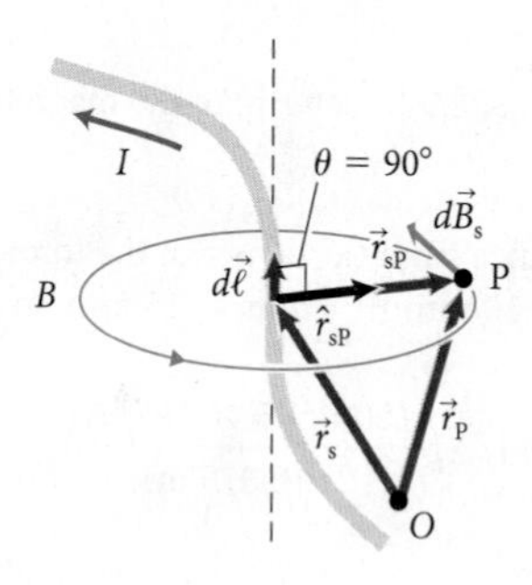

28.7 Magnetic fields due to currents

Ampère's law allows us to determine the magnetic field in only a few symmetrical situations involving current-carrying conductors. For any other situation, such as the one illustrated in **Figure 28.32*a***, Ampère's law is of little help. Suppose, for example, that we are interested in determining the magnetic field at point P due to the current of magnitude I through this conductor. To calculate this field, we develop a procedure that parallels the procedure we developed in Section 23.7 to calculate the electric field of continuous charge distributions.

We begin by treating the wire as a current path and dividing it into small segments, each of length $d\ell$. For each segment we can define a vector $d\vec{\ell}$ that has length $d\ell$ and points in the direction of the current. One such vector segment is shown in Figure 28.32*b*. Let the magnetic field at P due to this segment be $d\vec{B}_s$. If we obtain an expression for the magnetic field due to the segment $d\vec{\ell}$ at an arbitrary position, we can determine the contributions of all the segments that make up the wire to the magnetic field at P and sum these to obtain $\vec{B}$. In the limit of infinitesimally small segments, this summation becomes an integral:

$$\vec{B} = \int_{\substack{\text{current}\\ \text{path}}} d\vec{B}_s. \tag{28.10}$$

where the integration is to be taken along the path followed by the current—that is, a path in the shape of the wire and in the direction of the current.

Before we can carry out the integration, we need to obtain an expression for the magnetic field $d\vec{B}_s$ of a current-carrying segment $d\vec{\ell}$. Because the magnetic field caused by one small segment is too feeble to measure, we cannot determine this field experimentally. From the expression for the magnetic field of a long straight current-carrying wire, however, it is possible to deduce what the field of a segment such as $d\vec{\ell}$ should be.

Let us first examine the direction of the magnetic field contribution $d\vec{B}_s$ due to a segment $d\vec{\ell}$ that is located such that $d\vec{\ell}$ is perpendicular to the vector $\vec{r}_{sP}$ pointing from $d\vec{\ell}$ to P ($\theta = 90°$, Figure 28.32*c*). The field lines of this segment are circles centered on the line of motion of the charge carriers causing the current through $d\vec{\ell}$ (see Figure 28.32*a* and *c*). Therefore the magnetic field at P due to $d\vec{\ell}$ is tangent to the circular field line through P. The direction of $d\vec{B}_s$ can be determined by associating the direction of $d\vec{B}_s$ with a vector product. To this end, we denote the unit vector pointing from the segment to P by $\hat{r}_{sP}$, as shown in Figure 28.32*c*. The direction of the vector-product $d\vec{\ell} \times \hat{r}_{sP}$ is that of $d\vec{B}_S$: If you curl the fingers of your right hand from $d\vec{\ell}$ to $\hat{r}_{sP}$ in Figure 28.32*c*, your thumb points in the direction of $d\vec{B}_S$.

We expect the magnitude dB_s to be proportional to the current I through the segment and to the length $d\ell$ of the segment. The greater the current or the longer the segment, the stronger the magnetic field. We also expect dB_s to depend on the distance r_{sP} to the segment. The field line picture for magnetic fields suggests that the magnetic field should decrease with distance r_{sP} as $1/r_{sP}^2$, just like the electric field. Finally, for any other segment than the one shown in Figure 28.32*c* we expect the magnetic field to depend on the angle θ between $d\vec{\ell}$ and $\hat{r}_{sP}$: For $\theta = 0$ (which is along the direction of the current) the magnetic field is zero, and for $\theta = 90°$ the magnetic field is maximum. The trigonometric function that fits this behavior is $\sin\theta$. Putting all this information in mathematical form, we obtain

$$dB_s = \frac{\mu_0}{4\pi}\frac{I\,d\ell\,\sin\theta}{r_{sP}^2}\;(0 \leq \theta \leq \pi). \tag{28.11}$$

The proportionality factor $\mu_0/4\pi$ is obtained by deriving this expression from Ampère's law, but the mathematics is beyond the scope of this book. Instead, we use the reverse approach and show in Example 28.6 that Eq. 28.11 yields the correct result for a long straight current-carrying wire.

Incorporating what we know about the direction of $d\vec{B}_s$, we can also write Eq. 28.11 in vector form:

$$d\vec{B}_S = \frac{\mu_0}{4\pi}\frac{I\,d\vec{\ell} \times \hat{r}_{sP}}{r_{sP}^2} \quad \text{(constant current)}, \tag{28.12}$$

where $\hat{r}_{sP}$ is the unit vector pointing along $\vec{r}_{sP}$ from s to P. Equation 28.12 is known as the **Biot-Savart law**. By substituting Eq. 28.12 into Eq. 28.10, we have a prescription for calculating the magnetic field produced by any constant current.

Example 28.6 Another look at the magnetic field generated by a long straight current-carrying wire

A long straight wire carries a current of magnitude I. Use the Biot-Savart law to derive an expression for the magnetic field $\vec{B}$ produced at point P a radial distance r from the wire.

❶ GETTING STARTED I begin by making a sketch of the wire (**Figure 28.33**). I arbitrarily orient the wire vertically and then choose the x axis along the direction of the wire. Because the magnetic field produced by the current has cylindrical symmetry, I can set the origin anywhere along the axis without loss of generality. For simplicity, I let the origin be at the height of point P. I assume the wire is of infinite length.

Figure 28.33

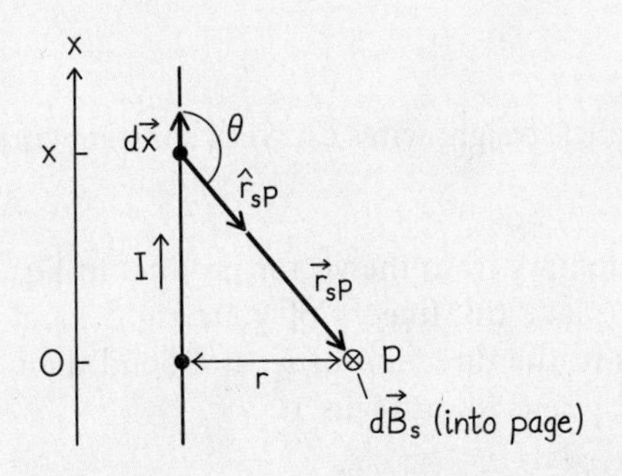

❷ DEVISE PLAN To use the Biot-Savart law (Eq. 28.12), I need to divide the wire into segments, determine the magnetic fields due to all segments, and then take the sum of the fields. In the limit of infinitesimally small segments, this sum becomes the integral given by Eq. 28.10 with the wire serving as the current path along which the integration is carried out.

I indicate one such segment in my sketch, calling it $d\vec{x}$. The magnetic field $d\vec{B}_s$ at P generated by this segment is given by Eq. 28.12. The unit vector $\hat{r}_{sP}$ in this equation points from $d\vec{x}$ to P, and the direction of $d\vec{B}_s$ is given by the right-hand current rule: I point my right thumb along the direction of the current and curl my fingers around the wire to determine the direction of $d\vec{B}_s$ (into the page at P in my sketch). Alternatively, I can use the vector product $d\vec{x} \times \hat{r}_{sP}$ to determine the direction of $d\vec{B}_s$: I line up the fingers of my right hand along $d\vec{x}$ in Figure 28.33 and curl them toward $\hat{r}_{sP}$. When I do this, my thumb points in the direction of the magnetic field. Both methods yield the same result: $d\vec{B}_s$ points into the page.

Note that all the segments $d\vec{x}$ along the wire produce a magnetic field in the same direction. This means I can take the algebraic sum of the *magnitudes* of $d\vec{B}_s$ to determine the magnitude of the magnetic field at P. Then I can use Eq. 28.11 to express dB_s in terms of dx and integrate the resulting expression from $x = -\infty$ to $x = +\infty$ to determine the magnitude of the magnetic field at P.

❸ EXECUTE PLAN Because r_{sP} and θ in Eq. 28.11 both depend on x, I need to express them in terms of x before I can carry out the integration. By the Pythagorean theorem,

$$r_{sP}^2 = x^2 + r^2,$$

and remembering that $\sin\theta = \sin(180° - \theta)$, I write

$$\sin\theta = \frac{r}{r_{sP}} = \frac{r}{\sqrt{x^2+r^2}}.$$

Substituting these last two results into Eq. 28.11 and using dx in place of $d\ell$, I get

$$dB_s = \frac{\mu_0 I}{4\pi}\frac{r}{\sqrt{x^2+r^2}}\frac{1}{x^2+r^2}dx = \frac{\mu_0 Ir}{4\pi}\frac{dx}{[x^2+r^2]^{3/2}},$$

and integrating this result over the length of the wire gives me

$$B = \frac{\mu_0 Ir}{4\pi}\int_{-\infty}^{+\infty}\frac{dx}{[x^2+r^2]^{3/2}} = \frac{\mu_0 Ir}{4\pi}\left[\frac{1}{r^2}\frac{x}{[x^2+r^2]^{1/2}}\right]_{x=-\infty}^{x=+\infty}$$

$$B = \frac{\mu_0 I}{2\pi r}. ✔$$

❹ EVALUATE RESULT This is identical to the result I obtained using Ampère's law in Example 28.3, a strong indication that my result here is correct.

28.15 Imagine a long straight wire of semi-infinite length, extending from $x = 0$ to $x = +\infty$, carrying a current of constant magnitude I. What is the magnitude of the magnetic field at a point P located a perpendicular distance d from the end of the wire that is at $x = 0$?

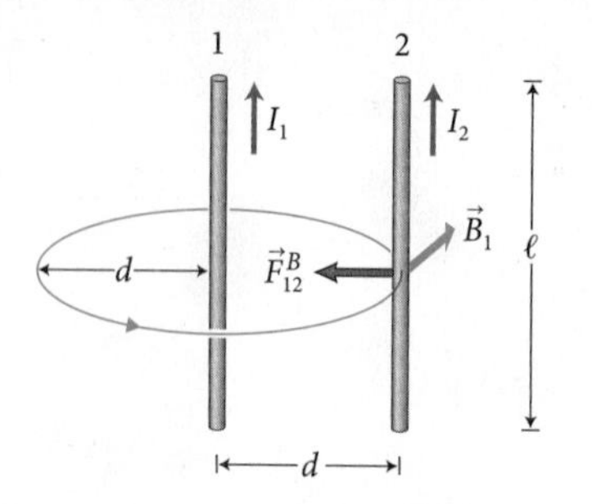

Figure 28.34 Calculating the magnetic force exerted by one current-carrying wire on another.

The expression for the magnitude of the magnetic field generated by a long straight wire obtained in Example 28.6 can lead us to an expression for the forces exerted by two current-carrying wires on each other. Consider the situation illustrated in **Figure 28.34**: Two parallel wires, of length ℓ and separated by a distance d, carry currents of magnitudes I_1 and I_2. To determine the magnetic force exerted by wire 1 on wire 2, we must first determine the magnitude and direction of the magnetic field generated by wire 1 at the location of wire 2 and then substitute this information into Eq. 27.8, which gives the magnetic force exerted on a straight current-carrying wire of length ℓ in a magnetic field $\vec{B}$:

$$\vec{F}^B = I\vec{\ell} \times \vec{B}, \tag{28.13}$$

where $\vec{\ell}$ is a vector whose magnitude is given by the length ℓ of the wire and whose direction is given by that of the current through the wire. Using the right-hand current rule, we see that the magnetic field $\vec{B}_1$ generated by wire 1 at the location of wire 2 points into the page (Figure 28.34). The result I obtained in Example 28.6 gives the magnitude of this field:

$$B_1 = \frac{\mu_0 I_1}{2\pi d}. \tag{28.14}$$

Because the magnetic field is perpendicular to $\vec{\ell}$, Eq. 28.13 yields for the magnitude of the magnetic force acting on wire 2

$$F^B_{12} = I_2 \ell B_1 \tag{28.15}$$

or, substituting Eq. 28.14,

$$F^B_{12} = \frac{\mu_0 \ell I_1 I_2}{2\pi d} \quad \text{(parallel straight wires carrying constant currents).} \tag{28.16}$$

The direction of this force follows from the vector product in Eq. 28.13: Using the right-hand force rule (place the fingers of your right hand along $\vec{\ell}$ in Figure 28.34, which points in the direction of I_2, and bend them toward $\vec{B}_1$), you can verify that the force points toward wire 1.

Example 28.7 The magnetic field generated by a circular arc of current-carrying wire

A wire bent into a circular arc of radius R subtending an angle ϕ carries a current of magnitude I (**Figure 28.35**). Use the Biot-Savart law to derive an expression for the magnitude of the magnetic field $\vec{B}$ produced at point P, located at the center of the arc.

Figure 28.35 Example 28.7

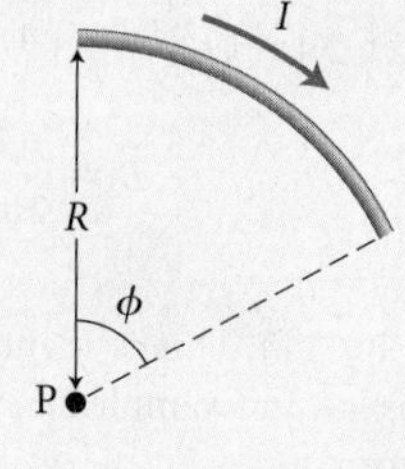

❶ GETTING STARTED I begin by evaluating what a small segment of length $d\ell$ along the arc contributes to the magnetic field. I therefore make a sketch showing one segment $d\vec{\ell}$ and the vector $\vec{r}_{sP}$ pointing from the segment to P (**Figure 28.36**). Using the right-hand vector product rule, I see that the direction of $d\vec{\ell} \times \vec{r}_{sP}$, and therefore the magnetic field $d\vec{B}_s$, are into the page at P.

Figure 28.36

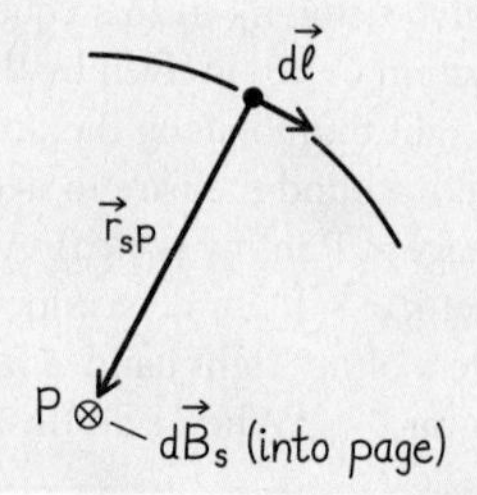

❷ **DEVISE PLAN** Because all segments contribute a magnetic field in the same direction, I can integrate Eq. 28.11 over the arc to obtain the magnitude of the magnetic field at P.

❸ **EXECUTE PLAN** Because $d\vec{\ell}$ and $\vec{r}_{sP}$ are always perpendicular to each other, I can write for Eq. 28.11

$$dB_s = \frac{\mu_0}{4\pi}\frac{I\,d\ell \sin 90°}{R^2} = \frac{\mu_0}{4\pi}\frac{I\,d\ell}{R^2}, \quad (1)$$

where I have substituted the radius R for the magnitude of $\vec{r}_{sP}$. To change the integration variable from $d\ell$ to the angular variable $d\phi$, I substitute $d\ell = R\,d\phi$ into Eq. 1:

$$dB_s = \frac{\mu_0 I}{4\pi R}\,d\phi$$

$$B = \int_{arc} dB_s = \frac{\mu_0 I}{4\pi R}\int_0^{\phi} d\phi$$

$$= \frac{\mu_0 I \phi}{4\pi R}. ✓$$

❹ **EVALUATE RESULT** My expression for B shows that at P the magnetic field magnitude is proportional to the current through the arc, as I expect. It is also proportional to the angle ϕ of the arc, also what I expect given that two such arcs should yield twice the magnetic field. Finally, B is inversely proportional to the arc radius R. That dependence on the radius makes sense because increasing the radius increases the distance between P and the arc, thus diminishing the magnetic field.

28.16 What is the magnitude of the magnetic field (*a*) at the center of a circular current loop of radius R and (*b*) at point P near the current loop in **Figure 28.37**? Both loops carry a current of constant magnitude I.

Figure 28.37 A current loop carrying a current of constant magnitude I (Checkpoint 28.16).

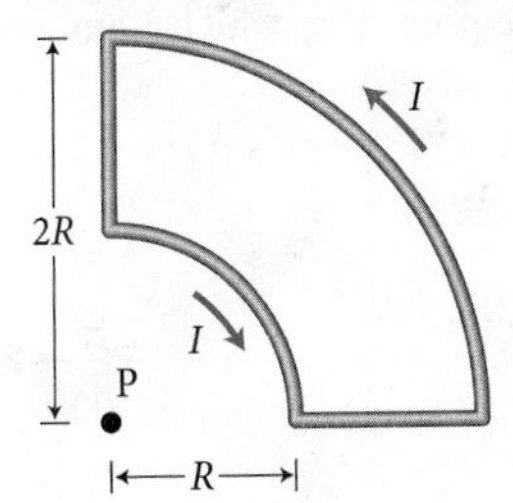

28.8 Magnetic field of a moving charged particle

Let us now use the Biot-Savart law to obtain an expression for the magnetic field caused by charged particles moving at constant velocity.* Consider first a straight wire carrying a current of magnitude I and aligned with the x axis, as in **Figure 28.38*a***. The magnetic field generated at point P by a small segment $d\vec{x}$ is given by Eq. 28.12:

$$d\vec{B}_s = \frac{\mu_0}{4\pi}\frac{I\,d\vec{x} \times \hat{r}_{sP}}{r_{sP}^2}, \quad (28.17)$$

where $\hat{r}_{sP}$ is a unit vector pointing from the segment $d\vec{x}$ to the point at which we wish to determine the magnetic field, and r_{sP} is the distance between the segment and the point P. Suppose the segment contains an amount of charge dq. Let the charge carriers responsible for the current take a time interval dt to have displacement $d\vec{x}$ (Figure 28.38*b*). According to the definition of current (Eq. 27.2), we have

$$I \equiv \frac{dq}{dt}, \quad (28.18)$$

and so

$$I d\vec{x} = \frac{dq}{dt}d\vec{x} = dq\,\frac{d\vec{x}}{dt} = dq\,\vec{v}, \quad (28.19)$$

where $\vec{v}$ is the velocity at which the charge carriers move down the wire. In the limiting case where the segment $d\vec{x}$ contains just a single charge carrier carrying a charge q (Figure 28.38*c*), dq becomes q and

$$I d\vec{x} = q\vec{v}. \quad (28.20)$$

Figure 28.38 We use the Biot-Savart law to obtain an expression for the magnetic field caused by charged particles moving at constant velocity.

(*a*) Small segment of current-carrying wire causes magnetic field at point P

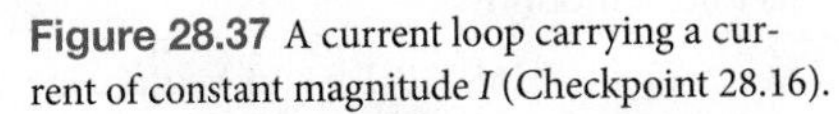
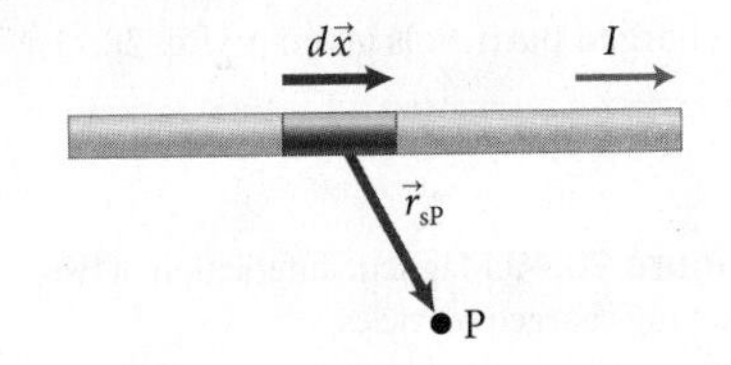

(*b*) Displacement of charge dq in time interval dt

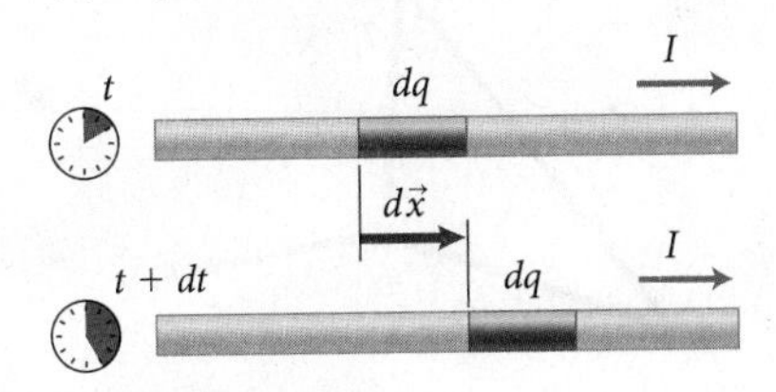

(*c*) Displacement of charged particle in time interval dt

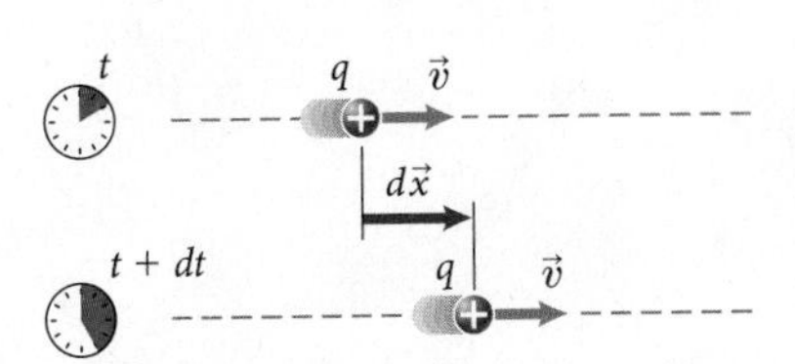

*In the derivation that follows, we assume $v \ll c_0$ and ignore any relativistic effects as described in Sections 27.4 and 27.8.

QUANTITATIVE TOOLS

Substituting this result into Eq. 28.17, we obtain an expression for the magnetic field of a single moving charged particle:

$$\vec{B} = \frac{\mu_0}{4\pi}\frac{q\vec{v} \times \hat{r}_{pP}}{r_{pP}^2} \quad \text{(single particle)}, \tag{28.21}$$

where r_{pP} is the distance between the particle and P, and $\hat{r}_{pP}$ is the unit vector pointing from the particle to P.

Example 28.8 Magnetic field generated by a moving electron

An electron carrying a charge $-e = -1.60 \times 10^{-19}$ C moves in a straight line at a speed $v = 3.0 \times 10^7$ m/s. What are the magnitude and direction of the magnetic field caused by the electron at a point P 10 mm ahead of the electron and 20 mm away from its line of motion?

❶ GETTING STARTED I begin by drawing the moving electron and the point P at which I am to determine the magnetic field (**Figure 28.39**).

Figure 28.39

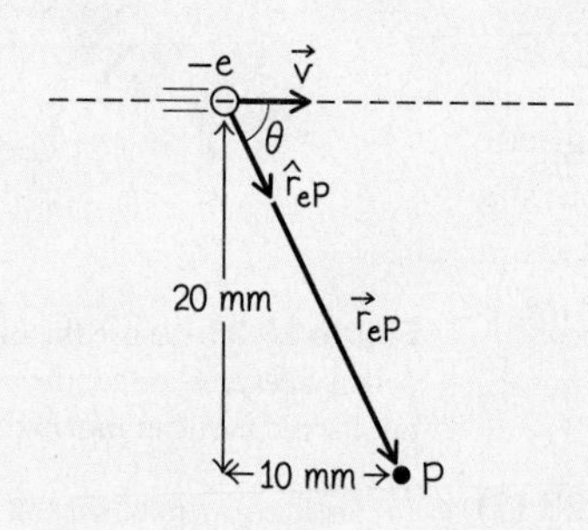

❷ DEVISE PLAN The magnetic field created by a moving charged particle is given by Eq. 28.21.

❸ EXECUTE PLAN The unit vector $\hat{r}_{eP}$ points from the electron to P, and so $\vec{v} \times \hat{r}_{eP}$ points into the page. Because the charge of the electron is negative, the magnetic field points in the opposite direction, out of the page. ✔

The magnitude of the vector $\vec{r}_{eP}$ in Figure 28.39 is $r_{eP} = \sqrt{(10 \text{ mm})^2 + (20 \text{ mm})^2} = 22.36$ mm, and $\sin\theta = (20 \text{ mm})/(22.36 \text{ mm}) = 0.8944$. Substituting these values into Eq. 28.11 thus yields for the magnitude of the magnetic field

$$B = \frac{(4\pi \times 10^{-7} \text{ T}\cdot\text{m/A})}{4\pi}\frac{(1.60 \times 10^{-19} \text{ C})(3.0 \times 10^7 \text{ m/s})(0.8944)}{(22.36 \times 10^{-3} \text{ m})^2}$$

$$= 8.6 \times 10^{-16} \text{ T.} ✔$$

❹ EVALUATE RESULT The magnetic field magnitude I obtained is much too small to be detected, but that's what I expect for a single electron. I can verify the direction of the magnetic field by applying the right-hand current rule to the current caused by the electron. Because the electron carries a negative charge, its motion to the right in Figure 28.39 causes a current to left. Pointing my right-hand thumb to the left in the figure, I observe that the fingers curl out of the page at P, in agreement with what I determined earlier for the direction of $\vec{B}$.

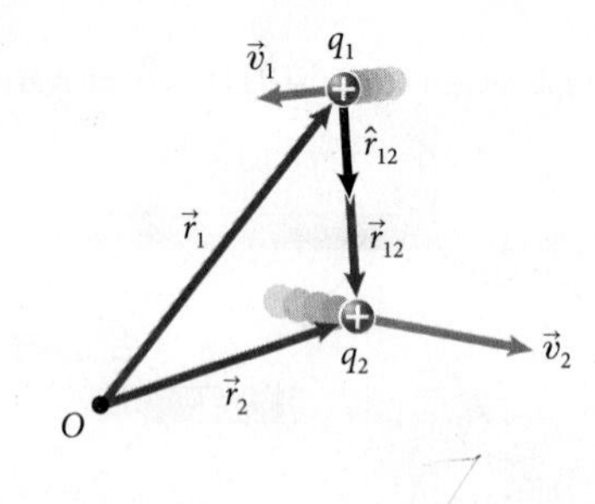

Figure 28.40 Magnetic interaction of two moving charged particles.

We can now combine the expression for the magnetic field caused by a moving charged particle (Eq. 28.21) with that for the magnetic force exerted on another moving charged particle (Eq. 27.19; $\vec{F}^B_p = q\vec{v} \times \vec{B}$) to determine the magnetic interaction between two moving charged particles. Consider the situation illustrated in **Figure 28.40**. The magnetic field caused by particle 1 at the location of particle 2 is given by Eq. 28.21:

$$\vec{B}_1(\vec{r}_2) = \frac{\mu_0}{4\pi}\frac{q_1\vec{v}_1 \times \hat{r}_{12}}{r_{12}^2}, \tag{28.22}$$

where r_{12} is the magnitude of the vector $\vec{r}_{12}$ pointing from particle 1 to particle 2 and $\hat{r}_{12}$ is a unit vector pointing along $\vec{r}_{12}$. Substituting this expression into Eq. 27.19, we obtain for the magnetic force exerted by particle 1 on particle 2

$$\vec{F}^B_{12} = q_2\vec{v}_2 \times \vec{B}_1(\vec{r}_2) = \frac{\mu_0}{4\pi}\frac{q_1 q_2}{r_{12}^2}\vec{v}_2 \times (\vec{v}_1 \times \hat{r}_{12}). \tag{28.23}$$

Notice the appearance of the double vector product: We must first take the vector product of $\vec{v}_1$ and $\hat{r}_{12}$ and then take the vector product of $\vec{v}_2$ and the result of the first vector product.

Equation 28.23 also shows that we need a moving charged particle in order to generate a magnetic field ($\vec{v}_1 \neq 0$) and another moving charged particle to "feel" that magnetic field ($\vec{v}_2 \neq 0$), in agreement with our discussion in the first part of this chapter.

In contrast to the electric force, the magnetic force does not satisfy Eq. 8.15, $\vec{F}_{12} = -\vec{F}_{21}$. To see this, consider the two positively charged moving particles shown in **Figure 28.41**. Particle 1, carrying a charge q_1, travels in the positive x direction, while particle 2, carrying charge q_2, travels in the positive y direction. The force exerted by 1 on 2 is given by Eq. 28.23. Applying the right-hand vector product rule, we determine that the vector product $\vec{v}_1 \times \hat{r}_{12}$ on the right side of Eq. 28.23 points out of the page. Applying the right-hand vector product rule again to the vector product of $\vec{v}_2$ and $\vec{v}_1 \times \hat{r}_{12}$, we see that $\vec{F}^B_{12}$ points in the positive x direction.* The force exerted by 2 on 1 is obtained by switching the subscripts 1 and 2 in Eq. 28.23:

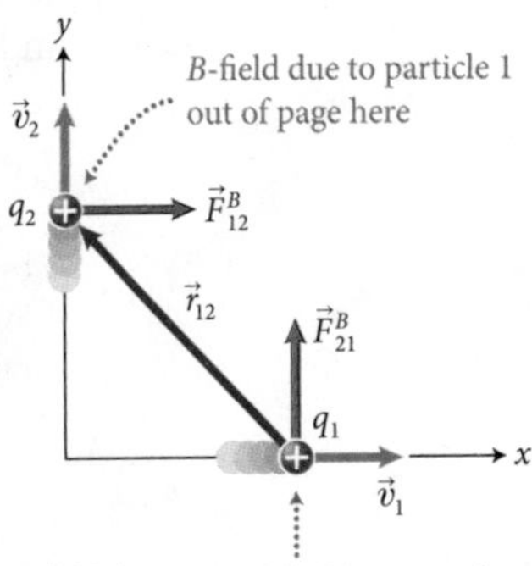

Figure 28.41 The forces that moving charged particles exert on each other do not necessarily point in opposite directions.

$$\vec{F}^B_{21} = \frac{\mu_0}{4\pi}\frac{q_1q_2}{r^2_{12}}\vec{v}_1 \times (\vec{v}_2 \times \hat{r}_{21}), \tag{28.24}$$

where we have used that $q_1q_2 = q_2q_1$ and $r^2_{12} = r^2_{21}$. Applying the right-hand vector product rule twice, we find that $\vec{F}^B_{21}$ points in the positive y direction, and so the magnetic forces that the two particles exert on each other, while equal in magnitude, do not point in opposite directions, as we would expect from Eq. 8.15.

We derived Eq. 8.15 from the fact that the momentum of an isolated system of two particles is constant, which in turn follows from conservation of momentum, one of the most fundamental laws of physics. The electric and magnetic fields of isolated moving charged particles, however, are not constant, and as we shall see in Chapter 30, we can associate a flow of both momentum and energy with changing electric and magnetic fields. Therefore we need to account not only for the momenta of the two charged particles in Figure 28.41, but also for the momentum carried by their fields—which goes beyond the scope of this book. Even though Eq. 8.15 is not satisfied by the magnetic force, the momentum of the system comprising the two particles and their fields is still constant.

Substituting Coulomb's law and Eq. 28.23 into Eq. 27.20, $\vec{F}^{EB}_{\text{p}} = q(\vec{E} + \vec{v} \times \vec{B})$, we obtain an expression for the electromagnetic force that two moving charged particles exert on each other:

$$\vec{F}^{EB}_{12} = \frac{1}{4\pi\epsilon_0}\frac{q_1q_2}{r^2_{12}}[\hat{r}_{12} + \mu_0\epsilon_0\vec{v}_2 \times (\vec{v}_1 \times \hat{r}_{12})]. \tag{28.25}$$

By comparing the result obtained in Example 28.6 with Eq. 27.38 and substituting $k = 1/(4\pi\epsilon_0)$ (Eq. 24.7), we see that $\mu_0 = 1/(\epsilon_0c^2_0)$ or $\mu_0\epsilon_0 = 1/c^2_0$, where c_0 is the speed of light. Using this information, we can write Eq. 28.25 as

$$\vec{F}^{EB}_{12} = \frac{1}{4\pi\epsilon_0}\frac{q_1q_2}{r^2_{12}}\left[\hat{r}_{12} + \frac{\vec{v}_2 \times (\vec{v}_1 \times \hat{r}_{12})}{c^2_0}\right] \quad (v \ll c_0). \tag{28.26}$$

*Here is another way you can determine the direction of the magnetic force exerted by particle 1 on particle 2. The motion of particle 1 corresponds to a current directed in the positive x direction. Using the right-hand current rule, it follows that the magnetic field of 1 points out of the xy plane at any location above the x axis. The magnetic field due to 1 at the location of 2 thus points out of the plane. The motion of particle 2 corresponds to a current directed in the positive y direction. To determine the direction of the magnetic force exerted on this "current," we can use the right-hand force rule. When you curl the fingers of your right hand from the positive y axis to the direction of $\vec{B}_1$ at particle 2 (out of the page), your thumb points in the positive x direction, as we found before.

This equation is the most general expression for the electromagnetic interaction between moving charged particles. Because in most applications the speeds of charged particles are significantly smaller than the speed of light, the second term in the square brackets in Eq. 28.26 is much smaller than the first term, which represents the electric contribution to the electromagnetic force from Coulomb's law.

28.17 Consider two protons 1 and 2, each carrying a charge $+e = 1.6 \times 10^{-19}$ C, separated by 1.0 mm moving at 3×10^5 m/s parallel to each other and perpendicular to their separation. *(a)* What is the direction of the magnetic force that each proton exerts on the other? *(b)* Determine the ratio of the magnitudes of the magnetic and electric forces that the two exert on each other.

Chapter Glossary

SI units of physical quantities are given in parentheses.

Ampère's law The line integral of the magnetic field along a closed path (called an *Ampèrian path*) is proportional to the current encircled by the path, I_{enc}:

$$\oint \vec{B} \cdot d\vec{\ell} = \mu_0 I_{enc}. \qquad (28.1)$$

In analogy to Gauss's law in electrostatics, Ampère's law allows us to determine the magnetic field for current distributions that exhibit planar, cylindrical, or toroidal symmetry.

Ampèrian path A closed path along which the magnetic field is integrated in Ampère's law.

Biot-Savart law An expression that gives the magnetic field at a point P due to a small segment $d\vec{\ell}$ of a wire carrying a current I:

$$d\vec{B}_s = \frac{\mu_0}{4\pi} \frac{I\, d\vec{\ell} \times \hat{r}_{sP}}{r_{sP}^2}, \qquad (28.12)$$

where $\hat{r}_{sP}$ is a unit vector pointing from the segment to the point at which the magnetic field is evaluated. The Biot-Savart law can be used to calculate the magnetic field of a current-carrying conductor of arbitrary shape by integration:

$$\vec{B} = \int_{\text{current path}} d\vec{B}_s. \qquad (28.10)$$

Current loop A current-carrying conductor in the shape of a loop. The magnetic field pattern of a current loop is similar to that of a magnetic dipole.

Magnetic constant μ_0 (T · m/A) A constant that relates the current encircled by an Ampèrian path and the line integral of the magnetic field along that path. In vacuum:

$$\mu_0 = 4\pi \times 10^{-7}\ \text{T} \cdot \text{m/A}.$$

Magnetic dipole moment $\vec{\mu}$ (A · m²) A vector that points from the S pole to the N pole for a bar magnet or along the axis of a planar current loop in the direction given by the right-hand thumb when the fingers of that hand are curled along the direction of the current through the loop. In an external magnetic field, the magnetic dipole moment tends to align in the direction of the external magnetic field.

Solenoid A long, tightly wound helical coil of wire. The magnetic field of a current-carrying solenoid is similar to that of a bar magnet.

Toroid A solenoid bent into a circle. The magnetic field of a toroid is completely contained within the windings of the toroid.

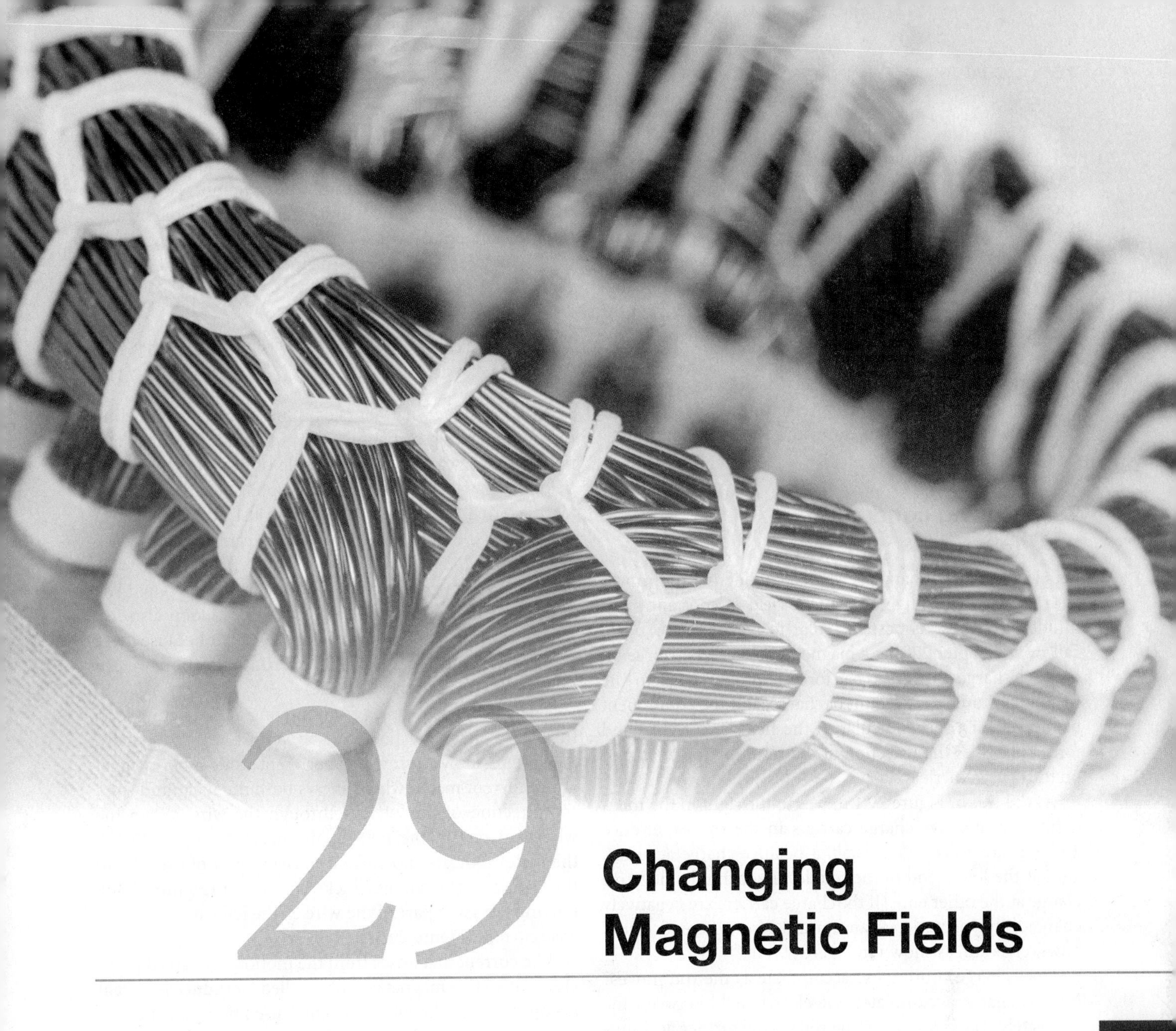

29 Changing Magnetic Fields

Up to now, the only electric fields we have encountered are those that arise from the presence of electrical charges. In this chapter we explore electric fields that accompany changing magnetic fields. As we saw in Chapter 27, constant magnetic fields exert forces on moving charge carriers. Now we discover that there is no fundamental difference between the interaction of changing magnetic fields with stationary charge carriers and the interaction of constant magnetic fields with moving charge carriers. We shall also learn that energy can be stored in magnetic fields. These ideas are harnessed in a wide variety of important applications, from electric motors and electrical power generation to electronic appliances and other everyday devices.

29.1 Moving conductors in magnetic fields

Let's examine what happens when a conducting rod moves with velocity $\vec{v}$ through a magnetic field of constant magnitude B, where $\vec{v}$ is perpendicular to $\vec{B}$ (**Figure 29.1**). Recall from Chapter 27 that a magnetic field exerts a force on moving charge carriers. That force is proportional to the field magnitude, the amount of charge, and the component of the velocity perpendicular to the field. The right-hand force rule gives the direction of the resulting force exerted on the charge carriers.

As the rod in Figure 29.1 moves to the right in the magnetic field, positive charge carriers in the rod experience a downward force. As a result, a positive charge accumulates at the lower end of the rod, leaving behind a negative charge at the other end. (If the charge carriers are negatively charged, they experience an upward force; the end result, however, is still as shown in Figure 29.1.)

Does charge continue to accumulate as the rod moves? No—as charge accumulates, the electric field between the oppositely charged ends of the rod resists further accumulation of charge. Once the opposing force due to the electric field is equal in magnitude to the force exerted by the magnetic field on the charge carriers, the vector sum of the forces exerted on the charge carriers is zero, and no further accumulation of charge takes place. The motion of the rod through the magnetic field thus establishes a charge separation, the amount of which is determined by the magnitude of the force exerted by the magnetic field on the charge carriers.

29.1 What happens if the rod in Figure 29.1 moves to the left?

Now consider what happens if, instead of a conducting rod, a conducting rectangular loop moves into a magnetic field (**Figure 29.2**). As the right side of the loop enters the field, its charge carriers move in response to the magnetic force exerted on them. Rather than just accumulating at the ends of the right side, however, the carriers now have a closed path along which to travel. Thus, the magnetic force exerted on the carriers creates a current in the loop. (The carriers in the top and bottom sides of the loop also feel magnetic forces, as shown in Figure 29.2, but because these carriers can't move very far in the vertical direction, we can ignore the effects of these forces.)

The current that arises from moving a conducting loop through a magnetic field can be verified experimentally. **Figure 29.3** shows a simple experiment using a bar magnet and a current meter, which shows both the magnitude and the direction of the current through the wire. When the wire is at rest in the field of the magnet (Figure 29.3*a*), the meter indicates no current. Moving part of the wire to the right causes a counterclockwise current (Figure 29.3*b*); moving the same part of the wire to the left causes a clockwise current (Figure 29.3*c*).

The current that arises from the motion of charged particles relative to a magnetic field is called an **induced current** (as opposed to a "regular" current caused by a potential difference between, for instance, the terminals of a battery).

Figure 29.1 When a conducting rod moves in a magnetic field, charge carriers in the rod experience a magnetic force. In the configuration shown, the right-hand force rule tells us that a positive charge accumulates at the bottom of the rod.

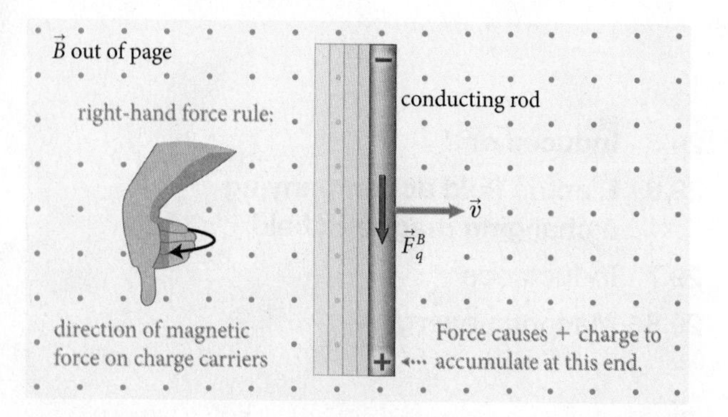

Figure 29.2 Direction of current in a rectangular loop moving into a magnetic field. Magnetic forces exerted on the charge carriers in the loop cause them to move around the loop.

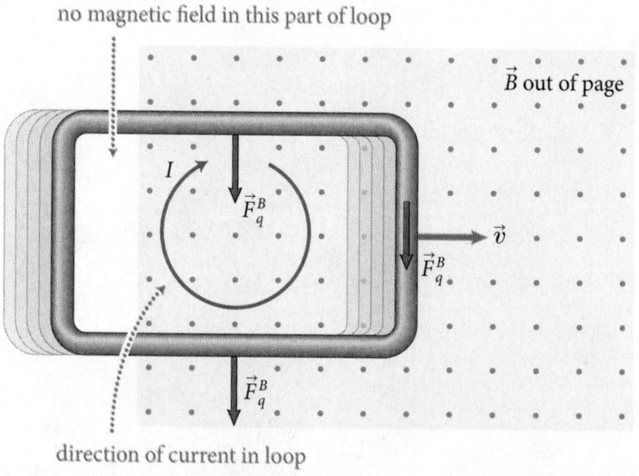

CONCEPTS

Figure 29.3 Experimental observation of induced current.

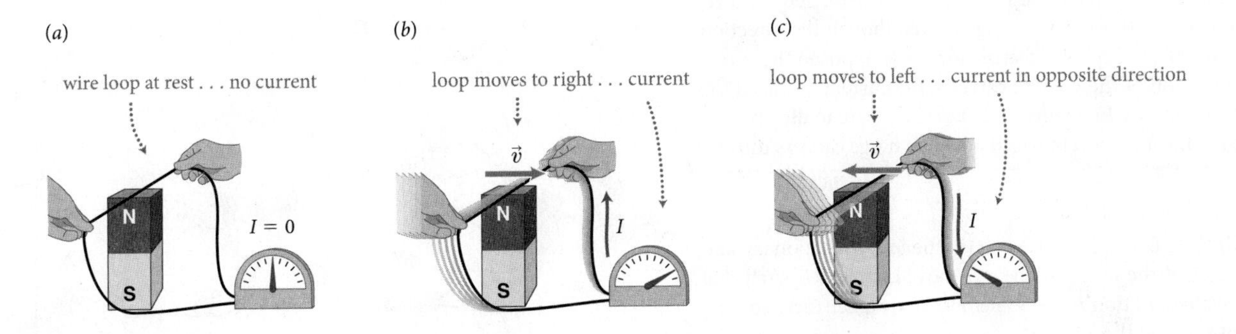

Example 29.1 A rectangular loop passing through a magnetic field

Consider a rectangular conducting loop traveling from left to right through a magnetic field as shown in **Figure 29.4**. During this motion, at what positions *a–e* is a current induced in the loop, and what is the current direction at each position?

❶ GETTING STARTED As the loop moves through the magnetic field, the charge carriers in the loop experience a magnetic force. For each of the five positions I need to determine whether or not the magnetic force causes the carriers to flow around the loop. To simplify the problem, I assume that only the positive charge carriers are free to move.

❷ DEVISE PLAN The direction of the magnetic force exerted on the charge carriers is given by the right-hand force rule. To see if there is a current in the loop, I'll sketch the loop in each position and determine the direction of the magnetic force exerted on the charge carriers in each side of the loop.

❸ EXECUTE PLAN No matter where along the loop perimeter the carriers are, they always move to the right with the loop, and so curling the fingers of my right hand from $\vec{v}$ to $\vec{B}$, I see that the magnetic force exerted on them is always toward the bottom of the page. I draw this force for each side of the loop that is inside the magnetic field (**Figure 29.5**). The magnetic forces exerted on the top and bottom sides of the loop cause no current through the loop because the charge carriers in these sides have essentially no mobility in the direction in which $\vec{F}^B$ is exerted.

In position (*a*), with no part of the loop in the magnetic field, no magnetic force is exerted on any of the carriers, and therefore no current is induced. When the right side of the loop passes into the magnetic field but the left edge has not yet entered (*b*), the magnetic force exerted on the charge carriers in the right side of the loop causes a clockwise current around the loop. Once the entire loop is in the magnetic field (*c*), the magnetic force exerted on the carriers in the left side drives a counterclockwise current equal in magnitude to the clockwise current driven by the magnetic force exerted on the carriers in the right side. The combined effect is that no current is induced. In position (*d*), only the left side of the loop is in the field, and the magnetic force exerted on those charge carriers induces a counterclockwise current around the loop. When the loop has passed out of the magnetic field (*e*), once again no magnetic forces are exerted on the carriers and therefore no current is induced. ✔

Figure 29.5

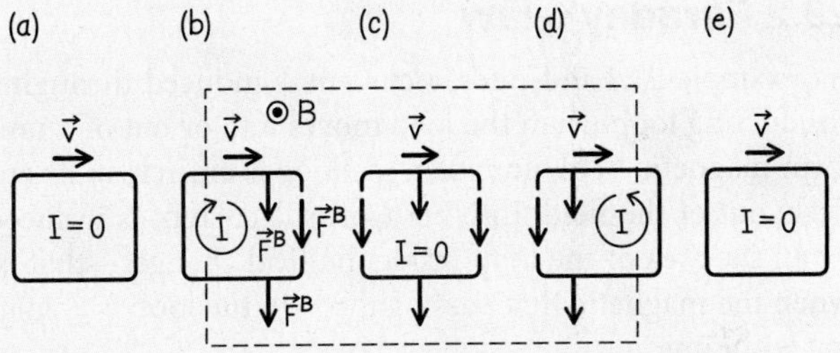

❹ EVALUATE RESULT In position (*b*) I have the same situation as in Figure 29.2, so it is reassuring that my answer is the same. If I ignore any dissipation of energy, the initial and final states are the same. Therefore it makes sense that the effect of moving the loop through the field is zero: The charge carriers first move clockwise as the loop moves into the field, then move counterclockwise as the loop moves out of the field.

Figure 29.4 Example 29.1.

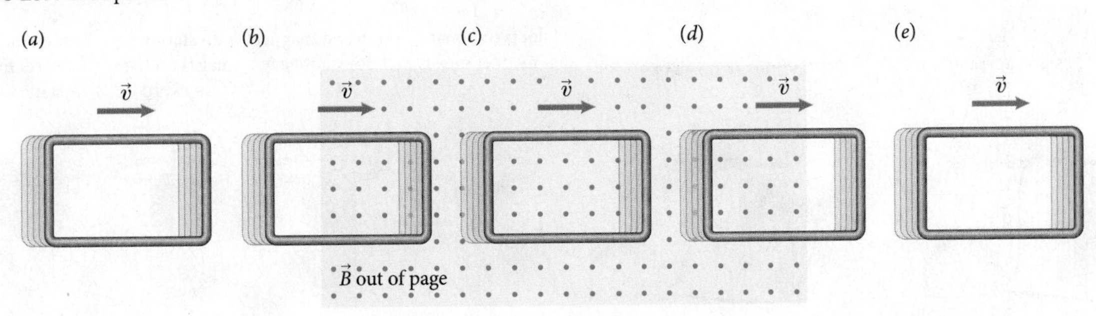

I assumed that only the positive charge carriers can move. Had I assumed that the negative charge carriers are free to move, my answer would not have changed. Even though the direction of motion of the negative charge carriers is opposite the direction of motion of the positive carriers, the current is still in the directions shown in Figure 29.5 because current direction is *defined* as the direction in which positive charge carriers move.

29.2 In Example 29.1, suppose the loop is stationary and the source of the magnetic field is moved to the left such that their relative motion is the same. Do you expect there to be a current through the loop?

29.2 Faraday's law

As Example 29.1 indicates, a current is induced through a conducting loop when the loop moves into or out of a uniform magnetic field, not when the loop is entirely in or entirely out of the field. In other words, a current is induced when the area of the loop inside the field changes—that is, when the magnetic flux passing through the loop is changing with time.

Experiments show that the rate at which the magnetic flux through the loop changes affects the magnitude of the induced current. As illustrated in **Figure 29.6**, moving the loop through the magnetic field faster, which causes a greater rate of change in the magnetic flux through the loop, produces a greater induced current.

As we learned in Chapters 6 and 14, physical phenomena do not depend on the reference frame in which we observe them. Indeed, if the change in magnetic flux through a conducting loop is responsible for inducing a current, it should not matter whether the loop is moving in a constant magnetic field or the source of the magnetic field is moving and the loop is stationary. Only the motion of one relative to the other matters. Experiments confirm that the same current is measured in the loop no matter whether the loop moves to the right with speed v or the source of the magnetic field moves to the left with speed v (**Figure 29.7**). In fact, current is induced in a stationary conducting loop in any region where a magnetic field is changing, even if the source of the magnetic field is stationary and the magnitude of the magnetic field is changing.

Figure 29.6 The speed at which a conducting loop is moved through a magnetic field affects the magnitude of the induced current.

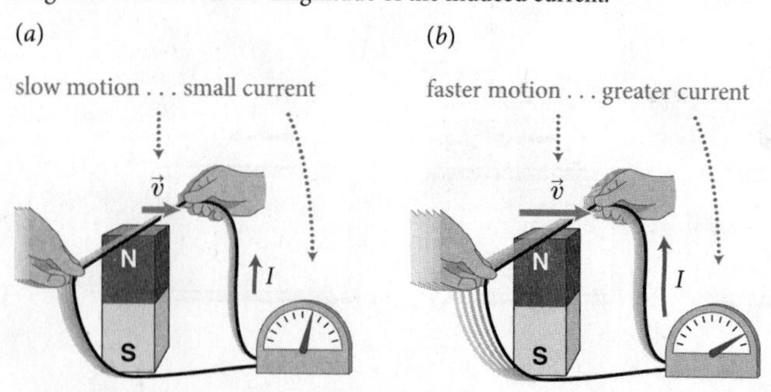

Figure 29.7 A current can be induced by moving either the loop or the magnet.

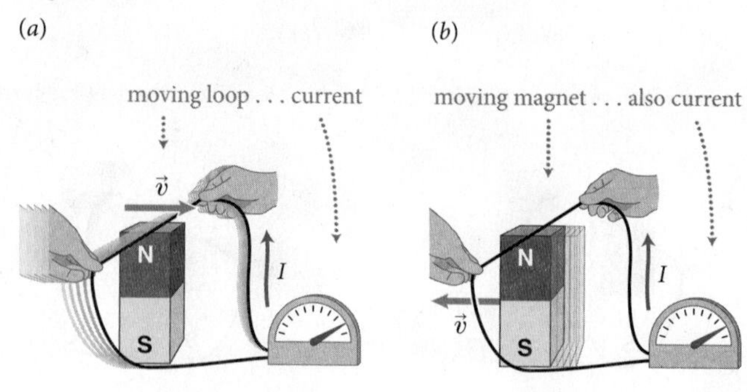

These experiments lead us to conclude:

A changing magnetic flux through a conducting loop induces a current in the loop.

This statement is called **Faraday's law.** The process by which a changing magnetic flux causes charge carriers to move, establishing a separation of charge or inducing a current, is called **electromagnetic induction.**

Example 29.2 Changing magnetic flux

(*a*) In which of the four loops shown in **Figure 29.8** is there a current that is caused by a magnetic force? (*b*) In which situation(s) is a current induced in the loop?

❶ GETTING STARTED I begin by making a sketch of the magnetic field of the bar magnet, based on what I learned in Chapter 27 (Figure 27.13). I show a cross section of the loop placed above the north pole of the magnet (**Figure 29.9**).

❷ DEVISE PLAN Magnetic forces are exerted on charge carriers moving relative to a magnetic field, provided there is a nonzero component of the magnetic field perpendicular to the carriers' velocity. To answer part *a*, therefore, I must determine which of the four situations satisfies those conditions. For part *b* I know that a changing magnetic flux induces a current, and so for each situation I must establish whether or not the magnetic flux through the loop is changing.

❸ EXECUTE PLAN (*a*) Because the loop is stationary in situations 1 through 3, there cannot be a magnetic force in those situations. For situation 4, I draw vectors in Figure 29.9 indicating the

Figure 29.8 Example 29.2.

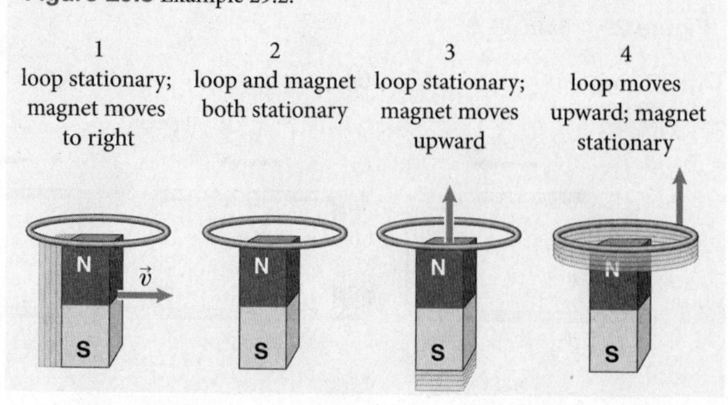

CONCEPTS

Figure 29.9

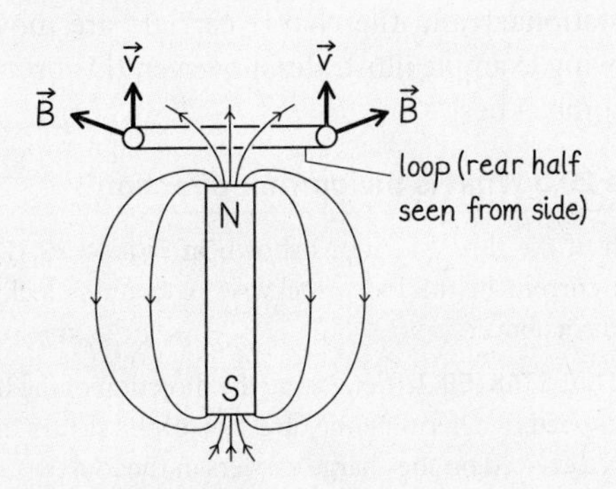

directions of the loop velocity and of the magnetic field at the position of the ring and the velocity of the loop. Because the magnetic field lines spread outward from the north pole, the magnetic field is not parallel to the velocity, and so a magnetic force is exerted on the charge carriers. According to the right-hand force rule, the force is in opposite directions on the two sides of the loop, so this magnetic force creates a current in the loop. ✔

(*b*) Because the field of a bar magnet is not uniform, any relative motion between the magnet and the loop causes the magnetic flux through the loop to change. Therefore, the magnetic flux is changing in situations 1, 3, and 4, which means current is induced in the loop in these three cases. ✔

❹ **EVALUATE RESULT** That there is a current created by a magnetic force only in situation 4 makes sense because a magnetic force requires a nonzero velocity of the charge carriers and only in situation 4 is the loop moving. That there is an induced current in every situation except 2 makes sense because Faraday's law tells me that a current is induced whenever there is a changing magnetic flux through the loop—that is, whenever the loop and magnet move relative to each other.

29.3 Is a magnetic force exerted on the (stationary) charge carriers in the loop of wire held above the magnet in Figure 29.7*b*?

29.3 Electric fields accompany changing magnetic fields

Example 29.2 and Checkpoint 29.3 lead to a surprising conclusion: Although no magnetic force is exerted on the charge carriers in a stationary loop, a current is still induced! **Figure 29.10** shows this situation in more detail. Experiments show that as a magnetic field moves past a stationary conducting rod, a charge separation and hence a potential difference develop between the ends of the rod even though no magnetic force is exerted on stationary charge carriers.

The potential difference that develops between the ends of the rod shown in Figure 29.10 is the same as that which would develop if the magnetic field were stationary and the rod were moving to the right (recall Figure 29.1). Any motion of the rod relative to the magnetic field produces

Figure 29.10 A stationary conducting rod in a moving magnetic field develops a charge separation, but no magnetic force is exerted on the charge carriers because their speed in the rod is zero, which means $\vec{v} \times \vec{B} = \vec{0}$.

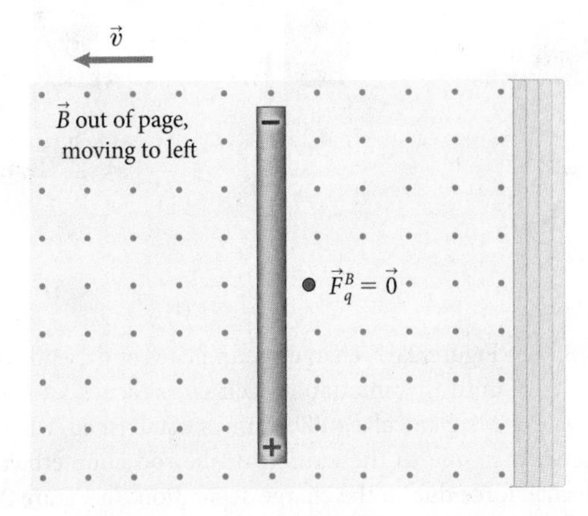

the same potential difference across the ends of the rod. By the same token, a current is induced in the stationary rectangular loop of **Figure 29.11** by the moving magnetic field, just as a current is induced in the rectangular loop moving through a constant magnetic field in Figure 29.2. In other words, whenever relative motion occurs between charge carriers and the source of a magnetic field, the resulting rearrangement of charge and potential difference is independent of the choice of inertial reference frame, as are any resulting currents.

What force causes the charge carriers to flow when the loop is stationary? We have encountered two types of forces that are exerted on charged particles: Electric and magnetic. Magnetic forces are exerted only on moving charged particles, so these forces cannot be responsible for the induced current in Figure 29.11. Only electric forces are exerted on stationary charged particles, so we must conclude that the force that causes charge to separate in Figure 29.10 is an electric force (because it behaves like one). Because electric forces are caused by electric fields, we are led to conclude:

A changing magnetic field is accompanied by an electric field.

Figure 29.11 A current is induced in a stationary conducting loop as a magnetic field moves past it when not all of the loop is in the field, even though no magnetic force is exerted on the charge carriers.

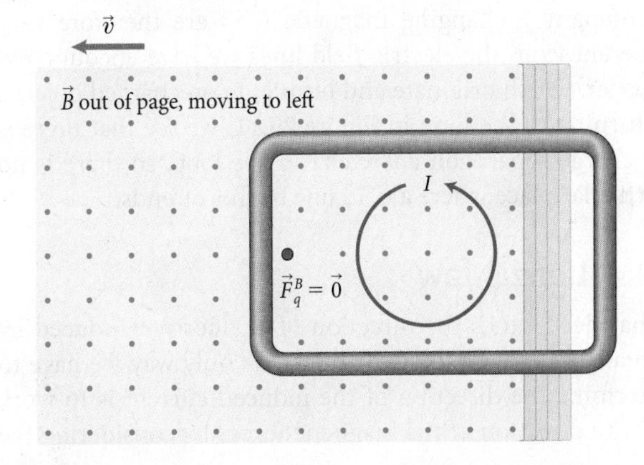

Figure 29.12 Examples 29.3 and 29.4.

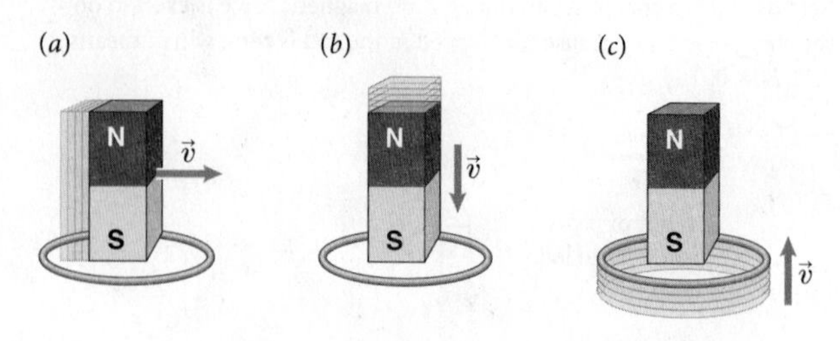

29.4 In Figure 29.1, charge accumulates at the ends of the moving rod until the amount at each end reaches an equilibrium value. Mechanical equilibrium is established when the magnetic force due to the motion of the rod counterbalances the electric force due to the charge separation. In Figure 29.10, what two forces determine the equilibrium state of the charge separation in the stationary rod?

We see that the nature of the force causing electromagnetic induction depends on the choice of reference frame. Forces that appear to be magnetic in one inertial reference frame are electric in another inertial reference frame. Indeed, the fact that our choice of reference frame affects whether a force is magnetic or electric is further evidence that electric and magnetic fields arise from fundamentally the same interaction (see Section 27.4). Because viewing the situation from a different inertial reference frame cannot alter the underlying interaction, magnetic and electric forces must be two manifestations of the same interaction.

29.5 As viewed from above, what is the direction of the induced current in situations 1 and 3 in Figure 29.8?

What do the field lines of the electric field that induces the current in Figure 29.11 look like? Electric field lines show the direction of the electric force exerted on a positively charged test particle. Because a counterclockwise current is induced in the loop (see position d in Example 29.1), we know that positive charge carriers in the loop experience a force directed counterclockwise all along the loop. This leads us to conclude that the field lines must form loops that close on themselves. The electric field lines that accompany a changing magnetic field are therefore very different from the electric field lines we have encountered thus far, which originate and terminate on charged objects. Returning to the loop in Figure 29.11, we see that there is no charge separation anywhere in the loop, so there is no particular place where a field line begins or ends.

29.4 Lenz's law

What determines the direction of the current induced by a changing magnetic flux? So far, the only way we have to determine the direction of the induced current is to work out the direction of the magnetic forces by considering the situation from a frame of reference in which the magnetic field is stationary and the charge carriers are moving. As the following example illustrates, however, this procedure is cumbersome at best.

Example 29.3 What is the current direction?

For each of the three situations shown in **Figure 29.12**, is the induced current in the loop clockwise or counterclockwise as viewed from above?

❶ GETTING STARTED To determine the direction of the induced current, I need to determine the direction of the electric or magnetic force exerted on the charge carriers in the loop.

❷ DEVISE PLAN I can use the right-hand force rule to determine the direction of the magnetic forces (assuming positive charge carriers), and the only way I know to do this is to consider each situation from a reference frame in which the magnetic field is stationary and the loop is in motion. For (*a*) I choose a reference frame moving along with the magnet; in this reference frame the loop moves to the left (**Figure 29.13*a***). Situations (*b*) and (*c*) are equivalent, so the direction of the induced current must be the same in both. I therefore need to consider only (*c*), where the magnetic field is stationary and the loop is in motion (Figure 29.13*b*). I can then use the right-hand force rule to determine the direction of the magnetic force at a couple of locations on the loop. Once I know these force directions, I can determine the direction of the induced current.

❸ EXECUTE PLAN I'll consider two locations on the loop: location P on the left and Q on the right. At each of these locations I draw arrows representing the velocity $\vec{v}$ of the (positive) charge carriers and the magnetic field $\vec{B}$. (All of the vectors I draw are in the vertical xz plane.)

(*a*) At location P on the loop in Figure 29.13*a*, the magnetic field points down and to the right, and the charge carriers in the loop are moving to the left at velocity $\vec{v}$. If I curl the fingers of my right hand from $\vec{v}$ to $\vec{B}$ at P, my thumb points out of the page, which means the magnetic force exerted on the charge carriers at P points out of the page. Thus at P the charge carriers are moving counterclockwise as viewed from above. At Q the magnetic field is up and to the left, and the charge carriers here are also moving to the left at velocity $\vec{v}$. When I curl the fingers of my right hand from $\vec{v}$ to $\vec{B}$, my thumb points into the page. This means the magnetic force at Q points into the page, moving the charge carriers in the loop counterclockwise as viewed from above. Thus the induced current is counterclockwise as seen from above. ✔

Figure 29.13

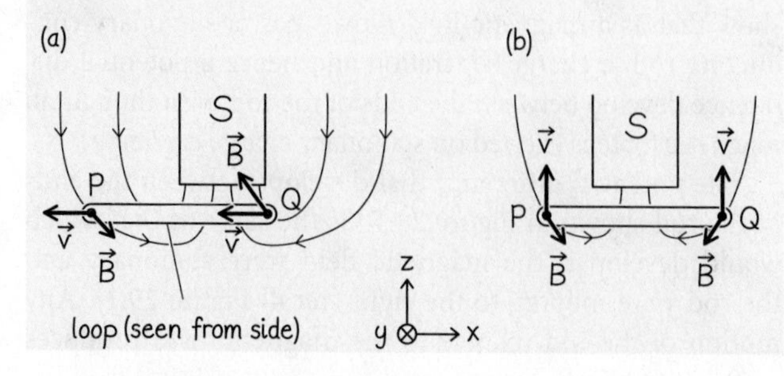

(b, c) At P in Figure 29.13b, the loop and its charge carriers are moving up at velocity $\vec{v}$, and the magnetic field has a vertical component and a component to the right in the plane of the page. Curling the fingers of my right hand from $\vec{v}$ to $\vec{B}$, I discover that the magnetic force exerted on the positive charge carriers points into the page. At Q it points out of the page, so the magnetic forces exerted on the charge carriers at P and Q induce a current that is clockwise as seen from above. ✔

❹ **EVALUATE RESULT** Comparing Figures 29.8 and 29.12, I note that Figure 29.12a is similar to situation 1 in Figure 29.8, and Figure 29.12b and c are similar to situations 3 and 4. In Example 29.2, I concluded that currents are induced in situations 1, 3, and 4, so in this example, too, I should expect currents in all three situations.

A simpler approach to determining the direction of an induced current follows from experimental observation:

The direction of an induced current through a conducting loop is always such that the magnetic flux produced by the induced current opposes the change in magnetic flux through the loop.

This principle is called **Lenz's law,** and the magnetic field resulting from the induced current is called the **induced magnetic field.**

Consider, for example, a conducting loop in a magnetic field that points up and is increasing in magnitude (Figure 29.14a). In order for the induced magnetic field $\vec{B}_{ind}$ to oppose the change in magnetic field, $\vec{B}_{ind}$ must point down, as shown in Figure 29.14b. What current direction produces a downward induced magnetic field? According to the right-hand dipole rule, the current must be clockwise as seen from above (Figure 29.14c). Indeed, this is what we found for a loop moving into a magnetic field in Example 29.1: As the magnetic flux through the loop increases, a clockwise current is induced.

Figure 29.14 Applying Lenz's law.

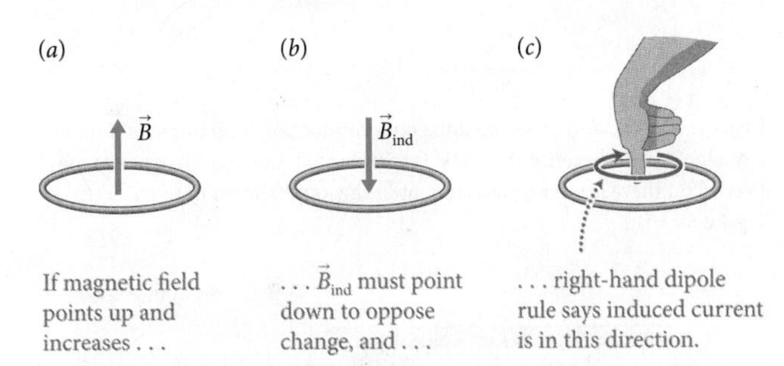
(a)
(b)
(c)
$\vec{B}$
$\vec{B}_{ind}$
If magnetic field points up and increases . . .
. . . $\vec{B}_{ind}$ must point down to oppose change, and . . .
. . . right-hand dipole rule says induced current is in this direction.

Example 29.4 Clockwise or counterclockwise?

Consider again the three situations of Figure 29.12. Use Lenz's law to determine the direction of the induced current in each loop.

❶ **GETTING STARTED** Using Lenz's law involves determining which direction for the induced current produces an induced magnetic field opposing the change in magnetic flux through the loop. The magnetic field that passes through the loop points primarily up (toward the south pole of the magnet) and is strongest along the long axis of the magnet.

❷ **DEVISE PLAN** According to Lenz's law, the induced current must be directed in such a way as to create an induced magnetic field $\vec{B}_{ind}$ that counteracts the change in magnetic flux. If the magnetic flux that causes the induced current decreases, $\vec{B}_{ind}$ must be in the same direction as the magnetic field of the magnet so that $\vec{B}_{ind}$ increases the magnetic flux through the loop. If the magnetic flux that causes the induced current increases, $\vec{B}_{ind}$ must be in the direction opposite the direction of the magnet's magnetic field. Once I know the direction of $\vec{B}_{ind}$, I can use the right-hand dipole rule to determine the direction of the current.

❸ **EXECUTE PLAN** (a) As the magnet moves to the right, the magnitude of the magnetic field at the loop, and therefore the magnetic flux through the loop, decrease. Therefore $\vec{B}_{ind}$ is upward, in the direction of the magnet's magnetic field. To use the right-hand dipole rule, I recall from Section 28.3 that the direction of a magnetic field near the poles of a bar magnet is given by the magnetic dipole moment vector, which points from south pole to north pole. Thus I point my right thumb in this direction in Figure 29.12a and see that my fingers curl counterclockwise as viewed from above, telling me that this is the direction of the induced current. ✔

(b) Moving the magnet toward the loop causes the magnetic flux through the loop to increase. The induced magnetic field $\vec{B}_{ind}$ associated with the induced current must therefore point downward, opposite the direction of the field created by the magnet, to oppose the increase in magnetic flux through the loop. The induced current is therefore clockwise as viewed from above. ✔

(c) Moving the loop toward the magnet results in an increase in magnetic flux through the loop, as in part b. The induced current therefore must be clockwise as viewed from above to produce a downward-pointing induced magnetic field $\vec{B}_{ind}$ that opposes the increase in flux. ✔

❹ **EVALUATE RESULT** My answers are the same as in Example 29.3, which gives me confidence that they are correct.

29.6 When current is induced in a conducting loop by the motion of a nearby magnet, the induced magnetic field $\vec{B}_{ind}$ exerts a force on the magnet. (a) In Figure 29.12b, what is the direction of the force exerted on the magnet by $\vec{B}_{ind}$? (b) Suppose Lenz's law stated that the induced current *adds to* the change that produced it instead of opposing it. What would be the direction of the force exerted by $\vec{B}_{ind}$ on the magnet?

Checkpoint 29.6 gives us a clue to the reason for Lenz's law. In Figure 29.12b, let's consider the system made up of magnet and loop. If the induced current further increased the magnetic flux through the loop, the induced current would exert an attractive force on the magnet. This attractive force would pull the magnet closer to the loop, increasing the magnetic flux further and inducing even more current. In the process, the kinetic energy of the magnet and the energy associated with the current in the loop would increase, increasing the energy of the magnet-loop system without any work being done on it, violating the law of conservation of energy.

Conservation of energy therefore *requires* that an induced current oppose the change that created it. In Figure 29.12*b*, for example, the force exerted by the induced magnetic field $\vec{B}_{ind}$ resists the magnet moving closer to the loop. An alternative statement of Lenz's law therefore is:

An induced current is always in such a direction as to oppose the motion or change that caused it.

Because of this opposing force, an agent moving the magnet at constant speed toward the loop must do work on the magnet. Where does this energy go? As work is done on the system, the current through the loop—and therefore the induced magnetic field—increases. Just as we can associate electric potential energy with an electric field, we can associate **magnetic potential energy** with a magnetic field. In the absence of dissipation, the work done on the magnet-loop system therefore increases the magnetic potential energy of the system. The energy diagram in **Figure 29.15** illustrates the conversion of mechanical work to magnetic potential energy.

Work must also be done to move a conducting loop into a magnetic field. When current is induced in the loop by its motion in the magnetic field, the magnetic field exerts a force on the current.

Figure 29.16 shows the directions of these magnetic forces exerted on each side of a rectangular conducting loop as it enters a magnetic field from the left. The vector sum of the forces points to the left. In order to move the loop through the field at constant speed, you must exert on the loop a constant force to the right that is equal in magnitude to the magnetic force (**Figure 29.17**). In doing so, you do work on the loop.

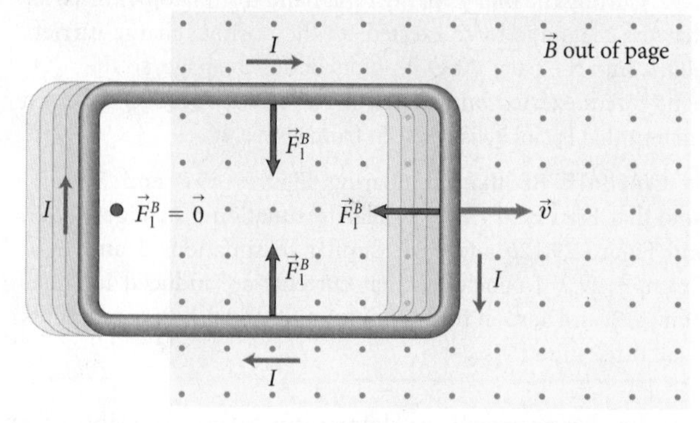

Figure 29.16 Direction of magnetic forces exerted by a magnetic field on each side of a rectangular conducting loop because of the current induced in the loop. As the loop moves into the field, the vector sum of the magnetic forces exerted on the loop resists its motion into the field.

29.7 (*a*) After the left edge of the loop in Figure 29.17 enters the magnetic field, is the work required to continue pulling the loop through the field at constant speed v positive, negative, or zero? (*b*) As the right edge emerges from the magnetic field, is the work required positive, negative, or zero?

As you move a conductor relative to a magnetic field (or vice versa), the work you do transfers energy to the conductor, setting the charge carriers in motion. This energy associated with the induced current can then be converted to other forms of energy by some device placed in the current path, such as a lamp, toaster, or electric motor. In **Figure 29.18**, for example, a lamp is inserted in the current path and the work done on the system is converted to light energy and thermal energy in the lamp. Doing work by moving a loop through a magnetic field and thereby setting charge carriers in motion is the basic idea behind electric generators.

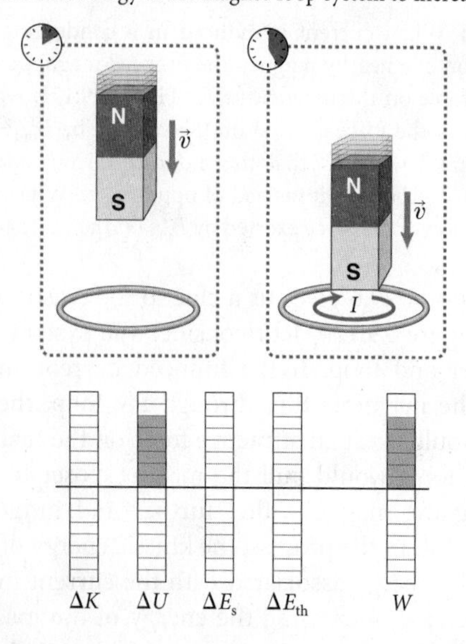

Figure 29.15 Energy diagram for a magnet-loop system when the magnet moves toward the conducting loop, inducing a current through the loop. Work must be done on the system to push the magnet at constant speed against the opposing induced magnetic field of the loop. This work causes the potential energy of the magnet-loop system to increase.

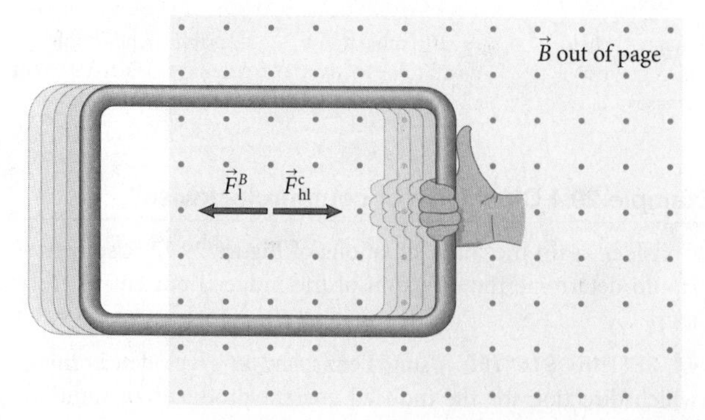

Figure 29.17 Work must be done on a conducting loop to move it into a magnetic field because magnetic forces exerted on the loop resist its motion when there is an induced current in the loop (shown in more detail in Figure 29.16).

Figure 29.18 An induced current can be used to light a lamp. The work done on the magnet is converted to light and thermal energy in the filament of the lamp.

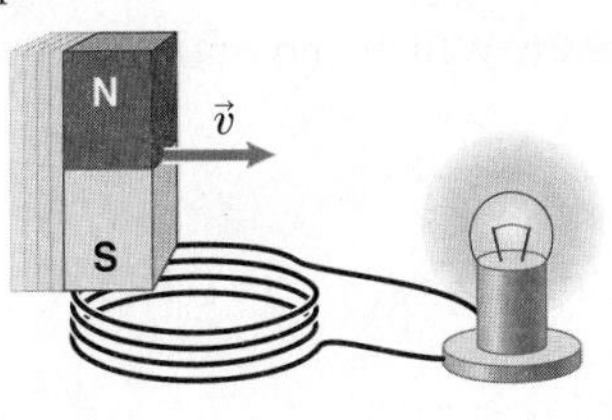

29.8 Which requires doing more work: moving a magnet toward a closed conducting loop or moving it toward a rod? Both motions are at constant speed.

It is not necessary to have a conducting object in the shape of a loop. If a bar magnet is moved toward an extended conducting object, such as a conducting sheet (**Figure 29.19**), circular currents are induced throughout the surface of the object. These currents are called **eddy currents** because they form eddy-like loops in the object's surface. (An *eddy* is the circular, whirlpool-like movement of water.)

Figure 29.19 The circular current loops induced in a conducting sheet by the motion of a nearby magnet are called eddy currents.

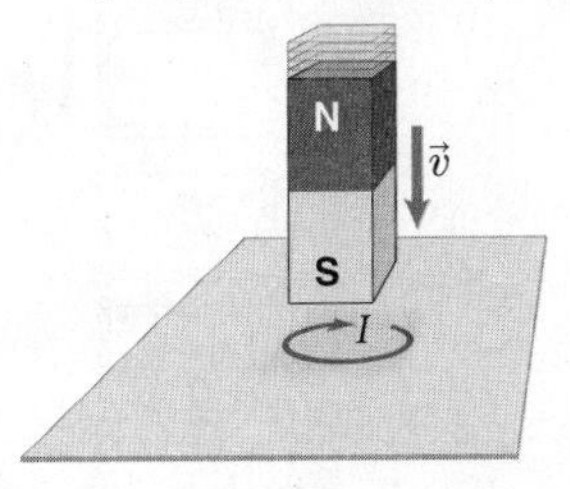

Because eddy currents dissipate energy, they can be used to convert kinetic energy (for example, the kinetic energy associated with the motion of a conductor relative to a magnet) to thermal energy. Dissipation of energy by eddy currents is used in braking systems, especially trains and roller coasters. In contrast to conventional friction brakes, eddy current brakes involve no physical contact and so there is no wear and tear and there are no brake pads to be replaced.

29.9 In **Figure 29.20**, a bar magnet moves parallel to a metal plate. (*a*) At the instant shown, does the magnitude of the magnetic flux increase, decrease, or stay the same through a small region around points P, Q, and R? (*b*) Are the eddy currents induced around these points clockwise, counterclockwise, or zero?

Figure 29.20 A bar magnet moves parallel to the surface of a conducting sheet, causing the local magnetic flux to change across the sheet surface and inducing eddy currents where the magnetic flux is changing (Checkpoint 29.9).

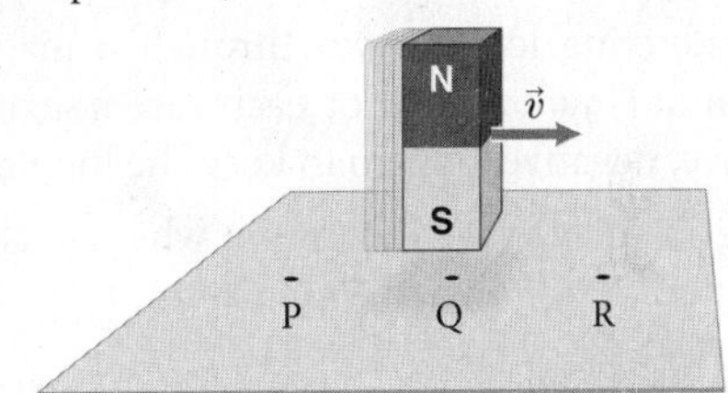

Self-quiz

1. A conducting rod moves through a magnetic field as shown in **Figure 29.21**. Which end of the bar, if any, becomes positively charged?

Figure 29.21 (*a*) (*b*) (*c*)

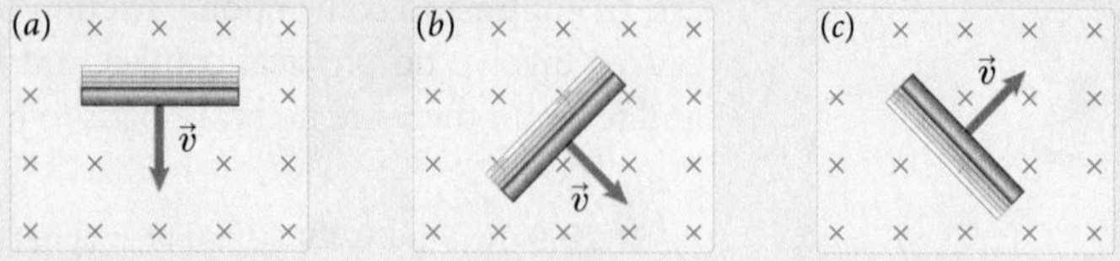

2. A conducting loop moves through a magnetic field as shown in **Figure 29.22*a*–*c***. Which way does the current run in the loop at the instant shown in each figure?

Figure 29.22

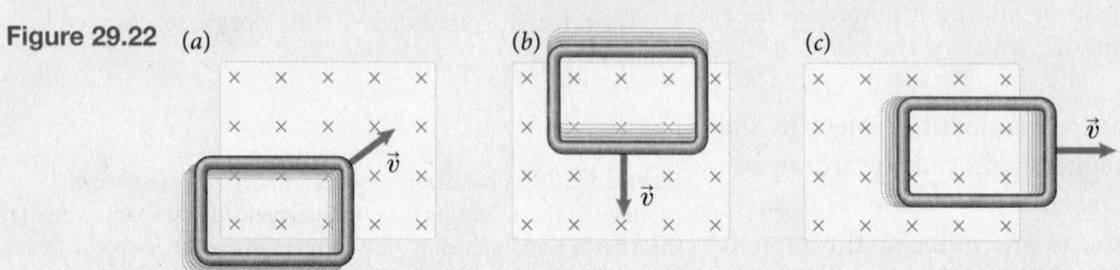

3. A conducting loop moves through a magnetic field at constant velocity as shown in **Figure 29.23**. For each case *a*–*e*, must the work done on the loop be positive, negative, or zero to keep the loop moving?

Figure 29.23 (*a*) (*b*) (*c*) (*d*) (*e*) $\vec{v}$

4. Using Faraday's law, determine whether charge carriers flow in the loop for each situation shown in **Figure 29.24**.

Figure 29.24

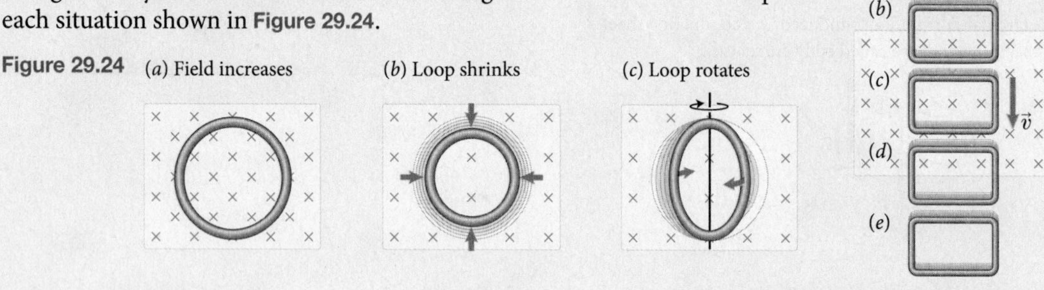

5. Using Lenz's law, determine the direction of the induced current, if any, at the instants shown in Figure 29.24.

Answers

1. The direction of the magnetic force on the positive charge carriers in the rod is given by the right-hand force rule: If you point the fingers of your right hand in the direction of $\vec{v}$ and then curl them toward $\vec{B}$ (into the page), your thumb points in the direction of the force. Applying this rule in the three situations shows that positive charge accumulates on the right end of the bar in (*a*) and (*b*) and on the left end of the bar in (*c*).
2. In (*a*) and (*b*), the magnitude of the magnetic flux through the loop increases into the page, so the induced magnetic field opposing this change must point out of the page. According to the right-hand dipole rule, an induced current must be counterclockwise to produce a magnetic field that points out of the page. In (*c*), the magnitude of the magnetic flux through the loop decreases into the page, so the induced current must be clockwise.
3. In (*a*), (*c*), and (*e*), the magnetic flux through the loop does not change and so no current is induced in the loop. The vector sum of the magnetic forces exerted on the loop is therefore zero, so no work is required to keep the loop moving at constant speed. In (*b*) and (*d*), the magnetic flux through the loop changes and a current is induced through the loop. According to Lenz's law the induced current is in such a direction as to oppose the motion or change that created it, so a positive amount of work is needed to keep the loop moving at constant speed.
4. Because the magnetic flux through the loop is changing in all three cases, charge carriers flow in all three cases.
5. (*a*) The magnetic field is directed into the page, and the magnitude of the magnetic flux is increasing into the page. The induced magnetic field opposing this change must therefore point out of the page, so according to the right-hand dipole rule, the induced current is counterclockwise. (*b*) Because the area of the loop decreases, the magnetic field is directed into the page and the magnitude of the magnetic flux decreases. The induced current is therefore clockwise. (*c*) At the instant shown in the figure, the magnitude of the magnetic flux through the loop decreases and so the induced current must be clockwise.

29.5 Induced emf

We first encountered the concept of emf in Chapter 26 in the context of batteries. Chemical reactions in a battery produce and maintain a separation of charge across the battery terminals; this separation results in a potential difference from one terminal to the other. If a light bulb is connected to the terminals, this potential difference drives a current through the light bulb. In Section 26.4, we defined the emf of a charge-separating device as the work done per unit charge by nonelectrostatic interactions in separating charge within the device. Because the charge separation across the ends of a rod caused by electromagnetic induction (Figure 29.10) is nonelectrostatic, we can associate an emf with this separation of charge. Emfs produced by electromagnetic induction are therefore called **induced emfs.**

Consider the setup illustrated in Figure 29.25. A conducting rod of length ℓ rests on two conducting rails connected to a light bulb. The rod is pulled to the right and moves at a constant speed v along the x axis. The motion of the rod changes the area of the magnetic field enclosed by the loop formed by the rod, rails, and wires attaching the bulb to the rails. Because of this area change, a current is induced through the loop. This current causes the light bulb to glow.

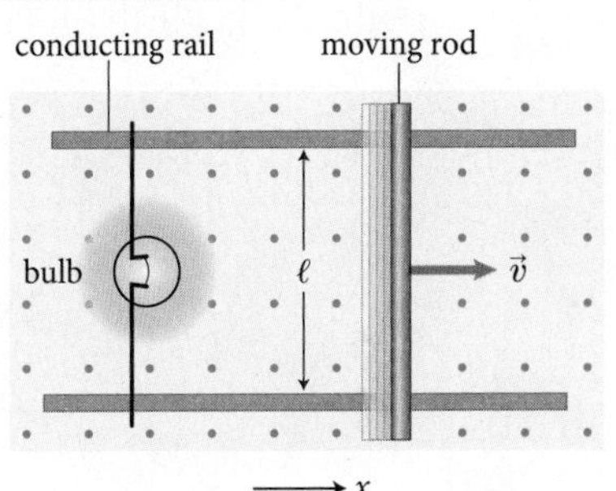

Figure 29.25 The emf that develops in a moving rod can be used to drive a current when that rod is part of a closed conducting loop. The rod is connected to the rest of the loop by sliding electrical contacts.

As we saw in Section 29.4, the magnetic field exerts a leftward force on the current-carrying rod. To move the rod at a constant speed v to the right, therefore, the agent pulling the rod must exert a rightward force of equal magnitude on it. In doing so, the external agent must do work on the rod, increasing the energy of the charge carriers in the loop. This energy is converted to light and thermal energy in the light bulb. We can obtain an expression for the induced emf in the loop by calculating the work done on the rod.

According to Eq. 27.4, the magnitude of the magnetic force exerted on a rod of length ℓ carrying a current I is $|I|\ell B$ when the rod is perpendicular to the magnetic field, so the contact force F^c_{ar} exerted by the agent on the rod required to pull it at constant speed must be of the same magnitude: $F^c_{\text{ar}} = |I|\ell B$. Because the work done on the rod is positive, it is equal to the product of the magnitude of the contact force exerted on it and the magnitude of the force displacement: $W_{\text{r}} \equiv F^c_{\text{ar}\,x}\Delta x_F = F^c_{\text{ar}}|\Delta x_F|$. In a time interval Δt, the rod is displaced a distance $\Delta x = v_x \Delta t$, so we have

$$W_{\text{r}} = F^c_{\text{ar}}|\Delta x_F| = B|I|\ell v\Delta t \tag{29.1}$$

or, using Eq. 27.1, $I \equiv q/\Delta t$, to simplify,

$$W_{\text{r}} = B\ell v|q|, \tag{29.2}$$

where q is the charge that passes through a given section of the rod in a time interval Δt. This result tells us how much energy is transferred to the current loop as the rod is pulled. Note that the force exerted by the magnetic field on the charge carriers in the rod transfers the mechanical work done on the rod to the charge carriers in the loop—the magnetic force itself does no work. (Because the direction of the magnetic force is always perpendicular to the motion of the charge carriers, the force cannot do any work on the carriers.)

If it bothers you that the external agent is doing the work and not the magnetic force that sets the charge carriers in motion, consider an analogy: Suppose you push yourself away from a wall starting from rest. The wall provides the force that accelerates you away from it. This force does *no work,* however, because the point of application does not move. In fact, no work at all is done on you. Internal (source) energy is converted to kinetic energy; the force exerted by the wall on you converts the internal energy to kinetic energy.

We can now calculate the magnitude of the induced emf, which is defined as the nonelectrostatic work per unit charge done on a charged particle (Eq. 26.7):

$$|\mathscr{E}_{\text{ind}}| \equiv \frac{W_{\text{r}}}{|q|} = B\ell v. \tag{29.3}$$

As shown in Figure 29.25, the induced emf in the closed conducting loop drives a current, just as a battery of the same emf would. This current has the same effect on the loop as a battery-driven current would have; in this case, it lights a light bulb. The magnitude of the resulting current—the **induced current**—depends on how much resistance the loop puts up to the motion of charge carriers. Experimentally we determine that the induced current through many conducting materials is proportional to the induced emf:

$$I_{\text{ind}} = \frac{\mathscr{E}_{\text{ind}}}{R}, \tag{29.4}$$

where the proportionality constant R is a measure of the *resistance* of the loop. The smaller the value of R, the greater the induced current for a given induced emf. We'll learn more about resistance in Chapter 31.

Equation 29.3 gives the magnitude of the induced emf in terms of v, B, and the length ℓ of the rod. In Section 29.2 we argued that the induced current—that is, the current that results from the induced emf—is due to a change in the magnetic flux enclosed by the loop. Let us therefore obtain an expression for the induced emf in terms of the magnetic flux Φ_B through the loop. In Section 27.6 we defined the magnetic flux through a loop by the surface integral (Eq. 27.10)

$$\Phi_B \equiv \int \vec{B} \cdot d\vec{A}, \tag{29.5}$$

where $d\vec{A}$ is the area vector of an infinitesimally small segment of a surface bounded by the loop.

Because the surface integral in Eq. 29.5 is independent of the choice of surface bounded by the loop (Eq. 27.15), we integrate over a flat surface bounded by the loop in Figure 29.25. Because the magnetic field is uniform, we can pull it out of the integral and so in Figure 29.25, if we let the vector $d\vec{A}$ point out of the page in the same direction as $\vec{B}$, $\Phi_B = BA$. When the moving side of the loop is displaced by Δx, the area enclosed by the expanding loop changes by $\Delta A = \ell \Delta x$. If this change occurs over a time interval Δt, the rate of change in the magnetic flux enclosed by the loop is thus

$$\frac{\Delta \Phi_B}{\Delta t} = \frac{B \Delta A}{\Delta t} = \frac{B\ell \Delta x}{\Delta t} = B\ell \frac{\Delta x}{\Delta t}, \tag{29.6}$$

so Eq. 29.3 can be written as

$$|\mathscr{E}_{\text{ind}}| = B\ell v = B\ell \left|\frac{\Delta x}{\Delta t}\right| = \left|\frac{\Delta \Phi_B}{\Delta t}\right|. \tag{29.7}$$

As we let the time interval Δt approach zero, this yields

$$\mathscr{E}_{\text{ind}} = -\frac{d\Phi_B}{dt}, \tag{29.8}$$

where we have added a negative sign to indicate that the direction of the induced emf is such that it drives a current that counteracts the change in magnetic flux, in accordance with Lenz's law. Equation 29.8, which relates a changing magnetic flux to an induced emf, is a quantitative statement of **Faraday's law.**

Note that Eq. 29.8 correctly gives a zero emf when applied to the situation shown in Figure 29.4*c*, in which the entire loop (not just one side) is moving and is in the magnetic field. We derived Eq. 29.8 for the special case of one side of a rectangular conducting loop moving in a magnetic field, but it is generally valid: It gives the value of the emf no matter what causes the magnetic flux through a conducting loop to change.

29.10 Sketch how the induced emf in the loop in Figure 29.4 varies as the loop moves through the five positions.

Equation 29.8 allows us to point out an important difference between potential difference and emf, both of which are related to the work per unit charge done on charge carriers—the former by electrostatic interactions, the latter by nonelectrostatic interactions. As we established in Chapter 25, potential difference depends only on starting and ending locations. It does not depend on the path followed to get from the starting location to the ending location. In other words, potential difference is path-independent. The work per unit charge done on charged particles by *nonelectrostatic* interactions, such as the induced emf in Figure 29.25, does depend on the path taken. Consider, for example, a rectangular conducting loop containing a light bulb and placed in a changing magnetic field (**Figure 29.26*a***). The light bulb glows because the changing magnetic flux through the area enclosed by the loop causes an induced emf throughout the loop. If the area enclosed by the loop is smaller (Figure 29.26*b*), then the light bulb is less bright. The emf induced in the loop is now smaller because the smaller area causes a smaller rate of change in the magnetic flux. Had we established a potential difference across the light bulb by connecting it via wires to, say, the terminals of a battery, how much area the wires enclose would not affect the brightness of the bulb (**Figure 29.27**). We shall discuss this difference in path-dependence between potential difference and emf in more detail in the next section.

We can use an induced emf to establish a potential difference between two locations. Consider, for example, the conducting rod of Figure 29.1. As we saw in Section 29.1, the magnetic force exerted on the negative charge carriers in the moving rod drives them upward. A surplus of negative charge accumulates at the top of the rod, leaving a surplus of positive charge at the bottom. This separation of charge produces an electric field inside the rod. The negative charge carriers flow until the magnitude of the downward electric force exerted on one of them due to the electric field inside the rod is equal to the magnitude of the upward magnetic force (**Figure 29.28** on the next page):

$$F_q^E = F_q^B. \tag{29.9}$$

The magnitude of the magnetic force exerted on a charged particle of charge q moving perpendicular to a magnetic field is given by Eq. 27.18, $F_q^B \equiv |q|vB$, and the electric force is given by Eq. 23.6, $F_q^E = |q|E$. With this information, we can rewrite Eq. 29.9 as

$$|q|E = |q|vB. \tag{29.10}$$

Thus E, the magnitude of the electric field inside the rod, is

$$E = vB. \tag{29.11}$$

Figure 29.26 Path-dependence of an induced emf.

(*a*)

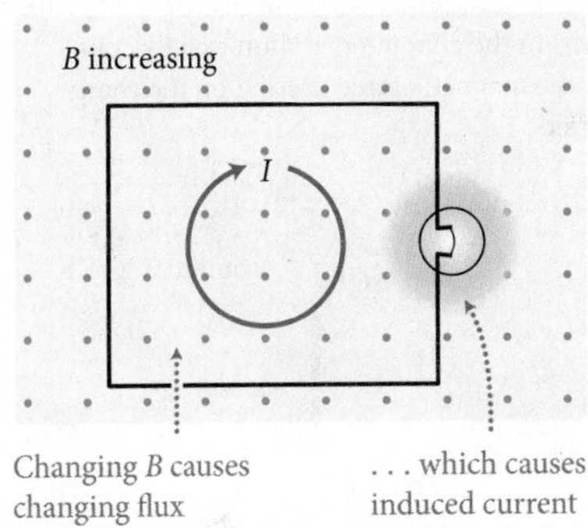
B increasing
I
Changing B causes changing flux through loop . . .
. . . which causes induced current that lights bulb.

(*b*)

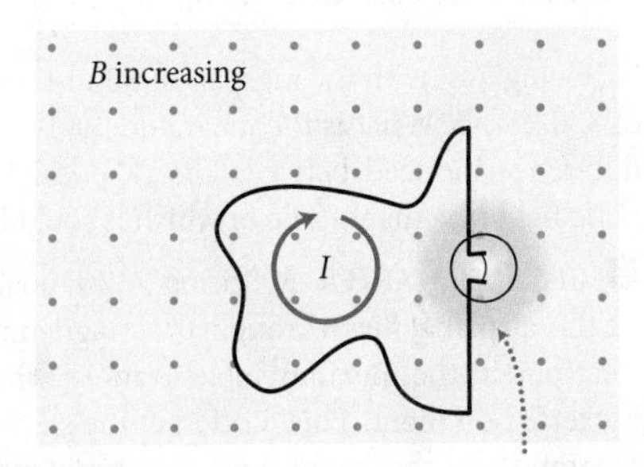
B increasing
I
If loop encloses less area, bulb is dimmer.

Figure 29.27 The emf produced by a battery does not depend on enclosed area.

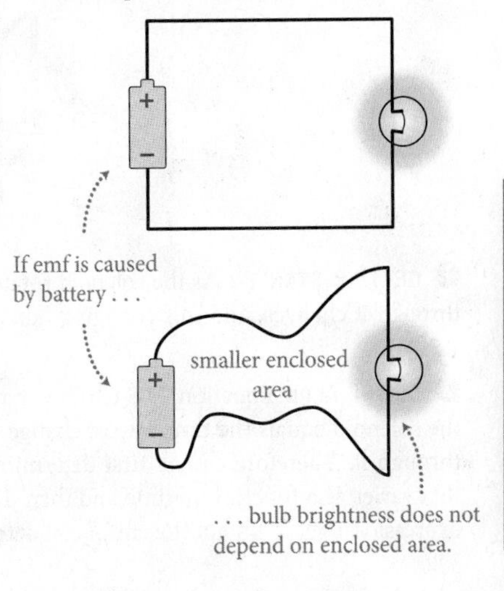
+
−
If emf is caused by battery . . .
smaller enclosed area
. . . bulb brightness does not depend on enclosed area.

QUANTITATIVE TOOLS

Figure 29.28 When a conducting rod moves in a magnetic field, the magnetic force exerted on the negative charge carriers in the rod causes a charge separation. This separation continues until the force exerted by the electric field resulting from the charge separation exactly counters the magnetic force exerted on the charge carriers.

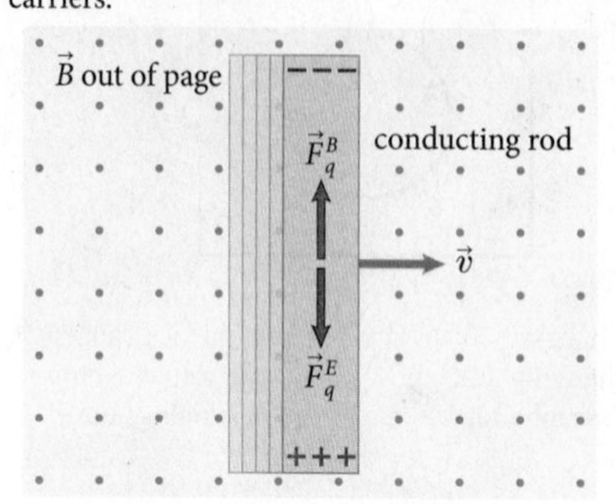

This electric field causes a potential difference between the top and the bottom of the rod. For a rod of length ℓ, the potential difference is (see Eq. 25.25)

$$V_{\text{tb}} = -\int_{\text{top}}^{\text{bottom}} \vec{E} \cdot d\vec{\ell} = \int_{\text{top}}^{\text{bottom}} E d\ell = E \int_{\text{top}}^{\text{bottom}} d\ell = E\ell, \quad (29.12)$$

where $\vec{E} \cdot d\vec{\ell} = -E d\ell$ because the electric field points from the bottom to the top of the rod whereas $d\vec{\ell}$ points in the direction of integration. In addition we have pulled E out of the integral because it is constant throughout the rod. Substituting the magnitude of the electric field from Eq. 29.11, we obtain

$$V_{\text{tb}} = vB\ell. \quad (29.13)$$

Thus we see that the electric field in the rod establishes across the ends of the rod a potential difference that is equal in magnitude to the induced emf (Eq. 29.3).

Example 29.5 Airplane wing "battery"

A Boeing 747 with a wingspan of 60 m flies at a cruising speed of 850 km/h. What is the magnitude of the maximum potential difference induced between the two wingtips by Earth's magnetic field (the magnitude of which is roughly 0.50×10^{-4} T)?

❶ **GETTING STARTED** A Boeing 747 is made of metal and so, as the airplane flies through the magnetic field of Earth, the electrons in the metal are able to move when a magnetic force is exerted on them. This motion of the electrons causes a charge separation in the wings, and thus a potential difference is induced. To solve this problem, I must make several simplifying assumptions. I take the wings to be a 60-m conducting rod and assume the plane is flying horizontally through a vertical magnetic field. As a result, the problem reduces to a rod moving through a perpendicular magnetic field.

❷ **DEVISE PLAN** To calculate the induced potential difference between the ends of the "rod" formed by the wings, I can use Eq. 29.13.

❸ **EXECUTE PLAN**

$$V_{\text{wings}} = (8.50 \times 10^5 \text{ m/h})(1 \text{ h}/3600 \text{ s})(0.50 \times 10^{-4} \text{ T})(60 \text{ m})$$
$$= 0.71 \text{ T} \cdot \text{m}^2/\text{s} = 0.71 \text{ V}. ✔$$

❹ **EVALUATE RESULT** My calculation yields a potential difference on the order of the one from an AA battery, but that should not cause any problems in the aircraft. Also, the magnetic field is unlikely to be perpendicular to the velocity of the plane, as I assumed, and therefore the actual value of the potential difference is smaller than what I obtained.

Example 29.6 Generator

In an electric generator a solenoid that contains N windings each of area A is rotated at constant rotational speed ω in a uniform magnetic field of magnitude B (**Figure 29.29**). What is the emf induced in the solenoid?

Figure 29.29 Example 29.6.

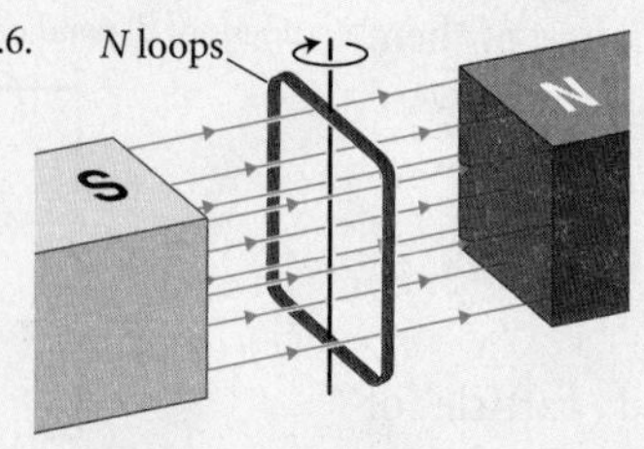

❶ **GETTING STARTED** As the solenoid rotates, the magnetic flux through it changes, and this changing flux causes an emf in the solenoid.

❷ **DEVISE PLAN** Equation 29.8 tells me that the emf induced in the solenoid equals the time rate of change of the magnetic flux through it. Therefore I must first determine how the magnetic flux varies as a function of time and then differentiate whatever expression I get to obtain the emf. I can determine the magnetic flux through the solenoid by multiplying the magnetic flux through a single winding, $\Phi_B \equiv \vec{B} \cdot \vec{A}$ (Eq. 27.9), by the number of windings N. To determine the magnetic flux through a single winding, I sketch a top view of a winding in the magnetic field, indicating the directions of the magnetic field $\vec{B}$ and the area vector $\vec{A}$ (**Figure 29.30**). I let the plane of the winding be perpendicular to the direction of the magnetic field at $t = 0$.

Figure 29.30

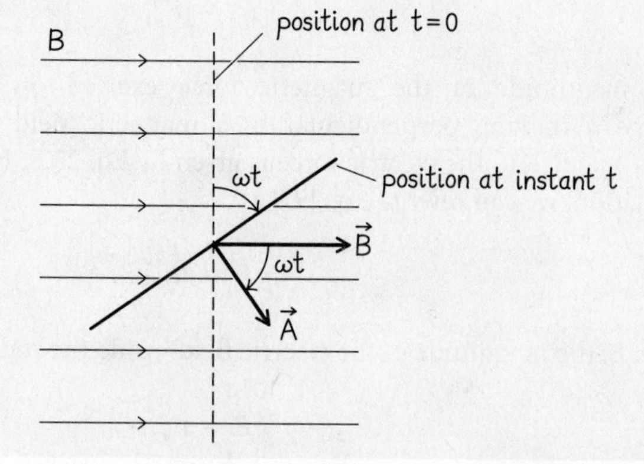

QUANTITATIVE TOOLS

❸ **EXECUTE PLAN** As the solenoid rotates, the scalar product $\vec{B}\cdot\vec{A}$ changes with time. At instant t shown in my sketch, the angle between $\vec{A}$ and $\vec{B}$ is ωt, and so the magnetic flux through a single winding is $\Phi_B = \vec{B}\cdot\vec{A} = BA\cos\omega t$. Through the N windings of the solenoid the magnetic flux is $\Phi_B = NBA\cos\omega t$. Substituting this value into Eq. 29.8, I get

$$\mathscr{E}_{\text{ind}} = -\frac{d}{dt}(NBA\cos\omega t) = \omega NBA\sin\omega t. \checkmark$$

❹ **EVALUATE RESULT** My result shows that the emf oscillates sinusoidally. It is zero when $\omega t = n\pi$ ($n = 0, 1, 2, \ldots$) and maximum when $\omega t = n\pi + \frac{\pi}{2}$ ($n = 0, 1, 2, \ldots$). That result makes sense because the rate of change of the magnetic flux through the solenoid is zero when the area vector of the windings is parallel to the magnetic field ($\omega t = n\pi$) and maximum when the area vector is perpendicular to the magnetic field ($\omega t = n\pi + \frac{\pi}{2}$).

29.11 The expression I derived in Example 29.6 indicates that the emf becomes negative after the solenoid has rotated 180° and remains negative through the next 180° of rotation. However, the solenoid orientation looks the same when the solenoid has rotated 180° as when it started. Why does the emf have a different sign for half of the rotation?

29.6 Electric field accompanying a changing magnetic field

We saw in Section 29.3 that when the magnetic flux through a conducting loop changes, an electric force is exerted on the initially stationary charge carriers in the loop. This electric force is caused by the electric field that accompanies the changing magnetic field. Let's explore the properties of this electric field in more detail.

Consider the case of a conducting circular loop in a uniform magnetic field that has a circular cross section (**Figure 29.31a**). Suppose the magnitude of the magnetic field increases steadily over time. Because the conducting loop encloses an increasing magnetic flux directed out of the page, a clockwise current is induced in the loop. What do the electric field lines that are responsible for this current look like?

We found in Section 29.3 that the electric field lines that accompany a changing magnetic field form loops because there are no isolated charge carriers on which the electric field lines can begin or end. These loops must be circular and centered on the axis of the magnetic field because the electric field that accompanies the changing magnetic field must have the same cylindrical symmetry as the magnetic field (Figure 29.31*b*). Lenz's law tells us that the induced current is clockwise, and so the electric field must also be pointing clockwise. When the magnetic field does not exhibit cylindrical symmetry, the electric field lines are not circular, and in general it is difficult to determine the shape of the electric field lines.

Figure 29.31 Electric field that accompanies an increasing cylindrical magnetic field.

(*a*) Conducting ring in cylindrical uniform magnetic field

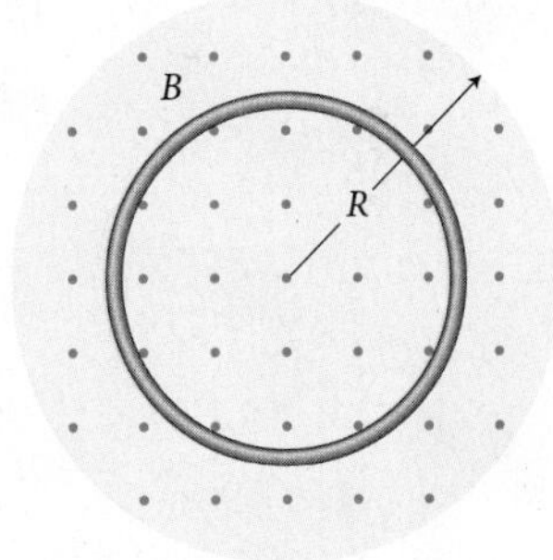

(*b*) Electric field that accompanies increasing magnetic field

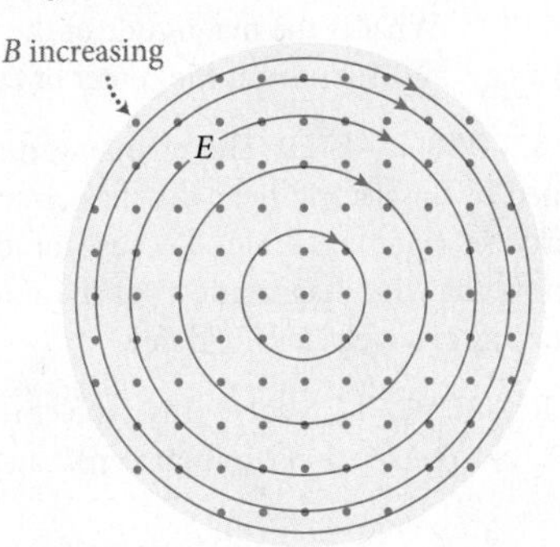

29.12 What do the electric field lines look like when the magnitude of the magnetic field in Figure 29.31*b* (*a*) is held constant and (*b*) decreases steadily?

To determine the magnitude of the electric field that accompanies the changing magnetic field, let's think about its effect on charge carriers in the conducting loop. A current is induced in the loop because the electric field does work on the charge carriers in the loop. We can calculate the work done by the electric field using Eq. 25.1, which gives the work done by an electric field $\vec{E}$ on a particle carrying a charge q in moving it from point A to point B:

$$W_q(\text{A} \to \text{B}) = \int_{\text{A}}^{\text{B}} \vec{F}_q^E \cdot d\vec{\ell} = q\int_{\text{A}}^{\text{B}} \vec{E}\cdot d\vec{\ell}. \quad (29.14)$$

(Note that the electric field accompanying the changing magnetic field is *not* an electrostatic field, and thus the quantity calculated in Eq. 29.14 is *not* electrostatic work.)

The work done on a particle carrying a charge q as it travels around the closed path formed by the conducting loop is then

$$W_q(\text{closed path}) = q \oint \vec{E} \cdot d\vec{\ell}. \tag{29.15}$$

The work done by the electric field per unit charge is thus

$$\frac{W_q}{q} = \oint \vec{E} \cdot d\vec{\ell}. \tag{29.16}$$

This work per unit charge is the induced emf (Eq. 26.7), and so combining Eq. 29.16 with Eq. 29.8 gives us an expression that contains the electric field accompanying the changing magnetic field:

$$\oint \vec{E} \cdot d\vec{\ell} = -\frac{d\Phi_B}{dt}. \tag{29.17}$$

Equation 29.17 is an alternative statement of Faraday's law. Note that the magnetic flux appearing in this equation is a signed quantity. In order to obtain the correct direction of the electric field in Eq. 29.17, the magnetic flux must be taken to be positive when the magnetic field points in the direction of your right-hand thumb as you curl the fingers of your right hand in the direction of the integration path.

Keep in mind that the electric field in Eq. 29.17 is not an electrostatic field. Electrostatic fields originate directly from static charge distributions. Rather than originating from static charged objects, the electric field in Eq. 29.17 appears with a changing magnetic field and the electric field lines close on themselves. As we saw in Chapter 25, the work done by an electrostatic field on a charged particle moving around a closed path is zero, which means the line integral around a closed path on the left in Eq. 29.17 is always zero for an electrostatic field (see Eq. 25.32). However, as Eq. 29.17 shows, in a region where the magnetic field is changing, the line integral of an electric field around a closed path is not necessarily zero. This means that the electric field accompanying the changing magnetic field can do work on charged particles that travel along a closed path and the amount of work depends on the choice of path. We cannot define a potential difference between two points for such an electric field because it would have different values for different paths between those points.

Example 29.7 Electric field magnitude

Let the uniform cylindrical magnetic field in Figure 29.31 have a radius $R = 0.20$ m and increase at a steady rate of 0.050 T/s. What is the magnitude of the electric field at a radial distance $r = 0.10$ m from the center of the magnetic field?

❶ **GETTING STARTED** The changing magnetic field is accompanied by an electric field that has circular field lines pointing clockwise (Figure 29.31). Because of the circular symmetry, I know that the magnitude E of the electric field cannot vary along any given electric field line.

❷ **DEVISE PLAN** To solve this problem, I can use Eq. 29.17 to work out the relationship between E and the rate of change of the magnetic field magnitude dB/dt. Because E is constant along any electric field line, it must be constant on *any* circular path centered on the axis of the cylindrical magnetic field. I therefore choose as my path of integration a clockwise circular path of radius $r = 0.10$ m centered on the long central axis of the magnetic field (**Figure 29.32**). Because $\vec{E}$ is always parallel to $d\vec{\ell}$, $\vec{E} \cdot d\vec{\ell} = E\,d\ell$ on the left side of Eq. 29.17, and because the magnitude of the electric field is the same everywhere on the path, I can take E out of the integral. To evaluate the right side, I must first calculate the magnetic flux through the circular integration path. Because the magnetic field is uniform, the magnetic flux is given by Eq. 27.9.

Figure 29.32

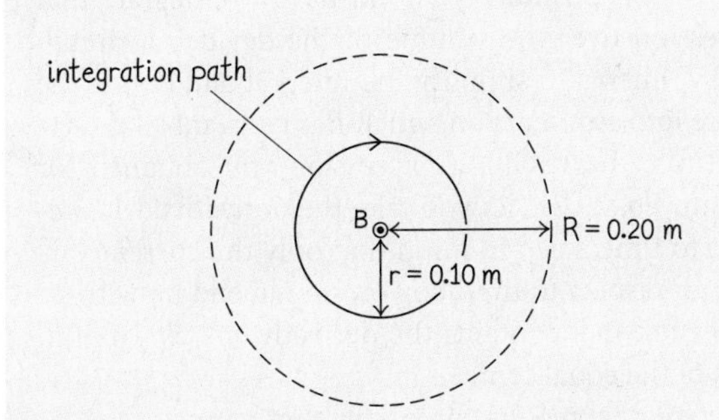

❸ EXECUTE PLAN The left side of Eq. 29.17 becomes

$$\oint \vec{E} \cdot d\vec{\ell} = 2\pi r E. \quad (1)$$

The magnetic flux through the area enclosed by the integration path is negative (curling the fingers of my right hand in the direction of the integration path makes my thumb point in the direction opposite the direction of $\vec{B}$), and so

$$\Phi_B = -BA = -\pi r^2 B.$$

Thus the right side of Eq. 29.17 becomes

$$-\frac{d\Phi_B}{dt} = +\pi r^2 \frac{dB}{dt}. \quad (2)$$

Substituting Eqs. 1 and 2 into Eq. 29.17 then yields

$$2\pi r E = \pi r^2 \frac{dB}{dt}$$

$$E = \tfrac{1}{2} r \frac{dB}{dt} = \frac{(0.10\ \text{m})(0.050\ \text{T/s})}{2}$$

$$= 2.5 \times 10^{-3}\ \text{T} \cdot \text{m/s} = 2.5 \times 10^{-3}\ \text{N/C}. ✔ \quad (3)$$

❹ EVALUATE RESULT Equation 3 shows that the magnitude of the electric field increases with radial distance r from the center of the magnetic field. Given the cylindrical symmetry of the situation, that's exactly what I expect.

Although the electric field lines and the conducting loop pictured in Figure 29.31 are circular, our derivation of Eq. 29.17 does not depend on the shape of the path of integration. This means we can use Eq. 29.17 to determine the line integral of the electric field around any closed path. For example, **Figure 29.33** shows a square path of integration that encloses the same area (and hence the same magnetic flux) as the circular path I used to solve Example 29.7. Because the enclosed magnetic flux is the same, we know that the integral of the electric field around the square path must yield the same result as the integral around the round path, even though the electric field is no longer tangent to the path and no longer of the same magnitude all along the path.

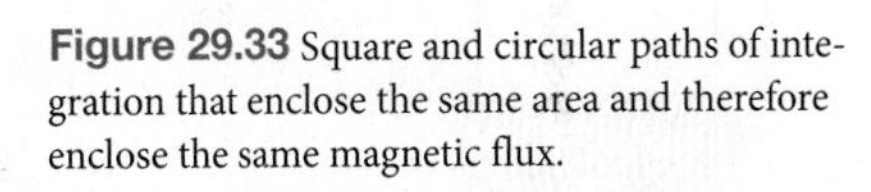
Figure 29.33 Square and circular paths of integration that enclose the same area and therefore enclose the same magnetic flux.

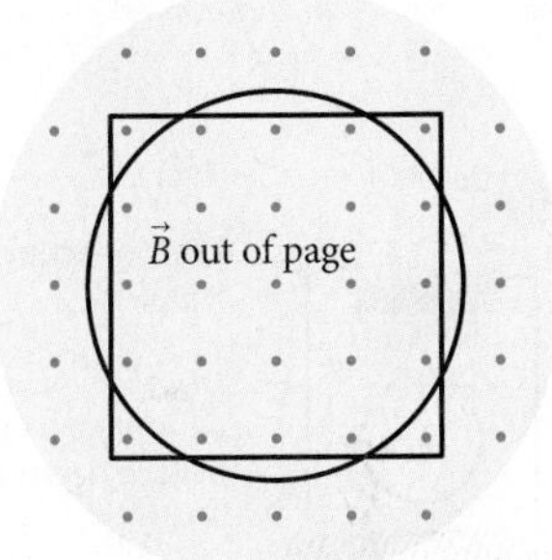

29.13 In Example 29.7 what is the magnitude of the electric field at a distance of 0.30 m from the center of the magnetic field?

29.7 Inductance

An interesting consequence of Faraday's law is that when a current through a conducting loop changes, the change induces an emf *in the loop itself*. To understand why this happens, consider **Figure 29.34a** (next page), in which a battery drives a current I through a solenoid. Because the current creates a magnetic field, a nonzero magnetic flux passes through the loops of the solenoid.

Now imagine the battery is suddenly disconnected, as in Figure 29.34*b*. Without the battery to supply an emf, the current, the magnitude of the magnetic field, and the magnitude of the magnetic flux all start to decrease. Faraday's law tells us, however, that the changing flux induces in the solenoid an emf that opposes the change in flux. Consequently, as the magnetic flux is decreasing, the induced emf causes an induced current that has the same direction as the original current. The magnetic flux associated with this induced current opposes the change in the magnetic flux.

The induced emf is proportional to the rate of change in the magnetic flux through the solenoid (Eq. 29.8). The magnetic flux is proportional to the magnitude B of the magnetic field, which is proportional to the current through the

Procedure: Calculating inductances

The inductance of a current-carrying device or current loop is a measure of the emf induced in the device or loop when current is changed. To determine the inductance of a particular device or current loop, follow these four steps.

1. Derive an expression for the magnitude of the magnetic field in the current-carrying device or current loop as a function of the current. Your expression should depend only on the current I and possibly—but not necessarily—the position within the device or current loop.
2. Calculate the magnetic flux Φ_B through the device or current loop. If the expression you derived in step 1 depends on position, you will have to integrate that expression over the volume of the device or circuit. Use symmetry to simplify the integral and divide the device into segments on which B is constant.
3. Substitute the resulting expression you obtained for Φ_B into Eq. 29.21. As you take the derivative with respect to time, keep in mind that only the current varies with respect to time, so you should end up with an expression that contains the derivative dI/dt on both sides of the equal sign.
4. Solve your expression for L after eliminating dI/dt.

Figure 29.34 Disconnecting a battery from a solenoid induces in the solenoid an emf that opposes the decrease in current through it.

(a)

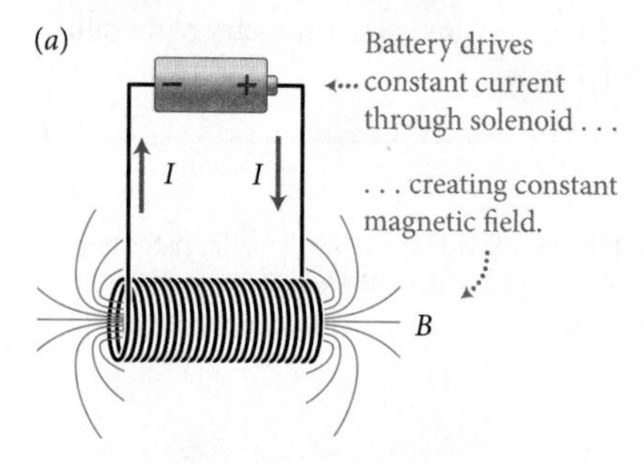
Battery drives constant current through solenoid . . .
I
I
. . . creating constant magnetic field.
B

(b)

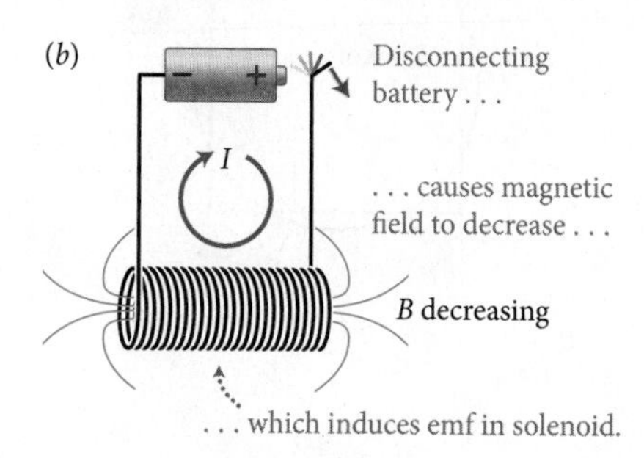
Disconnecting battery . . .
I
. . . causes magnetic field to decrease . . .
B decreasing
. . . which induces emf in solenoid.

loop (Eq. 28.6, $B = \mu_0 n I$). Thus the faster the current changes, the greater the magnitude of the induced emf:

$$\mathscr{E}_{\text{ind}} \propto \frac{d\Phi_B}{dt} \propto \frac{dB}{dt} \propto \frac{dI}{dt}. \tag{29.18}$$

We can define the **inductance** L of a loop or solenoid as the constant of proportionality between the emf and the rate of change of current:

$$\mathscr{E}_{\text{ind}} = -L\frac{dI}{dt}. \tag{29.19}$$

A large inductance leads to a large induced emf. The derived SI unit of inductance is the **henry:**

$$1\ \text{H} \equiv 1\ \text{V}\cdot\text{s/A} = 1\ \text{kg}\cdot\text{m}^2/\text{C}^2. \tag{29.20}$$

A device that has an appreciable inductance is called an **inductor.** Inductors are used widely in electric circuits to even out variations in current.

The inductance describes how much change in magnetic flux is associated with a change in current for a particular loop or solenoid, as we can see by substituting Eq. 29.8 in the left side of Eq. 29.19:

$$\frac{d\Phi_B}{dt} = L\frac{dI}{dt}. \tag{29.21}$$

The inductance of a current-carrying device or circuit depends only on its geometry, but for most real devices, calculating the inductance is not simple. The general procedure for determining the inductance is described in the Procedure box on this page. In a few particularly simple cases, it is possible to derive an algebraic expression for the inductance, as in the following example.

Example 29.8 Inductance of a solenoid

What is the inductance of a solenoid (Section 28.6) of length ℓ that has N windings, each of cross-sectional area A, when the current through the device is I?

❶ GETTING STARTED I begin by making a sketch of the solenoid (**Figure 29.35**). If I treat the solenoid as being infinitely long, I know from Section 28.6 that the magnetic field is uniform inside the solenoid and zero outside of it. To calculate the inductance of the solenoid, I follow the steps in the Procedure box on this page.

Figure 29.35

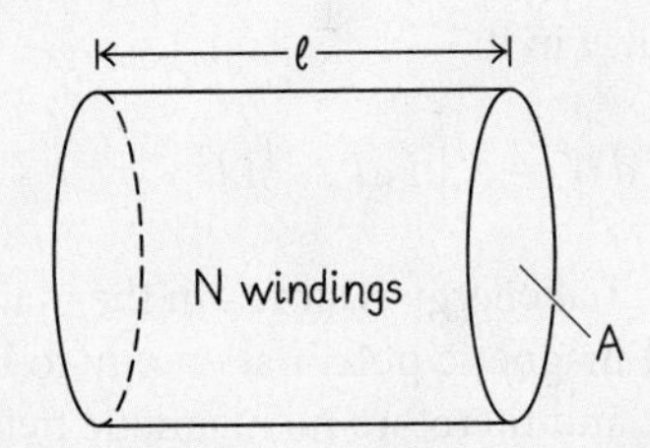

❷ DEVISE PLAN I first need to obtain an expression for the magnetic field magnitude B in terms of the current I. We derived this expression in Section 28.6 and found $B = \mu_0 nI$ (Eq. 28.6), where n is the number of windings per unit length $n = N/\ell$, and so $B = \mu_0 NI/\ell$. Knowing B, I can calculate the magnetic flux, substitute this result into Eq. 29.21, and then solve for the inductance L.

❸ EXECUTE PLAN The magnetic flux through one winding is the area A enclosed by the winding multiplied by the magnetic field B inside the solenoid: $\Phi_B = BA$. For the solenoid, the magnetic flux is then

$$\Phi = NBA = \frac{\mu_0 N^2 IA}{\ell}.$$

Substituting this expression for the magnetic flux into Eq. 29.21 yields

$$\frac{d}{dt}\frac{\mu_0 N^2 IA}{\ell} = L\frac{dI}{dt}.$$

The only part of the left side that is time-dependent is the current I, so differentiating with respect to time and simplifying give me

$$L = \frac{\mu_0 N^2 A}{\ell}. ✔$$

❹ EVALUATE RESULT My result shows that the inductance of a solenoid increases as the square of the number of windings N. This makes sense because both the magnetic field inside the solenoid and the magnetic flux increase with N. That inductance also depends on the area A enclosed by each winding also makes sense because increasing A increases the magnetic flux. Finally, that the inductance is inversely proportional to the length ℓ of the solenoid makes sense because the magnetic field is proportional to the number of turns per unit length.

29.14 A solenoid has 2760 windings of radius 50 mm and is 0.60 m long. If the current through the solenoid is increasing at a rate of 0.10 A/s, what is the magnitude of the induced emf?

29.8 Magnetic energy

We saw in Section 26.6 that work must be done to charge a capacitor and that this work increases the electric potential energy stored in the electric field of the capacitor. Likewise, work must be done on an inductor to establish the current through it because the change in current causes an induced emf that opposes this change; this work increases the **magnetic potential energy** stored in the magnetic field of the inductor.

In Chapter 26, we saw that electric potential energy can be attributed either to the configuration of charge in a system or to the electric field of this charge configuration. Likewise we can attribute the potential energy in an inductor either to the current through it or to the magnetic field caused by the current.

How much potential energy is stored in an inductor in which the current is I? To work this out, we calculate the work required to create a current I in an inductor. We begin by writing the work dW done on the inductor when an amount of charge dq moves through it. This work is the negative of the work done by the induced emf of the inductor:

$$dW = -\mathscr{E}_{\text{ind}}\, dq. \tag{29.22}$$

Using the definitions of current and Eq. 29.19, we obtain for the rate at which work is done

$$\frac{dW}{dt} = -\mathscr{E}_{\text{ind}}\frac{dq}{dt} = -\mathscr{E}_{\text{ind}} I = LI\frac{dI}{dt}, \tag{29.23}$$

and so $dW = LI\,dI$. We can then integrate both sides to obtain the work W done on the inductor to create a current I in it:

$$W = \int dW = L \int I\,dI = \tfrac{1}{2}LI^2. \tag{29.24}$$

To determine how much potential energy is stored in the magnetic field of an inductor, let's choose the zero of magnetic potential energy to be when there is no current through the inductor, and therefore no magnetic field. The magnetic potential energy stored in the inductor when there is a current through it is then equal to the work done to increase the current from zero to I:

$$U^B = \tfrac{1}{2}LI^2. \tag{29.25}$$

This is analogous to Eq. 26.4 for the electric potential energy stored in a capacitor, $U^E = \tfrac{1}{2}CV_{cap}^2$, with L taking the place of C and I taking the place of V_{cap}. Because this energy in the inductor is stored in the magnetic field, let's express the energy in terms of the magnetic field. We'll do this for the case of a long solenoid, but the result turns out to be generally applicable. Substituting the expression worked out in Example 29.8 for L into Eq. 29.25, we obtain

$$U^B = \tfrac{1}{2}\frac{\mu_0 N^2 A}{\ell}I^2, \tag{29.26}$$

where, as in Example 29.8, N is the number of windings, A is the area of each winding of the solenoid, and ℓ is the length of the solenoid.

In Chapter 26, we used the expression for the electric potential energy U^E to arrive at an expression for the energy density in an electric field: $u_E = \tfrac{1}{2}\epsilon_0 E^2$ (Eq. 26.6). Let us now use Eq. 29.26 to obtain an expression for the energy density in the magnetic field. Equation 28.6, $B = \mu_0 nI$, where $n = N/\ell$ is the number of windings per unit length, gives us the magnitude of the magnetic field inside a solenoid, and taking the square of this equation yields

$$B^2 = \frac{\mu_0^2 N^2 I^2}{\ell^2}. \tag{29.27}$$

Multiplying the right side of Eq. 29.26 by the factor $\mu_0\ell/(\mu_0\ell)$ gives

$$U^B = \tfrac{1}{2}\frac{\mu_0 N^2 A I^2}{\ell}\frac{\mu_0\ell}{\mu_0\ell} = \tfrac{1}{2}\frac{\mu_0^2 N^2 I^2}{\mu_0\ell^2}A\ell = \tfrac{1}{2}\frac{B^2}{\mu_0}A\ell. \tag{29.28}$$

Because $A\ell$ is the volume of the region of magnetic field inside the solenoid, dividing the energy U^B by this volume gives us the **energy density** of the magnetic field:

$$u_B \equiv \tfrac{1}{2}\frac{B^2}{\mu_0}. \tag{29.29}$$

This expression is analogous to the expression for the energy density u_E in the electric field (Eq. 26.6), with B appearing instead of E and $1/\mu_0$ replacing ϵ_0. If the magnetic field is nonuniform, we must subdivide the volume of interest into small enough segments that B can be considered uniform within each segment, then apply Eq. 29.29 to each segment and take the sum of all the contributions. This amounts to integrating the energy density of the magnetic field over the volume containing the magnetic field:

$$U^B = \int u_B dV. \tag{29.30}$$

Example 29.9 Magnetic energy stored in a square toroid

Consider a toroid with square windings (**Figure 29.36**). The inner radius is $R = 60$ mm, each winding has width $w = 30$ mm, and there are 200 windings, each carrying a current of 1.5 mA. What is the magnetic potential energy stored in this toroid?

Figure 29.36 Example 29.9.

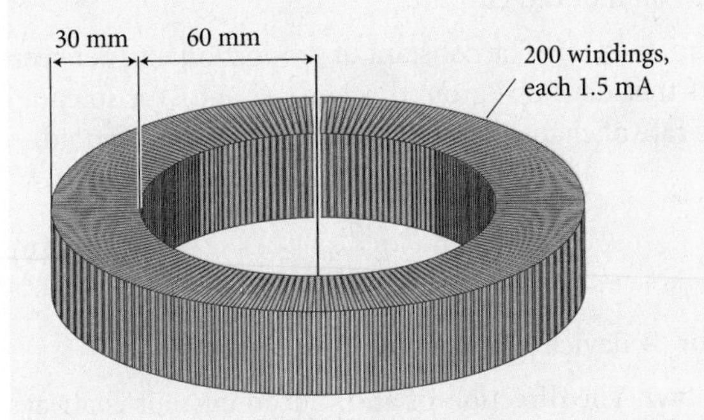

❶ **GETTING STARTED** The current through the toroid causes a magnetic field inside the toroid. This magnetic field stores magnetic potential energy. I note that the width of the windings is not negligible relative to the toroid radius, and this tells me that I must assume a nonuniform magnetic field magnitude across the width of the windings. I therefore must use Eq. 29.30 to determine the amount of magnetic potential energy stored.

❷ **DEVISE PLAN** Before I can use Eq. 29.30, I must obtain the energy density of the magnetic field, which is given by Eq. 29.29. To calculate this energy density, I need to know the magnitude B of the magnetic field. Because the magnetic field is nonuniform, I must determine B by using Eq. 28.9, $B = \mu_0 NI/2\pi r$, where r is the radial distance from the center of the ring formed by the toroid. So to determine the magnetic potential energy stored, I must integrate Eq. 29.29 over the volume of the space enclosed by the windings of the toroid.

❸ **EXECUTE PLAN** Equation 29.29 gives me

$$U^B = \int u_B dV = \frac{1}{2\mu_0}\int B^2 dV = \frac{\mu_0 N^2 I^2}{8\pi^2}\int \frac{dV}{r^2}. \qquad (1)$$

To integrate over the volume of the toroid, I divide the volume into ring-shaped segments (**Figure 29.37**) of height w, radius r, and thickness dr. The volume dV of each segment is equal to the product of the ring's circumference $2\pi r$, height w, and thickness dr: $dV = 2\pi rwdr$. Substituting this expression into Eq. 1 and integrating over r from the inner radius, $r = R$ of the toroid, to the outer radius, $r = R + w$, I get

$$U^B = \frac{\mu_0 N^2 I^2}{8\pi^2}\int_R^{R+w} \frac{2\pi wr\, dr}{r^2} = \frac{\mu_0 N^2 I^2 w}{4\pi}\int_R^{R+w} \frac{dr}{r}.$$

Figure 29.37

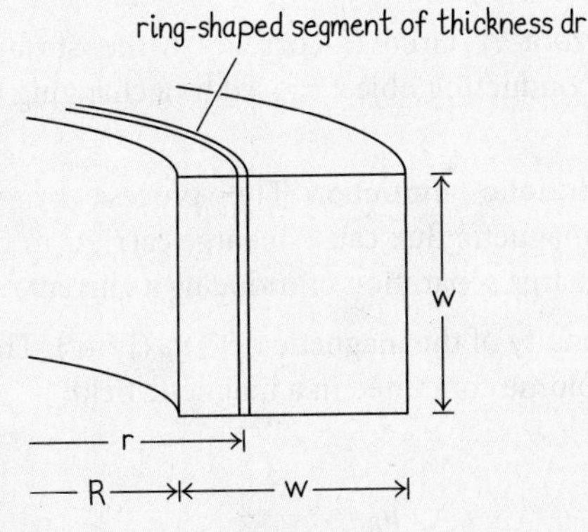

Working out the integration over r yields

$$U^B = \frac{\mu_0 N^2 I^2 w}{4\pi}\ln\left[1 + \frac{w}{R}\right]. \qquad (2)$$

Substituting in the values for μ_0, N, I, R, and w, I obtain for the magnetic potential energy stored in the toroid

$$U^B = (10^{-7}\,\text{T}\cdot\text{m/A})(200)^2(1.5\times10^{-3}\,\text{A})^2 \times (30\,\text{mm})\ln\left(1 + \frac{30\,\text{mm}}{60\,\text{mm}}\right)$$
$$= 1.1\times10^{-10}\,\text{T}\cdot\text{m}^2\cdot\text{A} = 1.1\times10^{-10}\,\text{J}, ✔$$

where I have used $1\,\text{T}\cdot\text{m}^2\cdot\text{A} = 1\,[\text{N}/(\text{A}\cdot\text{m})]\cdot\text{m}^2\cdot\text{A} = 1\,\text{N}\cdot\text{m} = 1\,\text{J}$ (Eq. 27.6).

❹ **EVALUATE RESULT** The result I obtained is a very small amount of energy, but I have no idea (yet) how much energy is stored in a magnetic field. I can, however, compare my result with the result I would obtain in the limit where the width of each winding is much less than the toroid radius, $w \ll R$. In that limit the magnetic field is uniform inside the toroid, and its magnitude is given by Eq. 28.9, $B = \mu_0 NI/2\pi R$. The volume inside the windings of the toroid is equal to the toroid circumference, $2\pi R$, times the area of a winding, w^2: $V = 2\pi Rw^2$. Substituting this expression for B in Eq. 29.29 and multiplying the energy density of the magnetic field by the volume V, I then get for the magnetic potential energy stored in the toroid

$$U^B = \frac{1}{2}\frac{\mu_0^2 N^2 I^2/(2\pi R)^2}{\mu_0}(2\pi Rw^2) = \frac{\mu_0 N^2 I^2 w^2}{4\pi R}$$
$$= \frac{\mu_0 N^2 I^2 w}{4\pi}\frac{w}{R}. \qquad (3)$$

Because $\ln(1+\epsilon) \approx \epsilon$ for $\epsilon \ll 1$, Eq. 2 is equal to Eq. 3 in the limit that $w \ll R$, giving me confidence in my integration result.

29.15 How does the energy density of a 1.0-T magnetic field compare with the energy density of an 1.0-V/m electric field?

Chapter Glossary

SI units of physical quantities are given in parentheses.

Eddy current A circular current at the surface of an extended conducting object caused by a changing magnetic field.

Electromagnetic induction The process by which a changing magnetic flux causes charge carriers to move, inducing a charge separation or inducing a current.

Energy density of the magnetic field u_B (J/m^3) The energy per unit volume contained in a magnetic field:

$$u_B \equiv \frac{1}{2}\frac{B^2}{\mu_0}. \quad (29.29)$$

Faraday's law A changing magnetic flux induces an emf:

$$\mathscr{E}_{\text{ind}} = -\frac{d\Phi_B}{dt}. \quad (29.8)$$

Henry The derived SI unit of inductance:

$$1\ \text{H} \equiv 1\ \text{V} \cdot \text{s/A}. \quad (29.20)$$

Induced current (A) The current caused by a changing magnetic flux.

Induced emf $\mathscr{E}_{\text{ind}}$ (V) The work per unit charge done by electromagnetic induction in separating positive and negative charge carriers.

Induced magnetic field $\vec{B}_{\text{ind}}$ (T) The magnetic field produced by an induced current.

Inductance L (H) The constant of proportionality between the emf that develops around a loop or across a solenoid and the rate of change of current in that loop or solenoid:

$$\mathscr{E}_{\text{ind}} = -L\frac{dI}{dt}. \quad (29.19)$$

Inductor A device with an appreciable inductance.

Lenz's law The direction of an induced current is always such that the magnetic flux produced by the induced current opposes the change in the magnetic flux that induces the current.

Magnetic potential energy U^B (J) The form of potential energy associated with magnetic fields:

$$U^B = \int u_B dV. \quad (29.30)$$

30 Changing Electric Fields

As we have seen in Section 29.3, electric fields accompany changing magnetic fields. Is the reverse true, too—do magnetic fields accompany changing electric fields? In this chapter we see that magnetic fields do indeed accompany changing electric fields. Consequently, a changing electric field can never occur without a magnetic field, and a changing magnetic field can never occur without an electric field. The interdependence of changing electric and magnetic fields gives rise to an oscillating form of changing fields called *electromagnetic waves.*

Electromagnetic waves are familiar to us as a wide range of phenomena: visible light, radio waves, and x-rays are all electromagnetic waves, the only difference being the frequency of oscillation of the electric and magnetic fields. We see our world by means of these waves, whether by using our eyes to observe our surroundings or by using x-ray diffraction to construct an image of a molecule or a material. Modern communications, from radio and television to mobile telephones, also make extensive use of electromagnetic waves. As we shall see, all these electromagnetic waves consist of changing electric and magnetic fields.

Figure 30.2 Checkpoint 30.1.

(*a*)

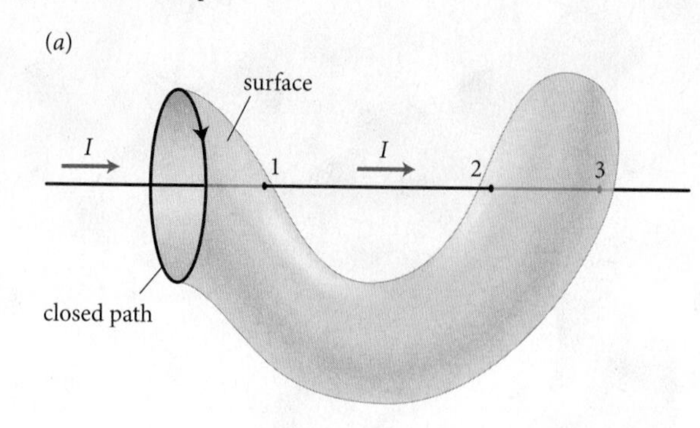

(*b*)

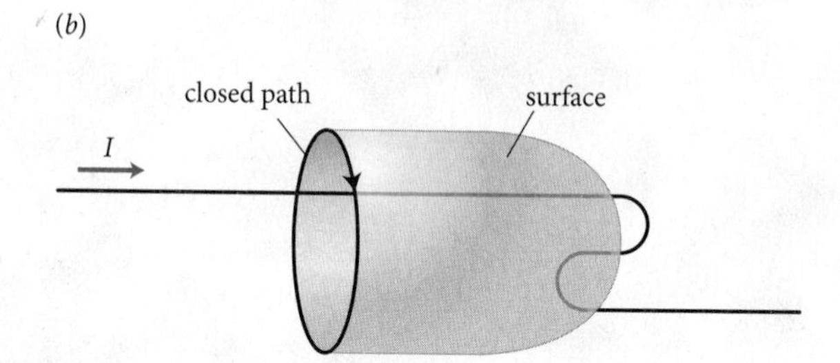

30.1 Magnetic fields accompany changing electric fields

In order to see that a magnetic field accompanies a changing electric field, let's revisit Ampère's law (see Section 28.5), which states that the line integral of the magnetic field along a closed path is proportional to the current encircled by the path (Eq. 28.1, $\oint \vec{B} \cdot d\vec{\ell} = \mu_0 I_{\text{enc}}$).

Figure 30.1 shows a current-carrying wire encircled by a closed path. The current encircled by the path is equal to the current through the wire, I. Another way to determine the encircled current is to consider any surface spanning the path and determine the current intercepted by that surface. For example, Figure 30.1 shows two different surfaces spanning the path. The current intercepted by either surface is I, the current encircled by the path.

30.1 Is the current intercepted by the surface equal to the current encircled by the closed path (*a*) in **Figure 30.2*a*** and (*b*) in Figure 30.2*b*?

Figure 30.1 Current-carrying wire encircled by a closed path. Surfaces A and B both span the path. Surface A lies completely in the plane of the path. Surface B extends as a hemisphere whose rim is the path.

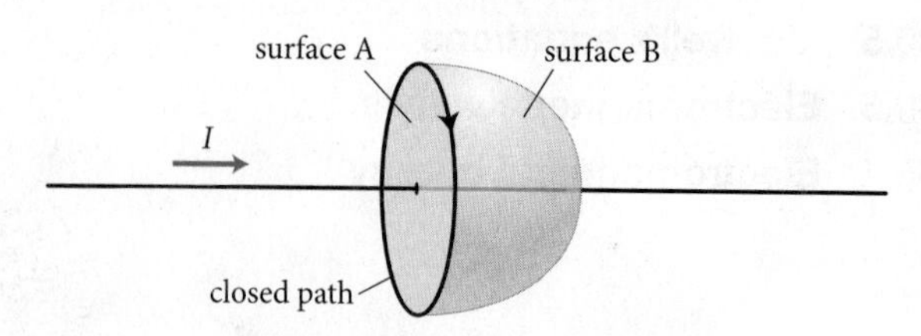

Checkpoint 30.1 shows that the current encircled by a closed path is equal to the current that is intercepted by any surface that spans the path, provided we keep track of the directions in which each interception takes place. Ampère's law can equally well be applied to the current encircled by a closed path and to the current intercepted by any surface spanning that closed path.

Now consider inserting a capacitor into our current-carrying wire while continuing to supply a constant current I to the wire. (That is, the capacitor is being charged.) **Figure 30.3*a*** again shows two surfaces A and B spanning the same closed path. The line integral of the magnetic field around the closed path does not depend on the choice of surface spanning the path. However, while the capacitor is charging, surface A is intercepted by a current I but

Figure 30.3 Capacitor being charged by a current-carrying wire. (*a*) The closed path of interest encircles the wire. Surface A intercepts the current, but surface B passes between the capacitor plates and does not intercept the current. (*b*) The closed path of interest lies between the capacitor plates. Surface A also lies between the plates and does not intercept the current, but surface B intercepts the current.

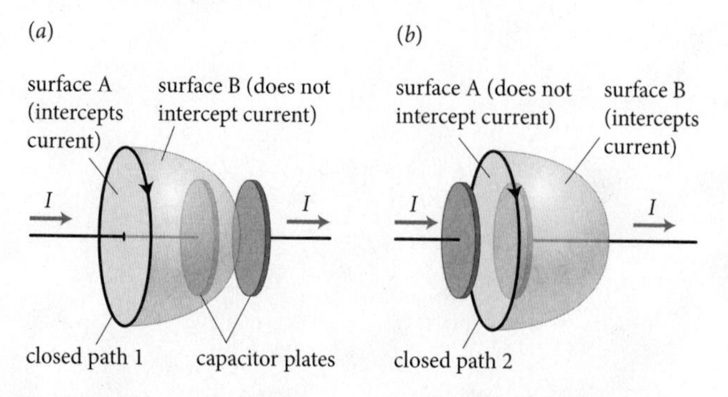

surface B, which passes between the capacitor plates, is not intercepted by any current. If we choose a closed path that lies between the capacitor plates (Figure 30.3*b*), a similar difficulty arises. Surface A intercepts no current, while surface B intercepts the current I.

In the case of a capacitor, therefore, the equivalence between encircled current and current intercepted by a surface spanning the encircling path doesn't hold. Surface B in Figure 30.3*a* would lead us to conclude that the line integral of the magnetic field around closed path 1 is zero. Because symmetry requires the magnetic field to always be tangent to the path and have the same magnitude all around the path, the line integral being zero means there is no magnetic field at the location of closed path 1 (even though the path encircles a current). Conversely, surface B in Figure 30.2*b* suggests there is a magnetic field at the location of closed path 2, even though that path encircles no current. Experiments do indeed confirm that there *is* a magnetic field in and around the gap between the plates of the charging capacitor. So only the surfaces that intersect the wires leading to the capacitor appear to provide the correct value of I_{enc} in Ampère's law for both closed paths in Figure 30.3.

Why must there be a magnetic field in and around the gap between the plates of the charging capacitor? Although there is no flow of charged particles between the plates of the capacitor, there *is* an electric field (**Figure 30.4**). Let us examine this electric field in more detail in the next checkpoint.

30.2 (*a*) While the capacitor of Figure 30.4 is being charged, is the current through the wire leading to or from the capacitor zero or nonzero? Is the electric field between the plates zero or nonzero? Is it constant or changing? (*b*) Answer the same questions for the capacitor fully charged.

The answers to Checkpoint 30.2 suggest that the magnetic field between the plates of the charging capacitor arises from the *changing* electric field. The current to the capacitor causes the electric field between the plates to

Figure 30.4 Capacitor being charged by a current-carrying wire. The electric field between the plates is shown. Closed path 1 encircles the current through the wire; closed path 2 encircles the electric field between the capacitor plates.

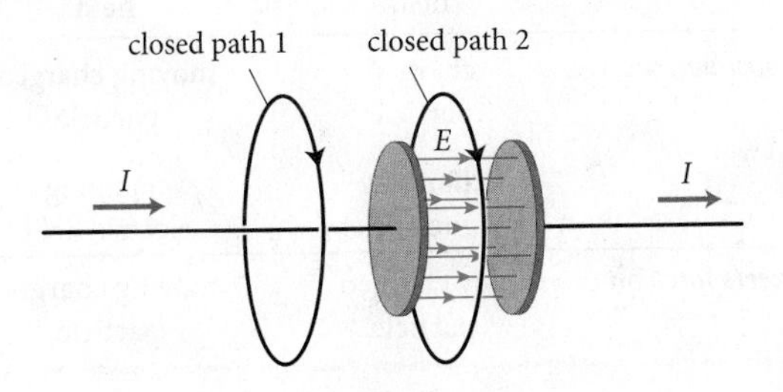

Figure 30.5 Parallels between (*a*) the electric field that accompanies a changing magnetic field and (*b*) the magnetic field that accompanies a changing electric field.

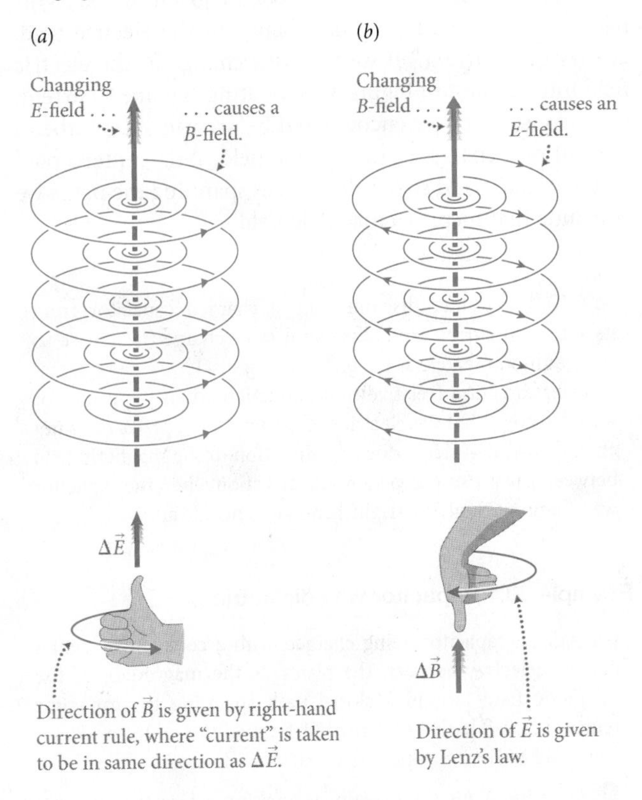

change, and the changing electric field between the capacitor plates acts in a way similar to the current that causes this change:

A changing electric field is accompanied by a magnetic field.

When the capacitor is fully charged, the current I into and out of the capacitor is zero, and there is no magnetic field surrounding the wires to the capacitor. Between the capacitor plates, the electric field is no longer changing, and the magnetic field is zero.

There are strong parallels between the electric field that accompanies a changing magnetic field and the magnetic field that accompanies a changing electric field, as **Figure 30.5** illustrates. Experiments show that the electric field lines that accompany a changing magnetic field form loops encircling the magnetic field, just as the magnetic field lines that accompany a changing electric field form loops encircling the electric field.

As we discussed in Section 27.3, the magnetic field surrounding a current-carrying wire forms loops that are clockwise when viewed looking along the direction of the current. The direction of these loops can be described by the right-hand current rule: Point the thumb of your right hand in the direction of the current, and your fingers curl in the direction of the magnetic field. Similarly, the direction of the loops formed by the magnetic field lines that accompany a changing electric field are given by the right-hand current rule, taking the change in the electric field, $\Delta\vec{E}$, as the "current." If we take this change in the electric field into account in Figure 30.3, treating $\Delta\vec{E}$ like a current, the inconsistency we encountered before vanishes: Either a current or a change in the electric field, $\Delta\vec{E}$, is intercepted by the surface, and so for all surfaces spanning the paths we conclude that there is a magnetic field.

30.3 Consider disconnecting a charged capacitor from its source of current and allowing it to discharge (to release its charge into an external circuit). During discharge, the current reverses direction (relative to its direction when the capacitor was charging), but the electric field between the plates does not change direction. How does the direction of the magnetic field between the plates compare to the direction when the capacitor was charging? Does the right-hand current rule apply?

Example 30.1 Capacitor with dielectric

Consider a capacitor being charged with a constant current I and a dielectric between the plates. Is the magnitude of the magnetic field around a closed path spanning the capacitor (such as closed path 2 in Figure 30.4) any different from what it would be without the dielectric? Why or why not?

❶ **GETTING STARTED** I begin by making a two-dimensional sketch of the capacitor, indicating the position of the closed path (**Figure 30.6**).

Figure 30.6

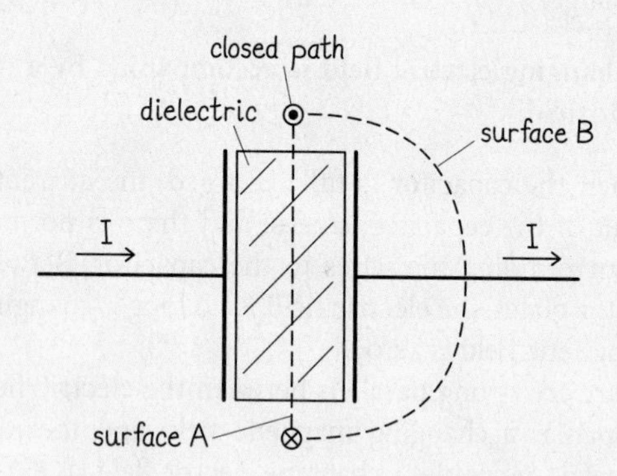

❷ **DEVISE PLAN** To determine the magnetic field magnitude at any position along the closed path, I need to examine the current and the changing electric field intercepted by a surface spanning the closed path.

❸ **EXECUTE PLAN** If I consider a flat surface through the closed path (surface A in Figure 30.6), the surface intersects the dielectric. While the capacitor is charging, the dielectric is being polarized: Negative charge carriers in the dielectric are displaced in one direction, and positive charge carriers are displaced in the opposite direction. This displacement of charge carriers corresponds to a current. Surface A also intercepts a changing electric field. However, without further information about the capacitor, I can determine neither the current nor the electric field between the capacitor plates, which is affected by the presence of the dielectric. Surface A therefore doesn't permit me to compare the magnetic field magnitude to what it would be without the dielectric. I therefore draw another surface, making this surface loop around one of the capacitor plates (surface B in Figure 30.6). This surface intercepts only the wire leading from the capacitor, and I know that the current through the wire is unchanged by the presence of the dielectric. The fact that the current through this wire is unchanged tells me that the effective current in the region containing the dielectric is also unchanged. Therefore, the magnetic field must be the same as it would be without the dielectric. ✔

❹ **EVALUATE RESULT** Intuitively I expect the magnetic field magnitude around my closed path to be unchanged when the magnetic field magnitude around the wires attached to the capacitor is unchanged. The electric field between the capacitor plates gives rise to a displacement of charge carriers within the dielectric and thus affects the electric field between the capacitor plates, but apparently everything adds up to yield, for a given current through the capacitor, the same magnetic field magnitude outside the capacitor for a given current to the capacitor regardless of the presence or absence of the dielectric.

We now have a complete picture of what gives rise to electric and magnetic fields and on what kind of charged particle these fields exert forces. **Table 30.1** summarizes the properties of electric and magnetic fields. Note the remarkable symmetry between the two. Each type of field is produced by charged particles and accompanies a changing field of the other type. Electric fields are produced by charged particles either at rest or in motion, but magnetic fields are produced only by charged particles in motion. Likewise, any charged particle—at rest or in motion—is subject to a force in the presence of an electric field, but only charged particles in motion are subject to forces in a magnetic field.

Table 30.2 summarizes what we know about the field lines for electric and magnetic fields. The most striking difference between electric and magnetic fields is that magnetic field lines always form loops but electric field lines do

Table 30.1 Properties of electric and magnetic fields

	Electric field	Magnetic field
associated with	charged particle	moving charged particle
	changing magnetic field	changing electric field
exerts force on	any charged particle	moving charged particle

CONCEPTS

Table 30.2 Electric and magnetic field lines

	Electric field	Magnetic field
lines emanate from or terminate on	charged particle	–
loops encircle	–	moving charged particle
	changing magnetic field	changing electric field

not always form loops. This is a direct consequence of the difference in the sources of these fields. Magnetic field lines must form loops because there is no magnetic equivalent of electrical charge—no magnetic monopole (see Section 27.1). Instead, magnetic fields arise from current loops that act as magnetic dipoles.

Electric and magnetic field lines that accompany changing fields both form loops around the changing field. When particles serve as the field sources, however, the difference between magnetic and electric fields is evident: Electric field lines emanate or terminate from charged particles, while magnetic field lines always form loops around moving charged particles (currents).

We shall return to these ideas quantitatively in Section 30.5.

30.4 The neutron is a neutral particle that has a magnetic dipole moment. What does this nonzero magnetic dipole moment tell you about the structure of the neutron?

30.2 Fields of moving charged particles

We have seen that capacitors generate changing electric fields when charging or discharging. What else produces changing electric fields? One answer to this question is: changes in the motion of charged particles.

Before examining the electric fields of accelerating charged particles, let's consider the electric fields generated by charged particles moving at constant velocity. **Figure 30.7** shows the electric field of a stationary charged particle and of the same particle moving at constant high speed. (By high speed, I mean a speed near enough the speed of light for relativistic effects to become important.)

The electric field of the stationary particle is spherically symmetrical; the electric field of the moving particle is still radial but definitely not spherically symmetrical. In this electric field, the field lines are sparse near the line along which the particle travels and are clustered together in the plane perpendicular to the motion. (This clustering is a relativistic effect and takes place for the same reason that objects moving at relativistic speeds appear shorter along the direction of motion, as discussed in Sections 14.3 and 14.6.) Consequently, the electric field created by the moving particle is strongest in that perpendicular plane. The faster the particle moves, the more the electric field lines bunch up in the transverse direction.

Keep in mind that as the particle moves at constant speed, the electric field lines move with it. At any instant, the electric field lines point directly away from the position of the particle *at that instant*. This means that as the particle moves, the electric field at a given position changes.

Because the particle in Figure 30.7*b* is moving, it is like a tiny current; it has a magnetic field that forms loops around its direction of travel, as shown in Figure 28.2*b*. The particle in Figure 30.7*a* does not have a magnetic field because it is at rest.

Now let's consider a particle that is initially at rest and then is suddenly set in motion. The electric field of this particle is shown at three successive instants in **Figure 30.8** on the next page. Figure 30.8*b* and *c* show something we have not seen before: electric field lines that do not point directly away from the charged particle that is their source but instead are disrupted by sharp kinks. What is more, these kinks, which appear when the particle accelerates (just after Figure 30.8*a*), do not go away once the particle

Figure 30.7 Electric field line pattern of a charged particle (*a*) at rest and (*b*) moving to the right with speed v (v is a significant fraction of the speed of light). For the moving particle, the electric field lines cluster around the plane perpendicular to the direction of motion.

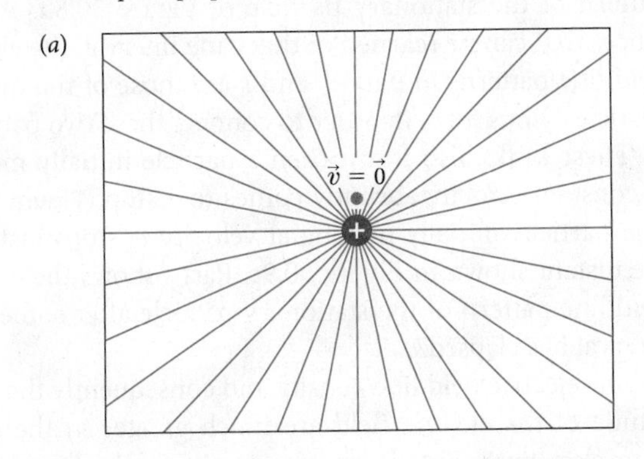

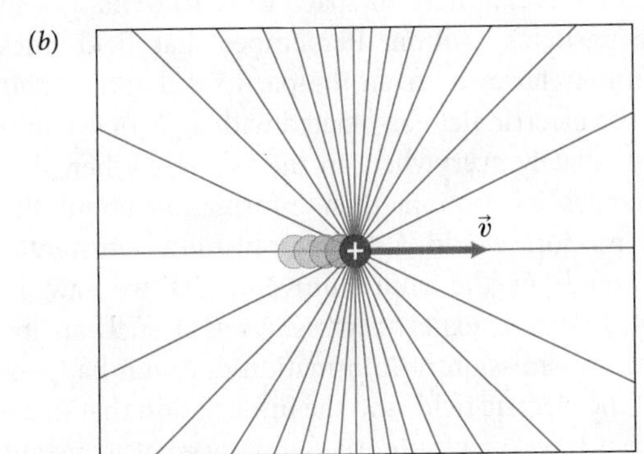

Figure 30.8 Electric field line pattern of a particle (*a*) initially at rest, (*b*) accelerating to speed v, and (*c*) moving at constant speed v. In (*b*) and (*c*), the ring of kinks in the electric field lines traveling outward from the particle corresponds to an electromagnetic wave pulse. Note that the speed v is smaller than the speed of the particle in Figure 30.7, indicated by the shorter arrow. Consequently, the electric field lines here are less sharply bunched around the vertical.

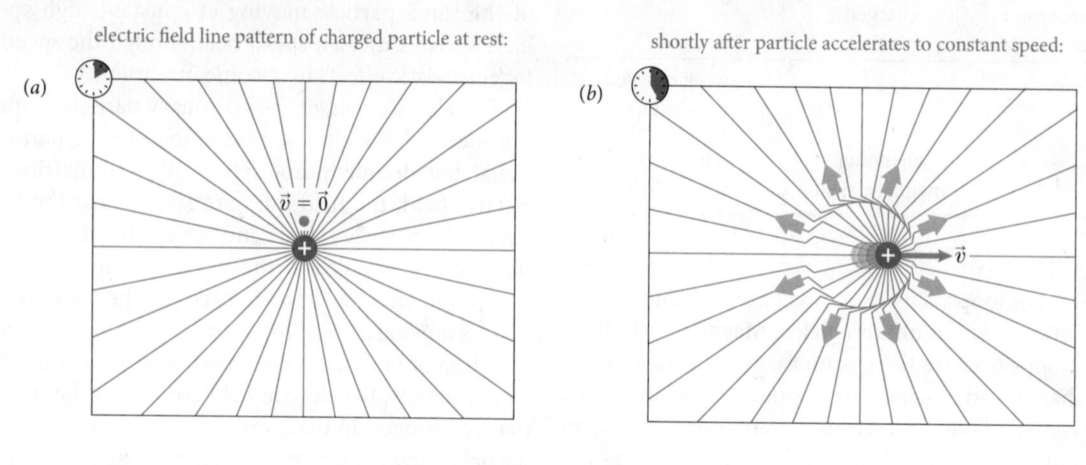

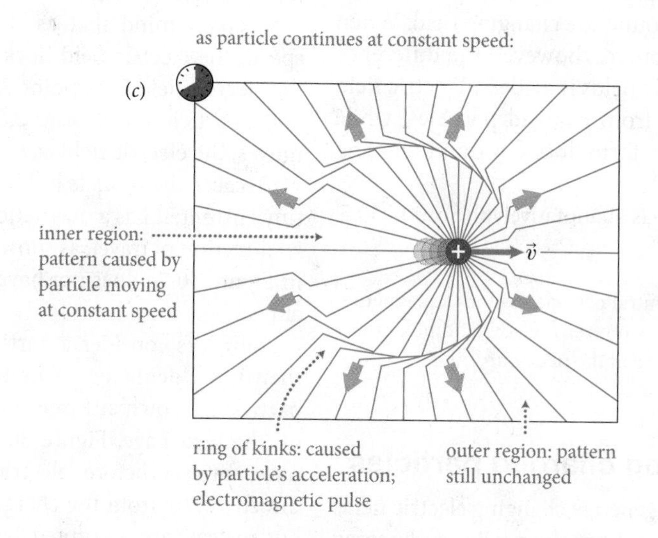

reaches its final constant speed. Instead, they travel radially out from the location where the particle was when it started moving.

Where do these kinks in the electric field pattern come from? They arise because the electric field cannot change instantaneously everywhere in space to reflect changes in the source particle's motion. Remember that field lines extend infinitely far away from the particles that are their sources. If the electric field associated with a particle could change immediately everywhere in the universe when that particle changes its motion, then information about the change in motion would also be transmitted instantaneously throughout the entire universe. As we saw in Chapter 14, however, experiments show that such an instantaneous transmission of information does not happen. Changes in the electric field, and the information that these changes carry, travel at a finite (though very great) constant speed. In fact, in vacuum such changes always travel at the same speed regardless of the details of the motion of the particles that produce them.

At distances that are too great for changes to reach in the time interval represented in Figure 30.8, the electric field line patterns in Figure 30.8*b* and *c* are still the same as the pattern of the stationary particle of Figure 30.8*a*. At distances that can be reached in that time interval, the electric field line patterns in parts *b* and *c* are those of the moving particle. Kinks form in order to connect these two patterns.

These kinks also form when a particle initially moving at constant velocity abruptly comes to a stop (**Figure 30.9**). The particle, initially moving at velocity v, stops just after the instant shown in Figure 30.9*e*. Part *f* shows the electric field line pattern of the stationary particle after some time interval has elapsed.

The electric field line density and consequently the magnitude of the electric field are much greater in the kinks than elsewhere. The energy density in the kinks region is

Figure 30.9 Electric field lines of a charged particle moving at some relativistic speed v. The upper diagrams show successive instants as the particle moves at constant velocity. The lower diagrams show the same instants, but the particle slows down to a stop between (*e*) and (*f*).

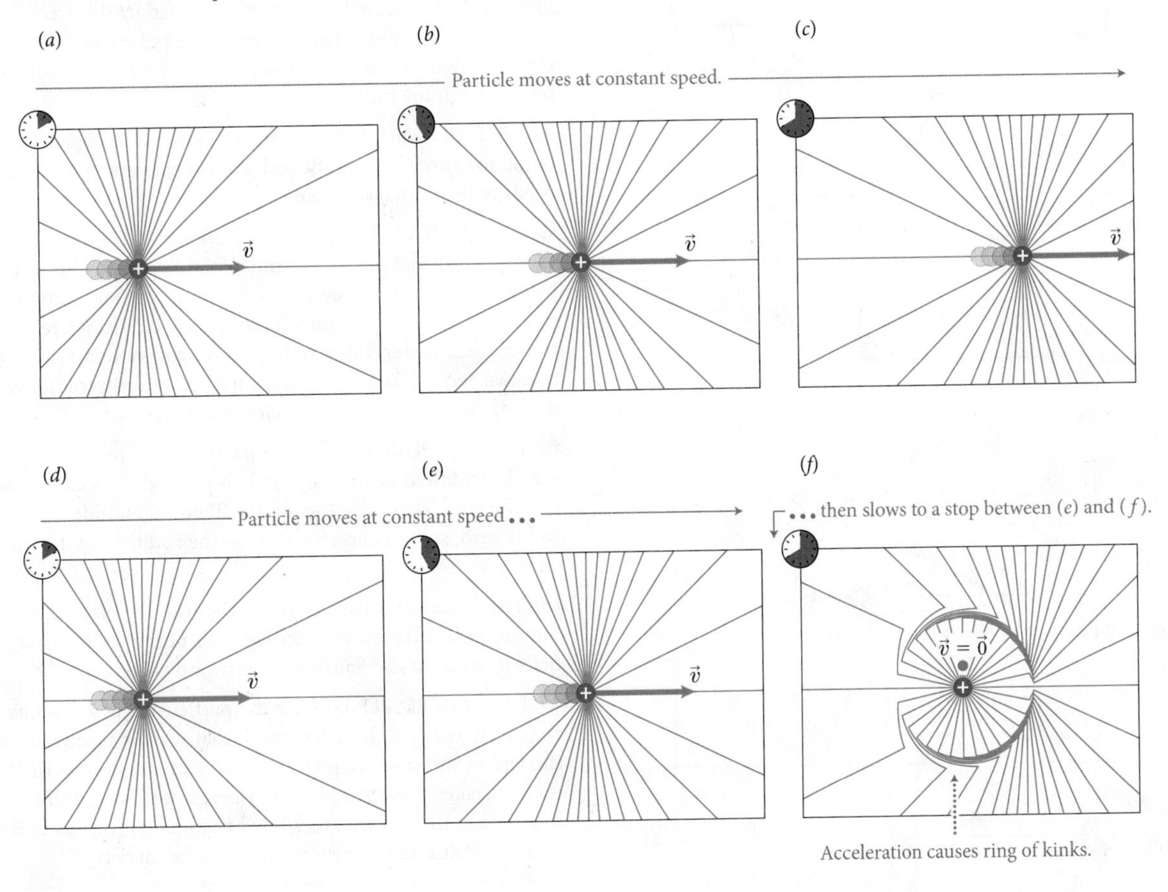

therefore greater than the energy density in other parts of the electric field. As the kinks move, they carry energy away from the particle. These kinks (and the energy carried by them) are one of the two parts of *electromagnetic waves*. As you might guess, kinks in magnetic field lines are the other part. Because changing electric fields are accompanied by changing magnetic fields (and vice versa), the two are always found together. An **electromagnetic wave** is thus a combined disturbance in an electric and a magnetic field that is propagating through space. Because a single isolated propagating disturbance is called a wave pulse (see Section 16.1), the kinks that appear in Figures 30.8 are 30.9 are *electromagnetic wave pulses*.

30.5 Estimate the final speed v of the charged particle in Figure 30.8 in terms of the speed of propagation c of the electromagnetic wave pulse produced by the particle's acceleration.

Let us now look at what effect an electromagnetic wave pulse has on a charged particle. **Figure 30.10** on the next page shows the force exerted on a stationary charged test particle by the electric field of an accelerated charged particle. At the first instant shown (Figure 30.10*a*), before the particle at the center of the panel is accelerated, the force exerted by the electric field on the test particle runs along the field line joining the two particles and points away from the center particle. At the second instant shown (Figure 30.10*b*), the center particle has been accelerated, and the wave pulse created by the acceleration has just reached the test particle. The force exerted on the test particle is no longer directed along the line joining the two particles but is directed along the kinks in the electric field lines. The force therefore has a component tangential to a circle centered on the original position of the accelerated particle at the center of the panel. (The exact direction of the force depends on the magnitude and duration of the acceleration of the accelerated particle.) Moreover, because the electric field line density is large in the region of the kinks, the force is large in magnitude.

At the final instant shown (Figure 30.10*c*), the wave pulse has traveled beyond the test particle and the force once again points away from the particle. The electric field line density is much smaller again, and so the magnitude of the force exerted on the test particle is again much smaller.

Figure 30.10 Force exerted on a stationary charged test particle by the electric field of an accelerated charged particle.

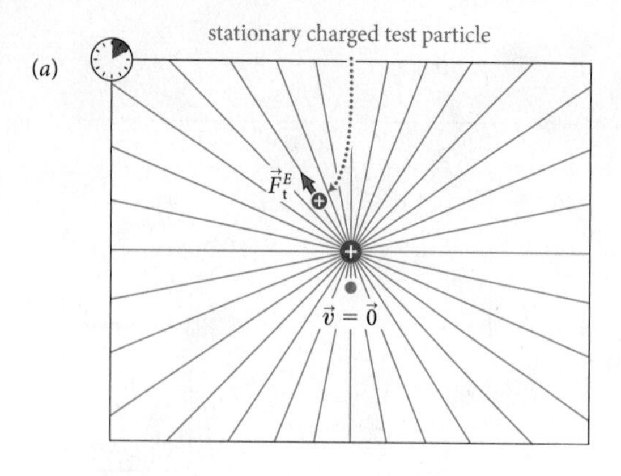

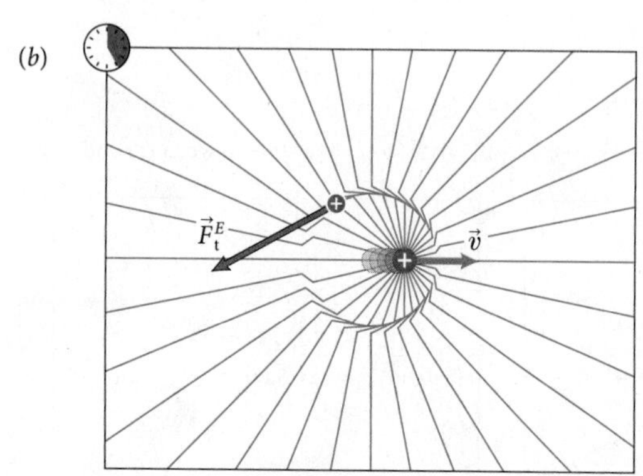

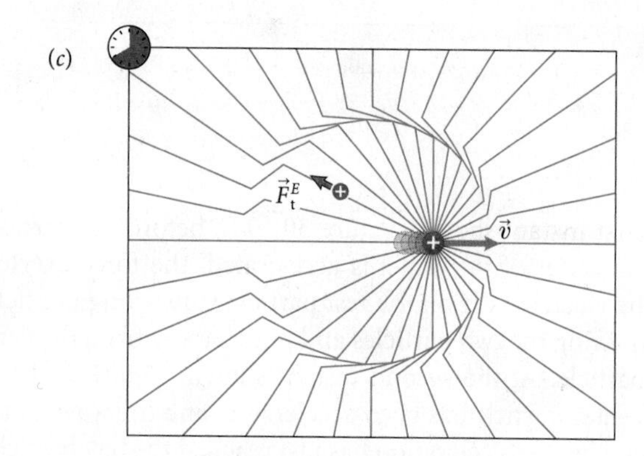

Example 30.2 Electromagnetic wave pulse

A particle carrying a negative charge is suddenly accelerated in a direction parallel to the long axis of a conducting rod, producing the electric field pattern shown in **Figure 30.11**. Does the electric field of the negatively charged particle create a current through the rod (*a*) at the instant shown in the figure, before the electromagnetic wave pulse created by the acceleration reaches the rod, and (*b*) at the instant the pulse reaches the rod? If you answer yes in either case, in which direction is the current through the rod?

Figure 30.11 Electric field of an accelerated particle near a conducting rod, before the wave pulse reaches the rod. (The electric field lines bend in near the conducting rod due to the rearrangement of charge carriers at the surface of the rod.)

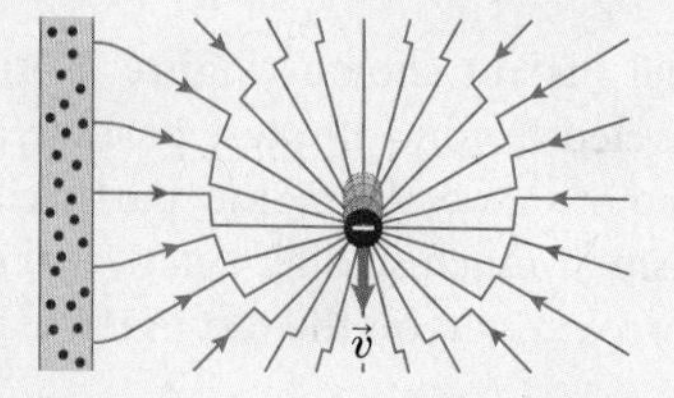

❶ **GETTING STARTED** A current is created when charge carriers in the rod flow through the rod. For the carriers to flow, a force needs to be exerted on them.

❷ **DEVISE PLAN** To determine whether there is a current through the rod, I must determine if the electric field is oriented in such a way as to cause a flow of charge carriers through the rod. Even though in a metallic rod only electrons are free to move, I can pretend that only positive charge carriers are free to move because as I saw in Section 27.3, my answer is independent of the sign of the mobile charge carriers.

❸ **EXECUTE PLAN** (*a*) No. Before the pulse reaches the rod, the electric field is constant and so the rod is in electrostatic equilibrium. Therefore the electric field magnitude inside the rod is zero, so no charge carriers in the rod flow at the instant shown. ✔

(*b*) Yes. Once the pulse arrives at the rod, the electric field in the rod points downward, accelerating positively charged particles downward and causing a downward current. ✔

❹ **EVALUATE RESULT** Because the particle being accelerated is negatively charged, it makes sense that it pulls positive charge carriers in the rod along (with a delay caused by the time interval it takes the wave pulse to travel to the rod). In practice, electrons in the rod are accelerated upward, but the result is the same as what I describe for positive charge carriers.

30.6 In Figure 30.10, in which regions of space surrounding the accelerating particle does a magnetic field occur?

30.3 Oscillating dipoles and antennas

The wave pulse we have just considered is a brief, one-time, propagating disturbance in the electric field, analogous to the disturbance created when the end of a taut rope is suddenly displaced (as in Figure 16.2, for instance). Just as a harmonic wave can be generated on a rope by shaking the end of the rope back and forth in a sinusoidal fashion, a harmonic electromagnetic wave can be generated when a charged particle oscillates sinusoidally. **Figure 30.12** shows the electric field of a charged particle undergoing sinusoidal oscillation. This electric field consists of periodic kinks traveling away from the particle in a wavelike fashion.

In practice, isolated charged particles are not common. More often, positive and negative charged particles are present together, whether in individual atoms or in solid or liquid materials. Thus, displacing a positive particle leaves a negative particle behind, forming an electric dipole. Let us therefore consider the electric field pattern of an oscillating dipole.

Figure 30.12 Electric field of a sinusoidally oscillating charged particle.

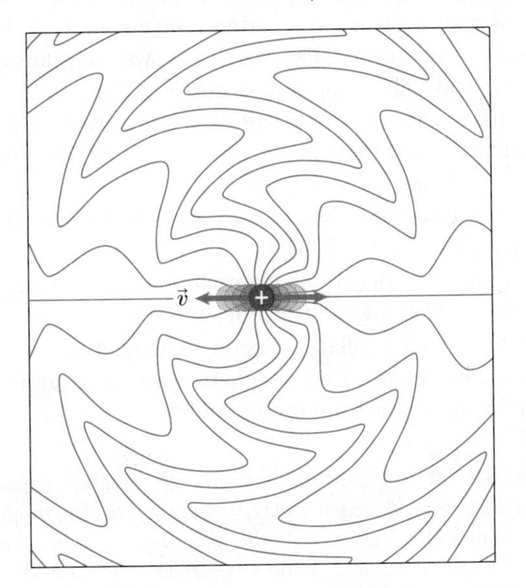

The electric field pattern of a stationary electric dipole with the positive charged particle above the negative charged particle is shown in **Figure 30.13a**. What about the electric field pattern of a stationary dipole made up of the same charged particles but with their positions switched, so that the dipole moment $\vec{p}$—which points from the negatively charged end to the positively charged end (see Section 23.4)—has reversed? The corresponding pattern of electric field lines has the same shape as shown in Figure 30.13*a*, but the directions of all the electric field lines are reversed.

Now let's work out the electric field pattern of an oscillating dipole, in which the two particles oscillate back and forth with a period T. We begin by considering the electric field of a dipole that undergoes only a single reversal of its dipole moment (that is, one-half of a single oscillation) rather than oscillating continuously. The dipole starts out as shown in Figure 30.13*a* at instant $t = 0$. The charged particles that constitute the dipole then switch places in a time interval $T/2$ (half a cycle) and remain there.

Figure 30.13*b* shows the electric field pattern at instant $t = T$, after the dipole has been at rest in its new orientation for a time interval $T/2$. We can divide the space surrounding the dipole into the three regions shown. First consider the region sufficiently close to the dipole that the electric field is just the electric field of the stationary dipole in its new orientation. If we denote the speed at which changes in the electric field travel outward by c and the dipole has been stationary for a time interval $T/2$, this innermost region occupies a circle of radius $R = cT/2$. (The origin of our coordinate system is the center of the dipole, midway between the two particles.) Inside this circle, the electric field is that of the stationary dipole, the same shape as shown in Figure 30.13*a* but with the electric field line directions reversed.

Now consider the region sufficiently far away that no information about the motion of the dipole has reached it yet. This region lies outside a circle of radius $R = cT$. In this region, the electric field pattern is identical to that shown in Figure 30.13*a*, the electric field of the original dipole before it flipped over.

In the highlighted region of Figure 30.13*b* between these two circles, the electric field pattern is not dipolar. Because there are no charged particles in this region, the electric field lines cannot begin or end here. Instead, they must be connected to the electric field lines in the inner and outer regions. Consequently the electric field lines split into two disconnected sets: a set that emanates from the ends of the dipole and a set of loops detached from the dipole.

Figure 30.13 (*a*) Electric field of a stationary electric dipole in which the positive particle lies above the negative particle. (*b*) Electric field of the same dipole after the dipole moment has reversed and the charged particles have returned to rest.

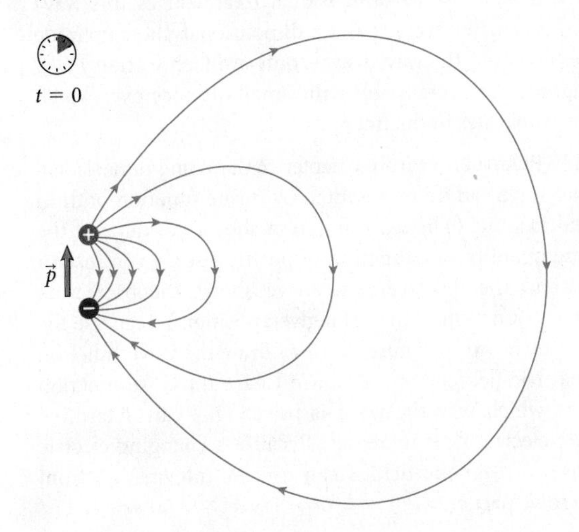

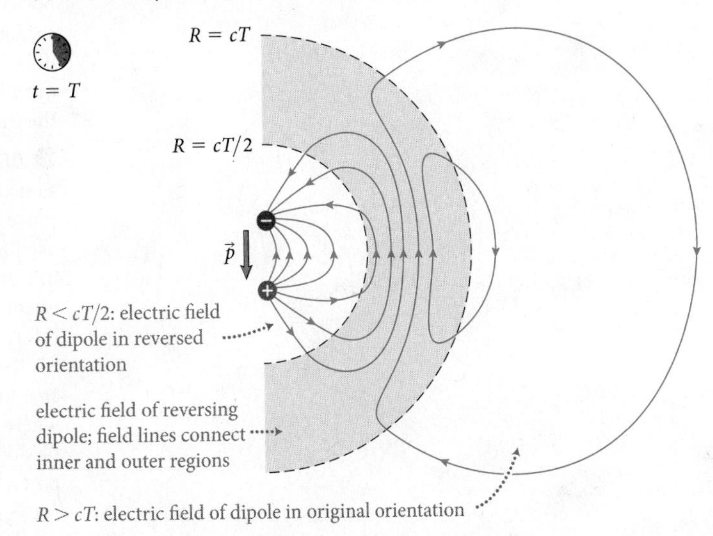

Figure 30.14 Snapshots of the electric field pattern of a sinusoidally oscillating dipole at time intervals of $T/8$ (where T is the period of oscillation).

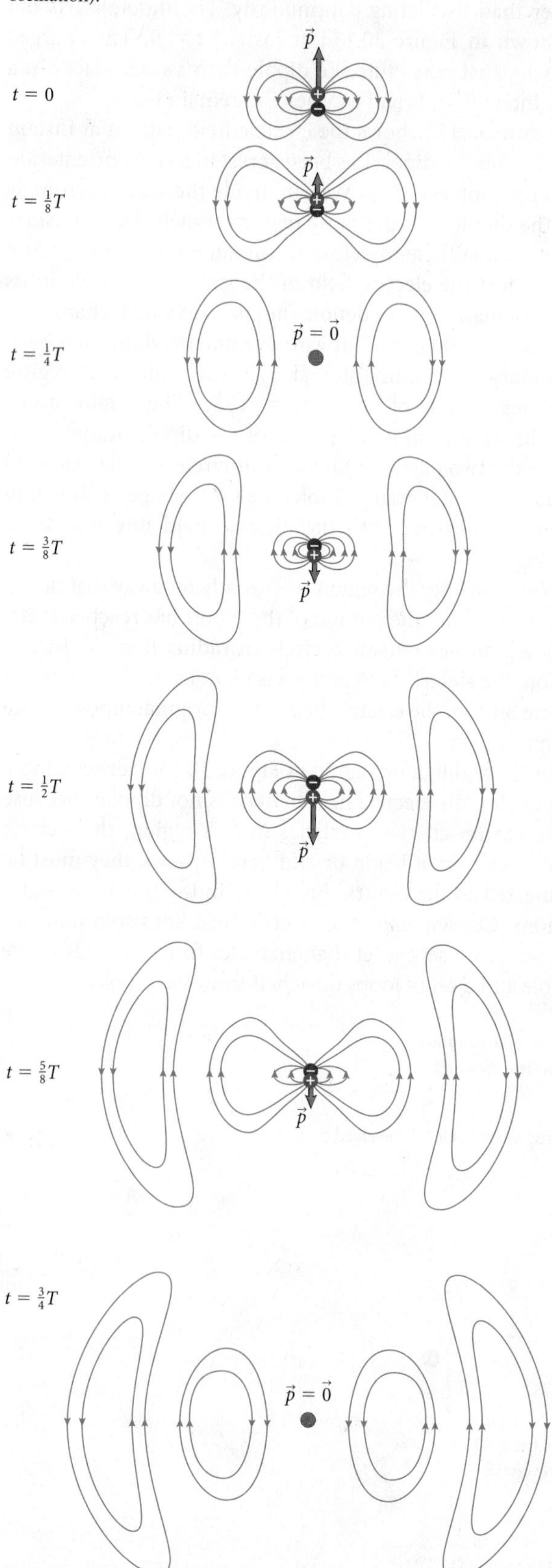

This electric field line pattern can be generalized to the case of a sinusoidally oscillating dipole (one that doesn't stop after half a cycle). **Figure 30.14** shows snapshots of the electric field pattern of such a dipole at time intervals of $T/8$. Just as with the single half-oscillation, we see a dipolar electric field near the dipole. Farther away, the electric field lines form loops.

Notice how these loops form every half-cycle (that is, at $t = T/4$ and $t = 3T/4$): As the charged particles of the dipole reach the origin during each oscillation, the electric field lines pinch off and the loops travel outward like puffs of smoke. This regular emission of looped electric field lines is a harmonic electromagnetic wave that travels away from the dipole horizontally left and right.

30.7 (*a*) If Figure 30.14 shows the oscillating electric field pattern at its actual size, estimate the wavelength of the electromagnetic wave. (*b*) If the wave is traveling at speed $c = 3 \times 10^8$ m/s, what is the wave frequency? (*c*) How long does one period last?

So far we have focused on the electric field pattern of this electromagnetic wave because it is natural to think about the electric field of a dipole. However, the changing electric field of the oscillating dipole is accompanied by a magnetic field. Consequently, the oscillation produces not only an electric field but also a magnetic field.

Example 30.3 Magnetic field pattern

Consider the electric field pattern of a sinusoidally oscillating dipole in Figure 30.14. (*a*) At $t = \frac{3}{4}T$, where along the horizontal axis bisecting the straight line connecting the two poles is the electric field increasing with time? Where is it decreasing? (*b*) Based on your answer to part *a*, what pattern of magnetic field lines do you expect in the horizontal plane that bisects the straight line connecting the two poles?

❶ **GETTING STARTED** Because the dipole oscillates sinusoidally, I expect the electric field to be a sinusoidally oscillating outward-traveling wave. The wave is three-dimensional, but the problem asks only about the electric field along the dipole's horizontal axis, so a one-dimensional treatment of this wave suffices. Because the wave is three-dimensional, the amplitude decreases as $1/r$ as the wave travels outward (see Section 17.1), but I'll ignore the decrease over the small distance over which the wave propagates in the figure.

❷ **DEVISE PLAN** I know from Chapter 16 that a one-dimensional sinusoidal wave can be represented by a sine function both in space and in time. (The wave function shows the value of the oscillating quantity as a function of position at a given instant in time, and the displacement curve shows the oscillating quantity as a function of time at a given position.) I can use the information shown in Figure 30.14 to draw the wave function for the electric field at $t = \frac{3}{4}T$. Once I have the wave function and know which way the wave is traveling, I can determine where the electric field increases. Because a changing electric field causes a magnetic field, I can use the information from part *a* to solve part *b*.

❸ **EXECUTE PLAN** (*a*) Because the problem asks for information at the instant $t = \frac{3}{4}T$, I begin by copying the right half of the bottom electric field pattern of Figure 30.14 (**Figure 30.15*a***). (The left half is simply the mirror image of the right half.) I draw a rightward-pointing horizontal axis through the center of the dipole and denote this as the *z* axis. I see that the electric field points downward parallel to the vertical axis (which I take to be the *x* axis) in the region between the dipole and the center of the first set of electric field loops. In the region between the centers of the first and the second set of loops, the electric field points upward. Because the electric field must vary sinusoidally, I can now sketch how its *x* component varies with position along the horizontal axis (Figure 30.15*b*).

Figure 30.15

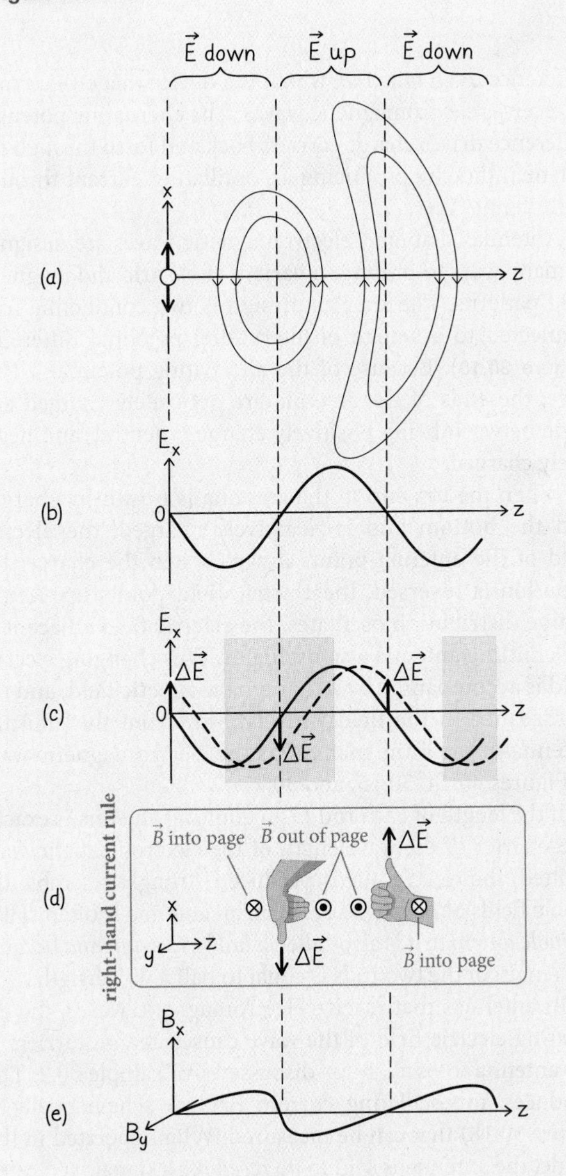

As the wave travels outward, the wave function of Figure 30.15*b* moves to the right (dashed curve in Figure 30.15*c*). The difference between the dashed and solid curves is the change in the electric field $\Delta\vec{E}$ (black arrows); $\Delta\vec{E}$ points down when the derivative of E_x with respect to *z* is positive (shaded regions) and up when it is negative (unshaded regions). ✔

(*b*) The direction of the magnetic field is determined by the right-hand current rule, taking the direction of the change in the electric field $\Delta\vec{E}$ as the "current." Pointing the thumb of my right hand down in the region where $\Delta\vec{E}$ points down and up where $\Delta\vec{E}$ points up (Figure 30.15*d*), I see from the way my fingers curl that the magnetic field lines form loops in the horizontal (*yz*) plane that are centered on the vertical black dashed lines, just like the electric field lines do. Consequently the magnetic field points out of the page when E_x is positive and into the page when E_x is negative. If I let the *y* axis point out of the page, the *y* component of the magnetic field must be positive when E_x is positive and negative when E_x is negative. I therefore draw a sinusoidally varying function for B_y as a function of position (Figure 30.15*e*). ✔

❹ **EVALUATE RESULT** My answer shows that the electric and magnetic fields have the same dependence on time but are oriented perpendicular to each other. Because the magnetic field changes, Faraday's law tells me that it is accompanied by an electric field. To analyze this electric field I can use an approach similar to the one I used to determine the magnetic field. **Figure 30.16*a*** shows the magnetic wave traveling outward. The difference between the dashed and solid curves is the change in the magnetic field $\Delta\vec{B}$. According to what I learned in Section 29.6, the direction of the electric field accompanying my changing magnetic field is given by Lenz's law and the right-hand dipole rule (Figure 30.16*b*). Consequently the electric field points into the page when B_y is positive and out of the page when B_y is negative. Because the *x* axis points into the page in this rendering (compare Figures 30.15*d* and 30.16*b*), the *x* component of the electric field must be positive when B_y is positive and negative when B_y is negative, as shown in Figure 30.16*c*. The electric field shown in Figure 30.16*c* is exactly the electric field I started out with in Figure 30.15*b*. In other words, the electric field yields the magnetic field and the magnetic field yields the electric field, and the two are entirely consistent with one another.

Figure 30.16

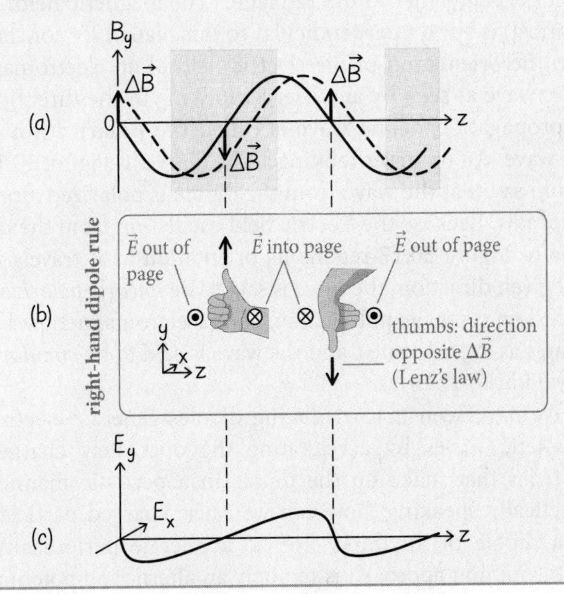

CONCEPTS

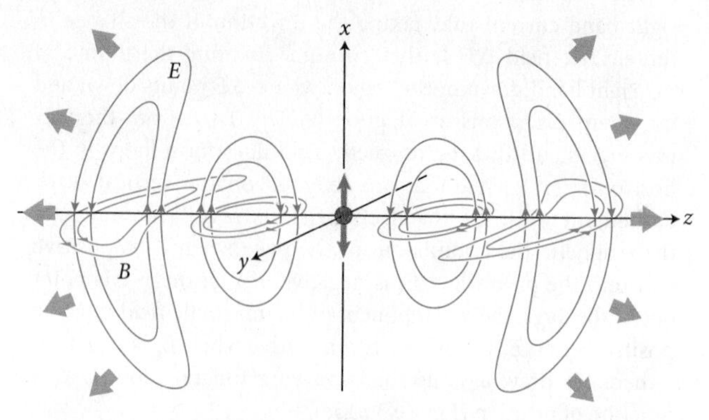

Figure 30.17 Electric and magnetic field pattern of oscillating dipole. The pink arrows indicate the direction of propagation of the electromagnetic wave pulse. For simplicity, only the fields in the xz and yz planes are shown.

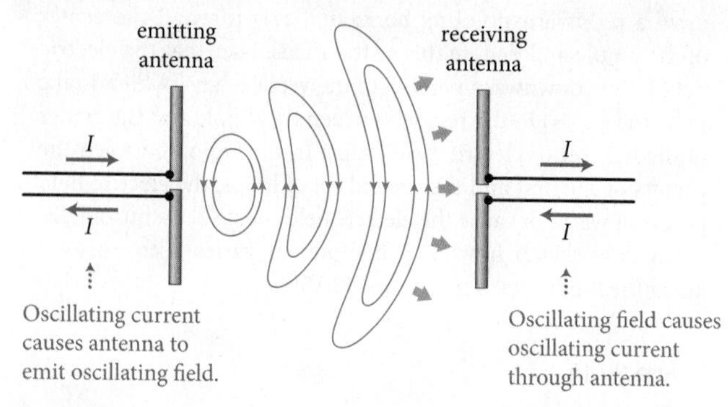

Figure 30.18 System of two antennas, one that emits electromagnetic waves and one that receives them. The emitting antenna is supplied with an oscillating current created by a source of alternating potential difference. An oscillating current is induced in the receiving antenna by the arriving electromagnetic wave.

The solution to Example 30.3 suggests that the magnetic field line pattern is similar to the electric field line pattern, but perpendicular to it. **Figure 30.17** shows the combined electric and magnetic field pattern of an oscillating dipole. Traveling electromagnetic waves, like the one shown in Figure 30.17 are transverse waves (see Section 16.1) because both the magnetic field and the electric field are perpendicular to the direction of propagation. Also, the electric and magnetic fields propagate at the same frequency, and both reach their maxima (or minima) simultaneously; the electric and magnetic fields are therefore in phase with each other.

30.8 (*a*) At the origin of the graphs in Figure 30.15, the electric field is zero, but there is a current due to the motion of the charged particles that constitute the dipole. Is this current upward, downward, or zero at the instant shown in Figure 30.15? (*b*) Is this current (or the absence thereof) consistent with the magnetic field pattern shown in Figure 30.17?

In the wave shown in Figure 30.17, not only are the electric and magnetic fields perpendicular to each other, but throughout the entire wave the electric field has no component perpendicular to the xz plane. (The magnetic field, in contrast, is always perpendicular to this plane.) By convention the orientation of the electric field of an electromagnetic wave as seen by an observer looking in the direction of propagation of the wave is called the **polarization** of the wave. An observer looking at the dipole in Figure 30.17 would say that the wave from the dipole is polarized along the x axis. Because the electric field oscillation from the dipole in Figure 30.17 retains its orientation as it travels in any given direction, the wave is said to be *linearly polarized.* In certain cases the polarization of an electromagnetic wave rotates as it propagates, and the wave is said to be *circularly* or *elliptically polarized.*

We have seen that oscillating dipoles generate electromagnetic waves by accelerating the oppositely charged particles that make up the dipole in a periodic manner. Practically speaking, how can we cause charged particles, in a dipole or anything else, to accelerate periodically? One common approach is to apply an alternating potential difference to an *antenna*, which is a device that either emits or receives electromagnetic waves. The alternating potential difference drives charge carriers back and forth through the antenna, thereby producing an oscillating current through the antenna.

Antennas that emit electromagnetic waves are designed in many ways to produce a variety of electric and magnetic field patterns. The simplest design is two conducting rods connected to a source of alternating potential difference (**Figure 30.18**). Because of the alternating potential difference, the ends of the antenna are oppositely charged and cycle between being positively charged, neutral, and negatively charged.

When the top end of the antenna is positively charged and the bottom end is negatively charged, the electric field of the antenna points down. When the charge distribution is reversed, the electric field points up. As the charge distribution oscillates, the electric field adjacent to the emitting antenna also oscillates. This changing electric field is accompanied by a changing magnetic field, and the disturbance in the fields travels away from the emitting antenna in the same manner as the electromagnetic wave of Figures 30.14, 30.15, and 30.17.

If the length of each rod in an emitting antenna is exactly one-quarter of the wavelength of the electromagnetic wave emitted, the electric fields produced strongly resemble the dipole fields of Figure 30.17. Such an antenna is often called a *dipole antenna;* it is also called a *half-wave antenna* because the length of the two rods is equal to half a wavelength.

In antennas that receive electromagnetic waves, the oscillating electric field of the wave causes charge carriers in the antenna to oscillate, as discussed in Example 30.2. This produces an oscillating current (shown schematically in Figure 30.18) that can be measured. When operated in this mode, the antenna is said to be *receiving* a signal.

30.9 To maximize the magnitude of the current induced in a receiving antenna, should the antenna be oriented parallel or perpendicular to the polarization of the electromagnetic wave?

Self-quiz

1. Suppose the current shown in **Figure 30.19** discharges the capacitor. What are the directions of $\vec{E}$, $\Delta\vec{E}$, and $\vec{B}$ between the plates of the discharging capacitor?

Figure 30.19

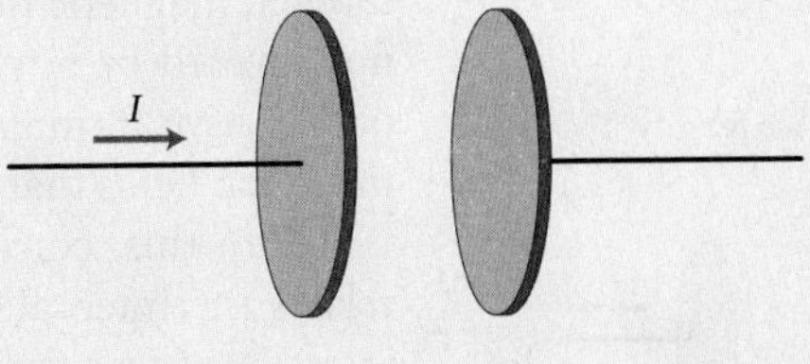

2. A positively charged particle creates the electric field shown in **Figure 30.20**. When the kinks in the electric field lines reach the rod, what is the direction of the current induced in the rod?

Figure 30.20

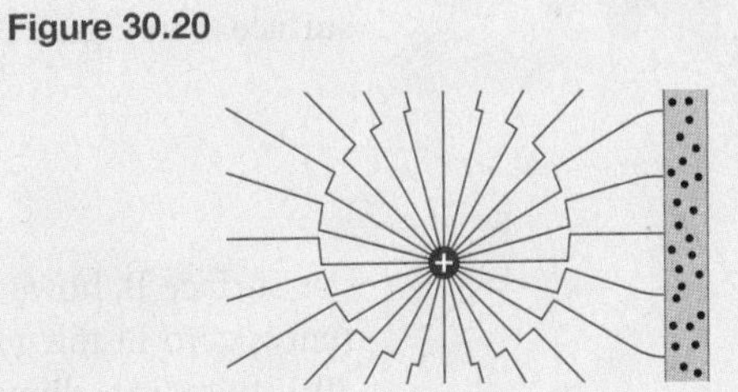

3. For the oscillating dipole of Figure 30.14, sketch the electric field pattern at $t = \frac{5}{4}T$.

4. In the electric field pattern for a sinusoidally oscillating dipole shown in **Figure 30.21**, what are (*a*) the direction of the change in the electric field $\Delta\vec{E}$ at point *C* as the electric field propagates and (*b*) the direction of the magnetic field loop near *C*?

Figure 30.21

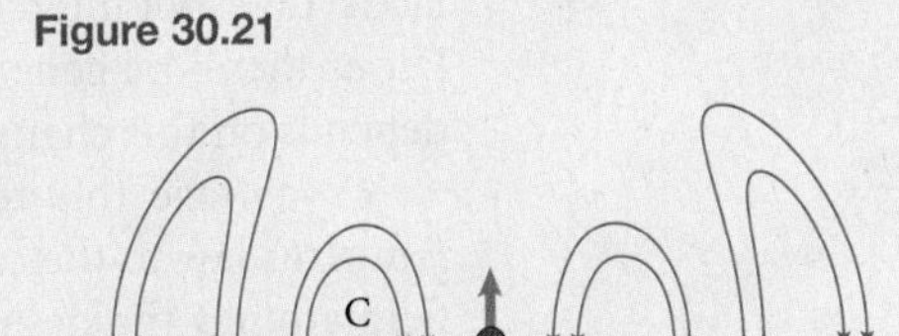

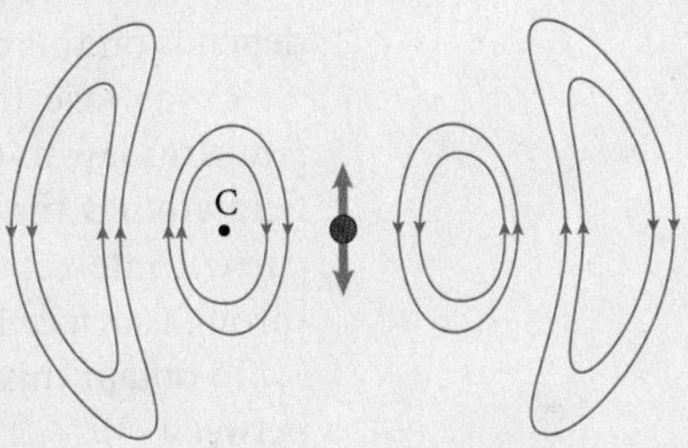

Answers

1. The current brings positive charge carriers to the left plate and removes them from the right plate. For the capacitor to discharge, the left plate must be negatively charged and the right one positively charged; this means $\vec{E}$ points left. The electric field decreases as the capacitor discharges, so $\Delta\vec{E}$ is to the right (just as the current is). The magnetic field lines are circular and centered on the axis of the capacitor in a direction given by the right-hand current rule with the thumb along the direction of $\Delta\vec{E}$. That is, the magnetic field lines are clockwise looking along the direction of the current.

2. Because the particle carries a positive charge, the electric field lines radiate outward. The electric field in the kinks therefore points up, and so the kinks induce an upward current through the rod.

3. See **Figure 30.22**. Because the loops move outward and a new pair of loops forms every half-period, the pattern now has three loops. Note in Figure 30.14 that the loops closest to the dipole at $t = \frac{1}{4}T$ and $t = \frac{3}{4}T$ have the same shape but opposite directions. Half a period after $t = \frac{3}{4}T$, at $t = \frac{5}{4}T$, the loops closest to the dipole again have the same shape (and the same direction as at $\frac{1}{4}T$). Likewise, at $t = \frac{5}{4}T$ the second closest loops curl in the direction opposite the direction at $\frac{3}{4}T$.

Figure 30.22

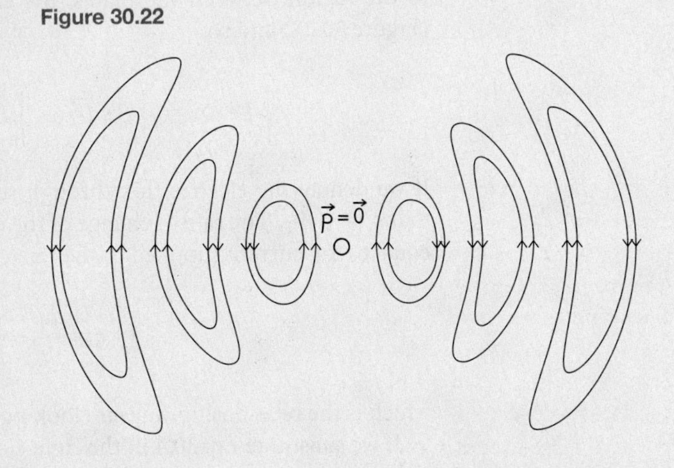

4. (*a*) The loops passing at *C* travel to the left. At the instant shown, the electric field is close to zero, but as the pattern moves to the left, the electric field lines point downward at *C*, so $\Delta\vec{E}$ is down. (*b*) The thumb of your right hand aligned in the direction of $\Delta\vec{E}$ makes your fingers curl in the direction of the magnetic field: clockwise viewed from the top.

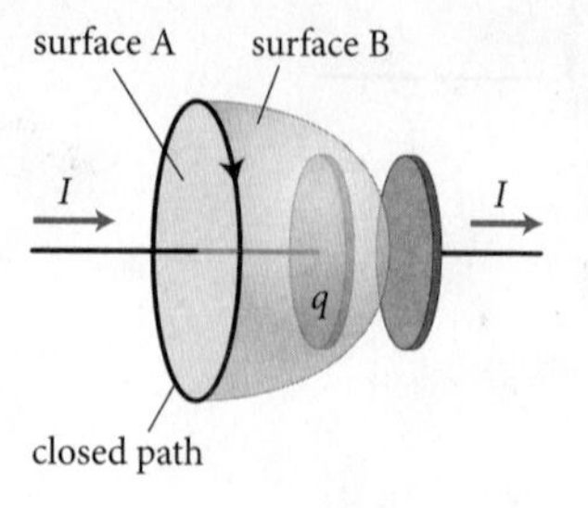

Figure 30.23 Capacitor being charged by a current-carrying wire. Because surfaces A and B both span the closed path shown, either surface can be used to calculate the magnetic field around the path. Ampère's law must be the same in either case.

30.4 Displacement current

The work we did with Ampère's law in Sections 28.4 and 28.5 dealt only with the magnetic field associated with an electric current. As we saw in Section 30.1, though, magnetic fields also accompany changing electric fields, a phenomenon not covered by our Chapter 28 form of Ampère's law. Let us now see how the quantitative formulation of Ampère's law must be modified to account for the magnetic fields that accompany changing electric fields.

To do this, consider the charging capacitor in **Figure 30.23**. Ampère's law relates the integral of the magnetic field around a closed path to the current intercepted by a surface spanning the path (Eq. 28.1). Applying Ampère's law to surface A in Figure 30.23, we have

$$\oint \vec{B} \cdot d\vec{\ell} = \mu_0 I. \tag{30.1}$$

For surface B, however, the right-hand side of Eq. 30.1 is zero because the current is zero in the gap between the capacitor plates. As we discussed in Section 30.1, there is a change in the electric field between the capacitor plates, so surface B does not intercept a current, but it does intercept a change in electric flux. Let us therefore generalize Ampère's law by adding to the right side a term that depends on this change in electric flux.

We choose this term so that when we apply the generalized version of Ampère's law to the capacitor shown in Figure 30.23, for example, the magnetic field around the designated path is the same whether we calculate it from the current intercepted by surface A or from the change in electric flux $d\Phi_E/dt$ through surface B.

To obtain this generalizing term, let's determine a mathematical relationship between $d\Phi_E/dt$ through surface B and the current to the plates. First, note that the change in electric flux is related to the change in the charge q on the plates, which, in turn, is related to the current I to the plate. Consider the closed surface surrounding the left capacitor plate in Figure 30.23 made up by surfaces A and B combined. Applying Gauss's law to this closed surface, we find that the electric flux through it is

$$\oint_{A+B} \vec{E} \cdot d\vec{A} = \frac{q}{\epsilon_0}, \tag{30.2}$$

where q is the charge on the capacitor plate. Because the electric field is confined to the region between the plates, the electric flux through surface A is zero (Figure 30.23) and so

$$\oint_{A+B} \vec{E} \cdot d\vec{A} = \int_A \vec{E} \cdot d\vec{A} + \int_B \vec{E} \cdot d\vec{A} = \int_B \vec{E} \cdot d\vec{A} = \frac{q}{\epsilon_0}. \tag{30.3}$$

If we denote the electric flux through surface B by Φ_E, we see from Eqs. 30.3 that $q = \epsilon_0 \Phi_E$. The rate of change of the charge on the capacitor plates, dq/dt, is equal to the current supplied to the capacitor, so

$$I \equiv \frac{dq}{dt} = \epsilon_0 \frac{d\Phi_E}{dt}, \tag{30.4}$$

which is the relationship we were looking for.

If we substitute Eq. 30.4 in the right side of Eq. 30.1, we obtain

$$\oint \vec{B} \cdot d\vec{\ell} = \mu_0 \epsilon_0 \frac{d\Phi_E}{dt}. \tag{30.5}$$

We can now use this expression to determine the line integral of the magnetic field around the closed path in Figure 30.23 by evaluating the change in electric flux through surface B. Because the right side of Eq. 30.5 is equal to $\mu_0 I$, we obtain for $\oint \vec{B} \cdot d\vec{\ell}$ the same value we found using the original form of Ampère's law with the current intercepting surface A.

To account for both a current and a changing electric flux, we generalize Ampère's law as follows:

$$\oint \vec{B} \cdot d\vec{\ell} = \mu_0 I_{\text{int}} + \mu_0 \epsilon_0 \frac{d\Phi_E}{dt}. \tag{30.6}$$

This equation holds for any surface spanning a closed path and is sometimes called the *Maxwell-Ampère law*, in honor of the Scottish physicist James Clerk Maxwell (1831–1879), who first introduced the additional term in Eq. 30.6. To reflect the fact that we must only include the current intercepted by the surface, not the current encircled by the integration path, we write I_{int} rather than I_{enc}.

The quantity on the right side of Eq. 30.4 is called the **displacement current:**

$$I_{\text{disp}} \equiv \epsilon_0 \frac{d\Phi_E}{dt}. \tag{30.7}$$

As you can see from Eq. 30.4, the SI units of the displacement current are indeed those of a current. The name is somewhat misleading because the derivation holds for a capacitor in vacuum where no charged particles are present in the space between the plates. Even if the term is somewhat of a misnomer, it is still useful to associate the change in the electric field with a "current" to determine the direction of the magnetic field accompanying a changing electric field. As we argued in Section 30.1, the direction of this displacement current is the same as that of the change in the electric field $\Delta\vec{E}$. We can then use the right-hand current rule to determine the direction of the magnetic field from the displacement current (see, for example, Figure 30.5).

Using Eq. 30.7, we can write Eq. 30.6 in the form

$$\oint \vec{B} \cdot d\vec{\ell} = \mu_0 (I_{\text{int}} + I_{\text{disp}}). \tag{30.8}$$

30.10 The parallel-plate capacitor in **Figure 30.24** is discharging so that the electric field between the plates *decreases*. What is the direction of the magnetic field (*a*) at point P above the plates and (*b*) at point S between the plates? Both P and S are on a line perpendicular to the axis of the capacitor.

Figure 30.24 Checkpoint 30.10.

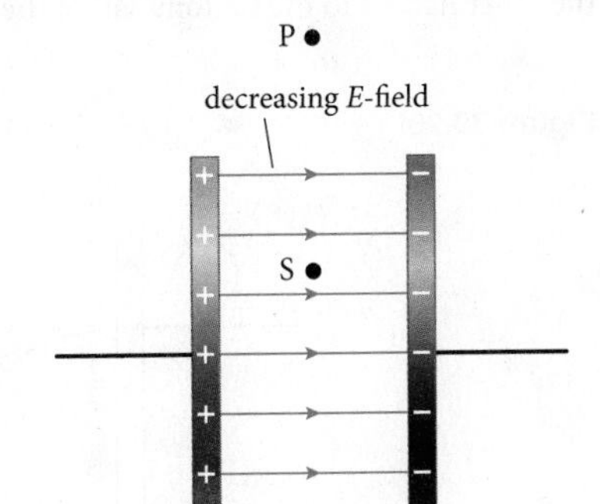

Example 30.4 A bit of both

The parallel-plate capacitor in **Figure 30.25** has circular plates of radius R and is charged with a current of constant magnitude I. The surface is bounded by a circle that passes through point P and is centered on the wire leading to the left plate and perpendicular to that wire. The surface crosses the left plate in the middle so that the top half of the plate is on one side of the surface and the bottom half is on the other side. Use this surface and Eq. 30.6 to determine the magnitude of the magnetic field at point P, which is a distance $r = R$ from the capacitor's horizontal axis.

Figure 30.25 Example 30.4.

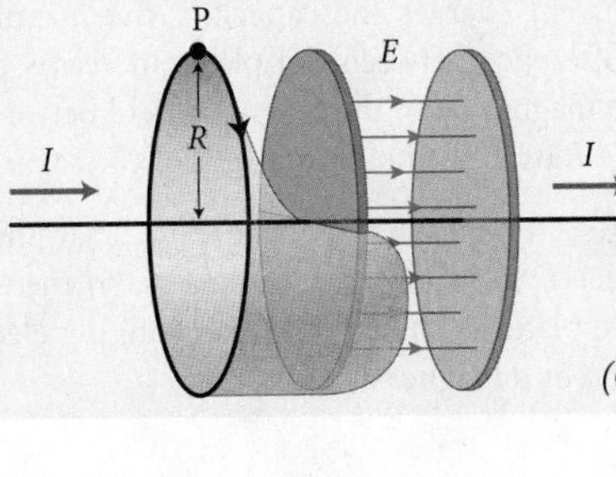

(Continued)

❶ GETTING STARTED The surface intercepts both a current (through the plate) and a changing electric field (between the plates). To apply Eq. 30.6, I must therefore determine both the current and the electric flux intercepted by the surface.

There also is a simple way to obtain the answer to this question: Ampère's law (Eq. 30.1). If I take the circle through point P centered on the wire and perpendicular to it as the integration path, then the integral on the left side of Eq. 30.1 is equal to the magnitude of the magnetic field times the circumference of the circle: $2\pi RB$. The path encircles the current I, so Ampère's law gives me $2\pi RB = \mu_0 I$ and $B = \mu_0 I/(2\pi R)$. Because the magnetic field magnitude cannot depend on the approach used to calculate it, I should obtain the same result using Eq. 30.6 and the surface in Figure 30.25.

❷ DEVISE PLAN The left side of Eq. 30.6 is identical to the left side of Eq. 30.1 and therefore equal to $2\pi RB$. To evaluate the right side of Eq. 30.6, I must determine both the ordinary current and the displacement current intercepted by the surface.

❸ EXECUTE PLAN If the surface intercepted the entire electric flux between the plates, the displacement current term on the right side of Eq. 30.6 would be equal to $\mu_0 I$. The surface intercepts only half of the electric flux, however, so the displacement current is

$$\epsilon_0 \frac{d\Phi_E}{dt} = \tfrac{1}{2}I.$$

Next I need to determine how much current is intercepted by the surface. To charge the plate uniformly, the current must carry charge carriers evenly to the two halves of the plate—half of the charge carriers go to the top half of the plate and the other half go to the bottom half of the plate (**Figure 30.26**).

Figure 30.26

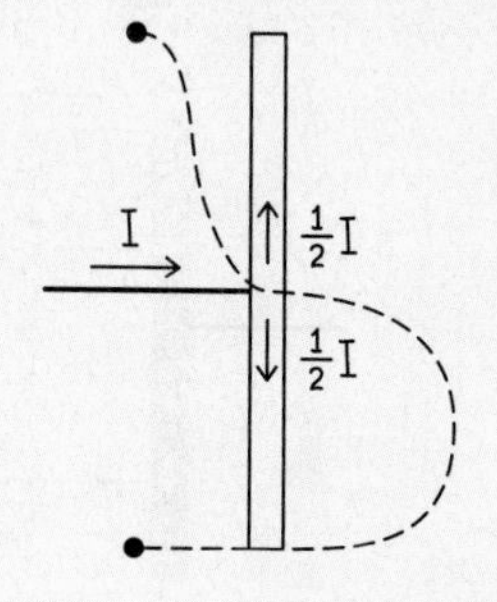

Because the top half of the plate is to the right of the surface, the current going to the top half of the plate must cross the surface. The current intercepted by the surface is thus $\frac{1}{2}I$, and the right side of Eq. 30.6 becomes

$$\mu_0 I_{\text{int}} + \mu_0 \epsilon_0 \frac{d\Phi_E}{dt} = \mu_0(\tfrac{1}{2}I) + \mu_0(\tfrac{1}{2}I) = \mu_0 I. \quad (1)$$

Substituting the right side of Eq. 1 into the right side of Eq. 30.6, I have $2\pi RB = \mu_0 I$ and so $B = \mu_0 I/(2\pi R)$, which is the same value I got using Eq. 30.1. ✔

❹ EVALUATE RESULT It's reassuring to see that the magnetic field magnitude at P does not depend on the choice of surface spanning the integration path. I can easily modify the argument above to show that any other surface that intercepts the left plate gives the same result. For example, a surface that intercepts one-quarter of the plate, as in **Figure 30.27**, intercepts one-quarter of the electric flux, and so the displacement current is only $I/4$. This surface intercepts the current twice: Where the surface intersects the wire, the current I crosses the surface from left to right, and where the surface intersects the plate, one-quarter of the current crosses the surface in the other direction, for a total contribution of $\frac{3}{4}I$. Again the sum of the ordinary and displacement currents intercepting the surface is I.

Figure 30.27

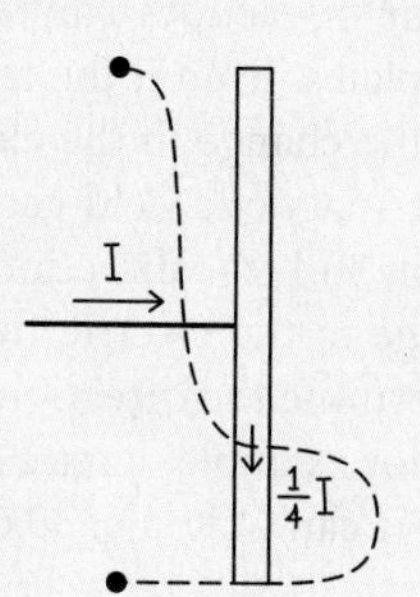

Example 30.5 Magnetic field in a capacitor

A parallel-plate capacitor has circular plates of radius $R = 0.10$ m and a plate separation distance $d = 0.10$ mm. While a current charges the capacitor, the magnitude of the potential difference between the plates increases by 10 V/μs. What is the magnitude of the magnetic field between the plates at a distance R from the horizontal axis of the capacitor?

❶ GETTING STARTED As the capacitor is charging, there is a changing electric flux between the plates, so the electric field between the plates is changing. This changing electric field is accompanied by a magnetic field.

❷ DEVISE PLAN Equation 30.6 relates the magnetic field to a changing electric flux. To work out the left side of Eq. 30.6, I chose a circular integration path centered on the horizontal axis of the capacitor so that I can exploit the circular symmetry of the problem. To work out the right side of Eq. 30.6, I need to determine the rate of change of the electric flux through an appropriate surface spanning the integration path. I choose the simplest possible surface: a flat surface parallel to the plates (**Figure 30.28**). To calculate the change in electric flux intercepted by this surface, I need to determine the magnitude of

Figure 30.28

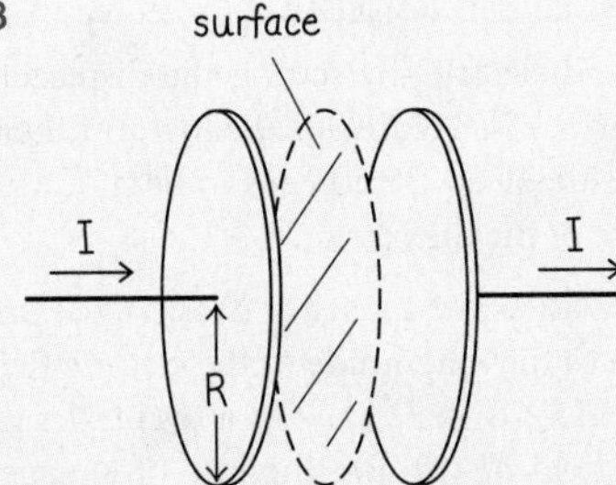

the (uniform) electric field between the plates. In Example 26.2 I determined that the magnitude of the electric field between the plates is related to the plate separation distance d and the magnitude of the potential difference between the plates: $V_{cap} = Ed$.

❸ **EXECUTE PLAN** Because the electric field between the plates is uniform and perpendicular to the plates, the electric flux Φ_E through the surface I chose is $\Phi_E = EA = E\pi R^2$, where $A = \pi R^2$ is also the area of the capacitor plates. The time rate of change of the electric flux is then given by

$$\frac{d\Phi_E}{dt} = \pi R^2 \frac{dE}{dt}.$$

Substituting this result in Eq. 30.6 and setting the current term equal to zero because no current is intercepted by the surface I've chosen, I get

$$\oint \vec{B} \cdot d\vec{\ell} = \mu_0 \epsilon_0 \pi R^2 \frac{dE}{dt}.$$

Around my integration path, the magnitude of the magnetic field is constant, and the left side of this expression simplifies to $2\pi RB$. Solving for B, I obtain

$$B = \frac{\mu_0 \epsilon_0 R}{2} \frac{dE}{dt}. \quad (1)$$

Because $V_{cap} = Ed$, I can write $E = V_{cap}/d$ and so

$$B = \frac{\mu_0 \epsilon_0 R}{2d} \frac{dV_{cap}}{dt} \quad (2)$$

$$B = (4\pi \times 10^{-7}\ \text{T} \cdot \text{m/A})(8.85 \times 10^{-12}\ \text{C}^2/\text{N} \cdot \text{m}^2)$$

$$\times \frac{0.10\ \text{m}}{2(0.10 \times 10^{-3}\ \text{m})} \frac{10\ \text{V}}{1.0 \times 10^{-6}\ \text{s}}$$

$$= 5.6 \times 10^{-8}\ \text{T},$$

where I have used the Eq. 25.16 definition of the volt, $1\ \text{V} \equiv 1\ \text{J/C} \equiv 1\ \text{N} \cdot \text{m/C}$ and the Eq. 27.3 definition of the ampere $1\ \text{C} \equiv 1\ \text{A} \cdot \text{s}$ to simplify the units. ✔

❹ **EVALUATE RESULT** The magnetic field magnitude I obtain is very small, in spite of the substantial rate at which the potential difference between the plates increases. I have no way of knowing whether my numerical result is reasonable or not, but what I can do to evaluate the result is use another method to obtain an expression for B. Because my flat surface intercepts all the electric flux, I know that the magnitude of the magnetic field should be the same at all positions a distance R from the current-carrying wire. I can obtain the current by solving Eq. 26.1, $q/V_{cap} = C$, for the charge q on the capacitor plate and then using the definition of current, $I \equiv dq/dt$:

$$I \equiv \frac{dq}{dt} = C\frac{dV_{cap}}{dt}.$$

Substituting the capacitance of a parallel-plate capacitor $C = \epsilon_0 A/d = \epsilon_0 \pi R^2/d$ (see Example 26.2) into this expression, and then substituting the result into the expression for the magnetic field around a current-carrying wire from Example 28.3, $B = \mu_0 I/(2\pi r)$ (setting $r = R$ for the distance to the wire), I get

$$B = \frac{\mu_0 I}{2\pi R} = \frac{\mu_0 \epsilon_0 R}{2d} \frac{dV_{cap}}{dt},$$

the same result I obtained in Eq. 2, as I expect.

Note that in Example 30.5 the rate of change of the electric field is very large (about 10^{11} V/(m · s)), but the accompanying magnetic field is small. This is not the case for electric fields that accompany changing magnetic fields, as substantial emfs can be induced by the motion of ordinary magnets.

30.11 Consider again the parallel-plate capacitor of Figure 30.23. For circular plates of radius R, calculate the magnitude of the magnetic field a distance $r < R$ from the horizontal axis of the capacitor (*a*) between the plates and (*b*) a short distance to the right of the right plate.

Example 30.6 Displacement current in the presence of a dielectric

Suppose a slab of dielectric with dielectric constant κ is inserted between the plates of the capacitor in Figure 30.23 and the capacitor is charged with a current I, as considered in Example 30.1. How does Eq. 30.6 have to be modified to account for the dielectric?

❶ GETTING STARTED For a given amount of charge on the capacitor plates, the presence of a dielectric decreases the magnitude of the electric field between them. As I concluded in Example 30.1, however, the magnetic field surrounding the wires that lead to the capacitor wires is determined only by the current I through the wires and therefore cannot be affected by the insertion of the dielectric.

❷ DEVISE PLAN Given that the magnetic field surrounding the wires cannot be affected by the presence of the dielectric, the displacement current intercepted by a surface spanning a circular path around the wire and passing between the capacitor plates (**Figure 30.29**) should be equal to I, regardless of the presence of the dielectric. By setting the displacement current through the surface in Figure 30.29 equal to I, I can determine how the right side of Eq. 30.6 needs to be modified to account for the presence of the dielectric.

Figure 30.29

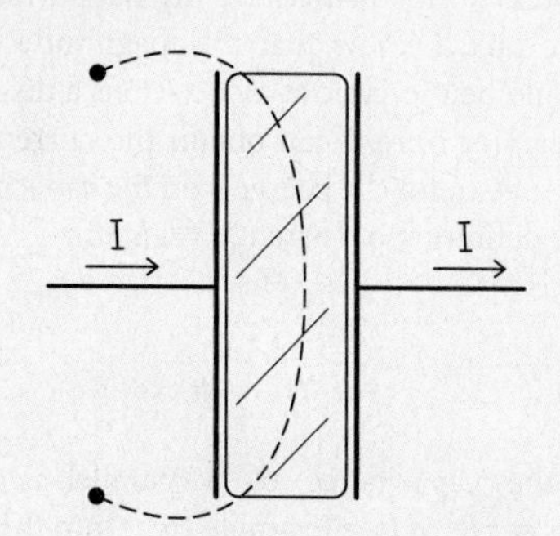

❸ EXECUTE PLAN As the capacitor charges, the presence of the dielectric reduces the magnitude of the electric field by a factor $1/\kappa$ (see Eq. 26.15): $E = E_{\text{free}}/\kappa$. As a result, the rate of change of the electric field dE/dt and the rate of change in the electric flux intercepted by the surface $d\Phi_E/dt$ are also reduced by a factor $1/\kappa$. To compensate for this reduction, I need to multiply $d\Phi_E/dt$ by κ in order to make the right side of Eq. 30.4 equal to I again. Therefore Eq. 30.6 becomes

$$\oint \vec{B} \cdot d\vec{\ell} = \mu_0 I + \mu_0 \epsilon_0 \kappa \frac{d\Phi_E}{dt}. \checkmark$$

❹ EVALUATE RESULT My modification to Eq. 30.6 is identical to the modification we made to make Gauss's law work in dielectrics (Eq. 26.25): In both cases the term containing the electric field or electric flux includes the factor κ.

30.5 Maxwell's equations

With Maxwell's addition of the displacement current $\epsilon_0 d\Phi_E/dt$ to Ampère's law, $\oint \vec{B} \cdot d\vec{\ell} = \mu_0 I$, we now have a complete mathematical description of electric and magnetic phenomena and the relationship between the two. Let us summarize this description in the absence of a dielectric.

Electric and magnetic fields are *defined* by Eq. 27.20, which gives the force exerted on a charged particle moving in an electric field and a magnetic field:

$$\vec{F} = q(\vec{E} + \vec{v} \times \vec{B}). \tag{30.9}$$

Charged particles are the source of electrostatic fields, and electrostatic field lines always begin or end on charged particles. We can calculate electric fields from each individual charged particle if we wish, but when dealing with a distribution of charge that exhibits a certain symmetry (Section 24.4), it is most convenient to use Gauss's law (Eq. 24.8) to work out the electric field. Gauss's law tells us that the electric flux Φ_E through a closed surface (a Gaussian surface) is proportional to the charge enclosed by the surface:

$$\Phi_E \equiv \oint \vec{E} \cdot d\vec{A} = \frac{q_{\text{enc}}}{\epsilon_0}. \tag{30.10}$$

Figure 30.30*a* shows the electric field created by a charged particle, along with a Gaussian surface enclosing that particle.

Figure 30.30 Graphical representation of the physics behind Maxwell's equations, together with their mathematical expressions. (*a*) Electric field surrounding a charged particle and a Gaussian surface enclosing that particle. Gaussian surfaces can be used to relate the electric field to the enclosed charge. (*b*) Magnetic field surrounding a current-carrying wire and a closed surface intercepted by the wire; the integral of the magnetic field over a closed surface is always zero. (*c*) Electrostatic field and two closed paths through that field; the path integral of the electric field around either path must be zero. (*d*) Steady magnetic field surrounding a current and two closed paths through that field; the path integral of the magnetic field is proportional to the encircled current. (*e*) Changing magnetic field and two closed paths through it; an electric field accompanies the changing magnetic field; the path integral of this electric field around either path is nonzero. (*f*) Changing electric field and two closed paths through it; a magnetic field accompanies the changing electric field; the path integral of this magnetic field around either path is nonzero.

(*a*) Surface integral of electric field (Gauss's law)

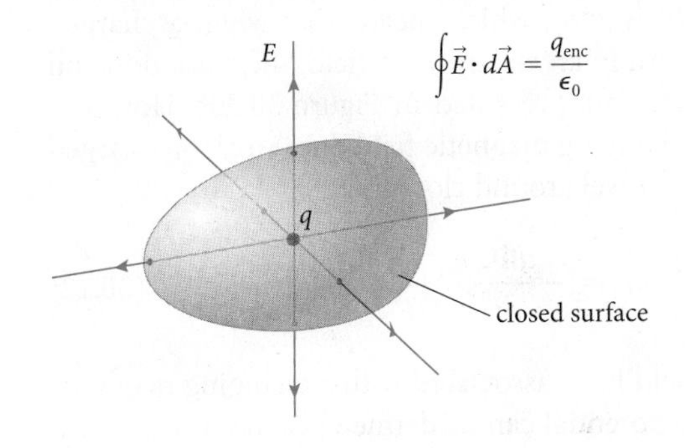

(*b*) Surface integral of magnetic field (Gauss's law for magnetism)

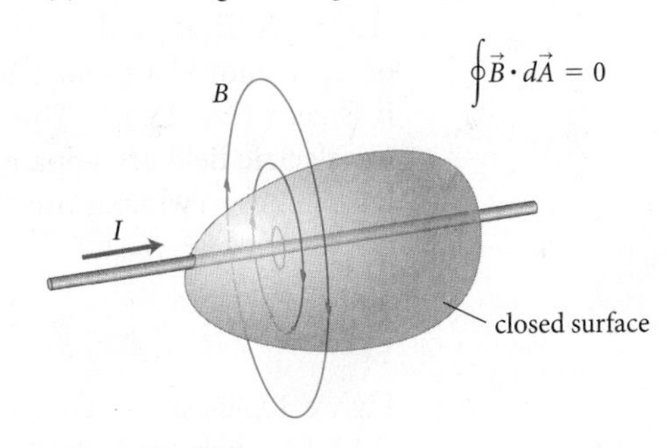

(*c*) Line integral of constant electric field

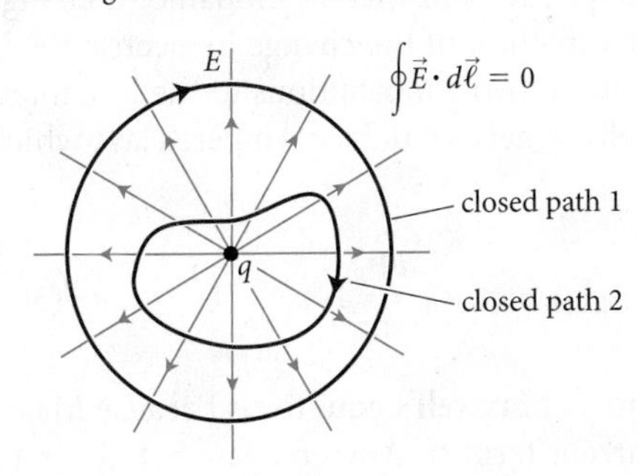

(*d*) Line integral of constant magnetic field (Ampère's law)

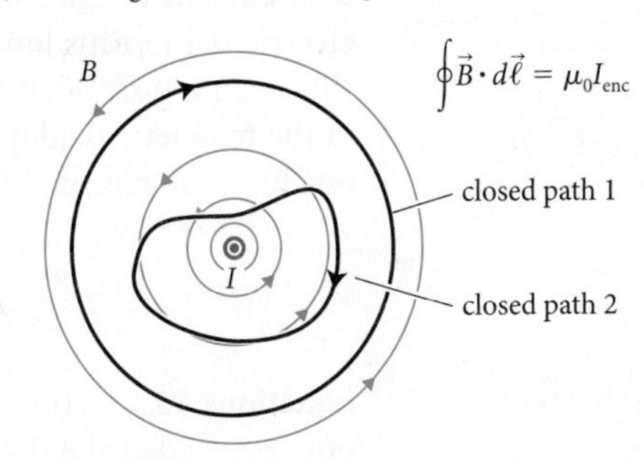

(*e*) Line integral of changing electric field (Faraday's law)

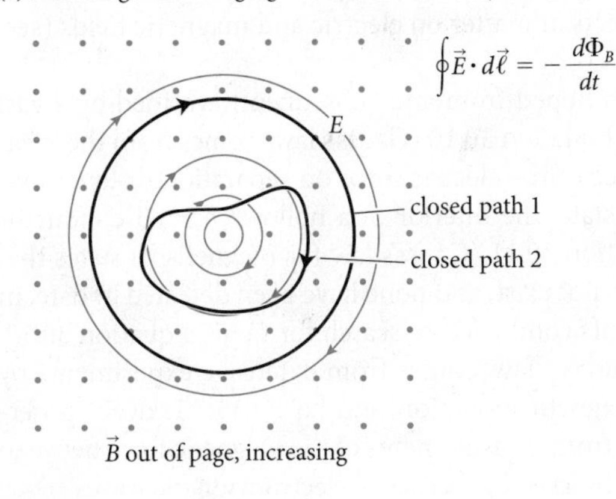

(*f*) Line integral of changing magnetic field (Maxwell's displacement current)

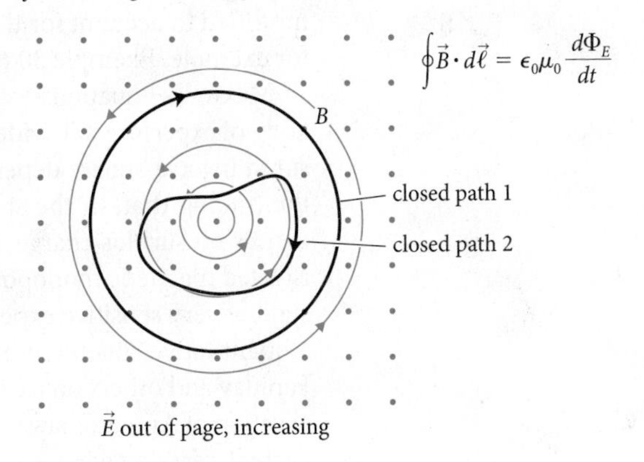

Magnetic fields are generated by moving charged particles, commonly in the form of currents. Unlike electric field lines, magnetic field lines always form loops. There are no isolated magnetic poles, only magnetic dipoles. Consequently, as we showed in Chapter 27, the magnetic flux through any closed surface is always zero (Eq. 27.11):

$$\Phi_B \equiv \oint \vec{B} \cdot d\vec{A} = 0. \tag{30.11}$$

Figure 30.30*b* shows the magnetic field surrounding a current-carrying wire, along with a closed surface intercepted by the wire.

For electrostatic fields, we showed in Chapter 25 that the path integral of the electric field around a closed path is zero, which means that when a charged object is moved around a closed path in an electrostatic field, the work done on it is zero (Eq. 25.32). This situation is represented in Figure 30.30*c*. However, the electric field accompanying a changing magnetic field does work on charged particles even when those particles travel around closed paths:

$$\oint \vec{E} \cdot d\vec{\ell} = -\frac{d\Phi_B}{dt}. \tag{30.12}$$

Figure 30.30*e* shows the electric field lines associated with a changing magnetic field. For such an electric field, no potential can be defined because the path integral of the electric field depends on the path chosen.

Finally, Ampère's law gives the line integral of the magnetic field produced by a current (Figure 30.30*d*). The magnetic field that accompanies a changing electric field forms loops around the direction of the change in electric field, as shown in Figure 30.30*f*. Combining these two contributions to the line integral of the magnetic field gives us Maxwell's generalization of Ampère's law, which is our Eq. 30.6, repeated here:

$$\oint \vec{B} \cdot d\vec{\ell} = \mu_0 I_{\text{int}} + \mu_0 \epsilon_0 \frac{d\Phi_E}{dt}. \tag{30.13}$$

Equations 30.10–30.13 are referred to as **Maxwell's equations** because Maxwell not only added the displacement current term to Ampère's law but also recognized the coherence and completeness of this set of equations. Together with conservation of charge, these four equations give a complete description of electromagnetic phenomena. In the presence of matter, these equations have to be modified to account for the effects of matter on electric and magnetic fields (see, for example, Example 30.6).

Maxwell's equations were developed from and subsequently verified by a vast body of experimental evidence. Equation 30.10 (Gauss's law) comes from the measured inverse-square dependence of the electric force on separation distance and the finding that, in the steady state, the interior of a hollow charged conductor carries no surplus charge. Equation 30.11 (Gauss's law for magnetism) states that isolated magnetic monopoles do not exist, and none have been detected to date, in spite of very sensitive experiments conducted to search for them. Equation 30.12, a quantitative statement of Faraday's law, comes from extensive experiments by Faraday and others on electromagnetic induction, and Eq. 30.13, Maxwell's generalization of Ampère's law, comes from measurements of the magnetic force between current-carrying wires and the observed properties of electromagnetic waves.

30.12 Suppose that isolated magnetic monopoles carrying a "magnetic charge" m did exist, and that the interaction between these monopoles depended on $1/r^2$, where r is the distance between two monopoles. How would you modify Maxwell's equations to account for these monopoles? Ignore any physical constants that may need to be added.

Example 30.7 Maxwell's equations in free space

What is the form of Maxwell's equations in a region of space that does not contain any charged particles?

❶ GETTING STARTED If there are no charged particles, there can be no accumulation of charge and no currents, which means $q_{enc} = 0$ and $I = 0$.

❷ DEVISE PLAN All I need to do is set q_{enc} and I equal to zero in Eqs. 30.10–30.13.

❸ EXECUTE PLAN Setting $q_{enc} = 0$ in Eq. 30.10 and $I = 0$ in Eq. 30.13, I obtain the following form of Maxwell's equations:

$$\oint \vec{E} \cdot d\vec{A} = 0 \quad (1)$$

$$\oint \vec{B} \cdot d\vec{A} = 0 \quad (2)$$

$$\oint \vec{E} \cdot d\vec{\ell} = -\frac{d\Phi_B}{dt} \quad (3)$$

$$\oint \vec{B} \cdot d\vec{\ell} = \mu_0 \epsilon_0 \frac{d\Phi_E}{dt}. \checkmark \quad (4)$$

❹ EVALUATE RESULT Maxwell's equations simplify greatly in the absence of charged particles (the only asymmetry is the sign difference between Eqs. 3 and 4, which comes from Lenz's law). Equations 1 and 2 state that both the electric and magnetic fluxes through a closed surface are zero in the absence of charged particles. Consequently, both electric and magnetic field lines must form loops. Equations 3 and 4 state that electric field line loops accompany changes in magnetic flux and magnetic field line loops accompany changes in electric flux.

30.13 As you saw in Section 30.3, the magnetic and electric fields in an electromagnetic wave are perpendicular to each other. How do Maxwell's equations in free space (Eqs. 1–4 of Example 30.7) express that perpendicular relationship?

30.6 Electromagnetic waves

From Maxwell's equations, we can derive the fundamental properties of electromagnetic waves. To begin, let's consider an electromagnetic wave pulse that arises from the sudden acceleration of a charged particle, as we discussed in the first part of this chapter (**Figure 30.31**). The magnitude of the electric field in the kinked part of the pulse in Figure 30.31 is essentially uniform and much greater than it is anywhere else. We shall consider the propagation of this wave pulse through a region of space containing no matter and no charged particles, so we can use the form of the Maxwell equations derived in Example 30.7. At great distances from the particle, only the transverse pulse is significant and we can ignore any other contributions to the electric field. This wave pulse is essentially a slab-like region of space that extends infinitely in the x and y directions and has a finite thickness in the z direction. Inside the slab, the electric field is uniform and has magnitude E; outside the slab, $E = 0$. We let the wave pulse move along the z axis (**Figure 30.32**) and denote its speed by c_0 (the subscript 0 indicates that

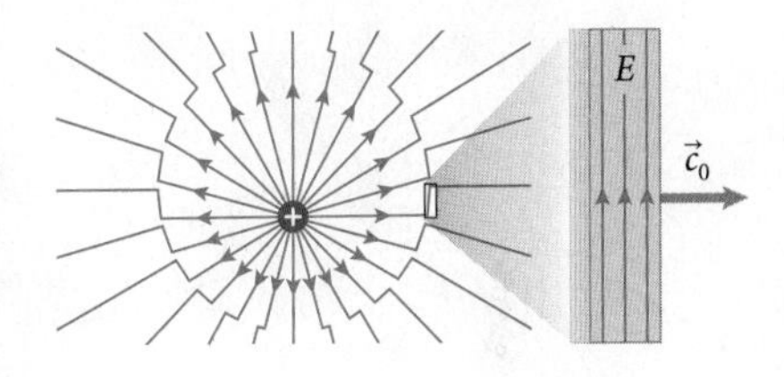

Figure 30.31 Electric field pattern of an accelerated charged particle. The kinks in the electric field pattern correspond to a transverse electric field pulse propagating away from the particle at speed c_0.

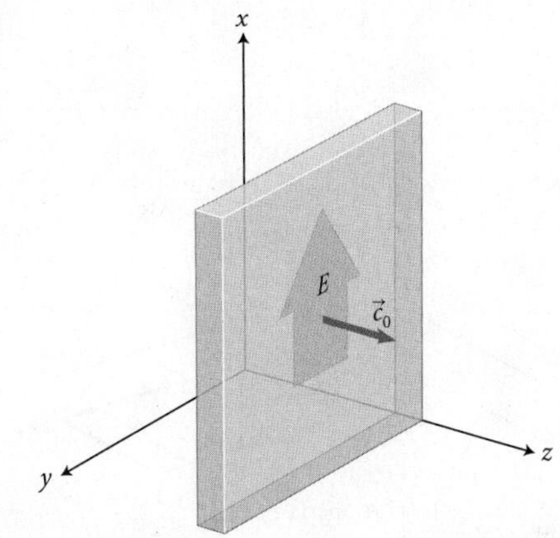

Figure 30.32 Perspective view of a planar electromagnetic wave pulse moving in the z direction. The electric field points in the x direction and has the same magnitude throughout an infinite plane parallel to the xy plane.

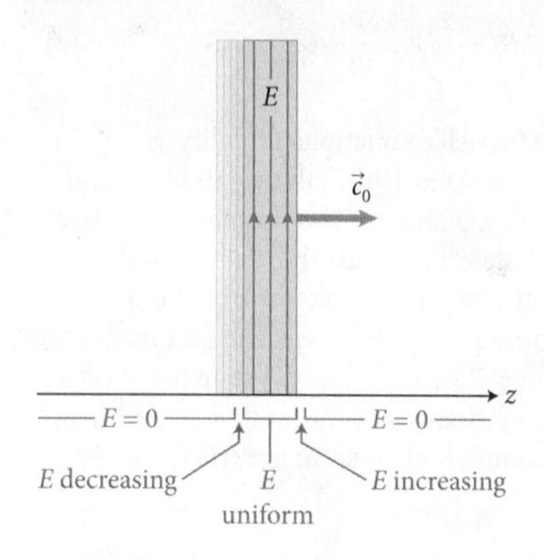

Figure 30.33 Side view of the planar electromagnetic wave pulse of Figure 30.32. The electric field inside the pulse is uniform except at the front and back surfaces, where it drops rapidly to zero.

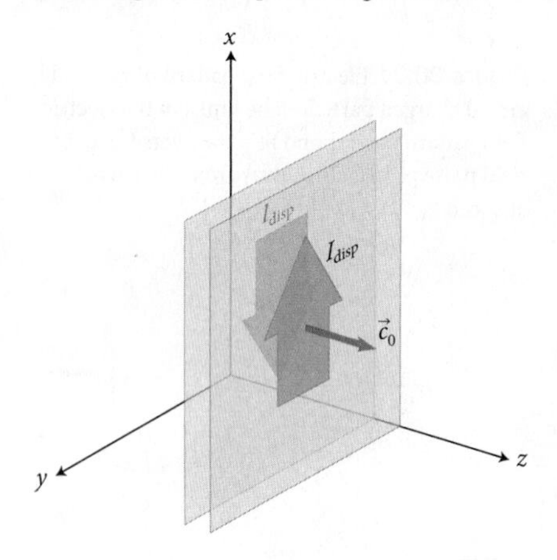

Figure 30.34 Displacement currents corresponding to the upwardly increasing electric field at the front surface and upwardly decreasing electric field at the back surface of the planar electromagnetic wave pulse of Figure 30.32.

this speed is in vacuum). The magnitude of the electric field depends only on z, not on x and y. The wave pulse is an example of a **planar electromagnetic wave** because of the constant magnitude of the electric field in a plane normal to the direction of propagation.

What magnetic field pattern is associated with the electric field in the planar electromagnetic wave pulse in Figure 30.32? Viewing Figure 30.32 from the side (**Figure 30.33**), we see that the electric field is zero in front and in back of the pulse, nonzero and uniform inside the pulse, and changing at the front and back surfaces of the pulse. At the front surface, the electric field increases in the upward direction, corresponding to an upward displacement current I_{disp} (**Figure 30.34**). At the back surface of the pulse, the displacement current points down.

In Checkpoint 28.13*b*, you determined the magnetic field of two infinite planar sheets of oppositely directed current. The electric field in the planar electromagnetic wave pulse gives a similar arrangement of oppositely directed displacement currents. The magnetic field associated with this current distribution is uniform and points in the $+y$ direction (**Figure 30.35**). We now see that the planar electromagnetic wave pulse consists of uniform electric and magnetic fields that are perpendicular to each other and to the direction of propagation of the pulse, as we already concluded in Section 30.3. In fact, for a planar electromagnetic wave pulse, there is a right-hand relationship among the directions of $\vec{E}$, $\vec{B}$, and $\vec{c}_0$. If you curl the fingers of your right hand from the direction of $\vec{E}$ to the direction of $\vec{B}$, in Figure 30.35, your thumb points in the direction of propagation of the pulse. (This means that the vector product $\vec{E} \times \vec{B}$ yields a vector pointing in the direction of propagation of the electromagnetic wave pulse.)

To calculate the magnitude of the magnetic field in the planar electromagnetic wave pulse, we can use the version of Eq. 30.13 valid in a region of space that does not contain any charged particles (see Example 30.7):

$$\oint \vec{B} \cdot d\vec{\ell} = \mu_0 \epsilon_0 \frac{d\Phi_E}{dt}. \tag{30.14}$$

Let's begin by evaluating the left side of this equation. To exploit the fact that the magnetic field points in the $+y$ direction in the pulse, we choose the Ampèrian path in Figure 30.35. This rectangular path lies in the yz plane and has width ℓ in the y direction; side ad is inside the pulse and side fg is far off to the right in the positive z direction. We let the direction of the path be such that it coincides with the direction of the magnetic field, so that $\vec{B} \cdot d\vec{\ell} = B\,d\ell$. Only

Figure 30.35 Magnetic field associated with the planar electromagnetic wave pulse of Figure 30.32. The Ampèrian path in the *yz* plane can be used to calculate the magnitude of the magnetic field.

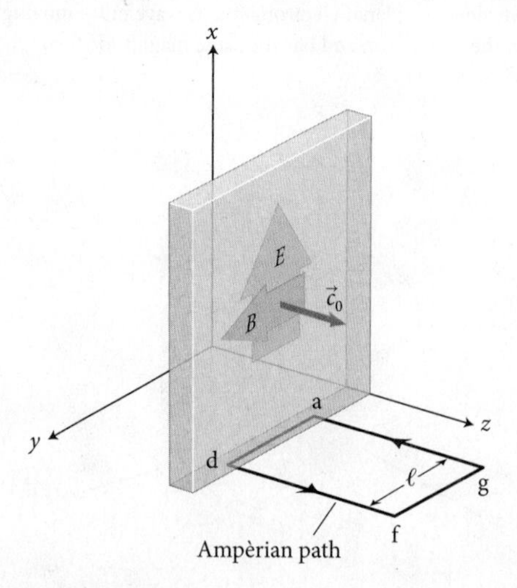

side ad of the rectangular path contributes to the line integral; the magnetic field is zero around side fg, and the two long sides are perpendicular to the magnetic field. Thus, the left side of Eq. 30.14 becomes

$$\oint \vec{B} \cdot d\vec{\ell} = B\ell. \tag{30.15}$$

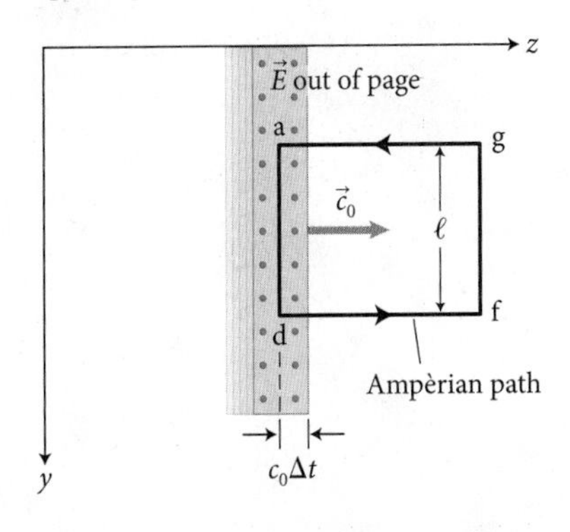

Figure 30.36 Top view of moving planar electromagnetic wave pulse of Figures 30.32–30.35, showing motion of the pulse through the Ampèrian path.

The electric flux through the path is given by $\Phi_E = \vec{E} \cdot \vec{A} = EA$, where $\vec{A}$ is a surface area vector pointing in the $+x$ direction, as dictated by the choice of direction of the integration path (Appendix B). To determine the rate of change of the electric flux through the path, note that the planar electromagnetic wave pulse is moving to the right with speed c_0. Before the front surface of the pulse reaches side ad of the Ampèrian path, the electric flux Φ_E through the path is zero. In a time interval Δt after the front surface of the pulse reaches side ad, the pulse travels a distance $c_0\Delta t$ into the rectangular path (**Figure 30.36**), and so at this instant the area over which the electric field is nonzero is $A = \ell c_0 \Delta t$. The electric flux through the path is then $\Phi_E = E\ell c_0 \Delta t$. The change in electric flux through the path during the interval Δt is thus

$$\Delta\Phi_E = E\ell c_0 \Delta t - 0, \tag{30.16}$$

and the rate of change in electric flux is

$$\frac{\Delta\Phi_E}{\Delta t} = E\ell c_0. \tag{30.17}$$

Substituting Eqs. 30.15 and 30.17 into Eq. 30.14 yields

$$B\ell = \mu_0\epsilon_0 E\ell c_0 \tag{30.18}$$

or, solving for B,

$$B = \mu_0\epsilon_0 E c_0. \tag{30.19}$$

We have thus obtained a relationship between the magnitudes of the magnetic and electric fields in the planar electromagnetic wave pulse.

We can now use Faraday's law (Eq. 30.12) to obtain an additional relationship between these two transverse fields:

$$\oint \vec{E} \cdot d\vec{\ell} = -\frac{d\Phi_B}{dt}. \tag{30.20}$$

Let's begin by evaluating the left side of this equation. To exploit the fact that the electric field points in the $+x$ direction in the pulse, we choose the rectangular integration path in **Figure 30.37**. We let the direction of the path be such that it coincides with the direction of the electric field in the pulse, so that $\vec{E} \cdot d\vec{\ell} = E\,d\ell$. As in our derivation of the magnetic field, the only contribution to the line integral of the electric field around the rectangular path comes from the left side of the path:

$$\oint \vec{E} \cdot d\vec{\ell} = Ew. \tag{30.21}$$

To evaluate the right side of Eq. 30.20, we note that the geometry of this situation is the same as in our treatment of the electric pulse, except that now $\Phi_B = \vec{B} \cdot \vec{A} = -BA$, where $\vec{A}$ is a surface area vector pointing in the $-y$ direction,

Figure 30.37 Top view of the planar electromagnetic wave pulse of Figures 30.32–30.35, showing motion of the pulse through an integration path lying in the xz plane.

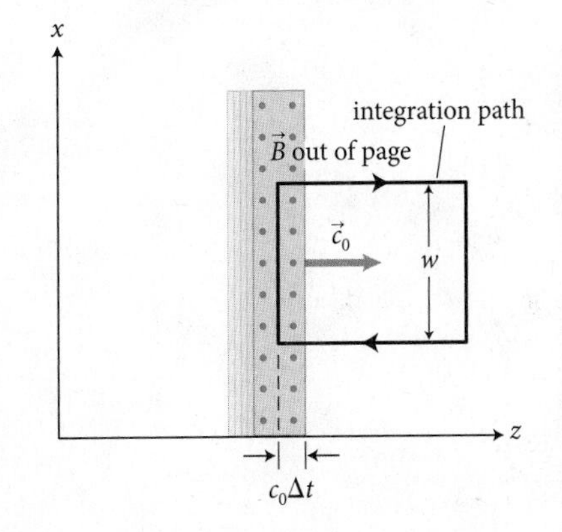

as dictated by the choice of direction of the integration path (see Appendix B). In analogy to Eqs. 30.16 and 30.17, the rate at which the magnetic flux through the path changes is then given by

$$\frac{\Delta\Phi_B}{\Delta t} = -Bwc_0. \tag{30.22}$$

Substituting Eqs. 30.21 and 30.22 into Eq. 30.20 gives us

$$Ew = Bwc_0 \tag{30.23}$$

or

$$B = \frac{E}{c_0}. \tag{30.24}$$

We now have two different relationships between the magnitudes of the electric and magnetic fields: Eq. 30.19 comes from Maxwell's generalization of Ampère's law, and Eq. 30.24 comes from Faraday's law. Setting the right sides of these two equations equal, we get

$$\frac{E}{c_0} = \mu_0\epsilon_0\, Ec_0. \tag{30.25}$$

This result implies that the speed of the planar electromagnetic wave pulse is

$$c_0 = \frac{1}{\sqrt{\epsilon_0\mu_0}}. \tag{30.26}$$

Equation 30.26 tells us something surprising: The speed of the planar electromagnetic wave pulse in vacuum is determined by two fundamental constants, ϵ_0 and μ_0. The first, ϵ_0, is introduced in Coulomb's law (see Eqs. 22.1 and 24.7). The second, μ_0 (see Eq. 28.1), is set by the definition of the ampere. In 1862, when Maxwell first worked out the relationship expressed in Eq. 30.26, no one knew that light and electromagnetic waves were related. To evaluate c_0 in Eq. 30.26, Maxwell used the results of experiments made with electric circuits and obtained a value of $c_0 = 3 \times 10^8$ m/s, in excellent agreement with values obtained for the speed of light in vacuum. This agreement led Maxwell to the remarkable conclusion that light is an electromagnetic wave.

Nowadays, the speed of light is set to be exactly 299,792,458 m/s (see Section 1.3) to define the meter. Likewise μ_0 is set by the definition of the ampere (see Eq. 28.1). The value of $\epsilon_0 = 1/(c_0^2\mu_0)$ as given in Eq. 24.7 is therefore also fixed.

Most electromagnetic waves have a more complex shape than the planar electromagnetic wave pulse we have used to arrive at Eq. 30.26. Through the superposition principle (see Section 16.3), however, we can superpose any number of planar electromagnetic wave pulses to obtain whatever planar electromagnetic wave shape interests us. The central property of these electromagnetic waves does not depend on shape: The electromagnetic wave pulse consists of electric and magnetic fields that are perpendicular to each other and to the direction of propagation of the pulse. The ratio of the magnitudes of the electric and magnetic fields is always given by Eq. 30.24. The field vectors $\vec{E}$ and $\vec{B}$ are always perpendicular, and they always travel at speed c_0 in a direction given by the vector product $\vec{E} \times \vec{B}$.

Mathematically, it is more convenient to build arbitrary wave shapes out of harmonic (sinusoidal) waves than out of rectangular wave pulses. **Figure 30.38** shows a planar electromagnetic wave for which the electric field varies sinusoidally in space. The field vectors for the electric field are shown embedded in rectangular slabs to emphasize that the electric field has the same magnitude

Figure 30.38 Perspective view of the electric field of a sinusoidal planar electromagnetic wave propagating in the z direction. The electric field vectors are embedded in rectangular slabs to emphasize that the electric field has the same magnitude everywhere in the plane of the slab. The magnitude of the electric field does vary from plane to plane along the z axis. The magnetic field (not shown) is uniform on planes parallel to the xy plane.

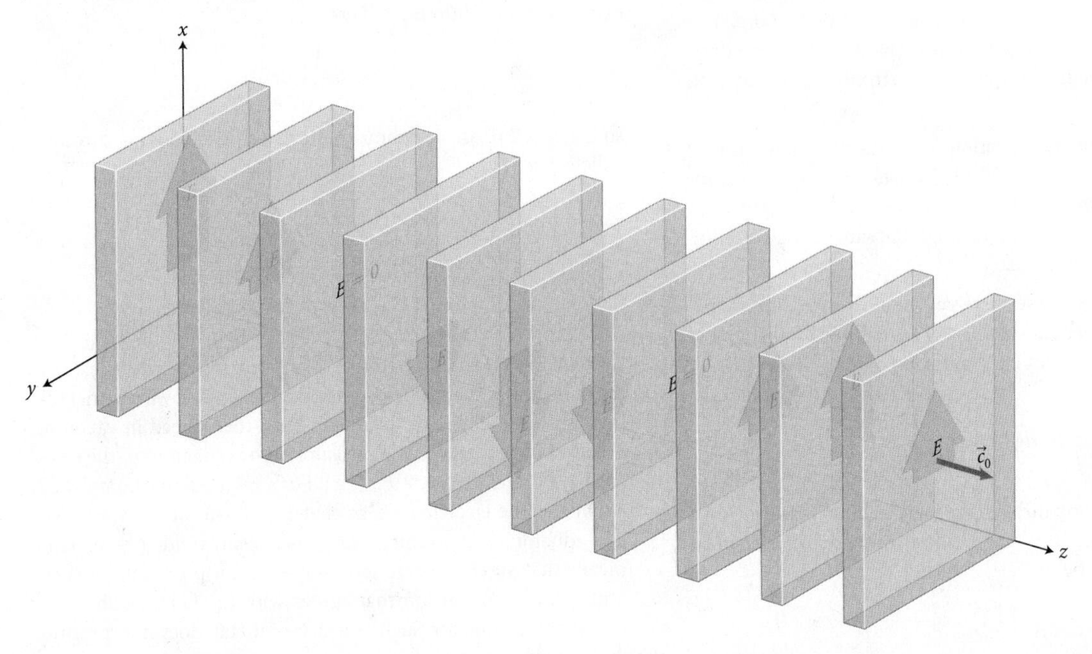

everywhere throughout the plane of a slab, not just on the z axis. The magnitude of this electric field depends only on z, not on x and y.

As we saw in Chapter 16, harmonic waves are characterized by a propagation speed c, a frequency f, and a wavelength λ, and these quantities are related by $c = f\lambda$. The remarkable thing about electromagnetic waves is that waves of all frequencies travel at the same constant speed c_0 in vacuum. Consequently, in vacuum, frequency and wavelength are inversely proportional to one another over a vast range of values.

Figure 30.39 shows the classification of electromagnetic waves as a function of wavelength and frequency. Extending over a span of nearly 20 orders of magnitude, the figure shows electromagnetic waves ranging from radio waves to gamma rays. Only a very small part of this range corresponds to what we are familiar with as "light." Our eyes are most sensitive to wavelengths between 430 nm and 690 nm, though we can see light somewhat outside this wavelength range if the light is sufficiently intense. However, waves outside the visible range are governed by exactly the same physics as visible light.

As we shall explore in more detail in Chapter 33, the frequency of an electromagnetic wave determines how the wave interacts with materials.

Figure 30.39 Classification of electromagnetic radiation as a function of frequency (top scale) and wavelength (bottom scale).

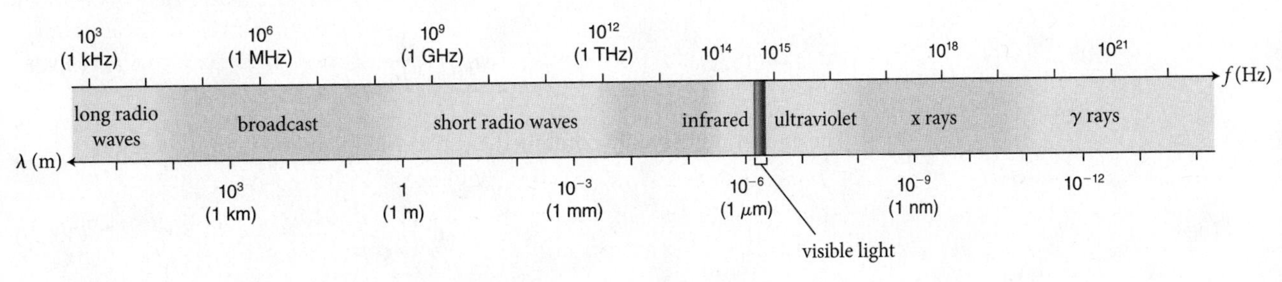

Example 30.8 Speed of light in a dielectric

At what speed does an electromagnetic wave pulse propagate through a dielectric for which the dielectric constant is κ?

❶ GETTING STARTED In the presence of a dielectric, both Gauss's law and the displacement current are modified (Eq. 26.25 and Example 30.6, respectively). These changes affect the derivation I used to obtain the speed of electromagnetic waves in vacuum, Eq. 30.26.

❷ DEVISE PLAN The modification of the displacement current by the dielectric changes Eq. 30.14. To obtain an expression for the speed at which the electromagnetic wave pulse propagates, I carry the modified expression through the same logic I used to go from Eq. 30.14 to Eq. 30.26.

❸ EXECUTE PLAN Because there can be no conventional currents through a dielectric, I in Eq. 30.13 is still zero. Substituting the modified displacement current from Example 30.6, I get

$$\oint \vec{B} \cdot d\vec{\ell} = \mu_0 \epsilon_0 \kappa \frac{d\Phi_E}{dt}.$$

Carrying this expression through the logic leading from Eq. 30.14 to Eq. 30.19, I obtain the magnitude of the magnetic field in terms of the electric field:

$$B = \mu_0 \epsilon_0 \kappa \, Ec, \tag{1}$$

where the κ comes from the modification of the displacement current and where I have written c for the speed in the dielectric rather than c_0, our symbol for the speed in vacuum.

Because Faraday's law is unaffected by the presence of the dielectric, the only modification required in Eq. 30.24 is replacing c_0 with c. Solving Eq. 30.24 for E and substituting that expression for E into Eq. 1, I get

$$B = \mu_0 \epsilon_0 \kappa \, Bc^2,$$

so the speed of an electromagnetic wave pulse moving through a dielectric is

$$c = \frac{1}{\sqrt{\mu_0 \epsilon_0 \kappa}} = \frac{c_0}{\sqrt{\kappa}}, \checkmark$$

where I have used Eq. 30.26, $c_0 = 1/\sqrt{\epsilon_0 \mu_0}$, to simplify.

❹ EVALUATE RESULT Because $\kappa > 1$ for most dielectrics, my result indicates that the speed of electromagnetic waves (including light) in a dielectric is smaller than their speed in vacuum. The dielectric constant is a measure of the reduction of the electric field (Eq. 26.15, $\kappa = E_{free}/E$). For a material that completely attenuates the electric field so that $E = 0$—in other words, for a conductor—κ is infinite and so my result yields $c = 0$. This means that an electromagnetic wave cannot propagate in such a material, a conclusion that agrees with the familiar observation that a conductor, such as a slab of metal, does not transmit visible light.

30.14 An electromagnetic wave with a wavelength of 600 nm in vacuum enters a dielectric for which $\kappa = 1.30$. What are the frequency and wavelength of the wave inside the dielectric?

30.7 Electromagnetic energy

Because electric and magnetic fields contain energy, energy is transported as an electromagnetic wave travels away from its source. Let us work out how much energy is transported by a planar electromagnetic wave. In Section 26.6, we calculated the energy density contained in an electric field in vacuum (Eq. 26.6):

$$u_E = \tfrac{1}{2}\epsilon_0 E^2. \tag{30.27}$$

Similarly, the energy density in a magnetic field in vacuum is given by Eq. 29.29:

$$u_B = \frac{1}{2}\frac{B^2}{\mu_0}. \tag{30.28}$$

The energy density in a combined electric and magnetic field is therefore

$$u = \tfrac{1}{2}\epsilon_0 E^2 + \frac{1}{2}\frac{B^2}{\mu_0}. \tag{30.29}$$

Because the magnitudes of the electric and magnetic fields in an electromagnetic wave are related by Eq. 30.24, we can rewrite Eq. 30.29 in terms of just the magnitude of the electric field. Using Eqs. 30.24 and 30.26, we get

$$u = \tfrac{1}{2}\epsilon_0 E^2 + \tfrac{1}{2}\frac{E^2}{c_0^2\mu_0} = \tfrac{1}{2}\epsilon_0 E^2 + \tfrac{1}{2}\epsilon_0 E^2 = \epsilon_0 E^2. \quad (30.30)$$

Comparing Eqs. 30.29 and 30.30, we see that in vacuum the electric and magnetic fields each contribute half of the energy density—the electric and magnetic energy densities are equal.

Alternatively, the energy density can be written in terms of only the magnitude of the magnetic field,

$$u = \frac{B^2}{\mu_0}, \quad (30.31)$$

or in terms of the magnitudes of both the electric and the magnetic fields,

$$u = \sqrt{\frac{\epsilon_0}{\mu_0}}EB. \quad (30.32)$$

Let us now calculate the rate at which energy flows through a certain area in an electromagnetic wave. Consider taking a slice of an electromagnetic wave normal to the direction of propagation (**Figure 30.40**). The slice has thickness dz and area A. The energy dU in this slice is the product of the energy density and the volume of the slice:

$$dU = uAdz. \quad (30.33)$$

From Section 17.5, you know that the intensity S of a wave is defined as the energy flow per unit time (the power) across a unit area perpendicular to the direction of wave propagation.* Using Eq. 30.33, we can express this relationship in the form

$$S = \frac{1}{A}\frac{dU}{dt}. \quad (30.34)$$

To determine the intensity of an electromagnetic wave, we substitute the expression for dU from Eq. 30.33 and recall that the wave travels at speed $dz/dt = c_0$. We can then rewrite Eq. 30.34 as

$$S = \frac{1}{A}uA\frac{dz}{dt} = uc_0 \quad (30.35)$$

or, after substituting Eqs. 30.32 and 30.26,

$$S = \frac{1}{\mu_0}EB. \quad (30.36)$$

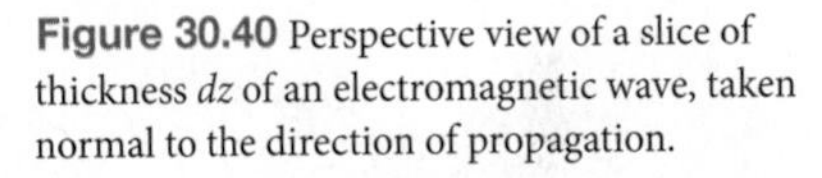

Figure 30.40 Perspective view of a slice of thickness dz of an electromagnetic wave, taken normal to the direction of propagation.

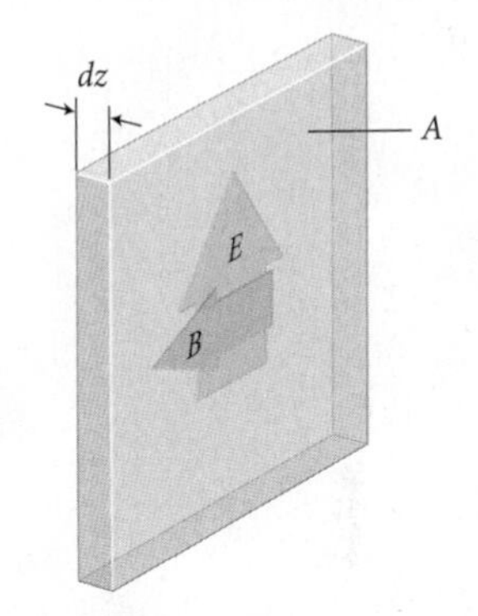

*In Chapter 17, we used I for intensity. That is fine in a discussion of mechanical waves, but now that we must deal with current I so frequently, we switch to the symbol S for intensity.

QUANTITATIVE TOOLS

As the electromagnetic wave travels, energy travels with it in the direction of wave propagation. As we saw in Section 30.6, the propagation direction is the same as that of the vector product $\vec{E} \times \vec{B}$. So with Eq. 30.36 we can define a vector that fully describes energy flow in the electromagnetic wave:

$$\vec{S} \equiv \frac{1}{\mu_0}\vec{E} \times \vec{B}. \quad (30.37)$$

This vector $\vec{S}$ is called the **Poynting vector,** after J. H. Poynting (1852–1914), the physicist who first defined this vector. The SI units of the Poynting vector are W/m^2 (Checkpoint 30.16). The Poynting vector represents the flow of energy in any combined electric and magnetic field, not just electromagnetic waves (Checkpoint 30.15), and the direction of $\vec{S}$ is the direction of energy flow.

When we describe electromagnetic waves, the magnitude S of the Poynting vector is called the *intensity of the electromagnetic wave* and, as noted in Eq. 30.34, is the instantaneous electromagnetic power (energy per unit time) crossing a unit area. For electromagnetic waves, this area is perpendicular to the direction of the vector product $\vec{E} \times \vec{B}$. To obtain the power P crossing a surface, we integrate the Poynting vector over the surface:

$$P = \int_{\text{surface}} \vec{S} \cdot d\vec{A}. \quad (30.38)$$

The results of this section give expressions for the *instantaneous* values of electromagnetic energy density (Eqs. 30.30–32), intensity (Eq. 30.36), and power (Eq. 30.38). Often, however, it is useful to consider average values over some time interval. Because the average of the square of a sine function is 1/2 (see Appendix B), the average value of $\sin^2 \omega t$ is 1/2. So, although sinusoidally oscillating electric and magnetic fields average to zero, their squares average to 1/2 of the square of their amplitudes: $(E^2)_{av} = \frac{1}{2}E^2_{max}$. The square roots of these average values are often referred to as the *root-mean-square values,* or rms values (see Eq. 19.21), and are represented by E_{rms} and B_{rms}:

$$E_{rms} \equiv \sqrt{(E^2)_{av}} \quad \text{and} \quad B_{rms} \equiv \sqrt{(B^2)_{av}}. \quad (30.39)$$

For sinusoidal electromagnetic waves, energy density, intensity, and power are proportional to the squares of a sine function, and so the average values of these quantities are related to the rms values of the fields in the same manner that the instantaneous values of energy density, intensity, and power are related to the instantaneous values of the fields. For example, Eq. 30.36 yields

$$S_{av} = \frac{1}{\mu_0}E_{rms}B_{rms} \quad \text{(sinusoidal electromagnetic wave).} \quad (30.40)$$

QUANTITATIVE TOOLS

Example 30.9 Tanning fields

The average intensity S of the Sun's radiation at Earth's surface is approximately $1.0\ \text{kW/m}^2$. Assuming sinusoidal electromagnetic waves, what are the root-mean-square values of the electric and magnetic fields?

❶ **GETTING STARTED** The electric and magnetic fields in an electromagnetic wave each contribute half of the energy density of the wave (Eq. 30.29). That energy density is related to the wave's Poynting vector $\vec{S}$, whose magnitude S is the intensity of an electromagnetic wave.

❷ **DEVISE PLAN** I can rewrite Eq. 30.35 to obtain the average energy density u_{av} in terms of the average intensity S_{av}. I can then use Eqs. 30.30 and 30.31 to relate the average energy density to the rms values of the electric and magnetic fields.

❸ **EXECUTE PLAN** From Eq. 30.35, I have

$$u_{av} = \frac{S_{av}}{c_0} = \frac{1.0 \times 10^3\ \text{W/m}^2}{3.0 \times 10^8\ \text{m/s}} = 3.3 \times 10^{-6}\ \text{J/m}^3.$$

To obtain the value of E_{rms}, I express the relationship between u_{av} and E_{rms} by analogy to Eq. 30.30:

$$u_{av} = \epsilon_0 E_{rms}^2$$

$$E_{rms} = \sqrt{\frac{u_{av}}{\epsilon_0}} = \sqrt{\frac{3.3 \times 10^{-6}\ \text{J/m}^3}{8.85 \times 10^{-12}\ \text{C}^2/\text{N}\cdot\text{m}^2}} = 6.1 \times 10^2\ \text{V/m}, \checkmark$$

where I have used the fact that $1\ \text{N/C} = 1\ \text{V/m}$.

For sinusoidal electromagnetic waves, the relationship between instantaneous values of the electric and magnetic field magnitudes holds for the rms values of the field magnitudes, so I can use Eq. 30.24 to determine the value of B_{rms}:

$$B_{rms} = \frac{E_{rms}}{c_0} = \frac{6.1 \times 10^2\ \text{N/C}}{3.0 \times 10^8\ \text{m/s}} = 2.0 \times 10^{-6}\ \text{T}. \checkmark$$

❹ **EVALUATE RESULT** As I expect based on the large value of c_0, the value of B_{rms} in teslas is much smaller than that of E_{rms} in volts per meter. This electric field magnitude is also not large, particularly compared with the electric fields often obtained in capacitors. (Applying a potential difference of just 1 V to the capacitor in the next example would produce an electric field magnitude of 10,000 V/m!) However, this electric field is greater than the typical atmospheric electric field magnitude of roughly 100 V/m (which is due to charged particles in the atmosphere rather than to the Sun's radiation).

30.15 Consider supplying a constant current to a parallel-plate capacitor in which the plates are circular. While the capacitor is charging, what is the direction of the Poynting vector at points that lie on the cylindrical surface surrounding the space between the capacitor plates? What does this Poynting vector represent?

Example 30.10 Capacitor power

A parallel-plate capacitor with circular plates of radius $R = 0.10$ m and separation distance $d = 0.10$ mm is charged by a constant current of 1.0 A. (*a*) What is the magnitude of the Poynting vector associated with the electric and magnetic fields at the edge of the space between the plates? (*b*) What is the rate at which electromagnetic energy is delivered to the cylindrical space between the plates?

❶ **GETTING STARTED** While the capacitor is charging, the charge on the plates and consequently the electric field between them and the energy stored in the capacitor are changing with time. The changing electric field is accompanied by a magnetic field. The electric field between the plates is uniform, and the magnetic field lines form circular loops centered on the horizontal axis of the capacitor, in a direction given by the right-hand current rule. At each point in the space between the plates, the electric and magnetic fields are perpendicular to each other, as shown in my sketch in **Figure 30.41**.

Figure 30.41

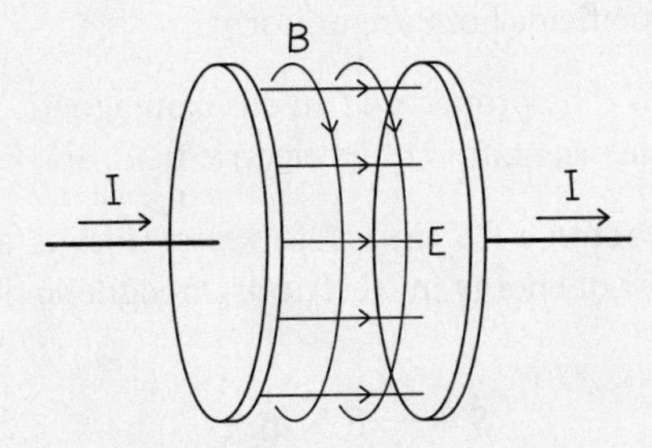

❷ **DEVISE PLAN** Because the electric and magnetic fields are perpendicular to each other, I can use Eq. 30.36 to calculate the magnitude of the Poynting vector. For this equation, I must know E and B. I calculated the magnitude of the electric field between two capacitor plates in Example 26.2, and from Checkpoint 30.11*a* I have an expression for the magnitude of the magnetic field between the plates.

To obtain the rate at which electromagnetic energy is delivered to the capacitor, I must integrate the Poynting vector over the cylindrical surface surrounding the space between the plates, as given in Eq. 30.38. From Checkpoint 30.15, I know that the Poynting vector points radially inward toward the capacitor axis, and I also know that at any given instant, the magnitudes of the electric and magnetic fields do not vary on the cylindrical surface surrounding the space between the plates. To integrate the Poynting vector over that surface, I therefore need to multiply the expression for the magnitude of the Poynting vector at any point on the surface by the area of the surface.

❸ **EXECUTE PLAN** (*a*) From Example 26.2 I know that the electric field between the capacitor plates is $E = q/(\epsilon_0 A)$, where $A = \pi R^2$ is the area of each plate. Because a constant current is supplied to the capacitor while the capacitor is charging, $q = I\Delta t$ (Eq. 27.1, where Δt is time interval that has elapsed since the capacitor began charging), I can therefore rewrite my expression for E in the form

$$E = \frac{I\Delta t}{\epsilon_0 A} = \frac{I\Delta t}{\pi\epsilon_0 R^2}.$$

From the solution to Checkpoint 30.11*a* I have $B = \mu_0 Ir/2\pi R^2$, where r is the radial distance from the axis of the capacitor to the position where the magnetic field is measured and R is the radius of each plate. For this case, $r = R$, so

$$B = \frac{\mu_0 I}{2\pi R}.$$

(*continued*)

QUANTITATIVE TOOLS

To obtain the magnitude of the Poynting vector, I substitute the above values for E and B into Eq. 30.36:

$$S = \frac{EB}{\mu_0} = \frac{I^2\Delta t}{2\pi^2\epsilon_0 R^3} = \frac{(1.0\text{ A})^2\Delta t}{2\pi^2[8.85\times10^{-12}\text{ C}^2/(\text{N}\cdot\text{m}^2)](0.10\text{ m})^3}$$
$$= [5.7\times10^{12}\text{ W}/(\text{m}^2\cdot\text{s})]\Delta t. ✔$$

(*b*) To calculate the rate at which electromagnetic energy is delivered to the cylindrical space between the capacitor plates, I multiply the expression for S by the area of the surface, which is the product of the circumference of the cylinder ($2\pi R$) and the height of the cylinder, which is the plate separation distance d:

$$P = 2\pi Rd\frac{I^2\Delta t}{2\pi^2\epsilon_0 R^3} = \frac{I^2 d\Delta t}{\pi\epsilon_0 R^2}$$
$$= \frac{(1.0\text{ A})^2(1.0\times10^{-4}\text{ m})\Delta t}{\pi[8.85\times10^{-12}\text{ C}^2/(\text{N}\cdot\text{m}^2)](0.10\text{ m})^2}$$
$$= (3.6\times10^8\text{ W/s})\Delta t. ✔$$

❹ EVALUATE RESULT I have no way of gauging the reasonableness of my numerical results, but I can check the reasonableness of my expression for the power (which incorporates my result for part *a*) by using an alternative method to obtain that expression. The rate at which energy is delivered to the capacitor is the power delivered to it by the current. I know from Eq. 26.2 that the electric potential energy of the capacitor changes by an amount $dU^E = V_{cap}dq$ during charging. To get the rate at which the amount of energy stored in the capacitor changes with time, I take the derivative

$$P \equiv \frac{dU_E}{dt} = V_{cap}\frac{dq}{dt} = V_{cap}I = \frac{qI}{C} = \frac{I^2\Delta t}{C}. \quad (1)$$

For a parallel-plate capacitor, $C = \epsilon_0 A/d$, so with $A = \pi R^2$,

$$P = \frac{I^2 d\Delta t}{\epsilon_0\pi R^2},$$

which is the same result I obtained for part *b*.

30.16 Use Eq. 30.36 to show that the SI units of the Poynting vector are W/m^2.

Chapter Glossary

SI units of physical quantities are given in parentheses.

Displacement current (A) A current-like quantity that contributes to Ampère's law caused by a changing electric flux:

$$I_{disp} \equiv \epsilon_0\frac{d\Phi_E}{dt}. \quad (30.7)$$

Electromagnetic wave A wave disturbance that consists of combined electric and magnetic fields. In vacuum and dielectrics, the electric and magnetic fields in an electromagnetic wave are always perpendicular to each other and the direction of propagation is given by the vector product $\vec{E}\times\vec{B}$. In vacuum, electromagnetic waves always propagate at speed

$$c_0 = \frac{1}{\sqrt{\epsilon_0\mu_0}}. \quad (30.26)$$

Maxwell's equations Equations that together provide a complete description of electric and magnetic fields:

$$\Phi_E \equiv \oint\vec{E}\cdot d\vec{A} = \frac{q_{enc}}{\epsilon_0} \quad (30.10)$$

$$\Phi_B \equiv \oint\vec{B}\cdot d\vec{A} = 0 \quad (30.11)$$

$$\oint\vec{E}\cdot d\vec{\ell} = -\frac{d\Phi_B}{dt} \quad (30.12)$$

$$\oint\vec{B}\cdot d\vec{\ell} = \mu_0 I + \mu_0\epsilon_0\frac{d\Phi_E}{dt}. \quad (30.13)$$

Planar electromagnetic wave An electromagnetic wave in which the wavefronts (the surfaces of constant phase) are planes perpendicular to the direction of propagation. On these planes, the instantaneous magnitudes of the electric and magnetic fields both are uniform.

Polarization The property of an electromagnetic wave that describes the orientation of its electric field.

Poynting vector $\vec{S}$ (W/m^2) The vector that is associated with the flow of energy in electric and magnetic fields:

$$\vec{S} \equiv \frac{1}{\mu_0}\vec{E}\times\vec{B}. \quad (30.37)$$

31 Electric Circuits

Electric circuits surround us. We light our work and living spaces, start our cars, communicate with each other, and cook our food with electric circuits. Almost everything that consumes energy in our offices and homes is powered by electricity, whether by batteries or by the electric circuitry of the building.

Electrical devices are ubiquitous because electric circuits offer an extremely versatile means of producing and distributing electrical energy for a variety of tasks. Circuits are designed to control the flow of charge carriers—the current—in a device. The current then provides energy to the device, allowing it to perform its function.

In this chapter we explore the basic principles of electric circuits powered by sources of electric potential energy that maintain a constant potential difference, such as batteries. Such circuits are known as *direct-current circuits* or *DC circuits*. In the next chapter, we'll learn about *alternating-current (AC)* circuits, which run on time-varying potential differences, such as the electricity delivered to buildings by means of electrical power lines.

31.1 The basic circuit

One familiar way to make electricity do something useful is to connect a battery to a light bulb. To do so, we connect each terminal of the battery to one of the two contacts on the light bulb (**Figure 31.1*a***). On standard light bulbs, the two contacts are the threaded metal casing and a metallic "foot" that is separated from the casing by an insulator (Figure 31.1*b*). Inside the bulb, each of these contacts is connected by a metal wire to one end of a tungsten *filament*, a very thin wire of tungsten wound in a tight coil. When connected as shown in Figure 31.1, the wires, contacts, and filament form a continuous conducting path from one terminal of the battery to the other, and charge carriers flow through the filament. The current in the filament causes the filament to become white hot and glow. (Tungsten is used for light bulb filaments because its very high melting temperature allows it to glow and not melt.)

Figure 31.1 A light bulb connected to a battery.

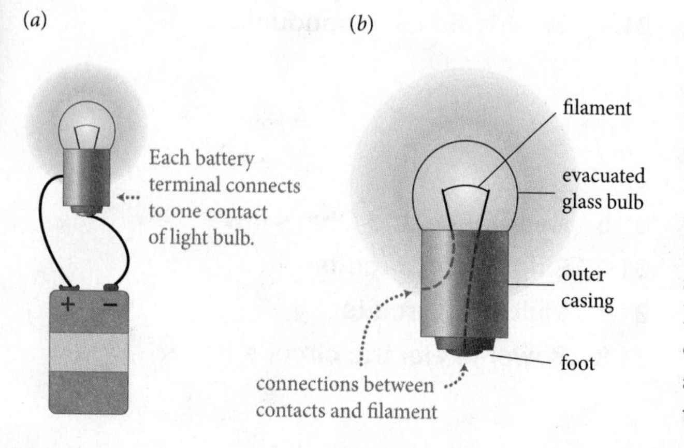

Experiments with light bulbs show that:

1. A bulb doesn't glow—not even briefly—if only one contact is connected to a battery terminal.
2. In order for a bulb to glow, both contacts must be connected through a continuous conducting path to the terminals of a battery. If the path is broken, the bulb goes out. It doesn't matter where in the path the break occurs.
3. A glowing light bulb generates light and thermal energy.
4. The wires connecting a light bulb to a battery usually do not get hot.
5. A light bulb left connected to a battery over a long time interval eventually goes out because the battery runs down. Once this happens, not enough chemical energy remains in the battery to maintain a large enough potential difference to light the bulb.

In the rest of this chapter we explore the reasons for these observations. The first two observations can be summarized as follows: In order for the light bulb to glow, one of the bulb's contacts must be connected to the positive terminal of the battery and the bulb's other contact must be connected to the negative terminal of the battery. Because the battery maintains a potential difference between its terminals, charge carriers move from one terminal toward the other when a conducting path is provided between the terminals.

The arrangement shown in Figure 31.1 is an example of an **electric circuit**—an interconnection of electrical components (called *circuit elements*). Any closed conducting path through the circuit is called a **loop.** We shall first study electric circuits that have a single loop.

31.1 (*a*) Consider the system comprising the single-loop circuit shown in Figure 31.1*a*, including the light and thermal energy generated by the bulb. (*a*) Is this system closed? (*b*) Is the energy of the system constant? (*c*) Where do the light energy and the thermal energy come from?

Checkpoint 31.1 demonstrates an important feature of circuits: Energy is converted from electric potential energy to some other form. With this in mind, we can draw a general single-loop circuit (**Figure 31.2*a***) consisting of a power source, a load, and wires that connect the load to the source. The **power source** provides electric potential energy to the rest of the circuit, usually by converting some form of energy to electric potential energy. The potential difference across the terminals of the power source drives a current in the circuit. The **load** in an electric circuit is all the circuit elements connected to the power source. In the load, the electric potential energy of the moving charge carriers is converted to other forms of energy, such as thermal or mechanical energy. The wires connecting the elements in a circuit are considered to be ideal; that is,

Figure 31.2 The energy conversions in a single-loop circuit and in a mechanical analog of the circuit.

(*a*) Schematic simple circuit

power source: $\Delta E_s \rightarrow \Delta U^E$

load: $\Delta U^E \rightarrow \Delta E_{other}$

wires carry current I

(*b*) Mechanical analog of circuit

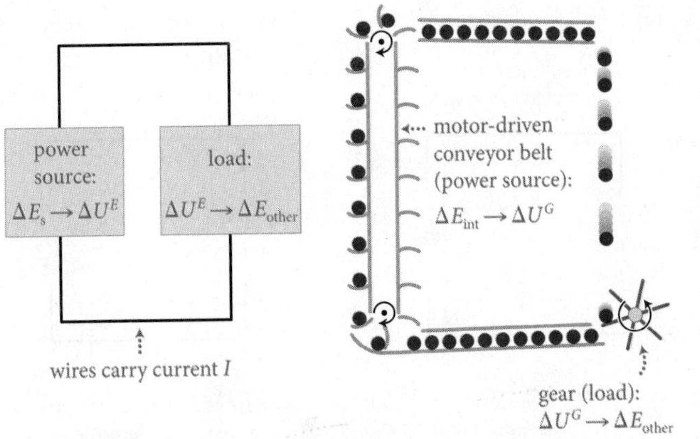

motor-driven conveyor belt (power source): $\Delta E_{int} \rightarrow \Delta U^G$

gear (load): $\Delta U^G \rightarrow \Delta E_{other}$

the wires serve to transport electric potential energy to the load, and we ignore the small amount of energy dissipated in the wires.

Figure 31.2*b* shows a mechanical analog of a single-loop electric circuit. The motor-driven conveyor belt is analogous to the power source in the electric circuit. The conveyor belt lifts the balls, converting source energy to gravitational potential energy. The balls then roll through a horizontal tube, losing only a tiny amount of gravitational potential energy in the process, in the same manner that charge carriers flow with very little loss of energy through wires. Next, the balls drop from the tube onto the gear, which corresponds to the load of an electric circuit. As the balls fall and make the gear turn, their gravitational potential energy is converted to mechanical energy. Finally, the balls roll down a second tube to the bottom of the conveyor belt, completing the cycle.

Notice two things about the mechanical circuit of Figure 31.2*b*. First, because the system is filled with balls, the balls have to move through the circuit one after the other. If the right end of the upper tube were blocked rather than open (the equivalent of the connecting wire in Figure 31.2*a* being disconnected from the load), no more balls could be pushed into it from the conveyor belt. There has to be a closed path through the system in order for the balls to move, just as charge carriers can flow through a circuit only when there is an unbroken conducting path.

The second thing to notice in Figure 31.2*b* is that there is no net transport of balls from the power source to the load. There are as many balls traveling from the power source to the load as the other way around. The balls travel around a closed path through the system, and if we watched one particular ball long enough, we would see it circle repeatedly around the circuit. The balls are simply vehicles transporting energy through the system.

We can therefore propose an operational description of a single-loop DC circuit in terms of energy conversion:

In a single-loop DC circuit, electric potential energy acquired by the carriers in the power source is converted to another form of energy in the load.

The power source in such an electric circuit doesn't have to be a battery—it can be anything that produces a constant potential difference, thereby driving a current around the circuit, such as a solar cell or an electric generator. Likewise, the load doesn't have to be a light bulb. A toaster acting as a circuit load converts electric potential energy to thermal energy, a loudspeaker acting as a load converts electric potential energy to mechanical energy in sound waves, and a motor load converts electric potential energy to mechanical energy.

Exercise 31.1 Solar fan

A solar cell, which converts solar energy (a form of source energy, see Section 7.4) to electric potential energy, is connected to a small fan. Represent this circuit by a diagram analogous to Figure 31.2*a* and describe the energy conversions taking place in this circuit.

SOLUTION The solar cell is the power source, and the fan is the load. **Figure 31.3** shows my diagram for this circuit. The solar cell converts solar energy to electric potential energy, and the fan converts electric potential energy to another form of energy by setting the air in motion. I add these conversions to my diagram. ✔

Figure 31.3

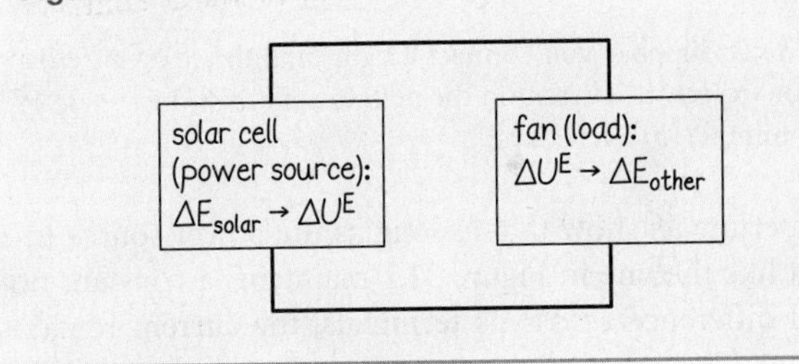

The various parts of electric circuits are commonly represented by graphical symbols rather than by such words as *power source* and *load*. **Figure 31.4** shows the symbols for some common circuit elements. In **Figure 31.5** these symbols are used to represent the bulb-and-battery circuit

Figure 31.4 Standard representations of common elements of electric circuits.

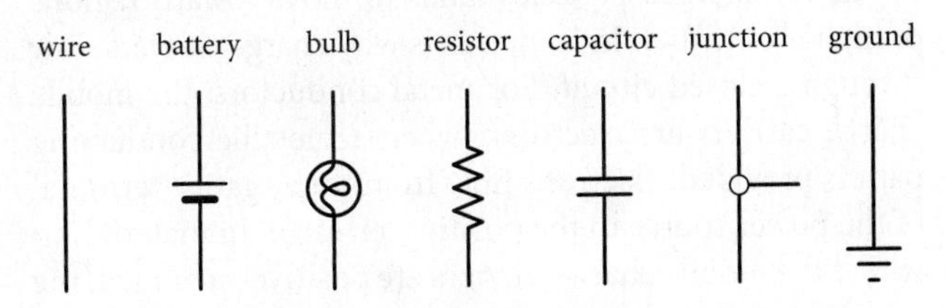

Figure 31.5 Circuit diagram of bulb-and-battery circuit of Figure 31.1.

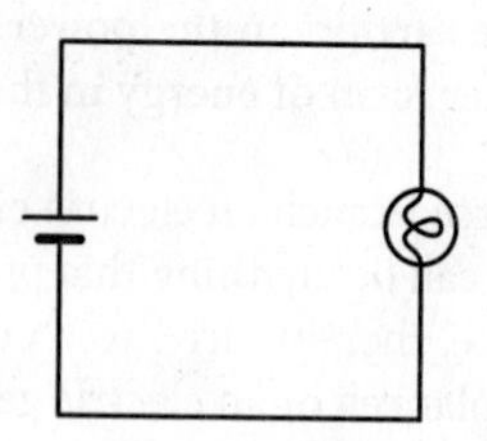

Figure 31.6 Because charge is conserved, in steady state it doesn't accumulate in the load or in any other part of the circuit. Hence, in steady state the current into any part of the circuit must be the same as the current out of that part.

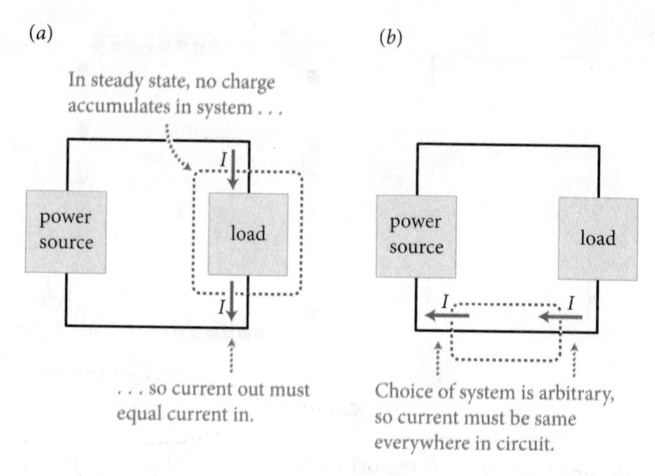

of Figure 31.1. Note that in the battery symbol the short, heavy line represents the negative terminal and the longer, thin line represents the positive terminal. A schematic representation of a circuit using the standard symbols shown in Figure 31.4 is called a **circuit diagram.**

31.2 Two wires connect the plates of a charged capacitor to the contacts of a light bulb. (*a*) Does this assembly constitute a circuit? If so, identify the power source and the load. (*b*) Does the bulb glow? (*c*) What energy conversions take place after the bulb and capacitor are connected?

31.2 Current and resistance

Why does the filament of a light bulb get hot and glow, but the wires connecting the bulb to the battery do not? And why does the battery run out of energy and the bulb eventually stop glowing? To answer these questions, we need to look at the motion of charge carriers through an electric circuit.

31.3 Suppose you connect a light bulb to a battery. How do you expect the current in the bulb to vary over the course of (*a*) a minute, (*b*) a few days?

Experiments show that as long as the power source in a circuit like the one in Figure 31.1 maintains a constant potential difference across its terminals, the current remains constant. Given that a battery connected to a bulb can maintain a constant potential difference for hours, the current in a circuit like the one in Figure 31.1 is constant over that same time interval. This current is established almost instantaneously after the circuit is completed (that is, after all the circuit elements are connected together), and it vanishes almost instantaneously when the circuit is broken. Other than the instants after the circuit is completed or broken, therefore, we have a **steady state** with constant current.

As we found in Section 25.3, positively charged particles tend to move toward regions of lower potential and negatively charged particles tends to move toward regions of higher potential. Indeed, this is why charge carriers flow through a closed circuit. For metal conductors, the mobile charge carriers are electrons; when a metallic conducting path is provided, electrons flow from the negative terminal of the power source to the positive terminal. In materials in which the mobile charge carriers are positively charged, the carriers travel in the opposite direction—from the positive terminal of the power source to the negative terminal. The rate at which the charges on the carriers cross a section in the conductor is the current, and the sign of the carriers and the direction of their motion determine the direction of the current (see Sections 27.3 and 27.5). Remember, however, that for a given potential difference between two points, the current direction is from points of higher potential to points of lower potential regardless of the sign of the mobile charge carriers (see Section 27.3).

In steady state we can draw an important conclusion regarding currents in circuits. Consider, for example, the system comprising just the load in the circuit shown in **Figure 31.6*a***. Because charge is conserved (see Section 22.3), the amount of charge inside this system can only change due to a flow of charge into or out of the system. In steady state the charge of the system is constant, and so the flow of charge carriers into the system must be equal to the flow of charge carriers out of it. Put differently, the current into the system must equal the current out of the system.

Because we can choose any system we like—as in Figure 31.6*b*, for example—we conclude:

In a steady state, the current is the same at all locations along a single-loop electric circuit.

We shall refer to this requirement as the **current continuity principle. Figure 31.7** illustrates the principle for balls flowing through a tube. Because the tube is filled, if the flow of balls through the tube is steady, there is nowhere for the balls to pile up without changing the rate of flow. As a result, when one ball is pushed into the left end of the tube, one ball must come out at the right end. Likewise, if two balls are pushed into the left end at the same instant, two must come out at the right end.

The current continuity principle tells us something else about the operation of circuits: As electrons flow through a light bulb, they are not accumulating, or "used up," in the bulb. For every electron that goes into one end of the

Figure 31.7 The continuity principle.

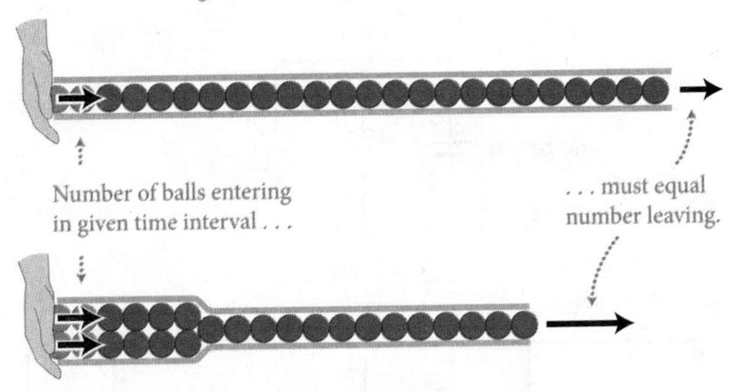
If flow of balls through tube is steady:
Number of balls entering in given time interval . . .
. . . must equal number leaving.

filament, one electron comes out the other end. What then is consumed (used up) by the bulb as it glows? In our discussion of the mechanical circuit of Figure 31.2*b*, we found that the balls moving through the circuit are not consumed anywhere. Instead, they act to transfer energy from the power source to the load. Before we begin our examination of the energy of the electrons in the electric circuit of Figure 31.2*a*, answer the next checkpoint.

31.4 Does an electron lose or gain electric potential energy (*a*) while moving inside a battery from the positive terminal to the negative terminal and (*b*) while moving through the rest of the circuit from the negative battery terminal to the positive terminal? (*c*) While flowing through the wire and load portions of the circuit, where do the electrons lose most of their energy?

As we saw in Section 26.4, *inside* a battery, chemical energy is converted to electric potential energy, and electrons at the negative terminal have greater potential energy than those at the positive terminal. In the rest of an electric circuit, electrons lose that same amount of electric potential energy as they move from the negative terminal through the load to the positive terminal, and this electric potential energy is converted to other forms of energy.

For a charged particle moving between two locations through the load, the difference in potential energy from one location to the other is equal to the potential difference between those two locations multiplied by the charge of the particle. Therefore, for each electron that moves through the load of a circuit, an amount of electric potential energy eV_{load} is converted to some other form of energy (light, say, or thermal energy, or mechanical energy), where V_{load} is the magnitude of the potential difference across the load.*

Thus, changes in the potential difference in an electric circuit show us where energy conversion is taking place. Experiments indicate that V_{load} is essentially equal to V_{batt}, while V_{wire} is negligible. In other words, essentially all of the energy conversion in the circuit takes place in the load and the source, and almost none takes place in the wires.

*When the subscript of potential refers to a device or circuit element (say, V_{device}), the symbol is taken to denote the *magnitude* of the potential difference across the device or circuit element.

Because we can ignore the energy dissipation that takes place in the wires:

Every point on any given wire is essentially at the same potential.

Therefore we can consider the load in a circuit to be connected directly to the battery if the two are connected to each other through wires.

Example 31.2 Current and potential difference

In **Figure 31.8**, two light bulbs are connected to each other by a wire and the combination is connected to a battery. In steady state, bulb A glows brightly and bulb B glows dimly. If the magnitude of the potential difference across the battery is 9 V, what can you say about the magnitude of the potential difference across A?

Figure 31.8 Example 31.2.

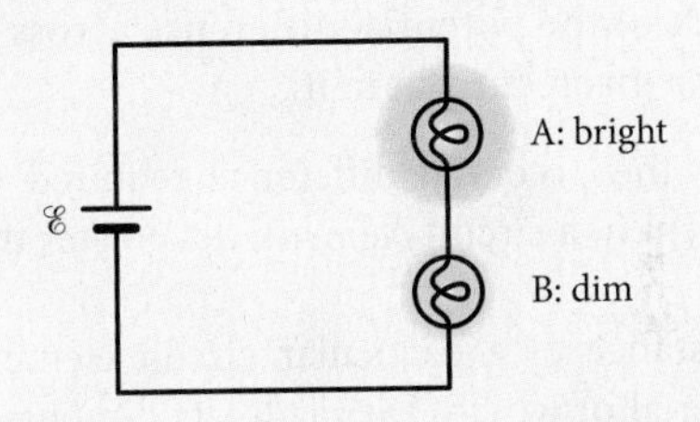
A: bright
B: dim
$\mathscr{E}$

❶ **GETTING STARTED** The battery and light bulbs constitute a circuit in steady state. The electrons in this circuit gain electric potential energy inside the battery, and this electric potential energy is converted to light (a form of energy) and thermal energy in the bulbs. I assume that the ratio of light to thermal energy is the same in the two bulbs. Because bulb A glows more brightly than bulb B, bulb A must convert more electric potential energy than bulb B.

❷ **DEVISE PLAN** Because this circuit is in steady state, I know that the current is the same throughout the circuit, and therefore during any given time interval, the same number of electrons pass through both bulbs. For each electron that moves through the load of the circuit, an amount of electric potential energy eV_{load} is converted to light and thermal energy. Therefore the magnitude of the potential difference across each bulb is proportional to the amount of electric potential energy converted to light and thermal energy.

❸ **EXECUTE PLAN** The fact that A glows more brightly than B tells me that A is producing more light and thermal energy. Bulb A must therefore be converting more electric potential energy to these other forms, and the electrons must be losing more energy in A than in B. Because the electrons lose more energy in A than in B, the magnitude of the potential difference across A, V_A, must be greater than the magnitude of the potential difference across B, V_B. Because $V_A > V_B$ and because $V_A + V_B$ must equal V_{batt}, it must be true that V_A is more than half of V_{batt}, which means $9.0 \text{ V} > V_A > 4.5 \text{ V}$. ✔

❹ **EVALUATE RESULT** It makes sense that the bulb that has the greater potential difference across it glows more brightly.

Example 31.2 illustrates an important point: The magnitudes of the potential difference across various circuit elements in an electric circuit need not be the same, even if the current in them is the same.

The two light bulbs in Figure 31.8 are said to be connected *in series:* There is only a single current path through them, and the charge carriers flow first through one and then through the other. As we have seen in Example 31.2:

The potential difference across circuit elements connected in series is equal to the sum of the individual potential differences across each circuit element.

Experiments show that to obtain a particular current in a circuit element, different elements require widely different potential differences. As noted in passing in Section 29.5, for a given circuit element, the constant of proportionality between the potential difference across the element and the current in it is called the **resistance** of the circuit element:

The resistance of any element in an electric circuit is a measure of the potential difference across that element for a given current in it.

The greater the potential difference required to obtain a certain current in a circuit element, the greater the element's resistance.

The resistance of a particular circuit element depends on the material of which it is made, its dimensions, and its temperature. The resistance of a combination of circuit elements connected in series is equal to the sum of the individual resistances of each element. Therefore, the more light bulbs we connect in series to a battery, the greater the resistance of the combination of light bulbs and the smaller the resulting current in the circuit.

31.5 (*a*) Which light bulb in Example 31.2 has the greater resistance? (*b*) Suppose you connect each bulb separately to the battery. Do you expect the current in bulb A to be greater than, equal to, or smaller than that in bulb B?

31.3 Junctions and multiple loops

Circuit elements can also be connected so that more than one conducting path is formed, as illustrated in Figure 31.9. Such circuits contain more than one loop and are called *multiloop circuits.* These circuits contain **junctions**—locations where more than two wires are connected together—and **branches**—conducting paths between two junctions that are not intercepted by another junction. The circuit in Figure 31.9, for example, contains two junctions (represented by open circles), three branches, and three loops.

The continuity principle permits us to draw some important conclusions about the currents in multiloop circuits. Let us begin by applying the current continuity principle to the circuit shown in Figure 31.10. Because we can select any system boundary along a branch, the current continuity principle requires the current to be the same throughout that branch. We can apply the same reasoning to the other two branches, so:

The current in each branch of a multiloop circuit is the same throughout that branch.

This statement is known as the **branch rule.** We shall label the current in each branch according to the branch. In Figure 31.10, for example, the currents in the three branches are I_1, I_2, and I_3.

To examine how charge carriers flow at a junction, we draw a system boundary around a junction, as in Figure 31.11*a*. Because the charge inside the system is not changing and because charge is conserved, it follows that the flow of charge carriers into the system must be equal to the flow of charge carriers out of the system. Specifically, if a current I_1 goes into the junction and the other two wires carry currents I_2 and I_3 out of the junction, then $I_2 + I_3$ must

Figure 31.9 Circuit diagram of two light bulbs connected to a battery.

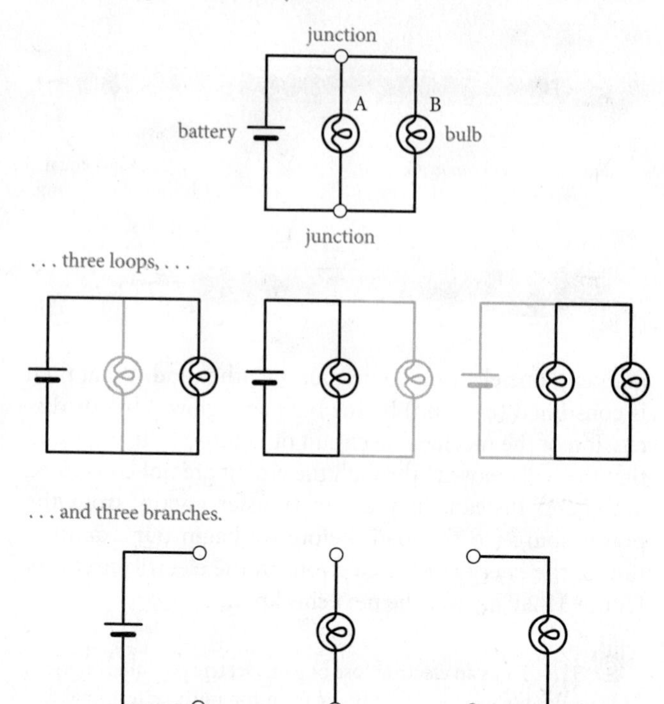

Figure 31.10 The branch rule: In each branch, the current must be the same throughout that branch.

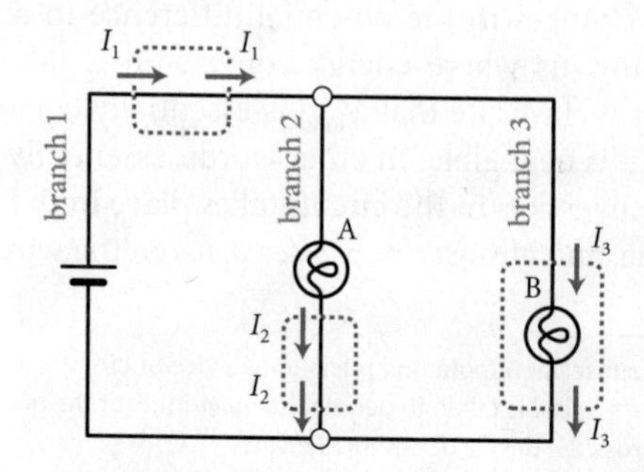

Figure 31.11 The continuity principle at a junction is illustrated by the flow of balls through a branched tube.

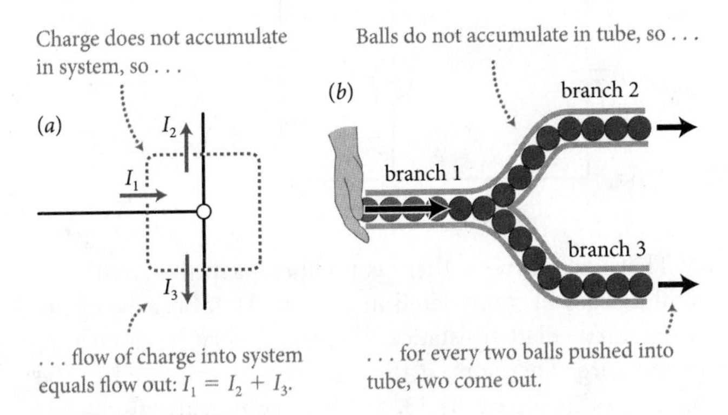
In steady state:
Charge does not accumulate in system, so . . .
Balls do not accumulate in tube, so . . .
(a)
(b)
branch 1
branch 2
branch 3
. . . flow of charge into system equals flow out: $I_1 = I_2 + I_3$.
. . . for every two balls pushed into tube, two come out.

equal I_1. The general statement of this principle is known as the **junction rule:**

> **The sum of the currents directed into a junction equals the sum of the currents directed out of the same junction.**

Figure 31.11*b* illustrates this principle for balls flowing through a junction of tubes. If we push four balls in at one end of branch 1, four balls have to come out from branches 2 and 3 combined. In the junction shown in the figure, branch 2 and branch 3 are equivalent and two balls are likely to come out from each. In general, however, the branches do not need to be equivalent, and in that case the number of balls going into branch 1 must be equal to the sum of the balls coming out of branches 2 and 3. Because the pushing in and the coming out occur during the same time interval, the "current" of balls through branch 1 equals the sum of the currents in branches 2 and 3, which is what we concluded for the currents in Figure 31.11*a*.

Let us return now to the circuit in Figure 31.9. Because the two bulbs are each connected to the same two junctions, they are said to be connected *in parallel* across the battery. Because each junction is at a given potential, we can conclude that for parallel circuit elements:

> **The potential differences across circuit elements connected in parallel are always equal.**

More specifically, for the circuit shown in Figure 31.9, the potential difference across the two light bulbs is equal to the emf of the battery.

In the next example we study the resistance of parallel combinations of circuit elements.

Example 31.3 Series versus parallel

Two identical light bulbs can be connected in parallel or in series to a battery to form a closed circuit. How do the magnitudes of the potential differences across each bulb and the current in the battery compare with those in a single-bulb circuit when the bulbs in the two-bulb circuit are connected in parallel? When they are connected in series?

❶ **GETTING STARTED** I begin by drawing circuit diagrams for the single-bulb circuit and for the parallel and series two-bulb circuits (**Figure 31.12**). To connect the bulbs in parallel to the battery, one contact of each bulb is connected directly to the positive terminal of the battery, and the other contact of each bulb is connected directly to the negative terminal of the battery. For the series circuit, one contact of bulb 1 is connected directly to the positive terminal of the battery, the other contact of bulb 1 is connected to one contact of bulb 2, and the other contact of bulb 2 is connected directly to the negative terminal of the battery.

Figure 31.12

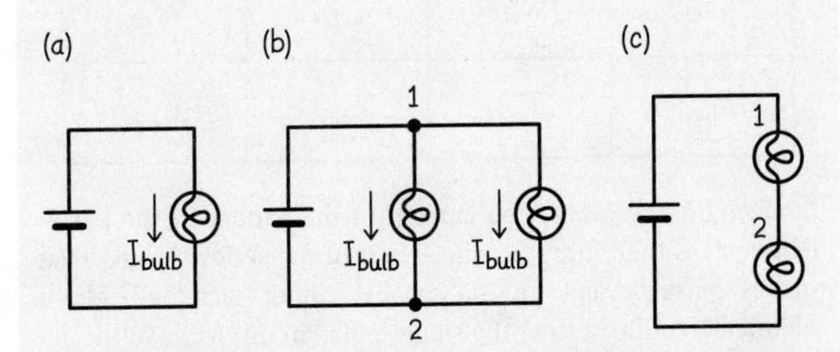
(a)
(b)
(c)
I_{bulb}
I_{bulb}
I_{bulb}
1
2
1
2

❷ **DEVISE PLAN** To determine the magnitude of the potential difference across each bulb, I need to analyze how the bulbs are connected to the battery. To determine the current in the battery, I must first determine the current through each bulb. To determine the current in each bulb, I use the fact that the resistance and potential difference determine the amount of current in each bulb. I can use the junction rule to determine the currents in the parallel circuit, and so I label the junctions 1 and 2 in Figure 31.12*b*. Because there is only one branch in the series circuit, I know that the current in the battery is the same as the current in the bulbs. Because the bulbs are identical, their resistances are identical too. I assume the wires have no resistance.

❸ **EXECUTE PLAN** Because in the parallel circuit one contact of each bulb is connected to the battery's positive terminal and the other contact of each bulb is connected to the negative terminal, the potential difference across each bulb is equal to the potential difference across the battery. By the same argument, the potential difference across the bulb in the single-bulb circuit is also equal to the potential difference across the battery. Therefore the potential difference across each bulb in the parallel circuit is the same as that across the bulb in the single-bulb circuit. ✔

Because the potential difference is the same across these three bulbs, the current must also be the same in all three bulbs. I'll denote this current by I_{bulb}. In the parallel circuit, the fact that the current in each bulb is I_{bulb} means that the current in the battery must be $2I_{bulb}$. (The current pathway at junction 1 of Figure 31.12*b* is just like the pathway shown in Figure 31.11, with $I_2 = I_3 = I_{bulb}$.) In the single-bulb circuit, the current must be the same at all locations in the circuit, so the current in the battery must be I_{bulb}. ✔

In the series circuit, the potential difference across the two-bulb combination is equal to the potential difference across the battery, which means the magnitude of the potential difference across each bulb must be half the magnitude of the potential difference across the battery. The potential difference across each bulb is thus equal to half the potential difference across the bulb in the single-bulb circuit. ✔

Because the potential difference across each bulb in series is half the potential difference across the battery, the current in each bulb must be half the current in the single-bulb circuit $I_{bulb}/2$. Because the series circuit in Figure 31.12*c* is a single-loop circuit, the current is the same at all locations. Therefore the current in the battery must also be $I_{bulb}/2$. The current in the battery in the single-bulb circuit is therefore twice that in the series two-bulb circuit. ✔

In tabular form my result are

	Parallel	Series	Single
V_{bulb}	V_{batt}	$V_{batt}/2$	V_{batt}
I_{batt}	$2I_{bulb}$	$I_{bulb}/2$	I_{bulb}

❹ **EVALUATE RESULT** The current in the battery in the parallel circuit is therefore four times that in the series circuit. That makes sense because in the parallel circuit each bulb glows identically to the bulb in the single-bulb circuit, while in the series circuit, the battery has to "push" the charge carriers through twice as much resistance. Therefore it makes sense that in the series circuit, both the potential difference across each bulb and the current in the battery are much smaller than they are in the parallel circuit.

31.6 In Figure 31.9, treat the parallel combination of two light bulbs as a single circuit element. Is the resistance of this element greater than, equal to, or smaller than the resistance of either bulb?

Checkpoint 31.6 highlights an important point about electric circuits: Adding circuit elements in parallel *lowers* the combined resistance and *increases* the current. How can adding elements with a certain resistance lower the combined resistance? The resolution to this apparent contradiction is that adding elements in parallel really amounts to adding paths through which charge carriers can flow, rather than adding resistance to existing paths in the circuit. If you are emptying the water out of a swimming pool, it empties faster if you have multiple hoses draining the water than if you drain the entire pool through a single hose.

Instead of connecting two light bulbs in parallel as in Figure 31.9, what if we connected a wire in parallel with a bulb as in **Figure 31.13**? Experimentally, we find that replacing bulb B of Figure 31.9 with a wire causes bulb A to stop glowing. Why does this happen? The potential difference across a branch is determined by the potential difference between the two junctions on either end of the branch, and therefore the potential difference across all branches between two junctions must be the same. Because the wire is made from a conducting material, the potential difference between its two ends is zero, and therefore the two junctions in Figure 31.13 are at the same potential. Consequently there is no current in the light bulb.

Figure 31.13 A wire that is connected to a battery *in parallel* to a light bulb constitutes a short circuit.

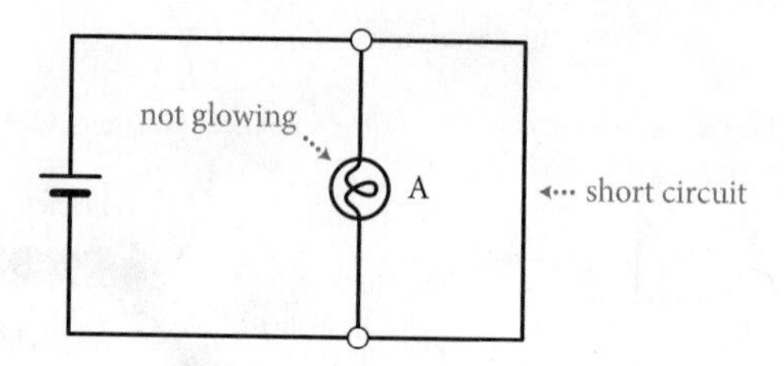

That doesn't mean there is no current in the circuit. As we'll discuss in more detail in Section 31.4, because of the wire's very small resistance, there is a very large current in the wire. Therefore, if the wire is left connected to the battery as in Figure 31.13, the battery quickly discharges through the wire and the wire heats up. A circuit branch with negligible resistance in parallel with an element is commonly called either a **short** or a **short circuit.**

The circuits in Figures 31.8, 31.10, and 31.12 are all drawn neatly with the circuit elements on a rectangular grid. Real circuits are rarely so neatly laid out, of course, and typically look more like the one shown in **Figure 31.14.** It takes time and concentration to look at the tangle of wires in Figure 31.14 and figure out whether or not the bulbs light up (they do) and identify the path taken by the charge carriers through the circuit.

To analyze such a circuit, therefore, it is helpful to draw a circuit diagram, with elements and wires arranged horizontally and vertically. The circuit diagram must accurately show the connections between elements that are present in the actual circuit, but wires that are not connected to each other should, as far as possible, be drawn so that they do not cross.

To draw a circuit diagram for the circuit in Figure 31.14, we begin by identifying the junctions and the branches connecting the junctions. Note that the two terminals of bulb A are connected to two wires each. Because the terminals of the bulb are also connected to each other via the filament inside the bulb, each of these terminals is a junction. There are three branches. One branch consists of the battery and the wires that connect the left and right contacts of light bulb A. The second branch connects the left contact of light bulb A through the filament to the right contact of light bulb A. The third branch connects the right contact of light bulb A through light bulbs B and C

Figure 31.14 Real circuits don't always look like circuit diagrams.

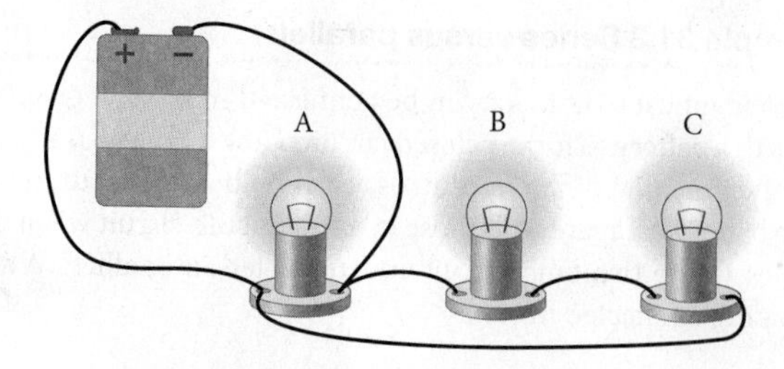

CONCEPTS

Figure 31.15 Circuit diagram for the circuit of Figure 31.14.

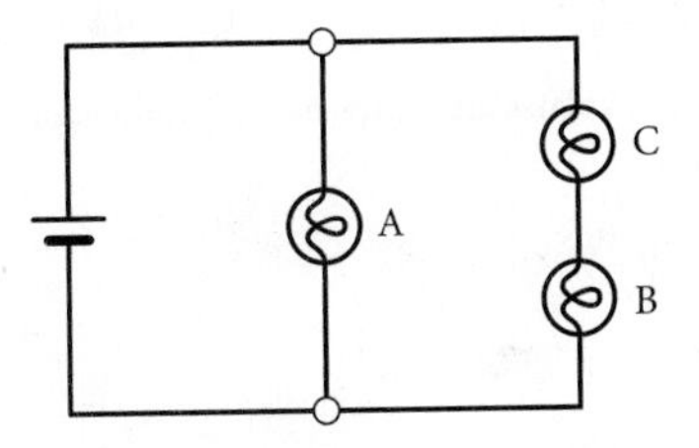

to the left contact of light bulb A. **Figure 31.15** shows a circuit diagram for the circuit in Figure 31.14. Junctions are marked by open circles. The battery and light bulbs have been replaced by the symbols introduced in Figure 31.4. The connecting wires between the circuit elements have been straightened out and replaced by lines.

It is not important that the branch of the circuit containing bulb A appears to the left of the branch containing bulbs B and C. The two-bulb branch could be shown to the left of the branch containing bulb A, and the diagram would still be correct. It is also not important that the branch that contains the battery is shown at the left. There is always more than one way to draw a circuit diagram for any given circuit; any diagram that correctly represents the connections between elements in the circuit is valid.

Exercise 31.4 Bulbs and batteries

Draw a circuit diagram for the arrangement shown in **Figure 31.16.**

Figure 31.16 Exercise 31.4.

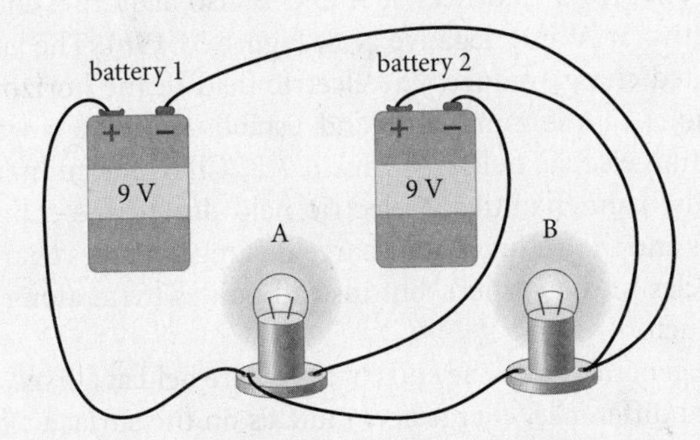

SOLUTION I begin by identifying the junctions and branches. I identify two junctions: one is the left terminal of bulb A and the other is the right terminal of bulb B. I can trace out three branches going from the left terminal of bulb A to the right terminal of bulb B. One branch includes battery 1. The second branch includes bulb A and battery 2, and the third branch includes just bulb B.

I begin by drawing the two junctions. Then I connect these two junctions by each of the three branches, making sure to get the directions of the batteries correct—positive terminal of battery 1 connected to the left terminals of bulbs A and B, and negative terminal of battery 1 connected to the positive terminal of battery 2 and bulb B. This yields the diagram shown in **Figure 31.17*a*.**

Figure 31.17

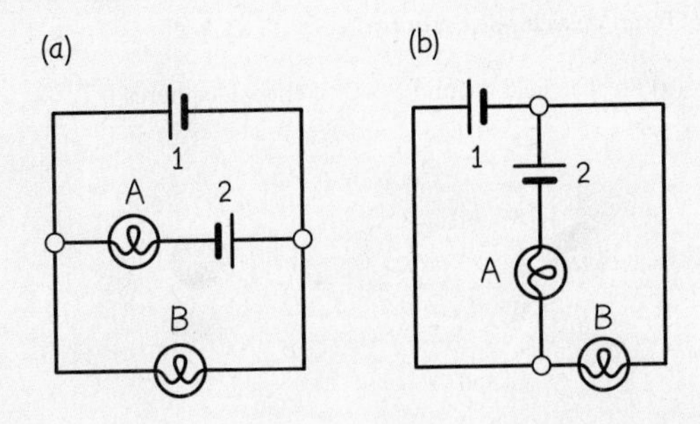

Note that the diagram I drew is not unique. Had I chosen to draw one junction above the other, I might have obtained the diagram shown in Figure 31.17*b*. By comparing the two diagrams, I see that by sliding battery 1 in Figure 31.17*b* to the leftmost vertical branch, sliding bulb B to the rightmost branch, and then rotating the diagram 90° clockwise, I obtain the diagram shown in Figure 31.17*a*.

31.7 If each battery in Figure 31.16 has an emf of 9 V, what is the magnitude of the potential difference across (*a*) bulb A and (*b*) bulb B? (*c*) If the two bulbs are identical, which one glows more brightly?

31.4 Electric fields in conductors

In Section 24.5 we found that the electric field is zero inside a conducting object in electrostatic equilibrium. A conductor through which charge carriers flow is *not* in electrostatic equilibrium, however. (Remember, *electrostatic* means that the arrangement of charged particles is fixed.) To keep charge carriers flowing through a conductor, we need an electric field inside it.

Let us examine more closely the electric field in current-carrying conductors. Consider connecting the two charged spheres of **Figure 31.18*a*** on the next page with a metal rod that is much thinner than the radius of the spheres. Charge carriers can now flow from one sphere to the other along this rod (Figure 31.18*b*). For simplicity, we assume that a power source (not shown in the figure) keeps the charge on each sphere and the potential difference between the two constant.

Initially, the electric field along the axis of the rod points from the positive sphere to the negative sphere. However, the electric field is stronger near the spheres, at A and C in Figure 31.18*b*, than in the middle, at B. Consequently, the charge carriers at A and C get pushed along horizontally more strongly than in the middle, causing positive charge carriers to accumulate between A and B and negative carriers to accumulate between B and C. This accumulation changes the electric field in the rod. For example, as charge carriers accumulate between A and B, the flow of carriers at A is reduced and that at B is enhanced. The carriers stop accumulating when the electric field is the same throughout

Figure 31.18 The source of the electric field in a current-carrying conductor. (For simplicity, we assume that the charge on each sphere is kept constant by an unseen power source.)

(*a*) Electric field around a pair of charged spheres

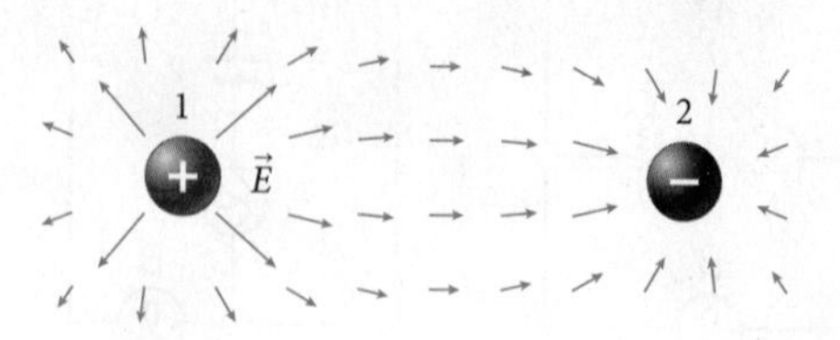

(*b*) Just after spheres are joined by conducting rod

Electric field in rod is smallest in middle (at B).

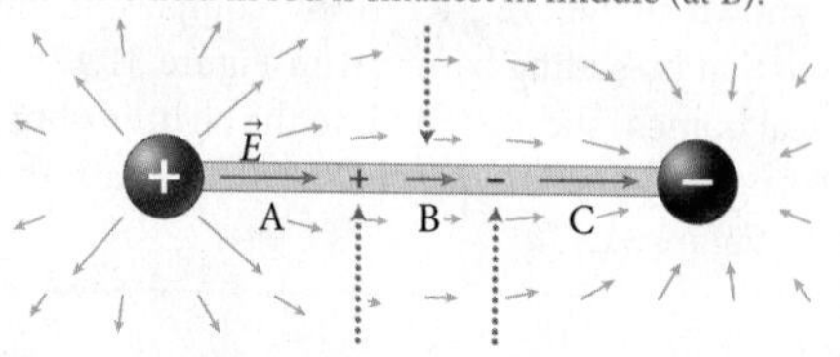

Unequal electric field in rod causes charge to accumulate.

(*c*) Steady state

Accumulation continues until field is equal throughout rod.

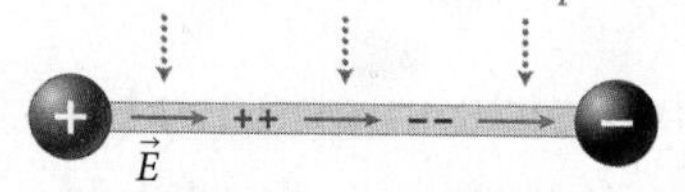

Figure 31.19 Electric field in a bent conductor that connects two charged plates.

(*a*) Just after charged plates are joined by bent conductor

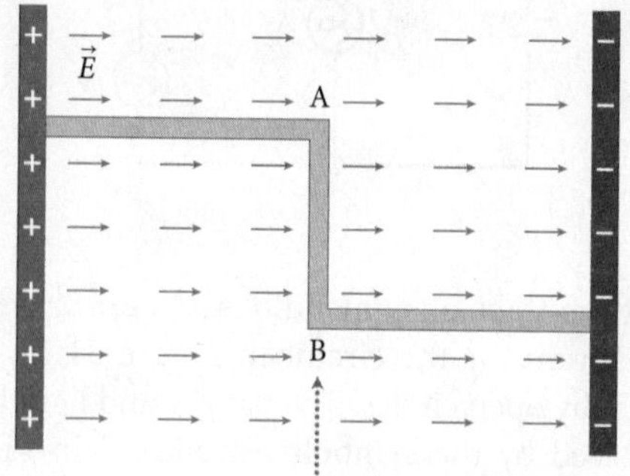

Electric field of capacitor pushes electrons in vertical segment to left.

(*b*) Steady state

Electric field causes charge to accumulate on surfaces of vertical segment.

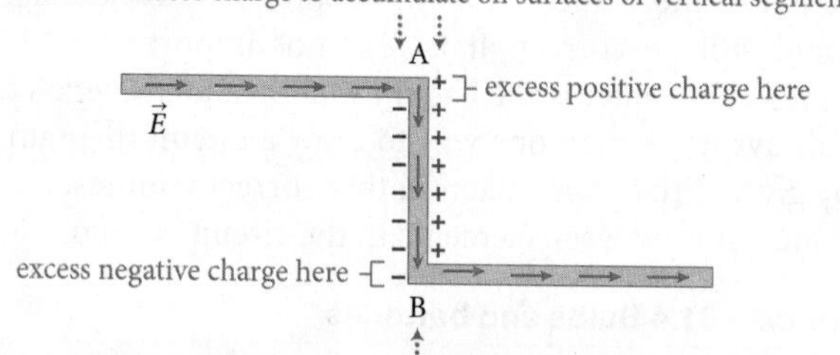

Excess charge at ends causes downward electric field in vertical segment.

the rod (Figure 31.18*c*). Thus, any unevenness in the electric field causes charge to accumulate, and the accumulation, in turn, affects the electric field. This feedback continues until the electric field is uniform and points along the conductor everywhere. In practice, this evening out takes place in an extremely short time interval (10^{-9} s).

31.8 Suppose the distance between the spheres in Figure 31.18 is ℓ and the potential difference between them is V_{12}. What is the magnitude of the electric field inside the connecting rod?

What is the electric field in a bent conductor, as opposed to the straight conductor of Figure 31.18? Consider the conductor connecting the charged plates in **Figure 31.19*a***. When the plates are first connected to each other via a long thin conductor, the only electric field present is that of the plates, and this field pushes electrons everywhere to the left. Because of this, electrons pile up on the left surface of vertical segment AB of the conductors, leaving behind positive ions on the right side (Figure 31.19*b*). The surface of the conductor accumulates charge in this manner until the horizontal component of the electric field is zero within segment AB. The electric field within AB then no longer pushes electrons to the left surface.

While charged particles accumulate on the two sides of segment AB, the corners at A and B also acquire a charge (positive at A and negative at B, Figure 31.19*b*). The accumulated charge reduces the electric field in the horizontal portions of the conductor and establishes a downward-pointing electric field in segment AB. Charge accumulates in this manner until the electric field due to the charged plates and the accumulated charge no longer pushes charged particles to the corners, but instead guides them along the conductor.

In general, when the ends of a wire are held at a fixed potential difference, charge accumulates on the surface of the conductor. The accumulation stops when the electric field due to the combined effects of the surface charge distribution and the applied potential difference has the same magnitude everywhere inside the conductor and points along a path through the conductor that is everywhere parallel to the sides of the conductor. Once this electric field is established, it no longer pushes charge carriers to the surface of the conductor but instead drives a steady flow of charge carriers through the conductor:

In a conductor of uniform cross section carrying a steady current, the electric field has the same magnitude everywhere in the conductor and is parallel to the walls of the conductor.

Keep in mind that because potential differences across conductors are generally small, the electric fields inside conductors are also generally small, even in the presence of substantial currents.

Example 31.5 Bending fields

Consider the three pieces of wire in **Figure 31.20**. All three are made of the same material and have identical circular cross sections. Conductor A has length ℓ; conductors B and C have length 2ℓ. With the wires kept in the configurations shown, a positively charged conducting plate is connected to the left end of each wire and a negatively charged conducting plate is connected to the right end of each wire. Rank the magnitudes of the steady-state electric field at P, Q, R, S, and T.

Figure 31.20 Examples 31.5.

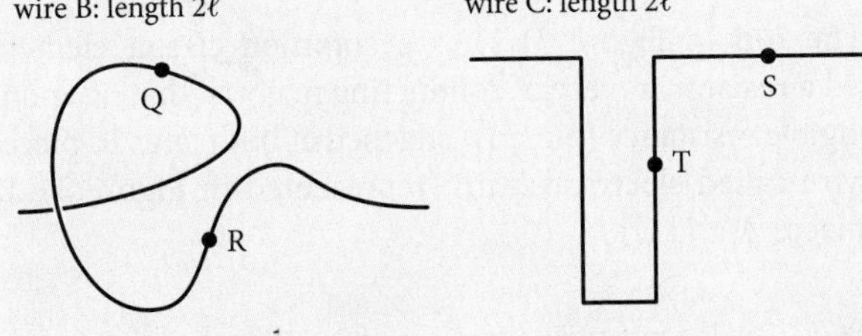

❶ **GETTING STARTED** Once a steady flow of charge carriers has been established in each wire, the electric field in each wire has the same magnitude everywhere in that wire and is always parallel to the walls of the wire.

❷ **DEVISE PLAN** Because the electric field is uniform along each wire, I can use the result I obtained in Example 25.5a (Eq. 1): the magnitude of a uniform electric field between two points is equal to the magnitude of the potential difference between those points divided by the distance between those points.

❸ **EXECUTE PLAN** In each case the magnitude of the potential difference between the ends of the conductor is equal to the potential difference between the plates, which I'll denote by V_{plates}. The electric field magnitude in A is then V_{plates}/ℓ, and that in B and C is $V_{plates}/2\ell$. Therefore, if I use E_P for the electric field magnitude at P, I can say $E_P > E_Q = E_R = E_S = E_T$. ✔

❹ **EVALUATE RESULT** My result is not surprising because I can think of a wire as a load with very small resistance. A wire of length 2ℓ is equivalent to two wires of length ℓ connected in series. Based on the result I obtained in Example 31.3, I expect the current in the wire of length ℓ (analogous to the single-bulb case of Example 31.3) to be twice the current in the wire of length 2ℓ (analogous to the two-bulb series case). To obtain this greater current, I need a greater electric field, as my result shows.

31.9 Sketch the electric field lines inside the conductors in Figures 31.18 and 31.19.

Figure 31.21 (*a*) Conductor of circular cross section that decreases in radius by a factor of 2 at the midpoint. (b) Electric field lines in the conductor when it is connected between two charged objects (not shown).

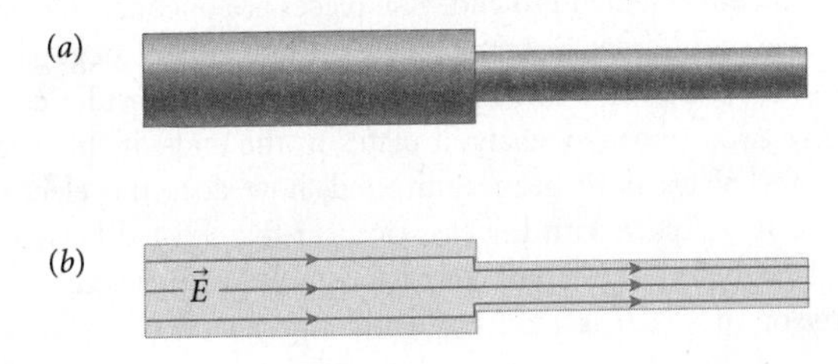

As Example 31.5 shows, the current in wire A is twice the current in wires B and C, which are twice as long. To get the same current in wires B and C as in A, the magnitude of the potential difference across wires B and C must be doubled. This means that the resistance of wires B and C is twice the resistance of wire A. Therefore, the resistance of a conductor is proportional to the length of the conductor.

So far, the only conductors we have considered have all had a uniform cross section. Suppose now that we have a conductor that has a circular cross section that decreases in radius by a factor of 2 in the middle (**Figure 31.21*a***). How do the current and the electric field in this conductor compare in the wide and the narrow parts?

If we think of this conductor as two circuit elements (the wide part and the narrow part) connected in series, we see that the current must be the same through the entire conductor. At steady state, electric field lines inside a conductor in a circuit must satisfy the continuity principle, just as the current must. Therefore the electric field lines in the wide part of our conductor must all continue into the narrow part (Figure 31.21*b*). The density of the electric field lines, and consequently the magnitude of the electric field, must therefore increase by the same factor by which the area decreases. Because the cross-sectional area decreases by a factor of 4, the magnitude of the electric field must increase by a factor of 4.

Now consider the magnitude of the potential difference across each part of the conductor. In each part, the magnitude of the electric field is equal to the magnitude of the potential difference across the part divided by the length of the part (see Example 31.5). Because the electric field is four times greater in the narrow part and because the two parts have the same length, the potential difference across the narrow part must also be four times greater than the potential difference across the wide part.

To get the same current in both parts, four times as much potential difference must be applied to the narrow part! This means that the resistance of the narrow part is four times greater than the resistance of the wide part, even though both are made of the same material. Therefore, the resistance of a conductor not only depends on the material from which it is made and the length of the conductor, but also is inversely proportional to the cross-sectional area of the conductor.

Example 31.6 Electric fields in conductors

Figure 31.22 shows a rod made of three pieces of conducting material connected end to end. The pieces are of equal size but are made of different materials; pieces A and C have negligible resistance, but piece B has significant resistance. Consider connecting two oppositely charged plates to the ends of this rod. After the plates have been connected, how does the electric field in B compare with the electric field that existed between the plates before they were connected to the rod? Assume the charge on the plates is maintained.

Figure 31.22 Example 31.6.

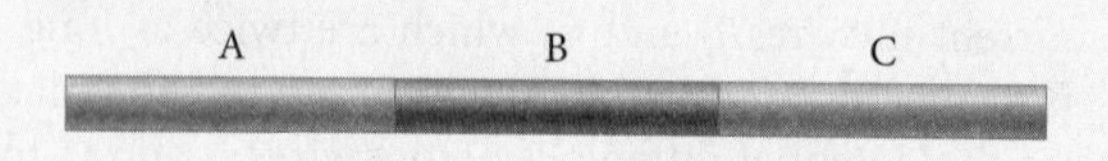

❶ **GETTING STARTED** I begin by making a sketch of the rod connected to the two charged plates (**Figure 31.23**).

Figure 31.23

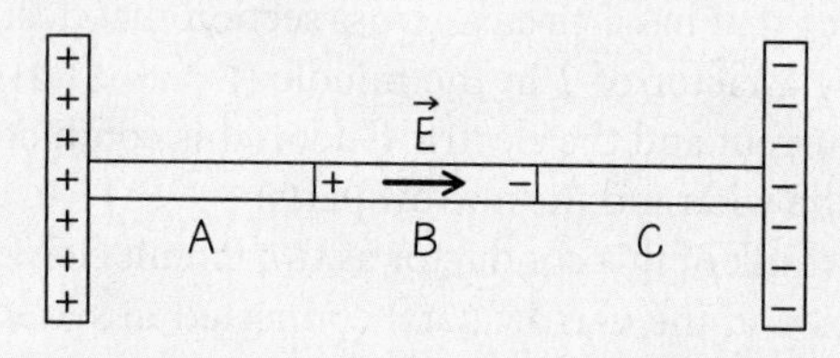

❷ **DEVISE PLAN** As I noted in Example 31.5 in using my result from Example 25.5*a*, the magnitude of the electric field in B is equal to the magnitude of the potential difference across B divided by the length of B. Likewise, the electric field magnitude between the plates is equal to the magnitude of the potential difference between the plates divided by the distance between them. To compare these two field magnitudes, I need to compare the magnitude of the potential difference across B with that across the plates.

❸ **EXECUTE PLAN** Because the resistance of A and C is negligible, the potential difference across A and C is negligibly small; the potential difference across B is therefore essentially the same as that between the plates. Because B is one-third as long as the distance between the plates, the electric field magnitude in B is roughly three times the electric field magnitude that existed between the plates before connection. ✔

❹ **EVALUATE RESULT** Because A and C have negligible resistance, I expect the charge to distribute itself along the rod as shown in Figure 31.23, with positive charge residing at the end of B nearer the positively charged plate and negative charge residing at the other end. This distribution of charge reduces the electric field in A and C and increases the electric field in B, as I concluded.

The rod in Figure 31.22 is a common circuit element called a **resistor,** a piece of conducting material that has non-negligible resistance (piece B) attached at both ends to pieces of wire called electrical *leads* (represented in Figure 31.22 by pieces A and C).

31.10 If a potential difference of 9 V is applied across the conductor in Figure 31.21, what is the magnitude of the potential difference (*a*) across the wide part and (*b*) across the narrow part?

Self-quiz

1. Why are none of the bulbs in **Figure 31.24** lit?

Figure 31.24

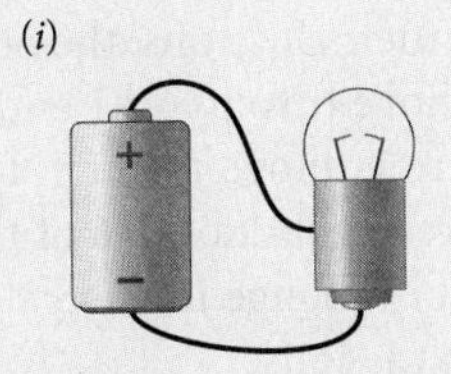

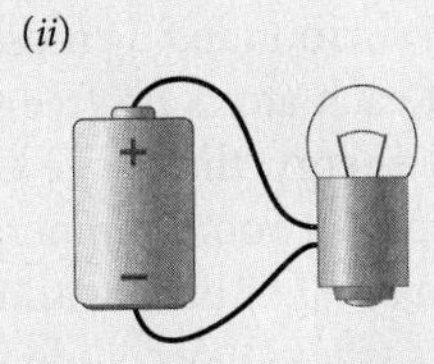

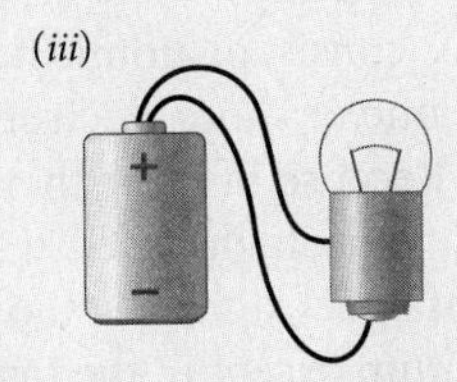

Figure 31.25

heating coil

A B D C

2. In **Figure 31.25**, identify the energy conversions that occur between points A and B, B and C, C and D, and D and A.
3. In **Figure 31.26**, bulb B is brighter than bulb C, which in turn is brighter than bulb A. Rank, largest first, (*a*) the magnitudes of the potential differences across the bulbs, (*b*) the currents in them, and (*c*) their resistances.

Figure 31.26

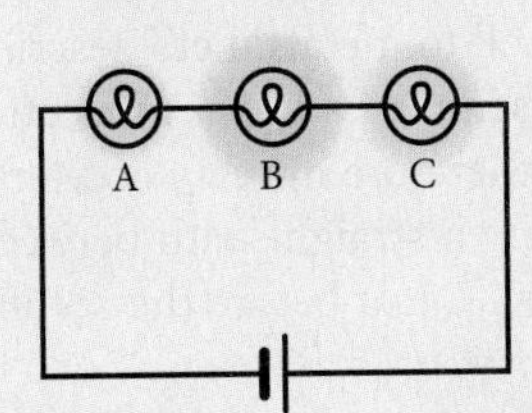

Figure 31.27

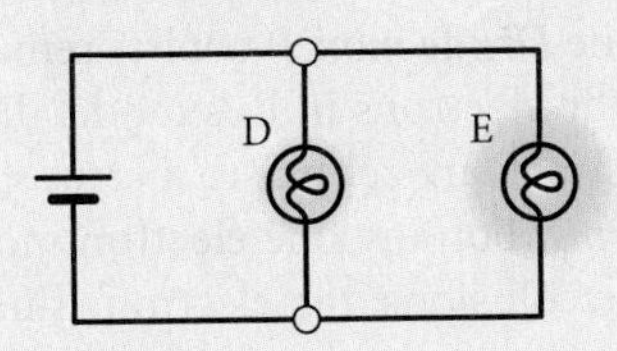

Figure 31.28

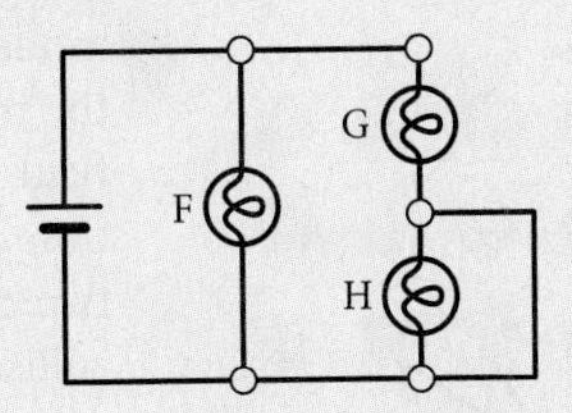

4. In **Figure 31.27**, bulb E is brighter than bulb D. Which bulb has (*a*) the greater potential difference across it, (*b*) more current in it, and (*c*) more resistance?
5. In **Figure 31.28**, which of the three bulbs F, G, and H light up?
6. A thick resistor and a thin resistor of the same length and material are connected in series, as shown in **Figure 31.29**. Which resistor has (*a*) the greater potential difference across it and (*b*) the greater resistance?

Figure 31.29

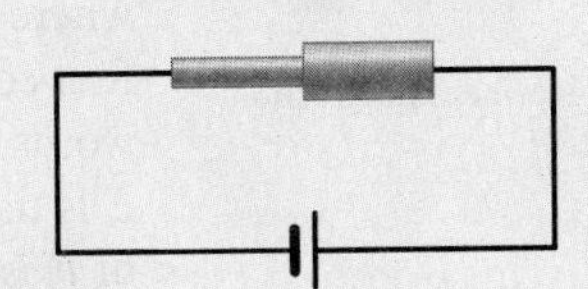

Answers

1. (*a*) Filament broken. (*b*) Battery connected to only one contact on bulb. (*c*) Bulb connected to only one battery terminal.
2. AB: electric potential energy converted to thermal energy; CD: chemical energy converted to electric potential energy; BC and DA: negligible conversion of energy.
3. (*a*) $V_B > V_C > V_A$ because the brightest bulb converts the most energy. (*b*) $I_B = I_C = I_A$ because the bulbs are in series and the current is the same throughout the circuit. (*c*) $R_B > R_C > R_A$ because, for a given current in a circuit element, R is proportional to V.
4. (*a*) $V_D = V_E$ because the bulbs are in parallel. (*b*) $I_E > I_D$ because E is brighter. (*c*) $R_D > R_E$ because less current passes through D.
5. F and G light up, but H does not because the wire in parallel with H carries all of the current through the right branch. (The wire "shorts out" H.)
6. (*a*) $V_{thin} > V_{thick}$ because the density of electric field lines through the thin resistor has to be greater, and if the lengths of the resistors are the same and E is greater, then V is greater. (*b*) $R_{thin} > R_{thick}$ because resistance is inversely proportional to cross-sectional area.

Figure 31.30 The effect of an applied electric field on the motion of a free electron through a lattice of ions.

(*a*) Motion in absence of an electric field

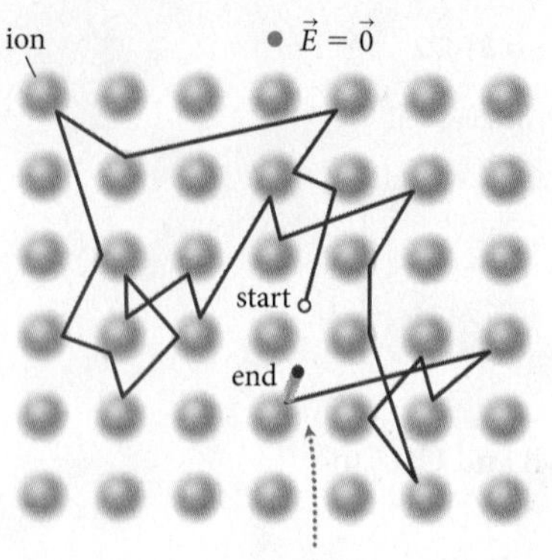

Electron's displacement is zero over long time interval.

(*b*) Motion with applied electric field

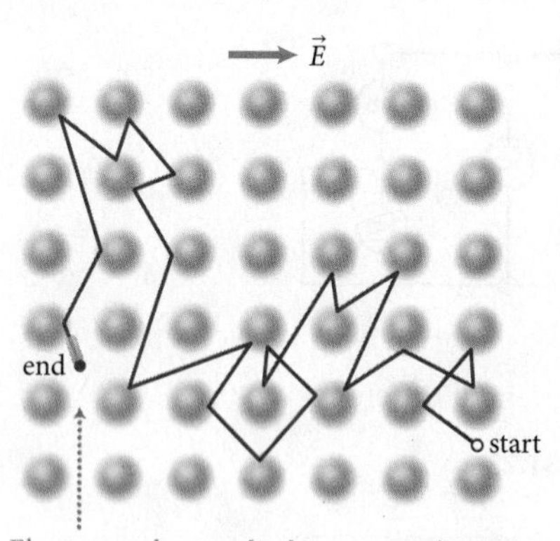

Electron undergoes displacement in direction opposite to electric field.

31.5 Resistance and Ohm's law

Recall from Section 22.3 that a metal consists of a lattice of positively charged ions through which electrons can move relatively freely. Although the lattice positions are fixed (meaning there is no motion of ions through the metal), the ions vibrate around their lattice positions, as if they were connected to those positions by springs. (The amplitude of the vibrations depends on temperature.) The ions consist of atomic nuclei surrounded by most of their electrons; the outermost one or more electrons of each atom are free to move through the entire lattice. Because of their thermal energy, these free electrons move through the lattice at very high speeds (10^5 m/s at room temperature, or about 0.1% of the speed of light), and they move in straight lines without any change in energy or momentum except in the instants when they collide with the ions.

In the absence of an electric field and over a time interval long enough for many collisions to take place, the displacement of an electron is very small in spite of its high speed (**Figure 31.30*a***). This is true because each collision changes the direction of the electron's motion, randomizing the direction of the electron's velocity. Consequently, the average velocity of all the electrons is zero. However, when an electric field is applied, as in Figure 31.30*b*, the electric field causes the electrons to accelerate in the direction opposite to the electric field.

A quantitative description of the motion of charge carriers in a conductor was given by P. K. Drude in 1900, shortly after the discovery of the electron. His model (called the *Drude model*) applies remarkably well to metals. Let's describe the motion of the electrons in this model. In the presence of a uniform electric field $\vec{E}$, the electrons are subject to a force $-e\vec{E}$ and therefore have an acceleration $\vec{a} = -e\vec{E}/m_e$. For any one electron moving along a straight path between two consecutive collisions, the electron's final velocity $\vec{v}_f$ just before the second collision is

$$\vec{v}_f = \vec{v}_i + \vec{a}\Delta t = \vec{v}_i - \frac{e\vec{E}}{m_e}\Delta t, \tag{31.1}$$

where $\vec{v}_i$ is the electron's initial velocity on that path (its velocity just after the first collision), Δt is the time interval the electron spends on that path (in other words, the time interval between collisions), e is the elementary charge (Eq. 22.3), $\vec{E}$ is the applied electric field, and m_e is the mass of the electron. The magnitude of $\vec{v}_i$ is roughly 10^5 m/s, as noted above, and the direction is determined by the first of the two collisions. Because of the high electron speed, the time interval Δt between collisions is extremely short—on the order of 10^{-14} s.

To calculate the average velocity of all the electrons, we take the average of Eq. 31.1 for all of the electrons:

$$(\vec{v}_f)_{av} = (\vec{v}_i)_{av} - \frac{e\vec{E}}{m_e}(\Delta t)_{av}. \tag{31.2}$$

(I have assumed here that the electric field is either constant over time or takes a time interval much longer than Δt to change significantly.)

Even though the magnitude of $\vec{v}_i$ is quite large, its average value for all electrons is zero because the collisions produce a random distribution of the directions of the initial velocities. The resulting average velocity, called the **drift velocity** $\vec{v}_d$ of the electrons, is thus

$$\vec{v}_d = -\frac{e\vec{E}}{m_e}\tau, \tag{31.3}$$

where $\tau \equiv (\Delta t)_{av}$ is the average time interval between collisions. (The value of τ depends on the number density, size, and charge of the lattice ions, and on temperature.) The magnitude of the drift velocity is called the *drift speed*. Equation 31.3 shows that the drift velocity of the electrons is in a direction opposite that of the electric field, and the drift speed is proportional to the electric field magnitude.

31.11 (*a*) Does the electric field do work on the electrons of Figure 31.30*b* as they accelerate between collisions? (*b*) On average, does the kinetic energy of the electrons increase as they drift through the lattice? (*c*) What do your answers to parts *a* and *b* imply about the energy in the lattice?

In Chapter 27 we found that the current in a conductor can be expressed in terms of the speed v of the charge carriers (Eq. 27.16). Because the speed of the charge carriers is what we are now calling the drift speed, we can write Eq. 27.16 in the form

$$I = nAqv_{\mathrm{d}}, \tag{31.4}$$

where A is the cross-sectional area of the conductor, q is the charge on each charge carrier in the conductor, and n is the number of carriers per unit volume in the conductor. Because of the current's dependence on cross-sectional area, it is convenient to introduce the current per unit area, called the **current density**, whose magnitude is given by the magnitude of the current per unit area:

$$J \equiv \frac{|I|}{A} = n|q|v_{\mathrm{d}}. \tag{31.5}$$

Because the drift velocity is a vector, the current density is a vector, too:

$$\vec{J} = nq\vec{v}_{\mathrm{d}}. \tag{31.6}$$

The direction of the current density is the same as that of the drift velocity for positive charge carriers and opposite the direction of the drift velocity for negative charge carriers. Therefore the current density is always in the same direction as the current. The SI unit of current density is $\mathrm{A/m^2}$.

Substituting the absolute value of the right side of Eq. 31.3 for v_{d} and e for $|q|$ in Eq. 31.5 yields

$$J = n(e)\left(\frac{eE}{m_{\mathrm{e}}}\tau\right) = \frac{ne^2\tau}{m_{\mathrm{e}}}E. \tag{31.7}$$

Equation 31.7 shows that the current density is proportional to the applied electric field. The proportionality constant σ is called the **conductivity** of the material of which the conductor is made:*

$$\sigma \equiv \frac{J}{E}. \tag{31.8}$$

*Note that conductivity is *not* the same as surface charge density, which is represented by the same symbol.

Table 31.1 Conductivities of various materials at room temperature $A/(V \cdot m)$

Conductors	Silver	6.3×10^7
	Copper	5.9×10^7
	Aluminum	3.6×10^7
	Tungsten	1.8×10^7
	Nichrome	6.7×10^5
	Carbon	7.3×10^4
Semiconductors	Silicon	4×10^{-4}
	Germanium	2
Poor conductors	Seawater	4
Insulators	Pure water	4.0×10^{-6}
	Glass	10^{-12}

The SI unit of conductivity is equal to $(A/m^2)/(V/m) = A/(V \cdot m)$. The conductivity is a measure of a material's ability to conduct a current for a given applied electric field. The conductivity is a property of the material and is therefore the same for any piece of that material you might choose.

Table 31.1 gives the conductivities of some common materials at room temperature. Note that the first four materials, all metals, have very similar conductivities. Nichrome is an alloy of nickel and chromium used in heating elements because of its relatively low conductivity. Silicon is a *semiconductor;* its conductivity is intermediate between that of an electrical conductor and that of an electrical insulator. The conductivities of insulators (such as pure water and glass) are many orders of magnitude smaller than those of metals. Seawater has a greater conductivity than pure water because the ions dissolved in seawater serve as charge carriers for a current, just as the ions in the electrolyte of a battery do.

Comparing Eqs. 31.7 and 31.8 gives us an expression for the conductivity of a metal in terms of the average time interval between collisions, the mass and charge of the electrons, and the number density of electrons present in the material:

$$\sigma = \frac{ne^2\tau}{m_e}. \tag{31.9}$$

Because of the temperature dependence of τ, the conductivity σ depends on temperature and, for some materials, on the magnitude of the current in the material, but it is independent of the shape of the piece of material in which the current density is measured.

We obtained Eq. 31.9 for a metal, in which the charge carriers are electrons, but it applies to any system in which the charge carriers in a conducting material move freely between collisions. To generalize Eq. 31.9 for charge carriers other than electrons, simply substitute q^2 for e^2, and m_q for m_e, where q is the charge on the carriers and m_q is the mass of one carrier.

The conductivity describes how large a current density and hence current are created by an external electric field. Such an electric field is produced by applying a potential difference across a material. In Section 31.2, we looked at resistance as a way of relating the applied potential difference V across a circuit element to the resulting current in the element. In general, the resistance of any element is defined to be

$$R \equiv \frac{V}{I}. \tag{31.10}$$

The derived SI unit of resistance is the **ohm** ($1\ \Omega \equiv 1\ \text{V/A}$). Resistance is always positive, so in Eq. 31.10 the direction in which V and I are measured must be such that they both have the same algebraic sign. The resistance of most circuit elements is typically in the range of 10 Ω to 100,000 Ω.

The concept of resistance is useful in describing conductors and other objects that provide a continuous conducting path for charge carriers, such as filaments, bulbs, and resistors. (It is not especially useful for other types of circuit elements, such as capacitors.) For some conducting materials, the resistance R of a piece of this material is fixed at a given temperature. For these materials, Eq. 31.10 indicates that the current in the conductor is proportional to the potential difference across it and inversely proportional to the resistance:

$$I = \frac{V}{R}. \quad (31.11)$$

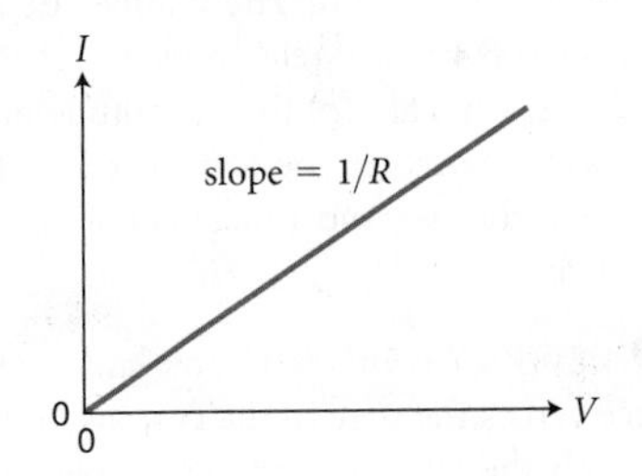

Figure 31.31 Current versus applied potential difference for an ohmic conductor. The current is proportional to the potential difference.

Such materials are said to be *ohmic*. If we plot the magnitude of the current in a piece of this material as a function of the potential difference across it, the result is a straight line whose slope is equal to the inverse of the resistance R of the piece (**Figure 31.31**).

Equation 31.11 is often referred to as **Ohm's law.** It is important to keep two things in mind about Eq. 31.11. First, many materials and many circuit elements are not ohmic. For such materials and elements, a plot of current as a function of applied potential difference is not a straight line, but has a more complicated shape, indicating that R depends on the potential difference. Second, Eq. 31.11 is really a definition, not a law. It simply amounts to the observation that in certain materials, the current is proportional to the applied potential difference. In this chapter, we concern ourselves only with ohmic materials and circuit elements.

Let's now relate the resistance of a conductor to the conductivity of the material of which it is made. From Checkpoint 31.8 we know that the magnitude of the potential difference across a wire of length ℓ is given by (Eq. 25.25 and Example 25.5):

$$|V| = E\ell. \quad (31.12)$$

Combining Eqs. 31.5 and 31.8, we see that $J \equiv |I|/A = \sigma E$, so $E = |I|/(\sigma A)$. Substituting this expression into Eq. 31.12 and dropping the absolute-value symbol, we obtain a relationship between V and I:

$$V = \frac{I}{\sigma A}\ell = I\frac{\ell}{\sigma A}. \quad (31.13)$$

Substituting this expression into the definition of resistance, Eq. 31.10, we obtain for ohmic conductors,

$$R = \frac{\ell}{\sigma A}. \quad (31.14)$$

Equation 31.14 shows that the resistance of a conductor not only depends on the material from which it is made (through the conductivity σ) but also is proportional to the length ℓ of the conductor and inversely proportional to the cross-sectional area, as we had concluded in Section 31.4.

Example 31.7 Drifting electrons

Consider a piece of copper wire that is 10 m long and has a diameter of 1.0 mm. The number density of free electrons in copper is 8.4×10^{28} electrons/m^3. If the wire carries a current of 2.0 A, what are (*a*) the magnitude of the potential difference across the wire, (*b*) the drift speed of the electrons in the wire, and (*c*) the average time interval between collisions for the electrons?

❶ GETTING STARTED Ohm's law relates the potential difference across the wire to the current in it and its resistance. Even though the wire is made of copper, it has a finite resistance, which depends on the conductivity of copper and on the length and cross-sectional area of the wire. This resistance is caused by collisions between the electrons and copper ions in the wire, and these collisions limit the drift speed of the electrons through the wire. The drift speed is related to the average time interval between collisions, which is one of the parameters I need to calculate and appears in the expression for conductivity in the Drude model, Eq. 31.3, $v_d = eE\tau/m_e$.

❷ DEVISE PLAN The potential difference across the wire is related to the current by Eq. 31.11, but in order to use this equation, I must determine the wire's resistance, which is given by Eq. 31.14. I can look up the conductivity of copper in Table 31.1 ($\sigma = 5.9 \times 10^7$ A/V · m). To obtain the drift speed of the electrons, I can use Eq. 31.4, which contains the charge on the electron ($q = e = 1.6 \times 10^{-19}$ C), the current (given), the number density of the electrons (given), and the cross-sectional area of the wire, which I can calculate from the given diameter. To obtain the average time interval between collisions, I can use Eq. 31.9, which contains the mass of the electron ($m_e = 9.11 \times 10^{-31}$ kg).

❸ EXECUTE PLAN (*a*) I obtain the resistance of the wire from Eq. 31.14 and the cross-sectional area of the wire, $A = \pi r^2 = \pi(5.0 \times 10^{-4}\text{ m})^2 = 7.9 \times 10^{-7}\text{ m}^2$:

$$R_{\text{wire}} = \frac{\ell}{\sigma A} = \frac{10\text{ m}}{(5.9 \times 10^7\text{ A/V}\cdot\text{m})(7.9 \times 10^{-7}\text{ m}^2)} = 0.21\ \Omega.$$

Now that I have R_{wire}, I can obtain V_{wire} from Eq. 31.11:

$$V_{\text{wire}} = IR_{\text{wire}} = (2.0\text{ A})(0.21\ \Omega) = 0.42\text{ V.} ✔$$

(*b*) I first obtain the drift speed of the electrons from Eq. 31.4:

$$v_d = \frac{I}{neA} = \frac{2.0\text{ C/s}}{(8.4 \times 10^{28}\text{ m}^{-3})(1.6 \times 10^{-19}\text{ C})(7.9 \times 10^{-7}\text{ m}^2)}$$
$$= 1.9 \times 10^{-4}\text{ m/s.} ✔$$

(*c*) Solving Eq. 31.9 for τ, I get

$$\tau = \frac{m_e\sigma}{ne^2} = \frac{(9.11 \times 10^{-31}\text{ kg})(5.9 \times 10^7\text{ A/V}\cdot\text{m})}{(8.4 \times 10^{28}\text{ m}^{-3})(1.6 \times 10^{-19}\text{ C})^2}$$
$$= 2.5 \times 10^{-14}\text{ s.} ✔$$

❹ EVALUATE RESULT Because the conductivity of copper is high, it makes sense that I obtain a small resistance and consequently a small potential difference across the wire even though it is 10 m long. My answer to part *b* shows that the drift speed of the electrons in the wire is very small. The drift speed indicates that it takes the electrons about 5 s to move just 1 mm, even though I know from experience that when I turn on a light switch, the light turns on instantly even if the bulb is meters away from the switch. The reason the light bulb comes on almost instantaneously, however, is that the current is the same throughout the circuit—all of the electrons throughout the circuit are set in motion almost simultaneously when I flip the switch.

How can my very small calculated value for drift speed be reasonable when the current is a significant 2.0 A = 2.0 C/s at any given location in the wire? The reason is that the number density of electrons in the wire is very high. Although it takes 5 s for an electron to move 1 mm, there are nearly 10^{20} electrons per cubic millimeter of the wire. This means that on the order of 10^{20} electrons, or 10 coulombs of charge, pass a given location in 5 s—hence the current is 2 C/s, or 2 A.

The very high number density of the electrons is also responsible for the extremely short time interval between collisions. If I imagine the 10^{20} electrons per cubic millimeter arranged on a cubic lattice where each cube of the lattice has a side length of 1.0 mm, there are about 10^7 electrons along each side of the cube, and so the average distance between them is about 10^{-10} m. As the electrons move at about 10^5 m/s (see Section 31.5), they cover this average distance in about 10^{-15} s, which is within an order of magnitude of what I obtained.

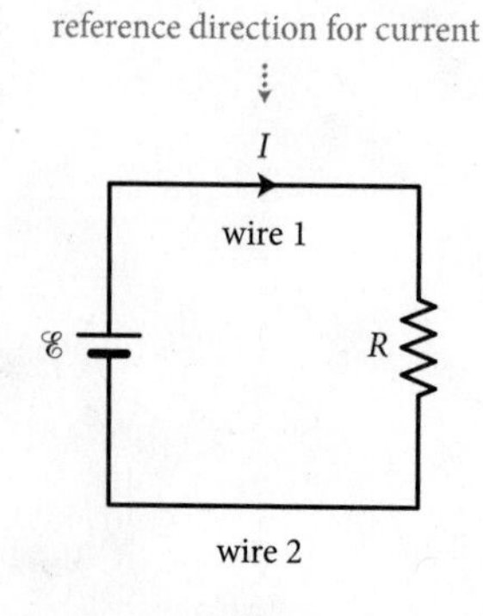

Figure 31.32 Circuit diagram for a battery connected in series with a single resistor.

31.12 If the temperature of a metal is raised, the amplitude of the vibrations of the metal-lattice ions increases. (*a*) What effect, if any, do you expect these greater vibrations to have on the resistance of a piece of that metal? (*b*) What effect does running a current in a metal have on the temperature of the metal? (*c*) Make a graph of current versus potential difference, taking into account the effect you described in part *b*.

31.6 Single-loop circuits

In this section, we quantitatively analyze single-loop circuits. We begin with the simplest possible circuit: a single battery connected to a single resistor, obeying Ohm's law (**Figure 31.32**). The emf of the battery is ℰ, and the resistance of the resistor is *R*.

QUANTITATIVE TOOLS

The emf of the battery establishes an electrostatic field that drives a current in the circuit. In steady state, this current is the same at all locations in the circuit, as we discussed in the first part of this chapter. For a single-loop circuit we can express this condition quantitatively as

$$I_{\text{battery}} = I_{\text{wire}} = I_{\text{resistor}} = I. \quad (31.15)$$

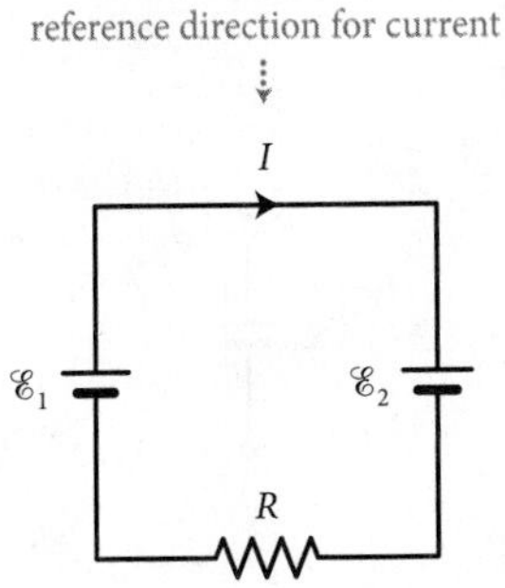

Figure 31.33 Checkpoint 31.13.

In Section 27.3 we called the direction of decreasing potential in a conducting object the *direction of current,* and we indicated this direction by an arrow labeled I next to the object carrying the current (Figure 27.18). When working with circuits, however, we generally won't know the direction of current in advance. To analyze the circuit we therefore need to choose a *reference direction for the current* in each circuit branch. We indicate this direction by an arrowhead on a wire in the circuit (Figure 31.32) and label that arrowhead with a symbol for the current in that branch. In a single-loop circuit, there are no junctions and therefore only one current. As the next checkpoint shows, the reference direction for the current and the direction of current need not be the same. As we shall see later, we obtain a positive value for I when the direction of current and the reference direction for the current are the same. When these two directions are opposite, we find that $I < 0$.

31.13 If $\mathscr{E}_1 < \mathscr{E}_2$ in **Figure 31.33**, is the direction of current the same as the reference direction for the current indicated in the diagram?

Let's next consider the energy transformations that occur in the circuit shown in **Figure 31.34**. Nonelectrostatic work done on the negative charge carriers as they travel from point a to point b through the source raises the electric potential energy of the charge carriers. In the load, this electric potential energy is converted to other forms of energy, such as thermal energy, mechanical energy, radiation energy:

$$\Delta E_{\text{other}} = W_{\text{nonelectrostatic}}(\text{a} \rightarrow \text{b}). \quad (31.16)$$

Figure 31.34 Choosing a direction of travel for analyzing a single-loop circuit.

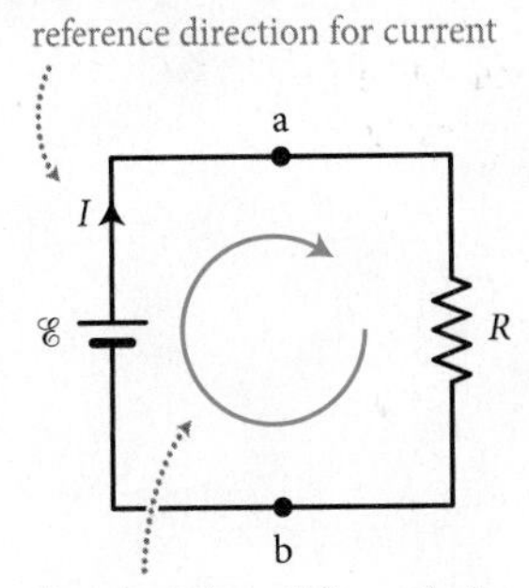

Considering just the load by itself, we also know that the amount of energy converted to other forms of energy must be equal to the work done by the electrostatic field on the charge carriers as they travel from point b to point a through the load:

$$\Delta E_{\text{other}} = W_{\text{electrostatic}}(\text{b} \rightarrow \text{a}), \quad (31.17)$$

so

$$W_{\text{nonelectrostatic}}(\text{a} \rightarrow \text{b}) = W_{\text{electrostatic}}(\text{b} \rightarrow \text{a}). \quad (31.18)$$

Using Eqs. 26.7 and 25.15 this becomes

$$q\mathscr{E} = -qV_{\text{ba}} \quad (31.19)$$

or

$$\mathscr{E} + V_{\text{ba}} = 0. \quad (31.20)$$

For circuits that contain many elements, Eq. 31.20 can be generalized by replacing each term by a sum. In that case the algebraic sum of the emfs and the potential differences around the loop is zero:

$$\Sigma\mathscr{E} + \Sigma V = 0 \quad \text{(steady state, around loop)}. \quad (31.21)$$

Equation 31.21 is called the **loop rule.**

When evaluating the sum on the left in Eq. 31.21, we need to pay close attention to the signs of the emfs and potential differences. We begin by choosing a *direction of travel* around the loop. We denote this direction of travel by a curved

Figure 31.35 Potential differences across batteries, resistors, and capacitors.

(*a*) Battery

direction of travel

$a \to b$ $\mathscr{E}$ $b \to a$

$V_{ab} = +\mathscr{E} < 0$ $V_{ba} = -\mathscr{E} > 0$

(*b*) Resistor

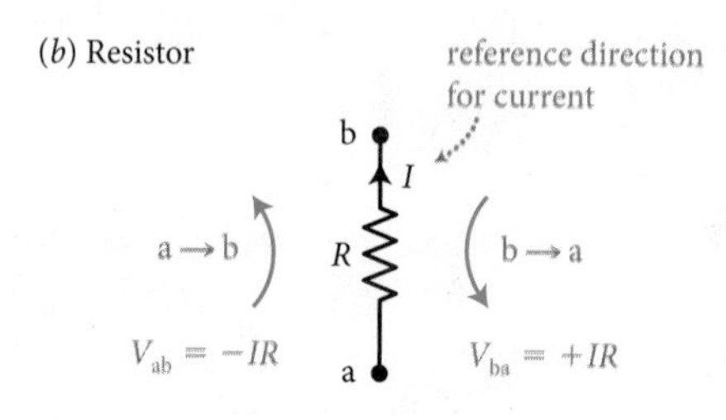

(*c*) Capacitor

$a \to b$ C $b \to a$

$V_{ab} = +q(t)/C > 0$ $V_{ba} = -q(t)/C < 0$

arrow at the center of the loop (Figure 31.34). As we shall see shortly, this choice does not affect the end result—we must just be sure to use the same direction for all elements in the loop. In single-loop circuits it makes sense to let the travel direction be the same as the reference direction for the current. In multiloop circuits, however, it is generally not possible to let both directions coincide.

Because the potential is the same everywhere along any of the wires, we only need to consider the potential differences across circuit elements: Each circuit element contributes one term to the sum. For a battery, the potential difference is positive when traveling from the negative terminal to the positive terminal because the potential of the positive terminal is higher than that of the negative terminal (**Figure 31.35*a***). Conversely, the potential difference is negative when traveling in the opposite direction. If the battery is ideal, the magnitude of the potential difference is equal to the emf $\mathscr{E}$ of the battery (Eq. 26.8).

The sign of the potential difference across a resistor depends on both the choice of current reference direction and the choice of travel direction (Figure 31.35*b*). When traveling in the same direction as the reference direction for the current, the potential difference is $-IR$. To see why this is so, let us assume that in Figure 31.35*b* the current is positive (that is, the current is in the reference direction for the current). Because the direction of current is from higher potential to lower potential, we see that the potential at b must be lower than at a, $V_b < V_a$, and therefore $V_{ab} = V_b - V_a < 0$. Indeed when $I > 0$ the potential difference $V_{ab} = -IR$ is negative. When traveling in the opposite direction, the potential difference is $+IR$.

Next we examine the potential difference across a capacitor (Figure 31.35*c*). The potential difference is positive when traveling from the negatively charged plate to the positively charged one because the potential of the positive terminal is higher than that of the negative terminal; the potential difference is negative when traveling in the opposite direction. The magnitude of the potential difference can be obtained from Eq. 26.1: $C \equiv q/V_{cap}$, where C is the capacitance and q is the magnitude of the charge on the plates. It is important to keep in mind that the charge on the plate and the potential difference typically vary as a function of time as the capacitor charges or discharges. At any given instant, the magnitude of the potential difference across the capacitor is

$$V_{cap}(t) = \frac{q(t)}{C}, \tag{31.22}$$

where $q(t)$ is the magnitude of the charge on the plates at that instant. Because of this time-dependence, the current also depends on time. When solving Eq. 31.21 for circuits that contain capacitors, remember that the time-dependent current $I(t)$ and $q(t)$ are related by Eq. 27.2: $I(t) \equiv dq/dt$.

The Procedure box and **Table 31.2** on the next page summarize how to apply the loop rule to a single-loop circuit. Let us return to the circuit in Figure 31.34 and apply this procedure to obtain a relationship between the current and the emf. The figure already indicates a reference direction for the current (the arrowhead labeled I) and a direction of travel (the clockwise circular arrow). We shall begin our analysis at the bottom left corner. If we go clockwise, the first circuit element is the battery; because we are traveling from the negative to the positive terminal, the potential difference is $+\mathscr{E}$ (Figure 31.35*a* and Table 31.2). Next is the resistor; because we are traveling in the same direction as the reference direction for the current, the potential difference is $-IR$. Substituting these values in Eq. 31.21, we get

$$+\mathscr{E} - IR = 0 \tag{31.23}$$

or

$$I = \frac{\mathscr{E}}{R}. \tag{31.24}$$

Procedure: Applying the loop rule in single-loop circuits

When applying the loop rule to a single-loop circuit consisting of resistors, batteries, and capacitors, we need to make several choices in order to calculate the current or the potential difference across each circuit element.

1. Choose a reference direction for the current in the loop. (This direction is arbitrary and may or may not be the direction of current, but don't worry, things sort themselves out in step 4.) Indicate your chosen reference direction by an arrowhead, and label the arrowhead with the symbol for the current (I).
2. Choose a direction of travel around the loop. This choice is arbitrary and separate from the choice of the reference direction for the current in step 1. (You may want to indicate the travel direction with a circular clockwise or counterclockwise arrow in the loop.)
3. Start traversing the loop in the direction chosen in step 2 from some arbitrary point on the loop. As you encounter circuit elements, each circuit element contributes a term to Eq. 31.21. Use Table 31.2 to determine the sign and value of each term. Add all terms to obtain the sum in Eq. 31.21. Make sure you traverse the loop completely.
4. Solve your expression for the desired quantity. If your solution indicates that $I < 0$, then the direction of current is opposite the reference direction you chose in step 1.

Table 31.2 Signs and values of potential differences across batteries and resistors (Figure 31.35)

Circuit element	Plus sign when traversing	Value
ideal battery	from − to +	$\mathscr{E}$
capacitor	from − to +	$q(t)/C$
resistor	opposite reference direction of current	IR

31.14 In the analysis of the circuit in Figure 31.34, we chose a clockwise reference direction for the current and a clockwise direction of travel. Redo the analysis using (*a*) a clockwise reference direction for the current and a counterclockwise direction of travel and (*b*) a counterclockwise reference direction for the current and a clockwise direction of travel.

Exercise 31.8 Series resistors

Consider the circuit shown in **Figure 31.36**, containing two resistors with resistances R_1 and R_2 and a battery with an emf $\mathscr{E}$. Determine the current in the circuit in terms of R_1, R_2, and $\mathscr{E}$.

Figure 31.36 Exercise 31.8.

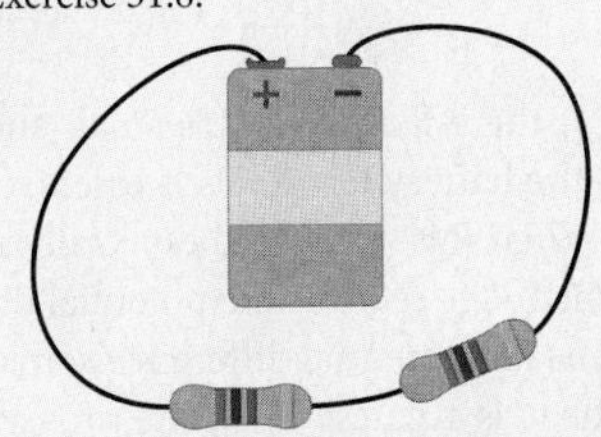

SOLUTION I begin by making a diagram, labeling the points a, b, and c, and choosing a clockwise reference direction for the current and a clockwise direction of travel (**Figure 31.37**). If I begin at point a, as I travel from the negative to the positive terminal, the potential difference is $+\mathscr{E}$. As I travel across resistor 1, I am traveling in the same direction as the reference direction for the current, so the potential difference is $-IR_1$. Next I travel across resistor 2 in the same direction as the reference direction

Figure 31.37

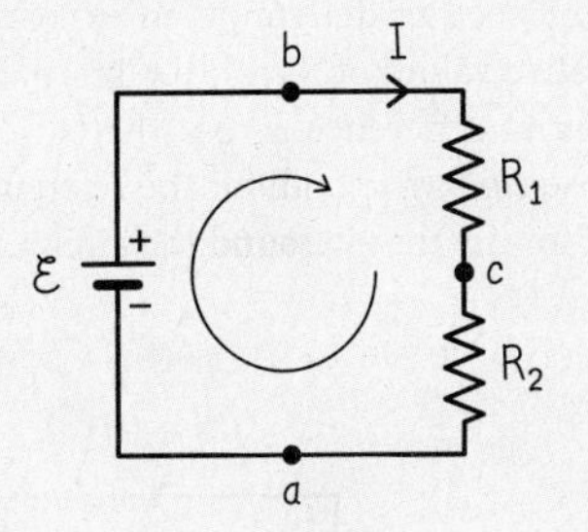

for the current, so the potential difference is $-IR_2$. I now add all these terms and set the sum to zero (Eq. 31.21):

$$+\mathscr{E} - IR_1 - IR_2 = 0.$$

Solving this equation for the current yields

$$I = \frac{\mathscr{E}}{R_1 + R_2}.$$

Comparing the result of Exercise 31.8 with Eq. 31.11, we see that the effect on the current of placing two (or more) resistors in series is equivalent to adding the resistances, as we had concluded in Section 31.2. In other words, the same current would exist in the circuit if we replaced the two resistors in series with a single resistor having resistance

$$R_{eq} = R_1 + R_2. \quad (31.25)$$

By following the same line of reasoning, I can show that for more than two resistors in series, this result generalizes to

$$R_{eq} = R_1 + R_2 + R_3 + \cdots \quad \text{(resistors in series)}. \quad (31.26)$$

The resistance that could be used to replace a combination of circuit elements without altering the current from the battery is often referred to as the **equivalent resistance** of that part of the circuit.

31.15 (*a*) In Figure 31.37, determine the potential difference between c and b by going counterclockwise from c to b. (*b*) For $\mathscr{E} = 9$ V, $R_1 = 10\ \Omega$, and $R_2 = 5\ \Omega$, calculate V_{cb} going counterclockwise from c to b.

Example 31.9 Internal resistance

Even the best batteries dissipate some energy, which means that not all the chemical energy is converted to electric potential energy. We can take this dissipation into account by modeling a nonideal battery as consisting of an ideal battery of emf $\mathscr{E}$ in series with a small resistor R_{batt}, often called the *internal resistance* of the battery. The effect of this internal resistance is that the potential difference across the battery terminals is smaller than $\mathscr{E}$ when there is a current. When a nonideal battery is used in a circuit where the load is a resistor of resistance R, is the potential difference across the battery terminals greatest when the resistance of the load is high or when it is low?

❶ GETTING STARTED I begin by drawing a circuit diagram for such a nonideal battery connected to a load of resistance R (**Figure 31.38**). I need to determine an expression that shows me how the relative values of R and the internal resistance R_{batt} affect the potential difference across the battery terminals. I arbitrarily choose clockwise both for the reference direction for the current and for my travel around the circuit.

Figure 31.38

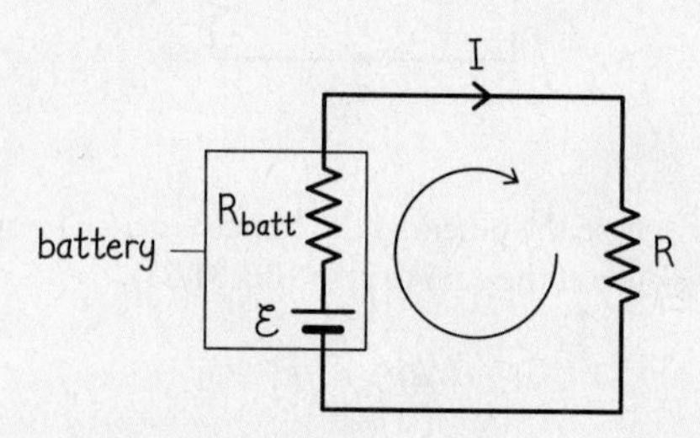

❷ DEVISE PLAN The potential difference across the nonideal battery terminals is the sum of the potential difference across the ideal battery, $\mathscr{E}$, and the potential difference across the internal resistance, $-IR_{batt}$. This latter term means I need to obtain a value for the current. I note that my circuit is equivalent to the two resistors in series in the circuit shown in Figure 31.37, and the final expression I obtained in Exercise 31.8 lets me express the current in terms of these two resistances. Thus I should be able to use this result to write an expression relating R and R_{batt} to V_{batt}.

❸ EXECUTE PLAN I use the solution I obtained in Exercise 31.8 to calculate the current in the circuit:

$$I = \frac{\mathscr{E}}{R_{batt} + R}.$$

I can now substitute this expression for I in the $-IR_{batt}$ term in my expression for V_{batt}:

$$V_{batt} = \mathscr{E} - IR_{batt} = \mathscr{E}\left(1 - \frac{R_{batt}}{R_{batt} + R}\right) = \mathscr{E}\left(\frac{R}{R_{batt} + R}\right).$$

When $R \gg R_{batt}$, the effect of R_{batt} is small and the potential difference across the battery terminals is essentially what it is in an ideal battery. When R is small, and especially when it is comparable to R_{batt}, then R_{batt} reduces the potential difference across the terminals. Thus the potential difference across the terminals of a nonideal battery is greatest when the resistance of the load is high. ✔

❹ EVALUATE RESULT If $R \gg R_{batt}$, the resistance in the circuit is great and hence the current is small. Because the current is small, the potential difference due to the internal resistance is small ($-IR_{batt}$) and hence the potential difference across the terminals does not differ significantly from its ideal value. This conclusion agrees with what my expression for V_{batt} shows.

According to the model of Example 31.9, an ideal battery is simply a battery with $R_{batt} = 0$. For a high-quality nonideal battery, R_{batt} can be extremely small—at most a few ohms. Thus, for ordinary loads, which are typically hundreds of ohms or more, the battery's internal resistance is not important.

Electrical measuring instruments

Current, potential difference, and resistance can be measured with the following three electrical measuring devices. Often the three functions are combined into one instrument, called a *multimeter*.

Ammeter: To measure the current in a single-loop circuit (or in a branch of a multiloop circuit), an ammeter must be inserted into the loop or branch, as shown in Figure 31.39*a* or *c*. The ammeter then indicates the current passing through it.

Voltmeter: To measure the potential difference between two points, a voltmeter is connected to those two points (Figure 31.39*b*). Unlike current measurement, a potential difference measurement does not require breaking the circuit.

Ohmmeter: A voltmeter and an ammeter can be combined to measure a resistance, as shown in Figure 31.39*c*. In practice resistance is measured using a device called an *ohmmeter*. Such a device puts a known potential difference across the resistor to be measured, measures the resulting current though the resistor, and then uses Ohm's law to determine the resistance. To measure the resistance of a circuit element, that element must be disconnected from the circuit.

Voltmeters and ammeters typically affect the circuit under observation. For example, if the resistance R_A of the ammeter in Figure 31.39*a* is nonzero, the current in the circuit is affected by the insertion of the ammeter. The resistance of ammeters must therefore be very small compared to the other resistances in the circuit; an *ideal ammeter* has zero resistance.

Similarly, the resistance R_V of the voltmeter in Figure 31.39*b* must be very great compared to other resistances in the circuit to prevent charge carriers from flowing through the voltmeter and changing the current in the circuit; an *ideal voltmeter* has infinite resistance.

In either case, if we know the resistance of the device, it is possible to correct for its effect on the circuit.

Figure 31.39 Electrical measuring instruments.

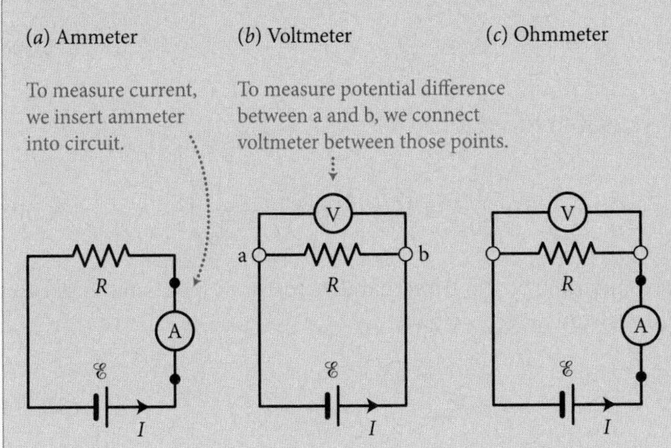
(*a*) Ammeter
(*b*) Voltmeter
(*c*) Ohmmeter
To measure current, we insert ammeter into circuit.
To measure potential difference between a and b, we connect voltmeter between those points.
V
V
a
b
R
R
R
A
A
ℰ
ℰ
ℰ
I
I
I

A number of measuring instruments exist to measure electrical quantities in electric circuits and circuit elements. The instrument to measure currents is called an *ammeter*; the instrument to measure potential differences is called a *voltmeter*. By measuring both the current in a resistor and the potential difference across it, one can determine the resistance using Ohm's law. An *ohmmeter* is used to accomplish this task: Connecting a resistor between the terminals of an ohmmeter yields a reading for the resistance. See the box above on Electrical measuring instruments for more information on these devices.

31.7 Multiloop circuits

The analysis of multiloop circuits follows the same principles we laid out in the preceding section, but we need to consider what happens in multiple branches, junctions, and loops. The first thing to note is that the branch rule requires that in steady state, the current in any branch is the same everywhere along that branch. Therefore, in a circuit containing M branches, there are M distinct currents. When analyzing multiloop circuits, you should therefore always begin by identifying all the branches and labeling the current in each of those branches.

As we saw in Section 31.3, the **junction rule** states that in steady state the sum of the currents going into a junction must be equal to the sum of the currents going out of that junction:

$$I_{\text{in}} = I_{\text{out}} \quad \text{(steady state)}. \tag{31.27}$$

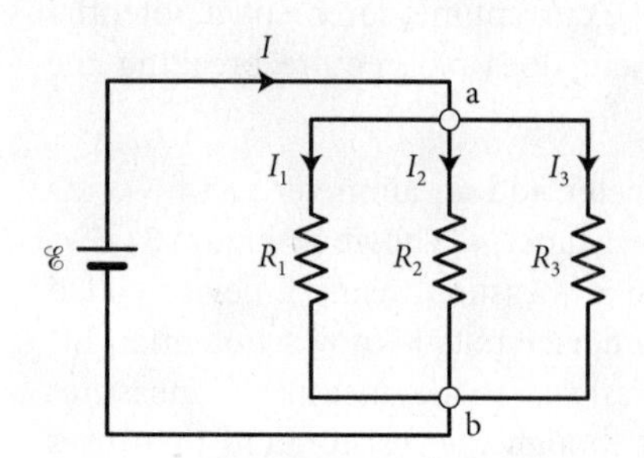

Figure 31.40 Circuit diagram for three resistors connected in parallel across an ideal battery.

As an example of a simple multiloop circuit, let us begin by considering three resistors connected in parallel across an ideal battery, as shown in **Figure 31.40**. Our goal will be to determine the equivalent resistance of the parallel combination of resistors. The circuit in Figure 31.40 has four branches, two junctions, and six loops (Can you identify them all?). Because there are four branches, we begin by labeling the currents in the four branches: the current in the battery, I, and the currents I_1, I_2, and I_3 through the three resistors. At junction a in Figure 31.40 we have $I_{in} = I$ and $I_{out} = I_1 + I_2 + I_3$, so the junction rule yields

$$I = I_1 + I_2 + I_3. \tag{31.28}$$

We can follow the same reasoning at junction b, but as you can easily verify, we would obtain exactly the same equation.

The potential difference V_{ba} is the same across each of the three resistors, so we can write for the currents in the resistors

$$I_1 = \frac{V_{ba}}{R_1}, \quad I_2 = \frac{V_{ba}}{R_2}, \quad I_3 = \frac{V_{ba}}{R_3}. \tag{31.29}$$

Substituting Eq. 31.29 into Eq. 31.28 gives

$$I = \frac{V_{ba}}{R_1} + \frac{V_{ba}}{R_2} + \frac{V_{ba}}{R_3} = V_{ba}\left(\frac{1}{R_1} + \frac{1}{R_2} + \frac{1}{R_3}\right). \tag{31.30}$$

If we replace the combination of the three parallel resistors by a single resistor having an equivalent resistance R_{eq}, we have

$$I = V_{ba}\left(\frac{1}{R_{eq}}\right). \tag{31.31}$$

The value of R_{eq} needed to obtain the same current I in the battery as with the three resistors connected in parallel is obtained by comparing Eqs. 31.30 and 31.31:

$$\frac{1}{R_{eq}} = \frac{1}{R_1} + \frac{1}{R_2} + \frac{1}{R_3}, \tag{31.32}$$

which tells us that the reciprocal of the equivalent resistance of resistors connected in parallel equals the sum of the reciprocals of the individual resistances. Because the line of argument I used in deriving Eqs. 31.28–31.32 can be followed for any number of resistors, this statement is true for any number of resistors in parallel:

$$\frac{1}{R_{eq}} = \frac{1}{R_1} + \frac{1}{R_2} + \frac{1}{R_3} + \cdots \quad \text{(resistors in parallel).} \tag{31.33}$$

For two resistors in parallel, the equivalent resistance can also be written as

$$R_{eq} = \frac{R_1 R_2}{R_1 + R_2} \quad \text{(two resistors in parallel).} \tag{31.34}$$

Note that the equivalent resistance is smaller than either of the two individual resistances ($R_{eq} < R_1$ and $R_{eq} < R_2$). This is always true for resistors in parallel. As we discussed in Section 31.3, although it may seem paradoxical that combining multiple resistors reduces the equivalent resistance, combining resistors in parallel amounts to providing multiple paths for the current to follow. Viewed from this perspective, it is not surprising that increasing the number of branches in the circuit reduces the equivalent resistance of the circuit.

31.16 Let $\mathscr{E} = 9$ V, $R_1 = 3\ \Omega$, $R_2 = 10\ \Omega$, and $R_3 = 5\ \Omega$ in Figure 31.40. (*a*) What is the equivalent resistance of the three resistors? (*b*) What is the current in the battery?

The circuit in Figure 31.40 has more than one loop, but it is still a relatively simple circuit. To analyze more complex multiloop circuits, we need to derive a set of mathematical relationships among the unknown quantities (typically the currents in the various branches). Specifically, we need to have as many independent mathematical relationships as there are unknown quantities in the circuit.

One way to obtain a suitable set of equations is to apply the junction rule and the loop rule as many times as necessary to obtain a suitable number of equations. The junction rule and the loop rule are sometimes referred to as either *Kirchhoff's circuit rules* or *Kirchhoff's laws,* after the German physicist Gustav Kirchhoff (1824–1887). These rules do not contain any new physical principles, but instead are simply the application of principles we have already encountered—continuity (the junction rule) and Eq. 25.32 representing conservation of energy (the loop rule)—to an electric circuit in steady state.

Suppose we are interested in determining the currents in the circuit shown in **Figure 31.41**. Specifically, we want to determine the currents I_1, I_2, and I_3 in the three resistors. (The chosen reference directions in Figure 31.41 are arbitrary.) We wish to determine values for these currents in terms of the resistances R_1, R_2, and R_3 and the emfs of the batteries.

Figure 31.41 Using the junction and loop rules to determine the currents in the resistors in this multiloop circuit.

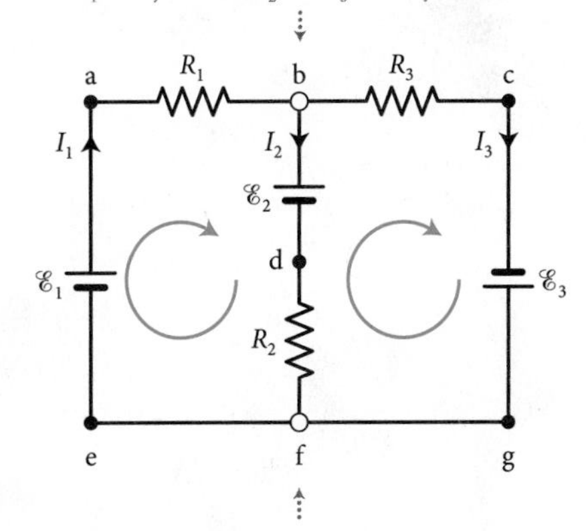

The circuit in Figure 31.41 has two junctions, three branches, and three loops. At junction b we have $I_{in} = I_1$ and $I_{out} = I_2 + I_3$, so, according to the junction rule,

$$I_1 = I_2 + I_3. \tag{31.35}$$

Applying the junction rule at f yields no new information because we would obtain the same equation. In general, it can be shown that in a circuit that contains N junctions, the junction rule yields $N - 1$ independent equations. This means that in a circuit with N junctions, we should apply the junction rule $N - 1$ times.

Next we need to use the loop rule. There are three unknown quantities in this problem (the three currents), so we need two equations in addition to Eq. 31.35 to determine the three currents. In other words, we need to apply the loop rule to two loops in the circuit.

First let us apply the loop rule to loop abdfea, choosing a clockwise travel direction around the loop:

$$V_{ab} + V_{bd} + V_{df} + V_{fe} + V_{ea} = 0. \tag{31.36}$$

Expressing the potential differences in this equation in terms of the emfs, resistances, and currents (see Table 32.2), we get

$$-I_1R_1 - \mathscr{E}_2 - I_2R_2 + 0 + \mathscr{E}_1 = 0. \tag{31.37}$$

Next we apply the loop rule to loop bcgfdb, going clockwise around the loop:

$$V_{bc} + V_{cg} + V_{gf} + V_{fd} + V_{db} = 0, \tag{31.38}$$

or

$$-I_3R_3 + \mathscr{E}_3 + 0 + I_2R_2 + \mathscr{E}_2 = 0. \tag{31.39}$$

Procedure: Analyzing multiloop circuits

Here is a series of steps for calculating currents or potential differences in multiloop circuits.

1. Identify and label the junctions in the circuit.
2. Label the current in each branch of the circuit, arbitrarily assigning a direction to each current.
3. Apply the junction rule to all but one of the junctions. (The choice of which junctions to analyze is arbitrary; choose junctions that involve the quantities you are interested in calculating.)
4. Identify the loops in the circuit and apply the loop rule (see the Procedure box on page 831) enough times to obtain a suitable number of simultaneous equations relating the unknowns in the problem. The choice of loops is arbitrary, but every branch must be in at least one of the loops. Traverse each loop in whichever direction you prefer, but be sure you traverse each loop completely and stick with the direction of travel and with the chosen directions of the currents.

There are several simplifications you can make during your analysis.

1. Multiloop circuits can sometimes be simplified by replacing parallel or series combinations of resistors by their equivalent resistances. If you can reduce the circuit to a single loop, you can solve for the current in the source. You may then need to "unsimplify" and undo the resistor simplification to calculate the current or potential difference across a particular resistor.
2. In general when solving problems, you should solve equations analytically before substituting known numerical values. When solving the simultaneous equations you obtain for multiloop circuits, however, you can often simplify the algebra if you substitute the known numerical values earlier on.

We now have three equations (Eqs. 31.35, 31.37, and 31.39) involving three unknowns (I_1, I_2, and I_3). (As stated earlier, the resistances and emfs are not unknowns. We consider R_1, R_2, R_3, $\mathscr{E}_1$, $\mathscr{E}_2$, and $\mathscr{E}_3$ to have certain values, although we are not given numerical values for them. Other problems might treat one or more of the resistances or emfs as unknowns, in which case some additional information, such as the values of some of the currents, would be needed.) Equations 31.35, 31.37, and 31.39 can be solved using standard algebraic techniques. Because every circuit is different, we shall not solve this particular problem. Complete solutions to such problems are provided in the examples below and in the Practice Volume. All the physics of the circuit is expressed in Eqs. 31.35–31.39; the remainder of the solution simply involves algebra. The steps for analyzing multiloop circuits are summarized in the Procedure box above.

Example 31.10 Multiloop circuit

Consider the circuit shown in **Figure 31.42**. Determine the magnitude of the potential difference across R_1 if $\mathscr{E}_1 = \mathscr{E}_2 = 9.0$ V and $R_1 = R_2 = R_3 = 300\ \Omega$.

Figure 31.42 Example 31.10.

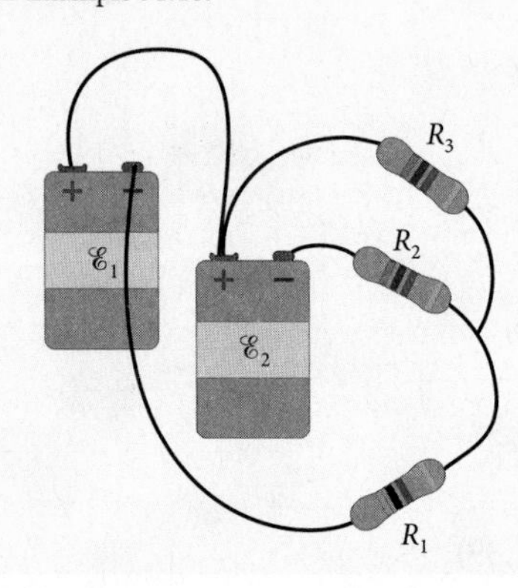

❶ GETTING STARTED I begin by drawing a circuit diagram for the circuit (**Figure 31.43**).

Figure 31.43

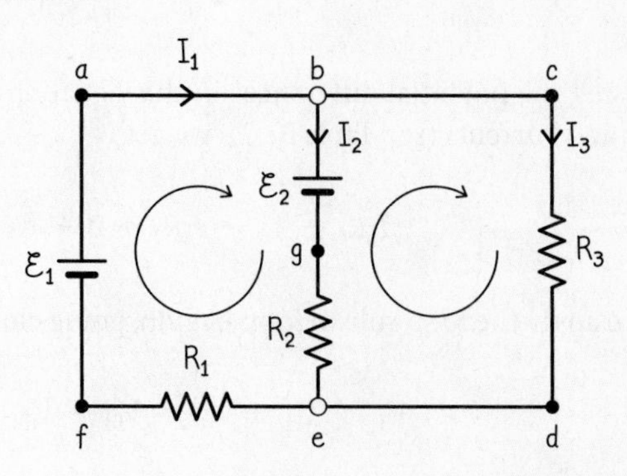

QUANTITATIVE TOOLS

❷ **DEVISE PLAN** To solve this problem I follow the steps in the Procedure box on page 836. There are two junctions in the circuit (b, e), three loops (abgefa, bcdegb, abcdefa), and three branches (efab, egb, edcb). I label the currents in the three branches I_1, I_2, and I_3. If I apply the junction rule to one of the junctions and the loop rule to two of the loops, I will obtain three equations that I can solve for the three unknowns I_1, I_2, and I_3. I choose loops abgefa and bcdegb and a clockwise travel direction in each of these loops. Once I know the current I_1, I can use Ohm's law (Eq. 31.11) to compute the potential difference across the resistor of resistance R_1. Because the values of the emfs and resistances are all the same, I set $\mathscr{E}_1 = \mathscr{E}_2 = \mathscr{E}$ and $R_1 = R_2 = R_3 = R$ to simplify the calculation.

❸ **EXECUTE PLAN** For junction b I have $I_{in} = I_1$ and $I_{out} = I_2 + I_3$, so applying the junction rule gives

$$I_1 = I_2 + I_3. \quad (1)$$

Applying the loop rule going clockwise around loop abgefa gives

$$0 - \mathscr{E} - I_2R - I_1R + \mathscr{E} = 0,$$

or

$$I_2 = -I_1. \quad (2)$$

Next I apply the loop rule going clockwise around loop bcdegb:

$$0 - I_3R + 0 + I_2R + \mathscr{E} = 0. \quad (3)$$

I now have three equations from which I can determine the three unknown currents.

Substituting Eq. 2 into Eq. 1 yields $I_1 = -I_1 + I_3$, or

$$I_3 = 2I_1. \quad (4)$$

Substituting this expression for I_3 and Eq. 2 into Eq. 3 yields

$$-2I_1R - I_1R + \mathscr{E} = 0.$$

Solving for I_1 then gives me

$$I_1 = \frac{\mathscr{E}}{3R}, \quad (5)$$

and substituting $\mathscr{E} = 9.0$ V and $R = 300\ \Omega$, I get

$$I_1 = \frac{9.0\text{ V}}{900\ \Omega} = 0.010\text{ A}.$$

The magnitude of the potential difference across the resistor with resistance R_1 is therefore $|-I_1R_1| = (0.010\text{ A})(300\ \Omega) = 3.0$ V. ✔

❹ **EVALUATE RESULT** I can use my expression for I_1 to determine the potential difference between positions b and e. Going through branch efab, I have $V_{eb} = V_{ef} + V_{fa} + V_{ab} = -I_1R + \mathscr{E} + 0$. Substituting my expression for I_1 (Eq. 5), I get $V_{eb} = \frac{2}{3}\mathscr{E} = +6.0$ V. Going through branch egb, I have $V_{eb} = V_{eg} + V_{gb} = +I_2R + \mathscr{E}$. Because $I_2 = -I_1$ (Eq. 2), this expression gives the same result, $V_{eb} = +6.0$ V. Going through branch edcb and using Eq. 4, I get $V_{eb} = 0 + I_3R + 0 = +2I_1R$. Substituting Eq. 5, I obtain $V_{eb} = \frac{2}{3}\mathscr{E} = +6.0$ V. Because I obtain the same value for the potential difference each time, I am confident that my result for I_1 is correct.

Example 31.11 Wheatstone bridge

The circuit shown in **Figure 31.44** includes a variable resistor, the resistance R_{var} of which can be adjusted, and a resistor of unknown value R. A circuit with such a network of resistors is called a *Wheatstone bridge* and can be used to determine the value of the unknown resistance R by adjusting the variable resistor so that the light bulb does not glow. The light bulb is initially glowing when R_{var} is set to 20 Ω, but it goes out when R_{var} is adjusted to 12 Ω. Determine the current in the battery when the light bulb is out.

Figure 31.44 Example 31.11.

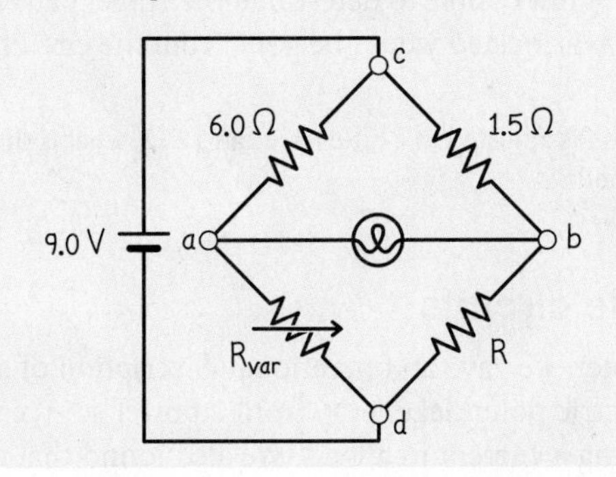

❶ **GETTING STARTED** The bulb glows when there is a current in it and stops glowing when the current in it is zero. In the latter case, it doesn't matter whether the bulb is connected or not, so I can omit it from the circuit and simplify the circuit diagram, as shown in **Figure 31.45a** on the next page. Because R_{var} is set to 12 Ω, I use that value rather than R_{var}. I use the label I_1 for the current in the branch containing the variable resistor, I_2 for the current in the branch containing R, and I_3 for the current in the battery, which is the current I need to determine.

❷ **DEVISE PLAN** In order for the current in the bulb to be zero, junctions a and b in Figure 31.44 must be at the same potential, and I must have $V_{ca} = V_{cb}$ and $V_{ad} = V_{bd}$. I can then use Ohm's law to express these two conditions in terms of the unknown resistance R and the currents I_1 and I_2, and solve the resulting set of equations for R. Once I know R, I can further simplify the circuit by replacing the sets of resistors connected in series in each branch by a single equivalent resistor. This yields the circuit shown in Figure 31.45b. To determine the current in the battery, I can then replace the two parallel resistors by a single equivalent resistor as in Figure 31.45c, and use Eq. 31.34 to determine its resistance.

Figure 31.45

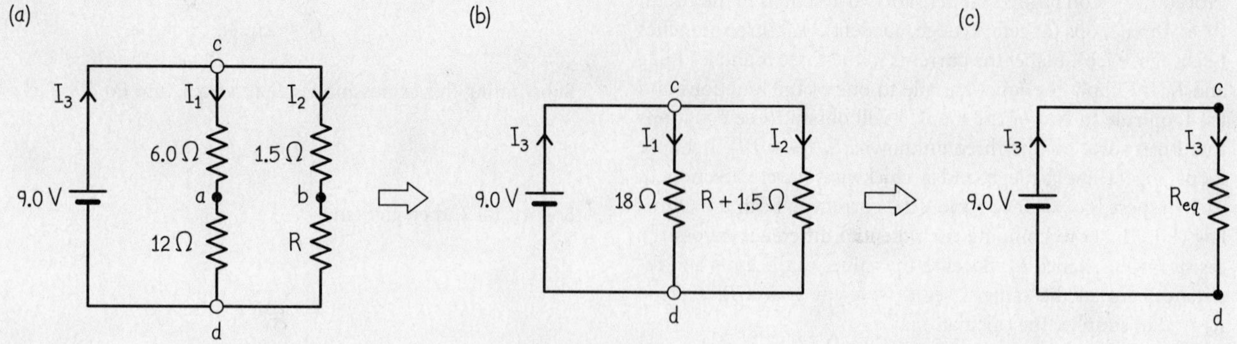

❸ **EXECUTE PLAN** Rearranging the Ohm's law equation (Eq. 31.11) to $V = IR$ and applying it to the two conditions $V_{ca} = V_{cb}$ and $V_{ad} = V_{bd}$, I have, when I travel clockwise in the circuit,

$$-(6.0\ \Omega)I_1 = -(1.5\ \Omega)I_2 \tag{1}$$

$$-(12\ \Omega)I_1 = -RI_2. \tag{2}$$

Solving Eq. 1 for I_1 gives me $I_1 = I_2/4$, and substituting this result into Eq. 2 and solving for R give me

$$R = 3.0\ \Omega.$$

With this value of R the resistance in the rightmost branch in Figure 31.45*b* becomes 4.5 Ω. The equivalent resistance of the two parallel resistors is, from Eq. 31.34,

$$R_{eq} = \frac{(18\ \Omega)(4.5\ \Omega)}{18\ \Omega + 4.5\ \Omega} = 3.6\ \Omega.$$

The current in the battery is then

$$I_3 = \frac{\mathscr{E}}{R_{eq}} = \frac{9.0\ \text{V}}{3.6\ \Omega} = 2.5\ \text{A.} ✔$$

❹ **EVALUATE RESULT** I can verify that $V_{ca} = V_{cb}$ by calculating I_1 and I_2 and substituting these values in Eq. 1. Applying the junction rule to junction c, I have $I_3 = I_1 + I_2$. I know that $I_1 = I_2/4$, so $I_2 = \frac{4}{5}I_3 = 2.0$ A and $I_1 = \frac{1}{5}I_3 = 0.5$ A. Using these values, I get $V_{ca} = -(6.0\ \Omega)(0.5\ \text{A}) = -3.0$ V and $V_{cb} = -(1.5\ \Omega)(2.0\ \text{A}) = -3.0$ V, so junctions a and b are at the same potential, as I expect when there is no current in the bulb.

Example 31.11 illustrates an alternative to our standard procedure for analyzing circuits. For circuits that contain many junctions and loops, it is sometimes easier to simplify the circuit by replacing part of it with a single equivalent resistance, as I did in Figure 31.45*b* of the example. When you do this, think about what question you ultimately want to answer about the circuit and about which parts of the circuit can be simplified without interfering with solving the problem. In Example 31.11, because what I wanted to determine was the current in the battery, I could replace everything outside the battery by a single equivalent resistor. However, if I had been asked to determine the currents at, say, junction b when the bulb was not glowing, I would have stopped simplifying the circuit at the stage shown in Figure 31.45*b*. Simplifying the circuit to just a single resistor and battery would have made it impossible to determine the current at b because the branch in which junction b is located would be gone from the circuit.

31.17 If R_{var} in Figure 31.44 is adjusted to a little less than 12 Ω, what is the direction of the current in the light bulb?

31.8 Power in electric circuits

At the beginning of this chapter, we gave an operational description of a single-loop circuit as converting electric potential energy from a power source to some other form of energy by driving a current in a load. We also found that the loop rule, which essentially embodies the idea of conservation of energy, is a powerful tool for analyzing circuits.

Let us examine how rapidly the power source in **Figure 31.46** can deliver energy to its load. From Chapter 25 we know that the electrostatic work done on a charge carrier carrying charge q in a time interval Δt as it passes through the load from point a to point b is the negative of the potential difference across the load multiplied by q (Eq. 25.17):

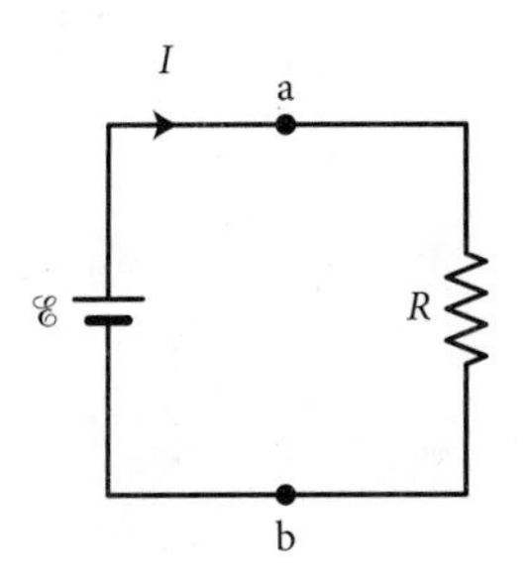

Figure 31.46 A battery delivers energy to a load connected to it.

$$W_q(\mathrm{a} \to \mathrm{b}) = -qV_{\mathrm{ab}}. \tag{31.40}$$

Because V_{ab} is negative, the work done by the power source on the charge carrier is positive, increasing its energy. While it travels through the load, however, this energy is converted to other forms, and so the amount of energy converted in the load is given by

$$\Delta E = -qV_{\mathrm{ab}}. \tag{31.41}$$

To determine the rate at which energy is converted, we substitute this expression into the expression for the average power (time rate of change of energy, Eq. 9.29). Because we are considering a steady-state situation, average and instantaneous power are the same, so we can drop the subscript av:

$$P \equiv \frac{\Delta E}{\Delta t} = \frac{-qV_{\mathrm{ab}}}{\Delta t} = -IV_{\mathrm{ab}}. \tag{31.42}$$

The form of energy to which the electric potential energy in a circuit is converted depends on the type of load. When the load is a light bulb, for example, the conversion is from electric potential energy to light and thermal energy. When the load is an electric motor, the conversion is to kinetic energy as, say, the blades of a fan start rotating. Here we're interested in circuits in which the load is a resistor, in which case the conversion is to thermal energy. Substituting $V_{\mathrm{ab}} = -IR$ in Eq. 31.42 gives us, for the rate at which energy is dissipated in a resistor,

$$P = I^2R. \tag{31.43}$$

It can also be useful to know the rate at which energy is dissipated in terms of the potential difference across the resistor. Substituting $-V_{\mathrm{ab}}/R$ for I in Eq. 31.43 yields

$$P = \frac{V_{\mathrm{ab}}^2}{R}. \tag{31.44}$$

Keep in mind that although Eq. 31.42 is valid for any electrical device, we have derived Eqs. 31.43 and 31.44 using Ohm's law for resistors, in which electric potential energy is converted to thermal energy.

We can apply a similar reasoning to the power source. Let's consider the case of a battery as the power source. When a charge carrier carrying charge q moves from the negative terminal to the positive terminal inside the battery, its electric potential energy increases at a rate given by

$$P = \frac{q\mathcal{E}}{\Delta t} = I\mathcal{E}. \tag{31.45}$$

For an ideal battery $\mathcal{E} = IR$ (Eq. 31.24), and so Eq. 31.45 becomes

$$P = I\mathcal{E} = I^2R. \tag{31.46}$$

Comparing Eqs. 31.46 and 31.43, we see that the rate at which electric potential energy increases in the battery is equal to the rate at which electric potential energy is dissipated in the load. For a nonideal battery, some energy is dissipated inside the battery. As we have seen in Example 31.9, we can account for this dissipation by attributing an internal resistance R_{batt} to the battery. This internal resistance decreases the potential difference across the terminals: $V_{ba} = \mathscr{E} - IR_{batt}$, so $\mathscr{E} = V_{ba} + IR_{batt}$. Substituting this expression and $V_{ba} = IR$ into Eq. 31.45 yields

$$P = I\mathscr{E} = IV_{ba} + I^2R_{batt} = I^2R + I^2R_{batt}. \quad (31.47)$$

So for a nonideal battery the rate at which chemical energy is converted is equal to the sum of the rates at which energy is dissipated in the load and inside the battery, as we would expect.

31.18 The SI units of power suggested by Eqs. 31.42 and 31.43 are $A \cdot V$ and $A^2 \cdot \Omega$, respectively. Show that these SI units are equivalent to the derived SI unit for power, the watt.

Example 31.12 Battery to battery

A 9.0-V and a 6.0-V battery are connected to each other (**Figure 31.47**). Each battery has an internal resistance of 0.25 Ω. At what rate is energy dissipated in the 6.0-V battery?

Figure 31.47 Example 31.12.

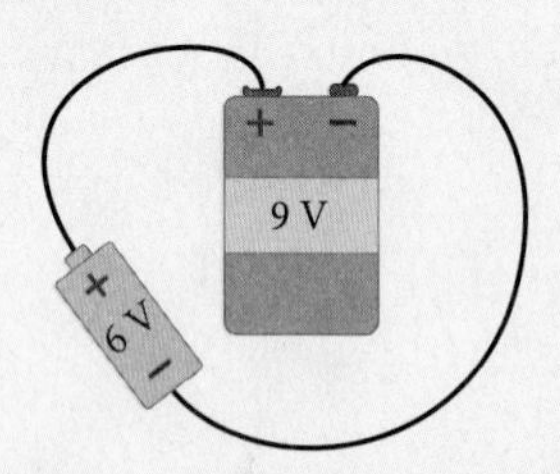

❶ **GETTING STARTED** I begin by drawing a circuit diagram for the two-battery combination, representing the internal resistance by two resistors in series with the batteries (**Figure 31.48**). I note that the negative terminals of the batteries are connected to each other, and the positive terminals are connected to each other. I arbitrarily choose clockwise both for the reference direction for the current and for my travel around the circuit.

Figure 31.48

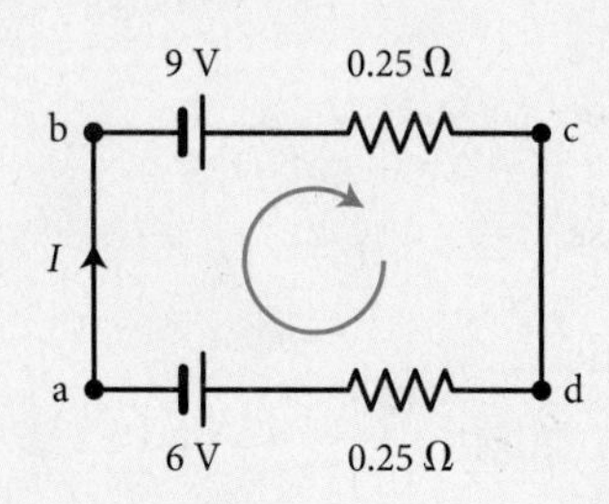

❷ **DEVISE PLAN** Because this circuit contains only one loop and no junctions, the current is the same everywhere in the circuit. To determine the current, I apply the loop rule and solve the resulting equation for the current. I can then apply Eq. 31.43 to determine the rate at which energy is dissipated in the 6.0-V battery.

❸ **EXECUTE PLAN** As I travel clockwise around the circuit starting at a, the loop rule yields

$$9.0 \text{ V} - I(0.25 \ \Omega) - I(0.25 \ \Omega) - 6.0 \text{ V} = 0$$

$$I = \frac{9.0 \text{ V} - 6.0 \text{ V}}{0.50 \ \Omega} = \frac{3.0 \text{ V}}{0.50 \ \Omega} = 6.0 \text{ A}.$$

The fact that I get a positive value for the current tells me that my assumed reference direction for the current (clockwise) is correct. Energy is dissipated in the 6.0-V battery as resistive losses in the internal resistance. Using Eq. 31.43 for the rate at which energy is dissipated in a resistor, I get

$$P = I^2R = (6.0 \text{ A})^2(0.25 \ \Omega) = 9.0 \text{ W}. \checkmark$$

❹ **EVALUATE RESULT** Even though the internal resistance is small, this power is substantial because the current is large. The current is large because there is very little resistance in the circuit. It's not a good idea to connect two batteries of different emf values in this manner!

31.19 (*a*) In Example 35.12, how would the answer change if we had chosen a counterclockwise travel direction around the circuit? (*b*) At what rate is energy dissipated in the 9-V battery?

Chapter Glossary

SI units of physical quantities are given in parentheses.

Branch The part of a circuit between two junctions that does not contain any junctions itself. In steady state, the current is the same at any location along a branch.

Branch rule The current in each branch of a multiloop circuit is the same throughout that branch.

Circuit diagram A schematic representation of an electric circuit, using standard symbols to represent circuit elements and straight lines to represent conducting connections between elements.

Conductivity σ (A/(V · m)) The proportionality constant relating current density to electric field in a conductor:

$$\sigma \equiv \frac{J}{E}. \quad (31.8)$$

The conductivity is a measure of a material's ability to conduct current. For a metal,

$$\sigma = \frac{ne^2\tau}{m_e}, \quad (31.9)$$

where n is the number density of the free electrons, e is the charge of the electron, m_e is the mass of the electron, and τ is the average time interval between collisions.

Current continuity principle In steady state, the current is the same at all locations along a single-loop electric circuit.

Current density $\vec{J}$ (A/m^2) A vector whose magnitude represents the current per unit area through a conductor of cross-sectional area A:

$$J \equiv \frac{|I|}{A}. \quad (31.5)$$

Drift velocity (m/s) The average velocity that an electron attains in a conductor due to an electric field:

$$\vec{v}_d = -\frac{e\vec{E}}{m_e}\tau. \quad (31.3)$$

Electric circuit An interconnection of electrical elements.

Equivalent resistance The resistance that can replace a combination of circuit elements without altering the current from the battery. For resistors connected in series,

$$R_{eq} = R_1 + R_2 + R_3 + \cdots \quad \text{(resistors in series)}, \quad (31.26)$$

and for resistors connected in parallel,

$$\frac{1}{R_{eq}} = \frac{1}{R_1} + \frac{1}{R_2} + \frac{1}{R_3} + \cdots \quad \text{(resistors in parallel)}. \quad (31.33)$$

Junction A location in a circuit where more than two wires or other circuit elements are connected together.

Junction rule The sum of the currents going into a junction is equal to the sum of the currents coming out of the same junction.

Load A combination of circuit elements connected to a power source where electric potential energy is converted to other forms of energy.

Loop A closed conducting path in an electric circuit.

Loop rule In steady state, the sum of the emfs and the potential differences around any loop in an electric circuit is zero:

$$\Sigma\mathscr{E} + \Sigma V = 0 \quad \text{(steady state, around loop)}. \quad (31.21)$$

Ohm (Ω) The derived SI unit of resistance:

$$1\ \Omega \equiv 1\ \text{V/A}.$$

Ohm's law The current in a conductor between two points is directly proportional to the potential difference across the two points and inversely proportional to the resistance between them:

$$I = \frac{V}{R}. \quad (31.11)$$

Power source A circuit element that provides electric potential energy to the elements in an electric circuit by maintaining a potential difference between the two locations in the circuit in which it is connected.

Resistance R (Ω) The resistance of a circuit element is proportional to the potential difference that must be applied across it to obtain a current of 1 A in the load.

$$R \equiv \frac{V}{I}. \quad (31.10)$$

Resistor A conducting object that has nonnegligible constant resistance, usually attached to wires at either end for easy incorporation into a circuit.

Short circuit A branch in a circuit in parallel with a load that consists of only wire. Because of its negligible resistance, a short circuit diverts all the current away from the load.

Steady state A circuit is in steady state when the current has a constant value at all points in the circuit; at the instants when the current is established or cut off, the current is changing throughout the circuit and is not in steady state.

32 Electronics

CONCEPTS

QUANTITATIVE TOOLS

In the preceding chapter, we discussed electric circuits in which the current is steady. As noted in that chapter, the steady flow of charge carriers in one direction only is called *direct current*. Batteries and other devices that produce static electrical charge, such as van de Graaff generators, are sources of direct current. Although direct current has many uses, it has several limitations as well. For example, in order to produce substantial currents, direct-current sources must be quite large and are therefore cumbersome. More important, steady currents do not generate any electromagnetic waves, which can be used to transmit information and energy through space, as we saw in Chapter 30.

Because of these and many other factors, most electric and electronic circuits operate with **alternating currents** (abbreviated AC)—currents that periodically change direction. The current provided by household outlets in the United States, for instance, alternates in direction, completing 60 cycles per second (that is, with a frequency of 60 Hz), and the currents in computer circuits change direction billions of times per second. It is no understatement to say that contemporary society *depends* on alternating currents.

In this chapter we discuss the basics of both household currents and the electronics that lie at the heart of computers.

32.1 Alternating currents

We have already encountered one example of an electrical device that produces a changing current: a capacitor that is either charging or discharging. Let's consider what happens when we connect an inductor to a charged capacitor (**Figure 32.1**). A circuit that consists of an inductor and a capacitor is called an *LC* circuit. As soon as the two circuit elements are connected, positive charge carriers begin to flow clockwise through the circuit. The magnitude of the current increases from its initial value of zero (**Figure 32.2a** on next page) to a nonzero value (Figure 32.2*b*–*d*). The capacitor discharges through the inductor, and the current causes a magnetic field in the inductor. As the current in the inductor increases, the magnetic field also increases, causing an induced emf (see Section 29.7) that opposes this increase and prevents the current from increasing rapidly. Consequently, the capacitor discharges more slowly than it would if we had connected it to a wire.

Figure 32.1 What happens when we connect an inductor to a charged capacitor?

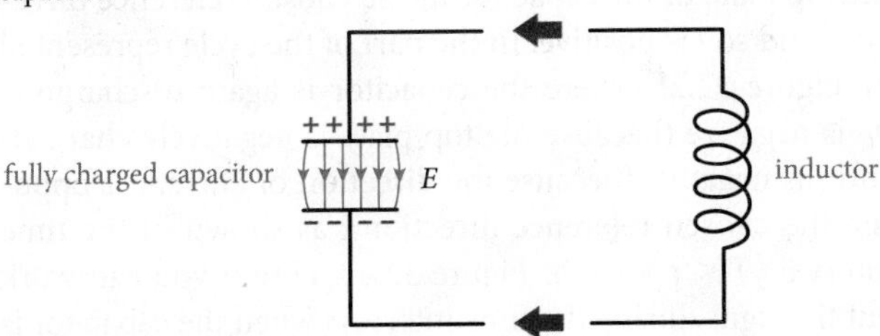

32.1 (*a*) Just before the inductor is connected to the charged capacitor, what type of energy is contained in the system comprising the two elements? (*b*) Once the two elements are connected to each other, what happens to that energy? (*c*) Once the capacitor is completely discharged, in what form is the energy in the circuit?

As you saw in Checkpoint 32.1, when the capacitor is completely discharged, all of the energy in an *LC* circuit is contained in the magnetic field and this field reaches its maximum magnitude (Figure 32.2*c*). Because the magnetic energy is proportional to the square of the current in the inductor (Eq. 29.25), the current, too, reaches its maximum value at this instant. Once the magnetic field and the current reach their maximum values, the current begins to charge the capacitor in the opposite direction (Figure 32.2*d*), and the charge on the capacitor increases as the magnetic field in the inductor decreases. When the magnetic field in the inductor is zero, the current is also zero and the capacitor has again maximum charge but with the opposite polarity (Figure 32.2*e*). The process then repeats itself with the current in the opposite direction (Figure 32.2*f*–*h*) until the capacitor is restored to its starting configuration. Then the cycle begins again.

Figure 32.3 on the next page shows the time dependence of the electric potential energy U^E stored in the capacitor and the magnetic potential energy U^B stored in the inductor. In the absence of dissipation, the energy in the circuit, $U^E + U^B$, must stay constant. Therefore, when the capacitor is not charged and U^E drops to zero, U^B must reach its maximum value, U_{max}.

There is always some dissipation in a circuit. Resistance in the connecting wires gradually converts electrical energy to thermal energy. Consequently, the oscillations decay in the same manner as the damped mechanical oscillations we considered in Section 15.8. Resistance therefore plays the same role in oscillating circuits as damping does in mechanical oscillators.

Throughout this chapter we work with time-dependent potential differences and currents. To make the notation as concise as possible, we represent time-dependent quantities with lowercase letters. In other words, v_C is short for $V_C(t)$ and i is short for $I(t)$. We also need a symbol for the maximum value of an oscillating quantity—its *amplitude* (see Section 15.1). For this we use a capital letter without the time-dependent marker (t); thus V_C is the maximum value of the potential difference across a capacitor, and I is the maximum value of the current in a circuit.

Unlike their counterparts in DC circuits, the potential difference across the capacitor, v_C, and the current in the *LC* circuit, i, change sign periodically. So, when analyzing AC circuits, we must carefully define what we mean by the sign of these quantities. To analyze the *LC* circuit in Figure 32.2, for example, we choose a reference direction for the current i and let the potential difference v_C be positive when the top capacitor plate is at a higher potential than the bottom plate (**Figure 32.4a** on page 845). Note that both of these choices are arbitrary.

Figure 32.2 A series of "snapshots" showing what happens when we connect an inductor to a charged capacitor.

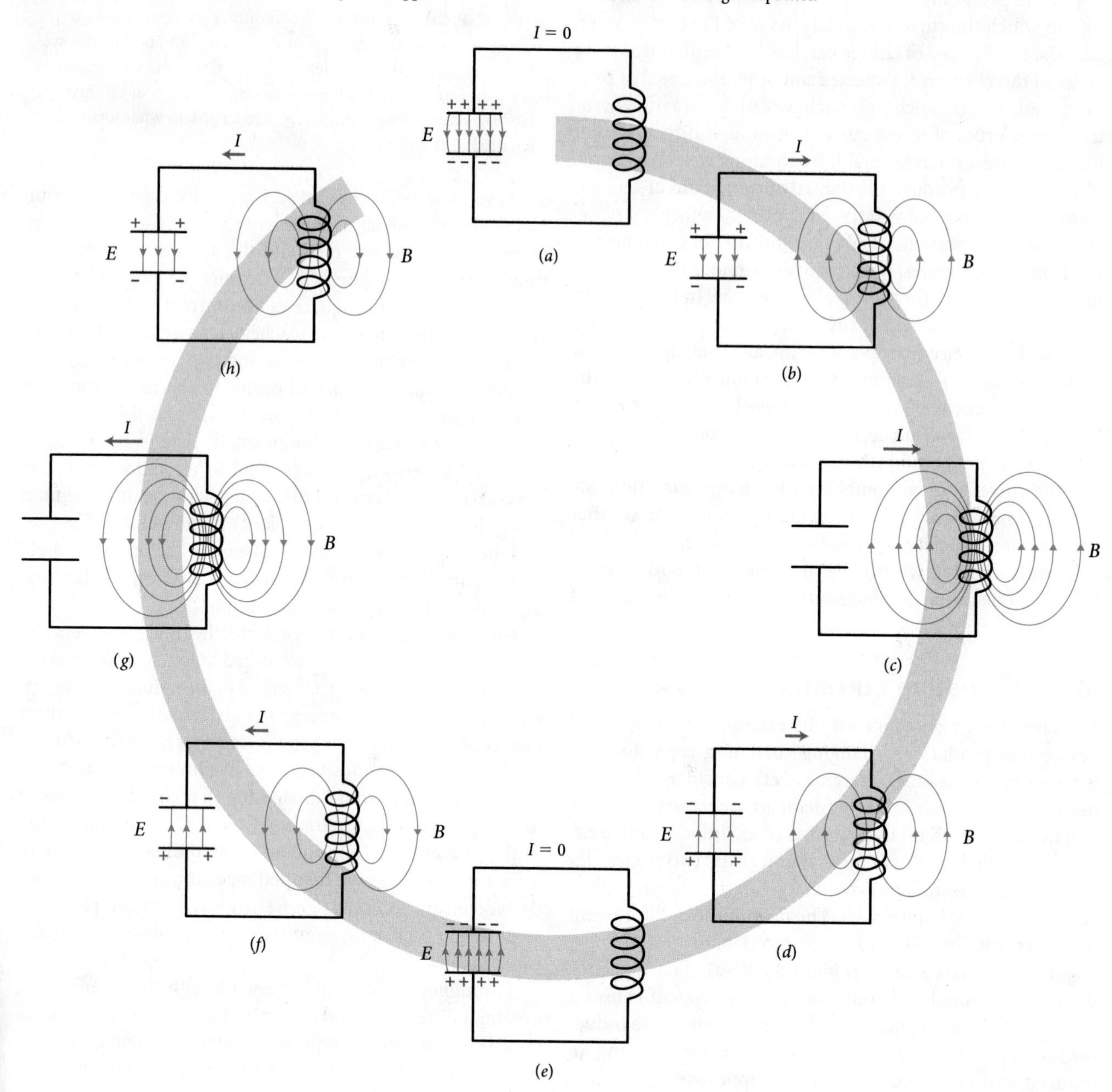

Figure 32.3 Time dependence of the electric potential energy U^E stored in the capacitor and the magnetic potential energy U^B stored in the inductor. In the absence of dissipation, the energy in the circuit, $U^E + U^B$, is a constant U_{max}.

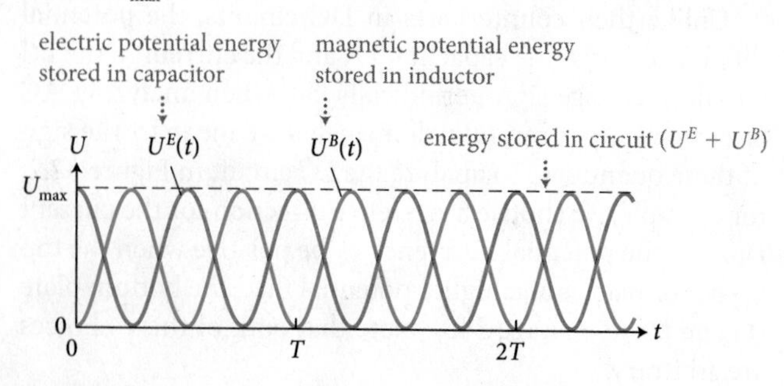

Figure 32.4*b* shows graphs for v_C and i with these choices. The potential difference across the capacitor v_C is initially positive, representing the situation at Figure 32.2*a*. During the first quarter cycle (Figure 32.2*b*), the capacitor is discharging and positive charge carriers travel away from the top plate of the capacitor in the chosen reference direction, and so i is positive. In the part of the cycle represented by Figure 32.2*f*, where the capacitor is again discharging, v_C is negative (because the top plate is negatively charged) and i is negative (because the direction of current is opposite the chosen reference direction), as shown in the time interval $\frac{1}{2}T < t < \frac{3}{4}T$ in Figure 32.4*b*. (See if you can work out the signs during the time intervals when the capacitor is

Figure 32.4 For the *LC* circuit shown in Figure 32.2, graphs of the time-dependent potential difference across the capacitor (defined to be positive when the top plate is at the higher potential) and the current in the circuit (defined to be positive when positive charge carriers travel away from top plate of the capacitor). One cycle is completed in a time interval T (the *period*).

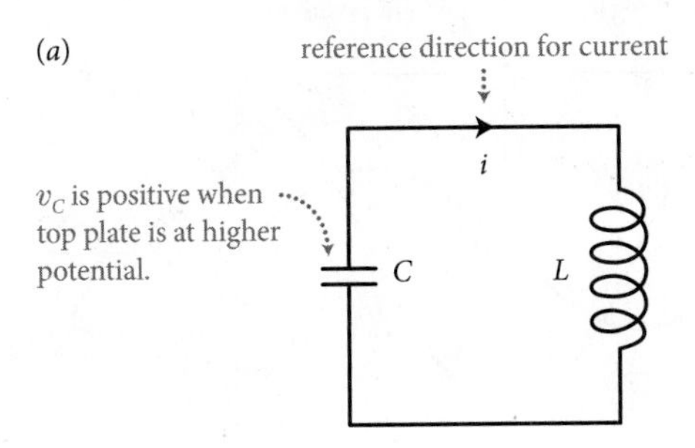

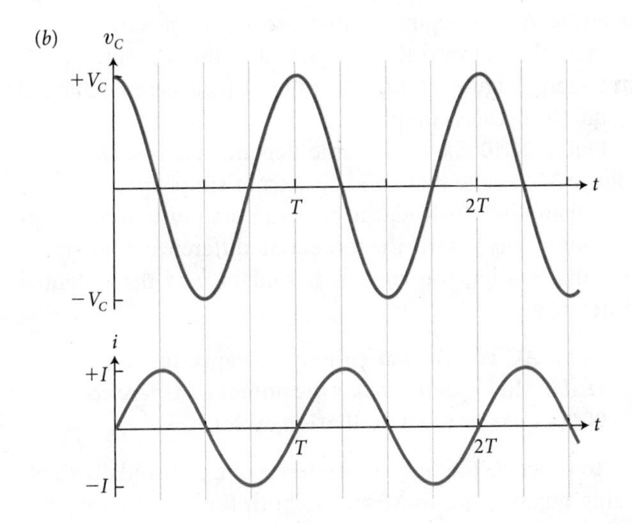

charging.) Both v_C and i vary sinusoidally in time, with v_C at its maximum when i is zero, and vice versa.

Because of dissipation, the *LC* circuit in Figure 32.1 is not a practical source of alternating current; instead, generators are widely used to produce sinusoidally alternating emfs in a circuit (see Example 29.6). The symbol for a source that generates a sinusoidally alternating potential difference or current is shown in **Figure 32.5**; such a source is called an **AC source**. The time-dependent emf an AC source produces across its terminals is designated $\mathscr{E}$, and its amplitude is designated $\mathscr{E}_{max}$.

Figure 32.5 Symbol that represents an AC source in an electric circuit. The AC source produces a sinusoidally varying emf $\mathscr{E}$ across its terminals.

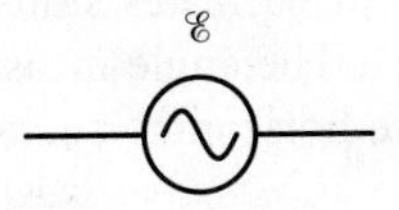

Exercise 32.1 AC source and resistor

Figure 32.6 shows a circuit consisting of an AC source and a resistor. The emf produced by the generator varies sinusoidally in time. Sketch the potential difference across the resistor as a function of time and the current in it as a function of time.

Figure 32.6 Exercise 32.1.

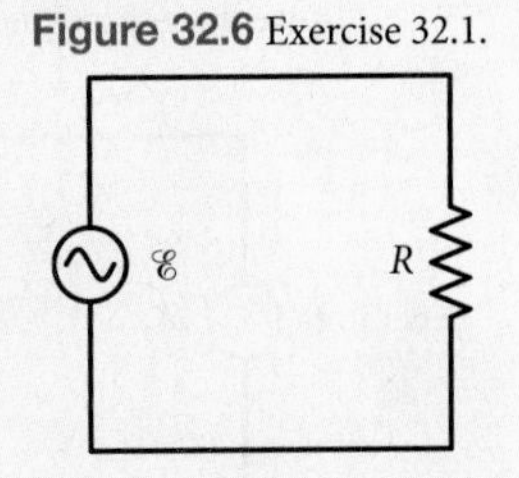

SOLUTION Ohm's law, the junction rule, and the loop rule (see Chapter 31) apply to alternating-current circuits just as they do to direct-current circuits. All I need to remember here is that the potential differences and currents are time dependent. Applying the loop rule to this circuit requires the time-dependent potential difference across the resistor v_R to equal the emf $\mathscr{E}$ of the AC source at every instant, so that the sum of the potential differences around the circuit is always zero. Consequently, v_R oscillates just as $\mathscr{E}$ oscillates, as shown in **Figure 32.7**; V_R is the maximum value of the potential difference across the resistor.

Figure 32.7

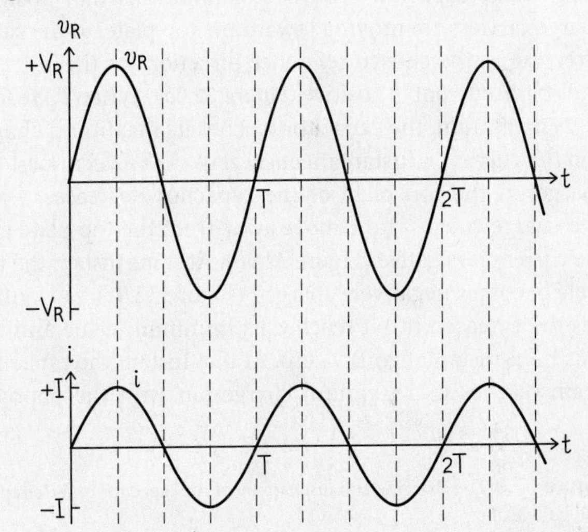

Ohm's law requires the time-dependent current i in the resistor to be proportional to v_R, which means that i also oscillates, with the current at its maximum when v_R is maximum. ✔

32.2 (*a*) Is energy dissipated in the resistor in the circuit of Figure 32.6? (*b*) If so, why doesn't the amplitude of the oscillations of v_R and i (shown in Figure 32.7) decrease with time?

32.2 AC circuits

The circuit discussed in Exercise 32.1 is an alternating current, or AC circuit. Such circuits exhibit more complex behavior when they contain elements that do not obey Ohm's law, so that the current is not proportional to the emf of the source. For example, let's consider the current in the circuit shown in **Figure 32.8** on the next page.

Figure 32.8 AC circuit with a capacitor connected to an AC source.

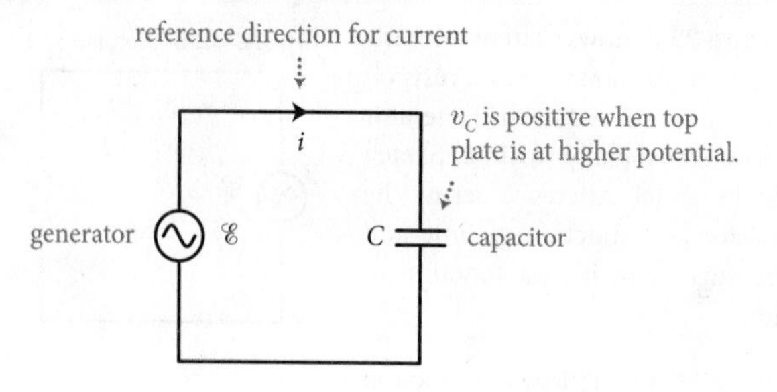

To analyze the circuit we choose a reference direction for the current i and let the potential difference v_C again be positive when the top capacitor plate is at a higher potential than the bottom plate (Figure 32.8). Because the capacitor is connected directly to the AC source, the time-dependent potential difference across the capacitor v_C equals the emf of the AC source at any instant. What is the current in the circuit? Let's begin considering what happens when the capacitor is uncharged. As v_C increases, the charge on the top plate of the capacitor increases. This means that positive charge carriers are moving toward the top plate, in the same direction as the chosen reference direction for the current, and so the current is positive (Figure 32.9*a*). When v_C reaches its maximum, the capacitor reaches its maximum charge and the current is instantaneously zero. As v_C decreases, the charge on the top plate of the capacitor decreases. Positive charge carriers now move away from the top plate and the current is negative (Figure 32.9*b*). At some instant the top plate becomes negatively charged (Figure 32.9*c*); v_C continues to decrease until it reaches its minimum value and the current is instantaneously zero. At that instant the capacitor again reaches its maximum charge but with the opposite

Figure 32.9 The charging and discharging of the capacitor in the circuit of Figure 32.8.

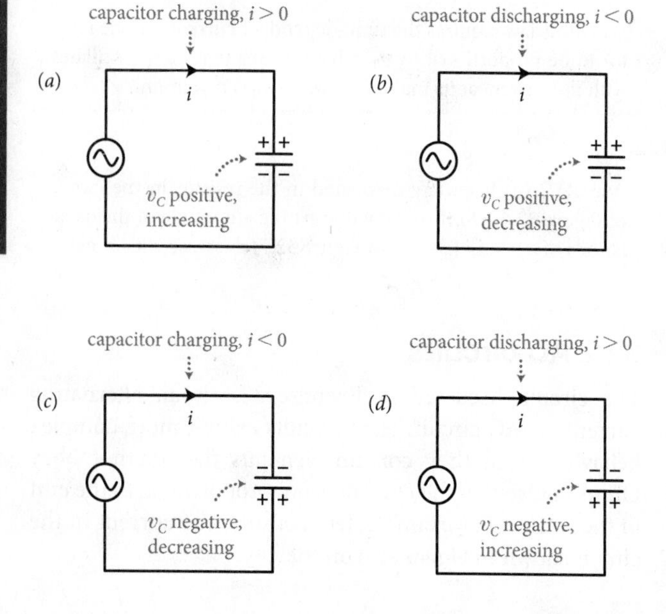

Figure 32.10 Time-dependent current in the circuit and potential difference across the capacitor for the circuit of Figure 32.9.

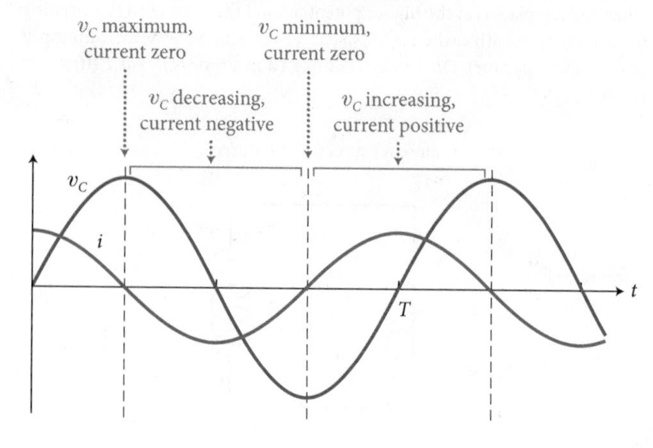

polarity. As v_C begins to increase again, positive charge carriers flow toward the top plate and the current is positive again (Figure 32.9*d*). When both plates are uncharged again, the cycle is complete.

Figure 32.10 shows the time dependence of i and v_C in Figure 32.9. Note that i and v_C are not simply proportional to one another. Instead, the current maximum occurs one-quarter cycle before the potential difference maximum. For this reason, the current is said to *lead* the potential difference:

In an AC circuit that contains a capacitor, the current in the capacitor leads the potential difference by 90° (a quarter of an oscillation cycle).

To describe the time dependence of a sinusoidally oscillating quantity, we must specify both the angular frequency of oscillation ω and the instant at which the oscillating quantity equals zero. As discussed in Chapter 15, a sinusoidally time-dependent quantity (such as the circuit potential difference we are looking at here) can be written in the form $v = V\sin(\omega t + \phi_i)$. The argument of the sine, $\omega t + \phi_i$, is the *phase*. At $t = 0$ the phase is equal to the *initial phase* ϕ_i (Chapter 15). When the phase of an oscillating quantity is zero, $\omega t + \phi_i = 0$, the quantity is zero as well because $\sin(0) = 0$.

We can analyze phase differences in AC circuits with lots of algebra, but the underlying physics is much clearer (and the analysis much simpler!) if we use the phasor notation developed in Chapter 15 to describe oscillatory motion. Following the approach of Section 15.5, we can represent an oscillating potential difference v by a phasor rotating in a reference circle (Figure 32.11). Because the length of the phasor equals the amplitude (maximum value) of v, the phasor is labeled V. The phasor rotates counterclockwise at angular frequency ω. The magnitude of v at any instant is given by the vertical component of the phasor; as the phasor rotates, that component oscillates sinusoidally in time, as shown in Figure 32.11. The angle measured counterclockwise from the positive horizontal axis to the phasor is the phase $\omega t + \phi_i$.

Figure 32.11 Phasor representation of a sinusoidally varying potential difference v. The phasor rotates counterclockwise at the same angular frequency at which v oscillates. The instantaneous value of v equals the length of the vertical component of the phasor.

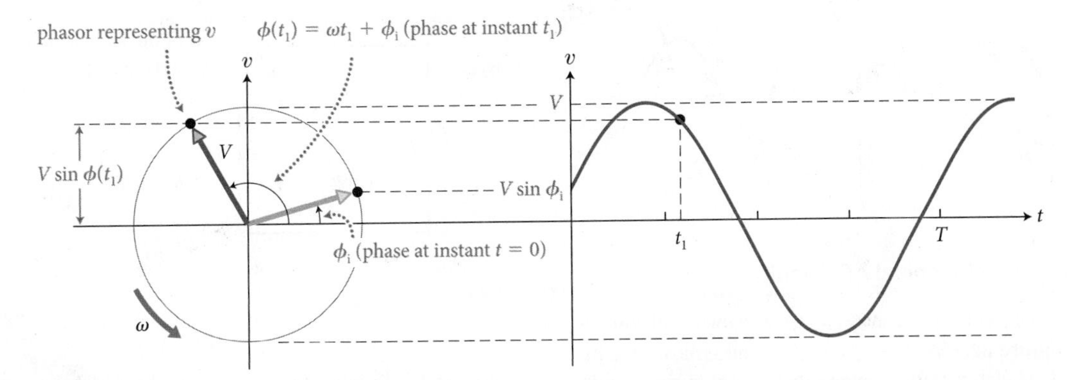

Example 32.2 Phasors

Consider the oscillating emf represented in the graph of **Figure 32.12**. Which of the phasors a–d, each shown at $t = 0$, correspond(s) to this oscillating emf?

Figure 32.12 Example 32.2.

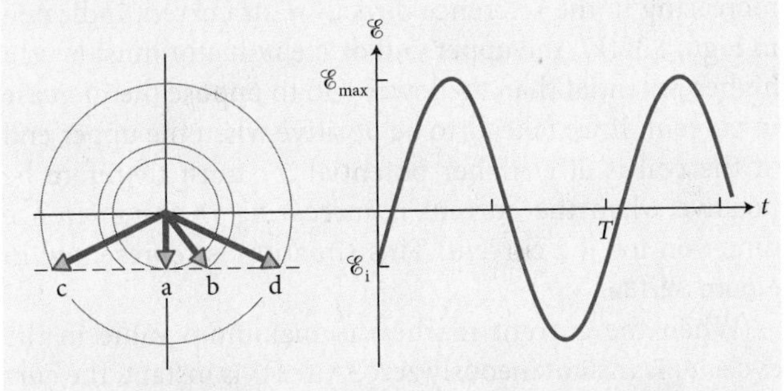

❶ **GETTING STARTED** I begin by observing from the graph that the emf is negative at instant $t = 0$ and increases until it reaches a maximum value $\mathscr{E}_{max}$.

❷ **DEVISE PLAN** To identify the correct phasor or phasors, I can use the following information: (1) the length of the phasor is equal to the amplitude of the oscillation, (2) the value of the emf at any instant corresponds to the vertical component of the phasor, and (3) the phasor rotates counterclockwise around the reference circle.

❸ **EXECUTE PLAN** The amplitudes of phasors a and b are too small and so I can rule these two out. The fact that the emf starts out negative at $t = 0$ and then increases tells me that the phasor representing it must be in the fourth quadrant (below the horizontal axis and to the right of the vertical axis), meaning the correct phasor must be d. ✔

❹ **EVALUATE RESULT** I can verify my answer by tracing out the projection of the phasor on the vertical axis as the phasors rotates counterclockwise. The initial value of the projection, initial phase, and amplitude all agree with the values of these variables represented in the graph.

32.3 Construct a phasor diagram for the time-dependent current and potential difference at $t = 0$ in the AC source-resistor circuit of Figure 32.6.

We can generalize the result of this checkpoint to represent i and v_R from Figure 32.7 at an arbitrary instant t_1. Because i and v_R are in phase for a resistor, the two phasors for i and v_R always have the same phase and so overlap (**Figure 32.13**). Note that the initial phase ϕ_i is zero because i and v_R are zero at $t = 0$ (at that instant both phasors point to the right along the horizontal axis).

The relative lengths of the I and V_R phasors are meaningless because the units of i and v_R are different. However, for circuits with multiple elements (resistors, inductors, or capacitors), the relative lengths of phasors showing the potential differences across different elements are meaningful and will prove very useful in analyzing the circuit.

Phasors are most useful when we need to represent quantities that are not in phase. **Figure 32.14** on the next page shows the phasor diagram that corresponds to Figure 32.10 (at the instant represented by Figure 32.9*a*). As the phasor diagram shows, the angle between V_C and I is 90°, and so the phase difference between the two phasors is $\pi/2$. Because the phasors rotate counterclockwise, we see that current phasor I is ahead of the potential difference phasor V_C, in agreement with our earlier conclusion that the current in a capacitor leads the potential difference across the capacitor.

Figure 32.13 Phasor diagram and graph showing time dependence of v_R and i from Figure 32.7.

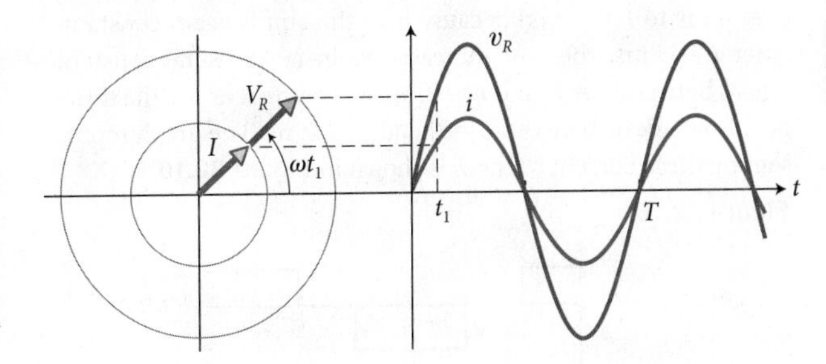

Figure 32.14 Phasor diagram and graph showing time dependence of i and v_C corresponding to Figure 32.10.

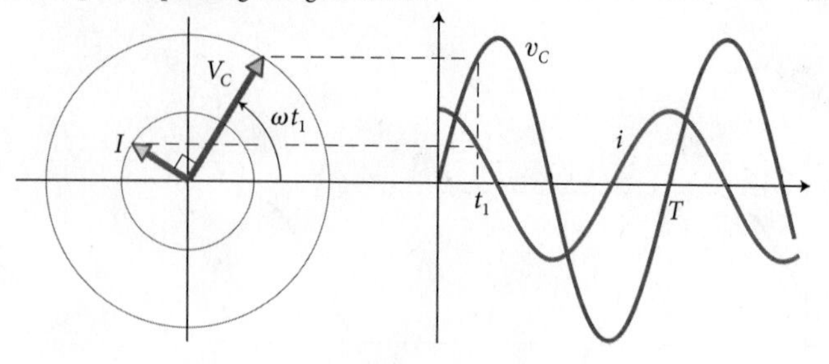

Example 32.3 Nonsinusoidal AC circuit

When a certain capacitor is connected to a nonsinusoidal source of emf as in **Figure 32.15a**, the emf varies in time as illustrated in Figure 32.15*b*. Sketch a graph showing the current in the circuit as a function of time.

Figure 32.15 Example 32.3.

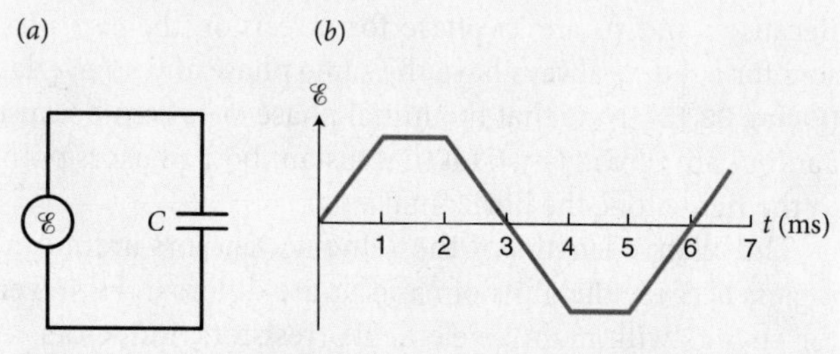

❶ **GETTING STARTED** From Figure 32.15*b* I see that the emf has five distinct parts during the time interval shown. During each part, the emf either is changing at a constant rate or is constant.

❷ **DEVISE PLAN** I know that the current is proportional to the rate at which the charge on the capacitor plates changes over time. I also know that the emf is proportional to the charge on the plates, and so the current is proportional to the derivative of the emf with respect to time.

❸ **EXECUTE PLAN** Between $t = 0$ and $t = 1$ ms, the emf increases at a constant rate, so $i = Cd\mathscr{E}/dt$ is constant and positive. Between $t = 1$ ms and $t = 2$ ms, the emf is constant, so $i = Cd\mathscr{E}/dt = 0$. Between $t = 2$ ms and $t = 4$ ms, the emf decreases at a constant rate, so $i = Cd\mathscr{E}/dt$ is constant and negative. Because the rate of decrease between $t = 2$ ms and $t = 4$ ms is the same as the rate of increase between $t = 0$ and $t = 1$ ms, the magnitude of the current between $t = 2$ ms and $t = 4$ ms should be the same as that between $t = 0$ and $t = 1$ ms. The current is zero again during the next millisecond ($t = 4$ ms to $t = 5$ ms) because here the emf is again constant. After $t = 5$ ms, the emf increases again at the same constant rate as between $t = 0$ and $t = 1$ ms, so the current has the same positive value as between $t = 0$ and $t = 1$ ms. The graph representing these current changes is shown in **Figure 32.16.** ✔

Figure 32.16

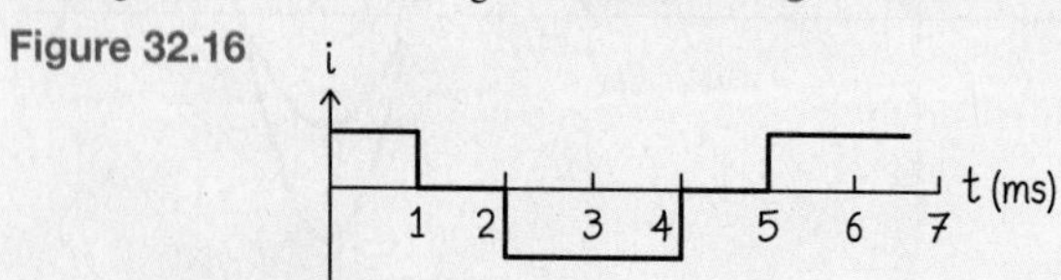

❹ **EVALUATE RESULT** When the current is positive, the emf is increasing; when the current is negative, the emf is decreasing; and when the current is zero, the emf is constant, as it should be.

Figure 32.17 AC circuit consisting of an inductor connected across the terminals of an AC source.

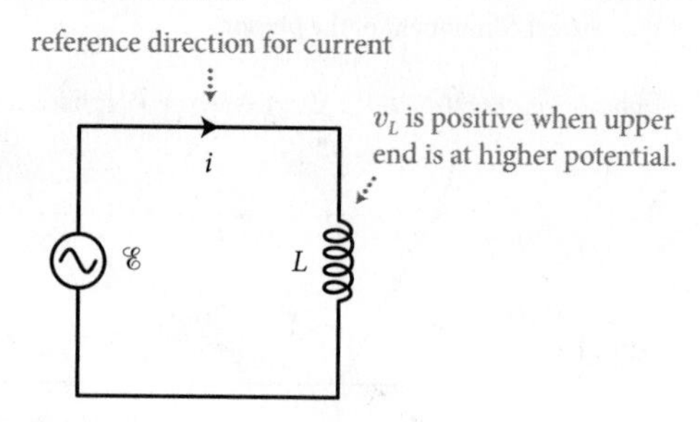

Now let's examine the behavior of an inductor connected to an AC generator (**Figure 32.17**). When the current in the circuit is changing, an emf is induced in the coil, in a direction to oppose this change (see Section 29.7). The potential difference between the ends of the inductor, which we'll denote by v_L, is proportional to the rate di/dt at which the current changes (Eq. 29.19). If the current is increasing in the reference direction for current indicated in Figure 32.17, the upper end of the inductor must be at a higher potential than the lower end to oppose the increase in current. If we take v_L to be positive when the upper end of the coil is at a higher potential, v_L must therefore be positive when the current is increasing in the reference direction for the current. This situation is represented in **Figure 32.18a**.

When the current reaches its maximum value in the cycle, v_L is instantaneously zero. After this instant, the current begins to decrease and the lower end of the inductor

Figure 32.18 Current and magnetic field oscillations through the inductor of Figure 32.17.

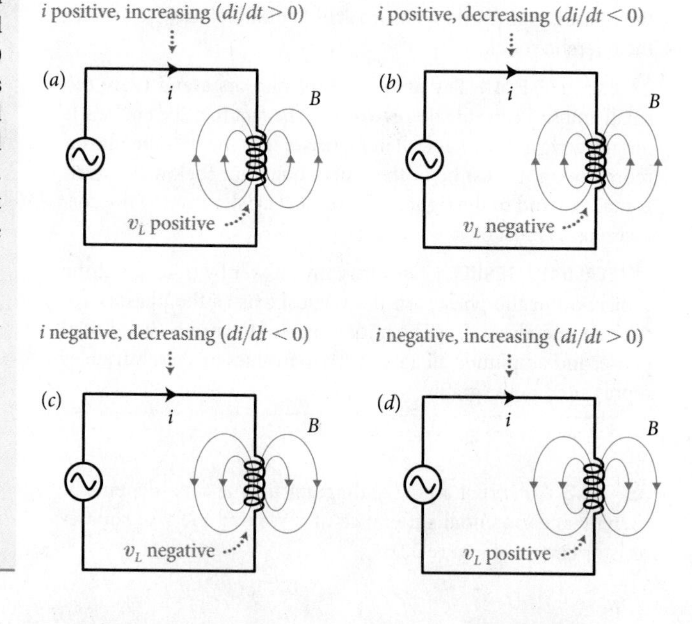

Figure 32.19 Graph of time-dependent current in the circuit and potential difference across the inductor for the circuit in Figure 32.17.

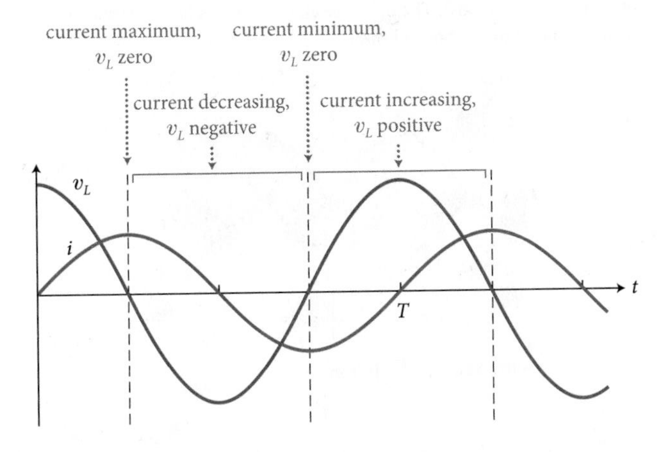

must be at a higher potential than the upper end to oppose this decrease in current. The potential difference v_L is now negative (Figure 32.18*b*). In the second half of the cycle, the current is in the opposite direction. As in the first part of the cycle, v_L has the same sign as di/dt (Figure 32.18*c* and *d*).

Figure 32.19 illustrates the time dependence of i and v_L in Figure 32.18. Note that the current maximum occurs one-quarter cycle after the potential difference maximum. For this reason, the current is said to *lag* the potential difference:

In an AC circuit that contains an inductor, the current in the inductor lags the potential difference by 90°.

Figure 32.20 shows the phasor diagram that corresponds to Figure 32.19 (at the instant represented by Figure 32.18*a*). Just as with the capacitor, the angle between V_L and I is 90° and so the phase difference is $\pi/2$, but in this case the current phasor I is behind the potential difference phasor V_L, in agreement with our earlier conclusion that the current in an inductor lags the potential difference across the inductor.

32.4 What are the initial phases for the phasors in Figures 32.13 and 32.20?

Figure 32.20 Phasor diagram and graph showing time dependence of i and v_L corresponding to Figure 32.19.

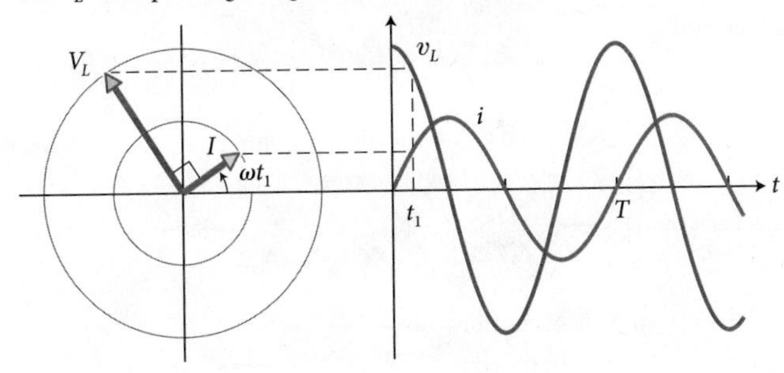

Figure 32.21 Schematic depiction of silicon, phosphorus, and boron atoms, shown as an inner core surrounded by valence electrons.

core: nucleus plus inner electrons

valence electron

	(*a*) silicon	(*b*) phosphorus	(*c*) boron
no. inner electrons:	10	10	2
no. valence electrons:	4	5	3

32.3 Semiconductors

Most modern electronic devices are made from a class of materials called **semiconductors**. Semiconductors have a limited supply of charge carriers that can move freely; consequently, their electrical conductivity is intermediate between that of conductors and that of insulators. Semiconductors are widely used in the manufacture of electronic devices such as transistors, diodes, and computer chips because their conductivity can be tailored chemically for particular applications layer by layer, even within a single piece of semiconductor.

Semiconductors are of two main types: intrinsic and extrinsic. *Intrinsic semiconductors* are chemically pure and have poor conductivity. *Extrinsic* or *doped semiconductors* are not chemically pure, have a conductivity that can be finely tuned, and are widely used in the microelectronics industry. The most widely used semiconductor is silicon, a nonmetallic element that makes up more than one-quarter of Earth's crust. **Figure 32.21*a*** shows a schematic of a silicon atom, which consists of a nucleus surrounded by fourteen electrons. Ten of these electrons are tightly bound to the nucleus—we'll refer to these electrons plus the nucleus as the *core* of the atom. The remaining outermost four electrons are called the atom's *valence electrons.* Each valence electron can form a covalent bond with a valence electron of another silicon atom. These bonds hold many identical silicon atoms together in a crystalline lattice (**Figure 32.22**).

Figure 32.22 Schematic of a crystalline lattice of silicon atoms, showing electrons participating in silicon-silicon bonds. (A real silicon crystal exists in three dimensions, and not all of the silicon-silicon bonds lie in a plane; this diagram illustrates only the essential idea that all of the valence electrons participate in covalent bonds.)

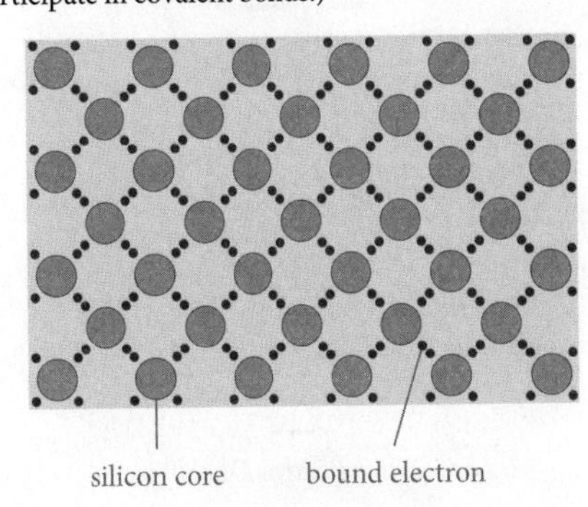

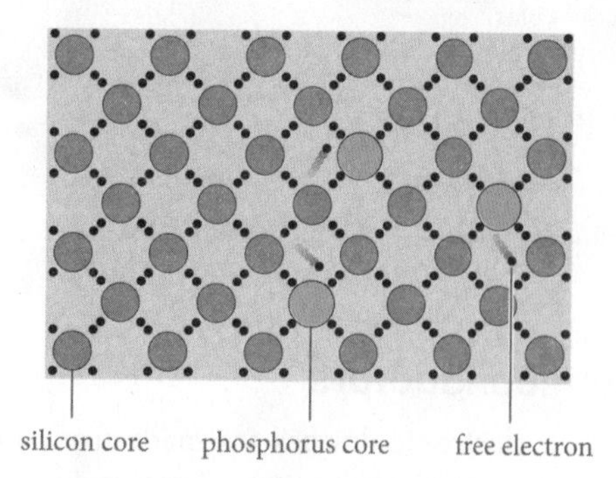

Figure 32.23 Schematic depiction of a crystalline lattice of silicon atoms doped with phosphorus atoms. The only charge carriers that are free to move in the crystal are the free electrons supplied by the phosphorus dopant atoms.

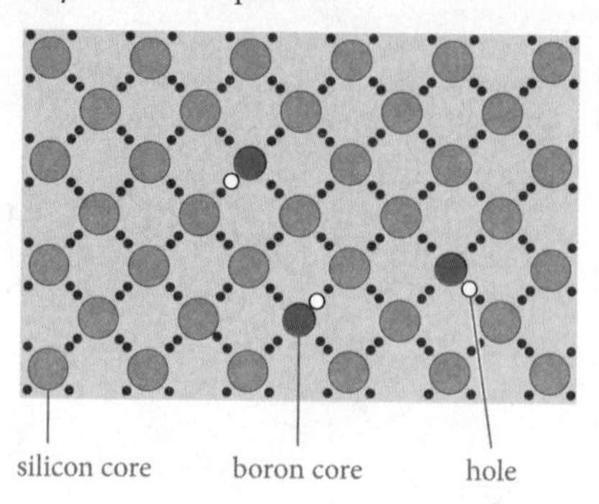

Figure 32.25 Schematic of crystalline lattice of silicon atoms with some boron atoms substituted for silicon, showing both bonding electrons and holes (missing electrons). The only free charge carriers in the crystal are the holes caused by the boron impurities.

The electrons in a covalent bond are not free to move; consequently, pure silicon has a very low electrical conductivity because all of its valence electrons form covalent bonds.

In extrinsic silicon, other types of atoms, such as boron or phosphorus, replace some of the atoms in the silicon lattice, introducing freely moving charge carriers into the lattice. The substituted atoms are called either *impurities* or *dopants*. For example, phosphorus has five valence electrons (Figure 32.21*b*). Because the silicon lattice structure requires only four bonds from each atom, the fifth electron from a phosphorus atom dopant is not involved in a bond and is free to move through the solid (**Figure 32.23**).

If an electric field is applied to the doped semiconductor of Figure 32.23, the free electrons move, creating a current in the semiconductor (**Figure 32.24**). As free electrons leave the semiconductor from one side, other free electrons enter it on the opposite side. Because the semiconductor must remain electrically neutral, the number of free electrons in the semiconductor at any given instant is always the same and it is equal to the number of phosphorus atoms in the material.

If boron atoms, which have three valence electrons (Figure 32.21*c*), are substituted for some silicon atoms in a silicon lattice, the "missing" fourth electron at each boron leaves behind what is called a **hole**—an incomplete bond (**Figure 32.25**). These holes behave like *positive* charge carriers and are free to move through the lattice (Figure 32.25). The holes therefore increase the ability of the silicon to conduct current, just as do the free electrons in phosphorus-doped silicon.

Keep in mind that the motion of holes involves electrons moving to fill existing holes, leaving new holes in the previous positions of the electrons (**Figure 32.26**). The boron

Figure 32.24 In an applied electric field, the free electrons in a phosphorous-doped semiconductor are free to move in the direction opposite the field direction. Free electrons leave the semiconductor at the left, travel through the circuit wire, and enter the semiconductor at the right.

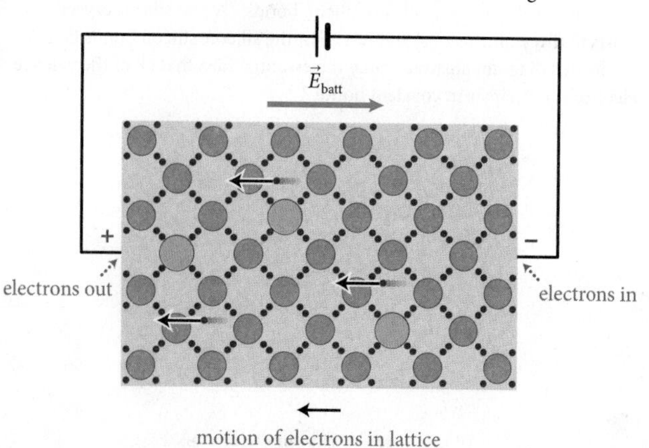

Figure 32.26 Sequence of four snapshots showing how holes "move" through a crystal by trading places with bonding electrons. In the presence of an electric field, holes move in the direction of the field (opposite to the directions in which the electrons move). To maintain continuity, free electrons from attached metal wires enter at the right, recombining with holes that accumulate there, and leave at the left.

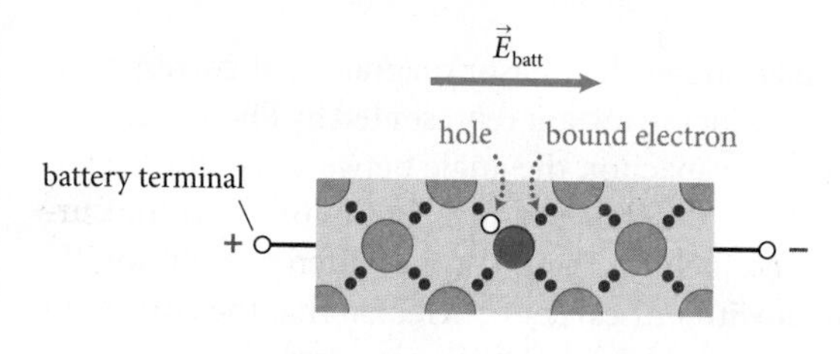

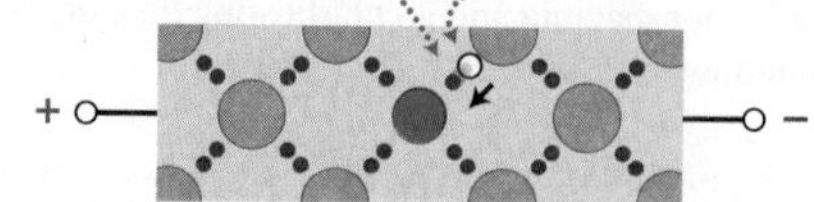

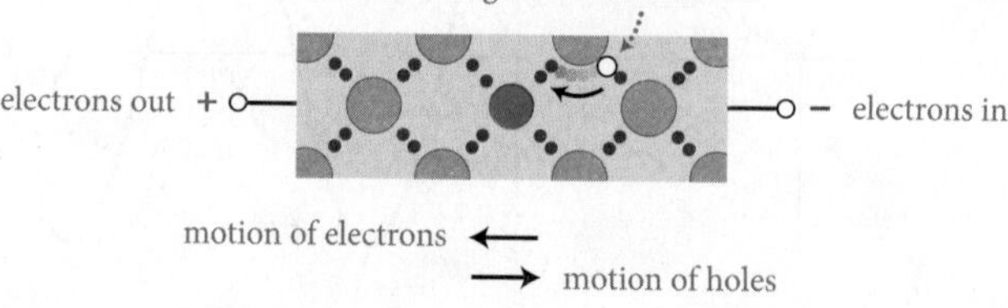

cores do not move! In the presence of an electric field, the positively charged holes move in the direction of the field as the negatively charged electrons move in the opposite direction. If the semiconductor is attached to metal wires on either side, as in Figure 32.26, free electrons travel into the semiconductor from the right (eliminating holes that reach the right edge) and travel out of the semiconductor on the left (producing holes on the left edge). Electrons thus flow from right to left, making holes travel in the opposite direction. Unlike the electrons, however, the holes never leave the semiconductor.

Doped semiconductors are classified according to the nature of the dopant. In a *p-type* semiconductor, the dopant has fewer valence electrons than the host atoms, contributing positively charged holes as the free charge carriers (thus the *p* in the name). In an *n-type* semiconductor, the dopant has more valence electrons than the host atoms, contributing negatively charged electrons as the free charge carriers (thus the *n* in the name). Substituting as few as ten dopant atoms per million silicon atoms produces conductivities appropriate for most electronic devices.

32.5 Is a piece of *n*-type silicon positively charged, negatively charged, or neutral?

32.4 Diodes, transistors, and logic gates

Although tailoring the conductivity of a single piece of semiconductor can be a useful procedure, the most versatile semiconductor devices combine doped layers that have different types of charge carriers. The simplest such device is a **diode,** made by bringing a piece of *p*-type silicon into contact with a piece of *n*-type silicon (**Figure 32.27*a***). Near the junction where the two pieces meet, free electrons from the *n*-type silicon wander into the *p*-type material, where they end up filling holes. This *recombination* process turns free electrons into bound electrons (that is, electrons not free to roam around in the material) and eliminates the holes. Likewise, some of the holes in the *p*-type silicon wander into the *n*-type silicon, where they recombine with free electrons.

As recombination events take place, a thin region containing no free charge carriers (neither free electrons nor holes), called the **depletion zone,** develops at the junction. Although there are no *free* charge carriers in this zone, the trapping of electrons on the *p*-side of the junction causes negative charge carriers that are nonmobile to accumulate there. Similarly, positive nonmobile charge carriers accumulate on the *n*-side of the junction. As a result, the depletion zone consists of a negatively charged region and a positively

(*a*) Pieces of *p*- and *n*-type doped silicon

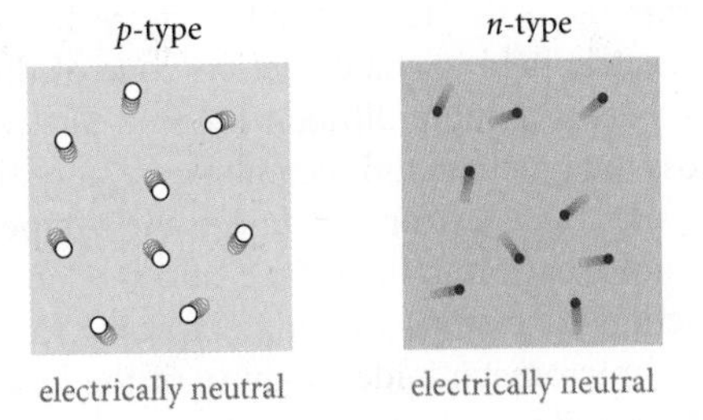

(*b*) When the two are put in contact, a diode is formed

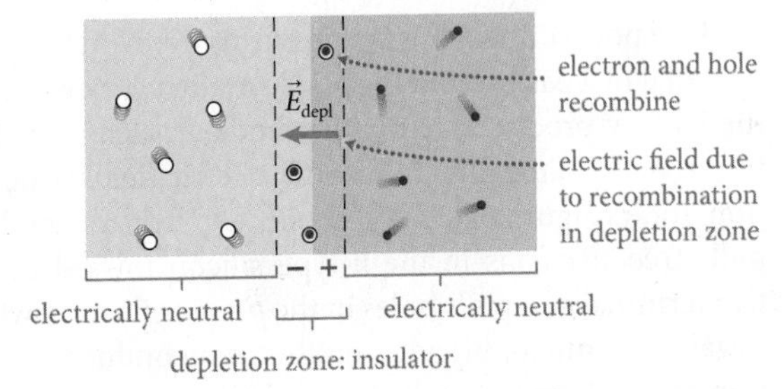

(*c*) Battery connected so as to produce electric field in *same* direction as electric field in depletion zone; diode blocks current

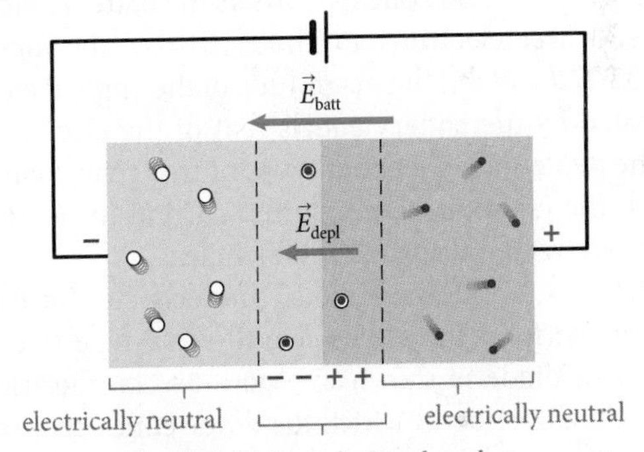

(*d*) Battery connected with the *opposite* polarity; diode conducts current

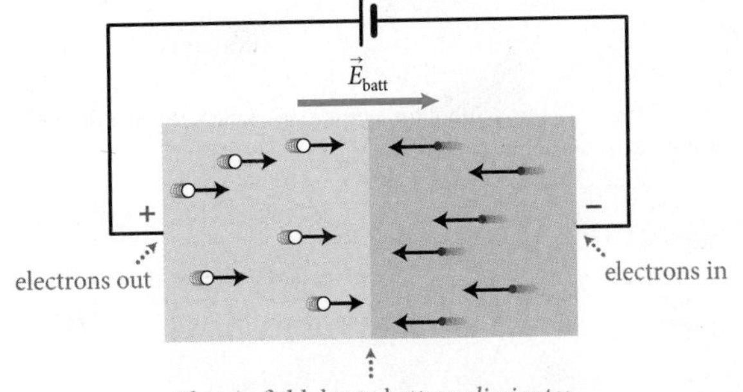

Figure 32.27 How a diode transmits current in one direction but blocks it in the other. If the battery is connected as shown in part *d* and produces a sufficiently strong electric field to compensate for the field of the depletion zone, there is a steady flow of both electrons and holes. (Remember, though: The holes never leave the semiconductor. Only the electrons enter and leave the semiconductor.)

charged region, and an electric field points across the depletion zone from the n-side to the p-side (Figure 32.27b).

As this electric field in the depletion zone of the diode increases, it becomes more difficult for free electrons and holes to cross the junction and recombine because the electric field pushes free electrons back into the n-type silicon and pushes holes back into the p-type silicon. Consequently, the depletion zone stops growing. Typically this region is less than a micrometer wide. Because of the lack of free charge carriers in it,

the depletion zone acts as an electrical insulator.

If we now connect the n-side of this diode to the positive terminal of a battery and the p-side to the negative terminal, the battery produces across the diode an electric field that points in the same direction as the electric field in the depletion zone (Figure 32.27c). The electric field of the battery pulls free electrons in the n-type silicon toward the positive terminal and pulls holes in the p-type silicon toward the negative terminal, broadening the (nonconducting) depletion zone. Connecting the battery in this manner therefore causes no flow of charge carriers in the diode.

When the battery is connected in the opposite direction, however, the depletion zone narrows as the battery's electric field pushes free electrons and holes toward the junction (Figure 32.27d). When the magnitude of the applied electric field created by the battery equals that of the electric field across the depletion zone, both types of free charge carriers can reach the junction, resulting in a current in the device carried both by free electrons and by holes.

As Figure 32.27 shows, a diode conducts current in one direction only: from the p-type side to the n-type side. The symbol for a diode is shown in **Figure 32.28a**; the triangle points in the direction in which the diode conducts current (from the p-side to the n-side).

32.6 In the diode of Figure 32.28a, which way do holes travel? Which way do electrons travel?

Figure 32.28 (a) Circuit symbol for a diode. (b) Schematic of a diode made using integrated-circuit technology.

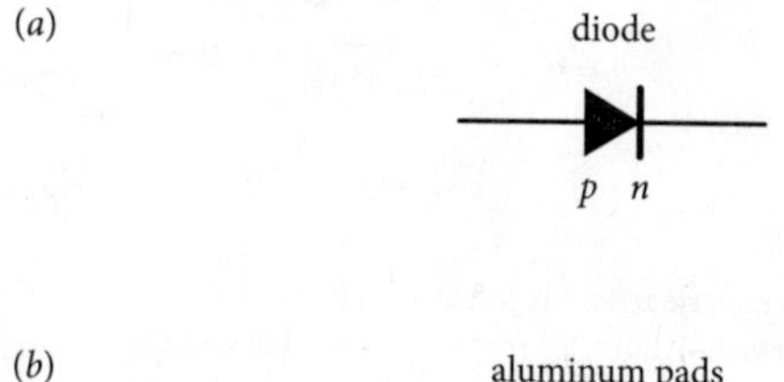

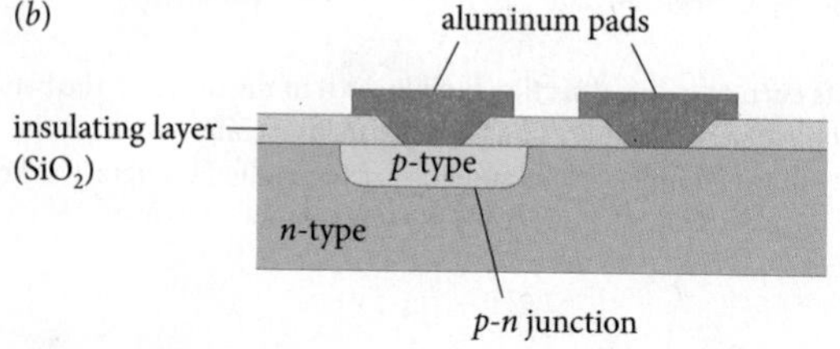

An *ideal diode* acts like a short circuit for current in the permitted direction and like an open circuit for current in the opposite direction. (That is not exactly how a diode behaves, but it's pretty close.)

32.7 Suppose a sinusoidally varying potential difference is applied across a diode connected in series with a resistor. Sketch the potential difference across the diode as a function of time, and then, on the same graph, sketch the current in the resistor as a function of time.

Example 32.4 Rectifier

Consider the arrangement of ideal diodes shown in **Figure 32.29**. This arrangement, called a *rectifier*, converts alternating current (AC) to direct current (DC). Sketch a graph showing, for a sinusoidally alternating source, the current in the resistor in the direction from b to c as a function of time.

Figure 32.29 Example 32.4.

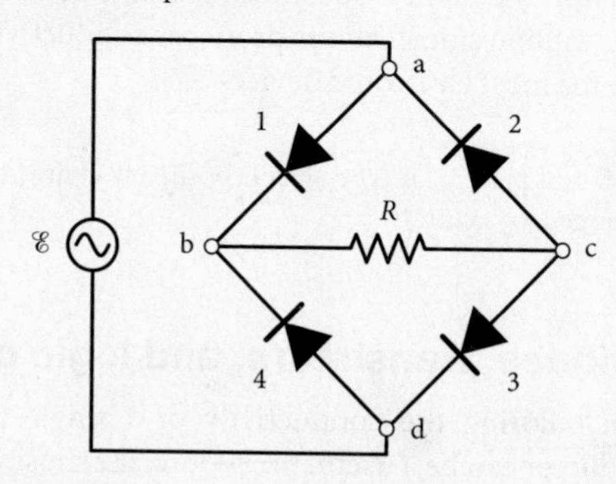

❶ **GETTING STARTED** Because the source is alternating, the current in the circuit periodically reverses direction. During part of the cycle the charge carriers creating the current flow clockwise through the source, and during another part of the cycle they flow counterclockwise. The diodes, however, conduct current in one direction only. I begin by making a sketch of the current between a and d, taking the direction from a to d to be positive (**Figure 32.30a**).

Figure 32.30

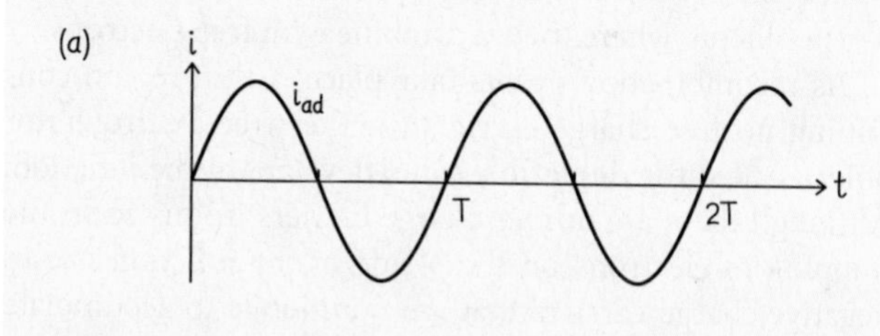

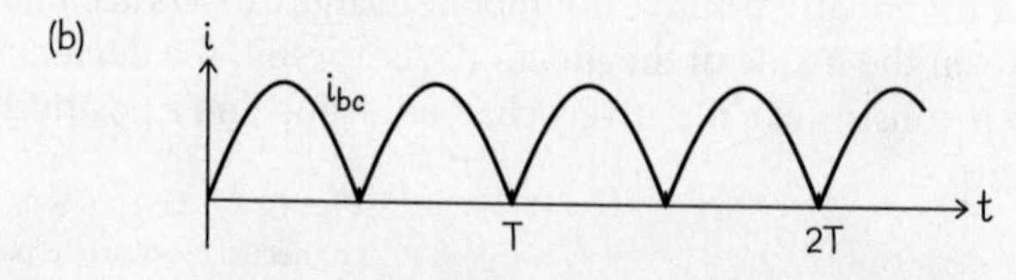

❷ **DEVISE PLAN** In an ideal diode, the charge carriers can flow only in the direction in which the triangle in the diode symbol points. I shall determine which diodes allow charge carriers to flow when the current direction is clockwise and when it is counterclockwise. I can then determine in each case which way the charge carriers flow through the resistor.

❸ **EXECUTE PLAN** When the current in the circuit is clockwise, only diodes 1 and 3 are conducting, so the current direction is abcd. When the current in the circuit is counterclockwise ($i_{ad} < 0$), only diodes 2 and 4 are conducting, so the current direction is dbca. At all instants, the current in the resistor points in the same direction: from b to c. This means that i_{bc} is positive regardless of whether i_{ad} is positive or negative. Whenever i_{ad} is negative, the diodes reverse the direction of the current in the resistor, so i_{bc} is always positive and my graph is as shown in Figure 32.30*b*. ✔

❹ **EVALUATE RESULT** The arrangement of diodes keeps the current from b to c always in the same direction, even though the current from a to d alternates in direction. It makes sense, then, that this arrangement of diodes is called a *rectifier*.

Figure 32.28*b* shows how a diode may be constructed as part of an integrated circuit (a computer chip, for example). An aluminum pad (part of the metal wire connecting the diode to the rest of the circuit) is in contact with a small *p*-type region of silicon, which is surrounded by a larger *n*-type region that is in contact with a second aluminum pad. The *p*-*n* junction forms at the interface between the *p*- and *n*-type regions. A thin layer of silicon oxide (SiO_2) insulates the aluminum from the underlying silicon except where electrical contact is needed. On a modern computer chip, the entire device is only a few micrometers wide.

Another important circuit element in modern electronics is the **transistor,** a device that allows current control that is more precise than the on/off control of a diode. A transistor consists of a thin layer of one type of doped semiconductor sandwiched between two layers of the opposite type of doped semiconductor. **Figure 32.31**, for example, shows an *npn-type bipolar transistor*—a thin layer of *p*-type silicon sandwiched between two thicker regions of *n*-type silicon.* If the *p*-type layer is thin, the depletion zone formed at the left *p*-*n* junction merges with the depletion zone formed at the right *p*-*n* junction. The merged depletion zones form one wide depletion zone.

When a potential difference is applied across such a transistor (**Figure 32.32*a***), the depletion zone across junction 1 disappears, but that across junction 2 grows, shifting the depleted region toward the positive terminal of the battery. While charge carriers can now cross junction 1 where the depletion zone has disappeared, the (shifted) depletion zone that still exists prohibits their movement, which means no current in the transistor. For historical reasons, the *n*-type region connected to the negative terminal is called the *emitter*, the *n*-type region connected to the positive terminal is called the *collector*, and the *p*-type layer is called the *base*. If the direction of the applied potential difference is reversed, the roles of the emitter and the collector are also reversed, and there is still no current in the transistor.

Figure 32.31 Schematic of an *npn*-type bipolar transistor, showing charge distribution and depletion zones for both *p*-*n* junctions.

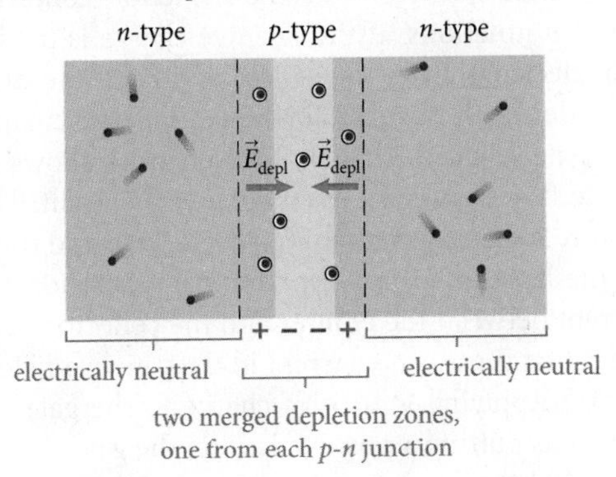

Figure 32.32 How an *npn*-type bipolar transistor works.

(*a*) Potential difference applied from collector to emitter only

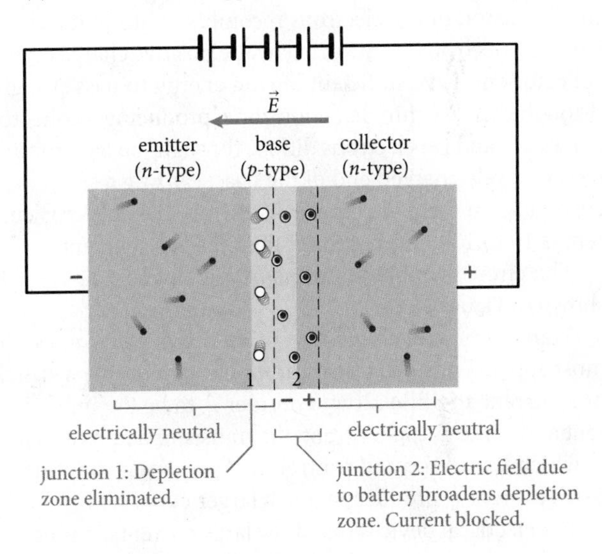

(*b*) Potential difference also applied from base to emitter

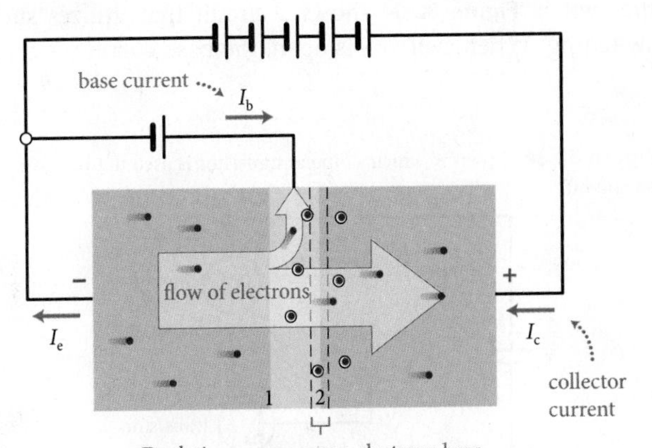

*Transistors in which a thin layer of *n*-type silicon is sandwiched between pieces of *p*-type silicon, called *pnp*-type bipolar transistors, are also used.

Figure 32.33 Circuit symbol for an *npn*-type bipolar transistor.

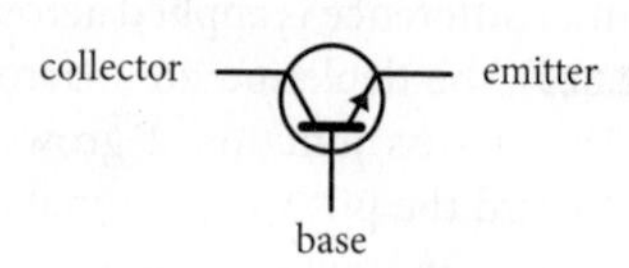

Figure 32.35 Schematic of an *npn*-type bipolar transistor made using integrated-circuit technology.

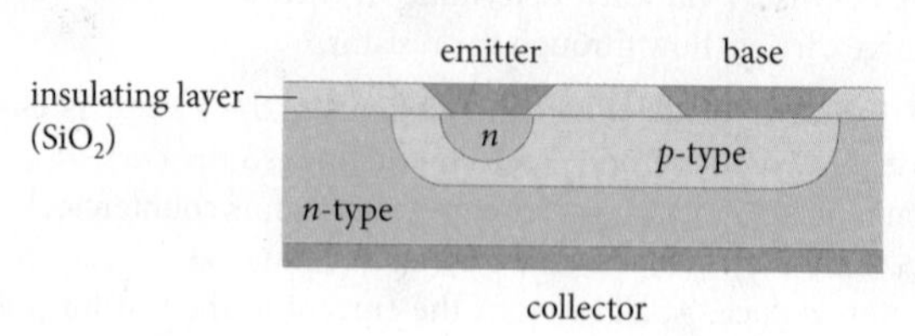

The situation changes drastically when, in addition to the potential difference between the emitter and the collector, a small potential difference is applied between the emitter and the base (Figure 32.32*b*). Adding this potential difference, called a *bias* or *bias potential difference,* makes the depletion zone much thinner than it is in Figure 32.32*a* because the formerly negatively charged region of this zone is brought to a positive potential, restoring mobile holes to that region. Because the emitter-base junction is conducting (remember, the depletion zone at junction 1 has disappeared), electrons now start flowing from the emitter toward the base. Once in the base, three things happen: (1) a small fraction of the electrons recombine with holes in the base, (2) electrons are attracted by the positive charge on the collector and have sufficient kinetic energy to pass straight through the very thin depletion zone, producing a collector current I_c, and (3) electrons diffuse through the base toward the positively charged end of the base, causing a small base current I_b. In a typical bipolar transistor, the collector current is 10 to 1000 times greater than the base current.

The circuit symbol for an *npn*-type bipolar transistor is shown in **Figure 32.33**.

Transistors are ubiquitous in modern electronics. In most applications, the transistor functions as either a switch or a current amplifier. If we consider I_b to be the input current and I_c the output current, the transistor acts as a switch in which I_b turns on and controls I_c. As a current amplifier, a small current I_b produces a much larger current I_c.

For electrical devices that draw large currents, it is useful to switch the device on and off with a mechanical switch wired in parallel with the device, rather than in series, so that the current in the device does not have to pass through the switch. **Figure 32.34** shows a circuit that utilizes such switching. When switch S is open, the base current is zero, and so the collector current (and therefore the current in the device) is zero. When switch S is closed, the small current from base to emitter causes a large current from collector to emitter that turns on the motor.

Figure 32.34 Circuit in which a bipolar transistor is used to turn a motor on and off.

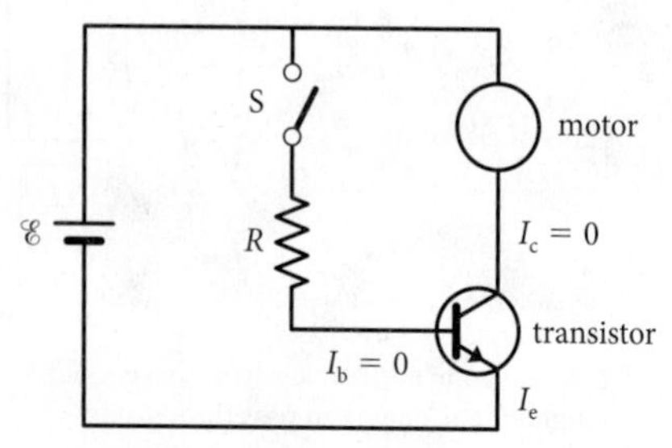

32.8 In a bipolar transistor, what relationship, if any, exists among I_b, I_c, and the emitter current I_e?

Figure 32.35 shows how an *npn*-type bipolar transistor can be fabricated. A drawback of this type of transistor, however, is that a continuous small current through the base is required to make the transistor conducting. For this reason, another type of transistor, called the *field-effect transistor,* is used much more frequently. **Figure 32.36*a*** shows the configuration of one. Two *n*-type wells are made in a piece of *p*-type material. The *p*-type material between the two wells is covered with a nonconducting oxide layer (typically SiO_2) and then with a metal layer called the *gate.* The two *n*-type wells are called the *source* and the *drain* (the *n*-type well that is kept at a higher potential is the drain).

Because of the depletion zones between the *p*-type and *n*-type materials, no charge carriers can flow from the source to the drain (or vice versa). The nonconducting layer between the gate and the *p*-type material prevents charge carriers from traveling between the gate and the rest of the device.

If the gate is given a positive charge, as in Figure 32.36*b*, the (positively charged) holes just underneath the gate are pushed away, forming underneath the gate an additional depletion zone that connects the depletion zones around the two *n*-*p* junctions. If the positive charge is made large enough, electrons from the source and from the drain are pulled underneath the gate, forming an *n*-type channel below the gate (Figure 32.36*c*). This channel allows charge carriers to flow between the source and the drain. The gate thus controls the current between the source and the drain, just as the base in an *npn*-type bipolar transistor controls the current between the emitter and the collector. (The difference is that there is no current in the gate in a field-effect transistor.) Applying a positive charge to the gate is often referred to as putting a positive *bias* on the gate.

Figure 32.37*a* shows the circuit symbol for a field-effect transistor, and Figure 32.37*b* shows how this type of transistor can be realized in an integrated circuit. This type of transistor has two advantages over the bipolar transistor

Figure 32.36 How a field-effect transistor works.

(*a*) Field-effect transistor with uncharged gate

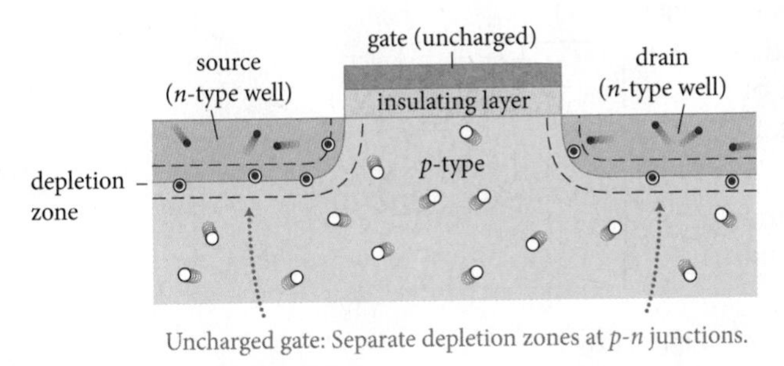

(*b*) Small positive charge on gate attracts electrons to gate and extends depletion zone below gate

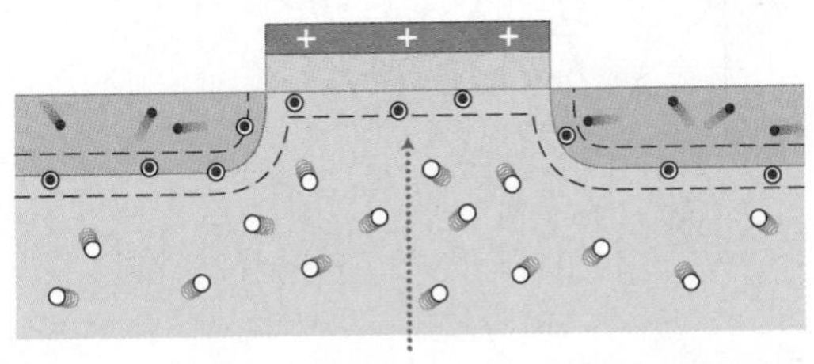

(c) Large positive charge on gate attracts more electrons to gate and causes *n*-type channel, which connects source and drain

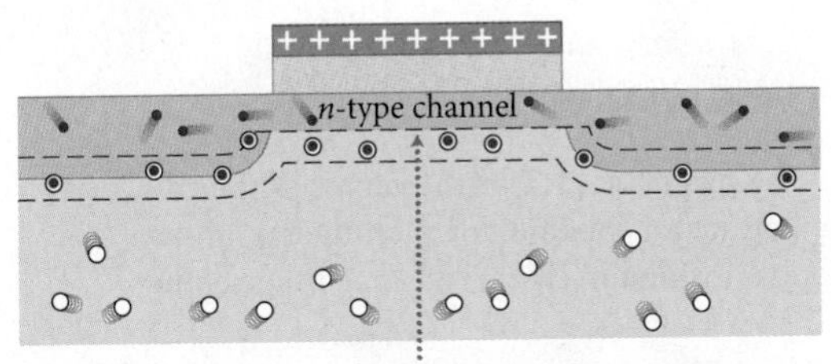

Figure 32.37 (*a*) Circuit symbol for a field-effect transistor. (*b*) Schematic of a field-effect transistor made using integrated-circuit technology.

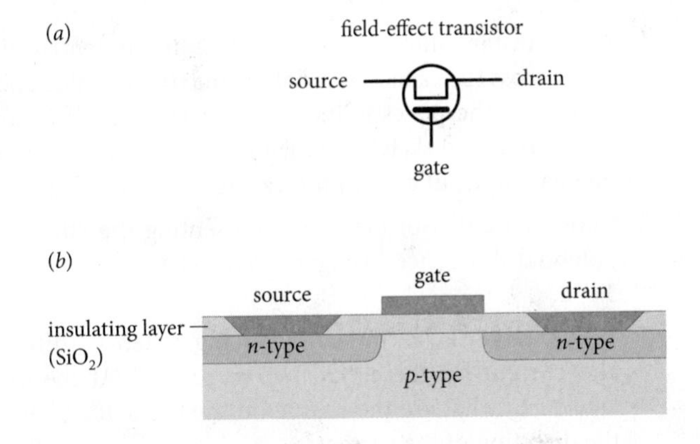

shown in Figure 32.35. First, all the terminals in the field-effect transistor are on the same side of the chip, making fabrication in integrated circuits much easier. Second, the current between the source and the drain is controlled by the charge on the gate, allowing a potential difference rather than a current to be used to control the source-drain current. Because no current is leaving the gate, no energy is required to keep current flowing from the source to the drain.

Field-effect transistors are widely used in devices called *logic gates,* which are the building blocks of computer processors and memory. A logic gate takes two input signals and provides an output after performing a logic operation on the input signals. For example, in a so-called AND gate, the output potential is nonzero with respect to ground only when both inputs are at positive potential with respect to ground. In an OR gate, the output potential is nonzero when either input potential is positive. The symbols used for these gates in circuit diagrams are shown in **Figure 32.38**; the inputs are on the left, and the output is on the right. In analyzing these circuits, we'll make the simplifying assumption that a transistor is just a switch that is open (off) when the potential of the gate is either at ground or negative with respect to ground and is closed (on) when the gate is at a positive potential.

Figure 32.38 Circuit symbols for AND and OR logic gates.

A, B — AND — $A \cap B$

A, B — OR — $A \cup B$

32.9 Circuit diagrams for two logic gates are shown in **Figure 32.39**. Which is the AND gate, and which is the OR gate? Explain briefly how each one works.

Figure 32.39 Checkpoint 32.9

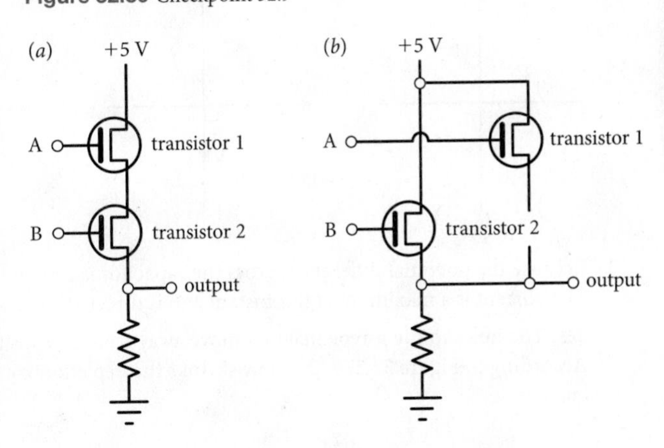

CONCEPTS

Self-quiz

1. At the instant shown in **Figure 32.40**, the potential difference across the capacitor is half its maximum value and the charge on the plates is increasing. Draw the direction of the current and sketch the magnetic field at this instant. Is the magnitude of current increasing or decreasing?

Figure 32.40

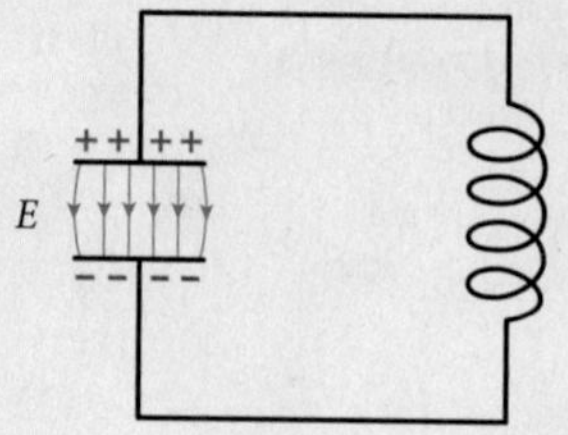

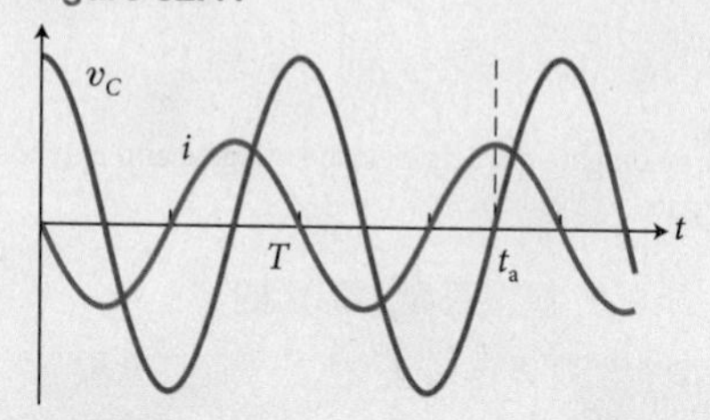

2. Construct a phasor diagram representing the current and potential difference in Figure 32.10 at $t = T/4$, $T/2$, and $3T/4$.
3. **Figure 32.41** shows the time-varying potential difference and current for the circuit of Figure 32.8. At the instant labeled t_a, what are the charge on the capacitor plates and the direction of the current?

Figure 32.41

4. Is there any current in a diode connected as shown in **Figure 32.42**?

Figure 32.42

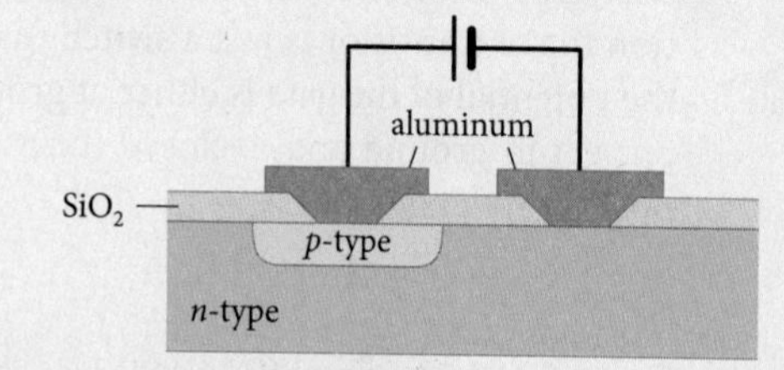

Answers:

1. Your sketch should show the current directed counterclockwise. The magnetic field in the center of the coil points up the page according to the right-hand dipole rule (assuming we are looking down on the top of the coil in Figure 32.40). Because the current is zero when the capacitor has maximum charge, the magnitude of the current is decreasing at the instant shown in Figure 32.40.
2. See **Figure 32.43.** At $t = T/4$ the potential difference phasor V_C points along the positive y axis because the potential difference reaches its maximum positive value at this instant, and the current phasor I points along the negative x axis because it leads the current by 90°. Each quarter cycle both phasors rotate 90° counterclockwise.

Figure 32.43

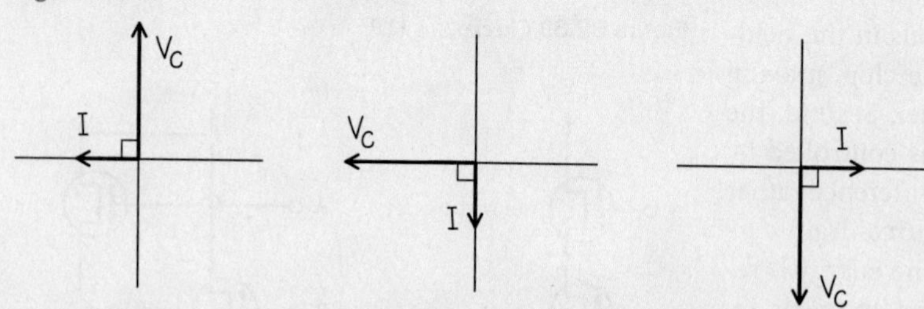

3. Because the potential difference across the capacitor is zero at instant t_a, the charge on the plates must be zero. The current is a maximum at this instant and is directed clockwise.
4. Yes. The holes in the p-type material move away from the positive terminal, and the electrons move toward it. According to Figure 32.27d, this flow shrinks the depletion zone, the charge carriers can flow, and so there is a current.

32.5 Reactance

Let us now develop a mathematical framework for analyzing alternating-current circuits. The instantaneous emf supplied by an AC source is customarily written as

$$\mathscr{E} = \mathscr{E}_{max} \sin \omega t, \qquad (32.1)$$

where $\mathscr{E}_{max}$ is the maximum value of the emf, typically called the *peak value* or *amplitude* (see Section 15.1), $\omega = 2\pi f$ is the angular frequency of oscillation in inverse seconds (Section 15.5), and f is the frequency in hertz. Most generators have frequencies of 50 Hz or 60 Hz. Audio circuits typically operate at kilohertz frequencies, radio transmitters at 10^8 Hz, for instance, and computer chips at 10^9 Hz. It's very important to remember to convert frequencies in hertz (cycles per second) to angular frequencies in s^{-1} when ω appears in the equations.

Note that the initial phase for the emf as written in Eq. 32.1 is zero. When we make this choice, the source emf serves as the reference for phase in the circuit.

Let's begin by revisiting the circuit from Exercise 32.1—a resistor connected to an AC source (**Figure 32.44**). At any instant, Ohm's law relates the potential difference across the resistor to the current in it, just as it does for DC circuits:

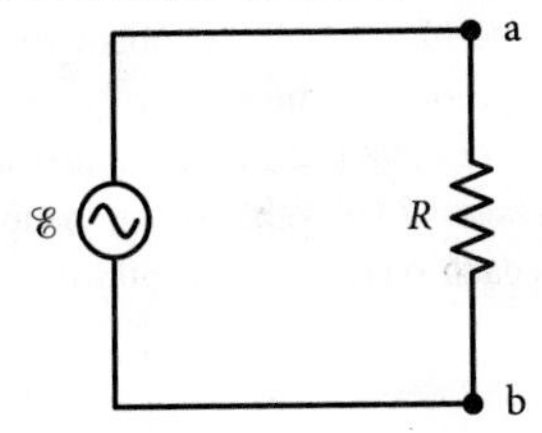

Figure 32.44 AC circuit consisting of a resistor connected across the terminals of an AC source.

$$v_R = iR. \qquad (32.2)$$

The only difference between Eq. 32.2 and Ohm's law for DC circuits (Eq. 31.11) is that the potential difference and the current in Eq. 32.2 oscillate in time.

Applying the loop rule to this circuit gives the AC version of Eq. 31.23:

$$\mathscr{E} - iR = 0. \qquad (32.3)$$

Equations 32.2 and 32.3 show that the potential difference across the load equals the emf supplied by the source (as we would expect):

$$v_R = \mathscr{E} = \mathscr{E}_{max} \sin \omega t. \qquad (32.4)$$

32.10 (*a*) In Figure 32.44, is the potential at point a higher or lower than the potential at b when the current direction is clockwise through the circuit? (*b*) If we define such a current to be positive, is $\mathscr{E}$ positive or negative? Express v_R in terms of the potential at a and the potential at b. (c) Half a cycle later, when the current is negative, is $\mathscr{E}$ positive or negative? Express v_R again in terms of the potential at a and the potential at b.

Using Eqs. 32.2 and 32.4, we can write the current in the resistor as

$$i = \frac{v_R}{R} = \frac{\mathscr{E}_{max} \sin \omega t}{R} = I \sin \omega t, \qquad (32.5)$$

where $I = \mathscr{E}_{max}/R$ is the amplitude of the current. Note that the current and the potential difference both oscillate at angular frequency ω and are in phase, as we concluded in Exercise 32.1. If we write $v_R = V_R \sin \omega t$, we see that the amplitudes of the current and the potential difference satisfy the relationship

$$V_R = IR. \qquad (32.6)$$

Figure 32.45 shows the corresponding phasor diagram and time dependence of v_R and i.

Figure 32.45 (*a*) Phasor diagram and (*b*) graph showing time dependence of i and v_R for the circuit shown in Figure 32.44.

(*a*)

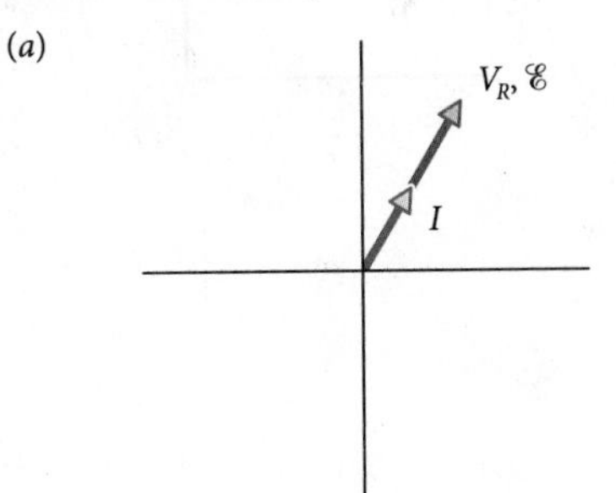

(*b*)

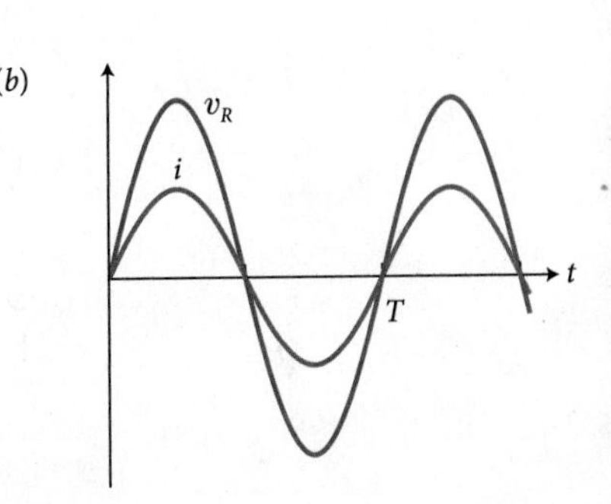

QUANTITATIVE TOOLS

Exercise 32.5 AC circuit with two resistors

In **Figure 32.46**, the resistances are $R_1 = 100\ \Omega$ and $R_2 = 60\ \Omega$, the amplitude of the emf is $\mathscr{E}_{max} = 160$ V, and its frequency is 60 Hz. (*a*) What is the amplitude of the potential difference across each resistor? (*b*) What is the instantaneous potential difference across each resistor at $t = 50$ ms?

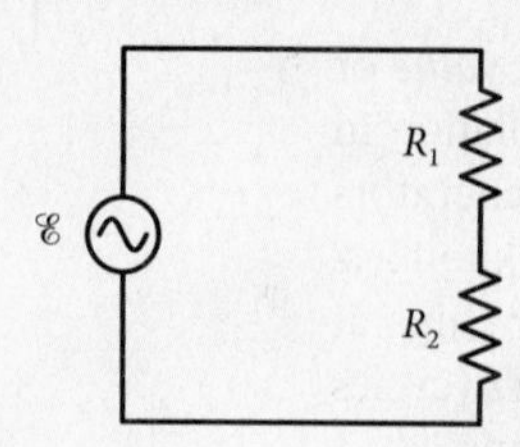

Figure 32.46 Exercise 32.5.

SOLUTION I analyze this circuit just as I would analyze a DC circuit containing two resistors, except now I must keep in mind that the current and potential differences are oscillating. The resistance of the load is

$$R_{load} = R_1 + R_2,$$

and the instantaneous current in the load is

$$i = \frac{\mathscr{E}}{R_{load}} = \frac{\mathscr{E}}{R_1 + R_2}.$$

(*a*) Because the current and the emf are in phase, they reach their maximum values at the same instant. As a result, the amplitude (maximum value) of the current is given by the amplitude of the emf divided by the resistance:

$$I = \frac{\mathscr{E}_{max}}{R_1 + R_2} = 1.0 \text{ A}.$$

The potential differences across the resistors are in phase with the current, and so I calculate the amplitude of the potential differences from the amplitude of the current using Eq. 32.6:

$$V_{R_1} = IR_1 = (1.0 \text{ A})(100\ \Omega) = 100 \text{ V}$$

$$V_{R_2} = IR_2 = (1.0 \text{ A})(60\ \Omega) = 60 \text{ V.} \checkmark$$

(*b*) I can use Eq. 32.5 to calculate the instantaneous value of the current:

$$i = (1.0 \text{ A}) \sin(2\pi \cdot 60 \text{ Hz} \cdot 0.050 \text{ s}) = 0.$$

(In 50 ms, three full cycles at 60 Hz take place.) Because the current is zero at 50 ms, the potential differences v_{R_1} and v_{R_2} at 50 ms are also zero. ✔

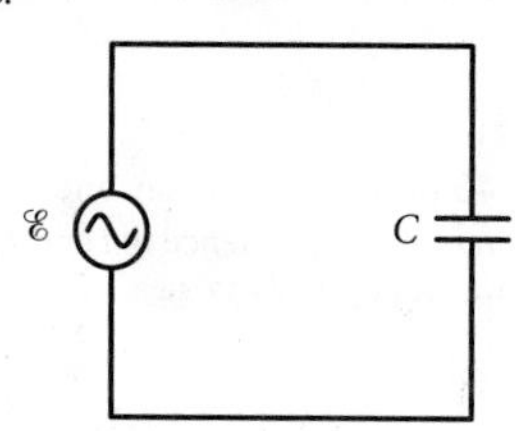

Figure 32.47 AC circuit consisting of a capacitor connected across the terminals of an AC source.

Next consider a capacitor connected to an AC source (**Figure 32.47**). Because the capacitor and the source are connected to each other, we have

$$\mathscr{E} = v_C, \tag{32.7}$$

and so the potential difference across the capacitor is

$$v_C = \mathscr{E}_{max} \sin \omega t = V_C \sin \omega t. \tag{32.8}$$

At any instant the potential difference across the capacitor and the charge on the upper plate are related by (see Eq. 26.1)

$$\frac{q}{v_C} = C, \tag{32.9}$$

where the potential difference v_C and the charge q on the plate oscillate in time. The charge on the upper capacitor plate is thus

$$q = Cv_C = CV_C \sin \omega t, \tag{32.10}$$

and the current is the rate of change of the charge on the plate:

$$i = \frac{dq}{dt} = \frac{d}{dt}(CV_C \sin \omega t) = \omega CV_C \cos \omega t. \tag{32.11}$$

Using the identity $\cos \alpha = \sin(\alpha + \frac{\pi}{2})$, we can rewrite this as

$$i = \omega CV_C \sin\left(\omega t + \frac{\pi}{2}\right) = I \sin\left(\omega t + \frac{\pi}{2}\right). \tag{32.12}$$

We now see that v_C and i are not in phase: i reaches its maximum value one-quarter period before v_C reaches its maximum value (**Figure 32.48**), as we found in Section 32.2.

The current in the capacitor of Figure 32.47 is not simply proportional to the potential difference across the capacitor because the two are out of phase. However, the *amplitude* of the current is proportional to the amplitude of the potential difference: $I = \omega C V_C$. Rewriting this to express V_C in terms of I gives

$$V_C = \frac{I}{\omega C}. \quad (32.13)$$

Note how this expression differs from the expression for a circuit that consists of only an AC source and a resistor, $V_R = IR$ (Eq. 32.6), where R is the proportionality constant between V and I. In Eq. 32.13, the proportionality constant is no longer a resistance (though it still has units of ohms). In circuits that contain capacitors and/or inductors, we use the general name **reactance** for the proportionality constant between the potential difference amplitude and the current amplitude. From Eq. 32.13 we see that this proportionality constant for a circuit that contains a capacitor is $1/\omega C$, and we call this constant the *capacitive reactance* X_C:

$$X_C \equiv \frac{1}{\omega C}, \quad (32.14)$$

so Eq. 32.13 becomes

$$V_C = IX_C. \quad (32.15)$$

Reactance is a measure of the opposition of a circuit element to a change in current. Unlike resistance, reactance is frequency dependent. At low frequency, the capacitive reactance X_C is large, which means that the amplitude of the current is small for a given value of V_C. At zero frequency, the current $I = \omega C V_C$ is zero, as it should be. (There is no direct current in a capacitor because the capacitor is just like an open circuit!) The higher the frequency of the source, the smaller the capacitive reactance and the greater the current (the less the capacitor opposes the alternating current).

Often, when analyzing AC circuits, the only things we are interested in are the amplitudes of the currents and potential differences. The capacitive reactance allows us to calculate the amplitude of the current in the capacitor directly from the amplitude of the potential difference across it—in this case, the emf of the source.

It is conventional to write the current in an AC circuit in the form

$$i = I\sin(\omega t - \phi), \quad (32.16)$$

where ϕ is called the **phase constant.** The negative sign in front of the phase constant is chosen so that a positive ϕ corresponds to shifting the curve for the current to the right, in the positive direction along the time axis, and a negative ϕ corresponds to shifting the curve to the left, in the negative direction along this axis (**Figure 32.49**).

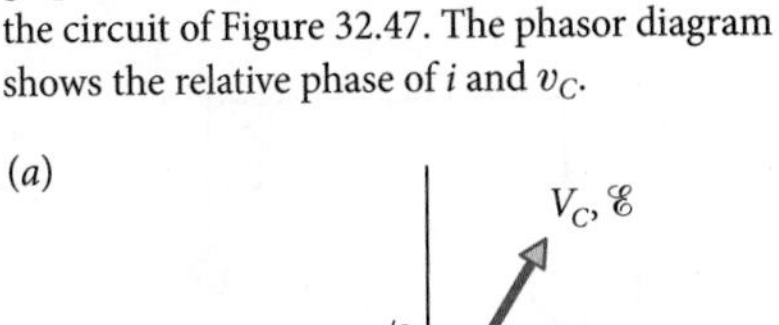
Figure 32.48 (*a*) Phasor diagram and (*b*) graph showing time dependence of i and v_C for the circuit of Figure 32.47. The phasor diagram shows the relative phase of i and v_C.

(*a*)

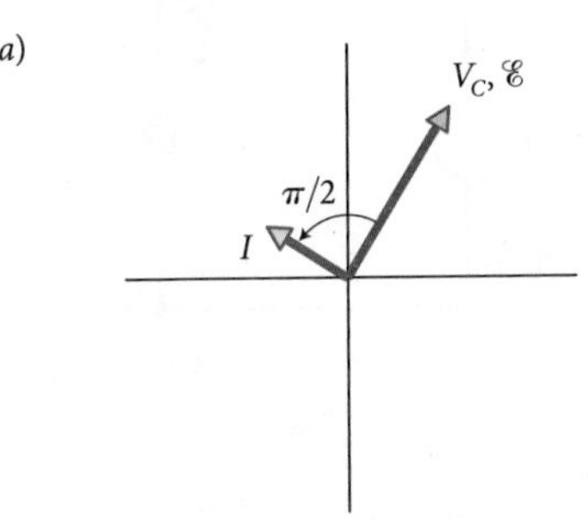

(*b*)

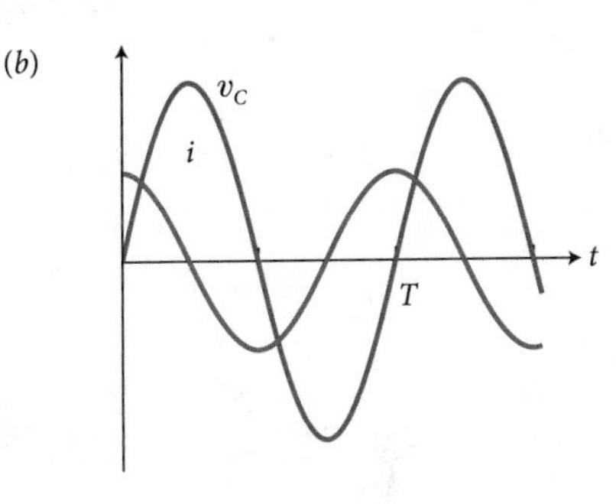

Figure 32.49 Positive and negative phase constant.

(*a*)

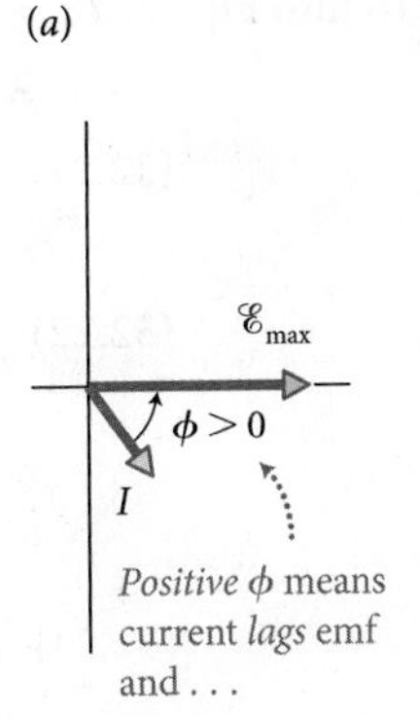

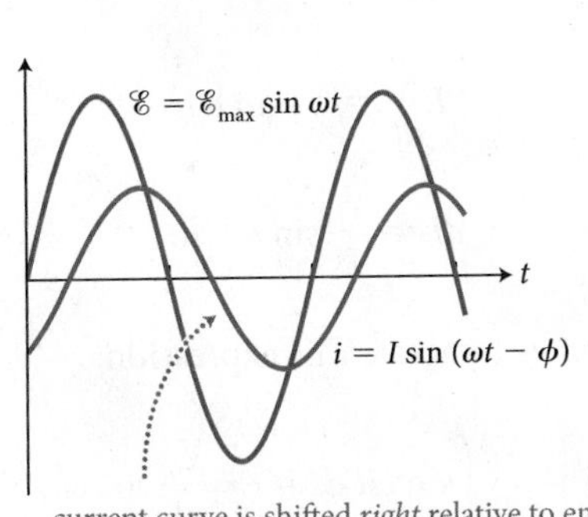

(*b*)

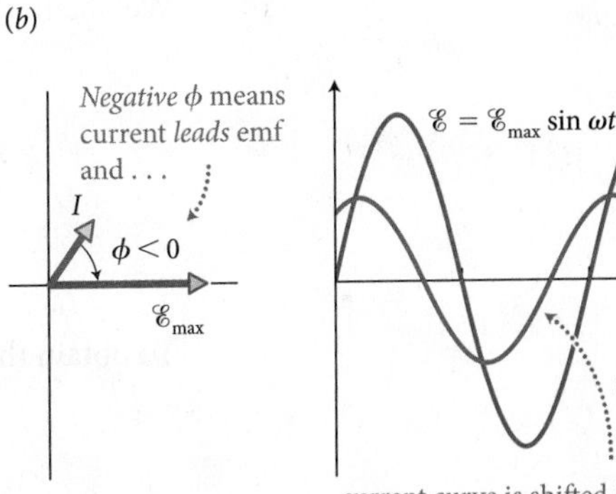

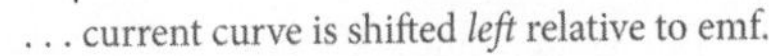

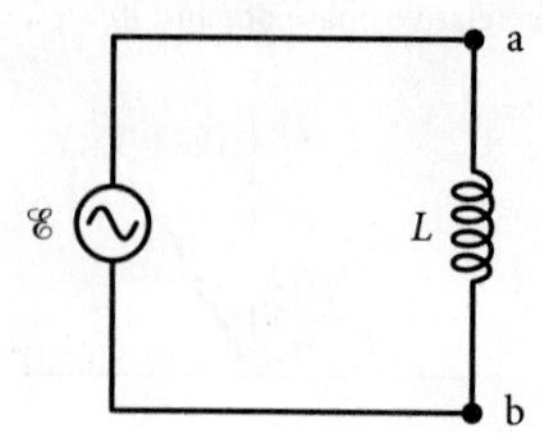

Figure 32.50 AC circuit consisting of an inductor connected across the terminals of an AC source.

The phase constant represents the phase difference between the source emf and the current. It is measured from the current phasor to the source emf phasor with the counterclockwise direction being positive (Figure 32.49). As a result, when the current leads the source emf, ϕ is negative; when the current lags the source emf, ϕ is positive.

Comparing Eqs. 32.12 and Eq. 32.16, we see that for the capacitor-AC source circuit of Figure 32.47, $\phi = -\pi/2$, as shown in Figure 32.48*a*. The negative phase constant means that the current leads the source emf. The curve for i is shifted to the left relative to the curve for v_C, as shown in Figure 32.48*b*. As you can see from the figure, when the capacitor has maximum charge (v_C maximum), the current is zero because at that instant the current reverses direction as the capacitor begins discharging. The current reaches its maximum value when the capacitor is completely discharged ($v_C = 0$).

32.11 As in the LC circuit discussed in Section 32.1, the current in the circuit of Figure 32.47 oscillates. If we think of v_C as corresponding to the position of the simple harmonic oscillator described in Section 15.5, what property of the circuit of Figure 32.47 corresponds to the velocity of the oscillator?

Finally, consider an inductor connected to an AC source (**Figure 32.50**). Because the inductor and the source are connected to each other, we have

$$\mathscr{E} = v_L, \tag{32.17}$$

so the potential difference across the inductor is

$$v_L = \mathscr{E}_{\text{max}} \sin \omega t = V_L \sin \omega t. \tag{32.18}$$

In Chapter 29 we saw that a changing current in an inductor causes an induced emf (Eq. 29.19):

$$\mathscr{E}_{\text{ind}} = -L\frac{di}{dt}. \tag{32.19}$$

The negative sign in this expression means that the potential decreases across the inductor in the direction of increasing current. Consequently, in Figure 32.50, the potential at b is lower than the potential at a when the current is increasing clockwise around the circuit. However, for consistency with Eq. 32.3, we always measure the potential difference v_L from b to a, just as we did with the AC source-resistor circuit of Figure 32.44. Therefore the sign of the potential difference across the inductor is the opposite of the sign in Eq. 32.19:

$$v_L = L\frac{di}{dt}. \tag{32.20}$$

We obtain the current in the circuit by substituting Eq. 32.18 into Eq. 32.20:

$$L\frac{di}{dt} = V_L \sin \omega t \tag{32.21}$$

$$di = \frac{V_L}{L} \sin \omega t \, dt. \tag{32.22}$$

To obtain the current, we integrate this expression:

$$i = \frac{V_L}{L}\int \sin \omega t \, dt = -\frac{V_L}{\omega L} \cos \omega t. \tag{32.23}$$

The amplitude of the current is thus

$$I = \frac{V_L}{\omega L}, \tag{32.24}$$

and using the identity $\cos \omega t = -\sin(\omega t - \frac{\pi}{2})$, we get

$$i = I \sin\left(\omega t - \frac{\pi}{2}\right). \tag{32.25}$$

The phase constant is $\phi = +\pi/2$, which means the current lags the source by 90°, as shown in **Figure 32.51**.

Just as we defined a capacitive reactance for a circuit that contains a capacitor, we define the *inductive reactance* X_L for a circuit that contains an inductor as the constant of proportionality between the amplitudes V_L and I in the circuit. From Eq. 32.24 we see that this proportionality constant is ωL:

$$X_L \equiv \omega L, \tag{32.26}$$

so that

$$V_L = IX_L. \tag{32.27}$$

Inductive reactance, like capacitive reactance, has units of ohms and depends on the frequency of the AC source. However, X_L increases with increasing frequency, so, at a given potential difference, the amplitude of the current is greatest at zero frequency and decreases as the frequency increases. This makes sense because for a constant current, an inductor is just a conducting wire and does not impede the current; as the frequency of the AC source increases, the emf induced across the inductor increases.

Figure 32.51 (*a*) Phasor diagram and (*b*) graph showing time dependence of i and v_L for the circuit of Figure 32.50. The phasor diagram shows the phase difference $\phi = \pi/2$ between i and v_L.

(*a*)

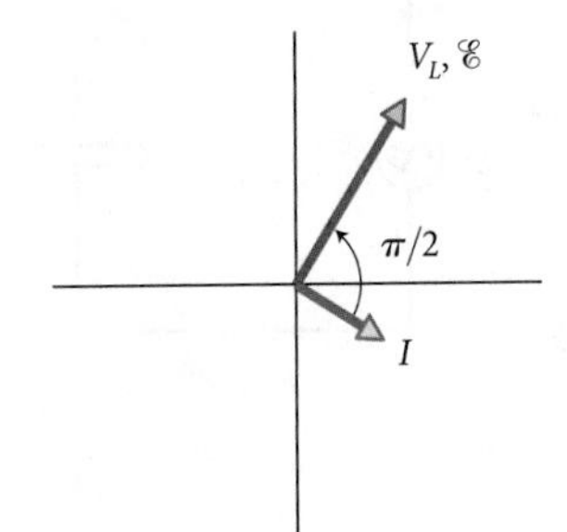

(*b*)

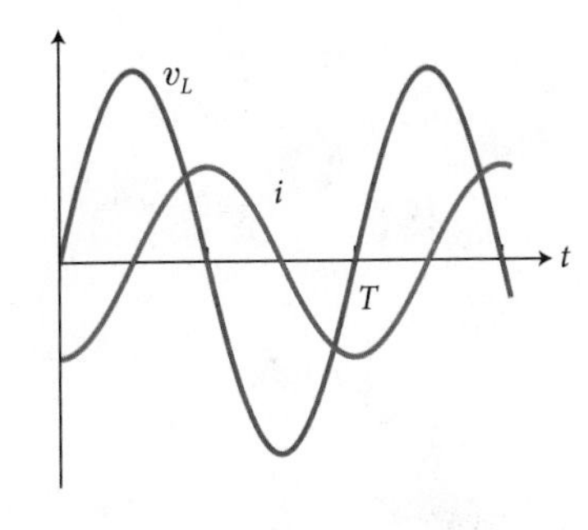

Example 32.6 Oscillating inductor

When a 3.0-H inductor is the only element in a circuit connected to a 60-Hz AC source that is delivering a maximum emf of 160 V, the current amplitude is I. When a capacitor is the only element in a circuit connected to the same source, what must the capacitance be in order to have the current amplitude again be I?

❶ **GETTING STARTED** I begin by identifying the information given in the problem statement: $\mathscr{E}_{max} = 160$ V, angular frequency $\omega = 2\pi(60$ Hz$)$, and inductance $L = 3.0$ H. The problem asks me to compare two circuits, one with an inductor connected to an AC source and the other with a capacitor connected to the same source. What I must determine is the capacitance value that makes the current amplitude the same in the two circuits.

❷ **DEVISE PLAN** For both circuits the potential difference across the load equals the source emf, so $\mathscr{E}_{max} = V_C = V_L$. I can use Eqs. 32.26 and 32.27 to get an expression for V_L in terms of I, from which I can express I in the inductor circuit in terms of V_L, ω, and L. Next I can use Eqs. 32.14 and 32.15 to get an expression for V_C in terms of I. I can then substitute this into my first expression for I and obtain an expression for C that contains only known quantities.

❸ **EXECUTE PLAN** Substituting the inductive reactance from Eq. 32.26, $X_L = \omega L$, into Eq. 32.27, $V_L = IX_L$, I get $V_L = I\omega L$, so the amplitude of the current is

$$I = \frac{V_L}{\omega L}. \tag{1}$$

Substituting the capacitive reactance from Eq. 32.14, $X_C = 1/\omega C$, into Eq. 32.15, $V_C = IX_C$, I get

$$V_C = \frac{I}{\omega C}. \tag{2}$$

Solving Eq. 2 for C and substituting Eq. 1 for I give

$$C = \frac{I}{\omega V_C} = \frac{V_L}{\omega^2 V_C L} = \frac{1}{\omega^2 L}$$

$$= \frac{1}{(2\pi \cdot 60 \text{ Hz})^2(3.0 \text{ H})} = 2.3 \times 10^{-6} \text{ F.} ✔$$

❹ **EVALUATE RESULT** To check my answer, I can calculate the inductive and capacitive reactances from Eqs. 32.26 and 32.14, respectively: $X_L = \omega L = (2\pi \cdot 60 \text{ Hz})(3.0 \text{ H}) = 1.1 \text{ k}\Omega$ and $X_C = 1/\omega C = 1/(2\pi \cdot 60 \text{ Hz})(2.3 \times 10^{-6} \text{ F}) = 1.1 \text{ k}\Omega$. The two are identical, as I expect given that they yield the same current amplitude for the same AC source.

Figure 32.52 An RC series circuit, consisting of a resistor and a capacitor in series across the terminals of an AC source.

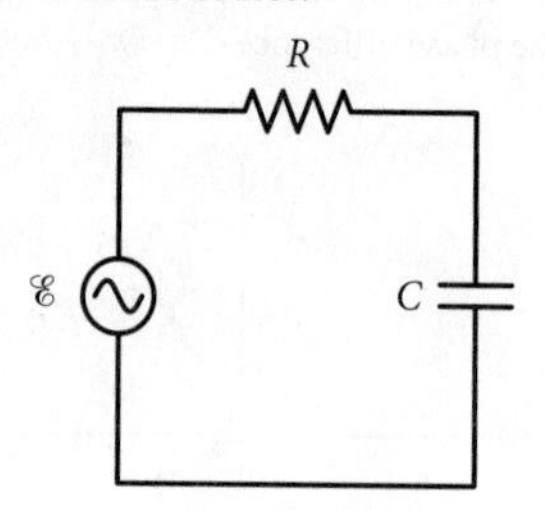

32.12 For the three circuits discussed in this section (AC source with resistor, capacitor, or inductor), sketch for a given emf amplitude (*a*) the resistance or reactance as a function of angular frequency ω and (*b*) the current amplitude in the circuit as a function of ω. Explain the meaning of each curve on your graphs.

32.6 *RC* and *RLC* series circuits

When an AC source is connected to multiple circuit elements, either in series or in parallel, applying the loop rule becomes more complicated than for DC circuits because we need to add several oscillating potential differences that may be out of phase with one another. For example, suppose we have a resistor and a capacitor in series with an AC source (**Figure 32.52**), known as an *RC series circuit*. The loop rule states that

$$\mathscr{E} = v_R + v_C. \quad (32.28)$$

To compute the sum on the right side of this equation, we must add potential differences that vary sinusoidally at the same angular frequency ω but are out of phase. The combined potential difference v of two potential differences v_1 and v_2 that oscillate at the same angular frequency is

$$v = V_1 \sin(\omega t + \phi_1) + V_2 \sin(\omega t + \phi_2), \quad (32.29)$$

where ϕ_1 and ϕ_2 are the initial phases of the two potential differences. Calculating this sum algebraically gets very messy, but using phasors to calculate it simplifies things greatly.

Figure 32.53*a* shows the phasors that correspond to the two terms on the right in Eq. 32.29. Recall that the instantaneous value of the quantity represented by a rotating phasor equals the vertical component of the phasor (see Figure 32.11). Therefore, v at any instant equals the sum of the vertical components of the phasors that represent v_1 and v_2. This sum is equal to the vertical component of the vector sum $V_1 + V_2$ of the phasors, as shown in Figure 32.53*b*.

Note that the combined potential difference v oscillates at the same angular frequency as v_1 and v_2. Consequently, the three phasors V_1, V_2, and $V_1 + V_2$ rotate as a unit at angular frequency ω, as shown in **Figure 32.54**. The phase relationship among the three phasors is constant, as is the phase relationship among the potential differences.

Figure 32.53 (*a*) Phasor diagram for a system of two oscillating potential differences v_1 and v_2. (b) Vector diagram indicating that the vertical component of the vector sum of the phasors equals the sum of the vertical components of the individual phasors.

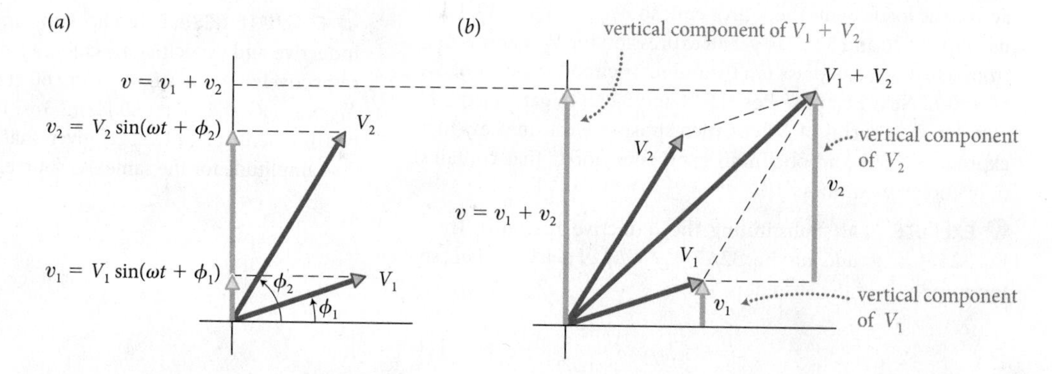

Figure 32.54 Phasor diagram and graph showing time dependence of v_1, v_2, and $v = v_1 + v_2$ from Figure 32.53. All three phasors rotate as a unit at angular frequency ω.

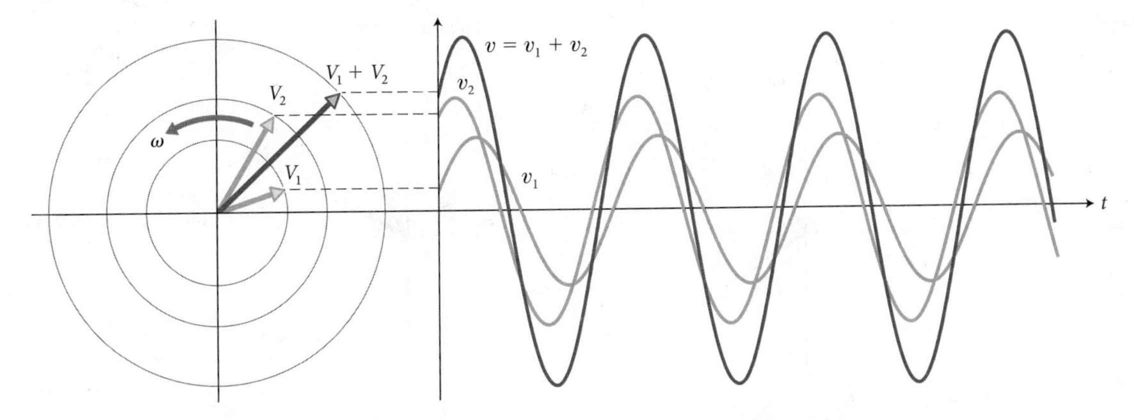

The next example shows how to apply these principles to a specific situation. To convince yourself that the phasor method is worthwhile, try adding the two original trigonometric functions algebraically after solving the problem using phasors!

Example 32.7 Adding phasors

Use phasors to determine the sum of the two oscillating potential differences $v_1 = (2.0 \text{ V}) \sin \omega t$ and $v_2 = (3.0 \text{ V}) \cos \omega t$.

❶ **GETTING STARTED** I begin by making a graph showing the time dependence of v_1 and v_2, and I draw the corresponding phasors V_1 and V_2 to the left of my graph (**Figure 32.55**). I add to my phasor diagram the phasor $V_1 + V_2$, which is the phasor that represents the potential difference sum $v_1 + v_2$ that I must determine. Using phasor $V_1 + V_2$, I can sketch the time dependence of the sum $v_1 + v_2$ by tracing out the projection of phasor $V_1 + V_2$ onto the vertical axis of my $V(\omega t)$ graph as this phasor rotates counterclockwise from the starting position I drew.

Figure 32.55

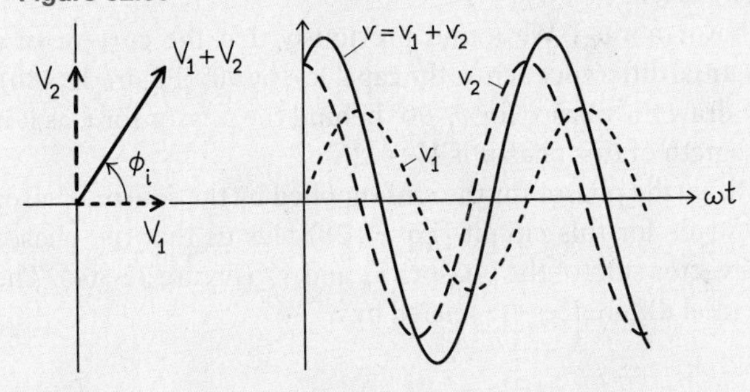

❷ **DEVISE PLAN** To obtain an algebraic expression for $v_1 + v_2$, I first write the oscillating potential differences in the form $v_1 = V_1 \sin(\omega t + \phi_1)$ and $v_2 = V_2 \sin(\omega t + \phi_2)$. Comparing these expressions with the given potential differences, I see that $V_1 = 2.0 \text{ V}$, $\phi_1 = 0$, and $V_2 = 3.0 \text{ V}$. In order to determine ϕ_2, I use the trigonometric identity $\cos(\omega t) = \sin(\omega t + \pi/2)$, and so my given information $v_2 = (3.0 \text{ V}) \cos \omega t = (3.0 \text{ V}) \sin(\omega t + \pi/2)$ tells me that $\phi_2 = \pi/2$. The sum $v_1 + v_2$ is a sinusoidally varying function that can be written as $v_1 + v_2 = A \sin(\omega t + \phi_i)$. The amplitude A is equal to the length of the phasor $V_1 + V_2$, and from my sketch I see that the initial phase ϕ_i is given by the angle between $V_1 + V_2$ and V_1.

❸ **EXECUTE PLAN** The length of the phasor $V_1 + V_2$ is given by the Pythagorean theorem applied to the right triangle containing ϕ_i in my phasor diagram:

$$A = \sqrt{V_1^2 + V_2^2} = \sqrt{(3.0 \text{ V})^2 + (2.0 \text{ V})^2}$$
$$= \sqrt{13 \text{ V}^2} = 3.6 \text{ V}.$$

The tangent of the angle between $V_1 + V_2$ and V_1 is then

$$\tan \phi_i = \frac{V_2}{V_1} = \frac{3.0 \text{ V}}{2.0 \text{ V}} = 1.5,$$

so $\phi_i = \tan^{-1}(1.5) = 56°$.

Now that I have determined A and ϕ_i, I can write the sum of the two potential differences as

$$v_1 + v_2 = (3.6 \text{ V}) \sin(\omega t + 56°). ✔$$

❹ **EVALUATE RESULT** The amplitude of the sinusoidal function I obtained is 3.6 V, which is greater than the larger of the two phasors I added, as I expect. My answer shows that the sum of the two potential differences reaches its maximum when $\omega t + \phi_i = 90°$, or when $\omega t = 90° - \phi_i = 34°$. This conclusion agrees with my phasor diagram: The phasor $V_1 + V_2$ reaches the vertical position after it rotates through an angle of $90° - \phi_i = 90° - 56° = 34°$. (I could also verify my answer by adding the two original sine functions algebraically, but the trigonometry needed in that approach is tedious.)

Figure 32.56 Steps involved in constructing a phasor diagram for the circuit in Figure 32.52. The diagram in part *d* indicates the phase of the current relative to the source emf.

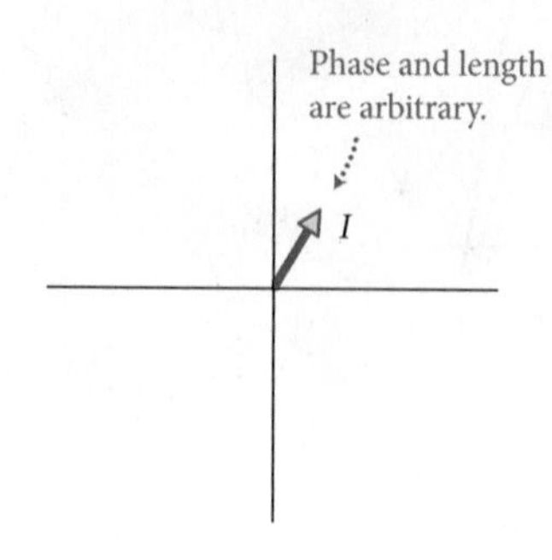

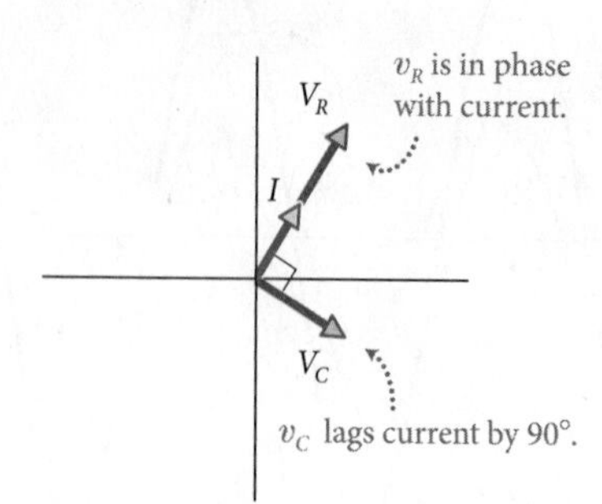

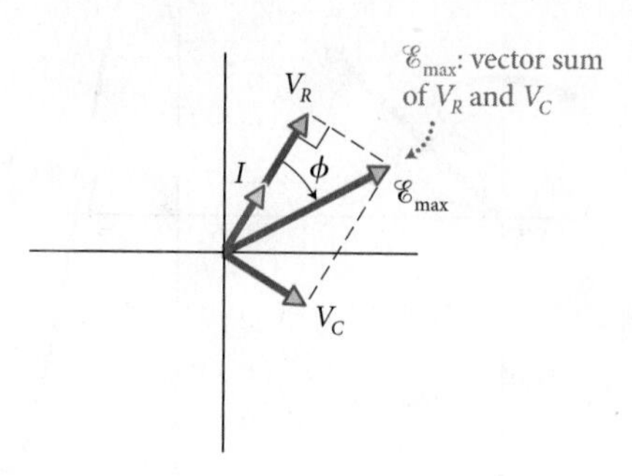

32.13 Suppose you need to add two potential differences that are oscillating at *different* angular frequencies—say, $2 \sin(\omega t)$ and $3 \cos(2\omega t)$. Can you use the phasor method described above to determine the sum? Why or why not?

Let us now return to the *RC* series circuit of Figure 32.52 and construct a phasor diagram in order to determine the amplitude and phase of the current in terms of the amplitude of the source emf and the resistance and capacitance of the circuit elements. From the current, we can calculate the potential differences across the circuit elements.

Because the circuit contains only one loop, the time-dependent current i is the same throughout. Therefore, we begin by drawing a phasor that represents i (**Figure 32.56*a***). We are free to choose the phase of this phasor because we have not yet specified the phase of any of the potential differences in the circuit. Also, the length we draw for phasor I is unimportant because it is the only current phasor for this circuit.

Next, we draw the phasors for v_R and v_C, the potential differences across the resistor and capacitor, respectively. We must get the relative phases right, and the lengths of the phasors must also be appropriately proportioned. Because the current is in phase with v_R (Figure 32.45*a*), we draw the corresponding phasor as shown in Figure 32.56*b*; its length is $V_R = IR$.

What about the phasor for v_C? We found previously that the current in a capacitor leads the potential difference across the capacitor by 90° (Figure 32.48*a*), which means we must draw the phasor for v_C 90° behind the phasor for i, as it is in Figure 32.56*b*. The length of this phasor is $V_C = IX_C$.

Finally, we need to draw the phasor for the emf supplied by the source. Phasor addition with the loop rule for this circuit (Eq. 32.28) tells us that the phasor $\mathscr{E}_{max}$ for the emf is the vector sum of the phasors V_R and V_C (Figure 32.56*c*). The amplitudes of the potential differences are related by

$$\mathscr{E}^2_{max} = V^2_R + V^2_C. \tag{32.30}$$

If we substitute $V_R = IR$ (Eq. 32.6) and $V_C = IX_C$ (Eq. 32.15), this becomes

$$\mathscr{E}^2_{max} = (IR)^2 + (IX_C)^2 = I^2(R^2 + X^2_C) = I^2\left(R^2 + \frac{1}{\omega^2 C^2}\right). \tag{32.31}$$

Solving for I gives

$$I = \frac{\mathscr{E}_{max}}{\sqrt{R^2 + 1/\omega^2 C^2}}. \tag{32.32}$$

Remembering that $\mathscr{E}_{max} = V_{load}$, we see that even though this load includes both resistive and reactive elements, I is still proportional to V_{load}! The constant of proportionality is called the **impedance** of the load and is denoted by Z:

$$I = \frac{\mathscr{E}_{max}}{Z}. \tag{32.33}$$

The impedance of the load is a property of the entire load. It is measured in ohms and depends on the frequency for any load that contains reactive elements.

Impedance plays the same role in AC circuits that resistance plays in DC circuits. In fact, Eq. 32.33 can be thought of as the equivalent of Ohm's law for AC circuits. Equation 32.32 shows that, for an RC series circuit, Z depends on both R and C:

$$Z_{RC} \equiv \sqrt{R^2 + 1/\omega^2C^2} \quad (RC \text{ series combination}). \tag{32.34}$$

To express V_R and V_C in terms of $\mathscr{E}_{max}$, R, C, and ω, we use Eq. 32.32:

$$V_R = IR = \frac{\mathscr{E}_{max}R}{\sqrt{R^2 + 1/\omega^2C^2}} \tag{32.35}$$

$$V_C = IX_C = \frac{\mathscr{E}_{max}/\omega C}{\sqrt{R^2 + 1/\omega^2C^2}}. \tag{32.36}$$

To calculate the phase constant ϕ, the geometry shown in Figure 32.56*c* gives us, with Eqs. 32.6 and 32.15,

$$\tan \phi = -\frac{V_C}{V_R} = -\frac{IX_C}{IR} = -\frac{1}{\omega RC} \tag{32.37}$$

or

$$\phi = \tan^{-1}\left(-\frac{1}{\omega RC}\right) \quad (RC \text{ series circuit}). \tag{32.38}$$

The negative value of ϕ indicates that the current in an RC series circuit leads the emf, just as it does in an AC circuit with only a capacitor. As you can see in Figure 32.56*c*, however, the phase difference between the emf and the current in the RC series circuit is less than 90°.

Example 32.8 High-pass filter

A circuit that allows emfs in one angular-frequency range to pass through essentially unchanged but prevents emfs in other angular-frequency ranges from passing through is called a *filter*. Such a circuit is useful in a variety of electronic devices, including audio electronics. An example of a filter, called a *high-pass filter*, is shown in **Figure 32.57**. Emfs that have angular frequencies above a certain angular frequency, called the *cutoff angular frequency* ω_c, pass through to the two output terminals marked v_{out}, but the filter attenuates the amplitudes of emfs that have frequencies below the cutoff value. (*a*) Determine an expression that gives, in terms of R and C, the cutoff angular frequency ω_c at which $V_R = V_C$. (*b*) Determine the potential difference amplitude v_{out} across the output terminals for $\omega \gg \omega_c$ and for $\omega \ll \omega_c$.

Figure 32.57 Example 32.8.

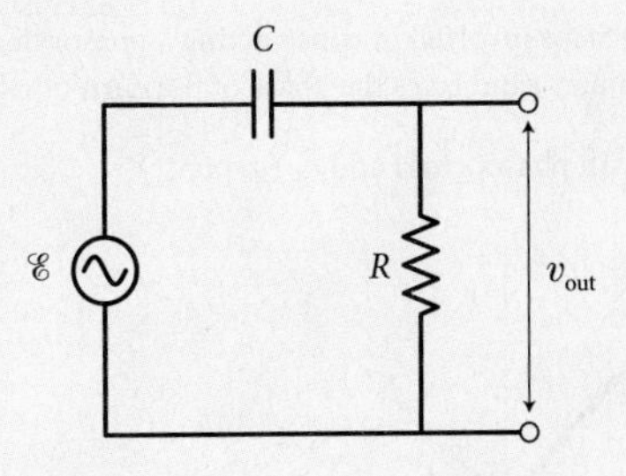

❶ GETTING STARTED This circuit is the same as the one in Figure 32.52, which I used to determine expressions for V_R (Eq. 32.35) and V_C (Eq. 32.36) in terms of R and C, so I can use

(Continued)

those results. From Figure 32.57 I see that the potential difference v_{out} is equal to the potential difference across the resistor, so $V_{out} = V_R$.

❷ **DEVISE PLAN** In order to determine the value of ω_c at which $V_R = V_C$, I equate the right sides of Eqs. 32.35 and 32.36. The resulting ω factor in my expression then is the cutoff value ω_c. For part *b*, I know that $V_{out} = V_R$. Therefore I can use Eq. 32.35 to determine V_{out} and then determine how V_{out} behaves in the limiting cases where $\omega \gg \omega_c$ and $\omega \ll \omega_c$.

❸ **EXECUTE PLAN** (*a*) Equating the right sides of Eqs. 32.35 and 32.36, I get $R = 1/\omega C$. Solving for ω yields the desired cutoff angular frequency ω_c:

$$\omega_c = \frac{1}{RC}. ✔$$

(*b*) To obtain the values of V_{out} for $\omega \gg \omega_c$ and for $\omega \ll \omega_c$, I first rewrite Eq. 32.35 in a form that contains ω_c:

$$V_{out} = V_R = \frac{\mathscr{E}_{max} R}{\sqrt{R^2 + 1/\omega^2 C^2}}$$

$$= \frac{\mathscr{E}_{max}}{\sqrt{1 + 1/R^2\omega^2 C^2}} = \frac{\mathscr{E}_{max}}{\sqrt{1 + \omega_c^2/\omega^2}}. \quad (1)$$

For $\omega \gg \omega_c$, the second term in the square root vanishes and Eq. 1 reduces to $V_{out} = \mathscr{E}_{max}$. ✔

For $\omega \ll \omega_c$, the second term in the square root dominates, so I can ignore the first term. Equation 1 then becomes

$$V_{out} = V_R = \frac{\mathscr{E}_{max}}{\sqrt{1 + \omega_c^2/\omega^2}} \approx \frac{\mathscr{E}_{max}}{\sqrt{\omega_c^2/\omega^2}}$$

$$= \frac{\mathscr{E}_{max}\omega}{\omega_c} = \mathscr{E}_{max}\omega RC.$$

In the limit that the angular frequency ω approaches zero, V_{out} approaches zero as well. ✔

❹ **EVALUATE RESULT** The name *high-pass filter* makes sense because this circuit allows emfs with an angular frequency higher than the cutoff angular frequency to pass through to the output but attenuates emfs of angular frequency lower than the cutoff angular frequency, preventing them from passing through to the output. It is the capacitor that does the actual passing or blocking. It blocks low-angular-frequency emfs because for these emfs the capacitive reactance, $X_C = 1/\omega C$, is very high. For high-angular-frequency emfs, X_C approaches zero, and so the capacitor passes the emf undiminished.

32.14 Interchange the resistor and the capacitor in Figure 32.57, and then show that the high-pass filter becomes a low-pass filter.

Filters can also be constructed by wiring an inductor and a resistor in series with an AC source. Such a circuit is called an *RL* series circuit and can be analyzed in exactly the manner we used to analyze an *RC* series circuit (see Example 32.9).

Finally, let's analyze an *RLC* series circuit: a resistor, a capacitor, and an inductor all in series with an AC source (**Figure 32.58**). As with the *RC* series circuit, the instantaneous current i is the same in all three elements, and the sum of all the potential differences equals the emf of the source:

$$\mathscr{E} = v_R + v_L + v_C. \quad (32.39)$$

The phasor diagram for this circuit is constructed in **Figure 32.59** for the case where $V_L > V_C$. As before, we begin with the phasors for i and v_R, and then note

Figure 32.58 An *RLC* series circuit, consisting of a resistor, an inductor, and a capacitor in series across the terminals of an AC source.

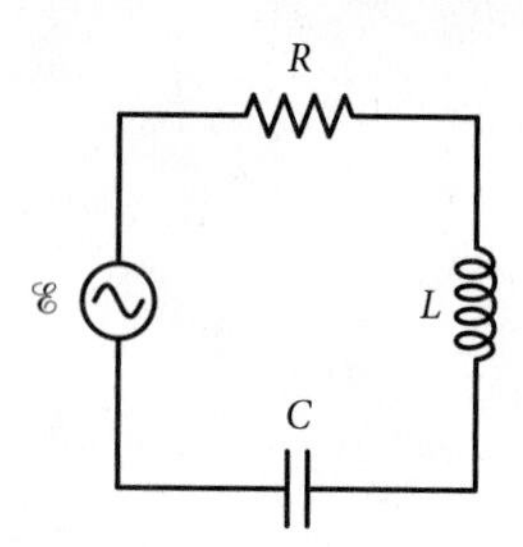

Figure 32.59 Steps involved in constructing a phasor diagram for the *RLC* series circuit in Figure 32.58. The diagram in part *c* indicates the phase of the current relative to the source emf.

(*a*) Begin with phasors for i and v_R (in phase)

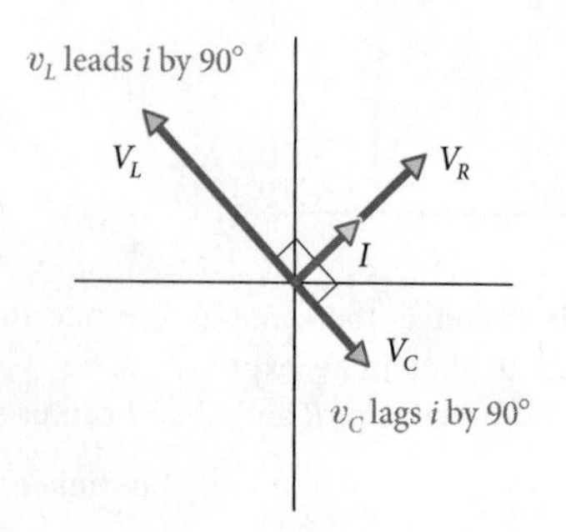

(*b*) Add V_C and V_L

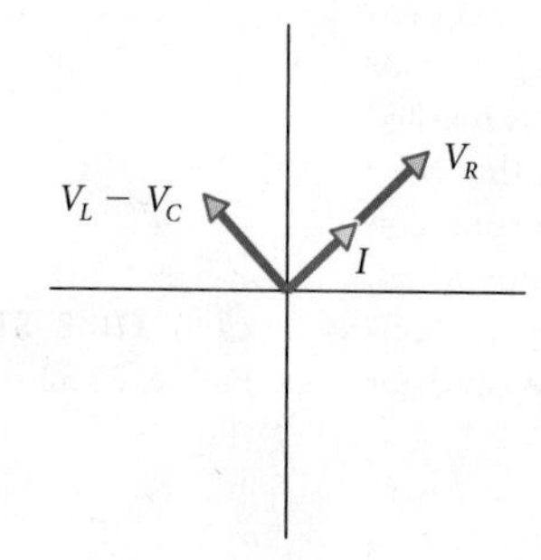

(*c*) Add $V_L - V_C$ and V_R to obtain $\mathscr{E}_{max}$

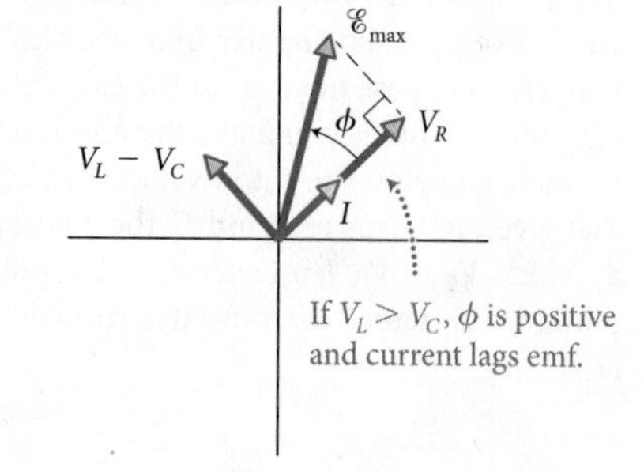

that v_C lags i by 90° and v_L leads i by 90° (Figure 32.59*a*). As a result, the phasors V_C and V_L can be added directly (Figure 32.59*b*). Finally, the loop rule (Eq. 32.39) requires the phasor for the emf to equal the vector sum of the phasors for the potential differences, as shown in Figure 32.59*c*. Consequently, the amplitudes V_R, V_L, and V_C must satisfy

$$\mathscr{E}_{max}^2 = V_R^2 + (V_L - V_C)^2. \quad (32.40)$$

Rewriting Eq. 32.40 in terms of I, R (from Eq. 32.6), X_L (from Eq. 32.27), and X_C (from Eq. 32.15) gives

$$\mathscr{E}_{max}^2 = I^2[R^2 + (X_L - X_C)^2] = I^2[R^2 + (\omega L - 1/\omega C)^2], \quad (32.41)$$

and thus

$$I = \frac{\mathscr{E}_{max}}{\sqrt{R^2 + (\omega L - 1/\omega C)^2}}. \quad (32.42)$$

The impedance of the *RLC* series combination (in other words, the constant of proportionality between I and $\mathscr{E}_{max}$) is therefore

$$Z_{RLC} \equiv \sqrt{R^2 + (\omega L - 1/\omega C)^2} \quad (RLC \text{ series combination}). \quad (32.43)$$

Table 32.1 lists the impedances of various loads.

Figure 32.59*c* shows that the phase relationship between the current and the source emf depends on the relative magnitudes of V_L and V_C. The phase of the current relative to the emf is given by

$$\tan\phi = \frac{V_L - V_C}{V_R} = \frac{X_L - X_C}{R}$$

$$= \frac{\omega L - 1/\omega C}{R} \quad (RLC \text{ series circuit}). \quad (32.44)$$

If $V_L > V_C$, as it is in Figure 32.59, ϕ is positive, meaning that the current lags the source emf. Here the inductor dominates the capacitor, and as a result the series combination of the inductor and capacitor behaves like an inductor. If $V_L < V_C$, ϕ is negative, the inductor-capacitor combination is dominated by the capacitor, and the current leads the source emf, just as in an *RC* series circuit.

In general, when analyzing AC series circuits, follow the procedure shown in the Procedure box on page 868.

Table 32.1 Impedances of various types of loads (all elements in series)

Load	Z
R	R
L	ωL
C	$1/\omega C$
RC	$\sqrt{R^2 + (1/\omega C)^2}$
RLC	$\sqrt{R^2 + (\omega L - 1/\omega C)^2}$

Note that impedances do not simply add the way resistances do. However, the impedance of any simpler load can be found from the impedance of the *RLC* combination; for example, $Z_{RC} = Z_{RLC}$ without the term containing L.

Procedure: Analyzing AC series circuits

When analyzing AC series circuits, we generally know the properties of the various circuit elements (such as R, L, C, and $\mathscr{E}$) but not the potential differences across them. To determine these, follow this procedure:

1. To develop a feel for the problem and to help you evaluate the answer, construct a phasor diagram for the circuit.
2. Determine the impedance of the load using Eq. 32.43. If there is no inductor, then ignore the term containing L; if there is no capacitor, ignore the term containing C; and so on.
3. To determine the amplitude of the current, in the circuit, you can now use Eq. 32.42; to determine the phase of the current relative to the emf, use Eq. 32.44.
4. Determine the amplitude of the potential difference across any reactive element using $V = XI$, where X is the reactance of that element. For a resistor, use $V = RI$.

Example 32.9 *RL* series circuit

Consider the circuit shown in **Figure 32.60**. (*a*) Determine the cutoff angular frequency ω_c and the phase constant at which $V_R = V_L$. (*b*) Can this circuit be used as a low-pass or high-pass filter?

Figure 32.60 Example 32.9.

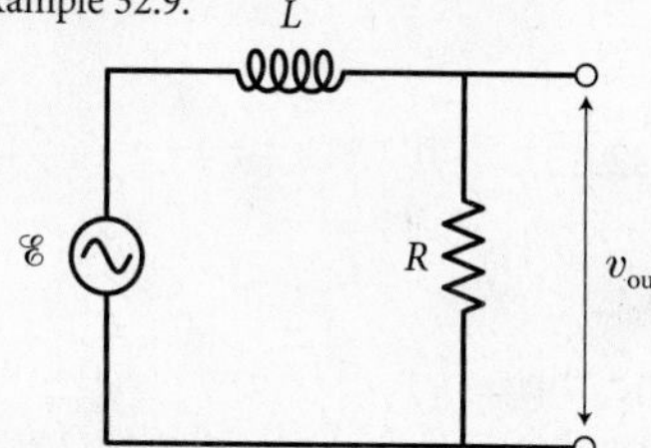

❶ GETTING STARTED This example is similar to Example 32.8, with the capacitor of that example replaced by an inductor here. As in Example 32.8, I see from the circuit diagram that the potential difference v_{out} is equal to the potential difference across the resistor, so $V_{out} = V_R$. I begin by drawing a phasor diagram for the circuit (**Figure 32.61**). I first draw phasors V_R and I, which I know from Figure 32.45*a* are in phase. I then add V_L, which leads I by 90° (Figure 32.51*a*). I make V_L have the same length as V_R because the problem asks about the circuit when $V_R = V_L$.

Figure 32.61

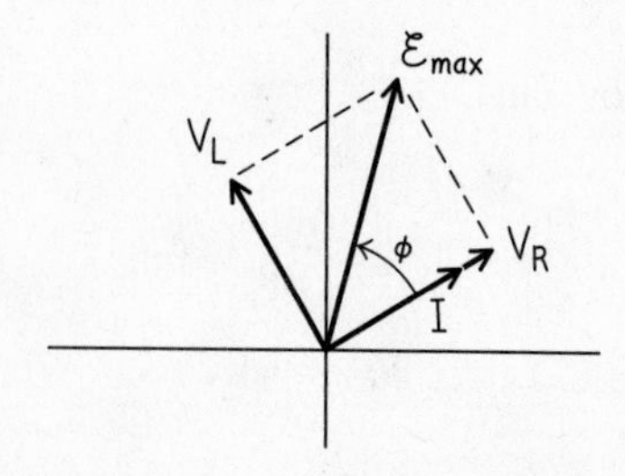

❷ DEVISE PLAN To determine the potential difference amplitudes V_L and V_R across the inductor and the resistor, I follow the procedure given in the Procedure box above. I then set these two amplitudes equal to each other in order to determine ω_c and the phase constant. To determine whether this circuit can be used as a low-pass or high-pass filter, I examine the behavior of V_{out} for $\omega \gg \omega_c$ and for $\omega \ll \omega_c$.

❸ EXECUTE PLAN (*a*) Ignoring the term containing C in Eq. 32.43 and substituting the result in Eq. 32.33, I get for the current amplitude

$$I = \frac{\mathscr{E}_{max}}{\sqrt{R^2 + (\omega L)^2}}.$$

I can now use Eq. 32.6 to calculate the amplitude of the potential difference across the resistor,

$$V_R = IR = \frac{\mathscr{E}_{max} R}{\sqrt{R^2 + (\omega L)^2}}, \quad (1)$$

and Eq. 32.27 to calculate the amplitude of the potential difference across the inductor,

$$V_L = IX_L = \frac{\mathscr{E}_{max} \omega L}{\sqrt{R^2 + (\omega L)^2}}, \quad (2)$$

where I have substituted ωL for X_L (Eq. 32.26). Equating the right sides of Eqs. 1 and 2 yields $R = \omega L$. Substituting ω_c for ω in this equation and solving for ω_c give me for the cutoff angular frequency value at which $V_R = V_L$:

$$\omega_c = \frac{R}{L}. ✔$$

To determine the phase constant for the condition $V_R = V_L$, I substitute V_R for V_L in Eq. 32.44 and set V_C equal to zero:

$$\tan\phi = \frac{V_L - V_C}{V_R} = \frac{V_R}{V_R} = 1,$$

so the phase constant is 45°. ✔

(*b*) Just as I did in Example 32.8, to obtain the limiting values of V_{out}, I first rewrite Eq. 1 in a form that contains ω_c:

$$V_{out} = V_R = \frac{\mathscr{E}_{max} R}{\sqrt{R^2 + (\omega L)^2}} = \frac{\mathscr{E}_{max}}{\sqrt{1 + (\omega L)^2/R^2}}$$

$$= \frac{\mathscr{E}_{max}}{\sqrt{1 + \omega^2/\omega_c^2}}. \quad (3)$$

QUANTITATIVE TOOLS

For $\omega \ll \omega_c$, the second term in the square root vanishes and Eq. 3 reduces to $V_{out} = \mathscr{E}_{max}$. For $\omega \gg \omega_c$ the second term in the square root dominates and we can ignore the first term. Equation 3 then becomes

$$V_{out} = V_R = \frac{\mathscr{E}_{max}}{\sqrt{1+\omega^2/\omega_c^2}} \approx \frac{\mathscr{E}_{max}}{\sqrt{\omega^2/\omega_c^2}} = \frac{\mathscr{E}_{max}\omega_c}{\omega} = \frac{\mathscr{E}_{max}R}{\omega L}.$$

In the limit that the angular frequency ω becomes very large, V_{out} approaches zero. The circuit thus blocks high-frequency emfs and allows low-frequency ones to pass through to the output. Therefore it can be used as a low-pass filter. ✔

❹ EVALUATE RESULT From my phasor diagram I see that the triangle that has V_R and V_L as two of its sides is an equilateral right-angle triangle, and so the phase constant ϕ must be 45°, as I obtained.

For part *b*, an emf is generated in the inductor whenever the current in it changes. This emf is proportional to the rate of change of the current in the inductor and it opposes the change in current (Eq. 29.19, $\mathscr{E}_{ind} = -L(di/dt)$). For a low-angular-frequency emf, di/dt is small, and the signal passes through the inductor essentially undiminished. For a high-angular-frequency signal, the inductive reactance $X_L = \omega L$ is high, and so the inductor essentially blocks the signal. It therefore makes sense that the arrangement in Figure 32.60 can serve as a low-pass filter.

Example 32.10 *RLC* series circuit

Consider an *RLC* circuit, such as the one shown in Figure 32.58. The source emf has amplitude 160 V and frequency 60 Hz. The resistance is $R = 50\ \Omega$ and the inductance is $L = 0.26$ H. If the amplitudes of the potential difference across the capacitor and the inductor are equal, what is the current in the circuit?

❶ GETTING STARTED I begin by drawing a phasor diagram for the circuit (**Figure 32.62**). I first draw phasors V_R and I, which are in phase, arbitrarily choosing the direction in which I draw them. I then add phasors V_C, which lags I by 90°, and V_L, which leads I by 90°.

Figure 32.62

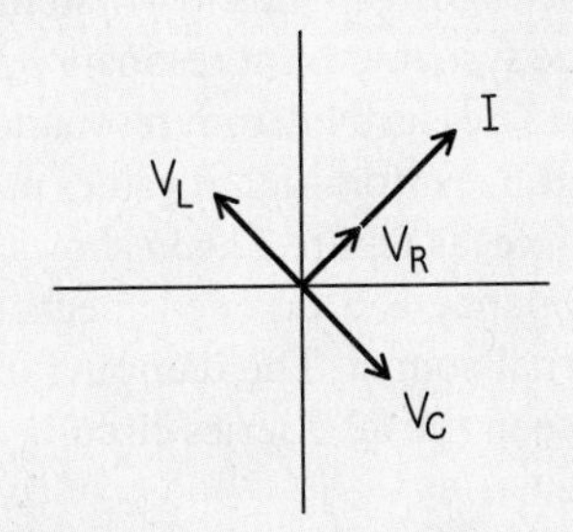

❷ DEVISE PLAN The current in the circuit depends on the impedance, which is given by Eq. 32.43. This equation contains C, however, and I am given no information about this variable. I am given, however, that $V_C = V_L$, and because both V_C and V_L are proportional to the current, I should be able to determine the current without knowing the capacitance in the circuit.

❸ EXECUTE PLAN I know from Eq. 32.24 that $V_L = I\omega L$, and I also know that $V_C = I/\omega C$ (Eq. 32.13). Because $V_L = V_C$ in this problem, I can equate the terms on the right in these two equations to obtain

$$\omega L = \frac{1}{\omega C}. \qquad (1)$$

Substituting ωL for $1/\omega C$ in Eq. 32.43 then yields $Z_{RLC} = R$. Now I can use Eq. 32.33 to determine the current in the circuit:

$$I = \frac{\mathscr{E}_{max}}{Z_{RLC}} = \frac{\mathscr{E}_{max}}{R} = \frac{160\ \text{V}}{50\ \Omega} = 3.2\ \text{A}. ✔$$

❹ EVALUATE RESULT When the amplitudes of the potential differences across the inductor and the capacitor are equal, the lengths of the phasors V_C and V_L are equal. Because the phasors point in opposite directions (Figure 32.62), they add to zero, and so the impedance in the circuit is due to the resistor only. This means the current is essentially given by Ohm's law, $I = V/R$, or, in the version I obtained here, $I = \mathscr{E}_{max}/R$.

32.15 (*a*) Calculate the maximum potential difference across each of the three circuit elements in Example 32.10. (*b*) Is the sum of the amplitudes V_R, V_L, and V_C equal to the amplitude of the source emf? Why or why not?

32.7 Resonance

Consider again the *RLC* series circuit of Figure 32.58. Suppose that the amplitude $\mathscr{E}_{max}$ of the source emf is held constant, but we vary its angular frequency. What happens to the amplitude of the current I and to the phase constant ϕ? Combining Eqs. 32.42 and Eq. 32.43, we can say

$$I = \frac{\mathscr{E}_{max}}{Z} = \frac{\mathscr{E}_{max}}{\sqrt{R^2 + (\omega L - 1/\omega C)^2}}. \qquad (32.45)$$

QUANTITATIVE TOOLS

Figure 32.63 Current and phase changes in the *RLC* circuit of Figure 32.58.

(*a*) Frequency dependence of the current at low, medium, and high *R*

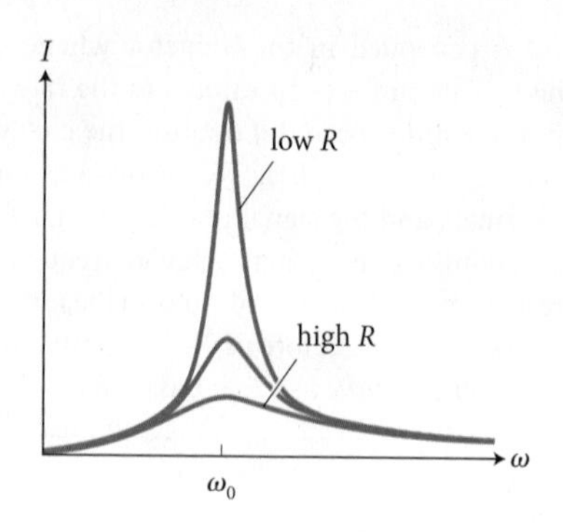

(*b*) Frequency dependence of current phase relative to source emf as function of angular frequency at low, medium, and high *R*

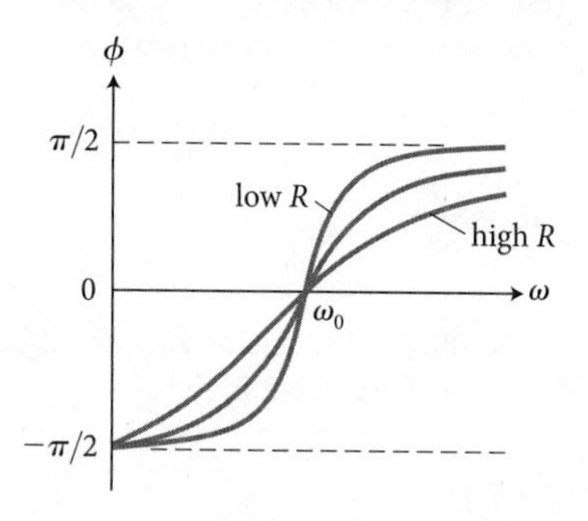

The current is at its maximum when the term in parentheses in the denominator is zero. When this term is zero, $V_L = V_C$ and the two potential differences are 180° out of phase. Therefore the effects of the inductor and the capacitor cancel each other and the circuit behaves as if only the resistor is present. The term in parentheses in Eq. 32.45 is zero when

$$\omega L = \frac{1}{\omega C}. \tag{32.46}$$

The angular frequency for which Eq. 32.46 is satisfied is called the **resonant angular frequency** ω_0 of the circuit:

$$\omega_0 = \frac{1}{\sqrt{LC}}. \tag{32.47}$$

The current amplitude and phase as a function of angular frequency are plotted in **Figure 32.63** for three values of R (with fixed values of $\mathscr{E}_{max}$, L, and C). Increasing or decreasing the angular frequency from ω_0 decreases the current amplitude. Changing R changes the maximum current that can be obtained and also changes how rapidly the current drops as the angular frequency increases or decreases from resonance.

Whenever an oscillating physical quantity has a peaked angular frequency dependence, the dependence is referred to as a *resonance curve*. The sharpness of the peak reflects the efficiency with which the source delivers energy to the system at or near resonance and depends on the amount of dissipation present in the system. A very tall, sharp peak corresponds to a system with low dissipation. In such a system, the source can pump an enormous amount of energy into the system at resonance. A short, broad peak corresponds to a system with high dissipation. Here, less energy goes into the system even at resonance, but that energy can be transferred in at angular frequencies farther from resonance. For the *RLC* series circuit, energy is dissipated via the resistor; high *R* values produce less current at ω_0 and a broader resonance curve, as Figure 32.63*a* shows.

Another system that exhibits resonance is a damped mechanical oscillator (see Section 15.8) driven by an external source. The damping in a mechanical oscillator is analogous to the resistance in the *RLC* series circuit.

32.16 How does the resonance curve in Figure 32.63 change if the value of C or L is changed?

In the *RLC* series circuit, the phase difference between the current and the driving emf also depends on the angular frequency of the AC source. The current can either lag or lead the emf (or be in phase with it), depending on the angular frequency. We found previously that the phase of the current relative to the source emf for an *RLC* series circuit is given by Eq. 32.44:

$$\tan\phi = \frac{\omega L - 1/\omega C}{R}. \tag{32.48}$$

Consider the limiting values of this expression for the relative phase by looking at the curves in Figure 32.63*b*. At resonance ($\omega = \omega_0$), $\phi = 0$ and the current and the source emf are in phase. When $\omega = 0$, $\tan\phi = -\infty$ and $\phi = -\pi/2$. When $\omega = \infty$, $\tan\phi = \infty$ and $\phi = +\pi/2$. Below resonance, $\phi < 0$, the capacitor provides the dominant contribution to the impedance, and the current leads the source emf. Above resonance, $\phi > 0$ and the inductor dominates, and the current lags the source emf.

32.17 In an *RLC* series circuit, you measure $V_R = 4.9$ V, $V_L = 6.7$ V, and $V_C = 2.5$ V. Is the angular frequency of the AC source above or below resonance?

32.8 Power in AC circuits

At the beginning of this chapter, we saw that in alternating-current circuits, the energy stored in capacitors and inductors can oscillate. Consequently, for part of each cycle, these elements put energy back into the source rather than taking up energy from the source. Thus, unlike what we see in DC circuits, the source in an AC circuit does not simply deliver energy steadily to the circuit. Let's take a closer look at how to determine the rate at which an AC source delivers energy to a load.

In general, the rate at which the source delivers energy to its load—in other words, the power of the source—is the time-dependent version of the result we found for DC circuits (Eq. 31.42):

$$p = iv_{\text{load}}. \tag{32.49}$$

Because the current and the emf oscillate, this power varies with time and in principle can be either positive or negative. Let's first consider a load that consists of just one resistor. Ohm's law tells us that the instantaneous energy delivered to the resistor is

$$p = iv_R = i^2R = I^2R\sin^2\omega t. \tag{32.50}$$

The time dependence of the potential difference, current, and power are shown in **Figure 32.64**. Because the current and potential difference are in phase, the power is always positive, and so the source always delivers energy to the resistor. This makes sense because the resistor dissipates energy regardless of the current direction. Consequently, the rate at which energy is dissipated in the resistor (the *power at the resistor*) is always positive and oscillates at *twice* the angular frequency of the emf.

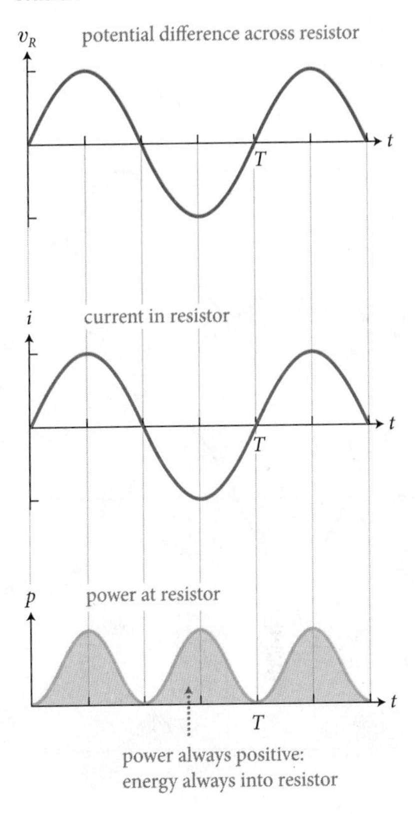

Figure 32.64 For an AC circuit consisting of a resistor connected across an AC source, time dependence of potential difference across the resistor, current in the resistor, and power at the resistor.

For most applications, we are interested in the time average of the power at the resistor. Using the trigonometric identity $\sin^2\alpha = \frac{1}{2}(1 - \cos 2\alpha)$, we can rewrite Eq. 32.50 as

$$p = I^2R[\tfrac{1}{2}(1 - \cos 2\,\omega t)] = \tfrac{1}{2}I^2R - \tfrac{1}{2}I^2R\cos 2\omega t. \tag{32.51}$$

The first term on the right is constant in time. The second term on the right averages to zero over a full cycle because the area under the positive half of the cosine is equal to the area under the negative half. As a result, for time intervals much longer than the period of oscillation, the time average of the power at the resistor is

$$P_{\text{av}} = \tfrac{1}{2}I^2R. \tag{32.52}$$

For a sinusoidally varying current, the *root-mean-square* or *rms* value of the current is (Eqs. 19.21 and 30.39)

$$I_{\text{rms}} \equiv \sqrt{(i^2)_{\text{av}}} = \sqrt{\tfrac{1}{2}I^2} = \frac{I}{\sqrt{2}}. \tag{32.53}$$

and so

$$P_{\text{av}} = I_{\text{rms}}^2R. \tag{32.54}$$

The advantage of writing the average power in terms of the rms current is that Eq. 32.54 is completely analogous to the expression for the energy dissipated by a resistor connected to a DC source (Eq. 31.43). Similarly, we can introduce rms values of potential difference and source emf:

$$V_{\text{rms}} = \frac{V_R}{\sqrt{2}} \quad \text{and} \quad \mathscr{E}_{\text{rms}} = \frac{\mathscr{E}_{\text{max}}}{\sqrt{2}}. \tag{32.55}$$

Figure 32.65 For an AC circuit consisting of a capacitor connected across an AC source, time dependence of potential difference across the capacitor, current in the capacitor, and power at the capacitor.

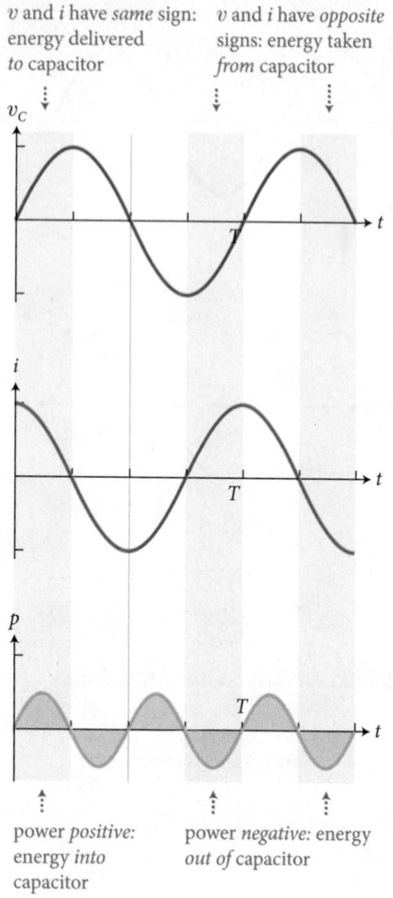

v and i have *same* sign: energy delivered *to* capacitor

v and i have *opposite* signs: energy taken *from* capacitor

power *positive:* energy *into* capacitor

power *negative:* energy *out of* capacitor

The rms value is a useful way to measure the average value of an oscillating current or emf because the strict time average of these quantities is zero. *Voltmeters* and *ammeters* typically measure the rms value of alternating potential differences and currents, respectively. Thus, for example, the wall potential difference in household electrical wiring in the United States is referred to as 120 V even though the amplitude is 170 V; the 120-V rating is the rms value.

Next let's look at a circuit made up of an AC source and a capacitor. How much power is delivered to the capacitor by the source? Now the current and the potential difference are out of phase (see Figure 32.48). Substituting from Eq. 32.8, $v_C = V_C \sin \omega t$, and Eq. 32.11, $i = \omega C V_C \cos \omega t$, into Eq. 32.49, we obtain

$$p = i v_C = (\omega C V_C \cos \omega t)(V_C \sin \omega t). \tag{32.56}$$

Using the trigonometric identity $\sin(2\alpha) = 2 \sin \alpha \ \cos \alpha$, we obtain

$$p = \tfrac{1}{2}\omega C V_C^2 \sin 2\omega t. \tag{32.57}$$

As in the resistor-only circuit, the power oscillates at twice the angular frequency of the source, but now the power is sometimes positive and sometimes negative, as shown in **Figure 32.65**. When v and i have the same sign, energy is transferred to the capacitor; when v and i have opposite signs, energy residing in the capacitor is transferred back to the source. The average power is zero, as it must be, because no energy is dissipated in a capacitor. The same is true for an inductor, except that in an inductor energy is stored as magnetic energy rather than electric energy; you can show this mathematically by converting Eqs. 32.56 and 32.57 to their V_L counterparts.

Finally, let's examine how power is delivered to the load in an *RLC* circuit. Although we could work out the power at each element, it's easier to consider the power for the entire load consisting of the *RLC* combination. The potential difference across the load is equal to the applied emf. Using Eq. 32.1, $\mathscr{E} = \mathscr{E}_{max} \sin \omega t$; and Eq. 32.16, $i = I \sin(\omega t - \phi)$; we can write the instantaneous power as

$$p = vi = \mathscr{E}_{max} I \sin \omega t \sin(\omega t - \phi). \tag{32.58}$$

Using the trigonometric identities $\sin(\alpha - \beta) = \sin \alpha \cos \beta - \cos \alpha \sin \beta$ to separate the ϕ and ω dependence and substituting $\sin \alpha \cos \alpha = \frac{1}{2} \sin 2\omega t$, we rewrite Eq. 32.58 in the form

$$\begin{aligned} p &= \mathscr{E}_{max} I (\cos \phi \sin^2 \omega t - \sin \phi \sin \omega t \cos \omega t) \\ &= \mathscr{E}_{max} I (\cos \phi \sin^2 \omega t - \sin \phi \tfrac{1}{2} \sin 2\omega t). \end{aligned} \tag{32.59}$$

The time average of $\sin^2 \omega t$ is $1/2$, and the second term inside the parentheses averages to zero, leaving us with

$$P_{av} = \tfrac{1}{2} \mathscr{E}_{max} I \cos \phi. \tag{32.60}$$

Writing this result using rms values (Eqs. 32.54 and 32.55) gives

$$P_{av} = \tfrac{1}{2}(\sqrt{2}\,\mathscr{E}_{rms})(\sqrt{2} I_{rms}) \cos \phi = \mathscr{E}_{rms} I_{rms} \cos \phi. \tag{32.61}$$

We can rewrite this in a more physically insightful way if we note that $\mathscr{E}_{rms} = I_{rms}Z$ (Eq. 32.33) and note from Figure 32.59*c* that

$$\cos\phi = \frac{V_R}{\mathscr{E}_{max}} = \frac{RI}{ZI} = \frac{R}{Z}. \tag{32.62}$$

With these substitutions, Eq. 32.61 becomes

$$P_{av} = I_{rms}Z\,I_{rms}\frac{R}{Z} = I_{rms}^2 R. \tag{32.63}$$

This result tells us that all of the energy delivered to the circuit is dissipated as thermal energy in the resistor—as it must be, because neither the capacitor nor the inductor dissipates energy. This energy is dissipated at the same average rate as in a circuit made up of a single resistor connected to an AC source (Eq. 32.54).

The factor $\cos\phi$ that appears in Eqs. 32.60–32.62 is called the **power factor** which is a measure of the efficiency with which the source delivers energy to the load. At resonance, when the current and the emf are in phase ($\phi = 0$), the current and the power factor are greatest, and the maximum power possible is delivered to the load. At angular frequencies away from resonance, less power is delivered to the load.

32.18 Calculate the rate P_{av} at which energy is dissipated in the *RLC* series circuit of Example 32.10.

Chapter Glossary

SI units of physical quantities are given in parentheses.

AC source A power source that generates a sinusoidally alternating emf.

alternating current (AC) Current that periodically changes direction. Circuits in which the current is alternating are called *AC circuits*.

depletion zone A thin nonconducting region at the junction between p-doped and n-doped pieces of a semiconductor where the charge carriers have recombined and become immobile.

diode A circuit element that behaves like a one-way valve for current.

hole An incomplete bond in a semiconductor that behaves like a freely moving positive charge carrier.

impedance Z (Ω) The proportionality constant between the amplitudes of the potential difference and the current in any load connected to an AC source. The impedance for the load in an RLC series circuit is

$$Z_{RLC} \equiv \sqrt{R^2 + (\omega L - 1/\omega C)^2} \qquad (32.43)$$

phase constant ϕ (unitless) A scalar that represents the phase difference between the source emf and the current. When the current leads the source emf, ϕ is negative; when the current lags the source emf, ϕ is positive.

power factor $\cos\phi$ (unitless) A scalar factor that is a measure of the efficiency with which an AC source delivers energy to a load:

$$\cos\phi = \frac{V_R}{\mathscr{E}_{\max}} = \frac{R}{Z}. \qquad (32.62)$$

reactance X (Ω) The proportionality constant between the amplitudes of the potential difference and the current in a capacitor or inductor connected to an AC source. The *capacitive reactance* is

$$X_C \equiv 1/\omega C, \qquad (32.14)$$

and the *inductive reactance* is

$$X_L \equiv \omega L. \qquad (32.26)$$

resonant angular frequency ω_0 (s^{-1}) In an RLC series circuit, the angular frequency at which the current is a maximum.

$$\omega_0 = \frac{1}{\sqrt{LC}}. \qquad (32.47)$$

In general, the resonant angular frequency of an oscillator of any kind is the angular frequency at which the maximum oscillation is obtained.

semiconductor A material that has a limited supply of charge carriers that can move freely and an electrical conductivity intermediate between that of conductors and that of insulators. An *intrinsic semiconductor* is made of atoms of one element only; a *doped/extrinsic semiconductor* contains trace amounts of atoms that alter the number of free electrons available and change the electronic properties. An *n-type* semiconductor has a surplus of valence electrons (relative to the number present in the original intrinsic semiconductor), which means it has some free electrons. A *p-type* semiconductor has a deficit of valence electrons, and so it has some free holes.

transistor A circuit element that behaves like a switch or a current amplifier.

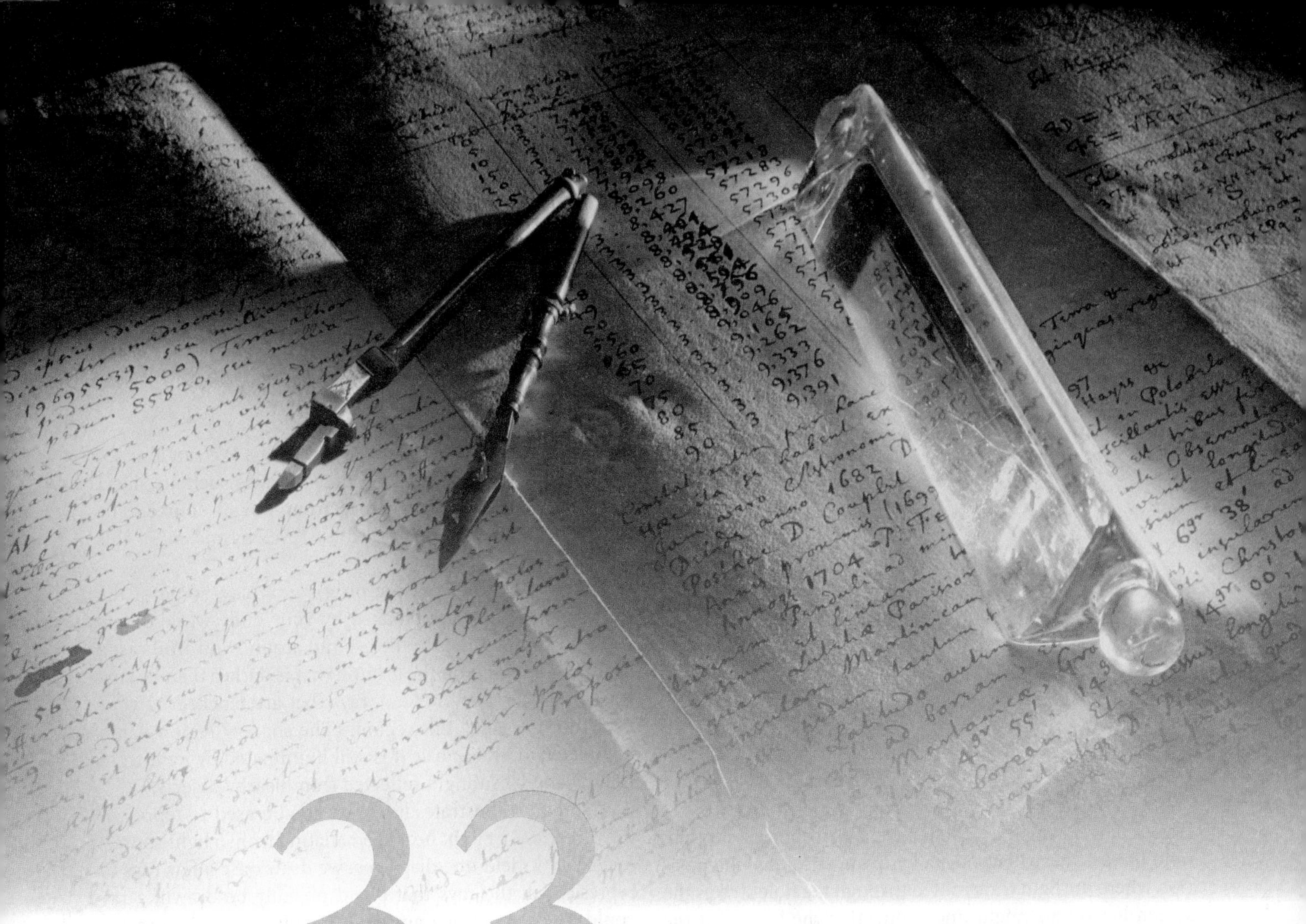

33 Ray Optics

CONCEPTS

QUANTITATIVE TOOLS

You can read these words because this page reflects light toward you; your eyes intercept some of the reflected light, and the lenses of your eyes redirect it, forming an image of the page on the retina. Where does the light reflected from the page come from? Our primary source of light during the day is the Sun, and our secondary source is the brightness of the sky. Indoors and at night, our light sources are flames in candles, white-hot filaments in light bulbs, and glowing gases in fluorescent bulbs. The light from all these sources comes from the accelerated motion of electrons as this motion produces electromagnetic waves.

In Chapter 30 we studied the propagating electric and magnetic fields that constitute electromagnetic waves, and we learned that a narrow frequency range of these waves corresponds to what we know as visible light. In this chapter we continue to study light, particularly its propagation and its interactions with materials. We shall not consider the electric and magnetic fields individually, but instead think of the behavior of rays of light. Such behavior, which is called *ray optics,* was understood long before it was known that light is an electromagnetic wave.

33.1 Rays

If you pierce a small hole in a piece of cardboard and then hold the cardboard between a lamp and a screen, the position where the light transmitted through the hole strikes the screen lies on a straight line connecting the lamp and the hole (Figure 33.1). This observation suggests that we can think of a light source as made up of many straight beams that spread out in three dimensions from the source. Each beam travels in a straight line until it interacts with an object. That interaction changes the beam's direction of travel.

We can represent the propagation of light by drawing **rays**:

> **A ray is a line that represents the direction in which light travels. A beam of light with a very small cross-sectional area approximately corresponds to a ray.**

In order to see an object, our eyes form an image by collecting light that comes from the object. If the object is a light source, we see it by the light it emits. We can also see an object that is not a light source because such an object interacts with light that comes from a light source. The light is then redirected toward our eyes by means of this interaction.

When you stand outside on a sunny day, some of the rays from the Sun are blocked by your body while others travel in straight lines to the ground around you. You cast a shadow—a region on the ground that is darker than its surroundings because the Sun's rays that are blocked by your body do not strike this region. (The shadow region is not completely dark because it is still illuminated by light from the sky and by sunlight reflected from nearby objects.)

Figure 33.2 illustrates how rays can be used to represent the directions of light beams emanating from a light source. Just as with field line diagrams, we draw only a few rays to represent all the rays that could possibly be drawn; a ray could be drawn along any line radially outward from the source. Although most sources of light—the Sun, a flame, a light bulb—are extended, when the distance to the source is much greater than the extent of the source, we can treat that source as a *point source* (See Section 17.1). That is, we can treat the source as if all the light were emitted from a single point in space. In the first part of this chapter, we develop a feel for which rays to draw in a given situation.

Figure 33.2 Rays emanating from a source of light.

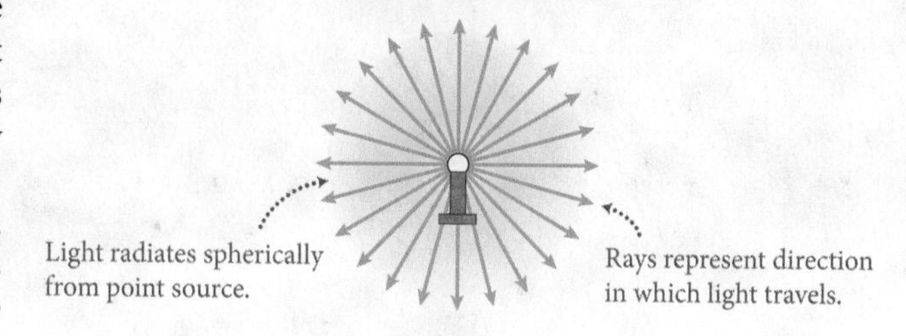
Light radiates spherically from point source.
Rays represent direction in which light travels.

> **33.1** Suppose a second bulb is added to the left of the one in Figure 33.1, as illustrated in Figure 33.3. What happens to (*a*) the brightness of the spot created on the screen by the first bulb and (*b*) the brightness at locations close to the vertical edges of the original shadow on the screen (points P, Q, R, and S)?

Figure 33.1 A light beam that is not disturbed travels in a straight line.

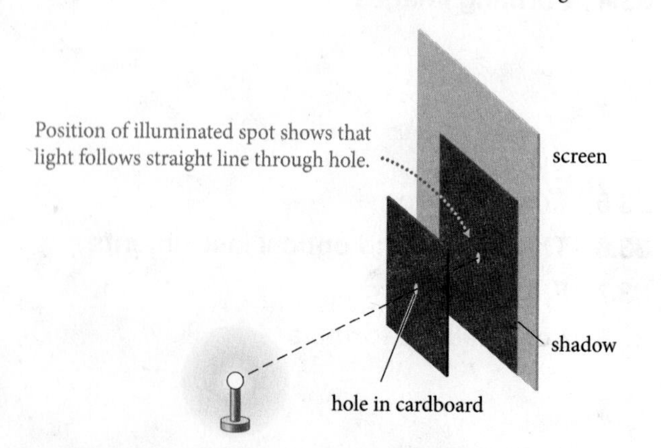
Position of illuminated spot shows that light follows straight line through hole.
screen
shadow
hole in cardboard

Figure 33.3 Checkpoint 33.1.

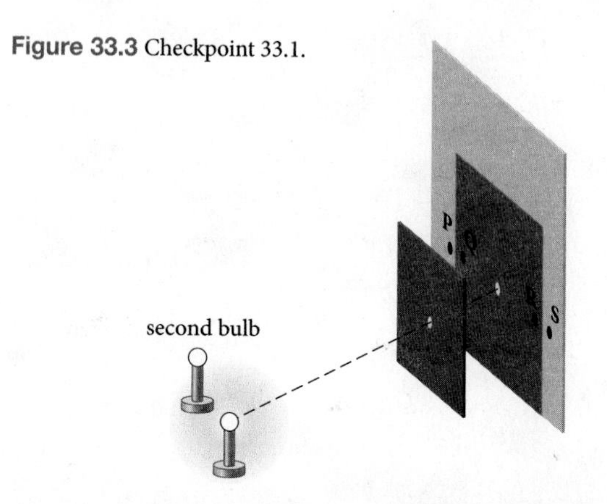
second bulb
P
Q
R
S

Example 33.1 Light and shadow

An object that has a small aperture is placed between a light source and a screen, as shown in **Figure 33.4**. Which parts of the screen are in the shadow?

Figure 33.4 Example 33.1.

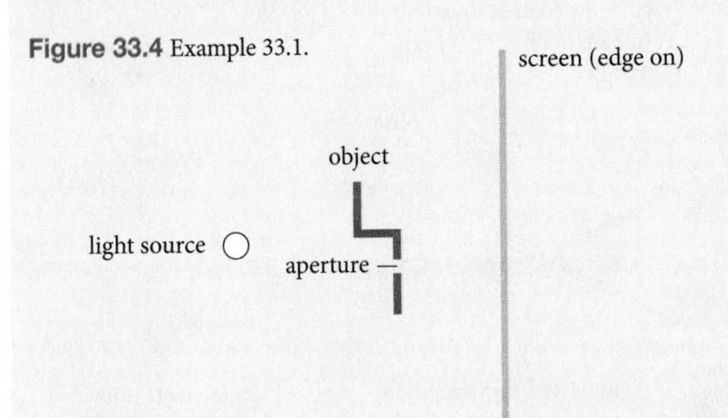

❶ **GETTING STARTED** The rays emitted by the source radiate outward in all directions following straight paths. The shadow is cast because the object prevents some of the rays from reaching the screen (except for the rays that make it through the aperture).

❷ **DEVISE PLAN** To locate the edges of the shadow, I draw straight lines from the source to the edges of the object (including the edges of the aperture) and extend these rays to the screen (**Figure 33.5**).

Figure 33.5

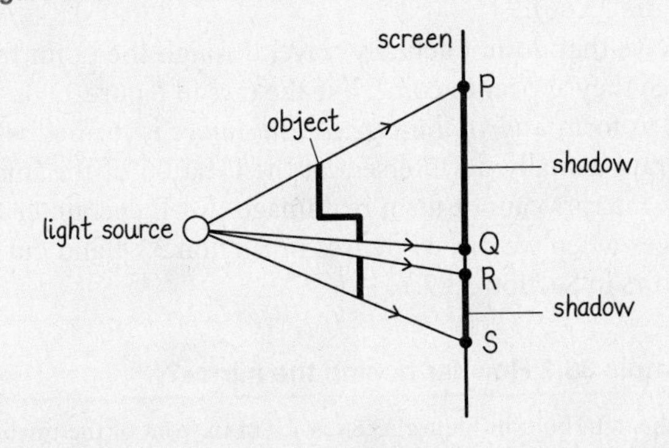

❸ **EXECUTE PLAN** The top and bottom edges of the shadow correspond to the highest and lowest screen locations (P and S) reached by light rays that are not blocked by the object. The gap in the shadow between locations Q and R corresponds to light rays that pass through the aperture, which means that this region of the screen is not in shadow. ✔

❹ **EVALUATE RESULT** A shadow that is taller than the object makes sense. Because the light rays from the source emanate in all directions, most of them reach the screen at angles other than 90°. This means that the distance from P to S must be greater than the object height. Indeed, I know from experience that the shadow cast by my hand gets larger as I move my hand closer to a lamp, increasing that angle.

33.2 Hold a piece of paper between your desk lamp (or any other source of light) and your desk or a wall. How does the sharpness of the edges of the shadow change as you move the paper closer to the bulb? Why does this happen?

33.2 Absorption, transmission, and reflection

Different materials interact differently with the light that strikes them, which is how you can visually distinguish wood from metal, fabric from skin, and a white piece of paper from a blue one. When light strikes an object, the light can be transmitted, absorbed, or reflected.

Transmitted light passes through a material. Objects that transmit light, such as a piece of glass, are said to be *transparent* (**Figure 33.6*a***). In *translucent* materials, such as frosted glass, light rays are *transmitted diffusely*—that is, they are redirected in random directions as they pass through, so that the transmitted light does not come from a definite direction (Figure 33.6*b*). Because translucent materials scatter light in this manner, we cannot see objects clearly through them.

Absorbed light enters a material but never exits again. Objects that absorb most of the light that strikes them, such as a piece of wood, are said to be *opaque*. When light strikes such materials, the energy carried by the light is converted to some other form (usually thermal energy) and the light propagation stops.

Reflected light is any light that is redirected away from the surface of the material (**Figure 33.7** on the next page). Smooth surfaces reflect light *specularly*—that is, each ray bounces off the surface in such a way that the angle between it and the normal to the surface doesn't change (Figure 33.7*a*). The angle between the incoming ray and the normal to the surface is called the **angle of incidence** θ_i; the angle between the outgoing ray and the normal is called the **angle of reflection** θ_r.

Figure 33.6 We see objects clearly through a sheet of clear glass but diffusely through frosted glass.

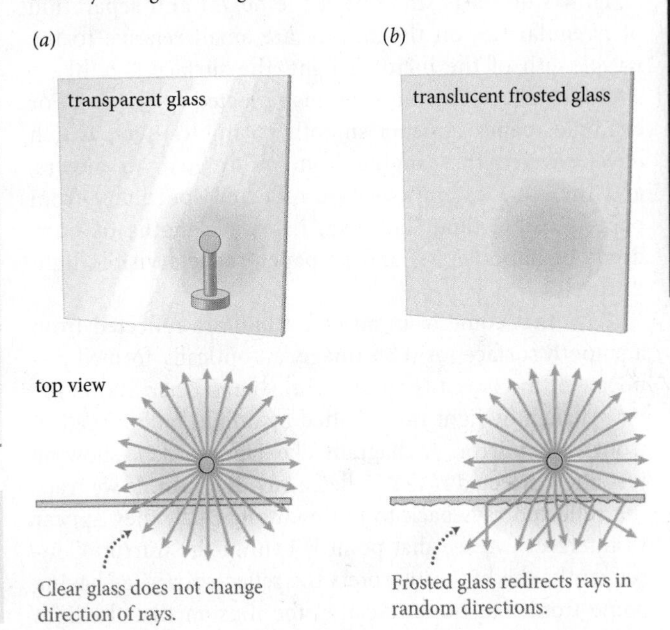

CONCEPTS

Figure 33.7 Light reflects specularly from a smooth surface, forming a mirror image. From a rough surface, it reflects diffusely (in random directions), so no image forms.

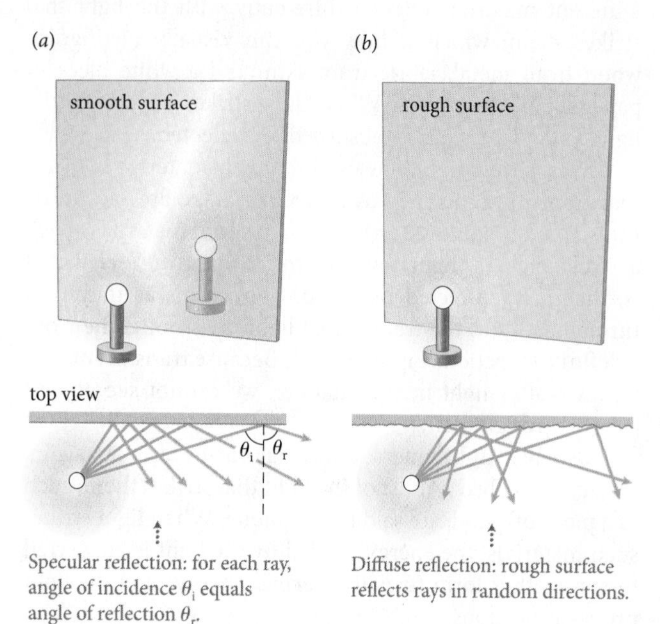
(a)
(b)
smooth surface
rough surface
top view
θ_i
θ_r
Specular reflection: for each ray, angle of incidence θ_i equals angle of reflection θ_r.
Diffuse reflection: rough surface reflects rays in random directions.

Figure 33.8 Diagrams showing the paths taken by light rays that are produced by a bulb and reflected by a mirror into an observer's eye. The reflected rays appear to come from behind the mirror, forming an image behind the mirror.

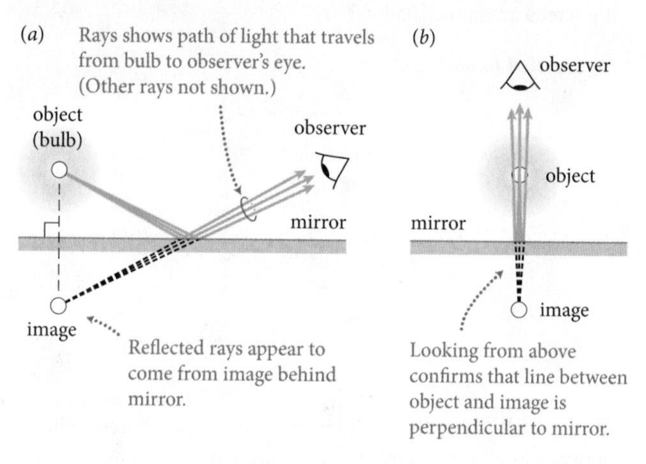
(a)
Rays shows path of light that travels from bulb to observer's eye. (Other rays not shown.)
object (bulb)
observer
mirror
image
Reflected rays appear to come from image behind mirror.
(b)
observer
object
mirror
image
Looking from above confirms that line between object and image is perpendicular to mirror.

Empirically we find:

For a ray striking a smooth surface, the angle of reflection is equal to the angle of incidence, and both angles are in the same plane.

This **law of reflection** holds at smooth surfaces for any angle of incidence.

Surfaces that are not smooth reflect light in many directions (Figure 33.7*b*). For such *diffuse reflection,* each ray obeys the law of reflection, but the direction of the surface normal varies over the surface and so the angle of reflection also varies.

How smooth is smooth? If the height and separation of irregularities on the surface are small relative to the wavelength of the incident light, the surface acts like a smooth surface and most light is reflected specularly. For example, paper appears smooth to microwaves, which have wavelengths ranging from millimeters to meters, and therefore microwaves are reflected specularly from paper. Visible light, however, has wavelengths of hundreds of nanometers, and so paper reflects visible light diffusely.

Rays that come from an object and are reflected from a smooth surface form an **image,** an optically formed duplicate of the object (Figure 33.7*a*). Figure 33.8*a* shows the paths taken by light rays emitted by a light bulb placed in front of a mirror. A diagram like Figure 33.8*a* showing just a few selected rays is called a **ray diagram.** If we trace the reflected rays back to the point at which they appear to intersect, we see that point is behind the mirror. Consequently, the brain interprets the reflected rays as having come from that point, creating the illusion that the light bulb is behind the mirror. The directions of the rays that reach the eyes of the observer are the same as if they had come from an object located behind the mirror.

Note that the image is located on the line through the object and perpendicular to the mirror, because if we look along that line, the image lies behind the object (Figure 33.8*b*).

Rays that do not actually travel through the point from which they appear to come, like the rays in Figure 33.8*a*, are said to form a *virtual image.* A *real image* is formed when the rays actually do intersect at the location of the image. (Flat mirrors cannot form real images; we'll encounter real images when we discuss lenses in Section 33.4 and curved mirrors in Section 33.7.)

Example 33.2 How far behind the mirror?

If the light bulb in Figure 33.8*a* is 1.0 m in front of the mirror, how far behind the mirror is the image?

❶ GETTING STARTED The location of the image is the location from which the rays reflected by the mirror appear to come—that is, the point at which they intersect. From Figure 33.8 I know that because the rays intersect directly behind the bulb, a line that passes through the bulb and is normal to the mirror passes through the image.

❷ DEVISE PLAN I can obtain the distance of the image behind the mirror by considering one ray that travels from the bulb to the observer and then tracing that ray back through the mirror to its intersection with the line that is perpendicular to the mirror and passes through the bulb.

❸ EXECUTE PLAN I begin by drawing a ray that travels from the bulb to the mirror and is reflected to the location of the observer (Figure 33.9). In my drawing, A denotes the bulb location, B denotes the point where the line connecting the bulb and its image intersects the mirror, and C denotes the point at which the ray that is reflected to the observer hits the mirror. According to the law of reflection, $\theta_r = \theta_i$.

Figure 33.9

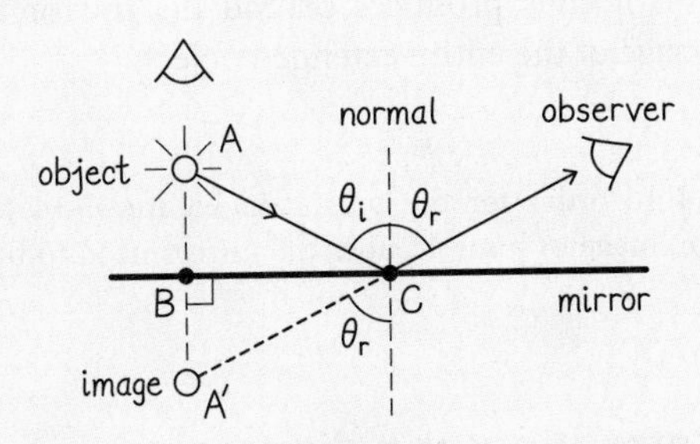

I now extend the reflected ray to behind the mirror (dashed line). I know that the image must lie somewhere along that dashed line and must also lie on the line that passes through the object and is perpendicular to the mirror. The image must therefore lie at the intersection of this line and the dashed ray extension; I denote that intersection point by A'.

To determine the distance BA', which is how far behind the mirror the image is, I note that angle A'CB is equal to $90° - \theta_r$ and angle ACB is equal to $90° - \theta_i$. Because $\theta_i = \theta_r$, angles A'CB and ACB are equal. Therefore triangles ABC and A'BC are congruent, and AB = BA'. That is, the image appears at the same distance behind the mirror as the object is in front of it: 1.0 m behind the mirror. ✔

❹ **EVALUATE RESULT** My result makes sense because I know from experience that as I walk toward a mirror, my image also approaches it.

33.3 If the observer in Figure 33.8 moves to a different position, does the location of the image change?

The colors of visible light we see correspond to different frequencies of electromagnetic waves. Red corresponds to the lowest frequency of the visible spectrum. As the frequency increases, the color changes to orange, yellow, green, blue, indigo, and finally violet (the highest frequency of the visible spectrum). The range of visible frequencies is quite small relative to the range of the complete electromagnetic spectrum, as **Figure 33.10** shows. Frequencies lower than the visible correspond to infrared radiation, and higher frequencies to ultraviolet. When a light source produces all the frequencies of the visible spectrum at roughly the same intensities, the emitted light appears white.

Figure 33.10 Visible light makes up only a small part of the electromagnetic spectrum.

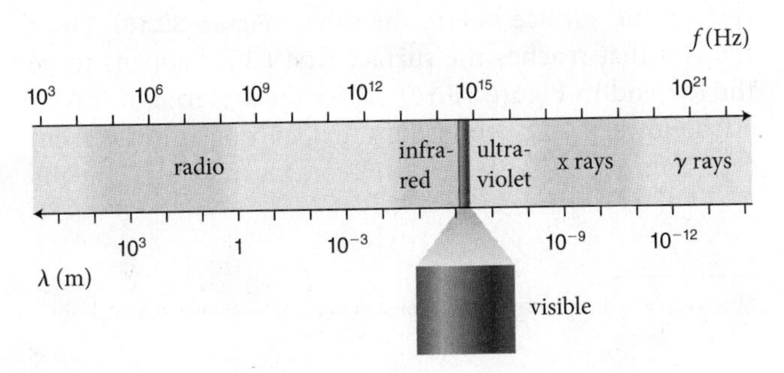

Figure 33.11 All colors of light pass through colorless glass (shown light blue for illustration purposes); orange glass transmits orange light and absorbs all colors of light except orange.

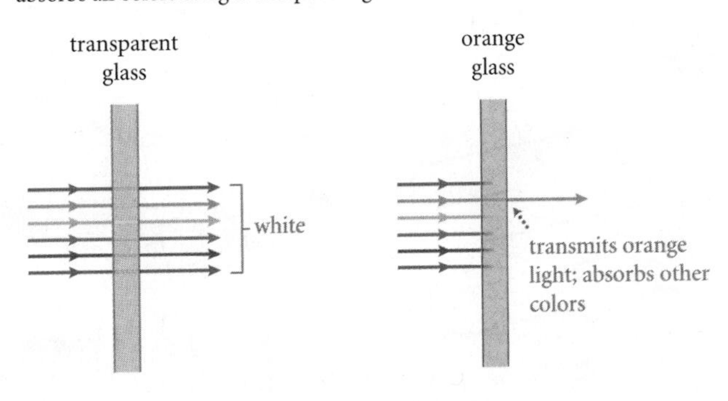

Different colors of light interact differently with different objects, affecting the color we perceive the object as being. Colorless materials, like a piece of ordinary window glass, transmit all colors of the visible spectrum. A piece of orange glass, on the other hand, transmits only the orange part of the visible spectrum. All other colors are absorbed in the glass (**Figure 33.11**). A red apple absorbs all colors of the visible spectrum except red, which is redirected to our eyes. Grass absorbs all colors except green, which is diffusively reflected at its surface.

Because light is a wave phenomenon, it is sometimes useful to represent the propagation of light with wavefronts, which we introduced in Section 17.1. Wavefronts are drawn perpendicular to the direction of propagation of the wave.* Because light rays point along the direction of propagation of the light, light wavefronts are perpendicular to light rays. **Figure 33.12*a*** shows the spherical

Figure 33.12 A point source of light produces spherical wavefronts; a beam of light contains planar wavefronts.

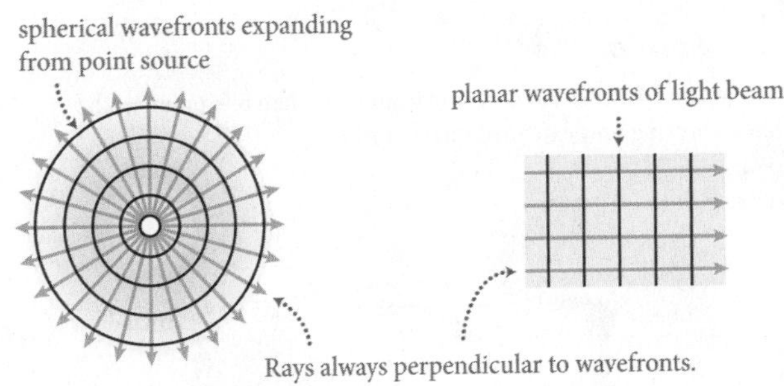

*For mechanical waves, the wavefronts are drawn at the locations of the wave crests, spaced by the wavelength of the wave. Because the wavelength of light is very short, the wavefronts of light cannot be represented to scale.

CONCEPTS

Figure 33.13 (*a*) The reflection of wavefronts from a smooth surface explains the law of reflection. (*b*) The corresponding rays and their angles of incidence θ_i and reflection θ_r.

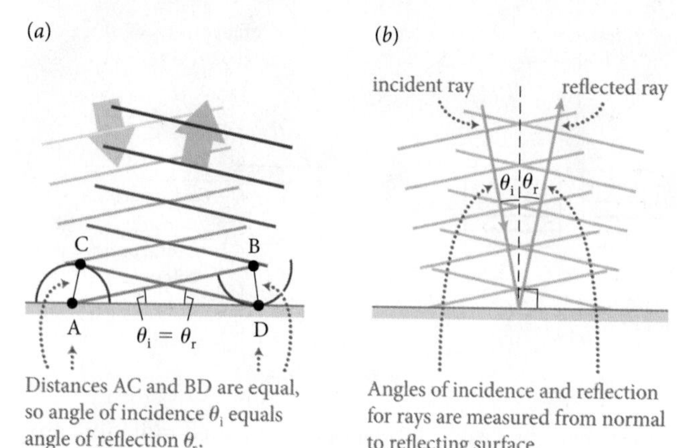
(a)
(b)
incident ray
reflected ray
θ_i θ_r
C
B
A
$\theta_i = \theta_r$
D
Distances AC and BD are equal, so angle of incidence θ_i equals angle of reflection θ_r.
Angles of incidence and reflection for rays are measured from normal to reflecting surface.

wavefronts for light coming from a point source, and Figure 33.12*b* shows the straight-line rays and wavefronts corresponding to a planar electromagnetic wave. Note that a planar wave is represented with rays that are parallel to one another because all the wavefronts are parallel to one another.

By looking at how wavefronts behave, we can understand the law of reflection. When a light ray strikes a smooth surface at an incidence angle $\theta_i \neq 0$ (**Figure 33.13**), the left end of the first wavefront to reach the surface gets there, at A, before the right end does. In the time interval it takes the right end to reach the surface at D, the left end has traveled back from the surface to C. The distance traveled by the right end toward the surface, BD, is the same as that traveled by the left end away from the surface, AC, so the angles BAD and CDA must be equal. The angle of incidence θ_i equals angle BAD. Likewise, the angle of reflection θ_r equals angle CDA. So $\theta_i = \theta_r$.

So far we have treated the object (and consequently the image) as a single point. **Figure 33.14** shows how images are formed of extended objects. Each point on the object reflects (or emits) light rays, and the reflections of these rays appear to come from a corresponding point on the image. A flat mirror thus produces behind the mirror an exact mirror image of the entire extended object.

Figure 33.14 Paths taken by rays from more than one point on the object, showing how extended images form.

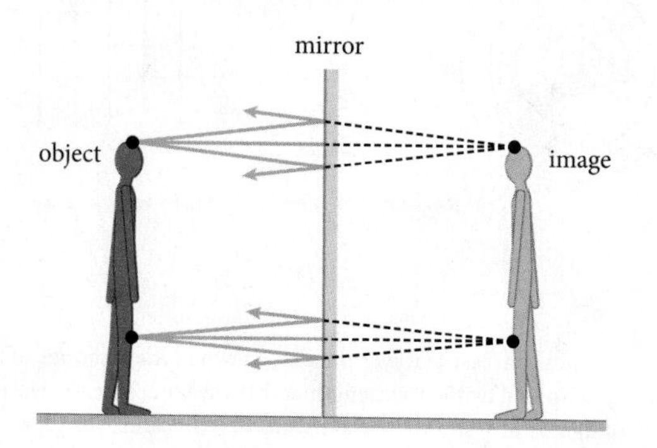
mirror
object
image

33.4 In order for the person in Figure 33.14 to see a complete image of himself, does the mirror need to be as tall as he is?

33.3 Refraction and dispersion

As we found in Chapter 30, light propagates with speed $c_0 = 3 \times 10^8$ m/s in vacuum. In air, the speed of light is almost the same as that in vacuum. In a solid or liquid medium, however, light propagates at a speed c that is generally less than c_0.* In glass, for example, visible light propagates at two-thirds of the speed of light in vacuum (see Example 30.8).

How does this change in speed affect the propagation of an electromagnetic wave? Recall from Chapter 16 that harmonic waves are characterized by both a wavelength λ and a frequency f and that the product of the wavelength and frequency equals the wave's speed of propagation (Eq. 16.10). The frequency of the wave must remain the same because the oscillation frequency of the electromagnetic field that makes up the wave is determined by the acceleration of charged particles at the wave's source. The acceleration of the source does not alter when the wave travels from one medium to another, and thus the frequency of the traveling wave also cannot change.

33.5 In vacuum, a particular light wave has a wavelength of 400 nm. It then travels into a piece of glass, where its speed decreases to two-thirds of its vacuum speed. What is the distance between the wavefronts in the glass?

As we found in Checkpoint 33.5, when rays of light pass through the interface between vacuum and a transparent material, the wavefronts inside the material are more closely spaced than they are in vacuum, due to the lower speed of the wavefronts. **Figure 33.15** illustrates this effect for wavefronts incident normal to the surface of the material.

What if the wavefronts strike the transparent material at an angle? In such a case, one end of the wavefront arrives at the surface before the other (**Figure 33.16**). Once the end that reaches the surface first (this happens to be the left end in Figure 33.16) enters the material, it travels at the lower speed while the other end of the wavefront (the right end in our example) continues to travel at the

*For visible light, c is less than c_0. For x rays, c can be greater than c_0.

Figure 33.15 Wavefronts for a ray traveling from vacuum into transparent glass in a direction normal to the glass surface.

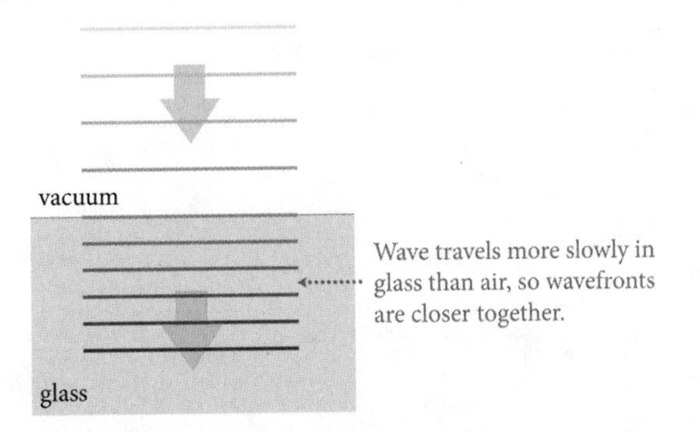

vacuum speed. This means the distance AC traveled by the left end is less than the distance BD traveled by the right end during the same time interval (Figure 33.16*a*). Consequently, the wavefront CD in the material is no longer parallel to the wavefront AB that has not yet entered the material.

The direction of the ray associated with these wavefronts therefore changes on entering the material. As shown in Figure 33.16*b*, the angle of incidence θ_1, between the ray in vacuum and the normal to the interface between the two materials, is greater than the angle θ_2, between the ray in the material and the normal to that interface. This bending of light as it moves from one material into another is referred to as **refraction,** and the angle θ_2 between the refracted ray and the normal to the interface between the materials is called the **angle of refraction.** Whenever light is refracted, the angle between the ray and the normal is always greater in the material in which the light travels faster, so:

When a light ray travels from one material into a second material where light travels more slowly, the ray bends toward the normal to the interface between the materials.

Generally, the speed of light decreases as the mass density of the material increases. Note also that, as shown in Figure 33.16*b*, both reflection and refraction take place at the interface between two media (or between vacuum and a medium).

The amount of bending depends on the angle of incidence and on the relative speeds in the two media. There is no bending for normal incidence (as we saw in Figure 33.15); the bending is less near normal incidence and becomes more pronounced as the angle of incidence increases. In Section 33.5, we'll work out a quantitative expression relating angles θ_1 and θ_2.

33.6 Suppose the ray in Figure 33.16 travels in the opposite direction—that is, from the denser medium to the less dense medium. If the angle of incidence is now θ_2, how does the angle of refraction compare with θ_1?

Because the relationship between the angles of incidence and refraction is completely determined by the speed of the wavefronts in the two media, the angles do not depend on which is the incident ray and which is the refracted ray. As shown in **Figure 33.17**, θ_1 and θ_2 have the same values whether θ_1 is the angle of incidence (Figure 33.17*a*) or the angle of refraction (Figure 33.17*b*). Keep in mind, however, that the reflected ray is always on the same side of the interface as the incident ray, and so the angle of reflection is *not* the same in Figure 33.17*a* and Figure 33.17*b*.

Figure 33.16 (*a*) Refraction is explained by the behavior of wavefronts that cross at an angle into a transparent medium in which they travel more slowly. (*b*) Incident, reflected, and refracted rays, showing the angles of incidence θ_1 and refraction θ_2 (measured from the normal to the surface).

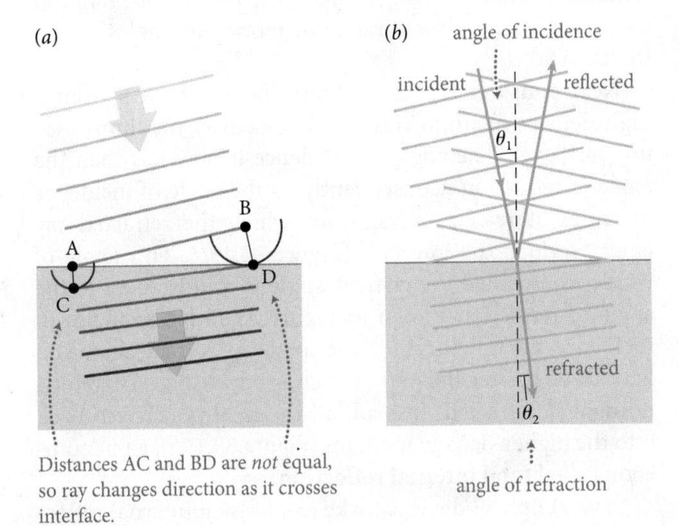

Figure 33.17 Because refraction is caused by the relative speeds of the wavefronts in two media, the angles of incidence and refraction do not depend on which is the incident ray and which the refracted ray.

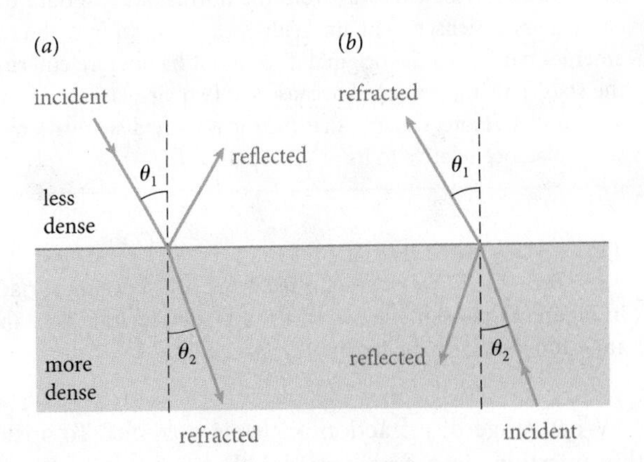

Example 33.3 Crossing a slab

Consider a light ray incident on a parallel-sided slab of glass surrounded by air, as shown in Figure 33.17*a*. The ray travels all the way through the slab and emerges into air on the other side. In what direction does the ray emerge?

❶ **GETTING STARTED** This problem involves two successive encounters of a light ray with interfaces between glass and air. At each interface, the ray is refracted. I need to determine the direction of the ray (its angle to the normal to the slab) after it crosses the lower interface of the slab represented in Figure 33.17*a*.

❷ **DEVISE PLAN** Because I want to know the direction of the emerging ray, I construct an appropriate ray diagram. Figure 33.17*a* shows the direction of the ray inside the slab. I extend the ray through the slab to the lower interface (**Figure 33.18**) and draw the emerging ray, labeling its angle to the normal θ_{lower}. To determine this angle, I need to consider the refraction that occurs at the lower interface.

Figure 33.18

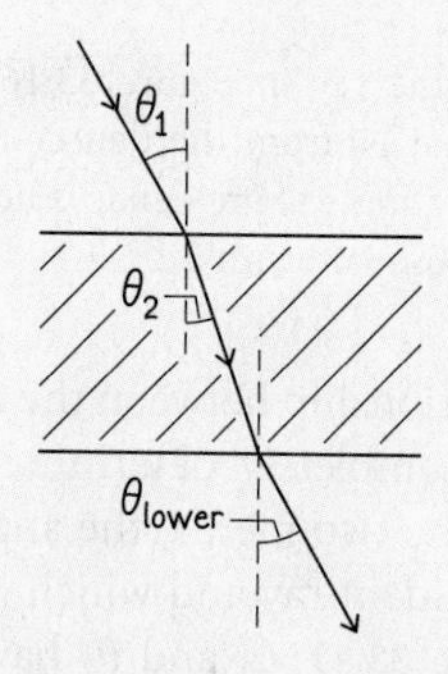

❸ **EXECUTE PLAN** Because the two interfaces are parallel, their normals are also parallel, and so the angle at which the ray is incident on the lower interface is equal to the angle θ_2 at which it is refracted at the upper interface. I saw in Figure 33.17*b* that if the angle between the ray and the normal in the slab is θ_2, it doesn't matter whether the ray in the slab is the incident ray or the refracted ray; either way, the angle between the ray in the air and the normal is θ_1. Therefore $\theta_{lower} = \theta_1$. ✔

❹ **EVALUATE RESULT** Crossing the lower interface from glass into air, the ray bends away from the normal, as it should because glass is denser than air. With $\theta_{lower} = \theta_1$, in fact, the ray emerges parallel to the original direction it had before entering the slab. This makes sense because the two air-glass interfaces are parallel. (Note, though, that the ray is shifted sideways by a small distance relative to its original path.)

33.7 When the ray reflected from the bottom surface in Figure 33.18 reemerges from the top surface, how does the angle it makes with the normal compare with θ_1?

What range of refraction angles is possible? To answer this question, let's first consider the case where the ray travels from a low-density medium into a denser medium (**Figure 33.19*a–d***). Because the angle of incidence is always greater than the angle of refraction in this situation, as the angle of incidence approaches 90°, the angle of refraction remains less than 90° (Figure 33.19*d*). The full 90° range of incidence angles gives a range of refraction angles that is less than 90°.

Figure 33.19 The range of possible refraction angles for a ray crossing into a medium of either higher or lower density.

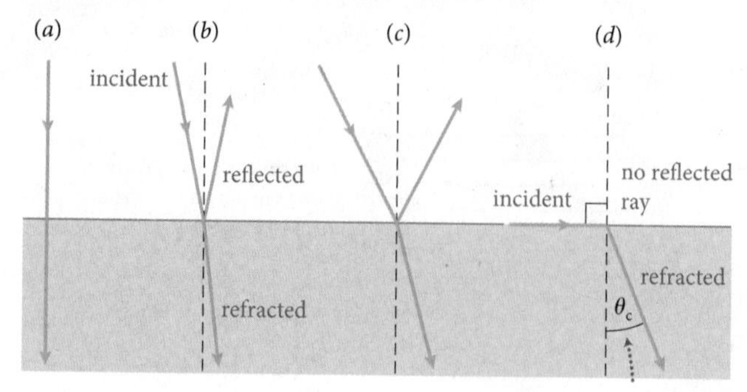

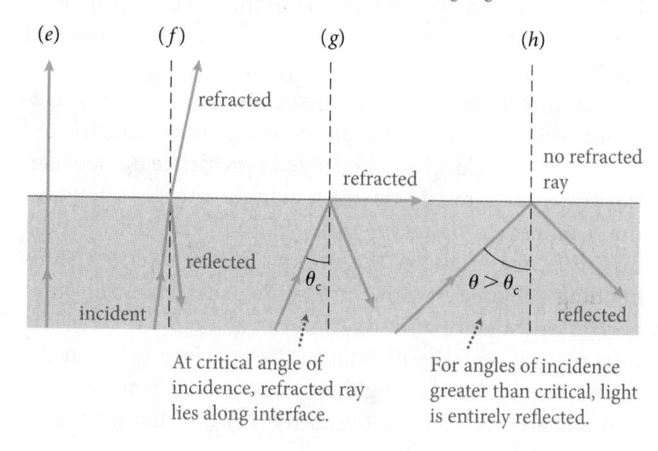

Next consider the case where the ray travels from a high-density medium into a lower-density medium (Figure 34.18*e–g*). The angle of incidence is now less than the angle of refraction. Consequently, as the angle of incidence increases, it reaches a value for which the refracted ray emerges along the interface (Figure 33.19*g*). This angle of incidence is called the **critical angle** θ_c and is equal to the angle of refraction shown in Figure 33.19*d*. For angles of incidence greater than θ_c, the angle of refraction would have to be greater than 90°, which is impossible. Therefore, no light is refracted. Instead, all the light is *reflected* back into the higher-density medium (Figure 33.19*h*), a phenomenon called **total internal reflection.**

Several optical devices make use of total internal reflection to direct light. The glass prism shown in **Figure 33.20**

Figure 33.20 A prism can act as a perfect mirror by means of total internal reflection.

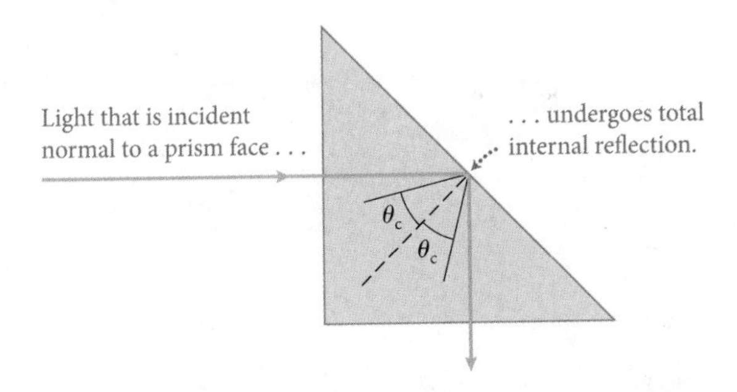

reflects light just as a mirror would. A light ray enters the prism's front surface at normal incidence. Because the back surface is slanted relative to the front surface, the angle at which the ray hits the back surface is less than 90°. This back-surface angle of incidence is greater than the critical angle for the glass, however, and so the light is totally reflected from the back surface. Such prisms are actually better mirrors than most regular mirrors; they reflect very close to 100% of the incident light, whereas mirrors are less reflective due to imperfections in the reflecting surface.

Optical fibers also guide light by means of total internal reflection. An optical fiber is a long, thin fiber made of a transparent material such as glass. If light shines into one end of the fiber at an angle greater than the critical angle, the light travels along the fiber through repeated total internal reflections, and essentially all of the light that entered the fiber emerges at the other end (**Figure 33.21**). Because very little light is lost as the light travels, only a faint glow comes from the rest of the fiber.

Figure 33.21 How optical fibers work.

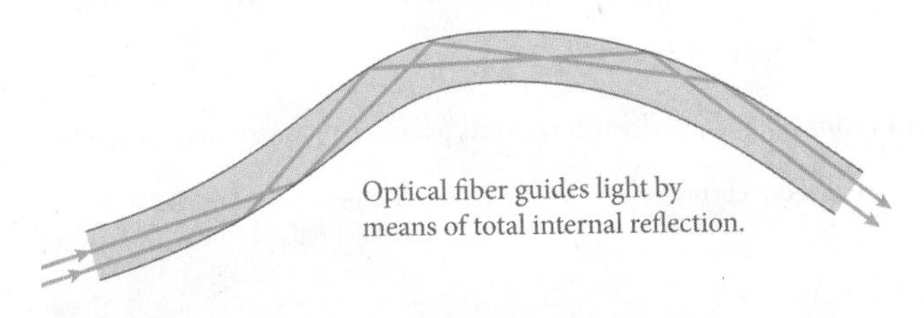

Figure 33.22 The phenomenon of dispersion, which results from the fact that the speed of light in a given medium (and hence the angle of refraction) depends slightly on the frequency of the light.

(*a*) Prism refracts light of single frequency

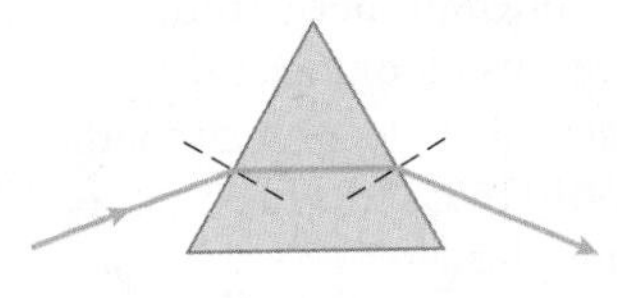

(*b*) Dispersion: different colors have different angles of refraction

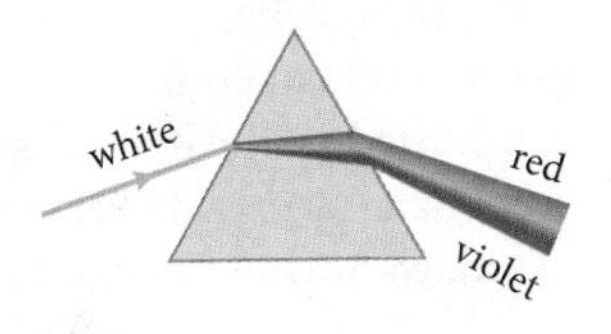

(*c*) Rainbows result from dispersion of sunlight by raindrops

Because the speed of light in any given medium depends slightly on the frequency of the light, the angle of refraction also depends on frequency. This phenomenon is called **dispersion** because it causes rays of different colors to separate—to be *dispersed*—when refracted. Prisms like the one shown in **Figure 33.22** are designed to separate colors by the frequency dependence of the angle of refraction. In most media, high-frequency light travels more slowly than low-frequency light, and so high-frequency light bends more strongly toward the normal. The lowest frequency of visible light is red and the highest is violet, which means violet light bends the most, as the rainbow of Figure 33.22*c* shows.

Both rainbows and the brilliance of gems result from a combination of total internal reflection and dispersion. In a rainbow (Figure 33.22*c*), the combination of total internal reflection and dispersion means that we see different colors coming from water droplets at different viewing angles. Gems are cut with many internal surfaces from which total internal reflection takes place. Because the light is also dispersed, colorless gems such as diamonds shine with many distinct colors.

33.8 Because of dispersion, the critical angle for total internal reflection in a given medium varies with frequency. Is the critical angle for a violet ray greater or less than that for a red ray?

Fermat's principle

Figure 33.23 shows four ways in which a light ray can travel between two locations A and B: directly, reflected from a mirror, refracted through a glass slab, and refracted through a prism.* You could say that in each case the ray reaches B because it is aimed properly from A. However, an entirely different way of looking at the path followed by the light was suggested by the French mathematician Pierre de Fermat (1601–1665) in a formulation today known as **Fermat's principle:**

> **The path taken by a light ray between two locations is the path for which the time interval needed to travel between those locations is a minimum.**

This principle may seem to imply that light always travels in a straight line. However, the *quickest path* between two locations is not necessarily the *shortest distance* when the speed of light differs in different regions.

Let's consider the four paths in Figure 33.23 using Fermat's principle. In Figure 33.23*a*, the ray does follow a straight path because the medium in which the ray travels is uniform. As a result, the quickest path is indeed the shortest distance: a straight line from A to B.

In Figure 33.23*b*, the fact that the straight-line path from A to B is blocked means that the ray must reflect somewhere off the mirror in order to travel from A to B. The path shown, which satisfies the law of reflection, is the shortest distance from A to B involving reflection from the mirror. Because the distance from A to the reflection location P equals the distance between the image location I and P, the straight line IB is equal in length to the path traveled by the ray from A to B. Moving the reflection location to either side of P, so that the angle of incidence does not equal the angle of reflection, increases the length of the path. Thus, Fermat's principle implies the law of reflection.

When the ray must travel through some air and some glass, as in Figure 33.22*c* and *d*, the quickest path is not a straight line because the ray's speed in the glass is only two-thirds of its speed in air. To minimize the time interval needed to travel from A to B, the ray bends on entering and exiting the glass. Such a bent path reduces the distance traveled through the glass without increasing the distance traveled in air so much that it offsets the amount of time saved. In Example 33.7, we shall see that calculating the bending angles with Fermat's principle gives the same result as with ray optics.

Figure 33.23 Ray diagrams illustrating the *quickest path* for a light ray traveling from A to B for four situations.

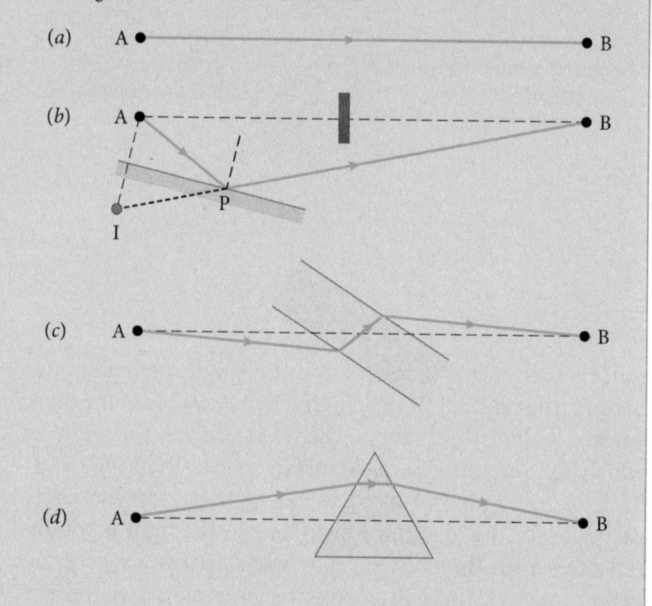

*Note that what distinguishes a glass slab from a glass prism is the way I use the terms: In a slab, the two opposite surfaces are parallel to each other; in a prism, they are not.

33.4 Forming images

As shown in **Figure 33.24**, by combining two prisms and a glass slab we can create a device that steers parallel light rays toward each other. The rays through the center of the device pass straight through, those through the top prism are refracted downward, and those through the bottom prism are refracted upward.

To bring all parallel incident rays to a single point, a structure called a **lens** is used. A lens is designed with curved surfaces so that the refraction of incident rays increases gradually as we move away from the center. To accomplish this, lenses are typically made with spherical surfaces, which are easy to manufacture.

Figure 33.25*a* shows a lens with *convex* spherical surfaces, where a *convex surface* is defined as one that curves like the outside of a sphere. Rays parallel to the lens *axis*—a line perpendicular to the lens through its center—converge through such a lens onto a single point called either the **focus** or the **focal point.** A lens with convex surfaces is therefore called a *converging lens.* The distance from the center of the lens to the focus is called the **focal length** f.

Figure 33.24 A device that redirects parallel light rays toward each other.

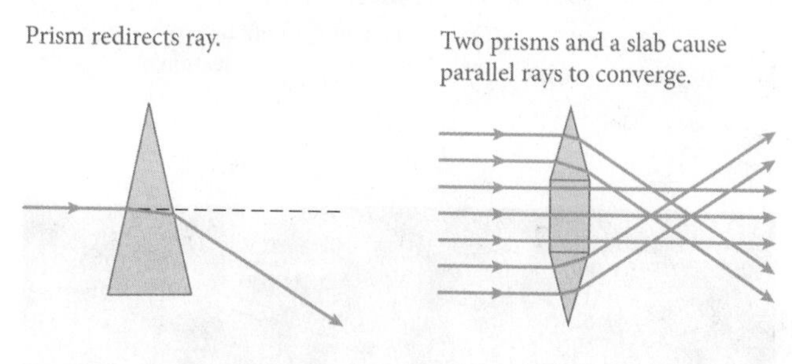

Prism redirects ray.
Two prisms and a slab cause parallel rays to converge.

Figure 33.25 Converging lens with convex spherical surfaces.

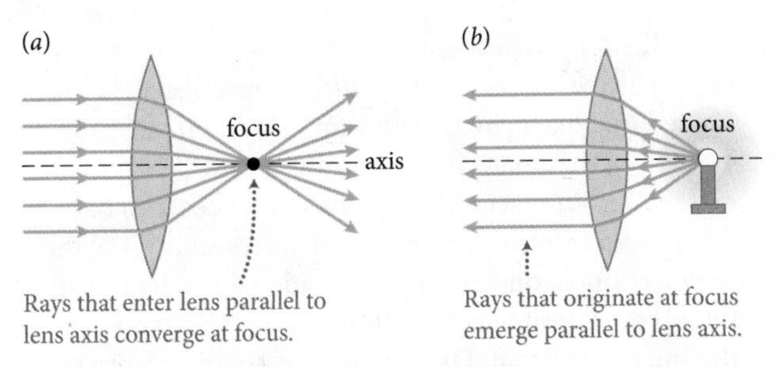
(a)
(b)
focus
axis
focus
Rays that enter lens parallel to lens axis converge at focus.
Rays that originate at focus emerge parallel to lens axis.

What if we place a light source at the focus of a lens, as in Figure 33.25*b*? As we saw in the preceding section, the path followed by a light ray is unaffected by reversing the direction of propagation of the ray, as long as the ray is not absorbed by the medium. So, if we place a light source at the focus of a lens, a beam of parallel rays emerges on the other side of the lens.

33.9 Sketch the wavefronts corresponding to all the rays in Figure 33.25*a*, both the parallel ones on the left and the refracted ones on the right.

We can reverse the direction of the rays through the lens of Figure 33.25, so that parallel rays enter from the right (**Figure 33.26**). The rays converge again at a focus that is the same distance f from the center of the lens. Thus, every lens has two foci, one on either side of the lens at the same distance f from it.

If we tilt the parallel rays a bit relative to the lens axis (**Figure 33.27**), the rays still converge at a distance f from the center, but the focus is no longer on the axis. Provided the parallel rays make only a small angle with the lens axis, they all converge at a point on a plane—called the *focal plane*—that is perpendicular to the axis a distance f from the lens. Rays that run near the lens axis—either parallel to it or at a small angle—are said to be *paraxial*.

Now that we know how parallel rays and rays that emanate from the focus of a lens are refracted by the lens, we can determine where images are formed. The image of a point on an object is formed where all the light rays emanating from that point converge. (These light rays then diverge from the location of the image; when they enter the eye, the brain interprets them as having come from the location of the image.) An image of the entire object is made up of the images of all the individual points on the object.

Figure 33.26 A lens has two equivalent foci, one on each side, at equal distances from the center of the lens.

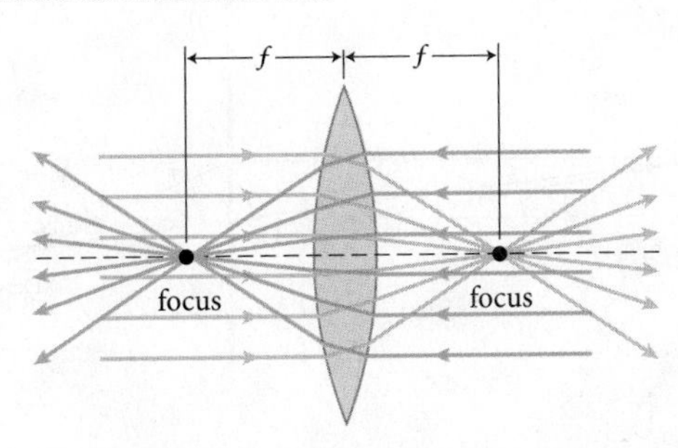
f
f
focus
focus

Figure 33.27 If the rays strike the lens at an angle, they no longer converge on the focus, but they still converge on a focal plane at the focal distance f.

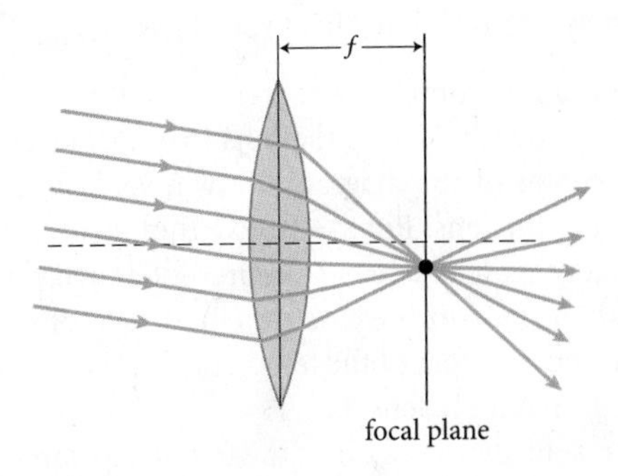
f
focal plane

To determine where the rays emanating from a point on an object converge, we don't need to draw all the rays. Instead, we draw three special ones, called **principal rays,** and see where they converge:

1. a ray that travels parallel to the lens axis before entering the lens,
2. a ray that passes through the center of the lens, and
3. a ray that passes through the focus that is on the same side of the lens as the object.

These three principal rays are shown in **Figure 33.28*a*** for the case where the object lies beyond the focus of the

Figure 33.28 The three principal rays for a spherical lens can be used to determine the location, size, and orientation of the image for a given object.

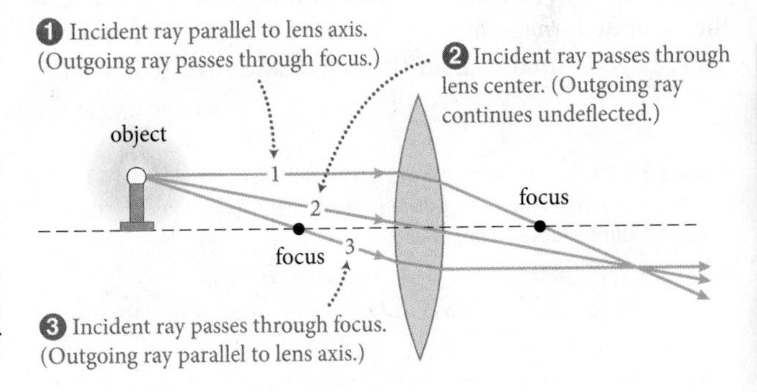
(a) The three principal rays
1 Incident ray parallel to lens axis. (Outgoing ray passes through focus.)
2 Incident ray passes through lens center. (Outgoing ray continues undeflected.)
object
1
2
3
focus
focus
3 Incident ray passes through focus. (Outgoing ray parallel to lens axis.)

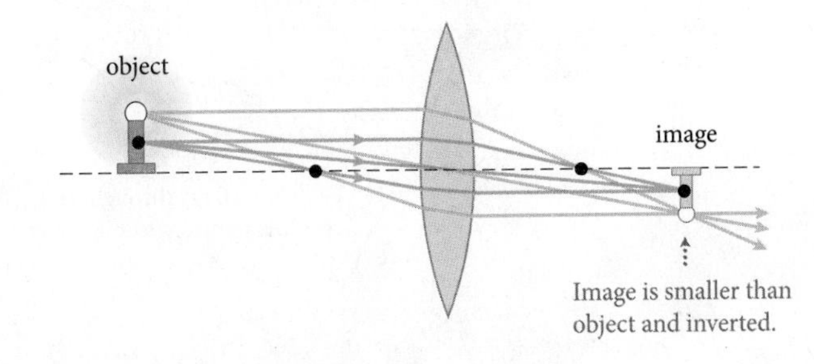
(b) Using principal rays to determine location and orientation of image
object
image
Image is smaller than object and inverted.

CONCEPTS

Procedure: Simplified ray diagrams for lenses

To determine the location and orientation of an image formed by a lens, follow this procedure:

1. Draw a horizontal line representing the lens axis (the line perpendicular to the lens through its center). In the center of the diagram, draw a vertical line representing the lens. Put a + above the line to represent a converging lens or a − to represent a diverging lens.
2. Put two dots on the axis on either side of the lens to represent the foci of the lens. The dots should be equidistant from the lens.
3. Represent the object by drawing an upward-pointing arrow from the axis at the appropriate relative distance from the lens. For example, if the distance from the object to the lens is twice the focal length of the lens, put the arrow twice as far from the lens as the dot you drew in step 2. The top of the arrow should be at about half the height of the lens.
4. From the top of the arrow representing the object draw two or three of the three *principal rays* listed in the Procedure box "Principal rays for lenses" on page 888.
5. The top of the image is at the point where the rays *that exit the lens* intersect (if they diverge, trace them backward to determine the point of intersection). If the intersection is on the opposite side of the lens from the object, the image is real; if it is on the same side, the image is virtual. Draw an arrow pointing from the axis to the intersection to represent the image (use a dashed arrow for a virtual image).

In general it is sufficient to draw two principal rays, but depending on the situation, some rays may be easier to draw than others. You can also use a third ray to verify that it, too, goes through the intersection. (If it doesn't, you have made a mistake.)

lens. We already know how rays 1 and 3 travel. Ray 1 passes through the focus on the other side of the lens, and ray 3 emerges from the lens parallel to the axis. As for ray 2, as long as it is paraxial, it passes straight through with negligible refraction (**Figure 33.29**). (Nonparaxial rays are shifted significantly; our treatment of lenses in this chapter is restricted to images formed by paraxial rays.) Ray 2 can therefore be drawn as traveling in a straight line through the center of the lens.

To determine the location and orientation of the image of an extended object, we work out the locations of the images of several points on the object. Figure 33.28*b* shows where the rays converge for two points on the object, and the extended image that can be inferred from these points. The image is smaller than the object and inverted. See the Procedure box "Simplified ray diagrams for lenses" on this page for a general description of how to draw simplified ray diagrams for lenses.

Figure 33.29 Paraxial and nonparaxial rays. For a paraxial ray or a lens that is not too thick, the refraction is so slight that we can consider this ray to be one uninterrupted straight line.

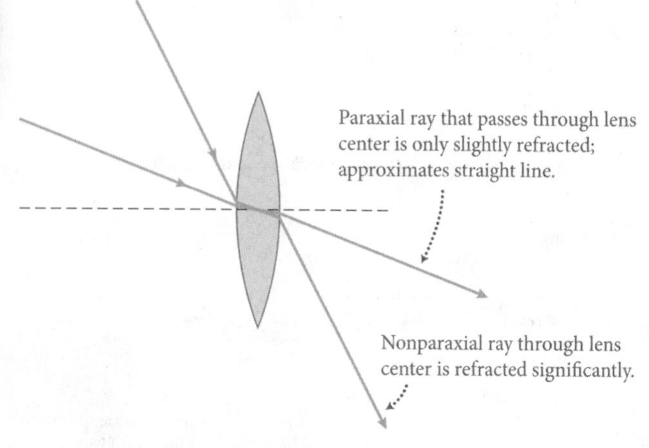

Paraxial ray that passes through lens center is only slightly refracted; approximates straight line.
Nonparaxial ray through lens center is refracted significantly.

33.10 Do you need to draw all three principal rays to determine the location of an image?

Example 33.4 Where is the image?

Consider the light bulb that is the object in Figure 33.28. If you move the bulb to the left, does the image shift left, shift right, or stay in the same place?

❶ GETTING STARTED Using the procedure for drawing simplified ray diagrams, I represent the lens, object, and image of Figure 33.28, and draw the bulb at its new position (**Figure 33.30*a***).

Figure 33.30

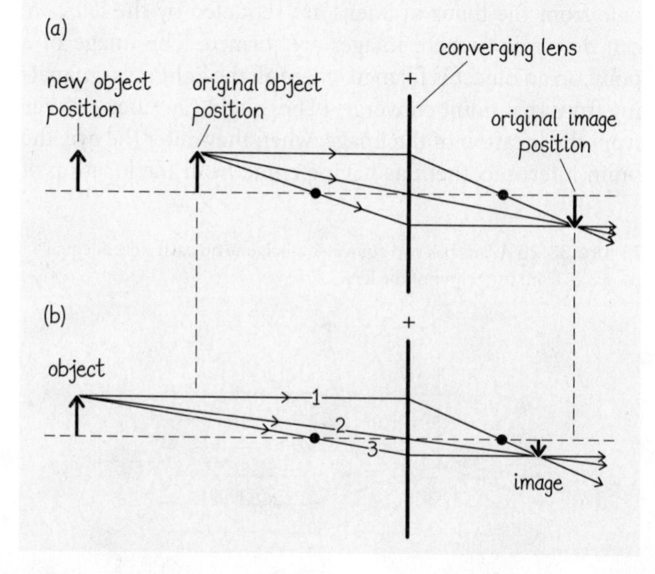

❷ **DEVISE PLAN** I can determine which way the image shifts by drawing the principal rays for the light bulb in its new position.

❸ **EXECUTE PLAN** With the rays drawn (Figure 33.30*b*), I see that the image shifts left. ✔

❹ **EVALUATE RESULT** As I move the light bulb to the left, principal ray 1 remains the same, but principal ray 2 makes less of an angle with the lens axis than before. Consequently, the location at which these two rays intersect is closer to the lens than before, shifting the image to the left, as I concluded from the diagram.

Notice in Figure 33.30 that the image gets smaller as the object moves farther away from the lens. Conversely, moving the object closer to the lens makes the image larger. Indeed, one of the most common uses of lenses is to enlarge images. If the object is placed at the focus, no image forms because the rays all emerge from the lens parallel to each other—in other words, the rays do not converge.

Consider placing an object *between* the lens and its focus, as in **Figure 33.31**. In this configuration, principal ray 3 does not pass through the focus. Instead, it lies on the line that joins the focus to the point of interest on the object.

The image formed in the configuration of Figure 33.28 (object beyond focus) is real because the rays really do converge at the point where the image is formed. In contrast, the image in Figure 33.31 is virtual because the rays do not actually converge at the point where the image is formed. (The extensions of these rays do cross the image point, however, and so an observer interprets the rays emerging from the lens as having traveled along straight lines from the location of the image, as indicated in Figure 33.31*a*.)

Figure 33.31 (*a*) When an object is located between the lens and the focus, the image is virtual and enlarged. (*b*) In a simplified ray diagram, the object and image are replaced with solid and dashed arrows, respectively, and the lens is replaced by a vertical line. (The + indicates a converging lens.)

(*a*) Ray diagram for an object located between the focus and the lens

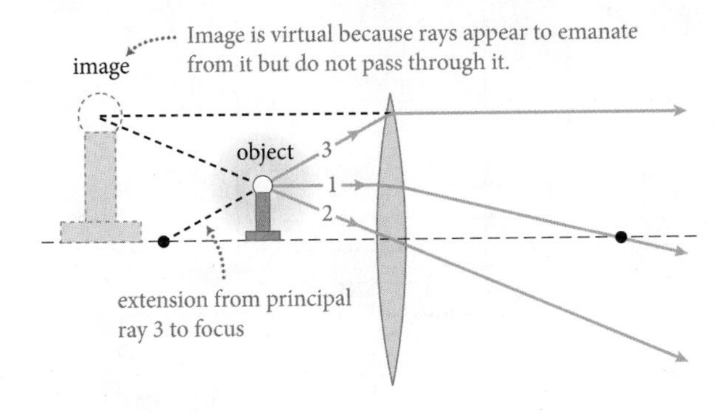

(*b*) Simplified version of ray diagram

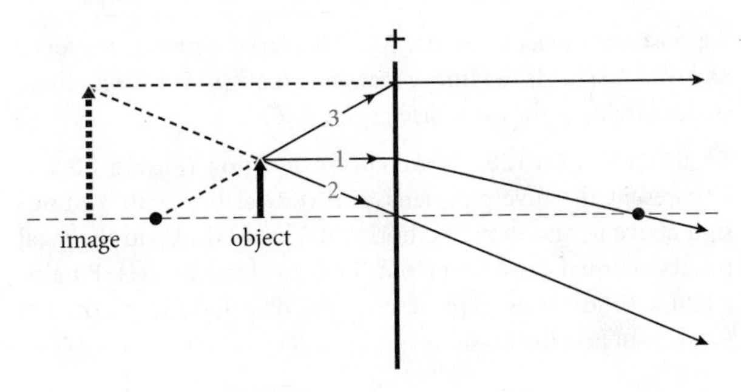

An important difference between real and virtual images is that if a screen is placed at the location of a real image, the image can be seen on the screen. Placing a screen at the location of a virtual image does not display the image because the light rays do not actually pass through the image location.

Figure 33.31*a* shows that, for this configuration of object and lens (object between focus and lens), the image is larger than the object and upright (unlike the image in Figure 33.28, which is inverted). A magnifying glass is designed to produce an enlarged, upright image of an object, which means that magnifying glasses are made with converging lenses and are held close to the object of interest (so that the object is between the lens and the focus).

33.11 As the object in Figure 33.31 is moved closer to the lens, does the size of the image increase, decrease, or stay the same?

Just as with electric circuits, it is convenient to use a simplified notation for ray diagrams. Figure 33.31*b* shows such a simplified version of the ray diagram of Figure 33.31*a*. Note that objects and real images are denoted by solid arrows and virtual images are denoted by dashed arrows.

Lenses can also be made with concave spherical surfaces rather than convex ones, where a *concave surface* is one that curves like the inside of a sphere. Such a lens is called a *diverging lens*, and **Figure 33.32** shows why: A series of parallel rays entering the lens are no longer parallel when they emerge. If we follow the path of the emerging rays back to the left side of the lens, we see that the diverging rays appear to all come from the same location on the left side. This location corresponds to the focus of a converging lens, but in a diverging lens it is a *virtual focus* rather than a real focus because the rays never actually travel through this location. (Just as for converging lenses, there is an equivalent focus on the other side of the lens.)

Figure 33.32 A diverging lens.

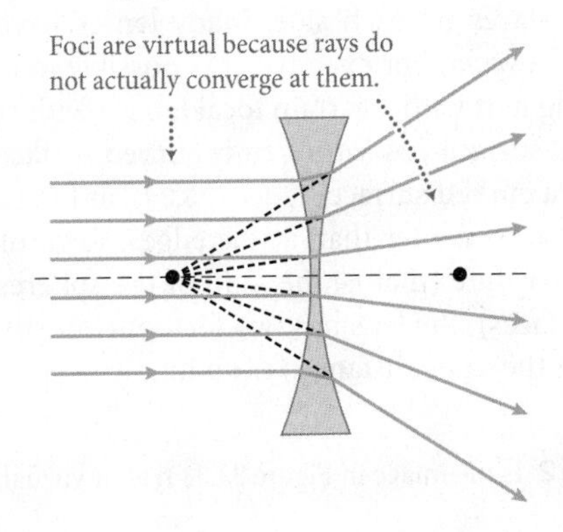

Procedure: Principal rays for lenses

The propagations of principal rays for converging and diverging lenses are very similar. The description below holds for rays that travel from left to right.

Converging lens

1. A ray that travels parallel to the lens axis before entering the lens goes through the right focus after exiting the lens.
2. A ray that passes through the center of the lens continues undeflected.
3. A ray that passes through the left focus travels parallel to the lens axis after exiting the lens. If the object is between the focus and the lens, this ray doesn't pass through the focus but lies on the line from the focus to the point where the ray originates.

Diverging lens

1. A ray that travels parallel to the lens axis before entering the lens continues along the line from the left focus to the point where the ray enters the lens.
2. A ray that passes through the center of the lens continues undeflected.
3. A ray that travels toward the right focus travels parallel to the lens axis after exiting the lens.

Figure 33.33 Ray diagram for an object outside the focus of a diverging lens.

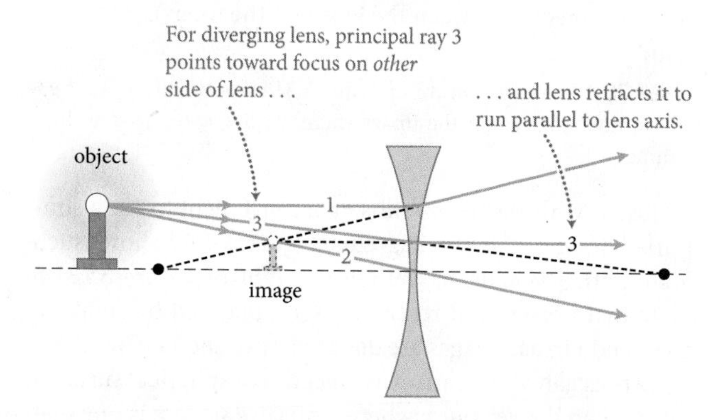

Figure 33.34 These lenses are all converging because each is thicker at the center than at the edges.

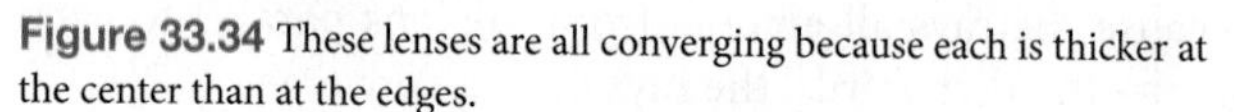

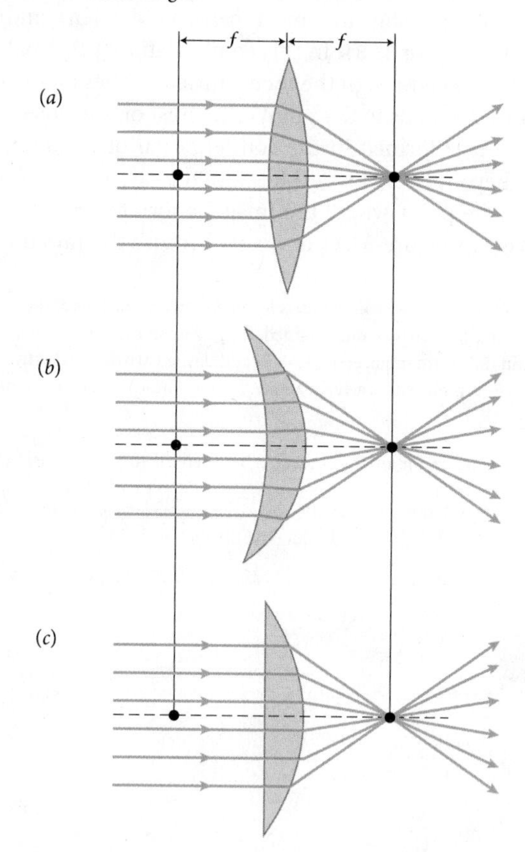

Figure 33.33 shows a ray diagram for a diverging lens. The same principal rays are drawn, but now ray 3 does not pass through the focus on the same side of the lens as the object. Instead, for diverging lenses ray 3 is drawn on the line that runs from the point where the ray originates to the focus on the other side of the lens; once refracted by the lens, this ray travels parallel to the lens axis. The general procedure for drawing ray diagrams for diverging lenses is still the same as the one for converging lenses (see the Procedure box "Simplified ray diagrams for lenses" on page 886), but the drawing of principal rays is a little bit different (see the Procedure box "Principal rays for lenses" above).

The lenses we have considered so far all have identical curved surfaces on each side. Many lenses have different surfaces, however. For example, it is possible to construct a converging lens with a certain focal length with two identical curved surfaces, two differently curved surfaces, or even a flat and a curved surface (**Figure 33.34**), as long as the lens is thicker at its center than at the edges. Regardless of the *radii of curvature* (that is, the radii of the spheres that best fit the surfaces), the lens has two foci, one on either side of the lens at the same distance f from it.

33.12 Is the image in Figure 33.33 real or virtual?

Example 33.5 Demagnifying glass

Suppose the object in Figure 33.33 is placed between the focus and the lens. (*a*) Is the image real or virtual? (*b*) Is it larger than, smaller than, or the same size as the object?

❶ GETTING STARTED To sketch the situation (**Figure 33.35**), I represent the diverging lens as a vertical line with a minus sign above it, and draw the horizontal lens axis. I add the focal points at equal distances from the lens along its axis. Finally, I add a solid arrow, representing the object, between the left focal point and the lens.

Figure 33.35

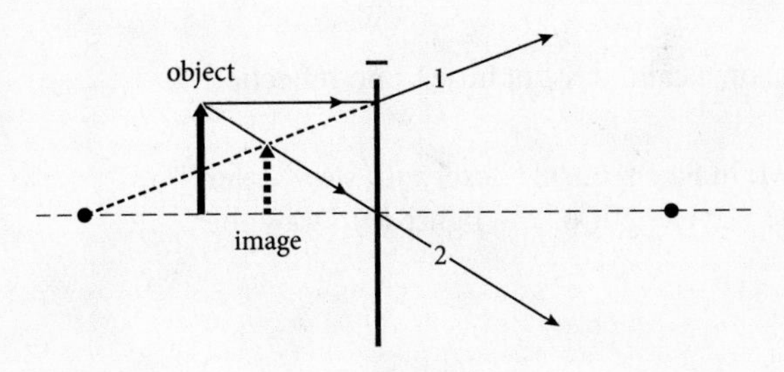

❷ **DEVISE PLAN** To locate the image and determine its size and whether it is real or virtual, I can draw the principal rays.

❸ **EXECUTE PLAN** (*a*) I add to my diagram the principal rays coming from the tip of my object arrow. All I need is rays 1 and 2; I do not need ray 3 because the intersection of ray 2 and the dashed extension of ray 1 unambiguously determines the location of the tip of the image arrow. The dashed extension I had to draw for ray 1 tells me that the rays do not actually intersect at this location; they only appear to intersect here. Therefore the image is virtual. ✔

(*b*) My diagram tells me that the image is smaller than the object. ✔

❹ **EVALUATE RESULT** Because a diverging lens spreads rays out rather than bringing them together, it makes sense that a virtual image will form. I also know from experience that in contrast to a converging lens, which magnifies images, diverging lenses create smaller images, as I found.

33.13 (*a*) Draw the third principal ray in **Figure 33.35**. Is there any position for the object in Figure 33.35 for which (*b*) the image is larger than the object and (*c*) the image is real?

Self-quiz

1. Why do you get a clear reflection from the surface of a lake on a calm day but little or no reflection from the surface on a windy day?
2. (*a*) As light travels from one medium into another, as shown in **Figure 33.36** ("fast" and "slow" refer to the wave speed in each medium), what happens to the wavelength of the light? (*b*) Draw the reflected and refracted rays at each surface.

Figure 33.36

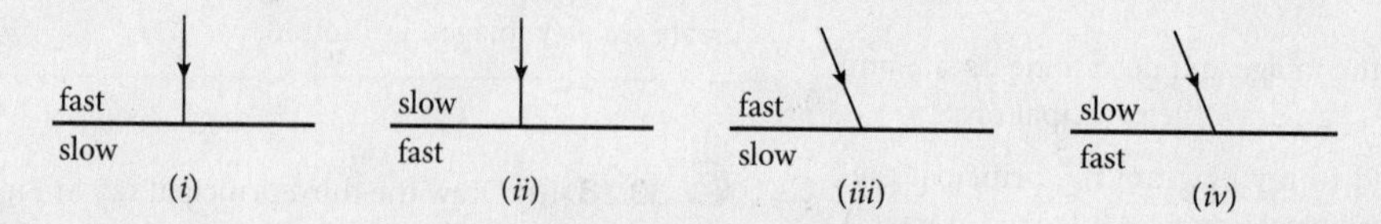

3. What is the difference between a real image and a virtual image?
4. In each situation in **Figure 33.37**, draw the three rays emanating from the top of the object and reflecting or refracting from the optical element shown. Show the image, and state whether it is real or virtual.

Figure 33.37

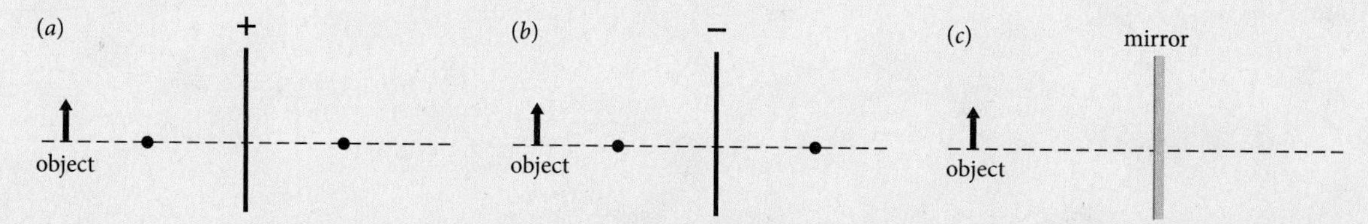

Answers:

1. On a calm day, the lake surface is smooth, and specular reflection is like that of a mirror. On a windy day, the surface is rough, which makes the reflection diffuse and prevents the formation of an image.
2. (*a*) The wavelength decreases when the wave travels more slowly in the second medium (*i* and *iii*) and increases when the wave travels faster in the second medium (*ii* and *iv*). (*b*) See **Figure 33.38**.

Figure 33.38

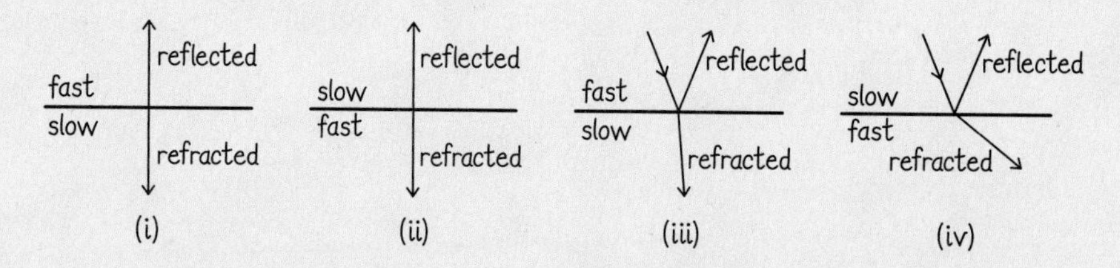

3. Real image: All rays actually pass through the location of the image, and the image can be seen on a screen placed at the image location. Virtual image: All rays do not pass through the location of the image (only the extensions of the rays do), and the image cannot be seen on a screen placed at the image location.
4. See **Figure 33.39**. For the lenses, the three principal rays can be used to locate the image; for the mirror, any rays and the law of reflection can be used to locate the image. The images are (*a*) real, (*b*) virtual, (*c*) virtual.

Figure 33.39

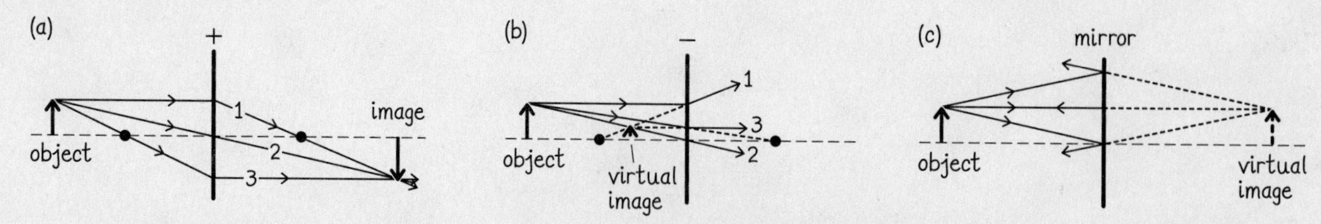

33.5 Snel's law

In the first part of this chapter, we saw that light refracts when it travels from one medium into another because the speed of light depends on the medium. The speed of light in a medium is specified by the **index of refraction:**

$$n \equiv \frac{c_0}{c}, \tag{33.1}$$

where c is the speed of light in the medium and c_0 is the speed of light in vacuum. (By definition, $n_{\text{vacuum}} = 1$; in air $n_{\text{air}} \approx 1$.) If a light wave of frequency f travels from one medium into another, the frequency doesn't change because the source determines the frequency (see also Checkpoint 16.10). The wavelength, however, does change; it is greater in the medium in which wave speed is greater.

The wavelength λ of the light is related to the wave speed and frequency, in the same manner that these quantities are related for harmonic waves (Eq. 16.10). In vacuum, for example,

$$\lambda = \frac{c_0}{f} \text{ (vacuum)}. \tag{33.2}$$

In a medium in which a wave has speed c_1, the wavelength λ_1 is given by

$$\lambda_1 = \frac{c_1}{f} = \frac{c_0/n_1}{f} = \frac{1}{n_1}\lambda, \tag{33.3}$$

where λ is the wavelength of the wave in vacuum and n_1 is the index of refraction of the medium. Thus, the wavelength decreases as the index of refraction increases. As discussed in Section 33.3, the amount of refraction a light wave undergoes varies somewhat with wavelength (see Figure 33.22) because different wavelengths of light travel at different speeds. Therefore the index of refraction depends on the wavelength. **Table 33.1** lists the indices of refraction for some common transparent materials at a wavelength of 589 nm.

Let us now work out the quantitative relationship between the angle of incidence and the angle of refraction. **Figure 33.40** shows wavefronts and one ray for a beam of light incident on the interface between medium 1 and medium 2 at angle θ_1 from the normal. The angle of refraction is θ_2. Using right triangles ABD and ACD, we can express angles θ_1 and θ_2 in terms of the wavelengths λ_1 and λ_2:

$$\sin\theta_1 = \frac{\text{BD}}{\text{AD}} = \frac{\lambda_1}{\text{AD}} \tag{33.4}$$

and

$$\sin\theta_2 = \frac{\text{AC}}{\text{AD}} = \frac{\lambda_2}{\text{AD}}. \tag{33.5}$$

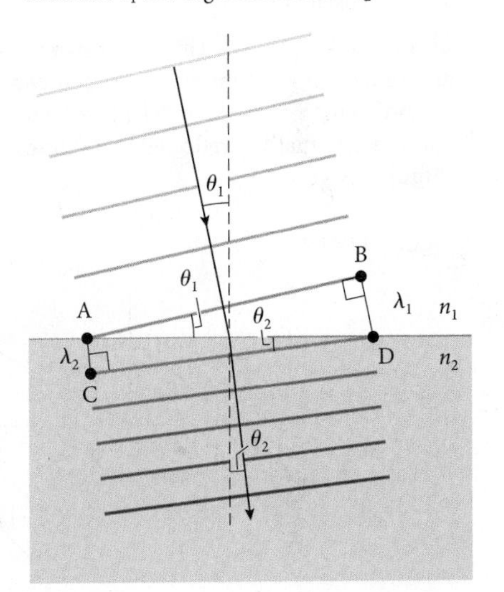

Figure 33.40 Relationship between angle of incidence θ_1 and angle of refraction θ_2.

Table 33.1 Indices of refraction for common transparent materials

Material	n (for λ = 589 nm)
Air (at standard temperature and pressure)	1.00029
Liquid water	1.33
Sugar solution (30%)	1.38
Sugar solution (80%)	1.49
Microscope cover slip glass	1.52
Sodium chloride (table salt)	1.54
Flint glass	1.65
Diamond	2.42

Combining these equations to eliminate AD and substituting $\lambda_1 = \lambda/n_1$ (Eq. 33.3), we get

$$\frac{\sin\theta_1}{\sin\theta_2} = \frac{\lambda_1}{\lambda_2} = \frac{\lambda/n_1}{\lambda/n_2} = \frac{n_2}{n_1}, \tag{33.6}$$

which can be written as

$$n_1 \sin\theta_1 = n_2 \sin\theta_2. \tag{33.7}$$

This relationship between the indices of refraction and the angles of incidence and refraction is called **Snel's law,** after the Dutch astronomer and mathematician Willebrord Snel van Royen (1580–1626).

Example 33.6 Bending 90°

A ray traveling through a medium for which the index of refraction is n_1 is incident on a medium for which the index of refraction is n_2. At what angle of incidence θ_1, expressed in terms of n_1 and n_2, must the ray strike the interface between the two media for the reflected and transmitted rays to be at right angles to each other?

❶ GETTING STARTED This problem involves both reflection and refraction at an interface between two media. To visualize the problem, I draw the incident, reflected, and refracted rays and indicate that the reflected and refracted rays are 90° apart (**Figure 33.41**).

Figure 33.41

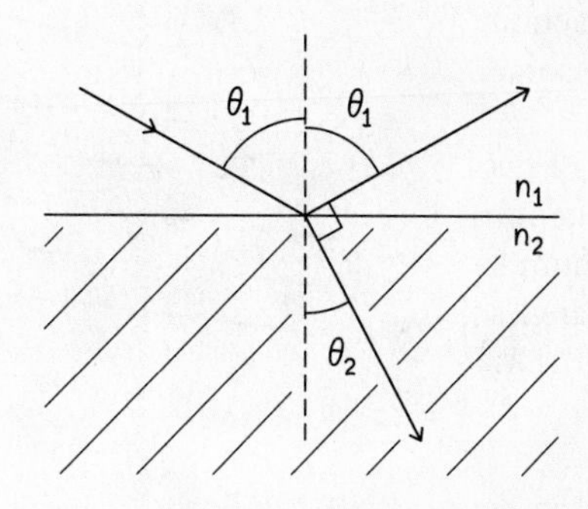

❷ DEVISE PLAN Snel's law (Eq. 33.7), the law of reflection, and the indices of refraction determine the paths taken by the reflected and refracted rays. Therefore I need to use those relationships to obtain an expression that tells me the value of θ_1 that produces reflected and refracted rays oriented 90° to each other. To obtain θ_1 in terms of n_1 and n_2, I need to eliminate θ_2 from Eq. 33.7. To do so, I use the fact that the angles on the right side of the normal to the interface must add to 180°. Thus, with reflected and refracted rays forming a 90° angle, I can say $180° = \theta_1 + 90° + \theta_2$. Solving this expression for θ_2 gives $\theta_2 = 90° - \theta_1$, which I can substitute into Eq. 33.7.

❸ EXECUTE PLAN Substituting $\theta_2 = 90° - \theta_1$ into Eq. 33.7, I get

$$n_1 \sin\theta_1 = n_2 \sin(90° - \theta_1) = n_2 \cos\theta_1,$$

and isolating the terms that contain θ_1 gives

$$\frac{\sin\theta_1}{\cos\theta_1} = \tan\theta_1 = \frac{n_2}{n_1}$$

$$\theta_1 = \tan^{-1}\left(\frac{n_2}{n_1}\right). \checkmark$$

❹ EVALUATE RESULT My result says that θ_1 increases as n_2 increases. This makes sense because as n_2 increases, the refracted ray bends more, meaning that θ_2 becomes smaller. To keep the reflected and refracted rays perpendicular to each other, the angle of reflection must increase, and so θ_1 must also increase.

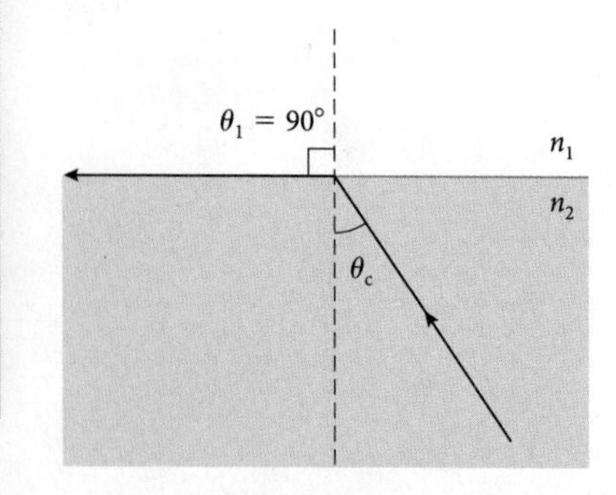

Figure 33.42 Critical angle for a ray traveling from a denser medium (n_2) to a less dense medium ($n_1 < n_2$).

Earlier in this chapter, we found that for rays traveling from a denser medium to a less dense medium, we can define a critical angle of incidence θ_c such that the angle of refraction θ_1 is equal to 90° (**Figure 33.42**); beyond this critical angle θ_c, total internal reflection occurs. We can calculate the critical angle θ_c for an interface between two media with indices of refraction n_1 and n_2 ($n_2 > n_1$) by applying Snel's law (Eq. 33.7) and setting $\theta_1 = 90°$:

$$\frac{\sin\theta_1}{\sin\theta_2} = \frac{1}{\sin\theta_c} = \frac{n_2}{n_1}. \tag{33.8}$$

Solving for θ_c gives

$$\theta_c = \sin^{-1}\left(\frac{n_1}{n_2}\right). \tag{33.9}$$

Example 33.7 Fermat's principle

For a light ray that crosses the interface between medium 1 having index of refraction n_1 and medium 2 having index of refraction n_2, what relationship between θ_1 and θ_2 follows from Fermat's principle (page 884)?

❶ **GETTING STARTED** I begin with a diagram that shows the two media and a ray traveling from an arbitrary point A in the n_1 medium to an arbitrary point C in the n_2 medium (**Figure 33.43**). Fermat's principle states that the path the ray takes from A to C is the path for which the time interval needed for the motion is a minimum. Therefore this ray must cross the interface at a point B that makes the time interval a minimum.

Figure 33.43

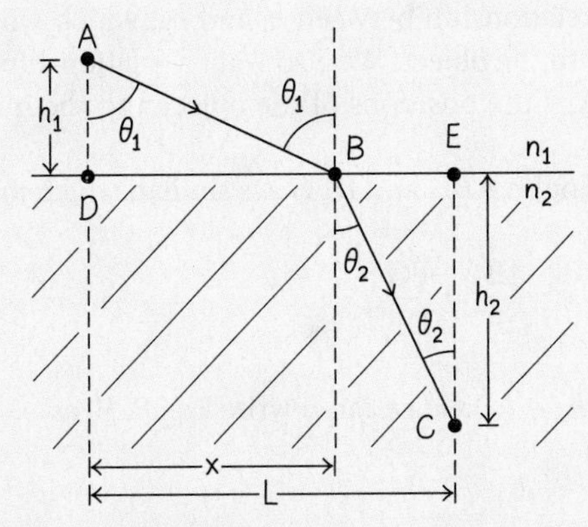

An alternative way to express this problem is: Given the locations of A and C, where must B lie so as to minimize the time interval needed to travel from A to C?

❷ **DEVISE PLAN** I add two more location labels to my drawing: D directly below A and lying on the interface, and E directly above C and lying on the interface. Doing so gives me two right-angle triangles that permit me to express the angles in terms of the distance traveled. I write h_1 for the distance AD and h_2 for the distance EC. I can think of the distance from D to B as unknown—I'll call it x—and the distance from D to E, which I'll call L, is fixed by the locations of A and C. My goal is to determine the value of x for which the travel time from A to C is minimized. Once I obtain x, I hope to obtain a relationship between θ_1 and θ_2.

❸ **EXECUTE PLAN** I begin by expressing the time interval Δt_{AC} the ray needs to travel from A to C in terms of the distances shown in Figure 33.43 and the speed of light in the two media:

$$\Delta t_{AC} = \Delta t_{AB} + \Delta t_{BC} = \frac{AB}{c_1} + \frac{BC}{c_2} = \frac{AB}{c_0/n_1} + \frac{BC}{c_0/n_2}. \quad (1)$$

Next I express AB and BC in terms of h_1, h_2, L, and x:

$$AB = \sqrt{h_1^2 + x^2}$$

$$BC = \sqrt{h_2^2 + (L - x)^2}.$$

Substituting these two expressions into Eq. 1 gives me

$$\Delta t_{AC} = \frac{\sqrt{h_1^2 + x^2}}{c_0/n_1} + \frac{\sqrt{h_2^2 + (L - x)^2}}{c_0/n_2}.$$

Except for x, all quantities in this expression are constants.

The path for which the time interval Δt_{AC} is a minimum—as it must be from Fermat's principle—is the path for which the derivative of Δt_{AC} with respect to x is zero:

$$\frac{d}{dx}(\Delta t_{AC}) = \frac{xn_1}{c_0\sqrt{h_1^2 + x^2}} - \frac{(L - x)n_2}{c_0\sqrt{h_2^2 + (L - x)^2}} = 0. \quad (2)$$

Solving this equation for x would tell me where the light ray crosses the interface, but I do not have values for L and x. However, the right triangles ADB and BEC in Figure 33.43 allow me to express these distances in terms of θ_1 and θ_2:

$$\sin\theta_1 = \frac{x}{\sqrt{h_1^2 + x^2}} \quad \text{and} \quad \sin\theta_2 = \frac{L - x}{\sqrt{h_2^2 + (L - x)^2}}. \quad (3)$$

I can now use these expressions to rewrite Eq. 2 in terms of θ_1 and θ_2. From Eq. 2 I obtain

$$\frac{xn_1}{c_0\sqrt{h_1^2 + x^2}} = \frac{(L - x)n_2}{c_0\sqrt{h_2^2 + (L - x)^2}}.$$

Canceling the c_0 factors that appear on both sides and substituting from Eq. 3, I get

$$n_1 \sin\theta_1 = n_2 \sin\theta_2. ✔$$

❹ **EVALUATE RESULT** My result is identical to Snel's law—which I derived by considering the effect of changing speed on the propagation of wavefronts. So Fermat's principle yields the same result as Snel's law, which I know to be correct.

Fermat's principle applies to all of ray optics, not only to refraction. As discussed in the box "Fermat's principle" on page 884, the law of reflection also follows from this principle.

33.14 We found in Example 33.3 that a light ray is refracted twice when it passes completely through a slab of transparent material (see Figure 33.18). The result of these two refractions is that the exiting ray is shifted sideways relative to the entering ray. Let the slab be in air with an index of refraction $n_1 = 1$. (*a*) Derive an expression for the distance (perpendicular to the ray) over which the ray is shifted sideways for an angle of incidence θ_1, slab thickness d, and slab index of refraction n_2. (*b*) Calculate the value of the shift for $\theta_1 = 30°$, $n_2 = 1.5$, $d = 0.010$ m.

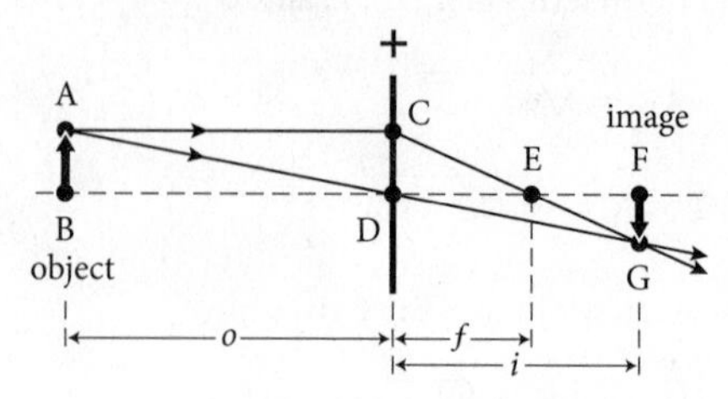

Figure 33.44 Simplified ray diagram for the formation of an image by a converging lens.

33.6 Thin lenses and optical instruments

In the first part of this chapter, we found that converging lenses form images of objects by focusing the light rays emanating from those objects. Let us now work out quantitatively the location and size of such images. We shall restrict our discussion to lenses that are thin enough that we can ignore the type of effects shown in Figure 33.29. Such lenses are called *thin lenses.*

A simplified ray diagram of the image formed by a converging lens is shown in **Figure 33.44**. The focal length f of the lens is DE, the distance o from the lens to the object (also called the *object distance*) is BD, and the distance i from the lens to the image (also called the *image distance*) is DF. The height of the object is AB, and the height of the image is FG. Let us denote the height of the object by h_o and the height of the image by h_i. We choose the values of h_o and h_i to be positive for upright objects and images and negative for inverted objects and images. We want to obtain a relationship between h_i and h_o, which will tell us how large the image is relative to the object. We also want a relationship among f, i, and o, which will tell us how the positions of the object and the image are related.

We begin by noting that triangles ABD and DFG are similar, which means

$$\frac{\text{AB}}{\text{DB}} = \frac{\text{FG}}{\text{DF}}. \tag{33.10}$$

Because the image is inverted, $h_i = -\text{FG}$, we can rewrite Eq. 33.10 as

$$\frac{h_o}{o} = \frac{-h_i}{i}. \tag{33.11}$$

Rearranging this expression gives

$$-\frac{h_o}{h_i} = \frac{o}{i}. \tag{33.12}$$

In this case, the absolute value of the ratio of the object height to the image height equals the ratio of the object distance to the image distance.

Triangles CDE and EFG are also similar, which means

$$\frac{\text{DE}}{\text{CD}} = \frac{\text{EF}}{\text{FG}}, \tag{33.13}$$

which can be written as

$$\frac{f}{h_o} = \frac{i - f}{-h_i}. \tag{33.14}$$

Using Eq. 33.12 to rewrite Eq. 33.14 in terms of f, o, and i gives us

$$\frac{f}{o} = \frac{i - f}{i} = 1 - \frac{f}{i}. \tag{33.15}$$

Dividing by f and rearranging terms yield

$$\frac{1}{f} = \frac{1}{o} + \frac{1}{i}. \tag{33.16}$$

This result is known as the **lens equation.**

QUANTITATIVE TOOLS

It can be shown that Eq. 33.16 is generally true for either real or virtual images formed by either converging or diverging lenses, as long as we choose the signs of *f*, *i*, and *o* properly. For a converging lens, *f* is positive and *o* is positive if the object is in front of the lens. (This is always true for a single lens and for the first lens in a lens combination. For situations involving multiple lenses, however, it is possible that the object imaged by a secondary lens is on the opposite side of the lens from the side where the rays enter it—that is, the object is "behind the lens." In that case *o* is negative.) If the image is on the same side of the lens as the emerging light, the image is real and *i* is positive; if the image is on the opposite side of the lens from the emerging light, the image is virtual and *i* is negative.

For a diverging lens (a lens with concave surfaces), the focal length *f* is negative because the focus is virtual rather than real—that is, parallel rays appear to come from the same side of the lens where the light source is rather than converging on the other side of the lens. The same sign convention applies to *o* as for converging lenses. A single diverging lens always produces a virtual image (see Example 33.5), so *i* is always negative for such lenses.

The sign conventions for *f*, *i*, and *o* are similar for images formed by spherical mirrors, which are discussed in the next section. **Table 33.2** summarizes these sign conventions.

The **magnification** of the image is defined as the ratio of the signed image height to the object height. Using Eq. 33.12, we get

$$M \equiv \frac{h_\text{i}}{h_\text{o}} = -\frac{i}{o}. \tag{33.17}$$

We define *M* this way so that the magnification of upright images is positive and that of inverted images is negative. Examining Figures 33.30, 33.31, and 33.33, we can see that for a single lens, when the image distance and object distance are both positive, as in Figure 33.30, we obtain an inverted image, whereas when the image distance is negative, as in Figures 33.31 and 33.33, we obtain an upright image.

33.15 If the diverging lens in Figure 33.33 has a focal length of 80 mm and the object is located 100 mm from the lens, (*a*) what is the image distance and (*b*) how tall is the image relative to the object?

Table 33.2 Sign conventions for *f*, *i*, and *o* (positive = real; negative = virtual)

Sign	Lens	Mirror
$f > 0$	converging lens	converging mirror
$f < 0$	diverging lens	diverging mirror
$o > 0$	object in front[b] of lens	object in front of mirror
$o < 0$[a]	object behind lens	object behind mirror
$i > 0$	image behind lens	image in front of mirror
$i < 0$	image in front of lens	image behind mirror
$h_\text{i} > 0$	image upright	image upright
$h_\text{i} < 0$	image inverted	image inverted
$\lvert M\rvert > 1$	image larger than object	image larger than object
$\lvert M\rvert < 1$	image smaller than object	image smaller than object

[a] Encountered only with lens or mirror combinations.
[b] For both lenses and mirrors, "in front" means on the side where the rays originate; "behind" refers to the opposite side.

Figure 33.45 An eye cannot focus on an object that is closer than its near point (which represents the limit of the biological lens's ability to change curvature). However, an external converging lens (such as a magnifying lens) makes it possible to see objects that are closer than the near point. It also enlarges them.

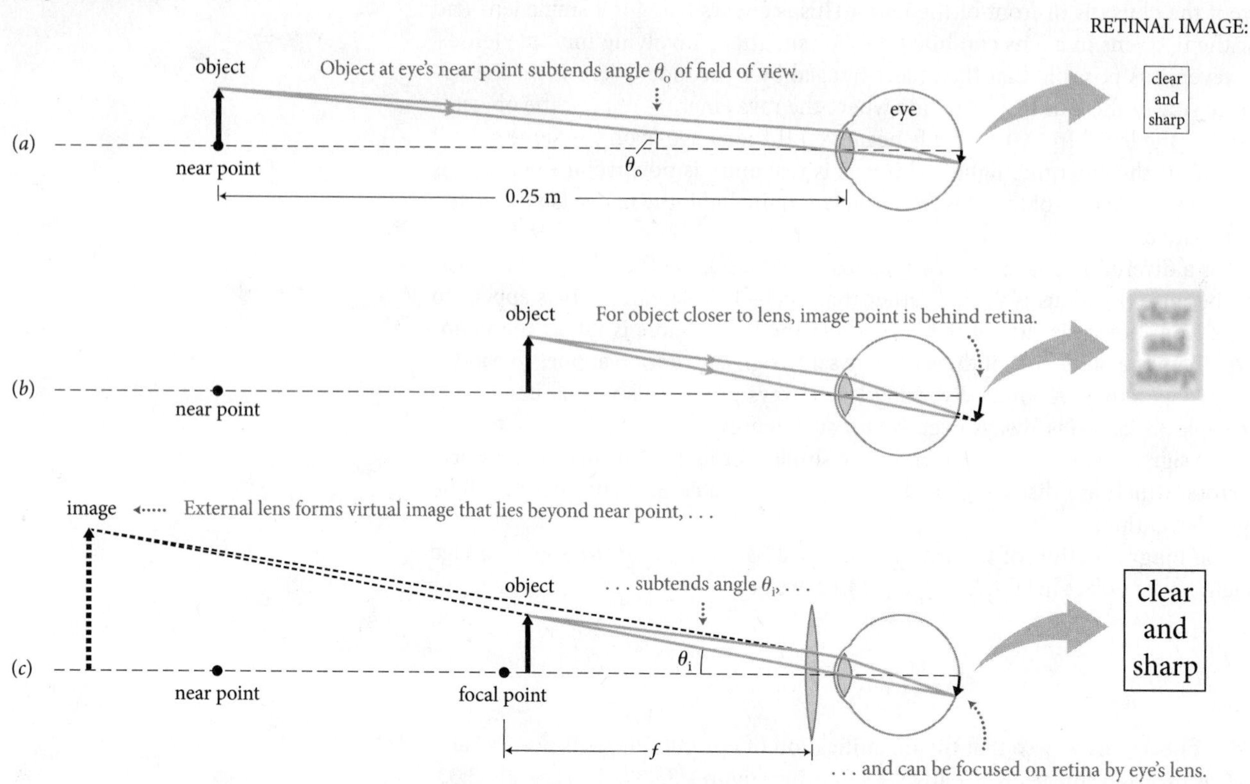

The human eye focuses incoming light rays, forming an image on the retina of the eye (**Figure 33.45*a***). One part of the eye is its lens, but unlike the lenses we have examined so far, the focal length of the eye's lens is variable, which allows us to see objects clearly over a wide range of distances. When the muscle around the lens is fully relaxed, the lens flattens out and the retina lies in the focal plane of the lens. Thus, light rays from distant objects focus onto the retina.

To the unaided eye, the largest (and thus most detailed) image of an object is observed when we bring the object as close as possible to the eye. However, there is a limit to how much the eye's lens can adjust. The *near point* is the closest object distance at which the eye can focus on the object comfortably. Typically, for an adult, the near point is about 0.25 m from the eye. An object positioned at the near point appears clear and sharp to the observer, as shown in Figure 33.45*a*. With age the distance between the near point and the eye tends to increase, and when an object is brought closer than the near point, the plane where the image is formed lies behind the retina and the image "seen" by the retina is blurry (Figure 33.45*b*). This situation can be corrected by an external lens that works in combination with your eye's lens to focus the image on the retina (Figure 33.45*c*).

An external converging lens properly placed between the object and the eye, as in Figure 33.45*c*, magnifies the object. To maximize the size of the image, the object is held near the focus of the external lens, and the lens is held as close as possible to the eye. The image formed by the external lens then serves as the object for the eye's lens. The image formed by the external lens is virtual and subtends an angle θ_i that is greater than the angle θ_o subtended by the object in Figure 33.45*a*, permitting the viewer to see finer details. The image is also outside the near point; if the object is placed exactly at the focus of the external lens, the image is at infinity and can be viewed comfortably. We can define the *angular magnification* produced by the lens as

$$M_\theta \equiv \left|\frac{\theta_\text{i}}{\theta_\text{o}}\right|. \quad (33.18)$$

For small angles and an object placed close to the focus of the external lens, as in Figure 33.45*c*, the angle θ_i subtended by the image can be expressed in terms of the object height h_o and the focal length f of the lens:

$$\theta_\text{i} \approx \tan\theta_\text{i} \approx \frac{h_\text{o}}{f} \quad \text{(object close to focus, small } \theta_\text{i}\text{).} \quad (33.19)$$

For small angles and an object placed at the eye's near point, as in Figure 33.45*a*, the angle subtended by the object is approximately

$$\theta_\text{o} \approx \tan\theta_\text{o} = \frac{h_\text{o}}{0.25\text{ m}} \quad \text{(object close to near point, small } \theta_\text{o}\text{).} \quad (33.20)$$

Substituting Eqs. 33.19 and 33.20 into Eq. 33.18 gives an angular magnification of

$$M_\theta \approx \frac{0.25\text{ m}}{f}. \quad (33.21)$$

This expression gives what is called either the *small-angle approximation* or the *paraxial approximation* to the angular magnification because it is obtained with the small-angle approximations of Eqs. 33.19 and 33.20. These approximations are good to within 1% for angles of 10° or less.

Lenses placed near the eye (in the form of eyeglasses) are used to correct vision for far-sighted or near-sighted eyes. The strength of eyeglass lenses (and of magnifying lenses, too) is commonly symbolized by d and measured in *diopters*:

$$d \equiv \frac{1\text{ m}}{f}. \quad (33.22)$$

The *lens strength d*, like the lens focal length f, is positive for converging lenses and negative for diverging lenses. For example, a +4-diopter lens is a converging lens with a focal length of 0.25 m. Diverging lenses are typically used to correct nearsightedness, with lens strengths ranging from −0.5 to −4 diopters.

33.16 A single-lens magnifying glass used to examine photographic slides produces eightfold angular magnification. (*a*) What is the lens strength in diopters? (*b*) What is the focal length of the lens?

Many optical instruments combine two or more lenses to increase magnification. To trace rays through a combination of lenses, use the following procedure: The image formed by the first lens serves as the object for the second lens, the image formed by the second lens serves as the object for the third lens, and so on. **Figure 33.46** on the next page shows a ray diagram constructed in two steps for a combination of two lenses. Figure 33.46*a* shows the object, image, and rays for lens 1, and Figure 33.46*b* shows these elements for lens 2. Note that the rays from object 2 (which is the image formed by lens 1) are *not* the continuation of those used to locate image 1.

Figure 33.46 Two-step process for tracing rays through a combination of two lenses. When lenses are combined, the image of each lens serves as an object for the next lens.

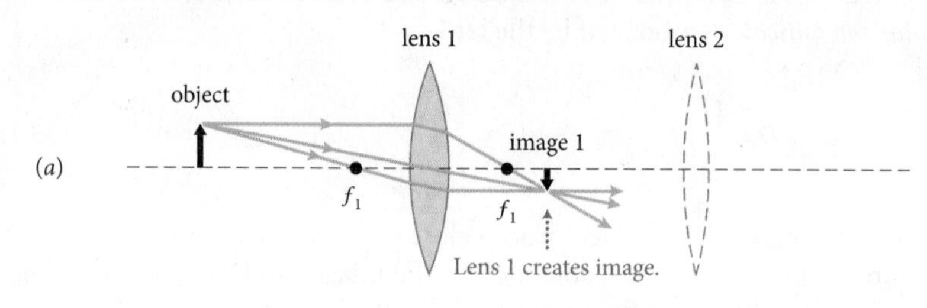

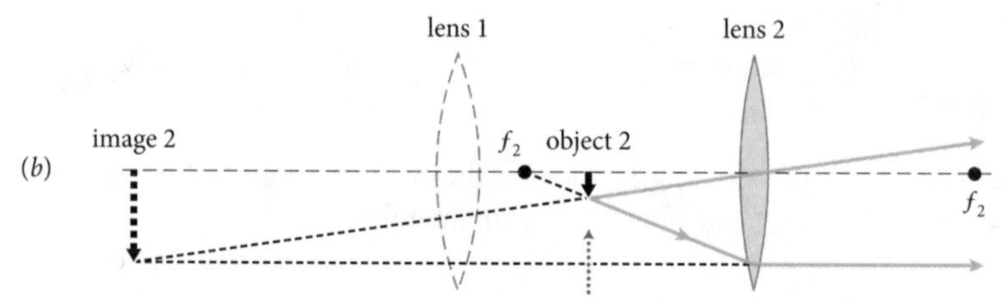

Example 33.8 Compound microscope

A compound microscope consists of two converging lenses, the *objective lens* and the *eyepiece lens*, positioned on a common optical axis (**Figure 33.47**). The objective lens is positioned to form a real, highly magnified image 1 of the sample being examined, and the eyepiece lens is positioned to form a virtual, further magnified image 2 of image 1. It is image 2 that the user sees. A knob on the microscope allows the user to move the objective lens upward and downward to change both the sample-objective lens distance and the distance between the two lenses. (*a*) How must the sample and the two lenses be positioned relative to one another so that the user sees a highly magnified, virtual image of the sample? (*b*) What is the overall magnification produced by the microscope?

Figure 33.47 Example 33.8.

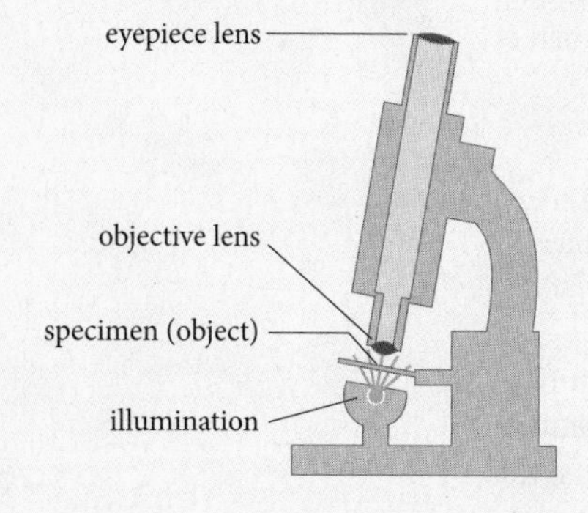

❶ GETTING STARTED I begin by examining Figure 33.46, which shows how, in a combination of two lenses 1 and 2, the image formed by lens 1 serves as the object for lens 2. The objective lens in a compound microscope corresponds to lens 1 in Figure 33.46, and the eyepiece lens corresponds to lens 2. Thus to keep things simple I refer to the objective lens as 1 and the eyepiece lens as 2.

❷ DEVISE PLAN To determine the relative positioning of the two lenses relative to each other, I must examine ray diagrams for various lens-sample distances and determine for which arrangement I get the greatest magnification. To determine the magnification M_1 of image 1, I can use the lens equation (Eq. 33.16) together with the relationship among magnification, image distance, and object distance (Eq. 33.17). The focal length of the lenses is fixed by their construction, and in operating a microscope, the observer can adjust the distance between the sample and the objective lens, so I can express this magnification in terms of f_1 and o_1. To determine the magnification of image 2, I recognize that lens 2 is used as a magnifying glass, so I can use Eq. 33.21, which gives the angular magnification $M_{\theta 2}$ produced by a simple magnifier. The overall magnification produced by the microscope is the product $M_1 M_{\theta 2}$.

❸ EXECUTE PLAN (*a*) In Figure 33.46, lens 1 produces an image that is smaller than the object. I am told that the image formed by lens 1 in a microscope is larger than the sample, and so I must choose a different sample position, one that yields an image 1 larger than the sample. I am also told that this image is real. Placing the sample just outside the focal point of lens 1 gives me an image 1 that is larger than the sample. My choices are to increase or decrease the sample-lens 1 distance. Drawing a ray diagram for each possibility, I see that moving the sample farther and farther from lens 1 makes the image smaller and smaller. Therefore I should position the sample closer to lens 1 than in Figure 33.46*a*. Should I choose a position inside or outside the lens focus? I know from the

QUANTITATIVE TOOLS

Figure 33.48

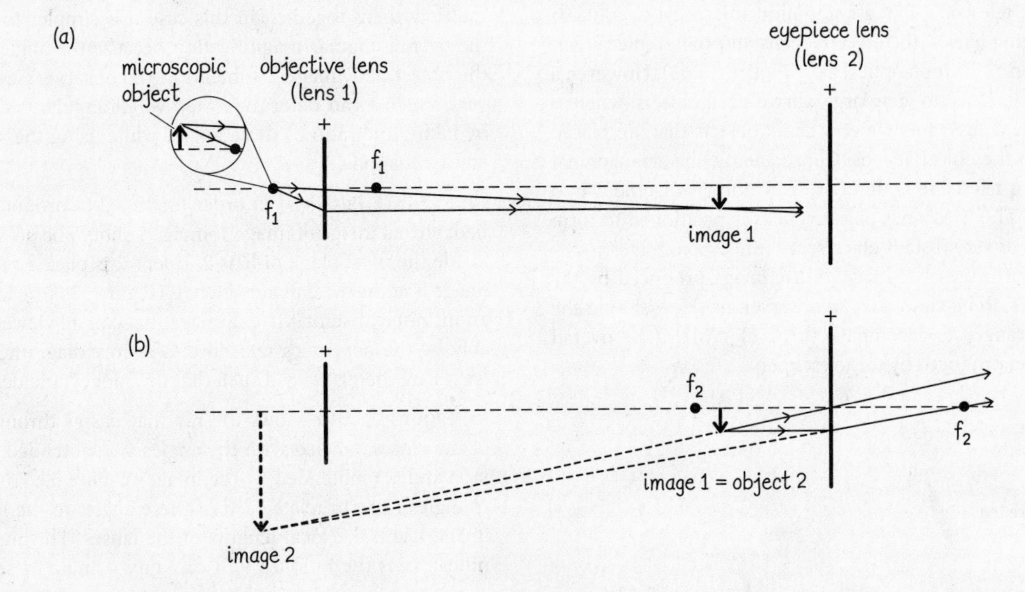

problem statement that this image is real, and I know from Figure 33.31 that an object inside the focus of a converging lens produces a virtual image. Thus my best choice is to adjust the sample-lens 1 distance so that the sample is just outside the lens focus (**Figure 33.48*a***).

I am told that image 2 is virtual and larger than image 1. I know from Figure 33.31 that a converging lens produces a virtual, magnified image when the object is inside the lens focus. I again draw ray diagrams for various positions inside the focus and see that the greatest magnification is obtained when I adjust the distance from lens 1 to lens 2 to make image 1 fall just inside the focal point of lens 2, as shown in Figure 33.48*b*. ✔

(*b*) To determine M_1, I use the lens equation to write i_1 in terms of f_1 and o_1:

$$i_1 = \frac{1}{\left(\frac{1}{f_1} - \frac{1}{o_1}\right)}.$$

I substitute this expression into Eq. 33.17:

$$M_1 = -\frac{i_1}{o_1} = -\frac{1}{o_1} \times \frac{1}{\left(\frac{1}{f_1} - \frac{1}{o_1}\right)} = -\frac{1}{\frac{o_1}{f_1} - 1},$$

which tells me that the magnification M_1 produced by lens 1 is determined by the ratio o_1/f_1. Because I have made o_1 slightly larger than f_1 in order to produce a real image 1, the denominator is positive and therefore M_1 is negative.

The angular magnification produced by lens 2 is approximately

$$M_{\theta 2} = \frac{0.25 \text{ m}}{f_2}.$$

The overall magnification produced by the microscope is thus

$$M = M_1 M_{\theta 2} = \frac{-0.25 \text{ m}}{f_2\left(\frac{o_1}{f_1} - 1\right)}. ✔$$

❹ **EVALUATE RESULT** Figure 33.48*a* indicates that image 1 is inverted, making M_1 negative and giving me confidence in my expression for M_1. Figure 33.48*b* tells me that image 2 is upright relative to its object, and so $M_{\theta 2}$ is positive, which agrees with my result. Because image 2 is inverted relative to the sample, the overall magnification is negative, as my result shows.

33.17 (*a*) Consider replacing the objective lens in Fig. 33.48*a* with one that has a greater focal length, and moving the sample in order to keep it just outside the focal point of the lens. Does the image formed by the objective lens move closer to the objective lens, stay in the same place, or move farther from the objective lens? (*b*) In practice it is desirable for a microscope to be fairly compact. To keep the microscope compact, should the focal length of the objective lens be chosen to be short or long, or does it matter?

Example 33.9 Refracting telescope

A refracting telescope, like a compound microscope, contains two converging lenses, the objective lens and the eyepiece lens, positioned on a common optical axis (**Figure 33.49**). However, a telescope is designed to view large, very distant objects, whereas a microscope is used to view very small objects that are placed very close to the objective lens. Consequently, the arrangement of lenses in a telescope is different from the arrangement in a microscope. The telescope's objective lens is positioned to form a real image of very distant objects, and the eyepiece lens is positioned to form a virtual image of the image produced by the objective lens, to be viewed by an observer. (*a*) How should the lenses be arranged to accomplish this? (*b*) What is the overall magnification produced by the telescope?

Figure 33.49 Example 33.9.

❶ **GETTING STARTED** I begin by examining Figure 33.46 and then construct a similar ray diagram with the object at a very great distance from the lenses. As in Example 33.8, I use lens 1 to refer to the objective lens and lens 2 to refer to the eyepiece lens. Because the object is very far away, light rays from it enter lens 1 as parallel rays. These rays form an image 1 in the focal plane of lens 1 (**Figure 33.50*a***). Because I know the location of image 1, I need to draw only one principal ray. As in the microscope, lens 2 is used as a simple magnifier to view image 1.

❷ **DEVISE PLAN** Because the original object is very distant and the final image is viewed by the observer's eye, I can calculate the angular magnification of this image. Although I could calculate the angular magnification produced by each lens and multiply them together, in this case it is simpler to determine the overall angular magnification because the angles θ_o and θ_i the object and image 2 subtend at the observer's eye are both very small. I can determine the overall angular magnification by taking the ratio of these angles while using the small-angle approximation.

❸ **EXECUTE PLAN** (*a*) In order for lens 2 to produce a magnified, virtual image of image 1, image 1 should be positioned just inside the focal plane of lens 2. If lens 2 is placed such that the image is at the focal plane of lens 2 (Figure 33.50*b*), lens 2 forms an infinitely distant, virtual image that can be viewed comfortably by the observer's relaxed eye. As my diagram shows, the lenses are then arranged such that their foci coincide. ✔

(*b*) **Figure 33.51*a*** shows the ray that passes through the foci of the lenses, labeled with the angles θ_o (subtended by the object) and θ_i (subtended by the image). Figure 33.51*b* shows the triangles I use to relate each of these angles to the height h_i of image 1 and the focal lengths of the lenses. The angular magnification is the ratio θ_o/θ_i. I can approximate these angles by

Figure 33.50

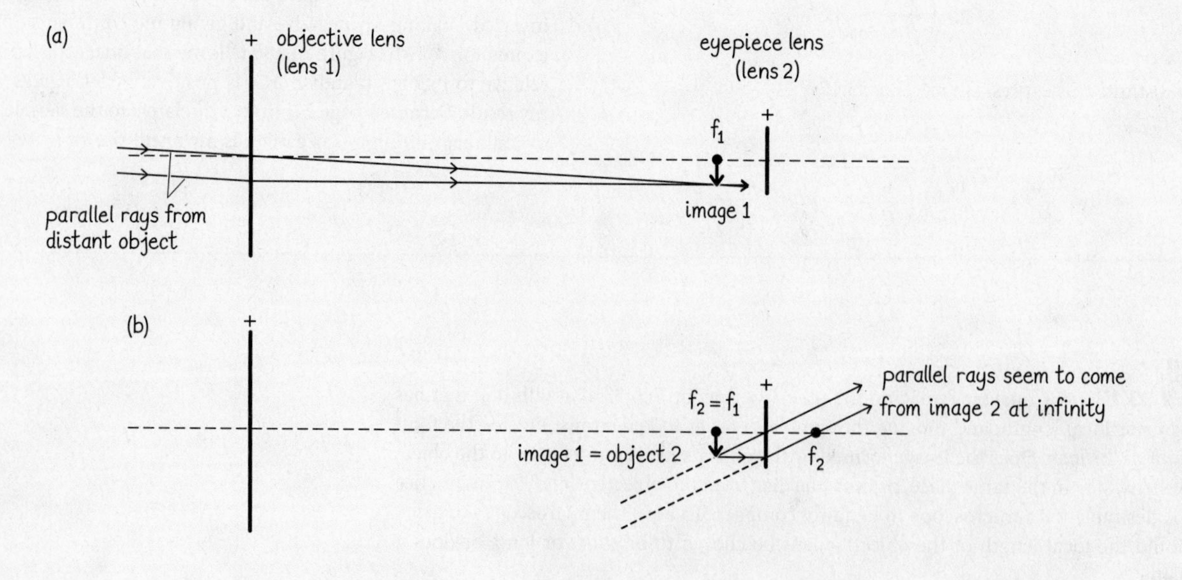

Figure 33.51

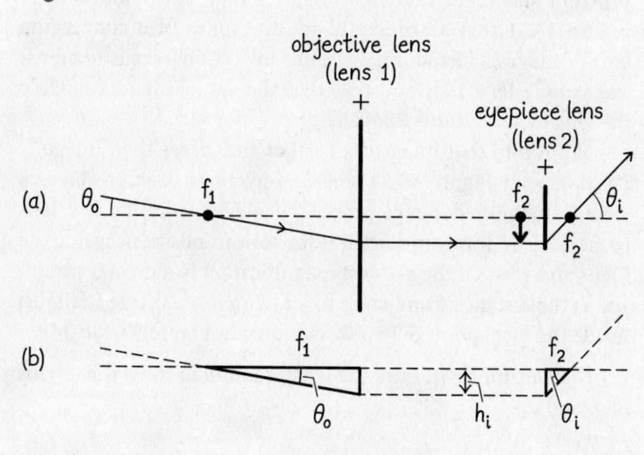

QUANTITATIVE TOOLS

their tangents; substitute for the tangents of the angles in terms of h_i, f_1, and f_2; and simplify:

$$M_\theta = \left|\frac{\theta_i}{\theta_o}\right| \approx \left|\frac{\tan\theta_i}{\tan\theta_o}\right| = \frac{|h_i/f_2|}{|h_i/f_1|} = \left|\frac{f_1}{f_2}\right|. ✔$$

❹ **EVALUATE RESULT** Figure 33.50 indicates that image 1 is inverted and that image 2 is upright relative to image 1, which means that image 2 is inverted relative to the distant object. This makes sense because the incoming ray in Figure 33.51 is angled downward but the outgoing ray is angled upward.

33.18 A telescope with a magnification of 22× has an eyepiece lens for which the focal length f_2 is 40.0 mm. (*a*) What is the focal length f_1 of the objective lens? (*b*) What is the length of the telescope?

33.7 Spherical mirrors

Just like lenses, spherical mirrors focus parallel rays (**Figure 33.52**). A concave mirror focuses rays to a point in front of the mirror, corresponding to a real focus; a diverging mirror makes rays diverge so that they appear to come from a point behind the mirror, corresponding to a virtual focus. Thus, just as with lenses, we can have both converging and diverging spherical mirrors. Unlike lenses, however, spherical mirrors have only a single focus.

To obtain the location of the focus of a converging mirror, we examine the reflection of the two rays shown in **Figure 33.53*a***. Ray 1 comes in parallel to the axis of the mirror, striking the mirror at A, and ray 2 comes in along the axis of the mirror. The focus of the mirror is at D, where the two reflected rays cross, and the center of the sphere on which the surface of the mirror lies is at C. Line CA is therefore a radius of the sphere and perpendicular to the mirror surface. We denote the length of the radius by R. This distance is called the *radius of curvature*.

Ray 2 strikes the mirror at normal incidence and so is reflected back along the axis. Ray 1 is reflected through the focus at D. Consequently, the distances CD and AD in Figure 33.53*a* are equal. Dividing triangle ACD into two congruent right triangles by drawing a line perpendicular to the base from D, we see that $\text{CD} = (R/2)/\cos\theta_i$. For small θ, $\cos\theta \approx 1$, and so $\text{CD} = R/2$ and $\text{BD} = R - \text{CD} = R/2$, independent of θ. Therefore, the focus, which is located at D because that is where the two reflected rays cross, lies halfway between the mirror and the center of the sphere, making the focal length

$$f = \frac{R}{2}. \tag{33.23}$$

As Figure 33.53*b* shows, the geometry and hence the position of the focus are exactly the same for diverging mirrors except that here the focus lies behind the mirror.

Figure 33.52 Spherical mirrors focus parallel rays just like lenses do.

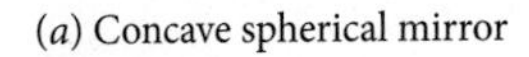
(*a*) Concave spherical mirror

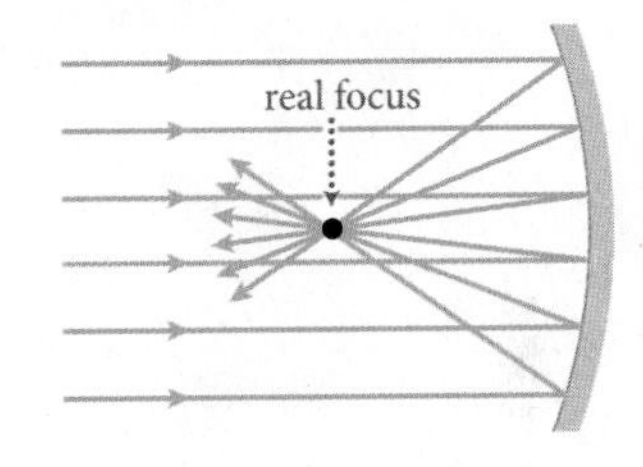

(*b*) Convex spherical mirror

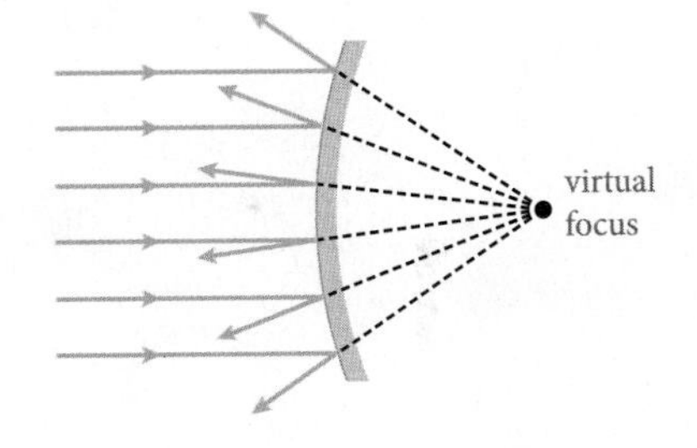

Figure 33.53 The two principal rays used to determine the focus of a spherical mirror.

(*a*) Concave spherical mirror

(*b*) Convex spherical mirror

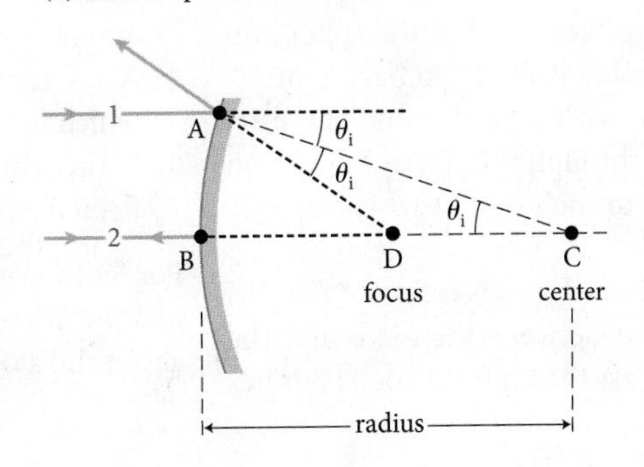

Procedure: Ray diagrams for spherical mirrors

Ray diagrams for spherical mirrors are very similar to those for lenses. This procedure is for rays traveling from the left to the right.

1. Draw a horizontal line representing the mirror axis. In the center of the diagram, draw a circular arc representing the mirror. A converging mirror curves toward the left; a diverging mirror curves toward the right.
2. Put a dot on the axis at the center of the circular arc and label it C. Add another dot on the axis, halfway between C and the mirror. This point is the focus. Label it *f*.
3. Represent the object by drawing an upward-pointing arrow from the axis at the appropriate relative distance to the left of the mirror. For example, if the distance from the object to a converging mirror is one-third the radius of curvature of the mirror, put the arrow a bit to the right of the focus. The top of the arrow should be at about half the height of the mirror.
4. From the top of the arrow representing the object draw two or three of the following three so-called *principal rays* listed in the Procedure box "Principal rays for spherical mirrors."
5. The top of the image is at the point where the rays that are reflected by the mirror intersect. If the intersection is on the left side of the lens, the image is real; if it is on the right side, the image is virtual. Draw an arrow pointing from the axis to the intersection to represent the image (use a dashed arrow for a virtual image).

To determine the location of images formed by mirrors, we follow the same ray-tracing procedure we used for lenses. The three principal rays emanating from a given point on the object are analogous: ray 1 approaching the mirror parallel to the mirror axis, ray 2 passing through the center C of the mirror, and ray 3 passing through the focus on its way to the mirror. Now, however, "center" refers to *the center of the sphere on which the mirror surface lies* rather than the center of the lens. **Figure 33.54** shows a ray diagram for an image formed by a converging mirror. Ray 1 is reflected through the focus, ray 2 strikes the mirror at normal incidence and thus reflects back on itself, and ray 3 is reflected parallel to the mirror axis. As Figure 33.54 shows, there is a fourth ray that can easily be drawn: A ray that hits the mirror on the axis is reflected back symmetrically about the axis. The procedures for drawing ray diagrams and principal rays for spherical mirrors are given in the Procedure boxes on this page.

Object distance, image distance, and focal length are measured from the surface of the mirror, and the relationship among o, i, and f is the same as that for lenses:

$$\frac{1}{f} = \frac{1}{o} + \frac{1}{i}. \qquad (33.24)$$

Procedure: Principal rays for spherical mirrors

This description holds for rays that travel from left to right.

Converging mirror

1. A ray that travels parallel to the mirror axis before reaching the mirror goes through the focus after being reflected.
2. A ray that passes through the center of the sphere on which the mirror surface lies is reflected back onto itself. If the object is between the center and the mirror, this ray doesn't pass through the center but lies on the line from the center to the point at which the ray originates.
3. A ray that passes through the focus is reflected parallel to the axis. If the object is between the focus and the mirror, this ray doesn't pass through the focus but lies on the line from the focus to the point at which the ray originates.

Diverging mirror

1. A ray that travels parallel to the mirror axis before reaching the mirror is reflected along the line that goes through the focus and the point where the ray strikes the surface.
2. A ray that passes through the center of the sphere on which the mirror surface lies is reflected back onto itself.
3. A ray whose extension passes through the focus is reflected parallel to the axis.

For both converging and diverging mirrors, a ray that hits the mirror on the axis is reflected back symmetrically about the axis.

Figure 33.54 Principal ray diagram for an object outside the focus of a concave spherical mirror. The image is real, inverted, and smaller than the object.

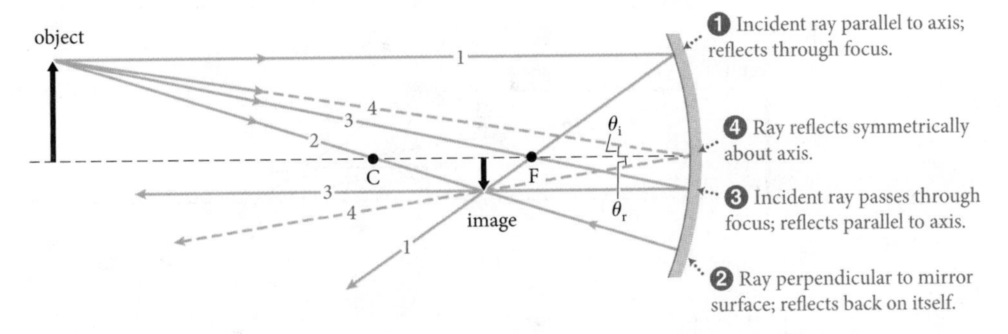
object
1
4
3
2
C
F
image
θi
θr
1 Incident ray parallel to axis; reflects through focus.
4 Ray reflects symmetrically about axis.
3 Incident ray passes through focus; reflects parallel to axis.
2 Ray perpendicular to mirror surface; reflects back on itself.

Just as for lenses, the focal length f for any spherical mirror is positive for a real focus and negative for a virtual focus, and o is positive when the object is in front of the mirror and negative when it is behind the mirror (that can happen only when the object is an image formed by another mirror or a lens). Similarly, i is positive for a real image and negative for a virtual image. With mirrors, however, a real image is located on the same side of the mirror as the object and a virtual image is located on the opposite side—the opposite of what happens with lenses. These sign conventions are summarized in Table 33.2. Finally, the relationship between object and image distances and heights for lenses (Eq. 33.12) also applies to mirrors, so that equation can be used to determine the size of the images formed by mirrors.

Example 33.10 Funny mirror

An object is placed 0.30 m in front of a converging mirror for which the radius of curvature is 1.0 m. (*a*) On which side of the mirror is the image? Is the image real or virtual? (*b*) If the object is 50 mm tall, what is the height of the image?

❶ **GETTING STARTED** To visualize the situation, I draw the mirror and its axis, and indicate its center of curvature C and its focal point f halfway between the mirror surface and C. I represent the object as a solid arrow (**Figure 33.55**).

Figure 33.55

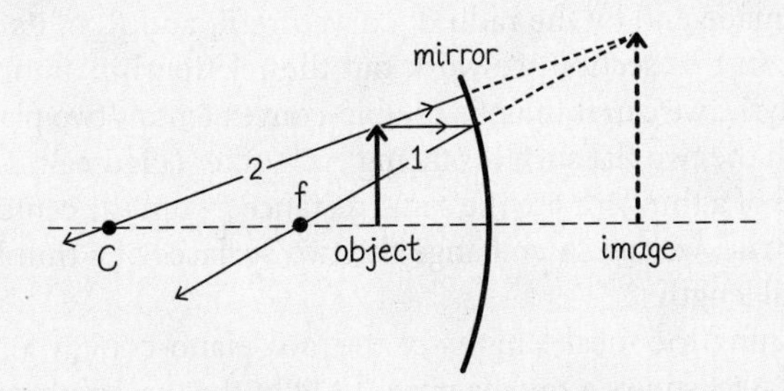
mirror
2
1
f
C
object
image

❷ **DEVISE PLAN** To locate the image and identify whether it is real or virtual, I can draw the principal rays. I can use Eq. 33.12 to obtain the image height from the object height and the image and object distances. I can determine the image distance from the focal length and the object distance using Eq. 33.24.

❸ **EXECUTE PLAN** (*a*) As Figure 33.55 shows, I need to draw only two principal rays because the intersection of two rays unambiguously determines the position of the image. I draw ray 1 parallel to the mirror axis and reflecting through the focus. Because the object is between the mirror and the center of curvature, I draw ray 2 along the line defined by C and the tip of the object. Ray 2 does not pass through C until after it is reflected from the mirror, however, and I indicate this by adding an arrowhead pointing toward the mirror on the part of the ray to the right of the object and an arrowhead pointing away from the mirror on the part to the left of the object. My diagram shows that the rays do not actually meet but appear to come from a point behind the mirror. Therefore the image is behind the mirror and virtual. ✔

(*b*) I begin by determining the image distance i. The focal length of the mirror is half the radius of curvature, $f = 0.50$ m. Substituting this value and $o = 0.30$ m into Eq. 33.24 and solving for i give me $i = -0.75$ m. The negative sign indicates that the image is virtual and therefore behind the mirror; this is consistent with my result from part *a*. I then solve Eq. 33.12 for h_i (the signed image height) and substitute the values from this problem:

$$h_i = \frac{-i\,h_o}{o} = \frac{-(-0.75\text{ m})(0.050\text{ m})}{(0.30\text{ m})} = 0.13\text{ m.} ✔$$

❹ **EVALUATE RESULT** My ray diagram (Figure 33.55) indicates that the image is enlarged and upright, so h_i should be positive and greater than h_o. This agrees with my result.

Figure 33.56 Ray diagram for an image formed by a convex spherical mirror (object distance greater than focal length). The image is virtual, upright, and smaller than the object.

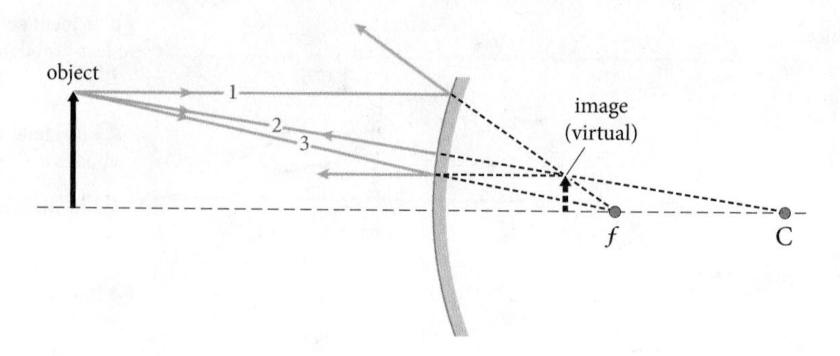
object
1
2
3
image
(virtual)
f
C

Example 33.10 shows that placing an object inside the focus of a converging mirror produces an upright, virtual image. Compare this with the situation in Figure 33.54, where a real, inverted image is formed for an object placed outside the mirror's focus. The same occurs with a converging lens: Placing the object inside the focus produces an upright, virtual image (Figure 33.31), while placing the object outside the focus produces a real, inverted image (Figure 33.30).

Diverging mirrors, like diverging lenses, always form virtual images when used alone because the light rays must diverge from the mirror surface. **Figure 33.56** shows a ray diagram for an image formed by a diverging mirror. The image is much smaller than the object, which allows a relatively large scene to be captured on a small mirror surface, and is upright. For these reasons, wide-angle rear-view mirrors on the passenger side of cars and trucks and wide-angle surveillance mirrors are typically convex. (A converging mirror also produces small images of distant objects, but the images are inverted, as Figure 33.54 shows.)

33.19 An object is placed 1.0 m in front of a diverging mirror for which the radius of curvature is 1.0 m. (*a*) Where is the image located relative to the mirror? Is the image real or virtual? (*b*) If the object is 0.30 m tall, what is the height of the image?

33.8 Lensmaker's formula

Figure 33.57 Analysis of a double-convex lens.

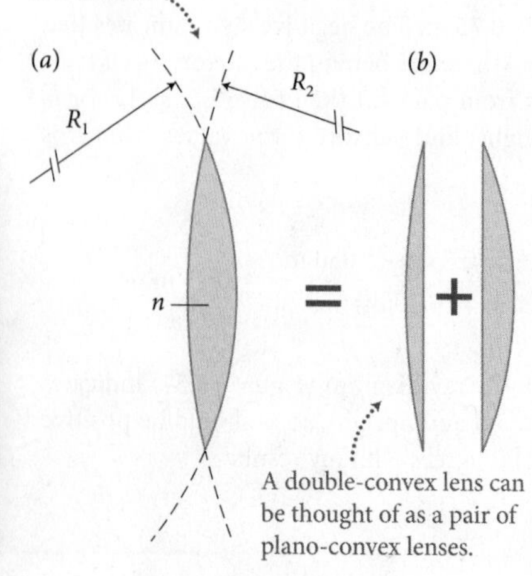
Focal length of lens depends on lens's index of refraction n and on radii of curvature R_1, R_2 of lens faces.
(a)
(b)
R_1
R_2
n
=
+
A double-convex lens can be thought of as a pair of plano-convex lenses.

The focal length of a lens is determined by the refractive index n of the material of which the lens is made and by the radii of curvature R_1 and R_2 of its two surfaces (**Figure 33.57*a***). In this section we work out the relationship among f, R_1, R_2, and n. In this analysis, we can think of a double-convex lens as two plano-convex lenses placed with the two flat surfaces facing each other (Figure 33.57*b*). Remember that both foci of a thin lens are the same distance f from the center of the lens. Because this is true, we can interchange the two surfaces of a thin lens without changing its focal length.

We begin by determining the focal lengths of the two plano-convex lenses in Figure 33.57*b*. **Figure 33.58** shows a ray diagram for light that passes through the right lens only. To calculate the focal length f_1 of this lens, consider a ray incident from the left that comes in parallel to the axis at a distance h above the axis. Because the ray is normal to the planar surface of the lens, it is not refracted at that surface. After passing through the lens, it strikes the curved surface at an angle θ_i measured from the normal to the curved surface and is refracted as it leaves the lens. The refracted ray emerges at an angle θ_r measured from the normal to the curved surface and crosses the lens axis a distance f_1 from the lens. Therefore the angle that the emerging ray makes with the lens axis is $\theta_r - \theta_i$.

QUANTITATIVE TOOLS

Figure 33.58 Ray diagram for light passing through the right-hand plano-convex lens of Figure 33.57*b*.

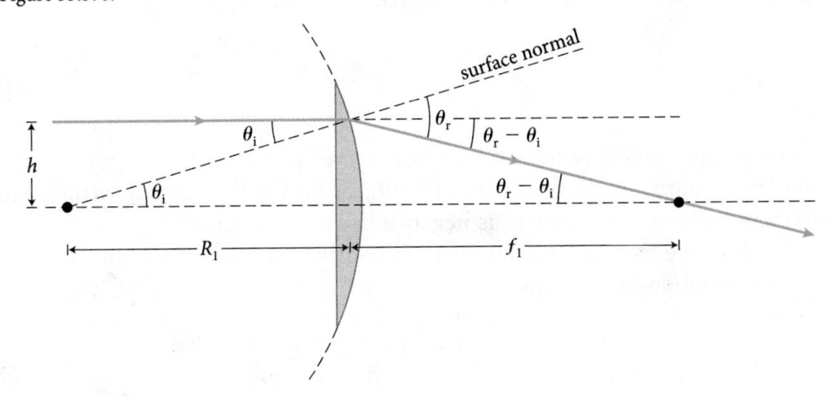

Applying Snel's law (Eq. 33.7) to this situation gives

$$n \sin \theta_i = \sin \theta_r. \tag{33.25}$$

(We do not need to show an n on the right because the medium is air and $n_{air} = 1$.) For paraxial rays, we can approximate the sines of the angles in Eq. 33.25 by the angles

$$n\theta_i = \theta_r \quad \text{(small angles)}. \tag{33.26}$$

Using this relationship, we can express the angle between the emerging ray and the lens axis as

$$\theta_r - \theta_i = n\theta_i - \theta_i = (n - 1)\theta_i. \tag{33.27}$$

For small angles, the angles are approximately equal to their tangents, which means this relationship can be expressed as

$$\theta_r - \theta_i = (n - 1)\theta_i \approx \frac{h}{f_1} \tag{33.28}$$

and

$$\theta_i \approx \frac{h}{R_1}. \tag{33.29}$$

Substituting Eq. 33.29 into Eq. 33.28 and dividing both sides by h, we get

$$\frac{1}{f_1} = \frac{n - 1}{R_1}. \tag{33.30}$$

We can follow the same procedure for the left lens in Figure 33.57*b*, using R_2 as our radius of curvature and a ray that originates at the left focus of that lens. The result analogous to Eq. 33.30 is

$$\frac{1}{f_2} = \frac{n - 1}{R_2}. \tag{33.31}$$

Now let us determine the focal length of the lens combination by working out the lens equation (Eq. 33.16) for the combination, just as we did for the microscope and telescope in Section 33.6. Consider both the object and light source to be on the left side of the lens combination in Figure 33.57. First, the light from the object strikes the left lens from the left and forms an image someplace to the right of the right lens. The lens equation that relates the location of this object and image, in terms of the focal length f_2 calculated in Eq. 33.31, is

$$\frac{1}{o_2} + \frac{1}{i_2} = \frac{1}{f_2} = \frac{n-1}{R_2}. \quad (33.32)$$

The image formed by the left lens now serves as the (virtual) object for the right lens. Consequently, $o_1 = -i_2$. (The object for the right lens is virtual, and therefore the object distance o_1 is negative because the object is located on the right side of the lens and the illumination comes from the left side of the lens.) The lens equation for the right lens is thus

$$-\frac{1}{i_2} + \frac{1}{i_1} = \frac{1}{f_1} = \frac{n-1}{R_1}, \quad (33.33)$$

where the rightmost equality comes from Eq. 33.30.

Adding Eqs. 33.32 and 33.33 yields

$$\frac{1}{o_2} + \frac{1}{i_1} = (n-1)\left(\frac{1}{R_1} + \frac{1}{R_2}\right). \quad (33.34)$$

The lens equation for the lens as a whole is simply

$$\frac{1}{o} + \frac{1}{i} = \frac{1}{f}, \quad (33.35)$$

where f, o, and i are the focal length, object distance, and image distance of the lens combination, respectively. The object of the left lens is the actual object, and the image formed by the right lens is the final image, which means that in Eq. 33.35 $o = o_2$ and $i = i_1$. Comparing Eqs. 33.34 and 33.35 gives us the **lensmaker's formula** for the focal length of our lens combination:

$$\frac{1}{f} = (n-1)\left(\frac{1}{R_1} + \frac{1}{R_2}\right). \quad (33.36)$$

Our derivation was for a double-convex lens, but it can be shown that the lensmaker's formula applies to any thin lens, not just a double-convex lens. The radii of curvature are positive for convex surfaces, negative for concave surfaces, and infinity for planar surfaces. For a double-convex lens, f is positive. For a double-concave lens, f is negative (because $n > 1$ for any material used for lenses).

33.20 How should the lensmaker's formula be modified if a lens for which the index of refraction is n_1 is submerged in a medium for which the index of refraction is n_2?

Chapter Glossary

SI units of physical quantities are given in parentheses.

absorbed, reflected, and **transmitted light** Light that enters a material but never exits again, light that is redirected away from the surface of the material, and light that passes through a material, respectively.

angle of incidence θ_i The angle between a ray that is incident on a surface and the normal to that surface

angle of reflection θ_r The angle between a ray that is reflected from a surface and the normal to that surface.

angle of refraction θ The angle between a ray that is refracted after crossing the surface between one medium and another and the normal to that surface.

critical angle θ_c (unitless) The angle of incidence for which the angle of refraction equals 90° when a ray travels from a medium with an index of refraction n_2 to one with an index of refraction $n_1 < n_2$:

$$\theta_c = \sin^{-1}\left(\frac{n_1}{n_2}\right). \quad (33.9)$$

dispersion The spatial separation of waves of different wavelength caused by a frequency dependence of the wave speed.

Fermat's principle The path taken by a light ray between any two locations is the path for which the time interval needed to travel between those locations is a minimum.

focal length f (m) The distance f from the center of the lens or the surface of the mirror to the focus. The value of f is positive for a converging lens or mirror and negative for a diverging lens or mirror.

focus (also called **focal point**) The location where parallel rays come together. If the rays cross at the focus, the focus is *real.* If only the extensions of the rays cross at the focus, it is *virtual.*

image An optical likeness of an object produced by a lens or mirror. The image is at the point from which the rays emanating from the surface of the lens or mirror appear to originate. If the rays travel through the point from which they appear to come, the image is *real;* if they do not travel through that point, the image is *virtual.*

index of refraction n (unitless) The ratio of the speed of light in vacuum to the speed of light in a medium:

$$n \equiv \frac{c_0}{c}. \quad (33.1)$$

law of reflection The angle of reflection for a ray striking a smooth surface is equal to the angle of incidence, and both angles are in the same plane.

lens An optical element that redirects light in order to form images. A *converging lens* directs parallel incident rays to a single point on the other side of the lens. A *diverging lens* separates parallel incident rays in such a manner that they appear to all come from a single point on the side of the lens where the rays came from.

lens equation The equation that relates the object distance o, the image distance i, and the focal length f of a lens or mirror:

$$\frac{1}{f} = \frac{1}{o} + \frac{1}{i}. \quad (33.16)$$

lensmaker's formula The relationship among the focal length f of a lens, the refractive index n of the material of which the lens is made, and the radii of curvature R_1 and R_2 of its two surfaces:

$$\frac{1}{f} = (n - 1)\left(\frac{1}{R_1} + \frac{1}{R_2}\right). \quad (33.36)$$

magnification M (unitless) The ratio of the signed image height h_i ($h_i > 0$ for upright image, $h_i < 0$ for inverted image) to the object height h_o:

$$M \equiv \frac{h_i}{h_o} = -\frac{i}{o}. \quad (33.17)$$

The *angular magnification* is defined as the ratio of the angle θ_i subtended by the image and the angle θ_o subtended by the object:

$$M_\theta \equiv \left|\frac{\theta_i}{\theta_o}\right|. \quad (33.18)$$

Provided these angles are small and for an object that is placed close to both the focus of the lens and the eye's near point, the angular magnification is $M_\theta \approx (0.25\ \text{m}/f)$.

principal rays a set of rays that can be used in ray diagrams to determine the location, size, and orientation of images formed by lenses or spherical mirrors.

ray A line that represents the direction in which light travels. A beam of light with a very small cross-sectional area approximately corresponds to a ray.

ray diagram A diagram that shows just a few selected rays, typically the so-called *principal rays* (see the Procedure boxes on pages 888 and 902).

refraction The changing in direction of a ray when it travels from one medium to another.

Snel's law The relationship among the indices of refraction n_1 and n_2 of two materials and the angle of incidence θ_1 and angle of refraction θ_2 at the interface of the materials

$$n_1 \sin\theta_1 = n_2 \sin\theta_2. \quad (33.7)$$

total internal reflection Mirrorlike reflection that occurs when a ray traveling in a medium strikes the medium boundary at an angle greater than the critical angle. The ray is completely reflected back into the medium.

34 Wave and Particle Optics

In Chapter 33, we considered the propagation of light along a straight path. The chapter title, "Ray Optics," reflects the fact that we considered propagating light only in the simplest way—as straight-line motion. You know from Chapter 30, however, that light is an electromagnetic wave. This means that it must undergo interference and diffraction, just like any mechanical wave. As you will learn in this chapter, light waves can interfere with one another and diffract when they pass through small openings.

Another fact about light you will learn in this chapter is that it has a dual nature: It is a wave, yes, but also has the properties of a particle!

34.1 Diffraction of light

As we saw in Chapter 17, when a water wave strikes a barrier that has a small opening, the wave diffracts (spreads out) after it passes through the opening. **Figure 34.1*a***, for example, shows surface water waves diffracting nearly circularly after they pass through an opening.

Given that light is a wave, as we discussed in Chapter 30, why don't we see light diffract in a similar fashion after it travels through, say, a window? As Figure 34.1*b* shows, after passing through a window, light continues to travel in a straight line, casting a sharp-edged shadow with no discernible diffraction.

The reason light does not diffract through a window is that the wavelength of the light is very much smaller than the size of the window. In Figure 34.1*a*, the wavelength of the water wave is about the same as the width of the opening, but the wavelength of the light in Figure 34.1*b* is about a million times smaller than the width of the window.

Diffraction is indeed observed with light waves but only when the width of the opening through which the light passes is not much greater in size than the wavelength of the light. Empirical evidence shows that diffraction occurs through openings approximately two orders of magnitude greater than the wavelength. Thus, visible light, with a wavelength on the order of 1 μm, diffracts through apertures up to hundreds of micrometers wide.

To understand diffraction, it is useful to consider the propagation of wavefronts. As discussed in Sections 17.1 and 33.2, a wavefront is a surface on which a wave spreading through space has constant phase. Wavefronts are everywhere perpendicular to the direction of propagation of the wave. By convention, wavefronts are drawn at the crests of the waves, which means the separation between adjacent wavefronts equals the wavelength (Figure 17.2). Although in principle wavefronts can take any shape, usually we consider only those that are either planar or spherical. Most sources of light, from light bulbs to stars, can be modeled as point sources—single points that produce concentric spherical wavefronts. As discussed in Section 17.1 (see especially Figure 17.7), far away from a point source, the radius of the spherical wavefronts is so great that the wavefronts are very nearly planar. As a result, distant point sources can be considered to be sources of planar waves. Lasers also produce planar waves, even very close to the source. For this reason, we can use a laser beam in seeing how electromagnetic waves behave. Keep in mind, however, that our analysis applies to any type of electromagnetic radiation, not just to laser beams.

Let us now determine under what conditions a planar wave spreads out as it propagates—in other words, under what conditions it undergoes diffraction. **Figure 34.2** shows planar wavefronts from a beam of electromagnetic waves propagating to the right, with point Q located at the upper end of the wavefronts and point P located outside the region reached by the wavefronts. Because a wavelet (see Section 17.4) centered on Q radiates toward P along the line QP, we expect the beam to spread out as it propagates. However, such spreading is not observed. The reason is that the wavelets centered on points below Q also radiate toward P,

Figure 34.1 Water waves diffract when they pass through a gap, whereas light coming in through a window seems not to diffract—it forms a sharp-edged shadow. Notice that the gap in the breakwater is roughly as wide as the wavelength of the water waves.

(*a*) (*b*)

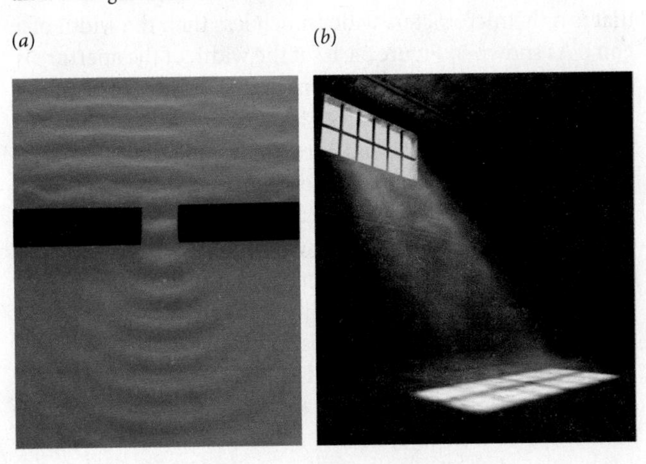

Figure 34.2 The reason we don't usually see diffraction for light beams, provided the beam is very much wider than the wavelength of the light waves.

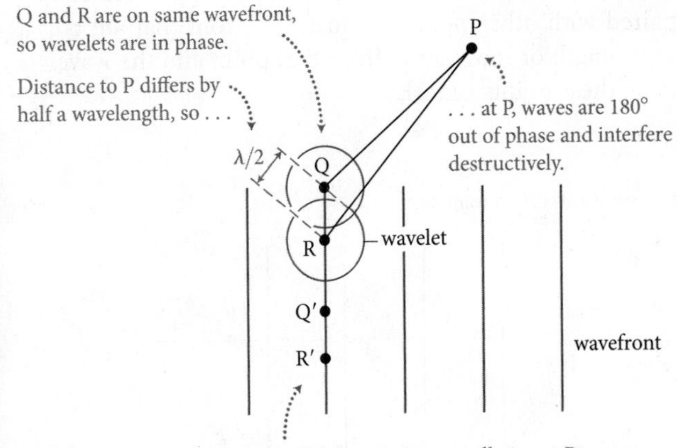

and we need to sum the contributions of all these wavelets. Consider, for example, the wavelet centered on R, for which the distance PR is exactly half a wavelength longer than the distance PQ. Because Q and R lie on the same wavefront, they produce coherent wavelets. (You should review Section 17.4 if you do not see why this is true.) This means that at P the electric field part of the wavelet from Q is 180° out of phase with the electric field part of the wavelet from R. Thus the two electric fields interfere destructively at P. Because the same is true for the magnetic field part of the wavelets, there is complete destructive interference (see Section 16.3) at P.

In the same manner, for points Q′ lying below Q on the wavefront, we can find on the same wavefront a point R′ for which the distance R′P is exactly half a wavelength longer than the distance Q′P. Thus the fields from the wavelet traveling along R′P cancel those from the wavelet traveling along Q′P. If the wavefronts extend far enough below Q, we can always find points that cancel the radiation from any other point. As a result, we conclude that the light does not spread outside the beam; in other words, there is no diffraction.

34.1 Consider the point P located ahead of the wavefronts shown in **Figure 34.3**. Following the line of reasoning used in the preceding discussion, what can you say about the intensity (power/area, as defined in Eq. 30.34 and Section 17.5) at P once the wave fronts reach it? (Hint: Consider separately points R above, below, and on the ray through P.)

Checkpoint 34 .1 demonstrates that the intensity of a planar wave is uniform as the wave propagates forward because the fields from individual wavelets cancel in any outward direction and reinforce only in the direction of propagation. Section 17.3 (especially Figure 17.20) shows that combining the waves from many adjacent point sources indeed produces a planar wave that propagates forward with very little spreading. This is true because the wavelets interfere destructively with one another in any direction other than the direction of propagation of the wavefronts.

The cancellation process we used in Figure 34.2 can be used only when the width of the laser beam, and hence the width of the wavefronts, is much greater than the wavelength of the light, so that all points on a wavefront can be paired with other points on that wavefront that are half a wavelength or more away from that point and the wavelets from these points cancel.

Figure 34.3 Checkpoint 34.1.

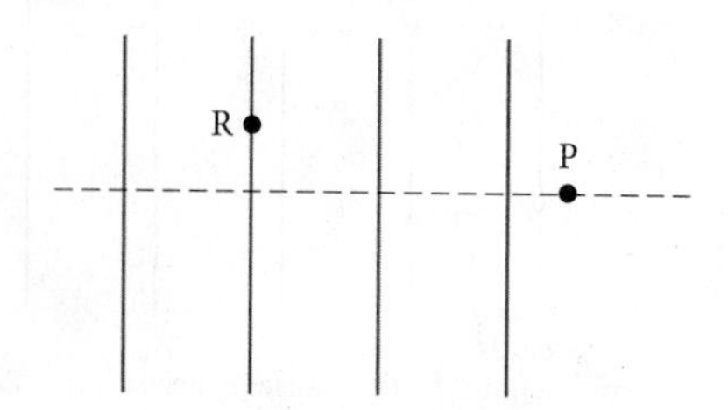

Figure 34.4 The reason we *do* see diffraction when a light beam is transmitted through a small aperture.

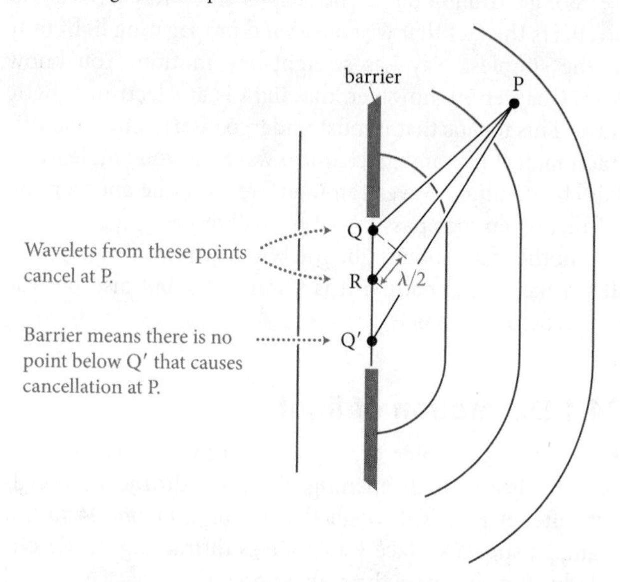

If the width of the beam is comparable to the wavelength of the light, not all points on the wavefronts can be paired in this manner. To see why this is so, let us place a barrier in front of our light source, as in **Figure 34.4**. You can think of this drawing as a bird's-eye view of a beam of light traveling to the right and running into a wall that has a gap in it. Light hitting the wall on either side of the gap cannot pass through. As each wavefront of the beam reaches the barrier, only the portion that hits the gap continues moving to the right. The width of each wavefront that passes through the gap is equal to the gap width.

Suppose the gap in Figure 34.4 has a width equal to 2λ. All radiation that reaches P from wavelets that originate above Q′ cancels, as indicated by the rays drawn from points Q and R in Figure 34.4; wavelets that originate at or below Q′ are not canceled at P because those wavelets lack corresponding wavelets at an appropriate distance below Q′. As a result, some of the light spreads out past the edges of the original path of the beam—the light is diffracted. Diffraction in this situation occurs for exactly the same reason as the diffraction of water waves pictured in Figure 34.1*a*. Diffraction of light occurs when a planar wave passes through an aperture that is only micrometers wide (much less than the width of a hair). As shown in Figure 34.1*a*, if the width of the aperture is equal to or less than the wavelength, the wavefronts coming from the aperture are spherical. If the aperture is a few wavelengths wide, the wavefronts are elongated right after they pass through the aperture, as shown in Figure 34.4.

An ordinary window, like that shown in Figure 34.1*b*, is effectively infinitely wide relative to the wavelength of the light, so only the light at the very edges of the window diffracts. In practice, not even diffraction from the edge of the window is observed because the edge is not perfectly smooth on a micrometer scale. However, it *is* possible to observe diffraction of light from the edge of a smooth razor blade, as shown in **Figure 34.5**.

Figure 34.5 Edge diffraction is not usually apparent for visible light because most edges are not smooth enough. However, it can be observed around the edge of a razor blade. The blade in this image is illuminated by a point source of monochromatic light.

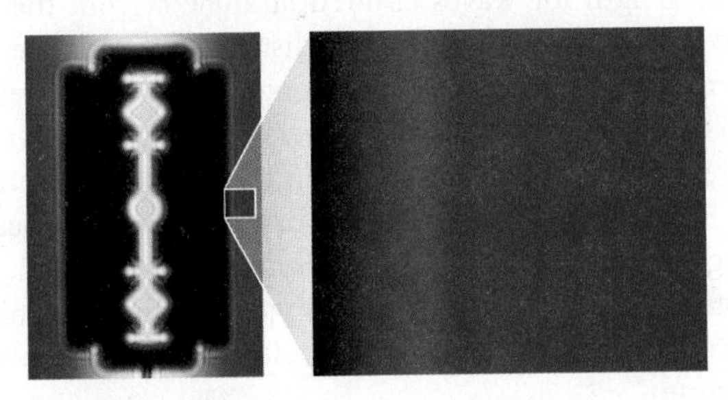

Example 34.1 Spreading out

Do you expect to be able to observe the diffraction of light through (*a*) the front door to your house; (*b*) the holes in a button; (*c*) the gaps between threads of the fabric of an umbrella?

❶ GETTING STARTED I expect to see noticeable diffraction through openings up to roughly two orders of magnitude times the wavelength of the light. I therefore need to estimate the width of each opening and determine the width-to-wavelength ratio.

❷ DEVISE PLAN To estimate the width of the front door and of the holes in the button, I can draw on my experience; to estimate the widths of the gaps between the threads of the fabric of an umbrella, I shall use an upper limit. Then I shall take the ratios of these widths to the wavelength of light in the middle of the visible range, 500 nm.

❸ EXECUTE PLAN (*a*) A door is about 1 m wide, so the ratio of the door's width to 500 nm is 2×10^6, much too great to see diffraction. ✔

(*b*) The holes in a button are about 1 mm in diameter, so the ratio of this width to 500 nm is 2×10^3, still too great to see diffraction. ✔

(*c*) I know that a human hair, which is less than 100 μm in diameter, cannot easily be threaded through a piece of fabric. Therefore I estimate the gaps between the threads in the fabric to be one-tenth of a hair diameter, or 10 μm at most. The ratio of a gap width to 500 nm is therefore 20 or less, and I expect to see diffraction through the gaps in the fabric. ✔

❹ EVALUATE RESULT I know from experience that I do not see diffraction through a doorway. I can check my answers to parts *b* and *c* by looking through the holes in a button and through an open umbrella. When I do so, I see no diffraction through the button but I can see diffraction through the umbrella if the fabric is dark. (This diffraction is particularly noticeable when I look at a streetlight at night through the umbrella.)

34.2 In discussing how a planar wave propagates, we could turn our earlier argument around and say that for each point Q in Figure 34.2 there is a point S somewhere on the wavefront that radiates toward P along a path exactly one wavelength longer than that from Q, and therefore there should be a nonzero intensity at P. What is wrong with this argument?

Figure 34.6 When the planar electromagnetic waves of a laser beam pass through a pair of narrow slits, what do we see on the screen?

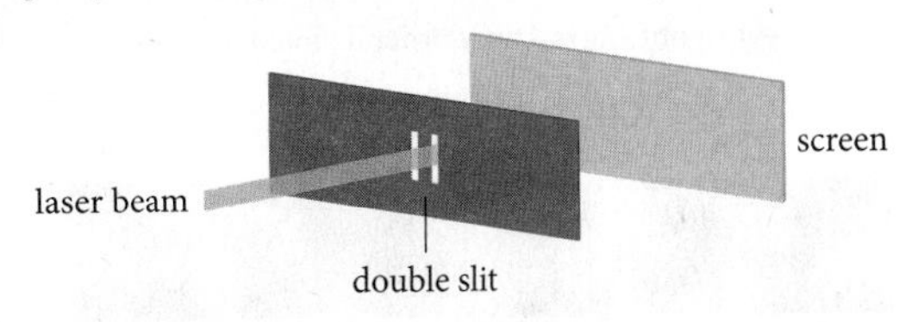

34.2 Diffraction gratings

What happens if instead of passing through a single small aperture, a planar electromagnetic wave strikes a barrier that contains two narrow slits at normal incidence, as shown in **Figure 34.6**? If the slit width is much less than the wavelength of the wave, the slits serve as two coherent point sources of electromagnetic waves of the same wavelength as the wave striking the barrier, and the waves from the two sources interfere with one another in the manner described in Section 17.3 for two adjacent sources of surface waves. The only difference is that electromagnetic waves are not confined to a planar surface but spread out in three dimensions. If the two slits are either round or square, the waves that emerge from them are spherical. If the slits are much taller than they are wide, as in Figure 34.6, the slits serve as lines of point sources and the waves that emerge from the slits have cylindrical wavefronts.

The crests of the waves from these two coherent point sources overlap in certain directions, as shown in **Figure 34.7** (and Figure 17.13). Between these directions where there is overlap are directions along which the waves from the two sources cancel. If we place a screen to the right of the slits, a pattern of alternating bright and dark bands appears on the screen, as shown in **Figure 34.8** on the next page. Such dark and bright bands are commonly called **interference fringes.** The bright fringes are labeled by a number called the **fringe order** *m*. The central bright fringe is the *zeroth-order bright fringe* ($m = 0$); around it are higher-order bright fringes ($m = 1, 2, \ldots$). Note that the pattern is symmetrical about this zeroth-order bright fringe.

Figure 34.7 Interference between the diffracted waves emerging from the two slits.

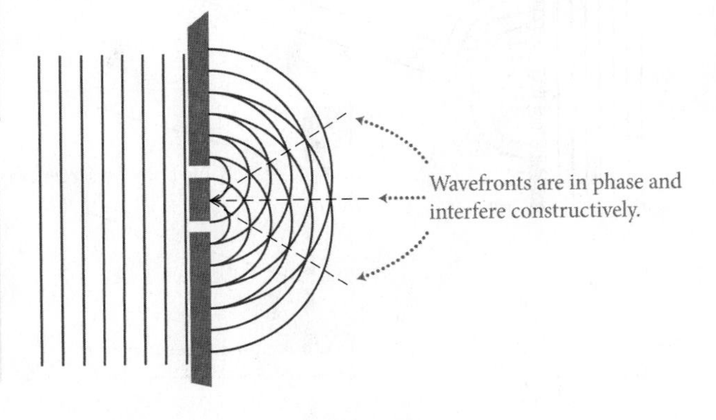

Figure 34.8 The interference pattern produced when the laser beam of Figure 34.6 passes through a pair of slits and strikes the screen.

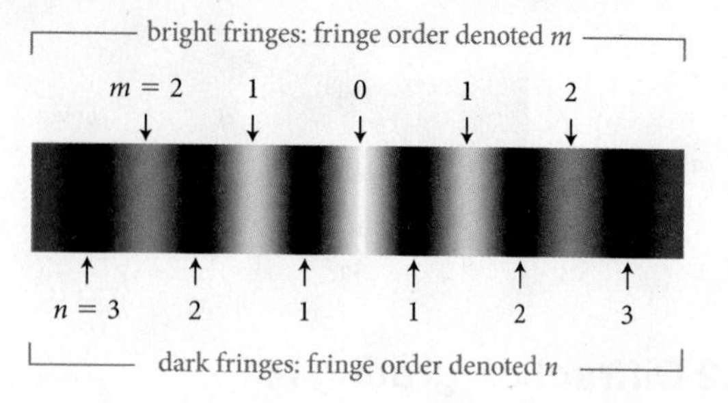

How do we determine the locations of the bright fringes? The central (zeroth-order, $m = 0$) bright fringe is simplest to locate. As shown in Figure 34.7, waves from the two sources interfere constructively along a perpendicular line running through the midpoint between the two slits. Because the waves coming from the two slits travel the same distance to reach any point along this perpendicular line, the waves arrive in phase with each other.

To locate the other bright interference fringes, we need to work out the directions in which the difference in path length between waves coming from the upper source and waves coming from the lower source is an integer number of wavelengths. In these directions, constructive interference between waves from the two sources produces bright fringes.

Let's consider two rays, one from each slit, that meet at the screen to form a fringe, as shown in **Figure 34.9a**. In general we shall take the distance from the sources to the screen to be much greater than the distance between the slits. Note from the figure how, even though the rays eventually meet at the screen, their paths are essentially parallel when they emerge from the slits (Figure 34.9*b*). If we denote the angle between the nearly parallel rays and the normal to the barrier as θ, we can say that the difference in path length for waves emitted at angle θ from the two sources is $d \sin \theta$, where d is the distance between the slits (Figure 34.9*b*). When this path-length difference is equal to an integer multiple of the wavelength, constructive interference occurs. The central bright fringe corresponds to $d \sin \theta = 0$ or $\theta = 0$. Subsequent bright fringes are located at angles given by $d \sin \theta_m = \pm m\lambda$, where $m = 1, 2, \ldots$ denotes the order of the bright fringe. The plus and minus signs give rise to bright fringes on either side of the central maximum.

Likewise, in directions that correspond to path-length differences of an odd number of half-wavelengths, waves from the two sources interfere destructively, resulting in dark fringes. For these fringes, we use n to denote fringe order. The smallest angle at which destructive interference occurs corresponds to $d \sin \theta = \frac{1}{2}\lambda$. More generally, the angles at which dark fringes occur are given by $d \sin \theta_n = \pm(n - \frac{1}{2})\lambda$, where $n = 1, 2, \ldots$ denotes the order of the dark fringe. The dark fringes around the zeroth-order bright fringe are the first-order dark fringes.

Example 34.2 Two-slit diffraction grating

Coherent green light of wavelength 530 nm passes through two very narrow slits that are separated by 1.00 μm. (*a*) Where is the first-order bright fringe? (*b*) What is the angular separation between the $n = 1$ and $n = 2$ dark fringes?

❶ **GETTING STARTED** This problem involves interference between light rays passing through two closely spaced, very narrow slits, as shown in Figure 34.9.

Figure 34.9 Determining the path-length difference between two rays traveling to a point on a distant screen in a two-slit interference setup.

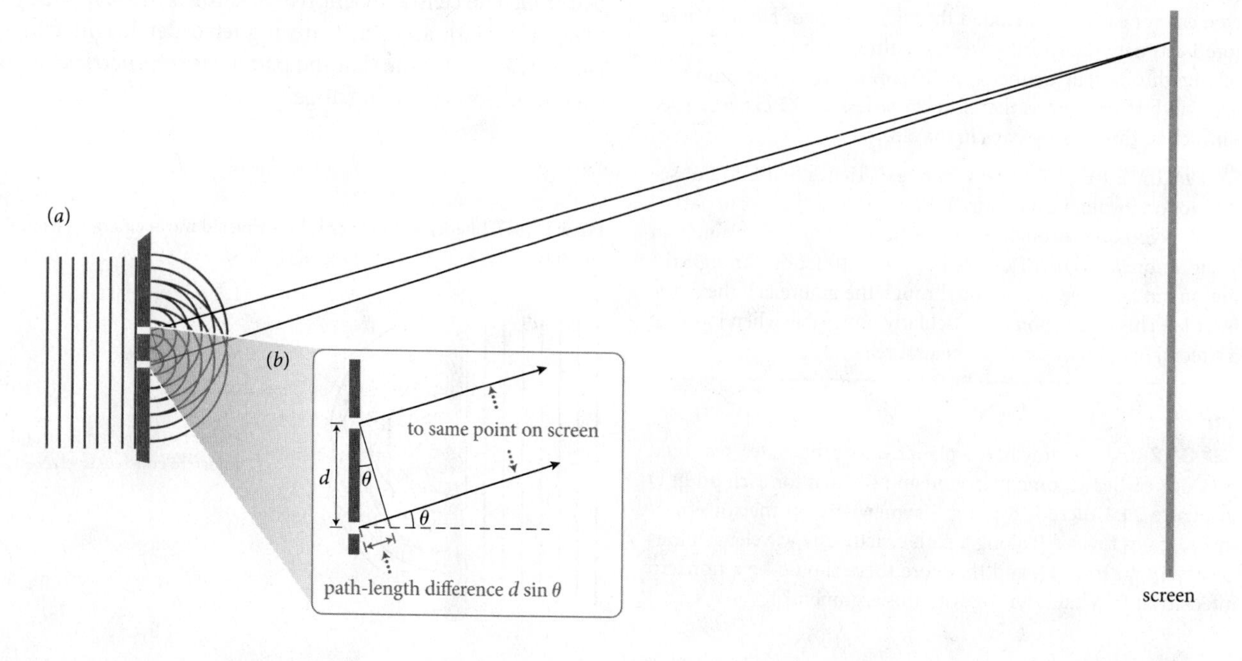

CONCEPTS

❷ DEVISE PLAN The angular locations of the centers of the bright fringes are given by the condition for constructive interference, $d \sin \theta_m = m\lambda$; the angular locations of the dark fringes are given by the condition for destructive interference, $d \sin \theta_n = (n - \frac{1}{2})\lambda$. (For both bright and dark fringes, I omit the $\pm$ signs because I'll only consider the fringes on one side of the central maximum.) I therefore need to use these relationships to calculate the angular locations of the fringes. To use the constructive and destructive interference conditions, I need to identify the appropriate values of m and n and then use them with the wavelength and the distance between slits to obtain the angular positions of the bright and dark fringes I am interested in.

❸ EXECUTE PLAN (*a*) The first-order bright fringe corresponds to $m = 1$, which means the center of the first-order bright fringe is located at the value θ_1 corresponding to $d \sin \theta_1 = \lambda$. Substituting for d and λ in this expression and solving for θ_1, I obtain

$$\theta_1 = \sin^{-1}\left(\frac{0.530\ \mu\text{m}}{1.00\ \mu\text{m}}\right) = 32.0°. ✔$$

(*b*) The two lowest-order dark fringes correspond to $n = 1$ and $n = 2$, and their centers occur at the angles corresponding to $d \sin \theta_1 = \lambda/2$ and $d \sin \theta_2 = 3\lambda/2$. Substituting and solving, I obtain

$$\theta_1 = \sin^{-1}\left(\frac{0.530\ \mu\text{m}}{2 \times 1.00\ \mu\text{m}}\right) = 15.4°$$

$$\theta_2 = \sin^{-1}\left(\frac{3 \times 0.530\ \mu\text{m}}{2 \times 1.00\ \mu\text{m}}\right) = 52.7°.$$

The angular separation is thus $52.7° - 15.4° = 37.3°$. ✔

❹ EVALUATE RESULT The center of the $m = 1$ bright fringe is roughly halfway between the centers of the $n = 1$ and $n = 2$ dark fringes, as I expect from Figure 34.8.

34.3 Does the spacing of the bright fringes in the two-slit arrangement in Figure 34.6 increase, decrease, or stay the same if we (*a*) increase the spacing d of the slits, or (*b*) increase the wavelength λ of the light incident on the arrangement?

Now consider the effect of many equally spaced narrow slits in a barrier on which planar waves are incident normally (**Figure 34.10**). Once again, we can determine the fringe pattern produced at a given location on a distant screen by combining all the waves (one from each slit) that travel to that location. The condition for constructive interference among all the waves is equivalent to that for two slits. As can be seen from Figure 34.10, the path-length difference between each pair of adjacent waves is the same. This means that if the waves from two of the slits interfere constructively, the waves from *all* of the slits do the same. Bright fringes therefore appear at angles corresponding to $d \sin \theta_m = \pm m\lambda$, with d the separation between adjacent slits. The location of bright fringes therefore does not depend on the number of slits as long as the separation d between adjacent slits is the same for all slits.

Figure 34.10 Path-length difference for planar waves striking a barrier with multiple slits.

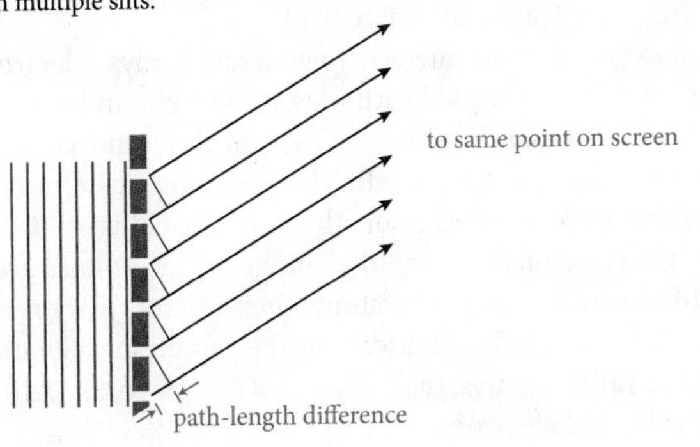

Figure 34.11 Three coherent waves interfere destructively when each is out of phase with the other two by one-third of a cycle.

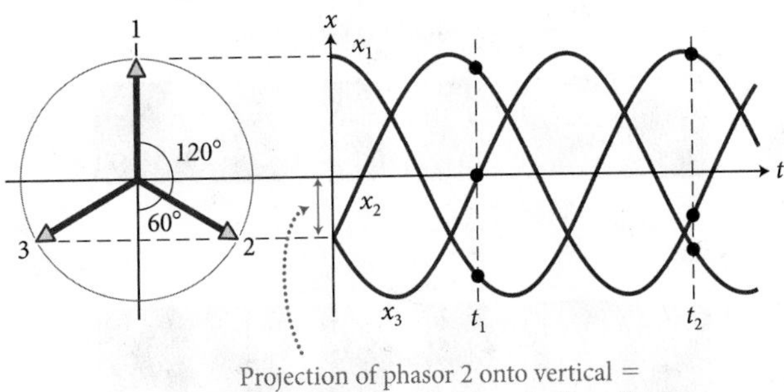

34.4 Suppose there are three slits in a barrier on which light is incident normally, with each slit separated from its neighbor by a distance d. Do the waves from all three slits cancel perfectly at angles given by $d \sin \theta = \pm(n - \frac{1}{2})\lambda$?

As Checkpoint 34.4 illustrates, the condition for complete destructive interference of all of the waves is not the same for three slits as for two. Instead, as **Figure 34.11** shows, three coherent waves cancel one another perfectly when each is out of phase with the other two by one-third of a cycle. The three waves also cancel one another perfectly when each is out of phase with the other two by two-thirds of a cycle. In that case, phasors 2 and 3 in Figure 34.11 are interchanged. So there are two dark fringes between each pair of bright fringes. In between these two dark fringes is a faint bright fringe (**Figure 34.12** on the next page). The brightest fringes are the *principal maxima* in the interference pattern. These correspond to constructive interference of the waves diffracted by all three slits. The fainter bright fringes are *secondary maxima*. At these locations the cancellation is not complete.

Four coherent waves cancel one another when adjacent waves are out of phase by one-fourth of a cycle. In the same manner, if there are N slits, the condition for complete destructive interference is that each wave must differ in phase by $1/N$ of a cycle from its immediate neighbors. Then the N waves are evenly distributed throughout one cycle of oscillation and add to zero. The condition for the path-length differences for the dark fringes is thus $d \sin \theta_k = \pm(k/N)\lambda$, where k is any integer that is *not* a whole-number multiple of N (because when $k/N = m$, we have constructive interference).

Figure 34.12 Interference pattern caused by the diffraction of a coherent beam of light through two, three, and eight narrow slits.

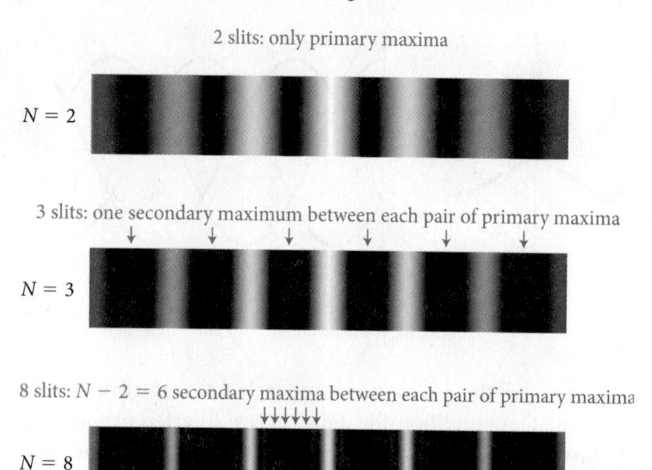

Although the bright fringes are in the same location regardless of the number of slits, there are now $N - 1$ dark fringes between the bright fringes and $N - 2$ secondary maxima between each pair of principal maxima. As a result, as N increases, the bright fringes become narrower and brighter, as shown in Figure 34.12. (The brightness of the pattern corresponds to the intensity of the light striking the screen.)

34.5 Why does the brightness of the fringes increase as the number of slits increases?

The interference of a planar electromagnetic wave as it passes through many closely spaced narrow slits is due to the diffraction that occurs at the slits. A barrier that contains a very large number of such slits is therefore called a **diffraction grating.** Diffraction gratings can be either transmissive (such as the one shown in Figure 34.10) or reflective. Reflective diffraction gratings are made by engraving grooves to reflect light from a surface, as shown in **Figure 34.13**. The grooves on a music compact disc (opening picture in this chapter) form a reflective diffraction grating.

Why is the light reflected from the compact disc surface so colorful? You found in Checkpoint 34.3 that the position of interference fringes produced by light of a single color depends on the wavelength. When white light, which contains many different colors and therefore many different wavelengths, falls on a diffraction grating, the fringes for each wavelength are displaced from each other, producing a series of rainbows.

Figure 34.13 Reflective diffraction grating.

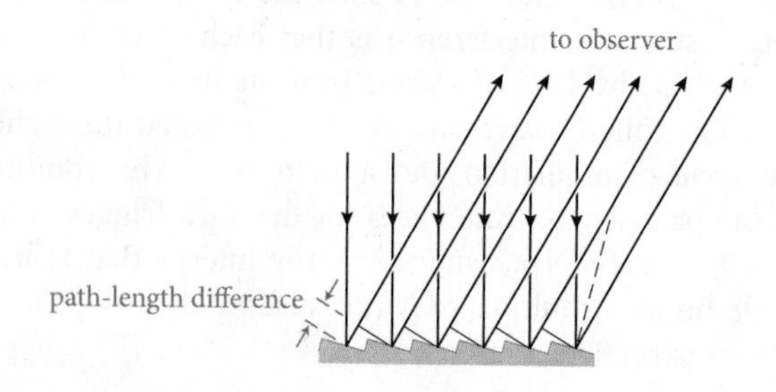

34.6 Suppose the light striking the reflective diffraction grating in Figure 34.13 is white light—that is, light consisting of all the colors of the rainbow. Red light has the longest wavelength of the colors that make up white light; violet has the shortest. (*a*) Is the angle at which first-order constructive interference occurs for violet light less than, equal to, or greater than the angle for red light? (*b*) Why are multiple rainbows visible in the reflected light?

Very precisely manufactured diffraction gratings have many uses in scientific equipment. They are most widely used to disperse visible, ultraviolet, or infrared light into its constituent wavelengths because the resulting spectrum provides information about the object that emitted the light. For example, astronomers often use a diffraction grating attached to a telescope to identify the wavelengths present in the light from stars, in order to understand the chemical composition of the stars or their distance from Earth.

Certain common objects can also function as diffraction gratings. For example, if you look through a piece of dark, finely woven, taut fabric at a point source of light (say, a distant street light through the fabric of a dark umbrella), you will see fringes.

34.7 Diffraction gratings used in astronomical instruments must be able to separate wavelengths that are quite close together. (*a*) To increase the ability to do this, should the separation between slits be made less or greater? (*b*) Does the width of the slits affect the diffraction pattern?

34.3 X-ray diffraction

The interference of very-short-wavelength electromagnetic radiation is widely used to study the structure of materials. **X rays** are electromagnetic waves that have wavelengths ranging from 0.01 nm to 10 nm, more than 100 times less than the wavelengths of visible light.

Figure 34.14 shows one way to generate X rays. Electrons are ejected from a heated cathode on the right and accelerated by a potential difference of several thousand volts toward a metal anode on the left. The electrons crash into the atoms that make up the anode; this inelastic collision decelerates the electrons very rapidly and gives the atoms a great deal of internal energy. The atoms then re-emit this energy in the form of x rays. In addition, the rapidly decelerating electrons radiate x rays, typically at a 90° angle to the path of the accelerated electrons.

CONCEPTS

Figure 34.14 Schematic diagram for a cathode ray tube x-ray emitter.

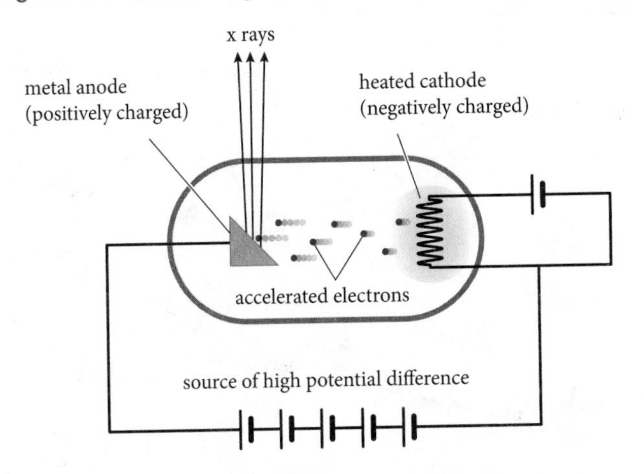

X rays can pass through many soft materials with low mass density that are opaque to visible light. For example, they pass through soft tissues in the human body but are strongly absorbed by bones and teeth. As a result, x rays are widely used to obtain photographic images of the skeleton (**Figure 34.15**). X-ray imaging of blood vessels or soft internal organs can be done by giving the patient a drug containing heavy atoms, such as iodine, because the heavy atoms absorb x rays. For example, one way cardiologists diagnose heart problems is to directly observe blood vessels in a patient's heart by injecting a heavy-atom drug into the patient's blood and taking x-ray movies of the beating heart as the blood is pumped through.

Because x-ray wavelengths are either shorter than or comparable to the typical distance between atoms in solid materials (0.1 nm to 1 nm), x-ray diffraction can be used to study atomic arrangements in solids. Many solids are *crystalline*, meaning that their atoms are arranged in a three-dimensional, regularly spaced grid called a *crystal lattice* (**Figure 34.16**). The lattice serves as a three-dimensional diffraction grating for x rays because the lattice spacing is comparable to the x-ray wavelength.

Figure 34.15 Bones and teeth absorb x rays, whereas soft tissues are nearly transparent to them.

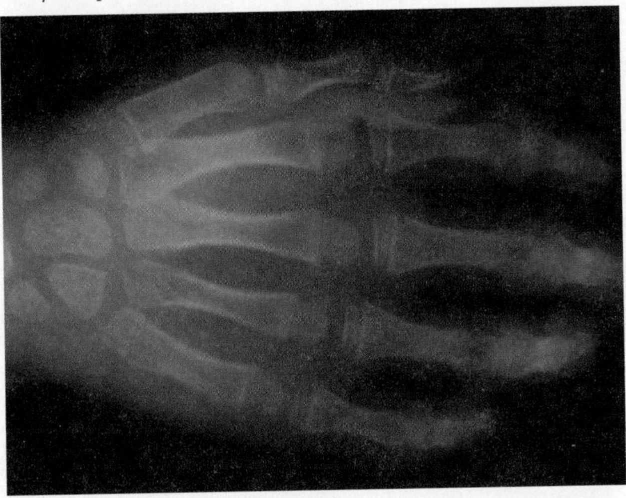

Figure 34.16 Two examples of crystal lattices.

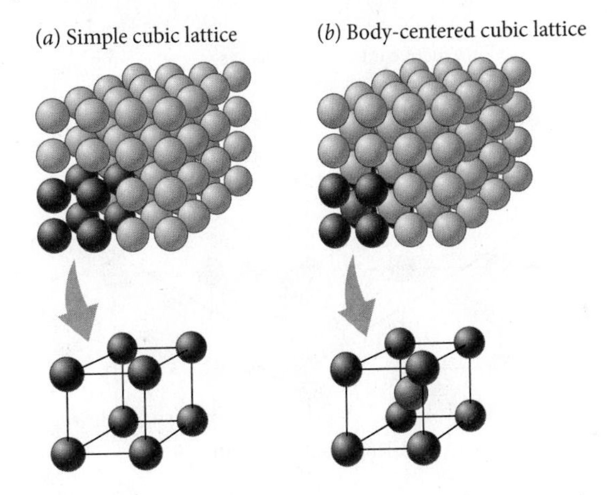

Consider what happens when x rays strike the top plane of atoms in a crystal lattice* at an angle θ (**Figure 34.17**). Each atom that is struck by the beam of x rays acts as the source of a wavelet emitting waves in all directions, much like the slits of the diffraction gratings we discussed in the preceding section. Waves emitted at $\theta' = \theta$ have the same path length and so they add constructively, yielding a strong reflected beam.

34.8 Considering only the top row of atoms in Figure 34.17, are there any other directions in which the x rays diffracted by the atoms interfere constructively?

As you saw in Checkpoint 34.8, x rays diffracted by the crystal in directions other than at angle $\theta' = \theta$ are much weaker than those diffracted at angle $\theta' = \theta$. We might therefore expect not to see any x-ray diffraction from crystals at other angles. However, a crystal consists of many planes of atoms, and some incident x rays penetrate into the

* The lattice shown is a so-called cubic lattice. Lattices can be much more complex than the cubic lattice, but the principles of diffraction are the same.

Figure 34.17 Diffraction of x rays by the atoms at the surface of a crystal lattice.

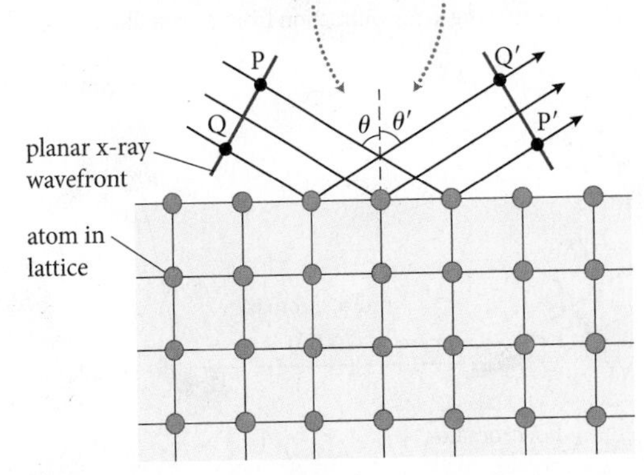

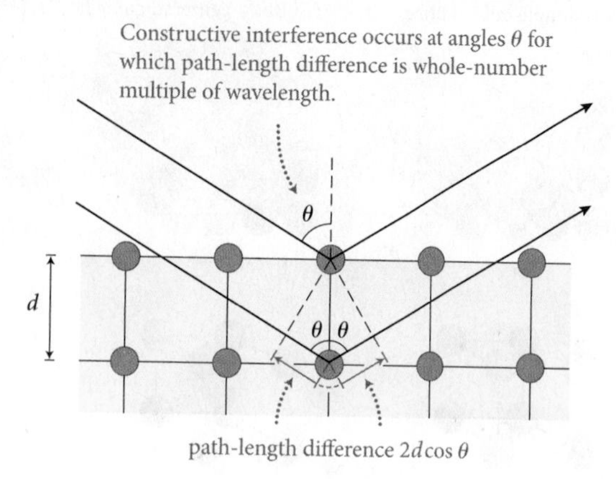

Figure 34.18 Interference of x rays diffracted by adjacent planes of a crystal.

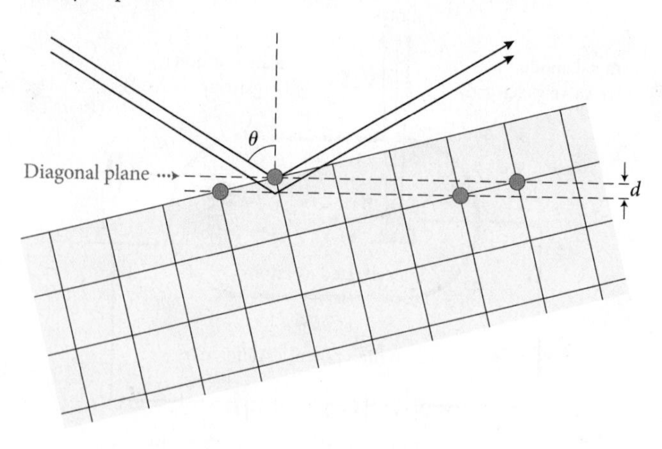

Figure 34.19 Constructive interference of x rays diffracted by two diagonal crystal planes.

crystal. We thus need to take into account the diffraction of the x rays by the atoms in multiple crystal planes to determine the diffraction of the crystal as a whole.

Figure 34.18 shows the x rays diffracted by atoms in two adjacent planes of a crystal. For most angles of incidence, the waves diffracted by atoms in different planes differ in phase and so interfere destructively. However, when the difference in path length between rays diffracted by atoms in different planes is a whole-number multiple of the x rays' wavelength, the rays are in phase and interfere constructively. The path-length difference equals $2d \cos \theta$, where d is the distance between adjacent planes. Therefore the condition for constructive interference is $2d \cos \theta = m\lambda$. This condition is called the **Bragg condition,** after the father-and-son team of physicists who formulated it. Because the atomic spacing and the x-ray wavelength are fixed, crystals reflect x rays only at those angles for which $2d \cos \theta$ is an integer multiple of the x-ray wavelength.

The atoms in the lattice of a crystal define many different lattice planes. **Figure 34.19** shows two sets of lattice planes in a cubic crystal, one indicated by dashed lines (planes parallel to the surface of the crystal) and the other indicated by solid lines (diagonal planes). If we tilt the crystal so that the angle the incident x rays make with its surface is different from the angle shown in Figure 34.18, a different set of planes with a different spacing can produce constructive interference of the diffracted waves.

By measuring the angles at which strong x-ray diffraction occurs, one can determine the arrangement of atoms in a crystalline solid. **Figure 34.20** shows how such a measurement is carried out. An x-ray source like that shown in Figure 34.14 is used to produce a beam of x rays of various wavelengths. The beam is then diffracted from a *crystal monochromator* (which is simply a crystal of known lattice spacing) positioned at an angle chosen so that the Bragg condition is satisfied for one desired wavelength of x rays. Because the other wavelengths in the original beam do not satisfy the Bragg condition, a *monochromatic* (single-wavelength) beam of x rays is diffracted from the monochromator to the sample.

The sample of crystalline material whose lattice is being studied is slowly rotated with respect to the monochromatic beam, and as this rotation takes place, the intensity of x rays diffracted from the sample is measured on a detector as a function of the angle α between the x rays and the sample surface (Figure 34.20*b*). This angle is often called the *Bragg angle* α. As this angle changes, the Bragg condition is

Figure 34.20 (*a*) Apparatus for studying x-ray diffraction from a crystalline solid. (*b*) Relationship between the incident angle θ and the Bragg angle α.

(*a*) Apparatus for studying x-ray diffraction from a crystalline solid

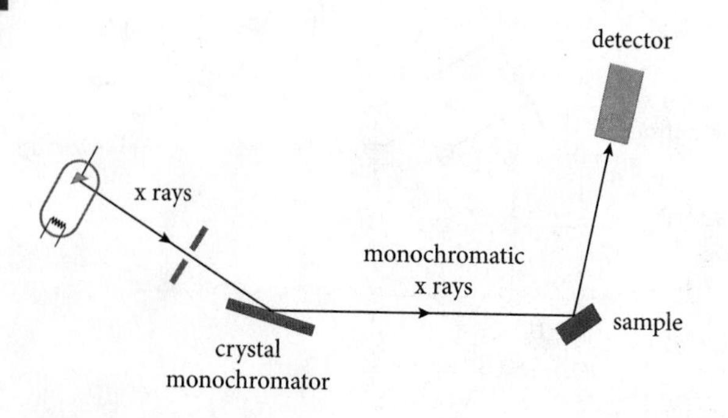

(*b*) Relationship between incident angle θ and Bragg angle α

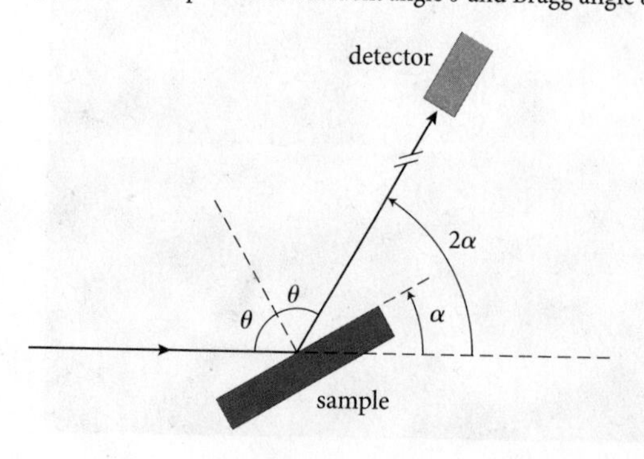

generally not satisfied. However, at specific Bragg angles, the various crystal planes in the sample satisfy the Bragg condition, producing a high intensity of x rays on the detector.

34.9 Express the Bragg condition $2d \cos \theta = m\lambda$ in terms of the Bragg angle α between the x rays and the surface of the sample rather than the angle θ between the x rays and the surface normal.

Example 34.3 X-ray diffraction

Figure 34.21 shows diffracted x-ray intensity as a function of the Bragg angle α, obtained using x rays having a wavelength of 0.11 nm. (*a*) Without calculating values for the lattice spacing d, identify which of the two peaks corresponds to a greater distance between adjacent planes in the sample being studied. (*b*) Calculate the distance between adjacent planes corresponding to each peak.

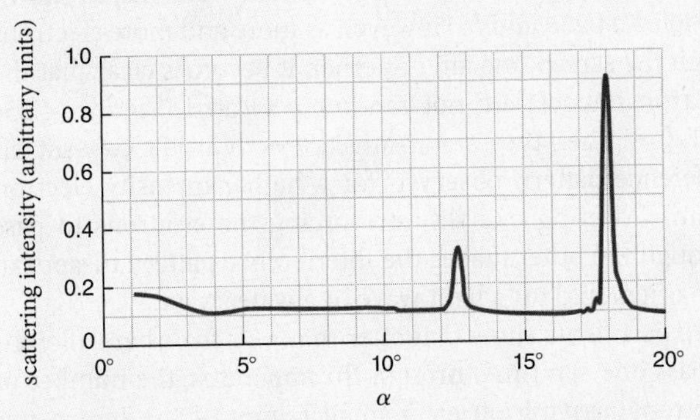

Figure 34.21 Example 34.3.

❶ **GETTING STARTED** The peaks in a graph of x ray intensity as a function of the Bragg angle α correspond to constructive interference and therefore values of α that satisfy the Bragg condition.

❷ **DEVISE PLAN** For part *a*, I can use the Bragg condition, together with the shape of the graph, to deduce which peak results from the greater d value. Because the graph gives intensity in terms of the Bragg angle, I shall need the Bragg condition expressed in terms of the Bragg angle.

For part *b*, I can solve this form of the Bragg condition for d and then insert my two given α values to determine the plane separation distance in each case. Looking at this form of the Bragg condition, I see that the Bragg angle α at which a peak occurs increases as m increases. Because this graph begins at $\theta = 0°$, the two peaks must correspond to $m = 1$ for the interference patterns when the crystal surface is oriented at the two Bragg angles I am working with.

❸ **EXECUTE PLAN** (*a*) From Checkpoint 34.9 I know that the Bragg condition in terms of the Bragg angle is $2d \sin \alpha = m\lambda$. In order for the product $2d \sin \alpha$ to remain constant, α must decrease as the distance d between adjacent planes increases. Therefore the peak at the smaller α value corresponds to the greater distance between planes. ✔

(*b*) To obtain d for each peak, I solve $2d \sin \alpha = m\lambda$ for d and then substitute $m = 1$ and the values for α and λ. For the short peak, $\alpha = 12.5°$, which gives $d = \lambda/(2 \sin \alpha) = (0.11 \text{ nm})/(2 \sin 12.5°) = 0.25$ nm. For the tall peak, $\alpha = 18°$, which gives $d = \lambda/2 \sin \alpha = (0.11 \text{ nm})/(2 \sin 18°) = 0.18$ nm. ✔

❹ **EVALUATE RESULT** The Bragg angle α at which constructive interference occurs decreases as the distance between planes increases. This is consistent with what I found previously for interference between two slits (Checkpoint 34.3), in which increasing the distance between slits also causes the angle between fringes to decrease. In general, the size of an interference pattern decreases as the distances between interfering sources increase. Finally, the smaller value of d multiplied by $\sqrt{2}$ gives the greater value of d, as it should for the distances between planes for the cubic lattice in Figure 34.19.

Many studies of crystal structure are done by passing x rays through the crystal rather than reflecting them from the various crystal planes. The crystal then acts like a three-dimensional transmissive diffraction grating for the x-ray beam. Instead of a single line of slits, the beam encounters many rows of slits. As a result, many rows of fringes usually called "spots" are formed. The experimental apparatus for such a measurement is shown in **Figure 34.22*a***. From the

Figure 34.22 X-ray crystallography.

(*a*) Schematic apparatus for x-ray crystallography

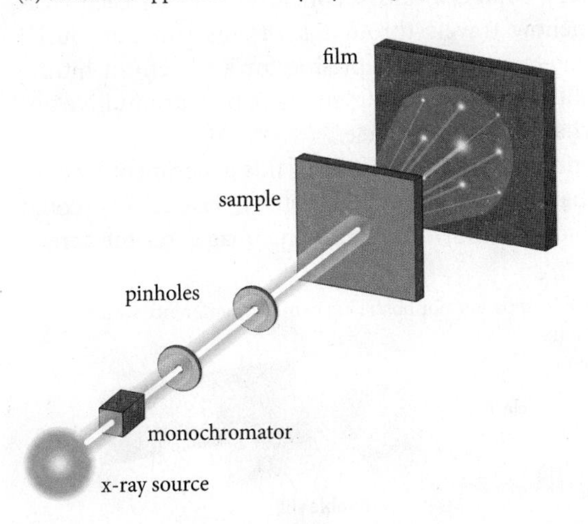

(*b*) X-ray diffraction pattern of diamond lattice

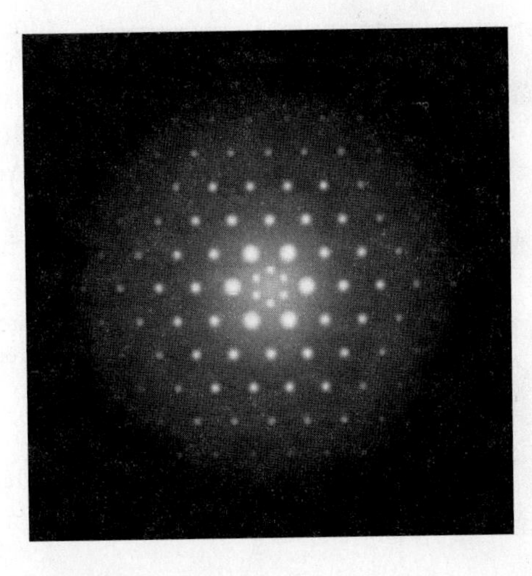

CONCEPTS

position and intensity of the spots in the resulting diffraction pattern (Figure 34.22*b*), we can deduce the arrangement of the atoms in the sample.

The earliest x-ray diffraction studies of crystals were done on simple crystals, such as metals that form cubic or other very simple lattices. However, x-ray diffraction has also been used to study the atomic structure of much more complicated molecules, by crystallizing solutions of such molecules. The double helical structure of DNA was determined using x-ray diffraction, also called *x-ray crystallography,* and crystallography is widely used today to determine the atomic structure of even more complicated biological molecules.

34.10 While comparing the x-ray diffraction patterns from two crystals, you determine that crystal A produces a pattern with more widely spaced diffraction spots than crystal B. Which crystal has the greater atomic spacing?

34.4 Matter waves

Patterns remarkably similar to x-ray diffraction patterns can be obtained by aiming a narrow beam of electrons at a crystal. **Figure 34.23** shows the pattern obtained by sending a beam of high-speed electrons through a solid crystalline sample in an instrument called an electron microscope. The shape of the pattern is similar to that obtained with x rays (Figure 34.22*b*), which tells us that electrons are also diffracted by crystals. This discovery, in turn, suggests that the electrons behave like waves because interference and diffraction are wave phenomena. Indeed, electrons have been found to exhibit interference in many other experiments. For example, a beam of electrons aimed at two very narrow slits produces an interference pattern similar to that of a beam of light aimed at two narrow slits (**Figure 34.24**).

Varying the speed of the electrons changes the spacing of the interference pattern, which indicates that the electron wavelength depends on speed.

Figure 34.23 Electron diffraction pattern for a diamond lattice. Notice the similarity to the x-ray pattern in Figure 34.22*b*.

34.11 Spots in an electron diffraction pattern, such as the one shown in Figure 34.23, move closer together as the speed of the electrons is increased. Does this mean the wavelength of the electrons increases or decreases with increasing speed?

Up to now we have considered electrons as being nearly pointlike particles, so it is surprising—to say the least—to discover that they also behave like waves. In fact, electrons can exhibit both particle behavior and wave behavior in a single experiment! This dual behavior was vividly demonstrated in an experiment done in 1989 in Japan using an apparatus similar to that shown in Figure 34.24. The number of electrons emitted by the source was kept very low, so as to ensure that at any given instant at most one electron was traveling from the source to the screen. The screen shows the place where each electron arrives as a bright dot, and at first these dots appear at seemingly random locations, as shown in **Figure 34.25*a*** and *b*. However, as more and more electrons reach the screen one after another, it becomes clear that the electron impacts are not randomly located (Figure 34.25*c* and *d*). Rather, they are arranged exactly in the two-slit interference pattern observed for a higher-intensity electron beam. Covering one slit and forcing the electrons to pass through the other makes the interference pattern disappear, just as it would for a light wave or a water wave.

Experiments show that after many electrons are allowed to pass one at a time through the apparatus, the number of electrons arriving on each small region of the screen per unit time is proportional to the intensity that would be observed at that region if light of the appropriate wavelength were shone on the slits. In other words, the *probability* of any single electron arriving at that small region in a fixed time interval corresponds to the intensity of the two-slit interference pattern at that location.

The observation of individual electrons arriving at the screen one after another indicates that the electrons are individual particles. However, if they are particles in the sense we think of for material objects, they must pass through either the right slit or the left slit, and so how can they produce an interference pattern?

The very counterintuitive conclusion is that each electron somehow travels through *both* slits simultaneously! Such a statement is not surprising for a wavefront hitting the two slits, but it goes completely against our intuitive notion of what we call a "particle."

We cannot explain the results of this experiment by concluding that electrons are *waves* because a classical wave could not produce the individual pinpoint images on the screen

Figure 34.24 Apparatus for observing two-slit interference with an electron beam.

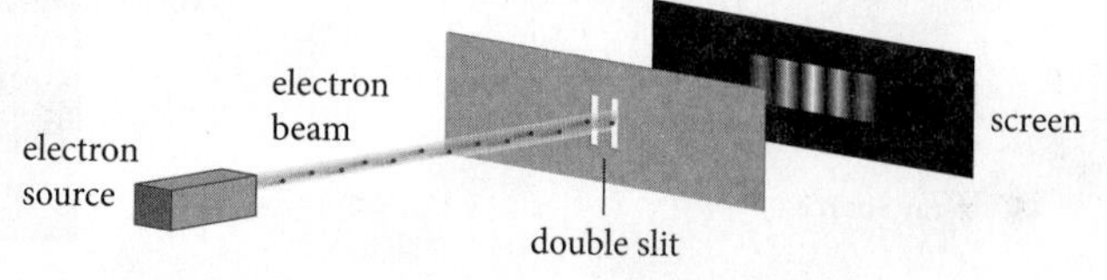

CONCEPTS

Figure 34.25 When we perform two-slit diffraction with a very weak electron beam, we can see the pattern build up over time. At first (*a*, *b*), the dots that mark electron impacts seem to be scattered randomly, but as more accumulate (*c*, *d*), the diffraction pattern becomes evident.

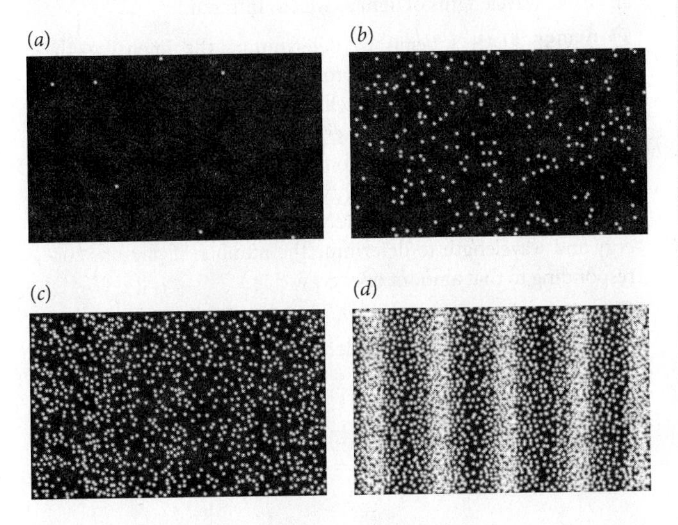

shown in, say, Figure 34.25*a*. If electrons were waves, the full diffraction pattern would be visible from the start (although it would be very faint). Scientists have concluded that electron behavior can be explained only if an electron has both particle properties and wave properties. This **wave-particle duality**—the possession of both wave properties and particle properties—has been observed not only for electrons but also for all other subatomic particles, for individual atoms, and, as we shall see in the next section, for light.

An expression for the wavelength of a particle, called the **de Broglie wavelength,** was proposed in 1924 by the French physicist Louis de Broglie* and confirmed experimentally a few years later. The de Broglie wavelength is inversely proportional to the momentum of the particle, $\lambda = h/p$, and the proportionality constant, called **Planck's constant,** is $h = 6.626 \times 10^{-34}\ \mathrm{J \cdot s}$. That this constant is a very small number indicates that the wavelength of macroscopic objects is extremely small. Because waves exhibit diffraction and interference only on length scales comparable to their wavelength, the wave nature of matter has been observed only with subatomic particles, atoms, and molecules. Exercise 34.4 illustrates this point.

Exercise 34.4 Electron versus baseball

Calculate the de Broglie wavelength associated with (*a*) a 0.14-kg baseball thrown at 20 m/s and (*b*) an electron of mass 9.1×10^{-31} kg moving at 5.0×10^{6} m/s.

SOLUTION I use the expression $\lambda = h/mv$ for the de Broglie wavelength.

(*a*)
$$\lambda_{\text{de Broglie, baseball}} = \frac{6.626 \times 10^{-34}\ \mathrm{J \cdot s}}{(0.14\ \mathrm{kg})(20\ \mathrm{m/s})} = 2.4 \times 10^{-34}\ \mathrm{m.}\ \checkmark$$

This is 24 orders of magnitude less than the diameter of an atom!

(*b*)
$$\lambda_{\text{de Broglie, electron}} = \frac{6.626 \times 10^{-34}\ \mathrm{J \cdot s}}{(9.1 \times 10^{-31}\ \mathrm{kg})(5.0 \times 10^{6}\ \mathrm{m/s})} = 1.5 \times 10^{-10}\ \mathrm{m} = 0.15\ \mathrm{nm.}\ \checkmark$$

Electron diffraction takes place on distances comparable to the spacing between atoms, whereas a baseball would diffract only through apertures more than 10^{24} times smaller than the typical spacing between atoms. (But the diameter of a baseball is about 0.10 m, so such an experiment is not possible.)

34.12 How would the electron diffraction pattern in Figure 34.23 change if the electrons were traveling more slowly?

34.5 Photons

We have now found that particles have wave properties, but these properties are observed only when the particle wavelength is comparable to the size of the objects the particles interact with. Up until now, you have most probably thought of light as being solely a wave. However, the dual wave-particle nature of particles may lead you to wonder whether light, too, is not just a wave but also a particle.

34.13 Compare the two-slit interference pattern obtained with electrons (Figure 34.25) with the two-slit interference pattern obtained with light (Figure 34.8). If light and electrons exhibit a similar wave-particle duality, how might you modify the two-slit experiment with light shown in Figure 34.6 to observe light behaving like a particle?

Figure 34.26*a* on the next page shows the image obtained by shining a beam of light onto the sensor chip of a digital camera. Pixels in the center of the beam register more light, and therefore the image of these pixels is brighter. The pixels at the edge of the beam are dimmer than those at the center, and outside the beam the pixels are black. If we place in front of the sensor chip a filter that lets only 50% of the light through, the image darkens because the brightness measured by each camera pixel is cut in half (Figure 34.26*b*).

Suppose we keep adding such filters, cutting the beam intensity in half with each addition. Does the image keep getting proportionally darker? If you carry out the experiment, you will discover that it does not, for one of two reasons. The first reason is mechanical: If you use the sensor chip of a digital camera, it stops detecting below a certain level. As you decrease the intensity of the beam by adding more and more filters, the image first gets grainy and then turns black.

The second reason has to do with the fundamental nature of light. Even if you use an extremely sensitive detector, you will still see that once the beam becomes very weak, adding another filter does not simply halve the image intensity. Instead, as shown in Figure 34.26*c* and *d*, the image of the beam breaks up into individual point-like flashes of equal

* "de Broglie" is pronounced "duh-Br-uh-y," the "y" sounding as in "yikes."

Figure 34.26 Images formed by using the sensor of a digital camera to record increasingly faint beams of light for the same exposure period. The separate dots recorded for the faintest beams reveal the particle-like behavior of light.

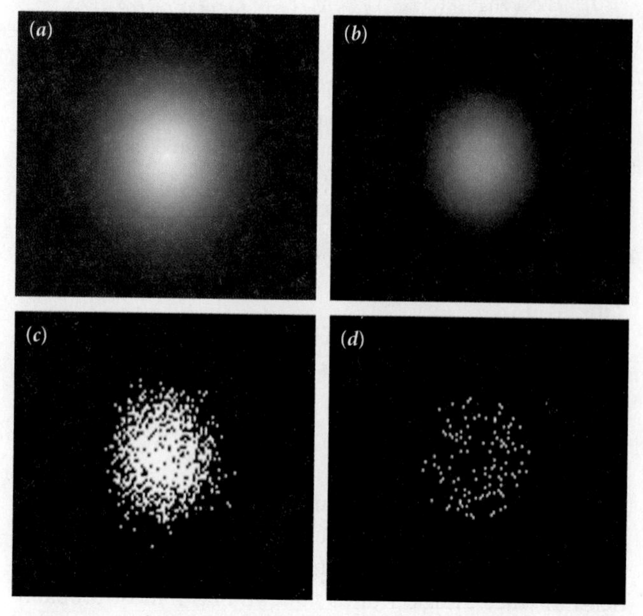

intensity, resembling the impacts of individual particles. The impacts may at first appear to be randomly distributed within the profile of the beam, but if you accumulate many of these impacts, you discover that the probability of observing a flash in a given location follows the intensity profile of the beam—the impacts are more likely to occur near the center of the beam. In fact, the probability of observing an impact in a particular location is proportional to the intensity of the beam at that location.

As the beam intensity is reduced, the individual impacts become separated in time. No two impacts occur at the same instant. Nor are any "half impacts" ever recorded. From these observations, we conclude that light indeed has particle properties as well as wave properties. The 'particles' of light are called **photons.** As we discuss further in Section 34.10, a photon represents the basic unit of a light wave and carries a certain amount of light energy. For a photon of frequency f, this energy equals hf, where h is again Planck's constant. Photons thus represent the quantum of electromagnetic energy—they cannot be subdivided.

CONCEPTS

Example 34.5 Photons from a light bulb

A 50-W incandescent light bulb emits about 5.0 W of visible light. (The rest is converted to thermal energy.) If a circular aperture 5.0 mm in diameter is placed 1.0 km away from the light bulb, approximately how many photons reach the aperture each second?

❶ **GETTING STARTED** This problem asks me to relate the power of light emitted by a light bulb to the rate at which photons pass through a certain area at a particular distance from the bulb. I can approximate the light bulb as a point source of light, meaning that its light is radiated uniformly in all directions. I can also simplify the problem by assuming that all of the light has the same wavelength, and I choose 500 nm for that wavelength. (This is a significant simplification because real light bulbs emit all visible wavelengths of light as well as infrared.)

❷ **DEVISE PLAN** I begin by determining the intensity—the power per unit area—of light produced by the light bulb over a sphere of radius 1 km, which tells me the intensity at the aperture. I can then multiply this intensity by the area of the aperture to obtain the power passing through the aperture, and I multiply that by 1 s to calculate the energy passing through the aperture in 1 s. Finally I shall use the relationship between photon energy and wavelength to determine the number of photons corresponding to that amount of energy.

❸ **EXECUTE PLAN** The intensity at the aperture is given by the power emitted by the bulb divided by the surface area of a sphere of radius 1.0 km:

$$\frac{(5.0\ \text{W})}{(4\pi)(1.0 \times 10^3\ \text{m})^2} = 4.0 \times 10^{-7}\ \text{J/m}^2 \cdot \text{s},$$

and therefore the amount of energy passing through a circular hole of radius 2.5 mm in 1.0 s is

$$(4.0 \times 10^{-7}\ \text{J/m}^2 \cdot \text{s})[\pi(2.5 \times 10^{-3}\ \text{m})^2] = 7.8 \times 10^{-12}\ \text{J}.$$

The energy of a single photon of wavelength 500 nm is

$$E = hf = \frac{hc_0}{\lambda}$$
$$= \frac{(6.626 \times 10^{-34}\ \text{J} \cdot \text{s})(3.00 \times 10^8\ \text{m/s})}{(500 \times 10^{-9}\ \text{m})}$$
$$= 3.98 \times 10^{-19}\ \text{J},$$

and so the number of photons that corresponds to the amount of energy passing through the hole is

$$\frac{7.8 \times 10^{-12}\ \text{J}}{3.98 \times 10^{-19}\ \text{J}} = 2.0 \times 10^7\ \text{photons.}\ ✔$$

❹ **EVALUATE RESULT** The number of photons I obtain is great even though I know from experience that the amount of light entering my eye 1.0 km from a 50-W bulb is exceedingly small. However, photons contain a vanishingly small amount of energy (about 10^{-19} J per photon), and so my answer is not implausible.

I assumed that all the photons have the same 500-nm wavelength. In reality the light bulb emits both longer- and shorter-wavelength photons. Suppose various wavelengths are equally distributed on both sides of 500 nm in the spectrum of the light emitted by the bulb. Photons with longer wavelengths have less energy and therefore there are more of them, but photons with shorter wavelengths have greater energy, and so there are fewer of them. The overall result is therefore about the same number of photons as the number I calculated using 500 nm as the only wavelength. To a first approximation, therefore, my result should be correct.

Figure 34.27 When we record a low-intensity beam of light with individual detectors, the beam acts like a stream of particles. Passing it through a double slit, however, causes an interference pattern to emerge.

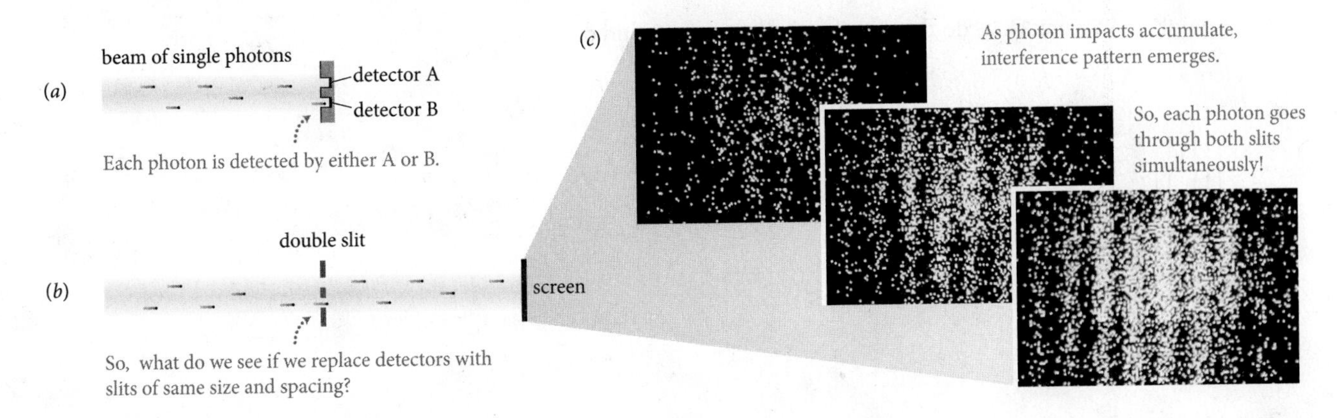
(a)
beam of single photons
detector A
detector B
Each photon is detected by either A or B.
(b)
double slit
screen
So, what do we see if we replace detectors with slits of same size and spacing?
(c)
As photon impacts accumulate, interference pattern emerges.
So, each photon goes through both slits simultaneously!

What are photons? How can light be a wave *and* consist of photons that behave like particles? I cannot answer this question because no one really knows the answer. I can, however, describe how photons behave. As you will see, their behavior defies common sense—it is unlike anything we ever experience.

If we reduce the intensity of a light beam (or a beam of any other type of electromagnetic radiation) so greatly that the flashes from the impacts of individual photons are well separated in time and then aim this beam at two very small, very closely spaced detectors (**Figure 34.27*a***), we see that each photon is detected by either one detector or the other. A simultaneous impact on both detectors is never recorded. (The detectors can be made as small as 1 μm wide and spaced by just a fraction of a micrometer.) This observation suggests that each photon takes a definite path—toward either detector A or detector B.

Now imagine replacing the two detectors in Figure 34.27*a* by two narrow slits of the same size as the detectors and placing a screen some distance back from the two slits (Figure 34.27*b*). The pattern on the screen initially looks like random impacts from photons that make it through either one slit or the other. As we accumulate the impacts of many photons, however, an interference pattern emerges (Figure 34.27*c*) that is identical to the one obtained by shining an intense beam of light on the two slits.

Notice that replacing the two detectors with two slits doesn't change anything about the beam or the photons contained in it. However, the experiment using the two detectors suggests that photons are particles detected by one or the other of the detectors; the experiment using the two slits indicates that each photon is a wave traveling through both slits simultaneously! So which is it? Do the photons travel through one slit only or through both? If we physically cover one of the slits (and therefore *force* the photons to go through the other slit), the interference pattern disappears. The pattern that emerges after accumulating many photons behind the slit corresponds to the diffraction pattern of light behind a single slit.

What this means is that photons behave as discrete particles when they are being detected. In transit, however, the wave nature of photons dictates their behavior. In other words, light behaves *both* like a wave and like a particle. It is impossible to explain the results of the above set of experiments by treating light as only a wave or only a particle; it must have qualities of both.

34.14 Figure 34.8 shows the interference pattern obtained by shining a strong laser beam on a pair of slits. If instead a very weak beam is shone on the same slits, so that the photons pass through the slits one photon at a time, at what angles is the probability of observing a particular photon the greatest?

Self-quiz

1. At point A in Figure 34.28, do the waves from the two slits add or cancel?

Figure 34.28

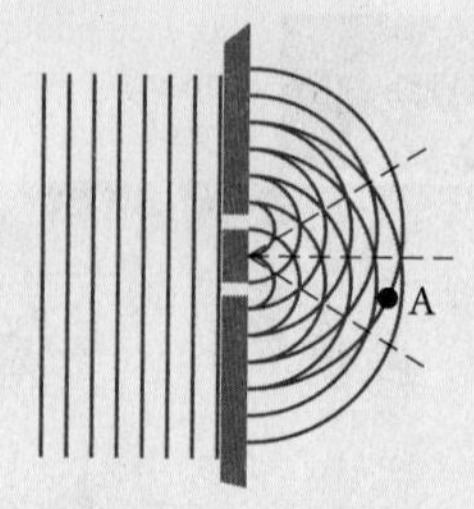

2. If the two sets of fringes shown in Figure 34.29 were produced by the same diffraction grating, which set is the product of the longer-wavelength radiation?

Figure 34.29

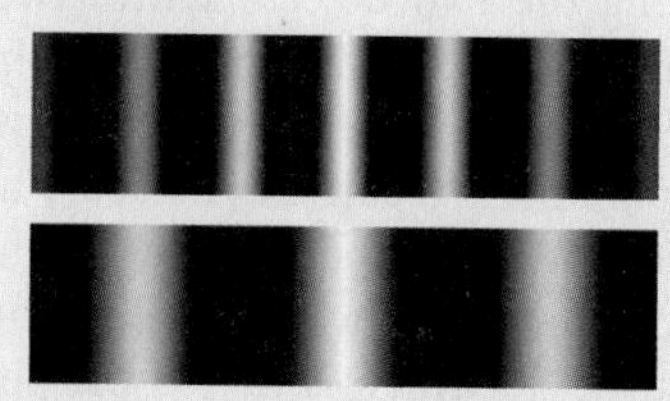

3. Coherent light of wavelength λ is normally incident on two slits separated by a distance d. What is the greatest possible fringe order?

4. Consider a proton and an electron moving at the same speed. Which has the longer wavelength?

5. Given the relationship between the energy E of a photon and its frequency f and the de Broglie expression relating momentum $p = mv$ and wavelength λ, determine the ratio E/p for a photon.

Answers:

1. At A, the crest of one wave overlaps the trough of the other wave, which means the waves cancel.
2. Because the wavelength is proportional to the sine of the angle the rays make with the normal to the diffraction grating, the fringes with the greater spacing were produced by the longer-wavelength radiation.
3. The fringe order is given by $d \sin\theta = \pm m\lambda$. Because the maximum value of $\sin\theta$ is 1, $d = m_{max}\lambda$ and so the maximum fringe order is $m_{max} = d/\lambda$. (Because m_{max} is an integer, you must truncate the value you obtain by dividing d by λ. For example, if $d/\lambda = 2.8$, then the greatest fringe order is 2.)
4. Because a proton has greater mass than an electron and because the de Broglie wavelength is inversely proportional to mass, $\lambda = h/mv$, the proton has a shorter wavelength than the electron.
5. The relationship between the energy of a photon and its frequency is $E = hf$ (see Section 34.5). The de Broglie wavelength is given by $\lambda = h/mv = h/p$, so $p = h/\lambda$. Therefore $E/p = hf/(h/\lambda) = f\lambda = c$, where c is the speed of light.

34.6 Multiple-slit interference

Let us now calculate the interference pattern produced by an electromagnetic wave normally incident on a barrier pierced by multiple closely spaced, very narrow slits. The width of each slit is much less than the wavelength of the radiation, so each slit serves as a point source of radiation. We begin by determining the pattern created when the barrier has just two very narrow slits.

As discussed earlier, the point sources corresponding to the two slits are in phase. Therefore the electric fields of the two waves that reach the screen differ in phase only due to any difference in the distance each wave travels from its slit to the screen. We shall refer to this difference as the *path-length difference* δs (**Figure 34.30a**). Taking the screen to be at position $x = 0$, we can write the electric fields of the waves traveling in the directions shown in Figure 34.30*a* as

$$E_1 = E_0 \sin \omega t \tag{34.1}$$

and

$$E_2 = E_0 \sin (\omega t + \phi), \tag{34.2}$$

where E_0 is the amplitude of the electric field and ϕ is a phase constant that is equal to the phase difference that results from the path-length difference between the two waves. The phase difference divided by 2π is the fraction of a cycle by which the two waves differ. This equals the path-length difference δs divided by the wavelength, or $\phi/2\pi = \delta s/\lambda$, and so

$$\phi = \frac{2\pi}{\lambda} d \sin \theta, \tag{34.3}$$

where d is the distance between the slits and θ is the angular position where the two rays meet on the screen.

We observe a bright fringe—a maximum in the intensity—when the two rays interfere constructively. This is the case when the phase difference ϕ equals an integer number times 2π:

$$\phi_m = \pm m(2\pi), \quad \text{for } m = 0, 1, 2, 3, \ldots. \tag{34.4}$$

Figure 34.30 (*a*) Interference of light diffracted by two very narrow slits. (*b*) We sum the phasors associated with the electric fields of the coherent light sources at slits 1 and 2.

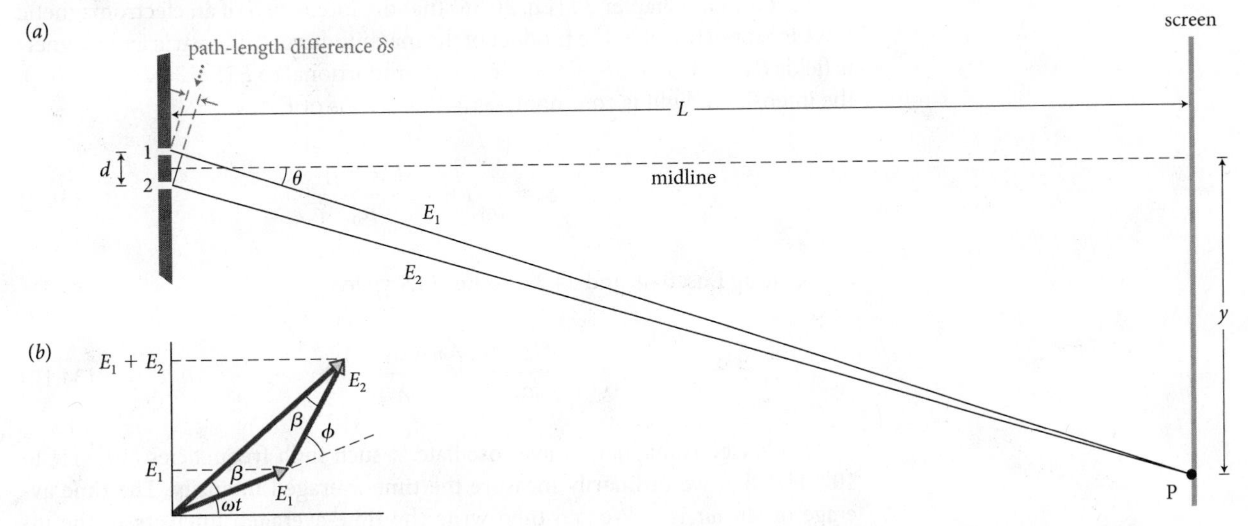

Combining Eqs. 34.3 and 34.4 and solving for $\sin\theta$ determine the angles θ_m for which bright fringes occur:

$$\sin\theta_m = \pm\frac{m\lambda}{d}, \quad \text{for } m = 0, 1, 2, 3, \ldots \quad \text{(bright interference fringes)}, \tag{34.5}$$

as we found in Section 34.2.

A dark fringe—a minimum in the intensity—occurs when the two rays interfere destructively. This is the case when the phase difference equals a odd number times π:

$$\phi_n = \pm(2n - 1)\pi, \quad \text{for } n = 1, 2, 3, \ldots. \tag{34.6}$$

Substituting ϕ into Eq. 34.3 and solving for $\sin\theta$ determine the angles θ_n for which dark fringes occur:

$$\sin\theta_n = \pm\frac{(n - \frac{1}{2})}{d}\lambda, \quad \text{for } n = 1, 2, 3, \ldots \quad \text{(dark interference fringes)}. \tag{34.7}$$

To calculate the light intensity as a function of θ, we start by determining the sum of the electric fields E_1 and E_2 (Eqs. 34.1 and 34.2) using phasors. The amplitude of the sum of two sinusoidal quantities is equal to the vector sum of the phasors representing those two quantities (see Section 32.6). The phasors that represent E_1 and E_2 and their sum phasor are shown in Figure 34.30*b*. For the isosceles triangle formed by the E_1 and E_2 phasors, the exterior angle ϕ is equal to the sum of the two opposite interior angles β, and so $\beta = \frac{1}{2}\phi$. From this, we calculate the amplitude E_{12} of the sum of the two electric fields:

$$E_{12} = 2(E_0 \cos\beta) = 2(E_0 \cos\tfrac{1}{2}\phi). \tag{34.8}$$

Because this combined electric field oscillates at the same frequency as the incident wave, the time-dependent electric field at the screen is $E = E_{12}\sin\omega t$. At the central bright fringe, $\theta = 0$ and the two beams are in phase, so $\phi = 0$. The amplitude of E_{12} at the central bright fringe is therefore twice the amplitude of the incident wave, as we would expect.

We found in Chapter 30 (Eq. 30.36) that the intensity S of an electromagnetic wave is proportional to the product of the magnitudes of the electric and magnetic fields E and B: $S = EB/\mu_0$. Because B is proportional to E (Eq. 30.24, $E = Bc_0$), the intensity of light is commonly written in terms of E^2:

$$S = \frac{1}{\mu_0}EB = \frac{E(E/c_0)}{\mu_0}. \tag{34.9}$$

Substituting Eqs. 34.8 and 34.3 into Eq. 34.9 yields

$$S = \frac{4E_0^2}{\mu_0 c_0}\cos^2\left(\frac{\pi d\sin\theta}{\lambda}\right)\sin^2\omega t. \tag{34.10}$$

Visible electromagnetic waves oscillate at such high frequencies (10^{14} Hz to 10^{15} Hz) that we ordinarily measure the time-averaged intensity. The time average of $\sin^2\omega t$ is $\frac{1}{2}$. We can then write the time-averaged intensity of the interference pattern in terms of the time-averaged intensity of the incident wave:

$S_{0,\text{av}} = E_0^2/(2\mu_0 c_0)$. Thus substituting $\frac{1}{2}$ for $\sin^2 \omega t$ and $S_{0,\text{av}}(2\mu_0 c_0)$ for E_0^2 in Eq. 34.10 gives us

$$S_{\text{av}} = 4S_{0,\text{av}} \cos^2\left(\frac{\pi d \sin \theta}{\lambda}\right). \quad (34.11)$$

The maximum intensity of the interference pattern is *not* just the sum of the intensities of the two interfering waves—it is twice the sum! Because intensity is proportional to the square of the electric field, we must add the electric fields and then calculate the intensity from the square of the combined electric field.

34.15 How can the energy of the closed system made up of the two interfering waves remain constant (as the energy law states it must) if the maximum time-averaged intensity is four times the individual time-averaged intensities of the two waves?

Now let us work out how this pattern looks on the screen. If we consider only small angles θ, we can approximate $\sin \theta \approx \tan \theta = y/L$, where y is the position on the screen corresponding to the angle θ measured from the midline between the slits and L is the distance between the screen and the barrier that contains the slits, as shown in Figure 34.30*a*. (Positive y corresponds to positions above the midline; negative y to positions below the midline.) We can then write the time-averaged intensity as

$$S_{\text{av}} = 4S_{0,\text{av}} \cos^2(\phi/2) \approx 4S_{0,\text{av}} \cos^2\left(\pi d \frac{y}{L\lambda}\right). \quad (34.12)$$

The intensity varies periodically with the phase difference ϕ. Near the central maximum, where θ is small, the intensity also varies periodically with y (**Figure 34.31**).

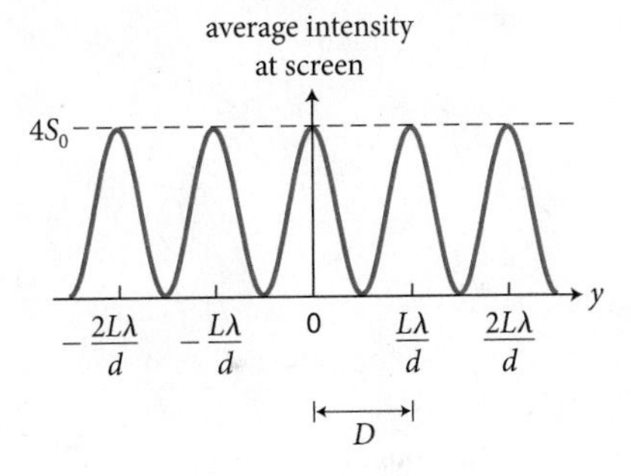

Figure 34.31 Average intensity produced on a screen by two-slit interference.

What is the distance D between adjacent intensity maxima of this pattern? Substituting y/L for $\sin \theta$ in Eq. 34.5, we obtain, for the positions of the maxima corresponding to any two values m and $m + 1$ of our order integer m,

$$m\lambda = d\frac{y_m}{L} \quad \text{and} \quad (m+1)\lambda = d\frac{y_{m+1}}{L}. \quad (34.13)$$

Subtracting the first of these equation from the second yields

$$\lambda = \frac{d}{L}(y_{m+1} - y_m). \quad (34.14)$$

The distance between adjacent maxima is $y_{m+1} - y_m = D$, so

$$D = \frac{L}{d}\lambda. \quad (34.15)$$

This expression tells us that the distance between the maxima is proportional to the wavelength of the light. The interference pattern "magnifies" the wavelength by the factor L/d.

If there are more than two slits in the barrier, the analysis proceeds in the same fashion as before. **Figure 34.32** on the next page shows a wave diffracting through six slits, each separated from its immediate neighbors by distance d. The condition for a maximum in the intensity pattern is the same regardless of the number of slits because if the difference in path length for waves from slits 1 and 2 is λ, then the difference in path length for any pair of adjacent slits is also λ. The principal maxima therefore appear at those locations where

$$d \sin \theta_m = \pm m\lambda, \quad \text{for } m = 0, 1, 2, 3, \ldots \quad \text{(principal maxima)}, \quad (34.16)$$

just as we found for a two-slit barrier (Eq. 34.5).

The principal maxima occur at the same angles regardless of the number of slits. However, as discussed in Checkpoint 34.5, the *intensity* at the maxima increases with the number of slits. As we found in Section 34.2, as the number of

Figure 34.32 Interference of light diffracted by six narrow slits.

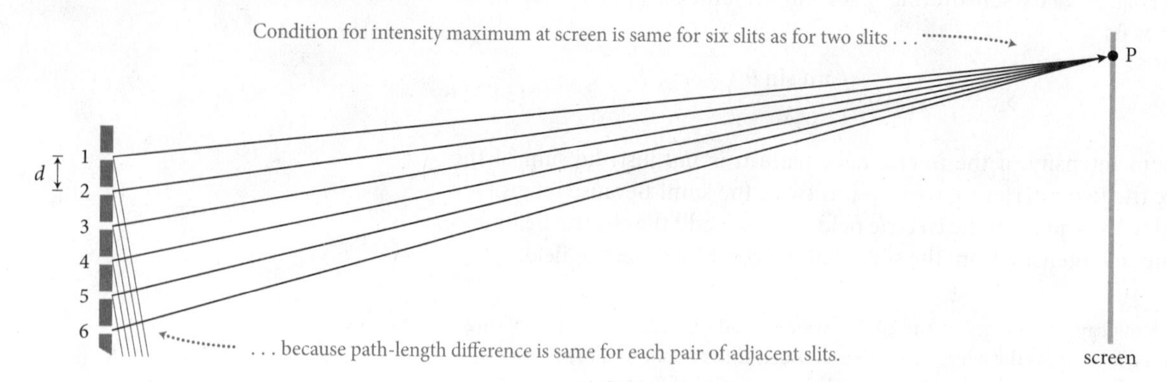

slits increases, the interference fringes also become narrower (Figure 34.33). This is so because the minima closest to the mth principal maximum occur for the ratio k/N that is as close as possible to m—namely, $(mN \pm 1)/N$. As the number of slits N becomes very great, the minima lie very close to m and so the interference pattern has extremely sharp maxima separated by broad dark regions with very faint secondary maxima.

Figure 34.33 Interference pattern produced by gratings with two, three, and eight slits.

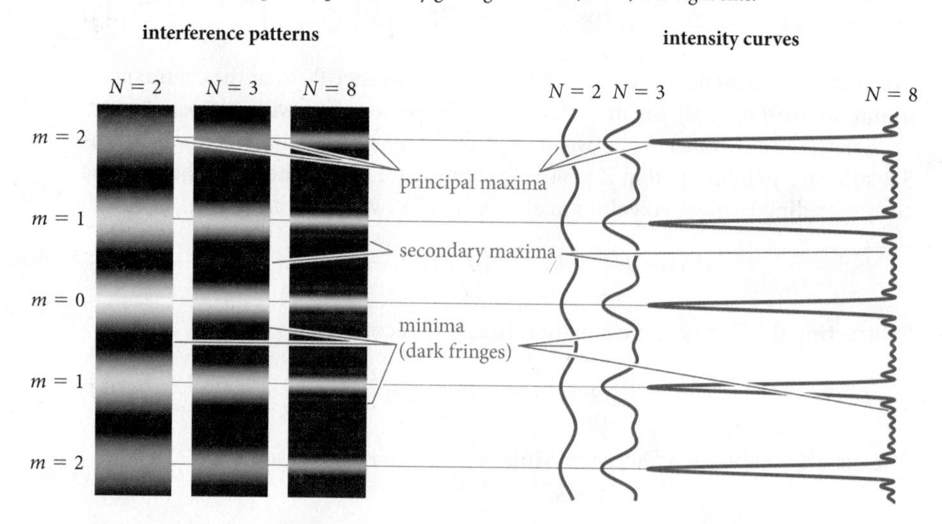

As discussed in Section 34.2, an important use of diffraction gratings is to disperse (separate) light into its constituent wavelengths in order to better understand the light source. The amount of information that can be obtained from a spectrum depends on whether the wavelengths of interest can be distinguished from one another in the spectrum (such wavelengths are said to be *resolved*).

Two wavelengths can be distinguished from each other if the principal maximum of one falls in the first dark region of the other, as shown in Figure 34.34. We found in Section 34.2 that, in general, a minimum occurs for interference through N slits when

$$d \sin \theta_{\text{min}} = \pm \frac{k}{N}\lambda, \quad \text{for integer } k \text{ that is not an integer multiple of } N \quad \text{(dark interference fringes).} \tag{34.17}$$

The angular position of the principal maxima increases with wavelength. Therefore, two wavelengths λ_1 and λ_2 can be distinguished from each other if the principal maximum for the longer wavelength falls at an angle greater than or equal to the angle for the $n = 1$ minimum for the shorter wavelength. As we

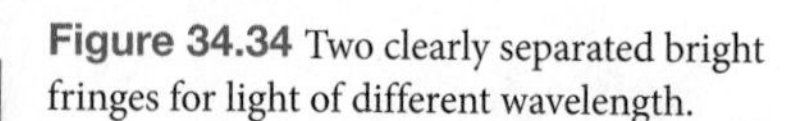
Figure 34.34 Two clearly separated bright fringes for light of different wavelength.

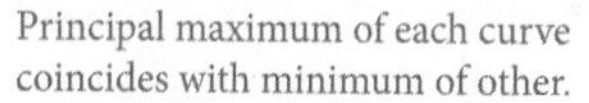

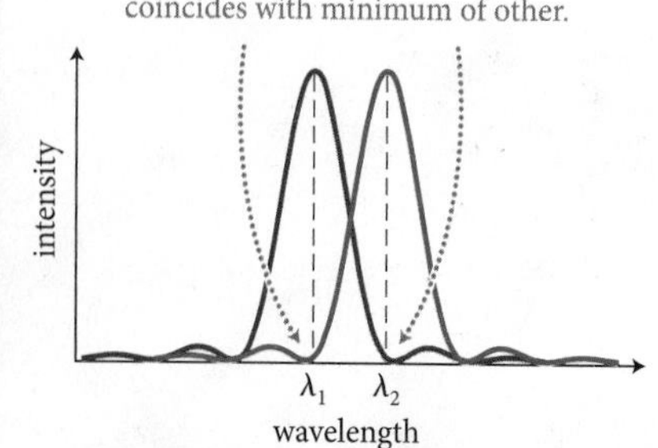

first explored in Checkpoint 34.7, the separation between the slits is critical in determining the smallest wavelength difference that can be distinguished (this wavelength difference is called the *resolution*).

Example 34.6 Resolution of wavelengths in a diffraction grating

An astronomer wishes to determine the relative heights of the intensity peaks for the bright fringes produced by two wavelengths of radiation emitted by sodium atoms. The wavelengths are 589.0 nm and 589.6 nm, and she uses a diffraction grating with 500.0 slits/mm to disperse the light collected by her telescope. (*a*) In which order are the intensity maxima for these two wavelengths farthest apart from each other: $m = 0$, $m = 1$, or $m = 2$? (*b*) If the part of the diffraction grating covered by the light is 4.000 mm wide, are the second-order principal maxima produced by these two spectral lines distinguishable from each other?

❶ **GETTING STARTED** This problem is about the overlapping diffraction patterns produced by two very similar wavelengths of light as the light passes through a diffraction grating. To answer the questions, I need to calculate the angular positions of the principal intensity maxima for each wavelength in three orders of diffraction. I also need to determine the angular position of the minimum adjacent to the second-order principal maximum for each wavelength, in order to determine whether the second-order principal maxima of the two wavelengths are distinguishable.

❷ **DEVISE PLAN** For part *a*, I can use Eq. 34.16 to locate the $m = 0$, 1, and 2 principal maxima for both wavelengths. For part *b*, I can locate the minimum adjacent to the second-order principal maximum for $\lambda = 589.0$ nm by applying the discussion following Eq. 34.17 and then compare that location with the location of the second-order principal maximum for $\lambda = 589.6$ nm.

❸ **EXECUTE PLAN** (*a*) I start by solving Eq. 34.16 for θ:

$$\theta_m = \pm\sin^{-1}\left(\frac{m\lambda}{d}\right). \quad (1)$$

This result indicates that the principal maxima are farthest apart for $m = 2$.

To calculate the positions of the principal maxima, I substitute the appropriate values of m, λ, and d in Eq. 1. The separation d between the slits is $1/(500 \text{ slits/mm}) = 2.000 \times 10^{-3}$ mm $=$ 2000 nm. The $m = 0$ principal maximum occurs at $\theta = 0$ for any wavelength. The first-order principal maxima are located at

$$\theta_{589.0} = \pm\sin^{-1}\left(\frac{589.0 \text{ nm}}{2000 \text{ nm}}\right) = \pm 17.13°$$

$$\theta_{589.6} = \pm\sin^{-1}\left(\frac{589.6 \text{ nm}}{2000 \text{ nm}}\right) = \pm 17.15°,$$

and the second-order principal maxima are located at

$$\theta_{589.0} = \pm\sin^{-1}\left(\frac{2 \times 589.0 \text{ nm}}{2000 \text{ nm}}\right) = \pm 36.09°$$

$$\theta_{589.6} = \pm\sin^{-1}\left(\frac{2 \times 589.6 \text{ nm}}{2000 \text{ nm}}\right) = \pm 36.13°.$$

For these two wavelengths, the second-order principal maxima are the ones farthest apart from each other. ✔

(*b*) Using the information given in the discussion following Eq. 34.17, I can write that the condition for the minimum adjacent to the second-order principal maximum for $\lambda = 589.0$ nm is

$$d \sin\theta_{\text{min},589.0} = \pm\lambda\frac{mN + 1}{N},$$

where I have replaced k in Eq. 34.17 by $mN + 1$ as explained in the discussion preceding that equation. Solving this expression for $\theta_{\text{min},589.0}$ gives

$$\theta_{\text{min},589.0} = \pm\sin^{-1}\left(\frac{\lambda}{d}\frac{mN + 1}{N}\right).$$

A region of the grating 4.000 mm wide contains 2000 slits. Substituting in the preceding expression, I get that, for $\lambda = 589.0$ nm, the minimum adjacent to the second-order principal maximum is at

$$\theta_{\text{min},589.0} = \pm\sin^{-1}\left(\frac{589.0 \text{ nm}}{2000 \text{ nm}} \times \frac{4001}{2000}\right) = \pm 36.10°. ✔$$

This minimum lies between 36.09°, the second-order principal maximum for $\lambda = 589.0$ nm, and 36.13°, the second-order principal maximum for $\lambda = 589.6$ nm, telling me that the second-order principal maxima for these two wavelengths are distinguishable from each other.

❹ **EVALUATE RESULT** Equation 34.16 tells me that the angles at which principal maxima occur are equal to the inverse sine of integer multiples of λ/d. For small angles $\sin\theta_m \approx \theta_m$, and so the angles θ_m are approximately equally spaced (**Figure 34.35**). As the curve for $\sin\theta$ bends toward the horizontal, however, the distance between adjacent θ_m values increases, in agreement with the result I obtained.

Figure 34.35

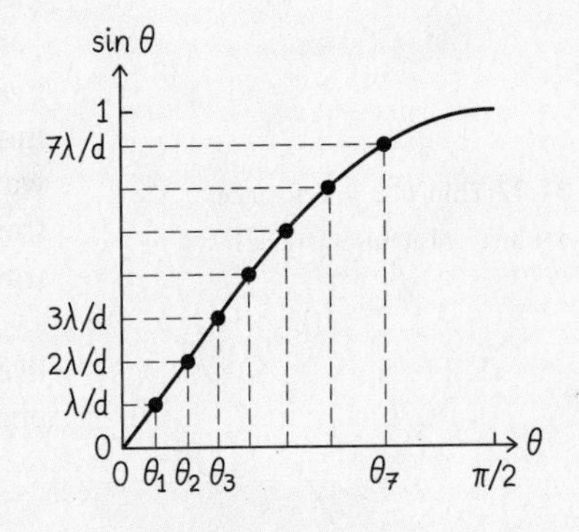

34.16 As the above discussion indicates, the separation distance between the principal maxima for different wavelengths increases as the fringe order m increases. Can you obtain an arbitrarily great separation distance by going to extremely high orders?

Figure 34.36 Soap bubble.

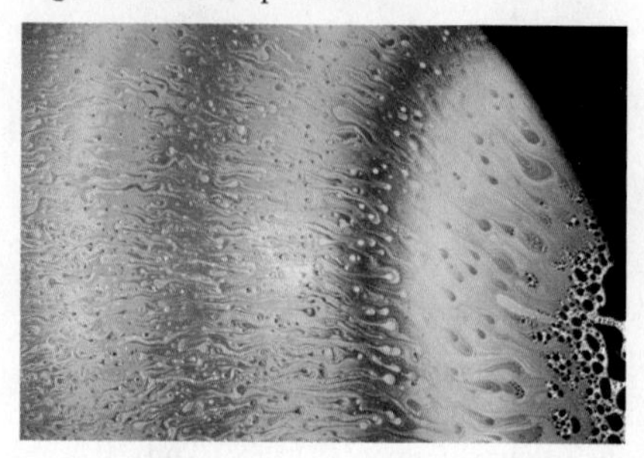

34.7 Thin-film interference

A familiar manifestation of the interference of light is the rainbow of colors reflected from thin films such as soap bubbles (**Figure 34.36**). This type of interference occurs in transparent materials whose thickness is comparable to the wavelengths of visible light. When such a thin material (we refer to it as a *film*) is either suspended in air, as in a soap bubble, or supported on a much thicker material with a different index of refraction, as in an oil slick on a puddle of water, white light reflecting from both the front and back surfaces of the film interferes, in the same manner as x rays reflecting from adjacent layers of atoms in a crystal. This is shown schematically in **Figure 34.37**. The film thickness t and index of refraction n_b, together with the angle of incidence, determine the path-length difference and corresponding phase difference between the reflected beams, and therefore determine which colors undergo constructive interference and which undergo destructive interference in any given direction.

To identify the conditions for constructive and destructive interference, we begin by expressing the electric fields of the waves reflected from the two surfaces, just as we did for waves passing through two slits in the preceding section:

$$E_1 = E_0 \sin(\omega t) \tag{34.18}$$

$$E_2 = E_0 \sin(\omega t + \phi). \tag{34.19}$$

The phase difference ϕ is due to the path-length difference $2\Delta s$ and the effect of the reflections on the phases of each wave. The path-length difference gives rise to a phase difference

$$\phi_{\text{path}} = \frac{2\pi\Delta s}{\lambda_b} = \frac{2\pi(2t\cos\theta_b)}{\lambda/n_b} = \frac{4\pi n_b t\cos\theta_b}{\lambda}, \tag{34.20}$$

where the path-length difference Δs and the angle θ_b of the ray relative to the surface normal inside the film are shown in Figure 34.37, and $\lambda_b = \lambda/n_b$ is the wavelength of light inside the film (see Eq. 33.3). Because of refraction, expressing Δs in terms of the angle of incidence θ, instead of the angle in the film θ_b, involves Snel's law and produces a rather complicated result. However, for normal incidence $\cos\theta_b = 1$, and so the expression is greatly simplified.

To determine the phase difference due to the reflections from the two film surfaces, recall the discussion of mechanical waves at boundaries from Section 16.4. We saw there that when a wave pulse is launched in a first medium and travels into a second medium, the pulse is partially reflected and partially transmitted at the boundary. The reflected pulse is inverted relative to the incident pulse if the wave speed c_1 in the first medium is greater than the speed c_2 in the second medium. When $c_1 < c_2$, the incident pulse is not inverted upon reflection.

Let us now extend these ideas to sinusoidal electromagnetic waves. When the incident wave in medium 1 reflects at the boundary with medium 2 of greater index of refraction ($n_1 < n_2$), the speed is less in medium 2 ($c_2 < c_1$), and so the wave is inverted upon reflection. This inverting of the wave is equivalent to the wave undergoing a phase shift of π upon reflection. If instead $n_1 > n_2$, the wave is not inverted and there is no phase shift upon reflection.

A phase shift of π can occur at each of the two interfaces, depending on the refractive indices n_a, n_b, and n_c of the three media, as indicated in Figure 34.37. Thus the phase difference associated just with the reflections for the two reflected waves is

$$\phi_r = \phi_{r2} - \phi_{r1}, \tag{34.21}$$

where ϕ_{r1} and ϕ_{r2} are either π or 0, depending on the values of the indices of refraction. Therefore the total phase difference between the two reflected waves is

$$\phi = \frac{4\pi n_b t\cos\theta_b}{\lambda} + \phi_{r2} - \phi_{r1}. \tag{34.22}$$

Figure 34.37 Thin-film interference.

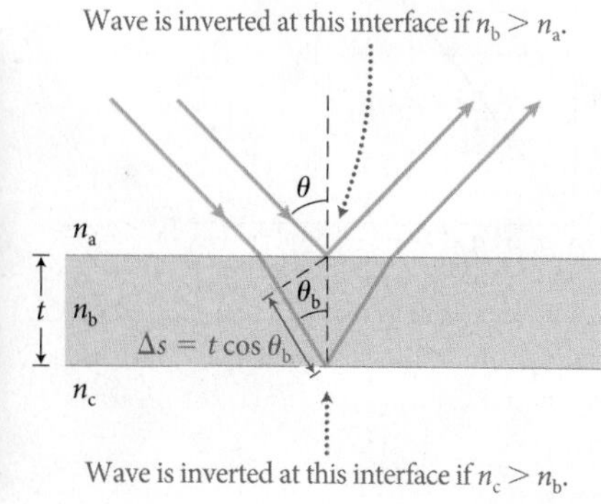
Wave is inverted at this interface if $n_b > n_a$.
θ
n_a
θ_b
t
n_b
$\Delta s = t\cos\theta_b$
n_c
Wave is inverted at this interface if $n_c > n_b$.

QUANTITATIVE TOOLS

Because ϕ_{r1} and ϕ_{r2} must each be π or 0, the effect of the two reflections on the phase cancels out when $\phi_{r1} = \phi_{r2}$, and the phase difference is due entirely to ϕ_{path}. If ϕ_{r1} is not equal to ϕ_{r2}, then ϕ differs from ϕ_{path} by π, causing the constructive and destructive interference conditions to be switched from what they would be due to just ϕ_{path}.

In our discussion of x-ray diffraction, the light source consisted of monochromatic x rays, and we found that constructive interference between adjacent planes takes place at only certain angles. In the current context—light reflections from thin films—we usually deal with light across the visible spectrum. Therefore ϕ depends on both wavelength and angle of incidence. For normal incidence, Eq. 34.22 simplifies to

$$\phi = \frac{4\pi n_b t}{\lambda} + \phi_{r2} - \phi_{r2} \quad \text{(normal incidence)}. \qquad (34.23)$$

Then as before, constructive interference occurs when the phase difference corresponds to an integer number times 2π, and destructive when it corresponds to an odd number times π.

Example 34.7 Antireflective coating

Eyeglass lenses made of crown glass ($n = 1.52$) are given a thin coating of magnesium fluoride ($n = 1.38$) to minimize reflection of light from the lens surface. What is the minimum coating thickness for which reflection from the lens surface is minimized?

❶ GETTING STARTED Reflection is minimized at all values of the coating thickness that cause the waves reflected from the front coating surface to interfere destructively with those reflected from the back coating surface. My task therefore is to obtain the minimum thickness needed for destructive interference.

❷ DEVISE PLAN I see from Eq. 34.22 that the condition for destructive interference depends not only on the coating thickness t and index of refraction n_b but also on two variables I have no values for: the angle of incidence and the wavelength of the light. Thus I must make some simplifying assumptions to solve this problem. As in the text discussion, I shall consider only light normally incident on the coated lens and shall assume the lens and coating surfaces are flat, as shown in my sketch (**Figure 34.38**). I choose a representative visible wavelength, 500 nm—in the middle of the visible spectral range.

Figure 34.38

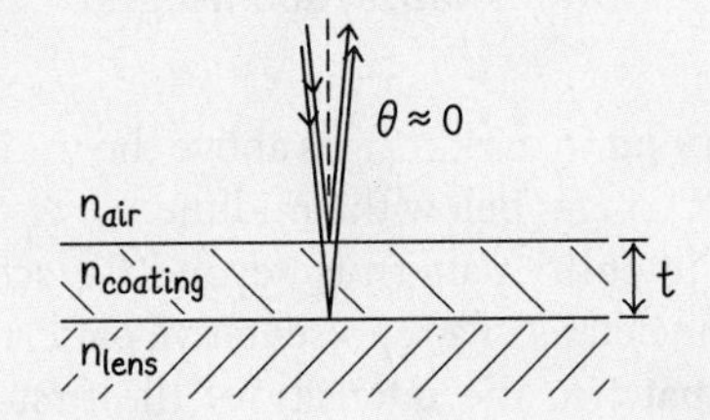

To obtain the condition for destructive interference, I can use Eq. 34.23 to determine the phase difference ϕ at normal incidence and equate it to an odd number times π radians. To work out the phase shifts ϕ_{r1} and ϕ_{r2} that occur at each reflection, I note that at the air-coating interface the light reflects from a medium for which the index of refraction is greater than that of the medium through which the incident wave travels, and so this wave undergoes a π phase shift upon reflection: $\phi_{r1} = \pi$. The wave that reflects from the coating-lens interface likewise reflects from a medium for which the index of refraction is greater than that of the medium through which the incident wave travels and thus also undergoes a π phase shift ($\phi_{r2} = \pi$). Consequently, ϕ_{r1} and ϕ_{r2} in Eq. 34.23 cancel, leaving only the phase difference due to the path-length difference. The minimum thickness produces a phase difference corresponding to the smallest number of cycles that gives destructive interference—namely, half a cycle.

❸ EXECUTE PLAN To determine the thickness that produces destructive interference for 500 nm light, I express the phase difference using Eq. 34.23 and equate it to a half-cycle phase difference:

$$\frac{4\pi n_b t}{\lambda} + \phi_{r2} - \phi_{r2} = \pi.$$

Substituting $\phi_{r1} = \phi_{r2} = \pi$, solving for the thickness t, and substituting values give

$$t = \frac{\lambda}{4n_b} = \frac{500\ \text{nm}}{4(1.38)} = 90.6\ \text{nm}. ✔$$

❹ EVALUATE RESULT The coating is thinner than the wavelength of the light traveling through the coating. That makes sense because, to create destructive interference, the wave that travels through the coating and reflects from the coating-lens interface must travel only half a wavelength farther than the wave that reflects from the air-coating interface.

34.17 When oil spreads on water, bands of different colors are visible. What causes the different colors?

Figure 34.39 Diffraction through a narrow slit.

(*a*) Wavefront in slit treated as a series of point sources

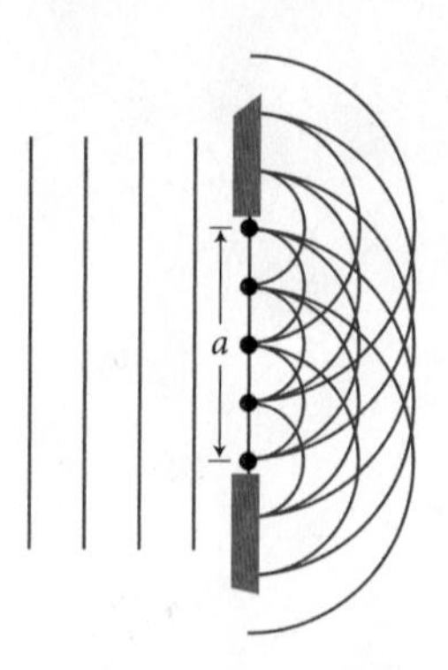

(*b*) Condition for first-order dark fringe: path lengths differ by $(\frac{1}{2}a)\sin\theta_1 = \lambda/2$

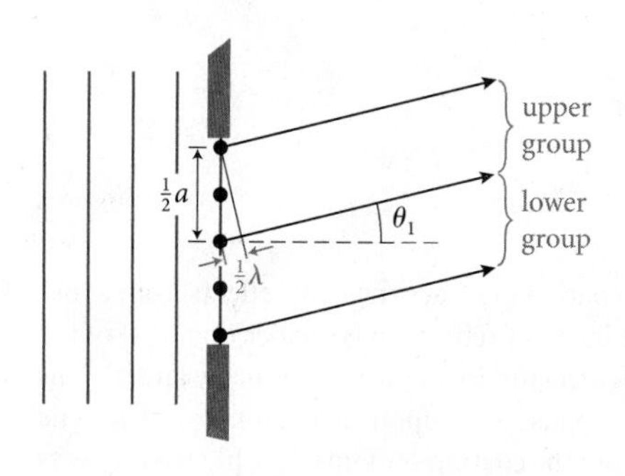

(*c*) Condition for second-order dark fringe: path lengths differ by $(\frac{1}{4}a)\sin\theta_2 = \lambda/2$

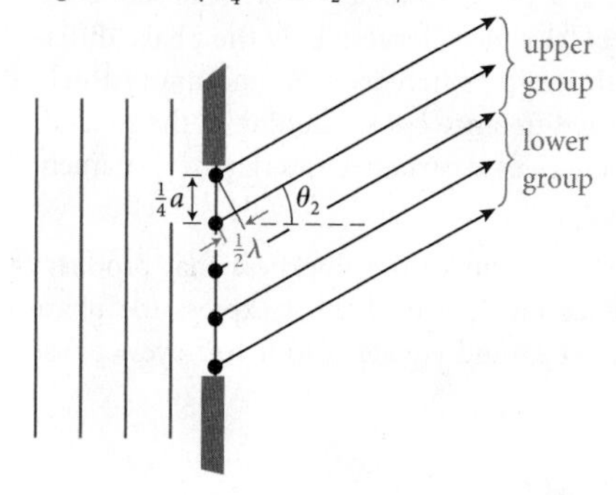

34.8 Diffraction at a single-slit barrier

Let us now describe quantitatively the diffraction by a single slit. Huygens' principle (see Section 17.4) allows us to describe the wavefront that reaches the opening as a series of point sources that emit spherical wavelets (**Figure 34.39*a***). The wave beyond the slit is the superposition of all these spherical wavelets. In the original direction of propagation, all of the waves add in phase, and as a result the amplitude of the transmitted wave is the maximum possible: the sum of all the individual wave amplitudes. If we divide the rays representing the wavelets into two equal groups, as shown in Figure 34.39*b*, we can pair each ray in the upper half with a corresponding ray in the lower half. In Figure 34.39*b*, for example, we can pair the top ray in the upper group with the top ray in the bottom group. As we discussed in Section 34.2, these rays are essentially parallel when they emerge from the slits; as they travel and intersect at some location on a screen placed far to the right of the slits, the rays must travel different distances to reach that location. For rays traveling at an angle θ to the original propagation direction, the difference in path length from two such corresponding rays is $(a/2)\sin\theta$, where a is the width of the aperture. When the path lengths differ by half a wavelength, $(a/2)\sin\theta_1 = \lambda/2$, the rays interfere destructively, which means that the direction for the first-order dark fringe in a single-slit diffraction pattern is given by

$$\sin\theta_1 = \frac{\lambda}{a} \quad \text{(first-order dark diffraction fringe).} \qquad (34.24)$$

Dividing the rays into four groups (Figure 34.39*c*) leads to a dark fringe (that is, a minimum in transmitted intensity) when $(a/4)\sin\theta = \frac{1}{2}\lambda$. Thus the direction for the second-order dark fringe is given by

$$\sin\theta_2 = 2\frac{\lambda}{a} \quad \text{(second-order dark diffraction fringe).} \qquad (34.25)$$

The general condition for a dark fringe is thus

$$\sin\theta_n = \pm n\frac{\lambda}{a}, \quad n = 1, 2, 3, \ldots \quad \text{(dark diffraction fringes).} \qquad (34.26)$$

Figure 34.40 Intensity plot for a single-slit diffraction pattern.

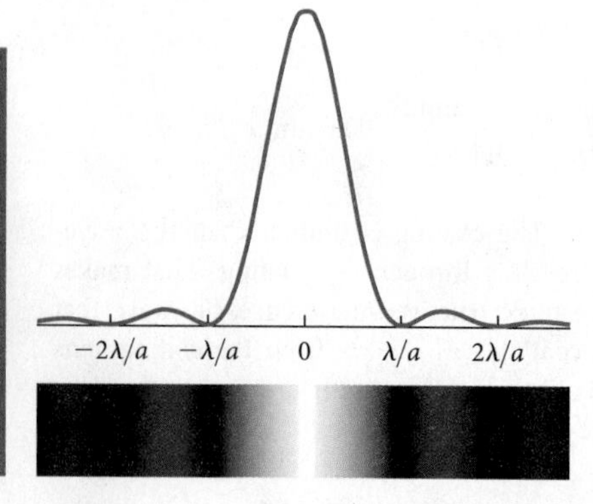

Positive values of $\sin\theta_n$ correspond to dark fringes above the midline, while negative values correspond to dark fringes below the midline.

Calculation of the detailed intensity pattern is beyond the scope of this text, but we can examine some of the details of a representative pattern (**Figure 34.40**). In this single-slit diffraction pattern, the intensity of the first-order ($m = 1$) bright fringe is less than 5% of the intensity of the central ($m = 0$) bright fringe; the intensity of the second-order ($m = 2$) bright fringe is less than 2% of the central bright fringe intensity. In other words, most of the transmitted energy falls within the central peak. In general, in a single-slit diffraction pattern, the intensity is greatest at the central bright fringe and decreases rapidly with distance from the center of the pattern.

What if we want to calculate the linear positions of the dark diffraction fringes on a screen located a distance L away from a barrier containing a single slit (**Figure 34.41**) rather than the angular positions? For dark fringes located at small angles θ_n from the original direction of wave propagation, we can approximate $\tan\theta_n$ as $\sin\theta_n$:

$$y_n = L\tan\theta_n \approx L\sin\theta_n, \quad (34.27)$$

and so, from Eq. 34.26,

$$y_n = \pm n\frac{\lambda L}{a} \text{ (dark diffraction fringes).} \quad (34.28)$$

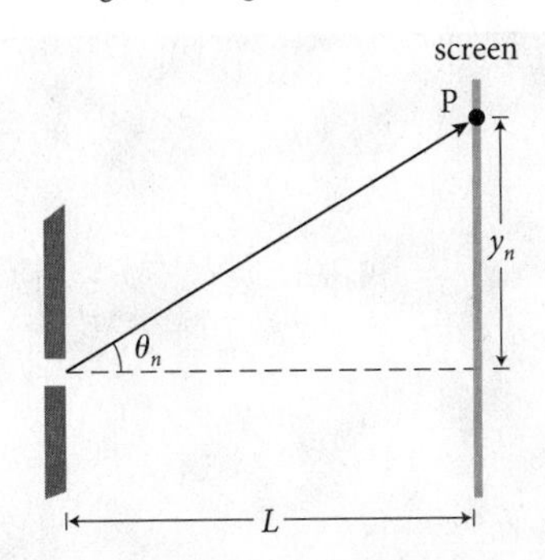

Figure 34.41 Calculating the positions of the dark fringes of a single-slit diffraction pattern.

Example 34.8 Spreading light

Consider the diffraction pattern shown actual size in Figure 34.40. If the pattern was formed by light from a 623-nm (red) laser passing through a single narrow slit and the screen on which the pattern was cast was 1.0 m away from the slit, what is the slit width?

❶ **GETTING STARTED** This problem asks me to relate the diffraction pattern in Figure 34.40, which was produced by a setup such as the one shown in Figure 34.41, to the width of the slit that produced it. Thus I need to relate the slit width to fringes whose position I can calculate from the parameters of the problem and can also measure on the image. Because the only variable I know how to calculate is the positions of minima in the diffraction pattern, I can measure the distance between the two first-order minima and relate that distance to the slit width and the geometry of the setup.

❷ **DEVISE PLAN** The positions of the two first-order minima, in terms of wavelength λ, slit-to-screen distance L, and slit width a, are given by Eq. 34.28 with $n = 1$. Subtracting the two values I obtain from each other gives me an expression for the distance between these two minima in terms of λ, L, and a. I am given the values of λ and L, and my task is to determine a. Thus if I know the distance between the n_1 minima, I can calculate a. Because the image in Figure 34.40 is actual size, I can measure this distance directly. Then I can solve Eq. 34.28 for a and insert my known values.

❸ **EXECUTE PLAN** Substituting $n = 1$ into Eq. 34.28 gives the linear positions of the two first-order minima:

$$y_1 = \pm\frac{\lambda L}{a},$$

so the distance w between the two minima is

$$w = \frac{2\lambda L}{a}. \quad (1)$$

I measure the distance between the centers of the two dark fringes on either side of the central bright fringe in Figure 34.40 to be 23 mm. Solving Eq. 1 for the slit width thus gives me

$$a = \frac{2\lambda L}{w} = \frac{2(623 \times 10^{-9}\text{ m})(1.0\text{ m})}{23 \times 10^{-3}\text{ m}}$$

$$= 5.4 \times 10^{-5}\text{ m} = 0.054\text{ mm} \checkmark$$

❹ **EVALUATE RESULT** The slit width is about a factor of 100 greater than the wavelength of the light, and that ratio is consistent with the general range of slit sizes that produce noticeable diffraction of visible light.

As long as λ is small relative to the slit width a, so that the small-angle approximation for θ_n is valid, most of the diffracted light intensity falls within this region defined by Eq. 1 in Example 34.8. Note that w increases with decreasing a: The narrower the slit, the more the wave spreads out after passing through the slit (**Figure 34.42**). If the slit width is equal to or less than the wavelength of the light, there are no dark fringes, as we found in Checkpoint 34.16. The wave simply spreads out in all directions behind the slit, which means the slit behaves as a point source.

Figure 34.42 The width of the central maximum in a diffraction pattern decreases as the slit is widened.

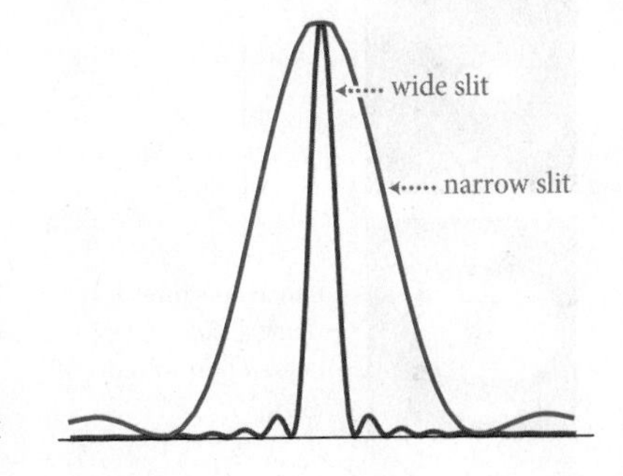

34.18 Using Eq. 34.24, calculate the angle at which the first dark fringe occurs when (*a*) $a < \lambda$ and (*b*) $a \gg \lambda$. Interpret your results.

34.9 Circular apertures and limits of resolution

When light passes through a circular aperture, the symmetry of the aperture causes the resulting diffraction pattern also to be circular (**Figure 34.43** on the next page). A circular central bright fringe is surrounded by circular dark diffraction fringes and additional diffraction bright fringes. The central bright

Figure 34.43 Diffraction pattern of light passing through a circular aperture.

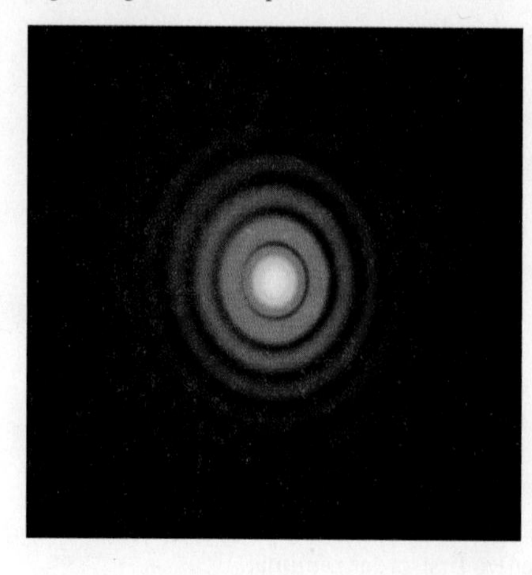

fringe is called the *Airy disk*, after the British astronomer and mathematician Sir George Airy, who developed the first detailed description of diffraction in 1835. Calculation of the circular diffraction pattern is rather involved, so all I shall do here is state the location of the first dark fringe. When light is diffracted by a circular aperture of diameter d, the first dark fringe occurs at angle θ_1 given by

$$\sin \theta_1 = 1.22 \frac{\lambda}{d}. \quad (34.29)$$

The result is similar to the one obtained in Eq. 34.24 for a slit of width a, except that now the sine of the angle is increased by a factor of 1.22. The increase in the angle for the same wavelength and aperture size can qualitatively be understood as follows: Equation 34.24 is obtained by considering the interference between wavelets coming from a slit whose width is a everywhere. A circular aperture has width d only across a diameter; the rest of the aperture is narrower. As the aperture gets narrower, diffraction through it becomes more pronounced. The factor of 1.22 quantitatively accounts for the varying horizontal width of the circular aperture.

The angular size of the Airy disk given in Eq. 34.29 determines the minimum angular separation of two point sources that can be distinguished by observing them with a (circular) lens—*regardless* of the magnification of the lens! To understand what this means, imagine imaging two distant, closely spaced point sources with a lens. These sources could be anything, from stars to organelles in a biological cell. The sources are not coherent, and so we can consider the Airy disks formed by the light from each source separately without considering interference between sources.

If there is overlap in the Airy disks of the images observed through the lens, it is difficult to tell whether there are two point sources or just one. Two objects being observed through a lens are just barely distinguishable when the center of one diffraction pattern is located at the first minimum of the other diffraction pattern. This happens when the angular separation between the two objects is at least the angle given in Eq. 34.29. If this is the case, we say that the two objects are *resolved*. This condition for distinguishability is called **Rayleigh's criterion.**

Figure 34.44 The resolution of these two stars improves as the aperture is made larger.

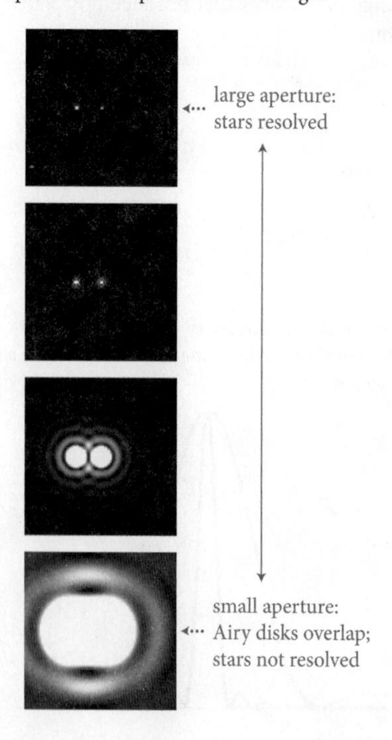

Because the diameter d of the lens is always much greater than the wavelength of the light, the angle in Eq. 34.29 is always small. Thus the minimum angular separation θ_r for which two sources can be resolved is approximately equal to the sine of the angle

$$\theta_r \approx \sin \theta_r = 1.22 \frac{\lambda}{d}. \quad (34.30)$$

Two objects that are separated by an angle equal to or greater than θ_r satisfy Rayleigh's criterion. For this reason, the closest two objects can be to each other and still be distinguished with an optical instrument such as a microscope or telescope depends not on the magnification but on the wavelength of the light and the size of the smallest aperture in the instrument.

Figure 34.44 shows the images of two stars obtained with a telescope. An aperture placed in front of the lens shows the effects of diffraction. When the opening of the aperture is small (bottom image in **Figure 34.44**), the images of the two stars are merged—the two stars cannot be resolved. As the aperture is opened, θ_r decreases and the two images separate cleanly.

Diffraction also determines the linear size of the images of point sources. In Chapter 33, we stated that the image of a point source formed on a screen by a lens is a point. **Figure 34.45a** shows parallel rays from a distant point source focused by a lens onto a screen placed in the focal plane of the lens. Without diffraction, the image formed by these rays would be an infinitesimally small point. In fact, because the aperture through which the light passes—the lens—has a

Figure 34.45 Analyzing the diffraction limit of a lens.

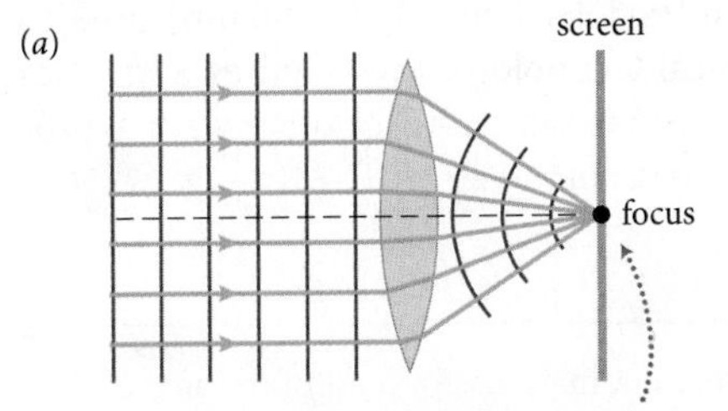

In absence of diffraction, lens would cause parallel rays to converge to point.

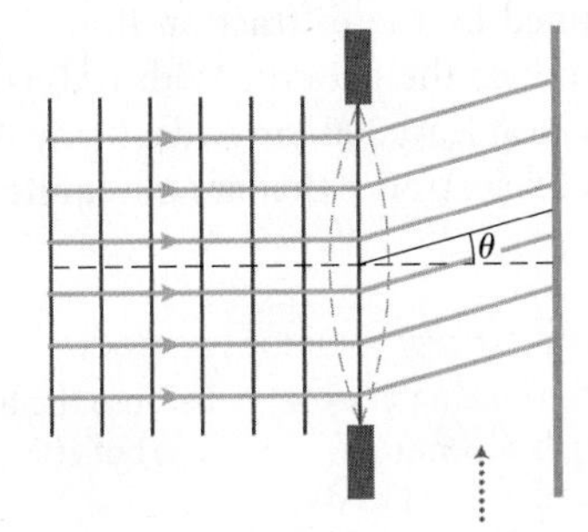

But lens is an aperture, so light is also diffracted.

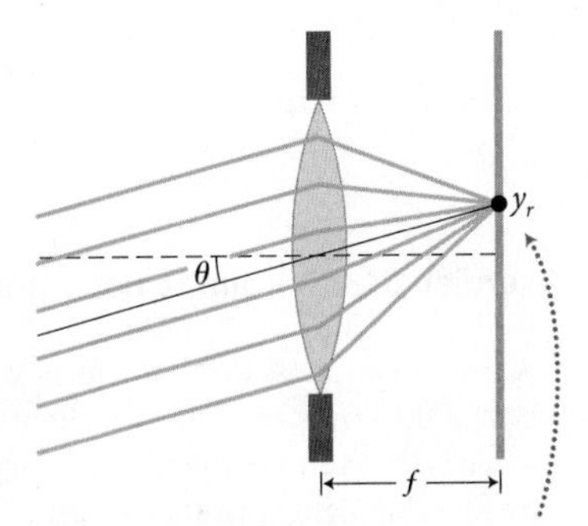

Focusing of diffracted light. Angular size of Airy disk depends on both diffraction and focusing.

finite diameter, such rays do not focus to an infinitely small point, and the image formed is a diffraction pattern just like that shown in Figure 34.43. The angular size of the central bright fringe of this diffraction pattern—in other words, the Airy disk—is given by Eq. 34.29.

To calculate the radius of the Airy disk formed in the focal plane of the lens,* we must account for both diffraction and focusing. Figure 34.45*b* shows the diffraction of light through an aperture of the same diameter as the lens. Light that originally traveled parallel to the lens axis is diffracted by an angle θ. Now consider how the lens focuses the diffracted light. Parallel rays that make an angle θ with the lens axis are focused at a point located a distance y above the axis (Figure 34.45*c*, see also Figure 33.27). This distance y is given by $y = f \tan\theta$, where f is the focal length of the lens.

Our next step in determining the Airy disk radius is to calculate the distance y_r from the center of the disk to the first dark fringe in the diffraction pattern, which is found at the angle given by Eq. 34.29. In the small-angle limit, $\sin\theta \approx \tan\theta$ and so we can substitute $\sin\theta_r = y_r/f$ into Eq. 34.29, giving

$$y_r = 1.22\frac{\lambda f}{d}. \qquad (34.31)$$

This expression gives the radius y_r of the Airy disk and the minimum size of the area to which light can be focused with light of wavelength λ by a lens of focal length f and diameter d. The best ratio of f/d that can be achieved with a lens is approximately unity, and so the smallest diameter to which light can be focused is about 2.5λ. This means that the smallest diameter of "points" in the resulting image is also 2.5λ.

This diffraction-determined minimum size of the features in an image is commonly called the *diffraction limit.* An industry in which the diffraction limit poses a serious problem is the manufacture of integrated circuits, such as computer chips. The transistors and logic gates described in Section 32.4 are produced by a series of processes known as *photolithography,* in which the semiconductor substrate is coated with a polymer that is sensitive to ultraviolet light. The polymer is then exposed to ultraviolet light in the pattern of the desired metal electrodes. This pattern of light is produced by illuminating a metal mask with holes in the shape of the electrodes and then imaging the resulting pattern of light onto the surface. Finally, the exposed polymer is dissolved with a chemical rinse, leaving the semiconductor surface exposed where metal is desired. The metal is then deposited in a subsequent step.

*The radius of the Airy disk is smallest when the screen is in the focal plane of the lens, and it increases as the screen is moved closer to or farther from the lens.

The smallest electrode that can be made by this process is therefore determined by the diffraction limit for the ultraviolet light ($\lambda \leq 150$ nm) used to produce the pattern. With ordinary optical technology, this requires electrodes to be at least 300 nm wide. Many researchers are searching for other ways to produce electrodes that are not limited by diffraction.

Exercise 34.9 A point is not a point

A magnifying glass that has a focal length of 0.25 m and a diameter of 0.10 m is used to focus light of wavelength 623 nm. (*a*) What is the radius of the smallest Airy disk that can be produced by focusing light with this lens? (*b*) How large is the Airy disk formed when this lens focuses a laser beam that has a 2.0-mm diameter?

SOLUTION (*a*) The radius of the smallest Airy disk is given by Eq. 34.31 with d equal to the diameter of the lens:

$$y = 1.22\frac{(623 \times 10^{-9}\ \text{m})(0.25\ \text{m})}{0.10\ \text{m}}$$

$$= 1.9 \times 10^{-6}\ \text{m} = 1.9\ \mu\text{m}. ✔$$

The minimum spot size is about three times greater than the wavelength of the light.

(*b*) In this case the radius is given by Eq. 34.31 with d equal to the diameter of the laser beam. I must use the beam diameter because the beam does not make use of most of the area of the lens but effectively defines its own aperture. I therefore have

$$y = 1.22\frac{(623 \times 10^{-9}\ \text{m})(0.25\ \text{m})}{0.0020\ \text{m}}$$

$$= 9.5 \times 10^{-5}\ \text{m} = 95\ \mu\text{m}. ✔$$

This Airy disk is much bigger than the one found in part *a*. This Airy disk radius means the central bright fringe has a diameter of 190 μm = 0.19 mm, which is smaller than the original beam diameter due to the focusing of the lens. Because the laser beam is much smaller than the diameter of the lens, it cannot be effectively focused by this lens. The diffraction partially cancels the focusing.

Example 34.10 Blurry images

The widths of one pixel in the sensor for a digital camera is about 2.0 μm. If the camera lens has a diameter of 40 mm and a focal length of 30 mm, is the resolution of the resulting image limited by the lens or the sensor?

❶ GETTING STARTED This problem involves comparing, for the image formed by a digital camera, the resolution limit due to diffraction and the limit due to the size of the pixels in the sensor. The limit due to diffraction is the size of the image of a point source; the limit due to the sensor is the width of a single pixel. The greater limit determines the image resolution. The problem does not specify the wavelength of the light involved, but because the problem is concerned with forming images with visible light, I choose $\lambda = 500$ nm, near the center of the visible spectrum.

❷ DEVISE PLAN The diffraction-limited image of a point source is the Airy disk at the center of the diffraction pattern, so I can use Eq. 34.31 to obtain the radius of the Airy disk formed by the camera. I can then compare the diameter (not the radius) of that disk with the width of a pixel. Whichever is greater limits the resolution possible for the image.

❸ EXECUTE PLAN Substituting the values given into Eq. 34.31, I obtain for the Airy disk radius

$$y = 1.22\frac{\lambda f}{d} = 1.22\frac{(0.500 \times 10^{-6}\ \text{m})(30 \times 10^{-3}\ \text{m})}{40 \times 10^{-3}\ \text{m}}$$

$$= 0.46 \times 10^{-6}\ \text{m} = 0.46\ \mu\text{m}.$$

The diameter of the Airy disk is thus $2y = 0.92\ \mu$m. This is significantly less than the pixel width, which means the resolution of the image is limited by pixel width rather than by diffraction. ✔

❹ EVALUATE RESULT My result for the radius of the Airy disk is reasonable because typically the diffraction-limited width of the image of a point source is comparable to the wavelength of light emitted by the source. (Note that if I had chosen $\lambda = 700$ nm, the longest visible wavelength, the Airy disk radius would increase only by a factor of $700/500 = 1.4$ and thus its diameter, 1.3 μm, would still not exceed the pixel width. If I had chosen a wavelength shorter than 500 nm, the Airy disk diameter would be less than my calculated value. So my conclusion is the same for any visible wavelength: The resolution is limited by the pixel width.).

34.19 Which of these three lenses offers (*a*) the highest resolution and (*b*) the lowest resolution: (*i*) $f = 10$ mm, $d = 8$ mm; (*ii*) $f = 15$ mm, $d = 10$ mm, (*iii*) $f = 20$ mm, $d = 18$ mm?

Figure 34.46 The photoelectric effect.

(*a*)

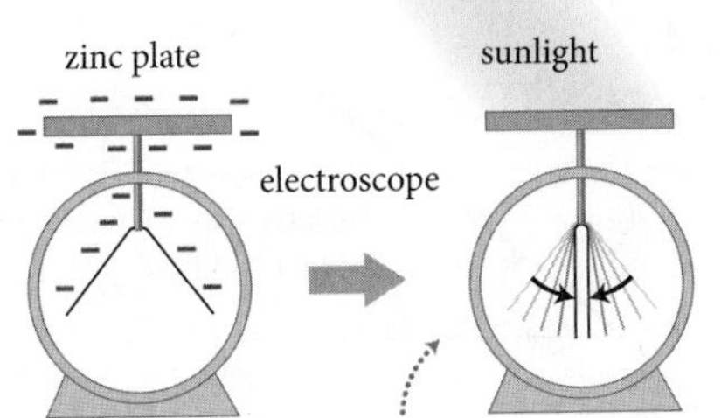

Sunlight discharges negatively charged zinc and electroscope.

(*b*)

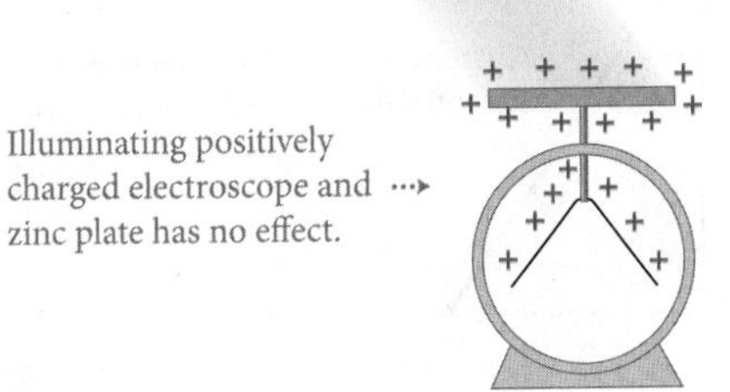

(*c*)

glass (blocks UV)

Passing light through glass slide prevents discharging of negatively charged electroscope and zinc plate.

34.10 Photon energy and momentum

In Chapter 30 we described light as an electromagnetic wave that has a wavelength λ and a frequency f and moves in vacuum at speed c_0, such that

$$c_0 = \lambda f. \tag{34.32}$$

We also found that the energy density in the electromagnetic wave is proportional to the square of the amplitude of the electric field oscillation. In addition, the experiment described in Section 34.4 suggests that light has particle properties. If light always propagates at speed c_0 in vacuum, what determines its energy? And if light is a particle that has energy, shouldn't it also have momentum?

The answer to the first question is provided by the **photoelectric effect,** a surprising phenomenon that cannot be explained by thinking of light as a wave (**Figure 34.46**). Place a piece of metal, such as zinc, on an electroscope (see Section 22.3) that is negatively charged, as shown in Figure 34.46*a*. Some of the charge immediately moves to the zinc so that it, too, is negatively charged. If you then shine sunlight on the metal, the light discharges the zinc and the electroscope. If the electroscope is positively charged, however, as in Figure 34.46*b*, nothing happens when light shines on it. If we place a piece of glass in the beam of light (Figure 34.46*c*), nothing happens even if the zinc plate is negatively charged and the light is very intense.

What is going on? In the situations of Figure 34.46*a* and *b*, the light knocks electrons out of the zinc plate. When the plate is initially negatively charged, like-charge repulsion causes the ejected electrons to accelerate away from the plate. When the plate is initially positively charged, opposite-charge attraction causes the ejected electrons to be attracted back to the plate, so that the charge on the plate does not change. Ultraviolet radiation cannot pass through ordinary glass, and thus we conclude from the situation in Figure 34.46*c* that ultraviolet light is essential in order for electrons to be ejected.

The apparatus illustrated in **Figure 34.47** is used to study the photoelectric effect. It allows us to measure the energy of the ejected electrons while separately controlling either the wavelength or the intensity of the light. A zinc target T is placed in an evacuated quartz bulb (quartz is transparent to ultraviolet light), along with another metal electrode called the collector (C). A power supply is used to maintain a constant potential difference V_{CT} between the target and the collector. The current from the target to the collector is measured with an ammeter. If the target is kept at a negative potential relative to the collector, so that V_{CT} is negative, any electrons ejected from the target by the light are accelerated by the electric field and move to the collector. With negative V_{CT}, the current measured is proportional to the intensity of the light source, suggesting that ejecting each electron requires a certain amount of light energy.

If the potential difference V_{CT} is made slightly positive (so that the target is positive relative to the collector), there is a small current detected, but the electric field

Figure 34.47 Apparatus to study the photoelectric effect. The potential difference V_{CT} is positive when $V_T > V_C$.

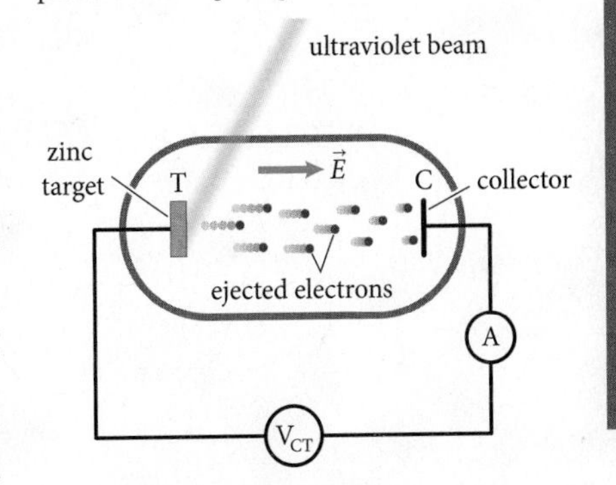

Figure 34.48 For the circuit in Figure 34.47, the current as a function of potential difference and the stopping potential as a function of the frequency of the incident light.

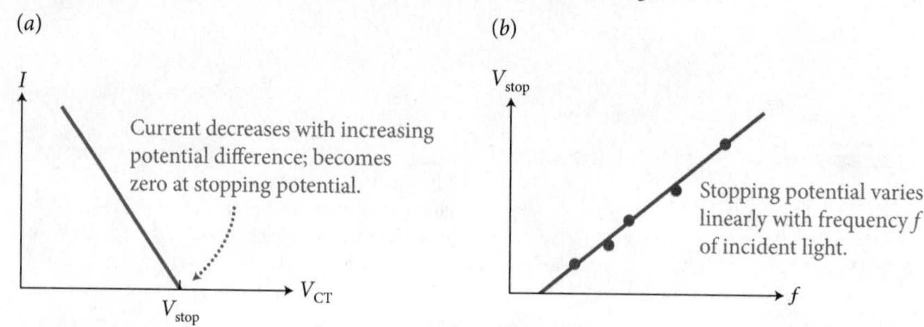

between T and C now *slows down* any ejected electrons that initially move toward the collector. As V_{CT} increases, there is a certain value of V_{CT} at which the flow of electron stops completely, as shown in the graph of current I versus V_{CT} in **Figure 34.48*a***. At this potential difference, called the *stopping potential difference,* the current is zero regardless of the intensity of the incident light. No matter how bright the light, there is no current between the target and the collector. This finding implies that the maximum kinetic energy with which the electrons leave the target does not depend on the intensity (and thus the incident power) of the light.

If we measure the kinetic energy of the ejected electrons (for $V_{CT} < V_{stop}$) directly, we discover that not all of them have the same kinetic energy. This happens because although the amount of energy absorbed from the photons is the same for all electrons, the energy required for the electron to make its way from its initial location in the target to the surface depends on the depth from which the electron is liberated. As a result, electrons released from the surface of the target have the maximum possible kinetic energy, which equals the amount of energy transferred to each electron by the light minus the energy required to liberate the electron from the metal.

For a given potential difference between the target and the collector, the electric field does work $-eV_{CT}$ on an electron as the electron moves from the target to the collector (see Eq. 25.17). The change in the electron's kinetic energy is thus

$$\Delta K = -eV_{CT}. \tag{34.33}$$

Given that the electrons just barely reach the collector at the stopping potential difference, we know that their final kinetic energy is zero, and so for these electrons $\Delta K = K_f - K_i = -K_i$. The maximum kinetic energy with which the electrons leave the target is thus

$$K_{max} = K_i = eV_{stop}. \tag{34.34}$$

Another clue to understanding the experiment in Figure 34.47 emerges when we change the frequency of the incident light and again measure the target-to-collector current as a function of V_{CT}. We observe that the stopping potential difference depends on the frequency; plotting this stopping potential difference as a function of the frequency of the light yields the results shown in Figure 34.48*b*.

34.20 (*a*) What does Figure 34.48*b* tell you about the relationship between the frequency of the incident light and the maximum kinetic energy of the ejected electrons? (*b*) What does the intercept of the line through the data points and the horizontal axis represent?

As Checkpoint 34.20, part *a* shows, the maximum kinetic energy of the ejected electrons depends on the frequency of the incident light, not its intensity. Electrons ejected by ultraviolet light, which has a higher frequency than visible light, have more kinetic energy than electrons ejected by visible light. (This is why putting glass in the beam of light in the experiment shown in Figure 34.46 essentially eliminates the effect. Although some electrons are ejected by the visible light, those electrons are liberated with less kinetic energy and are more likely to return to the target.) Furthermore, as you discovered in answering Checkpoint 34.20, part *b*, there is a certain minimum frequency of light below which electrons are not ejected at all, regardless of the intensity of the light.

Why does the photoelectric effect require us to think of light as a particle rather than as a wave? The stopping potential difference gives us the maximum kinetic energy with which electrons are released; the light must supply at least this much energy to the electrons in order to eject them. If light could be understood solely as a wave, the intensity of the wave, not its frequency, would determine the maximum amount of energy it could deliver to the electrons. Because the stopping potential difference depends not on light intensity but on frequency, we infer that light carries its energy in energy quanta and that the energy in each quantum is proportional to the frequency.

The photons described in Section 34.5 are these quanta. When an electron absorbs a photon, the electron acquires the photon's entire energy—the electron cannot absorb just part of a photon. The photon's energy frees the electron from the material and gives it additional kinetic energy. If we denote the minimum energy required to free the electron by E_0, we have

$$E_{\text{photon}} = hf = K_{\max} + E_0, \tag{34.35}$$

where $K_{\max}$ is the maximum kinetic energy of the electron as it is ejected. The energy E_0, called the **work function** of the target metal, is a property of the metal that measures how tightly electrons are bound to the metal.

The value of Planck's constant h can be determined by using the relationship between V_{stop} and f given in Figure 34.48*b*. Substituting Eq. 34.34 into Eq. 34.35 and solving the result for V_{stop}, we get

$$V_{\text{stop}} = \frac{h}{e}f - \frac{E_0}{e}. \tag{34.36}$$

This result shows that V_{stop} depends linearly on f and that the slope of the line in Figure 34.48*b* is h/e. By measuring the slope in Figure 34.48*b* and dividing that slope by the charge e of the electron, one obtains $h = 6.626 \times 10^{-34}\ \text{J}\cdot\text{s}$, the value given in Section 34.4.

As discussed in Section 34.5 and expressed in Eq. 34.35, Planck's constant relates the energy and frequency of a photon, $E_{\text{photon}} = hf_{\text{photon}}$, and in Section 34.4 we learned the relationship between the momentum and the wavelength of an electron (or anything else that is ordinarily thought of as a particle): $\lambda_{\text{electron}} = h/p_{\text{electron}}$. Because of the wave-particle duality, we can apply this expression for wavelength to photons as well as electrons: $\lambda_{\text{photon}} = h/p_{\text{photon}}$. If we calculate the momentum of a photon from its wavelength using this expression and then substitute $\lambda = c_0/f$, we see that the momentum of a photon is proportional to its energy:

$$p_{\text{photon}} = \frac{h}{\lambda_{\text{photon}}} = \frac{hf_{\text{photon}}}{c_0} = \frac{E_{\text{photon}}}{c_0}. \tag{34.37}$$

If we substitute this result into the equation relating energy and momentum derived in Chapter 14 (Eq. 14.57)

$$E^2 - (c_0 p)^2 = (mc_0^2)^2, \tag{34.38}$$

the left side of this equation becomes zero, and so we see that photons have zero mass ($m_{\text{photon}} = 0$). We derived Eq. 34.38 for particles that have nonzero mass. Now we see that we can treat photons as massless "particles of light." While these particles have no mass, they do have both momentum and energy:

$$E_{\text{photon}} = hf_{\text{photon}} \tag{34.39}$$

$$p_{\text{photon}} = \frac{hf_{\text{photon}}}{c_0}, \tag{34.40}$$

and, unlike ordinary particles, they always move at the speed of light c_0. Remember also that the mass of a particle is associated with the internal energy of that particle (Eq. 14.54), so $m_{\text{photon}} = 0$ means that photons have no internal energy and therefore no internal structure.

Example 34.11 Photoelectric effect

Light of wavelength 380 nm strikes the metal target in Figure 34.47. As long as the potential difference V_{CT} between the target and the collector is no greater than +1.2 V, there is a current in the circuit. Determine the longest wavelength of light that can eject electrons from this metal.

❶ **GETTING STARTED** To solve this problem, I recognize that the longest wavelength of light that can eject electrons corresponds to the lowest-energy photon that can eject an electron; this energy is equal to the work function. I therefore need to use the idea of stopping potential difference to determine the work function from the information given in the problem.

❷ **DEVISE PLAN** The fact that the current is zero when $V_{CT} > +1.2$ V tells me that the stopping potential difference is 1.2 V. Equation 34.36 gives the relationship between photon frequency and stopping potential difference. I can use the relationship between photon frequency and wavelength to rewrite Eq. 34.36 in terms of wavelength. Finally, Eq. 34.35 shows that the lowest photon energy comes when $K_{max} = 0$ so that the lower energy equals the work function E_0. Therefore I must determine the wavelength of a photon that has an energy equal to the work function, and for this I can use the expression I developed for the relationship between photon energy and wavelength.

❸ **EXECUTE PLAN** Solving Eq. 34.36 for E_0, then substituting c_0/λ for f and inserting numerical values, I obtain

$$E_0 = \frac{hc_0}{\lambda} - eV_{\text{stop}}$$

$$= \frac{(6.626 \times 10^{-34}\ \text{J}\cdot\text{s})(2.998 \times 10^{8}\ \text{m/s})}{380 \times 10^{-9}\ \text{m}}$$

$$- (1.602 \times 10^{-19}\ \text{C})(1.2\ \text{V})$$

$$= 3.3 \times 10^{-19}\ \text{J},$$

where I have used the equality $1\ \text{V} \equiv 1\ \text{J/C}$ (Eq. 25.16). Because the longest wavelength that can eject electrons has energy equal to the work function, I solve $E_0 = hc_0/\lambda$ for λ and substitute the value of E_0 I just calculated to obtain this maximum wavelength:

$$\lambda = \frac{hc_0}{E_0} = \frac{(6.626 \times 10^{-34}\ \text{J}\cdot\text{s})(2.998 \times 10^{8}\ \text{m/s})}{3.3 \times 10^{-19}\ \text{J}} = 0.60\ \mu\text{m}. \checkmark$$

❹ **EVALUATE RESULT** This value for the longest wavelength that can eject electrons is greater than 380 nm, the wavelength corresponding to the stopping potential difference of 1.2 V, as it should be.

34.21 A photon enters a piece of glass for which the index of refraction is about 1.5. What happens to the photon's (*a*) speed, (*b*) frequency, (*c*) wavelength, and (*d*) energy?

Chapter Glossary

SI units of physical quantities are given in parentheses.

Bragg condition The condition under which x rays diffracted by planes of atoms in a crystal lattice interfere constructively. For x rays of wavelength λ, diffracting from a crystal lattice with spacing d between adjacent planes of atoms at an angle θ between the incident rays and the normal to the scattering planes, the condition states that $2d \cos\theta = m\lambda$.

de Broglie wavelength λ (m) The wavelength associated with the wave behavior of a particle, $\lambda = h/p$.

diffraction grating An optical component with a periodic structure of equally spaced slits or grooves that diffracts and splits light into several beams that travel in different directions. When a diffraction grating is made up of slits, light passes through it and it is called a *transmission diffraction grating*; when a diffraction grating is made up of grooves, light reflects from it and it is called a *reflection grating*. The so-called *principal maxima* in the intensity pattern created by a diffraction grating occur at angles given by

$$d \sin\theta_m = \pm m\lambda, \quad \text{for } m = 0, 1, 2, 3, \ldots \tag{34.16}$$

and minima occur at angles give by

$$d \sin\theta_{\text{min}} = \pm\frac{k}{N}\lambda \tag{34.17}$$

for an integer k that is not an integer multiple of N.

fringe order m, or n (unitless) A number indexing interference fringes; the central bright fringe is called zeroth order ($m = 0$), and the index increases with distance from the central bright fringe. The dark fringes flanking the central bright fringe are first order ($n = 1$).

interference fringes A pattern of alternating bright and dark bands cast on a screen produced by coherent light passing through very small, closely spaced slits, apertures, or edges.

photoelectric effect The emission of electrons from matter as a consequence of their absorption of energy from electromagnetic radiation with photon energy greater than the work function.

photon The indivisible, discrete basic unit, or quantum, of light. A photon of frequency f_{photon} has energy

$$E_{\text{photon}} = hf_{\text{photon}} \tag{34.39}$$

and momentum

$$p_{\text{photon}} = \frac{hf_{\text{photon}}}{c_0}. \tag{34.40}$$

Planck's constant h (J · s) The fundamental constant that relates the energy of a photon to its frequency and also the de Broglie wavelength and momentum of a particle: $h = 6.626 \times 10^{-34}$ J · s.

Rayleigh's criterion Two features in the image formed by a lens can be visually separated (and are then said to be *resolved*) if they satisfy Rayleigh's criterion. For a lens of diameter d and light of wavelength λ, the minimum angular separation θ_r for which two sources can be resolved is

$$\theta_r \approx 1.22\frac{\lambda}{d}. \tag{34.30}$$

wave-particle duality The possession of both wave properties and particle properties, observed both for all atomic-scale material particles and for photons.

work function E_0 (J) The minimum energy required to free an electron from the surface of a metal. This energy measures how tightly the electron is bound to the metal.

x rays Electromagnetic waves that have wavelengths ranging from 0.01 nm to 10 nm.

Appendix A

Notation

Notation used in this text, listed alphabetically, Greek letters first.
For information concerning superscripts and subscripts, see the explanation at the end of this table.

Symbol	Name of Quantity	Definition	Where Defined	SI units
α (alpha)	polarizability	scalar measure of amount of charge separation occurring in material due to external electric field	Eq. 23.24	$C^2 \cdot m/N$
α	Bragg angle	in x-ray diffraction, angle between incident x rays and sample surface	Section 34.3	degree, radian, or revolution
α_ϑ	(ϑ component of) rotational acceleration	rate at which rotational velocity ω_ϑ increases	Eq. 11.12	s^{-2}
β (beta)	sound intensity level	logarithmic scale for sound intensity, proportional to $\log(I/I_{th})$	Eq. 17.5	dB (not an SI unit)
γ (gamma)	Lorentz factor	factor indicating how much relativistic values deviate from nonrelativistic ones	Eq. 14.6	unitless
γ	surface tension	force per unit length exerted parallel to surface of liquid; energy per unit area required to increase surface area of liquid	Eq. 18.48	N/m
γ	heat capacity ratio	ratio of heat capacity at constant pressure to heat capacity at constant volume	Eq. 20.26	unitless
Δ	delta	change in	Eq. 2.4	
$\Delta\vec{r}$	displacement	vector from object's initial to final position	Eq. 2.8	m
$\Delta\vec{r}_F$, Δx_F	force displacement	displacement of point of application of a force	Eq. 9.7	m
Δt	interval of time	difference between final and initial instants	Table 2.2	s
Δt_{proper}	proper time interval	time interval between two events occurring at same position	Section 14.1	s
Δt_v	interval of time	time interval measured by observer moving at speed v with respect to events	Eq. 14.13	s
Δx	x component of displacement	difference between final and initial positions along x axis	Eq. 2.4	m
δ (delta)	delta	infinitesimally small amount of	Eq. 3.24	
ϵ_0 (epsilon)	electric constant	constant relating units of electrical charge to mechanical units	Eq. 24.7	$C^2/(N \cdot m^2)$
η (eta)	viscosity	measure of fluid's resistance to shear deformation	Eq. 18.38	$Pa \cdot s$
η	efficiency	ratio of work done by heat engine to thermal input of energy	Eq. 21.21	unitless
θ (theta)	angular coordinate	polar coordinate measuring angle between position vector and x axis	Eq. 10.2	degree, radian, or revolution
θ_c	contact angle	angle between solid surface and tangent to liquid surface at meeting point measured within liquid	Section 18.4	degree, radian, or revolution
θ_c	critical angle	angle of incidence greater than which total internal reflection occurs	Eq. 33.9	degree, radian, or revolution
θ_i	angle of incidence	angle between incident ray of light and normal to surface	Section 33.1	degree, radian, or revolution

Symbol	Name of Quantity	Definition	Where Defined	SI units
θ_i	angle subtended by image	angle subtended by image	Section 33.6	degree, radian, or revolution
θ_o	angle subtended by object	angle subtended by object	Section 33.6	degree, radian, or revolution
θ_r	angle of reflection	angle between reflected ray of light and normal to surface	Section 33.1	degree, radian, or revolution
θ_r	minimum resolving angle	smallest angular separation between objects that can be resolved by optical instrument with given aperture	Eq. 34.30	degree, radian, or revolution
ϑ (script theta)	rotational coordinate	for object traveling along circular path, arc length traveled divided by circle radius	Eq. 11.1	unitless
κ (kappa)	torsional constant	ratio of torque required to twist object to rotational displacement	Eq. 15.25	$N \cdot m$
κ	dielectric constant	factor by which potential difference across isolated capacitor is reduced by insertion of dielectric	Eq. 26.9	unitless
λ (lambda)	inertia per unit length	for uniform one-dimensional object, amount of inertia in a given length	Eq. 11.44	kg/m
λ	wavelength	minimum distance over which periodic wave repeats itself	Eq. 16.9	m
λ	linear charge density	amount of charge per unit length	Eq. 23.16	C/m
μ (mu)	reduced mass	product of two interacting objects' inertias divided by their sum	Eq. 6.39	kg
μ	linear mass density	mass per unit length	Eq. 16.25	kg/m
$\vec{\mu}$	magnetic dipole moment	vector pointing along direction of magnetic field of current loop, with magnitude equal to current times area of loop	Section 28.3	$A \cdot m^2$
μ_0	magnetic constant	constant relating units of electric current to mechanical units	Eq. 28.1	$T \cdot m/A$
μ_k	coefficient of kinetic friction	proportionality constant relating magnitudes of force of kinetic friction and normal force between two surfaces	Eq. 10.55	unitless
μ_s	coefficient of static friction	proportionality constant relating magnitudes of force of static friction and normal force between two surfaces	Eq. 10.46	unitless
ρ (rho)	mass density	amount of mass per unit volume	Eq. 1.4	kg/m^3
ρ	inertia per unit volume	for uniform three-dimensional object, amount of inertia in a given volume divided by that volume	Eq. 11.46	kg/m^3
ρ	(volume) charge density	amount of charge per unit volume	Eq. 23.18	C/m^3
σ (sigma)	inertia per unit area	for uniform two-dimensional object, inertia divided by area	Eq. 11.45	kg/m^2
σ	surface charge density	amount of charge per unit area	Eq. 23.17	C/m^2
σ	conductivity	ratio of current density to applied electric field	Eq. 31.8	$A/(V \cdot m)$
τ (tau)	torque	magnitude of axial vector describing ability of forces to change objects' rotational motion	Eq. 12.1	$N \cdot m$
τ	time constant	for damped oscillation, time for energy of oscillator to decrease by factor e^{-1}	Eq. 15.39	s
τ_ϑ	(ϑ component of) torque	ϑ component of axial vector describing ability of forces to change objects' rotational motion	Eq. 12.3	$N \cdot m$
Φ_E (phi, upper case)	electric flux	scalar product of electric field and area through which it passes	Eq. 24.1	$N \cdot m^2/C$

Symbol	Name of Quantity	Definition	Where Defined	SI units
Φ_B	magnetic flux	scalar product of magnetic field and area through which it passes	Eq. 27.10	Wb
ϕ (phi)	phase constant	phase difference between source emf and current in circuit	Eq. 32.16	unitless
$\phi(t)$	phase	time-dependent argument of sine function describing simple harmonic motion	Eq. 15.5	unitless
Ω (omega, upper case)	number of basic states	number of basic states corresponding to macrostate	Section 19.4, Eq. 19.1	unitless
ω (omega)	rotational speed	magnitude of rotational velocity	Eq. 11.7	s^{-1}
ω	angular frequency	for oscillation with period T, $2\pi/T$	Eq. 15.4	s^{-1}
ω_0	resonant angular frequency	angular frequency at which current in circuit is maximal	Eq. 32.47	s^{-1}
ω_ϑ	(ϑ component of) rotational velocity	rate at which rotational coordinate ϑ changes	Eq. 11.6	s^{-1}
A	area	length $\times$ width	Eq. 11.45	m^2
A	amplitude	magnitude of maximum displacement of oscillating object from equilibrium position	Eq. 15.6	m (for linear mechanical oscillation; unitless for rotational oscillation; various units for nonmechanical oscillation)
$\vec{A}$	area vector	vector with magnitude equal to area and direction normal to plane of area	Section 24.6	m^2
$\vec{a}$	acceleration	time rate of change in velocity	Section 3.1	m/s^2
$\vec{a}_{Ao}$	relative acceleration	value observer in reference frame A records for acceleration of object o in reference frame A	Eq. 6.11	m/s^2
a_c	magnitude of centripetal acceleration	acceleration required to make object follow circular trajectory	Eq. 11.15	m/s^2
a_r	radial component of acceleration	component of acceleration in radial direction	Eq. 11.16	m/s^2
a_t	tangential component of acceleration	component of acceleration tangent to trajectory; for circular motion at constant speed $a_t = 0$	Eq. 11.17	m/s^2
a_x	x component of acceleration	component of acceleration directed along x axis	Eq. 3.21	m/s^2
$\vec{B}$	magnetic field	vector field providing measure of magnetic interactions	Eq. 27.5	T
$\vec{B}_{ind}$	induced magnetic field	magnetic field produced by induced current	Section 29.4	T
b	damping coefficient	ratio of drag force on moving object to its speed	Eq. 15.34	kg/s
C	heat capacity per particle	ratio of energy transferred thermally per particle to change in temperature	Section 20.3	J/K
C	capacitance	ratio of magnitude of charge on one of a pair of oppositely charged conductors to magnitude of potential difference between them	Eq. 26.1	F
C_P	heat capacity per particle at constant pressure	ratio of energy transferred thermally per particle to change in temperature, while holding pressure constant	Eq. 20.20	J/K
C_V	heat capacity per particle at constant volume	ratio of energy transferred thermally per particle to change in temperature, while holding volume constant	Eq. 20.13	J/K
$COP_{cooling}$	coefficient of performance of cooling	ratio of thermal input of energy to work done on a heat pump	Eq. 21.27	unitless

Symbol	Name of Quantity	Definition	Where Defined	SI units
$COP_{heating}$	coefficient of performance of heating	ratio of thermal output of energy to work done on a heat pump	Eq. 21.25	unitless
c	shape factor	ratio of object's rotational inertia to mR^2; function of distribution of inertia within object	Table 11.3, Eq. 12.25	unitless
c	wave speed	speed at which mechanical wave travels through medium	Eq. 16.3	m/s
c	specific heat capacity	ratio of energy transferred thermally per unit mass to change in temperature	Section 20.3	$J/(K \cdot kg)$
c_0	speed of light in vacuum	speed of light in vacuum	Section 14.2	m/s
c_V	specific heat capacity at constant volume	ratio of energy transferred thermally per unit mass to change in temperature, while holding volume constant	Eq. 20.48	$J/(K \cdot kg)$
$\vec{D}$	displacement (of particle in wave)	displacement of particle from its equilibrium position	Eq. 16.1	m
d	diameter	diameter	Section 1.9	m
d	distance	distance between two locations	Eq. 2.5	m
d	degrees of freedom	number of ways particle can store thermal energy	Eq. 20.4	unitless
d	lens strength	1 m divided by focal length	Eq. 33.22	diopters
E	energy of system	sum of kinetic and internal energies of system	Table 1.1, Eq. 5.21	J
$\vec{E}$	electric field	vector field representing electric force per unit charge	Eq. 23.1	N/C
E_0	work function	minimum energy required to free electron from surface of metal	Eq. 34.35	J
E_{chem}	chemical energy	internal energy associated with object's chemical state	Eq. 5.27	J
E_{int}	internal energy of system	energy associated with an object's state	Eqs. 5.20, 14.54	J
E_{mech}	mechanical energy	sum of system's kinetic and potential energies	Eq. 7.9	J
E_s	source energy	incoherent energy used to produce other forms of energy	Eq. 7.7	J
E_{th}	thermal energy	internal energy associated with object's temperature	Eq. 5.27	J
$\mathscr{E}$	emf	in charge-separating device, nonelectrostatic work per unit charge done in separating positive and negative charge carriers	Eq. 26.7	V
$\mathscr{E}_{ind}$	induced emf	emf resulting from changing magnetic flux	Eqs. 29.3, 29.8	V
$\mathscr{E}_{max}$	amplitude of emf	amplitude of time-dependent emf produced by AC source	Section 32.1, Eq. 32.1	V
$\mathscr{E}_{rms}$	rms emf	root-mean-square emf	Eq. 32.55	V
e	coefficient of restitution	measure of amount of initial relative speed recovered after collision	Eq. 5.18	unitless
e	eccentricity	measure of deviation of conic section from circular	Section 13.7	unitless
e	elementary charge	magnitude of charge on electron	Eq. 22.3	C
$\vec{F}$	force	time rate of change of object's momentum	Eq. 8.2	N
$\vec{F}^B$	magnetic force	force exerted on electric current or moving charged particle by magnetic field	Eqs. 27.8, 27.19	N
$\vec{F}^b$	buoyant force	upward force exerted by fluid on submerged object	Eq. 18.12	N
$\vec{F}^c$	contact force	force between objects in physical contact	Section 8.5	N

Symbol	Name of Quantity	Definition	Where Defined	SI units
$\vec{F}^{d}$	drag force	force exerted by medium on object moving through medium	Eq. 15.34	N
$\vec{F}^{E}$	electric force	force exerted between electrically charged objects or on electrically charged objects by electric field	Eq. 22.1	N
$\vec{F}^{EB}$	electromagnetic force	force exerted on electrically charged objects by electric and magnetic fields	Eq. 27.20	N
$\vec{F}^{f}$	frictional force	force exerted on object due to friction between it and a second object or surface	Eq. 9.26	N
$\vec{F}^{G}$	gravitational force	force exerted by Earth or any object having mass on any other object having mass	Eqs. 8.16, 13.1	N
$\vec{F}^{k}$	force of kinetic friction	frictional force between two objects in relative motion	Section 10.4, Eq. 10.55	N
$\vec{F}^{n}$	normal force	force directed perpendicular to a surface	Section 10.4, Eq. 10.46	N
$\vec{F}^{s}$	force of static friction	frictional force between two objects not in relative motion	Section 10.4, Eq. 10.46	N
f	frequency	number of cycles per second of periodic motion	Eq. 15.2	Hz
f	focal length	distance from center of lens to focus	Section 33.4, Eq. 33.16	m
f_{beat}	beat frequency	frequency at which beats occur when waves of different frequency interfere	Eq. 17.8	Hz
G	gravitational constant	proportionality constant relating gravitational force between two objects to their masses and separation	Eq. 13.1	$N \cdot m^2/kg^2$
g	magnitude of acceleration due to gravity	magnitude of acceleration of object in free fall near Earth's surface	Eq. 3.14	m/s^2
h	height	vertical distance	Eq. 10.26	m
h	Planck's constant	constant describing scale of quantum mechanics; relates photon energy to frequency and de Broglie wavelength to momentum of particle	Eq. 34.35	$J \cdot s$
I	rotational inertia	measure of object's resistance to change in its rotational velocity	Eq. 11.30	$kg \cdot m^2$
I	intensity	energy delivered by wave per unit time per unit area normal to direction of propagation	Eq. 17.1	W/m^2
I	(electric) current	rate at which charged particles cross a section of a conductor in a given direction	Eq. 27.2	A
I	amplitude of oscillating current	maximum value of oscillating current in circuit	Section 32.1, Eq. 32.5	A
I_{cm}	rotational inertia about center of mass	object's rotational inertia about an axis through its center of mass	Eq. 11.48	$kg \cdot m^2$
I_{disp}	displacement current	current-like quantity in Ampère's law caused by changing electric flux	Eq. 30.7	A
I_{enc}	enclosed current	current enclosed by Ampèrian path	Eq. 28.1	A
I_{ind}	induced current	current in loop caused by changing magnetic flux through loop	Eq. 29.4	A
I_{int}	intercepted current	current intercepted by surface spanning Ampèrian path	Eq. 30.6	A
I_{rms}	rms current	root-mean-square current	Eq. 32.53	A
I_{th}	intensity at threshold of hearing	minimum intensity audible to human ear	Eq. 17.4	W/m^2
i	time-dependent current	time-dependent current through circuit; $I(t)$	Section 32.1, Eq. 32.5	A
i	image distance	distance from lens to image	Section 33.6, Eq. 33.16	m

Symbol	Name of Quantity	Definition	Where Defined	SI units
$\hat{\imath}$	unit vector ("i hat")	vector for defining direction of x axis	Eq. 2.1	unitless
$\vec{J}$	impulse	amount of momentum transferred from environment to system	Eq. 4.18	$kg \cdot m/s$
$\vec{J}$	current density	current per unit area	Eq. 31.6	A/m^2
J_ϑ	rotational impulse	amount of angular momentum transferred from environment to system	Eq. 12.15	$kg \cdot m^2/s$
$\hat{\jmath}$	unit vector	vector for defining direction of y axis	Eq. 10.4	unitless
K	kinetic energy	energy object has because of its translational motion	Eqs. 5.12, 14.51	J
K	surface current density	current per unit of sheet width	Section 28.5	A/m
K_{cm}	translational kinetic energy	kinetic energy associated with motion of center of mass of system	Eq. 6.32	J
K_{conv}	convertible kinetic energy	kinetic energy that can be converted to internal energy without changing system's momentum	Eq. 6.33	J
K_{rot}	rotational kinetic energy	energy object has due to its rotational motion	Eq. 11.31	J
k	spring constant	ratio of force exerted on spring to displacement of free end of spring	Eq. 8.18	N/m
k	wave number	number of wavelengths in 2π units of distance; for wave with wavelength λ, $2\pi/\lambda$	Eqs. 16.7, 16.11	m^{-1}
k	Coulomb's law constant	constant relating electrostatic force to charges and their separation distance	Eq. 22.5	$N \cdot m^2/C^2$
k_B	Boltzmann constant	constant relating thermal energy to absolute temperature	Eq. 19.39	J/K
L	inductance	negative of ratio of induced emf around loop to rate of change of current in loop	Eq. 29.19	H
L_ϑ	(ϑ component of) angular momentum	capacity of object to make other objects rotate	Eq. 11.34	$kg \cdot m^2/s$
L_m	specific transformation energy for melting	energy transferred thermally per unit mass required to melt substance	Eq. 20.55	J/kg
L_v	specific transformation energy for vaporization	energy transferred thermally per unit mass required to vaporize substance	Eq. 20.55	J/kg
ℓ	length	distance or extent in space	Table 1.1	m
ℓ_{proper}	proper length	length measured by observer at rest relative to object	Section 14.3	m
ℓ_v	length	measured length of object moving at speed v relative to observer	Eq. 14.28	m
M	magnification	ratio of signed image height to object height	Eq. 33.17	unitless
M_θ	angular magnification	ratio of angle subtended by image to angle subtended by object	Eq. 33.18	unitless
m	mass	amount of substance	Table 1.1, Eq. 13.1	kg
m	inertia	measure of object's resistance to change in its velocity	Eq. 4.2	kg
m	fringe order	number indexing bright interference fringes, counting from central, zeroth-order bright fringe	Section 34.2, Eq. 34.5	unitless
m_v	inertia	inertia of object moving at speed v relative to observer	Eq. 14.41	kg
N	number of objects	number of objects in sample	Eq. 1.3	unitless
N_A	Avogadro's number	number of particles in 1 mol of a substance	Eq. 1.2	unitless
n	number density	number of objects per unit volume	Eq. 1.3	m^{-3}

Symbol	Name of Quantity	Definition	Where Defined	SI units
n	windings per unit length	in a solenoid, number of windings per unit length	Eq. 28.4	unitless
n	index of refraction	ratio of speed of light in vacuum to speed of light in a medium	Eq. 33.1	unitless
n	fringe order	number indexing dark interference fringes, counting from central, zeroth-order bright fringe	Section 34.2, Eq. 34.7	unitless
O	origin	origin of coordinate system	Section 10.2	
o	object distance	distance from lens to object	Section 33.6, Eq. 33.16	m
P	power	time rate at which energy is transferred or converted	Eq. 9.30	W
P	pressure	force per unit area exerted by fluid	Eq. 18.1	Pa
P_{atm}	atmospheric pressure	average pressure in Earth's atmosphere at sea level	Eq. 18.3	Pa
P_{gauge}	gauge pressure	pressure measured as difference between absolute pressure and atmospheric pressure	Eq. 18.16	Pa
p	time-dependent power	time-dependent rate at which source delivers energy to load; $P(t)$	Eq. 32.49	W
$\vec{p}$	momentum	vector that is product of an object's inertia and velocity	Eq. 4.6	$kg \cdot m/s$
$\vec{p}$	(electric) dipole moment	vector representing magnitude and direction of electric dipole, equal amounts of positive and negative charge separated by small distance	Eq. 23.9	$C \cdot m$
$\vec{p}_{ind}$	induced dipole moment	dipole moment induced in material by external electric field	Eq. 23.24	$C \cdot m$
p_x	x component of momentum	x component of momentum	Eq. 4.7	$kg \cdot m/s$
Q	quality factor	for damped oscillation, number of cycles for energy of oscillator to decrease by factor $e^{-2\pi}$	Eq. 15.41	unitless
Q	volume flow rate	rate at which volume of fluid crosses section of tube	Eq. 18.25	m^3/s
Q	energy transferred thermally	energy transferred into system by thermal interactions	Eq. 20.1	J
Q_{in}	thermal input of energy	positive amount of energy transferred into system by thermal interactions	Sections 21.1, 21.5	J
Q_{out}	thermal output of energy	positive amount of energy transferred out of system by thermal interactions	Sections 21.1, 21.5	J
q	electrical charge	attribute responsible for electromagnetic interactions	Eq. 22.1	C
q_{enc}	enclosed charge	sum of all charge within a closed surface	Eq. 24.8	C
q_p	dipole charge	charge of positively charged pole of dipole	Section 23.6	C
R	radius	radius of an object	Eq. 11.47	m
R	resistance	ratio of applied potential difference to resulting current	Eqs. 29.4, 31.10	Ω
R_{eq}	equivalent resistance	resistance that could be used to replace combination of circuit elements	Eqs. 31.26, 31.33	Ω
r	radial coordinate	polar coordinate measuring distance from origin of coordinate system	Eq. 10.1	m
$\vec{r}$	position	vector for determining position	Eqs. 2.9, 10.4	m
$\hat{r}_{12}$	unit vector ("r hat")	unit vector pointing from tip of $\vec{r}_1$ to tip of $\vec{r}_2$	Eq. 22.6	unitless
$\vec{r}_{AB}$	relative position	position of observer B in reference frame of observer A	Eq. 6.3	m
$\vec{r}_{Ae}$	relative position	value observer in reference frame A records for position at which event e occurs	Eq. 6.3	m

Symbol	Name of Quantity	Definition	Where Defined	SI units
$\vec{r}_{cm}$	position of a system's center of mass	a fixed position in a system that is independent of choice of reference frame	Eq. 6.24	m
$\vec{r}_p$	dipole separation	position of positively charged particle relative to negatively charged particle in dipole	Section 23.6	m
$r_\perp$	lever arm distance *or* lever arm	perpendicular distance between rotation axis and line of action of a vector	Eq. 11.36	m
$\Delta\vec{r}$	displacement	vector from object's initial to final position	Eq. 2.8	m
$\Delta\vec{r}_F$	force displacement	displacement of point of application of a force	Eq. 9.7	m
S	entropy	logarithm of number of basic states	Eq. 19.4	unitless
S	intensity	intensity of electromagnetic wave	Eq. 30.36	W/m^2
$\vec{S}$	Poynting vector	vector representing flow of energy in combined electric and magnetic fields	Eq. 30.37	W/m^2
s	arc length	distance along circular path	Eq. 11.1	m
s^2	space-time interval	invariant measure of separation of events in space-time	Eq. 14.18	m^2
T	period	time interval needed for object in circular motion to complete one revolution	Eq. 11.20	s
T	absolute temperature	quantity related to rate of change of entropy with respect to thermal energy	Eq. 19.38	K
$\mathcal{T}$	tension	stress in object subject to opposing forces stretching the object	Section 8.6	N
t	instant in time	physical quantity that allows us to determine the sequence of related events	Table 1.1	s
t_{Ae}	instant in time	value observer A measures for instant at which event e occurs	Eq. 6.1	s
Δt	interval of time	difference between final and initial instants	Table 2.2	s
Δt_{proper}	proper time interval	time interval between two events occurring at same position	Section 14.1	s
Δt_v	interval of time	time interval between two events measured by observer moving at speed v relative to an observer for whom the events occur at the same position	Eq. 14.13	s
U	potential energy	energy stored in reversible changes to system's configuration state	Eq. 7.7	J
U^B	magnetic potential energy	potential energy stored in magnetic field	Eqs. 29.25, 29.30	J
U^E	electric potential energy	potential energy due to relative position of charged objects	Eq. 25.8	J
U^G	gravitational potential energy	potential energy due to relative position of gravitationally interacting objects	Eqs. 7.13, 13.14	J
u_B	energy density of magnetic field	energy per unit volume stored in magnetic field	Eq. 29.29	J/m^3
u_E	energy density of electric field	energy per unit volume stored in electric field	Eq. 26.6	J/m^3
V	volume	amount of space occupied by an object	Table 1.1	m^3
V_{AB}	potential difference	negative of electrostatic work per unit charge done on charged particle as it is moved from point A to point B	Eq. 25.15	V
V_{batt}	battery potential difference	magnitude of potential difference between terminals of battery	Eq. 25.19	V
V_C	amplitude of oscillating potential	maximum magnitude of potential across circuit element C	Section 32.1, Eq. 32.8	V
V_{disp}	displaced volume	volume of fluid displaced by submerged object	Eq. 18.12	m^3

Symbol	Name of Quantity	Definition	Where Defined	SI units
V_P	(electrostatic) potential	potential difference between conveniently chosen reference point of potential zero and point P	Eq. 25.30	V
V_{rms}	rms potential	root-mean-square potential difference	Eq. 32.55	V
V_{stop}	stopping potential	minimum potential difference required to stop flow of electrons from photoelectric effect	Eq. 34.34	V
$\mathcal{V}$	"volume" in velocity space	measure of range of velocities in three dimensions	Eq. 19.20	$(m/s)^3$
v	speed	magnitude of velocity	Table 1.1	m/s
$\vec{v}$	velocity	time rate of change in position	Eq. 2.23	m/s
$\vec{v}_{12}$	relative velocity	velocity of object 2 relative to object 1	Eq. 5.1	m/s
$\vec{v}_{AB}$	relative velocity	velocity of observer B in reference frame of observer A	Eq. 6.3	m/s
v_C	time-dependent potential	time-dependent potential across circuit element C; $V_C(t)$	Section 32.1, Eq. 32.8	V
$\vec{v}_{cm}$	velocity, center of mass	velocity of the center of mass of a system, equal to the velocity of the zero-momentum reference frame of the system	Eq. 6.26	m/s
$\vec{v}_d$	drift velocity	average velocity of electrons in conductor in presence of electric field	Eq. 31.3	m/s
v_{esc}	escape speed	minimum launch speed required for object to reach infinity	Eq. 13.23	m/s
v_r	radial component of velocity	for object moving along circular path, always zero	Eq. 11.18	m/s
v_{rms}	root-mean-square speed	square root of average of square of speed	Eq. 19.21	m/s
v_t	tangential component of velocity	for object in circular motion, rate at which arc length is swept out	Eq. 11.9	m/s
v_x	x component of velocity	component of velocity directed along x axis	Eq. 2.21	m/s
W	work	change in system's energy due to external forces exerted on system	Eqs. 9.1, 10.35	J
$W_{P\to Q}$	work	work done along path from P to Q	Eq. 13.12	J
W_{in}	mechanical input of energy	positive amount of mechanical work done on system	Section 21.1	J
W_{out}	mechanical output of energy	positive amount of mechanical work done by system	Section 21.1	J
W_q	electrostatic work	work done by electrostatic field on charged particle moving through field	Section 25.2, Eq. 25.17	J
X_C	capacitive reactance	ratio of potential difference amplitude to current amplitude for capacitor	Eq. 32.14	Ω
X_L	inductive reactance	ratio of potential difference amplitude to current amplitude for inductor	Eq. 32.26	Ω
x	position	position along x axis	Eq. 2.4	m
$x(t)$	position as function of time	position x at instant t	Section 2.3	m
Δx	x component of displacement	difference between final and initial positions along x axis	Eq. 2.4	m
Δx_F	force displacement	displacement of point of application of a force	Eq. 9.7	m
Z	impedance	(frequency-dependent) ratio of potential difference to current through circuit	Eq. 32.33	Ω
Z	zero-momentum reference frame	reference frame in which system of interest has zero momentum	Eq. 6.23	

Math notation

Math notation	Name	Where introduced
$\equiv$	defined as	Eq. 1.3
$\approx$	approximately equal to	Section 1.9
Σ (sigma, upper case)	sum of	Eq. 3.25
$\int$	integral of	Eq. 3.27
$\parallel$	parallel	Section 10.2
$\perp$	perpendicular	Section 10.2
$\propto$	proportional to	Section 13.1
$\cdot$	scalar product of two vectors	Eq. 10.33
$\times$	vector product of two vectors	Eq. 12.35
$\frac{\partial f}{\partial x}$	partial derivative of f with respect to x	Eq. 16.47
$\vec{b}$	vector b	Eq. 2.2
$\lvert\vec{b}\rvert$ or b	magnitude of $\vec{b}$	Eq. 2.3
b_x	x component of $\vec{b}$	Eq. 2.2
$\vec{b}_x$	x component vector of $\vec{b}$	Eq. 10.5
$\hat{\imath}$	unit vector ("i hat")	Eq. 2.1
$\hat{r}_{12}$	unit vector ("r hat")	Eq. 22.6

Note concerning superscripts and subscripts

Superscripts are appended to forces and potential energies to indicate the type of force or energy. They may be found in the main list under F, for forces, and U, for potential energies. Uppercase superscripts are used for fundamental interactions.

Subscripts are used on many symbols to identify objects, reference frames, types (for example, of energy), and processes. Object identifiers may be numbers, letters, or groups of letters. Reference frames are indicated by capital letters. Object identifiers and reference frames can occur in pairs, indicating relative quantities. In this case, the main symbol describes a property of whatever is identified by the second subscript relative to that of the first. In the case of forces, the first subscript identifies the object that causes the force and the second identifies the object on which the force is exerted. Types and processes are identified in various ways; many are given in the main list. Here are some examples:

m_1	inertia of object 1
m_{ball}	inertia of ball
$\vec{v}_{\text{cm}}$	velocity of center of mass of system
$\vec{r}_{12}$	position of object 2 relative to object 1; $\vec{r}_{12} = \vec{r}_2 - \vec{r}_1$
$\vec{p}_1$	momentum of object 1
$\vec{p}_{\text{Z}2}$	momentum of object 2 as measured in zero-momentum reference frame
$\vec{v}_{\text{AB}}$	velocity of observer B as measured in reference frame of observer A
$\vec{v}_{\text{Ao}}$	velocity of object o as measured in reference frame A
$\vec{r}_{\text{Ee}}$	position of event e as measured in Earth reference frame
$\vec{F}^{\text{c}}_{\text{pw}}$	contact force exerted by person on wall
$\vec{F}^{\text{G}}_{\text{Eb}}$	gravitational force exerted by Earth on ball
E_{th}	thermal energy
K_{conv}	convertible kinetic energy
P_{av}	average power
a_{c}	centripetal acceleration
$W_{\text{P}\to\text{Q}}$	work done along path from P to Q

Initial and final conditions are identified by subscripts i and f, following other identifiers. For example:

$\vec{p}_{1\text{i}}$	initial momentum of object 1
$\vec{p}_{\text{Zball,f}}$	final momentum of ball as measured in zero-momentum reference frame

Italic subscripts are used to identify components of vectors. These include x, y, z, r (radial), t (tangential), and ϑ (angular, with respect to given axis). They are also used to enumerate collections, for example, as indices of summation, and to indicate that a subscript refers to another variable. Here are some examples:

r_x	x component of position
a_t	tangential component of acceleration
L_ϑ	ϑ component of angular momentum
$p_{\text{Zball}\,y,\text{f}}$	final y component of momentum of ball as measured in zero-momentum reference frame
$\delta m_n r_n^2$	contribution to rotational inertia of extended object of small segment n, with inertia δm_n at position r_n
c_P	specific heat capacity at constant pressure
W_q	electrostatic work

Appendix B

Mathematics Review

1 Algebra

Factors

$$ax + bx + cx = (a + b + c)x$$
$$(a + b)^2 = a^2 + 2ab + b^2$$
$$(a - b)^2 = a^2 - 2ab + b^2$$
$$(a + b)(a - b) = a^2 - b^2$$

Fractions

$$\left(\frac{a}{b}\right)\left(\frac{c}{d}\right) = \frac{ac}{bd}$$
$$\left(\frac{a/b}{c/d}\right) = \frac{a}{b} \div \frac{c}{d} = \frac{a}{b}\cdot\frac{d}{c} = \frac{ad}{bc}$$
$$\left(\frac{1}{1/a}\right) = a$$

Exponents

$$a^n = \underbrace{a \times a \times a \times \cdots \times a}_{n \text{ factors}}$$

Any real number can be used as an exponent:

$$a^{-x} = \frac{1}{a^x}$$
$$a^0 = 1$$
$$a^1 = a$$
$$a^{1/2} = \sqrt{a}$$
$$a^{1/n} = \sqrt[n]{a}$$
$$a^x a^y = a^{x+y}$$
$$\frac{a^x}{a^y} = a^{x-y}$$
$$(a^x)^y = a^{x \cdot y}$$
$$a^x b^x = (ab)^x$$
$$\frac{a^x}{b^x} = \left(\frac{a}{b}\right)^x$$

Logarithms

Logarithm is the inverse function of the exponential function:

$$y = a^x \Leftrightarrow \log_a y = \log_a a^x = x \quad \text{and} \quad x = \log_a (a^x) = a^{\log_a x}$$

The two most common values for the base a are 10 (the common logarithm base) and e (the natural logarithm base).

$$y = e^x \Leftrightarrow \log_e y = \ln y = \ln e^x = x \quad \text{and} \quad x = \ln e^x = e^{\ln x}$$

Logarithm rules (valid for any base):

$$\ln(ab) = \ln(a) + \ln(b)$$
$$\ln\left(\frac{a}{b}\right) = \ln(a) - \ln(b)$$
$$\ln(a^n) = n \ln(a)$$
$$\ln 1 = 0$$

The expression $\ln(a + b)$ cannot be simplified.

Linear equations

A linear equation has the form $y = ax + b$, where a and b are constants. A graph of y versus x is a straight line. The value of a equals the slope of the line, and the value of b equals the value of y when x equals zero.

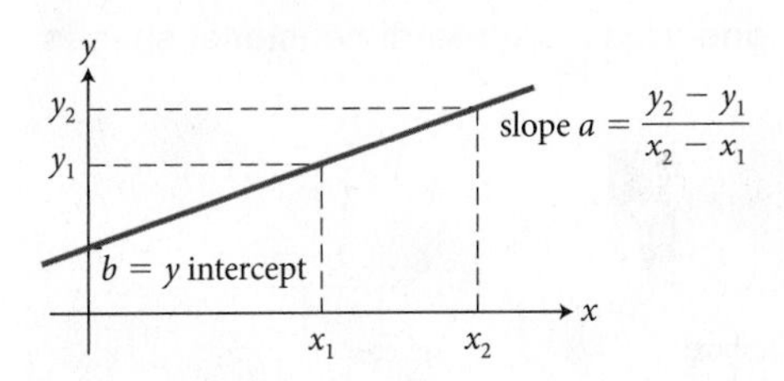

If $a = 0$, the line is horizontal. If $a > 0$, the line rises as x increases. If $a < 0$, the line falls as x increases. For any two values of x, say x_1 and x_2, the slope a can be calculated as

$$a = \frac{y_2 - y_1}{x_2 - x_1}$$

where y_1 and y_2 correspond to x_1 and x_2 (that is to say, $y_1 = ax_1 + b$ and $y_2 = ax_2 + b$).

Proportionality

If y is proportional to x (written $y \propto x$), then $y = ax$, where a is a constant. Proportionality is a subset of linearity. Because $y/x = a =$ constant for any corresponding x and y,

$$\frac{y_1}{x_1} = \frac{y_2}{x_2} \Leftrightarrow \frac{y_1}{y_2} = \frac{x_1}{x_2}.$$

Quadratic equation

The equation $ax^2 + bx + c = 0$ (the quadratic equation) has two solutions (called *roots*) for x:

$$x = \frac{-b \pm \sqrt{b^2 - 4ac}}{2a}$$

If $b^2 \geq 4ac$, the solutions are real numbers.

2 Geometry

Area and circumference for two-dimensional shapes

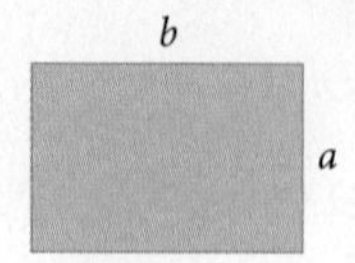

rectangle:
area = ab
circumference = $2(a + b)$

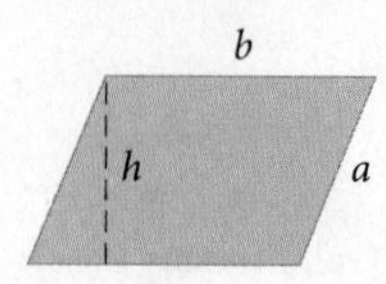

parallelogram:
area = bh
circumference = $2(a + b)$

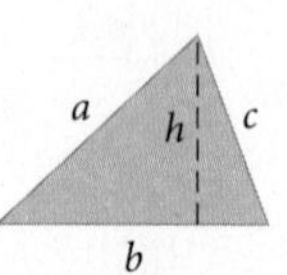

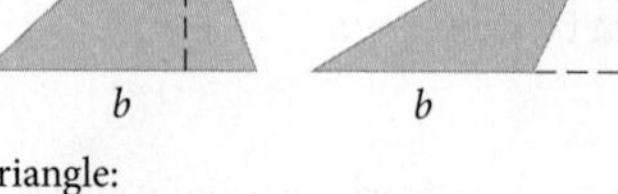

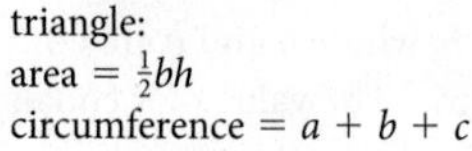

triangle:
area = $\frac{1}{2}bh$
circumference = $a + b + c$

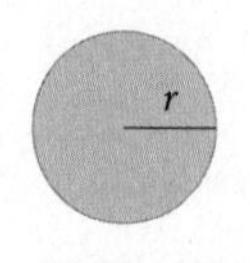

circle:
area = πr^2
circumference = $2\pi r$

Volume and area for three-dimensional shapes

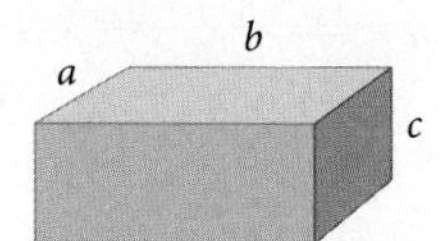

rectangular box:
volume = abc
area = $2(a^2 + b^2 + c^2)$

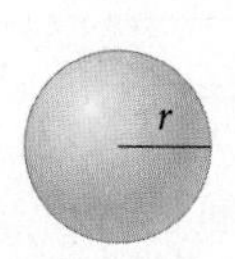

sphere:
volume = $\frac{4}{3}\pi r^3$
area = $4\pi r^2$

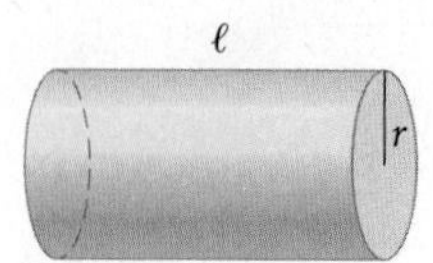

right circular cylinder:
volume = $\pi r^2 \ell$
area = $2\pi r\ell + 2\pi r^2$

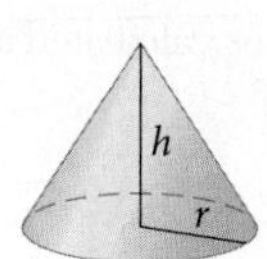

right circular cone:
volume = $\frac{1}{3}\pi r^2 h$
area = $\pi r^2 + \pi r\sqrt{r^2 + h^2}$

3 Trigonometry

Angle and arc length

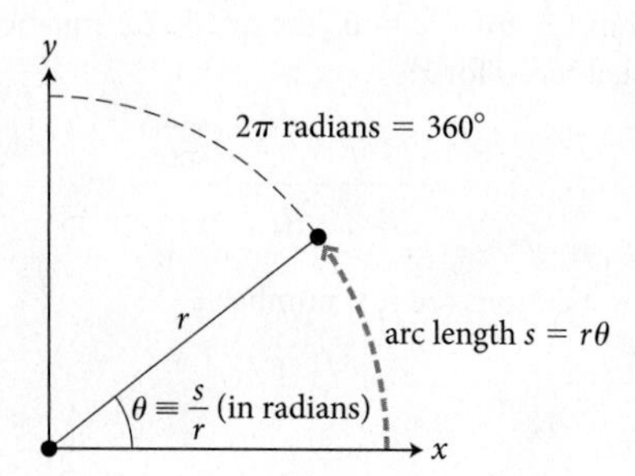

Right triangles

A right triangle is a triangle in which one of the angles is a right angle:

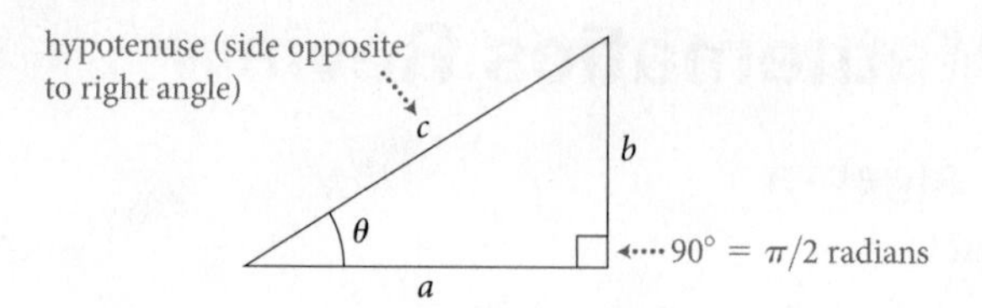

Pythagorean theorem: $a^2 + b^2 = c^2 \Leftrightarrow c = \sqrt{a^2 + b^2}$

Trigonometric functions:

$$\sin\theta = \frac{b}{c} = \frac{\text{opposite side}}{\text{hypotenuse}}, \quad \theta = \sin^{-1}\left(\frac{b}{c}\right) = \arcsin\left(\frac{b}{c}\right)$$

$$\cos\theta = \frac{a}{c} = \frac{\text{adjacent side}}{\text{hypotenuse}}, \quad \theta = \cos^{-1}\left(\frac{a}{c}\right) = \arccos\left(\frac{a}{c}\right)$$

$$\tan\theta = \frac{b}{a} = \frac{\text{opposite side}}{\text{adjacent side}}, \quad \theta = \tan^{-1}\left(\frac{b}{a}\right) = \arctan\left(\frac{b}{a}\right)$$

General triangles

For any triangle, the following relationships hold:

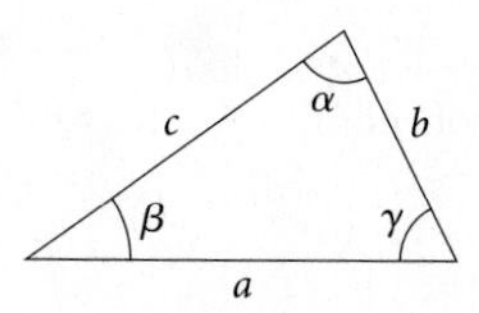

$$\alpha + \beta + \gamma = 180° = \pi \text{ rad}$$

Sine law: $\dfrac{\sin\alpha}{a} = \dfrac{\sin\beta}{b} = \dfrac{\sin\gamma}{c}$

Cosine law: $c^2 = a^2 + b^2 - 2ab\cos\gamma$

Identities

$$\tan\theta = \frac{\sin\theta}{\cos\theta}$$

$$\cot\theta = \frac{1}{\tan\theta} = \frac{\cos\theta}{\sin\theta}$$

$$\csc\theta = \frac{1}{\sin\theta}$$

$$\sec\theta = \frac{1}{\cos\theta}$$

Periodicity

$$\cos(\alpha + 2\pi) = \cos\alpha$$
$$\tan(\alpha + \pi) = \sin\alpha$$

Angle addition

$$\sin(\alpha \pm \beta) = \sin\alpha\cos\beta \pm \cos\alpha\sin\beta$$

$$\cos(\alpha \pm \beta) = \cos\alpha\cos\beta \mp \sin\alpha\sin\beta$$

Double angles

$\sin(2\alpha) = 2\sin\alpha\cos\alpha$

$\cos(2\alpha) = \cos^2\alpha - \sin^2\alpha = 1 - 2\sin^2\alpha = 2\cos^2\alpha - 1$

Other relations

$\sin^2\alpha + \cos^2\alpha = 1$

$\sin(-\alpha) = -\sin\alpha$

$\cos(-\alpha) = \cos\alpha$

$\sin(\alpha \pm \pi) = -\sin\alpha$

$\cos(\alpha \pm \pi) = -\cos\alpha$

$\sin(\alpha \pm \pi/2) = \pm\cos\alpha$

$\cos(\alpha \pm \pi/2) = \mp\sin\alpha$

The following graphs show $\sin\theta$, $\cos\theta$, and $\tan\theta$ as functions of θ:

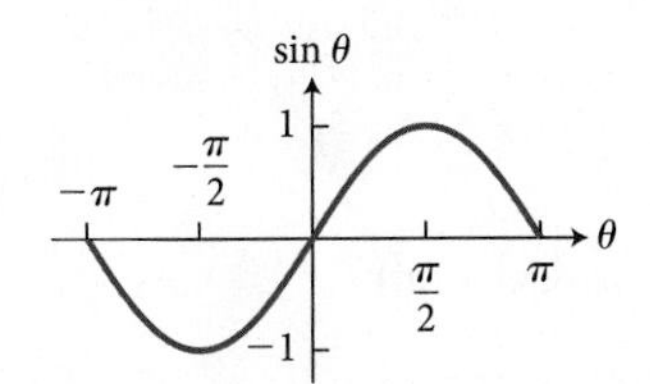

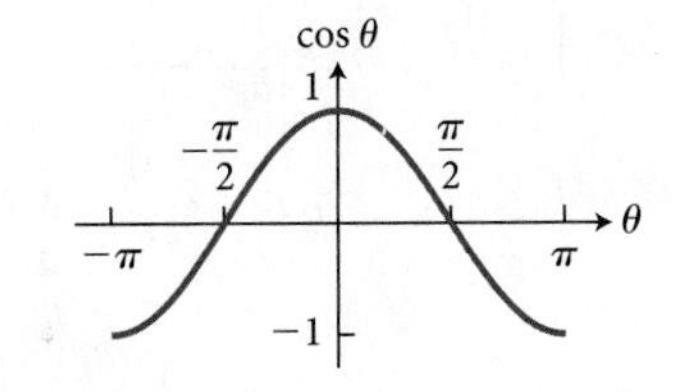

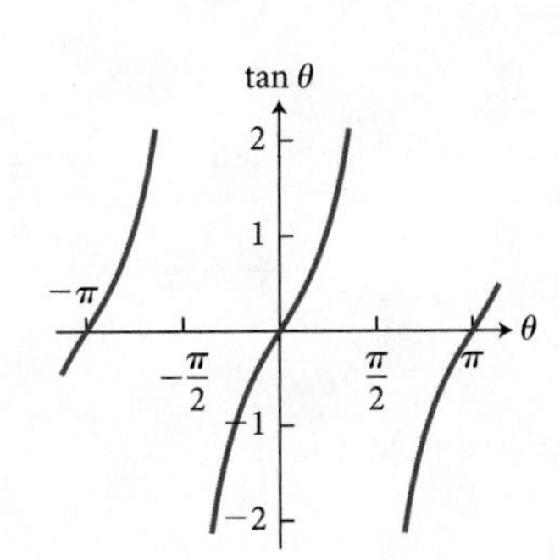

4 Vector algebra

A vector $\vec{A}$ in three-dimensional space can be written in terms of magnitudes A_x, A_y, and A_z of unit vectors $\hat{\imath}$, $\hat{\jmath}$, and $\hat{k}$, which have length 1 and lie along the x, y, and z axes:

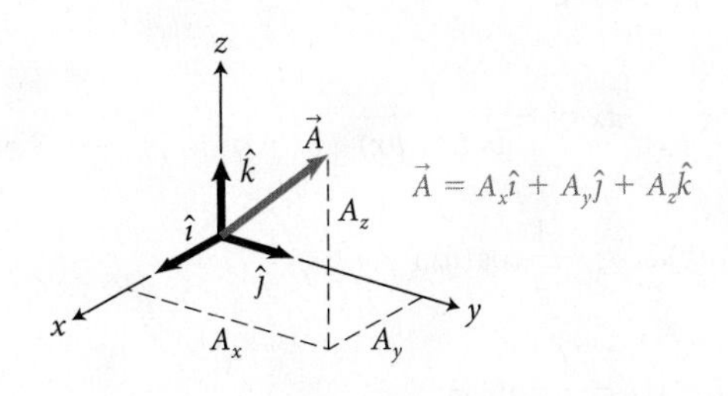

Dot products between vectors produce scalars:

$$\vec{A}\cdot\vec{B} = A_xB_x + A_yB_y + A_zB_z = |A||B|\cos\theta$$

(θ is the angle between vectors $\vec{A}$ and $\vec{B}$)

Cross products between vectors produce vectors:

$$\vec{A}\times\vec{B} = (A_yB_z - A_zB_y)\hat{\imath} + (A_zB_x - A_xB_z)\hat{\jmath} + (A_xB_y - A_yB_x)\hat{k}$$

$$|\vec{A}\times\vec{B}| = |\vec{A}||\vec{B}|\sin\theta \text{ (θ is the angle between vectors $\vec{A}$ and $\vec{B}$)}$$

The direction of $\vec{A}\times\vec{B}$ is given by the right-hand rule (see Figure 12.44).

5 Calculus

In this section, x is a variable, and a and n are constants.

Derivatives

Geometrically, the derivative of a function $f(x)$ at $x = x_1$ is the slope of $f(x)$ at x_1:

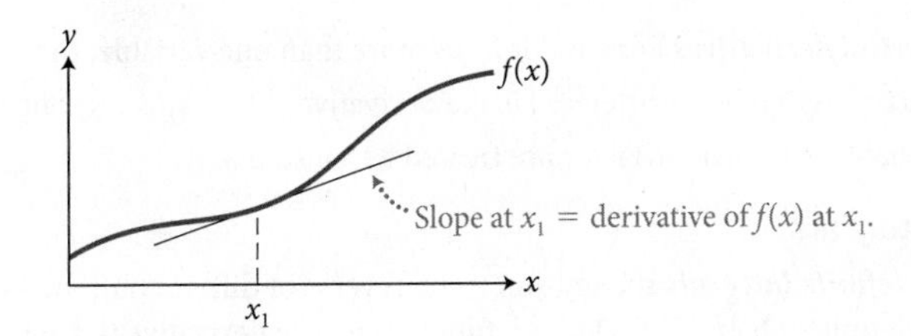

Derivatives of common functions

$$\frac{d}{dx}a = 0$$

$$\frac{d}{dx}x^n = nx^{n-1} \text{ (n need not be an integer)}$$

$$\frac{d}{dx}\sin x = \cos x$$

$$\frac{d}{dx}\cos x = -\sin x$$

$$\frac{d}{dx}\tan x = \frac{1}{\cos^2 x}$$

$$\frac{d}{dx}e^{ax} = ae^{ax}$$

$$\frac{d}{dx}\ln(ax) = \frac{1}{x}$$

$$\frac{d}{dx}a^x = a^x\ln a$$

Derivatives of sums, products, and functions of functions

Constant times a function: $\frac{d}{dx}[a\cdot f(x)] = a\cdot\frac{d}{dx}f(x)$

Sum of functions: $\frac{d}{dx}[f(x) + g(x)] = \frac{d}{dx}f(x) + \frac{d}{dx}g(x)$

Product of functions:

$$\frac{d}{dx}[f(x)\cdot g(x)] = g(x)\frac{d}{dx}f(x) + f(x)\frac{d}{dx}g(x)$$

Quotient of functions: $\frac{d}{dx}\left[\frac{f(x)}{g(x)}\right] = \frac{g(x)\frac{d}{dx}f(x) - f(x)\frac{d}{dx}g(x)}{[g(x)]^2}$

Functions of functions (the chain rule): If f is a function of u, and u is a function of x, then

$$\frac{d[f(u)]}{du}\cdot\frac{d[u(x)]}{dx} = \frac{d[f(x)]}{dx}$$

Second and higher derivatives The second derivative of a function f with respect to x is the derivative of the derivative:

$$\frac{d^2f(x)}{dx^2} = \frac{d}{dx}\left(\frac{d}{dx}f(x)\right)$$

Higher derivatives are defined similarly:

$$\frac{d^nf(x)}{dx^n} = \underbrace{\cdots\frac{d}{dx}\left(\frac{d}{dx}\left(\frac{d}{dx}f(x)\right)\right)}_{n \text{ uses of } \frac{d}{dx}}$$ (where n is a positive integer).

Partial derivatives For functions of more than one variable, the partial derivative, written $\frac{\partial}{\partial x}$, is the derivative with respect to one variable; all other variables are treated as constants.

Integrals

Indefinite integrals Integration is the reverse of differentiation. An indefinite integral $\int f(x)dx$ is a function whose derivative is $f(x)$.

That is to say, $\frac{d}{dx}[\int f(x)dx] = f(x)$.

If $A(x)$ is an indefinite integral of $f(x)$, then so is $A(x) + C$, where C is any constant. Thus, it is customary when evaluating indefinite integrals to add a "constant of integration" C.

Definite integrals The definite integral of $f(x)$, written as $\int_{x1}^{x2} f(x)dx$, represents the sum of the area of contiguous rectangles that each intersect $f(x)$ at some point along one base and that each have another base coincident with the x axis over some part of the range between x_1 and x_2; the indefinite integral evaluates the sum in the limit of arbitrarily small rectangle bases. In other words, the indefinite integral gives the net area that lies under $f(x)$ but above the x axis between the boundaries x_1 and x_2.

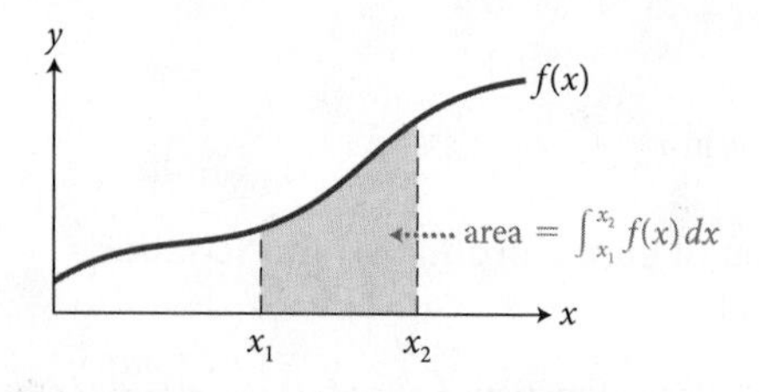

If $A(x)$ is any indefinite integral of $f(x)$, then the definite integral is given by $\int_{x1}^{x2} f(x)dx = A(x_2) - A(x_1) \equiv A(x)|_{x1}^{x_2}$. The constant of integration C does not affect the value of definite integrals and thus can be ignored (i.e., set to zero) during evaluation.

Integration by parts $\int_a^b udv$ is the area under the curve of $u(v)$. If $\int_a^b udv$ is difficult to evaluate directly, it is sometimes easier to express the area under the curve as the area within part of a rectangle minus the area under the curve of $v(u)$. In other words:

$$\int_a^b udv = uv|_a^b - \int_a^b vdu.$$

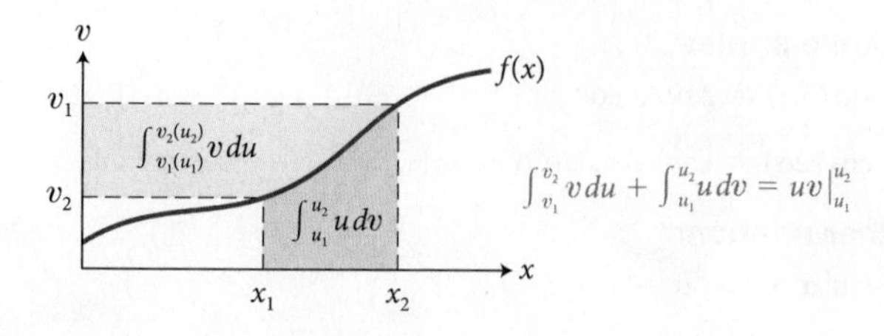

By choosing u and dv appropriately (both can be functions of x), this approach, called "integration by parts", can transform difficult integrals into easier ones.

Table of integrals In the following expressions, a and b are constants. An arbitrary constant of integration C can be added to the right-hand side.

$$\int x^n dx = \frac{1}{n+1}x^{n+1} \text{ (for } n \neq -1)$$

$$\int x^{-1}dx = \ln|x|$$

$$\int \frac{1}{a^2 + x^2}dx = \frac{1}{a}\tan^{-1}\frac{x}{a}$$

$$\int \frac{1}{(a^2 + x^2)^2}dx = \frac{1}{2a^3}\tan^{-1}\frac{x}{a} + \frac{x}{2a^2(x^2 + a^2)}$$

$$\int \frac{1}{\sqrt{\pm a^2 + x^2}}dx = \ln\left|x + \sqrt{\pm a^2 + x^2}\right|$$

$$\int \frac{1}{\sqrt{a^2 - x^2}}dx = \sin^{-1}\frac{x}{|a|} = \tan^{-1}\frac{x}{\sqrt{a^2 - x^2}}$$

$$\int \frac{x}{\sqrt{\pm a^2 - x^2}}dx = -\sqrt{\pm a^2 - x^2}$$

$$\int \frac{x}{\sqrt{\pm a^2 + x^2}}dx = \sqrt{\pm a^2 + x^2}$$

$$\int \frac{1}{(\pm a^2 + x^2)^{3/2}}dx = \frac{\pm x}{a^2\sqrt{\pm a^2 + x^2}}$$

$$\int \frac{x}{(a^2 + x^2)^{3/2}}dx = -\frac{1}{\sqrt{a^2 + x^2}}$$

$$\int \frac{1}{a + bx}dx = \frac{1}{b}\ln(a + bx)$$

$$\int \frac{1}{(a + bx)^2}dx = -\frac{1}{b(a + bx)}$$

$$\int \sin(ax)dx = -\frac{1}{a}\cos(ax)$$

$$\int \cos(ax)dx = \frac{1}{a}\sin(ax)$$

$$\int \tan(ax)dx = -\frac{1}{a}\ln(\cos ax)$$

$$\int \sin^2(ax)dx = \frac{x}{2} - \frac{\sin 2ax}{4a}$$

$$\int \cos^2(ax)dx = \frac{x}{2} + \frac{\sin 2ax}{4a}$$

$$\int x\sin(ax)dx = \frac{1}{a^2}\sin ax - \frac{1}{a}x\cos ax$$

$$\int x\cos(ax)dx = \frac{1}{a^2}\cos ax + \frac{1}{a}x\sin ax$$

$$\int e^{ax}dx = \frac{1}{a}e^{ax}$$

$$\int xe^{ax}dx = \frac{e^{ax}}{a^2}(ax - 1)$$

$$\int x^2e^{ax}dx = \frac{x^2e^{ax}}{a} - \frac{2}{a}\left[\frac{e^{ax}}{a^2}(ax - 1)\right]$$

$$\int \ln ax\,dx = x\ln(ax) - x$$

$$\int_0^\infty x^n e^{-ax}dx = \frac{n!}{a^{n+1}}$$

$$\int_0^\infty e^{-ax^2}dx = \frac{1}{2}\sqrt{\frac{\pi}{a}}$$

Line integrals. A *line integral* is an integral of a function that needs to be evaluated over a path (that is, a curve connecting two points in space). Consider, for example, the two-dimensional path C from point A to point B in the figure below. (The procedure described below is equally applicable in three dimensions.)

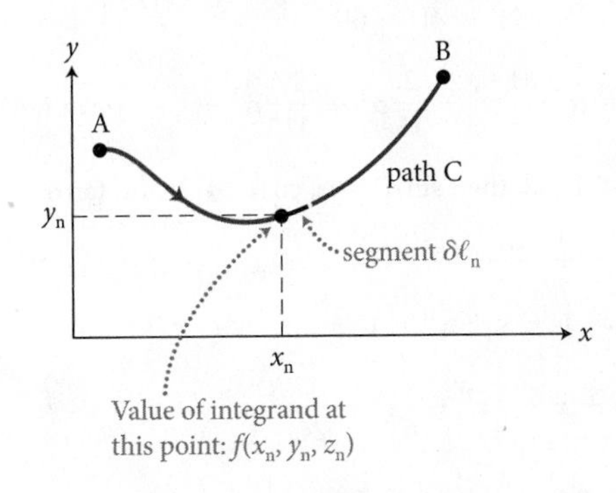

The path from A to B is *directed:* At any point the direction along the path away from A and toward B is forward (positive). Suppose we have a function $f(x, y, z)$ defined everywhere along the path. The function can be either a scalar or a vector; we will first discuss line integrals of scalar functions. We divide the path between A and B into small segments of length $\delta\ell_n$, each segment small enough that we can consider it essentially straight and small enough that the value of the function $f(x, y, z)$ can be considered constant over that segment. We then calculate the product $f(x_n, y_n, z_n)\delta\ell_n$ for each segment. The line integral of the function $f(x, y, z)$ along path C is then given by the sum of all those products along the path in the limit of infinitesimally small segments:

$$\int_C f(x, y, z)d\ell = \lim_{\delta\ell\to\infty} \sum_n f(x_n, y_n, z_n)\delta\ell_n.$$

To evaluate the integral on the right, we need to know the path C. Usually the path is specified in terms of the length parameter ℓ: $x = x(\ell), y = y(\ell), z = z(\ell)$. The line integral can then be written as an ordinary definite integral:

$$\int_C f(x, y, z)d\ell = \int_A^B f[x(\ell), y(\ell), z(\ell)]d\ell.$$

Next we consider the line integral of a vector function. We consider the same path C from A to B, but now we consider a vector function $\vec{F}(x, y, z)$. Instead of taking infinitesimally small scalar segments $d\ell_n$ along the path, we take small vector segments $d\vec{\ell}_n$ along the path, of length $d\ell_n$ and whose direction is tangent to the path in the direction of the path from A to B:

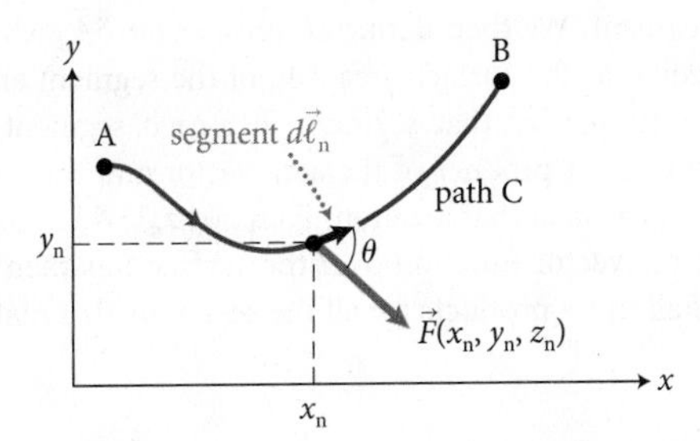

At each point we calculate the scalar product $\vec{F}(x_n, y_n, z_n) \cdot d\vec{\ell}_n$ and then sum these products over path C to obtain the line integral.

$$\int_C \vec{F}(x, y, z) \cdot d\vec{\ell}.$$

By writing out the scalar product, $\vec{F}(x, y, z) \cdot d\vec{\ell} = F(x, y, z)\cos\theta\, d\ell$, we can reduce the line integral of a vector function to that of a scalar function:

$$\int_C \vec{F}(x, y, z) \cdot d\vec{\ell} = \int_C F(x, y, z)\cos\theta\, d\ell.$$

In other words, we need to compute the line integral of the component of the vector $\vec{F}(x, y, z)$ along the tangent to the path.

If the path is closed—that is, the path returns to the starting point—we indicate that by putting a circle through the integration sign:

$$\oint_C \vec{F}(x, y, z) \cdot d\vec{\ell}.$$

Surface integrals. A *surface integral* is an integral of a function that needs to be evaluated over a surface. As with line integrals, the integrand of a surface integral can be a scalar or a vector function. We will only discuss the more general case of a vector function here.

The surface over which the integration is to be taken can be either *closed* or *open.* A closed surface, such as the surface of a sphere, divides space into two parts—an inside and an outside—and to get from one part to the other one has to go through the surface. An open surface does not have this property: For the surface S shown in the figure below, for example, one can go from one side of the surface to the other without passing through it.

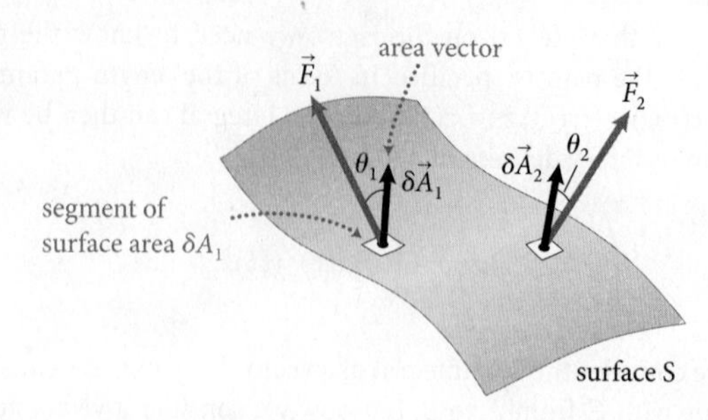

Consider a vector function $\vec{F}(x, y, z)$. To calculate the surface integral of this function over surface S, we begin by dividing the surface into small segments of surface area δA_n, each segment being small enough that we can consider it to be essentially flat and small enough so that the function $\vec{F}(x, y, z)$ can be considered constant over the segment. We then define an *area vector* $\delta\vec{A}_n$ whose magnitude is equal to the surface area δA_n of the segment and whose direction is normal to that segment. For each segment we then calculate the scalar product of the area vector and the value $\vec{F}_n$ of the vector function at that location: $\vec{F}_n(x_n, y_n, z_n) \cdot \delta\vec{A}_n$. The surface integral of the vector function over the surface S is then given by the sum of all those products for all the segments that make up the surface:

$$\int_S \vec{F}(x, y, z) \cdot d\vec{A} = \lim_{\delta A_n \to \infty} \sum_n \vec{F}_n(x_n, y_n, z_n) \cdot \delta\vec{A}_n.$$

If the surface is closed, we indicate that by putting a circle through the integration sign:

$$\oint_S \vec{F}(x, y, z) \cdot d\vec{A}$$

6 Complex numbers

A complex number $z = x + iy$ is defined in terms of its real part x and its imaginary part y. Both x and y are real numbers. i is Euler's constant, defined by the property $i^2 = -1$.

Each complex number z has a "complex conjugate" z^* which has the same real part but an imaginary part with opposite sign: $z = x + iy \Leftrightarrow z^* = x - iy$.

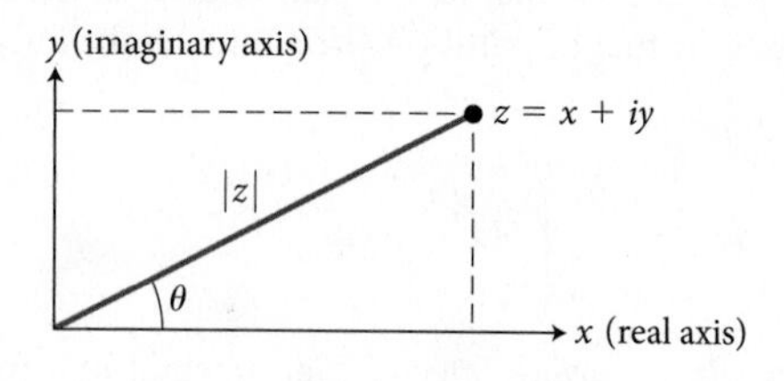

The real and imaginary parts can be expressed in terms of the complex number and its conjugate:

$$x = \tfrac{1}{2}(z + z^*)$$

$$y = \tfrac{1}{2}i(z - z^*)$$

A complex number is like a two-dimensional vector in a plane with a real axis and an imaginary axis. Thus, z can be described by a magnitude or length $|z|$ and an angle θ formed with the real axis (called the "phase angle"):

$z = |z|(\cos\theta + i\sin\theta)$, where $|z| = \sqrt{zz^*}$ and

$$\theta = \tan^{-1}\frac{y}{x} = \tan^{-1}\frac{i(z - z^*)}{(z + z^*)}.$$

Euler's formula says that $e^{i\theta} = \cos\theta + i\sin\theta$, allowing complex numbers to be written in the form $z = |z|e^{i\theta}$. This is a convenient form for expressing complex numbers. For example, it is easy to raise a complex number z to a power n: $z^n = |z|^n e^{in\theta}$.

7 Useful approximations

Binomial expansion

$$(1 + x)^n = 1 + nx + \frac{n(n-1)}{2}x^2 + \cdots$$

If $x \ll 1$, then $(1 + x)^n \approx 1 + nx$

Trigonometric expansions

$$\sin\alpha = \alpha - \frac{\alpha^3}{3!} + \frac{\alpha^5}{5!} - \frac{\alpha^7}{7!} + \cdots \quad (\alpha \text{ in rad})$$

$$\cos\alpha = 1 - \frac{\alpha^2}{2!} + \frac{\alpha^4}{4!} - \frac{\alpha^6}{6!} + \cdots \quad (\alpha \text{ in rad})$$

$$\tan\alpha = \alpha + \frac{1}{3}\alpha^3 + \frac{2}{15}\alpha^5 + \frac{17}{315}\alpha^7 + \cdots \quad (\alpha \text{ in rad})$$

If $\alpha \ll 1$ rad, then $\sin\alpha \approx \alpha$, $\cos\alpha \approx 1$, and $\tan\alpha \approx \alpha$.

Other useful expansions

$$\frac{1}{1 - x} = 1 + x + x^2 + x^3 + \cdots \quad \text{for } -1 < x < 1$$

$$e^x = 1 + x + \frac{1}{2}x^2 + \frac{1}{6}x^3 + \frac{1}{24}x^4 + \cdots$$

$$\ln(1 + x) = x - \frac{1}{2}x^2 + \frac{1}{3}x^3 - \frac{1}{4}x^4 + \cdots \quad \text{for } -1 < x < 1$$

$$\ln\left(\frac{1 + x}{1 - x}\right) = 2x + \frac{2}{3}x^3 + \frac{2}{5}x^5 - \frac{2}{7}x^7 + \cdots \quad \text{for } -1 < x < 1$$

Appendix C

SI Units, Useful Data, and Unit Conversion Factors

The seven base SI units

Unit	Abbreviation	Physical quantity
meter	m	length
kilogram	kg	mass
second	s	time
ampere	A	electric current
kelvin	K	thermodynamic temperature
mole	mol	amount of substance
candela	cd	luminous intensity

Some derived SI units

Unit	Abbreviation	Physical quantity	In terms of base units
newton	N	force	$kg \cdot m/s^2$
joule	J	energy	$kg \cdot m^2/s^2$
watt	W	power	$kg \cdot m^2/s^3$
pascal	Pa	pressure	$kg/m \cdot s^2$
hertz	Hz	frequency	s^{-1}
coulomb	C	electric charge	$A \cdot s$
volt	V	electric potential	$kg \cdot m^2/(A \cdot s^3)$
ohm	Ω	electric resistance	$kg \cdot m^2/(A^2 \cdot s^3)$
farad	F	capacitance	$A^2 \cdot s^4/(kg \cdot m^2)$
tesla	T	magnetic field	$kg/(A \cdot s^2)$
weber	Wb	magnetic flux	$kg \cdot m^2/(A \cdot s^2)$
henry	H	inductance	$kg \cdot m^2/(A^2 \cdot s^2)$

SI Prefixes

10^n	Prefix	Abbreviation	10^n	Prefix	Abbreviation
10^0	—	—			
10^3	kilo-	k	10^{-3}	milli-	m
10^6	mega-	M	10^{-6}	micro-	μ
10^9	giga-	G	10^{-9}	nano-	n
10^{12}	tera-	T	10^{-12}	pico-	p
10^{15}	peta-	P	10^{-15}	femto-	f
10^{18}	exa-	E	10^{-18}	atto-	a
10^{21}	zetta-	Z	10^{-21}	zepto-	z
10^{24}	yotta-	Y	10^{-24}	yocto-	y

Values of fundamental constants

Quantity	Symbol	Value
Speed of light in vacuum	c_0	3.00×10^8 m/s
Gravitational constant	G	6.6738×10^{-11} N · m²/kg²
Avogadro's number	N_A	6.0221413×10^{23} mol^{-1}
Boltzmann's constant	k_B	1.380×10^{-23} J/K
Charge on electron	e	1.60×10^{-19} C
Permittivity constant	ϵ_0	$8.85418782 \times 10^{-12}$ C²/(N · m²)
Permeability constant	μ_0	$4\pi \times 10^{-7}$ T · m/A
Planck's constant	h	6.626×10^{-34} J · s
Electron mass	m_e	9.11×10^{-31} kg
Proton mass	m_p	1.6726×10^{-27} kg
Neutron mass	m_n	1.6749×10^{-27} kg
Atomic mass unit	amu	1.6605×10^{-27} kg

Other useful numbers

Number or quantity	Value
π	3.1415927
e	2.7182818
1 radian	57.2957795°
Absolute zero ($T = 0$)	−273.15 °C
Average acceleration g due to gravity near Earth's surface	9.8 m/s²
Speed of sound in air at 20 °C	343 m/s
Density of dry air at atmospheric pressure and 20 °C	1.29 kg/m³
Earth's mass	5.97×10^{24} kg
Earth's radius (mean)	6.38×10^6 m
Earth–Moon distance (mean)	3.84×10^8 m

Unit conversion factors

Length

1 in. = 2.54 cm (defined)

1 cm = 0.3937 in.

1 ft = 30.48 cm

1 m = 39.37 in. = 3.281 ft

1 mi = 5280 ft = 1.609 km

1 km = 0.6214 mi

1 nautical mile (U.S.) = 1.151 mi = 6076 ft = 1.852 km

1 fermi = 1 femtometer (fm) = 10^{-15} m

1 angstrom (Å) = 10^{-10} m = 0.1 nm

1 light − year (ly) = 9.461×10^{15} m

1 parsec = 3.26 ly = 3.09×10^{16} m

Volume

1 liter (L) = 1000 mL = 1000 cm^3 = 1.0×10^{-3} m^3
= 1.057 qt (U.S.) = 61.02 $in.^3$

1 gal (U.S.) = 4 qt (U.S.) = 231 $in.^3$ = 3.785 L = 0.8327 gal (British)

1 quart (U.S.) = 2 pints (U.S.) = 946 mL

1 pint (British) = 1.20 pints (U.S.) = 568 mL

1 m^3 = 35.31 ft^3

Speed

1 mi/h = 1.4667 ft/s = 1.6093 km/h = 0.4470 m/s

1 km/h = 0.2778 m/s = 0.6214 mi/h

1 ft/s = 0.3048 m/s = 0.6818 mi/h = 1.0973 km/h

1 m/s = 3.281 ft/s = 3.600 km/h = 2.237 mi/h

1 knot = 1.151 mi/h = 0.5144 m/s

Angle

1 radian (rad) = 57.30° = 57°18'

1° = 0.01745 rad

1 rev/min (rpm) = 0.1047 rad/s

Time

1 day = 8.640×10^4 s

1 year = 365.242 days = 3.156×10^7 s

Mass

1 atomic mass unit (u) = 1.6605×10^{-27} kg

1 kg = 0.06852 slug

1 metric ton = 1000 kg

1 long ton = 2240 lbs = 1016 kg

1 short ton = 2000 lbs = 909.1 kg

1 kg has a weight of 2.20 lb where $g = 9.80\ m/s^2$

Force

1 lb = 4.44822 N

1 N = 10^5 dyne = 0.2248 lb

Energy and work

1 J = 10^7 ergs = 0.7376 ft · lb

1 ft · lb = 1.356 J = 1.29×10^{-3} Btu = 3.24×10^{-4} kcal

1 kcal = 4.19×10^3 J = 3.97 Btu

1 eV = 1.6022×10^{-19} J

1 kWh = 3.600×10^6 J = 860 kcal

1 Btu = 1.056×10^3 J

Power

1 W = 1 J/s = 0.7376 ft · lb/s = 3.41 Btu/h

1 hp = 550 ft · lb/s = 746 W

1 kWh/day = 41.667 W

Pressure

1 atm = 1.01325 bar = 1.01325×10^5 N/m^2 = 14.7 lb/$in.^2$ = 760 torr

1 lb/$in.^2$ = 6.895×10^3 N/m^2

1 Pa = 1 N/m^2 = 1.450×10^{-4} lb/$in.^2$

Periodic Table of the Elements

Number of protons → 29
Symbol for element → **Cu**
63.546 ← Average atomic mass in g/mol. For elements having no stable isotope, value in parentheses is approximate atomic mass of longest-lived isotope.

Period / Group	1	2	3	4	5	6	7	8	9	10	11	12	13	14	15	16	17	18
1	1 **H** 1.008																	2 **He** 4.003
2	3 **Li** 6.941	4 **Be** 9.012											5 **B** 10.811	6 **C** 12.011	7 **N** 14.007	8 **O** 15.999	9 **F** 18.998	10 **Ne** 20.180
3	11 **Na** 22.990	12 **Mg** 24.305											13 **Al** 26.982	14 **Si** 28.086	15 **P** 30.974	16 **S** 32.065	17 **Cl** 35.453	18 **Ar** 39.948
4	19 **K** 39.098	20 **Ca** 40.078	21 **Sc** 44.956	22 **Ti** 47.867	23 **V** 50.942	24 **Cr** 51.996	25 **Mn** 54.938	26 **Fe** 55.845	27 **Co** 58.933	28 **Ni** 58.693	29 **Cu** 63.546	30 **Zn** 65.409	31 **Ga** 69.723	32 **Ge** 72.64	33 **As** 74.922	34 **Se** 78.96	35 **Br** 79.904	36 **Kr** 83.798
5	37 **Rb** 85.468	38 **Sr** 87.62	39 **Y** 88.906	40 **Zr** 91.224	41 **Nb** 92.906	42 **Mo** 95.94	43 **Tc** (98)	44 **Ru** 101.07	45 **Rh** 102.906	46 **Pd** 106.42	47 **Ag** 107.868	48 **Cd** 112.411	49 **In** 114.818	50 **Sn** 118.710	51 **Sb** 121.760	52 **Te** 127.60	53 **I** 126.904	54 **Xe** 131.293
6	55 **Cs** 132.905	56 **Ba** 137.327	71 **Lu** 174.967	72 **Hf** 178.49	73 **Ta** 180.948	74 **W** 183.84	75 **Re** 186.207	76 **Os** 190.23	77 **Ir** 192.217	78 **Pt** 195.078	79 **Au** 196.967	80 **Hg** 200.59	81 **Tl** 204.383	82 **Pb** 207.2	83 **Bi** 208.980	84 **Po** (209)	85 **At** (210)	86 **Rn** (222)
7	87 **Fr** (223)	88 **Ra** (226)	103 **Lr** (262)	104 **Rf** (261)	105 **Db** (262)	106 **Sg** (266)	107 **Bh** (264)	108 **Hs** (269)	109 **Mt** (268)	110 **Ds** (271)	111 **Rg** (272)	112 **Uub** (285)	113 **Uut** (284)	114 **Uuq** (289)	115 **Uup** (288)	116 **Uuh** (292)	117 **Uus** (294)	118 **Uuo**

Lanthanoids	57 **La** 138.905	58 **Ce** 140.116	59 **Pr** 140.908	60 **Nd** 144.24	61 **Pm** (145)	62 **Sm** 150.36	63 **Eu** 151.964	64 **Gd** 157.25	65 **Tb** 158.925	66 **Dy** 162.500	67 **Ho** 164.930	68 **Er** 167.259	69 **Tm** 168.934	70 **Yb** 173.04
Actinoids	89 **Ac** (227)	90 **Th** (232)	91 **Pa** (231)	92 **U** (238)	93 **Np** (237)	94 **Pu** (244)	95 **Am** (243)	96 **Cm** (247)	97 **Bk** (247)	98 **Cf** (251)	99 **Es** (252)	100 **Fm** (257)	101 **Md** (258)	102 **No** (259)

Appendix D

Center of Mass of Extended Objects

We can apply the concept of center of mass to extended objects. Consider, for example, the object of inertia m in **Figure D.1**. If you imagine breaking down the object into many small segments of equal inertia δm, you can use Eq. 6.24 to compute the position of the center of mass:

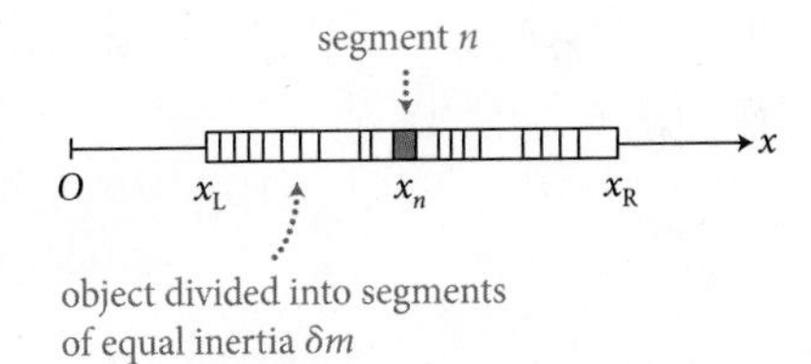

Figure D.1

$$x_{cm} = \frac{\delta m_1 x_1 + \delta m_2 x_2 + \cdots}{\delta m_1 + \delta m_2 + \cdots}, \tag{D.1}$$

where x_n is the position of segment δm_n. Because the sum of the inertias of all segments is equal to the inertia m of the extended object, $\delta m_1 + \delta m_2 + \cdots = m$, we can write Eq. D.1 as

$$x_{cm} = \frac{1}{m}(\delta m_1 x_1 + \delta m_2 x_2 + \cdots) = \frac{1}{m}\sum_n (\delta m_n x_n) \tag{D.2}$$

To evaluate this sum for the extended object, we take the limit of this expression as $\delta m \to 0$. In this limit, the sum becomes an integral:

$$x_{cm} = \frac{1}{m}\lim_{\delta m \to 0}\sum_n (\delta m_n x_n) \equiv \frac{1}{m}\int_{\text{object}} x\,dm. \tag{D.3}$$

To evaluate the integral we need to know how the inertia is distributed over the object. Let the *inertia per unit length* of the object be $\lambda \equiv dm/dx$. In general λ is a function of position—that is, the inertia per unit length need not be the same at different locations along the object—and so $\lambda = \lambda(x)$ and $dm = \lambda(x)dx$. Substituting this expression for dm into Eq. D.3, we obtain

$$x_{cm} = \frac{1}{m}\int_{\text{object}} x\lambda(x)dx. \tag{D.4}$$

In this expression, the limits of integration should be taken to be the positions of the left and right ends of the object (x_L and x_R, respectively).

For example, let the object in Figure D.1 have a uniformly distributed inertia so that $\lambda(x)$ has the same value (the inertia of the extended object divided by the length of the object) everywhere:

$$\lambda(x) = \frac{m}{x_R - x_L}. \tag{D.5}$$

Because $\lambda(x)$ does not depend on x, we can pull it out of the integral in Eq. D.4 and so, substituting Eq. D.5 into Eq. D.4, we get

$$\begin{aligned} x_{cm} &= \frac{\lambda(x)}{m}\int_{x_L}^{x_R} x\,dx = \frac{1}{x_R - x_L}\int_{x_L}^{x_R} x\,dx = \frac{1}{x_R - x_L}\left[\tfrac{1}{2}x^2\right]_{x_L}^{x_R} \\ &= \frac{x_R^2 - x_L^2}{2(x_R - x_L)} = \frac{x_R + x_L}{2}. \end{aligned} \tag{D.6}$$

In other words, the center of mass of the object is halfway between the ends of the object (at the center of the object), as we expect.

Appendix E

Derivation of the Lorentz Transformation Equations

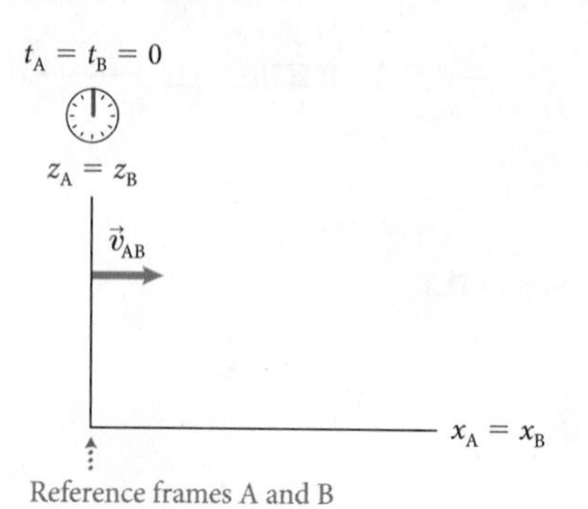

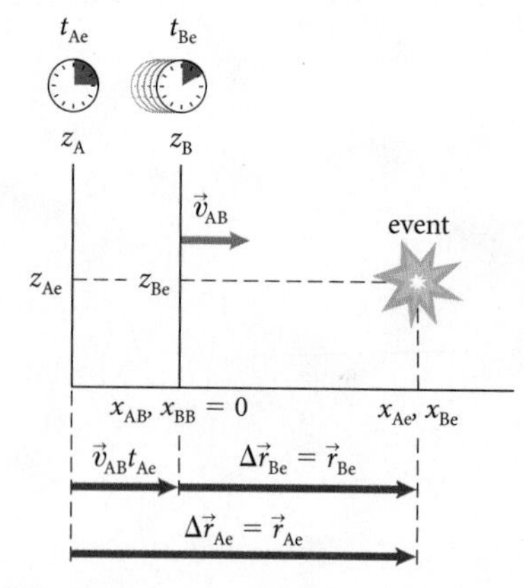

Figure E.1 Determining the space and time coordinates of an event from two reference frames that overlap at $t_A = t_B = 0$.

In this appendix we derive the Lorentz transformation equations (Eqs. 14.29–14.32), which relate position and time in two reference frames A and B. The origins of the reference frames coincide at $t_A = t_B = 0$, and the frames are moving at constant velocity $\vec{v}_{AB} = v_{ABx}\hat{\imath}$ in the x direction relative to each other (Figure E.1*a*).

Consider an event e that occurs at instant t_{Ae} and position x_{Ae} in reference frame A (Figure E.1*b*).* Our goal is to express t_{Be} and x_{Be}, the instant and the position at which the event occurs in reference frame B, in terms of t_{Ae} and x_{Ae}:

$$t_{Be} = f(x_{Ae}, t_{Ae}) \tag{E.1}$$

$$x_{Be} = g(x_{Ae}, t_{Ae}). \tag{E.2}$$

To obtain these relationships, we consider the space-time interval (Eq. 14.18) between the event in Figure E.1*b* and the event in Figure E.1*a* (this event being the overlapping of the two origins). The spatial separation of these two events is $\Delta x_{Ae} = x_{Ae} - 0 = x_{Ae}$ in reference frame A and $\Delta x_{Be} = x_{Be} - 0 = x_{Be}$ in reference frame B. The temporal separations are $\Delta t_{Ae} = t_{Ae} - 0 = t_{Ae}$ and $\Delta t_{Be} = t_{Be} - 0 = t_{Be}$. Because the space-time interval is an invariant, we have

$$(c_0 t_{Be})^2 - (x_{Be})^2 = (c_0 t_{Ae})^2 - (x_{Ae})^2. \tag{E.3}$$

To satisfy Eq. E.3, the relationships expressed in Eqs. E.1 and E.2 must be linear:

$$t_{Be} = Dt_{Ae} + Ex_{Ae} \tag{E.4}$$

$$x_{Be} = Ft_{Ae} + Gx_{Ae}, \tag{E.5}$$

where D, E, F, G are constants. To obtain relationships between these constants, consider the position of the origin of reference frame B at the instant illustrated in Figure E.1*b*. The origin of reference frame B moves at constant velocity $\vec{v}_{AB}$ relative to reference frame A, and so at t_{Ae} the position of B's origin in reference frame A is

$$x_{AB} = v_{ABx}t_{Ae}. \tag{E.6}$$

The position of B's origin in reference frame B is always zero, $x_{BB} = 0$.† According to Eq. E.5 we then have $x_{BB} = 0 = Ft_{Ae} + Gx_{AB}$, or substituting for x_{AB} from Eq. E.6, $0 = Ft_{Ae} + G(v_{ABx}t_{Ae})$. Therefore

$$F = -v_{ABx}G. \tag{E.7}$$

Conversely, the origin of reference frame A is located at $x_{BA} = -v_{ABx}t_{Be}$ in reference frame B, and so, with $x_{AA} = 0$, Eqs. E.4 and E.5 become

$$t_{Be} = Dt_{Ae} + 0 \tag{E.8}$$

$$-v_{ABx}t_{Be} = Ft_{Ae} + 0. \tag{E.9}$$

Solving Eqs. E.8 and E.9 for F gives $F = -v_{ABx}D$, and so, from Eq. E.7,

$$D = G. \tag{E.10}$$

---

*Remember our subscript format: The capital letter refers to the reference frame; the lower case e is for "event." Thus the vector $\vec{r}_{Ae}$ represents observer A's measurement of the position at which the event occurs.

†Do not be stymied by the unfamiliar subscript BB. It's the same format we've been using all along: The first B tells you the reference frame in which the measurement is made, the second B tells you what's being measured—in this case, the position of the origin of reference frame B.

Equations E.4 and E.5 then become

$$t_{\mathrm{Be}} = G t_{\mathrm{Ae}} + E x_{\mathrm{Ae}} \tag{E.11}$$

$$x_{\mathrm{Be}} = -v_{\mathrm{AB}x} G t_{\mathrm{Ae}} + G x_{\mathrm{Ae}}. \tag{E.12}$$

Substituting Eqs. E.11 and E.12 into the left side of Eq. E.3 and using $v_{\mathrm{AB}x}^2 = v^2$ yields

$$c_0^2 G^2 \left(1 - \frac{v^2}{c_0^2}\right) t_{\mathrm{Ae}}^2 + (c_0^2 EG + G^2 v_{\mathrm{AB}x}) 2 t_{\mathrm{Ae}} x_{\mathrm{Ae}} - (G^2 - c_0^2 E^2) x_{\mathrm{Ae}}^2$$
$$= (c_0 t_{\mathrm{Ae}})^2 - (x_{\mathrm{Ae}})^2, \tag{E.13}$$

and then gather all the $(t_{\mathrm{Ae}})^2$ terms, all the $x_{\mathrm{Ae}}\, t_{\mathrm{Ae}}$ terms, and all the $(t_{\mathrm{Ae}})^2$ terms:

$$\left[c_0^2 G^2 \left(1 - \frac{v^2}{c_0^2}\right) - c_0^2\right] t_{\mathrm{Ae}}^2 + 2(c_0^2 EG + G^2 v_{\mathrm{AB}x}) x_{\mathrm{Ae}} t_{\mathrm{Ae}}$$
$$- (G^2 - c_0^2 E^2 - 1) x_{\mathrm{Ae}}^2 = 0. \tag{E.14}$$

Because Eq. E.14 must hold for any value of t_{Ae} and x_{Ae}, the coefficient of each term must be zero. The coefficient of the t_{Ae}^2 term yields

$$c_0^2 G^2 \left(1 - \frac{v^2}{c_0^2}\right) = c_0^2. \tag{E.15}$$

Solving Eq. E.15 for G yields

$$G = \left(1 - \frac{v^2}{c_0^2}\right)^{-1/2} \equiv \gamma, \tag{E.16}$$

and so from Eq. E.7, $F = -v_{\mathrm{AB}x} G$, we see that $F = -\gamma v_{\mathrm{AB}x}$.

The coefficient of the $x_{\mathrm{Ae}} t_{\mathrm{Ae}}$ term in Eq. E.14 must also be zero. Substituting γ for G in that coefficient gives us

$$c_0^2 E \gamma + \gamma^2 v_{\mathrm{AB}x} = 0 \tag{E.17}$$

$$E = -\gamma v_{\mathrm{AB}x} / c_0^2. \tag{E.18}$$

Now we are ready to substitute for *D*, *E*, *F*, *G* in Eqs. E.4 and E.5—*D* from Eq. E.10, *E* from Eq. E.18, *F* from Eq. E.7, *G* from Eq. E.16:

$$t_{\mathrm{Be}} = \gamma\left(t_{\mathrm{Ae}} - \frac{1}{c_0^2} v_{\mathrm{AB}x} x_{\mathrm{Ae}}\right) \tag{E.19}$$

$$x_{\mathrm{Be}} = \gamma (x_{\mathrm{Ae}} - v_{\mathrm{AB}x} t_{\mathrm{Ae}}). \tag{E.20}$$

Because there is no length contraction in the directions perpendicular to the relative velocity $\vec{v}_{\mathrm{AB}}$ of the two reference frames, we have

$$y_{\mathrm{Be}} = y_{\mathrm{Ae}} \tag{E.21}$$

$$z_{\mathrm{Be}} = z_{\mathrm{Ae}}. \tag{E.22}$$

Equations E.19–E.22 are the Lorentz transformation equations we wanted to derive.

Velocities in two reference frames

We can use Eqs. E.19 and E.20 to derive Eq. 14.33, the relationship between the *x* components of the velocities of an object o measured in two reference frames A and B. Let the *x* component be $v_{\mathrm{Ao}x} = dx_{\mathrm{Ao}}/dt_{\mathrm{A}}$ in reference frame A and $v_{\mathrm{Bo}x} = dx_{\mathrm{Bo}}/dt_{\mathrm{B}}$ in reference frame B. We begin by writing Eqs. E.19 and E.20 in differential form:

$$dt_{\mathrm{B}} = \gamma\left(dt_{\mathrm{A}} - \frac{v_{\mathrm{AB}x}}{c_0^2} dx_{\mathrm{Ao}}\right) \tag{E.23}$$

$$dx_{\mathrm{Bo}} = \gamma (dx_{\mathrm{Ao}} - v_{\mathrm{AB}x}\, dt_{\mathrm{A}}), \tag{E.24}$$

$$dy_{\mathrm{Bo}} = dy_{\mathrm{Ao}}, \tag{E.25}$$

$$dz_{\mathrm{Bo}} = dz_{\mathrm{Ao}}, \tag{E.26}$$

where we have replaced the subscript e by o because we are considering an object rather than an event. Dividing Eq. E.24 by Eq. E.23 yields an expression for the x component of the velocity in reference frame B:

$$v_{\text{Box}} = \frac{dx_{\text{Bo}}}{dt_{\text{B}}} = \frac{\gamma(dx_{\text{Ao}} - v_{\text{AB}x}\,dt_{\text{A}})}{\gamma\left(dt_{\text{A}} - \frac{v_{\text{AB}x}}{c_0^2}dx_{\text{Ao}}\right)} = \frac{(dx_{\text{Ao}} - v_{\text{AB}x}\,dt_{\text{A}})}{\left(dt_{\text{A}} - \frac{v_{\text{AB}x}}{c_0^2}dx_{\text{Ao}}\right)}. \tag{E.27}$$

Finally, we divide the numerator and denominator of the rightmost term by dt_{A} to obtain Eq. 14.33:

$$v_{\text{Box}} = \frac{\left(\frac{dx_{\text{Ao}}}{dt_{\text{A}}} - v_{\text{AB}x}\right)}{\left(1 - \frac{v_{\text{AB}x}}{c_0^2}\frac{dx_{\text{Ao}}}{dt_{\text{A}}}\right)} = \frac{(v_{\text{Ao}x} - v_{\text{AB}x})}{\left(1 - \frac{v_{\text{AB}x}}{c_0^2}v_{\text{Ao}x}\right)}. \tag{E.28}$$

To find the y component of the velocity in reference frame B, we divide Eq. E.25 by Eq. E.23:

$$v_{\text{Bo}y} = \frac{dy_{\text{Bo}}}{dt_{\text{B}}} = \frac{dy_{\text{Ao}}}{\gamma\left(dt_{\text{A}} - \frac{v_{\text{AB}x}}{c_0^2}dx_{\text{Ao}}\right)}. \tag{E.29}$$

Dividing numerator and denominator by dt_{A} then yields

$$v_{\text{Bo}y} = \frac{v_{\text{Ao}y}}{\gamma\left(1 - \frac{v_{\text{AB}x}v_{\text{Ao}x}}{c_0^2}\right)}. \tag{E.30}$$

For the z component of the velocity we obtain a similar equation with y replaced by z

$$v_{\text{Bo}z} = \frac{v_{\text{Ao}z}}{\gamma\left(1 - \frac{v_{\text{AB}x}v_{\text{Ao}x}}{c_0^2}\right)}. \tag{E.31}$$

$$\Delta p_{sx} = +0.38\ \text{kg} \cdot \text{m/s} - 0$$

Solutions to Checkpoints

Chapter 22

22.1. (*a*) You should observe that the tape and battery attract each other. (If the battery repels the tape, you can get rid of the repulsion by wiping the entire battery surface with your hand.) It makes no difference how you orient the battery or whether the battery is fresh or spent. Any wooden object should also attract the tape in the same way a battery does. (*b*) The tape and the power cord attract each other. Turning the lamp on or off doesn't make any difference. It does not appear that the power cord has an effect different from any other object.

22.2. (*a*) You should see that the two strips repel each other quite strongly. (*b*) No. Regardless of how you orient the strips—sticky sides facing, nonsticky sides facing, one sticky side facing one nonsticky side—the two always repel. As you bring them closer, they twist to avoid contact.

22.3. Paper-paper, no interaction; tape-paper, interaction; tape-tape, interaction.

22.4. (*a*) The strip should be attracted to your hand and the strip of paper and be repelled by the second charged strip. (*b*) The uncharged strip should remain motionless when you hold your hand close to it. (*c*) You should see no interaction between the uncharged hanging strip and the strip of paper, and there should be an attractive interaction between the two tape strips. This tape-tape interaction is just like the attractive interaction between a charged tape strip and any other uncharged object.

22.5. The recharged strip should be attracted to your hand and to a strip of paper and repelled by another charged tape strip.

22.6. (*a*) There is an attractive interaction between both types of strips and uncharged objects. (*b*) A third charged strip attracts strip B and repels strip T. (*c*) Yes. Strip T must be charged because it interacts with other charged strips and with uncharged objects. (*d*) Yes. Strip B must be charged because it interacts with other charged strips and with uncharged objects. (*e*) After discharging, neither strip B nor strip T interacts with uncharged objects, and both strips are attracted to a charged strip. (See the box "Troubleshooting B and T strips" if your experimental results do not agree with these.)

Troubleshooting B and T strips

If your experiment with B and T strips doesn't work as expected, check the following:

1. You must pull off the combination in step 2 of Figure 22.5 *very* slowly. (The amount of charge that builds up on the strips is roughly proportional to the speed at which you separate them.) Be sure to remove *all* charge before proceeding.
2. Separating the B and T strips, on the other hand, must be done fairly rapidly. (If you do it too fast, however, so much charge may build up on your strips that it becomes hard to prevent them from being attracted to your hands. If they curl around and touch your hands, you must start over.)
3. Avoid any air currents on the suspended strips.
4. If the humidity of the air is high, the strips may lose their charge rapidly; you may need to repeat the experiment in a drier environment.

22.7. See Figure 22.6.

22.8. (*a*) The comb attracts the uncharged paper strip and repels the (charged) tape strip. (*b*) The comb repels one strip (usually the B strip) and attracts the other. (*c*) The charge on the comb behaves the same way as the charge on the strip it repels (usually B). The charge on the comb is therefore the same type as the charge on the strip it repels.

22.9. If in Checkpoint 22.8 your comb repelled the B strip, your brand of tape produces B strips that carry a negative charge. If the comb attracted the B strip, your brand of tape produces B strips that carry a positive charge.

22.10. (*a*) The red marbles, all carrying a positive charge, exert repulsive forces on one another and so move as far away as possible from one another. (*b*) The blue marbles, all carrying a negative charge, also exert repulsive forces on one another and so move as far as possible from one another. (*c*) The red and blue marbles are attracted to each other and so form red-blue pairs. The (sum of the) charge on any given pair is zero and so does not repel the other pairs. Consequently the pairs do not spread out. (The positive and negative charges do not overlap completely—a red marble and a blue one cannot be at the same place—and so, as we shall see later, some residual interaction is left.) (*d*) Each blue marble becomes part of a red-blue pair; the leftover red marbles repel one another and spread out. (*e*) Because of the surplus of positively charged red marbles, the entire collection carries a positive charge.

22.11. (*a*) Any charge deposited on the metal rod spreads out over the entire conducting system, which in this case is the metal rod plus your body. Most of the charge therefore ends up on you. (*b*) Because rubber is an electrical insulator, any charge on the rod that is not in contact with your hand stays on the rod.

22.12. When you rub together surfaces made of the same material, friction does occur, which tells you that bonds do form between the surfaces. Because the two surfaces are made of the same material, however, there is no preferred direction to transfer charge—the same numbers of electrons are transferred in each direction.

22.13. (*a*) Before the charged rod is brought nearby, the electroscope is neutral, meaning it has as many positive charge carriers as negative charge carriers. No charge is transferred to the electroscope, and so even with the rod nearby, the electroscope still has as many positive charge carriers as negative charge carriers (although they are now separated) and is still neutral. (*b*) The magnitudes must be equal because otherwise the electroscope would not be neutral. (*c*) The negative charge carriers on the rod attract the positive charge carriers on the electroscope ball and repel the negative charge carriers on the leaves. (*d*) The positive charge on the ball is equal in magnitude to the negative charge on the leaves. Therefore, all other things being equal, the magnitudes of the forces are the same. However, the distance between leaves and rod is greater than the distance between ball and rod. Because the electric force decreases with increasing distance, the magnitude of the attractive force exerted on the ball will be *greater* than the magnitude of the repulsive force exerted on the leaves.

22.14. The electron cloud is attracted not only to the external positive charge but also to the positive nucleus. The attractive electric interaction between electron cloud and nucleus prevents the electron cloud from leaving the atom entirely.

22.15. (*a*) Attractive. When the external negatively charged body in Figure 22.21 is replaced by a positively charged body, the atoms in the paper still get polarized. The polarization is now in the other direction, however, so that the negative charge appears on the top in Figure 22.21, meaning the force is again attractive. The vector sum of the forces exerted by the charged object on the piece of paper is still attractive. (*b*) It makes no difference whether positively charged protons move up or negatively charged electrons move down. The effect is the same: a surplus of positive charge on the ball of the electroscope and a deficit of positive charge on its leaves (Figure S22.15). (*c*) No, because the outcome does not depend on the type of charge that moves. The experiment that demonstrates that it is indeed the electrons that flow in a metal—as I asserted in Section 22.3—is beyond the scope of this chapter. We will discuss it in Chapter 27.

Figure S22.15

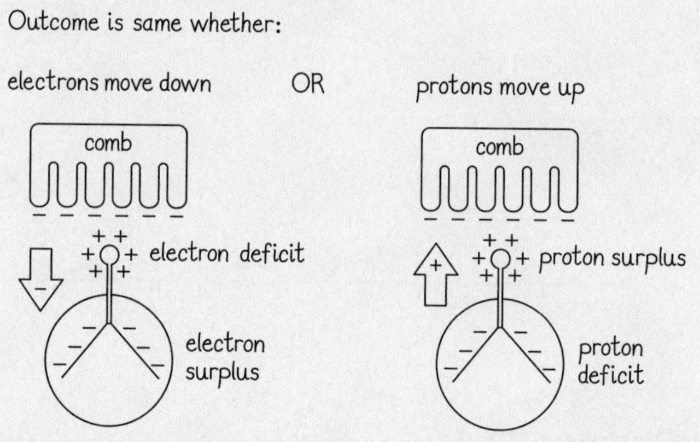

Outcome is same whether:

electrons move down OR protons move up

comb

electron deficit

electron surplus

comb

proton surplus

proton deficit

22.16. Greater. For simplicity we treat the spheres as particles so we can apply Coulomb's law. Before the spheres touch, the magnitude of the force exerted by them on each other is $k(q)(3q)/d^2 = 3kq^2/d^2$. When the spheres touch, the charge spreads out evenly over them, so each sphere carries a charge of $(q + 3q)/2 = 2q$. At separation d, therefore, the magnitude of the electric force between them is now $k(2q)(2q)/d^2 = 4kq^2/d^2$, which is greater than it was before.

22.17. Two particles carrying like charges repel each other. If you release them, they accelerate away from each other, gaining kinetic energy. Because the two particles form a closed system, this gain in kinetic energy must be due to a decrease in potential energy associated with the electric interaction. Thus the potential energy of a system consisting of two particles carrying like charges decreases as the separation of the two objects increases. Two particles carrying opposite charges attract each other and so accelerate toward each other. As with two particles carrying like charges, this acceleration means a gain in kinetic energy and a decrease in potential energy. The separation distance is decreasing, however, and we are asked about the change in the system's potential energy when the separation distance is *increasing*. We can reason that because the potential energy decreases as the separation distance between two oppositely charged particles decreases, it must increase as the separation distance increases.

22.18. (*a*) The positive charge carriers repel one another, making their separation greater than the distance between the centers of the spheres. Because Coulomb's law has a $1/r^2$ dependence, the greater separation results in a force of smaller magnitude. (*b*) If the two spheres carry opposite charges, the charge carriers attract one another, and so the distance between them becomes smaller than the distance between the centers of the spheres. Consequently, the magnitude of the attractive force is greater than that obtained from Coulomb's law.

22.19. Because particles 2 and 3 both interact attractively with particle 1, they must carry like charges, and so the interaction between 2 and 3 must be repulsive (Figure S22.19*a*). The vector sum of the forces exerted on 2 is the sum of a leftward horizontal force exerted by particle 1 and a force directed upward and to the right exerted by particle 3; see Figure S22.19*b*.

Figure S22.19

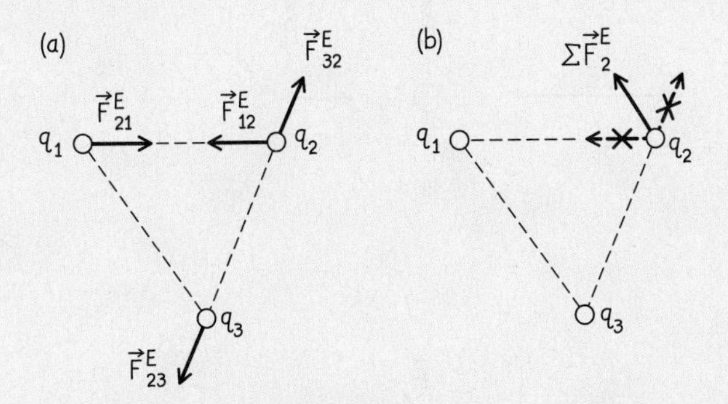

22.20. (*a*) The repulsive force $\vec{F}^E_{17}$ exerted by 1 on 7 points in the direction shown in Figure S22.20. To give 7 an acceleration to the right, the vector sum of the forces exerted on 7 must point to the right. This means that, in addition to $\vec{F}^E_{17}$, another sphere must exert a force $\vec{F}$ on 7 as indicated in the figure. Because the electric force is central (it always acts along the line connecting the two interacting objects), the only two spheres that can exert such a force are 2 and 5. (*b*) The magnitude F must be equal to F^E_{17} so that the components in the vertical direction add up to zero. So the magnitude of the charge on sphere 2 or 5 must be the same as that on 1 and 7. To give $\vec{F}$ the direction indicated in the figure, the charge must be negative if it is placed on 2 and positive if it is placed on 5.

Figure S22.20

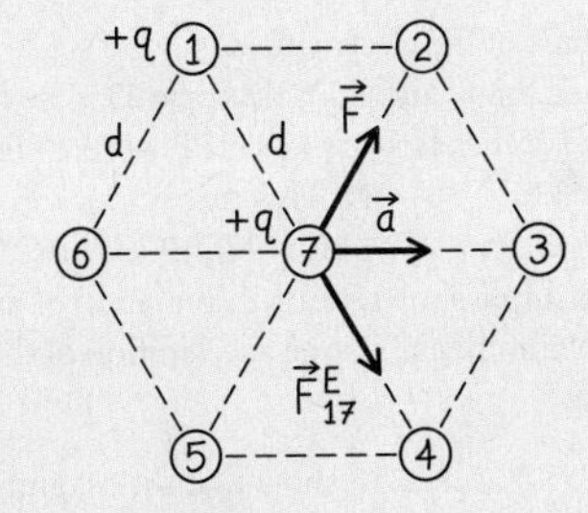

Chapter 23

23.1. (*a*) You know that the magnitude of the gravitational force exerted by Earth on an object of mass m_o is $F^G_{Eo} = Gm_Em_o/r^2_{Eo}$, where G is the gravitational constant and r_{Eo} is the Earth-object distance (Eq. 13.2). Because G and m_E are constants and r_{Eo} is the same for both objects, the magnitude of F^G_{Eo} is proportional to m_o, and thus $F^G_{E1}/F^G_{E2} = m_1/m_2$. (*b*) Yes, because mass is the only property of the objects in the gravitational force expression. Dividing by m_o removes that one property from the expression: $(Gm_Em_o/r^2_{Eo})/m_o = Gm_E/r^2_{Eo}$.

23.2. (*a*) The gravitational field at any location gives the gravitational force exerted on each kilogram of an object at that location. Because each kilogram of the satellite is subject to a force of 2.0 N, the satellite is subject to a force of magnitude $F^G_{Es} = (2.0\ \text{N/kg})(2000\ \text{kg}) = 4000\ \text{N}$. (*b*) $F^G_{Eb} = (2.0\ \text{N/kg})(0.20\ \text{kg}) = 0.40\ \text{N}$.

23.3. (*a*) The magnitudes are the same because the forces form an interaction pair. (*b*) Earth's gravitational field magnitude is greater than that of the ball because the gravitational pull exerted by Earth on objects is much greater than the pull of the ball on other ordinary objects. (*c*) The two forces are of equal magnitude because it is the *product* of the masses that determines the force magnitude: $F^G_{Eb} = Gm_Em_b/r^2_{Eb}$. The magnitude of the gravitational force exerted by a gravitational field on an object is equal to the product of the magnitude of the field at the location of the object and the mass of the object. Thus F^G_{Eb} is equal to the product of the magnitude g of Earth's gravitational field (proportional to m_E and large) and the (small) mass of the ball m_b. F^G_{bE} is equal to the product of the gravitational field of the ball (proportional to m_b and small) and the (large) mass of Earth m_E.

23.4. (*a*) No. $F^E_{pi} = kqq_i/d^2$ (Eq. 22.1) tells you that the magnitude of the force exerted by the particle on an object i is proportional to the charge on the object; because $q_1 \neq q_2$, the force magnitudes cannot be equal and therefore the forces cannot be equal. (*b*) The acceleration of the particle, from $\vec{a}_i = \vec{F}^E_{pi}/m_i$. (*c*) No, because the forces and mass of the particles are different (unless the difference in mass compensates for the difference in force). What is the same for both objects is $F^E_{pi}/q_i = kq/d^2$.

23.5. (*a*) Both forces point toward the particle because objects that carry charges of different types attract. (*b*) The electric field points in the same direction as the force exerted on a positively charged particle, so it points toward the particle. (*c*) Because objects that carry like charges repel, $\vec{F}^E_{p2}$ points away from the particle. The direction of the electric field, however, still points toward the particle. (*d*) The field points away from the particle because now the force exerted by the particle on a positively charged test particle is repulsive. (*e*) The two electric field magnitudes are the same because the magnitude of the force exerted on a test particle is proportional to the magnitude of the source charge and $|+q| = |-q|$.

23.6. The magnitude of the force is the product of the electric field magnitude $E = |\vec{E}\,|$ and the magnitude of the charge q. If q is positive, the direction of the force is the same as that of $\vec{E}$; if q is negative, the force points in the direction opposite the direction of $\vec{E}$.

23.7. (*a*) If q_2 is doubled, the force exerted by particle 2 doubles. Both the direction and the magnitude of $\Sigma\vec{F}^E$ change, and thus both the direction and the magnitude of $\vec{E}$ change (Figure S23.7a, next page). (*b*) See Figure S23.7*b*.

23.8. (*a*) In the same direction as the electric field at that point, that is, to the left. (*b*) The magnitude becomes three times as great; the direction is the same. (*c*) Yes, because the force exerted by object 3 changes. This change affects both the direction and the magnitude of $\Sigma\vec{F}^E$. (*d*) Opposite the direction of the electric field at that point, that is, down the page.

23.9. The droplet is subject to a downward force $\vec{F}^G_d$ and a horizontal force $\vec{F}^E_d$. The vector sum of these forces is downward at an angle, as illustrated in Figure S23.9. Both forces are constant in magnitude, so their vector sum is also constant. Consequently the droplet has a constant acceleration along the direction of the vector sum. Because the droplet begins at rest, its trajectory is a straight line along this direction.

Figure S23.7

Vector sum of forces on test particles

Electric field at points

Initial situation: $q_2 = q_1$

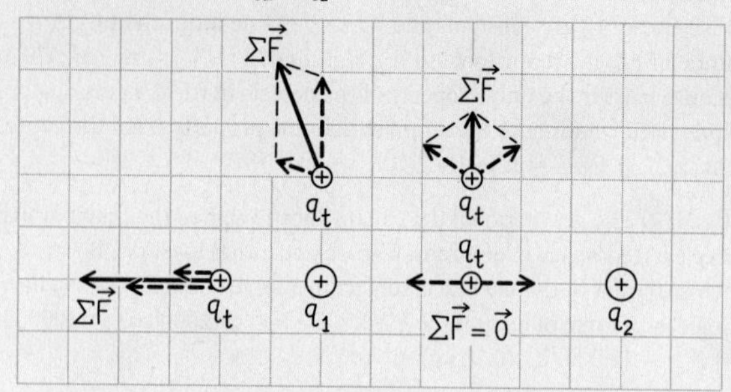

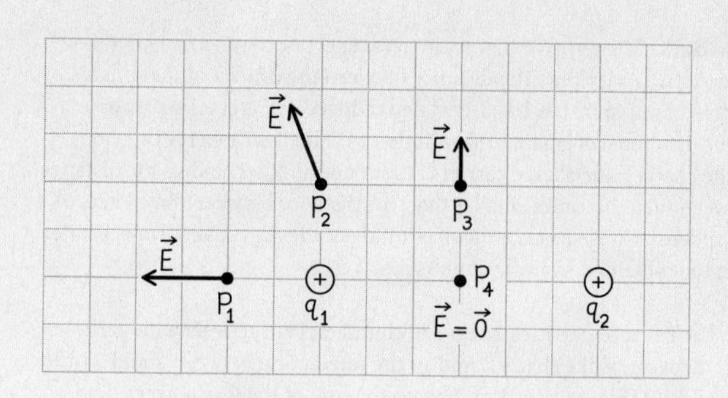

(a) q_2 is doubled: $q_2 = 2q_1$

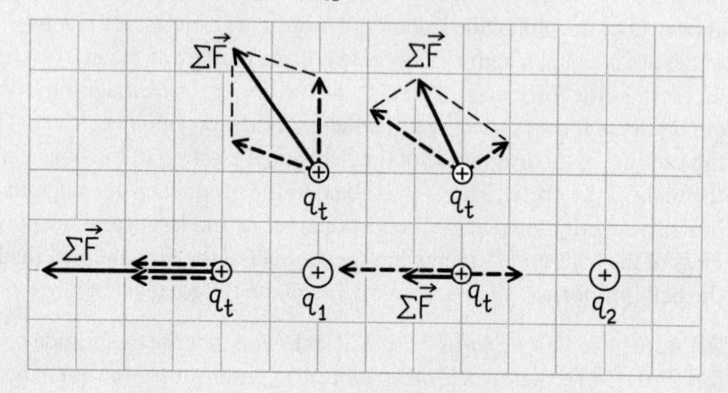

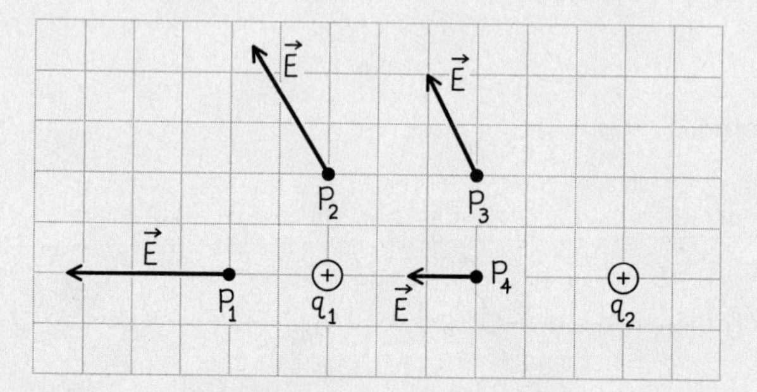

(b) Sign of q_2 is reversed: $q_2 = -q_1$

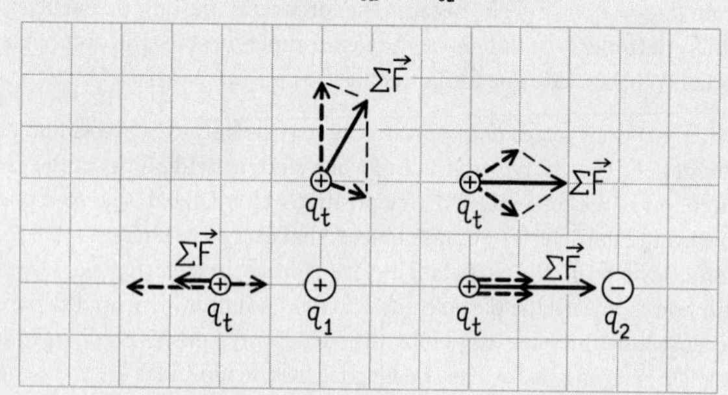

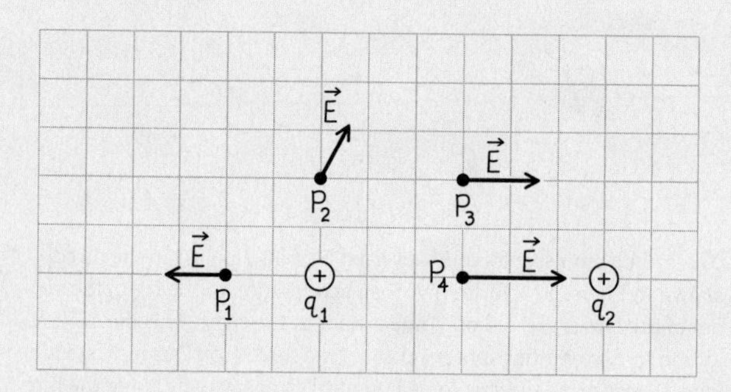

Figure S23.9

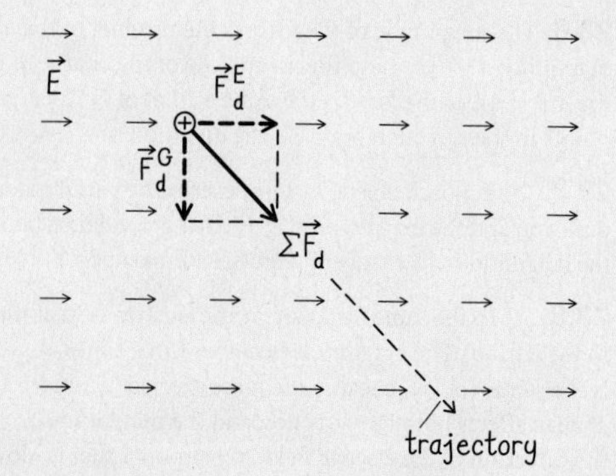

23.10. (*a*) The torque causes the molecule to rotate in the counterclockwise direction. (*b*) No. When the dipole axis that passes through the centers of the two particles is parallel to the electric field, the torque is zero because the two forces lie along the dipole axis and the lever arm of the torque is zero. When the axis is perpendicular to the electric field, the torque is a maximum.

23.11. (*a*) See Figure S23.11*a*. The vector sum of the forces—and therefore the center-of-mass acceleration—point to the left and down. (The force exerted on the negative end is greater than that on the positive end because the electric field is stronger there and both forces have a downward component.) (*b*) See Figure S23.11*b*. The acceleration now points to the right and down, and so the dipole begins by moving in that direction (*away* from the source of the electric field). The orientation of the dipole changes, however, because the forces cause a torque. As we saw in part *a*, as soon as the negative end of the dipole is more toward the left than the positive end, the dipole accelerates toward the left.

23.12. The magnitude of the electric field is $E = \sqrt{E_x^2 + E_y^2}$. Substituting the values we obtained for E_x and E_y in Example 23.3, we get $E = 1.9 \times 10^4$ N/C. The magnitude of the electric force is $F_e^E = eE = (1.6 \times 10^{-19}\ \text{C})(1.9 \times 10^4\ \text{N/C}) = 3.0 \times 10^{-15}$ N. This force causes an acceleration of magnitude $a_e = F_e^E/m_e = (3.0 \times 10^{-15}\ \text{N})/(9.1 \times 10^{-31}\ \text{kg}) = 3.3 \times 10^{15}\ \text{m/s}^2$. As this example shows, the accumulation of just tens of microcoulombs causes a phenomenally large acceleration of the electron even when it is meters away.

23.13. Equations 23.10 and 23.13 show that the magnitude of the electric field of a dipole is proportional to the dipole moment magnitude p. So the

Figure S23.11

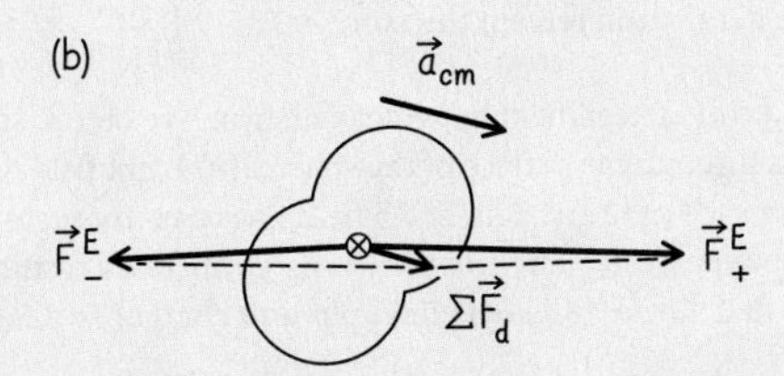

(b) $\vec{a}_{cm}$ $\vec{F}^E_-$ $\vec{F}^E_+$ $\Sigma\vec{F}_d$

dipole that creates the stronger field (dipole B) has the greater dipole moment. The magnitude of the dipole moment is the product of dipole separation d and dipole charge q_p, which means the greater dipole moment of B can be due either to greater d or to greater q_p. Because we cannot separate these two effects, we cannot say which dipole has the greater dipole charge.

23.14. To determine the electric field created by two parallel charged sheets, you must use the superposition principle. Figure S23.14*a* and *b* shows the electric fields of a positive and a negative sheet. In Figure S23.14*c* these two fields are added. (*a*) Between the sheets the electric fields are in the same direction, and so they reinforce, giving $E = 4k\pi\sigma$ between the sheets. (*b*) Outside the sheets, the electric fields are in opposite directions and therefore add to zero.

Figure S23.14

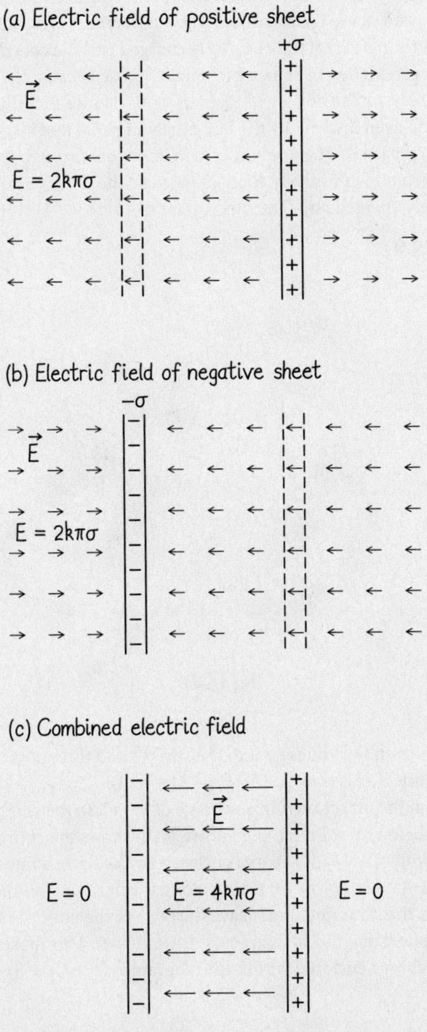

(a) Electric field of positive sheet
$+\sigma$
$\vec{E}$
$E = 2k\pi\sigma$

(b) Electric field of negative sheet
$-\sigma$
$\vec{E}$
$E = 2k\pi\sigma$

(c) Combined electric field
$E = 0$
$\vec{E}$
$E = 4k\pi\sigma$
$E = 0$

23.15. As you saw in Section 22.5, a direct consequence of the $1/r^2$ dependence in Coulomb's law is that the force exerted on a charged particle inside a hollow uniformly charged sphere is zero. This means that the electric field inside such a sphere is zero. For the electric field at some point P inside a uniformly charged sphere of radius R (Figure S23.15), the part of the sphere farther from the center than P does not contribute to the electric field at P. If the sphere carries a uniformly distributed charge q, the amount of charge on the part of the sphere closer to the center than P is $(r^3/R^3)q$, and the electric field due to this charge is

$$E_r = k\frac{r^3}{R^3}\frac{q}{r^2} = k\frac{q}{R^3}r.$$

In other words, the electric field *increases* in proportion to the distance r from the center. At the center, it is exactly zero.

Figure S23.15

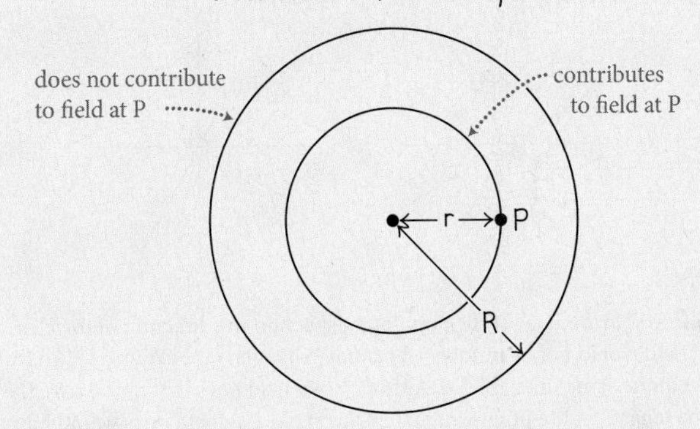

23.16. Yes. To see why, suppose the center of mass is off-center, as shown in Figure S23.16; everything else, including the magnitude of the charges and the dipole moment, is as before. Let the distance between the positive end and the center of mass be d_+. The force exerted on the positive end then causes a counterclockwise torque of magnitude $\tau_+ = r_\perp F^E_+ = (d_+ \sin\theta)(q_p E)$, and the force exerted on the negative end causes a counterclockwise torque of magnitude $\tau_- = r_\perp F^E_- = [(d - d_+)\sin\theta](q_p E)$, where $d = |\vec{r}_p|$. Adding the torques yields

$$\Sigma\tau_\vartheta = (d_+ \sin\theta)(q_p E) + [(d - d_+)\sin\theta](q_p E)$$
$$= d\sin\theta\,(q_p E) = (q_p d)E\sin\theta = pE\sin\theta,$$

identical to the result in Eq. 23.20.

Figure S23.16

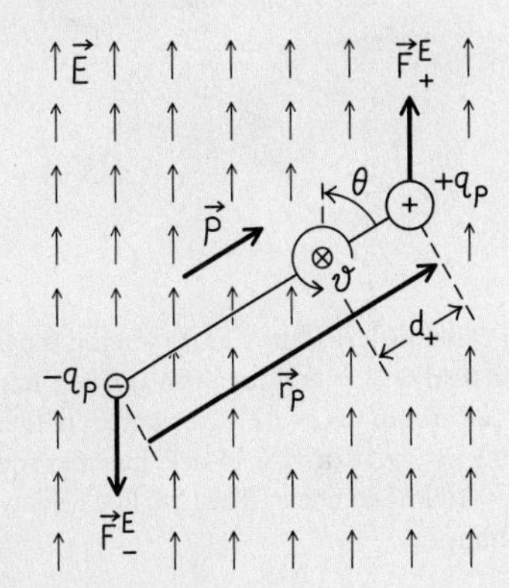

SOLUTIONS TO CHECKPOINTS

23.17. (*a*) Doubles the force (Eq. 23.23). (*b*) Doubles p and thus doubles the force. (*c*) Doubles p and thus doubles the force. (*d*) Reduces the force by a factor of $(\frac{1}{2})^3$ because of the $1/y^3$ dependence in Eq. 23.23.

23.18. The dipole moment $\vec{p}$ points in the same direction as the electric field that induces it, so α in Eq. 23.24 must be positive.

23.19. (*a*) Doubles the force (Eq. 22.1). (*b*) Quadruples the force because of the q^2 dependence in Eq. 23.26. (*c*) Doubling q_A in part *b* doubles the electric field created by A at the position of the induced dipole. This doubles the magnitude of p_{ind} (the stronger electric field increases the charge separation). This doubled dipole moment then interacts with a doubly strong electric field, increasing the force by a factor of 4. (*d*) No, because the polarization induced by the electric field is always along the electric field (Eq. 23.24). In other words, the induced dipole moment is parallel to the electric field, and the vector product of $\vec{p}$ and $\vec{E}$ is then zero.

Chapter 24

24.1 See Figure S24.1.

Figure S24.1

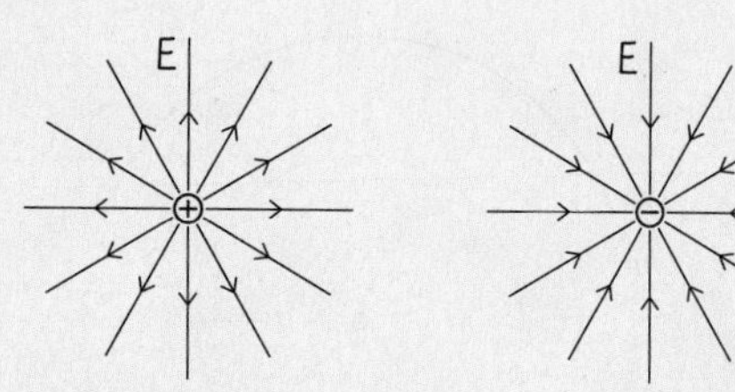

24.2 (*a*) No, because at the point of intersection the direction of the electric field would not be unique—it cannot be tangent to *both* intersecting lines at the same time. (*b*) No. Although two field lines that touch have the same tangent at the point where they touch, the two field lines would have different tangents on either side of that point, and so the direction of the electric field would again not be unique.

24.3 (*a*) As illustrated in Figure S24.3, the same 26 field lines pass through the surface of the hollow sphere. (*b*) The surface area of a sphere of radius R is $4\pi R^2$, so the number of field lines per unit area is $(26)/(4\pi R^2)$. (*c*) Again 26. (*d*) The number of field lines per unit area on the second sphere is $(26)/[4\pi(2R)^2] = (26)/(16\pi R^2)$, which is reduced by a factor of 4 from that for the sphere of radius R. (*e*) The electric field decreases as $1/r^2$, so doubling the distance reduces the electric field by a factor of 4.

Figure S24.3

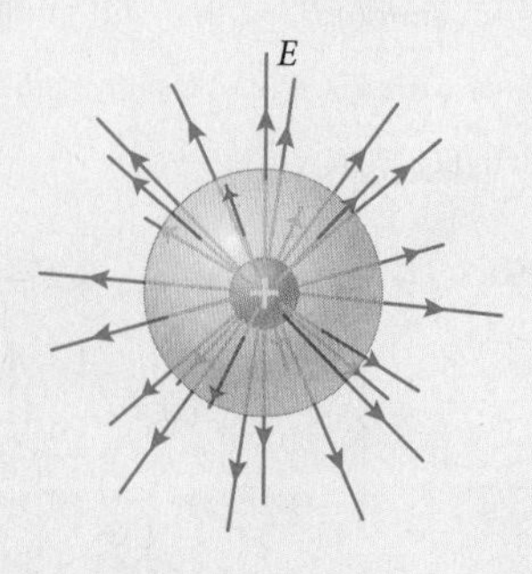

24.4 (*a*) No field lines cross the surface when it is parallel to the field lines. (*b*) If 16 field lines pass through an area of 1 m^2, then 8 field lines pass through an area of 0.5 m^2. (*c*) The number of field lines per unit area is $(8)/(0.5\ \text{m}^2) = 16\ \text{m}^{-2}$, or 16 field lines per square meter. (*d*) $(16)/(1\ \text{m}^2) = 16\ \text{m}^{-2}$, as above. The field line density is the same because the field is uniform.

24.5 (*a*) It remains the same—each field line still passes through the surface of the sphere. (*b*) Because neither the number of field lines nor the surface area of the sphere changes, the average number of field line crossings per unit surface area remains the same. (*c*) The electric field strength increases as the distance to the charged object decreases, so moving the sphere off-center increases the electric field strength on one side and decreases it on the opposite side. (*d*) No. The answer to part *b* gives the *average* field line density. Moving the sphere off-center increases the field line density on one side and decreases it on the other, so the average field line density can remain the same.

24.6 (*a*) Sixteen field lines emanate from the object, so 16 field lines pass through the surface of the sphere. (*b*) Eight field lines emanate from each of the objects, so 16 field lines pass through the surface of the sphere. (*c*) The amount of charge enclosed by the sphere is (20 field lines)/(8/q field lines per unit charge) $= 2.5q$.

24.7 (*a*) Each field line that reenters the donut must also exit, contributing a value of $(+1) + (-1) = 0$ to the field line flux. Thus, regardless of how many field lines reenter the donut, the field line flux remains 6. (*b*) No. Regardless of the shape of the surface, each field line contributes no more and no less than $+1$ to the field line flux, so the field line flux is always equal to 6.

24.8 (*a*) The field line flux is zero because each field line that enters the sphere also exits the sphere, and so does not contribute to the field line flux. (*b*) No. Moving the particle changes the number of field lines that enter the sphere, but regardless of how many field lines enter the sphere, each of them must also leave the sphere, so the field line flux remains zero.

24.9 Five field lines emanating from the positively charged particle contribute a flux of $+5$ (the field line that points from the positive end to the negative end doesn't pass through the surface; see Figure S24.9). Five field lines terminating on the negatively charged particle contribute a flux of -5. This gives a flux of zero, which makes sense because the amount of charge enclosed by the box is $+q + (-q) = 0$. [Notice that this answer remains valid even if we make the box much larger so that all the curved field lines fit inside it. Then only two field lines pass through the box—one emanating from the positively charged particle, the other terminating on the negatively charged one. The flux is then still zero: $(+1) + (-1) = 0$.]

Figure S24.9

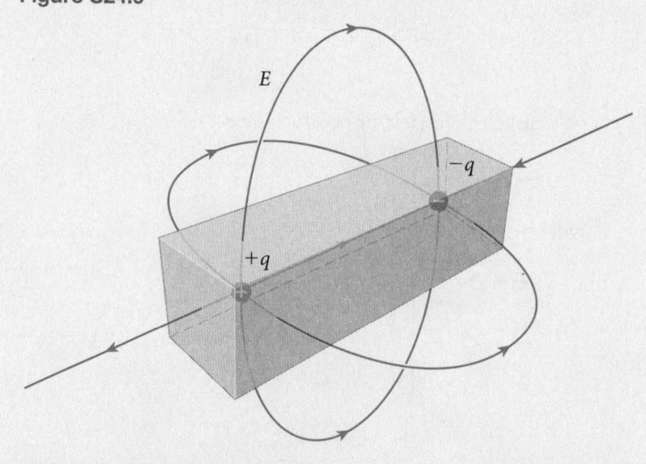

24.10 (*a*) Fifteen field lines go into the surface and three come out of it. The field line flux is thus $(-15) + (+3) = -12$. Twelve field lines emanate from the object carrying a charge of $+1$ C, so the charge inside the hidden region must be -1 C. Figure S24.10 shows the entire field line diagram; as you can see, the charge enclosed by the dashed line is indeed $(+3\ \text{C}) + (-4\ \text{C}) = -1$ C. (*b*) Going along the perimeter of the illustration, we note that four field lines leave the edge of the illustration and four enter it. The field line flux is thus $(+4) + (-4) = 0$. This makes sense because the charge enclosed within the diagram is $(+1\ \text{C}) + (-1\ \text{C}) = 0$.

Figure S24.10

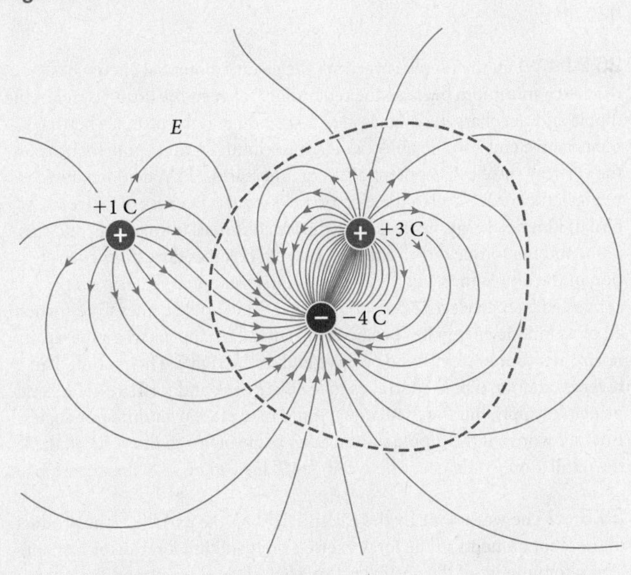

24.11 If there were a flux into the shell from the left and out of the shell on the right, then the situation would no longer have the required symmetry—if you were to rotate the sphere 180° about the vertical axis, the field line flux would be reversed and therefore the situation would be different. Because the situation is spherically symmetrical, this case is not possible.

24.12 (*a*) It stays the same because the field line density is the same everywhere on that surface. (*b*) It decreases because the field line spacing increases. (*c*) Zero. The field lines are all perpendicular to the wire, so no field lines pass through those surfaces.

24.13 (*a*) It stays the same because the field line density is the same everywhere along a plane parallel to the charged sheet. (*b*) It stays the same because the field line spacing doesn't change. (*c*) Positive, because the field lines cross from inside the surface to outside. (*d*) They are the same in magnitude and in sign. (*e*) Zero. The field lines are all perpendicular to the sheet, so no field lines pass through the curved surface.

24.14 (*a*) The field line flux is zero because the field is zero everywhere inside the conducting object. (*b*) According to the relationship between field line flux and enclosed charge derived in Section 24.3, the charge enclosed by the surface must be zero.

24.15 No. Imagine a Gaussian surface around the cavity, just inside the conducting object (Figure S24.15). Because $E = 0$ everywhere inside the object, the field line flux through this Gaussian surface is zero and so the charge enclosed by it must also be zero. So, the surplus charge on the object cannot reside on an inner surface.

Figure S24.15

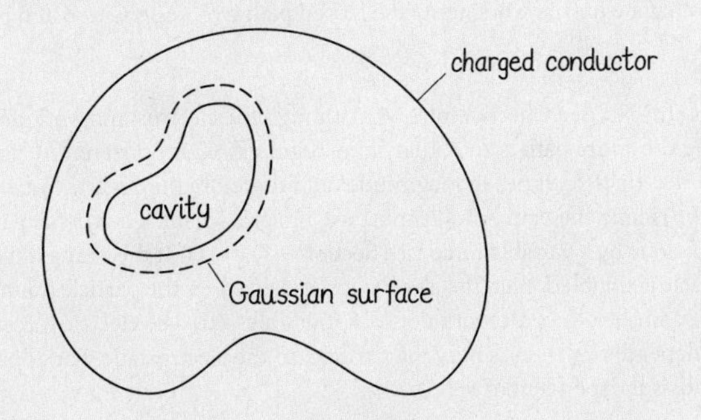

24.16 No. Field lines emanate from the positively charged particle and terminate on the negative charge carriers that line the cavity wall. Thus, "$E = 0$ everywhere inside a conducting object" means everywhere inside the bulk of the object. Cavities don't count!

24.17 (*a*) (*i*) Increases (more field lines intercepted); (*ii*) increases (greater field line density so more field lines intercepted); (*iii*) increases (more field lines intercepted). (*b*) (*i*) Increases (greater A yields greater Φ_E, by Eq. 24.1); (*ii*) increases (greater $|\vec{E}|$ yields greater Φ_E); (*iii*) increases (greater slope means smaller θ, which yields a greater $\cos\theta$).

24.18 (*a*) The area of the back surface is 1.0 m², so the magnitude of the corresponding area vector $\vec{A}_{back}$ is 1.0 m². From Figure S24.18 we see that $h/h_{front} = \cos\theta = \cos 30° = 0.87$, so the area of the front surface is larger by a factor of $h_{front}/h = 1.2$. The magnitude of the area vector $\vec{A}_{front}$ is thus 1.2 m². (*b*) We use Eq. 24.2 to calculate the electric fluxes. For the back surface we get $\Phi_E = EA_{back}\cos(180°) = (1.0\ \text{N/C})(1.0\ \text{m}^2)(-1) = -1.0\ \text{N}\cdot\text{m}^2/\text{C}$; for the front surface we have $\Phi_E = EA_{front}\cos\theta = (1.0\ \text{N/C})(1.2\ \text{m}^2)(0.87) = 1.0\ \text{N}\cdot\text{m}^2/\text{C}$.

Figure S24.18

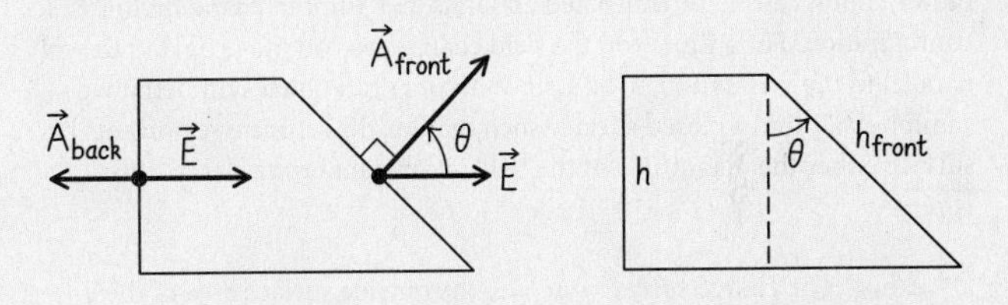

24.19 (*a*) At a distance r from a particle carrying a charge $+q$, the magnitude of the electric field is kq/r^2 (see Eq. 23.4). (*b*) Because the electric field is perpendicular to the sphere at all points and has the same constant magnitude, the electric flux is given by the product of the electric field and the area of the sphere: $\Phi_E = EA = E(4\pi r^2) = (kq/r^2)(4\pi r^2) = 4\pi kq$. (*c*) The enclosed charge is $+q$, so $\Phi_E = 4\pi kq_{enc}$. (*d*) No. If we double the radius r, the field decreases by a factor of 4 but the area of the sphere increases by a factor of 4. Thus the electric flux and its relationship to q_{enc} remain the same.

24.20 (*a*) With the modified Coulomb's law, the magnitude of the electric field at the surface of the sphere would be $kq/R^{2.00001}$. The electric flux thus becomes $\Phi_E = EA = E(4\pi R^2) = (kq/R^{2.00001})(4\pi R^2) = 4\pi kqR^{-0.00001}$. (*b*) $4\pi kqR^{-0.00001} = q_{enc}/\epsilon_0 = 4\pi kq_{enc}$ or $qR^{-0.00001} = q_{enc}$, which is not an equality. The cancellation of the radius R happens only when the dependence on r in Coulomb's law is *exactly* $1/r^2$.

24.21 Figure S24.21 (next page) shows how a wedge from q intersects the surface. The electric flux through intersection A_1 is equal to the electric flux through the intersection of the wedge with a sphere of radius r_1 centered on q. Likewise, the electric flux through A_2 is equal to the electric flux through the intersection of the wedge with a sphere of radius r_2 centered on q. As we saw in Checkpoint 24.19, the magnitudes of the electric flux through a sphere are independent of the radius. Because the intersections of the wedge with the two concentric spheres represent a fixed fraction of the total surface area of the spheres, the electric fluxes through these two intersections must also be the same. Consequently, the magnitudes of the electric fluxes through A_1 and A_2 are the same. Because the electric field points out of the sphere at A_1 and into the sphere at A_2, however, the algebraic signs of the electric fluxes are opposite: $\Phi_{E1} = -\Phi_{E2}$, so $\Phi_{E1} + \Phi_{E2} = 0$. Because this holds for *any* wedge, we learn that the electric flux through the surface due to q is zero when q is outside the surface.

Figure S24.21

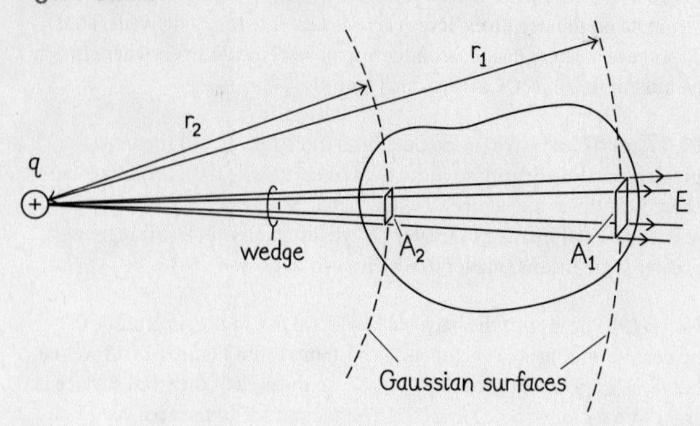

24.22 The electric field is also given by Eq. 24.15 because none of the arguments leading up to that result change: The field is still spherically symmetrical, and the enclosed charge is still $+q$.

24.23 No, because such a charge configuration does not have sufficient symmetry to make the calculation practical. A finite rod still has *rotational* symmetry (you can rotate it about its axis without changing the configuration), but it cannot be translated about its axis without changing the configuration. For a finite rod the field changes as you move up and down parallel to the y axis in Figure 24.30. With only rotational symmetry, we cannot easily find a closed surface such that for the separate regions of the surface either the magnitude of the field is constant or the electric flux is zero.

24.24 (*a*) The charge spreads out over the outside surface area of the plate. If we ignore the edges of the plate, $\frac{1}{2}q$ spreads out over each side of surface area A, and so the surface charge density (charge per unit surface area) is $\frac{1}{2}q/A$ on each side. The magnitude of the field is given by Eq. 24.17: $E = \sigma/\epsilon_0 = (\frac{1}{2}q/A)/\epsilon_0 = q/(2\epsilon_0 A)$. (*b*) The surface charge density is q/A; the magnitude of the electric field is given by the solution of Exercise 24.8: $E = \sigma/(2\epsilon_0) = (q/A)/(2\epsilon_0) = q/(2\epsilon_0 A)$, which is the same result as in part *a*. The point to remember therefore is that the difference between the solutions of Exercise 24.8 and Eq. 24.17 arises solely from the difference in the surface charge that should be used in each of them.

Chapter 25

25.1 (*a*) The kinetic energies are the same because the two particles are subject to the same force and the force displacements are the same, so the work done on them is the same. (*b*) Particle 2 has the greater momentum. As you may recall from Chapter 8, the change in momentum of an object is given by the product of the force exerted on it and the time interval during which the force is exerted. Because particle 2 has greater mass, its acceleration is smaller, and so it takes longer to fall the same distance as particle 1. (*c*) Both have the same momentum (because the product of force and time interval is the same for both), but particle 1 has greater kinetic energy (because particle 1 has smaller mass, its acceleration is greater and so it undergoes a greater displacement than particle 2; therefore the work done on it is greater and it gains more kinetic energy). (*d*) The electric force exerted on each particle is equal to the product of the electric field and the charge on the particle. Thus, their accelerations are given by

$$\vec{a}_1 = \vec{F}_1^E/m_1 = q_1\vec{E}/m_1$$

$$\vec{a}_2 = \vec{F}_2^E/m_2 = q_2\vec{E}/m_2.$$

Equating these two accelerations yields $q_1/m_1 = q_2/m_2$ or, rearranging terms, $q_1/q_2 = m_1/m_2$. In other words, if we make the ratio of the charges on the particles equal to the ratio of their masses, their accelerations are the same.

25.2 (*a*) When the dipole is vertical, the electric potential energy has reached a minimum because the separation between the positive end of the dipole and the charged object is at a maximum and the distance between the negative end and the object is at a minimum. As the dipole rotates past the vertical, the electric potential energy increases. (*b*) When the dipole is vertical, the torque due to the electric field is zero. Because of its (rotational) kinetic energy, however, the dipole continues to rotate beyond this point and the torque due to the electric field reverses direction. The rotation of the dipole therefore slows down and (rotational) kinetic energy is converted back to electric potential energy. The dipole comes to rest when all of its kinetic energy has been converted—this happens when it is again horizontal (but now with the positive end to the right). The motion then reverses and the dipole continues to oscillate back and forth. (*c*) The same principles apply, but the oscillation would take place over different angles. In other words, if the dipole started at an angle of 45° to the vertical, then the oscillation would occur between $\pm 45°$ instead of $\pm 90°$ as in part *b*.

25.3 (*a*) The work done by the electric field on the particle is the product of the x components of the force exerted on it and the force displacement. The x component of the particle's displacement is $x_B - x_A$ and the x component of the force exerted on it is qE_x, so $W_{Ep}(A \to B) = qE_x(x_B - x_A)$. (*b*) As the particle is moved back to its initial position, the force displacement is reversed, so the work done by the electric field on the particle is the negative of the amount done in part *a*: $W_{Ep}(B \to A) = -qE_x(x_B - x_A)$. (*c*) There is no change in the particle's kinetic energy and the particle possesses no internal energy, so its energy does not change as the agent moves it back. Therefore the agent must do an amount of work on the particle that is the negative of the work done by the electric field on it calculated in part *b*: $W_{ap}(B \to A) = -W_{Ep}(B \to A) = +qE_x(x_B - x_A)$. (*d*) The sum of the work done by the agent and by the electric field on the particle is zero. (*e*) See Figure S25.3. (There are no changes in the particle's energies and so the total work done on the particle is zero.)

Figure S25.3

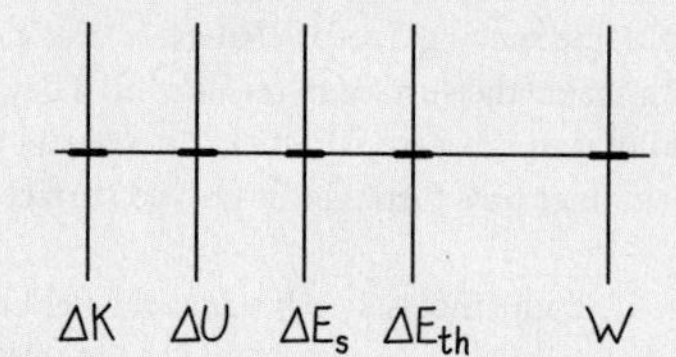

25.4 (*a*) The electrostatic work done on a particle as it moves along the gray path from A to C is equal to the electrostatic work done on it from A to B: $W_{Ep}(\text{gray path}) = W_{Ep}(A \to B)$. The work done from B to C is zero, as is that from C to B: $W_{Ep}(B \to C) = W_{Ep}(C \to B) = 0$. Thus, the electrostatic work done on the particle from C to B to A is just that done from B to A: $W_{Ep}(CBA) = W_{Ep}(B \to A)$. Because the electrostatic work done from B to A is the negative of that done from A to B (which is equal to W), the electrostatic work done from C to B to A must be $-W$. (*b*) Adding the result we found in part *a* to the electrostatic work done along the gray path, we calculate for the work along the closed path from A to C to B and back to A $W + (-W) = 0$.

25.5 (*a*) No. See Checkpoint 25.4. Although the electrostatic work done along the entire path is zero, that along segment AC is nonzero. (*b*) No. The electric field varies in magnitude (and therefore the electric force varies too) along the path AB. We must use integral calculus to calculate the work done by a variable force (see Section 9.7). (*c*) (*i*) If the charge on the particle is doubled, then the electric force exerted on the particle doubles and so the electrostatic work done on it doubles. (*ii*) The electric force does not depend on the mass m of the particle, so the electrostatic work done on it, too, is independent of m.

25.6 (*a*) Negative, because the electrostatic work done on a positively charged particle moving along any path from A to C is positive and the potential difference is the negative of the electrostatic work done per unit charge. (*b*) Zero, because the electrostatic work done on the particle is zero along any path from C to B. (*c*) Positive, because the electrostatic work done on the particle is negative (along the straight path from B to A the force and force displacement are in opposite directions). (*d*) The electrostatic work done on the particle is positive because the electric force and the force displacement are in the same direction. The potential difference is therefore negative. (*e*) The electrostatic work done on the particle is equal to the change in kinetic energy: $W_{Ep} = \Delta K$. This is the electrostatic work done on a particle carrying a charge q, and so the electrostatic work done per unit charge is $\Delta K/q$. The potential difference is the negative of this quantity: $-\Delta K/q$.

25.7 Yes. The surface of any sphere that is centered on the particle constitutes an equipotential surface. (The electric field is always perpendicular to such a surface, and so the electrostatic work done on a particle as it is moved along any path—regardless of its shape—that lies on such a surface is always zero.)

25.8 Zero. As we have seen in Section 24.5, the electric field inside a conducting object that is in electrostatic equilibrium is zero. The accumulation of positive and negative charge carriers on opposite ends of the sphere occurs precisely to cancel the effect of the external field of the charged rod anywhere inside the sphere: The electric field caused by the polarization of charge on the sphere exactly cancels the electric field of the rod. Therefore, because $\vec{E} = 0$ inside the conducting object, the electrostatic work done on a charged particle inside the object is zero, too. Consequently the potential difference between two points on or inside the object is zero.

25.9 (*a*) Along path CB, $r_i > r_f$, so the left side of Eq. 25.5 is negative. (*b*) Positive work must be done to push a positively charged particle toward another positively charged particle because the force the external agent doing the pushing must exert and the force displacement are in the same direction. Because the kinetic energy of particle 2 doesn't change, the total work done on 2 (the sum of the electrostatic work done by particle 1 on 2 and the work done by the external agent on 2) must be zero: $W_2 = W_{12} + W_{a2} = 0$. So if the electrostatic work done by particle 1 on 2 is negative ($W_{12} < 0$), then the work done by the external agent on 2 must be positive: $W_{a2} > 0$. (*c*) It we take both particles as our system, there is no longer any electrostatic work done on the system (the electric interaction is now internal). Consequently the work done by the external agent in moving 2 changes the electric potential energy of the two-particle system: $\Delta U^E = W_{a2}$. Using our answer to part *b*, we have $W_{a2} = -W_{12}(\text{C} \rightarrow \text{B})$ or, using Eq. 25.4,

$$\Delta U^E = -W_{12}(\text{C} \rightarrow \text{B}) = -\frac{q_1 q_2}{4\pi\epsilon_0}\left[\frac{1}{r_\text{C}} - \frac{1}{r_\text{B}}\right].$$

Because $r_\text{C} > r_\text{B}$, this change in energy is positive.

25.10 Yes. Figure S25.10 shows particle 1 moving instead of particle 2 (compare with Figure 25.15). We follow the same derivation as in Eq. 25.2, substituting $\vec{F}^E_{21}$ for $\vec{F}^E_{12}$. Because the magnitudes of these two forces are the same, the right-hand side of Eq. 25.2 remains the same. The initial and final separations are also still the same and so we obtain the same end result (Eq. 25.4). Alternatively, you can look at Eq. 25.5 and note that the expression is symmetrical in q_1 and q_2 and that r_{12i} and r_{12f} are independent of which of the two charged particles is moved.

25.11 Yes. The electrostatic work done on 3 as it is brought in first is zero: $W_3 = 0$; the electrostatic work done on 1 as it is brought in next is (compare with Eq. 25.9)

$$W_{31} = -\frac{q_3 q_1}{4\pi\epsilon_0}\frac{1}{r_{31}}.$$

Figure S25.10

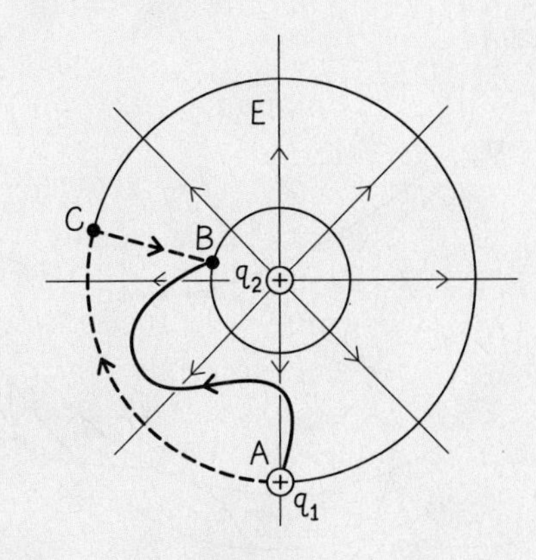

Likewise, the electrostatic work done on 2 when bringing in 2 is (compare with Eq. 25.12)

$$W_{12} + W_{32} = -\frac{q_1 q_2}{4\pi\epsilon_0}\frac{1}{r_{12}} - \frac{q_3 q_2}{4\pi\epsilon_0}\frac{1}{r_{32}},$$

The total electrostatic work $W = W_{31} + W_{12} + W_{32}$ done on the system is still the same.

25.12 (*a*) Positive. The potential difference between A and B is $V_\text{B} - V_\text{A}$, so A is the initial point and B is the final point: $r_i = r_\text{A}$ and $r_f = r_\text{B}$. Because $r_\text{A} > r_\text{B}$, the right side of Eq. 25.20 is positive. (*b*) Along the straight path from A to B the angle between the electric force and the force displacement is between 90° and 180°, so the electrostatic work done on the particle is negative. This is consistent with the answer to part *a*; see Eq. 25.15.

25.13 The electric field of a charged particle is given by Eq. 23.4, $\vec{E}_s(\text{P}) = (kq_s/r^2_{s\text{P}})\hat{r}_{s\text{P}}$ so Eq. 25.25 becomes

$$V_{\text{AB}} = -\int_\text{A}^\text{B} \vec{E} \cdot d\vec{\ell} = -\frac{q}{4\pi\epsilon_0}\int_\text{A}^\text{B} \frac{\hat{r}}{r^2} \cdot d\vec{\ell}.$$

The scalar product on the right-hand side is zero along a circular arc because $\hat{r}$ and $d\vec{\ell}$ are perpendicular. Along a radial line, $\hat{r}$ and $d\vec{\ell}$ are parallel, so $\hat{r} \cdot d\vec{\ell} = dr$. Therefore

$$V_{\text{AB}} = -\frac{q}{4\pi\epsilon_0}\int_{r_\text{A}}^{r_\text{B}} \frac{dr}{r^2} = \frac{q}{4\pi\epsilon_0}\left[\frac{1}{r}\right]_{r_\text{A}}^{r_\text{B}}$$

$$= \frac{q}{4\pi\epsilon_0}\left[\frac{1}{r_\text{B}} - \frac{1}{r_\text{A}}\right],$$

which is the same result we obtained in Eq. 25.20.

25.14 We begin by sketching some equipotentials for the electric field pattern shown in Figure 25.22. Because equipotentials are always perpendicular to field lines, we can sketch equipotential lines by always drawing them perpendicular to the field lines, as in Figure S25.14*a* (next page). Based on the equipotential lines I have drawn, I choose a set of representative points to evaluate how the potential varies. Moving clockwise around the path from P, I note that the displacement is upward, whereas the direction of the electric field (and therefore the direction of the electric force exerted on a positively charged test particle) is downward. Thus, the electrostatic work done on the particle is negative and the potential difference between a point above the equipotential passing through P and P is positive. In other words, the potential *increases*. At point 1 we reach maximum potential; moving beyond 1, the potential decreases again. At point 2 we cross the equipotential through point P again (the value of the potential

Figure S25.14

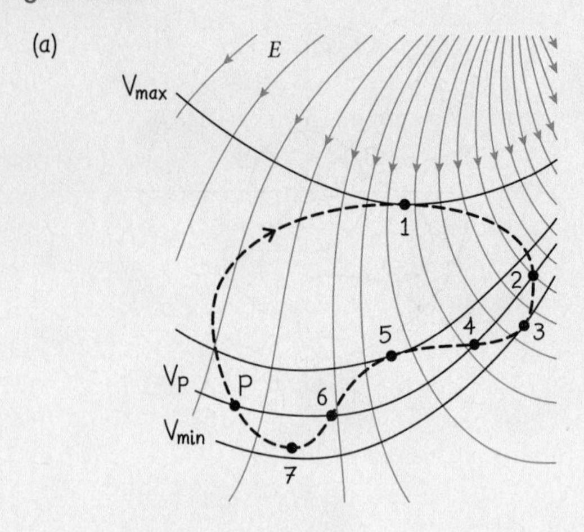

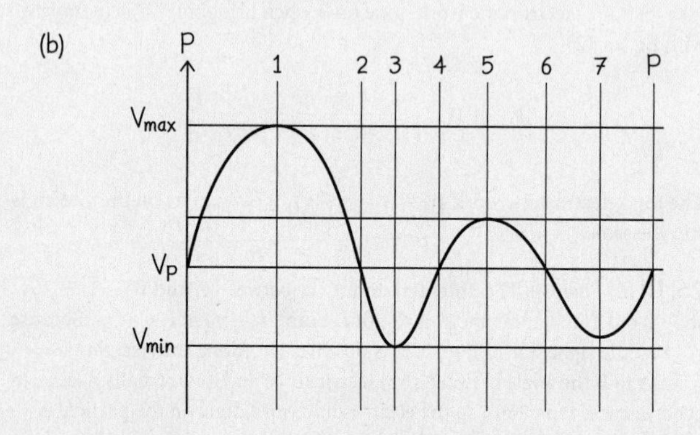

is again V_P). The potential continues to decrease until point 3, where it reaches a minimum value. The potential then increases again until a local maximum at point 5 is reached. As we return to P, the potential reaches another local minimum at 7 before increasing again to the initial value. See Figure S25.14*b*.

25.15 If the rod and the disk are positively charged and the zero of potential is at infinity, then both potentials should be *positive* (the electrostatic work done on a positively charged particle is negative as we bring it in from infinity). In the answer to Example 25.6, the factor in the logarithm is greater than 1 because $\sqrt{\ell^2 + d^2}$ is greater than d, so the numerator is always greater than the denominator. The logarithm is thus positive, yielding a positive V for positive q. Likewise, in the answer to Example 25.7, the factor in parentheses is always positive because $\sqrt{z^2 + R^2}$ is greater than $|z|$ so, for positive σ, the potential is positive.

25.16 In Example 25.5, we found for the potential between the plates, $V(a) = E(d - a)$, where a is the distance from the positive plate. Rewriting this result as a function of x, we have $V(x) = E(d - x)$. Because $V(x)$ is not a function of y and z, the partial derivatives of V with respect to y and z are zero. Thus $E_y = E_z = 0$. Therefore, the electric field must be in the positive x direction, and the component in that direction is

$$E_x = -\frac{\partial V}{\partial x} = -\frac{\partial}{\partial x}[E(d - x)]$$

$$= -\frac{\partial}{\partial x}(Ed) + \frac{\partial}{\partial x}(Ex) = 0 + E = E.$$

The electric field is in the x direction and of magnitude E, $\vec{E} = E\hat{\imath}$, in agreement with the situation shown in Figure 25.20.

25.17 The potential we obtained in Example 25.7 is a function of z only, so the x and y components of the electric field are zero. The z component is given by

$$E_z = -\frac{\partial V}{\partial z} = -\frac{\sigma}{2\epsilon_0}\frac{\partial}{\partial z}(\sqrt{z^2 + R^2} - |z|)$$

$$= -\frac{\sigma}{2\epsilon_0}\left(\frac{z}{\sqrt{z^2 + R^2}} - \frac{z}{|z|}\right).$$

Substituting $k = 1/(4\pi\epsilon_0)$, we get, for $z > 0$

$$E_z = 2k\pi\sigma\left[1 - \frac{z}{\sqrt{z^2 + R^2}}\right]$$

which is the same result we obtained by direct integration in Example 23.6 (see Eq. 2). Note how much easier and shorter the derivation of the electric field via the potential is compared to the direct integration of Section 23.7.

Chapter 26

26.1 (*i*) The force exerted on a unit positive charge at P is the vector sum of the electric forces due to all the charge carriers on the rod and the fur exerted on the unit charge. If the distribution of charge on the rod and the fur is the same as before, then doubling the charge doubles the magnitude of the individual electric forces without changing their direction. Consequently the direction of the electric field should be the same, but its magnitude should be twice as great. (*ii*) Because the electric field increases by a factor of 2, the potential difference (which is equal to the negative of the line integral of the electric field along any path between the two fixed points) increases by a factor of 2 as well. (*iii*) The potential energy stored in the system increases *by more than a factor of 2* because as charge is transferred from the fur to the rod, the charges already on the rod and the fur make the transfer more difficult because their field opposes the transfer. Thus, the transfer of each additional amount of charge requires more energy than the previous amount of charge. Put differently, the second half of the charge is transferred against a much greater potential difference than the first half.

26.2 The charged fur and rod attract each other, so you must do positive work on the rod-fur system to increase the separation of the rod and the fur. This work increases the electric potential energy of the rod-fur system.

26.3 See Figure S26.3. No external agent does any work on the system. The person, now part of the system, provides the energy from source energy.

Figure S26.3

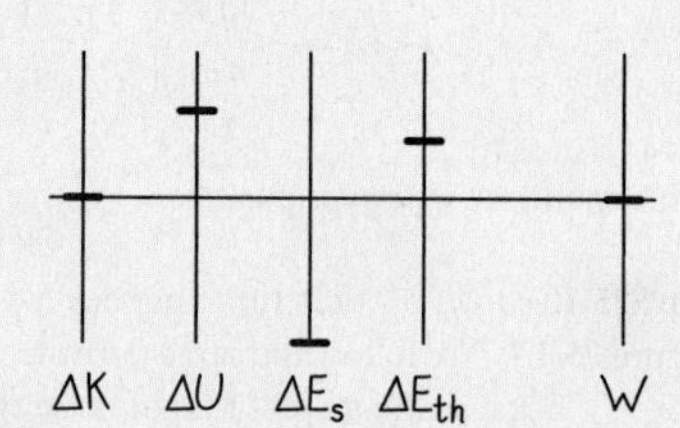

26.4 (*a*) When the wires are disconnected, any charge already on the plates remains there, so the potential difference between the plates remains the same. (*b*) The charging of the capacitor continues until the potential difference across the capacitor is equal to that across the battery. Thus, if the battery has a greater potential difference, then the potential difference between the plates is greater too. This greater potential difference requires more charge on each plate.

26.5 (*a*) Connecting the capacitors to a 9-V battery means that the potential difference across each capacitor is 9 V. As we saw in Section 25.5, the potential difference across a parallel-plate capacitor is the product of the electric field E and the plate separation d (see Example 25.5). Therefore, the capacitor with the smaller plate separation has the greater electric field. Because the electric field of a charged plate is proportional to its surface charge density (see Exercise 24.8), the plates of the capacitor with the smaller plate separation carry the greater charge. (*b*) The size and shape of the plates don't enter into the expression for the electric field, so the smaller plates must have the same surface charge density as the larger ones. Consequently, the capacitor with the smaller plate area stores less charge.

26.6 (*a*) The amount of charge remains the same; there are no other conductors to or from which charge carriers can go. (*b*) The charge carriers in the conducting slab always rearrange themselves so as to cancel the electric field inside the slab (see Section 24.5). To do so, they must set up an electric field of the same magnitude as that inside the capacitor, but in the opposite direction. The charge carriers in the slab rearrange themselves at the surfaces as shown in Figure 26.9*b*. For the electric field due to the charge carriers at the surfaces of the slab to be equal in magnitude to that caused by the capacitor plates, the magnitude of the surface charge density on the slab surfaces must be the same as that on the plates. Because the surface area of the slab (or at least the part of the slab that is in the electric field) is the same as that of the capacitor plates, the magnitude of the charge on each side of the slab is the same as that on the capacitor plates. (*c*) Zero, because the electric field inside the slab is zero. (*d*) It decreases. Between the capacitor plates and the surfaces of the slab, the electric field is the same as before, but inside the slab the electric field is now zero. The electrostatic work done on a unit charge as it is moved from one plate to the other is therefore less than it was before because it is zero inside the slab. Therefore the potential difference is decreased.

26.7 (*a*) No. Regardless of where the slab is inserted, the field inside it is always zero, while the field between the capacitor plates and the slab is equal to what it was before the slab was inserted. In the extreme case that the slab makes contact with one of the plates, the charge carriers on the plate move through the slab, all the way to its opposite side, as illustrated in Figure S26.7*a*, effectively reducing the plate separation. The magnitude of the electric field, however, remains the same. (*b*) See Figure 26.7*b*. Note that the slope of $V(x)$ on either side of the slab does not depend on the position of the slab.

Figure S26.7

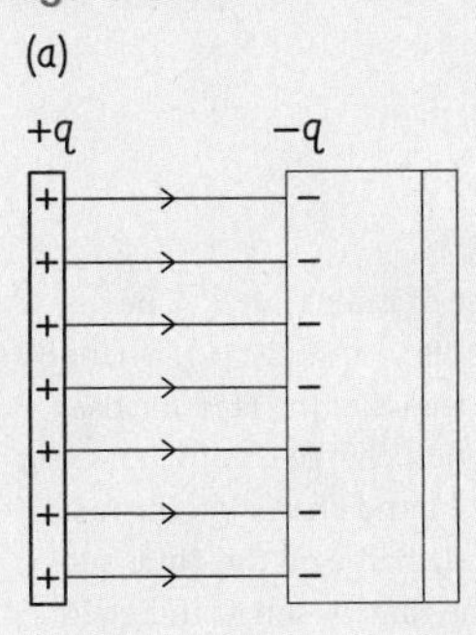

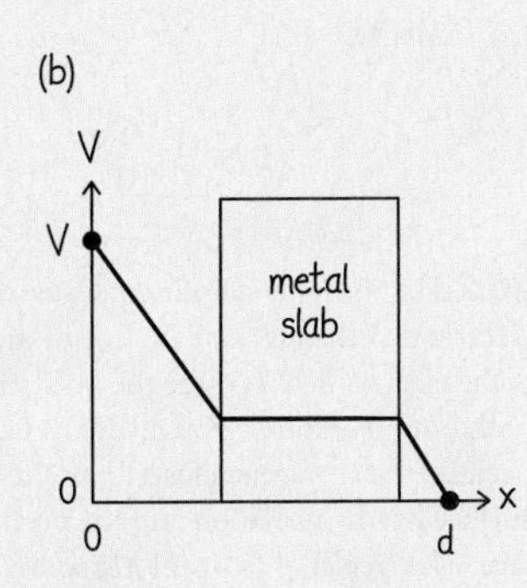

26.8 The direction of the electric field created by a given source is defined to coincide with the direction of the electric force exerted on a *positive* charge carrier (see Section 23.2). Because electrons carry a negative charge, the force exerted on them is in a direction opposite the direction of the electric field.

26.9 (*a*) The electric field is zero above the surface (so it has no direction) because the electric fields of the two surfaces cancel each other outside the dielectric, just like the fields of the two plates of a capacitor (see also Figure S23.14). (*b*) It points from the positively to the negatively charged surface—that is, opposite the direction of the electric field due to the capacitor plates.

26.10 (*a*) Zero, because the electric field due to the bound surface charge would be equal in magnitude and opposite in direction to that due to the free charge on the capacitor plates. In effect, the dielectric is like a conductor. (*b*) No. Suppose the bound charge accumulation at the surfaces becomes so great as to cancel the electric field due the free charge on the capacitor plates, as in part *a*. Because the electric field inside the dielectric is zero, there is nothing that would cause the material to polarize even further.

26.11 The capacitor with the dielectric, because the battery must do *additional* work on the charge carriers to increase the magnitude of the charge on each plate above that of the capacitor without the dielectric.

26.12 The electric field is in the direction of the flow of positive ions. (The electric force exerted on a positive charge carrier is in the same direction as the electric field.) Wait! Shouldn't the field be in the opposite direction, given the charge on the terminals and the fact that the electric field lines point away from positive charge carriers? The key to reconciling these two facts is to realize that the mechanism of the voltaic cell occurs in the small layers of electrolyte near the electrodes. There the field is in the opposite direction and chemical reactions push charge carriers against the electric field. The result is the pattern of charge and electric fields shown (not to scale) in Figure S26.12.

Figure S26.12

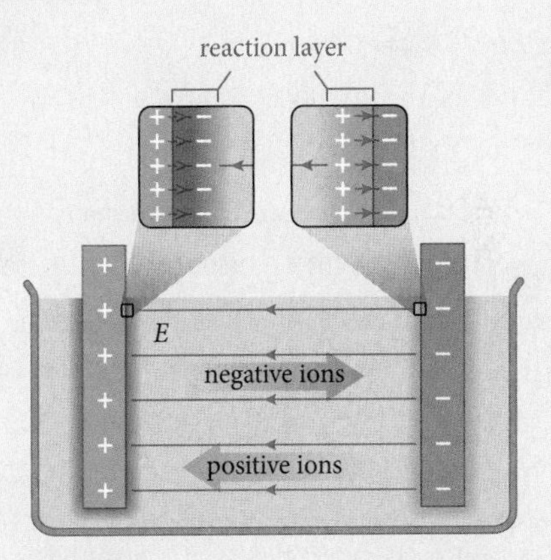

26.13 The work is negative because the system under consideration comprises the molecules undergoing chemical reactions, not the electrons. The molecules do positive work on the electrons, which are part of the molecules' environment, so the environment does *negative* work on the molecules.

26.14 A. The charge on each capacitor follows from Eq. 26.1: $q = C V_{cap}$. Thus, for the same V_{cap}, the capacitor with the greater capacitance C holds the greater charge q.

26.15 (*a*) Using the result of Example 26.2, we have

$$A = \frac{Cd}{\epsilon_0} = \frac{(1.0 \times 10^{-6}\ \text{F})(50 \times 10^{-6}\ \text{m})}{8.85 \times 10^{-12}\ \text{C}^2/(\text{N}\cdot\text{m}^2)} = 5.6\ \text{m}^2.$$

This is the area of a small room! A 1.0-F capacitor would therefore have a surface area of $5.6 \times 10^6\ \text{m}^2$—the size of a small town. (*b*) At the breakdown threshold, the potential difference across the capacitor is $V_{cap} = Ed = (3.0 \times 10^6\ \text{V/m})(50\ \mu\text{m}) = 150\ \text{V}$. The charge on the capacitor is then $q = C\,\Delta V = (1.0\ \mu\text{F})(150\ \text{V}) = 1.5 \times 10^{-4}\ \text{C}$. (*c*) This charge corresponds to $(1.5 \times 10^{-4}\ \text{C})/(1.6 \times 10^{-19}\ \text{C}) = 9.4 \times 10^{14}$ electrons.

26.16 The capacitance is given by the expression we obtained in Example 26.3. Substituting the values given in the checkpoint, we get

$$C = \frac{2\pi\epsilon_0\ell}{\ln(R_2/R_1)}$$

$$= \frac{2\pi[8.85 \times 10^{-12}\ \text{C}^2/(\text{N}\cdot\text{m}^2)](100\ \text{m})}{\ln[(2.0 \times 10^{-3}\ \text{m})/(2.0 \times 10^{-4}\ \text{m})]} = 2.4\ \text{nF}.$$

One nanofarad (1 nF) is 1×10^{-9} F, so this is a rather small capacitance. (A large capacitance is undesirable because it would allow charge carriers to "pile up" in the cable.)

26.17 (*a*) Example 26.4 gives an expression for the capacitance of a spherical capacitor. As R_2 approaches infinity, the R_1 in the denominator of this expression becomes negligible and so the capacitance reduces to

$$C = \lim_{R_2 \to \infty}\left[4\pi\epsilon_0 \frac{R_1 R_2}{R_2 - R_1}\right] = 4\pi\epsilon_0 \frac{R_1 R_2}{R_2} = 4\pi\epsilon_0 R_1.$$

(*b*) Substituting values for ϵ_0 and the radius R_1, we get

$$C = 4\pi[8.85 \times 10^{-12}\ \mathrm{C^2/(N\cdot m^2)}](2.5\ \mathrm{m})$$
$$= 2.8 \times 10^{-10}\ \mathrm{F}.$$

(*c*) The electric field is maximum at the surface of the dome. From Eq. 24.15 we know that the electric field at the surface of a charged sphere is

$$E = \frac{1}{4\pi\epsilon_0}\frac{q}{R^2} = \frac{1}{4\pi\epsilon_0}\frac{q}{R}\frac{1}{R} = V_R\frac{1}{R},$$

where V_R is the potential at the surface of the sphere and the potential is zero at infinity. Therefore, $V_R = ER$ and the charge stored on the dome is

$$q = CV_{\mathrm{cap}} = CER$$
$$= (2.8 \times 10^{-10}\ \mathrm{F})(3.0 \times 10^{6}\ \mathrm{V/m})(2.5\ \mathrm{m})$$
$$= 2.1 \times 10^{-3}\ \mathrm{C}.$$

26.18 (*a*) The potential difference across the capacitor is

$$V_{\mathrm{cap}} = Ed = (3.0 \times 10^{6}\ \mathrm{V/m})(50 \times 10^{-6}\ \mathrm{m}) = 150\ \mathrm{V}.$$

The energy stored in the capacitor then follows from Eq. 26.4:

$$U^E = \tfrac{1}{2}(1.0 \times 10^{-6}\ \mathrm{F})(150\ \mathrm{V})^2 = 1.1 \times 10^{-2}\ \mathrm{F\cdot V^2}.$$

Because $1\ \mathrm{F} = 1\ \mathrm{C/V}$ and $1\ \mathrm{V} = 1\ \mathrm{J/C}$, we have $(1\ \mathrm{F\cdot V^2}) = (1\ \mathrm{J})$, so the energy stored in the capacitor is 11 mJ.

(*b*) The increase in gravitational potential energy is given by $\Delta U^G = mg\,\Delta y$, so $\Delta y = \Delta U^G/(mg) = (11\ \mathrm{mJ})/[(2\ \mathrm{kg})(10\ \mathrm{m/s^2})] = 0.6\ \mathrm{mm}$.

26.19 (*a*) The surface charge density on each plate is given by $\sigma = q/A$, so the magnitude of the field due to one plate is

$$E_{\mathrm{plate}} = \frac{\sigma}{2\epsilon_0} = \frac{q}{2\epsilon_0 A}.$$

Note that this is not the electric field between the capacitor plates. The field between the plates is twice as great because each plate contributes to that field. The force exerted by the electric field of one plate on the other plate is thus

$$F_{Ep} = qE_{\mathrm{plate}} = \frac{q^2}{2\epsilon_0 A}.$$

(*b*) To increase the separation between the plates you must exert a force $\vec{F}$ of magnitude equal to the force exerted by the electric field on the plate calculated in part *a*. Because this force is constant, and because the force you exert must point in the same direction as the force displacement, we can write for the work you must do on the capacitor

$$W = F\Delta x_F = \frac{q^2\Delta x}{2\epsilon_0 A}.$$

(*c*) The change in the electric potential energy is equal to the work done on the capacitor, which we determined in part *b*.

(*d*) The volume of the additional space is $A\,\Delta x$, so the energy stored in the electric field in this additional space is $u_E\,A\,\Delta x$. Because the electric field between the plates of the capacitor is twice the electric field of a single plate calculated in part *a*, the energy density of the electric field is

$$u_E = \tfrac{1}{2}\epsilon_0 E^2 = \tfrac{1}{2}\epsilon_0\left(\frac{q}{\epsilon_0 A}\right)^2 = \frac{q^2}{2\epsilon_0 A^2},$$

so the additional energy is

$$\Delta U^E = u_E\,A\Delta x = \frac{q^2}{2\epsilon_0 A^2}A\Delta x = \frac{q^2\Delta x}{2\epsilon_0 A} = W.$$

26.20 (*a*) The energy stored in the capacitor is given by Eq. 26.4, so

$$U^E = \tfrac{1}{2}(10^{-4}\ \mathrm{F})(300\ \mathrm{V})^2 = 4.5\ \mathrm{J}.$$

(*b*) The average power is given by the change in energy divided by the time interval over which the change takes place (Eq. 9.29):

$$P_{\mathrm{av}} = \frac{4.5\ \mathrm{J}}{1 \times 10^{-3}\ \mathrm{s}} = 4.5 \times 10^{3}\ \mathrm{W}.$$

This is a phenomenally large power (but it is delivered for only a very short amount of time)—much greater than the few watts that can be delivered by a typical battery. Because it requires a great amount of power to illuminate a large space for a short amount of time, all flash units use a capacitor that is charged by a battery between flash firings.

26.21 (*a*) Because $q_0 = q_{\mathrm{free}} - q_{\mathrm{bound}}$ (see Figure 26.29), we have $q_{\mathrm{bound}} = q_{\mathrm{free}} - q_0$. If we substitute Eq. 26.21, this becomes

$$q_{\mathrm{bound}} = (\kappa - 1)q_0.$$

(*b*) The bound surface charge density on the dielectric is $\sigma_{\mathrm{bound}} = q_{\mathrm{bound}}/A$, where A is the area of the dielectric. Together with our answer to part *a*, this becomes

$$\sigma_{\mathrm{bound}} = (\kappa - 1)\frac{q_0}{A} = (\kappa - 1)\epsilon_0 E,$$

where $E = q_0/(\epsilon_0 A)$ (Eq. 24.17).

26.22 (*a*) Because $1\ \mathrm{V} = 1\ \mathrm{J/C}$ and $1\ \mathrm{J} = 1\ \mathrm{N\cdot m}$, we have

$$\frac{\mathrm{C^2}}{\mathrm{N\cdot m}} = \frac{\mathrm{C^2}}{\mathrm{J}} = \frac{\mathrm{C}}{\mathrm{V}} = \mathrm{F}.$$

(*b*) Similarly,

$$\mathrm{F\cdot V^2} = \frac{\mathrm{C}}{\mathrm{V}}\mathrm{V^2} = \mathrm{C\cdot V} = \mathrm{C}\left(\frac{\mathrm{J}}{\mathrm{C}}\right) = \mathrm{J}.$$

26.23 Consider a cylindrical Gaussian surface of radius $r > R$. The electric flux through the surface of the cylinder is $\Phi = E(2\pi rL)$, where E is the electric field outside the insulation. The Gaussian surface encloses both bound and free charges. The enclosed bound charge is zero, however, because the surface encloses both the negative bound charge on the inside surface of the insulation and the positive bound charge on the outer surface. The free charge is still λL, so we obtain the same result as in Eq. 26.25 because $\kappa = 1$ for $r > R$.

Chapter 27

27.1 No. Whether a pole is N or S is determined by the way the magnet orients itself with respect to Earth's North Pole. The experiment illustrated in Figure 27.3 then *shows* that like poles repel and unlike poles attract.

27.2 (*a*) Yes, because the charged object always pulls charge carriers of the opposite type toward itself, exerting an attractive force on the near side of the neutral object and a repulsive force on the far side. Because the electric force decreases with increasing distance, the magnitude of the attractive force is greater than that of the repulsive force and so the sum of the forces is attractive. (*b*) A north pole is induced in the clip on the left and a south pole on the right because the interaction between opposite magnetic poles is attractive.

27.3 (*a*) See Figure S27.3. (*b*) No poles because there are no faces where only one type of elementary magnet is exposed. (*c*) If you broke the ring in half and put the exposed faces near paper clips and other magnets, you would find that the faces of the cuts are magnetic poles.

Figure S27.3

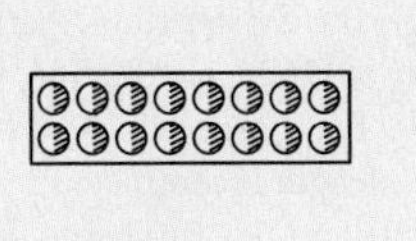

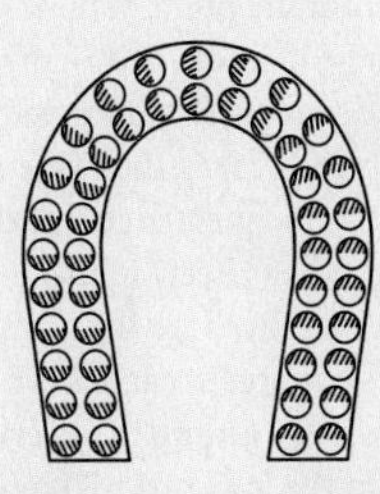

27.4 (*a*) By definition, the end that orients itself northward is the north pole of the needle. (*b*) The north pole of the magnet attracts the south pole of the needle and repels the north pole of the needle. Because the compass needle is free to rotate, the needle rotates so that its north-pointing end is as far away as possible from the north pole of the magnet. (*c*) No; if it were, it would repel the north pole of the compass needle. The earth's geographic North Pole is a magnetic south pole.

27.5 (*a*) The bar magnet's north pole attracts the needle's south pole and repels the needle's north pole. The two forces cause a torque on the needle that tends to align the needle with its north pole pointing away from the bar magnet's north pole. (*b*) Because the distance from the needle to the bar magnet's north pole is greater than the distance from the needle to the bar magnet's south pole, the magnitudes of the forces exerted by the north pole are smaller than those exerted by the south pole (Figure S27.5*a*). Consequently, the needle still aligns itself as in Figure 27.11*a*. (*c*) Once the needle has settled, we know that the torque on it must be zero. This means that the sum of the forces exerted by the poles of the bar magnet on each end of the needle must be aligned with the needle (Figure S27.5*b*).

Figure S27.5

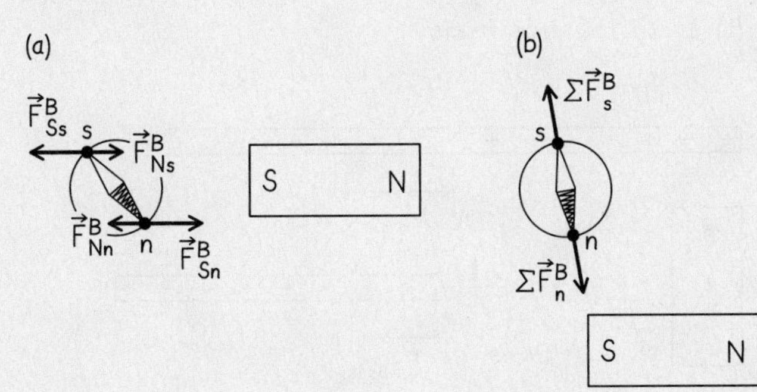

27.6 (*a*) Because elementary magnets have no physical extent, the magnetic field lines form loops that close on themselves (Figure S27.6). Because a particle cannot be cut in half, it must lie either inside or outside the closed surface. Consequently, each line that passes out through the closed surface must also pass back in through the surface. The straight field line entering at the bottom in Figure S27.6 is not an exception. It is balanced by the straight field line leaving at the top. The magnetic flux is zero. (*b*) No, because for this elementary magnet, too, each field line that leaves the surface must eventually reenter it. (*c*) Zero. Because the magnetic flux must be zero for each elementary magnet making up a bar magnet, the magnetic flux of the magnet must also be zero. (*d*) No. Because an elementary magnet cannot be divided, there is always a whole number of elementary magnets inside the closed surface, so the magnetic flux is still zero.

Figure S27.6

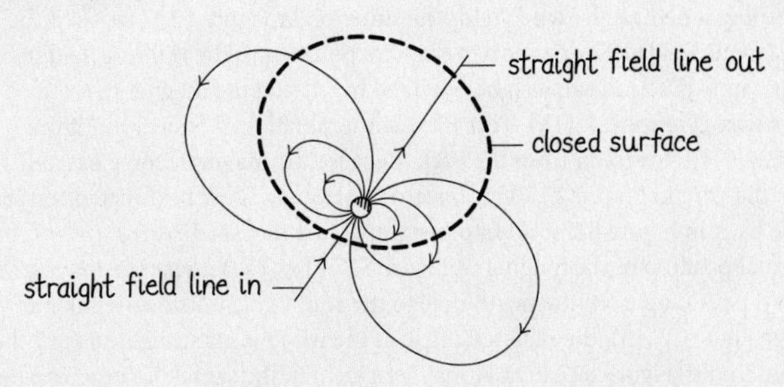

27.7 If we draw a closed surface around the north pole of the bar magnet (Figure S27.7), we see that all the field lines on the outside of the magnet are directed out of the closed surface. Because the magnetic flux through a closed surface must be zero, this means that the field lines inside the bar magnet point from the south pole to the north pole.

Figure S27.7

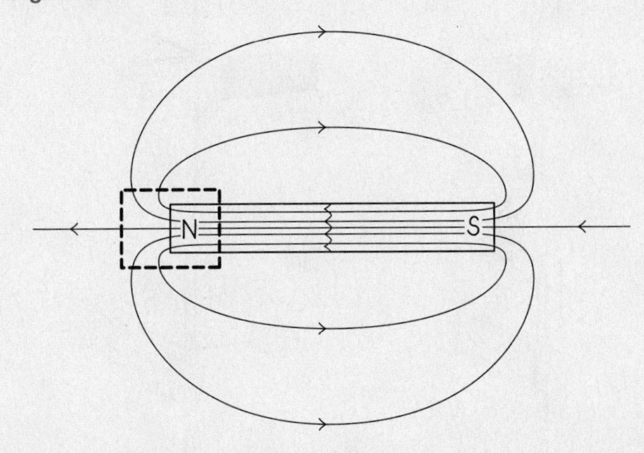

27.8 (*a*) No. If the rod were charged, the battery would have to carry an opposite charge, and charge carriers would start flowing back (see also Section 26.2). (*b*) No. Because it carries no surplus charge carriers, the rod cannot exert an electric force on another charged particle, and therefore there is no electric field due to the rod.

27.9 See Figure S27.9. If the needles align in a circular pattern, the magnetic field lines must be circular, curling in the direction that the north poles of the needles point.

Figure S27.9

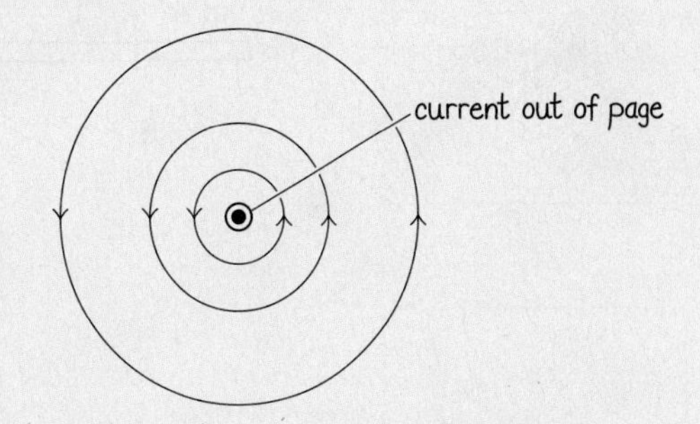

27.10 No. The circular magnetic field pattern generated by a current-carrying wire is unlike the field generated by any magnet or the elementary magnets of which any magnet is made. As we shall see in the next chapter, however, it is possible to use a current to generate a magnetic field pattern that approximates that of a bar magnet.

27.11 (*a*) Flipping the magnet in Figure 27.20*b* horizontally while keeping it behind the wire yields the situation in Figure S27.11*a*, with the magnetic field at the location of the wire pointing to the right. A field in the same direction can also be obtained by placing the magnet in front of the wire (Figure S27.11*b*). This is the same situation as shown in Figure 27.20*b*, but now seen from the back. Because the magnetic force exerted on the wire in Figure 27.20*b* is toward the front, it must be directed toward the back in Figure S27.11*b*. In other words, the force is directed *toward* the actual position of the magnet in Figure S27.11*a*. (*b*) Because the magnetic field lines loop from the north pole to the south pole outside the magnet (see Figure 27.13), the magnetic field at the wire points straight toward the south pole (Figure S27.11*c*). A magnetic field in the same direction can be obtained by placing the magnet to the right of the wire (Figure S27.11*d*). Comparing this situation with Figure 27.20*b*, you can conclude that the magnetic force exerted on the wire is now to the left. In other words, with the magnet in back of the wire as in Figure S27.11*c*, the magnet exerts a *sideways* force on the wire!

Figure S27.11

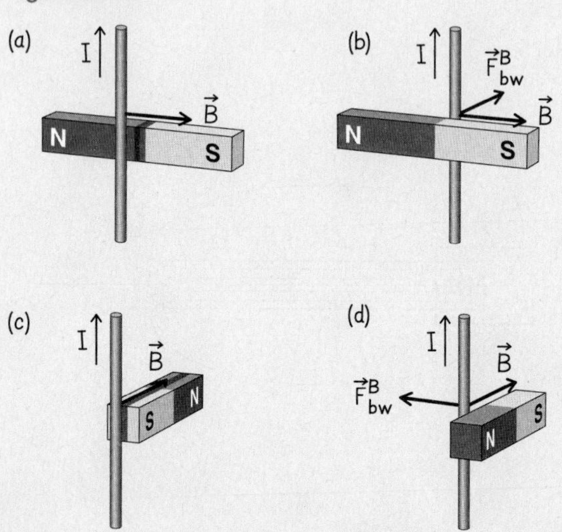

27.12 Reverse the direction of the current through rod 1 in Figure 27.23. As Figure S27.12*a* shows, at rod 1 the magnetic field due to the current through rod 2 is still out of the page. Placing the fingers of your right hand parallel to the current and curling them toward the magnetic field yield a downward-pointing thumb (Figure S27.12*b*), which means the magnetic force exerted by rod 2 on rod 1 is downward. The same reasoning tells you that the magnetic force exerted by rod 1 on rod 2 is upward. The two rods exert attractive forces on each other.

Figure S27.12

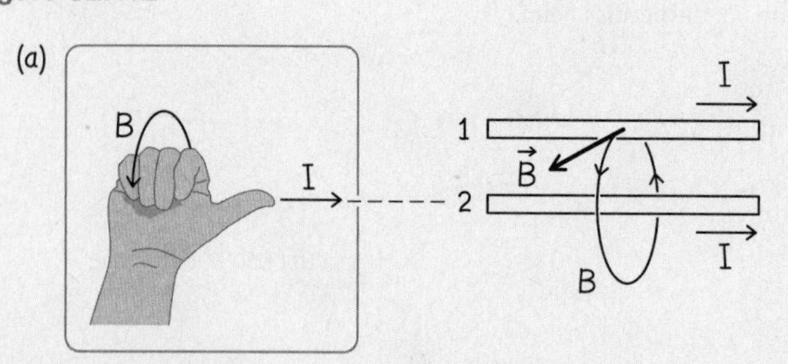

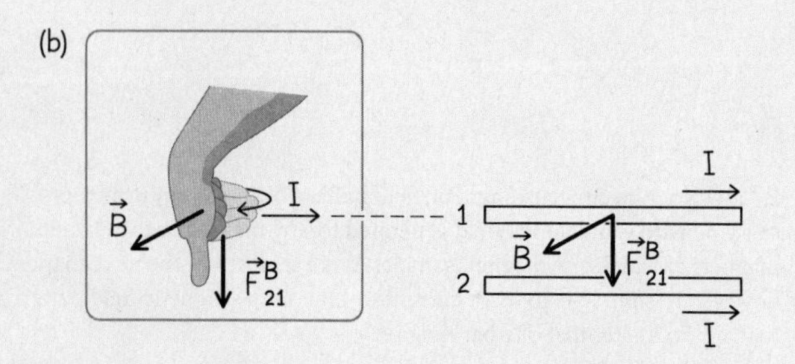

27.13 (*a*) From the point of view of observer M, the upper rod is moving to the left and the lower rod is at rest (Figure S27.13*a*.) Consequently, the upper rod's length is contracted and the lower one has length ℓ_{proper}. Because both rods carry the same charge, the shorter rod—that is, the upper one—has the greater charge density. (*b*) In Figure S27.13*b*, the charge on the lower rod has been adjusted so that the charge density matches that of the upper rod. (Because the lower rod's length is contracted according to observer E, the charge it carries is now smaller than the charge carried by the upper rod.) According to observer M, the upper rod, which is moving, is shorter than the lower rod. Thus observer M sees the shorter upper rod carrying more charge than the longer lower rod and says $\lambda_{\text{upper rod}} > \lambda_{\text{proper}}$ and $\lambda_{\text{lower rod}} < \lambda_{\text{proper}}$ (Figure S27.13*c*).

Figure S27.13

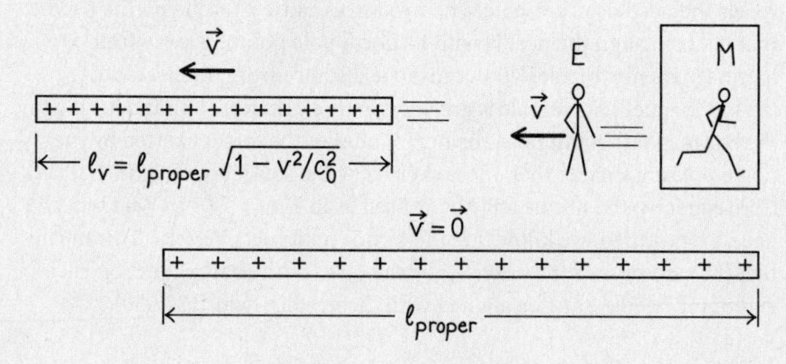

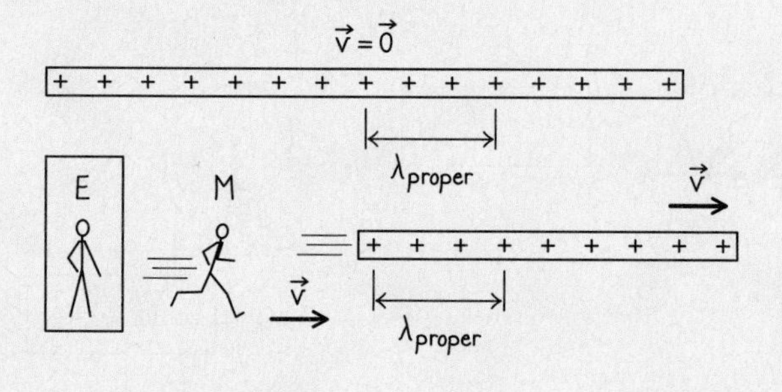

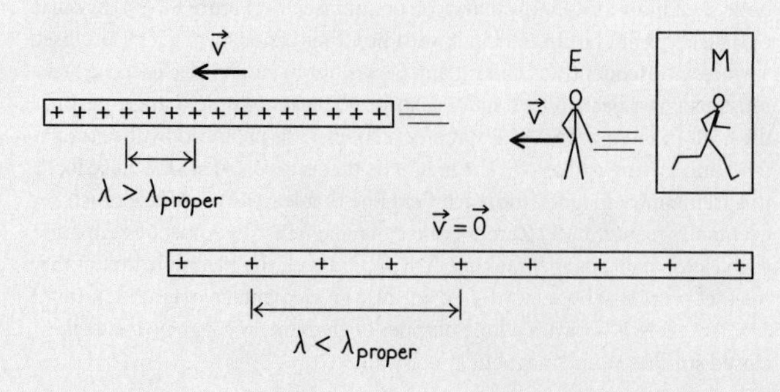

27.14 (*a*) The positive ions moving to the left correspond to a current to the left. (*b*) The negative electrons moving to the right correspond to a current to the left. (*c*) Yes. The positive particle moving to the right corresponds to a current to the right, and so the situation in Figure 27.27*c* corresponds to two parallel rods carrying currents in opposite directions. As you saw in Example 27.1, the force between two such rods is repulsive, in agreement with the repulsive force exerted by the rod on the particle. (*d*) The arguments for the wire remain unchanged: It appears electrically neutral to observer E and positively charged to observer M. The positively charged wire exerts an attractive electric force on the negatively charged particle. (*e*) Yes. Because a negative particle moving to the right corresponds to a current to the left, the situation is identical to that of two rods carrying currents in the same direction. As you saw in Checkpoint 27.12, the force between such rods is indeed attractive.

27.15 Example 27.2 shows that the current I is to the right in Figure 27.33. If this current is caused by negative charge carriers, the definition of current tells you that they must flow to the left.

27.16 (*a*) On the side through which the field enters, the field points into the cube but the area vector points outward, so $\theta = 180°$: $\Phi_B = AB\cos\theta = (1.0\text{ m}^2)(1.0\text{ T})\cos 180° = -1.0\text{ Wb}$. The minus sign reflects the fact that the magnetic field points *into* the cube. (*b*) Because as many field lines enter the cube as leave it, the magnetic flux is zero.

27.17 Because the magnitudes of the charge on the proton and the electron are the same, Eq. 27.23 tells you that the ratio of the radii of the paths is equal to the ratio of the masses: $R_p/R_e = m_p/m_e = (1.67 \times 10^{-27}\text{ kg})/(9.11 \times 10^{-31}\text{ kg}) = 1.83 \times 10^3$.

27.18 (*a*) See Figure S27.18*a* For negative charge carriers, $\vec{F}^E_p$ points in the direction opposite the direction of the electric field lines. Because q changes sign, $\vec{F}^B_p = q\vec{v} \times \vec{B}$ (Eq. 27.19) reverses direction as well. The magnitudes of the two forces are unchanged, and so they still add up to zero. (*b*) See Figure S27.18*b*. Reversing the direction of motion of the charge carriers does not affect $\vec{F}^E_p$, but it does reverse the direction of $\vec{v}$ and thus the direction of $\vec{F}^B_p = q\vec{v} \times \vec{B}$. Both forces therefore point in the same direction and so do not cancel. (*c*) See Figure S27.18*c*. Because $\vec{F}^B_p$ is always perpendicular to both the magnetic field and the velocity $\vec{v}$, the direction of $\vec{F}^B_p$ no longer lines up with that of $\vec{F}^E_p$, and so the two forces do not cancel.

Figure S27.18

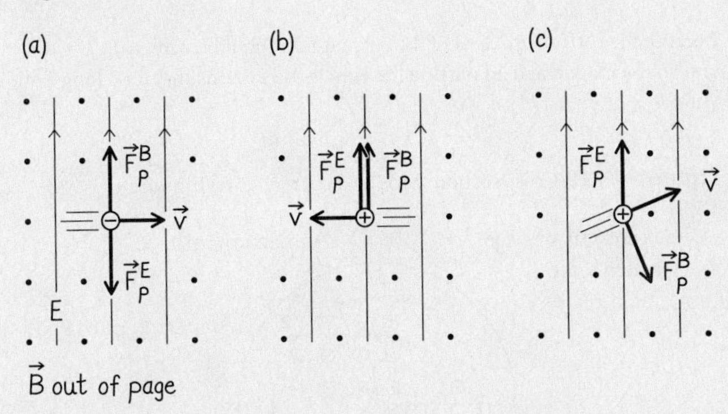

27.19 (*a*) The charge distribution is similar to that of a charged capacitor, so E is given by $E = |V_{RL}|/w$ (see Example 25.5), with the width w of the strip substituted for the separation between the positively and negatively charged sides. (*b*) Because the electric and magnetic forces exerted on the charge carriers are equal in magnitude, we have from Eq. 27.26 $v = E/B = |V_{RL}|/(wB)$. (*c*) From Eq. 27.16, $n = I/(A|q|v)$. Substituting the result from part *b*, we obtain $n = (IwB)/(A|qV_{RL}|)$. Because the charge carriers carry an elementary charge e and because $A = wh$, with h the height of the strip, we have $n = IB/(eh|V_{RL}|)$.

27.20 The wire has cylindrical symmetry and so the magnetic field spreads out only in two dimensions. The electric force seen by observer M must therefore decrease as $1/r$ (see Section 24.4). Because the electric force seen by observer M is the same as the magnetic force seen by observer E, that force must also decrease as $1/r$.

Chapter 28

28.1. (*a*) See Figure S28.1*a*. (*b*) The two particles exert no magnetic forces on each other because it takes a moving charged particle to detect the magnetic field of another moving charged particle. (*c*) See Figure S28.1*b*. The forces exerted on the horizontal wire cause a torque that tends to align that wire with the vertical one. Conversely, the forces exerted on the vertical wire cause a torque in the opposite direction, tending to align that wire with the horizontal one.

Figure S28.1

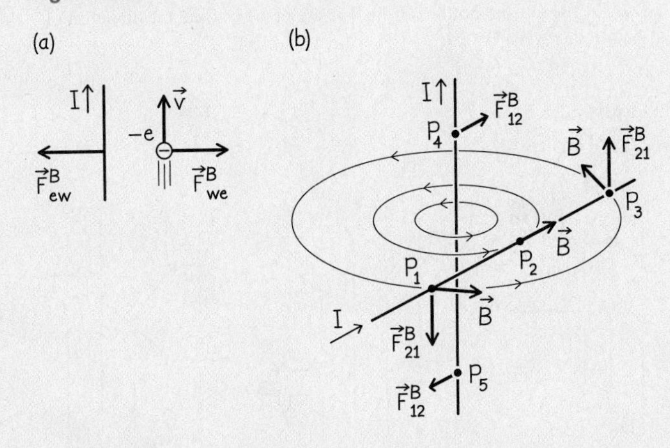

28.2. See Figure S28.2.

Figure S28.2

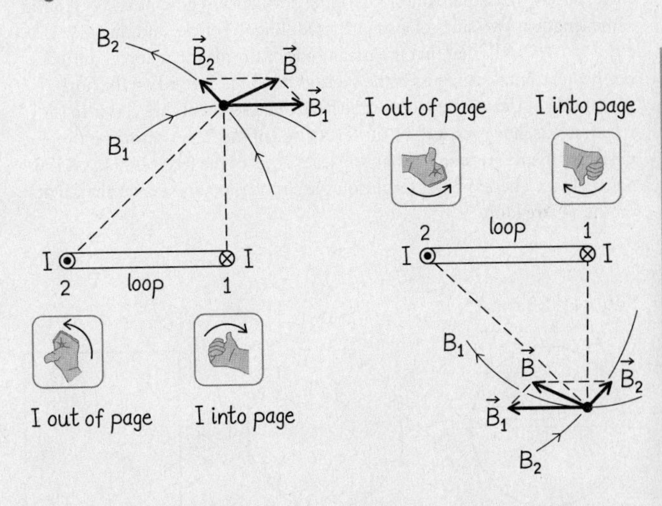

28.3. Comparing Figures 28.6*b* and 28.7, you see that the positively charged ring has a magnetic field similar to the field of a bar magnet, with its north pole up. The negatively charged ring thus has its north pole down, and this north pole is directly above the north pole of the positive ring. Because two north poles repel each other, the interaction is repulsive.

28.4. No. By definition the electric field points from positive charge carriers to negative charge carriers (see Section 23.2). The electric field along the axis passing through the poles of an electric dipole therefore points from the positive end to the negative end. The electric dipole moment, however, by definition points in the opposite direction (see Section 23.6).

28.5. (*a*) Because the magnetic field remains perpendicular to the horizontal sides of the loop, you know from Eq. 27.4 that the magnitude of the magnetic force exerted on each side is $F^B_{\max} = |I|\ell B$, where ℓ is the length of each side of the loop. Because none of these quantities changes, the magnitude of the force exerted on the horizontal sides stays the same. (*b*) See Figure S28.5*a* and *b*. The lever arm of the force exerted on the horizontal sides become smaller as the loop rotates, and so the torque caused by these forces decreases. (*c*) In the initial position, the vertical sides are parallel to the magnetic field, which means the magnetic force exerted on them is zero (Eq. 27.7, $F^B = |I|\ell B \sin\theta$ with $\theta = 0$). After the loop has rotated 90°, the vertical sides are perpendicular to $\vec{B}$ and thus subject to an outward magnetic force. For $0 < \theta < 90°$, the magnitude of the force is again given by Eq. 27.7. As you can verify using the right-hand force rule, the forces exerted on the vertical sides are directed outward along the rotation axis and so cause no torque. (*d*) See Figure S28.5*c*. Because the top and bottom of the loop are now reversed, the direction of the torque reverses.

Figure S28.5

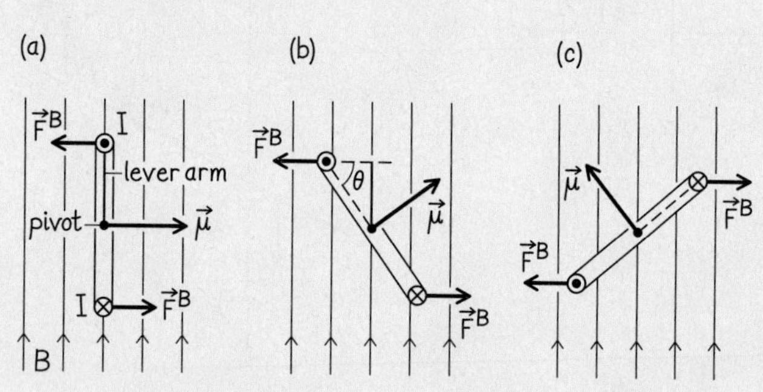

28.6. See Figure S28.6, where the circular loop is approximated by a series of vertical and horizontal segments. The vertical segments experience no force, but the horizontal ones do. All the horizontal segments add up to the same length as two sides of a square straddling the circle, and thus Eq. 27.4, $F^B_{\max} = |I|\ell B$, tells you that the magnitude of the magnetic force exerted on the horizontal segments is the same as the force exerted on the horizontal side of the square loop. Some horizontal segments are closer to the rotation axis, however, and so the lever arms of the forces acting on these closer segments are smaller than the lever arms of the forces acting on the square loop. Therefore the torque on the circular loop is smaller than that on the square loop.

Figure S28.6

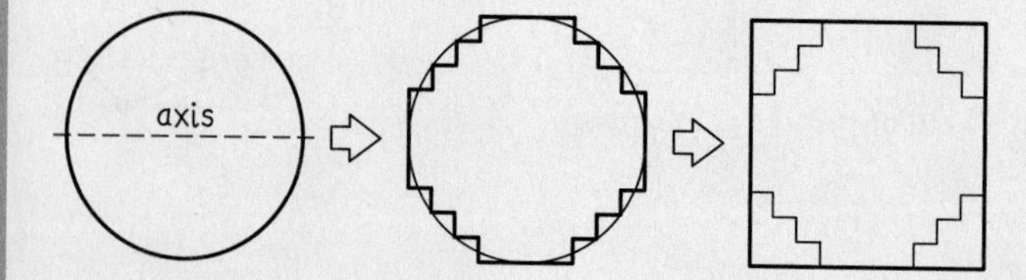

28.7. With the magnetic field on the left stronger than that on the right, $F^B_{\text{left}} > F^B_{\text{right}}$. Therefore the vector sum of the forces exerted on these two sides is nonzero and points upward. The effect is a clockwise rotation due to the torque and an upward acceleration due to the upward vector sum of the forces.

28.8. Increasing the current increases the magnitude of the magnetic field, and so the line integral increases.

28.9. (*a*) If the direction of the current is reversed, the direction of the magnetic field is reversed, and so the algebraic sign of the line integral is reversed. (*b*) Because the value of the line integral of the magnetic field around a closed path does not depend on the position of the wire inside the path, the second wire by itself gives rise to the same line integral as the first wire. Adding the second wire thus doubles the value of the line integral. (*c*) Reversing the current though the second wire flips the sign of the line integral. If its value was C, it is $-C$ after the current is reversed. The sum of the two line integrals is thus $C + (-C) = 0$.

28.10. No. If the path tilts, you can break it down into arcs, radial segments, and *axial* segments, which are segments parallel to the wire. Because the magnetic field is perpendicular to the axial segments, they don't contribute to the line integral. For the same reason, of course, the radial segments contribute nothing. So the line integral again is that along all the arcs, which is equivalent to a single complete circular path.

28.11. (*a*) Reversing the direction of the current through wire 1 changes the sign of its contribution from negative to positive, so adding the contributions of wires 1 and 3 yields a positive value. (*b*) The path does not encircle wire 2, so reversing the current through wire 2 does not affect the answer to Exercise 28.2. (*c*) Reversing the direction of the path inverts the signs of all the contributions to the line integral, but because the integral is zero, the answer doesn't change.

28.12. See Figure S28.12. Inside the wire, use Ampèrian path 1 of radius $r < R$. For this path,

$$\oint \vec{B} \cdot d\vec{\ell} = B_{\text{inside}} \oint d\ell = B_{\text{inside}}\, 2\pi r,$$

but the path encircles only part of the cross section and so encircles only part of the current I through the wire. The area of the cross section of the wire is πR^2, and the cross-sectional area of Ampèrian path 1 is πr^2, making the fraction of the wire cross section enclosed by the path $(\pi r^2)/(\pi R^2) = r^2/R^2$. Thus, the right side of Ampère's law is $\mu_0 I_{\text{enc}} = \mu_0(r^2/R^2)I$. Ampère's law thus yields $B_{\text{inside}}(2\pi r) = \mu_0(r^2/R^2)I$, or $B_{\text{inside}} = \mu_0 Ir/2\pi R^2$.
Outside the wire, use Ampèrian path 2 in Figure S28.12.
For this path,

$$\oint \vec{B} \cdot d\vec{\ell} = B_{\text{outside}} 2\pi r.$$

Because the path encircles all of the current I through the wire, $\mu_0 I_{\text{enc}} = \mu_0 I$, and so the magnetic field outside the wire is the same as that for a long thin wire: $B_{\text{outside}} = \mu_0 I/2\pi r$.

Figure S28.12 Cross section through a current-carrying wire

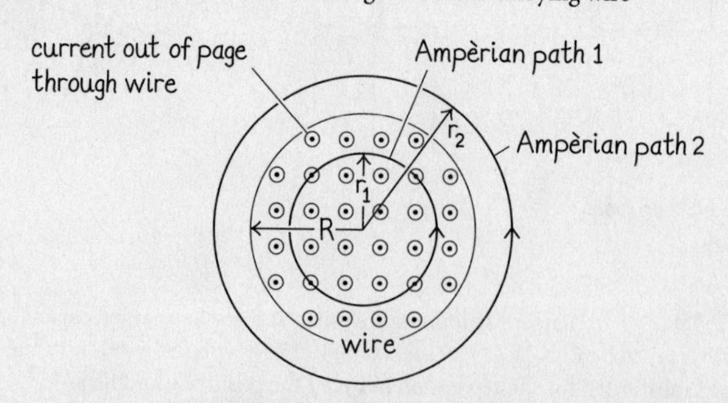

28.13. (*a*) Figure S28.13 shows a head-on view of the magnetic fields of the two sheets separately and their sum. The magnetic fields add up to zero between the sheets. Below and above the sheets, the magnetic field points to the left and its magnitude is twice that of a single sheet: $2(\frac{1}{2}\mu_o K) = \mu_o K$. (*b*) With the current through the lower sheet reversed, its magnetic field is also reversed. Now the magnetic fields add up to zero below and above the sheets. Between them the magnetic field is to the right and its magnitude is twice that of a single sheet: $\mu_0 K$.

Figure S28.13

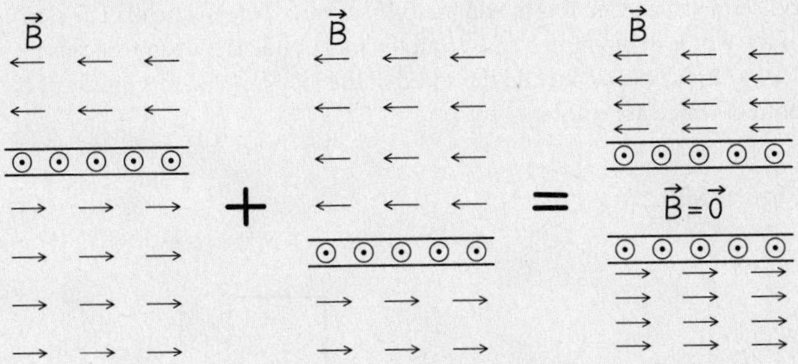

28.14. Outside the toroid, you use Ampèrian path 1 in Figure S28.14. The left side of Ampère's law is still given by Eq. 28.7, but the path now encircles both the current upward through the inside of each winding and the current downward through the outside of each winding. The encircled current is thus $NI - NI = 0$, making B outside the toroid zero. Inside the toroid's inner radius, use Ampèrian path 2. This path encloses no current, so here, too, $B = 0$.

Figure S28.14

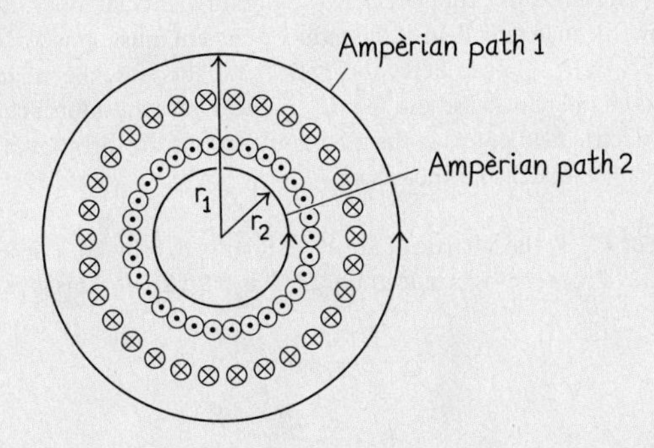

28.15. For this wire the integration in Example 28.6 extends from $x = 0$ to $x = +\infty$:

$$B = \frac{\mu_0 I d}{4\pi}\int_0^{+\infty} \frac{dx}{[x^2+d^2]^{3/2}}$$

$$= \frac{\mu_0 I d}{4\pi}\left[\frac{1}{d^2}\frac{x}{[x^2+d^2]^{1/2}}\right]_{x=0}^{x=+\infty} = \frac{\mu_0 I}{4\pi d}.$$

28.16. (*a*) Using our result from Example 28.7 with $\phi = 2\pi$:

$$B = \frac{\mu_0 I(2\pi)}{4\pi R} = \frac{\mu_0 I}{2R}.$$

(*b*) The outer arc spans an angle $\phi = \pi/2$ and has a radius $2R$. Using our result from Example 28.7 again, it contributes a magnetic field

$$B_{\text{outer}} = \frac{\mu_0 I(\pi/2)}{4\pi(2R)} = \frac{\mu_0 I}{16R}.$$

The inner arc also spans an angle of $\phi = \pi/2$, but it has radius R, and the current through it runs in the opposite direction so the magnetic field due to this arc is directed into the page. For this arc, the magnitude of the magnetic field is

$$B_{\text{inner}} = \frac{\mu_0 I(\pi/2)}{4\pi(R)} = \frac{\mu_0 I}{8R}.$$

The two straight segments do not contribute because they point straight toward P. Because B_{inner}, which points into the page, is greater than B_{outer}, which points out of the page, the magnetic field $\vec{B} = \vec{B}_{\text{inner}} + \vec{B}_{\text{outer}}$ is into the page rather than out of the page. The magnitude of the magnetic field is $B = \mu_0 I/8R - \mu_0 I/16R = \mu_0 I/16R$.

28.17. (*a*) The direction of the magnetic force is given by the vector products in Eq. 28.23. Figure S28.17*a* shows the vectors that appear in this expression. You must first evaluate $\vec{v}_1 \times \hat{r}_{12}$. The direction of the resulting vector is obtained by curling the fingers of your right hand from $\vec{v}_1$ to $\hat{r}_{12}$, which tells you $\vec{v}_1 \times \hat{r}_{12}$ points into the plane of the page (Figure S28.17*b*).

Figure S28.17

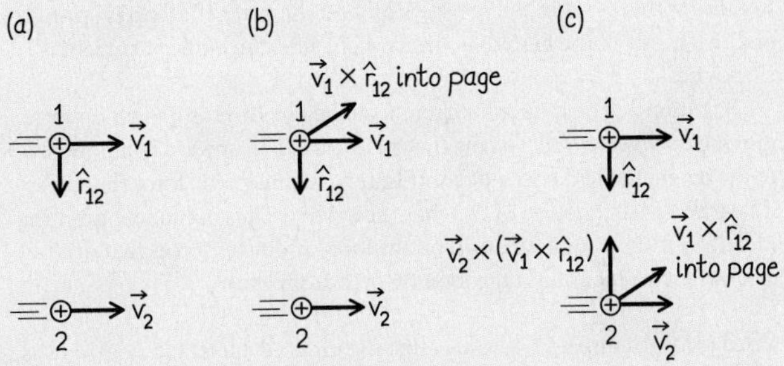

Next evaluate the vector product between $\vec{v}_2$ and the vector product you just found. The right-hand vector product rule tells you this yields a vector that points upward from proton 2 to proton 1. Thus, $\vec{F}^B_{12}$ points upward. Applying the same procedure to obtain $\vec{F}^B_{21}$, you find that it points downward, telling you that the magnetic interaction is attractive, like that of two parallel current-carrying wires.

(*b*) Because the angles involved in the vector products in Eq. 28.23 are all 90°,

$$F^B_{12} = \frac{\mu_0}{4\pi}\frac{q_1 q_2}{r^2_{12}} v_1 v_2.$$

The magnitude of the electric force is given by Coulomb's law:

$$F^E_{12} = \frac{1}{4\pi\epsilon_0}\frac{q_1 q_2}{r^2_{12}}.$$

The ratio is

$$\frac{F^B_{12}}{F^E_{12}} = \frac{\left(\frac{\mu_0}{4\pi}\right)\frac{q_1 q_2}{r^2_{12}} v_1 v_2}{\left(\frac{1}{4\pi\epsilon_0}\right)\frac{q_1 q_2}{r^2_{12}}} = \mu_0 \epsilon_0 v_1 v_2.$$

Substituting numerical values, you find that this ratio is 10^{-6}.

Chapter 29

29.1 The magnetic force exerted on the charge carriers points in the opposite direction—that is, upward rather than downward. As a result, positive charge accumulates at the upper end of the rod rather than the lower end.

29.2 Yes, in positions (*b*) and (*d*) in Figure 29.4. The only difference between moving the field and moving the loop is the observer's frame of reference. It would not make any sense for an observer in one reference frame to observe a current and an observer in another reference frame to see no current.

29.3 No, because magnetic forces are exerted only on moving charged particles, not on stationary ones.

29.4 There are two electric fields in Figure 29.10: the "regular" electrostatic field due to the charge separation and the electric field that accompanies the changing magnetic field due to the moving magnet. The charge distribution in the rod is in mechanical equilibrium when these two fields have the same magnitude and opposite directions, producing zero electric field inside the stationary rod, as we would expect.

29.5 Situation 1: The induced current must be the same as if the loop were moving to the left with the magnet stationary (the same relative motion). The upward component of the magnetic field exerts a force on a positive charge carrier that is directed into the page (toward the back of the loop). Because at this point in the magnet's motion the magnetic field is stronger on the left side of the loop than on the right side, the induced current is clockwise.

Situation 3: The induced current must be the same as if the loop were moving downward with the magnet stationary. The upward component of the magnetic field does not contribute to the magnetic force (because it's parallel to the motion of the charge carriers). The component pointing radially outward (looking down on the loop) produces forces that drive a clockwise current around the loop (viewed from above).

29.6 (*a*) In Example 29.4 we saw that the induced current is clockwise as viewed from above in Figure 29.12*b*. Inside the loop $\vec{B}_{\text{ind}}$ points down, and outside the loop $\vec{B}_{\text{ind}}$ points up. Remember that $\vec{B}_{\text{ind}}$ is stronger inside the loop than outside and that the direction of the force exerted by $\vec{B}_{\text{ind}}$ on the south pole is opposite the direction of $\vec{B}_{\text{ind}}$. Thus, the force exerted on the magnet by $\vec{B}_{\text{ind}}$ pushes the magnet upward, opposing the downward motion of the magnet that induced the current and field $\vec{B}_{\text{ind}}$.

(*b*) If Lenz's law indicated that the induced current and field should add to the change that produced them, the force exerted on the magnet by $\vec{B}_{\text{ind}}$ would accelerate the magnet downward more rapidly (and in the process increase the induced current).

29.7 (*a*) Zero. Once the loop is completely in the magnetic field, the magnetic flux enclosed by the loop is no longer changing. The loop's motion ceases to induce current, and the force that resists the motion vanishes, which means no further work is required to keep the loop moving at constant speed. (*b*) Positive. Once the right edge leaves the field, the magnetic flux enclosed by the loop begins to decrease. Lenz's law says that a force resists the magnetic flux decrease and therefore resists the loop's motion. Thus, the loop must be pulled out. The pulling force you exert points in the same direction as the motion, so the work done is positive.

29.8 More work must be done to move a magnet toward a closed conducting loop because the motion induces a current in the loop, and the work done on the magnet must provide the energy for this current. Because there cannot be a current through a rod, moving a magnet toward a rod induces only a static charge separation, which requires less work.

29.9 (*a*) The magnitude decreases through a small region around P because the magnet is moving away from P. It remains unchanged through a small region around Q because the magnet is directly overhead. It increases through a small region around R because the magnet is moving toward R. (*b*) The direction of any eddy current is such that the magnetic field associated with the current opposes the change in magnetic flux. A counterclockwise current (viewed from overhead) creates a field that points upward. Therefore the eddy current around P is counterclockwise and that around R is clockwise. The magnetic flux through a small region around Q remains unchanged, so there are no eddy currents around Q.

29.10 See Figure S29.10 (with magnetic flux out of the page in Figure 29.4 chosen to be positive). The induced emf is zero when the loop is completely outside (*a* and *e*) or completely inside (*c*) the field. The emf has the same constant value when it is moving into (*b*) or out of (*d*) the field. (The sign of the emf is different in *b* and *d*, but the magnitude is the same because $|\Delta\Phi_B/\Delta t|$ depends only on the speed of the loop and the magnitude of $\vec{B}$, both of which are constant.)

Figure S29.10

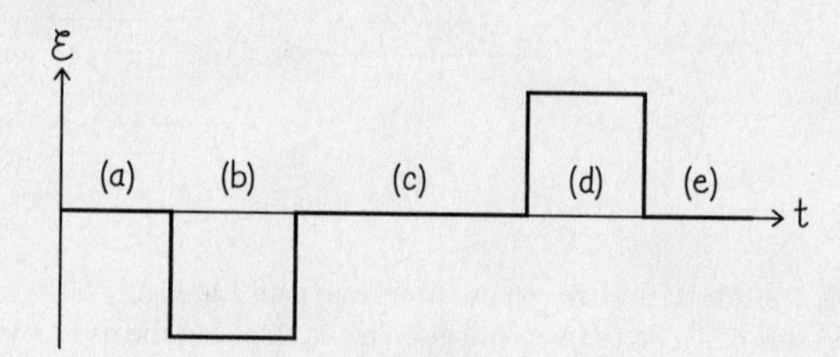

29.11 Although the solenoid orientation looks the same, the two sides have interchanged their positions. Because the two sides have changed positions, the sense of circulation has changed too. The emf thus has changed sign too.

29.12 (*a*) There is no change in the magnetic flux and thus no electric field. (*b*) They look like counterclockwise-pointing circular loops. By Lenz's law, the magnetic field of the induced current must oppose the decrease in the magnitude of the magnetic field. Therefore the induced current is counterclockwise and the electric field lines must form circles and the electric field points in the direction opposite the direction it has when the magnetic field is increasing.

29.13 For $r > R$, the left side of Eq. 29.17 is still given by Eq. 1 in Example 29.7, but the right side of Eq. 29.17 is $\pi R^2\, dB/dt$. This gives

$$2\pi r E = \pi R^2 \frac{dB}{dt}$$

$$E = \frac{R^2}{2r}\frac{dB}{dt}.$$

Substituting $dB/dt = 0.050$ T/s, $R = 0.20$ m, and $r = 0.30$ m into this expression yields

$$E = \frac{(0.20\text{ m})^2(0.050\text{ T/s})}{2(0.30\text{ m})} = 3.3 \times 10^{-3}\text{ N/C}.$$

29.14 You can calculate the induced emf from the inductance of the solenoid (Example 29.8) and Eq. 29.19:

$$L = \frac{(4\pi \times 10^{-7}\text{ T}\cdot\text{m/A})(2760)^2\pi(50 \times 10^{-3}\text{ m})^2}{0.60\text{ m}} = 0.13\text{ H}$$

$$|\mathscr{E}_{\text{ind}}| = L\frac{dI}{dt} = (0.13\text{ H})(1.0 \times 10^{-1}\text{ A/s}) = 13 \times 10^{-3}\text{ V}.$$

29.15

$$u_B = \frac{B^2}{2\mu_0} = \frac{(1.0\text{ T})^2}{2(4\pi \times 10^{-7}\text{ T}\cdot\text{m/A})} = 4.0 \times 10^5\text{ J/m}^3$$

$$u_E = \tfrac{1}{2}\epsilon_0 E^2 = \tfrac{1}{2}(8.85 \times 10^{-12}\text{ C}^2/\text{N}\cdot\text{m}^2)(1.0\text{ V/m})^2 = 4.4 \times 10^{-12}\text{ J/m}^3.$$

The ratio of u_B to u_E is 9.0×10^{16}.

Chapter 30

30.1 (*a*) Yes. The wire is intercepted by the surface three times, but the directions of the intercepts are different. Because the magnetic field points in the same direction as the integration direction along the closed path, we can use the right-hand current rule (see the Procedure box in Chapter 28, point 4) to determine that intercepts 1 and 3 are positive and intercept 2 is negative. The direction of the current at intercepts 1 and 3 is from inside to outside; at intercept 2 it is from outside to inside. We can expect current intercepted going from inside to outside to contribute oppositely to current intercepted going from outside to inside. From Figure 30.1 we see that current intercepted going from inside to outside through surface B is positive, so the current intercepted by the surface is $+I$ at 1 and 3 and $-I$ at 2, for a total of I, which is equal to the current encircled by the path. (*b*) Yes. The current intercepted by the surface is I: $+I$ for the top intercept, $-I$ for the middle intercept, and $+I$ for the bottom intercept. Again, the current intercepted by the surface is equal to the current encircled by the path.

30.2 (*a*) While the capacitor is charging, there is a current through the wire leading to or from the capacitor, and the electric field between the plates is nonzero and changing. (*b*) Once the capacitor is fully charged, the current through the wire drops to zero. There is still a nonzero electric field between the plates, but that field is no longer changing.

30.3 Although the electric field direction does not change, the direction of the *change* $\Delta\vec{E}$ in the electric field is reversed because the electric field is now decreasing. Taking the direction of $\Delta\vec{E}$ as the "current," you see that the direction of the magnetic field is reversed from what it was when the unit was charging, and the right-hand current rule applies.

30.4 The magnetic dipole moment of the neutron indicates that it must have an internal structure consisting of charged particles that form current loops. Indeed, in Chapter 7 we learned that the neutron consists of one up and two down quarks (Figure 7.17). These quarks do indeed carry charge.

30.5 You can estimate v by comparing the distance the particle moves in the time interval between parts *b* and *c* of Figure 30.8 with the distance the kinks have moved in that interval as measured from the point where they originate, which is at the center of each panel. The kinks move roughly twice as far as the particle, which means $v \approx 0.5c$.

30.6 The electric field is changing at the kinks and in the region between the kinks and the particle. In these regions there is a magnetic field. Beyond the kinks, the magnetic field is zero because the electric field is static.

30.7 (*a*) In a time interval equal to the period T, the wave travels a distance equal to its wavelength. Because the wave travels horizontally, you should measure the distance traveled along the horizontal axis through the electric field pattern. Between $t = 0$ and $t = T/2$ the wave advances half a wavelength. The distance that the rightmost field line advances in this time interval is 16 mm, so the wavelength is 32 mm. (*b*) From Eq. 16.10, $f = c/\lambda = 9.4 \times 10^9$ Hz. (*c*) From Eq. 15.2, $T = 1/f = 1.1 \times 10^{-10}$ s, or 0.11 ns.

30.8. (*a*) Figure 30.15*a* shows the bottom electric field pattern of Figure 30.14, so the motion of the particles must be the same in those two cases. From Figure 30.14 you see that at $t = \frac{3}{4}T$, the positively charged particle is moving up and the negatively charged particle is moving down, which means the current direction is up. (Although the dipole moment is zero at this instant because the particles are in the same location, the current is not zero because they are both moving. The current is instantaneously zero when the particles are at the two extremes of their travel paths and thus have zero velocity for an instant.) (*b*) Yes, because the magnetic field lines closest to the dipole are in the proper direction for the magnetic field of an upward current: Immediately to the right of the dipole, the magnetic field lines point into the page; immediately to the left of the dipole, they point out of the page.

30.9 Parallel, because parallel to the polarization means parallel to the electric field. From Section 23.4, you know that an electric field exerts a force on charged particles (see Figure 23.15, for instance). Only electric forces with a component along the axis of the antenna can accelerate the charge carriers in the antenna in a direction that produces current (along the antenna's length), so an electric field with a component parallel to the antenna is needed.

30.10 Whether between the plates or outside the capacitor, the magnetic field accompanies a change in the electric field that points left (the electric field is decreasing). From Section 30.1, you know that the "current" you must use with the right-hand current rule for determining the direction of $\vec{B}$ also points left. If you point the thumb of your right hand to the left along the axis in Figure 30.24, your curled fingers point into the page above the axis of the capacitor and out of the page below the axis. This means $\vec{B}$ is into the page at both P and S.

30.11 (*a*) Use a surface that lies between the plates, like the one in Figure 30.28 but with a radius $r < R$. Then follow the approach used in Example 30.5, substituting r for R in Eq. 1:

$$B = \frac{\mu_0\epsilon_0 r}{2}\frac{dE}{dt}.$$

To determine E, use Eq. 1 from Example 26.2 with $A = \pi R^2$ (because A is the area of each plate, not the area of the surface bounded by the path of integration). This substitution gives you

$$B = \frac{\mu_0\epsilon_0 r}{2}\frac{d}{dt}\left(\frac{q}{\epsilon_0\pi R^2}\right) = \frac{\mu_0 r}{2\pi R^2}\frac{dq}{dt}.$$

You know that dq/dt is the current, so you can write $B = \mu_0 Ir/(2\pi R^2)$. Thus, the magnetic field between the plates is smaller closer to the axis of the plates. (*b*) To the right of the right plate (and, of course, to the left of the left plate), the magnetic field is simply that of a current-carrying wire at a distance r from the wire. The result we obtained in Example 28.3 gives you $B = \mu_0 I/(2\pi r)$.

30.12 Equation 30.10 relates the electric field to its source charge distributions and thus needs no modification. The right side of Eq. 30.11 is zero because there are no magnetic monopoles. If monopoles having a $1/r^2$ relationship did exist, Eq. 30.11 would have to be analogous to Eq. 30.10. This means the right side must be proportional to m_{enc} rather than zero. You would expect electric fields to form loops around monopole currents, in the same way that magnetic fields form loops around currents, so the right side of Eq. 30.12 must gain a term proportional to dm/dt. Equation 30.13 needs no modification because $\oint \vec{B}\cdot d\vec{\ell}$ for a static distribution of monopoles should be zero, just as $\oint \vec{E}\cdot d\vec{\ell}$ for a static charge distribution is zero.

30.13 In vacuum, the line integral of the electric field is proportional to the time rate of change of magnetic flux through the surface bounded by the path (Eq. 3), and the line integral of the magnetic field is proportional to the time rate of change of electric flux through the surface bounded by the path (Eq. 4). To satisfy these relationships, as shown in Figure 30.30*e* and *f*, the magnetic field and electric field must be perpendicular to each other. The magnetic field must form loops around the electric field, and the electric field must form loops around the magnetic field.

30.14 When an electromagnetic wave enters a dielectric medium, the electric field of the wave accelerates charged particles in the medium back and forth with the frequency of the wave. The accelerated particles radiate electromagnetic waves that propagate with the same frequency as the incoming wave. The electromagnetic wave in the medium is the combination of all of these waves and hence has the same frequency, $f = c_0/\lambda = (3.00 \times 10^8\ \text{m/s})/(600\ \text{nm}) = 5.00 \times 10^{14}$ Hz. The wavelength in the dielectric medium decreases because the speed of the wave in the medium is smaller than in vacuum:

$$\lambda_{new} = \frac{c}{f} = \frac{c_0}{f\sqrt{\kappa}} = \frac{\lambda}{\sqrt{\kappa}} = \frac{600\ \text{nm}}{\sqrt{1.30}} = 526\ \text{nm}.$$

30.15 The electric field points from one plate to the other, and the magnetic field forms loops around the electric field. (Figure 30.4 represents this situation.) The vector product of these two vectors points radially inward, toward the capacitor axis (Figure S30.15). The Poynting vector thus represents the flow of energy into the region between the capacitor plates, where energy is being stored in the electric field. This makes sense because the energy density associated with the electric field inside the cylindrical surface defined by the capacitor plates is increasing. So there must be a flow of electromagnetic energy into this region.

Figure S30.15

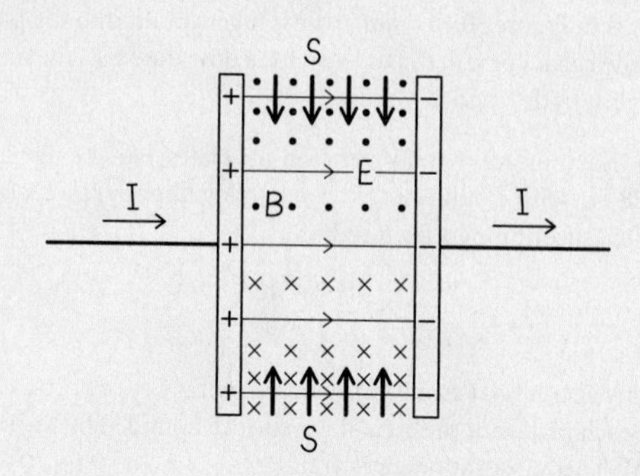

30.16 The SI units of μ_0 are $\text{T}\cdot\text{m/A}$; those of the electric field N/C and of the magnetic field T. Using Eq. 30.36, I thus get for the SI units of the Poynting vector:

$$\frac{[\text{N/C}][\text{T}]}{\text{T}\cdot\text{m/A}} = \frac{\text{N}\cdot\text{A}}{\text{C}\cdot\text{m}} = \frac{\text{N}\cdot\text{A}}{\text{A}\cdot\text{s}\cdot\text{m}} = \frac{\text{N}}{\text{m}\cdot\text{s}} = \frac{\text{J}}{\text{m}^2\cdot\text{s}} = \frac{\text{W}}{\text{m}^2}.$$

Chapter 31

31.1 (*a*) Yes. In practice, the light and thermal energy travel away from the bulb, so you need either a flexible definition of the system or a container capable of keeping the light and thermal energy in a well-defined volume. (*b*) Yes. (*c*) The energy to produce light and thermal energy comes from the electric potential energy associated with the potential difference between the battery terminals. This energy comes from chemical reactions taking place inside the battery.

31.2 (*a*) Yes. The charged capacitor is the source, and the bulb is the load. (*b*) The bulb glows during the brief time interval when the capacitor is discharging. Electrons flow from the negatively charged capacitor plate through the connecting wires and the bulb to the positively charged plate until the capacitor plates are no longer charged. (*c*) Electric potential energy stored in the charged capacitor is converted to light and thermal energy in the bulb.

31.3 These answers are appropriate for a flashlight bulb connected to four D batteries that are fresh. For different batteries and different bulbs, the times may vary somewhat. (*a*) The current should stay constant for many minutes. (*b*) After 24 hours, the batteries are depleted and the current is zero.

31.4 (*a*) Gains electric potential energy. (*b*) Loses the same amount of electric potential energy. (*c*) In the load (they lose very little energy when flowing through the wires).

31.5 (*a*) That resistance is proportional to potential difference tells you that A has the greater resistance. (*b*) The greater resistance of bulb A means that a greater potential difference is required to obtain a given current in A. Thus, the same potential difference across both bulbs produces a smaller current in A than in B.

31.6 Smaller than. When the circuit load consists of the parallel combination of two bulbs, the current out of the battery is greater than the current out of the battery when the load consists of a single bulb. (To be exact, the current is twice as great with the parallel combination of bulbs as with the single bulb, as shown in Example 31.3.) The potential difference across the load is the same in both cases. Therefore, the resistance of the parallel combination of two bulbs is half the resistance of the single bulb.

31.7 (*a*) One contact of A is connected to the positive terminal of battery 1, and the other contact is connected to the negative terminal of battery 2. The negative terminal of battery 1 is connected to the positive terminal of battery 2. Therefore the potential difference across A equals the potential difference across the two-battery combination, which is the sum of the potential differences across each battery, or 18 V. (*b*) One contact of bulb B is connected to the positive terminal of battery 1, and the other contact of bulb B is connected to the negative terminal of the same battery. Therefore, the magnitude of the potential difference across B must be 9 V. (*c*) Bulb A glows more brightly because there is a greater potential difference across it.

31.8 Because the electric field is uniform throughout the rod, the magnitude of the electric field is equal to the magnitude of the potential difference across the rod divided by the length of the rod, $E = |V_{12}|/\ell$ (see Eq. 1 in Example 25.5).

31.9 See Figure S31.9. The electric field lines are parallel to the sides of the conductors everywhere. The electric field line density is also uniform, representing the uniform magnitude of the electric field.

Figure S31.9

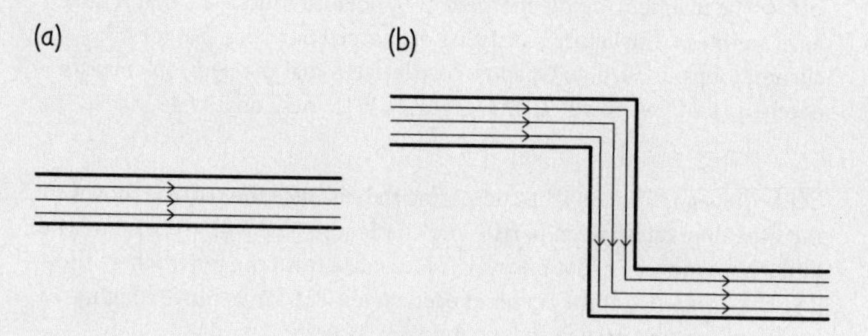

31.10 (*a*) You know from the text that $V_{\text{wide}} = V_{\text{narrow}}/4$. You also know that the magnitude of the potential difference across the entire conductor is 9 V. Therefore $V_{\text{wide}} + V_{\text{narrow}} = V_{\text{wide}} + 4V_{\text{wide}} = 9\text{ V}$, so $V_{\text{wide}} = 9\text{ V}/5 = 1.8\text{ V}$. (*b*) $V_{\text{narrow}} = 9\text{ V} - 1.8\text{ V} = 7.2\text{ V}$.

31.11 (*a*) Yes. The electric field does work on the electrons to accelerate them. (*b*) On average, the kinetic energy does not change over time because the average final velocity (drift velocity) does not depend on time. (*c*) Because the electric field does work on the electrons but their kinetic energy does not increase over time, some other form of energy in the system must increase. (As you'll see shortly, the increase takes the form of thermal energy.)

31.12 (*a*) Greater vibrations cause the ions to move around within a greater volume, increasing the probability of ion-electron collisions and decreasing the average time interval between collisions. Because conductivity is proportional to this time interval and resistance is inversely proportional to conductivity, decreasing the interval causes the resistance to increase. (*b*) A current causes the metal to heat up because the work done on the electrons by the applied electric field gets converted to thermal energy of the lattice ions. (*c*) See Figure S31.12. At low current, the heating is negligible. Because metals are ohmic at constant temperature, the curve is a line with slope $1/R$ at low current. At high current, which causes the metal to heat up, the resistance increases as the current increases, and so the line curves downward.

Figure S31.12

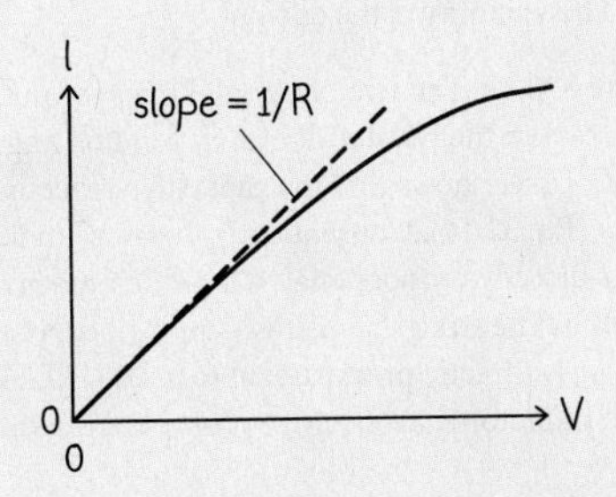

31.13 No. Because the positive terminals of the two batteries are connected to each other with a wire, the potential is the same at points a and b (Figure S31.13). So, if $\mathscr{E}_1 < \mathscr{E}_2$, then the potential at d must be greater than the potential at c. Because the direction of current is from high to low potential, the current is counterclockwise though the circuit, in the direction opposite the reference direction for the current indicated in the diagram.

Figure S31.13

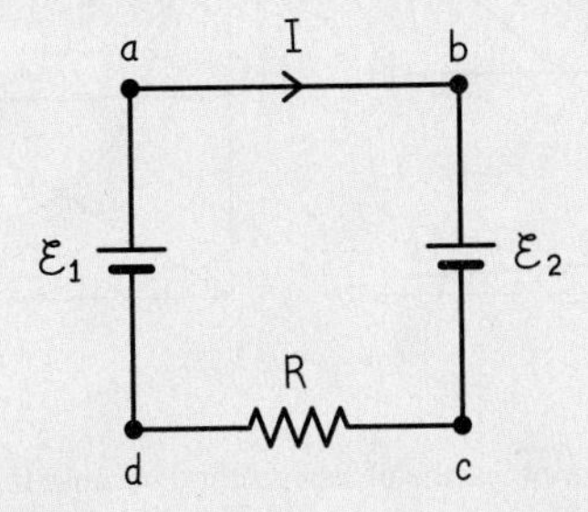

31.14 (*a*) See Figure S31.14*a*. If we go counterclockwise from the top, the first circuit element is the battery; because we are traveling from the positive to the negative terminal, the potential difference is $-\mathscr{E}$. Next is the resistor; because we are traveling in the opposite direction as the reference direction for the current, the potential difference is $+IR$. Substituting these values in Eq. 31.21, we get $-\mathscr{E} + IR = 0$ or $I = \mathscr{E}/R$, which is identical to the result we obtained in Eq. 31.24. We should get the same sign for the current because we have chosen the same reference direction for the current as in Figure 31.34. (*b*) See Figure S31.14*b*. If we go clockwise from the bottom, the first circuit element is the battery; because we are traveling from the negative to the positive terminal, the potential difference is again $\mathscr{E}$. Next is the resistor; because we are traveling in the opposite direction as the reference direction for the current, the potential difference is IR. Substituting these values in Eq. 31.21, we get $\mathscr{E} + IR = 0$ or $I = -\mathscr{E}/R$, which is opposite in sign to the result we obtained in Eq. 31.24. The negative sign means that the current direction is opposite the chosen reference direction for the current. That is, the current direction is not counterclockwise, but clockwise, in agreement with our earlier analysis.

Figure S31.14

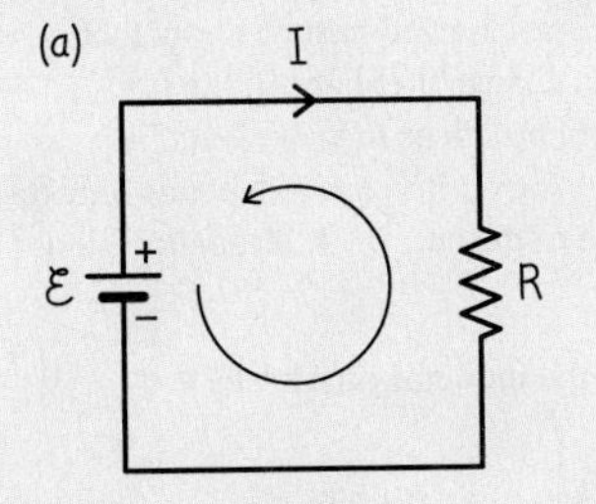

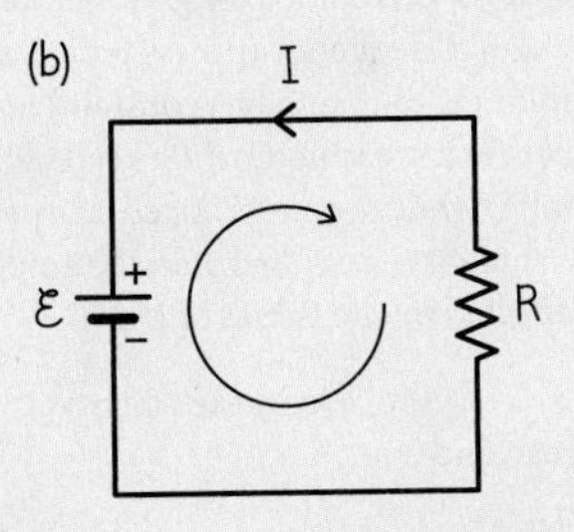

31.15 (*a*) The potential difference is now positive because you are moving closer to the positive terminal of the battery. The magnitudes of the current and resistance are the same, however, so the magnitude of the potential difference must be the same. Thus, $V_{cb} = IR_1$. (*b*) From the solution I obtained in Exercise 31.8, $I = (9\text{ V})/(10\ \Omega + 5\ \Omega) = 0.6$ A. Therefore, $V_{cb} = (0.6\text{ A})(10\ \Omega) = 6$ V. This result makes sense because the resistance of R_1 is two-thirds of the resistance of the combination of R_1 and R_2, which means you expect the potential difference across R_1 to be two-thirds of 9 V, the potential difference across the combination.

31.16 (*a*) From Eq. 31.32, $R_{eq} = [1/(3\ \Omega) + 1/(10\ \Omega) + 1/(5\ \Omega)]^{-1} = 1.6\ \Omega$. (*b*) From Eq. 31.31, $I = (9\text{ V})/(1.6\ \Omega) = 5.7$ A.

31.17 Now the resistance in the path from a to d is smaller than it is when $R_{var} = 12\ \Omega$. More charge carriers now flow from a to d, so a small number of carriers flow from b to a through the bulb to increase the current from a to d.

31.18 Because $1\text{ A} = 1\text{ C/s}$ and $1\text{ V} = 1\text{ J/C} = 1\text{ A}\cdot\Omega$,

$$1\text{ A}\cdot\text{V} = 1\frac{\text{C}}{\text{s}}\cdot\frac{\text{J}}{\text{C}} = 1\text{ J/s} = 1\text{ W}$$

$$1\text{ A}^2\cdot\Omega = 1\text{ A}\cdot\text{A}\cdot\Omega = 1\text{ A}\cdot\text{V} = 1\text{ W}.$$

31.19 (*a*) If I travel counterclockwise around the circuit starting at a, the loop rule yields $6.0\text{ V} + I(0.25\ \Omega) + I(0.25\ \Omega) - 9.0\text{ V} = 0$. Solving this expression for the current again yields $I = 6.0$ A, so the answer remains unchanged (as it should because nothing has physically changed). (*b*) Because the current is the same in both batteries and because they have the same internal resistance, the rate at which energy is dissipated is also the same: $P = 9.0$ W.

Chapter 32

32.1 (*a*) Electric potential energy in the electric field of the capacitor. (*b*) As the capacitor discharges, that energy is converted to magnetic energy stored in the magnetic field in the inductor. (*c*) Magnetic energy in the inductor.

32.2 (*a*) Yes (see Eq. 31.43). (*b*) Energy is supplied by the AC source, and so as long as the AC source delivers energy at the rate at which it is dissipated by the resistor, the amplitudes of the current and potential difference oscillations remain constant. (Because the power is proportional to the current *squared*, the changing sign of the current does *not* imply that sometimes the resistor delivers energy to the AC source. We discuss this point in Section 32.8.)

32.3 See Figure S32.3. Because the current and potential difference are in phase with each other, the phasors V_R and I overlap as in Figure 32.13. Because v_R and i are both zero at $t = 0$ and increase as time goes on, the phasors must lie on the horizontal axis.

Figure S32.3

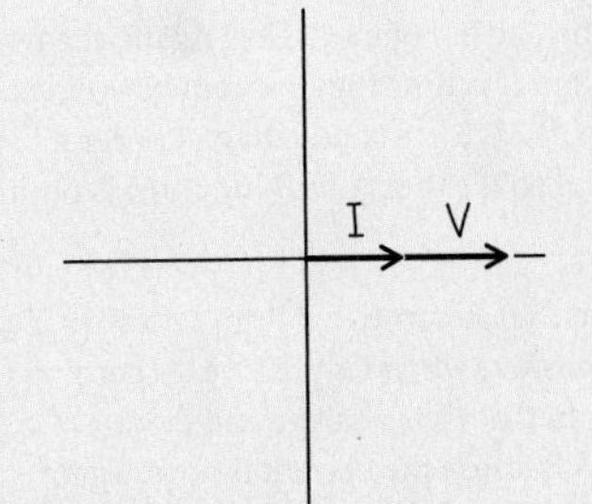

32.4 Figure 32.13: zero for both v_R and i. Figure 32.20: zero for i and $\pi/2$ for v_L.

32.5 A piece of doped silicon, whether *n*-type or *p*-type, is electrically neutral because the dopant atoms have the same number of protons as electrons. Although in a doped semiconductor there is either a surplus or a deficit of electrons relative to the number of electrons in a perfect silicon lattice, there is also an equal surplus or deficit of protons in the nuclei of the dopant atoms.

32.6 Holes travel from left to right (but only in the *p*-type region; in the *n*-type region they recombine with free electrons); simultaneously, electrons travel from right to left.

32.7 The current in the diode is in one direction only. If you assume the diode is attached in the circuit so that it conducts current only when $v_{\text{diode}} > 0$, your sketch looks like Figure S32.7*a*. If you assume the diode is attached so that it conducts current only when $v_{\text{diode}} < 0$, your sketch looks like Figure S32.7*b*.

Figure S32.7

(*a*)

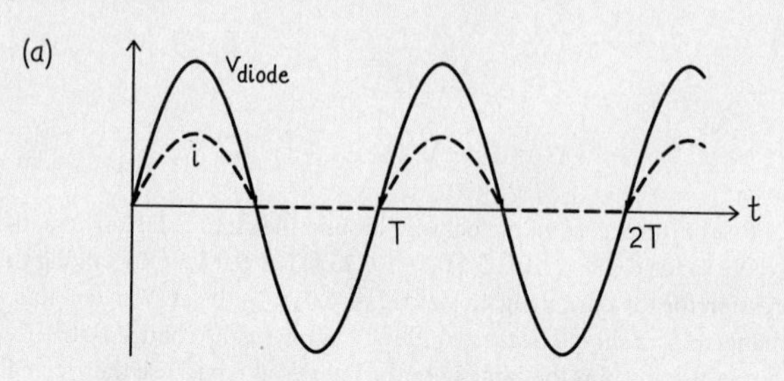

(b)

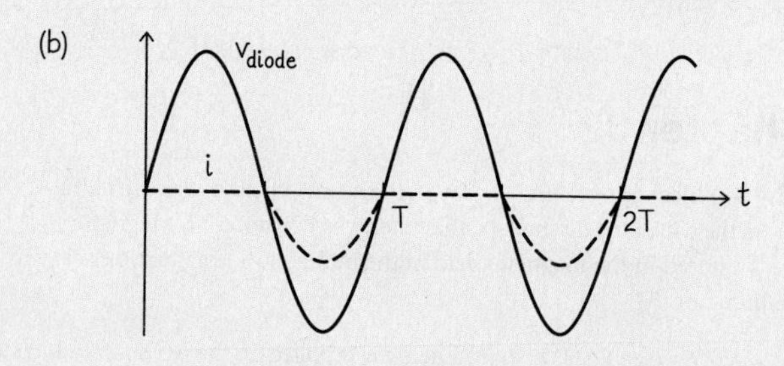

32.8 According to the junction rule, you must have $I_b + I_c = I_e$.

32.9 The AND gate is the one in Figure 32.39*a*. Both field-effect transistors function as switches and are closed (conducting) when their gates are positively charged. If the potential at A is positive with respect to ground, transistor 1 is conducting. If the potential at B is positive with respect to ground, transistor 2 is conducting. Because the transistors are connected in series, both must be conducting in order for the output potential to be different from ground, and so both A and B must be positive to obtain a nonzero output.

The OR gate is the one in Figure 32.39*b*. Again, the two transistors function as switches, but this time they are connected in parallel. Therefore only one of them needs to be conducting in order for the output potential to be nonzero, and so if either A or B (or both) is positive we obtain a nonzero output.

32.10 (*a*) Higher. (*b*) Because R is always positive, Eq. 32.2 tells you that v_R must be positive when i is positive. According to Eq. 32.3, $\mathscr{E}$ is positive, too. Given that v_R is positive and that $v_a > v_b$, you must have $v_R = v_a - v_b$. (*c*) Because the current is now negative, v_R and $\mathscr{E}$ must be negative, too. (Indeed, with the current now running counterclockwise, v_a is now smaller than v_b.) The potential difference across the resistor is still given by $v_R = v_a - v_b$.

32.11 The current in the circuit. The velocity of a simple harmonic oscillator is equal to the time derivative of its position. Because the potential difference v_C is proportional to the charge on the upper capacitor plate, the rate at which v_C changes—that is, the time derivative of v_C—is proportional to the rate at which the amount of charge on the plate changes, dq/dt. This quantity is the current.

32.12 See Figure S32.12. The two horizontal lines mean that the resistance and the current in the resistor do not depend on angular frequency. The hyperbolic X_C curve shows that the capacitive reactance is inversely proportional to ω (Eq. 32.14). Consequently the amplitude of the current in the capacitor is directly proportional to ω, which means the I_C curve has a constant slope. The straight, positive-slope X_L curve means that the inductive reactance is directly proportional to ω (Eq. 32.26); consequently the current in the inductor is inversely proportional to ω, as shown by the hyperbolic I_L curve.

Figure S32.12

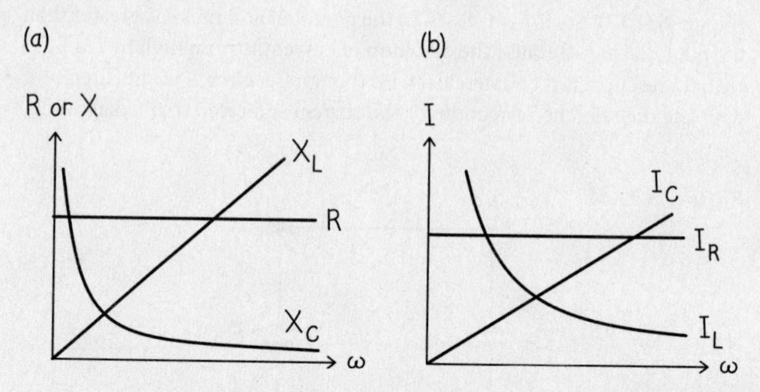

32.13 No. You cannot use the phasor method because the two phasors for the potential differences do not rotate as a unit with a constant phase difference.

32.14 We analyze this circuit as we did in Example 32.8, except that now $V_{\text{out}} = V_C$ instead of $V_{\text{out}} = V_R$. To take the high-ω and low-ω limits of V_{out}, we rewrite Eq. 32.36 as

$$V_{\text{out}} = V_C = \frac{\mathscr{E}_{\max}}{\sqrt{\omega^2 R^2 C^2 + 1}}.$$

The low-ω limit of V_{out} is then $V_{\text{out}} = \mathscr{E}_{\max}$, while the high-$\omega$ limit is $V_{\text{out}} = \mathscr{E}_{\max}/\omega RC$, which approaches zero for large ω. This circuit therefore passes low-frequency signals essentially unchanged and attenuates high-frequency signals.

32.15 (*a*) The maximum potential differences can be found from the maximum current (using Eqs. 32.6, 32.13, and 32.24). Rounding off to two significant digits, you have $V_R = IR = 1.6 \times 10^2$ V, $V_L = I\omega L = 3.1 \times 10^2$ V, and $V_C = I/\omega C = 3.1 \times 10^2$ V.

(*b*) No. $(V_R + V_L + V_C) \gg \mathscr{E}_{\max}$ because the potential differences are not in phase and are never simultaneously at their maxima.

32.16 Changing L or C while keeping R constant changes the resonant angular frequency (unless both L and C are changed in such a way that their product remains constant) and can change the shape of the $I(\omega)$ curve. Increasing C while keeping L fixed broadens the curve, and increasing L while keeping C fixed sharpens the curve. Because it depends only on R and on the applied emf, the amplitude of the current at resonance is not affected by the values of L or C.

32.17 Above, because when $V_L > V_C$, the inductor dominates the reactance.

32.18 We substitute the answers from Example 32.10 into Eq. 32.52:

$$P_{\text{av}} = I_{\text{rms}}^2 R = \frac{I^2}{(\sqrt{2})^2} R = \frac{(3.2\text{ A})^2(50\ \Omega)}{2} = 2.6 \times 10^2\text{ W}.$$

Chapter 33

33.1 (*a*) See Figure S33.1. In considering whether the brightness of any location on the screen has changed, note that the distribution of light from the first bulb has not changed, which means the brightness of a particular location changes only if additional light is cast on that location by the second bulb. If the second bulb does not cast light on a particular location, the brightness does not change. (The fact that a given location would be shadowed if it were illuminated by only the second bulb does not decrease the brightness of the light from the first bulb.) The brightness of the spot created by the first bulb does not change because no light from the second bulb strikes this spot. (*b*) Locations P and Q are now brighter than before because some light from the second bulb strikes them. Locations R and S are unaffected because no light from the second bulb reaches them.

Figure S33.1

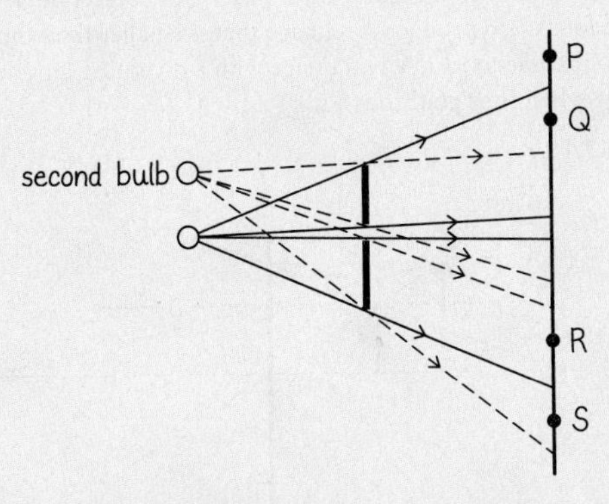

33.2 See Figure S33.2. The shadow edges become sharper as you move the paper farther from the bulb and fuzzier as you move the paper closer to the bulb. This happens because the bulb is not a point source of light. Rays from different parts of the bulb's surface pass by the edge of the paper at slightly different angles. When the paper is farther from the bulb, the difference in these angles is less and so the edge of the shadow is sharper. When the paper is close to the bulb, the angular size of the filament (as seen from the edge) is greater, which makes the edge of the shadow fuzzier.

Figure S33.2

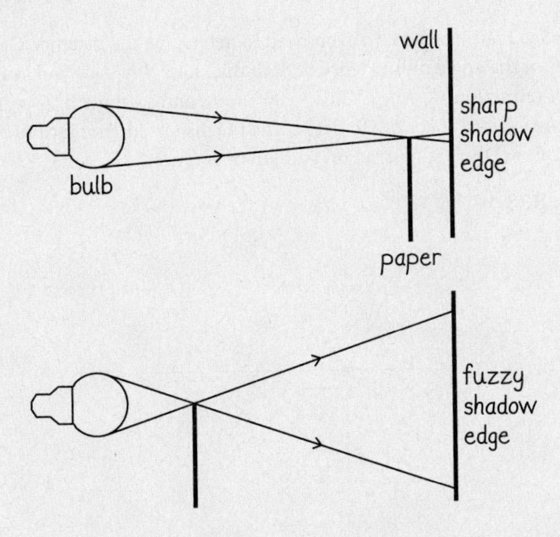

33.3 No. Figure S33.3 shows the reflected rays that reach the observer in the two locations. Because the angle of incidence always equals the angle of reflection when the reflecting surface is smooth, any ray from the object that reflects anywhere on the mirror can be traced back through the mirror to a location directly behind the object. Moving the observer changes only which subset of the reflected rays the observer sees and not where the rays appear to come from.

Figure S33.3

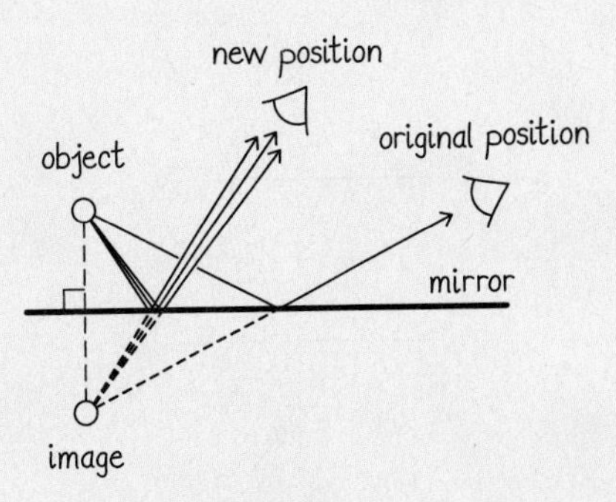

33.4 No. Figure S33.4 shows the rays that reflect into the person's eye from the highest and lowest points on his body. These rays strike the mirror at a height midway between where they originate on the person and the person's eye. So the mirror needs to extend only from halfway between the person's eye and the highest point on the person's body to halfway between the person's eye and the lowest point on the person's body. The ray from the highest point strikes the mirror at a height midway between the height h_1 in Figure S33.4 at which the ray originates and the height at which it strikes the eye, and the same is true for the ray from the lowest point. So the top of the mirror can be at a height that is half the height h_1 above the height of the eye, and the bottom of the mirror can be at a height that is half the height h_2 below the height of the eye. Because $h_1 + h_2$ is the person's height, this means the mirror needs to be only half this height. (However, the mirror must be positioned at the right height.)

Figure S33.4

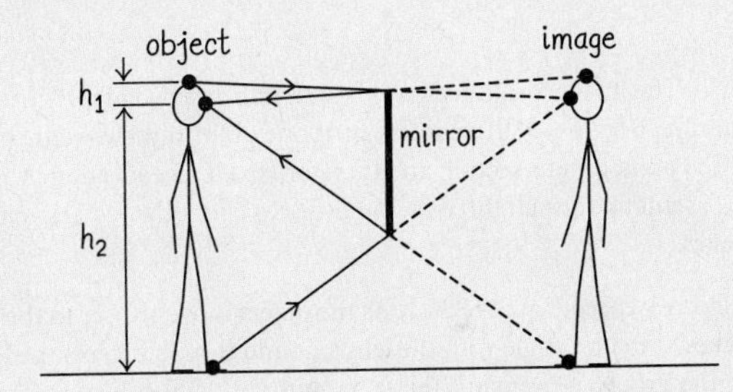

33.5 The wavefronts arrive at a given location in the glass at the same frequency that they arrive at any location in vacuum. However, they travel only two-thirds as fast in the glass as in vacuum. Sequential wavefronts arrive at the glass surface at instants separated by the period T of the wave ($T = 1/f$). If the wave traveled at the same speed in glass as it does in vacuum, then during one period, one wavefront would travel a distance into the glass equal to the vacuum wavelength (400 nm) before the next wavefront arrived at the surface of the glass, and the wavefronts would be separated by 400 nm. However, because the wavefronts travel at only two-thirds the speed of light in vacuum, the spacing between wavefronts in the glass must be two-thirds of the vacuum wavelength, or 267 nm.

33.6 The propagation of the wavefronts is exactly like that shown in Figure 33.16*a* with the direction of propagation reversed. Consequently, because the angle of incidence equals θ_2, the angle of refraction must equal θ_1.

33.7 The angle between the ray reflected from the bottom surface and the normal is θ_2, as shown in Figure S33.7 on the next page. This ray is now incident at the top surface and refracted as it emerges into the air. This is equivalent to the situation shown in Figure 33.17*b*, in which the ray originates in the glass and is refracted in the air. Therefore, as discussed in Example 33.3, in air the angle this refracted ray makes with the normal is θ_1. Note in Figure S33.7 that this refracted ray is shifted sideways relative to the ray *reflected* from the top surface.

Figure S33.7

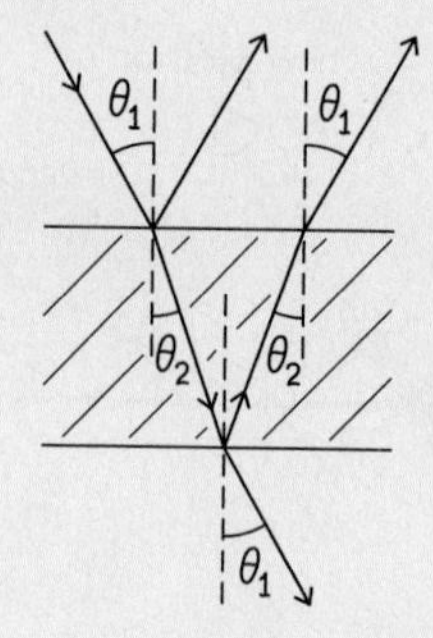

33.8 The critical angle corresponds to the angle of incidence for which the angle of refraction is 90°. When entering the medium, to obtain a given angle of refraction, the angle of incidence of a violet ray must be greater than that of a red ray; therefore the critical angle is smaller for a violet ray than for a red ray.

33.9 See Figure S33.9. The wavefronts are perpendicular to the light rays everywhere.

Figure S33.9

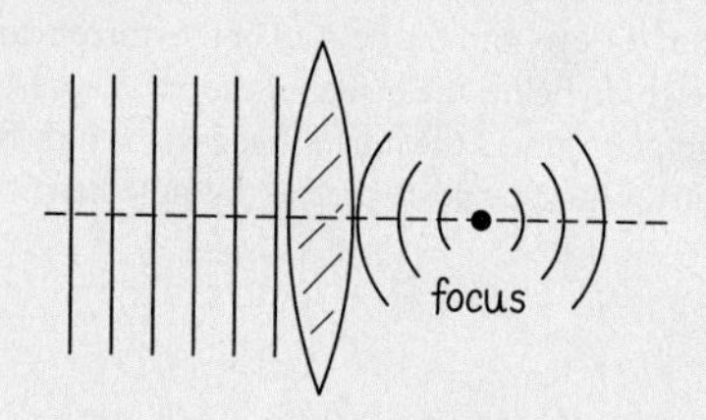

33.10 No. The image location can be found with any two of the three principal rays. Because all three intersect at one point, any two of them indicate the point of intersection and therefore the image location. It is, however, useful to draw all three rays in order to check that you have made no mistakes.

33.11 Decrease. See Figure S33.11. As the object moves closer to the lens, ray 3 makes a smaller angle with the lens axis and thus intercepts the lens closer to the axis. Ray 1 remains the same, and ray 2 makes a greater angle with the axis in order to pass through the center of the lens. As a result, the image point—the virtual intersection of the rays—is closer to the axis (as well as the lens), making the image smaller.

Figure S33.11

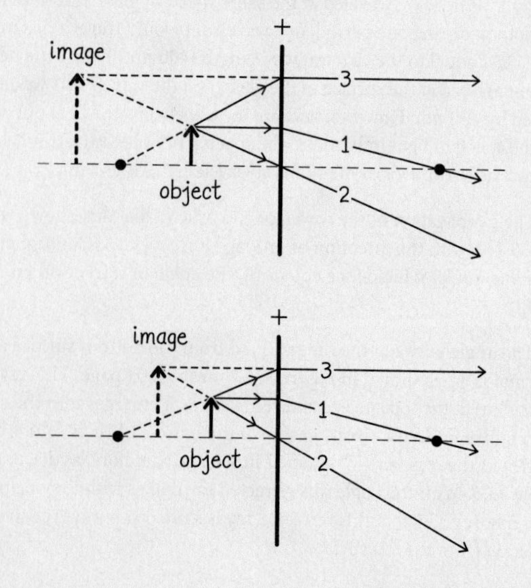

33.12 Virtual, because the rays do not actually intersect anywhere on the image. They merely appear to originate from a common point on the virtual image.

33.13 (*a*) See Figure S33.13*a*. You draw ray 3 by first drawing a line connecting the point of interest on the object with the focus on the side of the lens opposite the object. Run ray 3 along this line from the object to the lens surface; then change the ray direction so that it emerges parallel to the lens axis. (*b*) There is no position for which the image is larger than the object. To see why, all you need to consider is principal rays 1 and 2. When the object is moved, the direction of ray 1 does not change—only ray 2 changes direction. When the object is moved to the left, as in Figure S33.13*b*, ray 2 now intersects ray 1 closer to the lens axis and the image is smaller. Moving the object right, toward the lens, makes the image larger, but it cannot become larger than the object because on the left side of the lens where the image forms, ray 1 is always below the tip of the object. As shown in Figure 33.33, placing the object beyond the object-side focus also produces an image that is smaller than the object. (*c*) It is not possible to form a real image with a diverging lens because rays diverge when they go through it.

Figure S33.13

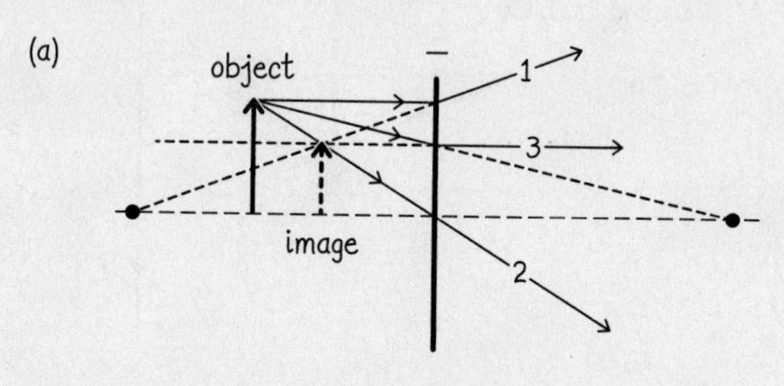

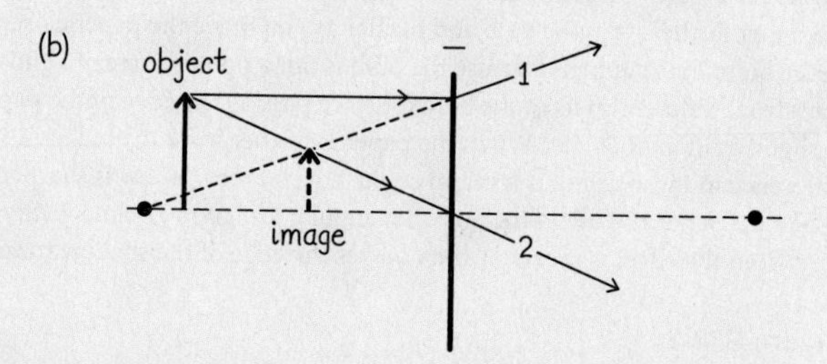

33.14 See Figure S33.14. (*a*) You need to determine the distance CE in terms of the angle of incidence θ_1, slab thickness d = AD, and slab index of refraction n_2. Angle CBE = 90° − θ_1, and so angle BEC = θ_1. Therefore CE = BE cos θ_1. To determine the distance BE, you can say DB + BE = DE, DE = $d \tan \theta_1$, and DB = $d \tan \theta_2$.

Figure S33.14

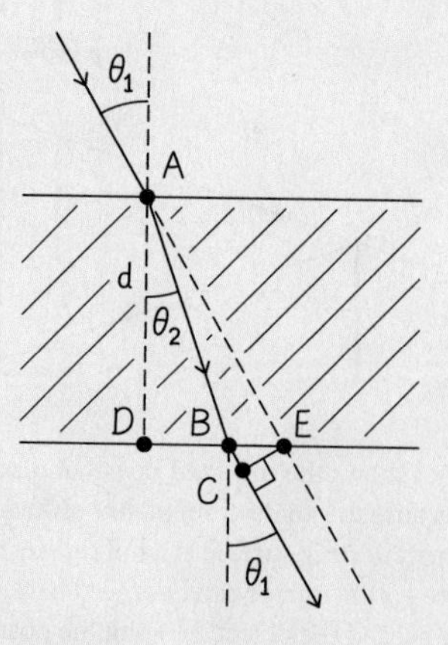

Substituting the second and third expressions in the first, you obtain $d \tan \theta_2$ + BE = $d \tan \theta_1$, which can be solved for BE:

$$\text{BE} = d(\tan\theta_1 - \tan\theta_2) = d\left(\frac{\sin\theta_1}{\cos\theta_1} - \frac{\sin\theta_2}{\cos\theta_2}\right). \quad (1)$$

Applying Snel's law (Eq. 33.7) to this case and solving Eq. 33.7 for $\sin\theta_2$ give you $\sin\theta_2 = n_1\sin\theta_1/n_2$. Next you can use the identity $\sin^2\theta + \cos^2\theta = 1$ to get an expression for $\cos\theta_2$:

$$\cos\theta_2 = \sqrt{1 - \sin^2\theta_2} = \sqrt{1 - \left(\frac{n_1\sin\theta_1}{n_2}\right)^2}.$$

You can substitute these expressions for $\sin\theta_2$ and $\cos\theta_2$ into Eq. 1:

$$\text{BE} = d\left[\frac{\sin\theta_1}{\cos\theta_1} - \frac{n_1\sin\theta_1}{n_2\sqrt{1 - \left(\frac{n_1\sin\theta_1}{n_2}\right)^2}}\right],$$

which simplifies to

$$\text{BE} = d\sin\theta_1\left(\frac{1}{\cos\theta_1} - \frac{n_1}{\sqrt{n_2^2 - n_1^2\sin^2\theta_1}}\right).$$

From this you can determine the distance CE:

$$\text{CE} = \text{BE}\cos\theta_1 = d\sin\theta_1\left(1 - \frac{n_1\cos\theta_1}{\sqrt{n_2^2 - n_1^2\sin^2\theta_1}}\right).$$

(*b*) $$\text{CE} = (0.010\ \text{m})(0.50)\left(1 - \frac{(1)(0.87)}{\sqrt{(1.5)^2 - (1)^2(0.5)^2}}\right)$$

$$= 0.0019\ \text{m}.$$

Thus, at an angle of incidence of 30°, in passing through a glass slab that is 10 mm thick (about three times the thickness of a typical windowpane), a light ray is shifted sideways by 2 mm (20% of the thickness of the slab).

33.15 (*a*) f is negative (lens is diverging), so $f = -80$ mm; o is positive (object is on same side as illumination), so $o = 100$ mm. Substituting these values into Eq. 33.16 gives

$$\frac{1}{i} = \frac{1}{-80\ \text{mm}} - \frac{1}{100\ \text{mm}} = -0.0225\ \text{mm}^{-1},$$

and solving for i gives $i = -44$ mm. The value of i is negative, as it should be for a virtual image, and the absolute value of i is less than o, which is consistent with Figure 33.33.

(*b*) Equation 33.17 gives

$$M = \frac{h_i}{h_o} = \frac{-i}{o} = \frac{-(-44\ \text{mm})}{100\ \text{mm}} = 0.44.$$

The image height is 44% of the object height.

33.16 (*a*) Solving Eq. 33.21 for f gives $f = (0.25\ \text{m})/M_\theta$, and substituting this result into Eq. 33.22 gives $d = 4M_\theta$. For $M_\theta = 8$, you have $d = +32$ diopters.

(*b*) $f = (0.25\ \text{m})/M_\theta = (0.25\ \text{m})/8 = 0.031\ \text{m} = 31$ mm.

33.17 (*a*) The image moves farther from the objective lens. This can be seen either by constructing a ray diagram (Figure S33.17) or by considering the relationship among object distance, image distance, and focal length (Eq. 33.16). Solving Eq. 33.16 for i_1 gives

$$\frac{1}{i_1} = \frac{1}{f_1} - \frac{1}{o_1} = \frac{o_1 - f_1}{o_1 f_1} \quad \text{or} \quad i_1 = \frac{o_1 f_1}{o_1 - f_1}.$$

Increasing both o_1 and f_1 will increase the numerator of the expression for i_1; keeping the sample "just outside the focal point" of the lens implies that the distance between the sample and the lens, $o_1 - f_1$, is kept at least roughly the same, so increasing the numerator while keeping the denominator constant increases i_1. (Practical limitations on the construction of lenses actually allow the distance $o_1 - f_1$ to be smaller for shorter-focal-length lenses, so in fact, the denominator also increases as the focal length increases. However, this last effect is not a consequence of the simple thin-lens treatment but of ways that lenses are not ideal that are beyond the scope of this text.)

Figure S33.17

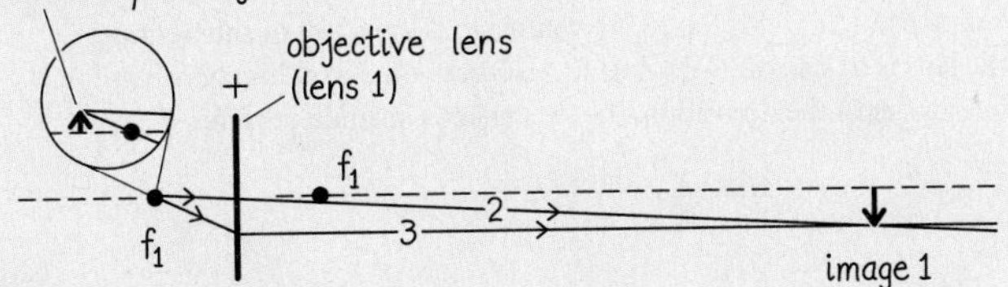

(*b*) Shorter focal lengths give a more compact microscope, because the first image must fall close to the focal point of the eyepiece lens, and thus the distance between the two lenses is roughly $i_1 + f_2$. From part *a* we know that i_1 depends on f_1, and so decreasing f_1 decreases i_1.

33.18 (*a*) Using the result of Example 33.9 gives you $f_1 = M_\theta f_2 = (22)\ (0.0400\ \text{m}) = 0.88$ m.

(*b*) The length of the telescope is roughly the sum of the focal lengths of the two lenses (see Figure 33.50): 0.88 m + 0.04 m = 0.92 m. Because the eyepiece lens is by design a short-focal-length lens, the length of the telescope is determined primarily by the focal length of the objective lens.

33.19 (*a*) The ray diagram in Figure S33.19 shows that the image is behind the mirror and virtual. You calculate the image distance with Eq. 33.24. The focal length is half the radius of curvature, C = 1.0 m, and is negative because it is a virtual focus:

$$i = \left(\frac{1}{f} - \frac{1}{o}\right)^{-1} = \left(\frac{1}{-0.50} - \frac{1}{1.0}\right)^{-1} = -0.33\ \text{m}.$$

The negative value for i tells you that the image is located 0.33 m behind the mirror.

Figure S33.19

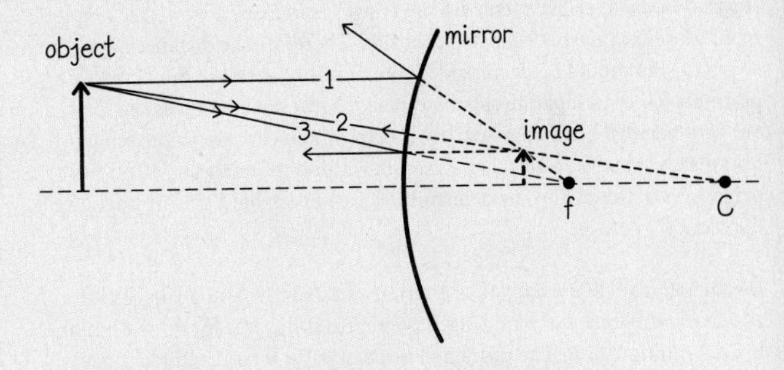

(*b*) $$h_i = -\frac{h_o i}{o} = -\frac{(0.30\ \text{m})(-0.33\ \text{m})}{(1.0\ \text{m})} = 0.10\ \text{m}.$$

The image is smaller than the object, and the positive value for h tells you that the image is upright.

33.20 The only change to the physics behind the derivation of the lensmaker's formula is that Eq. 33.25 becomes $n_1 \sin \theta_i = n_2 \sin \theta_r$. In the paraxial approximation, this result can be written as $(n_1/n_2)\theta_i = \theta_r$. Comparing this with Eq. 33.26, you see that the effect of submerging the lens is to substitute (n_1/n_2) for n. Once you make this change in the remainder of the derivation, the lensmaker's formula becomes

$$\frac{1}{f} = \left(\frac{n_1}{n_2} - 1\right)\left(\frac{1}{R_1} + \frac{1}{R_2}\right).$$

Chapter 34

34.1. Once the fronts reach P, the intensity at P is the same as the intensity at any point on any of the wavefronts. For any point R above or below the ray through P, you can locate a corresponding point R′ for which the distance R′P is exactly half a wavelength greater than the distance RP. Those points therefore do not contribute to the intensity at P. Only the point exactly on the ray (dashed line) contributes.

34.2. If you identify pairs of points that radiate in phase at P, then for each pair, there exists another pair that is exactly 180° out of phase with the first pair and thus cancels the radiation from the first pair.

34.3. (*a*) The spacing between fringes decreases as the separation between the slits increases because $\sin \theta$ is proportional to $1/d$. (*b*) The spacing between fringes increases as the wavelength increases because $\sin \theta$ is proportional to λ.

34.4. No. Two of the waves cancel each other perfectly, but the third wave is not canceled. The intensity observed on the screen, at angles given by $d \sin \theta = \pm(n - \frac{1}{2})\lambda$, is the intensity of one of the three waves.

34.5. Fringe brightness is determined by the intensity (power/area) that strikes the screen. Increasing the number of slits increases the amount of power that travels through the slits to the screen. In addition, increasing the number of slits causes the fringes to sharpen, with the result that the power that reaches the screen is concentrated into narrower fringes, decreasing the area and thus increasing the brightness further.

34.6. (*a*) As you found in Checkpoint 34.3, the angle of the fringes increases with wavelength, and so the angle is less for violet light than for red. White light is thus dispersed into a rainbow, with the violet end of the spectrum at smaller angles than the red. (*b*) Each rainbow corresponds to the fringes of a particular order for all colors, or said another way, each order produces its own rainbow.

34.7. (*a*) The spacing between fringes in a two-slit interference pattern increases with increasing wavelength and with decreasing distance between slits (see Checkpoint 34.3). The same is true for diffraction gratings: The separation distance between adjacent fringes is increased by decreasing the separation distance between adjacent slits. Therefore the distance between adjacent slits should be decreased if your aim is to increase the diffraction grating's ability to separate close wavelengths. (*b*) No, as long as the slits are very narrow. The fringe spacing is determined entirely by the separation distance between adjacent slits. (Slit width does, however, affect the brightness of the pattern by determining how much light gets through the diffraction grating.)

34.8. Yes. The diffracted x rays interfere constructively at all angles for which the difference in path length is either zero (as it is for $\theta' = \theta$) or an integer multiple of λ. The diffracted beam at $\theta' = \theta$ corresponds to the zeroth-order maximum; other beams correspond to higher orders and are much weaker.

34.9. From Figure 34.20*b* you can see that $2\theta + 2\alpha = \pi$, so $\theta = -\alpha + \frac{\pi}{2}$. Using trigonometry, I get $\cos \theta = \cos(-\alpha + \frac{\pi}{2}) = -\sin(-\alpha) = \sin \alpha$. Substituting this into the Bragg condition yields $2d \sin \alpha = m\lambda$.

34.10. Crystal B. Spots closer together correspond to greater distances between crystal planes and hence greater atomic spacing.

34.11. It decreases, because the spacing between spots is proportional to the wavelength.

34.12. If the electrons travel more slowly, their wavelength increases, and the diffraction spots spread farther apart.

34.13. You could perform the experiment shown in Figure 34.6 with a light source so weak that the light "particles" pass through the pair of slits only one at a time. If light indeed has particle properties, a source this weak would give you a pattern that initially is like the one in Figure 34.25*a*, showing where individual particles of light hit the screen.

34.14. The maximum probability is at the locations of greatest intensity in the interference pattern; these locations are at angles θ for which $\sin \theta$ is a multiple of λ/d.

34.15. Although the maximum time-averaged intensity is greater than the sum of the intensities of the original two beams, averaging this time-averaged intensity over the entire area filled by the bright and dark bands of the interference pattern gives just the sum of the intensities of the original two beams, $2S_{0,av}$ (because the average of the cosine squared is one-half).

34.16. No, because the higher-order bright fringes exist only as long as $m\lambda/d < 1$. For values of m such that $m\lambda/d > 1$, there are no bright fringes because these correspond to $\sin \theta_m > 1$, for which there is no angle θ_m.

34.17. The oil forms a thin film that causes thin-film interference. The thickness of the oil layer determines for which wavelength constructive interference occurs. If the thickness of the film varies spatially, different wavelengths interfere constructively at different locations.

34.18. (*a*) If $a < \lambda$, there are no dark fringes because that would require $\sin \theta > 1$. This indicates that no minimum exists, and the light coming through an aperture with a diameter less than the wavelength acts as a true point source as assumed in our original discussion of multiple-slit interference. When $a < \lambda$, the slit acts as a point source with the wavelets spreading out spherically from the slit. (*b*) For $a \gg \lambda$, the first dark fringe occurs very close to $\theta = 0$. This means there is essentially no diffraction, and the beam just propagates straight ahead. This illustrates that single-slit diffraction is observed primarily with slits from a few wavelengths wide to tens of wavelengths wide.

34.19. Resolution is determined by wavelength and the ratio f/d: $y_r = 1.22\,\lambda f/d$. For a given wavelength, the greater f/d is, the larger the Airy disk. For the three lenses, f/d is (*i*) 1.3, (*ii*) 1.5, and (*iii*) 1.1. (*a*) The highest resolution is obtained with the smallest Airy disk, produced by lens *iii*. (*b*) The lowest resolution is obtained with the largest Airy disk, produced by lens *ii*. (The wavelength determines the exact size of the Airy disk; the comparisons made here assume the same wavelength for all three lenses.)

34.20. (*a*) The stopping potential difference V_{stop} is proportional to the frequency (Figure 34.48*b*) of the incident photons, and the maximum kinetic energy K_{max} of the ejected electrons is proportional to V_{stop} (Eq. 34.34). Therefore, K_{max} must be proportional to the frequency of the incident photons. (*b*) The minimum frequency the light can have and be able to eject electrons. Lower-frequency light does not eject electrons.

34.21. (*a*) The photon slows down because the speed of light is less in a medium that has an index of refraction of 1.5 than in air (index of refraction 1). (*b*) The frequency remains unchanged, as discussed in Chapter 33. (*c*) The wavelength decreases because the wavefronts travel less far in a given time interval. (*d*) The photon's energy does not change because the medium does not take away any energy. This is why photon energy must be expressed in terms of frequency rather than wavelength, because neither frequency nor energy depends on the medium, whereas wavelength does.

Credits

COVER PHOTO Franklin Kappa

CHAPTER 22 **P. 593** valdezrl/Fotolia; **p. 594** Sean McGrath. Used under a Creative Commons License. http://creativecommons.org/licenses/by/2.0/deed.en; **p. 595** Eric Schrader/Pearson Education; **p. 600** Paul A. Souders/Corbis; **p. 604** GIPhotoStock/Science Source.

CHAPTER 23 **P. 614** John W. Belcher; **p. 619** John W. Belcher; **p. 621** John W. Belcher.

CHAPTER 24 **P. 639** Fermilab; **p. 641** John W. Belcher.

CHAPTER 25 **P. 639** Scott Sinklier/Alamy; **p. 668** both USGS.

CHAPTER 26 **P. 685** Dave Lindblom; **p. 688** MIT Museum; **p. 698** yurazaga/Shutterstock.

CHAPTER 27 **P. 710** Yamada Taro/Getty Images; **p. 711** bottom Eric Mazur; **p. 711** right Alex Kosev/Shutterstock; **p. 715** Cordelia Molloy/Science Source.

CHAPTER 28 **P. 735** Mircea RUBA/Shutterstock; **p. 750** Custom Medical Stock Photography/Newscom.

CHAPTER 29 **P. 759** Bosca78/Getty Images.

CHAPTER 30 P. **781** ALMA (ESO/NAOJ/NRAO)/L. Calçada (ESO)/H. Heyer (ESO)/H. Zodet (ESO).

CHAPTER 31 **P. 811** Powered by Light/Alan Spencer/Alamy.

CHAPTER 32 **P. 842** kubais/Shutterstock.

CHAPTER 33 **P. 875** Erich Lessing/Art Resource; **p. 883** left Steve Hamblin/Alamy; **p. 883** right Jan Mika/New Zealand photos/Alamy.

CHAPTER 34 **P. 908** David Palau/Shutterstock; **p. 909** left Richard Megna/Fundamental Photographs; **p. 909** right GualtieroBoffi/Shutterstock; **p. 911** Ken Kay/Fundamental Photographs; **p. 912** Eric Mazur; **p. 914** Eric Mazur; **p. 915** Custom Medical Stock Photography/Newscom; **p. 918** Omikron/Science Source; **p. 920** Eric Mazur; **p. 921** Wolfgang Rueckner; **p. 922** Eric Mazur; **p. 926** Eric Mazur **p. 928** Anyka/Fotolia; **p. 930** Eric Mazur; **p. 932** top GIP Photo Stock/Science Source; **p. 932** bottom Nick Strobel/www.astronomynotes.com.

Index

D

E

J

K

Q

R

S

T

U

V

Unit conversion factors

Length

1 in. = 2.54 cm (defined)

1 cm = 0.3937 in.

1 ft = 30.48 cm

1 m = 39.37 in. = 3.281 ft

1 mi = 5280 ft = 1.609 km

1 km = 0.6214 mi

1 nautical mile (U.S.) = 1.151 mi = 6076 ft = 1.852 km

1 fermi = 1 femtometer (fm) = 10^{-15} m

1 angstrom (Å) = 10^{-10} m = 0.1 nm

1 light − year (ly) = 9.461×10^{15} m

1 parsec = 3.26 ly = 3.09×10^{16} m

Volume

1 liter (L) = 1000 mL = 1000 cm^3 = 1.0×10^{-3} m^3
= 1.057 qt (U.S.) = 61.02 $in.^3$

1 gal (U.S.) = 4 qt (U.S.) = 231 $in.^3$ = 3.785 L = 0.8327 gal (British)

1 quart (U.S.) = 2 pints (U.S.) = 946 mL

1 pint (British) = 1.20 pints (U.S.) = 568 mL

1 m^3 = 35.31 ft^3

Speed

1 mi/h = 1.4667 ft/s = 1.6093 km/h = 0.4470 m/s

1 km/h = 0.2778 m/s = 0.6214 mi/h

1 ft/s = 0.3048 m/s = 0.6818 mi/h = 1.0973 km/h

1 m/s = 3.281 ft/s = 3.600 km/h = 2.237 mi/h

1 knot = 1.151 mi/h = 0.5144 m/s

Angle

1 radian (rad) = 57.30° = 57°18'

1° = 0.01745 rad

1 rev/min (rpm) = 0.1047 rad/s

Time

1 day = 8.640×10^4 s

1 year = 365.242 days = 3.156×10^7 s

Mass

1 atomic mass unit (u) = 1.6605×10^{-27} kg

1 kg = 0.06852 slug

1 metric ton = 1000 kg

1 long ton = 2240 lbs = 1016 kg

1 short ton = 2000 lbs = 909.1 kg

1 kg has a weight of 2.20 lb where $g = 9.80\ m/s^2$

Force

1 lb = 4.44822 N

1 N = 10^5 dyne = 0.2248 lb

Energy and work

1 J = 10^7 ergs = 0.7376 ft · lb

1 ft · lb = 1.356 J = 1.29×10^{-3} Btu = 3.24×10^{-4} kcal

1 kcal = 4.19×10^3 J = 3.97 Btu

1 eV = 1.6022×10^{-19} J

1 kWh = 3.600×10^6 J = 860 kcal

1 Btu = 1.056×10^3 J

Power

1 W = 1 J/s = 0.7376 ft · lb/s = 3.41 Btu/h

1 hp = 550 ft · lb/s = 746 W

1 kWh/day = 41.667 W

Pressure

1 atm = 1.01325 bar = 1.01325×10^5 N/m^2 = 14.7 lb/$in.^2$ = 760 torr

1 lb/$in.^2$ = 6.895×10^3 N/m^2

1 Pa = 1 N/m^2 = 1.450×10^{-4} lb/$in.^2$

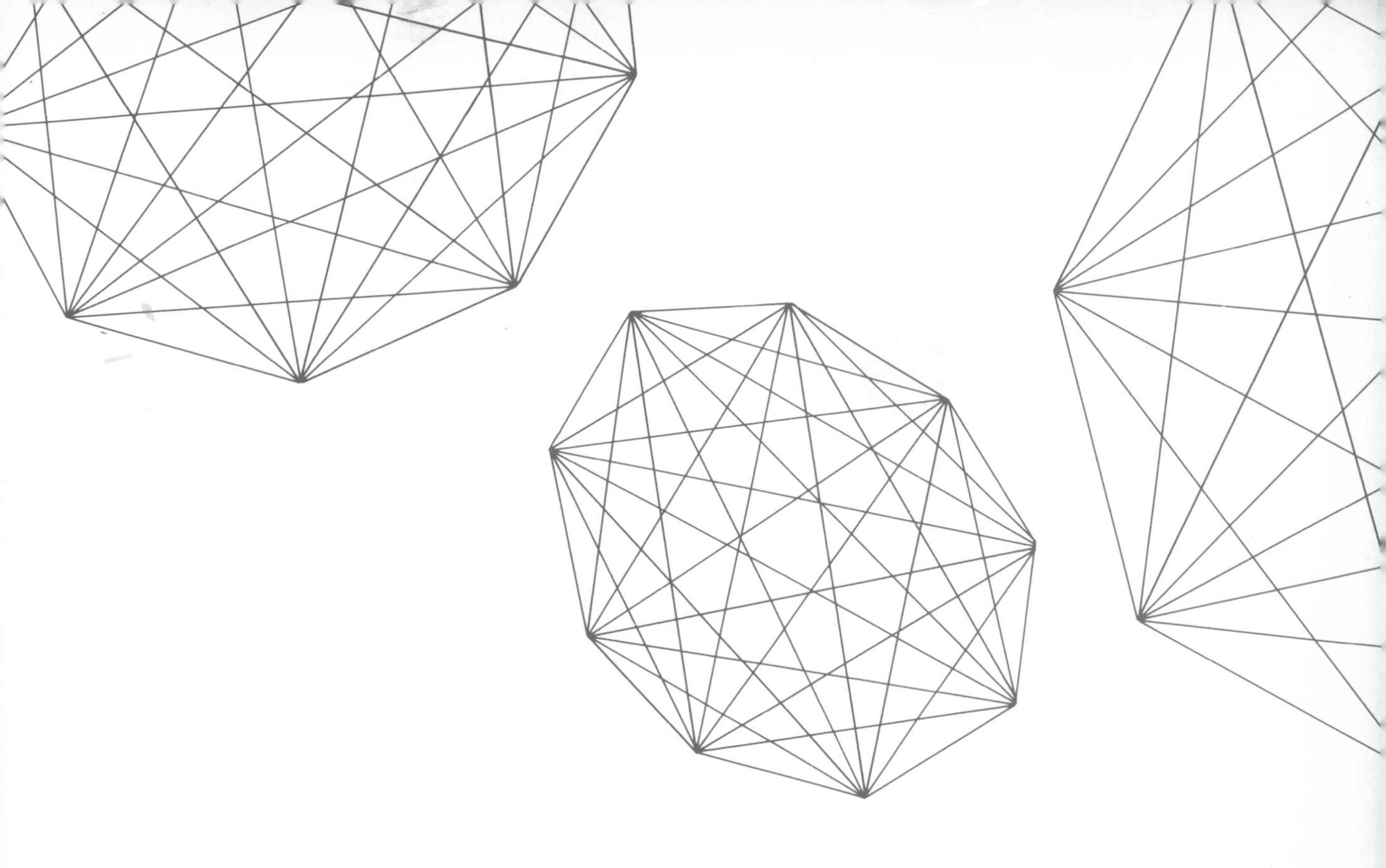

COMPUTER GRAPHICS

Donald Hearn
Department of Computer Science, University of Illinois

M. Pauline Baker
Department of Computer Science, Western Illinois University

PRENTICE-HALL, Englewood Cliffs, New Jersey 07632

Library of Congress Cataloging-in-Publication Data

Hearn, Donald.
Computer graphics.

Bibliography: p.
Includes index.
1. Computer graphics. I. Baker,
M. Pauline. II. Title.
T385.H38 1986 006.6 85-19137
ISBN 0-13-165382-2

Editorial/production supervision: Tracey Orbine and Kathryn Gollin Marshak
Interior design: Lee Cohen
Cover design: Lee Cohen
Manufacturing buyer: Gordon Osbourne
Page layout: Meg Van Arsdale
Cover art: Courtesy Melvin L. Prueitt, Los Alamos National Laboratory

Printed in the United States of America

10 9 8 7 6 5 4 3 2 1

ISBN 0-13-165382-2 025

PRENTICE-HALL INTERNATIONAL (UK) LIMITED, *London*
PRENTICE-HALL OF AUSTRALIA PTY. LIMITED, *Sydney*
PRENTICE-HALL CANADA INC., *Toronto*
PRENTICE-HALL HISPANOAMERICANA, S.A., *Mexico*
PRENTICE-HALL OF INDIA PRIVATE LIMITED, *New Delhi*
PRENTICE-HALL OF JAPAN, INC., *Tokyo*
PRENTICE-HALL OF SOUTHEAST ASIA PTE. LTD., *Singapore*
EDITORA PRENTICE-HALL DO BRASIL, LTDA., *Rio de Janeiro*
WHITEHALL BOOKS LIMITED, *Wellington, New Zealand*

TO OUR FOLKS

Rose, John, Millie, and J. Osborne

CONTENTS

PREFACE

Computer graphics is one of the most exciting and rapidly growing fields in computer science. Some of the most sophisticated computer systems in use today are designed for the generation of graphics displays. We all know the value of a picture as an effective means for communication, and the ability to converse pictorially with a computer is revolutionizing the way computers are being used in all areas.

This book presents the basic principles for the design, use, and understanding of graphics systems. We assume that the reader has no prior background in computer graphics but is familiar with fundamental computer science concepts and methods. The hardware and software components of graphics systems are examined, with a major emphasis throughout on methods for the design of graphics packages. We discuss the algorithms for creating and manipulating graphics displays, techniques for implementing the algorithms, and their use in diverse applications. Programming examples are given in Pascal to demonstrate the implementation and application of graphics algorithms. We also introduce the reader to the Graphical Kernel System (GKS), which is now both the United States and the international graphics-programming language standard. GKS formats for graphics-routine calls are used in the Pascal programs illustrating graphics applications.

The material presented in this text was developed from notes used in graduate and undergraduate graphics courses over the past several years. All of this material could be covered in a one-semester course, but this requires a very hasty treatment of many topics. A better approach is to select a subset of topics, depending on the level of the course. For the self-study reader, early chapters can be used to provide an understanding of graphics concepts, with individual topics selected from the later chapters according to the interests of the reader.

Chapter 1 is a survey of computer graphics, illustrating the diversity of applications areas. Following an introduction to the hardware and software components of graphics systems in Chapter 2, fundamental algorithms for the generation of two-dimensional graphics displays are presented in Chapters 3 and 4. These two chapters examine methods for producing basic picture components and techniques for handling color, shading, and other attributes. This introduces students to the programming techniques necessary for implementing graphics routines. Chapters 5 and 6 treat transformations and viewing algorithms. Methods for organizing picture components in segments and for interactive input are given in Chapters 7 and 8.

Three-dimensional techniques are introduced in Chapter 9. We then discuss the different ways that solid objects can be represented (Chapter 10) and manipulated (Chapter 11). Methods for forming three-dimensional views on a graphics display device are detailed in Chapter 12. The various algorithms for removing hidden surfaces of objects are discussed in Chapter 13, and models for shading and color are taken up in Chapter 14. These five chapters treat both the standard graphics methods and newer techniques, such as fractals, octrees, and ray tracing.

In Chapter 15, we explore techniques for modeling different systems. Modeling packages provide the structure for simulating systems, which is then passed to the graphics routines for display. Finally, methods for interfacing a graphics package to the user are examined in Chapter 16.

At the undergraduate level, an introductory course can be organized with a detailed treatment of fundamental topics from Chapters 2 through 8 plus an introduction to three-dimensional concepts and methods. Selected topics from the later chapters can be used as supplemental material. For a graduate course, the material on two-dimensional methods can be covered at a faster pace, with greater emphasis on the later chapters. In particular, methods for three-dimensional representations, three-dimensional viewing, hidden-surface removal, and shading and color models, can be covered in greater depth.

A great many people have contributed to this project in a variety of ways. To the many organizations and individuals who furnished photographs and other materials, we again express our appreciation. We are also grateful to our graphics students for their comments on the presentation of this material in the classroom. We thank the many people who provided comments on the manuscript, and we are especially indebted to Norman Badler, Brian Barsky, and Steve Cunningham for their helpful suggestions for improving the presentation of material. And a very special thanks goes to our editor, Jim Fegen, for his patience and encouragement during the preparation of this book, and to our production editors, Tracey Orbine and Kathy Marshak. Thanks also to our designer, Lee Cohen, and the Prentice-Hall staff for an outstanding production job.

Donald Hearn
M. Pauline Baker

1 A SURVEY OF COMPUTER GRAPHICS

Computers have become a powerful tool for the rapid and economical production of pictures. There is virtually no area in which graphical displays cannot be used to some advantage, and so it is not surprising to find the use of computer graphics so widespread. Although early applications in engineering and science had to rely on expensive and cumbersome equipment, advances in computer technology have made interactive computer graphics a practical tool. Today, we find computer graphics used routinely in such diverse areas as business, industry, government, art, entertainment, advertising, education, research, training, and medicine. Figure 1–1 shows a few of the many ways that graphics is put to use. Our introduction to the field of computer graphics begins with a tour through a gallery of graphics applications.

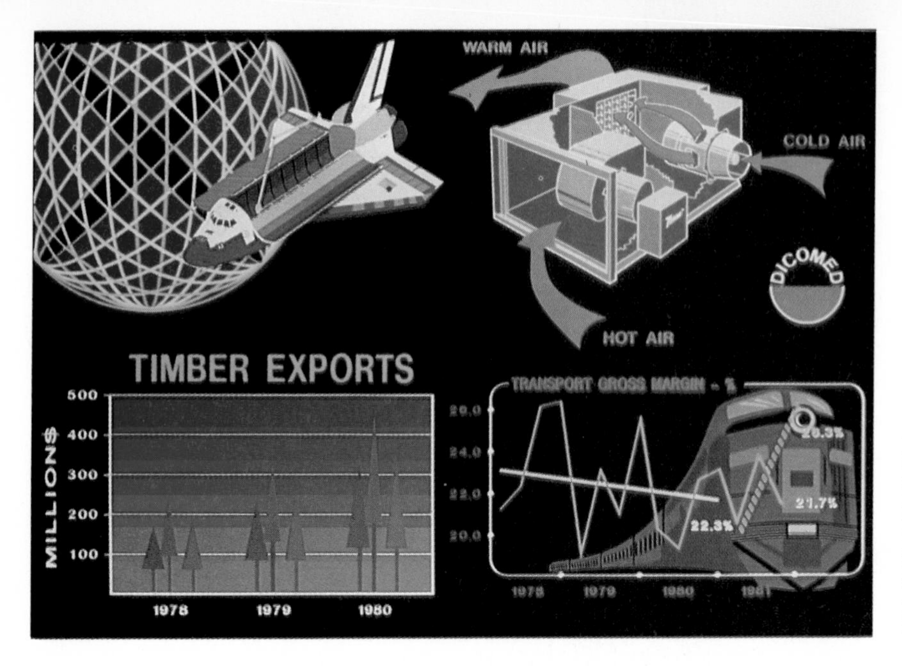

FIGURE 1–1
Examples of computer-generated displays used in modeling and simulation applications, training demonstrations, and the graphing of information. Courtesy DICOMED Corp.

1–1 Computer-Aided Design

For a number of years, the biggest use of computer graphics has been as an aid to design. Generally referred to as **CAD, computer-aided design** methods provide powerful tools. Parts design and drafting are done interactively, producing outlines (Fig. 1–2) or more realistic renderings (Fig. 1–3). When an object's dimensions have been specified to the computer system, designers can view any side of the object to see how it will look after construction. Experimental changes can be made

FIGURE 1–2
Drafting layout produced with a CAD system. Courtesy Evans & Sutherland.

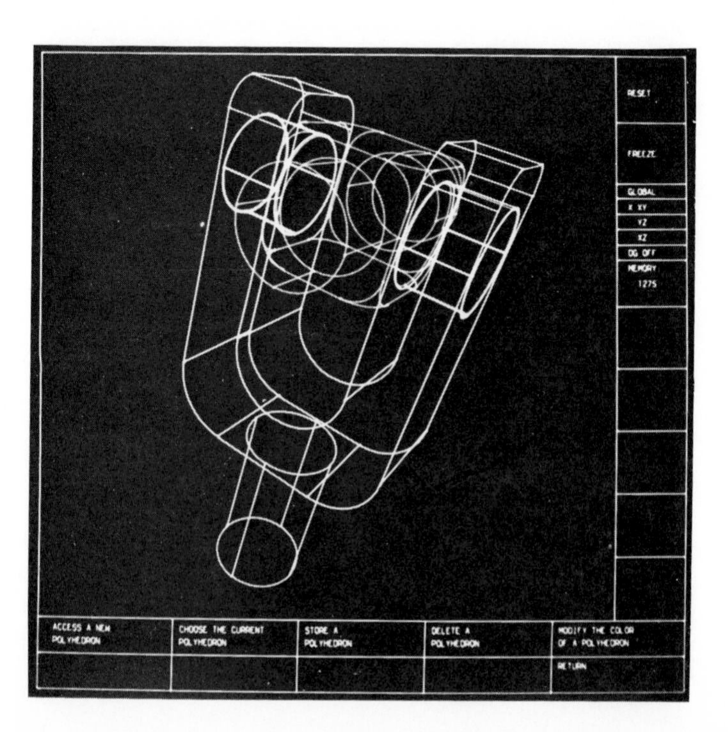

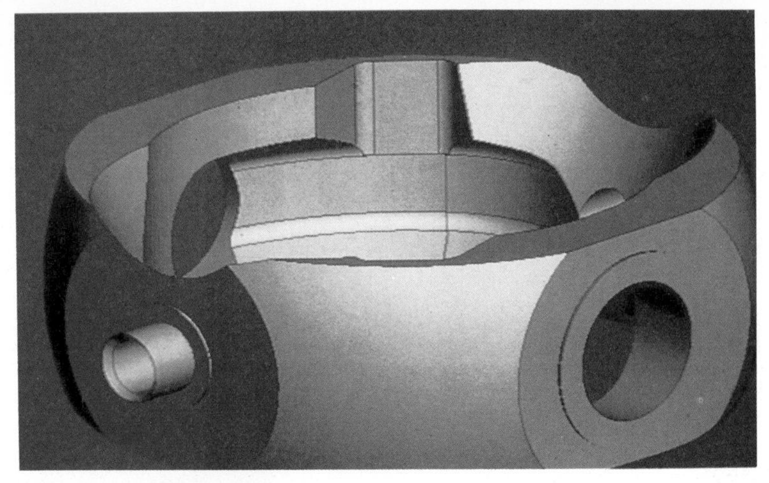

(a)

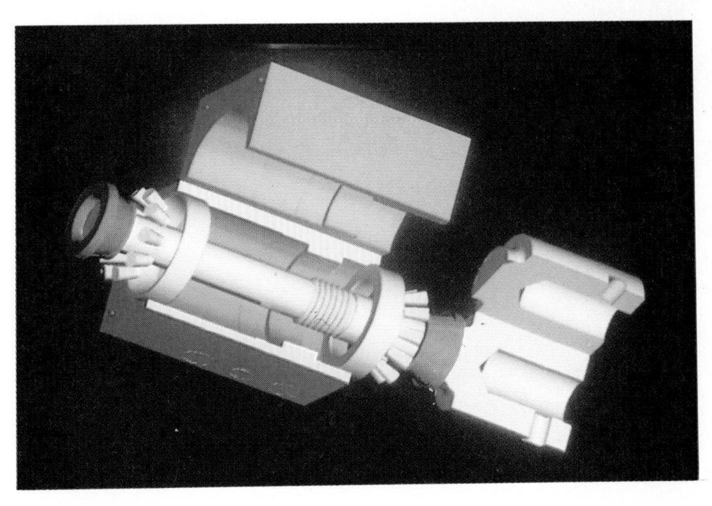

(b)

FIGURE 1–3
Three-dimensional rendering of machine parts using CAD systems. (a) Courtesy Lexidata Corp. (b) Bearing housing assembly, Courtesy Intergraph Corp.

freely since, unlike hand drafting, the CAD system quickly incorporates modifications into the display of the object. The manufacturing process also benefits in that layouts show precisely how the object is to be constructed. Figure 1–4 shows the path to be taken by machine tools over the surfaces of an object during its construc-

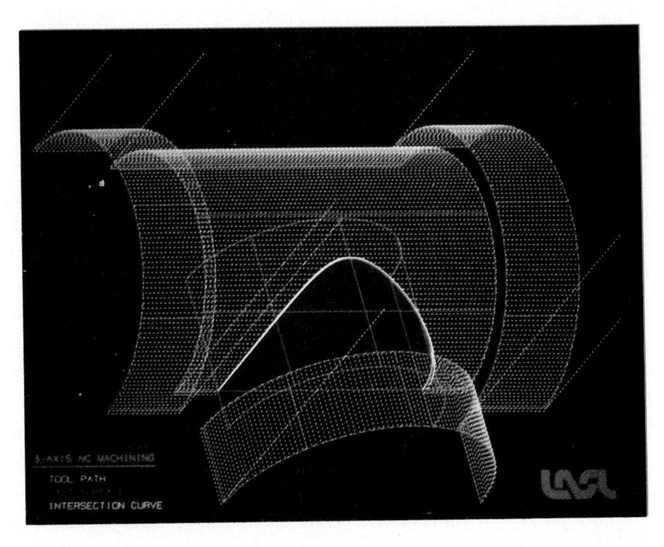

FIGURE 1–4
CAD layout for three-dimensional numerically controlled machining of a part. The part surface is displayed in one color and the tool path in another color. Courtesy Los Alamos National Laboratory.

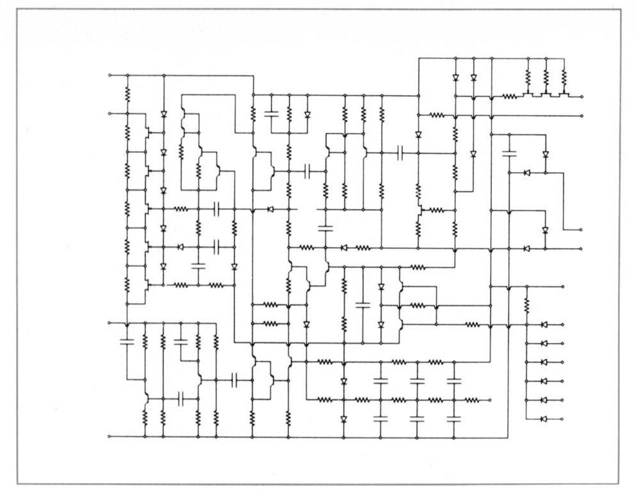

FIGURE 1–5
Circuit layout generated using CAD methods. Courtesy Precision Visuals, Inc., Boulder, Colorado.

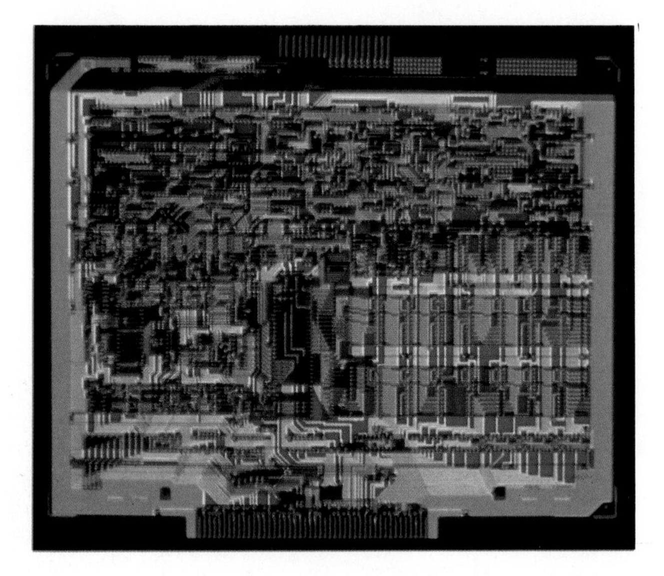

FIGURE 1–6
Computer board produced with a CAD system. Courtesy Lexidata Corp.

tion. Numerically controlled machine tools are then set up to manufacture the part according to these construction layouts.

Electrical and electronics engineers rely heavily on CAD methods. Electronic circuits, for instance, are typically designed with interactive computer graphics systems. Using pictorial symbols to represent various components, a designer can build up a circuit on a video monitor (Fig. 1–5) by successively adding components to the circuit layout. The graphics display can be used to try out alternate circuit schematics as the designer tries to minimize the number of components or the space required for the circuit. Figure 1–6 displays a completed circuit board generated on a video monitor. Similar techniques are used to design communications networks and water and electricity supply systems.

Automobile, aircraft, aerospace, and ship designers use CAD techniques in the design of various types of vehicles. Wire-frame drawings, such as those shown in Fig. 1–7, are used to model individual components and plan surface contours for automobiles, airplanes, spacecraft, and ships. Individual surface sections and vehicle components (Fig. 1–8) can be designed separately and fitted together to display the total object. Simulations of the operation of a vehicle are often run to test the

FIGURE 1–7

CAD systems are used by engineers to generate wire-frame layouts for overall body designs. (a) Courtesy Megatek Corp. (b) Courtesy Evans & Sutherland.

(a)

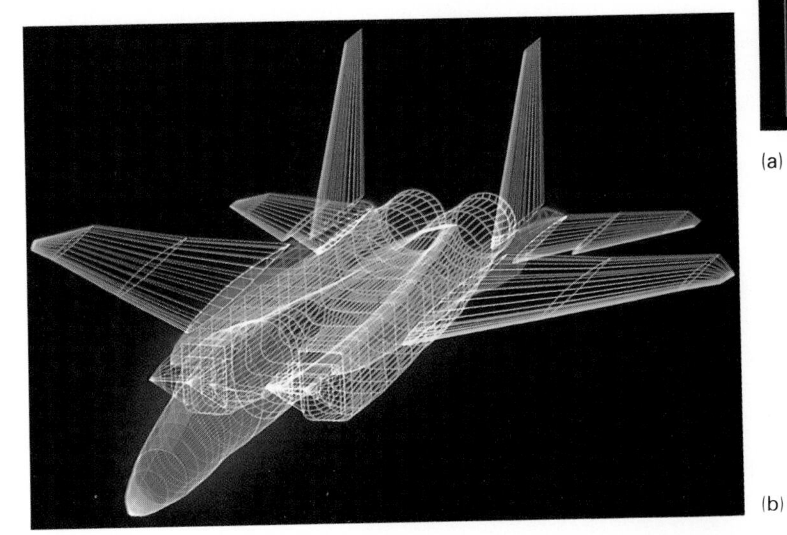

(b)

FIGURE 1–8

Surface sections and vehicle components are typically designed with CAD systems. Courtesy Evans & Sutherland.

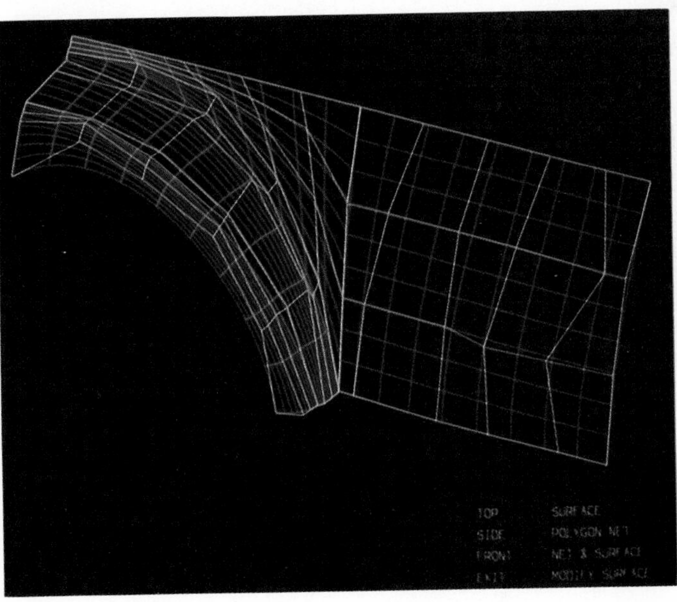

(a)

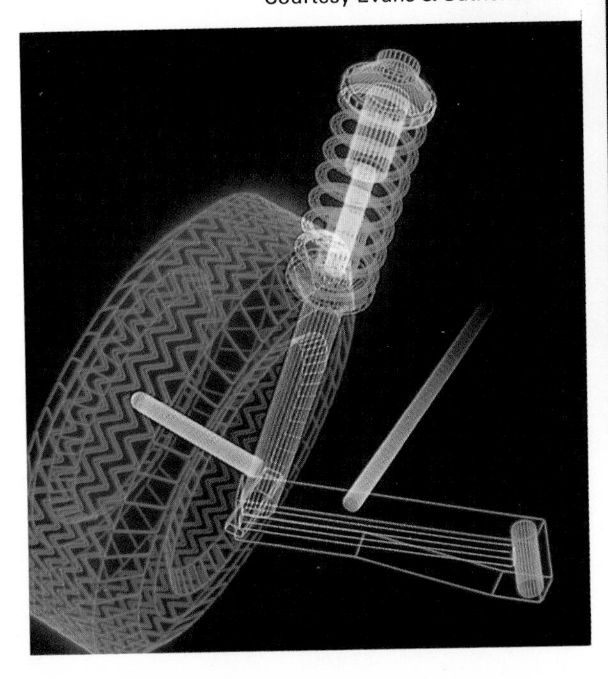

(b)

vehicle performance, as in the example of Fig. 1–9. Realistic renderings, such as those shown in Fig. 1–10, allow the designer to see how the finished product will appear.

Building designs are also created with computer graphics systems. Architects interactively design floor plans (Fig. 1–11), arrangements of doors and windows, and the overall appearance of a building. Working from the display of a building layout, the electrical designer can try out arrangements for wiring, electrical outlets, and fire warning systems. Utilization of space in the office or on the manufacturing floor is worked out using specially designed graphics packages.

Three-dimensional building models (Fig. 1–12) permit architects to study the appearance of a single building or a group of buildings, such as a campus or industrial complex. Using sophisticated graphics packages, designers can go for a simulated "walk" through the rooms or around the outsides of buildings to better appreciate the overall effect of a particular design.

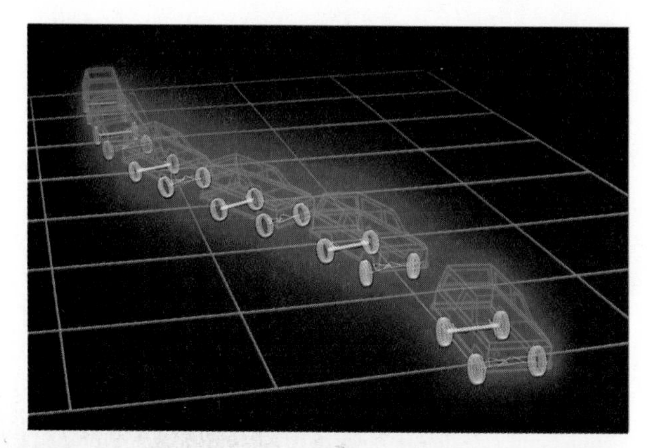

FIGURE 1–9
Simulation of Ford Bronco II performance during lane changes using a graphics display system. Courtesy Evans & Sutherland and Mechanical Dynamics, Inc.

(a)

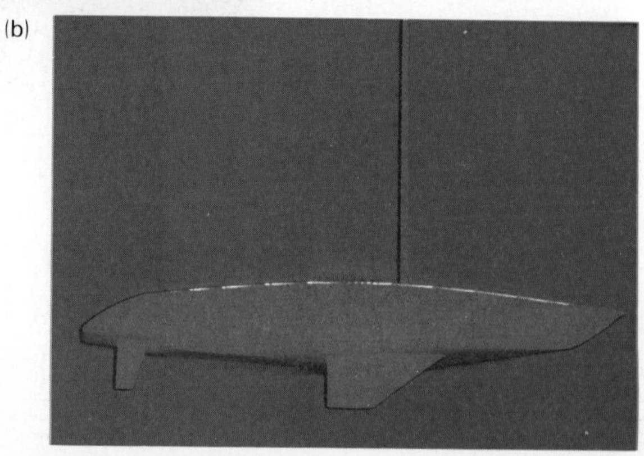

(b)

FIGURE 1–10
Solid modeling of an automobile and ship hull using graphics design systems. (a) Courtesy DICOMED Corp. (b) Courtesy Intergraph Corp.

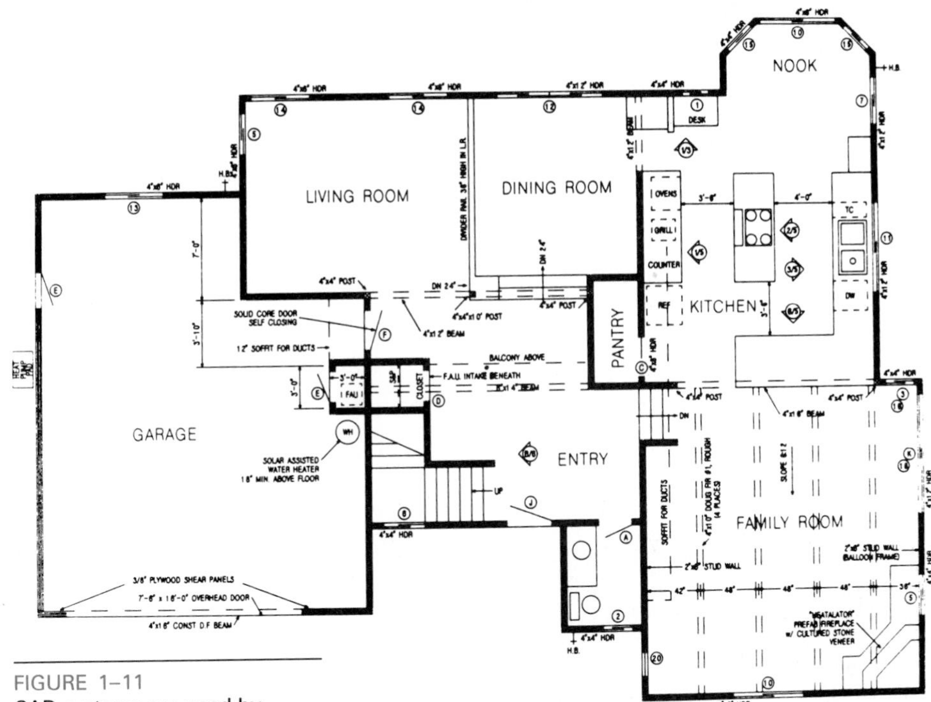

FIGURE 1–11
CAD systems are used by architects in the design of building layouts. Courtesy Precision Visuals, Inc., Boulder, Colorado.

(a)

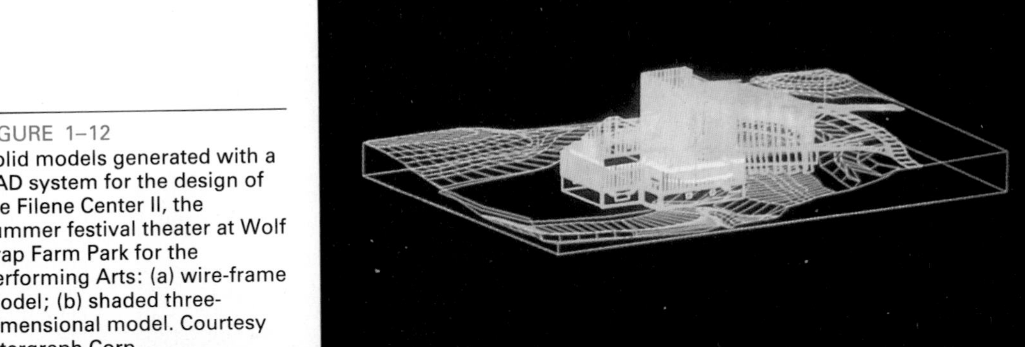

FIGURE 1–12
Solid models generated with a CAD system for the design of the Filene Center II, the summer festival theater at Wolf Trap Farm Park for the Performing Arts: (a) wire-frame model; (b) shaded three-dimensional model. Courtesy Intergraph Corp.

(b)

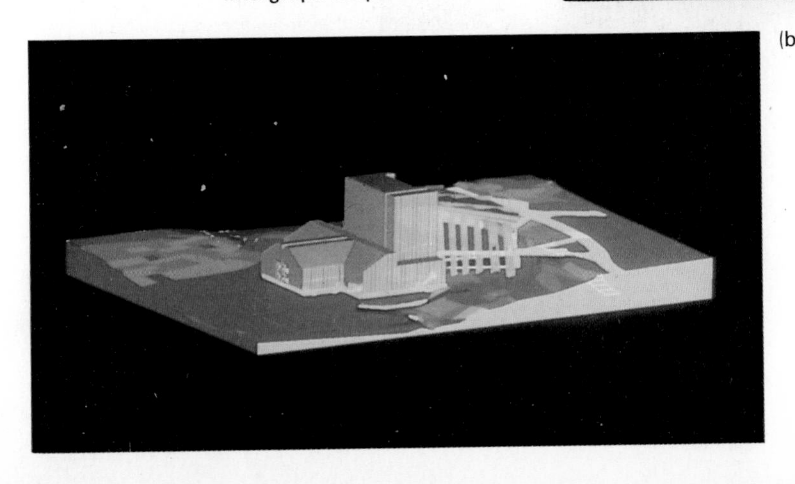

1–2 Graphs, Charts, and Models

A number of commercially available graphics programs are designed specifically for the generation of graphs and charts. Often a graph-plotting program will have the capability of generating a variety of graph types, such as bar charts, line graphs, surface graphs, or pie charts. Many programs are capable of summarizing data in either two-dimensional or three-dimensional form. Three-dimensional plots are typically used to illustrate multiple relationships, as in the graphs of Figs. 1–13 and 1–14. In some cases (Fig. 1–15), three-dimensional graphs are used to provide a more dramatic or more attractive presentation of data.

Business graphics, one of the most rapidly growing areas of application, makes extensive use of visual displays as a means for rapidly communicating the vast amounts of information that are compiled for managers and other individuals in an organization. Graphs and charts are typically used to summarize financial, statisti-

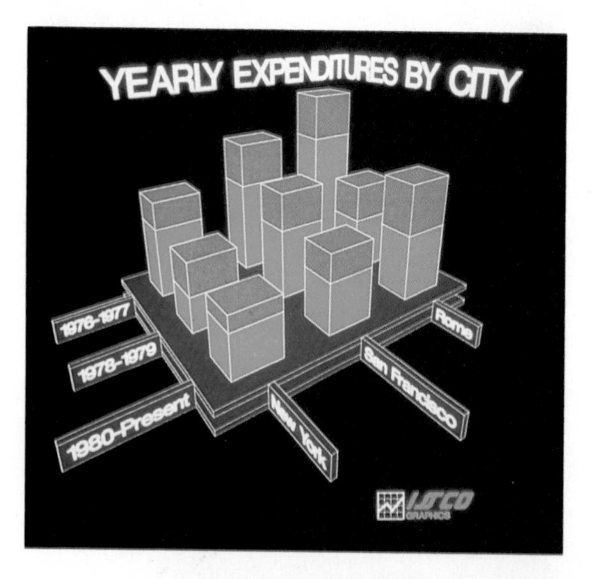

FIGURE 1–13
Three-dimensional bar chart used to illustrate several relationships within one graph. Reprinted with permission from ISSCO Graphics, San Diego, California.

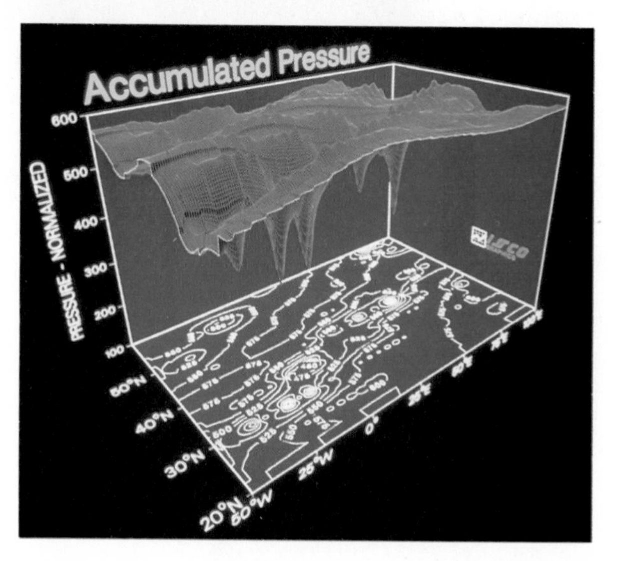

FIGURE 1–14
Three-dimensional pressure chart. Reprinted with permission from ISSCO Graphics, San Diego, California.

FIGURE 1–15
Three-dimensional graphs designed for dramatic effect. Reprinted with permission from ISSCO Graphics, San Diego, California.

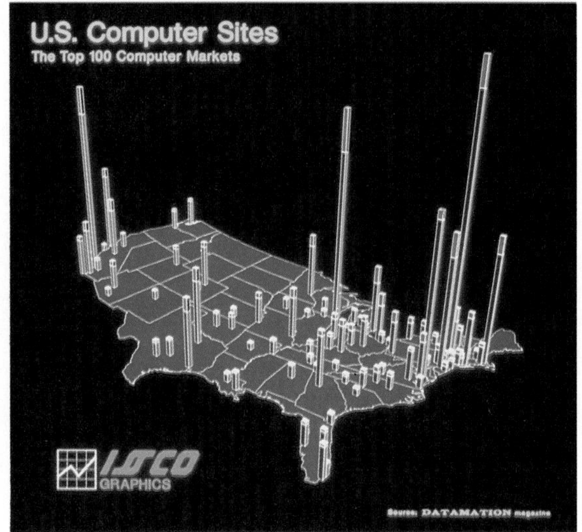

(a)

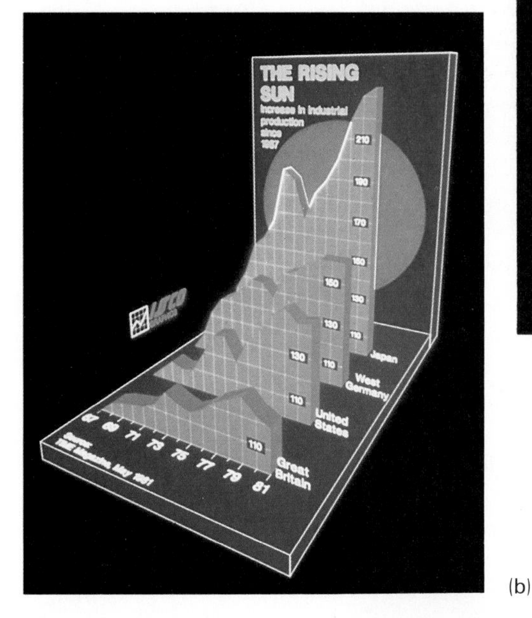

(b)

FIGURE 1–16
Several types of graphs are often combined within one display as an effective means for presenting related information. Reprinted with permission from ISSCO Graphics, San Diego, California.

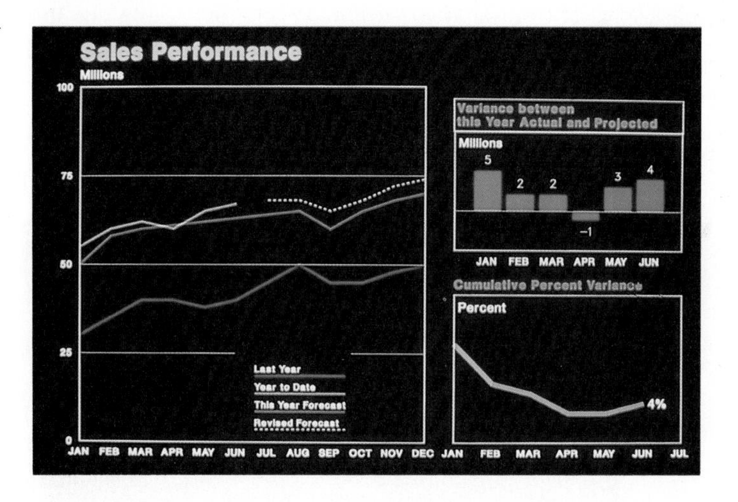

cal, mathematical, scientific, or economic data, and several graphs are often combined in one presentation (Fig. 1–16). These graphs are generated for research reports, managerial reports, consumer information bulletins, and visual aids used during presentations. Figure 1–17 illustrates a time chart used in task planning. Project management techniques make use of time charts and task network layouts to schedule and monitor projects. Some graphics systems include the capability of generating 35mm slides or overhead transparencies from the graphs displayed on a video monitor.

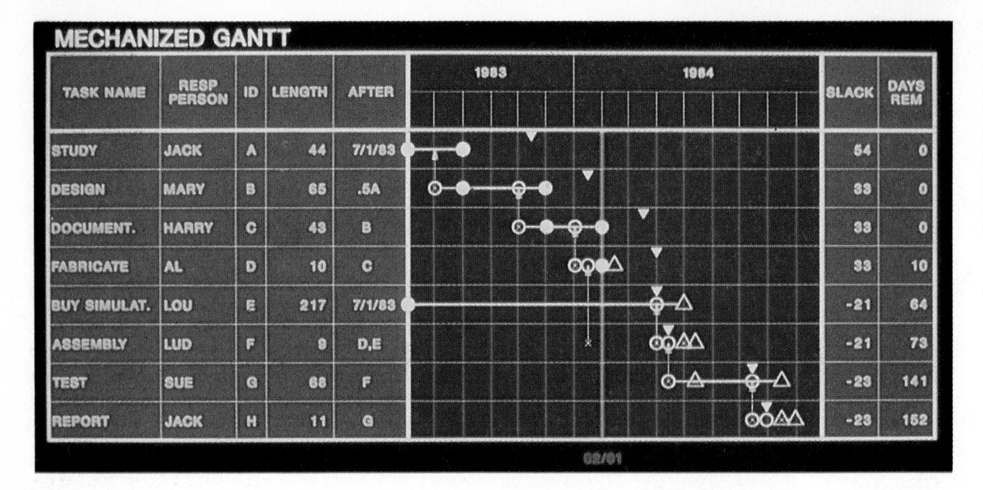

FIGURE 1–17
Time chart displaying relevant information about project tasks. Reprinted with permission from ISSCO Graphics, San Diego, California.

The behavior of physical systems is often studied by constructing graphs and models. When large amounts of data are to be analyzed, a color-coded graph, such as that in Fig. 1–18, can help researchers understand the structure of a system. Without the aid of such plots, it would be difficult for a researcher to interpret data tables containing millions of entries. In a similar way, computer-generated models (Fig. 1–19) are used to study the behavior of systems.

Computer-generated models of physical, financial, and economic systems are often used as educational aids. Models of physiological systems, population trends, or physical equipment, such as the color-coded diagram in Fig. 1–20, can help trainees to understand the operation of a system.

Many data-plotting applications are concerned with graphing some type of geographical information. Often such plots are used to display different types of regional or global statistics, such as plotting sales data in different districts. Programs designed to produce weather maps can take data from individual observation stations to produce plots such as those in Fig. 1–21. Cartographic programs are available for generating maps for selected geographic areas or the entire world.

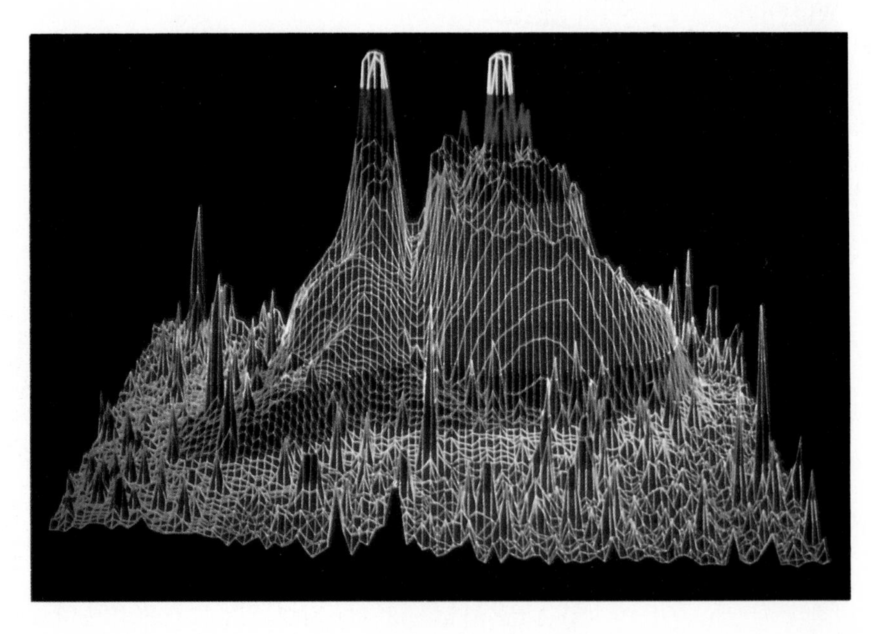

FIGURE 1–18
A color-coded plot with 16 million density points of relative brightness observed for the Whirlpool Nebula reveals two distinct galaxies. Courtesy Los Alamos National Laboratory.

FIGURE 1–19
Graphics model of a test surface used in the study of atomic and nuclear collisions. Courtesy Los Alamos National Laboratory.

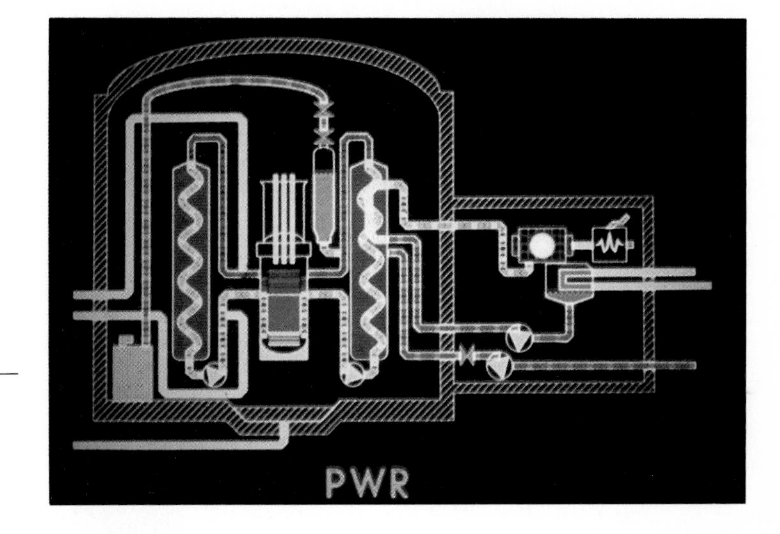

FIGURE 1–20
Color-coded diagram used to explain the operation of a nuclear reactor. Courtesy Los Alamos National Laboratory.

(a)

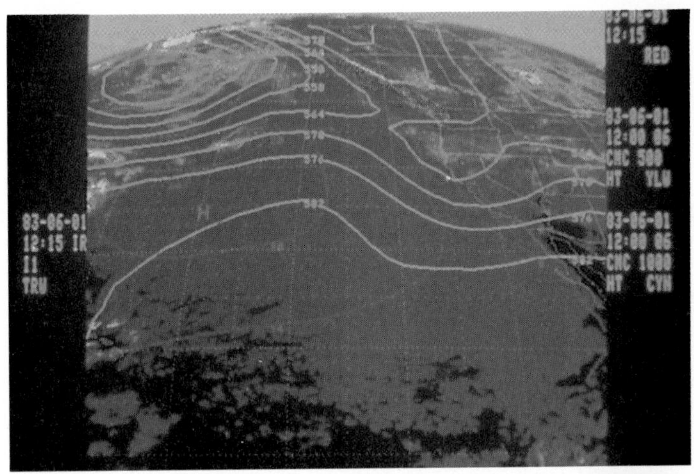

FIGURE 1–21
Weather charts generated with the aid of computer graphics for display on a newscast. Courtesy Gould Inc., Imaging & Graphics Division and GOES-WEST.

(b)

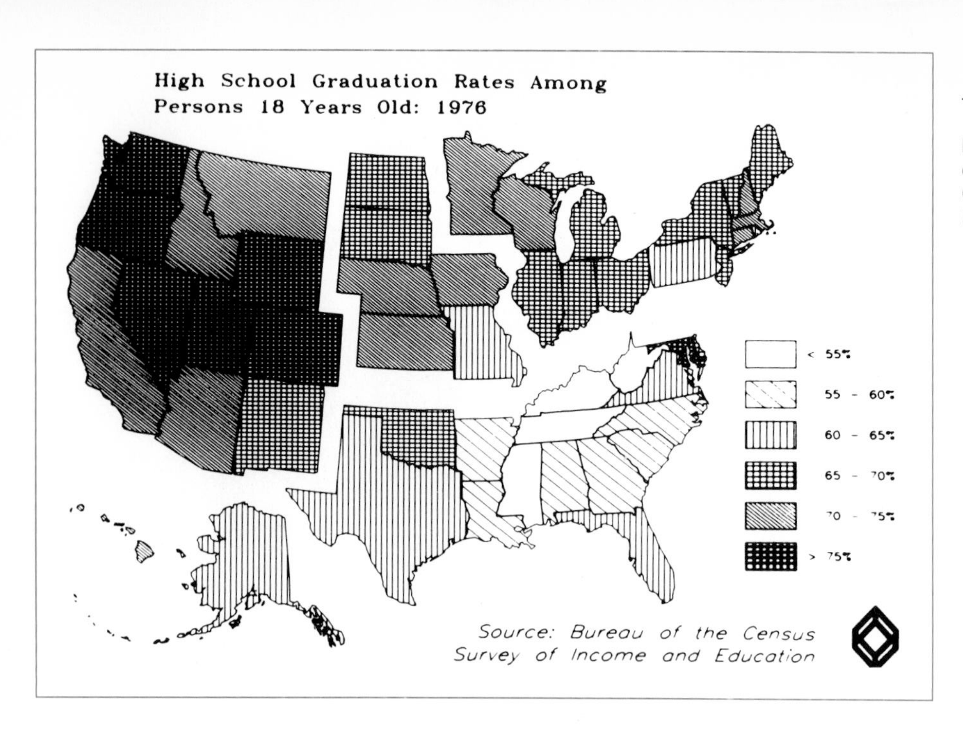

FIGURE 1–22
Exploded map produced with a cartographic program. Courtesy Precision Visuals, Inc., Boulder, Colorado.

Some of these programs allow shading or color for the various geographic areas to be chosen by the user, and some allow sections to be separated (or "exploded") for emphasis (Fig. 1–22).

1–3 Computer Art

Both creative and commercial art applications make extensive use of computer graphics. The abstract display in Fig. 1–23 is created by plotting a series of mathematical functions in varying colors. Figure 1–24 illustrates another type of art design drawn with a pen plotter.

"Paintbrush" programs allow artists to create pictures on the screen of a video monitor, somewhat like the character in Fig. 1–25. Actually, the artist might draw the picture on a "graphics tablet" using a stylus as input. A paintbrush program was also used to create the characters in Fig. 1–26, who seem now to be busy on

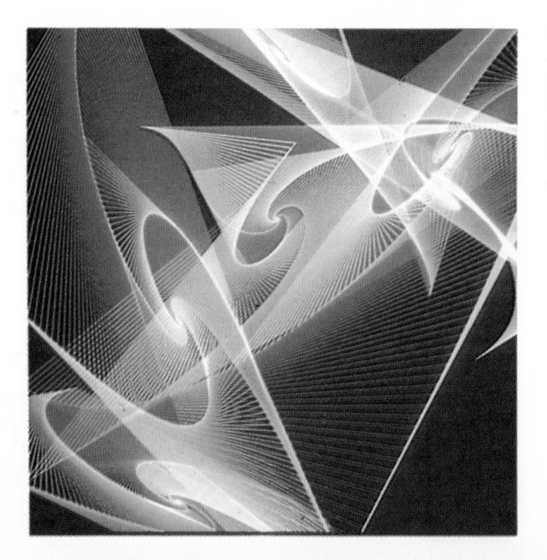

FIGURE 1–23
Abstract design can be created by plotting mathematical functions in various colors. Courtesy Melvin L. Prueitt, Los Alamos National Laboratory.

FIGURE 1–24
Computer "art" produced on a pen plotter. Courtesy Computer Services, Western Illinois University.

FIGURE 1–25
Cartoon drawing produced with a paintbrush program, symbolically illustrating a creative artist at work on a video monitor. Courtesy Gould Inc., Imaging & Graphics Division and Aurora Imaging.

(a)

FIGURE 1–26
Cartoon demonstrations of an "artist" creating a picture with a paintbrush program. The picture, drawn on a graphics tablet, is displayed on the video monitor as the elves look on. In (b), the cartoon is superimposed on the famous Thomas Nast drawing of Saint Nicholas, which was input to the system with a video camera, then scaled and positioned. Courtesy Gould Inc., Imaging & Graphics Division and Aurora Imaging.

(b)

FIGURE 1–27
Examples of creative art generated with an Aycon 16 series computer using a paintbrush program. Courtesy Aydin Controls, a division of Aydin Corp.

(a)

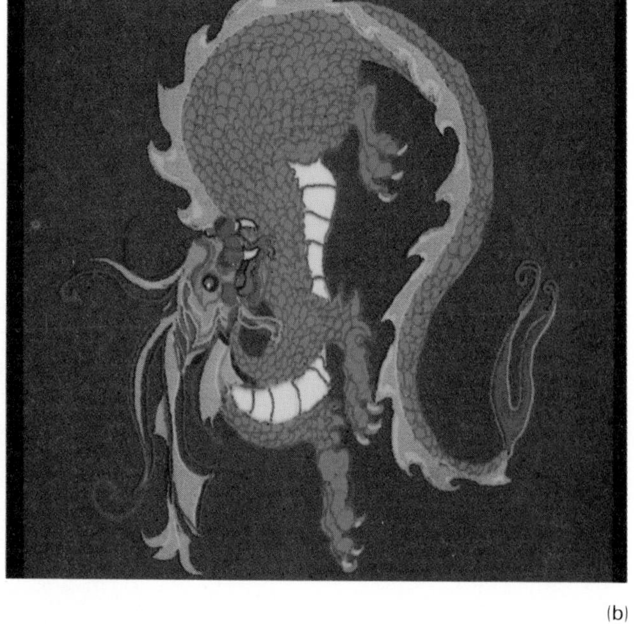

(b)

(c)

(d)

a creation of their own. Paintbrush programs are not limited to cartoon characters; they are also used to produce the sort of art seen in Figs. 1–27 and 1–28.

Computer-generated art is widely used in commercial applications. Figure 1–29 illustrates a rug pattern that was designed with a computer graphics system. Logos and advertising designs for TV messages are now commonly produced with graphics systems. Figure 1–30 shows two examples of computer-generated art for use in TV advertising spots. In addition, graphics programs have been developed for applications in publishing and word processing, which allow graphics and text-editing operations to be combined. Figure 1–31 is an example of the type of output that can be generated with such systems.

FIGURE 1–28
This piece of art took a Japanese artist 15 minutes to create on an Aurora 100 paintbrush system. Courtesy Gould Inc., Imaging & Graphics Division.

FIGURE 1–29
Oriental rug pattern produced with computer graphics design methods. Courtesy Lexidata Corp.

FIGURE 1–30
Graphics applications in commercial art for TV advertising: (a) Panasonic "Glider"; (b) TRW "Line." Courtesy Robert Abel & Associates.

(b)

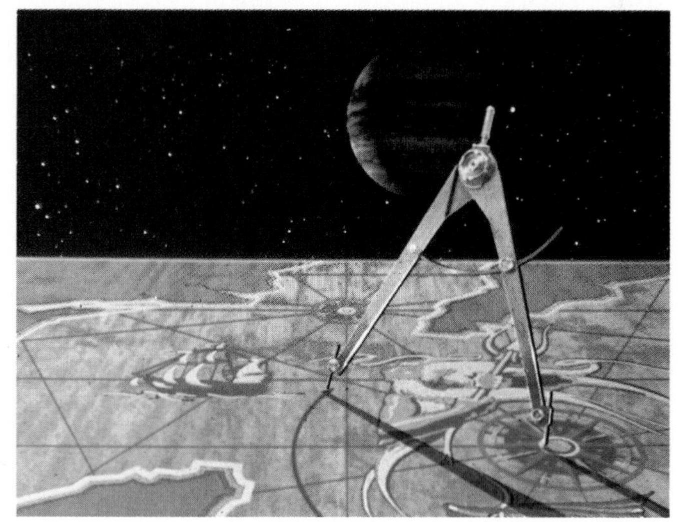

FIGURE 1–31
An "author's workstation," combining graphics and word processing. Courtesy Information Displays, Inc.

1–4 Computer Animation

Each of the photos in Fig. 1–30 represents one frame of an animation sequence. The glider in Fig. 1–30 (a) is moved slightly from one frame to the next to simulate motion about the room. Similarly, the compass in Fig. 1–30 (b) is displayed at a different position in each frame to simulate drawing of the curved line. Such frame techniques are used also in creating cartoons and science-fiction movies. Each frame is drawn with a graphics system and recorded on film, with only slight changes in the positions of objects from one frame to the next. When the frames are displayed in rapid succession, we have an animated movie sequence. Figure 1–32 shows artists creating a cartoon frame, and Fig. 1–33 shows some scenes that were generated for the movie *Star Trek—The Wrath of Khan.*

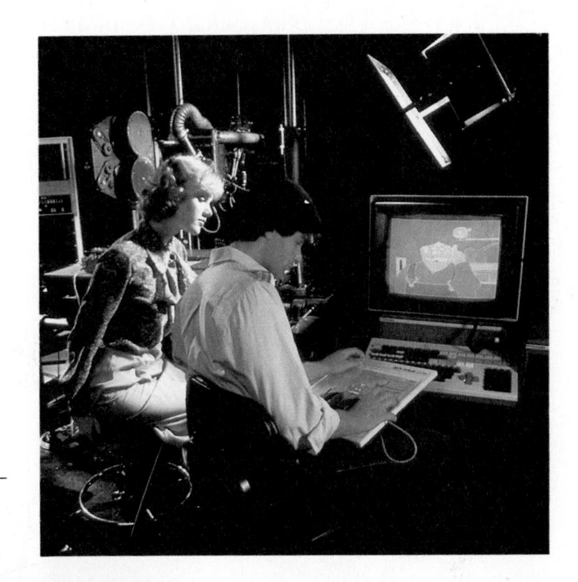

FIGURE 1–32
Artists designing a cartoon frame. Courtesy Lexidata Corp.

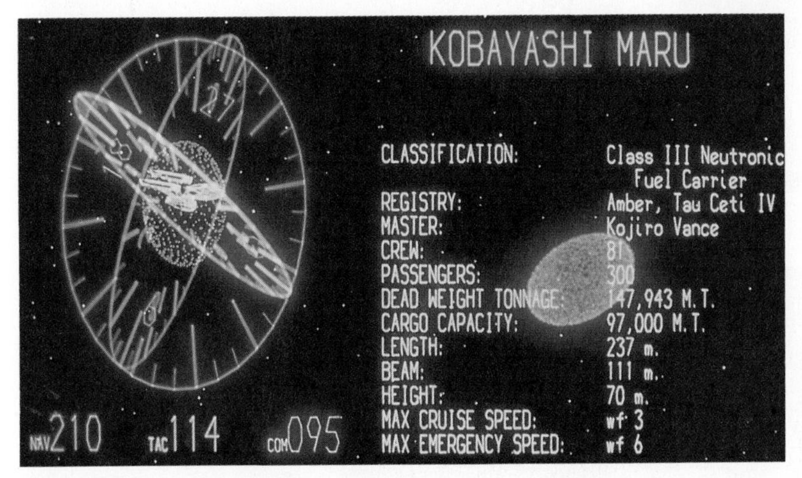

(a)

FIGURE 1–33
Graphics developed for the Paramount Pictures movie *Star Trek: The Wrath of Khan.* Courtesy Evans & Sutherland.

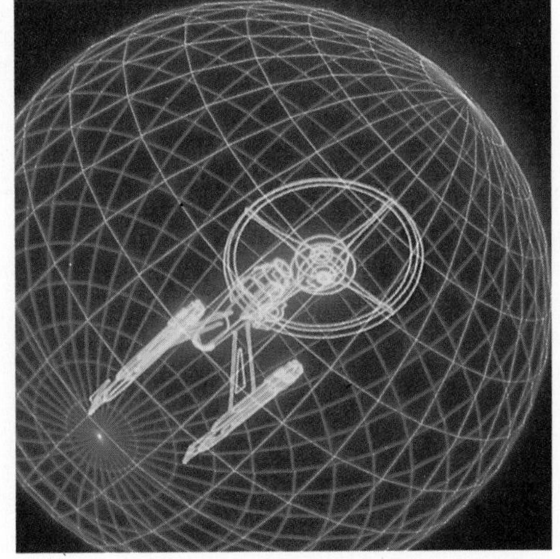

(b)

Animation methods are also used in education, training, and research applications. Figure 1–34, for example, is an animated simulation for the *Voyager* space probe. Such simulations can be used to study the behavior of physical systems or as an aid to instruction.

For some training applications, special systems are designed. Examples of such specialized systems are the simulators for training ship captains and aircraft pilots. One type of aircraft simulator is shown in Fig. 1–35. Figures 1–36 through 1–38 show some scenes that can be presented to a pilot operating a flight simulator. An output from an automobile-driving simulator is given in Fig. 1–39. This simulator is used to investigate the behavior of drivers in critical situations. The drivers' reactions are then used as a basis for optimizing vehicle design to maximize traffic safety.

(a)

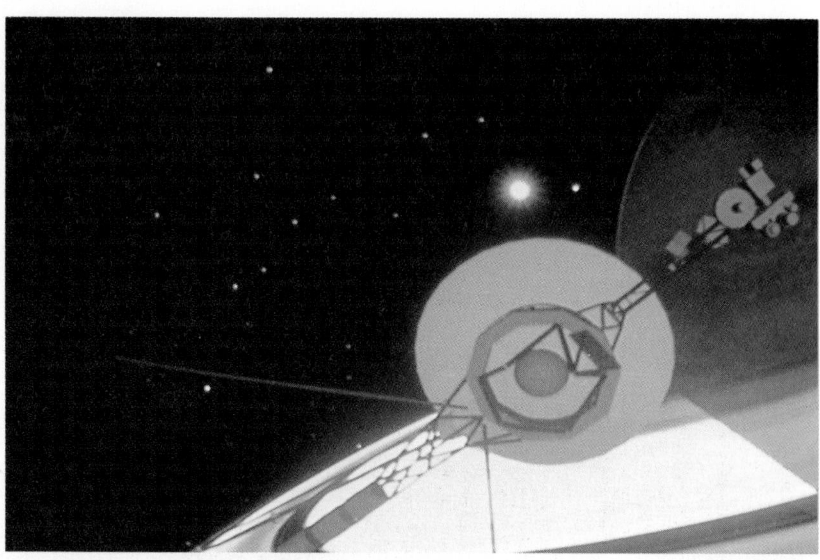

(b)

(c)

FIGURE 1–34
Graphics simulation of the trek of the *Voyager* space probe. (a) *Voyager* passing the surface of the planet Saturn. (b) The dark side of Saturn with Earth in the distance. (c) Opposite view of (b) from the Earth. Courtesy Gould Inc., Imaging & Graphics Division and Computer Graphics Lab of the Jet Propulsion Laboratory.

FIGURE 1–35
The GE F-5 flight simulator allows the instructor (in the rear) to enter test conditions for the pilot at the controls in front. Courtesy General Electric.

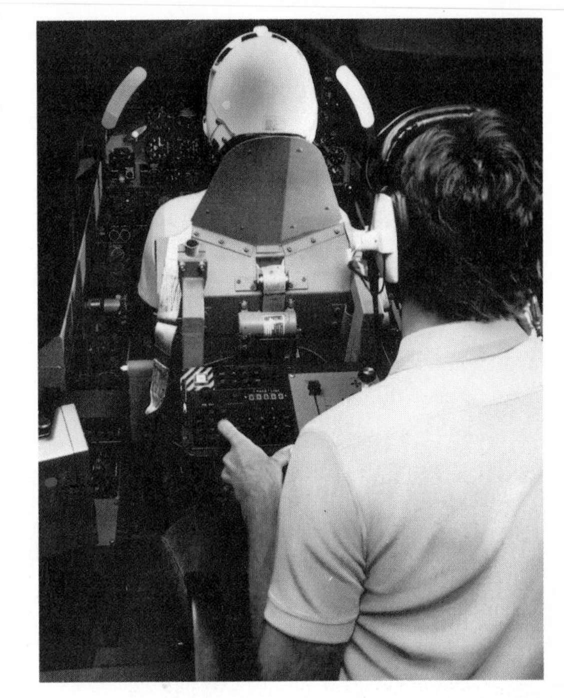

FIGURE 1–36
A computer-generated scene for the GE F-5 flight simulator shown in Fig. 1–34. Courtesy General Electric.

FIGURE 1–37
Computer-generated view of a runway used by pilots operating a flight simulator to practice landings. Courtesy Evans & Sutherland and Rediffusion Simulation.

FIGURE 1–38
Graphics display of terrain, buildings, and planes used as part of an animation sequence allowing practice in formation flying with a flight simulator. Courtesy Evans & Sutherland.

FIGURE 1–39
Scene generated by a driving simulator, used to study driver reactions in critical situations. Courtesy Evans & Sutherland and Daimler-Benz AG, Stuttgart, West Germany.

1–5 Graphical User Interfaces

Input options to many computer programs are designed as a set of **icons,** which are graphic symbols that look like the processing option they are meant to represent. Users of such programs select processing options by pointing to the appropriate icon. The advantage of such systems is that the icons can take up less screen space than the corresponding textual description of the functions, and they can be understood more quickly if well designed.

Figure 1–40 illustrates typical icons used in word processing programs. In these programs, the process for adding an item to a file is represented by a picture of a filing cabinet, a file drawer, or a file folder. Deleting a file can be represented by a wastebasket icon. A stop sign can stand for an exit operation, and a picture of a ruler can be used to adjust text margins. Icons can be useful in many applications.

FIGURE 1–40
Some icons that are useful in word-processing applications.

1–6 Graphics for Home Use

A major use of computer graphics in home applications is in video games. All of these games employ graphics methods in one form or another. Some systems display graphics on built-in screens, but many are designed to be attached to a TV set.

With the increasing popularity of personal computers and the ever-increasing graphics capabilities of these systems, graphics applications in the home have been steadily expanding. Personal computers allow the home user to generate financial or calorie-counting graphs or to create designs, such as those in Fig. 1–41, for customized greeting cards or stationery.

FIGURE 1–41
Graphics designs can be created on personal computers for a variety of uses.

1–7 Image Processing

The graphics technique used for producing visual displays from photographs or TV scans is called **image processing.** Although computers are used with these displays, image processing methods differ from conventional computer graphics methods. In traditional computer graphics, a computer is used to create the picture. Image processing techniques, on the other hand, use a computer to digitize the shading and color patterns from an already existing picture. This digitized information is then transferred to the screen of a video monitor. Such methods are useful for viewing many systems or objects that we cannot see directly, such as TV scans from spacecraft or views from the eye of an industrial robot. Figure 1–42 displays an image-processed picture of one of the moons of the planet Jupiter.

Once a picture has been digitized, additional processing techniques can be applied to rearrange picture parts, to enhance color separations, or to improve the quality of shading. An example of the application of image processing methods to the enhancement of picture quality is shown in Fig. 1–43.

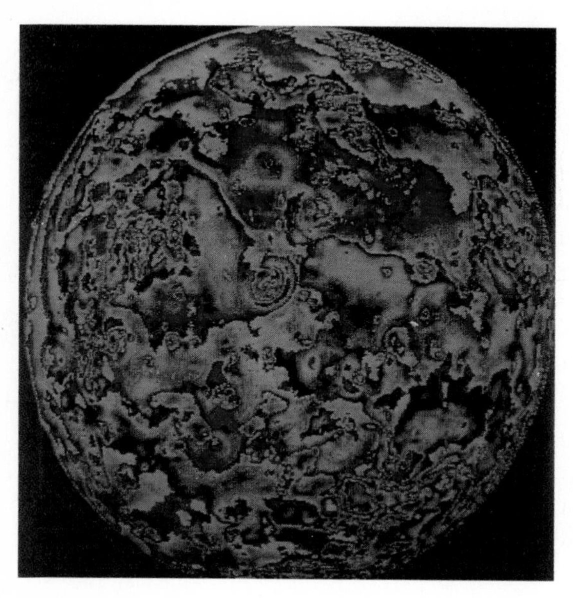

FIGURE 1–42
An enhanced display of Jupiter's moon Io. Courtesy Gould Inc., Imaging & Graphics Division.

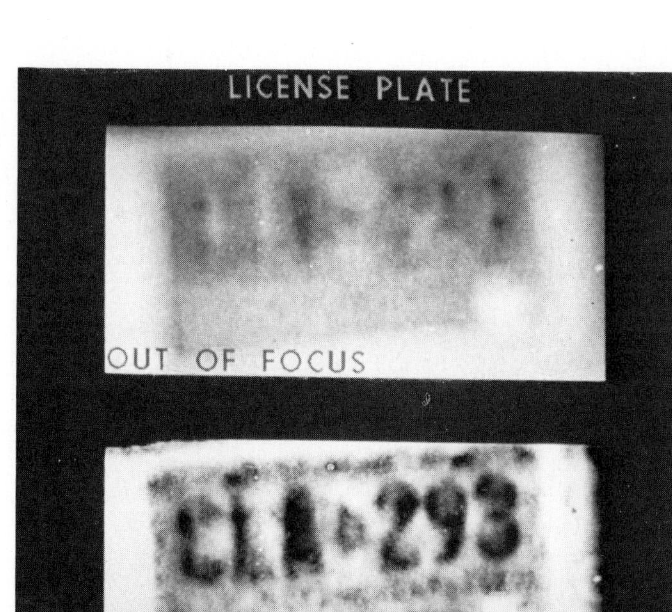

FIGURE 1–43
A blurred photograph of a license plate becomes legible after the application of image processing methods. Courtesy Los Alamos National Laboratory.

Image processing is used extensively in commercial art applications involving the retouching and rearranging of sections of photographs and other artwork. An image processing workstation for setting up advertising layouts is shown in Fig. 1–44.

Medical applications make use of image processing techniques both for picture enhancements and in tomography. Figure 1–45 shows a cancer cell enhanced through image processing. Tomography is a technique of X-ray photography that allows cross-sectional views of physiological systems to be displayed. Both computed X-ray tomography (CT) and position emission tomography (PET) use projec-

FIGURE 1–44
Image processing workstation used by commercial artists to produce advertising layouts. Courtesy COMTAL/3M Corp.

(a)

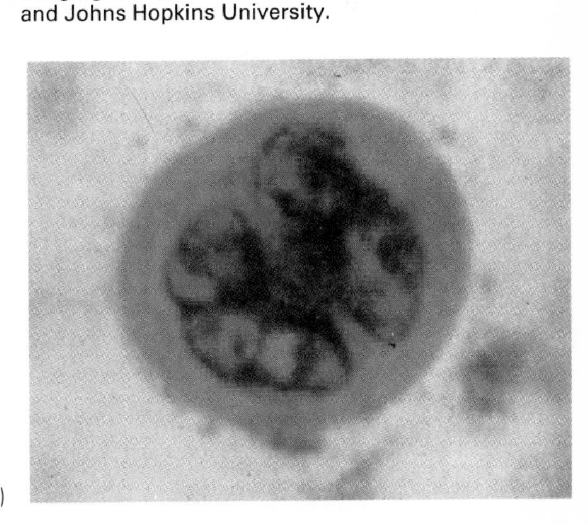

(b)

FIGURE 1–45
Image processing techniques applied to a picture of a cancerous cell, marked within a cell culture (a), produce the enlarged and enhanced picture (b). Courtesy Gould Inc., Imaging & Graphics Division and Johns Hopkins University.

tion methods to reconstruct cross sections from digital data. An example of a CT scan is given in Fig. 1–46, while Figs. 1–47 and 1–48 show PET scans. These techniques are also used to monitor internal functions and show cross sections during surgery. Other medical imaging techniques include ultrasonics and nuclear medicine scanners. With ultrasonics, high-frequency sound waves, instead of X-rays, are used to generate digital data. Nuclear medicine scanners collect digital data from radiation emitted from ingested radionuclides and plot color-coded images. Conventional graphics methods are also used in medical applications to model and study physical functions and in the design of artificial limbs.

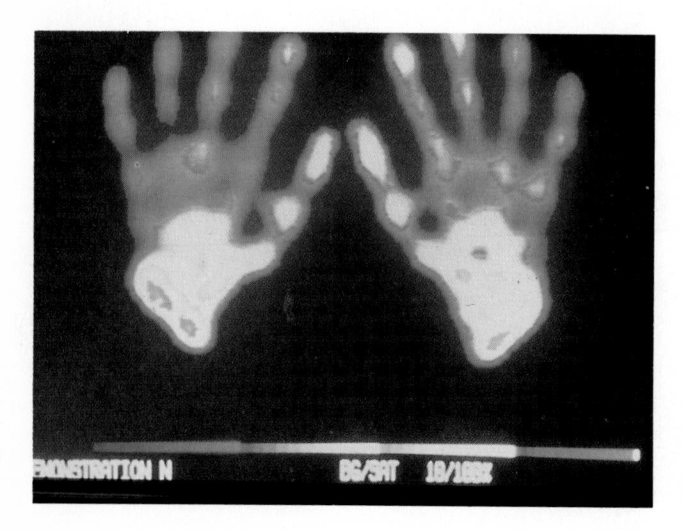

FIGURE 1–46
CT scan used to view physiological functions. Courtesy Gould Inc., Imaging & Graphics Division.

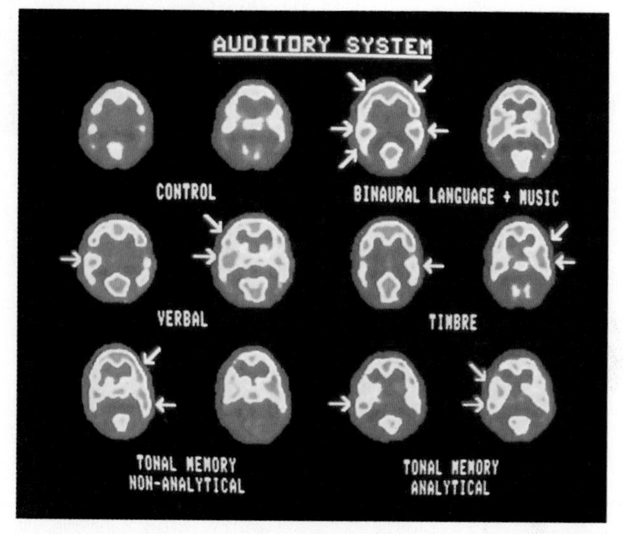

FIGURE 1–47
PET scans showing brain activation from sound stimulus. The scans show that verbal stimulus produces left-brain activation and music produces right-brain activation. Scans on the lower right came from a musician who created mental "histograms" of note frequencies, producing left-brain activation because his perceptions were analytical. Courtesy Gould Inc., Imaging & Graphics Division and UCLA.

FIGURE 1–48
PET scans, displaying blood flow in a horizontal slice across the beating heart muscle, used to measure chest pain as it occurred. Courtesy Gould Inc., Imaging & Graphics Division and UCLA.

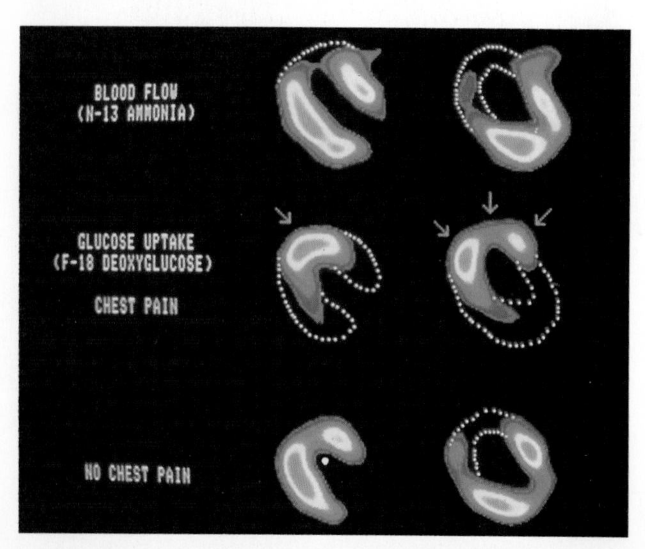

Numerous other fields make use of imaging techniques to generate pictures and analyze collected data. Figure 1–49 shows satellite photos used to analyze terrain features. The display of Fig. 1–50 was used to discover an oil field in the North Sea. Data from solar flares is plotted in Fig. 1–51, and galaxies are reconstructed in Fig. 1–52 from astronomical observations collected and arranged in large databases.

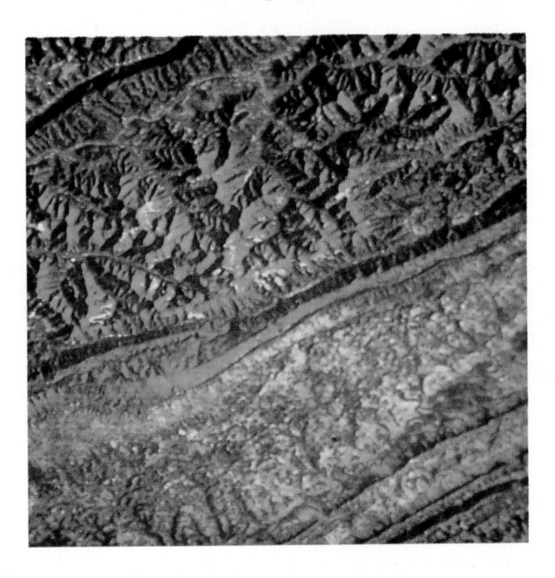

(b)

(a)

FIGURE 1–49
Landsat photos: (a) Kentucky mountain range color-coded to show areas destroyed by strip mining; (b) caldera on the border between Argentina and Chili. Courtesy Gould Inc., Imaging & Graphics Division.

FIGURE 1–50
Image showing oil deposits on either side of a salt plug (vertical red column) in the North Sea. Courtesy Gould Inc., Imaging & Graphics Division.

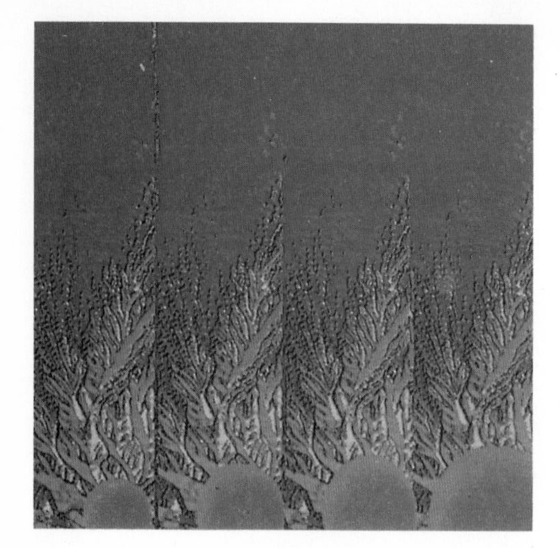

FIGURE 1–51
Display plotted from solar flare data. Courtesy Gould Inc., Imaging & Graphics Division and Science Applications, Inc.

(a)

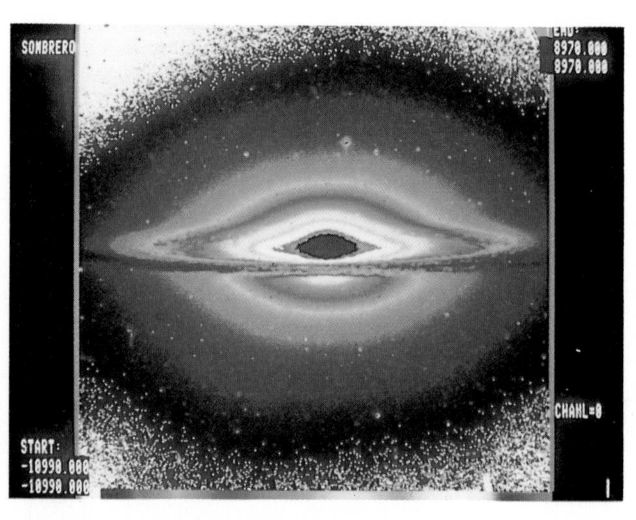

(b)

FIGURE 1–52
Galaxies plotted from astronomical observation data. Courtesy Gould Inc., Imaging & Graphics Division and European Southern Observatory.

REFERENCES

A detailed discussion of computer-aided design and manufacturing (CAD/CAM) is given in Pao (1984). This book discusses applications in different industries and various design methods. Other sources for CAD applications include Bouquet (1978) and Yessios (1979).

Graphics techniques for flight simulators are presented in Schachter (1983). Fu and Rosenfeld (1984) discuss simulation of vision, and Weinberg (1978) gives an account of space shuttle simulation.

Sources for graphics applications in mathematics, science, and technology include Gardner and Nelson (1983), Grotch (1983), and Wolfram (1984).

A discussion of graphics methods for visualizing music is given in Mitroo, Herman, and Badler (1979). Graphics icon and symbol concepts are presented in Lodding (1983) and in Loomas (1983).

For additional information on medical applications see Hawrylyshyn, Tasker, and Organ (1977); Preston, et. al. (1984); and Rhodes (1983).

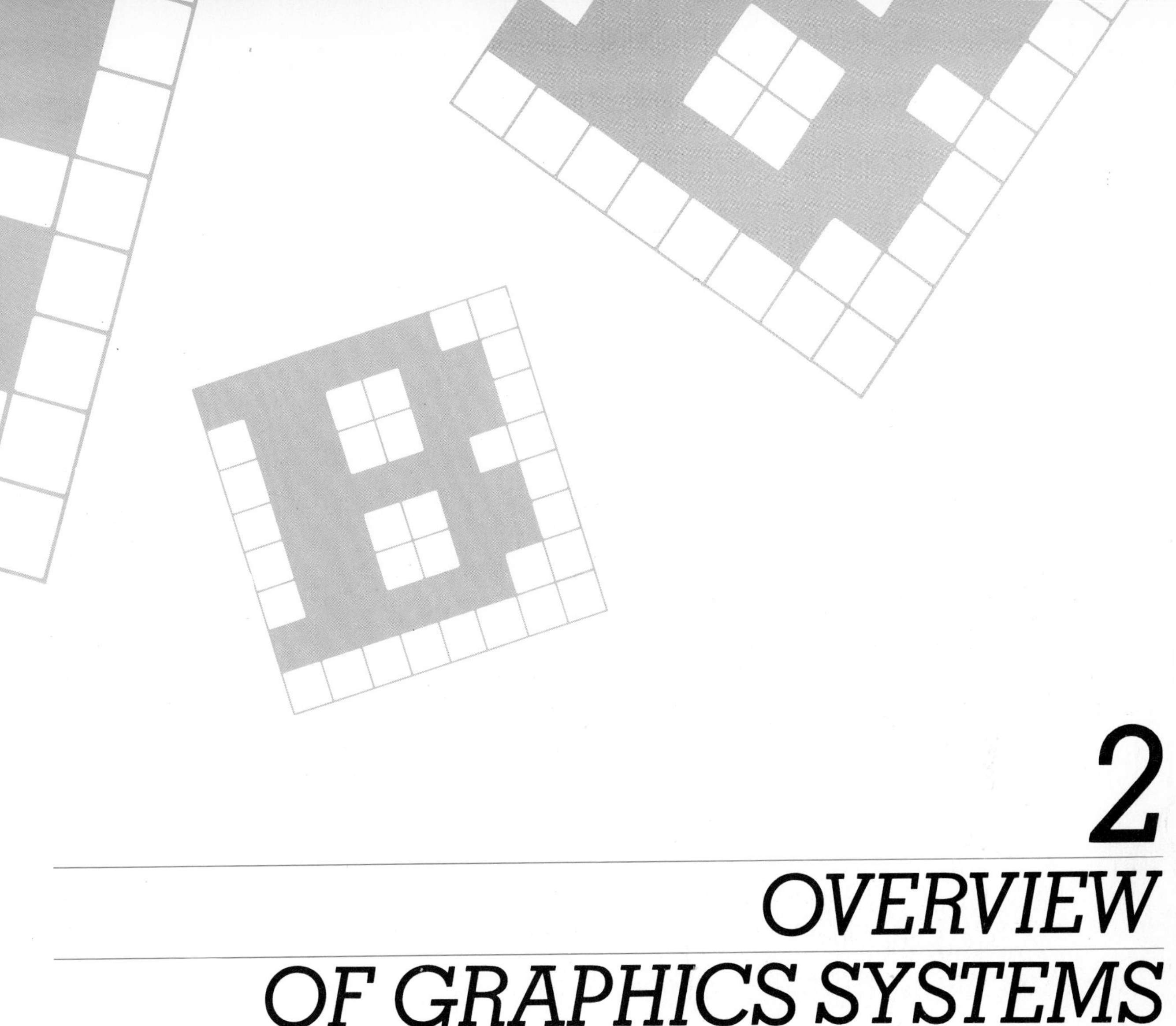

2 OVERVIEW OF GRAPHICS SYSTEMS

Computer systems can be adapted to graphics applications in various ways, depending on the hardware and software capabilities. Any general-purpose computer can be used to do character graphics by using elements of the system's character set to form patterns. Output statements, such as PRINT and WRITE, are used to produce the appropriate character string across each print line of a printer or video monitor. Though simple in technique, character graphics can produce intricate designs (Fig. 2–1). With the addition of a plotter, line drawings such as those in Fig. 2–2 can be obtained. If a video monitor and line-drawing commands are part of our system, we can interactively create and manipulate these drawings on the monitor screen. For more sophisticated applications, a variety of software packages and hardware devices are available.

FIGURE 2–1
Pictures can be produced with printers using various combinations of characters and overprinting to produce shading patterns. Courtesy Computer Services, Western Illinois University.

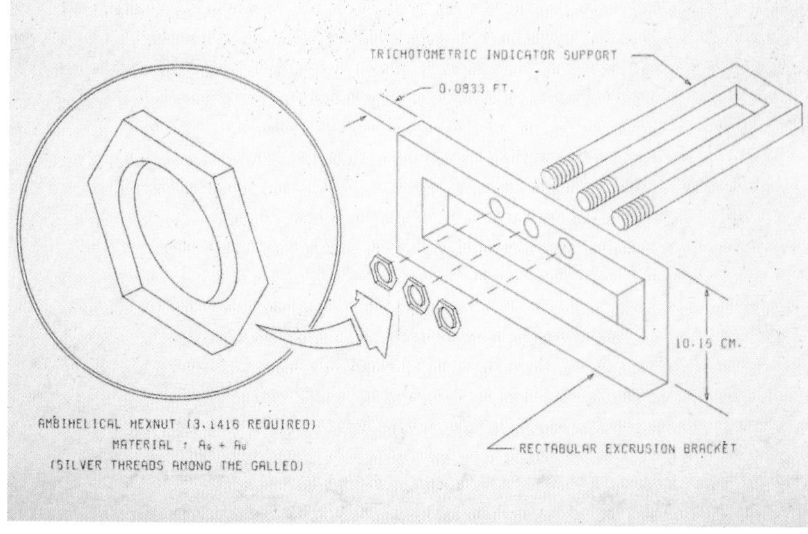

(a)

(b)

FIGURE 2–2
Examples of pen plotter output. Courtesy Computer Services, Western Illinois University and CalComp Group; Sanders Associates, Inc.

2–1 Display Devices

Interactive systems use some type of video monitor as the primary output device (Fig. 2–3). The operation of most video monitors is based on the standard **cathode-ray tube (CRT)** design, but several other technologies exist.

FIGURE 2–3
An interactive graphics system. Courtesy CalComp Group; Sanders Associates, Inc.

Refresh Cathode-Ray Tubes

Figure 2–4 illustrates the basic operation of a CRT. A beam of electrons (cathode rays), emitted by an electron gun, passes through focusing and deflection systems that direct the beam toward specified points on the phosphor-coated screen. The phosphor then emits a small spot of light at each point contacted by the electron beam. Since the light emitted by the phosphor fades very rapidly, some method is needed for maintaining the screen picture. One way to keep the phosphor glowing is to redraw the picture repeatedly by quickly directing the electron beam back over the same points. This type of display is called a **refresh CRT.**

Different types of phosphors are available for use in a CRT. Besides color, a major difference between phosphors is their persistence: how long they continue to emit light after the electron beam is removed. Persistence is defined as the time it takes emitted light to decay to one-tenth of its original intensity. Lower-persis-

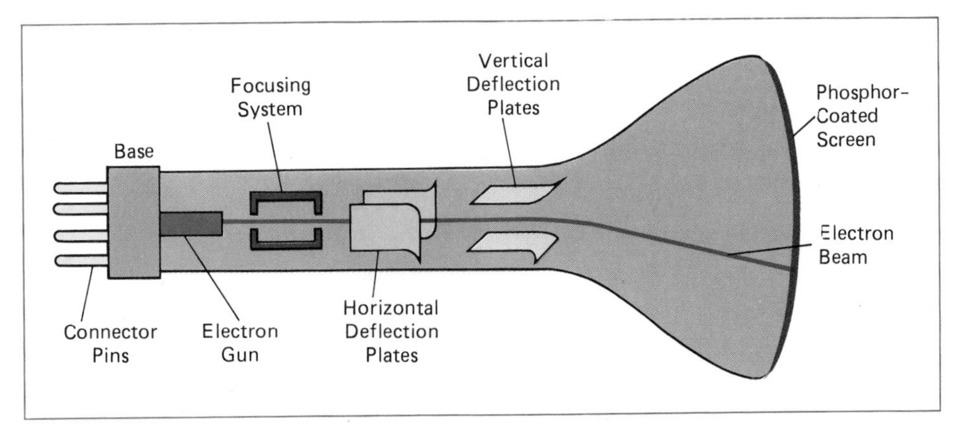

FIGURE 2–4
Basic design of a CRT, using electrostatic deflection fields.

tence phosphors require higher refresh rates to maintain a picture on the screen without flicker. A phosphor with low persistence is useful for animation, while high-persistence phosphors are better suited for displaying highly complex, static pictures.

Deflection of the electron beam is done either with electric fields or with magnetic fields. The electrostatic method is illustrated in Fig. 2–4. Here the beam passes between two pairs of metal plates: one pair vertical, the other pair horizontal. A voltage difference is applied to each pair of plates according to the amount that the beam is to be deflected in each direction. As the electron beam passes between each pair of plates, it is bent toward the plate with the higher positive voltage. In Fig. 2–4, the beam is first deflected toward one side of the screen. Then, as the beam passes through the horizontal plates, it is deflected toward the top or bottom of the screen. Magnetic fields provide another method for deflecting the electron beam. The proper deflections can be attained by adjusting the current through coils placed around the outside of the CRT envelope.

The basic components of an electron gun in a CRT are the heated metal cathode and a control grid (Fig. 2–5). Heat is supplied to the cathode by directing a current through a coil of wire, called the filament, inside the cylindrical cathode structure. This causes electrons to be "boiled off" the hot cathode surface. In the vacuum inside the CRT envelope, the free, negatively charged electrons are then accelerated toward the phosphor coating by a high positive voltage. The accelerating voltage can be generated with a positively charged metal coating on the inside of the CRT envelope near the phosphor screen, or an accelerating anode can be used, as in Fig. 2–5. Sometimes the electron gun is built to contain the accelerating anode and focusing system within the same unit.

Intensity of the electron beam is controlled by setting voltage levels on the control grid, which is a metal cylinder that fits over the cathode. A high negative voltage applied to the control grid will shut off the beam by repelling electrons and stopping them from passing through the small hole at the end of the control grid structure. A smaller negative voltage on the control grid simply decreases the number of electrons passing through. Since the amount of light emitted by the phosphor coating depends on the number of electrons striking the screen, we control the brightness of a display by varying the voltage on the control grid. A control knob is available on video monitors to set the brightness for the entire screen. We also can set the intensity level of individual positions on the screen using program commands.

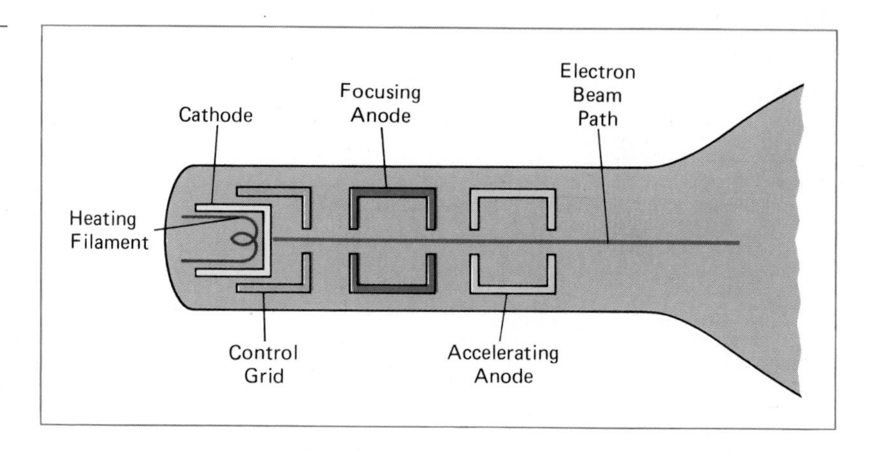

FIGURE 2–5
Operation of an electron gun with an accelerating anode.

The focusing system in a CRT is needed to force the electron beam to converge into a small spot as it strikes the phosphor. Otherwise, the electrons would repel each other, and the beam would spread out as it approaches the screen. Focusing is accomplished with either electric or magnetic fields. For electrostatic focusing, the electron beam passes through a metal cylinder with a positive voltage, as shown in Fig. 2–5. The positive voltage forces the electrons to stay along the axis of the beam. Similar focusing forces can be applied to the electron beam with electromagnetic fields set up by coils mounted around the outside of the CRT envelope.

Another type of focusing is used in high-precision systems to keep the beam in focus at all screen points. The distance that the electron beam must travel to different points on the screen varies because the radius of curvature for most CRTs is greater than the distance from the focusing system to the screen center. Therefore, the electron beam will be focused properly only at the center of the screen. As the beam moves to the outer edges of the screen, displayed images become blurred. To compensate for this, the system can adjust the focusing according to the screen position of the beam.

The maximum number of points that can be displayed without overlap on a CRT is referred to as the **resolution.** A more precise definition of resolution is the number of points per centimeter that can be plotted horizontally and vertically, although it is often simply stated as the total number of points in each direction. Resolution of a CRT is dependent on the type of phosphor used and the focusing and deflection systems. High-precision systems can display a maximum of about 4000 points in each direction, for a total of 16 million addressable screen points. Since a CRT monitor can be attached to different computer systems, the number of screen points that are utilized depends on the capabilities of the system to which it is attached.

An important property of video monitors is their **aspect ratio.** This number gives the ratio of vertical points to horizontal points necessary to produce equal-length lines in both directions on the screen. (Sometimes aspect ratio is stated in terms of the ratio of horizontal to vertical points.) An aspect ratio of 3/4 means that a vertical line plotted with three points has the same length as a horizontal line plotted with four points.

Random-Scan and Raster-Scan Monitors

Refresh CRTs can be operated either as random-scan or as raster-scan monitors. When operated as a **random-scan** display unit, a CRT has the electron beam directed only to the parts of the screen where a picture is to be drawn. Random-scan monitors draw a picture one line at a time and, for this reason, are also referred to as **vector** displays (or **stroke-writing** or **calligraphic** displays). The component lines of a picture can be drawn and refreshed by a random-scan system in any order specified (Fig. 2–6). A pen plotter operates in a similar way and is an example of a random-scan, hard-copy device.

Raster-scan video monitors shoot the electron beam over all parts of the screen, turning the beam intensity on and off to coincide with the picture definition. The picture is created on the screen as a set of points (Fig. 2–7), starting from the top of the screen. Definition for a picture is now stored as a set of intensity values for all the screen points, and these stored values are "painted" on the screen one row (scan line) at a time. The capability of a raster-scan system to store intensity information for each screen point makes it well suited for displaying shading

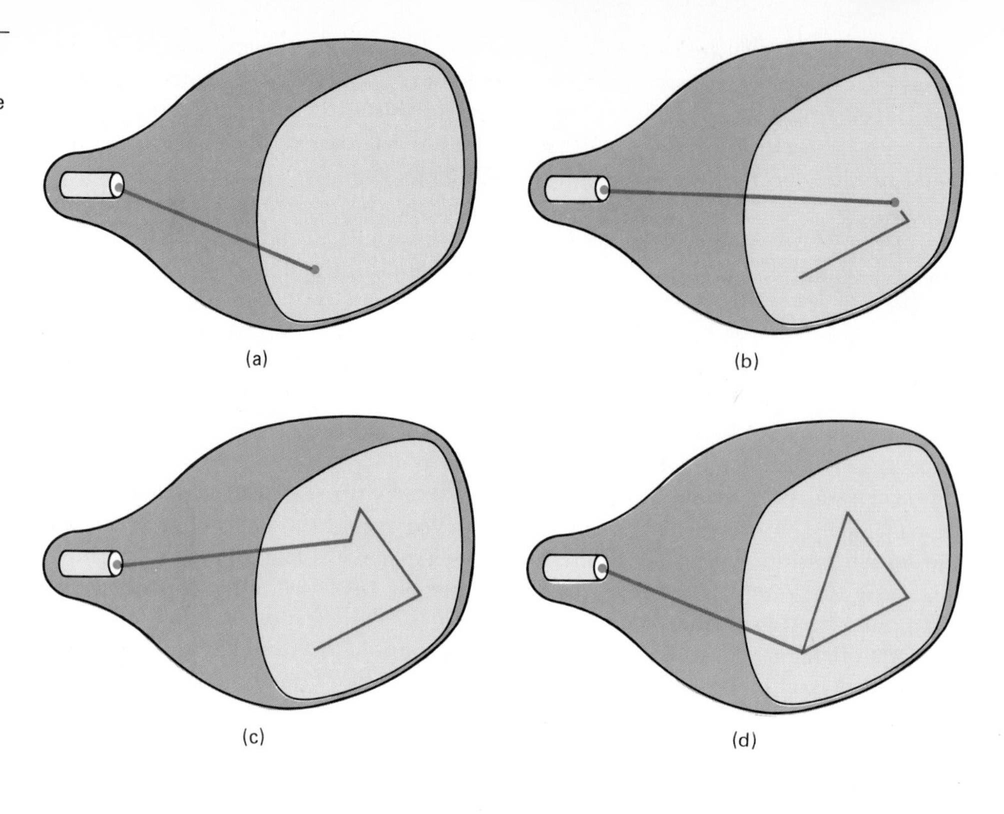

FIGURE 2–6
A random-scan system draws the component lines of a figure in any order specified.

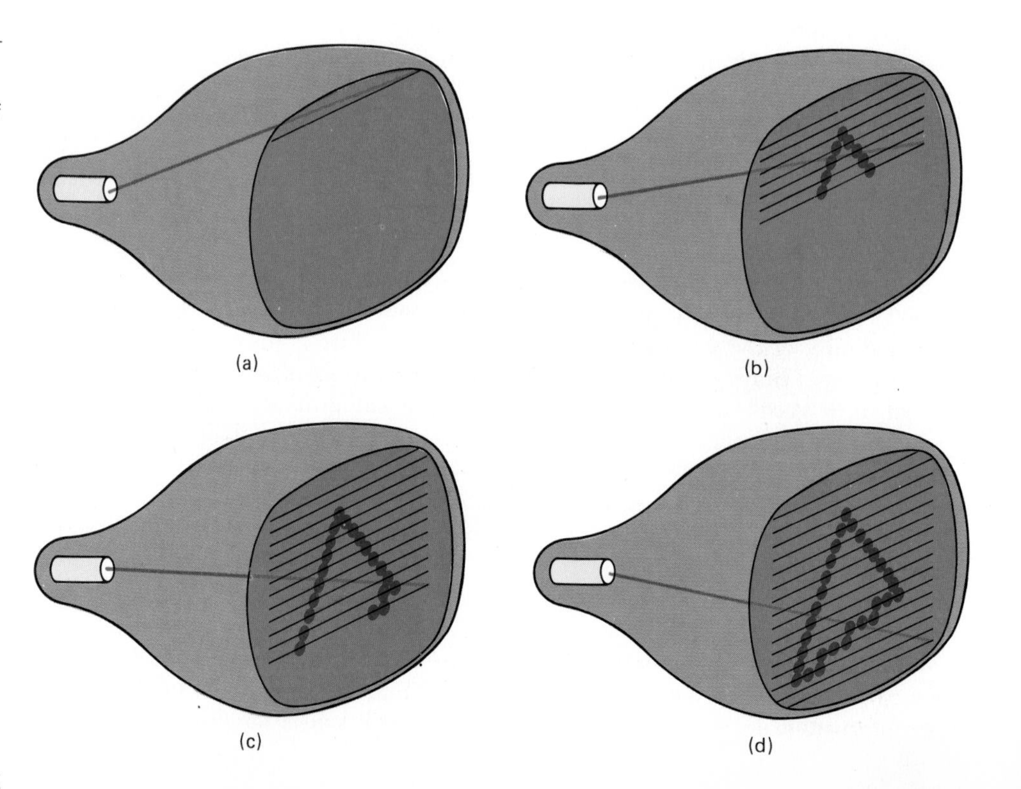

FIGURE 2–7
A raster-scan system displays the object of Fig. 2–6 as a set of points across each screen scan line.

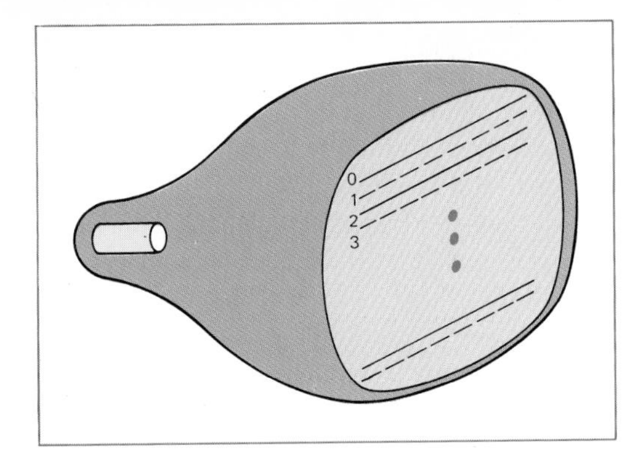

FIGURE 2–8
A raster-scan monitor can be set up to interlace scan lines: First, all points on the even-numbered (solid) lines are displayed; then all points along the odd-numbered (dashed) lines are displayed.

and color areas, whereas vector systems are restricted to line-drawing applications. Employing highly developed television technology, raster-scan monitors are now commonly used in the design of efficient and inexpensive computer graphics systems. Home television sets and printers are examples of other systems using raster-scan methods.

Often the refresh cycle in a raster-scan monitor (and in TV sets) is carried out by sweeping the beam across every other line on one pass from top to bottom, then returning (**vertical retrace**) to sweep across the remaining screen lines on the next pass down the screen (Fig. 2–8). This interlacing of the scan lines helps to reduce flicker at slower refresh rates. We essentially see the entire screen display in one-half the time it would have taken to sweep across all the lines at once from top to bottom.

Refreshing on raster-scan systems is done at rates of 30 to 60 frames per second. At the rate of 30 frames per second, the electron beam traces over all screen lines from top to bottom 30 times each second. A refresh rate below about 25 frames per second causes the picture to flicker. For systems with a large number of scan lines (1000 or more), higher refresh rates are usually needed. This is often accomplished by interlacing. In a 30-frames-per-second interlaced system (Fig. 2–8), each half of the set of screen lines is displayed in 1/60 second. This puts the overall refresh rate closer to 60 frames per second. This same refresh rate is accomplished with a 60-frames-per-second noninterlaced system employing faster processors.

Random-scan monitors are also often designed to draw the component lines of a picture 30 to 60 times each second, although some are designed to begin the refresh cycle immediately after all lines are drawn. Since random-scan systems are designed for line-drawing applications and do not store intensity values for all screen points, they generally have higher resolutions than raster systems.

Color CRT Monitors

A CRT monitor displays color pictures by using a combination of phosphors that emit different-colored light. By combining the emitted light from the different phosphors, a range of colors can be generated. The two basic techniques for producing color displays with a CRT are the beam-penetration method and the shadow-mask method.

The **beam-penetration** method for displaying color pictures has been used with random-scan monitors. Two layers of phosphor, usually red and green, are coated onto the screen, and the displayed color depends on how far the electron beam penetrates into the phosphor layers. A beam of slow electrons excites only the outer red layer. A beam of very fast electrons penetrates through the red layer and excites the inner green layer. At intermediate beam speeds, combinations of red and green light are emitted to show two additional colors, orange and yellow. The speed of the electrons, and hence the screen color at any point, is controlled by the beam acceleration voltage. Beam penetration has been an inexpensive way to produce color in random-scan monitors, but only four colors are possible, and the quality of pictures is not as good as with other methods.

Shadow-mask methods are commonly used in raster-scan systems (including color TV) since they produce a much wider range of colors than the beam-penetration method. A shadow-mask CRT has the screen coated with tiny triangular patterns, each containing three different closely spaced phosphor dots. One phosphor dot of each triangle emits a red light, another emits a green light, and the third emits a blue light. This type of CRT has three electron guns, one for each color dot, and a shadow-mask grid just behind the phosphor-coated screen (Fig. 2–9). The three electron beams are deflected and focused as a group onto the shadow mask, which contains a series of holes aligned with the phosphor-dot patterns. When the three beams pass through a hole in the shadow mask, they activate a dot triangle, which appears as a small color spot on the screen. The phosphor dots in the triangles are arranged so that each electron beam can activate only its corresponding color dot when it passes through the shadow mask.

Color variations in a shadow-mask CRT are obtained by combining different intensity levels of the three electron beams. By turning off the red and green guns, we get only the color coming from the blue phosphor. Other combinations of beam intensities produce a small light spot for each triangle, whose color depends on the amount of excitation of the red, green, and blue phosphors in the triangle. A white (or gray) area is the result of activating all three dots with equal intensity. Yellow is produced with the green and red dots only, magenta is produced with the blue and red dots, and cyan shows up when blue and green are activated equally. In low-cost systems, the electron beam can only be set to on or off, limiting displays to eight colors. More sophisticated systems can set intermediate intensity levels for the electron beams, allowing as many as several million different colors to be generated.

Personal computer systems with color graphics capabilities are often designed

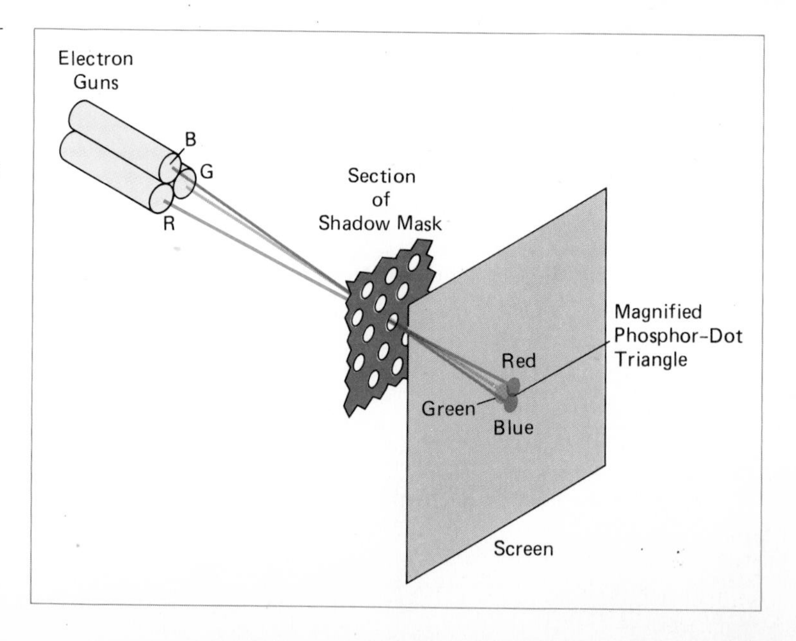

FIGURE 2–9
Operation of a shadow-mask CRT. Three electron guns, arranged to coincide with the color dot patterns on the screen, are directed to each dot triangle by a shadow mask.

to be used with several types of CRT display devices. These devices include TV sets, composite monitors, and RGB (red-green-blue) monitors. An inexpensive system typically makes use of a color TV and an RF (radio-frequency) modulator. The purpose of the RF modulator is to simulate the signal from a broadcast TV station. This means that the color and intensity information of the picture must be combined and superimposed on the broadcast-frequency carrier signal that the TV needs to have as input. Then the circuitry in the TV takes this signal from the RF modulator, extracts the picture information, and paints it on the screen. As we might expect, this extra handling of the picture information by the RF modulator and TV circuitry decreases the quality of displayed images.

Composite monitors are adaptations of TV sets that allow bypass of the broadcast circuitry. These display devices still require that the picture information be combined, but no carrier signal is needed. Since picture information is combined into a composite signal and then separated by the monitor, the resulting picture quality is still not the best attainable.

High-quality color CRTs are designed as **RGB monitors.** These monitors take the intensity level for each electron gun (red, green, and blue) directly from the computer system without any intermediate processing. In this way, fewer signal distortions are generated.

Direct-View Storage Tubes

An alternative method for maintaining a screen image is to store the picture information inside the CRT instead of refreshing the screen. A **direct-view storage tube (DVST)** stores the picture information as a charge distribution just behind the phosphor-coated screen. Two electron guns are used in a DVST. One, the primary gun, is used to store the picture pattern; the second, the flood gun, maintains the picture display.

A simplified cross section of a DVST is shown in Fig. 2–10. In this device, the primary electron gun is used to draw the picture definition on the storage grid, a nonconducting material. High-speed electrons from the primary gun strike the storage grid, knocking out electrons, which are attracted to the control grid. Since the storage grid is nonconducting, the areas where electrons have been removed will keep a net positive charge. This stored positive charge pattern on the storage grid is the picture definition. The flood gun produces a continuous stream of low-speed electrons that pass through the control grid and are attracted to the positive areas of the storage grid. These low-speed electrons penetrate through the storage grid to the phosphor coating, without appreciably affecting the charge pattern on the storage surface.

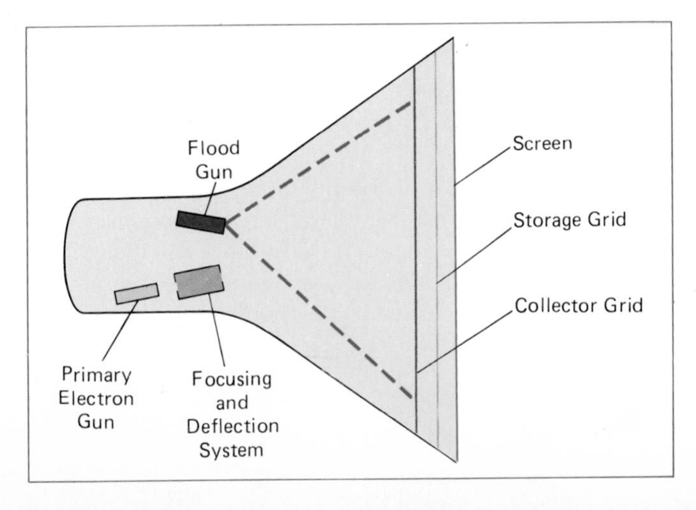

FIGURE 2–10
Operation of a DVST.

FIGURE 2–11
The Tektronix 4054 direct-view storage tube features a 19-inch screen and a resolution of 4096 by 3125. Courtesy Tektronix, Inc.

A DVST monitor has both disadvantages and advantages compared to the refresh CRT. Since no refreshing is needed, very complex pictures can be displayed clearly without flicker. As illustrated by the system in Fig. 2–11, very high resolution is possible, and the cost of DVST systems has been lower than many random-scan systems. Disadvantages of DVST systems are that they ordinarily do not display color and that selected parts of a picture cannot be erased. To eliminate a picture section, the entire screen must be erased by storing a positive charge over all parts of the storage grid. Then the flood gun electrons strike the phosphor coating at all points on the screen, erasing the picture in a flash of light. The entire picture is then redrawn, minus the parts to be omitted.

Plasma-Panel Displays

Although most graphics monitors contain CRTs, other technologies are used in some systems. **Plasma-panel displays** (Fig. 2–12) are constructed by filling the region between two glass plates with neon gas. A series of vertical and horizontal electrodes, placed on the front and rear glass panels, respectively, are used to light up individual points in the neon. One way to provide the individual points is to separate the gas into small "bulbs" with a center glass plate containing a number of closely spaced holes, as shown in Fig. 2–12.

An individual neon point in a plasma panel is turned on by applying a "firing voltage" of about 120 volts to the appropriate pair of horizontal and vertical electrodes. Once the point is turned on, the voltage on these electrodes is then lowered to a "sustaining voltage" level (about 90 volts) that keeps the neon cell glowing. A constant 90 volts can be applied across all electrode pairs, and those points that are to be lit have their electrode potentials momentarily raised to the firing voltage. This ensures that the points to be displayed continue to glow while all other points remain off. Thus the plasma panel is an inherent memory device that requires no refreshing. Erasing the screen is accomplished by lowering the voltage on each electrode below the sustaining voltage level.

The number of points that can be displayed by a plasma panel is limited, and its cost is somewhat higher than that of a refresh CRT. Advantages of plasma panels are that they are very rugged devices that require no refreshing. They also have flat screens and are transparent, so displayed images can be superimposed with pictures from slides or other media projected through the rear panel. Examples of plasma-panel systems are shown in Fig. 2–13.

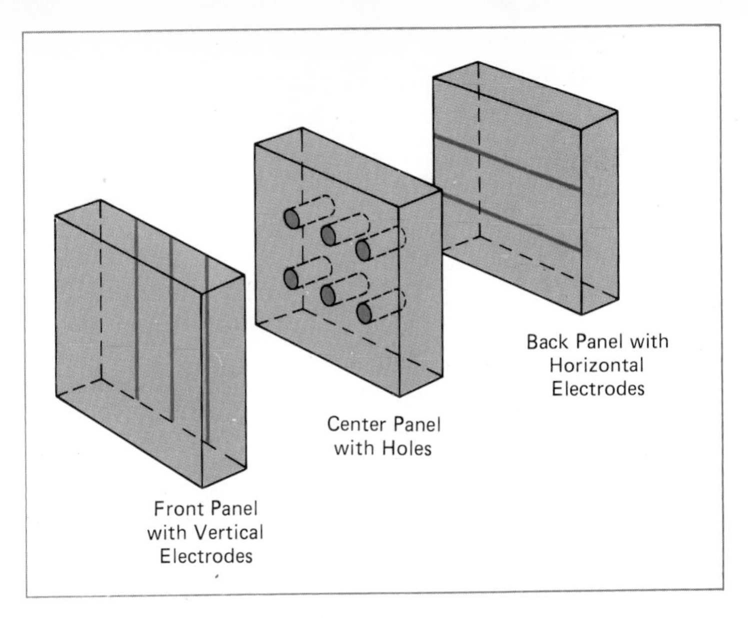

FIGURE 2–12
Plate construction in a plasma-panel display device.

FIGURE 2–13
Plasma-panel display systems: (a) large flat panel display with a resolution of about 1200 by 1600; (b) smaller 1024 by 768 portable terminal, showing graphics output superimposed on a map inserted behind the panels. Courtesy Magnavox Electronic Systems Co.

FIGURE 2–14
Liquid crystal display screens used with a "pocket" version of the Donkey Kong game. Each of the two small screens measures 5.5 cm by 3.5 cm.

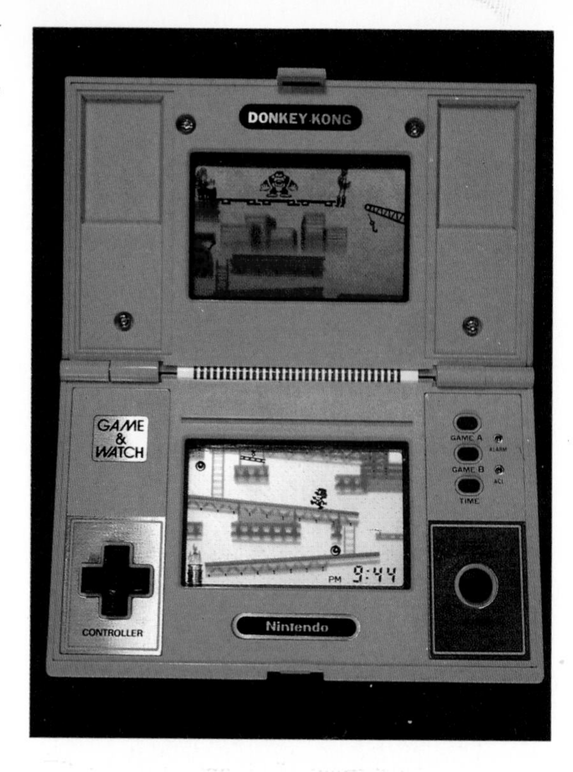

LED and LCD Monitors

Two other technologies used in the design of graphics monitors are light-emitting diodes (LEDs) and liquid-crystal displays (LCDs). These devices use light emitted from diodes or crystals instead of phosphors or neon gas to display a picture. LEDs and LCDs are particularly useful in the design of the miniscreens used with some graphics games. A "pocket" video game with two 1 1/4-inch by 2-inch screens is shown in Fig. 2–14. Screens for some wristwatch games are even smaller.

An auxiliary memory, similar to a frame buffer, is used to store the screen patterns for a display. The system repeatedly cycles through the memory area, turning on the appropriate LED or LCD positions in succession by applying a firing voltage to appropriate horizontal and vertical wire pairs. Each LED or LCD fired produces a very short pulse of light. However, the firing rate is rapid enough so that the picture is perceived as a set of steadily glowing points.

Laser Devices

An additional non-CRT technique for generating graphics output is to trace patterns on photochromic film, which is temporarily darkened by exposure to light. The patterns are formed with a laser beam, deflected by electromechanically controlled mirrors. Another light source is then used to project the images on a screen. A change in screen displays is obtained by winding the roll of film to the next blank frame and repeating the process. Highly complex patterns can be displayed in a very short time with these systems, but no selective erasure is possible. Changes to a picture can be made only by completely redrawing the patterns on the next film frame.

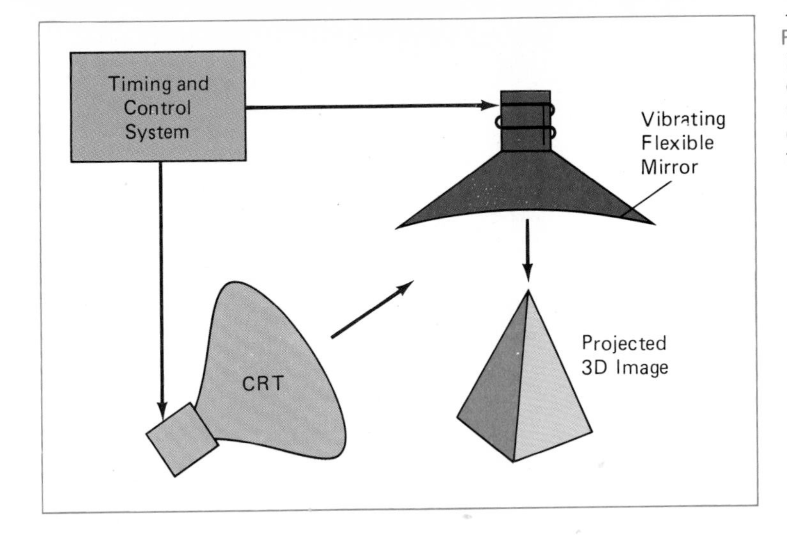

FIGURE 2–15
Basic operation of a three-dimensional display system using a vibrating mirror that changes focal length to match the depth of points in a scene.

Three-Dimensional Monitors

Graphics monitors for the display of three-dimensional scenes have been devised using a technique that reflects a CRT image from a vibrating, flexible mirror. The operation of such a system is demonstrated in Fig. 2–15. As the mirror (called a varifocal mirror) vibrates, it changes focal length. These vibrations are synchronized with the display of an object on a CRT so that each point on the object is reflected from the mirror at a position corresponding to the depth of that point. A viewer can see under, around, or over the top of the object. Figure 2–16 shows the Genisco SpaceGraph system, which uses a vibrating mirror to project three-

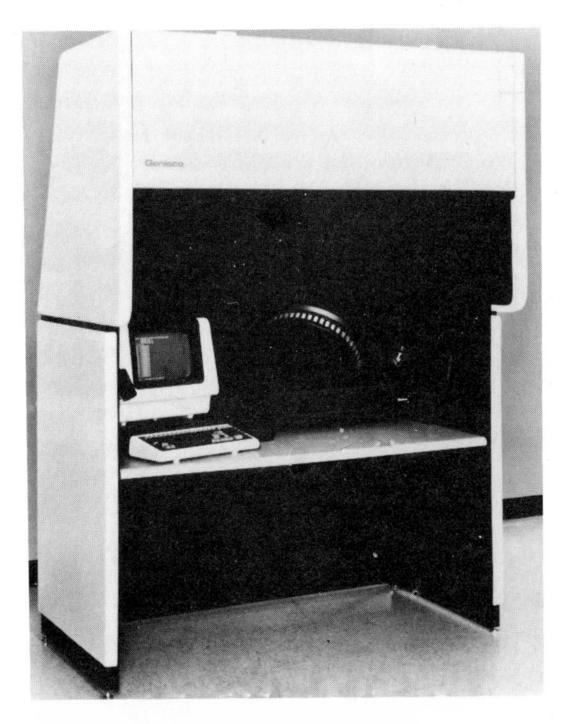

FIGURE 2–16
The SpaceGraph interactive graphics system displays objects in three dimensions using a vibrating, flexible mirror. Courtesy Genisco Computers Corp.

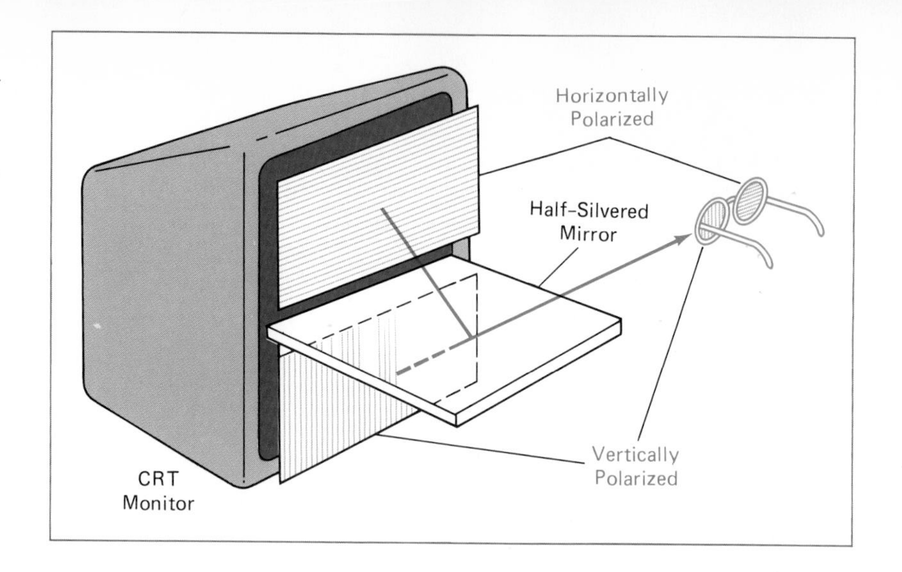

FIGURE 2–17
Stereoscopic viewing system, employing polarizing viewing filters and a half-silvered mirror to combine the images from two halves of a display screen.

dimensional objects into a 25cm by 25cm by 25cm volume. This system is also capable of displaying two-dimensional cross-sectional "slices" of objects selected at different depths. Such systems are used in medical applications to analyze data from ultrasonography and CAT scan devices, in geological applications to analyze topological and seismic data, in design applications involving three-dimensional objects, and in three-dimensional simulations of systems, such as molecules and terrain.

Another technique for representing three-dimensional objects is displaying stereoscopic views. This method does not produce true three-dimensional images, but it does provide a three-dimensional effect by presenting a different view to each eye of an observer. One way to do this is to display the two views of a scene on different halves of a CRT screen. Each view is projected onto the screen from a viewing direction that the corresponding eye (left or right) would normally see. A polarizing filter is placed over each half of the screen, and the user views the screen through a pair of polarized glasses (Fig. 2–17). Each lens of the glasses is polarized to admit the light from only one half of the screen, so the user sees a stereoscopic view of the three-dimensional scene. A stereoscopic effect can also be achieved by displaying each of the two views across the entire screen on alternate refresh cycles. The screen is again viewed through glasses, but now each lens is designed to act as a rapidly alternating shutter that is synchronized to block out one of the views. Electric fields can be used to cause the material in each lens to alternate between opaque and clear. Other types of shutter devices could also be used to alternate the two views.

Stereoscopic views have also been generated using a headset arrangement that contains two small video monitors and an optical system. One of the video monitors displays a view for the left eye, and the other displays a view for the right eye. A sensing system in the headset keeps track of the viewer's position, so that the front and back of objects can be seen as the viewer "walks through" the display.

2–2 Hard-Copy Devices

Many graphics systems are equipped to produce hard-copy output directly from a video monitor in the form of 35mm slides or overhead transparencies. Hard-copy pictures can also be obtained by directing graphics output to a printer or plotter.

Printers

Although originally designed for producing text pages, standard printers are acceptable graphics devices for applications not requiring high-quality output. They are generally accessible, so that even systems without special graphics capabilities can be used to generate graphics output on an attached printer. Some printers may be specifically adapted for graphics applications. Also, printers provide one of the least expensive devices for producing hard-copy pictures.

Printers produce output by either impact or nonimpact methods. **Impact printers** press formed character faces against an inked ribbon onto the paper. The familiar line printer is an example of an impact device, with the typefaces mounted on bands, chains, drums, or wheels. The daisy-wheel printer uses the impact method to print characters one at a time. A dot-matrix print head, containing a rectangular array of tiny pins, is often used in impact character printers to form individual characters by activating selected pin patterns. **Nonimpact printers** are faster and quieter and often use a dot-matrix method to print characters or draw lines. Ink-jet sprays, laser techniques, xerographic processes (as used in photocopying machines), electrostatic methods, and electrothermal methods are all used in the design of nonimpact printers.

Dot-matrix methods offer extended possibilities for graphics output. Besides printing preset character patterns, these printers can be adapted to print any dot combination selected by a graphics program. A printed page can now be thought of as a grid of closely spaced points, rather than simply as an array of character positions. Figure 2–18 is an example of the type of output that can be produced with a dot-matrix impact printer.

Ink-jet methods produce output by squirting ink in horizontal rows across a roll of paper wrapped on a drum. The ink stream, which is electrically charged, is deflected by an electric field to produce dot-matrix patterns. Electrostatic methods place a negative charge on a flat sheet of paper, one complete row at a time along the length of the paper. Then the paper is exposed to a black toner. The toner is positively charged and so is attracted to the negatively charged areas, where it adheres to produce the specified output. Electrothermal methods use heat in the dot-matrix print head to output patterns on heat-sensitive paper. A laser printer

FIGURE 2–18
A picture generated on a dot-matrix impact printer, which permits shading control by varying the density of the dot patterns. Courtesy Apple Computer, Inc.

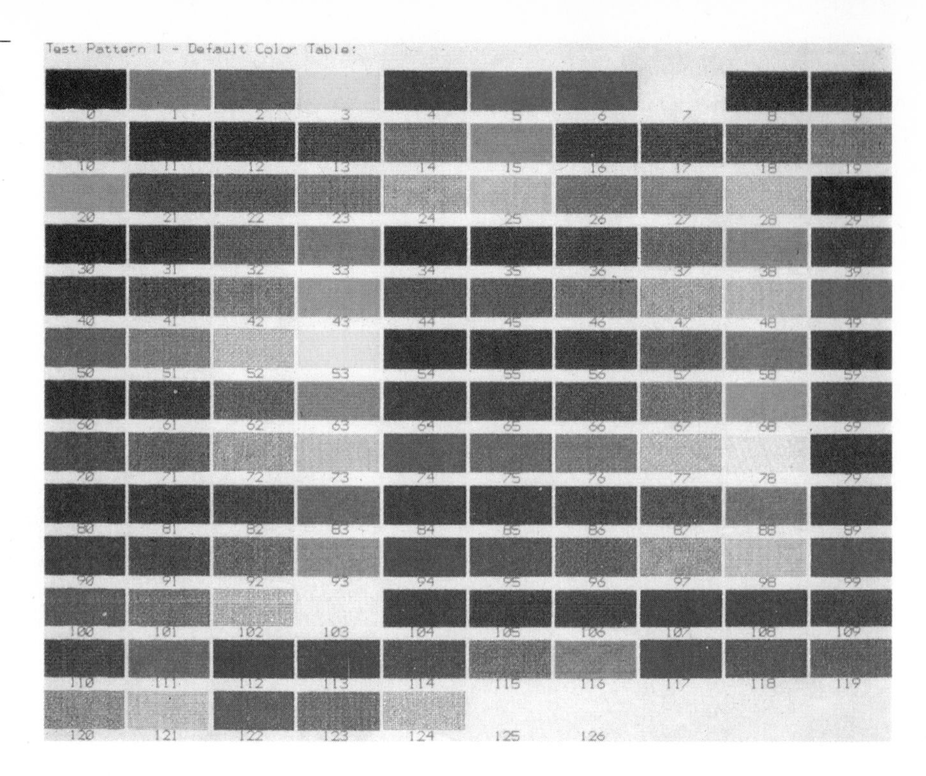

FIGURE 2–19
Color patterns displayed on an ink-jet printer. Courtesy PrintaColor Corp.

operates similarly to a Xerox copier. The laser beam creates a charge distribution on a drum coated with a photoelectric material, such as selenium. Toner is applied to the drum and then transferred to paper.

Both black-and-white and color printers are available. Different-colored ribbons have been used in impact printers to get color variations, but the nonimpact printers are better suited for color output. The nonimpact devices use various techniques to combine three color pigments (cyan, magenta, and yellow) to produce a range of color patterns. Laser and xerographic printers deposit the three pigments on separate passes; ink-jet printers shoot the three colors simultaneously on a single pass along each print line on the paper. Examples of color patterns produced with an ink-jet printer are given in Fig. 2–19.

Plotters

These devices produce hard-copy line drawings. The most commonly encountered plotters use ink pens to generate the drawings, but many plotting devices now employ laser beams, ink-jet sprays, and electrostatic methods. Unlike standard printers, plotters require additional software commands to direct the plotter output from an applications program.

Pen plotters normally use one or more ink pens mounted on a carriage, or crossbar, to draw lines on a sheet of paper. Wet ink, ball-point, and felt-tip pens are all possible choices for use with a pen plotter. The plotter paper is usually either stretched flat or rolled onto a drum or belt. Examples of **flatbed plotters** are shown in Fig. 2–20. The crossbar can move from one end of the plotter to the other, while the pen moves back and forth along the bar. Either clamps, a vacuum,

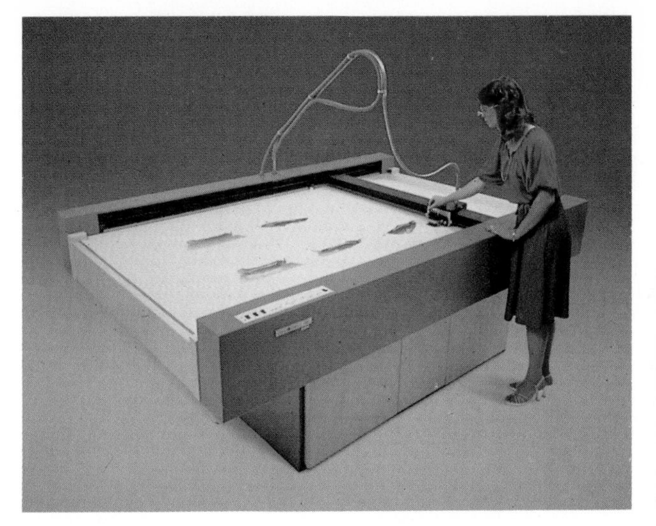

(a)

FIGURE 2–20
Flatbed plotters with ink pens mounted on a movable crossbar are available in a variety of sizes. (a) Courtesy CalComp Group; Sanders Associates, Inc. (b) Courtesy Tektronix, Inc.

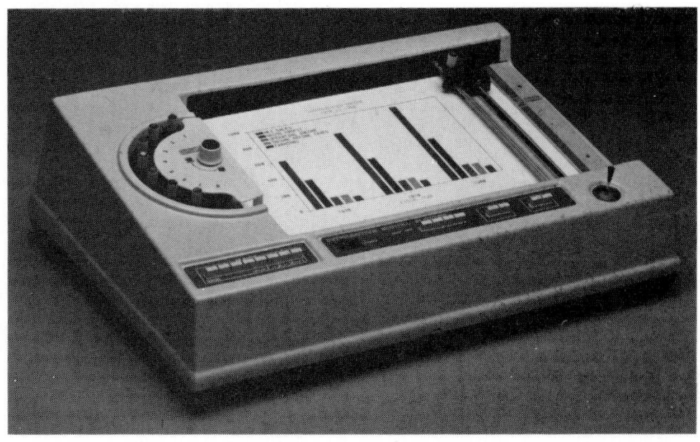

(b)

or an electrostatic charge hold the paper flat as the carriage and pen move across the paper. These plotters can vary in size from about 12 by 18 inches to over 6 by 10 feet. Figure 2–21 shows one type of **drum plotter.** Here the carriage is stationary, and the paper is made to move forward and backward on the drum as the pen slides across the carriage. Drum plotters are available in sizes that vary from about 1 foot to 3 feet in width. A cross between a flatbed and a drum is the **beltbed plotter** (Fig. 2–22). This plotter moves the paper on a wide, continuous belt over a flat surface as the pen carriage moves across the paper to produce the drawing.

Typical commands to a pen plotter from an application program include those for raising and lowering the pen and for moving the pen to a specified position. An electromagnet is used to raise and lower the pen, and servomotors move the pens and either the carriage or the drum. Some plotters allow pen movement in unit steps only, while others are capable of accepting commands for moves of more than one unit in various directions. Depending on the capabilities of a plotter, from 4 to 16 different directions can be chosen for the pen movement. Microprocessors are

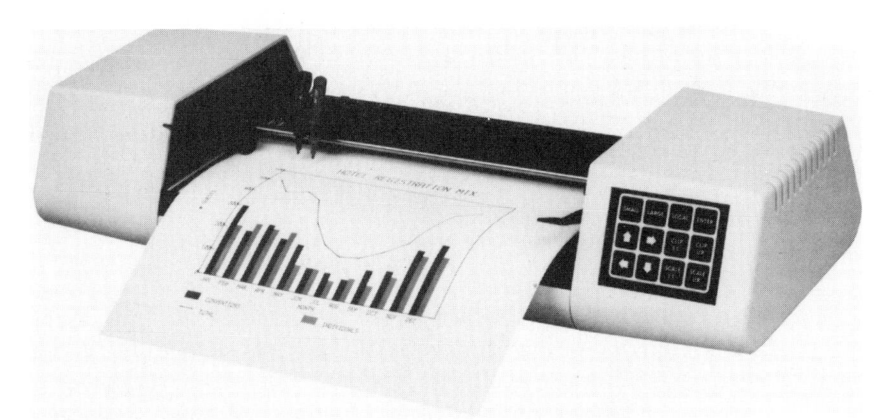

FIGURE 2–21
A small drum plotter, with ink pens mounted in a pen cartridge that can move laterally. The paper is attached to a rotating drum. Courtesy Houston Instrument.

FIGURE 2–22
A beltbed plotter. The paper is attached to a movable belt, and the pens move along the stationary crossbar. Courtesy CalComp Group; Sanders Associates, Inc.

often placed in plotters to allow automatic generation of common figures, such as lines, circles, and ellipses, and to output automatically different character patterns. Plotters with microprocessors are sometimes referred to as "vector" plotters.

Color drawings are produced with pen plotters by mounting different-colored pens on the carriage. Some plotters are designed to hold more than one pen on the carriage so that the different colors can be selected with program commands.

Plotters can also use pens of different widths to produce a variety of shadings and line styles. Sometimes plotters are equipped with a "pen stylus" to cut lines into a soft material or with a "pen light" to get the graphics output onto a photographic negative.

Line drawings can be produced by devices other than pen plotters. Printers employing ink-jet, laser, electrostatic, and other dot-matrix methods can be used interchangeably as either printers or plotters. These printer-plotter devices typically operate much faster than pen plotters, although the quality of the electrostatic devices is somewhat lower since the toner will spread slightly into uncharged areas. All of these methods require that the information for a drawing be stored in a buffer and delivered to the output device so that the drawing can be produced one row at a time from the top of the page to the bottom.

2–3 Interactive Input Devices

Graphics workstations, such as those shown in Figs. 2–23 and 2–24, typically include several types of input and output devices. Many systems, such as the one in Fig. 2–24, provide two monitors so that both high-resolution line drawings and raster color pictures can be displayed. In addition, a number of input devices are

FIGURE 2–23
A menu-driven artist's workstation, the Dicomed Imaginator, featuring a color raster monitor, a recessed keyboard, and a graphics tablet with hand cursor. Additional features are a light table for transparency analysis, disk drives, and a telephone with modem for voice and data communication. Courtesy DICOMED Corp.

FIGURE 2–24
A graphics workstation featuring both a vector and a raster display. Input devices include light pen and graphics tablet. Magnetic tape and hard disk units are also included. Courtesy Information Displays, Inc.

made available for interaction with the displays. A complete discussion of the operational features of input devices and the various interactive input methods appears in Chapter 8. Here, we briefly mention a few of the commonly used input devices and their primary function.

Keyboards are included on most graphics systems for the input of character strings and data values. Various types of keys, dials, and switches can be included on a keyboard to handle different applications. Some special-purpose keyboards contain only a set of dials or switches.

Coordinate values specifying screen positions are most often input with a graphics tablet, light pen, mouse, or a joystick. These devices typically are used to sketch pictures or make menu selections. A tablet inputs coordinate positions by activating a hand cursor or a stylus at selected positions on the tablet surface. A light pen pointed at a video monitor records coordinate positions by responding to the light emitted from phosphors on the screen. And a mouse or joystick is used to position the screen cursor at the coordinate locations to be selected.

Several other types of input devices are useful in graphics applications. They include touch panels, voice systems, and devices for entering three-dimensional coordinate information.

2–4 Display Processors

Interactive graphics systems typically employ two or more processing units. In addition to the central processing unit, or CPU, a special-purpose **display processor** is used to interact with the CPU and control the operation of the display device (Fig. 2–25). Stand-alone systems, such as microcomputers with graphics capabilities, contain both processors, while graphics terminals connecting to a host computer might only contain the display processor. Basically, the display processor is used to convert digital information from the CPU into corresponding voltage values needed by the display device. The manner in which this digital-to-analog conversion takes place depends on the type of display device used and the particular graphics functions that are to be hardware-implemented. In some systems, more than one processor is used to implement the graphics display functions.

Application programs for interactive graphics systems provide picture information to the display processor in terms of light-intensity levels for coordinate points on the screen. For many graphics monitors, the coordinate origin is defined at the lower left screen corner (Fig. 2–26). The screen surface is then represented as the first quadrant of a two-dimensional coordinate system, with positive x values increasing to the right and positive y values increasing from bottom to top. On some microcomputers, the coordinate origin is referenced at the upper left corner of the screen, so the y values are inverted. Graphics systems often allow applications programs to define picture points using any coordinate reference that is convenient for the user. A transformation is then performed by the system to convert user coordinates to screen values.

A fundamental task for the display processor is the display of line segments. Intensity levels (or color values) to be used in plotting coordinate positions along a line are supplied by the application program and converted to voltage levels, which

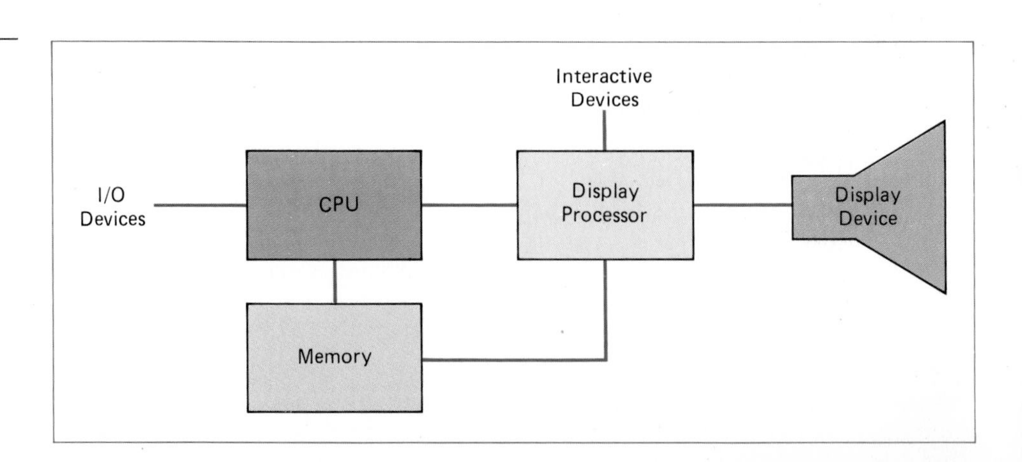

FIGURE 2–25
Simplified hardware diagram of an interactive graphics system.

are then applied to the display device. For simple black-and-white systems, no intensity information need be specified in the program, since any point is either on or off. Higher-quality systems allow the intensity of screen points to be varied so that shades of gray can be displayed.

Another typical function of a display processor is character generation. A standard character set is available on all systems, but some systems also allow user-generated character patterns to be stored and reproduced by the display processor.

Advanced display processors are designed to perform a number of additional operations. These functions include generating various line styles (dashed, dotted, or solid), displaying color areas, producing curved lines, and performing certain transformations and manipulations on displayed objects. Also, display processors are typically designed to interface with interactive input devices, such as a light pen.

With refresh CRT systems, the display processor may also be required to cycle through the picture definition, refreshing the screen often enough to eliminate flicker. The picture definition is kept in a refresh storage area. On many systems, this screen-refreshing task can be assigned to an additional processor, called the **display controller.** This allows the display processor to devote full time to the other functions.

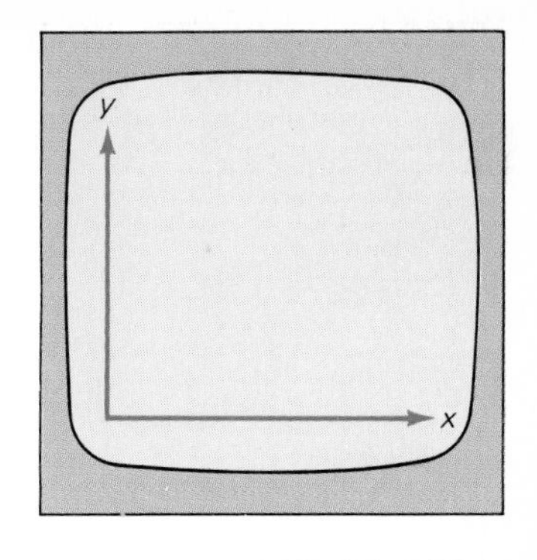

FIGURE 2–26
The origin of the coordinate system for identifying screen positions is usually specified in the lower left corner.

Random-Scan Systems

Figure 2–27 is a simplified diagram of the logical operations performed by a random-scan system. Graphics commands in an application program are translated into a **display file program,** which is accessed by the display processor to refresh the screen. The display processor cycles through each command in the display file program once during every refresh cycle.

When a display controller is included in a random-scan system, two files can be used. The translated graphics commands are first stored in a display file, as shown in Fig. 2–28. Then the display processor copies commands into a **refresh display file** for access by the display controller. This refresh display file is created by applying the viewing operations that select the particular view to be displayed on the screen. During the refresh process, which is now carried out by the display controller, the display processor may be updating the refresh file as interactive commands are input. These updates must be synchronized with the refreshing process so as not to distort the picture while it is in the process of being refreshed.

Graphics patterns are drawn on a random-scan system by directing the electron beam along the component lines of the picture. Lines are defined by the values for their coordinate endpoints, and these input coordinate values are converted to x and y deflection voltages. A scene is then drawn one line at a time by positioning the beam to fill in the line between specified endpoints.

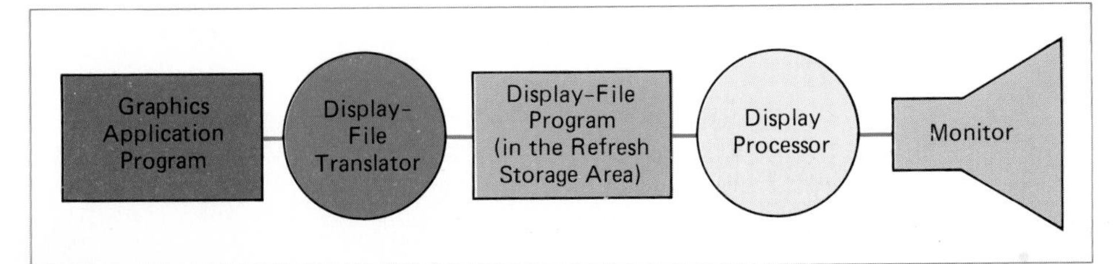

FIGURE 2–27
Simplified block diagram of the functions performed by a random-scan system.

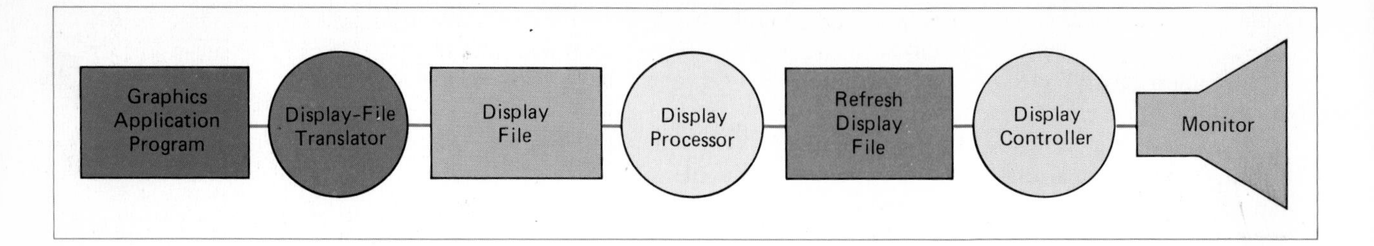

FIGURE 2–28
Block diagram of the functions performed by a random-scan system with a display controller.

Straight lines are drawn in a random-scan system with a **vector generator,** the hardware component of the display processor (or display controller), which produces deflection voltages for the electron beam. These deflection voltages can be generated in one of two ways. An **analog vector generator** sends the electron beam directly from one line endpoint to the other by linearly varying the deflection voltages. This produces a smooth straight line between the two points. A **digital vector generator** calculates successive points along the line, starting at one end, and converts these coordinate values to voltages. In this way, a straight line is constructed as a set of points. Although the digital method does not produce as smooth a line as the analog method, it is generally faster and less expensive. Line styles, such as a dashed line, are handled by turning the electron beam alternately on and off as the line is drawn.

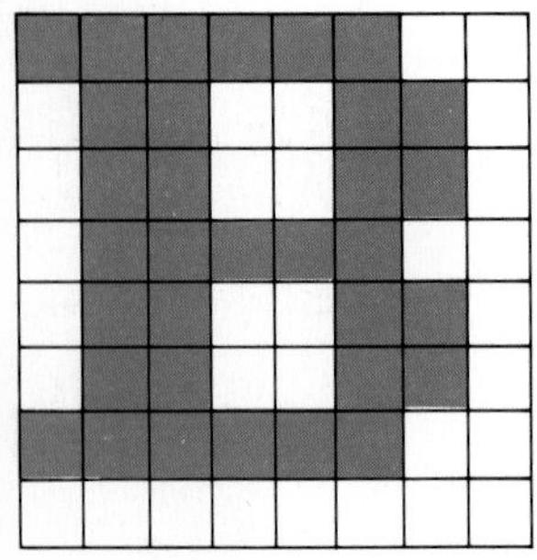

FIGURE 2–29
Rectangular grids can be used to define characters as a set of points.

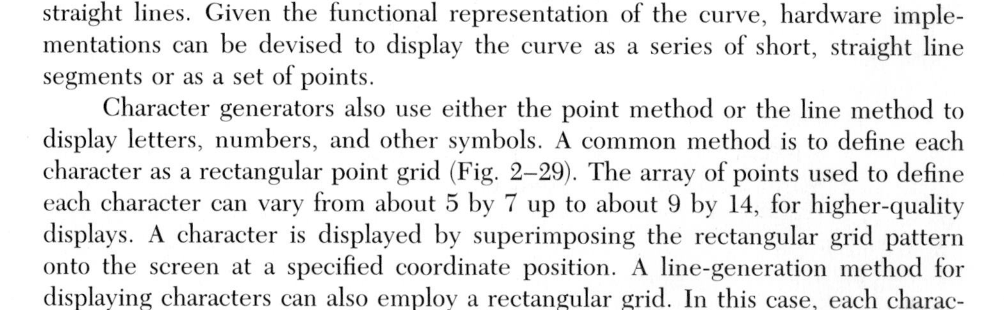

Curved lines can be generated by methods similar to those for drawing straight lines. Given the functional representation of the curve, hardware implementations can be devised to display the curve as a series of short, straight line segments or as a set of points.

Character generators also use either the point method or the line method to display letters, numbers, and other symbols. A common method is to define each character as a rectangular point grid (Fig. 2–29). The array of points used to define each character can vary from about 5 by 7 up to about 9 by 14, for higher-quality displays. A character is displayed by superimposing the rectangular grid pattern onto the screen at a specified coordinate position. A line-generation method for displaying characters can also employ a rectangular grid. In this case, each character is defined as a set of line segments within the grid rather than a point pattern.

DVST Systems

Display processors for a DVST can be designed to operate with or without a refresh storage area. A processor would employ a refresh area whenever a "write-through" capability is desired. This feature allows images to be superimposed on the stored picture as a refresh process. Without refresh storage, the display processor can generate lines and characters for direct storage with the same methods used in a random-scan system. Additional functions needed for a DVST display processor include commands for storing the picture information and for erasing the screen.

Raster-Scan Systems

The operation of a raster-scan system differs from that of a random-scan system in that the refresh storage area is used to store intensity information for each screen position, instead of graphics commands. For raster-scan systems, the refresh

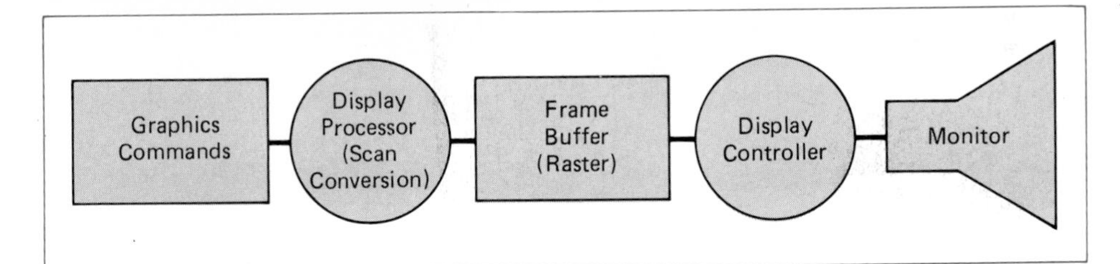

FIGURE 2–30
Simplified block diagram of a raster-scan system.

storage is usually referred to as the **frame buffer** or **refresh buffer** (Fig. 2–30). Other names for the frame buffer are **raster** and **bit map.** Each position in the frame buffer is called a **picture element,** or **pixel.** Pictures are painted on the screen from the frame buffer, one row at a time, from top to bottom. Each horizontal line of pixels is referred to as a **scan line,** and the process of generating pixel information into the frame buffer from the application program is called **scan conversion.** Intensity values are entered into the raster during the vertical-retrace time.

Pixel positions in the frame buffer are organized as a two-dimensional array of intensity values, corresponding to coordinate screen positions. The number of pixel positions in the raster is called the **resolution** of the display processor (or resolution of the frame buffer). Good-quality raster systems have a resolution of about 1024 by 1024, although higher-resolution systems are available. Since the resolution of a CRT monitor depends on the size of the phosphor dot that can be produced, a graphics system may have two resolutions, one for the monitor used with the system and one for the frame buffer. To generate the best-quality pictures, the resolution for the video monitor should be equal to or higher than the resolution of the frame buffer.

Graphics commands are translated by the scan-conversion process into intensity values for storage in the frame buffer. In a simple black-and-white system, each screen point is either on or off, so only one bit per pixel is needed to control the intensity of screen positions. A bit value of 1, for example, would mean that the electron beam is to be turned on at that position, while a value of 0 indicates that the beam intensity is to be off. Additional bits are needed when color and intensity variations can be displayed. Up to 24 or more bits per pixel are included in high-quality systems, although storage requirements for the frame buffer then get quite high. A system employing 24 bits per pixel with a screen resolution of 1024 by 1024 would require a frame buffer with 3 megabytes of storage.

Straight lines, curves, and characters are represented by the digital techniques discussed for random-scan systems. Each line is stored as a set of points, and characters are stored as dot-matrix patterns. A major advantage of the storage of intensity values in frame buffers is in the representation of areas that are to be filled with color or shading patterns. Once an area has been scan-converted, intensity values for all points within this area are stored in the raster for immediate use by the display controller.

In an effort to reduce memory requirements in random-scan systems, methods have been devised for organizing the raster as a linked list and encoding the intensity information. One way to do this is to store each scan line as a set of integer pairs. One number of each pair indicates an intensity value, and the second number specifies the number of adjacent pixels on the scan line that are to have that intensity. This technique, called **run-length encoding,** can result in a considerable saving in storage space if a picture is to be constructed mostly with long runs of a single color each. A similar approach can be taken when pixel intensities change linearly. Another approach is to encode the raster as a set of rectangular areas (**cell encoding**). The disadvantages of encoding runs are that intensity changes are difficult to make and storage requirements actually increase as the length of the runs decreases. In addition, it is difficult for the display controller to process the raster when many short runs are involved.

2–5 Graphics Software

Programming commands for displaying and manipulating graphics output are designed as extensions to existing languages. An example of such a graphics package is the PLOT 10 system developed by Tektronix, Inc., for use with FORTRAN on their graphics terminals. Basic functions available in a package designed for the graphics programmer include those for generating picture components (straight lines, polygons, circles, and other figures), setting color and intensity values, selecting views, and applying transformations. By contrast, application graphics packages designed for nonprogrammers are set up so that users can produce graphics without worrying about how they do it. The interface to the graphics routines in such packages allows users to communicate with the programs in their own terms. Examples of such applications packages are the artist's painting programs and various business, medical, and CAD systems.

Coordinate Representations

Most graphics packages are designed to use Cartesian coordinate systems. More than one Cartesian system may be referenced by a package, since different output devices can require different coordinate systems. In addition, packages usually allow picture definitions to be set up in any Cartesian reference system convenient to the application at hand. The coordinates referenced by a user are called **world coordinates,** and the coordinates used by a particular output device are called **device coordinates,** or **screen coordinates** in the case of a video monitor. World coordinate definitions allow a user to set any convenient dimensions without being hampered by the constraints of a particular output device. Architectural layouts might be specified in fractions of a foot, while other applications might define coordinate scales in terms of millimeters, kilometers, or light-years. Once the world coordinate definitions are given, the graphics system converts these coordinates to the appropriate device coordinates for display.

A typical procedure used in graphics packages is first to convert world coordinate definitions to **normalized device coordinates** before final conversion to specific device coordinates. This makes the system flexible enough to accommodate a number of output devices (Fig. 2–31). Normalized x and y coordinates are each assigned values in the interval from 0 to 1. These normalized coordinates are then transformed to device coordinates (integers) within the range (0, 0) to (x_{max}, y_{max}) for a particular device. To accommodate differences in scales and aspect ratios, normalized coordinates can be mapped into a square area of the output device so that proper proportions are maintained. On a video monitor, the remaining area of the screen is often used to display messages or list interactive program options.

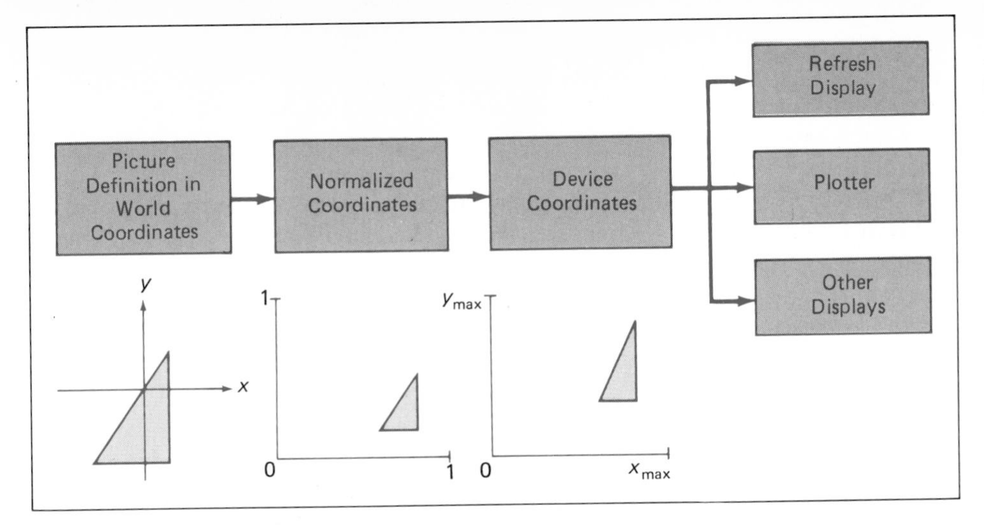

FIGURE 2–31
A transformation from a user's picture definition (world coordinates) to normalized coordinates is often performed by a graphics system in order to provide an interface to different types of output devices.

Graphics Functions

A general-purpose graphics package provides users with a variety of functions for creating and manipulating pictures. These routines can be categorized according to whether they deal with output, input, attributes, segment transformations, viewing, or general control.

The basic building blocks for pictures are referred to as **output primitives.** They include character strings and geometric entities, such as points, straight lines, polygons, and circles. Routines for generating output primitives provide the basic tools for constructing pictures.

Attributes are the properties of the output primitives. They include intensity and color specifications, line styles, text styles, and area-filling patterns. Functions within this category can be used to set attributes for groups of output primitives.

Given the primitive and attribute definition of a picture in world coordinates, a graphics package projects a selected view of the picture on an output device. **Viewing transformations** are used to specify the view that is to be presented and the portion of the output display area that is to be used.

Pictures can be subdivided into component parts, or **segments.** Each segment defines one logical unit of the picture. A scene with several objects could define the construction of each object in a separate named segment. Routines for processing segments carry out operations such as the creation, deletion, and transformation of segments.

Interactive graphics applications make use of various types of input devices, such as light pens, tablets, and joysticks. **Input operations** are used to control and process the data flow from these interactive devices.

Finally, a graphics package typically contains a number of housekeeping tasks, such as clearing a display screen or initializing parameters. We can lump the functions for carrying out these chores under the heading **control operations.**

Software Standards

The main goal of standardized graphics software is portability. When packages are designed with standard graphics functions, software can be moved easily to different types of hardware systems and used in different implementations and applications. Without standards, programs designed for one hardware system often cannot be transferred to another system without rewriting the software.

International and national standards-planning organizations in many countries have cooperated in an effort to develop a generally accepted standard for computer graphics. After considerable effort, this work on standards led to the development of the **Graphical Kernel System (GKS).** This system has been adopted as the graphics software standard by the International Standards Organization (ISO) and by various national standards organizations, such as the American National Standards Institute (ANSI). Although GKS was originally designed as a two-dimensional graphics package, a three-dimensional GKS extension was subsequently developed.

The final GKS functions, adopted as standards, were influenced by several earlier proposed graphics standards. Particularly important among these earlier proposals is the **Core Graphics System** (or simply **Core**), developed by the Graphics Standards Planning Committee of SIGGRAPH, the Special Interest Group on Computer Graphics of the Association for Computing Machinery (ACM).

Standard graphics functions are defined as a set of abstract specifications, independent of any programming language. To implement a graphics standard in a particular programming language, a **language binding** must be defined. This binding defines the syntax for accessing the various graphics functions specified within the standard. For example, GKS specifies a function to generate a sequence of connected straight line segments with the descriptive title

```
polyline (n, x, y)
```

In FORTRAN 77, this procedure is implemented as a subroutine with the name GPL. A graphics programmer, using FORTRAN, would invoke this procedure with the subroutine call statement

```
CALL GPL (N, X, Y)
```

GKS language bindings have been defined for FORTRAN, Pascal, Ada, C, PL/I, and COBOL. Each language binding is defined to make best use of the corresponding language capabilities and to handle various syntax issues, such as data types, parameter passing, and errors.

In the following chapters, we use the standard functions defined in GKS as a framework for discussing basic graphics concepts and the design and application of graphics packages. Example programs are presented in Pascal to illustrate the algorithms for implementation of the graphics functions and to illustrate also some applications of the functions. Descriptive names for functions, based on the GKS definitions, are used whenever a graphics function is referenced in a program.

Although GKS presents a specification for basic graphics functions, it does not provide a standard methodology for a graphics interface to output devices. Nor does it specify methods for real-time modeling or for storing and transmitting pictures. Separate standards have been developed for each of these three areas. Standardization for device interface methods is given in the **Computer Graphics Interface (CGI)** system. The **Computer Graphics Metafile (CGM)** system specifies standards for archiving and transporting pictures. And the **Programmer's Hierarchical Interactive Graphics Standard (PHIGS)** defines standard methods for real-time modeling and other higher-level programming capabilities not considered by GKS.

REFERENCES

A general treatment of display devices is available in Sherr (1979). The conceptual design of display devices is discussed in Haber and Wilkinson (1982) and in Myers (1984). Storage tubes are surveyed in Preiss (1978), and flat panel devices are discussed in Margolin (1985), Slottow (1976), and Tannas (1978). Woodsford (1976) describes the operation of a laser system.

Three-dimensional terminals are discussed in Ikedo (1984) and in Vickers (1970). Stereoscopic methods are presented in Roese and McCleary (1979).

Further information on graphics processors and system architecture is available in Carson (1983), Clark (1982), Foley and van Dam (1982), Levy (1984), Matherat (1978), and Niimi (1984).

For additional discussions of software standards see Graphics Standards Planning Committee (1977 and 1979), Hatfield and Herzog (1982), and Warner (1981).

Hopgood, et al. (1983) present an excellent introduction to the two-dimensional GKS graphics standard. A comparison of GKS and CORE concepts is given in Encarnaçao (1980). Other sources of information on GKS include Bono, et al. (1982), Enderle, Kansy, and Pfaff (1984), and Mehl and Noll (1984).

EXERCISES

2–1. List the relative advantages and disadvantages of the major display technologies for video monitors: vector refresh systems, raster refresh systems, DVST systems, and plasma panels.

2–2. For each of the various display technologies used in video monitors, list some applications in which that type of monitor might be appropriate.

2–3. Compare the advantages and disadvantages of a three-dimensional monitor using a varifocal mirror with conventional "flat" monitors. What techniques could be used by a two-dimensional monitor to provide three-dimensional views of objects?

2–4. Determine the resolution (pixels per centimeter) in the x and y directions for the video monitor in use on your system. Calculate the aspect ratio for this monitor, and explain how relative proportions of objects can be maintained on this system.

2–5. Consider a raster system with a resolution of 1024 by 1024. What is the size of the raster (in bytes) needed to store 4 bits per pixel? How much storage is required if 8 bits per pixel are to be stored?

2–6. For each of the rasters in Ex. 2–5, how long would it take to load the raster if 10^5 bytes can be transferred per second?

2–7. For each raster size in Ex. 2–5, how many pixels are accessed per second by a display controller that refreshes the screen at a rate of 30 frames per second? What is the access time per pixel?

2–8. Video monitor screen size is often specified by the length of the screen diagonal. A 19-inch screen is one with a diagonal length of 19 inches. What is the diameter of each point on a 19-inch screen that displays 1024 by 1024 pixels with equal resolution in each direction?

2–9. Refresh rates for video monitors are sometimes expressed as scan rates in units of Hertz (Hz), which specify the number of scan lines that can be displayed per second. (A television scan rate of 15.75 kHz means that a screen with 525 scan lines can be refreshed 30 times per second.) If the vertical retrace time of a system is eight percent of the total refresh cycle time, what scan rate in kHz would be required to refresh a 1024 by 1024 monitor 60 times each second?

2–10. Compare the functions performed by display processors in random-scan and in vector-scan systems. How is the erasure of a selected section of a screen accomplished in each of these systems?

2–11. Define the world coordinate system that would be appropriate for each of several application areas. For each type of application, state the units to be used for the coordinate axes and the range of coordinate values appropriate for describing an object or scene (group of objects).

2–12. Outline the procedures that would be necessary to convert normalized device coordinates to the device coordinates used by various types of output devices.

2–13. Illustrate the distinction between a graphics package designed for a programmer and one intended for a nonprogrammer by describing the structure of "commands" available to a user of each package.

2–14. What are some of the important considerations in designing a standard graphics package?

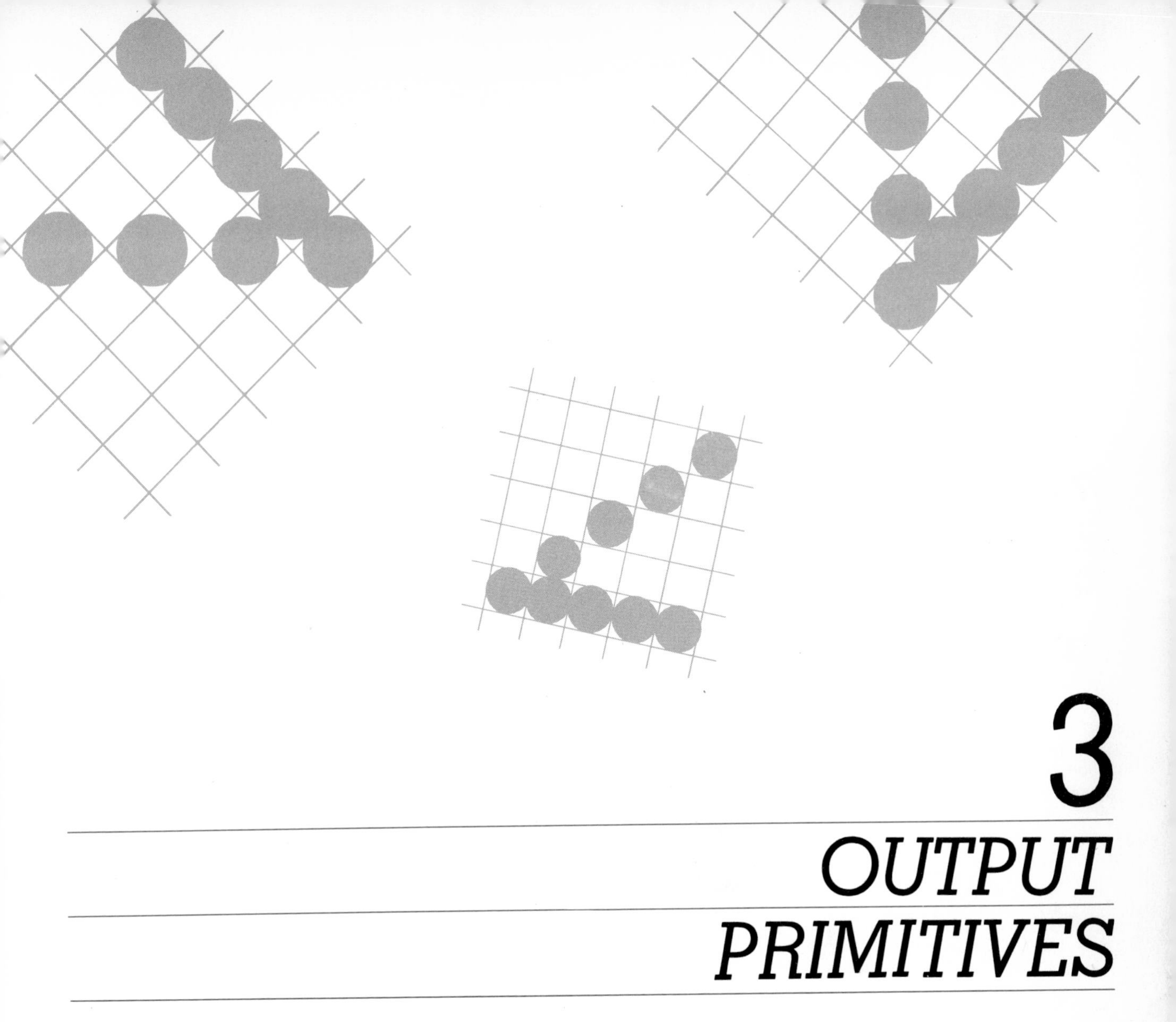

3 OUTPUT PRIMITIVES

Procedures that display output primitives direct an output device to produce specified geometric structures at designated locations. Such procedures take coordinate input and invoke display algorithms to construct a geometric shape on a selected output device. The simplest geometric components of a picture are points and lines. Other types of output primitives are polygon areas, curved figures, and character strings. For each type of output primitive, we consider the basic techniques and algorithms for displaying the primitive on different types of graphics systems, such as raster and vector systems. The major emphasis, however, is on methods appropriate to interactive graphics systems. Output primitive functions in GKS are examined, following the discussion of the display-generation algorithms.

3–1 Points and Lines

Point plotting is implemented in a graphics package by converting the coordinate information from an application program into appropriate instructions for the output device in use. With a CRT monitor, for example, the electron beam is turned on to illuminate a phosphor dot at the screen location specified. This is accomplished with a black-and-white raster display by setting the bit value at the specified coordinate position within the frame buffer to 1. Then, as the electron beam sweeps across each horizontal scan line, it emits a burst of electrons (plots a point) whenever a value of 1 is encountered in the frame buffer. For random-scan monitors, the point-plotting instruction is stored in the display file, and coordinates are converted to voltage deflections that move the electron beam to that position during each refresh cycle.

Line-drawing instructions in an application program define component lines of a picture by specifying endpoint coordinates for each line. The output device is directed to fill in the straight-line path between each pair of endpoints. For analog devices, such as a pen plotter or a random-scan display with an analog vector generator, a straight line is drawn smoothly from one endpoint to the other. Linearly varying horizontal and vertical deflection voltages are generated that are proportional to the required changes in the x and y directions to produce the smooth line.

Digital devices, such as a raster-scan display, produce a line by plotting pixels between the two endpoints. Pixel positions are computed from the equation of the line, and the appropriate bits are set in the frame buffer. Reading from the frame buffer, the display controller then activates corresponding positions on the screen. Since pixels are plotted at integer positions, the plotted line may only approximate actual line positions between the specified endpoints. For example, if position (10.33, 20.72) is calculated to be on the line, the pixel position (10, 21) is plotted. This rounding of coordinate values to integers causes lines to be displayed with a stairstep appearance ("the jaggies"), which can be quite noticeable on lower-resolution systems (Fig. 3–1). The appearance of raster lines can be improved by using high-resolution systems and also by applying techniques that have been specially developed for smoothing point-generated lines.

FIGURE 3–1
Stairstep effect ("jaggies") produced when a line is generated as a series of pixel positions.

3–2 Line-Drawing Algorithms

The equation for a straight line can be stated in the form

$$y = m \cdot x + b \tag{3–1}$$

with m as the slope of the line and b as the y intercept. Given that the two endpoints of a line segment are specified as (x_1, y_1) and (x_2, y_2), as shown in Fig. 3–2, we can determine values for the slope m and y intercept b with the following calculations:

$$m = \frac{y_2 - y_1}{x_2 - x_1} \tag{3–2}$$

$$b = y_1 - m \cdot x_1 \tag{3–3}$$

Algorithms for displaying straight lines are based on the line equation 3–1 and the calculations given in Eqs. 3–2 and 3–3.

FIGURE 3–2
Line segment, specified by coordinate endpoints (x_1, y_1) and (x_2, y_2).

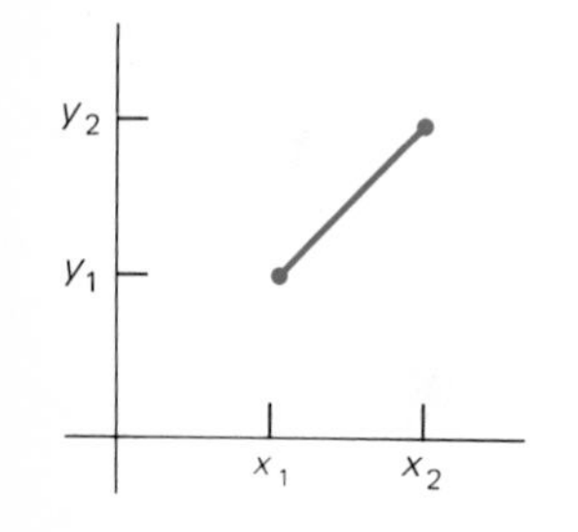

For any given x interval Δx along a line, we can compute the corresponding y interval Δy from Eq. 3–2 as

$$\Delta y = m \cdot \Delta x \tag{3–4}$$

This equation forms the basis for determining deflection voltages in analog devices. The change in the horizontal deflection voltage is set proportional to Δx, and the change in the vertical deflection voltage is set proportional to the value of Δy calculated from Eq. 3–4. These deflections are then used to generate a line with slope m between the specified endpoints.

DDA Algorithm

The digital differential analyzer (DDA) is an algorithm for calculating pixel positions along a line, using Eq. 3–4. This is accomplished by taking unit steps with one coordinate and calculating corresponding values for the other coordinate.

We first consider a line with positive slope, as shown in Fig. 3–2. If the slope is less than or equal to 1, we take the change in the x-coordinate values to be 1 and compute each succeeding y-coordinate value as

$$y_{i+1} = y_i + m \tag{3–5}$$

Subscript i takes integer values starting from 1, for the first point, and increases by 1 until the final endpoint is reached. Since m can be any real number, the calculated y values must be rounded to the nearest integer.

For lines with a positive slope greater than 1, we reverse the roles of x and y. That is, we move in unit y steps and calculate each succeeding x value as

$$x_{i+1} = x_i + \frac{1}{m} \tag{3–6}$$

Equations 3–5 and 3–6 assume that we are proceeding along the line from the left endpoint to the right endpoint (Fig. 3–2). If these endpoints are reversed, so that the starting endpoint is at the right, then either we have $\Delta x = -1$ and

$$y_{i+1} = y_i - m \tag{3–7}$$

or (when the slope is greater than 1) we have $\Delta y = -1$ with

$$x_{i+1} = x_i - \frac{1}{m} \tag{3–8}$$

Equations 3–5 through 3–8 can also be used to calculate points along a line with a negative slope. If the absolute value of the slope is less than 1 and the start endpoint is at the left, we set $\Delta x = 1$ and calculate y values with Eq. 3–5. When the start endpoint is at the right (for the same slope), we set $\Delta x = -1$ and obtain y positions from Eq. 3–7. Similarly, when the absolute value of a negative slope is greater than 1, we use $\Delta y = -1$ and Eq. 3–8 or we use $\Delta y = 1$ and Eq. 3–6.

This algorithm is summarized in the following procedure, which accepts as input the line endpoints ($x1$, $y1$) and ($x2$, $y2$). Differences in the input coordinate values in each direction are calculated as parameters dx and dy. The difference with the greater magnitude determines the value of parameter *steps*, which specifies the number of points to be plotted along the line. Starting at position ($x1$, $y1$), an amount is added to each coordinate to generate the next coordinate position. This is repeated *steps* times. If the magnitude of dx is greater than the magnitude of dy and $x1$ is less than $x2$, the values of the increments in the x and y direction are 1 and m, respectively. If the greater change is in the x direction, but $x1$ is greater than $x2$, then the values -1 and $-m$ are added to generate each new point on the line. Otherwise, we use a unit increment (or decrement) in the y direction and an x increment (or decrement) of $1/m$. We assume that points are to be plotted on a single-intensity system so that the command *set_pixel* is a call to the procedure for storing a pixel value of 1 ("on") in the frame buffer at the position specified by coordinate parameters x and y.

```
procedure dda (x1, y1, x2, y2 : integer);
  var
    dx, dy, steps, k : integer;
    x_increment, y_increment, x, y : real;
  begin
    dx := x2 - x1;
    dy := y2 - y1;
    if abs(dx) > abs(dy) then steps := abs(dx)
       else steps := abs(dy);
    x_increment := dx / steps;
    y_increment := dy / steps;
    x := x1; y := y1;
    set_pixel (round(x), round(y));
    for k := 1 to steps do begin
        x := x + x_increment;
        y := y + y_increment;
        set_pixel (round(x), round(y))
      end {for k}
  end; {dda}
```

The DDA algorithm is a faster method for calculating pixel positions than the direct use of Eq. 3–1. It eliminates the multiplication in Eq. 3–1 by taking advantage of raster characteristics, so that unit steps are taken in either the x or y direction to the next pixel location along the line. However, the calculations are slowed by the divisions needed to set increment values, the use of floating-point arithmetic, and the rounding operations.

Bresenham's Line Algorithm

A more efficient line algorithm to determine pixel positions, developed by Bresenham, finds the closest integer coordinates to the actual line path using only integer arithmetic. Figures 3–3 and 3–4 illustrate sections of a display screen where straight line segments are to be drawn. Pixel positions on the screen are represented by the rectangular areas between grid lines. In each of these examples, we need to decide between two pixel choices at each x position. Starting from the left endpoint of the line in Fig. 3–3, we need to determine whether the next point along the line is to be plotted at position (11, 10) or at (11, 11). Similarly, Fig. 3–4 shows a line path with a negative slope. Here, we need to decide between the points (51, 50) and (51, 49) as the next pixel position to be turned on. The next pixel to be plotted in each of these two examples is the one whose y value is closer to the actual y position on the line.

Again, we start with a line whose slope is positive and less than 1. Pixel positions along the line path can then be plotted by taking unit steps in the x

FIGURE 3–3
Section of a display screen where a straight line segment is to be displayed, starting from position (10, 10). Pixel positions are represented by the numbered rectangular areas.

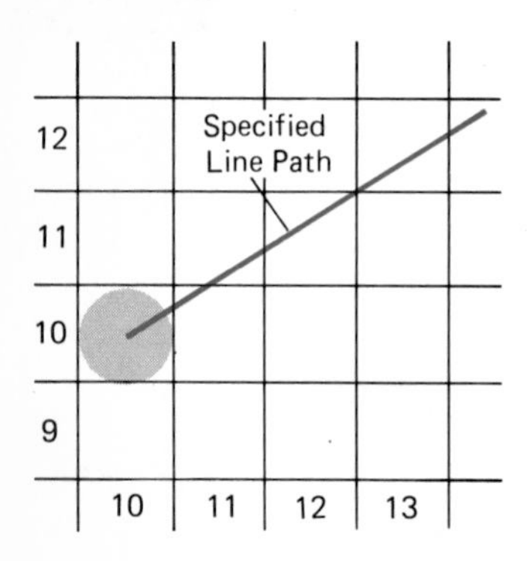

direction and determining the y-coordinate value of the nearest pixel to the line at each step. To establish the calculations needed in the algorithm, we consider the situation shown in Fig. 3–5. In this figure, we assume that pixel position (x_i, y_i) has been plotted and we now need to decide which is the next pixel to plot. The two choices for the next pixel position are at coordinates $(x_i + 1, y_i)$ and $(x_i + 1, y_i + 1)$.

In Fig. 3–6, the coordinate differences between the center of the two pixels and the line coordinate y are labeled d_1 and d_2. Position y can be calculated as

$$y = m(x_i + 1) + b$$

Then

$$\begin{aligned} d_1 &= y - y_i \\ &= m(x_i + 1) + b - y_i \end{aligned}$$

and

$$\begin{aligned} d_2 &= (y_i + 1) - y \\ &= y_i + 1 - m(x_i + 1) - b \end{aligned}$$

The difference between these two distances is

$$d_1 - d_2 = 2m(x_i + 1) - 2y_i + 2b - 1 \tag{3–9}$$

We now define a parameter that provides a measure of the relative distances of two pixels from the actual position on a given line. Substituting $m = \Delta y/\Delta x$, we can rewrite Eq. 3–9 so that it involves only integer arithmetic:

$$\begin{aligned} p_i &= \Delta x(d_1 - d_2) \\ &= 2\Delta y \cdot x_i - 2\Delta x \cdot y_i + c \end{aligned} \tag{3–10}$$

The constant c has the value $2\Delta y + \Delta x(2b - 1)$ and could be calculated once for all points, but we will see that Eq. 3–10 can be revised to eliminate this constant. The parameter p_i has a negative value if the pixel at position y_i is closer to the line than the upper pixel. In that case, we select the lower pixel; otherwise the upper pixel is chosen.

Equation 3–10 is simplified by relating parameters for successive x intervals. Then the value for each succeeding parameter is obtained from the previously calculated parameter. We can rewrite Eq. 3–10 in the form

$$p_{i+1} = 2\Delta y \cdot x_{i+1} - 2\Delta x \cdot y_{i+1} + c$$

Subtracting Eq. 3–10 from this expression, we have

$$p_{i+1} - p_i = 2\Delta y(x_{i+1} - x_i) - 2\Delta x(y_{i+1} - y_i)$$

But $x_{i+1} = x_i + 1$, so that

$$p_{i+1} = p_i + 2\Delta y - 2\Delta x(y_{i+1} - y_i) \tag{3–11}$$

This equation gives us a way to calculate the value of each successive parameter from the previous one. The first parameter, p_1, is obtained by evaluating Eq. 3–10 with (x_1, y_1) as the starting endpoint and $m = \Delta y/\Delta x$:

$$p_1 = 2\Delta y - \Delta x \tag{3–12}$$

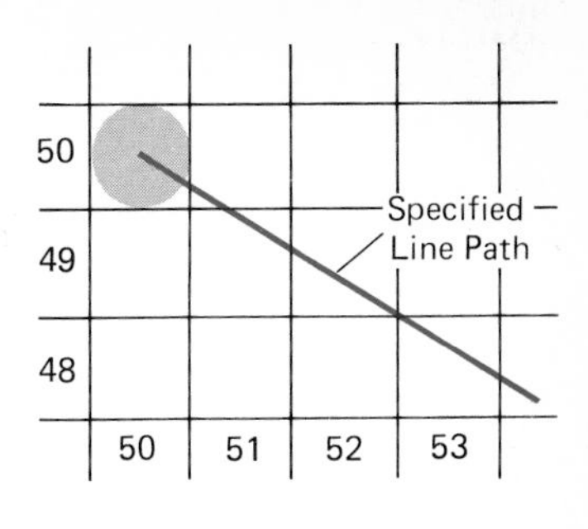

FIGURE 3–4
Section of a display screen where a line segment with negative slope is to be drawn. The left endpoint of the line is at position (50, 50).

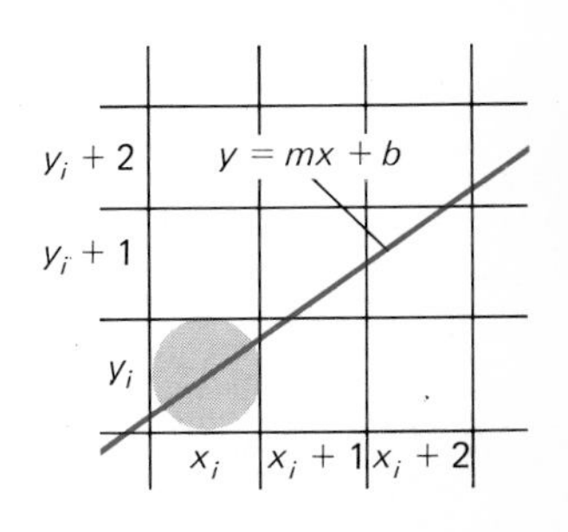

FIGURE 3–5
Section of the screen grid where a line passing through (x_i, y_i) is to be displayed.

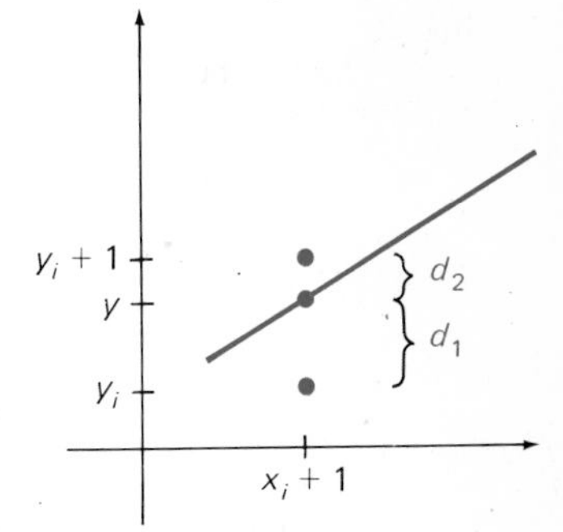

FIGURE 3–6
Coordinate differences between pixel centers and the y position on the line path at $x_i + 1$.

FIGURE 3–7
Bresenham's line algorithm.

1. Input line endpoints. Store left endpoint in (x_1, y_1). Store right endpoint in (x_2, y_2).
2. The first point to be selected for display is the left endpoint (x_1, y_1).
3. Calculate $\Delta x = x_2 - x_1$, $\Delta y = y_2 - y_1$, and $p_1 = 2\Delta y - \Delta x$. If $p_1 < 0$, the next point to be set is $(x_1 + 1, y_1)$. Otherwise, the next point is $(x_1 + 1, y_1 + 1)$.
4. Continue to increment the x coordinate by unit steps. At position $x_i + 1$, the coordinate to be selected, y_{i+1}, is either y_i or $y_i + 1$, depending on whether $p_i < 0$ or $p_i \geq 0$. The calculations for each parameter p depend on the last one. If $p_i < 0$, the form for the next parameter is
$$p_{i+1} = p_i + 2\Delta y$$
But if $p_i \geq 0$, the next parameter is
$$p_{i+1} = p_i + 2(\Delta y - \Delta x)$$
Then, if $p_{i+1} < 0$, the next y coordinate to be selected is y_{i+1}. Otherwise, select $y_{i+1} + 1$. (Coordinate y_{i+1} was determined to be either y_i or y_{i+1} by the parameter p_i in step 3.)
5. Repeat the procedures in step 4 until the x coordinate reaches x_2.

We summarize the steps for Bresenham's algorithm in Fig. 3–7, for a line with a positive slope less than 1. Since the constants $2\Delta y$, Δx, and $2(\Delta y - \Delta x)$ need be evaluated and stored only once, the arithmetic involves only integer addition and subtraction.

A procedure for implementing the algorithm of Fig. 3–7 is given in the following program. Endpoint coordinates for the line are input to this procedure through parameters $x1$, $y1$, $x2$, and $y2$. The call to *set_pixel* sets the position in the frame buffer for the point selected.

```
procedure bres_line (x1, y1, x2, y2 : integer);
  var
    dx, dy, x, y, x_end, p, const1, const2 : integer;
  begin
    dx := abs(x1 - x2);
    dy := abs(y1 - y2);
    p := 2 * dy - dx;
    const1 := 2 * dy;
    const2 := 2 * (dy - dx);
    {determine which point to use as start, which as end}
    if x1 > x2 then begin
        x := x2; y := y2;
        x_end := x1
      end {if x1 > x2}
    else begin
        x := x1; y := y1;
        x_end := x2
      end; {if x1 <= x2}
    set_pixel (x, y);
    while x < x_end do begin
        x := x + 1;
        if p < 0 then p := p + const1
```

we can change while_do statement to for-loop statement.

```
      else begin
        y := y + 1;
        p := p + const2
      end; {else begin}
      set_pixel (x, y)
    end {while x < x_end}
  end; {bres_line}
```

So far we have limited the discussion to lines with a positive slope between 0 and 1. We can extend the algorithm to positive slopes greater than 1 by interchanging the roles of the x and y coordinates. That is, we step along the y direction in unit steps and calculate successive x positions. For negative slopes, the procedures are similar, except that now one coordinate decreases as the other increases.

Loading the Frame Buffer

Whenever points and lines are to be displayed with a raster system, the frame buffer contents must be modified to contain appropriate intensity values for the specified coordinate positions. We have assumed that this is accomplished with the *set_pixel* procedure. This procedure converts coordinate values to corresponding addresses within the raster and stores intensity values at these positions in the frame buffer array.

As a specific example, suppose that the frame buffer array is addressed in row-major order for a display monitor with coordinate locations varying from (0, 0) at the lower left corner to (x_{max}, y_{max}) at the top right corner (Fig. 3–8). For a bilevel monitor (requiring one bit of storage per pixel), the bit address within the raster for a screen coordinate position (x, y) is calculated as

$$\text{ADDR}(x, y) = \text{ADDR}(0, 0) + y(x_{max} + 1) + x \tag{3–13}$$

This calculation is implemented in the *set_pixel* procedure when a single point is to be plotted and when the intensity value for the starting endpoint of a line is to be set.

For a line-drawing algorithm, we can simplify the calculations in Eq. 3–13 for intermediate points along the line by taking advantage of the fact that we are making unit steps in the x and y directions. The calculation of addresses within the frame buffer for these points can be carried out by incrementing previously calcu-

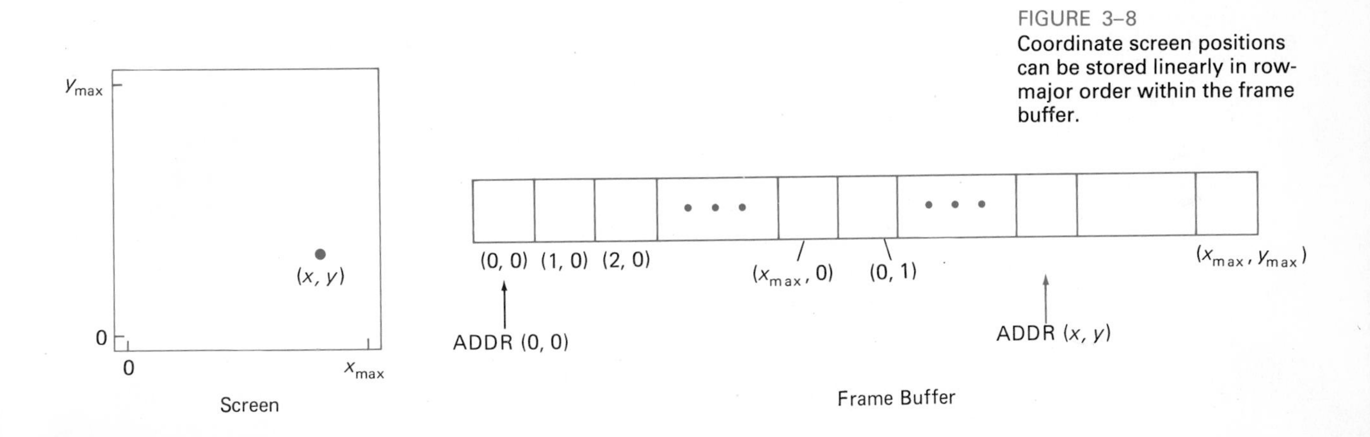

FIGURE 3–8
Coordinate screen positions can be stored linearly in row-major order within the frame buffer.

lated addresses. For example, from any position (x, y), the next address to be loaded in the raster (for a line with positive slope less than 1) can be calculated as one of the following two possibilities:

$$\mathrm{ADDR}(x + 1, y) = \mathrm{ADDR}(x, y) + 1 \tag{3–14}$$

$$\mathrm{ADDR}(x + 1, y + 1) = \mathrm{ADDR}(x, y) + x_{max} + 2 \tag{3–15}$$

Similar recursive calculations for other coordinate changes can be obtained from Eq. 3–13. These relations provide an efficient method for addressing since the calculations involve only integer addition.

Methods for storing the frame buffer and implementing the *set_pixel* procedure depend on the capabilities of a particular system and the design requirements of the software package. With systems that make use of color or a range of intensity values, an additional input parameter for *set_pixel* would specify the intensity value to be stored in the frame buffer.

3–3 Antialiasing Lines

FIGURE 3–9
Representing a line as a rectangle of finite width on a pixel grid. The left endpoint of the line is at position (10, 10).

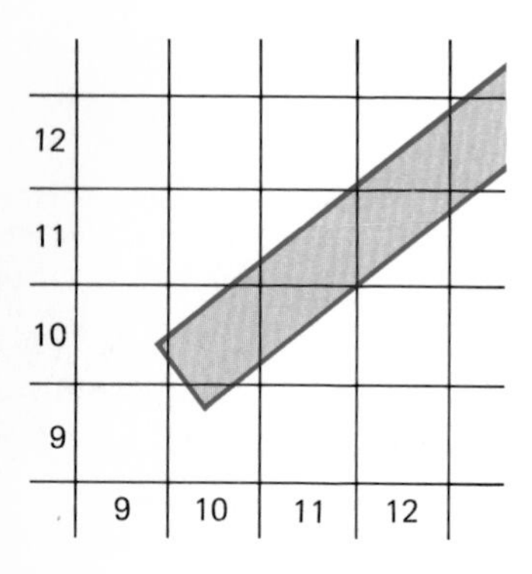

FIGURE 3–10
Unequal length lines displayed with the same number of pixels in each line.

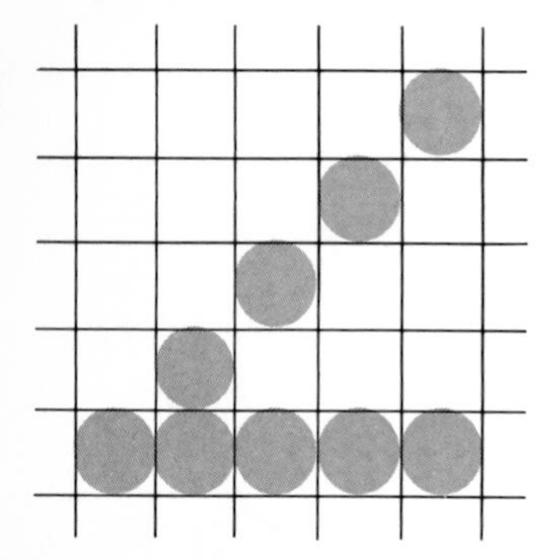

The raster algorithms we have discussed so far generate lines that have a jagged, or stairstep, appearance. Higher resolutions can be used to improve the appearance of lines, but this requires increased refresh buffer sizes, and this approach does not completely remove the stairstep effect. Object representations mapped onto a raster are subject to distortions because of **aliasing.** The digitization process rounds coordinate points on the object to discrete integer pixel positions in the raster. We can modify line-drawing algorithms to compensate for this raster effect by adding **antialiasing** routines that smooth out the display of a line on a video monitor. Antialiasing techniques remove the stairstep appearance by adjusting pixel intensities along the line path.

One method for developing an antialiasing routine is based on **sampling theory.** The idea behind this method is that natural geometric entities, such as points and lines plotted on a display screen, have finite dimensions. A pixel is not an infinitesimal mathematical point but a spot of light covering a small area of the screen. And lines have a width approximately equal to that of a pixel. When natural objects are digitized onto a rectangular grid (raster), the grid areas are "sampled" to determine appropriate light-intensity values. Raster systems that can display more than two intensity levels can use this method to adjust pixels so that each grid area is assigned the proper intensity.

Figure 3–9 shows a line represented with finite width on a pixel grid. Pixel areas are assumed to be square, and the width of the line is set equal to the width of a pixel. Instead of plotting the line with a single pixel at each x position, all pixels that are overlapped by the line area are displayed with an intensity proportional to the area of overlap. In the example shown, pixels at positions (11, 10) and (11, 11) are about half covered by the line. So each of these pixels is set to an intensity level of approximately 50 percent of the maximum. Similarly, the pixel at location (9, 10) is set to an intensity of about 10 percent of maximum. Although this method for antialiasing can improve the appearance of lines, the calculations are time-consuming.

Adjusting pixel intensities along the length of a line also compensates for another raster effect, illustrated in Fig. 3–10. Both lines are plotted with the same number of pixels, yet the diagonal line is longer than the horizontal line by a factor

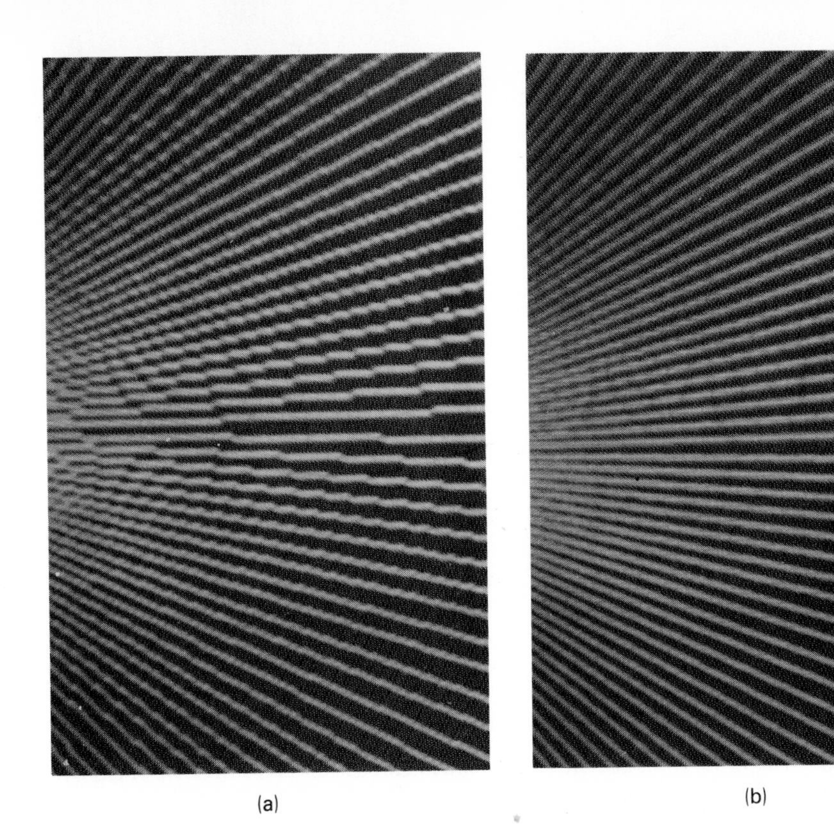

FIGURE 3–11
Jagged lines (a), plotted on the Merlin 9200 system, are smoothed (b) with an antialiasing technique called pixel phasing. This technique increases the number of addressable points on the system from 768 × 576 to 3072 × 2304. Courtesy Megatek Corp.

of $\sqrt{2}$. The visual effect of this is that the diagonal line appears less bright than the horizontal line, since the diagonal line is displayed with a lower intensity per unit length. A line-drawing algorithm could be adapted to compensate for this effect by adjusting the intensity of each line according to its slope. Horizontal and vertical lines would be displayed with the lowest intensity, while 45° lines would be given the highest intensity. However, if antialiasing techniques are applied to a display, intensities are automatically compensated. When the finite width of lines is taken into account, pixel intensities are adjusted so that lines display a total intensity proportional to their length.

Another technique for antialiasing lines is the **pixel phasing** approach developed by the Megatek Corporation. Intensities of line edges are adjusted by "micropositioning" of the electron beam. The systems incorporating this technique are designed so that individual pixel positions can be shifted by a fraction of a pixel diameter. By allowing shifts of 1/4, 1/2, and 3/4 of a pixel diameter, a line can be displayed by plotting points closer to the true path of the line. These systems also allow the size of individual pixels to be altered. Figure 3–11 illustrates antialiasing of lines using the pixel phasing technique.

3–4 Line Command

Graphics packages can be designed to include one command for basic line drawing and another for plotting individual points. However, the usefulness of a separate command to plot a single point is limited. Most graphics applications involve the construction of figures with straight line segments. Moreover, a single command can serve both purposes, since a point can be considered as a very short line segment.

A command for plotting both points and lines can be defined in the form:

```
polyline (n, x, y)
```

For our purposes, we will assume that the *polyline* function is to be provided as the basic output primitive command in a graphics package. This command is used to specify a single point, a single straight line segment, or a series of connected line segments, depending on the value assigned to parameter n. Coordinate values for the line endpoints (or single point) are stored in arrays x and y.

Implementation of the *polyline* function is through the *set_pixel* and line-drawing procedures, such as the Bresenham algorithm. When a single point is to be plotted, the *set_pixel* routine is invoked. When a line, or series of lines, is to be drawn, *polyline* causes the line-drawing procedure to be executed.

For point generation, a user would set the value of n at 1 and give the (x, y) coordinate values in $x[1]$ and $y[1]$. As an example, the statements

```
x[1] := 150;
y[1] := 100;
polyline (1, x, y);
```

specify that a single point is to be plotted at coordinate position (150, 100). We assume that coordinate references in the *polyline* command are stated as **absolute coordinate** values. This means that the values specified are the actual point positions in the coordinate system in use.

Some graphics systems employ line (and point) commands with **relative coordinate** specifications. In this case, coordinate values are stated as offsets from the last position referenced (called the **current position**). For example, if location (3, 2) is the last position that has been referenced in an application program, a relative coordinate specification of (2, −1) corresponds to an absolute position of (5, 1).

A straight line segment is obtained with a command such as *polyline* by setting n to 2 and assigning values for the two coordinate endpoints of the line to the x and y arrays. The following example program segment specifies a line with endpoints at (50, 100) and (250, 25).

```
x[1] := 50;
y[1] := 100;
x[2] := 250;
y[2] := 25;
polyline (2, x, y);
```

With graphics packages employing the concept of current position, a user need only give one coordinate point in a line-drawing statement. This signals the system to display a line from the current position to the given coordinates. The current position is then updated to the coordinate location stated in the line command. A series of connected lines is produced with such packages by a sequence of line commands, one for each line to be drawn.

Since any number of points can be specified with the *polyline* function, a user has the capability to generate a sequence of connected line segments with one statement. This is done by setting n equal to the number of line endpoints and storing the endpoint coordinates in arrays x and y. The graphics package then displays a series of $n - 1$ line segments that connect the n adjacent coordinates from position $(x[1], y[1])$ to position $(x[n], y[n])$.

We can implement the *polyline* command for $n > 2$ with a procedure that makes repeated calls to the line-drawing algorithm. Each successive call to the line-drawing algorithm passes the coordinate pair needed to plot the next line segment. The line-drawing algorithm is accessed by this procedure a total of $n - 1$ times.

3–5 Fill Areas

When shading or color patterns are to be applied to areas of a scene or graph, it is convenient for a user to be able to specify the area that is to be filled. Although fill patterns could be applied to the interiors of polygon borders defined with a line command, processing is simplified if a separate procedure is used to define a fill area. This approach allows a designated area to be immediately flagged as one that is to be displayed with a specified interior.

We introduce the following command to define a polygon fill area:

```
fill_area (n, x, y)
```

The area to be filled is inside the boundary defined by the series of n connected line segments from $(x[1], y[1])$ to $(x[n], y[n])$ and back to $(x[1], y[1])$.

Implementation of the *fill_area* command in a graphics package depends on the type of fill that is to be used to display the area. A user might want the area left blank or filled with a solid color or some pattern. When the interior of the area is left blank, *fill_area* simply produces the boundary outline. This is analogous to using the *polyline* procedure with the starting and ending coordinates set to the same values.

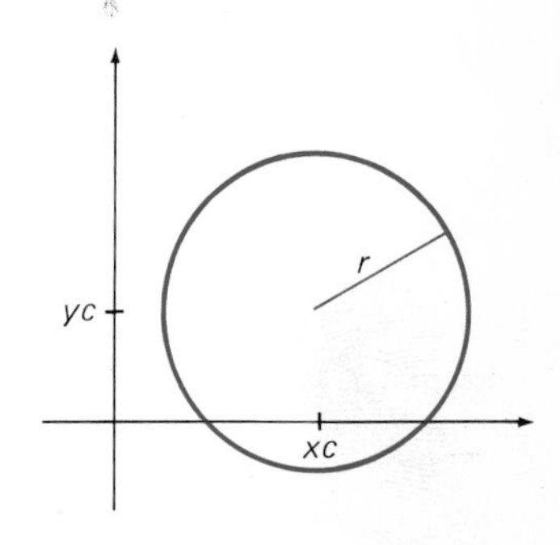

FIGURE 3–12
Circle with center coordinates *(xc, yc)* and radius *r*.

3–6 Circle-Generating Algorithms

Since the circle is a common component in many types of pictures and graphs, procedures for generating circles (and ellipses) are often included in graphics packages. The basic parameters that define a circle are the center coordinates (xc, yc) and the radius r (Fig. 3–12). We can express the equation of a circle in several forms, using either Cartesian or polar coordinate parameters. Figure 3–13 shows the relationship between Cartesian and polar parameters.

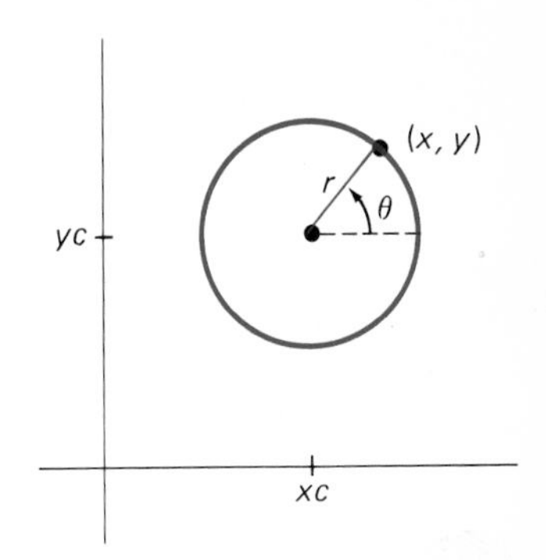

FIGURE 3–13
Relationship between Cartesian and polar coordinates.

Circle Equations

A standard form for the circle equation is the Pythagorean theorem:

$$(x - xc)^2 + (y - yc)^2 = r^2 \tag{3–16}$$

This equation could be used to draw a circle by stepping along the x axis in unit steps from $xc - r$ to $xc + r$ and calculating the corresponding y values at each position as

$$y = yc \pm \sqrt{r^2 - (x - xc)^2} \tag{3–17}$$

Obviously, this approach involves considerable computation at each step, and the

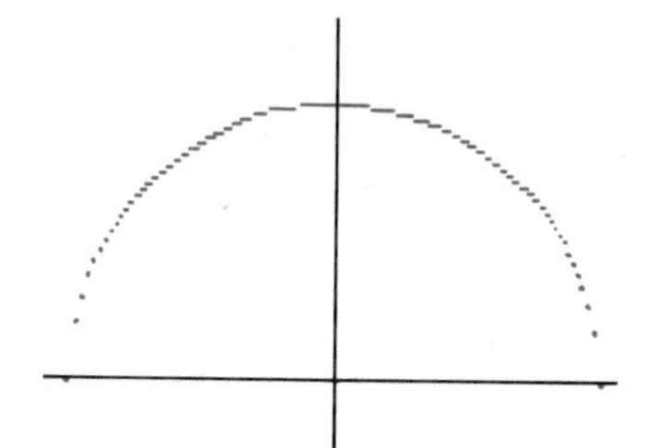

FIGURE 3–14
Positive half of a circle plotted with Eq. 3–17 and $(xc, yc) = (0, 0)$.

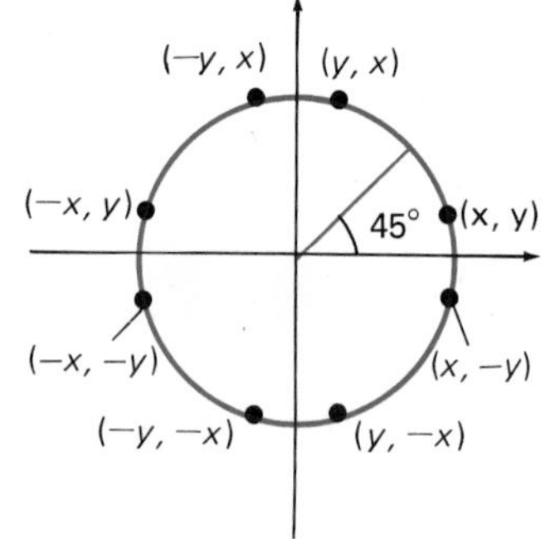

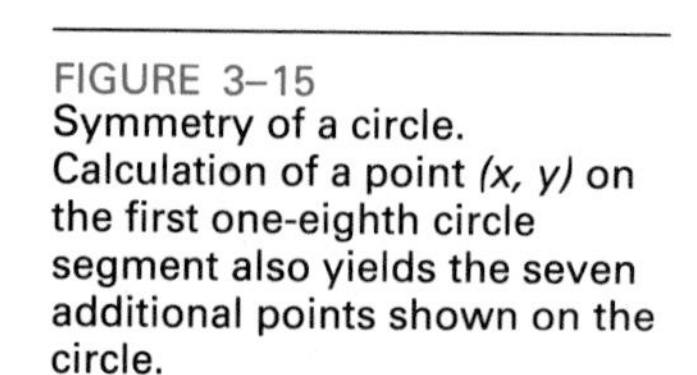

FIGURE 3–15
Symmetry of a circle. Calculation of a point (x, y) on the first one-eighth circle segment also yields the seven additional points shown on the circle.

spacing between plotted pixel positions is not uniform, as demonstrated in Fig. 3–14. We could adjust the spacing by interchanging x and y (stepping through y values and calculating x values) whenever the absolute value of the slope of the circle became greater than 1. But this adds computations and checking to the algorithm.

One way to eliminate the unequal spacing associated with Eq. 3–17 is to calculate points along the circular boundary using polar coordinates. Expressing the circle equation in parametric polar form yields the pair of equations

$$\begin{aligned} x &= xc + r \cdot \cos\theta \\ y &= yc + r \cdot \sin\theta \end{aligned} \tag{3–18}$$

When a display is generated with these equations using a fixed angular value for θ, a circle is plotted with equally spaced points along the circumference. The step size chosen for θ depends on the application. For circle generation with a raster-scan system, we can set the step size at $1/r$ and calculate closely spaced pixel positions. This step size gives us pixels that are approximately one unit apart.

An improvement in these methods can be made by taking advantage of the symmetry of circles. A given point on the circumference can be mapped into several other circle points by interchanging coordinates and alternating the sign of the coordinate values. As illustrated in Fig. 3–15, a point at position (x, y) on a one-eighth circle sector can be used to plot the other seven points shown. Using this approach, we could generate all pixel positions around a circle for a raster display by calculating only the points within the sector from $x = 0$ to $x = y$.

Determining pixel positions along a circle circumference using either Eq. 3–17 or Eq. 3–18 requires a good deal of computation time. The Pythagorean theorem approach involves multiplications and square root calculations, while the parametric equations contain multiplications and trigonometric calculations. We can improve the efficiency of circle generation by using a method that reduces the computations as much as possible to integer arithmetic.

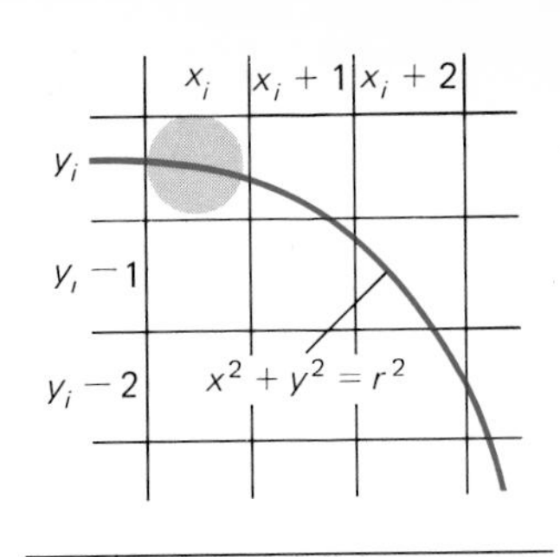

FIGURE 3–16
Section of the screen grid where a circle passing through the point (x_i, y_i) is to be displayed.

As in the line-generating algorithm, integer positions along a circular path can be obtained by determining which of two pixels is nearer the circle at each step. To simplify the algorithm statements, we first consider a circle centered at the coordinate origin ($xc = 0$ and $yc = 0$). We also calculate the points for a one-eighth circle segment, assuming that we are going to get the remaining points by symmetry for storage in a raster. (A random-scan system with a vector generator could extend the calculations through one complete cycle.) Unit steps are taken in the x direction, starting from $x = 0$ and ending when $x = y$. The starting coordinate in our algorithm is then $(0, r)$.

The situation at some arbitrary step in the algorithm is shown in Fig. 3–16. We assume that position (x_i, y_i) has been determined to be closer to the circle path. The next position is then either $(x_i + 1, y_i)$ or $(x_i + 1, y_i - 1)$.

From Eq. 3–16, the actual y value on the circle path is determined as

$$y^2 = r^2 - (x_i + 1)^2$$

Figure 3–17 illustrates the relationship between y and the integer coordinate values, y_i and $y_i - 1$. A measure of the difference in coordinate positions can be defined in terms of the square of the y values as

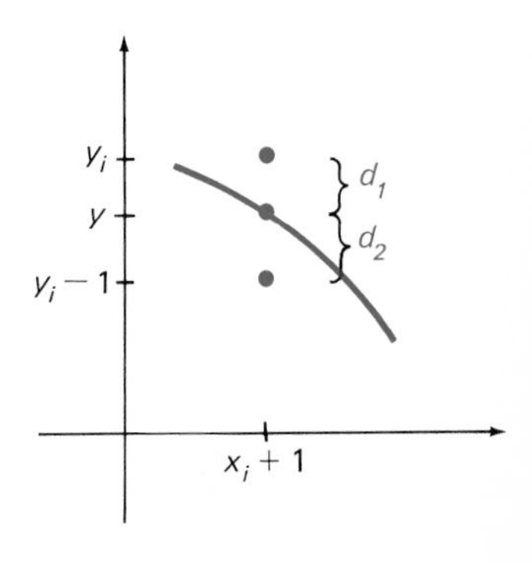

FIGURE 3–17
Coordinate differences between pixel centers and the y position on a circle at $x_i + 1$.

$$\begin{aligned} d_1 &= y_i^2 - y^2 \\ &= y_i^2 - r^2 + (x_i + 1)^2 \end{aligned} \tag{3–19}$$

and

$$\begin{aligned} d_2 &= y^2 - (y_i - 1)^2 \\ &= r^2 - (x_i + 1)^2 - (y_i - 1)^2 \end{aligned} \tag{3–20}$$

We now set up a parameter for determining the next coordinate position as the difference between d_1 and d_2:

$$\begin{aligned} p_i &= d_1 - d_2 \\ &= 2(x_i + 1)^2 + y_i^2 + (y_i - 1)^2 - 2r^2 \end{aligned} \tag{3–21}$$

If p_i is negative, we select the pixel at position y_i. Otherwise, we select the pixel at location $y_i - 1$.

The test for selecting the next pixel holds whether the actual path passes above y_i or below $y_i - 1$, as shown in Fig. 3–18. For the first case, Fig. 3–18 (a),

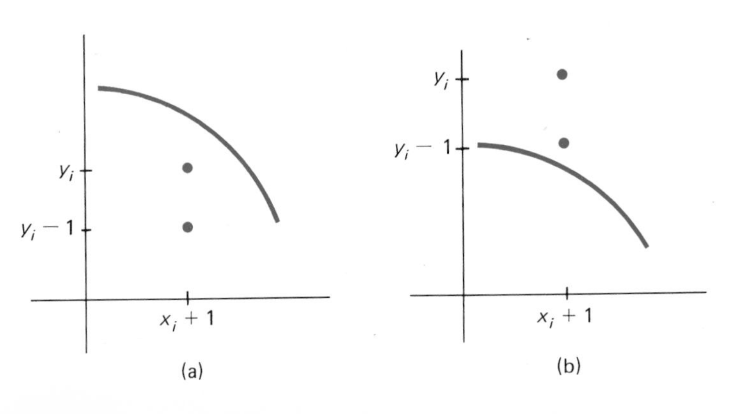

FIGURE 3–18
Possible pixel positions: (a) Both pixel centers are below the circle path, and (b) both pixel centers are above the circle path.

we have $d_1 < 0$, $d_2 > 0$, and $p_i < 0$, so that the point at y_i would be selected. In the second case, Fig. 3–18 (b), $d_1 > 0$ and $d_2 < 0$. Now $p_i > 0$, and the point at $y_i - 1$ is selected.

A recursive form for the parameter p is obtained by evaluating p_{i+1} in terms of p_i:

$$p_{i+1} = 2[(x_i + 1) + 1]^2 + y_{i+1}^2 + (y_{i+1} - 1)^2 - 2r^2$$

This expression can be written in terms of Eq. 3–21 as

$$p_{i+1} = p_i + 4x_i + 6 + 2(y_{i+1}^2 - y_i^2) - 2(y_{i+1} - y_i) \qquad (3\text{–}22)$$

The y position y_{i+1} is either the same as y_i or the same as $y_i - 1$, depending on the value of p_i. Starting from p_1, the algorithm determines each successive p parameter from the preceding one. We obtain p_1 by setting $(x_1, y_1) = (0, r)$ in Eq. 3–21:

$$p_1 = 3 - 2r \qquad (3\text{–}23)$$

Figure 3–19 summarizes the steps for calculating integer coordinates closest to the defined circle. To generalize the algorithm so that a circle with an arbitrary center position can be plotted, we simply add xc to each successive x value and add yc to each calculated y value.

FIGURE 3–19
Bresenham's circle algorithm.

1. Select the first position for display as
$$(x_1, y_1) = (0, r)$$
2. Calculate the first parameter as
$$p_1 = 3 - 2r$$
If $p_1 < 0$, the next position is $(x_1 + 1, y_1)$. Otherwise, the next position is $(x_1 + 1, y_1 - 1)$.
3. Continue to increment the x coordinate by unit steps, and calculate each succeeding parameter p from the preceding one. If for the previous parameter we found that $p_i < 0$, then
$$p_{i+1} = p_i + 4x_i + 6.$$
Otherwise (for $p_i \geq 0$),
$$p_{i+1} = p_i + 4(x_i - y_i) + 10$$
Then, if $p_{i+1} < 0$, the next point selected is $(x_i + 2, y_{i+1})$. Otherwise, the next point is $(x_i + 2, y_{i+1} - 1)$. The y coordinate is $y_{i+1} = y_i$, if $p_i < 0$ or $y_{i+1} = y_i - 1$, if $p_i \geq 0$.
4. Repeat the procedures in step 3 until the x and y coordinates are equal.

Although a multiplication is required in the calculation of each parameter, the multiplier is a power of 2, so the multiplication can be implemented as a logical shift operation. All other operations are simply integer additions or subtractions. The following procedure is a coding for this circle algorithm. Input to the procedure are the coordinates for the circle center and the radius. The procedure loads the frame buffer array with points along the circle circumference by calls to the *set_pixel* operation.

```
procedure bres_circle (x_center, y_center, radius : integer);
  var
    p, x, y : integer;
```

```
procedure plot_circle_points;
  begin
    set_pixel (x_center + x, y_center + y);
    set_pixel (x_center - x, y_center + y);
    set_pixel (x_center + x, y_center - y);
    set_pixel (x_center - x, y_center - y);
    set_pixel (x_center + y, y_center + x);
    set_pixel (x_center - y, y_center + x);
    set_pixel (x_center + y, y_center - x);
    set_pixel (x_center - y, y_center - x)
  end; {plot_circle_points}

begin {bres_circle}
  x := 0;
  y := radius;
  p := 3 - 2 * radius;
  while x < y do begin
      plot_circle_points;
      if p < 0 then p := p + 4 * x + 6
      else begin
          p := p + 4 * (x - y) + 10;
          y := y - 1
        end; {if p not < 0}
      x := x + 1
    end; {while x < y}
    if x = y then plot_circle_points
end; {bres_circle}
```

Ellipses

A circle-drawing algorithm can be extended to plot either circles or ellipses. In Fig. 3–20 we show one orientation for an ellipse, with r_1 labeling the semimajor axis and r_2 as the semiminor axis. The standard form for the elliptical equation is

$$\left(\frac{x - xc}{r_1}\right)^2 + \left(\frac{y - yc}{r_2}\right)^2 = 1 \qquad (3\text{–}24)$$

Using polar coordinates r and θ, we can also write the elliptical equations in parametric form:

$$\begin{aligned} x &= xc + r_1 \cdot \cos\theta \\ y &= yc + r_2 \cdot \sin\theta \end{aligned} \qquad (3\text{–}25)$$

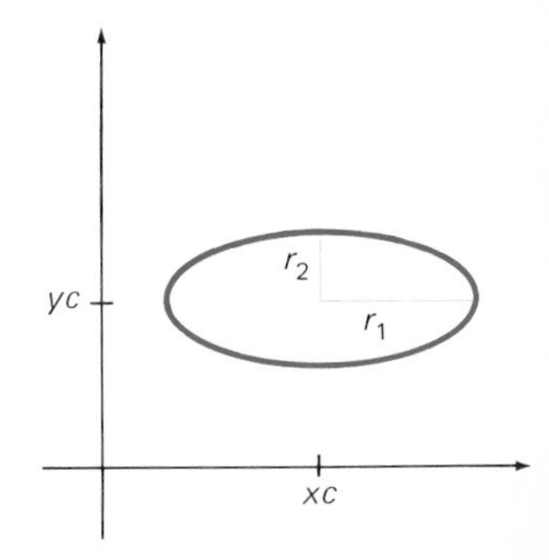

FIGURE 3–20
Ellipse centered at (xc, yc) with semimajor axis r_1 and semiminor axis r_2.

Bresenham's algorithm can be modified to generate elliptical shapes by using Eq. 3–24, instead of the circle equation, in the evaluation of parameter p_i. That is, for an ellipse centered at the origin, we can express y values in the form

$$y^2 = r_2^{\,2}\left(1 - \frac{x^2}{r_1^{\,2}}\right) \qquad (3\text{–}26)$$

The only difference in the algorithm is in the form of the p parameters. An ellipse is plotted at an arbitrary position by adding offsets to the output x and y values, as in the generation of circle positions.

To provide users with the capability for generating ellipses (and circles), a graphics package could include a command of the form

```
ellipse (xc, yc, r1, r2)
```

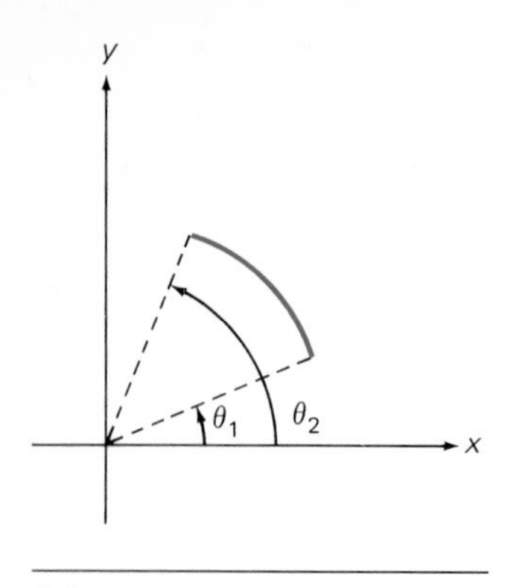

FIGURE 3–21
Circular arc specified by beginning and ending angles. Circle enter is at the coordinate origin.

Coordinates for the ellipse center are assigned to parameters *xc* and *yc*, and the semimajor and semiminor axes are specified in *r*1 and *r*2. If the values assigned to *r*1 and *r*2 are equal, the system displays a circle. These parameters are passed to the ellipse-generating algorithm, which plots the specified figure.

Commands to generate circles and ellipses often include the capability of drawing curve sections by specifying parameters for the line endpoints. Expanding the parameter list allows specification of the beginning and ending angular values for an arc, as illustrated in Fig. 3–21.

3–7 Other Curves

Procedures to display various curves use methods similar to those that generate circles and ellipses. Commonly encountered curves include sine functions, exponential functions, polynomials, probability distributions, and spline functions.

If we can express a curve in functional form, $y = f(x)$, values for y can be calculated and plotted over a specified interval for x. The coordinate values must be rounded to the nearest integer, and the curve path can be filled in with either individual points or straight line segments. For most applications, curves approximated with line segments are more convenient. A point-plotting method leaves gaps in the curve in areas where the magnitude of the slope is greater than 1. To avoid the gaps with a point-plotting method means that we must obtain the inverse function, $x = f^{-1}(y)$, and calculate values of x for given y values whenever the magnitude of the slope becomes large.

Symmetry considerations improve the efficiency of some curve-generating algorithms. Many curves have repeated patterns, so it may be possible to obtain more than one point on the curve with a single calculation. Parabolas and the normal probability distribution are symmetric about a center point, while all points within one cycle of a sine curve can be generated from the points in a 90° interval.

For a curve defined by a data set of discrete coordinate points, we must graph the curve in other ways. One method is simply to plot the individual data points and connect them with straight line segments. Another approach is to use curve-fitting techniques to obtain a curve approximation to the data points. Special curve-fitting methods have been devised for design applications, and we return to this topic in Chapter 10.

FIGURE 3–22
Possible bit pattern for the letter *B*, using an 8 by 8 rectangular grid.

1	1	1	1	1	1	0	0
0	1	1	0	0	1	1	0
0	1	1	0	0	1	1	0
0	1	1	1	1	1	0	0
0	1	1	0	0	1	1	0
0	1	1	0	0	1	1	0
1	1	1	1	1	1	0	0
0	0	0	0	0	0	0	0

3–8 Character Generation

Rectangular-grid patterns are typically used to define and plot characters. Figure 3–22 illustrates a bit pattern for the letter *B*, defined on an 8 by 8 dot matrix for use with a bilevel raster system. When this pattern is copied to some area of the frame buffer, the 1 bits designate which pixel positions are to be displayed on the monitor. Rectangular grids for character definitions vary from about 5 by 7 to 9 by 14 or more and are used with both raster and vector systems, although some vector systems generate characters with line segments. Standard character patterns for letters, numbers, and other symbols are predefined and stored in read-only memory. With some systems, additional user-defined character patterns can be accommodated, allowing specialized fonts.

In addition to allowing users to define special characters, graphics packages can provide options for various types of character manipulation. Characters can be shifted relative to each other to provide special effects or spacing, and character patterns can be rotated or scaled to vary their size.

User commands to output character strings are provided with all graphics packages. A basic character-string command can be defined as

```
text (x, y, string)
```

Parameter *string* is assigned any character sequence, which is displayed starting at **text position** (x, y). For example, the statement

```
text (100, 450, "population distribution")
```

could be used as a label on a distribution graph.

Text position (x, y) sets the coordinate location for the lower left corner of the first character of the horizontal string to be displayed. This provides the reference for copying the character grid definitions onto the frame buffer. The *text* command is implemented by a routine that places the character bit patterns into the raster array, one at a time from left to right starting at the text position. Graphics packages often allow for other string orientations, such as vertical lettering. For these options, the text position (x, y) may be interpreted differently by the *text* routine.

Another convenient character command is one that places a designated character (marker symbol) at one or more selected positions. This command can be defined similarly to the line command:

```
polymarker (n, x, y)
```

Parameters n, x, and y in *polymarker* are given the same meaning as in the *polyline* command. The difference is that *polymarker* causes a predefined character to be placed at each of the n coordinate positions specified in arrays x and y. The type of symbol used in *polymarker* depends on the particular implementation, but we assume for now that an asterisk is to be used. Figure 3–23 illustrates plotting of a data set with the statement

```
polymarker (6, x, y)
```

The implementation of *polymarker* is carried out by a routine that repeatedly copies the marker grid definition into the frame buffer at the designated coordinate position. As with the *text* command, the coordinate specifications set the position for the lower left corner of the character grid.

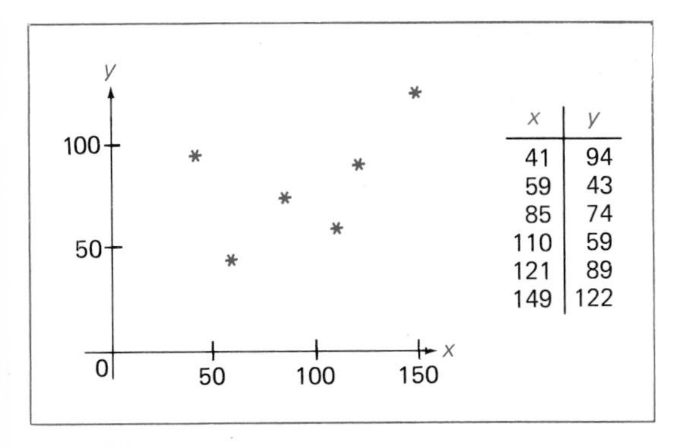

FIGURE 3–23
Sequence of data points specified with the POLYMARKER command.

3–9 Instruction Sets for Display Processors

Output primitive functions stated in an application program are converted by the graphics system into a form suitable for the display processor to generate the picture on an output device. The specific instruction set used by the display processor to generate output depends on the type of device in use.

Raster-Scan Systems

A number of registers are made available to the processors in a raster system to store coordinate positions and various instructions. Four registers can be used to hold coordinate values for line endpoints, circle and ellipse parameters, and positions for character strings or markers. The size of these registers is determined by the resolution of the system. If the raster is designed to display 1024 by 1024 pixels, coordinate registers need 10 bits to store the coordinate values from 0 to 1023. Additional registers are used to store output operations and the instructions for processing the output primitive commands. Typical size of the instruction registers is from 12 to 24 bits.

Instruction formats for the display processor are organized into opcode and address fields. The opcode field determines the type of operation to be performed, such as loading a register, drawing a line, or displaying a character. Address fields are used to specify register or memory locations.

Each output primitive command in an application program is compiled into a corresponding instruction format. The scan-conversion routine then loads the instruction into a register and processes it using an appropriate hardware-implemented procedure, such as the Bresenham line-drawing algorithm or retrieval of a character pattern. Pixel intensity values generated by the scan-conversion procedure are then loaded into the frame buffer at the specified positions.

Intensity values in the raster are used by a refresh procedure to display the picture on the video monitor. Two registers are used to store the coordinates of a pixel. Initially, the x register is set to 0 and the y register is set to the coordinate value for the top of the screen (say 1023). The value stored in the raster for this pixel position is then used to adjust the intensity of the electron beam of a refresh CRT. Then the x register is incremented by 1, and the process repeated for the next pixel on the top scan line. This procedure is repeated for each pixel along the scan line. After the last pixel on the top scan line has been processed, the x register is reset to 0 and the y register is decremented by 1. Pixels along this scan line are processed in turn, and the procedure is repeated for each scan line. When the pixels along the bottom scan line ($y = 0$) have been displayed, the y register is reset to the top value and the refresh process starts over. This refresh process is carried out at a rate of 30 to 60 frames per second.

Random-Scan Systems

Instruction formats for vector systems are similar to those for raster systems. The major difference is that a display file of instructions is created for the refresh process instead of loading a frame buffer.

To carry out the refresh process, the first instruction in the display file is loaded into a register. Appropriate subroutines are then referenced to display a line or character string on the output device, using either analog or digital vector-generation techniques. Next, an instruction counter is incremented, and the next instruction is retrieved from the display file and processed. After all instructions in the display file have been processed, the instruction counter is reset, and the refresh procedure is repeated from the beginning of the file. Depending on the type of vector generator used, a line can be refreshed by a random-scan system at rates varying from a few microseconds to a tenth of a microsecond.

3–10 Summary

Table 3–1 lists the output primitives discussed in this chapter and the command formats for displaying them from an application program. These primitives provide the basic tools for constructing displays with straight lines, areas, curves, and text.

TABLE 3–1.
Summary of Output Primitives

Output Primitive Function	Description
polyline (n, x, y)	Defines a connected sequence of n − 1 line segments. The n endpoints of the lines are specified in the arrays x and y. If n = 1, a point is specified.
fill_area (n, x, y)	Defines a polygon fill area with the coordinates for the n polygon vertices specified in arrays x and y.
text (x, y, string)	Displays a character string starting at coordinate position (x, y). The string to be displayed is specified in parameter string.
polymarker (n, x, y)	Displays a series of predefined characters at each of the n coordinate positions specified in arrays x and y.
ellipse (xc, yc, r1, r2)	Defines an ellipse centered at position (xc, yc) with axes r1 and r2. A circle is displayed if r1 = r2.

3–11 Applications

Following are some example programs illustrating the use of output primitives. The first program produces a line graph of a household's natural gas consumption for a year. Output of this procedure is drawn in Fig. 3–24.

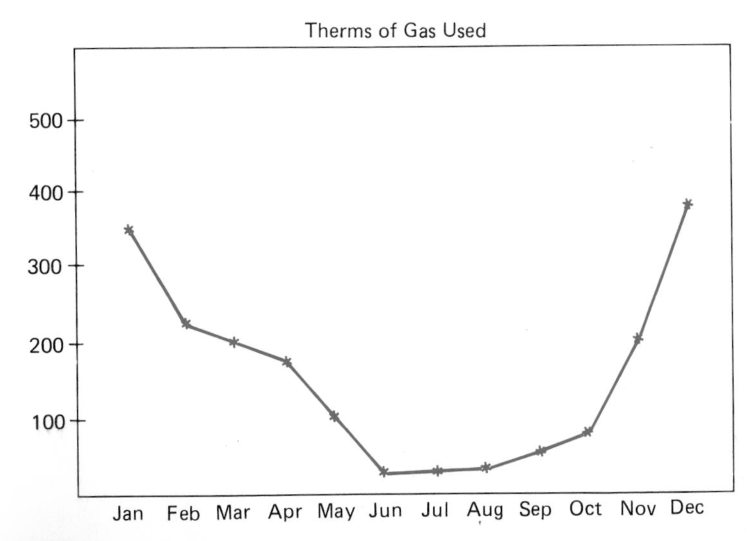

FIGURE 3–24
A plot of data points output by PROCEDURE LINE_GRAPH.

```
{The following are already defined as constants—
  max_points, grid_left, grid_right, grid_top, grid_bottom,
  grid_height, grid_center, first_mark, and interval.
  Convert_to_string is a predefined function that returns
  string equivalent of an integer; Max returns the maximum
  value found in the array of data values}

type
  data = array [1..12] of integer;
  points = array [1..max_points] of integer;

procedure line_graph (therms : data);
  var  x, y : points;

  procedure axes_and_labels;
    var  k : integer;
    begin
      x[1] := grid_left;     y[1] := grid_bottom;
      x[2] := grid_right;    y[2] := grid_bottom;
      x[3] := grid_right;    y[3] := grid_top;
      x[4] := grid_left;     y[4] := grid_top;
      fill_area (4, x, y);  {make outer box}
      {put title and bottom labels}
      text (grid_center - 72, grid_top + 6, 'Therms of Gas Used');
      text (grid_left, 0,
          'Jan Feb Mar Apr May Jun Jul Aug Sep Oct Nov Dec');
      {make tic marks and labels on left}
      x[1]:= grid_left - 3;
      x[2]:= grid_left + 3;
      for k := 1 to 5 do begin
          y[1]:= grid_bottom + k * 32;
          y[2]:= y[1];
          polyline (2, x, y);
          text (1, y[1] - 4, convert_to_string(k*100))
        end
    end; {axes_and_labels}

  procedure graph_data;
    var
      marker_placement : real;
      k, data_range : integer;

    begin
      data_range := max(therms)
      for k := 1 to 12 do begin
          marker_placement:= grid_height * (therms[k]/data_range);
          y[k]:= grid_bottom + round(marker_placement);
          x[k]:= first_mark + interval * (k - 1)
        end;
      polymarker (12, x, y);
      polyline (12, x, y)
    end; {graph_data}

  begin {line_graph}
    axes_and_labels;
    graph_data
  end; {line_graph}
```

Pie charts are used to show the percentage contribution of individual parts to the whole. The next procedure constructs a pie chart, with the number and relative size of the slices determined by input. A sample output from this procedure appears in Fig. 3–25.

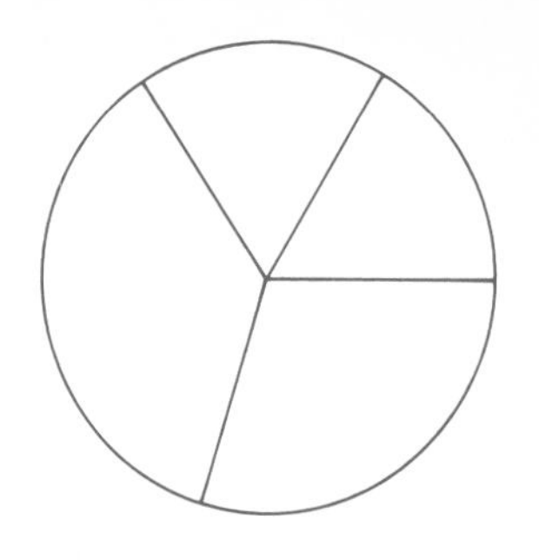

FIGURE 3–25
Output generated from PROCEDURE PIE_CHART.

```
const
  max_slice  =  8;
type
  data  =  array [1..max_slice] of integer;

procedure pie_chart (xc,yc,radius,slices : integer; data_values : data);
  var
    total, k, : integer;
    last_slice, new_slice : real;
  begin
    ellipse (xc, yc, radius, radius);
    total := 0;
    for k := 1 to slices do total := total + data_values[k];
    x[1] := xc;         {every slice line will start at pie center}
    y[1] := yc;
    last_slice := 0;
    for k := 1 to slices do begin
        new_slice := 6.28 * data_values[k] / total + last_slice;
        x[2] := round(xc + radius * cos(new_slice));
        y[2] := round(yc + radius * sin(new_slice));
        polyline (2, x, y);
        last_slice := new_slice    {update last slice}
      end
  end; {pie_chart}
```

Some variations on the circle equations are output by this next procedure. The shapes shown in Fig. 3–26 are generated by varying the radius r of a circle. Depending on how we vary r, we can produce a spiral, cardioid, limacon, or other similar figure.

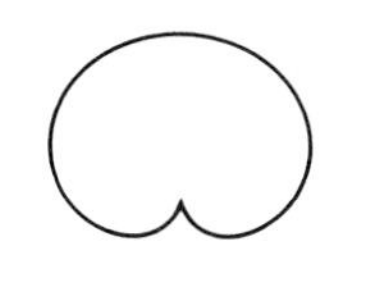
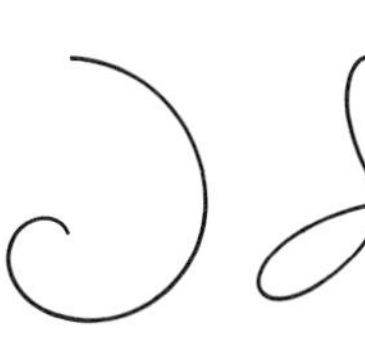

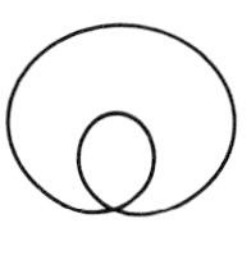

FIGURE 3–26
Curved figures produced with PROCEDURE DRAW_SHAPE.

```
type
  shape  =  (cardioid, three_leaf, spiral, four_leaf, limacon);

procedure draw_shape (figure : shape; xc, yc, a, b : integer);
  {xc, yc is figure center. a and b are used to compute radius}
  var
    theta, dtheta, r : real;
  begin
    theta  :=  0;
    dtheta  :=  1 / a;  {angle increment}
    {set first point for figure}
    y[1]  :=  yc;
    case figure of
      cardioid   : x[1] := round(xc + a * 2);
```

```
        three_leaf : x[1] := round(xc + a);
        spiral     : x[1] := xc;
        four_leaf  : x[1] := round(xc + a);
        limacon    : x[1] := round(xc + b + a)
      end;
      {compute other points}
      while theta < 6.28 do begin
        case figure of
          cardioid   : r := a * (1 + cos(theta));
          three_leaf : r := a * cos(3 * theta);
          spiral     : r := a * theta;
          four_leaf  : r := a * cos(2 * theta);
          limacon    : r := b + a * cos(theta)
        end; {case}
        x[2] := round(xc + r * cos(theta));
        y[2] := round(yc + r * sin(theta));
        polyline (2, x, y);
        x[1] := x[2];           {save the current point}
        y[1] := y[2];
        theta := theta + dtheta   {update the angle}
      end {while theta}
    end; {draw_shape}
```

REFERENCES

Sources of information on algorithms for generating output primitives include Bresenham (1965 and 1977), McIlroy (1983), and Pavlidis (1982).

Antialiasing techniques are discussed in Crow (1981), Fujimoto and Iwata (1983), Korein and Badler (1983), Pitteway and Watkinson (1980), and Turkowski (1982).

For additional information on output primitive functions in GKS see Enderle, Kansy, and Pfaff (1984) and Hopgood, et al. (1983).

EXERCISES

3–1. Implement the *polyline* command using the DDA algorithm, given any number (n) of input points. A single point is to be plotted when $n = 1$.

3–2. Extend Bresenham's line algorithm to generate lines with any slope. Implement the *polyline* command using this algorithm as a routine that displays the set of straight lines between the n input points. For $n = 1$, the routine displays a single point.

3–3. Implement the *set_pixel* routine in Bresenham's line algorithm of Ex. 3–2 so that the frame buffer is loaded using efficient address-calculation methods, such as those in Eqs. (3–14) and (3–15).

3–4. Modify a line-drawing algorithm so that the intensity of the output line is set according to its slope. That is, by adjusting pixel intensities according to the value of the slope, all lines are displayed with the same intensity per unit length.

3–5. Implement an antialiasing procedure by extending Bresenham's line algorithm to adjust pixel intensities in the vicinity of a line path.

3–6. Modify Bresenham's circle algorithm to display any ellipse, as specified by input values for the ellipse center and the major and minor axes. Write a program to implement this algorithm on your system.

3–7. Implement the *set_pixel* operation in the program of Ex. 3–6 using efficient methods for loading the frame buffer, such as those in Eqs. (3–14) and (3–15).

3–8. Outline a method for antialiasing a circle boundary. How would this method be modified to antialias elliptical boundaries?

3–9. Write a routine to implement the *text* function.

3–10. Write a routine to implement the *polymarker* function.

3–11. Devise an efficient algorithm that takes advantage of symmetry properties to display a sine function.

3–12. Extend the algorithm of Ex. 3–11 to display the function describing damped harmonic motion:

$$y = A\,e^{-kx}\sin(\omega s + \theta)$$

where ω is the angular frequency and θ is the phase of the sine function. Plot y as a function of x for several cycles of the sine function or until the maximum amplitude is reduced to $A/10$.

3–13. Choose any three functions that possess some form of symmetry, and explain how efficient methods could be devised to minimize the calculations needed to plot each function.

3–14. Write a program to display a bar graph using the *polyline* command. Input to the program is to include the data points and the labeling required for the x and y axes. The data points are to be scaled by the program so that the graph is displayed across the full screen area.

3–15. Extend the program of Ex. 3–14 to allow the graph to be displayed in any selected screen area.

3–16. Write a procedure to display a line graph for any input set of data points in any selected area of the screen. The input data set is then to be scaled to fit the selected screen area. Data points are to be displayed as asterisks joined with straight line segments, and the x and y axes are to be labeled according to input specifications.

3–17. Modify the procedure in Ex. 3–16 to display the data points as small circles.

3–18. Using the *ellipse* function, write a routine to display any specified pie chart with appropriate labeling. Input to the routine is to include the relative percentage of each section, the name of the piechart, and the section names. Each section label is to be displayed outside the boundary of the pie chart near the corresponding pie section.

3–19. Design an instruction set for a raster-scan system using 16-bit words.

3–20. Design an instruction set for a random-scan display processor using 16-bit word lengths.

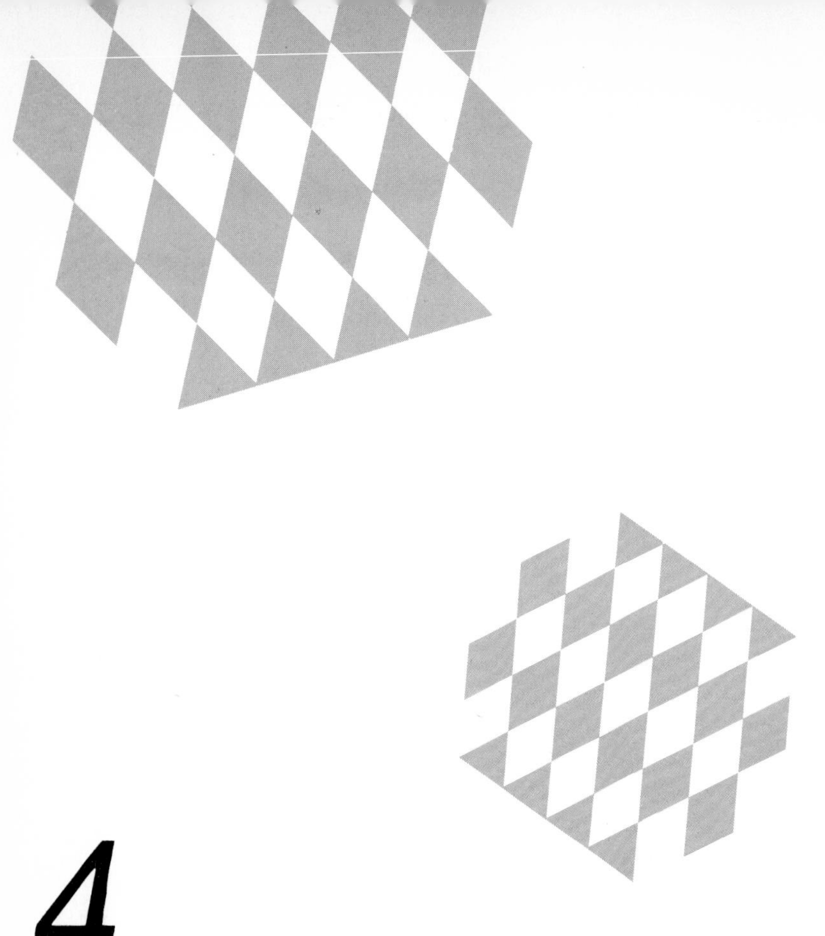

4
ATTRIBUTES OF OUTPUT PRIMITIVES

We can display output primitives with a variety of attributes. Lines can be dotted or dashed, fat or thin. Areas might be filled with a particular color or with a combination of colors used in a pattern. Text can appear reading from left to right, slanted diagonally across the screen, or in vertical columns. And individual characters can be displayed in different colors and sizes.

One way to incorporate attribute options into a graphics package is to extend the parameter list associated with each output primitive command to include the appropriate attributes. A point-plotting command, for example, could contain a color parameter in addition to coordinate parameters. This attribute parameter is passed directly to the output primitive routine.

Another approach is to maintain a system list of attributes and their current values. Separate commands are then included in the graphics package for setting the attribute values in the system attribute list. To display an output primitive, the system checks the relevant attributes and draws the primitive according to the current attribute settings. Some packages provide users with both attribute functions and attribute parameters in the output primitive commands. Our approach is to consider a package designed with functions that modify a system attribute list, and we discuss how a standard set of attributes can be defined and implemented in this fashion.

4–1 Line Styles

Attributes for line style determine how a line is to be displayed by a line-drawing routine. Typical attributes of a line are its type, its width, and its color. Routines for line drawing must be structured to produce lines with the specified characteristics.

Line Type

The line type attributes include solid lines, dashed lines, and dotted lines. They are implemented in a graphics package through modifications to the line-drawing algorithms to accommodate the type of line requested by a user. When a dashed line is to be displayed, the line-drawing algorithm outputs short solid sections along the line while leaving intervening sections blank. A dotted line could be generated by plotting every other point, or every third point, along the line path. Any number of line type variations can be built into the basic line-generating algorithms.

To set line type attributes in an applications graphics package, we introduce the function

```
set_linetype (lt)
```

We assume here that the line type parameter *lt* is assigned values such as *solid*, *dashed*, and *dotted*. Other values for *lt* could be used to provide a combination of dots and dashes or for variations in the dot-dash patterns.

With the *linetype* command, a user sets the current value of the line type attribute. All subsequent line-drawing commands produce lines with this line type. The following program segment illustrates use of a *linetype* command to display the data plots in Fig. 4–1.

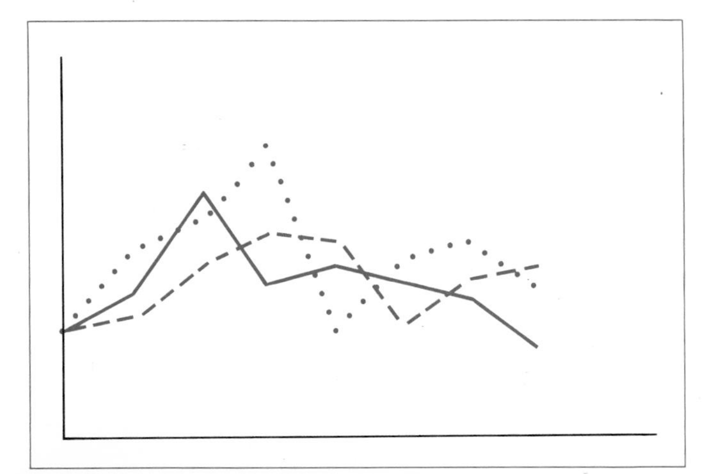

FIGURE 4–1
Plotting three data sets with three different line styles, as output by PROCEDURE LINE_TYPE_CHART.

```
{Draw_axes and get_data are predefined procedures.
Get_data returns arrays x and y, specifying
12 coordinate positions to be plotted.}

type
  points = array [1..8] of integer;

procedure line_type_chart;
  var k : integer; x, y : points;
  begin
    draw_axes;
    for k := 1 to 3 do begin
        get_data (x, y);
        case k of
          1 : set_linetype (solid);     {set line type}
          2 : set_linetype (dashed);
          3 : set_linetype (dotted)
        end; {case}
        polyline (8, x, y)              {draw line}
      end {for k}
  end; {line_type_chart}
```

Line Width

Implementation of line width options depends on the type of output device used. A heavy line on a video monitor could be drawn as adjacent parallel lines, while a pen plotter might require pen changes. As with other attributes, a line width command can be used to set a current line width value in the attribute list. This value is used by the line-drawing algorithms to control the width of lines generated with subsequent output primitive commands.

A command for setting the line width attribute is defined as

```
set_linewidth_scale_factor (lw)
```

Line width parameter *lw* is assigned a positive number to indicate the relative width of the line to be displayed. Assuming that a value of 1 specifies a standard-width line, a user could set *lw* to a value of 0.5 to generate a line whose width is one-half that of the standard line. Values greater than 1 would be used to produce lines thicker than the standard.

Line Color

When a system provides color (or intensity) options, a parameter giving the current color index is included in the list of system attribute values. This color index is then stored in the frame buffer at the appropriate locations by the *set_pixel* procedure. The number of color choices depends on the number of bits of storage available per pixel in the raster.

A graphics package can provide color options for lines with the command

```
set_line_color_index (lc)
```

Nonnegative integer values, corresponding to various color choices, can be assigned to the line color parameter *lc*. This parameter is used as an index into a color table, which lists the available colors for the output device in use. Any line drawn in the background color is invisible, and a user can erase previously displayed lines on a video monitor by respecifying them in the background color.

An example of the use of the various line attribute commands in an applications program is given by the following sequence of statements:

```
set_linetype (dashed);
set_linewidth_scale_factor (2);
set_line_color_index (5);
polyline (n1, x1, y1);

set_line_color_index (6);
polyline (n2, x2, y2);
```

This program segment would display two figures, drawn with double-wide, dashed lines. The first is displayed in a color corresponding to code 5, and the second in color 6.

In a color raster system, the number of bits of storage needed per pixel depends on the number of color choices available and the method used for storing color values. Color tables can be structured to provide extended color capabilities to a user without requiring huge frame buffers.

4–2 Color and Intensity

Various color and intensity-level options can be made available to a user, depending on the capabilities and design objectives of a particular system. Some systems provide for a wide choice of colors; others have only a few options. Raster-scan systems, for example, are often designed to display a great many colors, while random-scan monitors typically offer only a few color choices, if any. The color codes are usually assigned numeric values ranging from 0 through the positive integers. For a CRT, these color codes are converted to intensity-level settings for the electron beams in either shadow-mask or beam-penetration monitors. With a color plotter, the codes could control ink-jet deposits or pen changes.

Color Tables

A simple scheme for storing color code selections in the frame buffer of a raster system is shown in Fig. 4–2. When a particular color code is specified in an application program, the corresponding binary value is stored in the frame buffer

COLOR CODE	STORED COLOR VALUES IN FRAME BUFFER			DISPLAYED COLOR
	RED	GREEN	BLUE	
0	0	0	0	Black
1	0	0	1	Blue
2	0	1	0	Green
3	0	1	1	Cyan
4	1	0	0	Red
5	1	0	1	Magenta
6	1	1	0	Yellow
7	1	1	1	White

FIGURE 4–2
Color codes stored in a frame buffer with three bits per pixel.

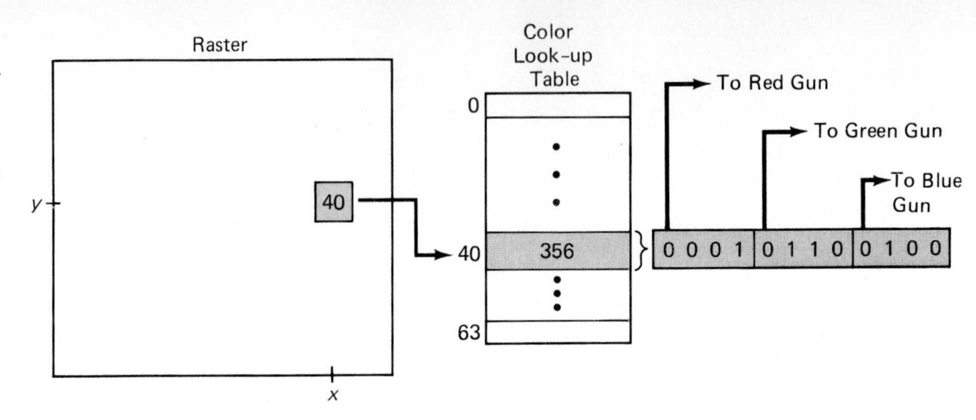

FIGURE 4–3
A color lookup table, providing 12 bits per entry, used with a raster containing 6 bits for each pixel position. A value of 40 stored in raster position *(x, y)* references the location in this lookup table containing the value 356. Each 4-bit segment of this entry controls the intensity level of one of the three electron guns in an RGB monitor, as shown.

for each component pixel in the output primitives to be displayed in that color. The scheme given in Fig. 4–2 allows eight color choices with 3 bits per pixel of storage. Each of the three bit positions is used to control the intensity level (either on or off) of the corresponding electron gun in an RGB monitor. The leftmost bit controls the red gun, the middle bit controls the green gun, and the rightmost bit controls the blue gun. Adding more bits per pixel to the frame buffer increases the number of color choices. In a raster with 6 bits per pixel, 2 bits can be used for each gun. This provides four intensity levels for each color gun (red, green, and blue), and 64 different color codes could be stored.

Another method for storing color codes is diagrammed in Fig. 4–3. In this scheme, the electron gun intensities are controlled by values stored in a **color lookup table** instead of by the values stored in the raster. The raster values are used as indices into the lookup table. For the example shown in this figure, a raster with 6 bits per pixel can reference any one of the 64 positions in the lookup table. Each entry in the table uses 12 bits to specify a color, so that a total of 4096 different colors is now available. Four bits of intensity information is provided for each electron gun. Systems employing this lookup table would permit a user to select any combination of 64 colors from a 4096-color palette. Also, the lookup table entries could be changed at any time, allowing designs, scenes, or graphs displayed on the screen to take on new color combinations.

Color table entries could be set by a user with the command

```
set_color_table (ct, c)
```

Parameter *ct* is used as a color table position number (0 to 64 for the example in Fig. 4–3), and parameter *c* is the code for one of the possible color choices.

Use of a color lookup table can dramatically increase the color options without corresponding increases in raster size. Some high-quality color systems use as many as 24 bits for each position in the color lookup table and 9 bits per pixel in the frame buffer. This allows 512 colors to be used in each display, with over 16 million color choices for the lookup table entries.

Gray Scale

With monitors that have no color capability, the *line_color* command can be used in an application program to set the intensity level, or **gray scale,** for points along displayed lines. Many packages use numeric values within the range of 0 to 1 to set gray scale levels. This allows the package to be adapted to hardware with differing gray scale capabilities.

INTENSITY CODES	STORED INTENSITY VALUES IN THE FRAME BUFFER (Binary Code)		DISPLAYED GRAY SCALE
0.0	0	(00)	Black
0.33	1	(01)	Dark Gray
0.67	2	(10)	Light Gray
1.0	3	(11)	White

FIGURE 4–4
Conversion of intensity values to integer codes for storage in a frame buffer accommodating a gray scale with four levels. Two bits of storage for each pixel position are needed in the frame buffer.

Storing intensity levels in a raster is similar to storing color codes. If only one bit per pixel is provided in the raster, on (white) and off (black) are the only possibilities for the gray scale. Three bits per pixel can accommodate eight different intensity levels. Higher-quality systems might provide 8 or more bits per pixel in the frame buffer for the intensity levels.

Intensity values specified in an application program are converted to appropriate binary codes for storage in the raster. Figure 4–4 illustrates conversion of user specifications to codes for a four-level gray scale. In this example, any intensity input value near 0.33 would store the binary code 01 in the frame buffer and result in a dark gray shading for these pixels. In an alternative scheme, the user specification might be converted directly to the voltage value that produces this gray scale level on the output device in use.

4–3 Area Filling

An advantage of raster systems is their ability to easily store and display areas filled with a color or shading pattern. Fill patterns for such areas are stored as color or intensity values in a frame buffer. Displaying shaded areas on a vector system is considerably more difficult, since area fill requires drawing line segments within the area boundary during each refresh cycle. Various algorithms have been developed for displaying filled areas on raster systems. One method uses the boundary definition to identify which pixels belong to the interior of an area. Other methods start from a position within the area and paint outward from this point.

Scan-Line Algorithm

A scan-line algorithm uses the intersections between area boundaries and scan lines to identify pixels that are inside the area. Figure 4–5 illustrates an area outline and an individual scan line passing through a polygon. Pixel positions along the

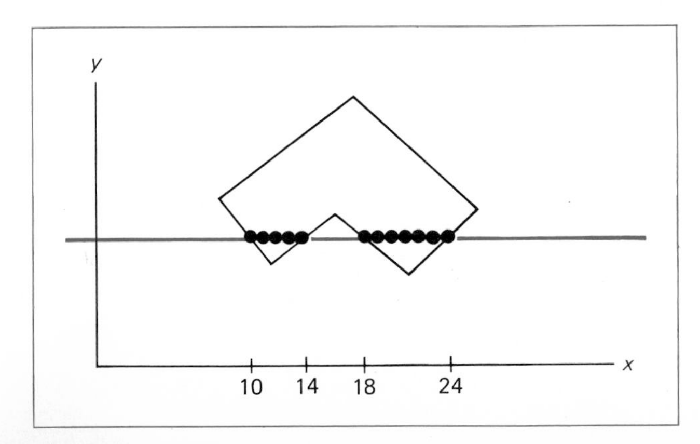

FIGURE 4–5
Interior pixels along a scan line passing through an area to be filled.

scan line that are within the polygon definition are set to the intensity or color values specified in the application program. We first consider how a scan-conversion algorithm can be set up for polygons. The algorithm can then be adapted to other figures, such as circles, by replacing the straight-line equations with the equations defining the figure boundary to be filled.

Taking each scan line in turn, a scan-conversion algorithm locates the intersection points of the scan line with each edge of the area to be filled. Proceeding from left to right, intersections are paired, and the intervening pixels are set to the specified fill intensity or color. In the example of Fig. 4–5, the four intersection points with the polygon boundaries define two stretches of interior pixels.

When a scan line intersects a polygon vertex, it may require special handling. A scan line passing through a vertex intersects two polygon edges at that position, adding two points to the list of intersections for the scan line. In Fig. 4–6, scan line 1 intersects a polygon boundary four times. Two interior stretches are defined: one from the left boundary to the vertex, and a second from the vertex to the right edge of the polygon. But scan line 2 generates five intersections with polygon edges, and resultant pairs do not correspond to the polygon interior.

To fill a polygon correctly, its overall topology must be considered. If the vertices of the polygon are specified in clockwise order, scan line 2 in Fig. 4–6 intersects a vertex whose connecting edges are monotonically decreasing in the y direction. When successive edges of the polygon are monotonically increasing or decreasing, a correct determination of interior points along a scan line is obtained by recording only one intersection point for the vertex. The vertex intersection on scan line 1 in this figure connects two lines with opposite y directions. One line has decreasing y-coordinate values, and the other has increasing y-coordinate values. When such a local minimum (or a local maximum) is encountered by a scan line, two intersection points should be generated to correctly identify interior pixels along the scan line.

Scan-conversion algorithms typically process a polygon from the top of the screen to the bottom and from left to right across each scan line. Calculations performed in such algorithms can be dramatically reduced by making use of various **coherence** properties of the objects being processed. Very often, we can expect properties of pixels along a scan line to be related, so that the properties of one pixel can be determined from those of the preceding pixel. Similarly, we can expect the properties of each scan line to be quite like those of the preceding scan line.

We can take advantage of coherence in calculating scan-line intersections with a polygon by noting that each successive scan line has a y value that is only one

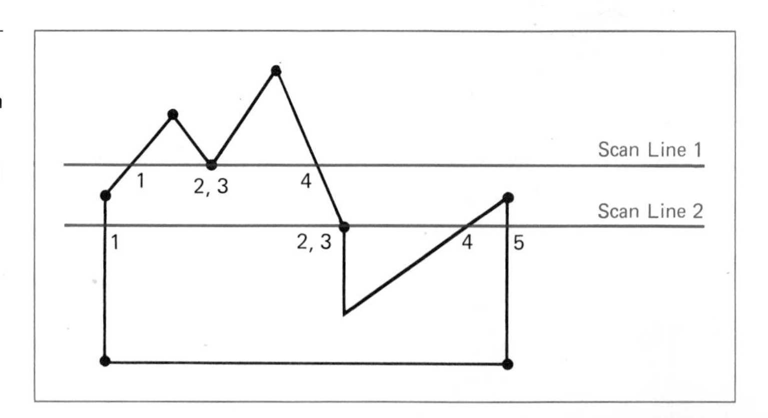

FIGURE 4–6
Intersection points along scan lines that pass through polygon vertices. Scan line 1 generates an even number of intersections that can be paired to correctly identify interior pixels. Scan line 2, however, generates an odd number of intersections.

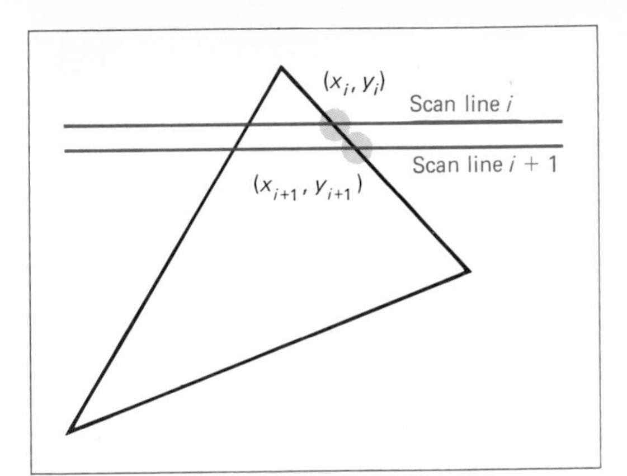

FIGURE 4–7
Two successive scan lines intersecting a polygon boundary.

unit smaller than the previous line. Figure 4–7 shows two successive scan lines crossing a side of a polygon. The slope of this polygon boundary line is

$$m = \frac{y_{i+1} - y_i}{x_{i+1} - x_i}$$

Since the changes in y coordinates is simply

$$y_{i+1} - y_i = -1$$

the x-intersection value x_{i+1} on the lower scan line can be determined from the x-intersection value x_i on the preceding scan line as

$$x_{i+1} = x_i - \frac{1}{m} \tag{4–1}$$

Once the first x-coordinate intersection value x_1 for a polygon side is found for one scan line, we can obtain the x-coordinate values for intersection points on each successive scan line by subtracting the inverse of the slope.

In many cases, we can expect an individual scan line to intersect only some of the total number of sides defining a fill area. To avoid unnecessary checking for intersection points, we can maintain a list of the polygon sides that are crossed by the current scan line. To do this, we create a list of all the polygon sides, sorted on the larger y coordinate of each side. Pointers into this list define the active list of sides for each scan line. Figure 4–8 illustrates the specification of an active list with pointers into the total list of sorted sides. As we move to the next scan line, the pointers are updated to define a new active list.

Procedure *fill_area_solid* incorporates these ideas to fill a defined polygon with a solid color. It accepts parameters specifying the number of vertices in the polygon and the coordinates of the vertices, given in clockwise order. The array of records *sides* stores information about each edge in the polygon. An entry in the array includes the larger y coordinate (*y_top*) of the edge, the length of the edge in the y direction (*delta_y*), and the inverse slope (*x_change_per_scan*). Initially, *x_int* contains the x coordinate of the vertex with the larger y value. When the edge is on the active list, this entry is updated to contain the x coordinate of the edge intersection with the current scan line.

A number of routines are defined within the scope of *fill_area_solid*. Routine *sort_on_bigger_y* creates the edge table and also returns, as parameter *bottomscan*,

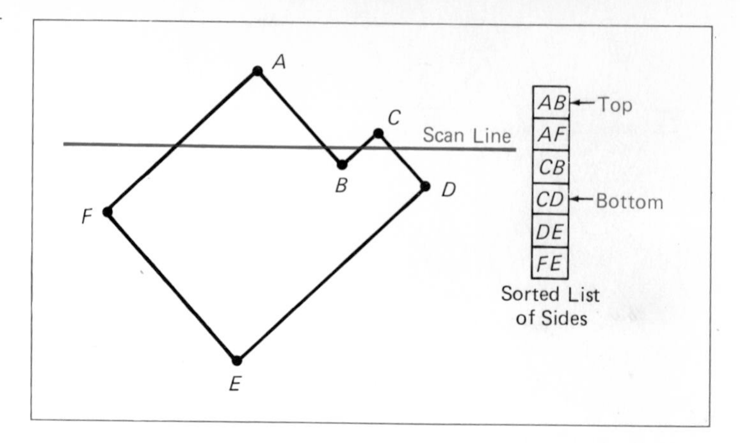

FIGURE 4-8
The active list of sides for a scan line passing through a polygon can be set up as pointers into the sorted list of all polygon sides. Pointers for the scan line shown are labeled Top and Bottom.

the *y* coordinate of the lowermost vertex. The polygon is filled from the top scan line to *bottomscan*. Routine *update_first_&_last* modifies the pointers *first_s* and *last_s* that define the active list for the current scan line. The number of intersection points for this scan line is determined by *process_x_intersections*, and *draw_lines* pairs the intersections and sets the interior pixels to the fill color. Array *sides* is then updated and made ready for the next scan line.

```
type
  points = array [1..max_points] of integer;

procedure fill_area_solid (count : integer; x, y : points);
  type
    each_entry = record
      y_top : integer;                {larger y coordinate for line}
      x_int : real;                   {x that goes with larger y}
      delta_y : integer;              {difference in y coordinates}
      x_change_per_scan : real        {x change per unit change in y}
      end;  {each_entry}
    list = array [0..max_points] of each_entry;
  var
    sides : list;
    side_count, first_s, last_s, scan, bottomscan, x_int_count, r :
        integer;

  begin {fill_area_solid}
    sort_on_bigger_y (count, x, y, sides, side_count, bottomscan);
    first_s := 1; last_s := 1;  {initialize pointers into sorted list}
    for scan := sides[1].y_top downto bottomscan do begin
        update_first_&_last (sides, count, scan, first_s, last_s);
        process_x_intersections (sides, scan, first_s, last_s,
            x_int_count);
        draw_lines (sides, scan, x_int_count, first_s);
        update_sides_list (sides)
      end  {for scan}
  end;  {fill_area_solid}
```

The first of the second-level routines, *sort_on_bigger_y*, enters edge information in *sides*. Endpoints of each edge are passed to *put_in_sides_list*. In this routine,

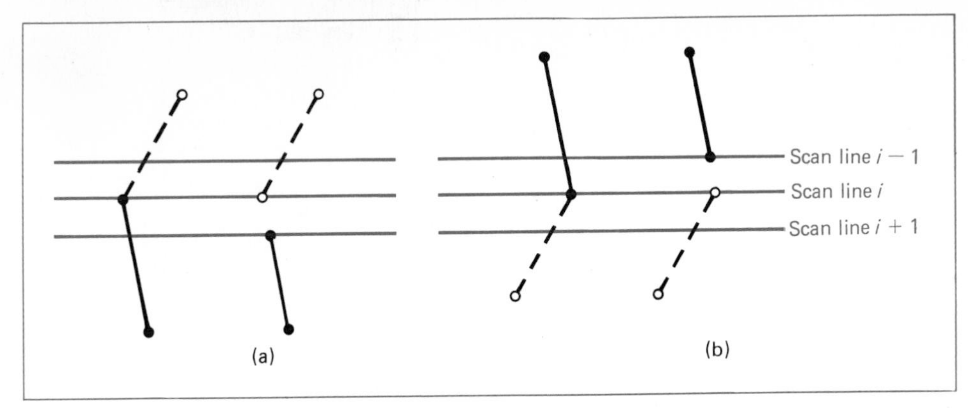

FIGURE 4–9
Adjusting endpoint y values for a polygon edge (solid line). In (a), vertex y coordinates are increasing, so the edge is lowered one unit. In (b), vertex y coordinates are decreasing, so the edge is raised one unit.

a check is made to determine if this edge and the next nonhorizontal edge are monotonically increasing or decreasing. If so, the edge being processed is shortened, to ensure that only one intersection point is generated for the scan line going through the vertex joining the edges. Figure 4–9 illustrates shortening the edge. When the y coordinates of the edges are increasing, the y value of the endpoint is decreased by 1, as in Fig. 4–9 (a). The y value is increased, as in Fig. 4–9 (b), if the y coordinates of the edges are monotonically decreasing. In both cases, the x value of the edge at the adjusted y coordinate is found.

Only nonhorizontal edges are entered into *sides*. An insertion sort based on the larger y coordinate of the edge's endpoints determines placement of the edge in the table. Information stored for the edge are the larger y coordinate and its corresponding x value, the difference between y coordinates of the edge endpoints, and the inverse slope.

```
procedure sort_on_bigger_y (n : integer; x, y : points;
            var sides : list; var side_count, bottomscan : integer);
  var  k, x1, y1, : integer;

  function next_y (k : integer) : integer;
    begin
      {returns y value of the next vertex whose
       y coordinate is not equal to y[k]}
    end;  {next_y}

  procedure put_in_sides_list (var sides : list;
              entry, x1, y1, x2, y2, next_y : integer);
    var
      maxy : integer;
      x2_temp, x_change_temp : real;

    begin
      {make adjustments for problem vertices}
      x_change_temp := (x2 - x1) / (y2 - y1);
      x2_temp := x2;
      if (y2 > y1) and (y2 < next_y) then begin
          y2 := y2 - 1;
          x2_temp := x2_temp - x_change_temp
        end
      else
```

```
        if (y2 < y1) and (y2 > next_y) then begin
          y2 := y2 + 1;
          x2_temp := x2_temp + x_change_temp
        end;
      {insert into sides list}
      if y1 > y2 then maxy := y1 else maxy := y2;
      while (entry > 1) and (maxy > sides[entry - 1].y_top) do begin
        sides[entry] := sides[entry - 1];
        entry := entry - 1
      end; {while}
      with sides[entry] do begin
        y_top := maxy;
        delta_y := abs(y2 - y1) + 1;
        if y1 > y2 then x_int := x1 else x_int := x2_temp;
        x_change_per_scan := x_change_temp
      end {with}
    end; {put_in_sides_list}

  begin {sort_on_bigger_y}
    side_count := 0;
    y1 := y[n]; x1 := x[n];            {initialize}
    bottomscan := y[n];
    for k := 1 to n do begin
      if y1 <> y[k] then begin  {put nonhorizontal edges in table}
        side_count := side_count + 1;
        {pass old point, current point, and}
        {y of next nonhorizontal point}
        put_in_sides_list (sides, side_count, x1, y1, x[k], y[k],
          next_y[k])
      end {if}
      else {horizontal}
        {draw x1,y1 to x[k],y1 with fill_color}
      if y[k] < bottomscan then bottomscan := y[k];
      y1 := y[k]; x1 := x[k]   {save for next side}
    end {for k}
  end; {sort_on_bigger_y}
```

Procedure *update_first_&_last* is called once for each scan line, and it updates the pointers defining the beginning and end of the active list. Figure 4–10 shows a polygon and its list of sorted edges. The pointers *first_s* and *last_s* remain positioned as shown for processing of all lines between the polygon's top scan line and scan line i. When processing of scan line i is completed, the active list is redefined for the next scan line ($i + 1$). Pointer *last_s* is moved to the table's fifth entry (GA), but *first_s* stays where it is since edge AB intersects scan line $i + 1$. Edge FG is now no longer an active side, but it is embedded within the range of the active-list pointers.

One way to eliminate edge FG from further processing is to shift the entries for edges AB, BC, and EF down one spot in *sides* to maintain a contiguous active list. To avoid shifting edges around in the table, we use the *delta_y* parameter of each edge to indicate the status of that edge in the active list. When this entry has the value 0, the edge is no longer considered active even though it is included within the range of the pointers. This entry is originally set to the difference in y values of the two endpoints, and its value is equal to the number of times this edge will intersect

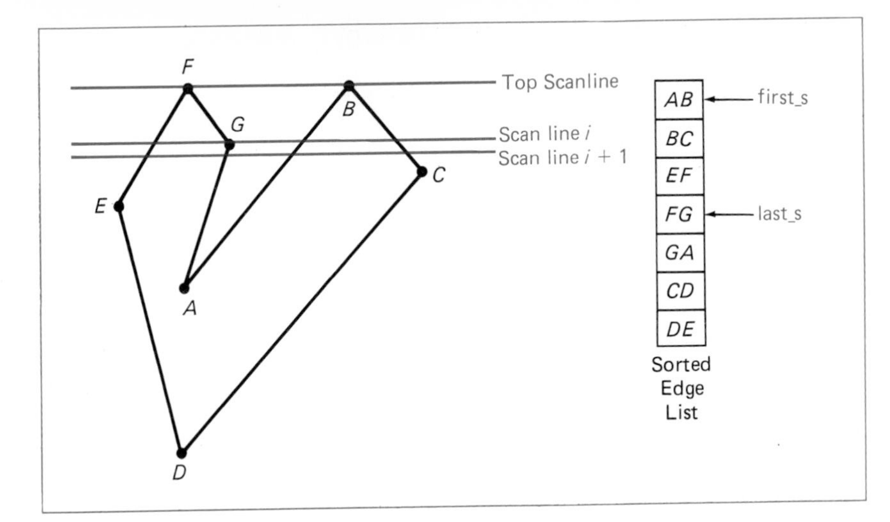

FIGURE 4–10
Processing of pointers for the active list within the sorted edge list for a specified fill area. The sorted edge list is obtained from the input vertices, specified here in the order *A* through *G*. Positions of the pointers FIRST_s and LAST_s show the active list for all scan lines from the top of the polygon through scan line *i*.

with subsequent scan lines. Each time we process a scan line, *delta_y* is decremented by 1. Active edges, then, are those that have a *delta_y* value greater than 0 and that reside between the *first_s* and *last_s* pointers. The pointer *first_s* is updated only when the *delta_s* value of the entry it points to has become 0.

```
procedure update_first_&_last (sides : list; count, scan : integer;
          var first_s, last_s : integer);
  begin
    while (sides[last_s + 1].y_top >= scan) and
          (last_s < count) do
      last_s := last_s + 1;
    while sides[first_s].delta_y = 0 do
      first_s := first_s + 1
  end;  {update_first_&_last}
```

Taking note of the *delta_y* value for each edge, *process_x_intersections* identifies elements of the active list that intersect the current scan line. These edges are shuffled within the active list to facilitate pairing the intersections that define the polygon interior. The list is reordered on increasing *x_int* values. For example, the active list shown in Fig. 4–10 would assume the new ordering EF, FG, AB, BC. Movement needed within the active list is minimized by the fact that an established ordering will remain constant across a number of scan lines.

The task of identifying pairs of intersections is left to the routine *draw_lines*. This routine takes two successive *x_int* values and sets the intervening pixels on this scan line to the specified fill color.

```
procedure process_x_intersections (var sides : list;
                                   scan, first_s, last_s : integer;
                                   var x_int_count : integer);
  var k : integer;

    {swap is predefined and reverses placement
     of two entries within the table sides}
```

```
procedure sort_on_x (entry, first_s : integer);
  begin
    while (entry > first_s) and
          (sides[entry].x_int < sides[entry-1].x_int) do begin
      swap (sides[entry], sides[entry - 1]);
      entry := entry - 1
    end {while}
  end; {sort_on_x}

begin {process_x_intersections}
  x_int_count := 0;
  for k := first_s to last_s do
    if sides[k].delta_y > 0 then begin
      x_int_count := x_int_count + 1;
      sort_on_x (k, first_s)
    end
end; {process_x_intersections}

procedure draw_lines (sides : list; scan, x_int_count, index : integer);
  var k, x, x1, x2 : integer;
  begin
    for k := 1 to round(x_int_count / 2) do begin
      while sides[index].delta_y = 0 do index := index + 1;
      x1 := round(sides[index].x_int);
      index := index + 1;
      while sides[index].delta_y = 0 do index := index + 1;
      x2 := round(sides[index].x_int);
      for x := x1 to x2 do
        set_pixel (x, scan, fill_color);
      index := index + 1
    end {for k}
  end; {draw_lines}
```

Procedure *update_sides* is the last of the second-level routines. Its job is to ready the entries in *sides* for processing of the next scan line. For each edge on the active list, *delta_y* is decremented, and *x_change_per_scan* is subtracted from *x_int*.

```
procedure update_sides_list (var sides : list);
  var k : integer;
  begin
    for k := first_s to last_s do with sides[k] do
      if delta_y > 0 then begin
        {determine next x_int, decrease delta_y}
        delta_y := delta_y - 1;
        x_int := x_int - x_change_per_scan
      end {if delta_y > 0}
  end; {update_list}
```

To produce a patterned fill, we modify these scan-line procedures so that a selected pattern is superimposed onto the scan lines. Beginning from a specified start position for a pattern fill, the rectangular patterns would be positioned horizontally across corresponding scan lines and vertically across groups of scan lines. Figure 4–11 illustrates the method for positioning a pattern across a defined area.

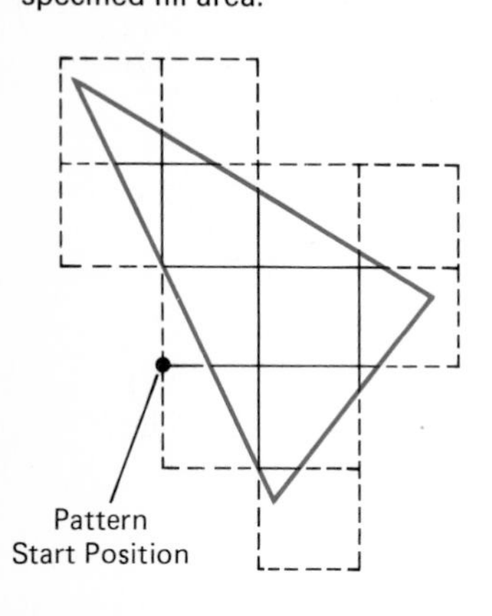

FIGURE 4–11
From the start position shown, patterns are laid out so as to cover all scan lines passing through the interior of the specified fill area.

Antialiasing Area Boundaries

The antialiasing concepts we have discussed for lines can be applied to the boundaries of areas to remove the jagged appearance generated on raster systems. We can implement these procedures into a scan-line algorithm to smooth the area outline as the area is generated.

If system capabilities permit the repositioning of pixels, area boundaries can be smoothed by adjusting boundary pixel positions so that they are along the line defining an area boundary. Other methods adjust each pixel intensity at a boundary position according to the percent of pixel area that is inside the boundary. This situation is depicted in Fig. 4–12, where each pixel area is represented as a small rectangle.

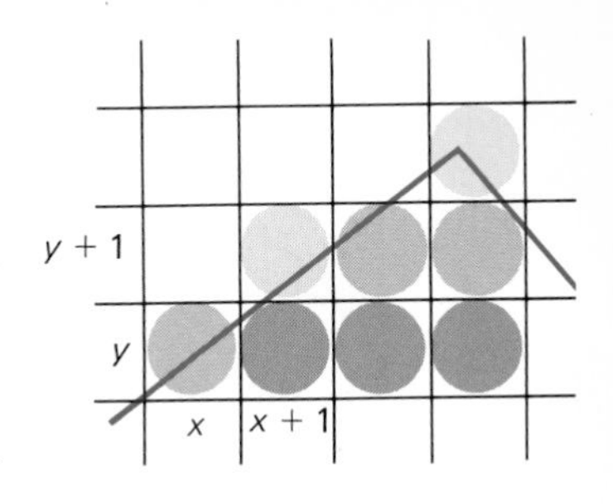

FIGURE 4–12
Adjusting pixel intensities along an area boundary.

In this example, the pixel at position (x, y) has about half its area inside the polygon boundary. Therefore, the intensity at that position would be adjusted to one-half its assigned value. At the next position $(x + 1, y + 1)$ along the boundary, the intensity is adjusted to about one-third the assigned value for that point. Similar adjustments, based on the percent of pixel area coverage, are applied to the other intensity values around the boundary.

Various techniques can be used to estimate the amount of each pixel inside the boundary of an area. One way to estimate this interior area is to subdivide the total area and determine how many subdivisions are inside the boundary. In Fig. 4–13, each pixel area is partitioned into four parts so that the original 5 by 5 grid of pixels is subdivided into a 10 by 10 grid. This turns the original five scan lines covering these pixels into ten scan lines for the subdivisions. A scan-line method can then be used to determine which subdivisions are on or inside the boundary line, as shown in Fig. 4–14. In this example, the two scan lines are processed to determine that three subdivisions are inside the boundary. The pixel intensity, therefore, is adjusted to 75 percent of its assigned value. With this method, the accuracy of intensity setting depends on the number of pixel subdivisions used.

FIGURE 4–13
A 5 by 5 pixel section of a raster display subdivided into a 10 by 10 grid.

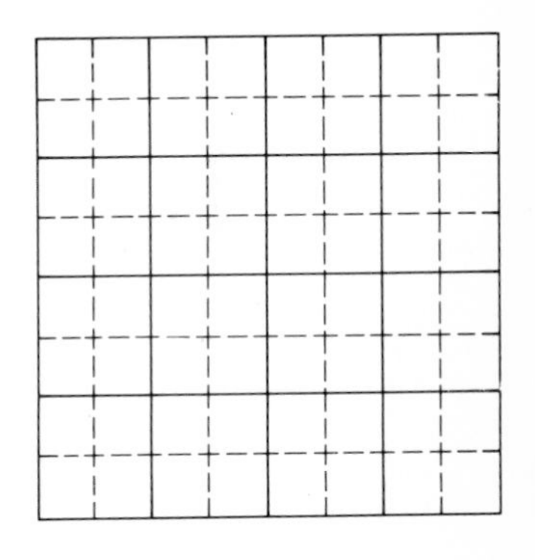

Another method for determining the percent of pixel area within a boundary, developed by Pitteway and Watkinson, is based on Bresenham's line algorithm. This algorithm selects the next pixel along a line by determining which of two pixels is closer to the line, as determined by the sign of a parameter p that measures relative distances of the two pixels from the line. By slightly modifying the form of p, we obtain a quantity that gives the percent of pixel area that is covered by a surface.

We consider the method for a line with slope m in the range from 0 to 1. In Fig. 4–15, a line with equation $y = mx + b$ is shown on a pixel grid. Assuming that the pixel at position (x_i, y_i) has been plotted, the next pixel nearest the line at $x = x_i + 1$ is either the pixel at y_i or the one at $y_i + 1$. We can determine which pixel is nearer with the calculation

$$y - y_{\text{mid}} = (mx_i + b) - (y_i + 0.5) \tag{4–2}$$

This gives the distance from the y coordinate on the line to the halfway point between pixels at positions y_i and $y_i + 1$. If this difference calculation is negative, the pixel at y_i is closer to the line. If the difference is positive, the pixel at $y_i + 1$ is closer. We can adjust this calculation so that it produces a positive number in the range from 0 to 1 by adding the quantity $1 - m$:

$$p = (mx_i + b) - (y_i + 0.5) + (1 - m) \tag{4–3}$$

FIGURE 4–14
A subdivided pixel area with three subdivisions inside the boundary line.

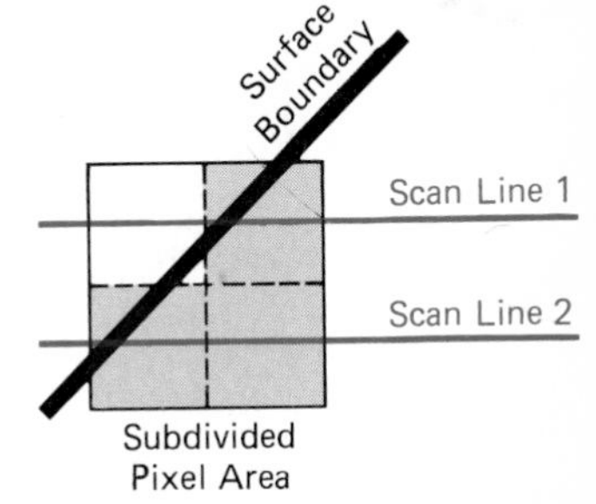

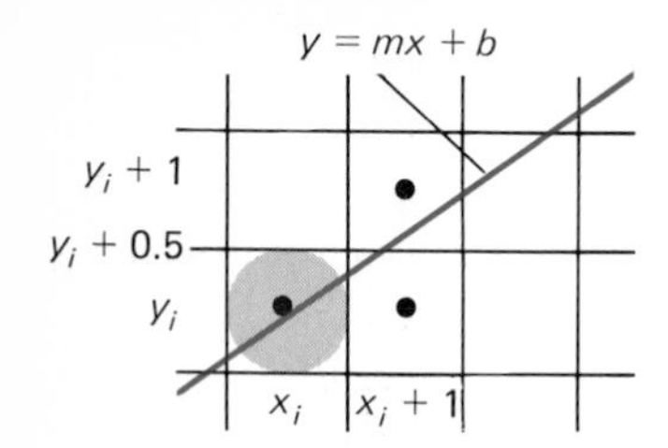

FIGURE 4–15
Boundary line of an area passing through a pixel grid section.

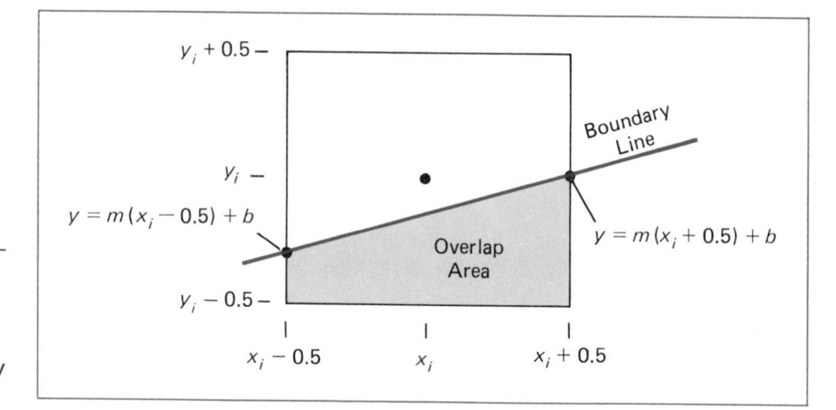

FIGURE 4–16
A pixel represented as a rectangular area with center at (x_i, y_i) has an overlap with a bounded region as indicated by the shaded area.

Now the pixel at y_i is nearer if $p < 1 - m$, and the pixel at $y_i + 1$ is nearer if $p > 1 - m$.

Parameter p also measures the amount of a pixel that is overlapped by an area. For the rectangular pixel in Fig. 4–16, the part of the pixel that is inside the boundary line has an area that can be calculated as

$$area = mx_i + b - y_i + 0.5 \tag{4-4}$$

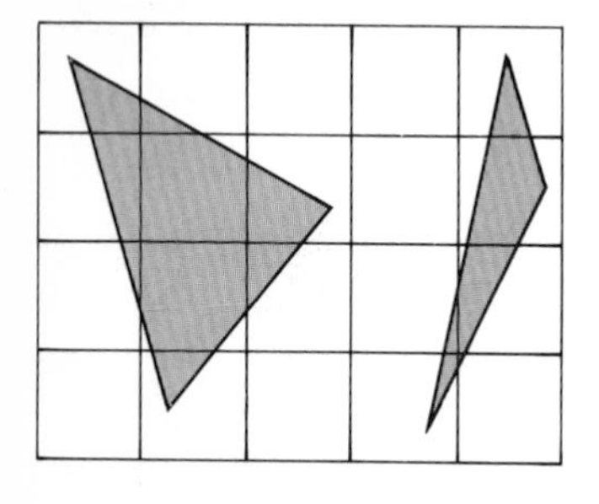

FIGURE 4–17
Areas on a pixel grid that have more than one boundary line passing through a pixel region.

We get the same result from parameter p when we substitute the coordinates $(x_i + 1, y_i + 1)$ for (x_i, y_i) in Eq. 4–3. Therefore, by evaluating p at each successive pixel position, we can determine the percent of area coverage for the preceding pixel. This value then sets the intensity level for the pixel.

We can incorporate this method into a scan-line algorithm by adjusting the pixel intensities at the points on each scan line that intersect the area boundaries. Also, the parameters in p can be adjusted to accommodate lines with negative slopes and slopes greater than 1. This provides an efficient method for antialiasing boundaries when only one edge of an area passes through a pixel. More than one side of a boundary can cross a single pixel at polygon vertices and when very skinny areas are to be displayed (Fig. 4–17). For these cases, we can apply other methods, such as pixel subdivision.

The various antialiasing methods can be applied to polygon areas or to regions with curved boundaries. In either case, boundary equations are used to determine the position of the boundaries relative to pixel positions. Using a scan-line method, coherence techniques are used to simplify the calculations from one scan line to the next.

Boundary-Fill Algorithm

As an alternative to the scan-line method, an area can be filled by starting at a point inside the figure and painting the interior in a specified color or intensity. Painting proceeds until the figure's boundary is encountered. This method, called the **boundary-fill algorithm,** is useful in interactive sketching and painting packages. Using a graphics tablet or other interactive device, a user sketches a figure outline, picks an interior point, and selects a color or pattern from a color palette. The system then paints the figure interior.

A boundary-fill algorithm accepts as input the coordinates of the interior point (x, y), a fill color, and a boundary color. Starting from (x, y), neighboring points are tested to determine whether they are of the boundary color. If not, they are painted with the fill color. In this way, all points are tested up to the area boundary.

Figure 4–18 shows two methods for proceeding to neighboring points from the interior start point (x, y). In Fig. 4–18 (a), four neighboring points are tested. These are the points that are above, below, to the right, and to the left of the starting point. Areas filled by this method are called **4-connected.** Another approach, shown in Fig. 4–18 (b), is used to fill more complex figures. Here the set of neighboring points also includes the four diagonal pixels. Fill methods using this approach are called **8-connected.** An 8-connected boundary-fill algorithm would correctly fill the interiors of the areas defined in Fig. 4–19, but a 4-connected boundary-fill algorithm produces the partial fill shown.

The following procedure illustrates a recursive method for filling a 4-connected area with an intensity specified in parameter *fill_color* up to a border color specified by *boundary*.

```
{inquire_color is predefined function that
 returns the current color of point (x, y)}

procedure boundary_fill (x, y, fill_color, boundary : integer);
  var present_color : integer;
  begin
    present_color := inquire_color (x, y);
    if (present_color <> boundary) and
       (present_color <> fill_color) then begin
        set_pixel (x, y, fill_color);
        boundary_fill (x + 1, y    , fill_color, boundary);
        boundary_fill (x - 1, y    , fill_color, boundary);
        boundary_fill (x    , y + 1, fill_color, boundary);
        boundary_fill (x    , y - 1, fill_color, boundary);
      end {if present color}
  end;  {boundary_fill}
```

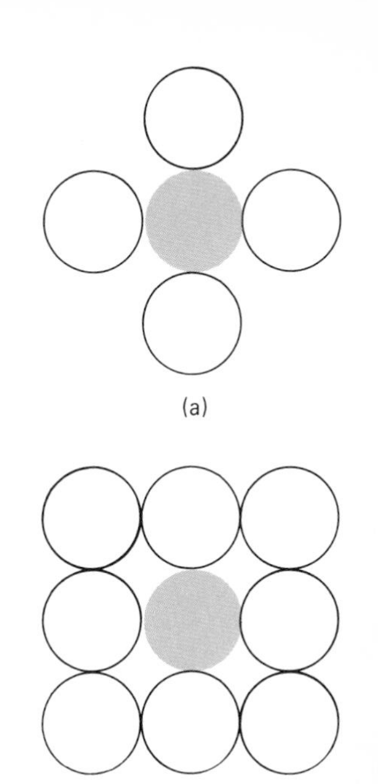

FIGURE 4–18
Boundary-fill methods applied to (a) a 4-connected area and (b) an 8-connected area. Hollow circles represent pixels to be tested from the interior start point (solid).

FIGURE 4–19
The defined areas (a) are only partially filled (b) by a 4-connected boundary-fill algorithm.

Since this procedure requires considerable stacking of neighboring points, more efficient methods are generally employed. These methods fill in all points along each horizontal line, one at a time, instead of testing 4-connected or 8-connected neighboring points. The general approach is first to completely fill in the scan line containing the starting point. Then process all lines below the start line, down to the bottom boundaries. Finally, all lines above the start line are processed, up to the top boundaries.

Flood-Fill Algorithm

Another way to fill an area from some start point (x, y) is to specify an interior color value that is to be replaced by the fill color. Instead of looking for a specified boundary color, this **flood-fill algorithm** looks for the interior color that is to be replaced. Using either the 4-connected or 8-connected approach, this procedure stops when no neighboring points have the specified interior color. In a scan-line approach, adjacent pixels with the specified interior color would be set to the fill color until a different color is encountered.

FIGURE 4–20
Interior polygon fill styles.

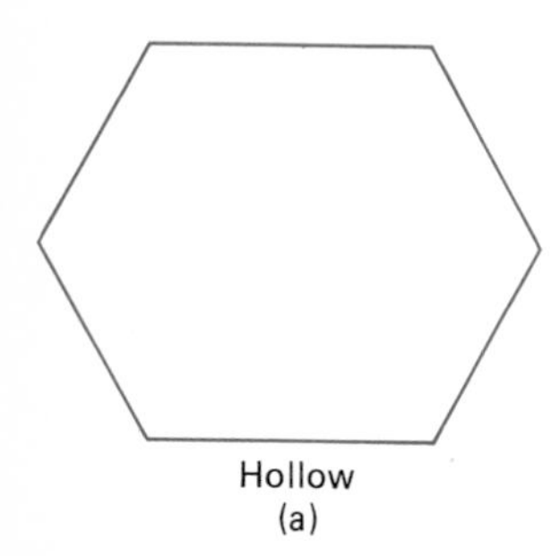

Hollow
(a)

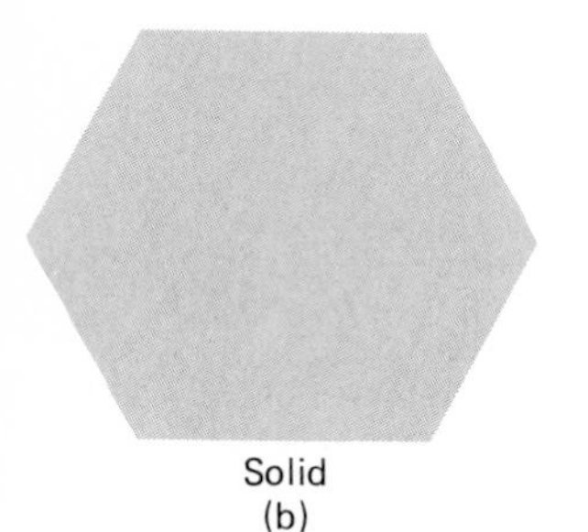

Solid
(b)

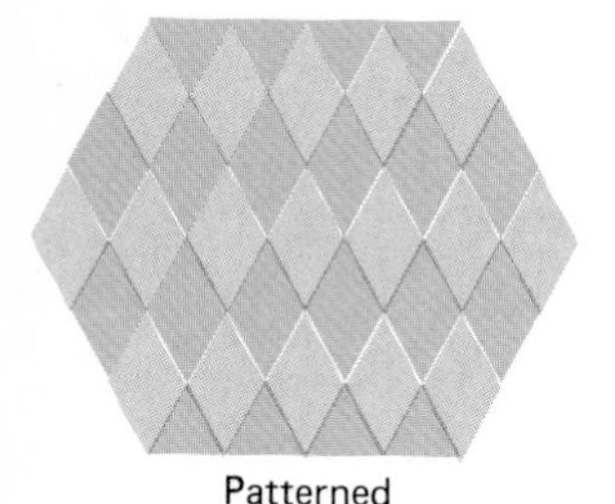

Patterned
(c)

Area-Filling Commands

Several attribute options for fill areas can be provided to a user of a graphics package. These attributes include fill style, fill color, and fill pattern. Fill style refers to the general type of interior, such as hollow or patterned, and fill pattern references the particular pattern code to be used. Parameters specifying these options can be added to the system attribute list and used by the scan-line algorithm to generate fill areas.

A user might select a particular type of interior fill with a command such as

```
set_fill_area_interior_style (fs)
```

Possible types of fill that could be chosen for the fillstyle parameter *fs* are *hollow*, *solid*, and *patterned* (Fig. 4–20). As with line attributes, the fill style parameter would be recorded in the list of system attributes so that areas defined by subsequent commands (such as *fill_area* and *ellipse* would be displayed with this fill style.

Hollow areas are displayed with only a boundary outline. The color for an area outline, or for a solid interior, is chosen with an area color command:

```
set_fill_area_color_index (fc)
```

where fill color *fc* is set to the desired color code.

Patterns can be specified with the command

```
set_fill_area_pattern_index (pi)
```

where the pattern index parameter *pi* is assigned a pattern code. If integer codes are used to select patterns, the following set of statements would fill the area defined in the *fill_area* command with the second pattern type stored in the pattern table:

```
set_fill_area_interior_style (patterned);
set_fill_area_pattern_index (2);
fill_area (n, x, y);
```

A set of standard patterns can be predefined for a user, or a pattern table could be created with user-defined patterns. Commands to create patterns can be made available to a user in the form

```
set_pattern_representation (pi, nx, ny, cp)
```

Parameter *pi* sets the pattern index number, *nx* and *ny* specify the number of points in the *x* and *y* directions to be defined in the pattern, and *cp* is a two-dimensional color pattern array of size *nx* by *ny*. The program segment

```
cp[1,1] := 4; cp[2,2] := 4;
cp[1,2] := 0; cp[2,1] := 0;
set_pattern_representation (1, 2, 2, cp);
```

could be used to set the first entry in the pattern table of Fig. 4–21. This pattern, when applied to a defined area, produces alternate red and black diagonal lines.

FIGURE 4–21
A pattern table with two entries, using the color codes of Fig. 4–2.

INDEX (PI)	PATTERN (CP)
1	$\begin{pmatrix} 4 & 0 \\ 0 & 4 \end{pmatrix}$
2	$\begin{pmatrix} 2 & 1 & 2 \\ 1 & 2 & 1 \\ 2 & 1 & 2 \end{pmatrix}$

Positioning patterns on the interior of a fill area is specified with the statement

```
set_pattern_reference_point (xp, yp)
```

The scan-line algorithm uses the coordinates (*xp*, *yp*) to fix the lower left corner of the rectangular pattern. From this starting position, the pattern is then mapped onto all scan lines covering the area to be filled. To illustrate the use of the pattern commands, the following program example creates a black-and-white pattern on the interior of the area in Fig. 4–22.

```
type
  points = array [1..max_points] of integer;

procedure fill_triangle;
  var
    pattern : array [1..3,1..3] of integer;
    x, y : points;
```

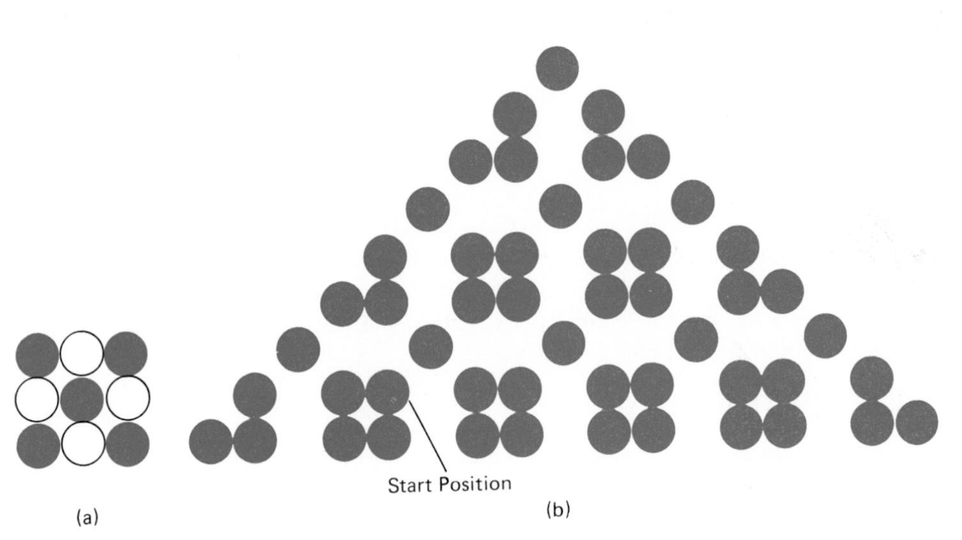

FIGURE 4–22
A defined fill pattern (a) is superimposed on a triangular fill area to produce the patterned display (b).

```
begin
  pattern[1,1] := 1; pattern[1,2] := 0; pattern[1,3] := 1;
  pattern[2,1] := 0; pattern[2,2] := 1; pattern[2,3] := 0;
  pattern[3,1] := 1; pattern[3,2] := 0; pattern[3,3] := 1;
  set_pattern_representation (8, 2, 2, pattern);
  x[1] := 10; y[1] := 10;
  x[2] := 26; y[2] := 10;
  x[3] := 18; y[3] := 18;
  set_fill_area_interior_style (patterned);
  set_fill_area_pattern_index (8);  {use pattern from above}
  set_pattern_reference_point (14, 11);
  fill_area (3, x, y);
end;  {fill_triangle}
```

Another option that can be made available for patterned fill is modification of the pattern size. Graphics packages providing this option include a command to change the dimensions of previously defined pattern arrays. Pattern size can be changed relative to the lower left corner of the array. If a pattern is to be reduced, an appropriate portion of the lower left corner of the array is maintained. For enlargement, the pattern is expanded out from the lower left corner.

4–4 Character Attributes

The appearance of displayed characters is controlled by attributes such as color, style, and orientation. Procedures for displaying character attributes operate on the grid definitions of characters and their placement in the raster. Attributes can be set both for character strings (text) and for additional characters defined as marker symbols.

Text Attributes

Some packages provide a number of choices for text style. The different styles can be made available to the package as predefined sets of grid patterns. Each style can be assigned a text font code, and the character style chosen for display is determined by the current text font setting in the system attribute list. An applications programmer could select a particular code with the text font parameter *tf* in the command

```
set_text_font (tf)
```

Color settings for displayed text are stored in the system attribute list and used by the procedures that load character definitions into the raster array. When a character string is to be displayed, the character grid patterns are entered into the raster with the current color values. Control of text color (or intensity) can be managed from an application program with

```
set_text_color_index (tc)
```

where text color parameter *tc* specifies an allowable color code.

Text size can be adjusted by scaling the height and width of characters. This scaling is applied to the character grid definitions in the horizontal and vertical directions before storage in the frame buffer. Graphics packages can provide separate size adjustment for height and width, or a single attribute parameter can be

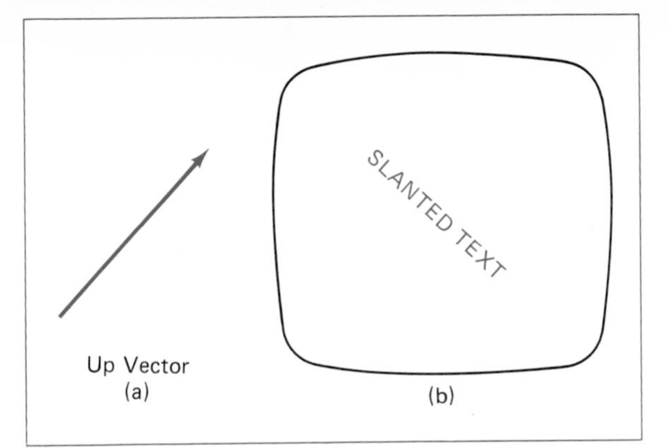

FIGURE 4–23
The direction for the up vector (a) controls the orientation of text (b).

used. When a single parameter is used, a standard ratio of width to height is maintained. A command for specifying text size is

```
set_character_height (ch)
```

This command determines both the height and width of characters. A value of 1 for the character height parameter *ch* displays characters in the standard size. Values greater than 1 enlarge text, and values smaller than 1 scale down the character sizes.

The orientation for a displayed character string can be set according to the direction of a **character up vector.** Two parameters can be used to specify the slope of this vector, and text is then aligned so that the tops of characters are in the direction of the up vector. A procedure for orienting text would rotate each character string so that the string is stored in the raster in a direction that is perpendicular to the direction of the up vector. Figure 4–23 illustrates the appearance of text oriented by an up vector at 45°. The direction of the up vector is set with the command

```
set_character_up_vector (dx, dy)
```

where the slope of the vector is equal to the ratio *dy/dx*.

It is useful in many applications to be able to arrange character strings vertically or horizontally (Fig. 4–24). An attribute parameter for this option could be used to direct a routine to load character patterns into the raster horizontally from the start position or vertically down from the start position. These options could be set by a user with the statement

```
set_text_path (tp)
```

where the text path parameter *tp* is assigned a value of *right* or *down*. Other possible options for text path are *left* and *up*.

Character strings can also be oriented by using a combination of up vector and text path specifications to produce slanted horizontal and vertical text. Figure 4–25 shows the directions of character strings generated by the various text path settings for a 45° up vector. Examples of text generated for *down* and *right* text path values with a 45° up vector are illustrated in Fig. 4–26.

Another handy attribute for character strings is alignment. Figure 4–27 shows various possibilities for text alignment with horizontal and vertical strings. The characters in a string can be aligned on the left, on the right, at the top or bottom, or they can be centered on some position. Once the alignment parameters have been given, procedures for displaying text load the characters into the frame buffer accordingly. A package can provide for alignment with the command

FIGURE 4–24
Text path attributes can be set to produce horizontal or vertical arrangements of character strings.

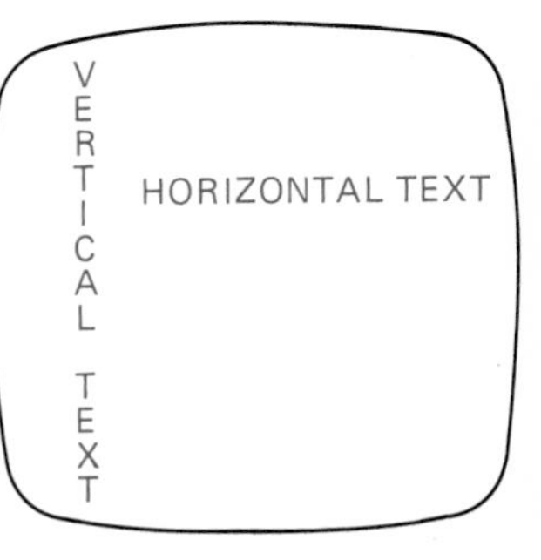

FIGURE 4–25
An up-vector direction (a) can be used to control directions for text path (b).

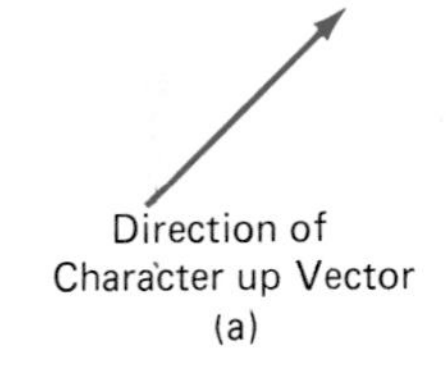

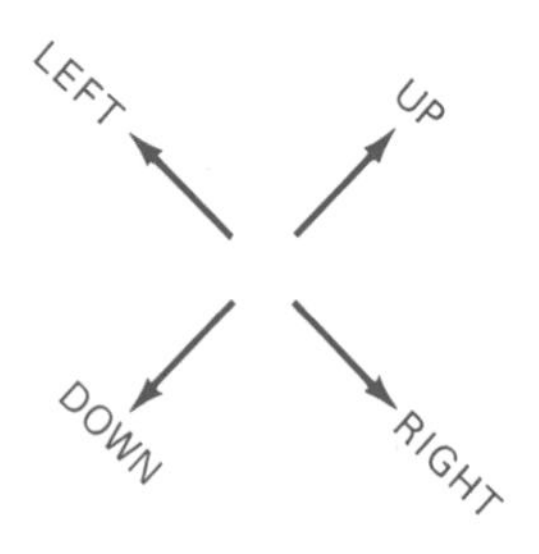

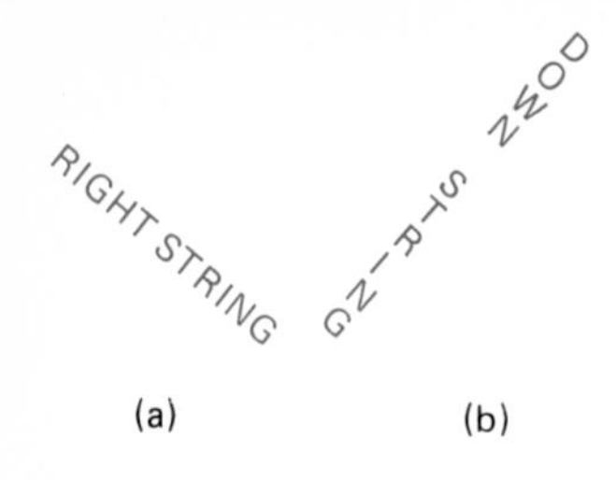

FIGURE 4–26
The 45° up vector in Fig. 4–25 produces the string (a) for a right path and the string (b) for a down path.

FIGURE 4–27
Possible character string alignments.

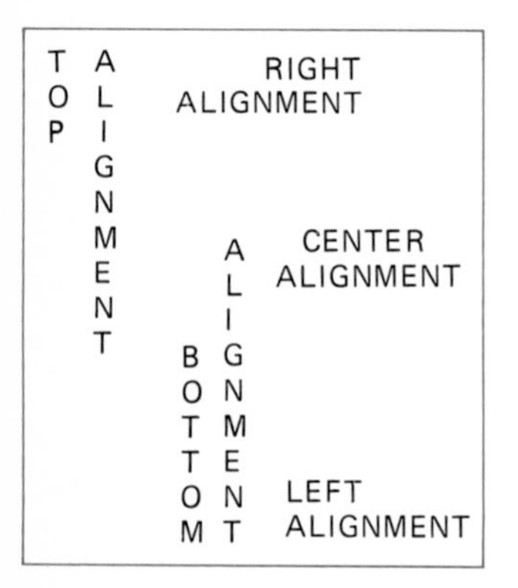


```
set_text_alignment (h, v)
```

Parameters *h* and *v* control horizontal and vertical alignment, respectively. Value *right*, *left*, or *center* is assigned to *h*, and *v* can be assigned *top* or *bottom*. These values specify which part of the string is to be placed at the starting coordinates given in the *text* command.

Marker Attributes

Basic marker attributes are the same as those for text. Markers can be different colors or sizes, and different marker symbols can be chosen. These attributes are implemented by procedures that load a selected symbol into the raster at the defined positions with the specified color and size.

Packages can provide for user settings of these attributes with *set_marker* commands. A command to set the marker type parameter is

```
set_marker_type (mt)
```

Similarly, different colors can be chosen with

```
set_marker_color_index (mc)
```

and marker size is adjusted with

```
set_marker_size_scale_factor (ms)
```

where the marker scale parameter *ms* specifies character enlargement when assigned values greater than 1 and size reductions for values less than 1.

4–5 Inquiry Functions

When attributes are repeatedly changed in an application program, it can be helpful to a user to be able to check current settings of the attribute values. Inquiry functions are used for this purpose. They allow current attribute values to be copied into specified parameters. These attribute values can then be saved for later use, or they could be used to provide for program checks in case of errors.

Examples of inquiry functions that can be included in a graphics package are

```
inquire_linetype (ilt);
```

and

```
inquire_fill_area_color_index (ifc)
```

Current values for line type and fill color indices are then copied into parameters *ilt* and *ifc*. Similar functions would be used for the other attribute options.

An example of the use of inquiry functions to save an attribute value for later use is given in the following program segment. Here, the user saves the current line type value while drawing with some new line style.

```
inquire_linetype (old_style);
set_linetype (new_style);
        .
        .
        .
set_linetype (old_style);
```

4–6 Bundled Attributes

The attribute-setting commands we have described are designed to specify exactly how an output primitive is to be displayed. Each line type parameter value, for instance, is defined as one explicit line style. These attributes are called **unbundled attributes,** and they are meant to be used with an output device that is capable of displaying primitives with the specified attributes. If an application program, employing unbundled attributes, is interfaced to several types of output devices, some of the devices may not have the capability to reproduce the intended attributes. A program using unbundled color attributes may have to be modified to produce acceptable output on a monochromatic monitor.

Unbundled attribute commands provide a simple and direct method for specifying attributes when a single type of output device is to be used. When several types of output devices are available at a graphics installation, it is convenient for a user to be able to say how a particular attribute setting is to be interpreted on each of the different devices. In this case, an application program could specify an attribute index that maps into a table defining which attributes, corresponding to the input index, are to be generated on the different output devices. Attributes specified in this manner are called **bundled attributes.** The table defining specific attributes for each output device is called the **bundle table.** An application program can set up its own bundle table for the various attributes, or the bundle tables can be established by the installation. Graphics packages can be designed to accommodate both types of attribute specifications, with one type chosen as the default mode.

With bundle attributes, an application program must specify the type of output device to be used, as well as the attribute index for the output primitives. As a means for identifying various devices, a user **workstation** is defined to be the set of input/output devices available for a particular application. Figure 4–28 illustrates some categories of workstations that might be defined at an installation. Some of these categories include both input and output devices; others are output devices only. To use a particular workstation, a command such as

```
activate_workstation (ws)
```

is issued in the application program. All output is then directed to the devices defined by the workstation code *ws*. When the output devices are to be changed, a new workstation is activated. Attributes for each workstation type can be set by a user with separate commands for each of the attribute categories: line, area fill, text, and marker.

WORKSTATION TYPE CODE	WORKSTATION DESCRIPTION
1	Raster Color Monitor with Keyboard
2	Raster Color Monitor with Keyboard and Graphics Tablet
3	DVST with Keyboard and Thumb Wheels
4	Vector Refresh Monitor with Keyboard
5	Color Plotter

FIGURE 4–28 Possible categories of workstations used with bundled attribute specifications.

Line Attributes

To specify attributes for lines, a single command,

```
set_line_index (li)
```

is used in place of the three separate commands for line type, line width, and line color. Line index parameter *li* points into the bundle table, which defines possible attribute combinations for the workstation specified.

The bundle table for line attributes is established for the various workstations with

```
set_line_representation (ws, li, lt, lw, lc)
```

Parameter *ws* is the workstation code, and *li* is the line index. Attribute parameters *lt*, *lw*, and *lc* set the line type, line width, and line color specifications. For example, the following commands set line attributes for index number 3 on two different workstations:

```
set_line_representation (1, 3, 2, 0.5, 1);
set_line_representation (4, 3, 1, 1, 7);
```

These statements might be used to display dashed lines at half thickness in a blue color on workstation number 1, while on workstation 4, this same index generates solid, standard-sized white lines.

Each workstation code may have several line indices, defining a range of possible line attributes. For example, line indices 1 through 8 might define solid lines of standard width in eight different colors. Line indices 9 through 16 could be used to specify dashed lines, with standard widths, in the same eight colors. In this way, an extensive table of line indices can be defined for each workstation.

Color and Intensity Attributes

Color indices specified for line, fill area, and text primitives are pointers into color and intensity lookup tables. These lookup tables can be set up for each workstation independently with the command

```
set_color_representation (ws, ci, r, g, b)
```

Again, *ws* identifies the workstation, and *ci* is the color index parameter. Floating-point values for parameters *r*, *g*, and *b* could be assigned in the range from 0 to 1, establishing the amount of red, green, and blue components in the color defined by index *ci*.

An illustration of how a table could be set up for a color monitor is given in Fig. 4–29. Figure 4–30 shows a possible intensity table specification.

Area-Filling Attributes

For a bundled attribute specification, the command

```
set_fill_area_index (fi)
```

selects attributes on each workstation for a given fill index *fi*. This single statement replaces the three commands for setting fill style, fill, color, and pattern index. Integer values are assigned to *fi*, which map into a bundle table for fill attributes.

COLOR INDEX CI	RED COMPONENT (R)	GREEN COMPONENT (G)	BLUE COMPONENT (B)	COLOR DESCRIPTION
0	0	0	0	Black
1	0.25	0	0	Shades of Red
2	0.50	0	0	
3	0.75	0	0	
4	1.0	0	0	
5	0	0.25	0	Shades of Green
6	0	0.50	0	
7	0	0.75	0	
8	0	1.0	0	
⋮	⋮	⋮	⋮	⋮

FIGURE 4–29
Color table defined for a particular workstation specification.

COLOR INDEX CI	RED COMPONENT (R)	GREEN COMPONENT (G)	BLUE COMPONENT (B)	GRAY-SCALE DESCRIPTION
0	0	0	0	Black
1	0.33	0.33	0.33	Dark Gray
2	0.67	0.67	0.67	Light Gray
3	1.0	1.0	1.0	White

FIGURE 4–30
A gray-scale intensity table defined for a particular workstation category.

Bundle tables for fill areas are set with

```
set_fill_area_representation (ws, fi, fs, fc, pi)
```

which defines the attribute list for fill index *fi* on workstation *ws*. Parameters *fs*, *fc*, and *pi* are assigned values for the fill style, color code, and pattern index, respectively.

Commands for setting pattern attributes are similar to those used in the unbundled mode. The difference is that a workstation must be named for each pattern defined. A user could set pattern representation, pattern reference point, and pattern size with

```
set_pattern_representation (ws, fsi, nx, ny, cp);
set_pattern_reference_point (xp, yp);
set_pattern_size (sx, sy);
```

Color codes are set in the *nx* by *ny* array *cp*. Position (*xp*, *yp*) establishes the starting point for the left corner of the rectangular area *cp*, and the size of the pattern to be used is defined by *sx* and *sy*.

Text Attributes

An index for text attributes is chosen with

```
set_text_index (ti)
```

and a user no longer employs individual commands for setting text font, color, or spacing. The text index *ti* maps into the text bundle table, which is established with

```
set_text_representation (ws, ti, tf, te, ts, tc)
```

Parameters *tf* and *tc* set the text font and text color for workstation *ws*. Additional parameters, *te* and *ts*, are used to adjust character size and spacing. The text expansion parameter *te* defines the ratio of width to height for characters, and *ts* specifies the spacing to be used between characters.

Other text attributes are set by the same commands used with unbundled attributes. They include setting the character up vector, text path, character height, and text alignment.

Marker Attributes

The marker index setting is accomplished with

```
set_marker_index (mi)
```

and the marker bundle table for this index is defined with the command

```
set_marker_representation (ws, mi, mt, ms, mc)
```

This defines the marker type *mt*, marker scale factor *ms*, and marker color *mc* for index *mi* on workstation *ws*. Values for the attributes parameters *mt*, *ms*, and *mc* can be set with the same commands used with unbundled attributes.

4–7 Summary

In this chapter, we have explored the various types of attributes that affect the appearance of output primitives. Procedures for displaying primitives use attribute settings to adjust the output of line-, area-, and character-generating algorithms.

The basic line attributes are those for line type, line color, and line width. These attributes are generated with a line-drawing algorithm modified to accommodate attribute choices by a user.

Area attributes include the type of interior fill, the area color, and the fill pattern. The basic area-filling procedure is the scan-line algorithm, which can be adapted to generate patterned fill. Two other methods for filling areas are the boundary-fill and flood-fill algorithms.

Characters can be displayed in different colors, sizes, and orientations. Procedures for controlling the display of character attributes use parameter settings to modify character grid definitions and store the altered grid patterns in the raster.

Graphics packages can be devised to handle both unbundled and bundled attribute specifications. Unbundled attributes are those that are defined for only one type of output device. Bundled attribute specifications allow different sets of attributes to be used on different devices, as defined in a bundle table. Bundle tables may be installation-defined, user-defined, or both. Commands to set the bundle table specify the type of workstation to be used and the attributes for a given attribute index.

Table 4–1 lists the attribute discussed in this chapter for the output primitive classifications: line, fill area, text, and marker. The attribute commands that can be used in graphics packages are listed for each category.

TABLE 4–1.
Summary of Attributes

Output Primitive Type	Associated Attributes	Attribute-Setting Commands	Bundled-Attribute Commands
Line	Type Color Width	set_linetype set_line_color_index set_linewidth_scale_factor	set_line_index set_line_representation
Fill Area	Interior Style Fill Color Pattern	set_fill_area_interior_style set_fill_area_color_index set_fill_area_pattern_index set_pattern_representation set_pattern_reference_point set_pattern_size	set_fill_area_index set_fill_area_representation
Text	Font Color Size Orientation	set_text_font set_text_color_index set_character_height set_character_up_vector set_text_path set_text_alignment	set_text_index set_text_representation
Marker	Type Color Size	set_marker_type set_marker_color_index set_marker_size_scale_factor	set_marker_index set_marker_representation

REFERENCES

Algorithms for filling polygons are treated in Barrett and Jordan (1974), Dunlavey (1983), Jordan and Barrett (1973), and Pavlidis (1978 and 1982).

Color and gray scale considerations are discussed in Crow (1978) and in Heckbert (1982).

Additional information on GKS workstations and attribute operations can be found in Enderle, Kansy, and Pfaff (1984) and in Hopgood, et. al. (1983).

EXERCISES

4–1. Implement the *linetype* command by modifying Bresenham's line-drawing algorithm to display either solid, dashed, or dotted lines.

4–2. A line specified by two endpoints and a width can be converted to a rectangular polygon with four vertices and then displayed using a scan-line method. Develop an efficient algorithm for computing the four vertices needed to define such a rectangle using the line endpoints and line width.

4–3. Implement the *linewidth* command in a line-drawing program so that any one of three line widths can be displayed.

4–4. Write a program to output a line graph of three data sets defined over the same x coordinate range. Input to the program is to include the three sets of data values, labeling for the axes, and the coordinates for the display area on the screen. The data sets are to be scaled to fit the specified area, each plotted line is to be displayed in a different line type (solid, dashed, dotted), and labeled axes are to be provided.

4–5. Write the program of Ex. 4–4 so that the coordinate axes and each plotted line is in a different color. If possible, harmonious color combinations should be used.

4–6. Define and implement a function for controlling the line type (solid, dashed, dotted) of displayed ellipses.

4–7. Define and implement a function for setting the width of displayed ellipses.

4–8. Devise an algorithm for implementing a color lookup table and the *set_color_table* operation.

4–9. Write a routine to display a bar graph in any specified screen area. Input is to include the data set, labeling for the coordinate axes, and the coordinates for the screen area. The data set is to be scaled to fit the designated screen area, and the bars are to be displayed in a solid color.

4–10. Write the routine of Ex. 4–9 to display two data sets defined over the same x coordinate range. The bars for one of the data sets are to be displaced horizontally to produce an overlapping bar pattern for easy comparison of the two sets of data. Use a different color or a different fill pattern for the two sets of bars.

4–11. Modify the scan-line algorithm to apply any specified rectangular fill pattern to a polygon interior, starting from a designated pattern position.

4–12. Write a program to scan-convert the interior of an ellipse into a solid color.

4–13. Extend the program of Ex. 4–12 to generate either a solid or a patterned fill.

4–14. Develop an algorithm for antialiasing elliptical boundaries.

4–15. Extend the Pitteway-Watkinson algorithm for antialiasing an area so that the intensity for boundary lines with any slope can be adjusted.

4–16. Write a program to implement the algorithm of Ex. 4–15 as a scan-line procedure to fill a polygon interior, using the routine *set_pixel* (x, y, p) to load intensity level p into the frame buffer at location (x, y).

4–17. Modify the scan-line algorithm for area fill to incorporate antialiasing. Use coherence techniques to reduce calculations on successive scan lines.

4–18. Modify the boundary-fill algorithm for a 4-connected region to avoid excessive stacking by incorporating scan-line methods.

4–19. Write a boundary-fill procedure to fill an 8-connected region.

4–20. Develop and implement a flood-fill algorithm to fill the interior of any specified area.

4–21. Write a procedure to implement the *set_pattern_representation* function.

4–22. Define and implement a command for changing the size of an existing rectangular fill pattern.

4–23. Devise an algorithm for adjusting the height and width of characters defined as rectangular bit grids.

4–24. Implement routines for setting the character up vector and the text path for controlling the display of character strings.

4–25. Write a program to align text as specified by input values for the alignment parameters.

4–26. Develop procedures for implementing the marker attribute functions.

4–27. Compare attribute-implementation procedures needed by systems that employ bundled attributes to those needed by systems using unbundled attributes.

4–28. Develop procedures for storing and accessing attributes in unbundled system attribute tables. The procedures are to be designed to store designated attribute values in the system tables, to pass attributes to the appropriate output routines, and to pass attributes to memory locations specified in inquiry commands.

4–29. Set up the procedures in Ex. 4–28 to handle bundled system attribute tables.

5 TWO-DIMENSIONAL TRANSFORMATIONS

With the procedures for displaying output primitives and their attributes, we can create a variety of picture and graph forms. In many applications, there is also a need for altering or manipulating displays. Sometimes we need to reduce the size of an object or graph to place it into a larger display. Or we might want to test the appearance of design patterns by rearranging the relative positions and sizes of the pattern parts. For animation applications, we need to produce continuous motion of displayed objects about the screen. These various manipulations are carried out by applying appropriate geometric transformations to the coordinate points in a display. The basic transformations are translation, scaling, and rotation. We first discuss methods for performing these transformations and then consider how transformation functions can be incorporated into a graphics package.

5–1 Basic Transformations

Displayed objects are defined by sets of coordinate points. Geometric transformations are procedures for calculating new coordinate positions for these points, as required by a specified change in size and orientation for the object.

Translation

A **translation** is a straight-line movement of an object from one position to another. We translate a point from coordinate position (x, y) to a new position (x', y') by adding **translation distances,** Tx and Ty, to the original coordinates:

$$x' = x + Tx, \qquad y' = y + Ty \tag{5–1}$$

The translation distance pair (Tx, Ty) is also called a **translation vector** or **shift vector.**

Polygons are translated by adding the specified translation distances to the coordinates of each line endpoint in the object. Figure 5–1 illustrates movement of a polygon to a new position as determined by the translation vector $(-20, 50)$.

Objects drawn with curves are translated by changing the defining coordinates of the object. To change the position of a circle or ellipse, we translate the center coordinates and redraw the figure in the new location.

Translation distances can be specified as any real numbers (positive, negative, or zero). If an object is translated beyond the display limits in device coordinates, the system might return an error message, clip off the parts of the object beyond the display limits, or present a distorted picture. Systems that contain no provision for handling coordinates beyond the display limits will distort shapes because coordinate values overflow the memory locations. This produces an effect known as **wraparound,** where points beyond the coordinate limits in one direction will be displayed on the other side of the display device (Fig. 5–2).

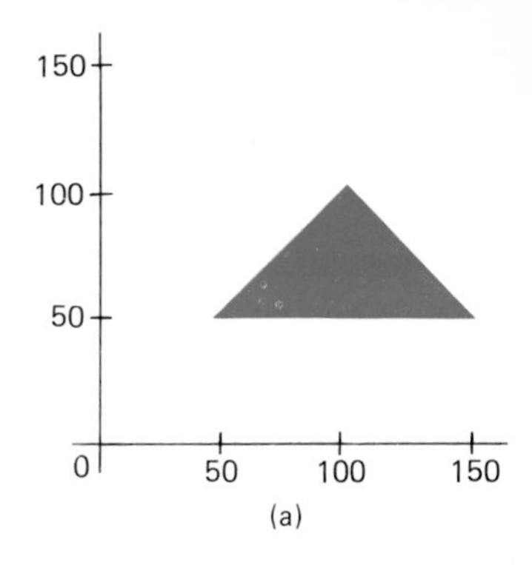

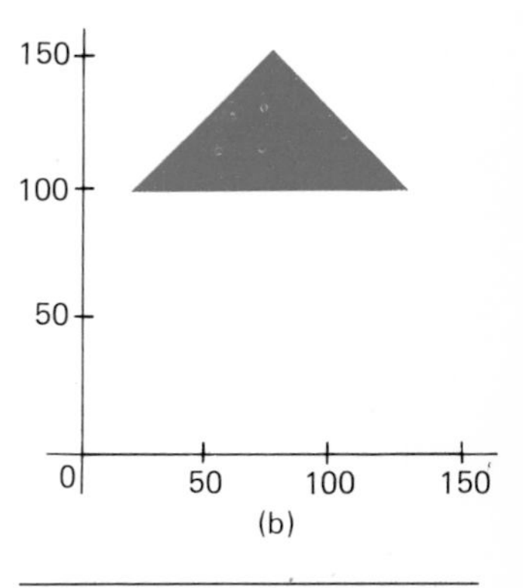

FIGURE 5–1
Translation of an object from position (a) to position (b) with translation distances $(-20, 50)$.

FIGURE 5–2
Possible effects of wraparound on the display of a polygon with one vertex *(P)* specified beyond the display coordinate limits. Point *P* is wrapped around to position *P′*, with a corresponding distortion (solid lines) in the display of the figure.

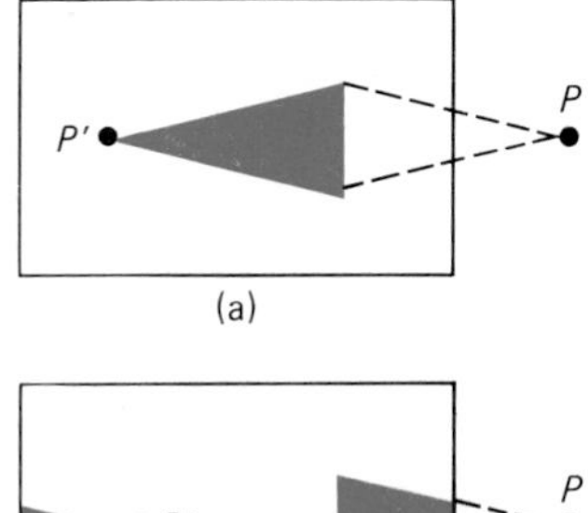

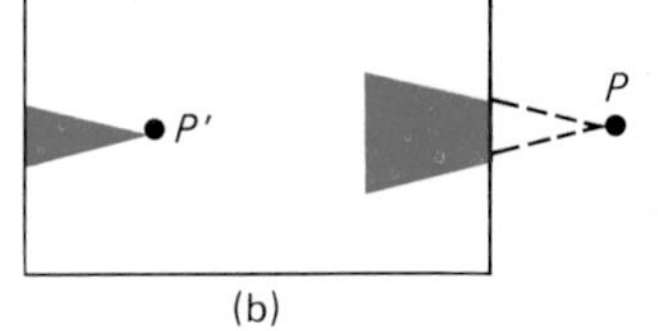

Scaling

A transformation to alter the size of an object is called **scaling.** This operation can be carried out for polygons by multiplying the coordinate values (x, y) for each boundary vertex by **scaling factors** Sx and Sy to produce transformed coordinates (x', y'):

$$x' = x \cdot Sx, \qquad y' = y \cdot Sy \tag{5–2}$$

Scaling factor Sx scales objects in the x direction, while Sy scales in the y direction.

Any positive numeric values can be assigned to the scaling factors Sx and Sy. Values less than 1 reduce the size of objects; values greater than 1 produce an enlargement. Specifying a value of 1 for both Sx and Sy leaves the size of objects unchanged. When Sx and Sy are assigned the same value, a **uniform scaling** is produced, which maintains relative proportions of the scaled object. Unequal values for Sx and Sy are often used in design applications, where pictures are constructed from a few basic shapes that can be modified by scaling transformations (Fig. 5–3).

When an object is redrawn with scaling equations 5–2, the length of each line in the figure is scaled according to the values assigned to Sx and Sy. In addition, the distance from each vertex to the origin of the coordinate system is also scaled. Figure 5–4 illustrates scaling a line by setting Sx and Sy to 1/2 in Eqs. 5–2. Both the line length and the distance to the origin are reduced by a factor of 1/2. Enlarged objects are moved away from the coordinate origin.

We can control the location of a scaled object by choosing a position, called the **fixed point,** which is to remain unchanged after the scaling transformation. Coordinates for the fixed point, (x_F, y_F) can be chosen as one of the vertices, the center of the object, or any other position (Fig. 5–5). A polygon is then scaled relative to the fixed point by scaling the distance from each vertex to the fixed point. For a vertex with coordinates (x, y), the scaled coordinates (x', y') are calculated as

$$x' = x_F + (x - x_F)Sx, \qquad y' = y_F + (y - y_F)Sy \tag{5–3}$$

We can rearrange the terms in these equations to obtain scaling transformations relative to a selected fixed point as

$$\begin{aligned} x' &= x \cdot Sx + (1 - Sx)x_F \\ y' &= y \cdot Sy + (1 - Sy)y_F \end{aligned} \tag{5–4}$$

where the terms $(1 - Sx)x_F$ and $(1 - Sy)y_F$ are constant for all points in the object.

As in translation, a scaling operation might extend objects beyond display coordinate limits. Transformed lines beyond these limits could be either clipped or distorted, depending on the system in use.

Scaling transformations 5–4 are applied to the vertices of a polygon. Other types of objects could be scaled with these equations by applying the calculations to each point along the defining boundary. For standard figures, such as circles and ellipses, these transformations can be carried out more efficiently by modifying distance parameters in the defining equations. We scale a circle by adjusting the radius and possibly repositioning the circle center.

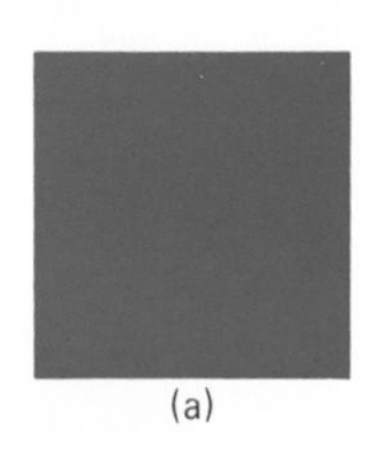

(a)

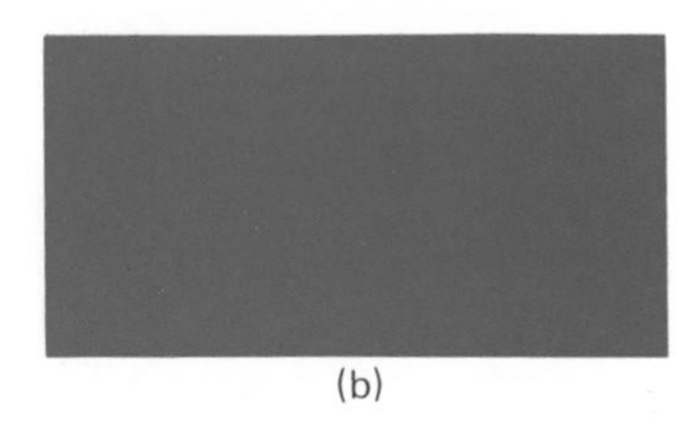

(b)

FIGURE 5–3
Turning a square (a) into a rectangle (b) by setting $S_x = 2$ and $S_y = 1$.

FIGURE 5–4
A line scaled with Eqs. (5–2) and $S_x = S_y = \frac{1}{2}$ is reduced in size and moved closer to the coordinate origin.

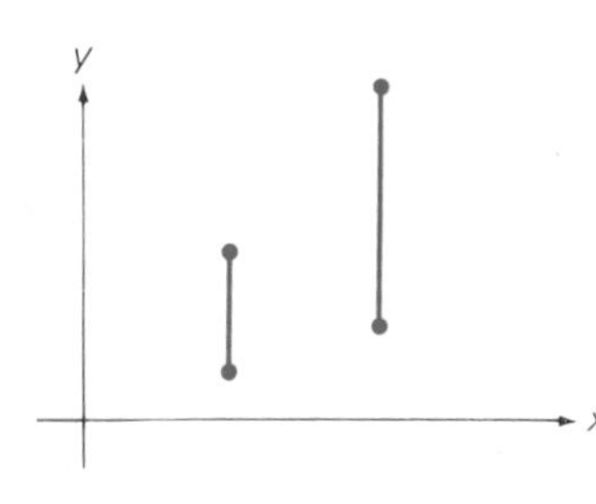

FIGURE 5–5
Scaling relative to a chosen fixed point (x_F, y_F). Distances from each polygon vertex to the fixed point are scaled by transformation equations (5–3).

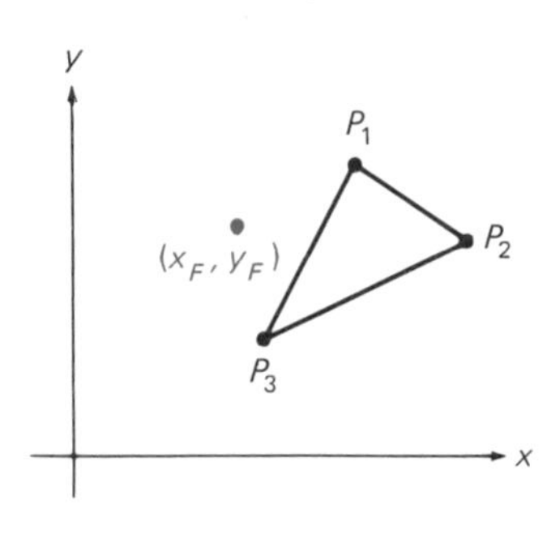

Rotation

Transformation of object points along circular paths is called **rotation.** We specify this type of transformation with a **rotation angle**, which determines the amount of rotation for each vertex of a polygon.

Figure 5–6 illustrates displacement of a point from position (x, y) to position (x', y'), as determined by a specified rotation angle θ relative to the coordinate origin. In this figure, angle ϕ is the original angular position of the point from the horizontal. We can determine the transformation equations for rotation of the point from the relationships between the sides of the right triangles shown and the associated angles. Using these triangles and standard trigonometric identities, we can write

$$\begin{aligned} x' &= r\cos(\phi + \theta) = r\cos\phi\cos\theta - r\sin\phi\sin\theta \\ y' &= r\sin(\phi + \theta) = r\sin\phi\cos\theta + r\cos\phi\sin\theta \end{aligned} \tag{5–5}$$

where r is the distance of the point from the origin. We also have

$$x = r\cos\phi, \qquad y = r\sin\phi \tag{5–6}$$

so that Eqs. 5–5 can be restated in terms of x and y as

$$x' = x\cos\theta - y\sin\theta$$
$$y' = y\cos\theta + x\sin\theta \qquad (5\text{–}7)$$

Positive values for θ in these equations indicate a counterclockwise rotation, and negative values for θ rotate objects in a clockwise direction.

Objects can be rotated about an arbitrary point by modifying Eqs. 5–7 to include the coordinates (x_R, y_R) for the selected **rotation point** (or **pivot point**). Rotation with respect to an arbitrary rotation point is shown in Fig. 5–7. The transformation equations for the rotated coordinates can be obtained from the trigonometric relationships in this figure as

$$x' = x_R + (x - x_R)\cos\theta - (y - y_R)\sin\theta$$
$$y' = y_R + (y - y_R)\cos\theta + (x - x_R)\sin\theta \qquad (5\text{–}8)$$

The pivot point for the rotation transformation can be set anywhere inside or beyond the outer boundary of an object. When the pivot point is specified within the object boundary, the effect of the rotation is to spin the object about this internal point. With an external pivot point, all points of the object are displaced along circular paths about the pivot point.

Since the rotation calculations involve trigonometric functions and several arithmetic operations for each point, computation time can become excessive. This is of particular concern in applications that require transformation of a large number of points or many repeated rotations. For animation and other applications that may involve small rotation angles, we can make some improvements in the efficiency of the rotation calculations. When the rotation angle is small (less than 10°), the trigonometric functions can be replaced with approximation values. For small angles, cosθ is approximately 1 and sinθ has a value that is very close to the value of θ in radians. The error introduced by these approximations will decrease as the rotation angle decreases.

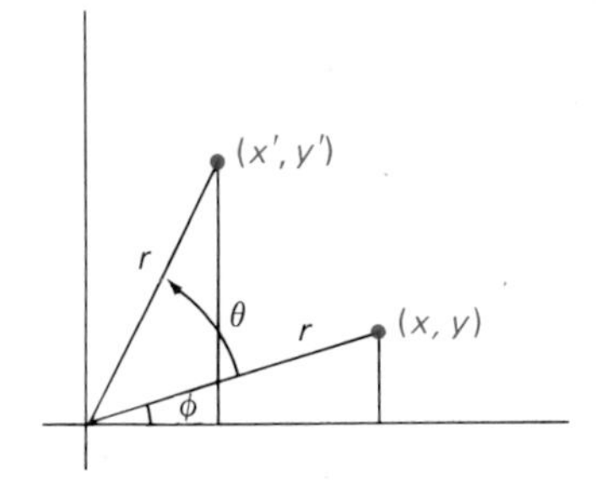

FIGURE 5–6
Rotation of a point from position *(x, y)* to position *(x′, y′)* through a rotation angle θ, specified relative to the coordinate origin. The original angular position of the point from the *x* axis is φ.

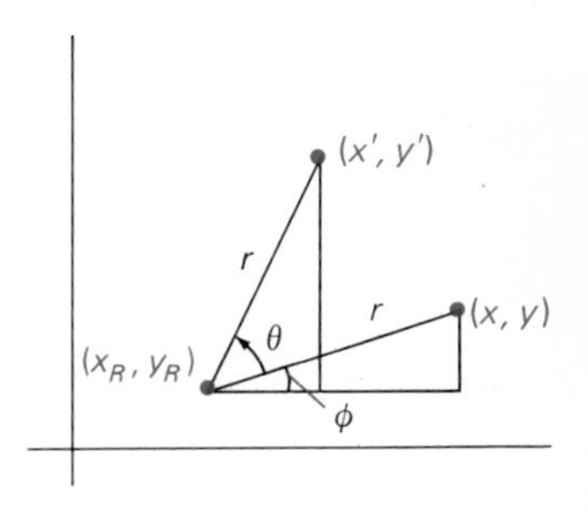

FIGURE 5–7
Rotation of a point from *(x, y)* to *(x′, y′)* through an angle θ, specified relative to a pivot point at (x_R, y_R).

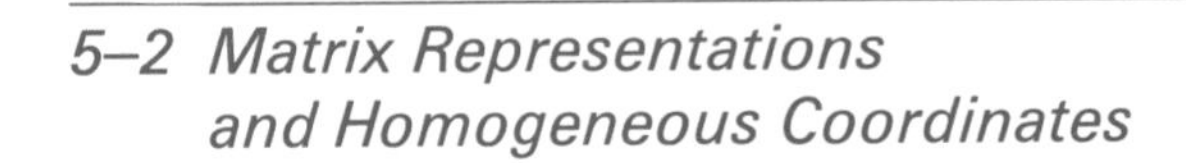

5–2 Matrix Representations and Homogeneous Coordinates

There are many applications that make use of the basic transformations in various combinations. A picture, built up from a set of defined shapes, typically requires each shape to be scaled, rotated, and translated to fit into the proper picture position. This sequence of transformations could be carried out one step at a time. First, the coordinates defining the object could be scaled, then these scaled coordinates could be rotated, and finally the rotated coordinates could be translated to the required location. A more efficient approach is to calculate the final coordinates directly from the initial coordinates using matrix methods, with each of the basic transformations expressed in matrix form.

We can write transformation equations in a consistent matrix form by first expressing points as **homogeneous coordinates.** This means that we represent a two-dimensional coordinate position (x, y) as the triple [x_h y_h w], where

$$x_h = x \cdot w, \qquad y_h = y \cdot w \qquad (5\text{–}9)$$

The parameter w is assigned a nonzero value in accordance with the class of transformations to be represented. For the two-dimensional transformations discussed in the last section, we can set $w = 1$. Each two-dimensional coordinate position then has the homogeneous coordinate form $[x\ y\ 1]$. Other values for w are useful with certain three-dimensional viewing transformations.

With coordinate positions expressed in homogeneous form, the basic transformation equations can be represented as matrix multiplications employing 3 by 3 **transformation matrices.** Equations 5–1, for translation, become

$$[x'\ y'\ 1] = [x\ y\ 1]\begin{bmatrix} 1 & 0 & 0 \\ 0 & 1 & 0 \\ Tx & Ty & 1 \end{bmatrix} \tag{5–10}$$

We also introduce the abbreviated notation $T(Tx, Ty)$ for the 3 by 3 transformation matrix with translation distances Tx and Ty

$$T(Tx, Ty) = \begin{bmatrix} 1 & 0 & 0 \\ 0 & 1 & 0 \\ Tx & Ty & 1 \end{bmatrix} \tag{5–11}$$

Using this notation, we can write the matrix form for the translation equations more compactly as

$$P' = P \cdot T(Tx, Ty) \tag{5–12}$$

where $P' = [x'\ y'\ 1]$ and $P = [x\ y\ 1]$ are 1 by 3 matrices (three-element row vectors) in the matrix calculations.

Similarly, the scaling equations 5–2 are now written as

$$[x'\ y'\ 1] = [x\ y\ 1]\begin{bmatrix} Sx & 0 & 0 \\ 0 & Sy & 0 \\ 0 & 0 & 1 \end{bmatrix} \tag{5–13}$$

or as

$$P' = P \cdot S(Sx, Sy) \tag{5–14}$$

with

$$S(Sx, Sy) = \begin{bmatrix} Sx & 0 & 0 \\ 0 & Sy & 0 \\ 0 & 0 & 1 \end{bmatrix} \tag{5–15}$$

as the 3 by 3 transformation matrix for scaling with parameters Sx and Sy.

Equations 5–7, for rotation, are written in matrix form as

$$[x'\ y'\ 1] = [x\ y\ 1]\begin{bmatrix} \cos\theta & \sin\theta & 0 \\ -\sin\theta & \cos\theta & 0 \\ 0 & 0 & 1 \end{bmatrix} \tag{5–16}$$

or as

$$P' = P \cdot R(\theta) \tag{5–17}$$

where

$$R(\theta) = \begin{bmatrix} \cos\theta & \sin\theta & 0 \\ -\sin\theta & \cos\theta & 0 \\ 0 & 0 & 1 \end{bmatrix} \tag{5–18}$$

is the 3 by 3 transformation matrix for rotation with parameter θ.

Matrix representations are standard methods for implementing the basic transformations in graphics systems. In many systems, the scaling and rotation transformations are always stated relative to the coordinate origin, as in Eqs. 5–13 and 5–16. Rotations and scalings relative to other points are handled as a sequence of transformations. An alternate approach is to state the transformation matrix for scaling in terms of the fixed-point coordinates and to specify the transformation matrix for rotation in terms of the pivot-point coordinates.

5–3 Composite Transformations

Any sequence of transformations can be represented as a **composite transformation matrix** by calculating the product of the individual transformation matrices. Forming products of transformation matrices is usually referred to as a **concatenation,** or **composition,** of matrices.

Translations

Two successive translations of an object can be carried out by first concatenating the translation matrices, then applying the composite matrix to the coordinate points. Specifying the two successive translation distances as (Tx_1, Ty_1) and (Tx_2, Ty_2), we calculate the composite matrix as

$$\begin{bmatrix} 1 & 0 & 0 \\ 0 & 1 & 0 \\ Tx_1 & Ty_1 & 1 \end{bmatrix} \cdot \begin{bmatrix} 1 & 0 & 0 \\ 0 & 1 & 0 \\ Tx_2 & Ty_2 & 1 \end{bmatrix} = \begin{bmatrix} 1 & 0 & 0 \\ 0 & 1 & 0 \\ Tx_1 + Tx_2 & Ty_1 + Ty_2 & 1 \end{bmatrix} \qquad (5\text{–}19)$$

which demonstrates that two successive translations are additive. Equation 5–19 can be written as

$$T(Tx_1, Ty_1) \cdot T(Tx_2, Ty_2) = T(Tx_1 + Tx_2, Ty_1 + Ty_2) \qquad (5\text{–}20)$$

The transformation of coordinate points for a composite translation is then expressed in matrix form as

$$P' = P \cdot T(Tx_1 + Tx_2, Ty_1 + Ty_2) \qquad (5\text{–}21)$$

Scalings

Concatenating transformation matrices for two successive scaling operations produces the following composite scaling matrix:

$$S(Sx_1, Sy_1) \cdot S(Sx_2, Sy_2) = S(Sx_1 \cdot Sx_2, Sy_1 \cdot Sy_2) \qquad (5\text{–}22)$$

The resulting matrix in this case indicates that successive scaling operations are multiplicative. That is, if we were to triple the size of an object twice in succession, the final size would be nine times that of the original.

Rotations

The composite matrix for two successive rotations is calculated as

$$R(\theta_1) \cdot R(\theta_2) = R(\theta_1 + \theta_2) \qquad (5\text{–}23)$$

As is the case with translations, successive rotations are additive.

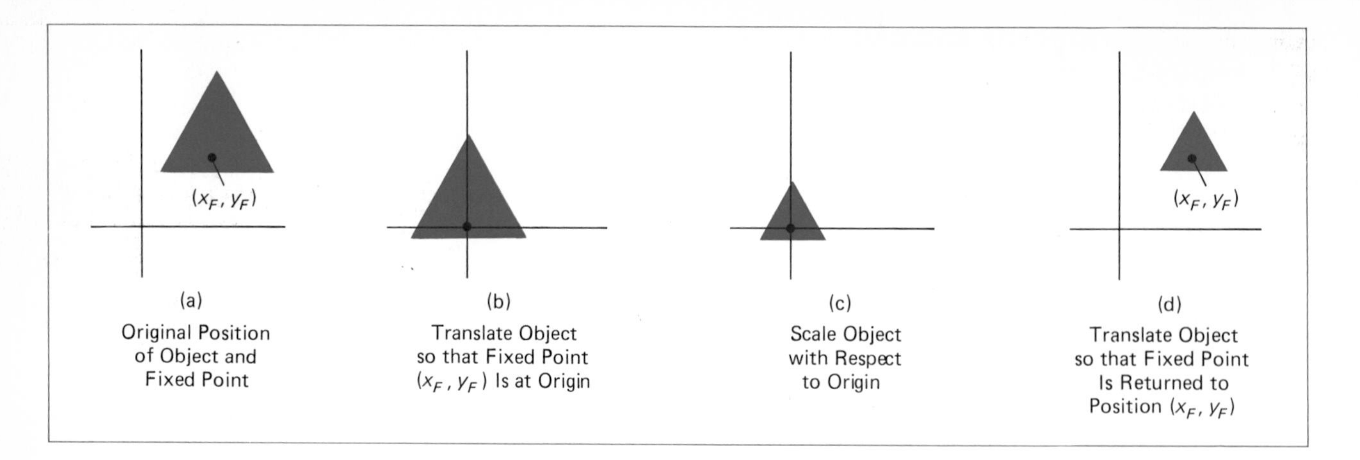

FIGURE 5–8
Sequence of transformations necessary to scale an object with respect to a fixed point using transformation matrices (5–11) and (5–15).

Scaling Relative to a Fixed Point

Using the transformation matrices for translation (Eq. 5–11) and scaling (Eq. 5–15), we can obtain the composite matrix for scaling with respect to a fixed point (x_F, y_F) by considering a sequence of three transformations. This transformation sequence is illustrated in Fig. 5–8. First, all coordinates are translated so that the fixed point is moved to the coordinate origin. Second, coordinates are scaled with respect to the origin. Third, the coordinates are translated so that the fixed point is returned to its original position. The matrix multiplications for this sequence yield

$$\begin{bmatrix} 1 & 0 & 0 \\ 0 & 1 & 0 \\ -x_F & -y_F & 1 \end{bmatrix} \cdot \begin{bmatrix} Sx & 0 & 0 \\ 0 & Sy & 0 \\ 0 & 0 & 1 \end{bmatrix} \cdot \begin{bmatrix} 1 & 0 & 0 \\ 0 & 1 & 0 \\ x_F & y_F & 1 \end{bmatrix} = \begin{bmatrix} Sx & 0 & 0 \\ 0 & Sy & 0 \\ (1 - Sx)x_F & (1 - Sy)y_F & 1 \end{bmatrix} \tag{5–24}$$

Rotation About a Pivot Point

Figure 5–9 illustrates a transformation sequence for obtaining the composite matrix for rotation about a specified pivot point (x_R, y_R). First, the object is translated so that the pivot point coincides with the coordinated origin. Second, the object is rotated about the origin. Third, the object is translated so that the pivot point returns to its original position. This sequence is represented by the matrix product:

$$\begin{bmatrix} 1 & 0 & 0 \\ 0 & 1 & 0 \\ -x_R & -y_R & 1 \end{bmatrix} \cdot \begin{bmatrix} \cos\theta & \sin\theta & 0 \\ -\sin\theta & \cos\theta & 0 \\ 0 & 0 & 1 \end{bmatrix} \cdot \begin{bmatrix} 1 & 0 & 0 \\ 0 & 1 & 0 \\ x_R & y_R & 1 \end{bmatrix}$$

$$= \begin{bmatrix} \cos\theta & \sin\theta & 0 \\ -\sin\theta & \cos\theta & 0 \\ (1 - \cos\theta)x_R + y_R \cdot \sin\theta & (1 - \cos\theta)y_R - x_R \cdot \sin\theta & 1 \end{bmatrix} \tag{5–25}$$

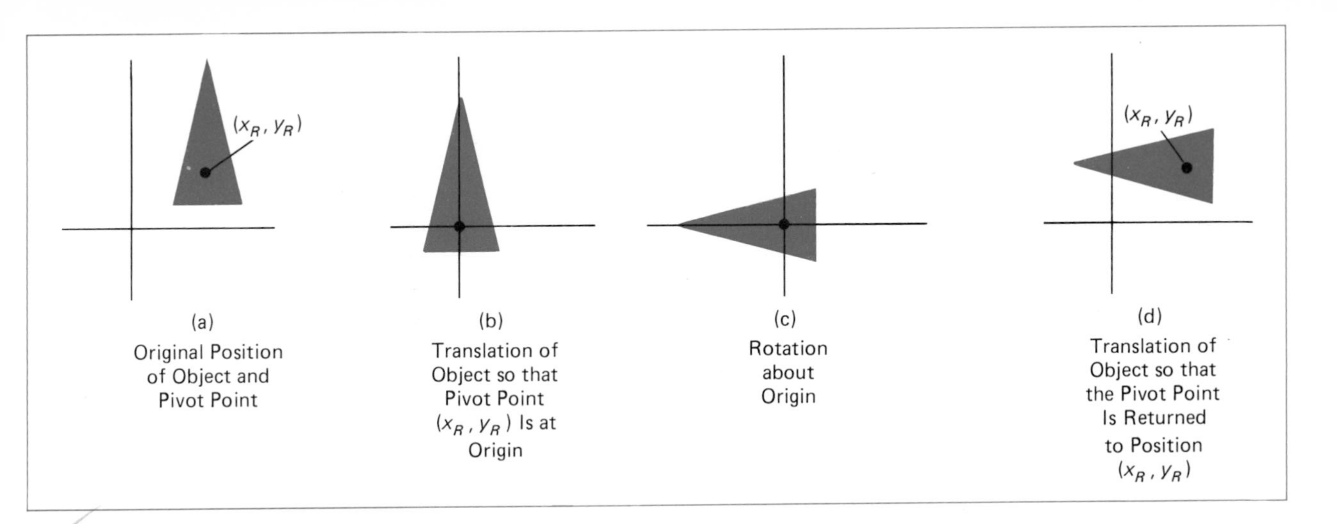

FIGURE 5–9
Sequence of transformations necessary to rotate an object about a pivot point using transformation matrices (5–11) and (5–18).

Arbitrary Scaling Directions

Scaling parameters Sx and Sy affect only the x and y directions. We can scale objects in other directions by performing a combination of rotations and scaling transformations.

Suppose we wanted to apply scaling factors with values specified by S_1 and S_2 in the directions shown in Fig. 5–10. To accomplish this, we first perform a rotation so that the directions for S_1 and S_2 coincide with the x and y axes, respectively. Then the scaling transformation is applied, followed by an opposite rotation to return points to their original orientations. The composite matrix resulting from the product of these three transformations is

$$\begin{bmatrix} S_1 \cdot \cos^2\theta + S_2 \cdot \sin^2\theta & (-S_1 + S_2)\sin\theta\cos\theta & 0 \\ (-S_1 + S_2)\sin\theta\cos\theta & S_1 \cdot \sin^2\theta + S_2 \cdot \cos^2\theta & 0 \\ 0 & 0 & 1 \end{bmatrix} \tag{5–26}$$

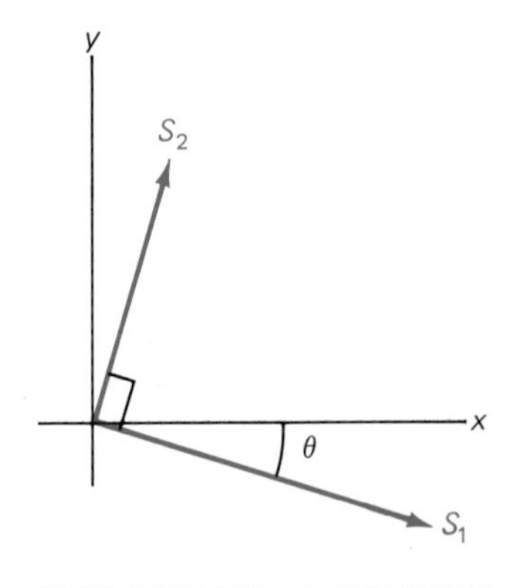

FIGURE 5–10
Scaling parameters S_1 and S_2 are applied in directions determined by an angular displacement θ from the x axis.

As an example of the application of this scaling transformation, we turn a square into a parallelogram (Fig. 5–11) by stretching it along the diagonal from (0, 0) to (1, 1). Our transformation rotates this diagonal onto the y axis and applies a scaling factor to double its length. We do this by setting $\theta = 45°$, $S_1 = 1$, and $S_2 = 2$ in the matrix 5–26.

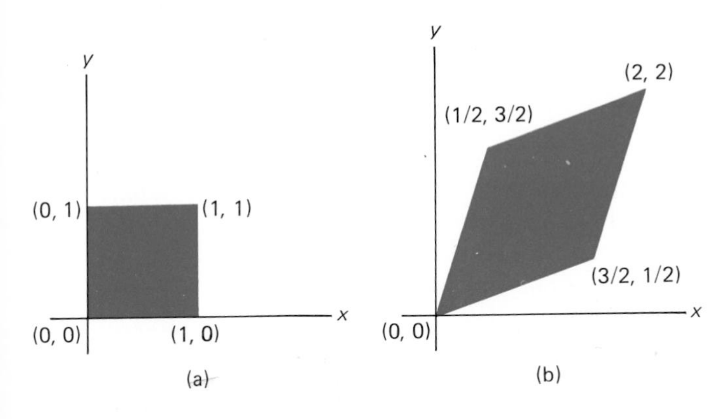

FIGURE 5–11
A square (a) is converted to a parallelogram (b) using the composite transformation matrix (5–26), with $S_1 = 1$, $S_2 = 2$, and $\theta = 45°$.

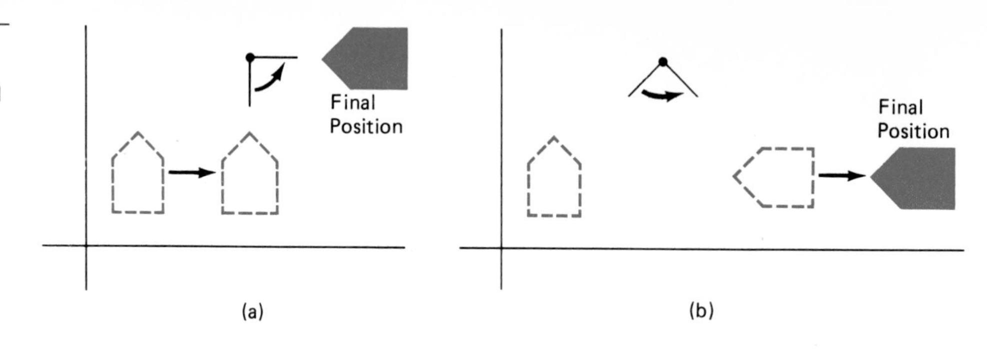

FIGURE 5–12
The order in which transformations are performed affect the final position of an object. In (a), the object is first translated, then rotated. In (b), the object is rotated first, then translated.

Concatenation Properties

Matrix multiplication is associative. That is, for any three matrices, A, B, and C, the matrix product $A \cdot B \cdot C$ can be performed by first multiplying A and B or by first multiplying B and C:

$$A \cdot B \cdot C = (A \cdot B) \cdot C = A \cdot (B \cdot C) \tag{5–27}$$

Therefore, we can evaluate composite matrices in either order.

On the other hand, composite transformations may not be commutative: The matrix product $A \cdot B$ is not equal to $B \cdot A$, in general. This means that if we want to translate and rotate an object, we must be careful about the order in which the composite matrix is evaluated (Fig. 5–12). For some special cases, the multiplication of transformation matrices is commutative. Two successive transformations of the same kind are commutative. For example, two successive rotations could be performed in either order and the final position would be the same. This commutative property holds also for two successive translations or two successive scalings. Another commutative pair of operations is a uniform scaling ($Sx = Sy$) and a rotation.

General Transformation Equations

Any general transformation, representing a combination of translations, scalings, and rotations, can be expressed as

$$[x' \; y' \; 1] = [x \; y \; 1] \begin{bmatrix} a & d & 0 \\ b & e & 0 \\ c & f & 1 \end{bmatrix} \tag{5–28}$$

Explicit equations for calculating the transformed coordinates are then

$$x' = ax + by + c, \qquad y' = dx + ey + f \tag{5–29}$$

These calculations involve four multiplications and four additions for each coordinate point in an object. This is the maximum number of computations required for the determination of a coordinate pair for any transformation sequence, once the individual matrices have been concatenated. Without concatenation, the individual transformations would be applied one at a time and the number of calculations could be significantly increased. An efficient implementation for the transformation operations, therefore, is to formulate transformation matrices, concatenate any transformation sequence, and calculate transformed coordinates by Eqs. 5–29.

The following procedure implements composite transformations. A transformation matrix T is initialized to the identity matrix. As each individual transformation is specified, it is concatenated with the total transformation matrix T. When all transformations have been specified, this composite transformation is applied to a given object. For this example, a polygon is scaled, rotated, and translated. Figure 5–13 shows the original and final positions of a polygon transformed by this sequence.

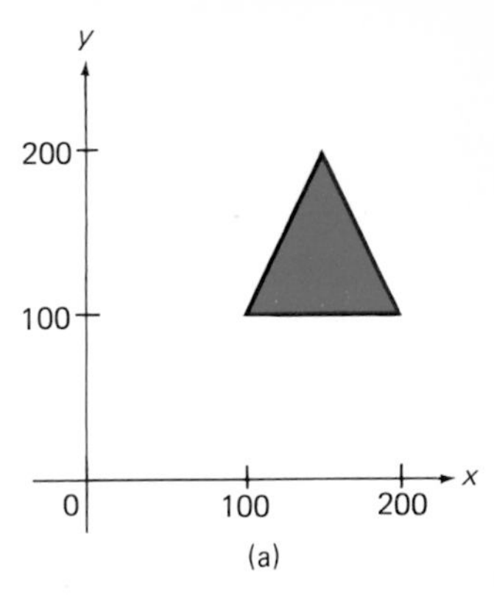

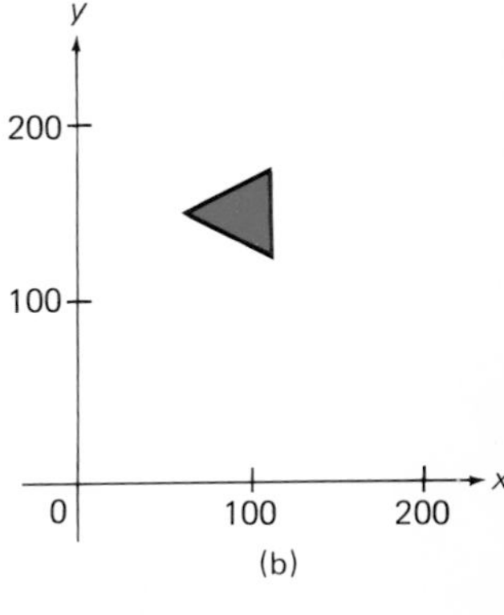

FIGURE 5–13
A polygon (a) is transformed into (b) by the composite operations in PROCEDURE TRANSFORM_OBJECT.

```
procedure transform_object;
  type
    matrix = array [1..3,1..3] of real;
    points = array [1..10] of real;
  var
    t : matrix;
    x, y : points;
    xc, yc : integer;

  procedure transform_points (n : integer; var x, y : points);
    var k : integer; tempx : real;
    begin
      for k := 1 to n do begin
          tempx := x[k] * t[1,1] + y[k] * t[2,1] + t[3,1];
          y[k] :=x[k] * t[1,2] + y[k] * t[2,2] + t[3,2];
          x[k] := tempx
        end
    end; {transform_points}

  procedure fill_area (n : integer; x, y : points);
    begin
        {remainder of fill_area steps}
    end; {fill_area}

  procedure get_vertices_and_center (n : integer; var x, y: points;
                                     var xc, yc : integer);

    begin
      {get vertices and center point}
    end;

  procedure make_identity (var m : matrix);
    var  r, c : integer;
    begin
      for r := 1 to 3 do
        for c := 1 to 3 do
          if r = c then m[r,c] := 1 else m[r,c] :=0
    end; {make_identity}

  procedure combine_transformations (var t : matrix; m : matrix);
    var r, c : integer; temp : matrix;
    begin
      for r := 1 to 3 do
        for c:= 1 to 3 do
          temp[r,c] := t[r,1]*m[1,c] + t[r,2]*m[2,c] + t[r,3]*m[3,c];
      for r := 1 to 3 do
        for c := 1 to 3 do
          t[r,c] := temp[r,c]
    end; {combine_transformations}

  procedure scale (sx, sy : real; xf, yf : integer);
    var  m : matrix;
```

```
  begin
    make_identity (m);
    m[1,1] := sx;  m[2,2] := sy;
    m[3,1] := (1 - sx) * xf;
    m[3,2] := (1 - sy) * yf;
    combine_transformations (t, m) {multiply m into t}
  end; {scale}

procedure rotate (a : real; xr, yr : integer);
  var  ca, sa : real;  m : matrix;
  function radian_equivalent (a : real) : real;
    begin
      radian_equivalent := a * 3.14159 / 180
    end;  {radian_equivalent}
  begin  {rotate}
    make_identity (m);
    a := radian_equivalent (a);
    ca := cos(a);  sa := sin(a);
    m[1,1] := ca; m[1,2] := sa;
    m[2,1] := -sa; m[2,2] := ca;
    m[3,1] := xr * (1 - ca) + yr * sa;
    m[3,2] := yr * (1 - ca) - xr * sa;
    combine_transformations (t, m) {multiply m into t}
  end; {rotate}

procedure translate (tx, ty : integer);
  var  m : matrix;
  begin
    make_identity (m);
    m[3,1] := tx;  m[3,2] := ty;
    combine_transformations (t, m) {multiply m into t}
  end; {translate}

begin  {transform_object}
  get_vertices_and_center (3, x, y, xc, yc);
  make_identity (t);
  set_fill_area_interior_style (hollow);
  set_fill_area_color_index (1);
  fill_area (3, x, y);
  scale (0.5, 0.5, xc, yc);
  rotate (90, xc, yc);
  translate(-60, 20);
  fill_area (3, x, y)
end;  {transform_object}
```

5-4 Other Transformations

Basic transformations such as translation, scaling, and rotation are included in most graphics packages. Some packages provide a few additional transformations that are useful in certain applications. Two such transformations are reflection and shear.

Reflection

A **reflection** is a transformation that produces a mirror image of an object. The mirror image is generated relative to an **axis of reflection.**

Objects can be reflected about the x axis using the following transformation matrix

$$\begin{bmatrix} 1 & 0 & 0 \\ 0 & -1 & 0 \\ 0 & 0 & 1 \end{bmatrix} \tag{5-30}$$

This transformation keeps x values the same, but "flips" the y values. The resulting orientation of an object after it has been reflected about the x axis is shown in Fig. 5–14.

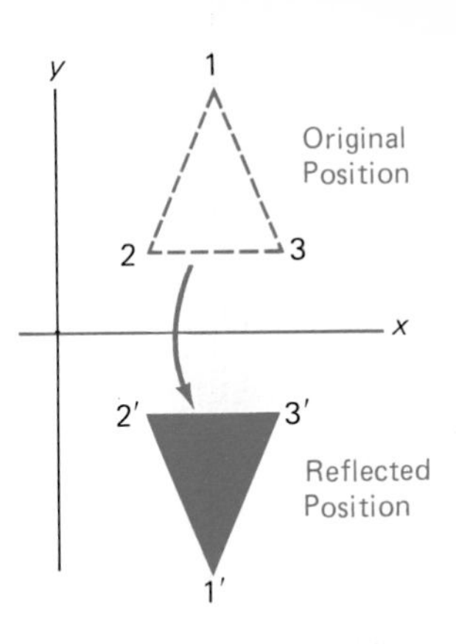

FIGURE 5–14
Reflection of an object about the x axis.

A reflection about the y axis flips the x coordinates while keeping y coordinates the same. We can perform this reflection with the transformation matrix

$$\begin{bmatrix} -1 & 0 & 0 \\ 0 & 1 & 0 \\ 0 & 0 & 1 \end{bmatrix} \tag{5-31}$$

Figure 5–15 illustrates the change in position of an object that has been reflected about the y axis.

FIGURE 5–15
Reflection of an object about the y axis.

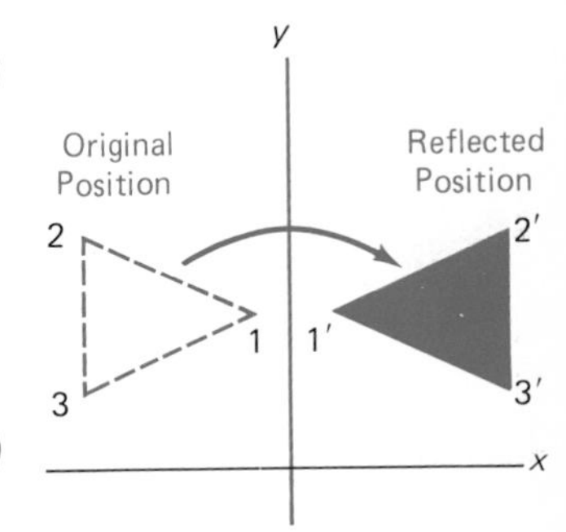

Another type of reflection is one that flips both the x and y coordinates by reflecting relative to the coordinate origin. (The axis of reflection in this case is the line perpendicular to the xy plane and passing through the origin.) We write the transformation matrix for this type of reflection as

$$\begin{bmatrix} -1 & 0 & 0 \\ 0 & -1 & 0 \\ 0 & 0 & 1 \end{bmatrix} \tag{5-32}$$

An example of reflection about the origin is shown in Fig. 5–16.

The following matrix performs a reflection transformation about the line $y = x$ (Fig. 5–17).

$$\begin{bmatrix} 0 & 1 & 0 \\ 1 & 0 & 0 \\ 0 & 0 & 1 \end{bmatrix} \tag{5-33}$$

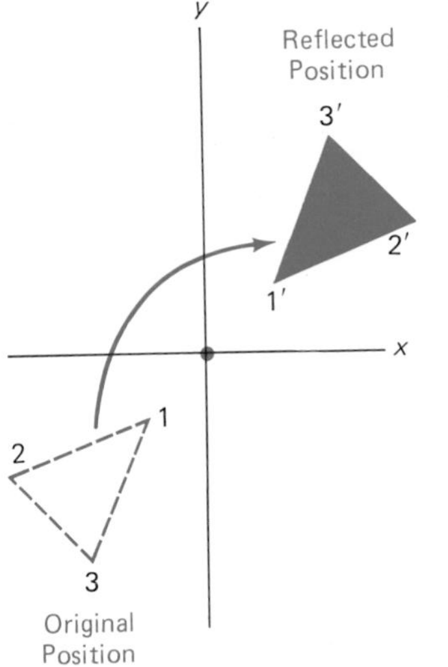

FIGURE 5–16
Reflection of an object relative to the coordinate origin.

FIGURE 5–17
Reflection of an object with respect to the line $y = x$.

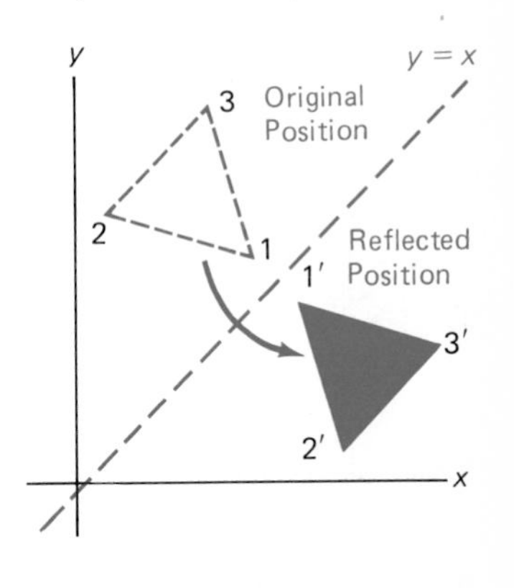

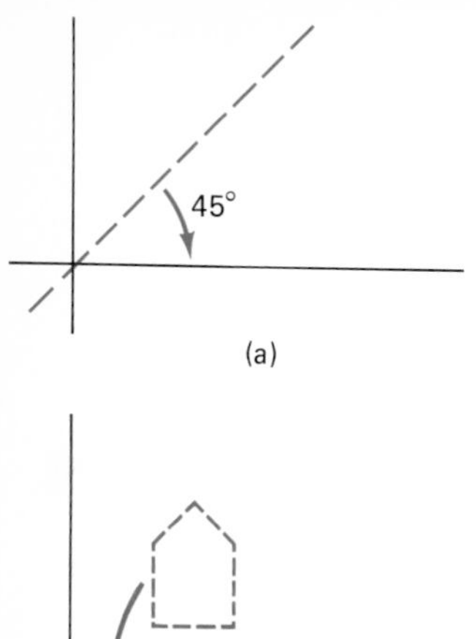

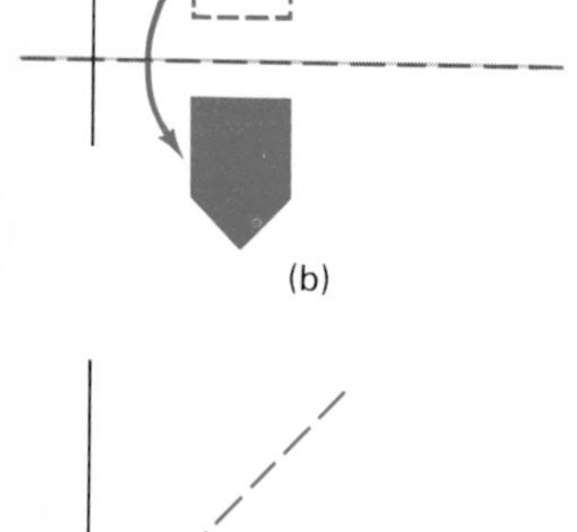

(c)

FIGURE 5–18
Sequence of basic transformations to produce reflection about the line $y = x$: (a) clockwise rotation of 45°; (b) reflection about the x axis; (c) counterclockwise rotation by 45°.

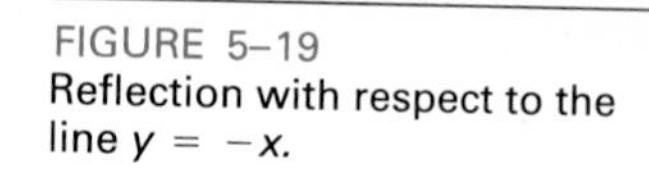

FIGURE 5–19
Reflection with respect to the line $y = -x$.

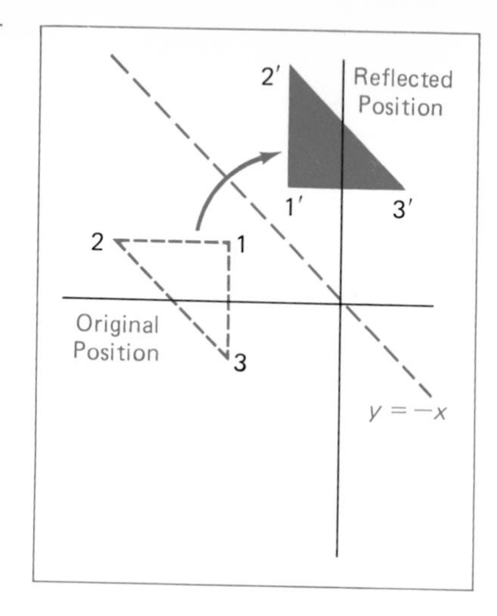

This transformation could be obtained from a sequence of rotations and coordinate-axis reflections. One possible sequence is shown in Fig. 5–18. Here, we first perform a clockwise rotation through a 45° angle, which rotates the line $y = x$ onto the x axis. The final step is to rotate the line $y = x$ back to its original position with a counterclockwise rotation through 45°. Concatenating this sequence of three transformations yields the matrix 5–33. An equivalent sequence of transformations is first to reflect the object about the x axis, then rotate counterclockwise 90°.

To obtain a reflection about the line $y = -x$, we perform the transformation sequence: (1) clockwise rotation by 45°, (2) reflection about the y axis, and (3) counterclockwise rotation by 45°. This sequence produces the transformation matrix

$$\begin{bmatrix} 0 & -1 & 0 \\ -1 & 0 & 0 \\ 0 & 0 & 1 \end{bmatrix} \tag{5–34}$$

Figure 5–19 shows the original and final positions for an object reflected about the line $y = -x$.

Reflections about other lines can be accomplished with similar sequences of transformations. For a reflection line in the xy plane passing through the origin, we can rotate the line onto a coordinate axis, reflect about that axis, and rotate the line back to its original position. For reflection lines that do not pass through the origin, we first translate the line so that it does pass through the origin, then carry out the rotate, reflect, rotate sequence.

Shear

These transformations produce shape distortions that represent a twisting, or **shear,** effect, as if an object were composed of layers that are caused to slide over each other. Two common shearing transformations are those for x-direction shear and y-direction shear.

An x-direction shear is produced with the transformation matrix

$$\begin{bmatrix} 1 & 0 & 0 \\ SHx & 1 & 0 \\ 0 & 0 & 1 \end{bmatrix} \tag{5–35}$$

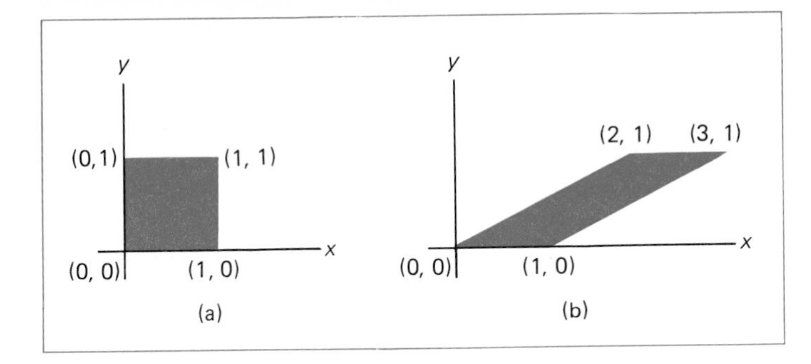

FIGURE 5–20
A unit square (a) converted to a parallelogram (b) using the x-direction shear matrix (5–35) with $SHx = 2$.

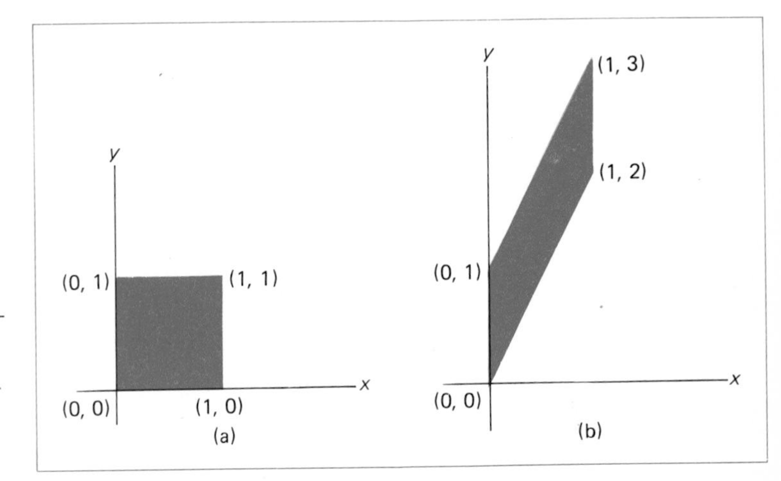

FIGURE 5–21
A unit square (a) is turned into a parallelogram (b) using the y-direction shearing transformation matrix (5–36) with $SHy = 2$.

Any real number can be assigned to the parameter SHx. This transformation affects only the x coordinates; y-coordinate values remain unchanged. Each point in an object is displaced horizontally by a distance that is proportional to the y coordinate. Setting SHx to 2 changes the square in Fig. 5–20 into a parallelogram. Negative values for SHx would produce horizontal shifts to the left.

A similar shearing transformation matrix can be set up to generate y-direction shearing:

$$\begin{bmatrix} 1 & SHy & 0 \\ 0 & 1 & 0 \\ 0 & 0 & 1 \end{bmatrix} \tag{5–36}$$

This transformation generates vertical shifts in coordinate positions. Parameter SHy can be set to any real number, which alters the y component of a coordinate position by an amount that is proportional to the x value. Figure 5–21 illustrates the conversion of a square into a parallelogram with $SHy = 2$.

5–5 Transformation Commands

Graphics packages can be structured so that separate commands are provided to a user for each of the transformation operations, as in *procedure transform_object*. A combination of transformations is then performed by referencing each individual function. An alternate formulation is to use one transformation command that can be used for a combination of transformations or a single operation. Since transformations are often performed in combination, a composite transformation command can provide a more convenient method for applying transformations.

We introduce the following command to perform composite transformations involving translation, scaling, and rotation:

```
create_transformation_matrix (xf,yf,sx,sy,xr,yr,a,tx,ty,matrix)
```

Parameters in this command are the scaling fixed point (*xf, yf*), scaling parameters *sx* and *sy*, rotation pivot point (*xr, yr*), rotation angle *a*, translation vector (*tx, ty*), and the output *matrix*. We assume that this command evaluates the transformation sequence in the fixed order: first scale, then rotate, then translate. The resulting composite transformation for this sequence is then stored in the parameter *matrix*. A procedure for implementing this command performs a concatenation of transformation matrices, using the values specified for the input parameters.

A single transformation, or a sequence of two or three transformations (in the order stated), can be carried out with this transformation command. To translate an object, a user sets $sx = sy = 1$, $a = 0$, and assigns translation values to *tx* and *ty*. The fixed-point and pivot-point coordinates could be set to any values, since they do not affect the transformation calculations when no scaling or rotation takes place. Similarly, rotations are specified by setting $sx = sy = 1$, $tx = ty = 0$, and giving appropriate rotation-angle and pivot-point values to parameters *a*, *xr*, and *yr*.

Since the transformation command can carry out only one fixed sequence of transformations, we can provide for alternative sequences by defining an additional command:

```
accumulate_transformation_matrix (matrix_in, xf, yf, sx, sy,
    xr, yr, a, tx, ty, matrix_out)
```

Parameters *xf, yf, sx, sy, xr, yr, a, tx,* and *ty* are the same as in the *create_transformation_matrix* command. This accumulate operation will take any previously defined transformation matrix (*matrix_in*) and concatenate it with the transformation operations defined by the parameter list, in the order *matrix_in*, scale, rotate, translate. The resulting transformation matrix is stored in *matrix_out*.

Using the *accumulate_transformation_matrix* command in conjunction with *create_transformation_matrix* allows a user to perform transformations in any order. For instance, a translation followed by a rotation cannot be carried out with the *create* command alone. But this transformation sequence could be accomplished with the following program statements:

```
create_transformation_matrix (0, 0, 1, 1, 0, 0, 0, tx, ty, m1);
accumulate_transformation_matrix (m1, 0, 0, 1, 1, xr, yr, a, 0, 0,
    m2);
```

The composite matrix *m2* is then applied to the points defining the object to be translated and rotated.

Several transformation matrices could be constructed in an application program. The particular matrix that is to be applied to subsequent output primitives could be selected with a function such as

```
set_transformation (matrix)
```

Parameter *matrix* stores the matrix elements that are to be applied to all subsequent output primitive commands until the transformation is reset. A method for turning off the transformation operations is to set *matrix* to the identity matrix.

Another implementation method for applying a particular matrix to a defined object is to group and label related output primitives (picture components). A transformation command could then be structured so that it is applied to a selected object by referencing the label assigned to the group of primitives defining the object. We return to this topic in Chapter 7.

5-6 Raster Methods for Transformations

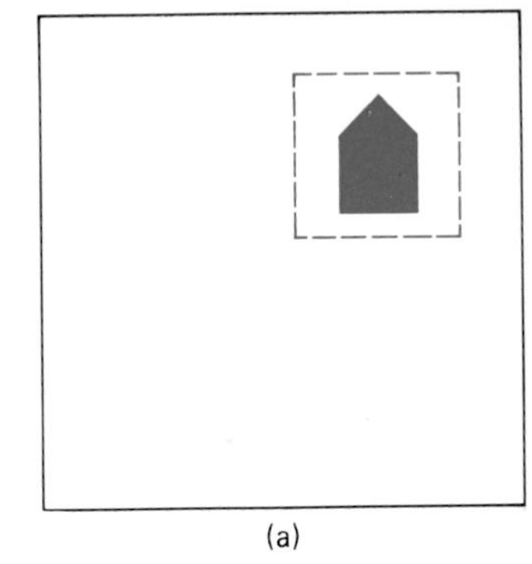

(a)

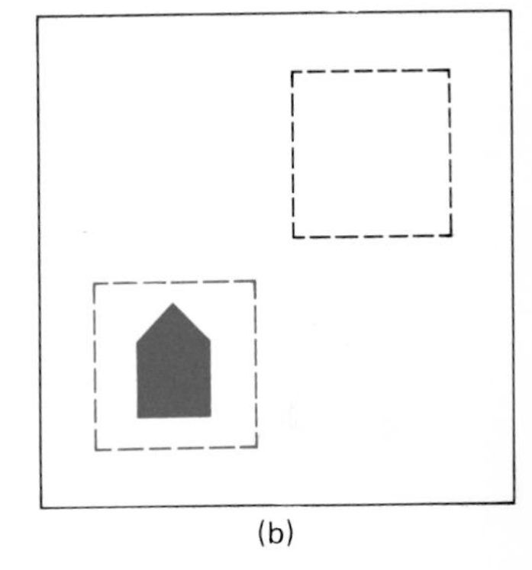

(b)

FIGURE 5–22
Block transfer of a raster area can be used to translate an object from one screen position to another.

The particular capabilities of raster systems suggest an alternate way to approach some transformations. Raster systems store picture information by setting bits in the frame buffer. Some simple transformations can be carried out by manipulating the frame buffer contents directly. No arithmetic operations are needed, so the transformations are particularly efficient.

Figure 5–22 illustrates a translation performed as a block transfer of a raster area. All bit settings in the rectangular area shown are copied as a block into another part of the raster. This translation is accomplished by reading pixel intensities from a specified rectangular area of the raster into an array, then copying the array back into the raster at the new location. The original object could be erased by filling its rectangular area with the background intensity.

Two functions can be provided to a user for carrying out these translation operations. One function is used for reading a rectangular area of the raster into a specified array, and the other is used to copy the pixel values in the array back into the frame buffer. Parameters for a read function are the name of the array and the size and location of the raster area. Parameters for a copy function are the name of the array and the copy position within the raster.

Some implementations provide options for the copy function so that bit values in the array can be combined with the raster values in various ways. Depending on the mode selected, the copy function could simply replace bit settings in the frame buffer with those in the array, or a Boolean or binary arithmetic operation could be applied. For example, bit settings to be introduced into some area of the frame buffer might be combined with the existing contents of the buffer using an AND or an OR operation. The Boolean *exclusive-or* can be particularly useful. With the *exclusive-or* mode, two successive copies of a block to the same raster area restores the values that were originally present in that area. This technique can be used to move an object across a scene without destroying the background.

In addition to translation, rotations in 90° increments could be done using block transfers. A 90° rotation is accomplished by copying each row of the block into a column in the new frame buffer location. Reversing the order of bits within each row rotates the block 180°.

Block transfers of raster areas, sometimes referred to as bit-block transfers or bit-blt, are quick. These techniques form the basis for many animation implementations. Once an array of bit settings has been saved for an object, it can be repeatedly placed at different positions in the raster to simulate motion.

REFERENCES

Two-dimensional transformations are discussed in Pavlidis (1982). Additional information on transformation operations in GKS can be found in Enderle, Kansy, and Pfaff (1984) and in Hopgood, et. al. (1983).

EXERCISES

5–1. Write a program to continuously rotate an object about a pivot point. Small angles are to be used for each successive rotation, and approximations to the sine and cosine function are to be used to speed up the calculations. The rotation angle for each step is to be chosen so that the object makes one complete revolution in less then 30 seconds. Because of the accumulation of coordinate errors, the original coordinate values for the object are to be reused at the start of each new revolution.

5–2. Write a set of procedures to implement the *create_transformation_matrix* and the *accumulate_transformation_matrix* operations to produce a composite transformation matrix.

5–3. Write a program that utilizes the procedures of Ex. 5–2 to apply any specified sequence of transformations to a displayed object. The program is to be designed so that a user selects the transformation sequence and associated parameters from displayed menus, and the composite transformation is used to transform the object. Display the original object and the transformed object in different colors or different fill patterns.

5–4. Modify transformation matrix (5–26), for scaling in an arbitrary direction, to include coordinates for any specific scaling fixed point (x_F, y_F).

5–5. Prove that the multiplication of transformation matrices for the following sequence of operations is commutative:
a. Two successive rotations
b. Two successive translations
c. Two successive scalings

5–6. Prove that a uniform scaling ($Sx = Sy$) and a rotation form a commutative pair of operations but that, in general, scaling and rotation are not commutative operations.

5–7. Show that transformation matrix (5–33), for a reflection about the line $y = x$, is equivalent to a reflection relative to the x axis followed by a counterclockwise rotation of 90°.

5–8. Show that transformation matrix (5–34), for a reflection about the line $y = -x$, is equivalent to a reflection relative to the y axis followed by a counterclockwise rotation of 90°.

5–9. Determine the form of the transformation matrix for a reflection about an arbitrary line with equation $y = mx + b$.

5–10. Determine a sequence of basic transformations that are equivalent to the x-direction shearing matrix (5–35).

5–11. Determine a sequence of basic transformations that are equivalent to the y-direction shearing matrix (5–36).

5–12. Set up procedures for implementing a block transfer of a rectangular area of a raster, using one function to read the area into an array and another function to copy the array into the designated transfer area.

5–13. Determine the results of performing two successive block transfers into the same area of a raster using the various Boolean operations.

5–14. What are the results of performing two successive block transfers into the same area of a raster using the binary arithmetic operations?

5–15. Implement a routine to perform block transfers in a raster using any specified Boolean operation or a replacement (copy) operation.

5–16. Write a routine to implement rotations in raster block transfers.

6 WINDOWING AND CLIPPING

Applications programs define pictures in a world coordinate system. This can be any Cartesian coordinate system that a user finds convenient. Pictures defined in world coordinates are then mapped by the graphics system into device coordinates. Typically, a graphics package allows a user to specify which area of the picture definition is to be displayed and where it is to be placed on the display device. A single area could be chosen for display, or several areas could be selected. These areas can be placed in separate display locations, or one area can serve as a small insert into a larger area. This transformation process involves operations for translating and scaling selected areas and for deleting picture parts outside the areas. These operations are referred to as windowing and clipping.

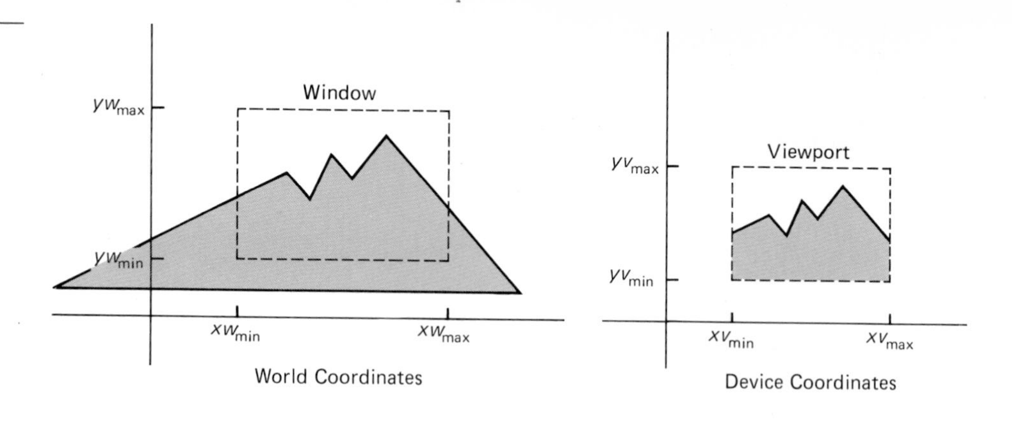

FIGURE 6–1
A window-to-viewport mapping.

6–1 Windowing Concepts

A rectangular area specified in world coordinates is called a **window.** The rectangular area on the display device to which a window is mapped is called a **viewport.** Figure 6–1 illustrates the mapping of a picture section that falls within the window area onto a designated viewport. This mapping is called a **viewing transformation, windowing transformation,** or **normalization transformation.**

Commands to set the window and viewport areas from an application program can be defined as

```
set_window (xw_min, xw_max, yw_min, yw_max)
set_viewport (xv_min, xv_max, yv_min, yv_max)
```

Parameters in each function are used to define the coordinate limits of the rectangular areas. The limits of the *window* are specified in world coordinates. Normalized device coordinates are most often used for the *viewport* specification, although device coordinates may be used if there is only one output device in the system. When normalized device coordinates are used, the programmer considers the output device to have coordinate values within the range 0 to 1. A viewport specification is given by values within this interval. The following specifications place a portion of a world coordinate definition into the upper right corner of a display area, as illustrated in Fig. 6–2.

```
set_window (-60.5, 41.25, -20.75, 82.5);
set_viewport (0.5, 0.8, 0.7, 1.0);
```

If a window is to be mapped onto the full display area, the viewport specification is given as

```
set_viewport (0, 1, 0, 1)
```

Coordinate positions expressed in normalized device coordinates must be converted to device coordinates before display by a specific output device. A device-specific routine is included in graphics packages for this purpose. The advantage of using normalized device coordinates is that the graphics package is largely device-independent. Different output devices can be used by providing the appropriate device drivers.

All coordinate points referenced in graphics packages must be specified relative to a Cartesian coordinate system. Any picture definitions that might originally

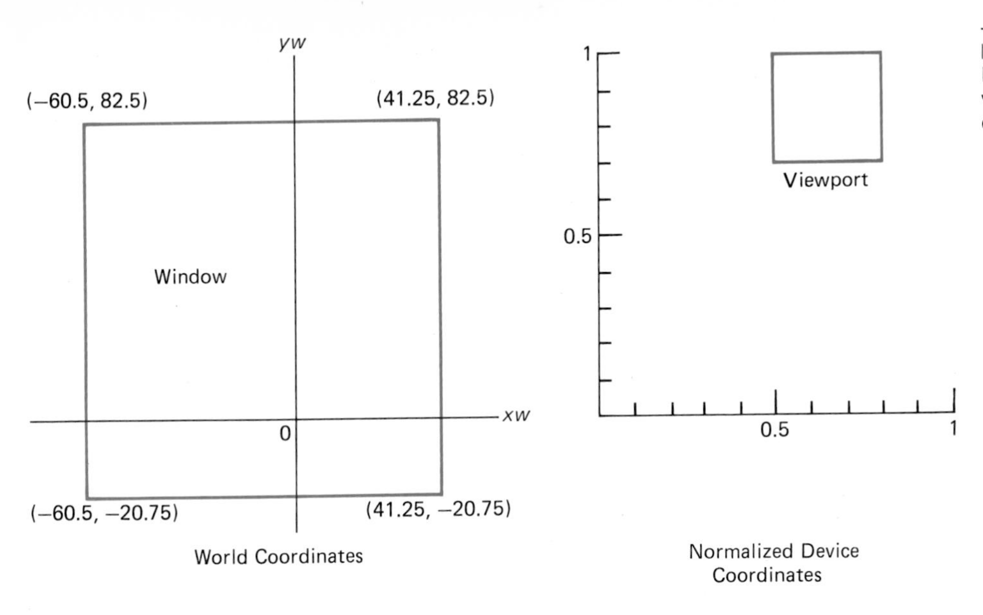

FIGURE 6–2
Mapping a window to a viewport in normalized device coordinates.

be defined in another system, such as polar coordinates, must first be converted by the user to a Cartesian world coordinate specification. These Cartesian coordinates are then used in window commands to identify the picture parts for display.

Window and viewport commands are stated before the procedures to specify a picture definition. Specifications for the window and viewport remain in effect for any subsequent output commands until new specifications are given.

By changing the position of the viewport, objects can be displayed at different positions on an output device. Also, by varying the size of viewports, the size and proportions of objects can be changed. When different-sized windows are successively mapped onto a viewport, **zooming** effects can be achieved. As the windows are made smaller, a user can zoom in on some part of a scene to view details that are not shown with the larger windows. Similarly, more overview is obtained by zooming out from a section of a scene with successively larger windows. **Panning** effects are produced by moving a fixed-size window across a large picture.

An example of the use of multiple window and viewpoint commands is given in the following procedure. Two graphs are displayed on different halves of a display device (Fig. 6–3).

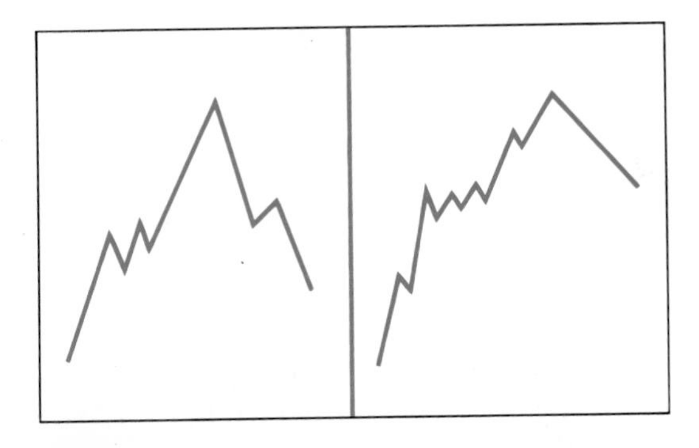

FIGURE 6–3
Simultaneous display of two graphs, using multiple window and viewport specifications.

```
type
  points = array [1..max_points] of real;
procedure two_graphs;
  var x, y. : points; k : integer;
  begin
    set_window (0, 1, 0, 1);              {draw center dividing line}
    set_viewport (0, 1, 0, 1);
    x[1] := 0.5; y[1] := 0; x[2] := 0.5; y[2] := 1;
    polyline (2, x, y);

    for k := 1 to 9 do begin              {read data for first graph}
        x[k] := k;
        readln (y[k])                     {data values between 300 and 700}
      end; {for k}
    set_window (1, 9, 300, 700);
    set_viewport (0.1, 0.4, 0.2, 0.8);       {put in left part of screen}
    polyline (9, x, y);

    for k := 1 to 13 do begin             {read data for second graph}
        x[k] := k;
        readln (y[k])                     {data values between 10 and 100}
      end; {for k}
    set_window (1, 13, 10, 100);
    set_viewport (0.6, 0.9, 0.2, 0.8);       {put in right part of screen}
    polyline (13, x, y)
  end; {two graphs}
```

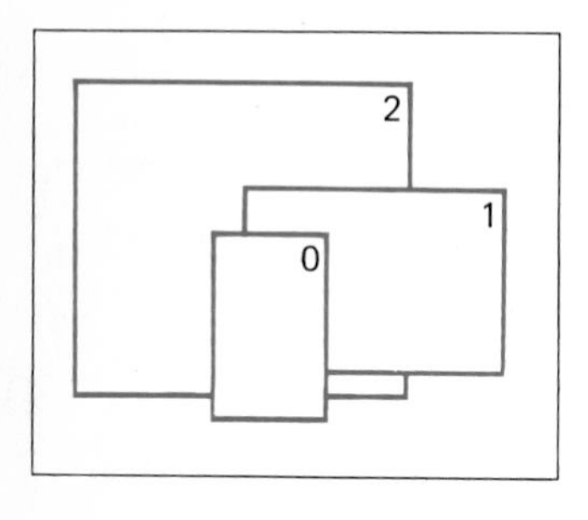

FIGURE 6–4
Displaying viewports in priority order. Lower-numbered viewports are given higher priority.

Another method for establishing multiple window and viewport areas in a graphics package is to assign a label to each specification. This could be done by including a fifth parameter in the window and viewport commands to identify each defined area. The parameter can be an integer index (0, 1, 2, . . .) that numbers the window or viewport definition. An additional command is then needed to indicate which window-to-viewport transformation number is to be applied to a set of output statements. This numbering scheme could also be used to associate a priority with each viewport, so that the visibility of overlapping viewports is decided on a priority basis. Viewports displayed according to priority are shown in Fig. 6–4.

For implementations that include multiple workstations, an additional set of window and viewport commands might be defined. These commands include a workstation number to establish different window and viewport areas on different workstations. This would allow a user to display various parts of the final image on different output devices. An architect, for example, might display the whole of a house plan on one monitor and just the second story on a second monitor.

The window and viewport commands we have introduced are used with rectangular areas, whose boundaries are parallel to the cordinate axes. Some graphics packages allow users to select other types of windows and viewports. A rotated window, as in Fig. 6–5, could be specified with an additional angle parameter A in a window command. Another possibility is to designate any type of polygon as the window by giving the sequence of vertices that define the polygon boundary. We begin by presenting algorithms for implementing rectangular windows and viewports whose boundaries are parallel to the x and y axes. The special handling required for arbitrarily shaped windows is then discussed as extensions to these algorithms.

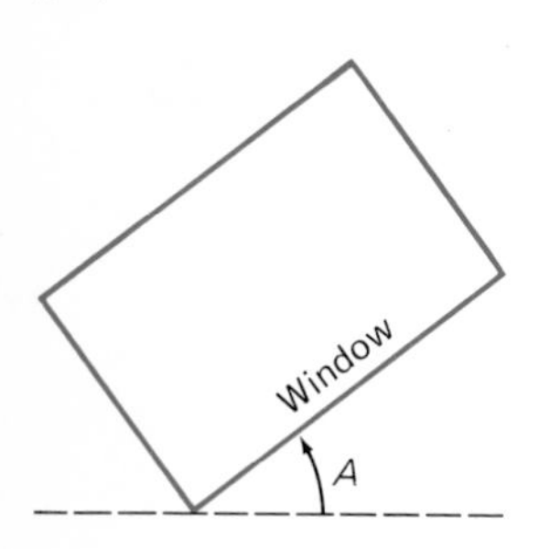

FIGURE 6–5
Rotated window, specified with angle parameter *a*.

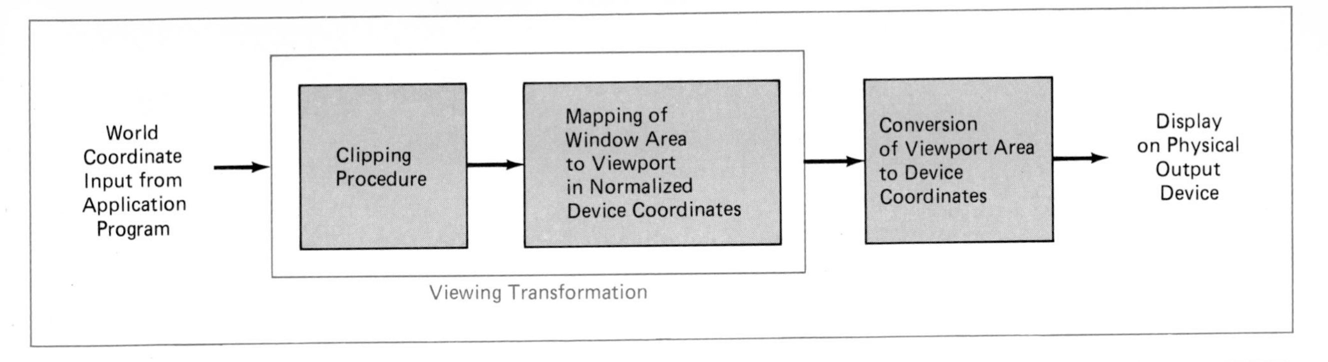

FIGURE 6–6
Processing input window specifications into displayed viewports.

6–2 Clipping Algorithms

Mapping a window area onto a viewport results in the display of only the picture parts within the window. Everything outside the window is discarded. Procedures for eliminating all parts of a defined picture outside of specified boundaries are referred to as **clipping algorithms** or simply **clipping.**

Implementation of a windowing transformation is often performed by clipping against the window, then mapping the window interior into the viewport (Fig. 6–6). Alternatively, some packages map the world coordinate definition into normalized device coordinates first and then clip against the viewport boundaries. In the following discussion, we assume that clipping is to be performed relative to the window boundaries in world coordinates. After clipping is completed, points within the window are mapped to the viewport.

Point clipping against a window specification simply means that we test coordinate values to determine whether or not they are within the boundaries. A point at position (x, y) is saved for transformation to a viewport if it satisfies the following inequalities:

$$xw_{\min} \leqslant x \leqslant xw_{\max}, \; yw_{\min} \leqslant y \leqslant yw_{\max} \tag{6–1}$$

If any one of these four inequalities is not satisfied, the point is clipped. In Fig. 6–7, point P_1 is saved, and point P_2 is clipped.

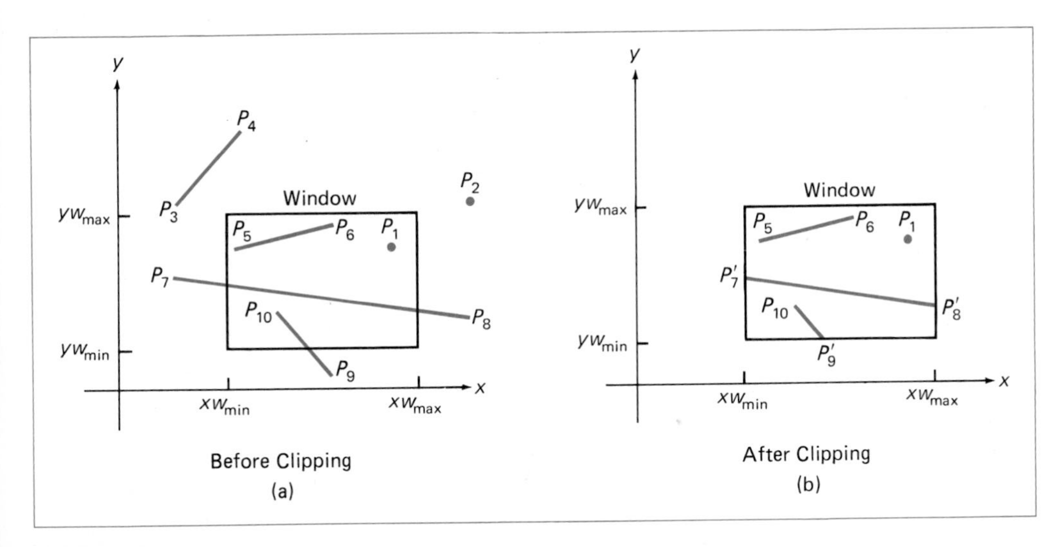

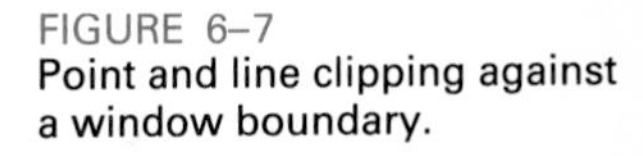
FIGURE 6–7
Point and line clipping against a window boundary.

Line Clipping

Figure 6–7 illustrates possible relationships between line positions and window boundaries. We test a line for clipping by determining whether the endpoints are inside or outside the window. A line with both endpoints inside the window boundaries, such as the line from P_5 to P_6, is saved. A line with one endpoint outside (P_9) and one endpoint inside (P_{10}) is clipped off at the boundary intersection point (P'_9). Lines that have both endpoints beyond the boundaries are either totally outside the window or cross two window boundaries. The line from P_3 to P_4 is completely clipped. But the line from P_7 to P_8 extends beyond the window boundaries on both sides, and so the section of the line from P'_7 to P'_8 is saved.

A line-clipping algorithm determines which lines are wholly within the window boundaries and which lines are to be totally or partially clipped. For lines that are to be partially clipped, the intersection points with the window boundaries must be calculated. Since a picture definition can contain thousands of line segments, the clipping process should be performed as efficiently as possible. Before performing intersection calculations, an algorithm should identify all lines that are to be completely saved or completely clipped. With these lines eliminated from consideration, the determination of intersection points for the remaining lines should be carried out with a minimum of calculations.

FIGURE 6–8
Binary region codes for line endpoints, used to define coordinate areas relative to a window.

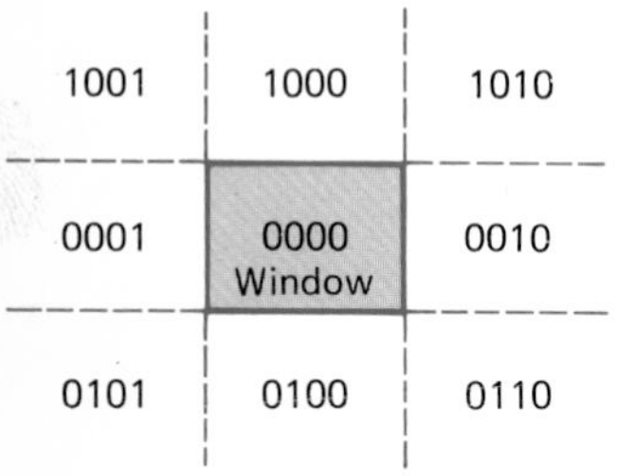

1001	1000	1010
0001	0000 Window	0010
0101	0100	0110

One approach to line clipping is based on a coding scheme developed by Cohen and Sutherland. Every line endpoint in a picture is assigned a four-digit binary code, called a **region code,** that identifies the coordinate region of the point. Regions are set up in reference to the window boundaries, as shown in Fig. 6–8. Each bit position in the region code is used to indicate one of the four relative coordinate positions of the point with respect to the window: to the left, right, top, or bottom. Numbering the bit positions in the region code as 1 through 4 from right to left, the coordinate regions can be correlated with the bit positions as

bit 1 – left
bit 2 – right
bit 3 – below
bit 4 – above

A value of 1 in any bit position indicates that the point is in that relative position; otherwise, the bit position is set to 0. If a point is within the window, the region code is 0000. A point that is below and to the left of the window has a region code of 0101.

Bit values in the region code are determined by comparing endpoint coordinate values (x, y) to the window boundaries. Bit 1 is set to 1 if $x < xw_{\min}$. The other three bit values can be determined using similar comparisons. For languages in which bit manipulation is possible, region-code bit values can be determined by these steps: (1) Calculate differences between endpoint coordinates and window boundaries. (2) Use the resultant sign bit of each difference calculation to set the corresponding value in the region code. Bit 1 is the sign bit of $x - xw_{\min}$; bit 2 is the sign bit of $xw_{\max} - x$; bit 3 is the sign bit of $y - yw_{\min}$; and bit 4 is the sign bit of $yw_{\max} - y$.

Once we have established region codes for all line endpoints, we can quickly determine which lines are completely inside the window and which are clearly outside. Any lines that are completely contained within the window boundaries have a region code of 0000 for both endpoints, and we trivially accept these lines. Any lines that have a 1 in the same bit position in the region codes for each endpoint are completely outside the window, and we trivially reject these lines. We would discard the line that has a region code of 1001 for one endpoint and a code of 0101 for the other endpoint. Both endpoints of this line are left of the window, as indicated by the 1 in the first bit position of each region code. A method that can be used to test lines for total clipping is to perform the logical AND operation with both region codes. If the result is not 0000, the line is outside the window.

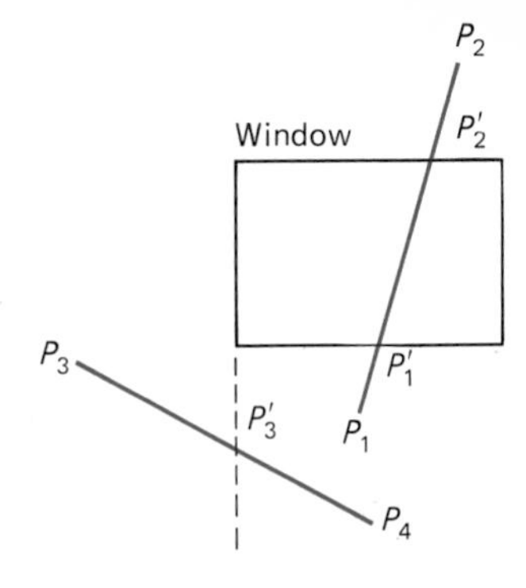

FIGURE 6–9
Lines extending from one coordinate region to another may pass through the window, or they may intersect window boundaries without entering the window.

Lines that cannot be identified as completely inside or completely outside a window by these tests are checked for intersection with the window boundaries. As shown in Fig. 6–9, such lines may or may not cross into the window interior. We can process these lines by comparing an endpoint that is outside the window to a window boundary to determine how much of the line can be discarded. Then the remaining part of the line is checked against the other boundaries, and we continue until either the line is totally discarded or a section is found inside the window. We set up our algorithm to check line endpoints against window boundaries in the order left, right, bottom, top.

To illustrate the specific steps in clipping lines against window boundaries using the Cohen-Sutherland algorithm, we show how the lines in Fig. 6–9 could be processed. Starting with the bottom endpoint of the line from P_1 to P_2, we check P_1 against the left, right, and bottom boundaries in turn and find that this point is below the window. We then find the intersection point P'_1 with the bottom boundary and discard the line section from P_1 to P'_1. The line now has been reduced to the section from P'_1 to P_2. Since P_2 is outside the window, we check this endpoint against the boundaries and find that it is above the window. Intersection point P'_2 is calculated, and the line section for P'_1 to P'_2 is saved. This completes processing for this line, so we save this part and go on to the next line. Point P_3 in the next line is to the left of the window, so we determine the intersection P'_3 and eliminate the line section from P_3 to P'_3. By checking region codes for the line section from P'_3 to P_4, we find that the remainder of the line is below the window and can be discarded also.

Intersection points with the window boundary are calculated using the line-equation parameters. For a line with endpoint coordinates (x_1, y_1) and (x_2, y_2), the y coordinate of the intersection point with a vertical window boundary can be obtained with the calculation

$$y = y_1 + m\,(x - x_1) \tag{6–2}$$

where the x value is set either to xw_{min} or to xw_{max}, and the slope m is calculated as $m = (y_2 - y_1) / (x_2 - x_1)$. Similarly, if we are looking for the intersection with a horizontal boundary, the x coordinate can be calculated as

$$x = x_1 + \frac{y - y_1}{m} \tag{6–3}$$

with y set either to yw_{min} or to yw_{max}.

The following procedure demonstrates the Cohen-Sutherland line-clipping algorithm. Codes for each endpoint are maintained as four-element arrays of Boolean values.

```
var
  xw_min, xw_max, yw_min, yw_max : real;

procedure clip_a_line (x1, y1, x2, y2 : real);
  type
    boundaries = (left, right, bottom, top);
    code = array [boundaries] of boolean;
  var
    code1, code2 : code;
    done, display : boolean;
    m : real;

  procedure encode (x, y : real; var c : code);
    begin
      if x < xw_min then c[left] := true
         else c[left] := false;
      if x > xw_max then c[right] := true
         else c[right] := false;
      if y < yw_min then c[bottom] := true
         else c[bottom] := false;
      if y > yw_max then c[top] := true
         else c[top] := false
      end; {encode}

  function accept (c1, c2 : code) : boolean;
    var k : boundaries;
    begin
      {if either point has 'true' in its code,
       a trivial accept is not possible}
      accept := true;
      for k := left to top do
         if c1[k] or c2[k] then accept := false
    end; {accept}

  function reject (c1, c2 : code) : boolean;
    var k : boundaries;
    begin
      {if endpoints have matching 'trues',
       line can be rejected}
      reject := false;
      for k := left to top do
         if c1[k] and c2[k] then reject := true
    end; {reject}

  procedure swap_if_needed (var x1, y1, x2, y2: real;
                             var c1, c2 : code);
    begin
      {insures that x1, y1 is a point outside
       of window and c1 contains its code}
    end; {swap_if_needed}

  begin {clip_a_line}
    done := false;
    display := false;
    while not done do begin
        encode (x1, y1, code1);
        encode (x2, y2, code2);
```

combine first two if statements and last two together

the only finding of a true in c1[k] or c2[k], we can get out of the loop

same as above

```
      if accept (code1, code2) then begin
          done := true;
          display := true
        end {if accept}
      else
        if reject (code1, code2) then done := true
        else begin {find intersection}
            {make sure that x1, y1 is outside window}
            swap_if_needed (x1, y1, x2, y2, code1, code2);
            m := (y2–y1) / (x2–x1);
            if code1[left] then begin
              y1 := y1 + (xw_min - x1) * m;
              x1 := xw_min
              end {crosses left}
            else
              if code1[right] then begin
                y1 := y1 + (xw_max - x1) * m;
                x1 := xw_max
                end {crosses right}
              else
                if code1[bottom] then begin
                  x1 := x1 + (yw_min- y1) / m;
                  y1 := yw_min
                  end {crosses bottom}
                else
                  if code1[top] then begin
                    x1 := x1 + (yw_max - y1) / m;
                    y1 := yw_max
                    end {crosses top}
          end {else find intersection}
      end; {while not done}
    if display then {draw x1, y1, to x2, y2}
  end; {clip_a_line}
```

A technique for locating window intersections without the line-equation calculations is a binary search procedure, called **midpoint subdivision.** Initial testing of lines is again carried out using region codes. Any lines that are not completely accepted or rejected with the region code checks are tested for window intersections by examining line midpoint coordinates.

This approach is illustrated in Fig. 6–10. For any line with endpoints (x_1, y_1) and (x_2, y_2), the midpoint of the line is calculated as

$$x_m = \frac{x_1 + x_2}{2}, \quad y_m = \frac{y_1 + y_2}{2} \tag{6–4}$$

Each calculation for a midpoint coordinate involves only an addition and a division by 2 (a shift operation). Once the midpoint coordinates have been determined, each half of the line can be tested for total acceptance or rejection. If half of the line can be accepted or rejected, then the other half is processed in the same way. This continues until an intersection point is found. If one half of the line cannot be trivially accepted or rejected, each half is processed until either the line is totally rejected or a visible section is found. Hardware implementation of this method can provide fast line clipping against the viewport boundaries after object descriptions have been transformed to device coordinates.

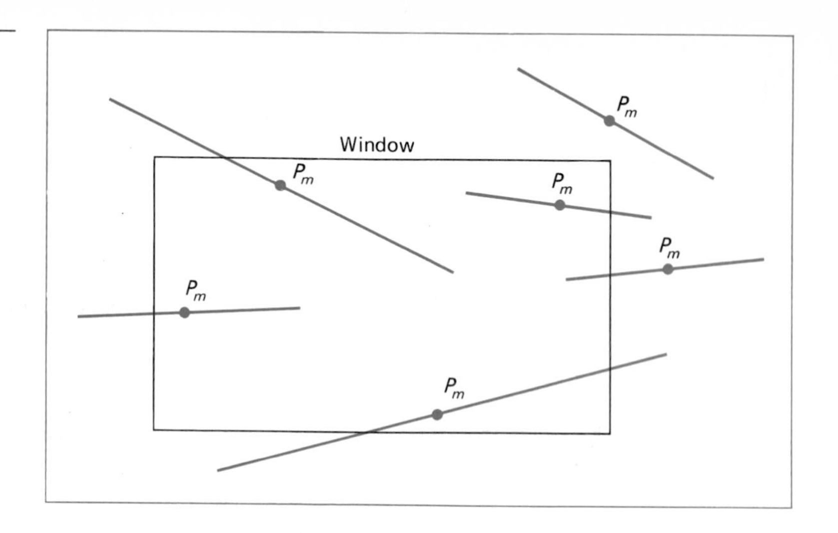

FIGURE 6–10
Line midpoints, P_m, used in a clipping algorithm.

Other techniques for line clipping have been devised that use a parametric form of the line equation. We can write the equation of a line segment defined between endpoints (x_1, y_1) and (x_2, y_2) in the parametric form

$$\begin{aligned} x &= x_1 + (x_2 - x_1)u = x_1 + \Delta x\, u \\ y &= y_1 + (y_2 - y_1)u = y_1 + \Delta y\, u \end{aligned} \tag{6–5}$$

where $\Delta x = x_2 - x_1$ and $\Delta y = y_2 - y_1$. Parameter u is assigned values between 0 and 1, and coordinates (x, y) represent a point on the line specified by a value of u in this range. When $u = 0$, $(x, y) = (x_1, y_1)$. At the other end of the line, $u = 1$ and $(x, y) = (x_2, y_2)$.

An efficient line-clipping algorithm using these parametric equations has been developed by Liang and Barsky. They note that if a point (x, y) along the line is inside a window defined by coordinates (xw_{min}, yw_{min}) and (xw_{max}, yw_{max}), the following conditions are satisfied:

$$\begin{aligned} xw_{min} &\leq x_1 + \Delta x\, u \leq xw_{max} \\ yw_{min} &\leq y_1 + \Delta y\, u \leq yw_{max} \end{aligned} \tag{6–6}$$

These four inequalities can be rewritten in the form

$$p_k\, u \leq q_k, \quad k = 1, 2, 3, 4 \tag{6–7}$$

where p and q are defined as

$$\begin{aligned} p_1 &= -\Delta x, & q_1 &= x_1 - xw_{min} \\ p_2 &= \Delta x, & q_2 &= xw_{max} - x_1 \\ p_3 &= -\Delta y, & q_3 &= y_1 - yw_{min} \\ p_4 &= \Delta y, & q_4 &= yw_{max} - y_1 \end{aligned} \tag{6–8}$$

Any line that is parallel to one of the window boundaries has $p_k = 0$ for the value of k corresponding to that boundary (k = 1, 2, 3, and 4 correspond to the left, right, bottom, and top boundaries, respectively). If, for that value of k, we also find $q_k < 0$, then the line is completely outside the boundary and can be eliminated from further consideration. If $q_k \geq 0$, the parallel line is inside the boundary.

When $p_k < 0$, the infinite extension of the line proceeds from the outside to the inside of the infinite extension of this particular window boundary. If $p_k > 0$, the line proceeds from the inside to the outside. For a nonzero p_k, we can calculate the value of u that corresponds to the point where the infinitely extended line intersects the extension of the window boundary k as

$$u = q_k/p_k \tag{6-9}$$

For each line, we can calculate values for parameters u_1 and u_2 that define that part of the line that lies within the window. The value of u_1 is determined by looking at the window edges for which the line proceeds from the outside to the inside ($p < 0$). For these window edges, we calculate $r_k = q_k / p_k$. The value of u_1 is taken as the largest of the set consisting of 0 and the various values of r. Conversely, the value of u_2 is determined by examining the boundaries for which the line proceeds from inside to outside ($p > 0$). A value of r_k is calculated for each of these window boundaries, and the value of u_2 is the minimum of the set consisting of 1 and the calculated r values. If $u_1 > u_2$, the line is completely outside the window and it can be rejected. Otherwise, the endpoints of the clipped line are calculated from the two values of parameter u.

This algorithm is presented in the following procedure. Line intersection parameters are initialized to the values $u1 = 0$ and $u2 = 1$. For each window boundary, the appropriate values for p and q are calculated and used by the function *cliptest* to determine whether the line can be rejected or whether the intersection parameters are to be adjusted. When $p < 0$, the parameter r is used to update $u1$; when $p > 0$, parameter r is used to update $u2$. If updating $u1$ or $u2$ results in $u1 > u2$, we reject the line. Otherwise, we update the appropriate u parameter only if the new value results in a shortening of the line. When $p = 0$ and $q < 0$, we can discard the line since it is parallel to and on the outside of this boundary. If the line has not been rejected after all four values of p and q have been tested, the endpoints of the clipped line are determined from values of $u1$ and $u2$.

```
var
  xwmin, xwmax, ywmin, ywmax : real;

procedure clipper (var x1, y1, x2, y2 : real);
  var
    u1, u2, dx, dy : real;

  function cliptest (p, q : real; var u1, u2 : real);
    var
      r : real;
      result : boolean;
    begin {cliptest}
      result := true;
      if p < 0 then begin                       {line from outside to inside
                                                 of this boundary}
        r := q / p;
        if r > u2 then result := false           {reject line or}
        else if r > u1 then u1 := r              {update u1 if appropriate}
      end {if p < 0}
      else
        if p > 0 then begin                     {line from inside to outside
                                                 of this boundary}
```

```
        r := q / p;
        if r < u1 then result := false        {reject line or}
        else if r < u2 then u2 := r            {update u2 if appropriate}
      end {if p > 0}
    else                                       {p = 0; line parallel to
                                               boundary}
        if q < 0 then result := false;         {outside of boundary}
    cliptest := result
  end; {cliptest}

begin {clipper}
  u1 := 0;
  u2 := 1;
  dx := x2 - x1;
  if cliptest (-dx, x1 - xwmin, u1, u2) then
    if cliptest (dx, xwmax - x1, u1, u2) then begin
      dy := y2 - y1;
      if cliptest (-dy, y1 - ywmin, u1, u2) then
        if cliptest (dy, ywmax - y1, u1, u2) then begin
          {if u1 and u2 are within range of 0 to 1,
           use to calculate new line endpoints}
          if u1 > 0 then begin
              x1 := x1 + u1 * dx;
              y1 := y1 + u1 * dy
            end; {if u1 > 0}
          if u2 < 1 then begin
              x2 := x1 + u2 * dx;
              y2 := y1 + u2 * dy
            end {if u2 < 1}
        end {if cliptest}
    end {if cliptest}
end; {clipper}
```

plot the line segment (x_1, y_1) to (x_2, y_2)

The Liang and Barsky line-clipping algorithm reduces the computations that are needed to clip lines. Each update of u_1 and u_2 requires only one division, and window intersections of the line are computed only once, when values of u_1 and u_2 have been finalized. In contrast, the Cohen and Sutherland algorithm repeatedly calculates points of intersection between the line and window boundaries, and each intersection calculation requires both a division and a multiplication.

When rotated windows or arbitrarily shaped polygons are used for windows and viewports, the line-clipping algorithms discussed would require some modification. It is still possible to do preliminary screening of the lines. A rotated window, or any other polygon shape, can be enclosed within a larger rectangle whose sides are parallel to the coordinate axes (Fig. 6–11). Any lines outside the larger bounding rectangle are outside the window also. Inside tests are not so easily done, and intersection points must be calculated using line equations for the window boundaries as well as for the lines to be clipped.

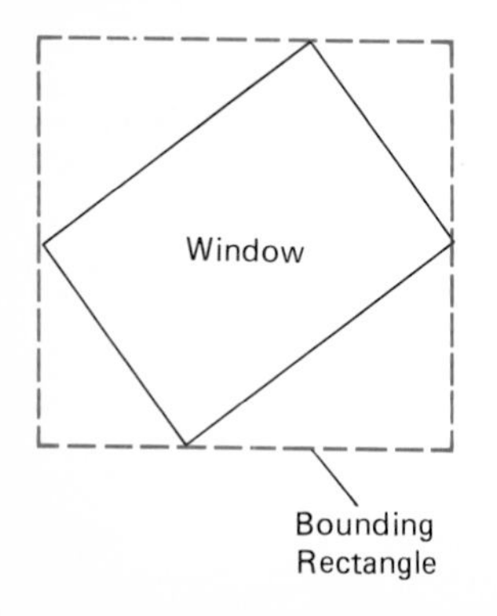

FIGURE 6–11
Rotated window enclosed by a larger bounding rectangle whose boundaries are parallel to coordinate axes.

Area Clipping

Hollow polygons used in line-drawing applications can be clipped by processing each component line through the line-clipping algorithms discussed. A polygon processed in this way is reduced to a series of clipped lines (Fig. 6–12).

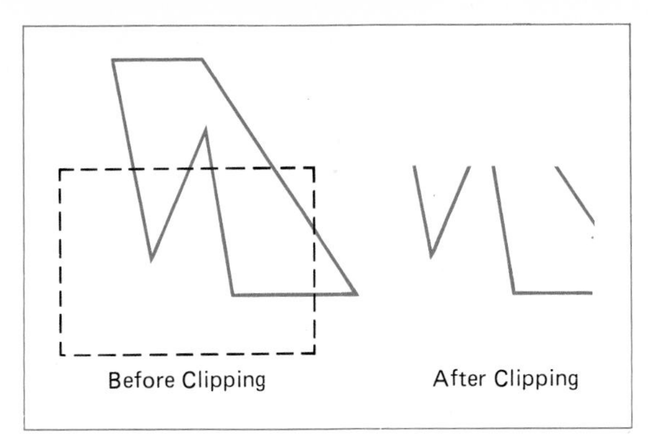

FIGURE 6–12
Polygon clipped by a line-clipping algorithm.

When a polygon boundary defines a fill area, as in Fig. 6–13, a modified version of the line-clipping algorithm is needed. In this case, one or more closed areas must be produced to define the boundaries for area fill.

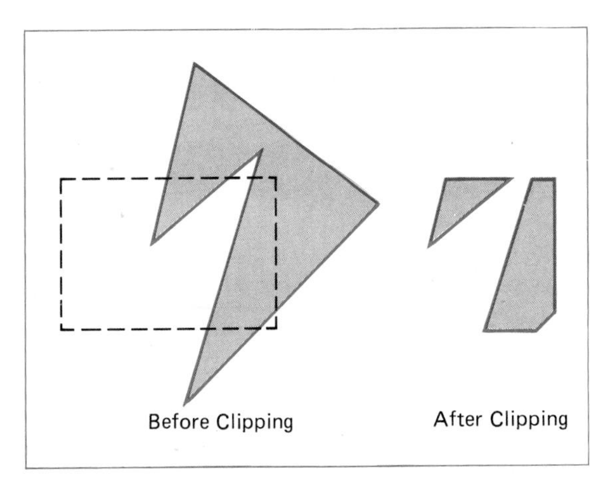

FIGURE 6–13
A shaded area, before and after clipping.

A technique for polygon clipping, developed by Sutherland and Hodgman, performs clipping by comparing a polygon to each window boundary in turn. The output of the algorithm is a set of vertices defining the clipped area that is to be filled with a color or shading pattern. The basic method is illustrated in Fig. 6–14.

Polygon areas are defined by specifying an ordered sequence of vertices. To clip a polygon, we compare each of the vertices in turn against a window boundary. Vertices inside this window edge are saved for clipping against the next boundary;

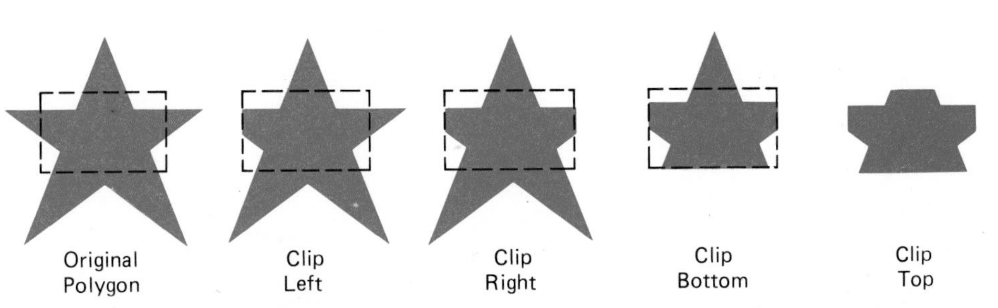

FIGURE 6–14
Clipping a polygon area against successive window boundaries.

FIGURE 6–15
Processing vertices of a polygon relative to a window boundary (dashed lines). From vertex *S*, the next vertex point processed *(P)* may generate one point, no points, or two points to be saved by a clipping algorithm.

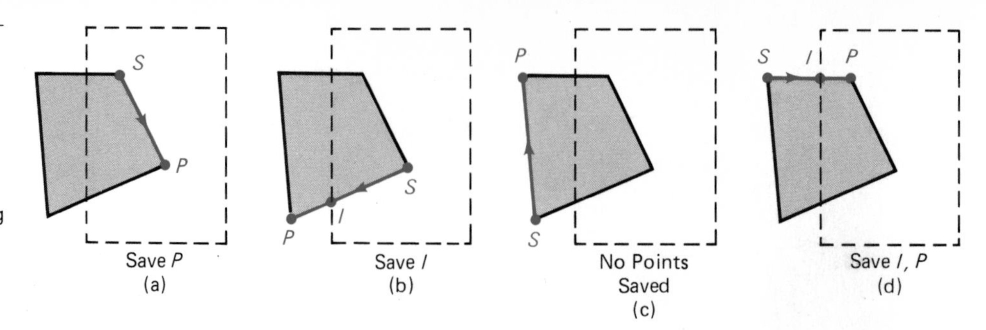

vertices outside the window edge are discarded. If we proceed from a point inside the window edge to an outside point, we save the intersection of the line with the window boundary. Both the intersection and the vertex are saved if we cross from the outside of a window edge to the inside. The four possible situations that can occur as we process a point *(P)* and the previous point *(S)* against a window boundary are illustrated in Fig. 6–15. A point inside the window boundary is saved (case a), while an outside point is not (case c). If a point *P* and the previous point *S* are on opposite sides of a boundary, the intersection *I* is calculated and saved (cases b and d). In case d, the point *P* is inside and the previous point *S* is outside so both the intersection *I* and the point *P* are saved. Once all vertices have been processed for the left window boundary, the set of saved points is clipped against the next window boundary.

We illustrate this method by processing the area in Fig. 6–16 against the left window boundary. Vertices 1 and 2 are found to be on the outside of the boundary. Passing to vertex 3, which is inside, we calculate the intersection and save both the intersection point and vertex 3. Vertices 4 and 5 are determined to be inside, and they also are saved. The sixth and final vertex is outside, so we find and save the intersection point. Using the five saved points, we repeat the process for the next window boundary.

Implementing the algorithm as described necessitates the use of extra storage space for the saved points. This can be avoided if we take each point that would be saved and immediately pass it to the clipping routine, along with instructions to clip it against the next boundary. We save a point (either an original vertex or a calculated intersection) only after it has been processed against all boundaries. It is as if we have a pipeline of clipping routines, with each stage in the pipeline clipping against a different window boundary. A point that is inside or on the window boundary at one stage is passed along to the next stage. A point that is outside at some stage simply does not continue in the pipeline.

The following procedure demonstrates this approach. An array, *s*, records the most recent point that was clipped for each window edge. The main routine passes each vertex *p* to the *clip_this* routine for clipping against the first window edge. If the line defined by endpoints *p* and *s*[*edge*] crosses this window edge, the intersection is found and is passed to the next clipping stage. If *p* is inside the window, it is passed to the next clipping stage. Any point that survives clipping against all window edges is then entered into the output arrays *x_out* and *y_out*. The array *first_point* stores for each window edge the first point that is clipped against that edge. After all polygon vertices have been processed, a closing routine clips lines defined by the first and last points clipped against each edge.

FIGURE 6–16
Clipping a polygon against the left edge of a window, starting with vertex 1. Primed numbers are used to label the points saved by the clipping algorithm.

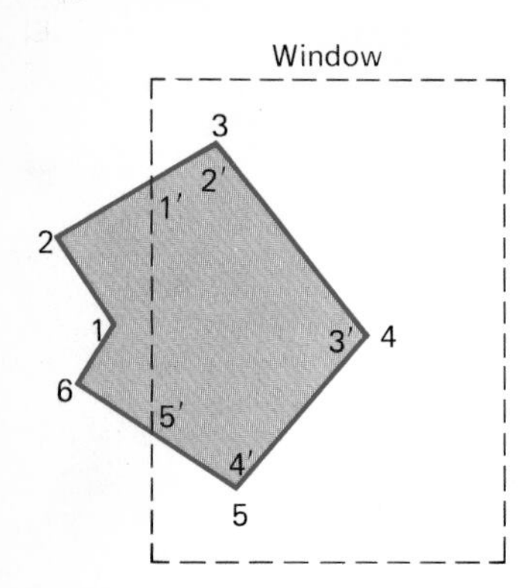

```
type
  points = array [1..max_points] of real;

procedure polygon_clip (n : integer; x, y : points; var m : integer;
                        var x_out, y_out : points);
  const
    boundary_count = 4;
  type
    vertex = array [1..2] of real;
    boundary_range = 1..boundary_count;
  var
    k : integer;
    p : vertex;
    s, first_point : array [1..boundary_count] of vertex;
    new_edge : array [1..boundary_count] of boolean;

  function inside (p : vertex; edge : boundary_range) : boolean;
    begin
      {returns true if vertex p is inside of window edge}
    end; {inside}

  function cross (p, s : vertex; edge : integer) : boolean;
    begin                                    boundary_range
      {returns true if polygon side ps intersects window edge}
    end; {cross}

  procedure output_vertex (p : vertex);
    begin
      m := m + 1;
      x_out[m] := p[1]; y_out[m] := p[2]
    end; {output_vertex}

  procedure find_intersection (p, s : vertex;
                               edge : boundary_range; var i : vertex);
    begin
      {returns in parameter i the intersection of ps with window edge}
    end; {intersection}

  procedure clip_this (p : vertex; edge : boundary_range);
    var i : vertex;
    begin {clip_this}
      {save the first point clipped against a window edge}
      if new_edge[edge] then begin
          first_point[edge] := p;
          new_edge[edge] := false
        end {new_edge}
      else
        {if ps crosses window edge, find intersection,
         clip intersection against next window edge}
        if cross (p, s[edge], edge) then begin
            find_intersection (p, s[edge], edge, i);
            if edge < boundary_count then clip_this (i, edge + 1)
            else output_vertex (i)
            end; {if p & s cross edge}
      {update saved vertex}
```

```
    s[edge] := p;
    {if p is inside this window edge,
     clip it against next window edge}
    if inside (p, edge) then
      if edge < boundary_count then clip_this (p, edge + 1)
      else output_vertex (p)
  end; {clip_this}

procedure clip_closer;
    {closing routine. For each window edge, clips the
     line connecting the last saved vertex and the first_point
     processed against the edge}
    var
      i : vertex;
      edge : integer;
    begin
      for edge := 1 to boundary_count do
          if cross (s[edge], first_point[edge], edge) then begin
               find_intersection (s[edge], first_point[edge], edge, i);
               if edge < boundary_count then clip_this (i, edge + 1)
               else output_vertex (i)
               end  {if s and first_point cross edge}
               end;  {clip_closer}

begin  {polygon_clip}
  m := 0;                                 {number of output vertices}
  for k := 1 to boundary_count do
    new_edge[k] := true;
  for k := 1 to n do begin                {puts each vertex into pipeline}
      p[1] := x[k]; p[2] := y[k];
      clip_this (p, 1)                    {clip against first window edge}
    end;  {for k}
  clip_closer                             {close polygon}
end;  {polygon_clip}
```

When a concave polygon is clipped against a rectangular window, the final clipped area may actually represent two or more distinct polygons. Since this area-clipping algorithm produces only one list of vertices, these separate areas will be joined with connecting lines. An example of this effect is shown in Fig. 6–17. Special considerations can be given to such cases to remove the extra lines, or more general clipping algorithms can be employed.

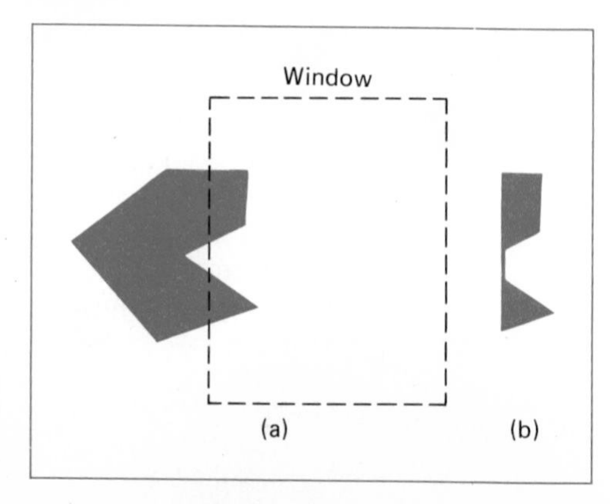

FIGURE 6–17
Clipping the concave polygon in (a) against the window generates the two connected areas in (b).

Although we have limited our discussion to rectangular windows aligned with the x and y axes, we could implement this algorithm to clip against a window of any polygon shape. We would need to store information about each of the window boundaries, and we would need to modify the routines *inside* and *find_intersection* to handle arbitrary boundaries.

Another approach to polygon-area clipping is to employ parametric-equation methods. Arbitrarily shaped windows would then be processed by using parametric line equations to describe both the window boundaries and the boundaries of the areas to be clipped.

Clipping areas other than polygons requires a little more work, since the area boundaries are not defined with straight-line equations. In Fig. 6–18, for example, circle equations are needed to find the two intersection points on the window boundary.

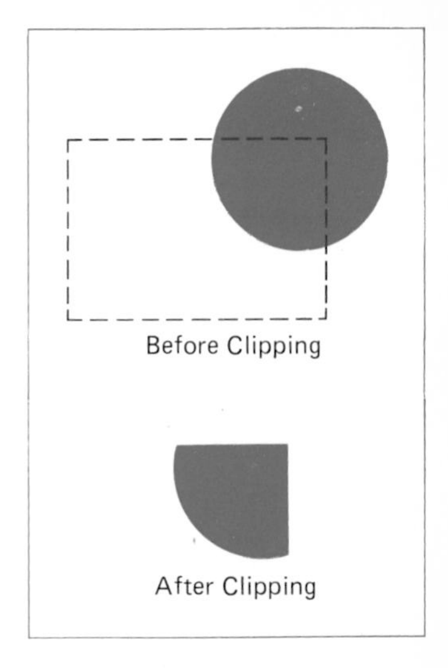

FIGURE 6–18
Clipping a circular area.

Text Clipping

There are several techniques that can be used to provide text clipping in a graphics package. The particular implementation chosen will depend on the methods used to generate characters and the amount of sophistication needed by a user of the package in the processing of text.

The simplest method for processing character strings relative to a window boundary is to use the "all-or-none text-clipping" strategy shown in Fig. 6–19. If all of the string is inside a window, we keep it. Otherwise, the string is discarded. This procedure can be implemented by considering a bounding rectangle around the text pattern. The boundary positions of the rectangle are then compared to the window boundaries, and the string is rejected if there is any overlap. This method produces the fastest text clipping.

An alternative to rejecting an entire character string that overlaps a window boundary is to use the "all-or-none character-clipping" strategy. Here we discard only those characters that are not completely inside the window (Fig. 6–20). In this case, the boundary limits of individual characters are compared to the window. Any character that either overlaps or is outside a window boundary is clipped.

A final method for handling text clipping is to clip individual characters. We now treat characters in much the same way that we treated lines. If an individual character overlaps a window boundary, we clip off the parts outside the window (Fig. 6–21). Characters formed with line segments can be processed in this way using a line-clipping algorithm. Processing characters formed with bit maps requires clipping individual pixels by comparing the relative position of the grid patterns to the window boundaries.

FIGURE 6–19
Text clipping using bounding rectangles. Any rectangles overlapping the boundary are entirely discarded.

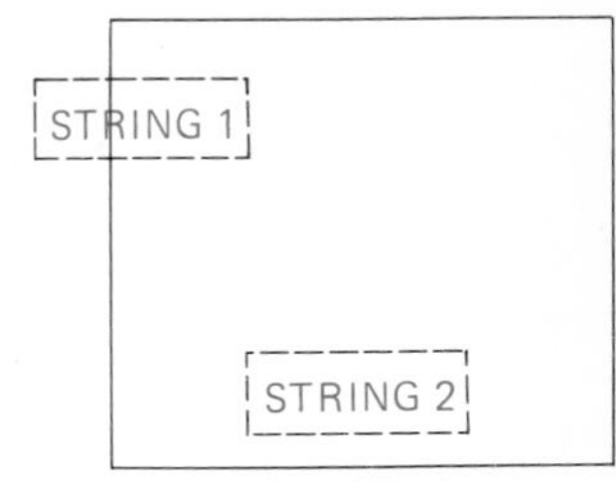

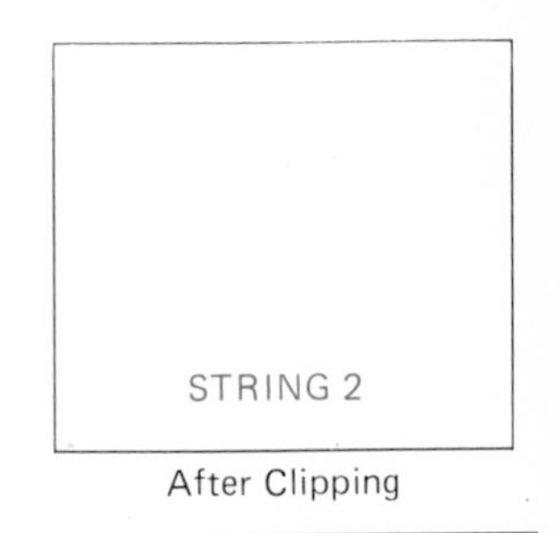

Blanking

Instead of saving information inside a defined region, a window area can be used for **blanking** (erasing) anything within its boundaries. What is saved is outside.

Blanking all output primitives within a defined area is a convenient means for overlapping different pictures. This technique is often used for designing page layouts in advertising or publishing applications or for adding labels or design patterns to a picture. The technique can also be used for combining graphs, maps, or schematics. Figure 6–22 illustrates some applications of blanking.

When two displays are to be overlaid using blanking methods, one display can be thought of as the foreground and the other as the background. A blanking window, encompassing the foreground display area, is superimposed on the back-

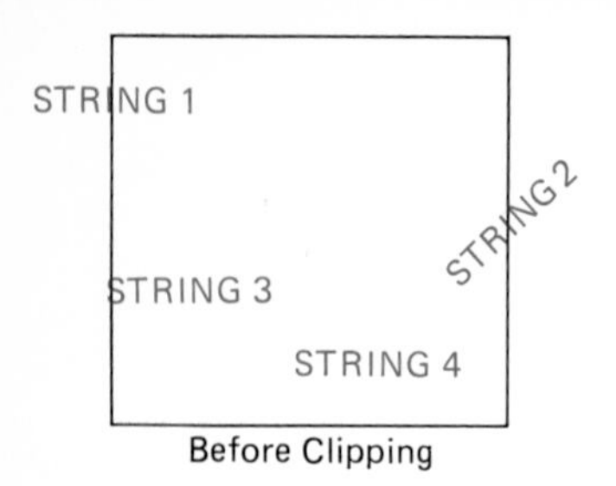

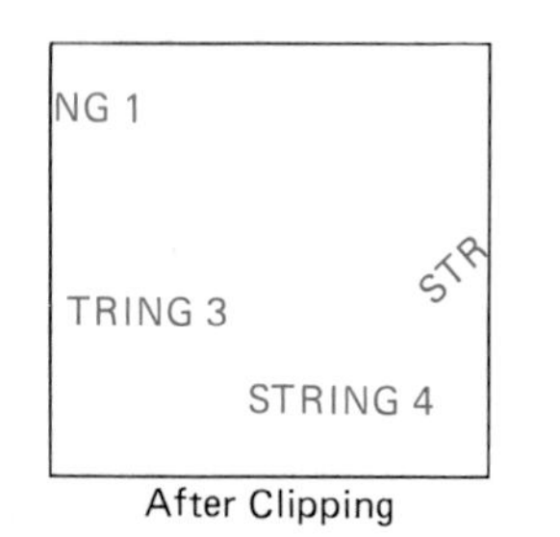

FIGURE 6–20
Character strings can be clipped so that only characters that are completely inside a window are retained.

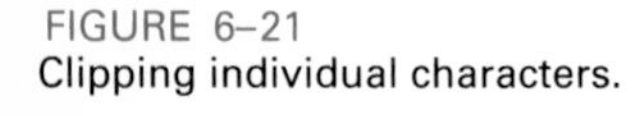

FIGURE 6–21
Clipping individual characters.

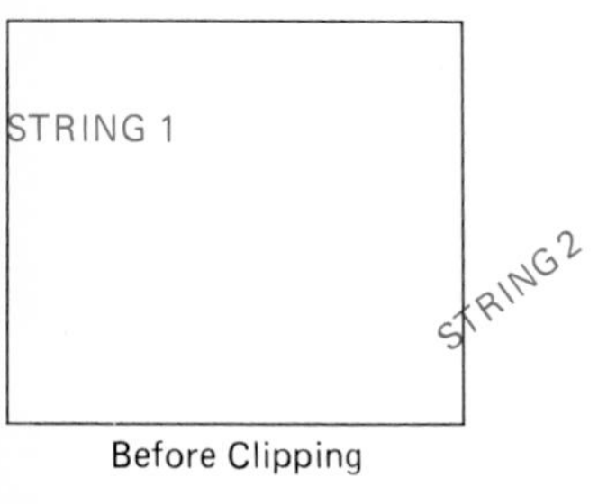

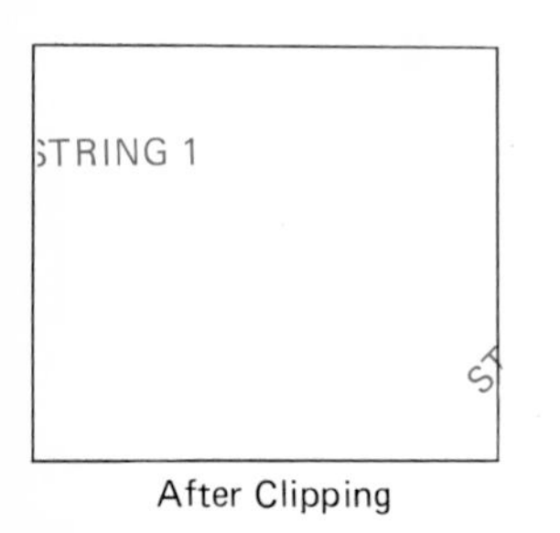

ground picture, and the picture parts within the window area are blanked. The two displays are then combined, with the foreground information placed within the blanked window area.

6–3 Window-to-Viewport Transformation

Once all points, lines, polygons, and text have been clipped, they are mapped onto the viewport area for display. This transformation to the viewport is carried out so that relative proportions are maintained.

In Fig. 6–23, a point at position (xw, yw) in a window is mapped into position (xv, yv) in a viewport. To maintain the same relative placement in the viewport as in the window, we require that

$$\frac{xw - xw_{min}}{xw_{max} - xw_{min}} = \frac{xv - xv_{min}}{xv_{max} - xv_{min}} \tag{6–10}$$

and

$$\frac{yw - yw_{min}}{yw_{max} - yw_{min}} = \frac{yv - yv_{min}}{yv_{max} - yv_{min}} \tag{6–11}$$

We can rewrite Eqs. 6–10 and 6–11 as explicit transformation calculations for the coordinates xv and yv:

$$\begin{aligned} xv &= \frac{xv_{max} - xv_{min}}{xw_{max} - xw_{min}}(xw - xw_{min}) + xv_{min} \\ yv &= \frac{yv_{max} - yv_{min}}{yw_{max} - yw_{max}}(yw - yw_{min}) + yv_{min} \end{aligned} \tag{6–12}$$

These window-to-viewport transformation calculations can be written more compactly as

$$\begin{aligned} xv &= sx(xw - xw_{min}) + xv_{min} \\ yv &= sy(yw - yw_{min}) + yv_{min} \end{aligned} \tag{6–13}$$

which include both scaling and translation factors. The ratios represented by sx and sy scale objects according to the relative sizes of the window and viewport. These ratios must be equal if objects are to maintain the same proportions when they are mapped into the viewport. When the window and viewport are the same size ($sx = sy = 1$), there is no change in the size of transformed objects. Values for xv_{min} and yv_{min} provide the translation factors for moving objects into the viewport area.

Character strings can be handled in two ways when they are mapped to a viewport. The simplest mapping maintains a constant character size, even though the viewport area may be enlarged or reduced relative to the window. This method would be employed when text is formed with standard character fonts that cannot be changed. In systems that allow for changes in character size, string definitions can be windowed the same as other primitives. For characters formed with line segments, the mapping to the viewport could be carried out as a sequence of line transformations.

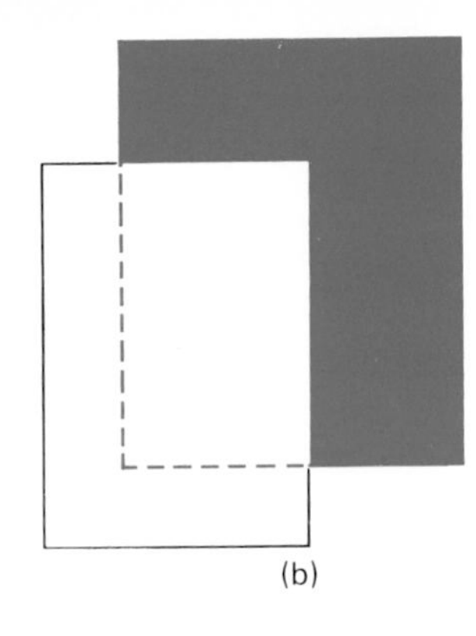

FIGURE 6–22
Examples of blanking: (a) An area is provided for labeling; (b) an area used to erase part of a previous display to provide a blank area for overlays.

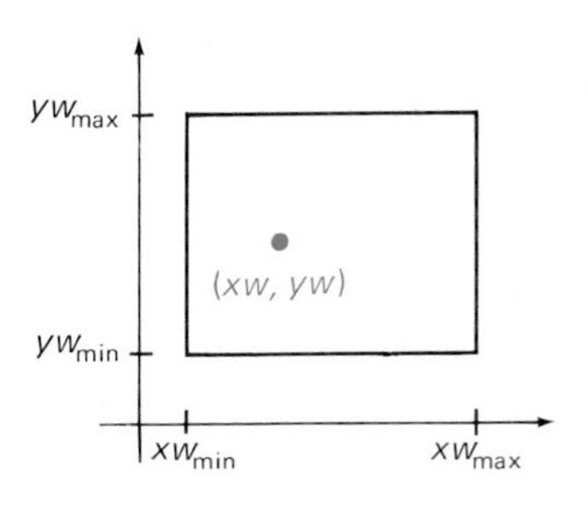

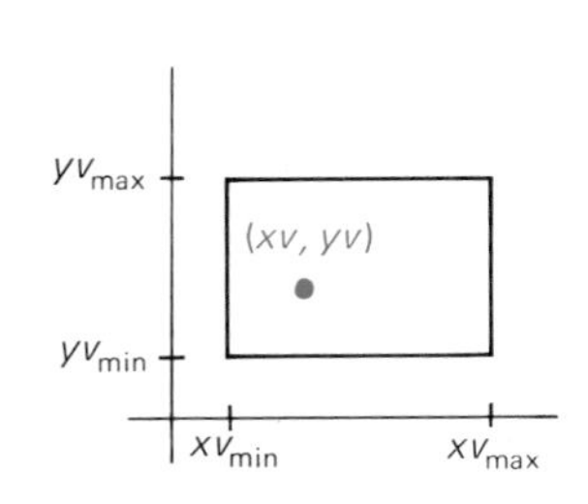

FIGURE 6–23
A point at position *(xw, yw)* in a window is mapped to the viewport point *(xv, yv)*. The mapping is performed so that relative proportions in the two areas are the same.

REFERENCES

Line-clipping algorithms are discussed in Liang and Barsky (1984), Pavlidis (1982), and Sproull and Sutherland (1968). Cyrus and Beck (1978) describe an algorithm that is similar to the Liang-Barsky clipping procedures and that can be used to clip against an arbitrarily shaped window.

The reentrant polygon clipping procedure is presented in Sutherland and Hodgman (1974). A method for polygon clipping using parametric line equations is given in Liang and Barsky (1983). And Weiler and Atherton (1977) discuss general techniques for clipping arbitrarily shaped polygons against each other.

Sources of information on windowing and clipping functions in GKS are Enderle, Kansy, and Pfaff (1984) and Hopgood, et. al. (1983).

EXERCISES

6–1. Implement the Cohen-Sutherland clipping algorithm with Pascal sets used to represent region codes. That is, the region code for a line endpoint is represented as a subset of the set *region = set of (left, right, bottom, top).*

6–2. Carefully discuss the rationale behind the various tests and methods for calculating the intersection parameters u_1 and u_2 in the Liang-Barsky line-clipping algorithm.

6–3. Compare the number of arithmetic operations performed in the Cohen-Sutherland and the Liang-Barsky line-clipping algorithms for several different line orientations relative to a clipping window.

6–4. Implement the Liang-Barsky line-clipping algorithm on your system.

6–5. Devise an algorithm to perform line clipping using the midpoint-subdivision method. Would a software implementation of this algorithm offer any benefits over the other two line-clipping algorithms discussed in this chapter?

6–6. Set up an algorithm to clip lines against a rotated window, defined with minimum and maximum coordinate values and a rotation angle as shown in Fig. 6–5.

6–7. Modify the polygon-clipping algorithm to clip concave polygon areas correctly. (One method for accomplishing this is to divide the concave polygon into a number of convex polygons.)

6–8. Adapt the Liang-Barsky line-clipping algorithm to polygon clipping.

6–9. Write a routine to clip an ellipse against a rectangular window.

6–10. Assuming that characters are defined in a pixel grid, develop a text-clipping algorithm that clips individual characters according to the "all-or-none" strategy.

6–11. Develop a text-clipping algorithm that clips individual characters, assuming that the characters are defined in a pixel grid.

6–12. Write a routine to perform blanking on any part of a defined picture using any specified blanking window size.

6–13. Set up procedures to implement the *window* and *viewport* commands. That is, the procedures are to take the coordinate parameters in these commands and perform the complete viewing transformation for a specified scene: clipping in world coordinates, transformation to normalized coordinates, then transformation to device coordinates.

6–14. Expand the procedures in Ex. 6–13 to allow multiple, numbered windows and viewports to be used in an application program. The parameter list in the *window* and *viewport* commands will now include a transformation number, specified as a nonnegative integer. In addition, the application program must state which window and viewport are to be used. This is accomplished with the command: *select_normalization_transformation (n),* where integer *n* identifies the window-viewport pair to be used with subsequent output primitive statements.

6–15. Compare the relative advantages and disadvantages of clipping in world coordinates to clipping in device coordinates (either normalized or physical device coordinates). Factors to be considered include hardware versus software implementations, application of geometric transformations (allowing possible composition of matrices), use of rotated windows or viewports, and the structure of a picture relative to the window boundaries.

7 SEGMENTS

For many applications, graphics displays are handled most efficiently by defining and modifying a picture as a set of subpictures. Graphics packages that are designed to store and manipulate pictures in terms of their component parts are more complicated, but they allow users greater flexibility. By defining each object in a picture as a separate module, a user can make modifications to the picture more easily. In design applications, different positions and orientations for a component of a picture can be tried out without disturbing other parts of the picture. Or a component can be taken out of a version of the picture, then later put back into the display. An animation program can apply transformations to an individual object definition so that the object is moved around while the other objects in the scene remain stationary.

7–1 Segment Concepts

A set of output primitives (and their associated attributes) that are arranged and labeled as a group is called a **segment.** To allow for segment operations, a graphics package must include capabilities for creating and deleting segments. When a segment is created by a user, the coordinate positions and attribute values specified in the segment are stored as a labeled group in a system segment list. As segments are deleted, they are removed from this list. Routines must then be provided in the package for allocating and reclaiming the segment storage areas.

Packages can provide for the creation of a segment with the command

```
create_segment (id)
```

The label for the segment is the positive integer assigned to parameter *id.* After all primitives and attributes have been listed, the end of the segment is signaled with a *close_segment* statement. For example, the following program statements define segment number 6 as the line sequence specified in *polyline:*

```
create_segment (6);
  polyline (n, x, y);
close_segment;
```

Any number of segments can be defined for a display using these commands. In most systems, only one segment can be open at a time. The open segment must be closed before a new segment can be created. This method eliminates the need for a segment identification number in the *close_segment* statement.

A user deletes a segment with the command

```
delete_segment (id)
```

Once a segment has been deleted, its name can be reused for another sequence of primitive statements.

In some situations, a user may want to modify a segment after it has been closed. Another command could be defined for reopening segments or for appending to segments, but the processing required to implement such functions is complex and costly. Generally, packages do not allow closed segments to be modified. The user can, of course, delete an existing segment and then re-create it with the desired changes. For example, a new segment 6 is created with the following statements:

```
delete_segment (6);
create_segment (6);
  polyline (n, x, y);
  text (xt, yt, "figure label");
close_segment;
```

Some systems do provide a function for copying the contents of an existing segment into a newly opened segment. While not strictly an append operation, this can be used to produce the same effect.

It is sometimes useful to rename a segment. For this purpose, we define

```
rename_segment (id_old, id_new)
```

One use for the rename operation is to consolidate the numbering of the segments after several segments have been deleted.

7–2 Segment Files

We will refer to any list of segments maintained by a graphics system with the generic term **segment file.** A segment file can take on a number of forms, depending on the type of graphics system. The display file program of a simple vector system can be structured according to a user's definition of segments. In this case, the display file program is a type of segment file, referred to as a **segmented display file.** On vector systems with both a display processor and a display controller, more than one segment file may be maintained by the system (Fig. 7–1).

Raster systems display a picture from the definition in a frame buffer, not from a display file program as in vector systems. However, graphics packages to be used on raster-scan equipment can be designed to maintain a **pseudo display file** as an internal data structure. The pseudo display file is a type of segment file that contains the segment definitions and other information regarding the structure of a picture. Scan-conversion routines work from this file and set appropriate bits in the frame buffer, as shown in Fig. 7–2. Maintaining such a segment file on a raster system requires extra memory, but it allows for greater compatibility with vector systems. Pseudo display files are also used because they provide a fast and efficient method for regeneration of a picture after modifications have been made.

Segment files can be physically stored in a number of ways. A linked-list structure, as in Fig. 7–3, offers the most flexible arrangement for handling segment creation, deletion, and manipulation. Segments are added to the linked list by setting appropriate pointer fields. If segments are assigned priorities, the linked list is organized in priority order, with each new segment inserted into the list at the proper position.

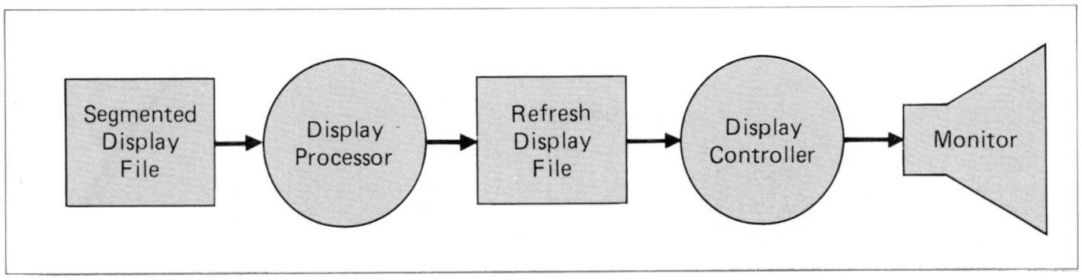

Segmented Display File → Display Processor → Refresh Display File → Display Controller → Monitor

FIGURE 7–1
Segment file organizations for a random-scan system.

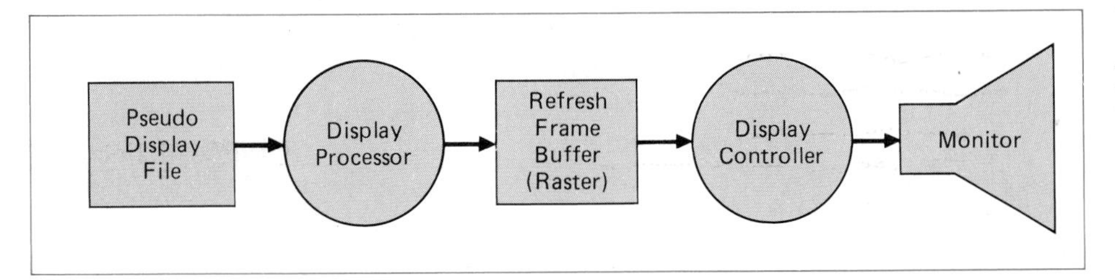

FIGURE 7–2
A segmented display file is used to update the frame buffer in a raster system.

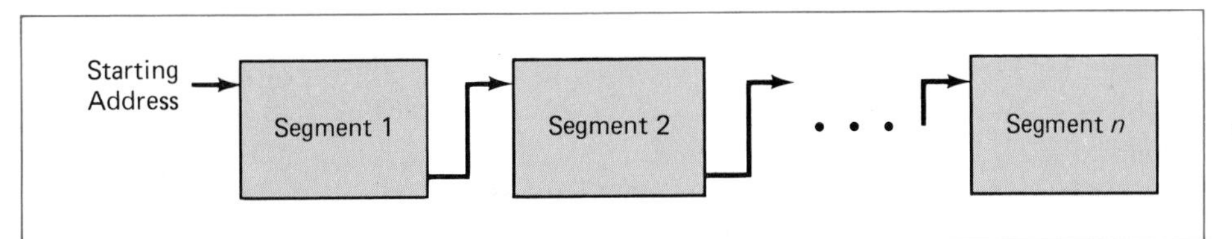

FIGURE 7–3
A segment file organized as a linked list.

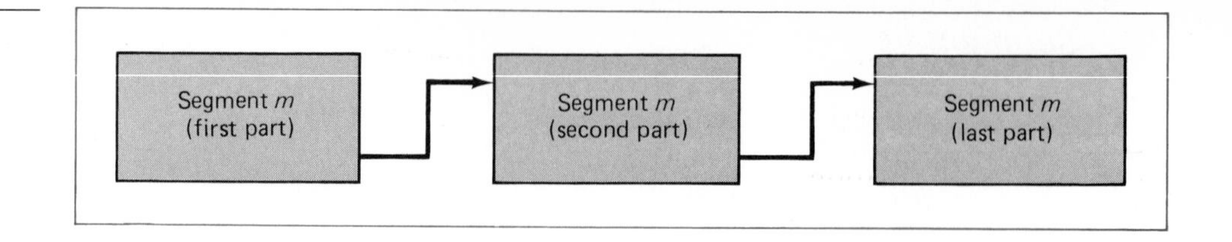

FIGURE 7–4
A segment fragmented for storage in three fixed-size memory blocks.

A memory management routine is used to handle segment storage assignments. Blocks of storage are assigned to segments as they are created, and storage blocks are returned to the storage pool when a segment is deleted. A procedure for deleting a segment sets pointer fields in the list to bypass the segment and then returns the segment block address to the storage pool. Different strategies adopted for assigning storage include fixed-size block assignments and variable-length block assignments to accommodate the different lengths of segments. Also, different memory compaction schemes are possible with variable-sized blocks to collect unused storage and minimize wasted space.

Fixed-size blocks are easy to manage, but they can result in fragmented segments (Fig. 7–4). If a segment is too big for the assigned block, it is overflowed into another block. The overflow process continues into successive blocks until all of the segment is stored. This approach requires some threading through pointer fields to locate all parts of large segments, and the refresh process is slowed if this method is used as the storage scheme for the refresh display file of a vector system.

Variable-sized blocks avoid fragmented segments, but they involve more processing by the memory management routines. Since the size of a segment is not known until it is closed, the correct-sized block cannot be assigned in advance. One way to accommodate the different-sized segments is simply to begin storing a segment in the first available storage block. If a segment overflows the assigned block, a larger block is requested immediately. In this way, a large segment is continually being transferred to larger blocks as it grows. Another possibility is to fragment a large segment over more than one block as it is being created and then compact it into a larger block after the segment is closed.

There often will be unused space at the end of a segment block. To simplify processing, this unused space could be left in the block rather than returned to the memory pool. If the memory pool should run out of blocks, a compaction method can be used to collect some of the unused space.

To insert and delete segments efficiently, we need a quick method for locating segments within the segment file. One method is to store segment names and related information in a separate segment directory. Alternatively, these data could be stored in segment headers at the beginning of each segment block.

Figure 7–5 shows one arrangement for a segment directory. When a segment is to be located, the directory is searched to find the address for the given segment name. A binary search procedure can be used to locate segments when entries are sorted on the name field. Using a directory to locate segments can be useful when a large number of segments are to be stored and manipulated. However, this method does require sorting and searching procedures for the directory and some method, such as an expandable directory size, for handling arbitrary numbers of segments.

A scheme for storing information in segment headers is illustrated in Fig. 7–6. Following the name field is the "visibility pointer field." Normally, this field

Name	Address	Length	Visibility	Other Parameters
N1	ADDR1	L1	V1	• • •
N2	ADDR2	L2	V2	• • •
⋮	⋮	⋮	⋮	⋮

FIGURE 7–5
Organization of a segment directory.

points to the beginning of the output primitive entries. Segments are made invisible by setting this field to the address of the next segment. The "next-segment pointer field" is provided at the beginning of the block to speed up searches through the linked list. Without this entry, a segment search would need to find the pointer field at the end of the block to proceed to the next segment. Finally, the "end-of-segment pointer field" can mark the segment end by pointing back to the "next-segment pointer field," as in Fig. 7–6. Such headers have advantages over storing the segment addresses in two places. The last field of each segment is always set to point back to the third field within each segment, and only one pointer field in a segment needs to be set during a deletion or insertion operation.

Another possible organization for a segment file is shown in Fig. 7–7. Here, segments are stored as adjacent blocks within a one-dimensional array. This scheme simplifies the setting of pointer fields, and it is particularly suitable as a simple

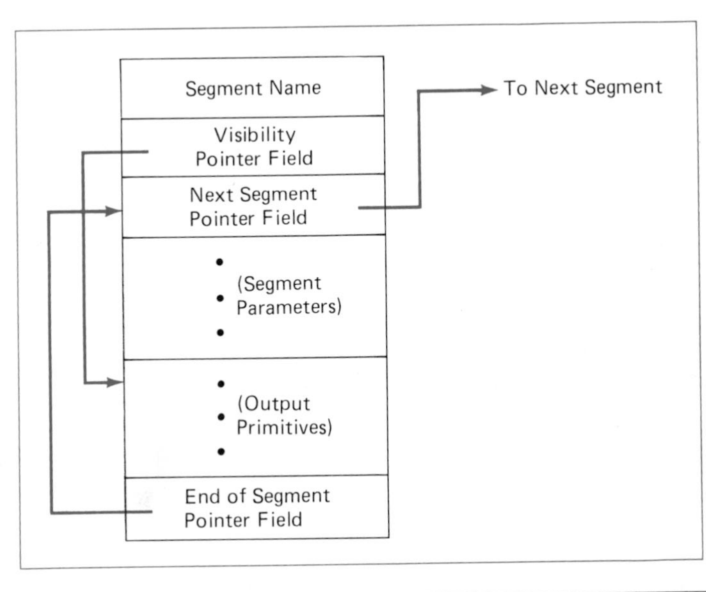

FIGURE 7–6
Organization of a segment block with a header. The visibility pointer field is used to set the visibility condition. A segment is invisible if this field points to the next segment.

Segment 1	Segment 2	• • •	Segment *n*

FIGURE 7–7
Organization of a segment file as a linear array.

organization for a vector display file program. As with linked lists, searches through this segment file can be done with a segment directory or with headers. Deletion is accomplished by removing segments from the array and compacting the blocks back into contiguous locations. Similarly, insertions are carried out by shifting blocks to make room for new segments. Visibility changes are again implemented by setting pointer fields. A segment is made invisible by setting the visibility pointer to the next segment address. It is made visible again by setting the pointer field to the starting address of the primitive entries.

On a vector system, the refresh display file is accessed by both the display processor running the graphics package and the display controller. The two processors operate concurrently and share the segment file data structure. Synchronization between the two processors is important. If the operations of the processors are not synchronized, updates to the file could cause the display controller to produce a distorted picture on the display device. Usually, the display processor locks the file while making an update, denying access to the display controller.

7–3 Segment Attributes

A number of parameters, or attributes, can be assigned to a created segment. Most graphics packages permit users to set the values for a segment's visibility, its priority, and whether or not it is to be highlighted. Some packages also allow a transformation matrix to be associated with a segment. With this transformation matrix attribute, a user can vary the size, position, and orientation of a segment.

Visibility of segments is established by a user with the command

```
set_visibility (id, v)
```

Parameter v is used to control the display of segments. If v is set to the value *invisible* for a specified segment *id*, that segment is maintained in the segment list but is not displayed on the output device in use. Setting v to *visible* causes the segment to be displayed.

The ability to turn segments on and off is useful in design applications where the effects of different segments are to be tested. It is also useful for setting up menu and message segments that can be displayed at certain times to direct interactive input to a program.

Priority can be set for segments with the command

```
set_segment_priority (id, p)
```

A numeric value between 0 and 1 is assigned to priority parameter p. These numbers determine the order for displaying segments, and they are mostly used on raster-scan systems. Lower-priority segments (values near 0) are displayed first up to the highest-priority segment, which is displayed last. Higher-priority segments "paint" over the lower-priority segments in areas of overlap. This gives the user a convenient way to set up background and foreground segments in applications such as animation, where objects move past each other.

In any implementation of the priority command, only a finite number of priority levels will be allowed. This number could be quite large on some systems, while other systems might restrict the priority levels to a very few values, such as 0, 0.5, and 1. Any other numbers assigned as priorities are rounded to the nearest level. When two segments are assigned the same priority number, the system must use another method to decide which to display first. A first-come, first-served scheme could be used, or some other property such as segment number could be used to break the priority tie.

Highlighting a segment is done with

```
set_highlighting (id, h)
```

where parameter h can take values *highlighted* and *normal*. By setting h to *highlighted*, a user specifies that some mechanism is to be applied to accent the designated segment. The type of highlighting displayed depends on the type of output device in use. With a video monitor, a highlighted segment could be displayed in a higher intensity, or it could be set to blink on and off.

When segment transformations are allowed in a graphics package, a user can specify the transformation matrices that are to be applied to individual segments. Typically, the identity matrix is linked to a segment when it is first created. Subsequently, the application program can change the associated matrix with the command

```
set_segment_transformation (id, matrix)
```

Parameter *matrix* contains the elements of the transformation matrix, as set by the methods discussed in Chapter 5. During the picture-generation process, all output primitives within the given segment are transformed according to the operations specified in *matrix*. The user can vary the appearance of the segment by associating a new transformation matrix with the segment.

Clipping and the window-to-viewport transformation must also be applied to output primitives making up a segment. Usually the window and viewport specifications that are current when the segment is opened are used to generate the display of that particular segment.

In our previous discussions, clipping was performed in world coordinates using the window boundaries. The interior of the window was then mapped to the viewport. This is not the most efficient arrangement for packages that allow transformations of segment definitions. Each time a transformation matrix is associated with a segment, the window-to-viewport mapping would be repeated. Since this mapping doesn't change, it is advantageous to omit it from the repeated reprocessing of the segment.

To do this, we can convert world coordinates to normalized device coordinates first, before the output primitives are even stored in the segment file. The current window and viewport specifications are used to supply the normalization transformation factors. All world coordinate values, regardless of their relationship to the window, are now normalized. As seen in Fig. 7–8, it is possible for some normalized points to extend beyond the normalized coordinate range from 0 to 1. Segment transformations are then applied to the normalized description of the object, and then clipping is performed against the viewport boundaries. This sequence of operations is demonstrated in Fig.7–9.

Since segment primitives are stored in normalized coordinates, segment transformation parameters must be applied in normalized coordinates. However, it is usually more convenient for the user to describe transformation matrices in world coordinates. In this case, the graphics package first applies the window-to-viewport normalization to the transformation parameters (fixed point, pivot point, and translation distances) before constructing the transformation matrix.

FIGURE 7–8
When the normalization transformation is performed before clipping, some world coordinate points can be mapped into positions outside the normalized coordinate range.

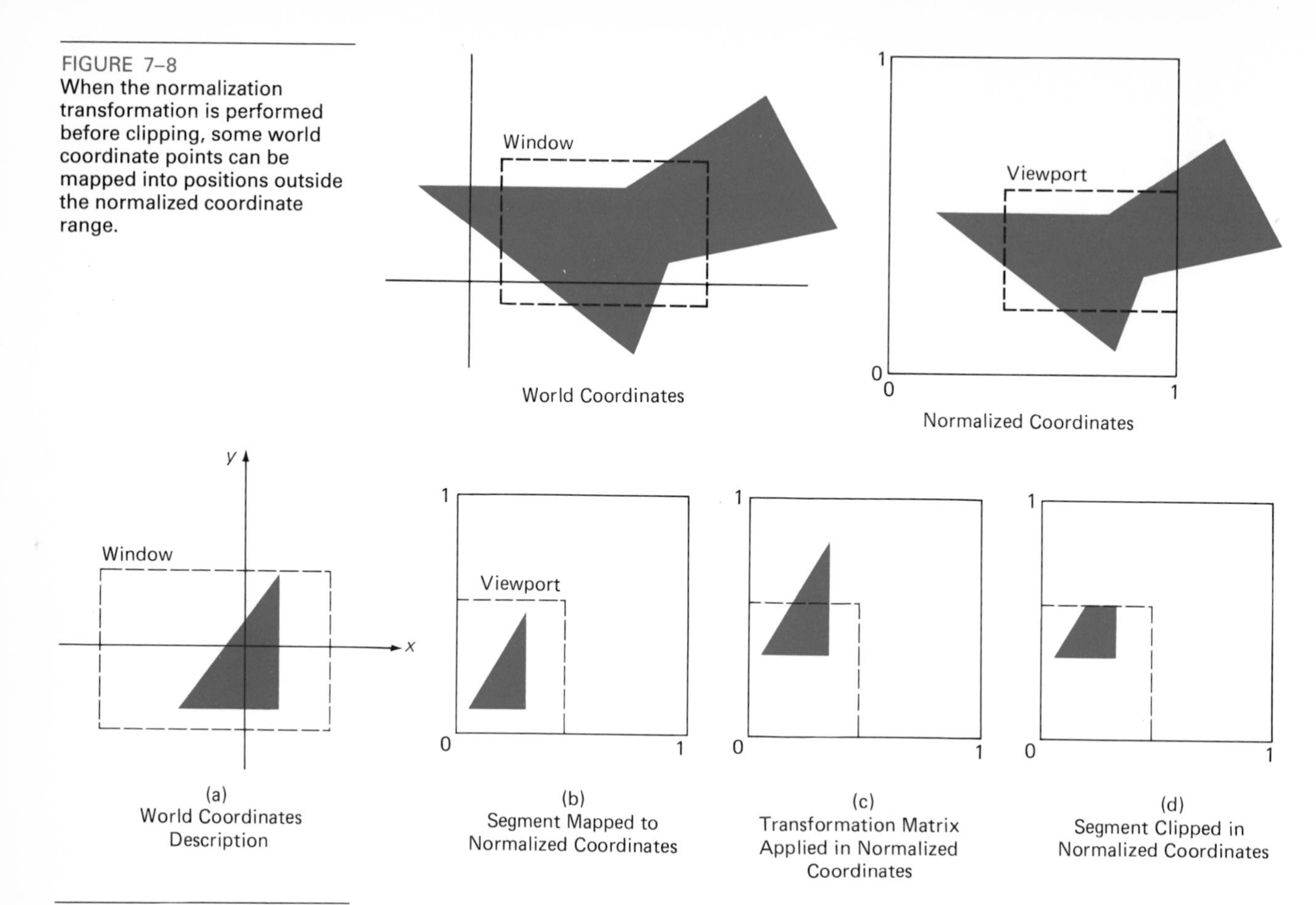

FIGURE 7–9
Sequence of operations performed when a normalization transformation is applied to a segment before clipping.

7–4 Multiple Workstations

Graphics systems often utilize several output devices, each identified with a unique workstation number. When multiple output devices are available, a user may be provided with the means to control the display of segments on the various workstations. This can be accomplished by allowing the user to activate and deactivate workstations. When a segment is created, it is associated with only those workstations that are active. For example,

```
activate_workstation (5);
create_segment (12);
      .
      .
      .
```

```
close_segment;
activate_workstation (2);
create_segment (13);
      .
      .
      .
close_segment;
deactivate_workstation (5);
create_segment (14);
      .
      .
      .
close_segment;
deactivate_workstation (2);
```

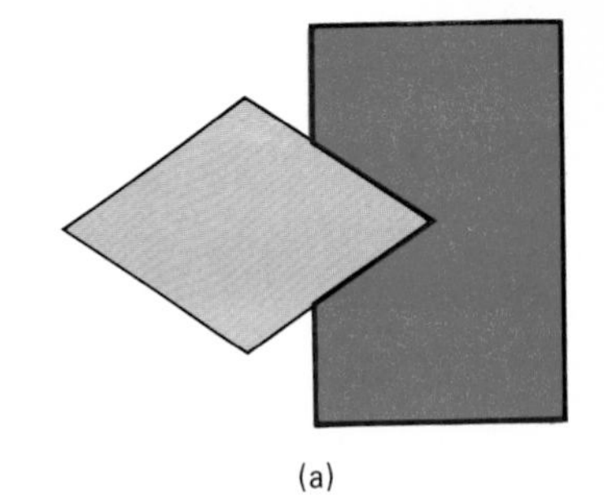

(a)

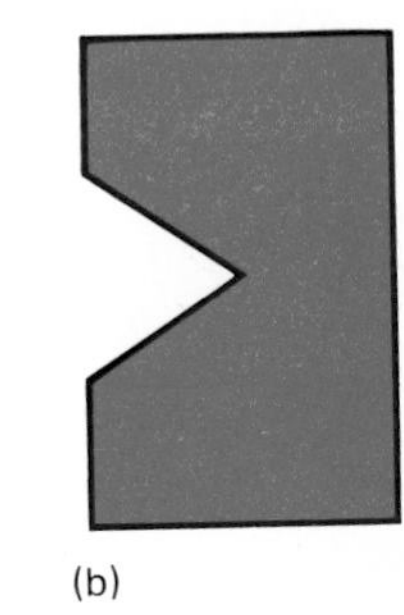

(b)

FIGURE 7–10
Erasing an object (segment) that overlaps another object (a) by redrawing the object in a background color also erases part of the background object (b).

attaches segments 12 and 13 to workstation 5, while segments 13 and 14 are associated with workstation 2.

Additional commands are provided to manage segments on the individual workstations. All segments are deleted from workstation *ws* with *clear_workstation (ws)*, while the command *delete_segment_from_workstation (ws, id)* removes a single segment.

Deleting a segment from a specific workstation can be handled in various ways. With a vector system, segments are deleted simply by removing them from the refresh display file. A fast method for deleting a segment with a raster system is to reset all of the segment pixels in the frame buffer to the background color. This method for erasing a segment also erases any overlapping parts of other segments, leaving gaps in a displayed picture (Fig. 7–10). To redraw the picture so that such gaps are eliminated, a user could issue the command

```
redraw_segments_on_workstation (ws)
```

Alternatively, a graphics package could be structured to redraw segments automatically after a deletion. Since many segments may be involved in a picture, it could be more efficient for a package to find and redraw only those objects that overlap a deleted segment. One way to locate these segments is to check each segment for overlap with the bounding rectangle of the deleted object, as shown in Fig. 7–11. The area within the bounding rectangle could be set to the background color, and all overlapping segments could be redrawn.

Greater flexibility in making segment assignments to workstations can be attained using a central, device-independent segment file. Segments are initially stored in the central file and later selectively assigned to various workstations, using some form of copy function. Construction and manipulation of segments could be monitored on a video display. When the user is satisfied with the appearance of all segments, the final picture could be directed to a plotter or other hard-copy device.

So far, we have discussed only files for the temporary storage of segments. These files exist only during the current session with a graphics package. A file used for long-term storage of graphical information is referred to as a **metafile.** Segment definitions in the metafile are retained and used during subsequent sessions with a graphics package. An important feature of metafiles is that they can be used by different graphics packages. Segments created in the metafile with one package can be read and used by other packages.

FIGURE 7–11
Locating overlapping objects using a bounding rectangle.

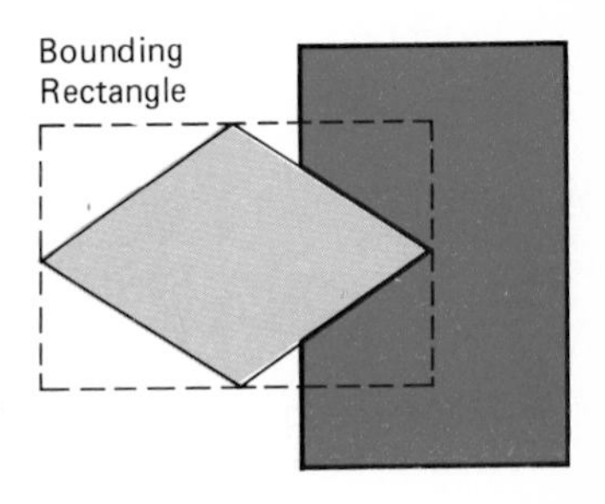

7–5 Summary

A segment defined in an application program is a group of output statements specifying a picture component. By creating pictures with individual segments, a user can easily add, delete, or manipulate picture components independently of one another. As segments are created by a user, they are entered by the graphics package into a segment file.

Segment files are usually organized as linked lists, to facilitate addition, deletion, and other operations on segments. One segment list may be maintained by a system and used by all output devices. Separate segments lists might also be maintained for each workstation available on the system.

Attribute parameters determine how segments are to be displayed. With the visibility parameter, a user can turn off the display of a segment while retaining it in the segment list. Priority specifies the display order among segments, so that foreground objects are displayed over background objects. A highlighting parameter is used to emphasize a displayed segment with blinking, color, or high-intensity patterns. And transformation matrices can be applied to each segment, allowing manipulation of individual picture components.

Table 7–1 lists the categories of segment operations. The various functions for manipulating segments are those for creating, deleting, and renaming segments and those for setting segment attributes.

TABLE 7–1.
Summary of Segment Functions

Function	Description
Creating and Deleting Segments	
create_segment (id)/close_segment	Creates segment with name specified in parameter *id*
delete_segment (id)	Deletes a segment from segment file
rename_segment (id_old, id_new)	Changes segment name
Workstation Operations	
delete_segment_from_workstation (ws, id)	Deletes segment *id* from workstation ws
clear_workstation (ws)	Deletes all segments from workstation *ws*
redraw_segments_on_workstation (ws)	Clears the device and redraws all segments
copy_segment_to_workstation (ws, id)	Copies segment *id* from a central, workstation-independent file to workstation *ws*
Segment Attributes	
set_segment_transformation (id, matrix)	Transformation elements in *matrix* applied to segment in normalized coordinates
set_visibility (id, v)	Parameter *v* set to *visible* or *invisible*
set_segment_priority (id, p)	Parameter *p* set to a real number in the range 0 to 1
set_highlighting (id, h)	Parameter *h* set to *highlighted* or *normal*

REFERENCES

For additional information on segment operations within GKS see Enderle, Kansy, and Pfaff (1984) or Hopgood, et. al. (1983)

EXERCISES

7–1. List the relative advantages and disadvantages of allowing more than one segment to be open at one time.

7–2. Consider how a graphics package could be designed to include a function to append information to a previously defined segment. For what types of applications would such an operation be preferable to deleting the segment and re-creating it with the new information? In general, what are the trade-offs in the two approaches?

7–3. Write a procedure to implement a segment directory table.

7–4. Write a procedure to manage a segment display file organized as a linked list with segment headers.

7–5. Consider the relative advantages and disadvantages of storing segments in doubly linked lists.

7–6. What are the relative advantages and disadvantages of using a binary tree structure or hash coding to store segments?

7–7. Discuss procedures for implementing the visibility attribute for the different segment file organizations.

7–8. Discuss various implementation procedures for the priority attribute, depending on the organization of the segment file.

7–9. Write a procedure to implement the highlighting function with a blinking operation.

7–10. Implement the segment-transformation command with a procedure for transforming all output primitives within the segment according to values in the transformation matrix.

7–11. Implement a viewing transformation for segments that first converts world coordinates to normalized coordinates. Any specified segment transformation matrix is also to be converted to normalized coordinates and applied to the segment. Finally, the segment is to be clipped against the viewport boundaries.

7–12. Develop an algorithm for locating any segments that overlap a segment that is to be deleted from a segment file. These overlapping segments are to be passed to a routine that redraws them after the unwanted segment has been removed from the segment file.

7–13. Discuss implementation details for maintaining separate workstation segment files and for maintaining a central segment file accessed by multiple workstations.

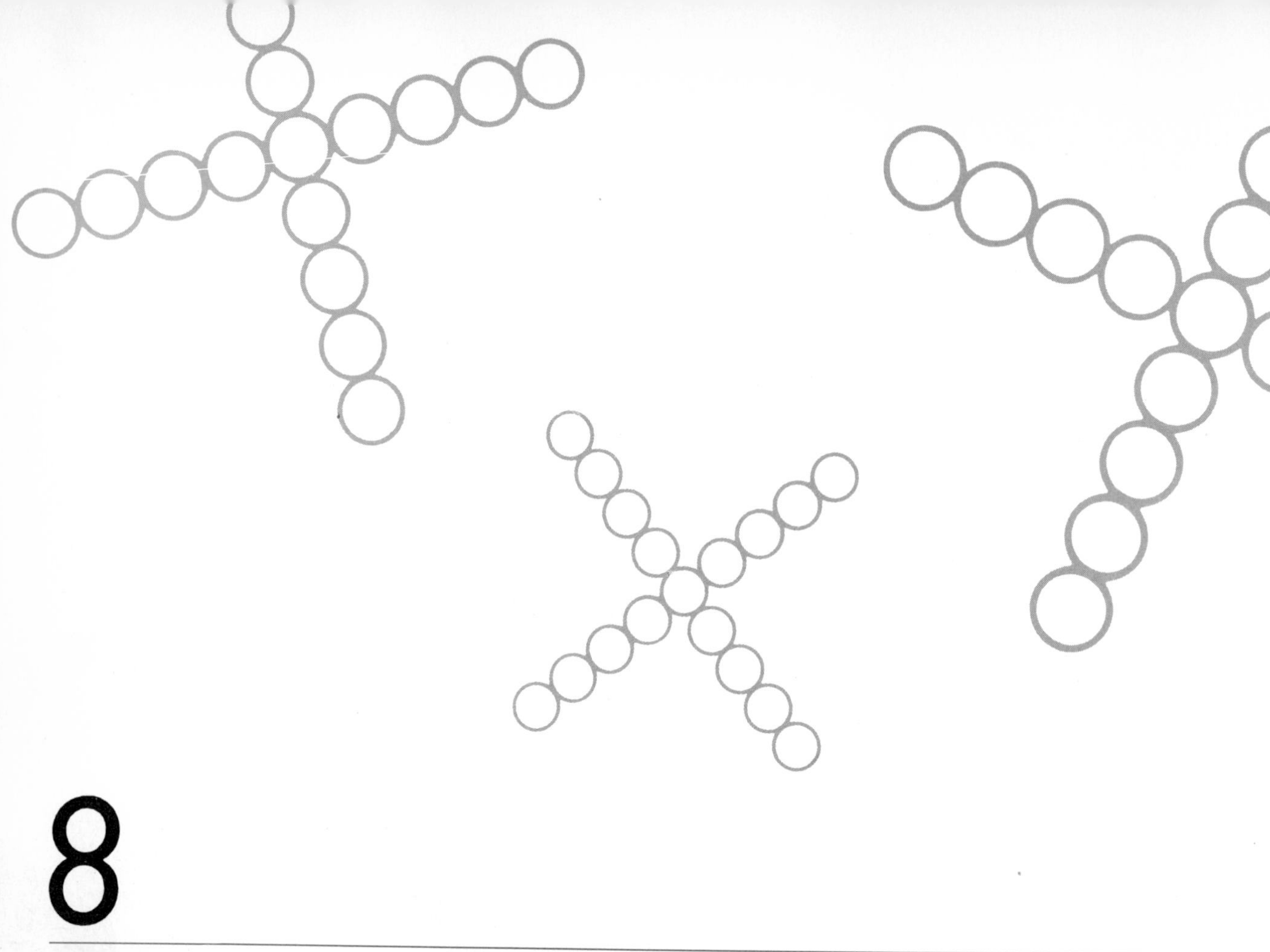

8 INTERACTIVE INPUT METHODS

Our discussion of graphics methods has concentrated, until now, on generating output in its various forms. We have seen how output primitives can be defined, how attributes are set, how picture segments can be managed, and how to apply transformations. However, we have yet to consider how a graphics package can be made interactive. To handle interactive input, a system must allow the user dynamically to input coordinate positions, select functions, or specify transformation parameters. A variety of input devices exists, and we consider now how graphics packages can be designed to interface with the different types of input and to carry on a dialogue with users.

8–1 Physical Input Devices

Various devices are available for data input on a graphics workstation. Most interactive systems have a keyboard and one or more additional devices specially designed for interactive input. These include dials, touch panels, light pens, tablets, voice systems, joysticks, trackballs, and mice.

Keyboards

An alphanumeric keyboard on a graphics system is used primarily as a device for entering text. The keyboard is an efficient device for inputting such nongraphic data as picture labels associated with a graphics display. Keyboards can also be provided with features to facilitate entry of screen coordinates, menu selections, or graphics functions.

Cursor-control keys and function keys are typical features found on general-purpose keyboards. Function keys allow users to enter commonly used operations in a single keystroke, and cursor-control keys select coordinate positions by positioning the screen cursor on a video monitor. Additionally, a numeric keypad is often included on the keyboard for fast entry of numeric data. Figure 8–1 shows two types of general-purpose keyboard layouts providing alphanumeric keys, cursor-control keys, programmable function keys, and a numeric keypad. The "joydisk" in the upper left corner of the keyboard in Fig. 8–1(a) is used to position the screen cursor.

(a)

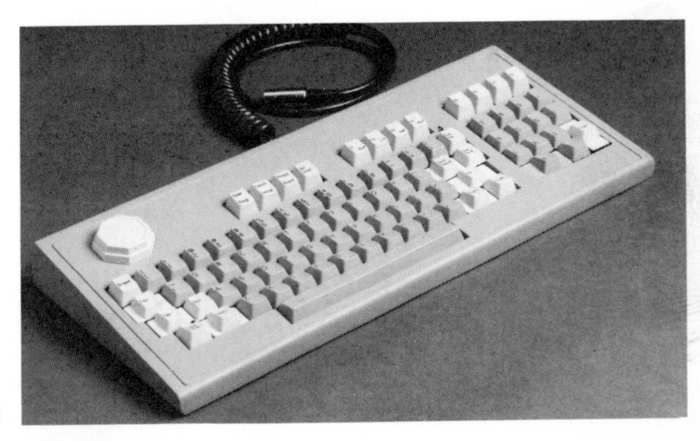

(b)

FIGURE 8–1
Examples of general-purpose display terminal keyboards. (a) Courtesy Tektronix, Inc. (b) Courtesy Information Displays, Inc.

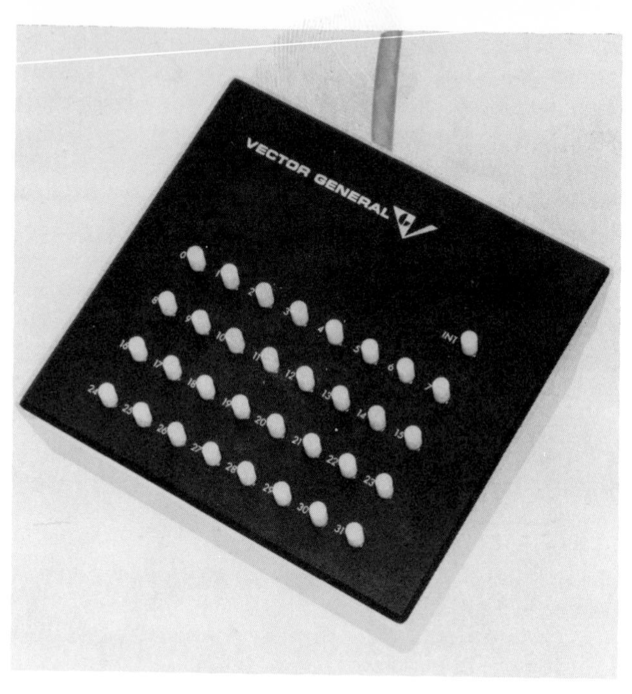

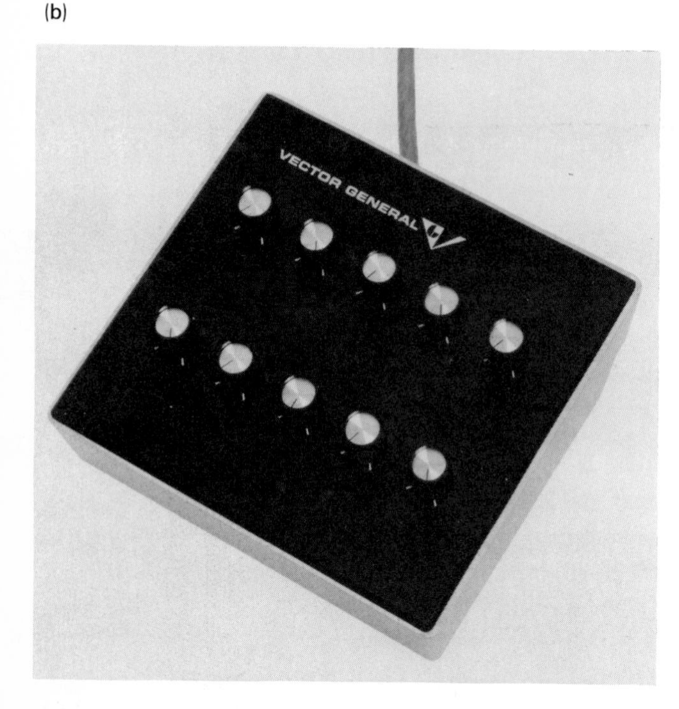

FIGURE 8–2
Special-purpose keyboards using buttons (a) or dials (b) can be used to select predefined graphics functions. Courtesy Vector General.

For specialized applications, a keyboard may contain only the set of buttons, dials, or switches that select the customized graphics operations needed in a particular application (Fig. 8–2). Buttons and switches are often used to input predefined functions, and dials are common devices for entering scalar values. Real numbers within some defined range are selected for input with dial rotations.

Dials can be used as fast input devices for moving the screen cursor. One dial can control horizontal movement of the cursor, and a second dial can control vertical movement. Potentiometers are used to measure dial rotations, which are then converted to deflection voltages for cursor movement. The two dials are sometimes arranged as "thumb wheels" on a keyboard, as shown in Fig. 8–3. The thumb wheels are placed at right angles to each other and have serrated edges so that they

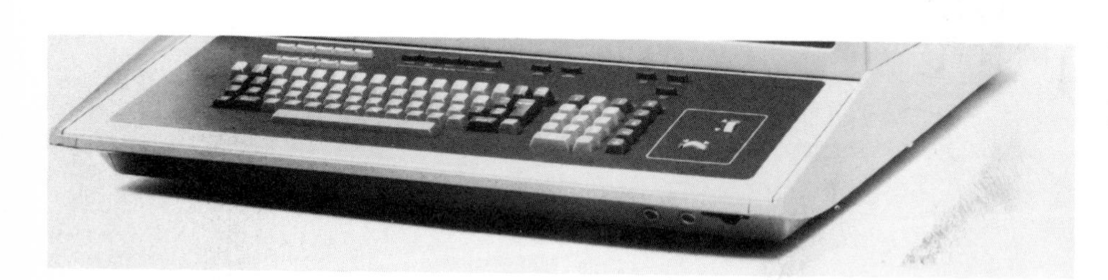

FIGURE 8–3
A keyboard with thumbwheels. Courtesy Tektronix, Inc.

can be easily rotated with the operator's thumb. Dials are convenient for horizontal and vertical movement of the cursor, but they are less effective than a joystick or trackball for cursor movement along diagonal lines or curves.

Touch Panels

As the name implies, these devices allow screen positions to be selected with the touch of a finger, as shown in Fig. 8–4. A **touch panel** is a transparent plate that fits over the CRT screen. When the plate is touched the contact position is recorded by an optical, electrical, or acoustical method.

Optical touch panels employ a line of light-emitting diodes (LEDs) along one vertical edge and along one horizontal edge of the frame. The opposite vertical and horizontal edges contain light detectors. These detectors are used to record which beams are interrupted when the panel is touched. The two crossing beams that are interrupted identify the horizontal and vertical coordinates of the screen position selected. Positions can be selected with an accuracy of about 1/4 inch. With closely spaced LEDs, it is possible to break two horizontal or two vertical beams simultaneously. In this case, an average position between the two interrupted beams is recorded. The LEDs operate at infrared frequencies so that the light is not visible to a user. Figure 8–5 illustrates the arrangement of LEDs in an optical touch panel.

FIGURE 8–4
A touch panel allows operators quickly and easily to select processing options by pointing to their graphical representations. Courtesy Carroll Touch Technology.

FIGURE 8–5
An optical touch panel, showing the arrangement of infrared LED units and detectors around the edges of the frame. Courtesy Carroll Touch Technology.

An electrical touch panel is constructed with two transparent plates separated by a small distance. One of the plates is coated with a conducting material, and the other plate is coated with a resistive material. When the outer plate is touched, it is forced into contact with the inner plate. This contact creates a voltage drop across the resistive plate that is converted to the coordinate values of the selected screen position.

In acoustical touch panels, high-frequency sound waves are generated in the horizontal and vertical directions across a glass plate. Touching the screen causes part of each wave to be reflected from the finger to the emitters. The screen position at the point of contact is calculated from a measurement of the time interval between the transmission of each wave and its reflection to the emitter.

Light Pens

Figure 8–6 shows two styles of **light pens.** These pencil-shaped devices are used to select screen positions by detecting the light coming from points on the CRT screen. They are sensitive to the short burst of light emitted from the phosphor coating at the instant the electron beam strikes a particular point. Other light sources, such as the background light in the room, are not detected by a light pen. An activated light pen, pointed at a spot on the screen as the electron beam lights up that spot, generates an electrical pulse that signals the computer to record the coordinate position of the electron beam. Since the electron beam sweeps across the screen 30 to 60 times every second, the detection of a lighted spot is essentially instantaneous. Coordinates entered with a light pen are often used by graphics programs for selecting or positioning objects on the screen.

Activating a light pen is accomplished with a mechanical or capacitive switch. Pens with a mechanical switch, such as those in Fig. 8–6, are activated either with a push tip or with a side button. Capacitive switches are formed with a metal band near the tip that must be touched with a finger to activate the pen.

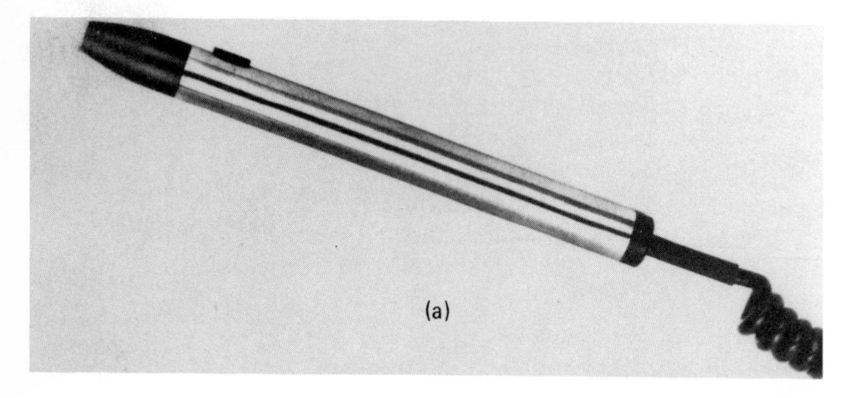

FIGURE 8–6
Light pens can be activated with button switches that are (a) mounted on the side or (b) installed as a push tip. Courtesy Interactive Computer Products.

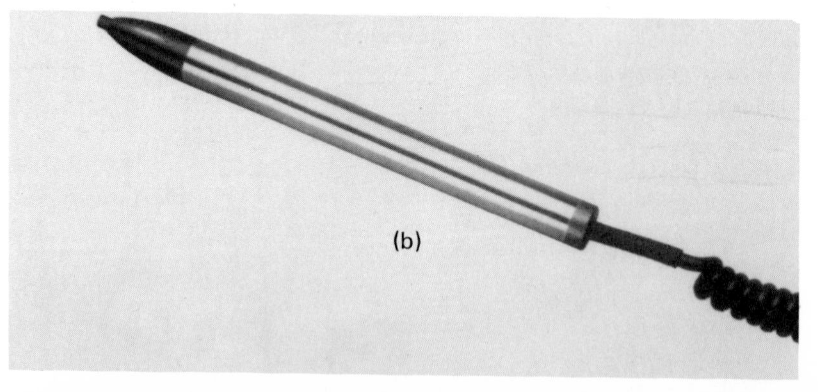

(a)

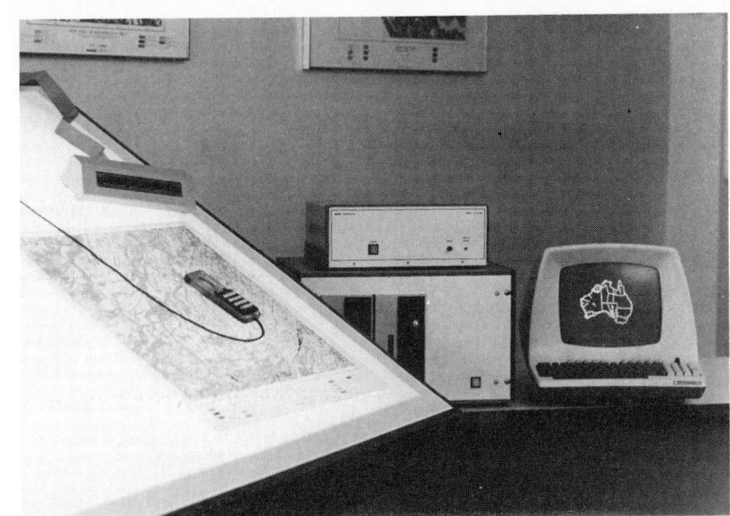

(b)

FIGURE 8–7
Graphics tablets used with (a) hand-cursor input and (b) stylus input. (a) Courtesy Altek Corp. (b) Courtesy Calcomp Group; Sanders Associates, Inc.

Graphics Tablets

Another method for selecting screen positions is by activating a hand cursor or a stylus at corresponding positions on a flat surface, called a **graphics tablet** or **digitizer.** A hand cursor contains cross hairs for sighting positions, while a stylus is a pencil-shaped device that is pointed at positions on the tablet. The two methods for entering data with a graphics tablet are illustrated in Figure 8–7. Graphics tablets provide one of the most accurate methods for selecting coordinate positions. They also have some other advantages. Movement of a stylus over the tablet surface does not block the user's view of any part of the screen, as does a light pen as it is moved over the screen surface. When a selected position is to be input, the hand cursor or stylus is activated by pressing a button. Several buttons are often available to provide options, such as storing a single point or a stream of points as the cursor is moved across the tablet surface.

Many graphics tablets are constructed with a rectangular grid of wires embedded in the tablet surface. Each wire has a slightly different voltage, which is correlated with the coordinate position of the wire. Voltage differences between wires in the horizontal and vertical directions correspond to coordinate differences in these directions on the screen. By activating a stylus or hand cursor at some point on the tablet, a user causes the voltages at that point to be recorded. These voltages are then converted to screen positions for use in the graphics routines. Some tablets employ electromagnetic fields instead of voltages to record positions. With these systems, the stylus is used to detect coded pulses or phase shifts in the wire grid.

Another type of graphics tablet uses sound waves. One such system is constructed with two perpendicular strip microphones (L-frame assembly) to detect the sound emitted by an electrical spark from a stylus tip. The position of the stylus is calculated by timing the arrival of the generated sound at the two microphones. An example of a graphics tablet employing an L-frame microphone assembly is shown in Fig. 8–8 (a). Point microphones, as in Fig. 8–8 (b), are used in some systems to replace the larger strip microphones. The compact size of the point microphone systems make them more portable, but the working area provided can be smaller than that available with an L-frame microphone assembly. A table top, or any other surface, is used as the "tablet" work area with these sonic devices.

Three-dimensional sonic digitizers are constructed with three or more microphones to record positions in space. Figure 8–9 shows such a three-dimensional system using four point microphones to digitize a surface. As the stylus is moved over the solid object, the Cartesian x, y, and z coordinates are calculated at each point by timing the arrival of the sound pulse at three of the microphones. The object is then projected onto the screen, as shown in the figure. A clear line of sight to at least three of the microphones is required to digitize positions on the surface of an object. The fourth microphone is provided for convenience in tracing over surfaces that might obscure the stylus from one of the microphones.

(a)

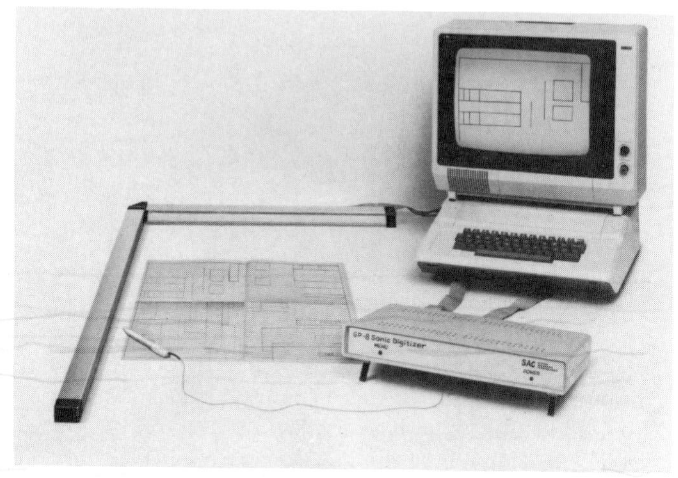

(b)

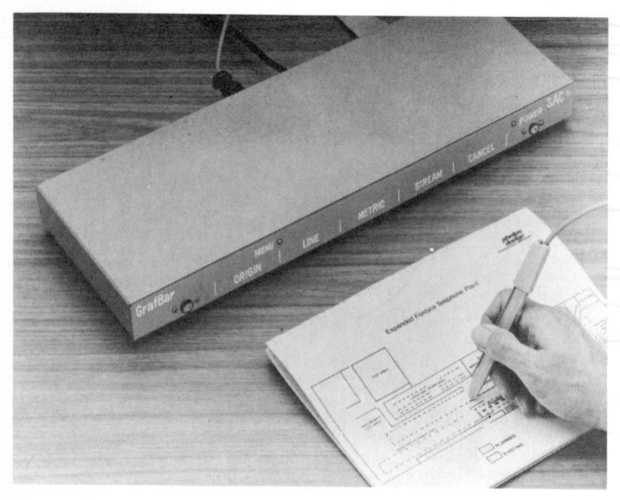

FIGURE 8–8
Acoustic "tablets": (a) an L-frame microphone assembly; (b) point microphones mounted on the front panel of the cabinet. Courtesy Science Accessories Corp.

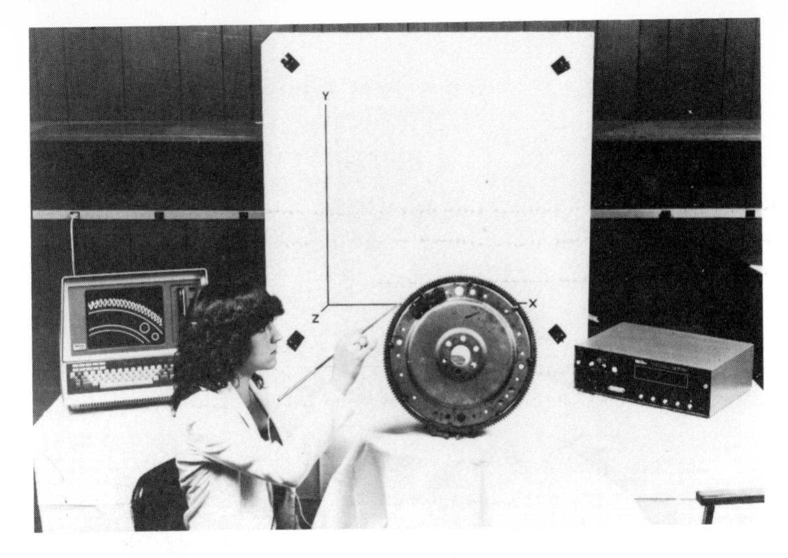

FIGURE 8–9
A three-dimensional sonic digitizer, using four point microphones (at the corners of the large white panel). Courtesy Science Accessories Corp.

Joysticks

A **joystick** consists of a small, vertical lever (called the stick) mounted on a base that is used to steer the screen cursor around. Most joysticks select screen positions with actual stick movement; others respond to pressure on the stick. Figure 8–10 shows a movable joystick. The distance that the stick is moved in any direction from its center position corresponds to screen cursor movement in that direction. Potentiometers are used in the joystick to measure the amount of movement, and springs return the stick to the center position when it is released. Sometimes joysticks are constructed with one or more buttons that can be programmed as input switches to signal some type of action once a screen position has been selected.

In another type of movable joystick, the stick is used to activate switches that cause the screen cursor to move at a constant rate in the direction selected. Eight switches, arranged in a circle, are usually provided, so that the stick can select any one of eight directions for cursor movement.

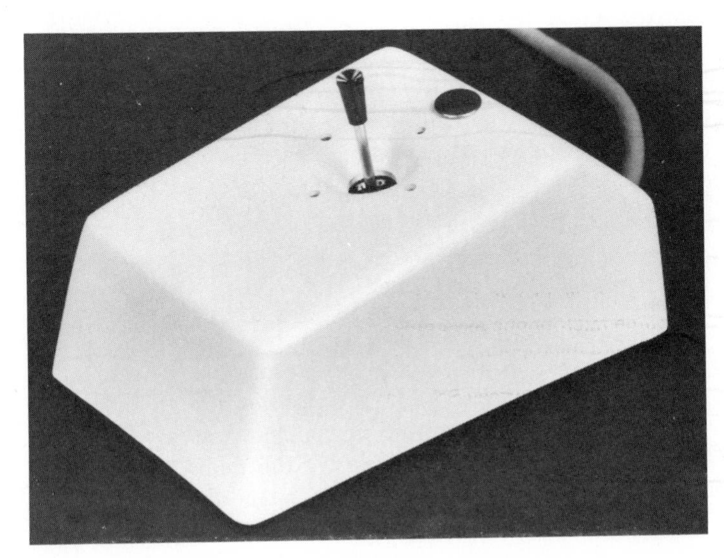

FIGURE 8–10
A movable joystick. Courtesy CalComp Group; Sanders Associates, Inc.

FIGURE 8–11
A pressure-sensitive joystick. Courtesy Measurement Systems, Inc., Norwalk, Connecticut.

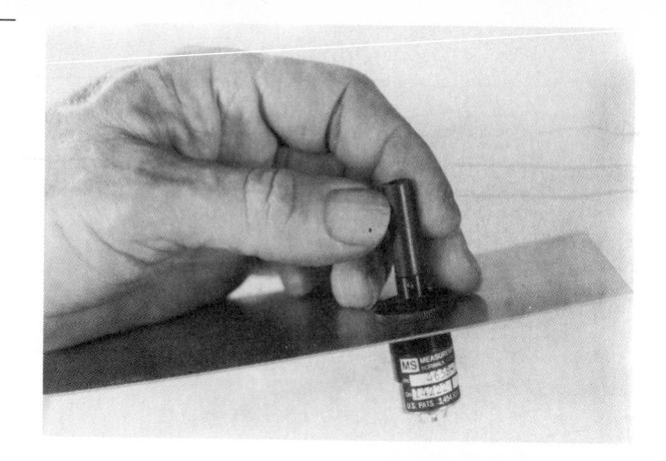

FIGURE 8–12
A trackball. Courtesy Measurement Systems, Inc., Norwalk, Connecticut.

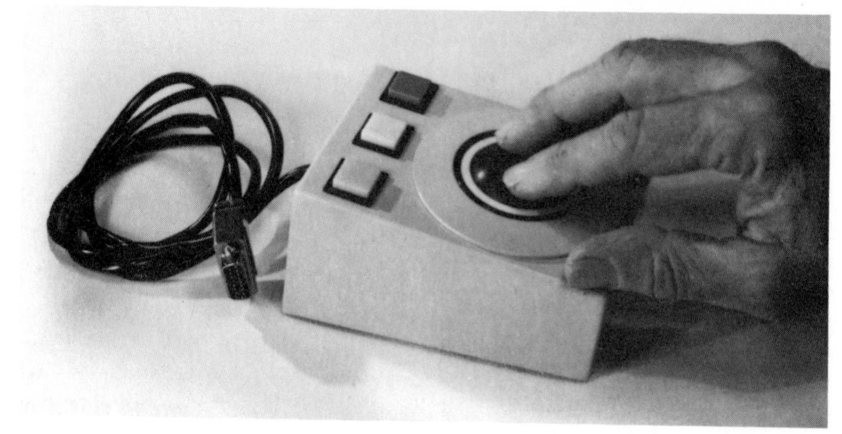

A pressure-sensitive joystick is shown in Fig. 8–11. This device, also called an isometric joystick, has a nonmovable stick. Pressure on the stick is measured with strain gauges and converted to movement of the cursor in the direction specified.

Trackball

The operation of a **trackball** (Fig. 8–12) is similar to that of a joystick. In this case, cursor movement is obtained by moving a sphere instead of a stick. The sphere can be rotated in any direction, and potentiometers measure the amount and direction of rotation.

Mouse

A **mouse** is small hand-held box with wheels or rollers on the bottom, as shown in Fig. 8–13. As the mouse is pushed across a flat surface, the rollers record the amount and direction of motion for conversion to a corresponding movement of the screen cursor. Since a mouse can be picked up and put down at another position without change in cursor movement, it is used for making relative changes in the position of the screen cursor. Buttons on the top of the mouse are used as switches to signal the execution of some operation, such as recording cursor position.

FIGURE 8–13
A mouse. Courtesy Summagraphics Corp.

Voice Systems

Speech recognizers are used in some graphics workstations as input devices to accept voice commands. The voice input can be used to initiate graphics operations or to enter data. These systems operate by matching an input against a predefined dictionary of words.

A dictionary is set up for a particular operator by having the operator speak the command words to be used into the system. Each word is spoken several times, and the system analyzes the word and establishes a frequency pattern for that word in the dictionary along with the corresponding function to be performed. Later, when a voice command is given, the system searches the dictionary for a frequency-pattern match. Voice input is typically spoken into a microphone on a headset, as in Fig. 8–14. The microphone is designed to minimize input of other background sounds. If a different operator is to use the system, the dictionary must be reestablished with that operator's voice patterns. Voice systems have some advantage over other input devices, since the attention of the operator does not have to be switched from one device to another to enter a command.

FIGURE 8–14
A speech-recognition system. Courtesy Threshold Technology, Inc.

8–2 Logical Classification of Input Devices

Graphics programs use a number of types of input data. They include values for coordinate positions, character strings for labeling, scalar values for the transformation parameters, values specifying menu options, and values for segment identification. Any of the devices discussed in the foregoing section can be used to input these various data types. To make graphics packages independent of the particular hardware devices used, input commands can be structured according to the data description to be handled by each command. This approach provides a logical classification for input devices in terms of the kind of data to be input.

The various types of input data are summarized in the following six logical device classifications:

LOCATOR — a device for specifying a coordinate position (x, y)
STROKE — a device for specifying a series of coordinate positions
STRING — a device for specifying text input
VALUATOR — a device for specifying scalar values
CHOICE — a device for selecting menu options
PICK — a device for selecting picture components

In some packages, a single logical device is used for both locator and stroke operations. Some other mechanism, such as a switch, can then be used to indicate whether one coordinate position or a "stream"of positions is to be input.

Each of the six logical input device classifications can be implemented with any of the hardware devices, but some hardware devices are more convenient for certain kinds of data than others. A device that can be pointed at a screen position is more convenient for entering coordinate data than a keyboard, for example. In the following sections, we discuss how the various physical devices are used to provide input within each of the logical classifications.

8–3 Locator Devices

A typical method for establishing interactive coordinate input is by positioning the screen cursor. This can be accomplished with movement of thumbwheels, dials, a trackball, a joystick, a mouse, or a tablet stylus or hand cursor. Movement of each of these physical devices is translated into a corresponding screen cursor movement (Fig. 8–15). When the screen cursor is at the desired location, a button is activated to input the coordinates of that point.

Light pens are also used to input coordinate positions, but some special implementation considerations are necessary. Since light pens operate by detecting light emitted from the screen phosphors, some nonzero intensity level must be present at the coordinate position to be selected. With a raster system, we can paint a color background onto the screen. As long as no black areas are present, a light pen can be used to select any screen position.

When a background color or intensity cannot be provided, a light pen can be used as a locator by creating a small light pattern for the pen to detect. The pattern could be a single point, a single character, or a line pattern that is moved around the screen until it finds the light pen. A straightforward method for carrying out this search is to move the pattern across each scan line in turn until the pattern

FIGURE 8–15
Relative position changes of a mouse can be used to control screen cursor movement.

moves into the light pen's field of view. At that instant, the activated pen signals that it has seen the pattern. This light pen interrupt signal causes the coordinate position of the pattern at that instant to be recorded as the input. Since the field of view of a light pen is large compared to the size of a screen pixel, a larger pattern produces more accurate coordinate selections than a small pattern. With a point or other small pattern, a coordinate position could be recorded at the edge of the pen's field of view instead of at the center position, as demonstrated in Fig. 8–16.

FIGURE 8–16
A small pattern detected by a light pen at the edge of its field of view causes a displaced coordinate position to be input.

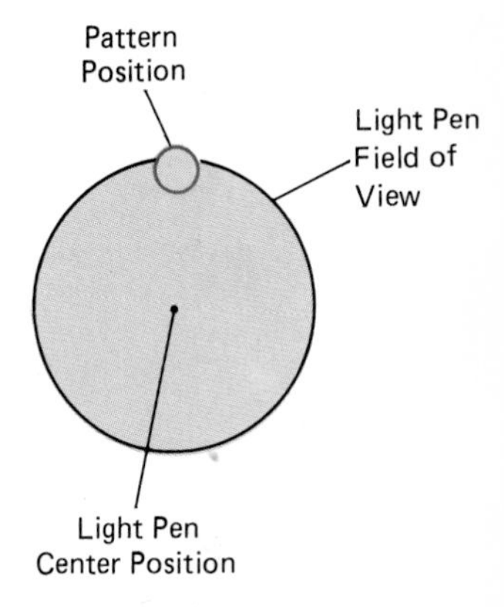

A technique used on some vector systems is to have the light pen move a tracking-cross pattern (Fig. 8–17) to the desired coordinate position. The cross is initially displayed at an arbitrary screen location, such as the screen center. To select a coordinate position, the pen first must be pointed at the cross, then moved slowly to the destination position. As the pen moves, each new cross position is calculated from a knowledge of how much of the cross the pen last saw. This is determined by drawing the cross from the outside toward the cross center, as shown in Fig. 8–18. Points along the arms of the cross are "turned on," one at a time, starting with the endpoints. As soon as a point on each of the four arms is detected by the pen, its position is recorded. The average x and y positions of these four points determine the new center position for the cross. When the cross finally reaches its destination, button activation can be used to signal storage of the coordinate position.

If the light pen is moved too rapidly, it could lose the tracking cross. In this case, some mechanism for restarting the tracking procedure could be initiated by the system. A small pattern could be scanned across the screen to find the pen. Once the light pen has been located, the cross can be moved to that position. If the pen has moved off screen and cannot be found, tracking is terminated. Other procedures, such as restarting the tracking process from the beginning, could be used when tracking is lost. To reduce the possibility of losing the tracking cross, a tracking procedure can be set up that attempts to move the cross slightly ahead of the pen. A straight-line prediction of the probable next pen position can be calculated and the cross moved to that position at each step.

FIGURE 8–17
A light-pen tracking cross, formed by two perpendicular arms drawn with adjacent screen points.

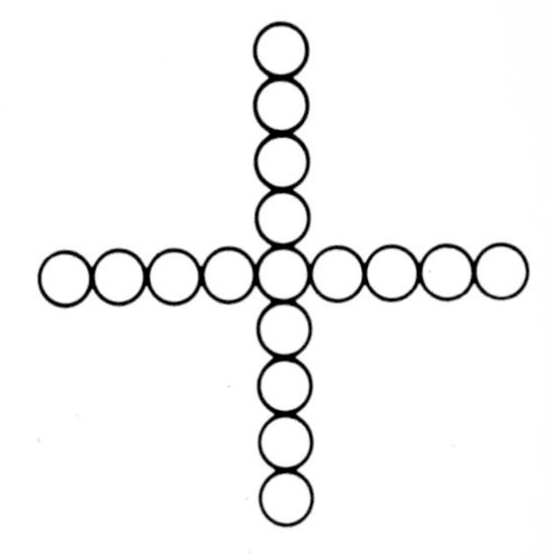

Keyboards can be used for locator input in two ways. Coordinate values could be typed in, but this is a slow process in which the user is required to know exact coordinate values. As an alternative, cursor keys can be used to move the screen cursor to the required position. Four keys provide for relative horizontal and vertical cursor movement (up, down, right, left). With an additional four keys, the cursor can be moved diagonally as well. Rapid cursor movement can often be accomplished by holding down one of the cursor keys. When the cursor has been positioned properly, the input key is pressed.

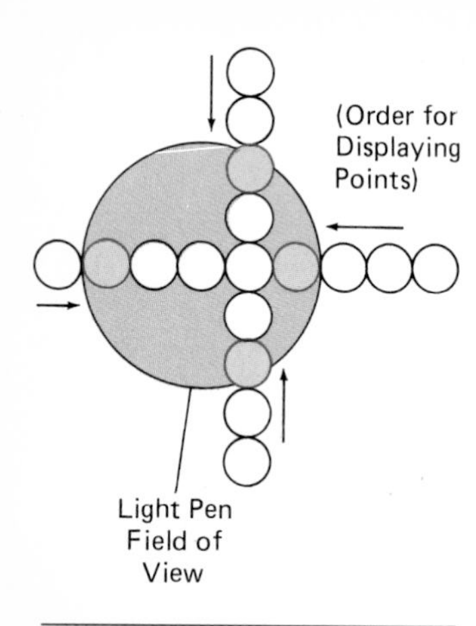

FIGURE 8–18
As a tracking cross is displayed from the outer points to the center, the average position of the first points detected by the light pen (color dots) determine the new tracking-cross center position.

8–4 Stroke Devices

This class of logical devices is used to input a sequence of coordinate positions. Stroke-device input is equivalent to multiple calls to a locator device. The set of input points is usually used to display a line section.

Many of the physical devices used for generating locator input can be used as stroke devices. Continuous movement of a mouse, trackball, or joystick is translated into a series of input coordinate values.

Two of the more common stroke devices are the tablet and the light pen. Button activation can be used to place a tablet hand cursor into "continuous" mode. As the cursor is moved across the tablet surface, a stream of coordinate values is generated. This process is used in paintbrush systems that allow artists to draw scenes on the screen and in engineering systems where layouts can be traced and digitized for storage. A similar procedure is used for stroke input with a light pen. By moving an activated pen across a CRT screen, a sequence of input points can be generated.

8–5 String Devices

The primary physical device used for string input is the keyboard. Input character strings are typically used for labels or program commands.

Other physical devices are sometimes used for generating character patterns in a "text-writing" mode. In these applications, individual characters are drawn on the screen with a stroke or locator-type device. A pattern recognition program then interprets the characters using a stored dictionary of predefined patterns. Figure 8–19 illustrates character patterns input in this way.

FIGURE 8–19
(a) A character drawn by plotting points along the path of a moving cursor. (b) A series of drawn letters input as a label.

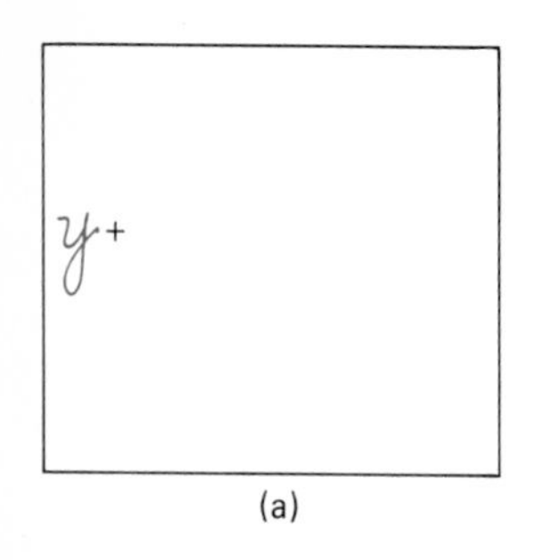

(a)

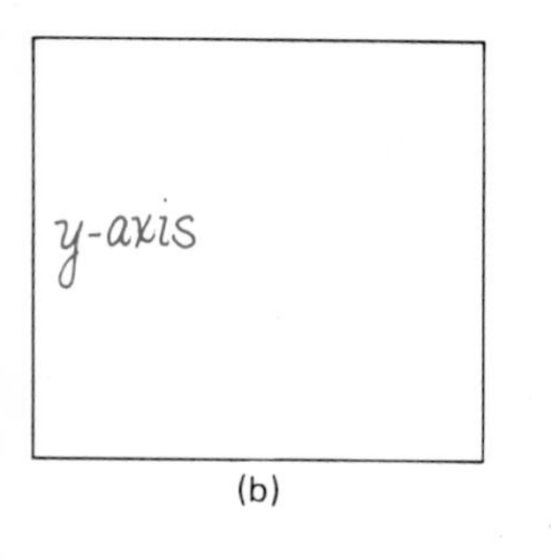

(b)

8–6 Valuator Devices

This logical class of devices is employed in graphics systems to input scalar values. Valuators are used for setting various graphics parameters, such as rotation angle and scale factors, and for setting physical parameters associated with a particular application (temperature settings, voltage levels, stress factors, etc.).

A typical physical device used to provide valuator input is a set of control dials. Floating-point numbers within any predefined range are input by rotating the dials. Dial rotations in one direction increase the numeric input value, and opposite rotations decrease the numeric value. Rotary potentiometers convert dial rotation into a corresponding voltage. This voltage is then translated into a real number within a defined scalar range, such as -10.5 to 25.5. Instead of dials, slide potentiometers are sometimes used to convert linear movements into scalar values.

Any keyboard with a set of numeric keys can be used as a valuator device. A user simply types the numbers directly in floating-point format. This is a slower method than using dials or slide potentiometers.

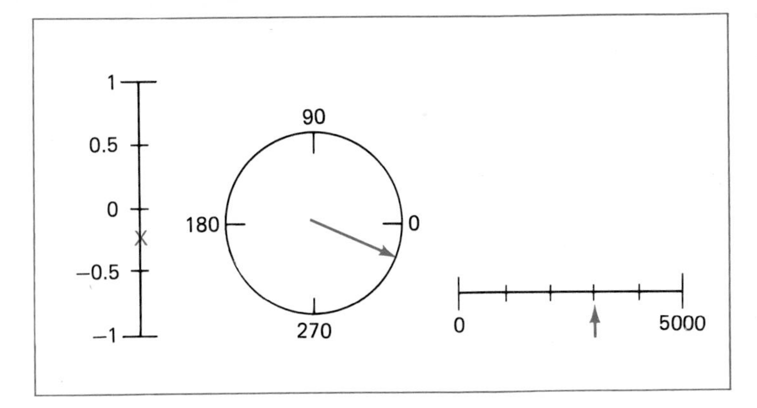

FIGURE 8–20
Scales displayed on a screen and used for valuator input.

Joysticks, trackballs, tablets, and other such devices can be adapted for valuator input by interpreting movement of the device relative to a scalar range. For one direction of movement, say left to right, increasing scalar values can be input. Movement in the opposite direction decreases the scalar input value.

Another technique for providing valuator input is to display the scale ranges in pictorial form on a video monitor. Figure 8–20 illustrates some possibilities for scale representations. A light pen, joystick, mouse, tablet, or other locator device is used to select a coordinate position on the scale representation. This coordinate value is then converted to a numeric input value. As a feedback mechanism for the user, the selected position can be marked with some symbol. In Fig. 8–20, the X symbol marks the value of -0.25 on the vertical scale, while an arrow is used on the horizontal scale at the value 3000. An arrow is used also on the circular scale to indicate a position at 330°. As new positions are selected, these markers are moved to show the new values. The numeric values may also be displayed somewhere on the screen to confirm the selection value.

8–7 Choice Devices

Menu selections are common inputs in many graphics applications. A choice device is defined as one that enters a selection from a list (menu) of alternatives. Packages employ menus to input programming options, parameters, and figures to be used in the construction of a picture. Commonly used choice devices are a set of buttons, a touch panel, and the light pen.

A function keyboard, or "button box," designed as a stand-alone unit, is often used to enter menu selections. Usually each button is programmable, so that its function can be altered to suit different applications. Single-purpose systems have a fixed, predefined function for each button. Programmable function keys and fixed-function buttons are often included with other standard keys on a keyboard.

A touch panel or a light pen is used with many graphics systems for selecting menu options listed on the screen. With a vector system, the activation of a light pen or touch panel triggers the system to record which particular menu item was in the process of being refreshed at that instant. On a raster system, a coordinate position (x, y) is recorded by the light pen or touch panel. This screen location is compared to the area boundaries for each listed menu item to determine which option has been selected. Each menu option can be considered to be bounded by a rectangular area. A menu item with vertical and horizontal boundaries at the coordinate values x_{min}, x_{max}, y_{min}, and y_{max} is selected if the input coordinates (x, y) satisfy the inequalities

$$x_{min} < x < x_{max}, \quad y_{min} < y < y_{max} \tag{8-1}$$

Other devices, such as a tablet, joystick, or mouse, are sometimes used to make menu selections with coordinate input. These devices can move the screen cursor, a character, or some special pattern to the location of an item to be selected. Button activation then signals the input of coordinates at that position.

Alternate methods for choice input include keyboard and voice entry. A standard keyboard can be used to type in commands or menu options. For this method of choice input, some abbreviated format is useful. Menu listings can be numbered or given short identifying names. Similar codings can be used with voice input systems. Voice input is particularly useful when a small number of options (20 or less) are to be used.

8–8 Pick Devices

Segment selection is the function of this logical class of devices. Typical uses for pick input are the application of transformations and manipulation procedures to the selected segments.

The light pen is the standard pick device used with many graphics systems. On a vector system, activation of a light pen causes the display processor to note which segment was being refreshed at that instant. Each segment is refreshed using the set of output primitives defining that segment. When the pen interrupt occurs, a check is made to determine which set of commands was being processed. The easiest way to do this is for the system to use a register to store the name of each segment in turn as it is being refreshed. Then a check of the current contents of the register identifies the segment selected.

With a raster system, refreshing is done from the frame buffer, and light-pen input is a screen coordinate position. This coordinate information must be mapped onto the frame buffer to determine which segment area encompasses that point. Several levels of search may be necessary to identify the correct segment. First, a rectangular boundary area can be established for each segment (Fig. 8–21). If the input coordinates from the pen fall within this area and no segment areas overlap, the desired segment has been identified. Should two or more segment areas overlap, the process is repeated for individual primitives. That is, bounding rectangles

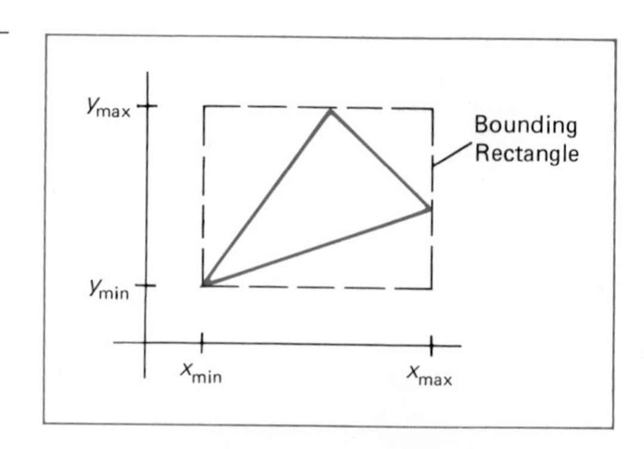

FIGURE 8–21
A picture component (triangle) and corresponding bounding rectangle.

for the component lines in a segment (Fig. 8–22) are used to narrow the search to individual primitives. If the pen coordinates are found to be in the primitive bounding areas of only one segment, the search is over. If the pen coordinates are found to be in overlapping primitive areas from two different segments, calculation of the shortest distance from the input point to each line segment (Fig. 8–23) finally settles the issue.

Other physical devices can be used for segment selection by identifying the segment position with a screen cursor. A coordinate position is selected by moving the screen cursor with a mouse, trackball, joystick, or tablet to the position of a segment and activating a button. This coordinate input is used in the same way as pen coordinates to identify the selected segment.

Keyboards and buttons also are used as segment selection devices in some applications. With a keyboard, the segment name can be typed. A technique for button input is to use one button to highlight successive segments and another button to stop the process when the desired segment is highlighted. If very many segments are to be searched in this way, the process can be speeded up and additional buttons used to help identify the segment. The first button can initiate a rapid successive highlighting of segments. A second button can again be used to stop the process, and a third button can be used to back up more slowly if the desired segment passed before the operator pressed the stop button.

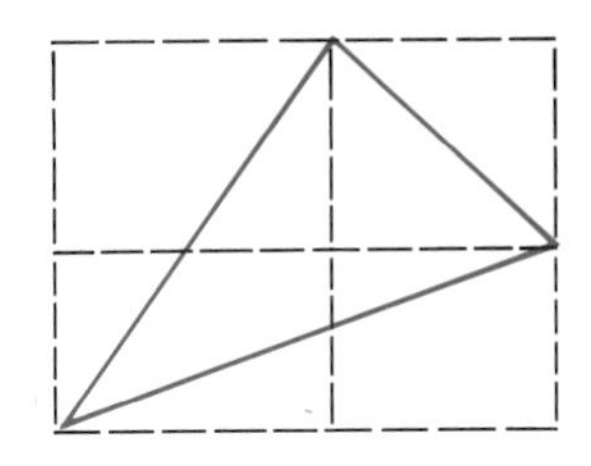

FIGURE 8–22
Bounding rectangles for component lines in an object (triangle).

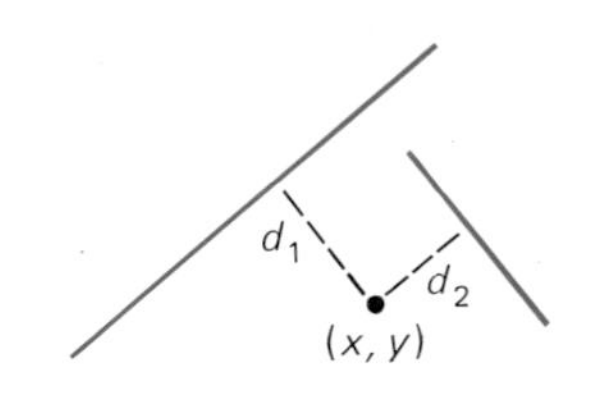

FIGURE 8–23
Distances d_1 and d_2 from a point at (x, y) to two line segments.

8–9 Interactive Picture-Construction Techniques

There are several techniques that we can incorporate into graphics packages to aid users in interactively constructing pictures. Various input options can be provided to a user, so that coordinate information entered with locator and stroke devices can be interpreted according to the selected option. Input coordinates can establish the position or boundaries for objects to be drawn, or they can be used to rearrange previously displayed objects.

Basic Positioning Methods

Coordinate values supplied by locator input are often used with **positioning** methods to specify a location for displaying an object or a character string. Screen positions can be input with a light pen or by moving the screen cursor to the desired position. Figure 8–24 illustrates object positioning with a cursor. The cursor is moved to the desired screen location; then a button is pressed to signal that the object should be placed at this coordinate position. In a similar way, a text position can be specified. Depending on the options available, the text position can specify the coordinates for the beginning, ending, or middle position of the character string. Figure 8–25 demonstrates text positioning by selecting the center position of a string.

Line drawing with locator input is shown in Fig. 8–26. The first coordinate position selects the starting endpoint, and the second input position indicates the opposite endpoint. A line is then displayed between the two selected endpoints.

As an aid in positioning objects, we can design the package to display coordinate information for the user. Figure 8–27 illustrates how coordinates might be displayed near the cursor position. With this technique, a user can move the cursor until the correct coordinates are shown, then press the button to select this position.

FIGURE 8–24
Positioning an object with the screen cursor.

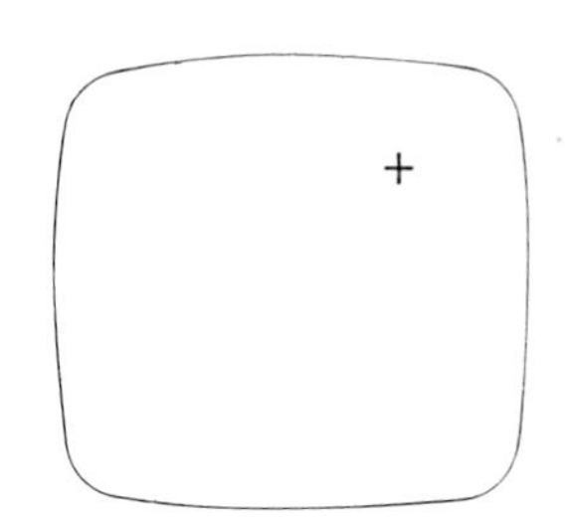

Position Cursor and Press Button

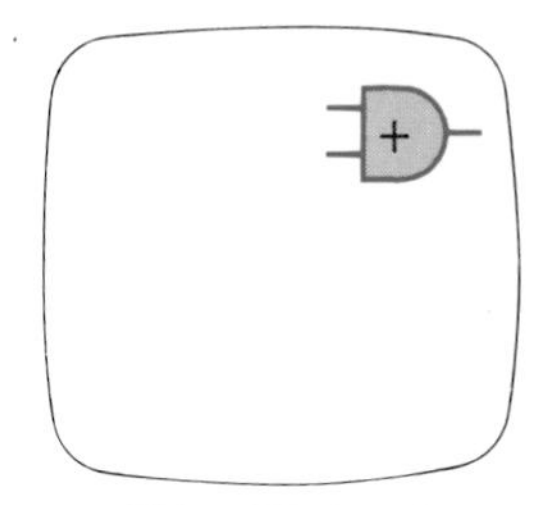

Object Displayed at Cursor Position

FIGURE 8–25
Displaying a character string by selecting the center position of the string with a screen cursor.

Press Button to Select Text Position

Text Displayed, Centered on Selected Position

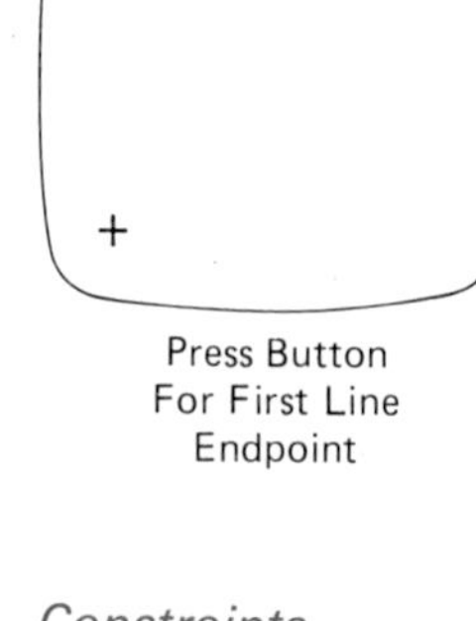
Press Button
For First Line
Endpoint

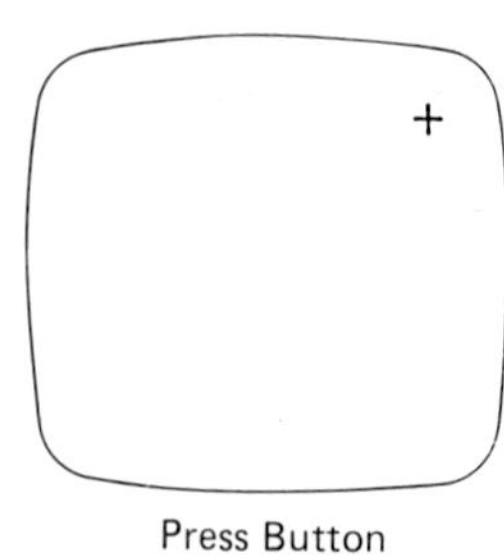
Press Button
at Second
Line Endpoint

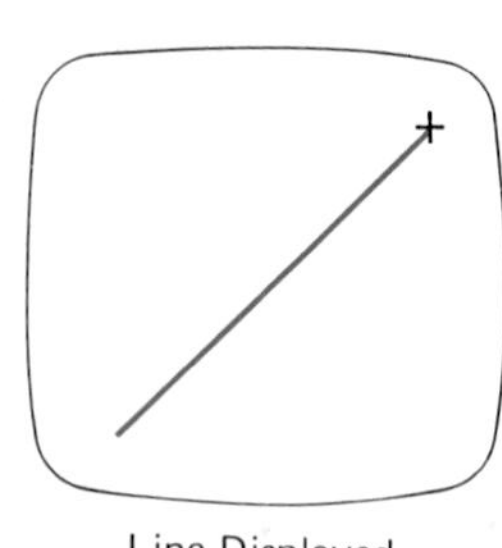
Line Displayed
Between the Two
Chosen Endpoints

FIGURE 8–26
Line drawing by positioning a screen cursor.

Constraints

With some applications, certain types of prescribed orientations or object alignments are useful. A **constraint** is a rule for altering input coordinate values to produce a specified orientation or alignment of the displayed coordinates. There are many kinds of constraint functions that can be specified, but the most common constraint is a horizontal or vertical alignment of straight lines. This type of constraint, shown in Figs. 8–28, and 8–29, is useful in forming network layouts. With this constraint, a user can create horizontal and vertical lines without worrying about precise specification of endpoint coordinates.

One method for implementing the line constraints demonstrated in Figs. 8–28 and 8–29 is to allow the package to determine whether any two input coordinate endpoints are more nearly horizontal or more nearly vertical. If the difference in the y values of the two endpoints is smaller than the difference in x values, a horizontal line is displayed. Otherwise, a vertical line is drawn.

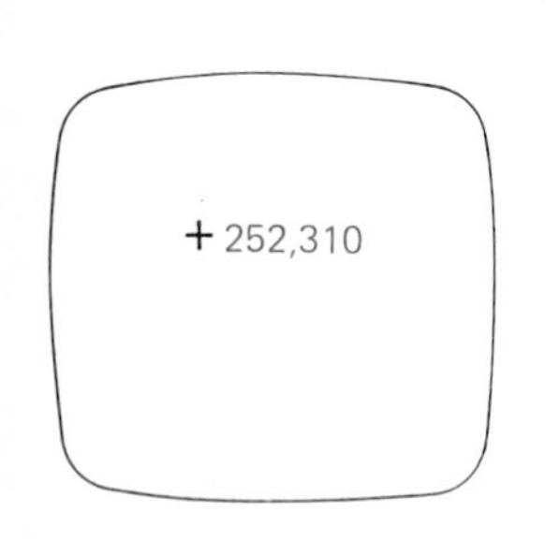
+ 252,310

FIGURE 8–27
Coordinates displayed near a cursor position, as an aid in selecting screen locations.

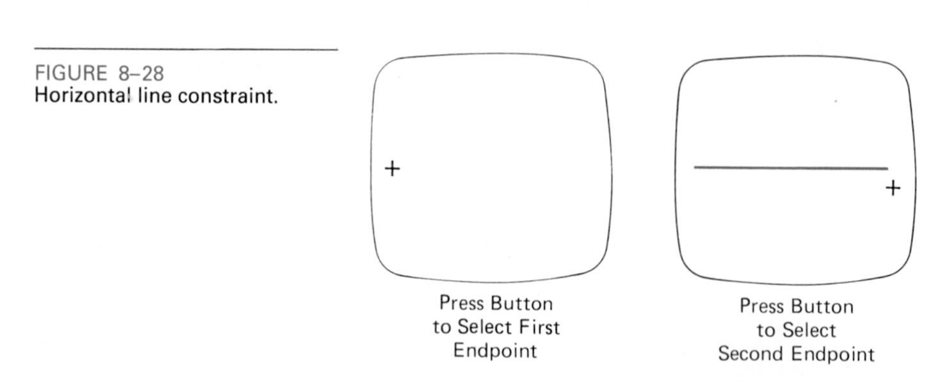
Press Button
to Select First
Endpoint
Press Button
to Select
Second Endpoint

FIGURE 8–28
Horizontal line constraint.

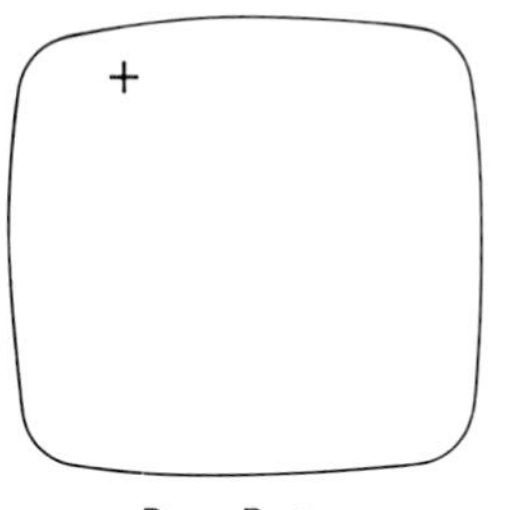
Press Button
to Select
First Endpoint

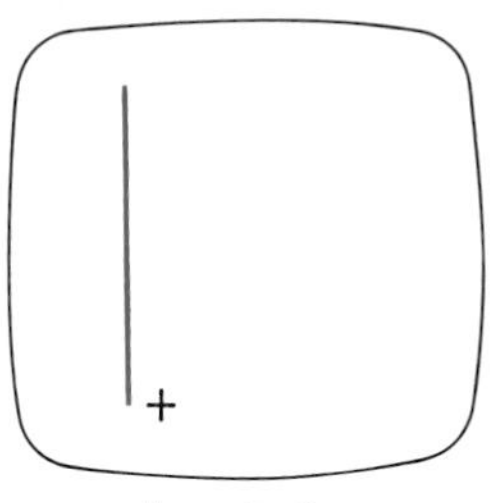
Press Button
to Select
Second Endpoint

FIGURE 8–29
Vertical line constraint.

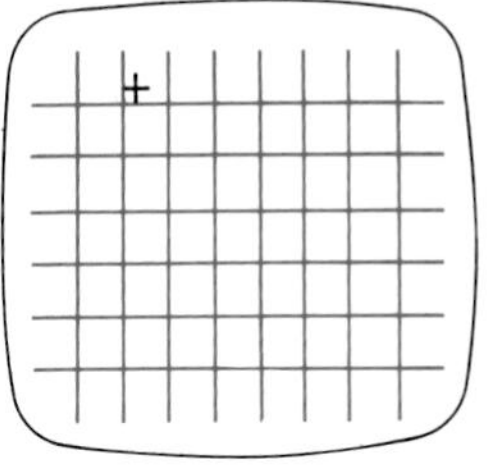
Press Button to
Select First Endpoint

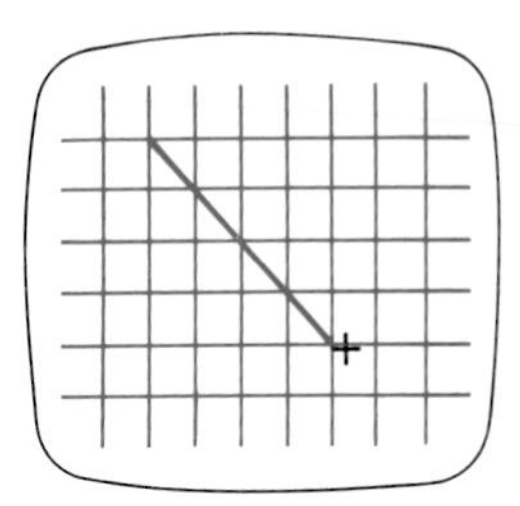
Press Button
to Select
Second Endpoint

FIGURE 8–30
Line drawing using a grid.

Other kinds of constraints can be applied to input coordinates to produce a variety of alignments. Lines could be constrained to have a particular slant, such as 45°, and input coordinates could be constrained to lie along predefined paths, such as circular arcs.

Grids

Another type of constraint is a **grid** of rectangular lines superimposed on the screen coordinate system. When a grid is used, any input coordinate position is rounded to the nearest intersection of two grid lines. Figure 8–30 illustrates line drawing with a grid. Each of the two cursor positions is converted to a grid intersection point, and the line is drawn between these grid points. This technique can help users to draw lines and figures more accurately and neatly. Grids also make the construction of objects easier, since positioning a cursor at an exact coordinate position can be a difficult process.

Grid lines can be used also for positioning and aligning objects or text. The spacing between grid lines is often left as an option for the user, and the grid pattern may or may not be displayed. It is also possible to use partial grids and grids of different sizes in different screen areas.

Gravity Field

In the construction of figures, a user often needs to connect a new line to a previously drawn line. If a grid is available, the grid points could be used for all line intersections. But sometimes a user may want to connect two lines at a position between grid points. Since exact positioning of the screen cursor at the connecting point can be difficult, graphics packages can be designed to convert any input position near a line to a position on the line.

This conversion of input position is accomplished by creating a **gravity field** area around the line. Any selected position within the gravity field of a line is moved ("gravitated") to the nearest position on the line. A gravity field area around a line is illustrated with a dashed boundary in Fig. 8–31. Areas around the endpoints are enlarged to make it easier for a user to connect lines at their endpoints. The size of gravity fields is chosen large enough to aid the user but small enough to reduce chances of overlap with other lines. If many lines are displayed, gravity areas can overlap, and it may be difficult for a user to specify points correctly. Normally, the boundary for the gravity field is not displayed; the user simply selects points as near to the line as possible.

FIGURE 8–31
Gravity field around a line.

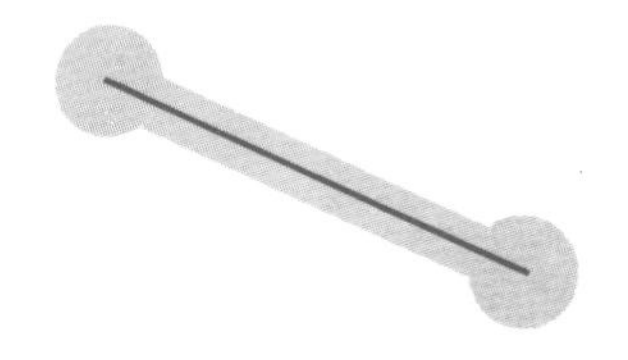

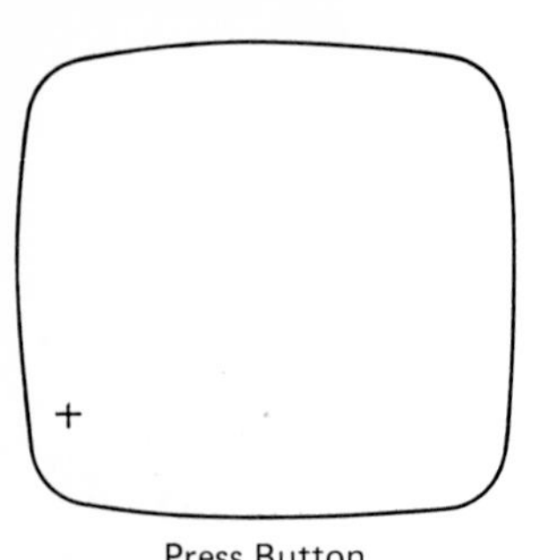
Press Button
to Start
Rubber-Band
Line Drawing

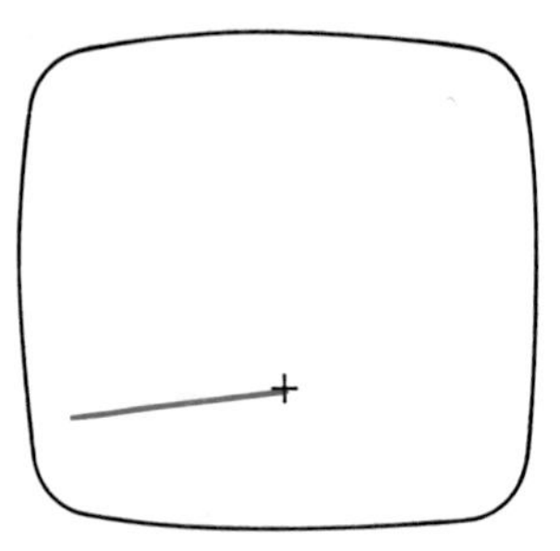
As the Cursor
Moves, A Line
Stretches out
From Initial
Point

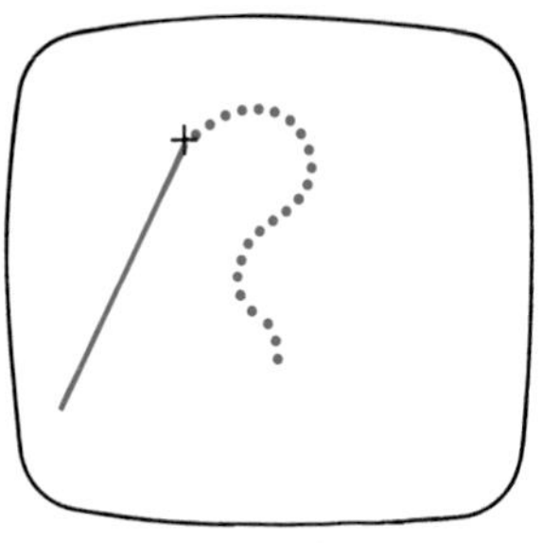
Line is
Positioned and a
Button Pressed to
End Process

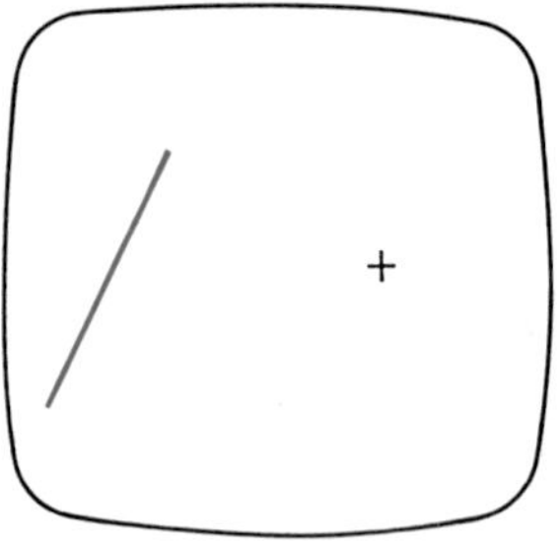
After Stop Button is
Pressed, Cursor Can
be Moved Without
Rubber-Band
Effects

FIGURE 8–32
Rubber-band method for drawing and positioning a line.

Rubber-Band Methods

Straight lines can be constructed and positioned by **rubber-band methods,** which stretch out a line from a starting position as the screen cursor is moved. Figure 8–32 demonstrates the rubber-band method. The start of the process is signaled by pressing a start button. As the cursor moves around, the line is displayed from the start position to the current position of the cursor. A stop button is used to signal the termination of the operation and fix the position of the line. On some systems, the rubber-band drawing of a line starts when a button is pressed and stops when the button is released.

Rubber-band methods can be used to construct and position various types of objects. In Fig. 8–33 a circular arc is produced, and Fig. 8–34 demonstrates rubber-band construction of a rectangle.

FIGURE 8–33
Drawing a circular arc using a rubber-band method.

FIGURE 8–34
Rubber-band method for constructing a rectangle.

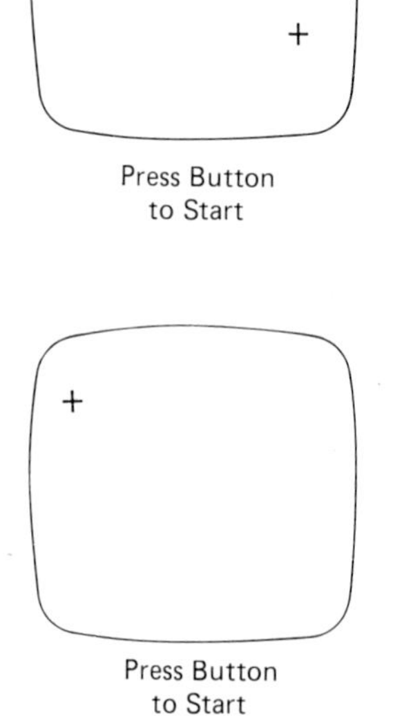
Press Button
to Start
Press Button
to Start

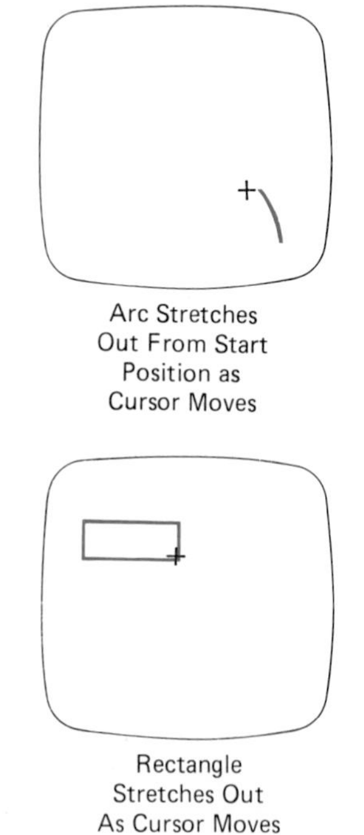
Arc Stretches
Out From Start
Position as
Cursor Moves
Rectangle
Stretches Out
As Cursor Moves

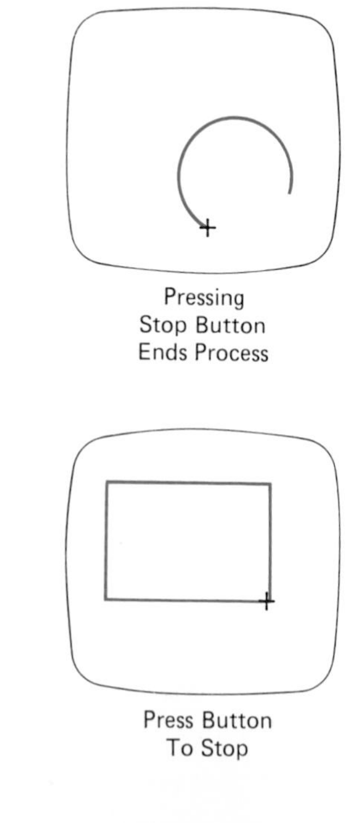
Pressing
Stop Button
Ends Process
Press Button
To Stop

Sketching

For many applications it is convenient to be able to **sketch** objects by tracing their outline on the screen. An extension of the rubber-band method can be used to create objects formed with straight line segments. Instead of drawing a single line, a series of connected lines can be drawn before the rubber-band procedure is terminated. As illustrated in Fig. 8–35, two buttons can be used. One button controls the beginning and ending of the overall procedure, and the second button is used to mark the intermediate line endpoints. This type of sketching is useful for facility layouts, circuit diagrams, and various network layouts.

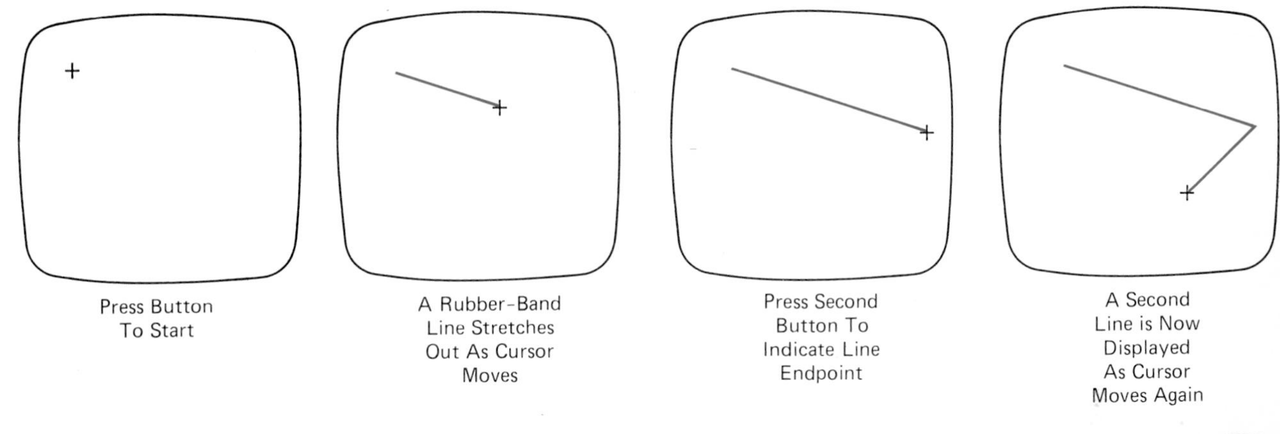
Press Button
To Start
A Rubber-Band
Line Stretches
Out As Cursor
Moves
Press Second
Button To
Indicate Line
Endpoint
A Second
Line is Now
Displayed
As Cursor
Moves Again

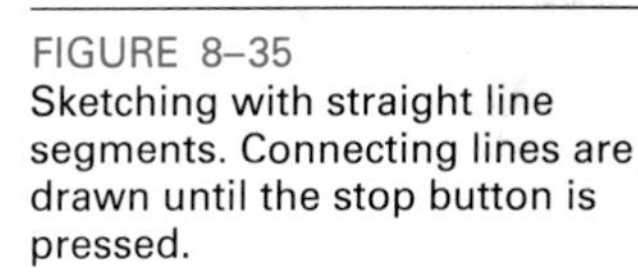
FIGURE 8–35
Sketching with straight line segments. Connecting lines are drawn until the stop button is pressed.

Instead of displaying straight line segments, a sketching program could plot individual points along the path of the cursor as it is moved around the screen. Various curved figures can then be displayed, as illustrated in Fig. 8–36. This type of sketching routine is used with paintbrush programs. The artist is furnished with a number of "brushes," which are used to generate lines of varying thickness and style. A color palette provides a menu of colors that can be displayed with the brushes. Lines are erased by retracing them in the background color.

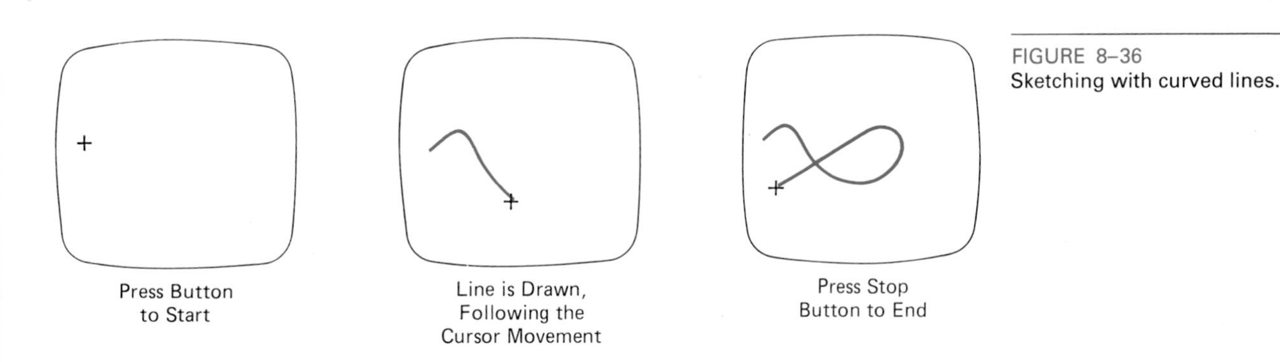
Press Button
to Start
Line is Drawn,
Following the
Cursor Movement
Press Stop
Button to End

FIGURE 8–36
Sketching with curved lines.

Dragging

A technique that is often used in interactive picture construction is to move objects into position by **dragging** them with the screen cursor. Usually, a menu of shapes is displayed, and the user selects an item, then drags it into position. This

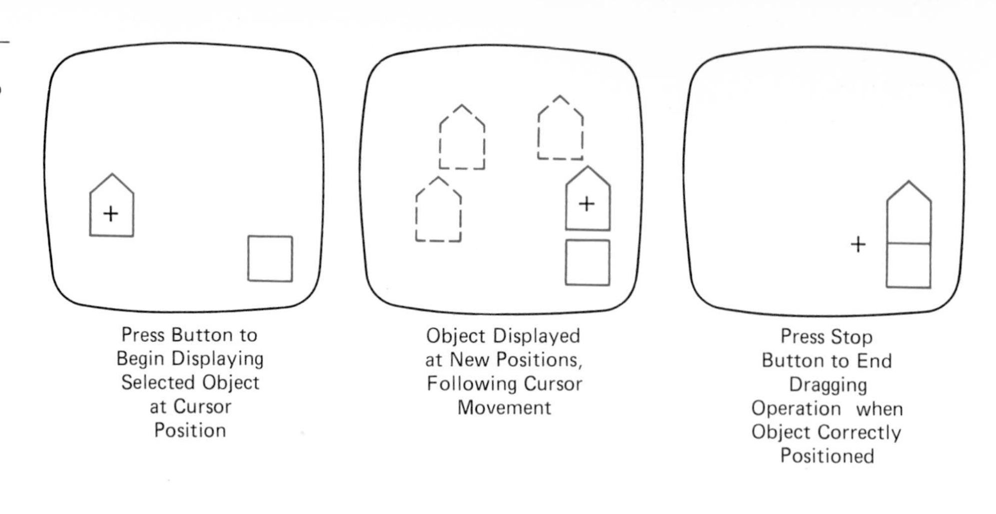
+
+
+
Press Button to Begin Displaying Selected Object at Cursor Position
Object Displayed at New Positions, Following Cursor Movement
Press Stop Button to End Dragging Operation when Object Correctly Positioned

FIGURE 8–37
Dragging a selected shape into position.

technique is demonstrated in Fig. 8–37. After an object has been selected from the menu, a button can be pressed to initiate the dragging process. When the user is satisfied with the object's position, the button is again pressed to release the object from the cursor.

8–10 Input Functions

We can design a graphics package using the six logical input classifications in various ways. An input statement in an application program could specify both the logical classification and the type of physical device employed to input the data. The other parameters needed depend on the type of data to be input. In addition, there are alternative ways to structure the interaction between an application program and the input devices.

Input Modes

Commands to provide input can be structured to operate in various **input modes,** which specify how the program and input devices interact. Input could be initiated by the program, or the program and input devices both could be operating simultaneously, or data input could be initiated by the devices. These three input modes are referred to as request mode, sample mode, and event mode.

In **request mode,** the application program requests input and suspends processing until the required data are input. The program and the input devices operate alternately. Devices are put into a wait state until an input request is made; then the program waits until the data are delivered.

In **sample mode,** the application program and input devices operate at the same time. Input data are stored and continually updated to current values as the input devices operate. The program samples these current data as it needs input.

In **event mode,** the input devices initiate data input to the application program. A stream of input values is provided to the program, and these data values are processed one at a time by the program. In this mode, the processing is directed by the input data.

Input commands can be defined for each of the three input modes, and a package can allow a user to specify the mode of operation for each logical device class. For example, the command

```
set_locator_mode (ws, device_code, input_mode)
```

specifies that a locator device is to be used in the particular mode set in parameter *input_mode*. The three choices for this parameter are *request, sample,* and *event.* Parameter *ws* identifies the workstation to be used, and parameter *device_code* is assigned a value that identifies which particular physical device is to be used as a locator. One possible assignment of device codes is given in Table 8–1. Using the assignments in this table, the command

```
set_stroke_mode (4, 2, event)
```

designates the graphics tablet as a stroke device in *event* mode on workstation 4.

TABLE 8–1.
Assignment of Device Codes

Device Code	Physical Device Type
1	Keyboard
2	Graphics Tablet
3	Light Pen
4	Touch Panel
5	Thumbwheels
6	Dial
7	Button
8	Joystick
9	Mouse
10	Trackball
11	Voice Entry

Similar commands are issued by a user to set the input mode for the other logical input classes. Devices can operate in only one mode at a time, but several devices could be operating simultaneously in different modes. Once the mode has been set, commands in that mode can be used.

Request Mode

Input commands used in this mode correspond to input commands such as the *read* statement in a programming language. An input request is made, and further processing waits for the input data to be received. After a device has been assigned a request mode, as discussed in the preceding section, input requests can be made to that device. For example, to request input from a locator device, the following command could be used in an application program:

```
request_locator (ws, device_code, x, y)
```

where parameters x and y are used to store the input coordinates for a single point.

Parameters included in the input commands for each logical classification depend on the type of data to be input. Stroke input is requested with

```
request_stroke (ws, device_code, n, xa, ya)
```

where the coordinates of the *n* input points are to be stored in arrays *xa* and *ya*. Similarly, text could be input in request mode with

```
request_string (ws, device_code, nc, text)
```

where parameter *text* specifies an input string of length *nc* (number of characters). To select a segment, an application program could include the following command, which identifies the segment with parameter *segment_id:*

```
request_pick (ws, device_code, segment_id)
```

Other parameters can be included in the input commands for request mode, depending on the options offered at a particular facility. For instance, it is useful in some applications to be able to identify component parts of segments, such as individual lines. An additional parameter could be included in the pick command to select labeled segment parts. This means, of course, that the segment parts must be labeled as the segment is being created. An activated light pen, pointed at the segment, would then input both the name of the segment and the name of a segment part.

Sample Mode

Once sample mode has been set for one or more physical devices, data input begins without waiting for a program input statement. If a joystick has been designated as a locator device in sample mode, coordinate values for the current position of the activated joystick are immediately stored. As the activated stick position changes, the stored values are continually replaced with the coordinates of the current stick position.

Sampling of the current values from a physical device in this mode begins when a *sample* command is encountered in the application program. A locator device is sampled with a call to

```
sample_locator (ws, device_code, x, y)
```

Sampling the other logical devices is performed with similar commands.

Event mode

When an input device is placed in event mode, the program and device operate simultaneously. Data input from the device can be accumulated in an **event queue,** or **input queue.** All input devices active in event mode can enter data (**events**) into this single event queue, with each device entering data values as they are generated. At any one time, the event queue can contain a mixture of data types, in the order they were input. Data entered into the queue are identified according to logical class, workstation number, and physical device code.

An application program can be directed to check the event queue for any input with the command

```
await_event (time, device_class, ws, device_code)
```

Parameter *time* is used to set a maximum waiting time for the application program. If the queue happens to be empty, processing is suspended until either the number of seconds specified in *time* has elapsed or an input arrives. Should the waiting time run out before data values are input, the parameter *device_class* can be used to return the value *none*. When *time* is assigned the value 0, the program checks the queue and immediately returns to other processing if the queue is empty.

When processing is directed to the event queue with the *await_event* command and the queue is not empty, the first event in the queue is transferred to a **current event record.** The particular logical device class, such as locator or stroke, that made this input is stored in parameter *device_class*. Codes, identifying the particular workstation and physical device that made the input, are stored in parameters *ws* and *device_code*, respectively.

To retrieve a data input from the current event record, an event-mode input command is used. The form of the commands in event mode is similar to those in request and sample modes. However, no workstation and device code parameters are necessary in the commands, since the values for these parameters are stored in the data record. A user could ask for locator data from the current event record with the command

```
get_locator (x, y)
```

As an example of the use of *await_event* and *get* commands, the following program section inputs a set of points from a tablet on workstation 1 and plots a series of straight line segments connecting the input coordinates:

```
set_stroke_mode (1, 2, event); {set tablet to stroke device, event
                                mode}
repeat
  await_event (3600, device_class, ws, device_code)
until device_class = stroke;
get_stroke (n, x, y);
polyline (n, x, y);
```

The *repeat–until* loop bypasses any data from other devices that might be in the queue. Also, the waiting time is set to one hour, which is long enough to ensure that the input data will be received. If the tablet is the only active input device in event mode, this loop is not necessary.

A number of devices can be used at the same time in event mode for rapid interactive processing of displays. The following statements plot input lines from a tablet with attributes specified by a button box:

```
set_polyline_index (1);
set_stroke_mode (1, 2, event); {set tablet to stroke device, event
                                mode}
set_choice_mode (1, 7, event); {set buttons to choice device, event
                                mode}
repeat
  await_event (120, device_class, ws, device_code);
  if device_class = choice then begin
      get_choice (option);
      set_polyline_index (option)
    end
  else
    if device_class = stroke then begin
        get_stroke (n, x, y);
        polyline (n, x, y)
      end
until device_class = none;
```

In this example, we have assumed that only the buttons and tablet are in use. Termination is signaled by a value of *none* for logical device parameter *device_class*. This occurs after a waiting time of two minutes if no event arrives in the queue during this time.

Some additional housekeeping commands can be used in event mode. Commands for clearing the event queue are useful when a process is terminated and a new application is to begin. These commands can be set to clear the entire queue or to clear only data associated with specified input devices and workstations.

Concurrent Use of Input Modes

An example of the simultaneous use of input devices in different modes is given in the following procedure. An object is dragged around the screen with a light pen. When a final position has been decided on, a button is pressed to terminate any further movement of the object. The pen positions are obtained in sample mode, and the button input is sent to the event queue.

```
{drags object in response to light pen input}
{button is used to terminate processing}

begin
  set_locator_mode (1, 3, sample);      {set pen to locator device,
                                         sample mode}
  set_choice_mode (1, 7, event);        {set button to choice device,
                                         event mode}
  repeat
    sample_locator (1, 3, x, y);        {read from pen}

      {translate object to x, y and draw}

    await_event (0, class, ws, code)    {check event queue for input}
  until class = choice                  {stop if button has been used}
end;
```

8–11 Summary

Many types of hardware devices can be used to input data to a graphics program. They include keyboards, buttons and dials, touch panels, light pens, two-dimensional and three-dimensional tablets, joysticks, and trackball, mouse, and voice devices. The type of device used in a particular system depends on the type of application and the data type to be input.

A logical classification of input describes devices in terms of the type of input data. This provides a basis for structuring input routines in a graphics package to make the package independent of the type of physical devices that may be used. Locator devices are any devices used by a program to input a single coordinate position. Stroke devices input a stream of coordinates. String devices are used to input text. Valuator devices are any input devices used to enter a scalar value. Choice devices enter menu selections. Pick devices input a segment name.

Interactive picture-construction methods are used in many design and painting packages. These methods provide users with the capability to position objects, to constrain figures to predefined orientations or alignments, to sketch figures, and to drag objects around the screen. Grids, gravity fields, and rubber-band methods are also used to aid users in constructing pictures.

Commands available in a package can be defined in three input modes. Request mode places input under the control of the application program. Sample mode allows the input devices and program to operate concurrently. Event mode allows input devices to initiate data entry and control processing of data. Once a mode has been chosen for a logical device class and the particular physical device to be used to enter this class of data, input commands in the program are used to enter data values into the program. An application program can make simultaneous use of several physical input devices operating in different modes.

REFERENCES

The evolution of the concept of logical (or virtual) input devices is discussed in Rosenthal, et al. (1982) and in Wallace (1976). An early discussion of input-device classifications is to be found in Newman (1968).

Additional information on GKS operations with input devices is to be found in Enderle, Kansy, and Pfaff (1984) and in Hopgood, et. al. (1983).

EXERCISES

8–1. Suppose the positions of the four microphones for a three-dimensional sonic digitizer have coordinates (x_k, y_k, z_k), where $k = 1, 2, 3, 4$. Solve three of the following four equations to find the input position (x, y, z) of the stylus:

$$(x - x_k)^2 + (y - y_k)^2 + (z - z_k)^2 = r^2_k, \quad k = 1, 2, 3, 4$$

where each r_k is the measured distance to one microphone from the input position (as determined by the time of arrival of the sound from the stylus).

8–2. How can all four equations in Ex. 8–1 be used to provide cross checks on the calculated values for the input coordinates?

8–3. Develop a program to allow a user to position objects on the screen using a locator device. An object menu of geometric shapes is to be presented to a user who is to select an object and a placement position. The program should allow any number of objects to be positioned until a "terminate" signal is given.

8–4. Extend the program of Ex. 8–3 so that selected objects can be scaled and rotated before positioning. The transformation choices and transformation parameters are to be presented to the user as menu options.

8–5. Write a program that allows a user to interactively sketch pictures using a stroke device.

8–6. Discuss the methods that could be employed in a pattern-recognition procedure to match input characters against a stored library of shapes.

8–7. Write a routine that allows a user to select a numeric value from a displayed linear scale. Values are to be chosen by positioning the cursor (or other symbol) along the scale line. Both the position on the line and the scalar value selected are to be displayed after a choice has been made.

8–8. Write the routine in Ex. 8–7 so that angles (in degrees) can be selected from a circular scale. Mark the position selected with a radial arrow.

8–9. Design a program that will produce a picture as a set of line segments drawn between specified endpoints. The coordinates of the individual line segments are to be selected with a locator device.

8–10. Extend the program of Ex. 8–9 so that a gravity field is used to aid the designer in joining line segments.

8–11. Write a procedure to allow a user to design a pattern with a locator device so that all lines are constrained to be horizontal or vertical.

8–12. Develop a program that displays a grid pattern and rounds input screen positions to grid intersections. The program can be used to design patterns or position objects, using a locator device to input positions. Once a picture is complete, the grid pattern can be removed before transferring the picture to a hardcopy device.

8–13. Write a routine that allows a designer to create a picture by sketching straight lines with a rubber-band method.

8–14. Extend the program of Ex. 8–13 so that designers can sketch either straight lines or circular arcs.

8–15. Write a program that allows a user to design a picture from a menu of basic shapes by dragging each selected shape into position with a pick device.

8–16. Design an implementation of the input commands for request mode.

8–17. Design an implementation of the sample mode input commands.

8–18. Design an implementation of the input commands for event mode.

8–19. Combine the implementations of Exs. 8–16, 8–17, and 8–18 into one package.

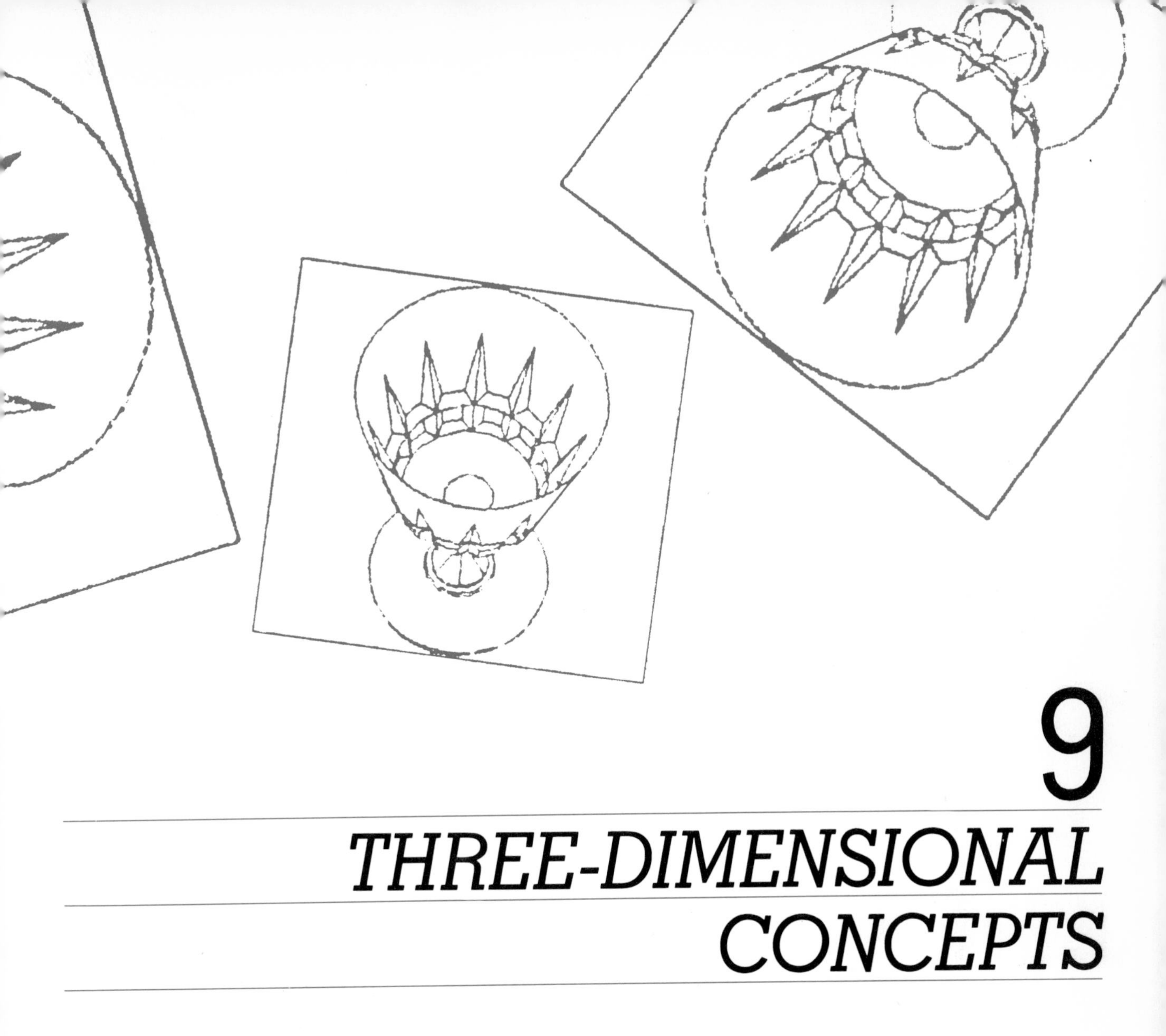

9 THREE-DIMENSIONAL CONCEPTS

We can classify three-dimensional graphics applications according to whether we are dealing with representations of existing ("real") objects or whether we are designing new shapes. A representation for an existing three-dimensional object is an approximate description that we use to construct the display. For example, we could describe a solid object as a framework of lines or as a set of flat surfaces (Fig. 9–1). In another application, we might need to specify a representation that includes curved lines and surfaces, as in Fig. 9–2. In computer-aided design applications, on the other hand, the goal is to create objects by constructing and manipulating patterns to form new three-dimensional shapes. Automobile and aircraft bodies are fashioned by rearranging surface patterns until certain design criteria are satisfied. For either type of application, descriptions of solid objects are specified in a three-dimensional world coordinate system and mapped onto the two-dimensional reference of a video monitor or other output device.

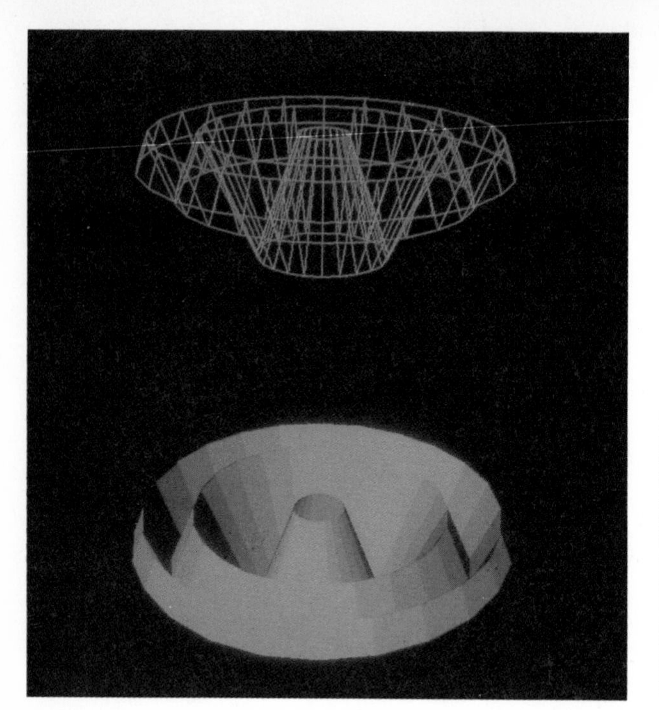

FIGURE 9–1
A three-dimensional object can be represented as a line framework or as a set of plane surfaces. Courtesy Aydin Controls, a division of Aydin Corp.

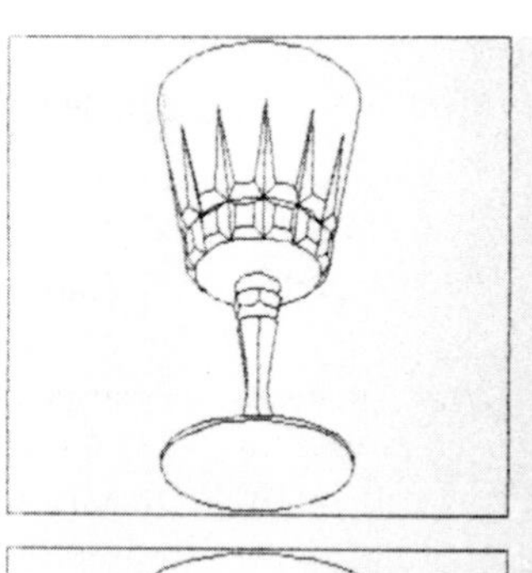

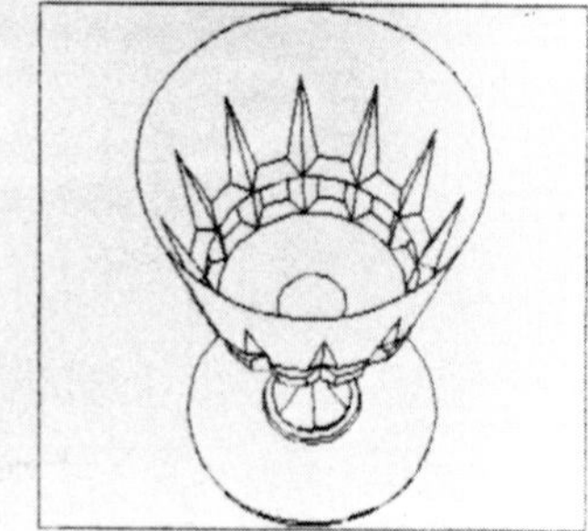

FIGURE 9–2
Representation of a solid object as a combination of plane and curved surfaces. Courtesy Selanar Corp.

9–1 Three-Dimensional Coordinate Systems

Figure 9–3 (a) shows the conventional orientation of coordinate axes in a three-dimensional Cartesian reference system. This is called a right-handed system because the right-hand thumb points in the positive z direction when we imagine grasping the z axis with the fingers curling from the positive x axis to the positive

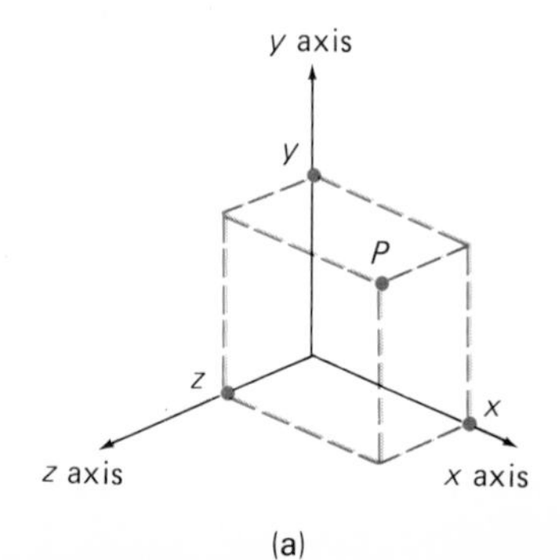

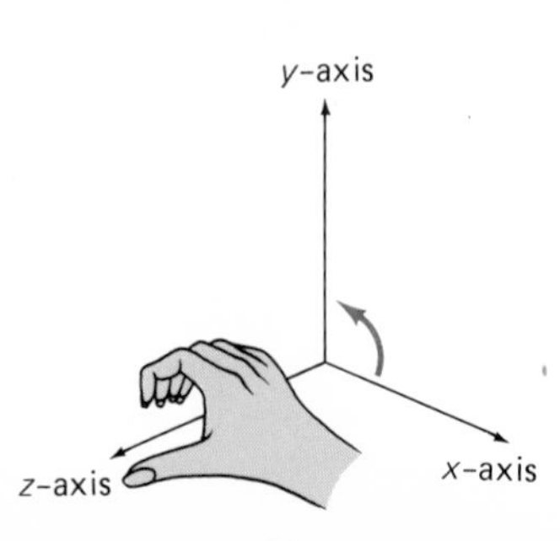

FIGURE 9–3
Coordinate representation of a point P at position (x, y, z) in a right-handed reference system.

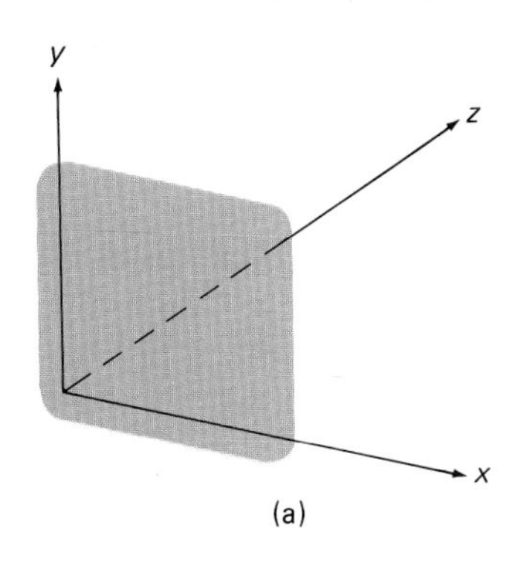

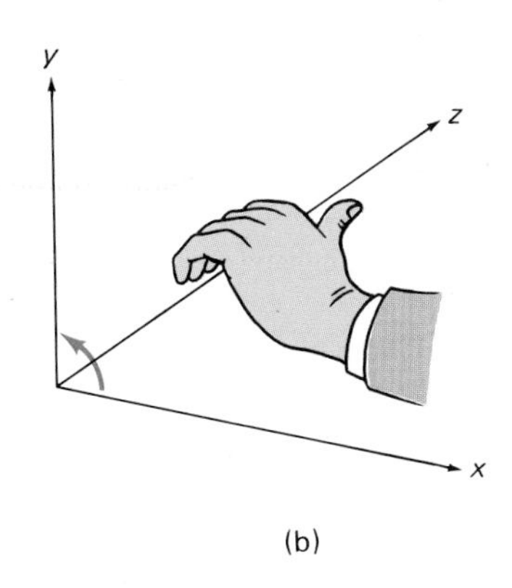

FIGURE 9–4
Left-handed coordinate axes superimposed on the surface of a display screen.

y axis (through 90°), as illustrated in Fig. 9–3 (b). We will assume that all world coordinate descriptions of objects are to be stated relative to a right-handed reference system.

Another possible arrangement of coordinate axes is the left-handed system shown in Fig. 9–4. For this system, the fingers of the left hand curl from the positive x axis to the positive y axis through 90° when we imagine grasping the z axis so that the left thumb points in the positive z direction. This orientation of axes is convenient in many graphics applications. With the coordinate origin of a left-handed system placed in the lower left screen corner, the first quadrant of the xy plane maps directly to screen coordinates, and positive z values indicate positions behind the screen as in Fig 9–4 (a). This coordinate reference can be used conveniently to specify device coordinates, with the screen in the $z = 0$ plane. Larger values along the positive z axis are then interpreted as being farther from the viewer.

Other coordinate references also can be useful in describing three-dimensional scenes. For some objects, it may be convenient to define coordinate positions using a spherical, cylindrical, or other system that takes advantage of any naturally occurring symmetries. However, for graphics packages that allow only Cartesian world coordinate representations, any non-Cartesian coordinate descriptions must first be converted by the user to corresponding Cartesian coordinates.

Like two-dimensional descriptions, three-dimensional world coordinate descriptions are converted to normalized device coordinates and then displayed in the coordinate reference for a particular output device. Unlike two-dimensional descriptions, a three-dimensional object can be viewed from any position. We can think of this as analogous to photographing the object, where the photographer walks around and chooses a position and orientation for the camera.

9–2 Three-Dimensional Display Techniques

Representations of a solid object on a viewing surface usually contain depth information so that a veiwer can easily identify, for the particular view selected, which is the back and which is the front of the object. Figure 9–5 illustrates the ambiguity that can result when an object is displayed without depth information. There are several techniques that can be used to include depth information in the two-dimensional representation of solid objects; the choice of display technique depends on the requirements of a particular application.

Parallel Projection

A solid object can be represented in two dimensions by projecting points on the object surface along parallel lines onto a plane viewing surface. By selecting different viewing positions, we can project visible points onto the viewing surface to obtain different two-dimensional views of the object, as in Fig. 9–6. In a parallel projection, parallel lines on the surface of the object project into parallel lines on the two-dimensional viewing plane. This technique is used in engineering and architectural drawings to represent the object with a set of views that maintain relative proportions of the object. The appearance of the solid object can then be reconstructed from the major views.

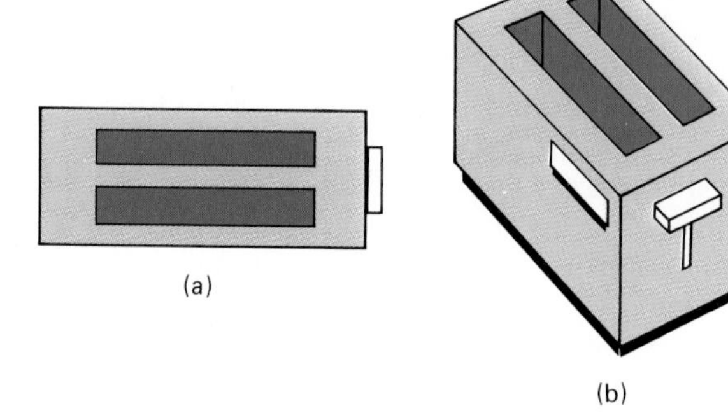

FIGURE 9–6
Two parallel projection views of an object, used to show relative proportions. Lines on the object that are parallel remain parallel in the projection views.

Perspective Projection

Another method for representing depth on a two-dimensional viewing surface is perspective projection. Instead of projecting points along parallel lines, perspective projections change the sizes of objects so that those farther from the viewing position are displayed smaller than those nearer the viewing position. Parallel lines on the surface of an object are now projected into lines that tend to converge. Objects displayed as perspective projections appear more natural, since this is the way that the eye and a camera lens form images. In Fig. 9–7 a building and its interior are represented in perspective, with the back lines shorter than the front lines.

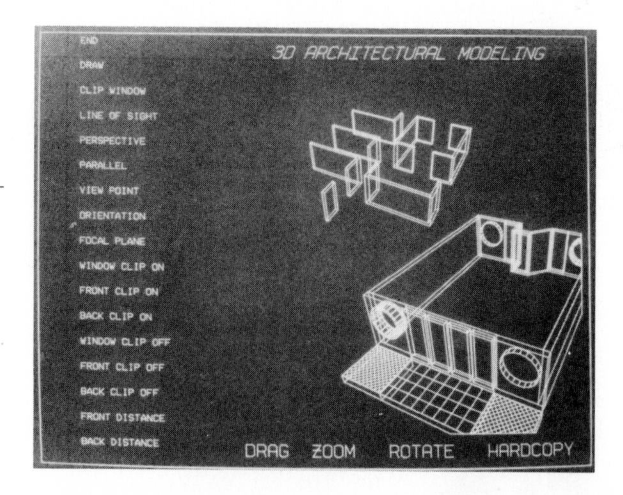

FIGURE 9–7
A perspective layout of a building. Parallel lines on the building are projected onto the screen as converging lines. Courtesy Precision Visuals, Inc., Boulder, Colorado.

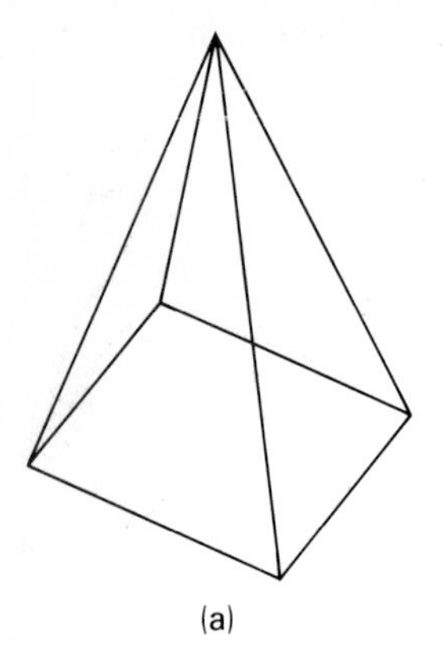

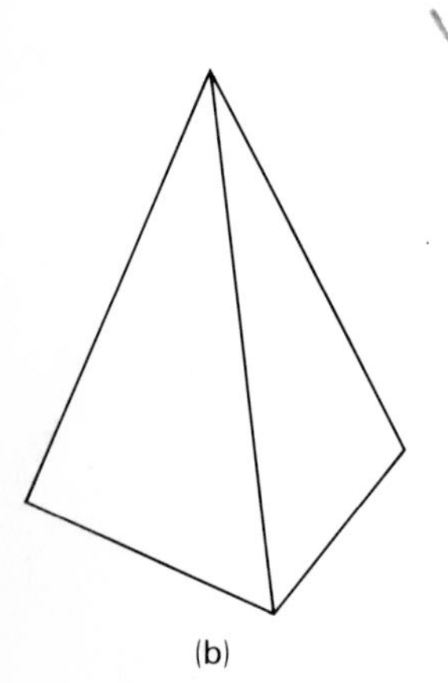

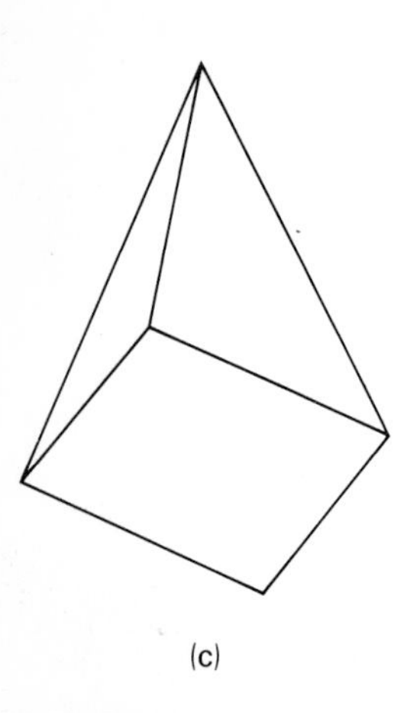

FIGURE 9–5
The wire-frame representation of a pyramid (a) contains no depth information to indicate whether the viewing direction is from above (b) or below the base (c).

Intensity Cuing

A simple method for indicating depth in a display is to vary the intensity of lines according to their distance from the viewing position. Figure 9–8 shows an object with front lines highlighted and drawn wider than back lines. More sophisticated intensity-cuing techniques employ a range of intensities, so that the brightness gradually decreases for lines farther from the viewer.

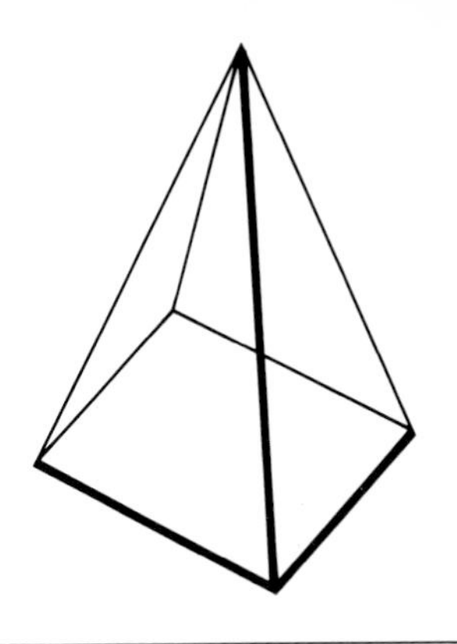

FIGURE 9–8
Highlighting, one form of intensity cuing, can be used to mark the front edges of objects.

Hidden-Line Removal

This technique is useful when objects are defined as a set of lines representing the edges of the surfaces of the objects. For a selected view, any lines that are hidden by front surfaces are removed before the object is displayed. Figure 9–2 displays an object with hidden lines removed. Although extensive computation may be required to locate all hidden line segments in a scene, this method gives a more realistic display of depth information in line drawings. For some applications it is helpful to replace hidden edges of objects with dashed lines (Fig. 9–9) rather than remove them all together. In this way more information about back surfaces is provided.

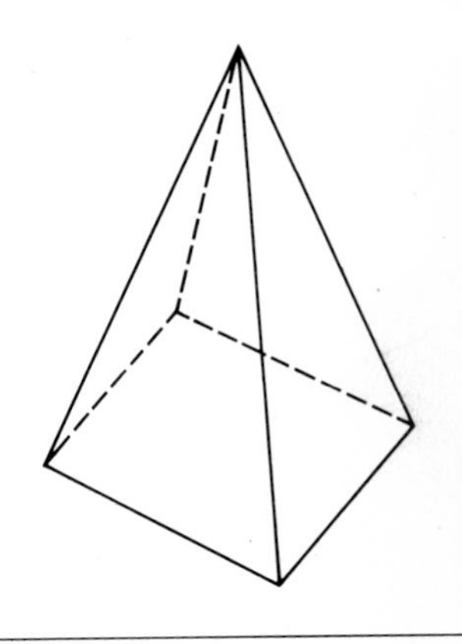

FIGURE 9–9
Displaying hidden edges as dashed lines provides more information about the back surfaces.

Hidden-Surface Removal and Shading

When objects are defined as surfaces filled by color or shading patterns, hidden-surface removal techniques are used to take out any back surfaces that are hidden by visible surfaces, as in the display of Fig. 9–1. Removing hidden surfaces is generally a complicated and time-consuming process, but it provides a highly realistic method for displaying objects. Added realism is attained by combining hidden-surface removal with perspective projections and by including shadows and surface texture. Figure 9–10 is an example of a three-dimensional scene modeled in this way.

FIGURE 9–10
Scene generated using hidden-surface removal, perspective, shading, and surface texture techniques. Courtesy Evans & Sutherland and Daimler-Benz AG, Stuttgart, West Germany.

Exploded and Cutaway Views

Hidden-line and hidden-surface removal techniques can be combined with exploded and cutaway views to give additional information about the structure of three-dimensional objects. The exploded line drawing of Fig. 9–11 clearly shows the structure of component parts, including those that are invisible in the assembled view. An alternative to exploding an object into its component parts is the cutaway view, which removes part of the visible surfaces to show internal structure. Figure 9–12 illustrates the cutaway technique.

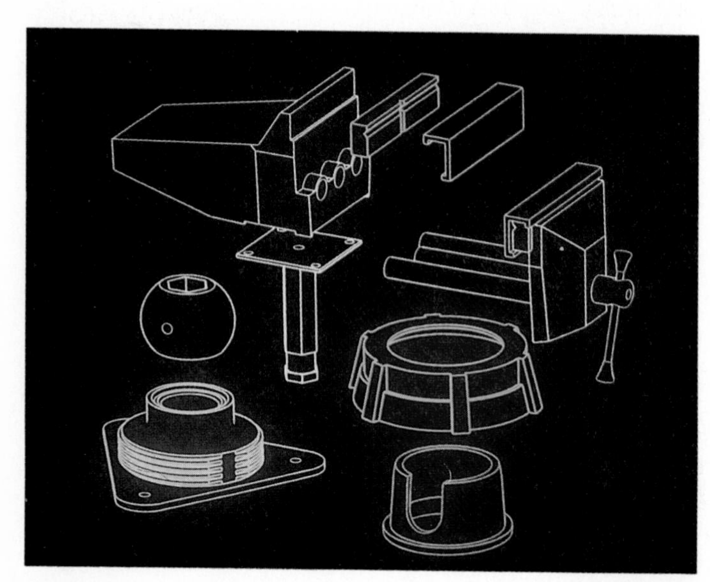

(a)

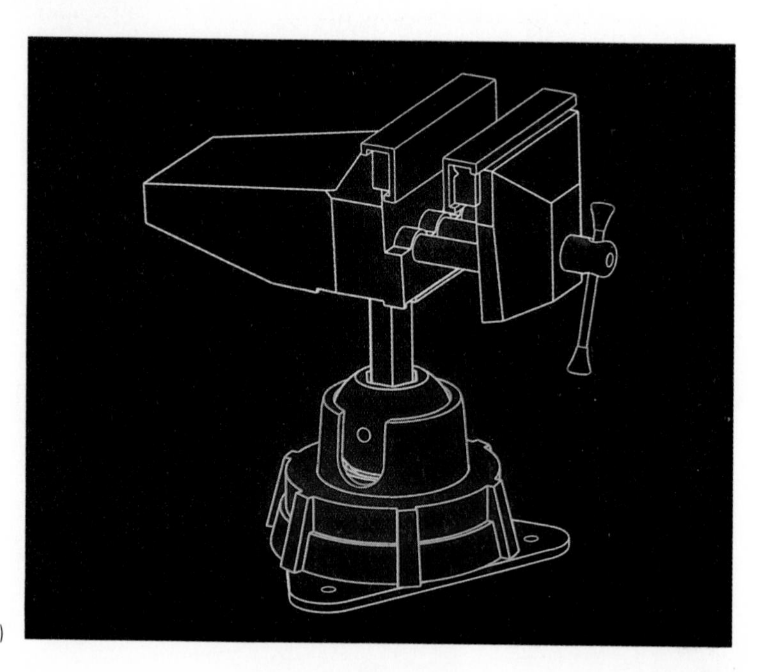

(b)

FIGURE 9–11
Exploded view (a) of an object (b) composed of several parts. Courtesy Evans & Sutherland and Shape Data, Ltd.

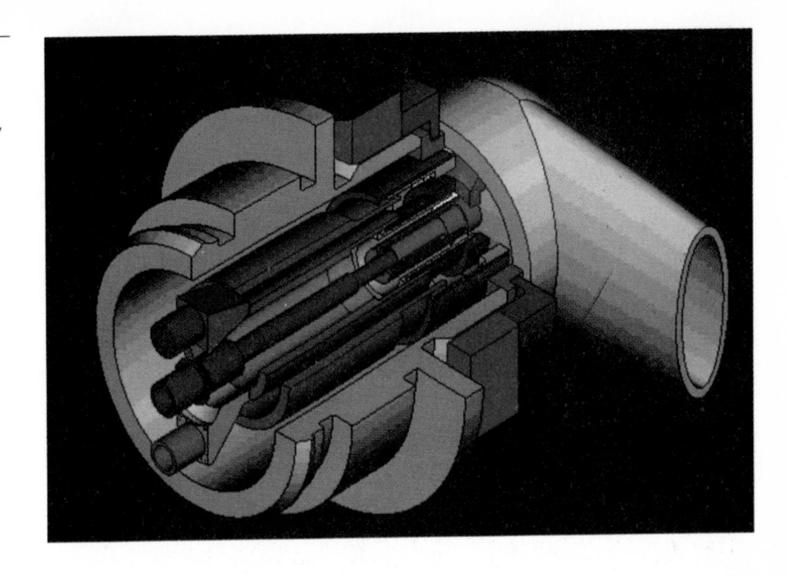

FIGURE 9–12
Cutaway view of an object showing the structure of internal components. Courtesy Chromatics, Inc.

Three-Dimensional and Stereoscopic Views

Instead of presenting a two-dimensional projection to a viewer, three-dimensional and stereoscopic views are sometimes used to give more depth information. As we have seen in Chapter 2, video monitors can be adapted to display such views.

A method for displaying true three-dimensional views is to reflect a CRT image off a vibrating, flexible mirror. The vibrations of the mirror are synchronized with the display of the scene on the CRT. As the mirror vibrates, the focal length varies so that each point in the scene is reflected to a position corresponding to its depth.

Stereoscopic devices present two simultaneous two-dimensional displays to a viewer, one for the left eye and the other for the right eye. The two views are generated by selecting viewing positions that correspond to those for each eye. These two views can be displayed on a single monitor using shutters or polarized filters to present each view to the correct eye. Another method for creating stereoscopic views is to use two small CRTs on a headset, so that each eye of the viewer sees only the display from one of the CRTs.

9–3 Three-Dimensional Graphics Packages

Design of three-dimensional packages requires some considerations that are not necessary with two-dimensional packages. A significant difference between the two packages is that a three-dimensional package must include methods for mapping three-dimensional descriptions onto a flat viewing surface. We need to consider implementation procedures for selecting different views and for using different projection techniques. We also need to consider how surfaces of solid objects are to be modeled, how hidden surfaces can be removed, how transformations of objects are performed in space, and how to describe the additional spatial properties introduced by three dimensions. Later chapters explore each of these considerations in detail.

Other considerations for three-dimensional packages are straightforward extensions from two-dimensional methods. World coordinate descriptions are extended to three dimensions, and users are provided with output and input routines accessed with specifications such as

```
polyline_3 (n, x, y, z)
fill_area_3 (n, x, y, z)
text_3 (x, y, z, string)
get_locator_3 (x, y, z)
```

Algorithms for displaying three-dimensional output primitives on a flat viewing surface are basically the same as for two-dimensional packages, since these procedures are applied after the projection transformations have been accomplished.

Three-dimensional input can be entered by a user with a three-dimensional device, such as the three-dimensional sonic digitizer, but most often two-dimensional input devices, such as a light pen or mouse, are used. When two-dimensional devices are used to input three-dimensional descriptions, coordinate input must be transformed from the device coordinates to three-dimensional world coordinates for processing. For example, if a three-dimensional position is to be input by pointing a light pen at a screen, the selected screen coordinates must be mapped through the inverse transformation back to world coordinates to obtain the "actual" position selected.

Two-dimensional attribute functions that are independent of geometric considerations can be used for both two-dimensional and three-dimensional descriptions. No new attribute functions need be defined for colors, line styles, marker attributes, or text fonts. However, a graphics package with three-dimensional capabilities would need to extend attribute procedures for orienting character strings and for filling defined areas. Text attribute routines associated with the up vector require expansion to include z-coordinate data so that strings can be given any spatial orientation. Area-filling routines, such as those for positioning the pattern reference point and for mapping patterns onto a fill area, would be expanded to accommodate various orientations of the fill-area plane and the pattern plane. Except for transformations, all two-dimensional segment operations can also be carried over to a three-dimensional package.

REFERENCES

Applications of three-dimensional and stereoscopic methods are discussed by Grotch (1983) and by Roese and McCleary (1979).

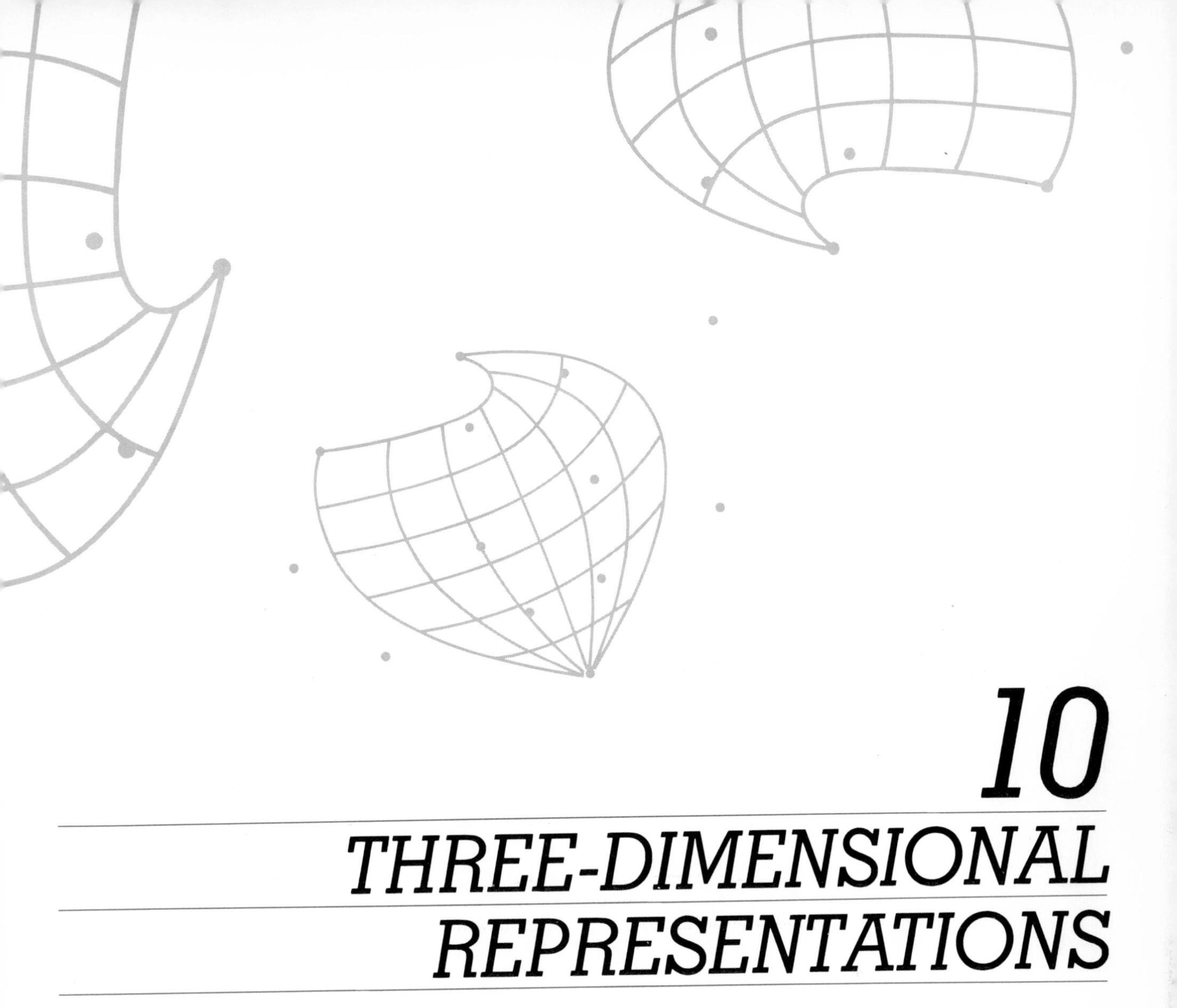

10 THREE-DIMENSIONAL REPRESENTATIONS

Solid objects can be represented for graphics display in a number of ways. In some cases, we may be able to give a precise description of an object in terms of well-defined surfaces. A cube is constructed with six plane faces; a cylinder is a combination of one curved surface and two plane surfaces; and a sphere is formed with a single curved surface. When more complex objects are to be represented, approximation methods are used to define the objects. In design applications, the exact appearance of an object is often not known in advance, and the designer wants to be able to experiment with surface features.

One way for a user to represent an arbitrary object to a graphics package is to approximate its shape with a set of plane, polygon faces. Surfaces of solids could also be described using parametric curve equations or fractal representations. For some applications, definitions of solids are given as construction methods that build the solids from simpler shapes. One technique for building a solid is to sweep a two-dimensional pattern through some region of space, creating a volume. Another construction technique uses solid-geometry methods to combine a basic set of three-dimensional objects, such as blocks, pyramids, and cylinders. Finally, solids can be represented using octree encoding, which defines the properties of each elementary volume of the three-dimensional space containing the solids.

10–1 Polygon Surfaces

Any three-dimensional object can be represented as a set of plane, polygon surfaces. For some objects, such as a polyhedron, this precisely defines the surface features. In other cases, a polygon representation provides an approximate description of the object. Figure 10–1 displays a solid object modeled as a mesh of polygon surfaces. This polygon-mesh representation can be displayed quickly to give a general indication of the object's structure, and the approximation can be improved by dividing object surfaces into smaller polygon faces.

Each polygon in an object can be specified to a graphics package using line or area-fill commands to define the vertex coordinates. CAD packages often allow users to enter vertex positions along polygon boundaries with interactive methods. These vertices can represent the result of digitizing a drawing, or they can be input by a designer who is creating a new shape.

Polygon Tables

Once each polygon surface has been defined by a user, the graphics package organizes the input data into tables that are to be used in the processing and display of the surfaces. The data tables contain the geometric and attribute properties of the object, organized to facilitate processing. Geometric data tables contain boundary coordinates and parameters to identify the spatial orientation of the polygon surfaces. Attribute information for the object includes designations for any color and shading patterns that are to be applied to the surfaces.

Geometric data can be organized in several ways. A convenient method for storing coordinate information is to create three lists: a vertex table, an edge table,

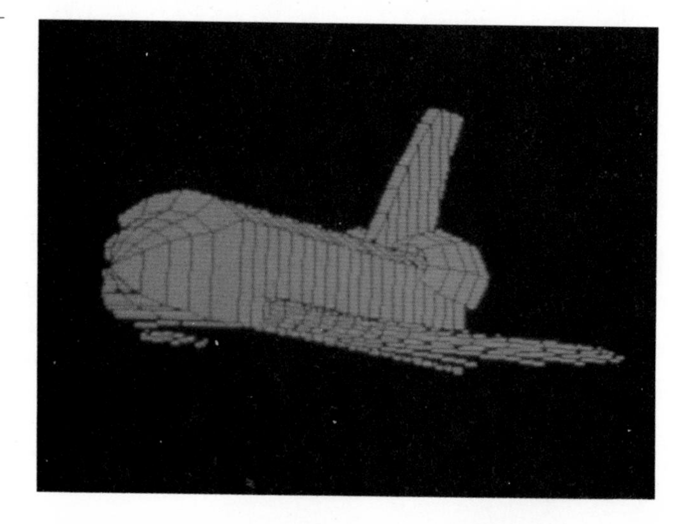

FIGURE 10–1
Solid object represented as a set of polygon surfaces. Courtesy Megatek Corp.

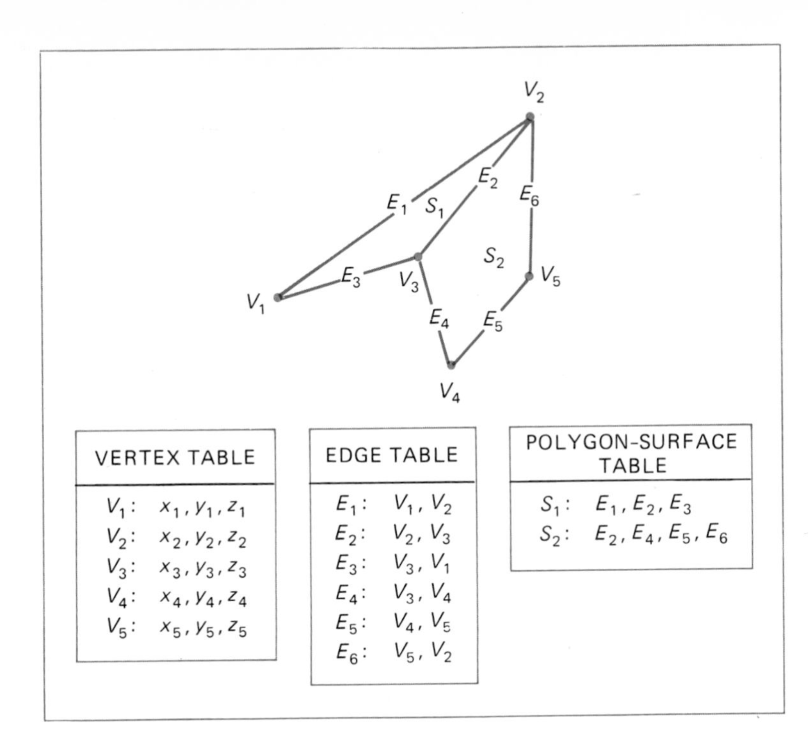

VERTEX TABLE
V_1: x_1, y_1, z_1
V_2: x_2, y_2, z_2
V_3: x_3, y_3, z_3
V_4: x_4, y_4, z_4
V_5: x_5, y_5, z_5

EDGE TABLE
E_1: V_1, V_2
E_2: V_2, V_3
E_3: V_3, V_1
E_4: V_3, V_4
E_5: V_4, V_5
E_6: V_5, V_2

POLYGON-SURFACE TABLE
S_1: E_1, E_2, E_3
S_2: E_2, E_4, E_5, E_6

FIGURE 10–2
Geometric data tables for a three-dimensional object represented by two plane polygon surfaces, formed with six edges and five vertices.

and a polygon table. Coordinate values for each vertex in the object are stored in the vertex table. The edge table lists the endpoint vertices defining each edge. Each polygon is defined in the polygon table as a list of component edges. This scheme is illustrated in Fig. 10–2 for an object consisting of two polygon surfaces. The polygon table contains pointers back to the edge table, which, in turn, contains pointers back to the coordinate values in the vertex table. When more than one object is to be represented in a scene, each object can be identified in an object table by the set of polygon surfaces in the polygon table defining that object.

Listing the geometric data in three tables, as in Fig. 10–2, provides a convenient reference to the components (vertices, edges, and polygons) of an object. Also, the object can be displayed efficiently by using data from the edge table to draw the component lines. Without the edge table, the object would be displayed using data in the polygon table, and this means that some lines would be drawn twice. If the vertex table were also absent, the polygon table would have to list explicit coordinates for each vertex in each polygon. All information regarding vertices and edges would have to be reconstructed from the polygon table, and coordinate positions would be duplicated for vertices on the border of two or more polygons.

Additional information could be incorporated into the data tables of Fig. 10–2 for faster information extraction. For instance, we could expand the edge table to include pointers into the polygon table so that common edges between polygons could be identified more rapidly (Fig. 10–3). This is particularly useful when shading models are applied to the surfaces, with the shading patterns varying smoothly from one polygon to the next. We could also expand the vertex table by cross-referencing vertices to corresponding edges.

Since the geometric data tables may contain extensive listings of vertices and edges for complex objects, it is important that the data be checked for consistency and completeness. When vertex, line, and polygon definitions are specified, it is

FIGURE 10–3
Expanded edge table for the surface of Fig. 10–2, with pointers to the polygon table.

E_1:	V_1, V_2, S_1
E_2:	V_2, V_3, S_1, S_2
E_3:	V_3, V_1, S_1
E_4:	V_3, V_4, S_2
E_5:	V_4, V_5, S_2
E_6:	V_5, V_2, S_2

possible that certain input errors could be made that would distort the display of the object. The more information included in the data tables, the easier it is to check. Some of the tests that could be performed by a graphics package are (1) that every vertex is listed as an endpoint for at least two lines, (2) that every line is part of at least one polygon, (3) that every polygon is closed, (4) that each polygon has at least one shared edge, and that (5) if the edge table contains pointers to polygons, every edge referenced by a polygon pointer has a reciprocal pointer back to the polygon.

Plane Equations

Parameters specifying the spatial orientation of each polygon are obtained from the vertex coordinate values and the equations that define the polygon planes. These plane parameters are used in viewing transformations, shading models, and hidden-surface algorithms that determine which lines and planes overlap along the line of sight.

The equation for a plane surface can be expressed in the form

$$Ax + By + Cz + D = 0 \tag{10–1}$$

where (x, y, z) is any point on the plane. The coefficients A, B, C, D are constants that can be calculated using the coordinate values of three noncollinear points in the plane. Typically, we use the coordinates of three successive vertices on a polygon boundary to find values for these coefficients. Denoting coordinates for three vertices of a polygon as (x_1, y_1, z_1), (x_2, y_2, z_2), and (x_3, y_3, z_3), we can solve the following set of simultaneous plane equations for the ratios A/D, B/D, and C/D:

$$(A/D)x_i + (B/D)y_i + (C/D)z_i = -1; \quad i = 1, 2, 3 \tag{10–2}$$

Using a solution method such as Cramer's rule, we can write the solution for the plane parameters in determinant form:

$$\begin{aligned} A &= \begin{vmatrix} 1 & y_1 & z_1 \\ 1 & y_2 & z_2 \\ 1 & y_3 & z_3 \end{vmatrix} \\ B &= \begin{vmatrix} x_1 & 1 & z_1 \\ x_2 & 1 & z_2 \\ x_3 & 1 & z_3 \end{vmatrix} \\ C &= \begin{vmatrix} x_1 & y_1 & 1 \\ x_2 & y_2 & 1 \\ x_3 & y_3 & 1 \end{vmatrix} \\ D &= - \begin{vmatrix} x_1 & y_1 & z_1 \\ x_2 & y_2 & z_2 \\ x_3 & y_3 & z_3 \end{vmatrix} \end{aligned} \tag{10–3}$$

We can expand the determinants and write the calculations for the plane coefficients in the explicit form:

$$\begin{aligned} A &= y_1 (z_2 - z_3) + y_2 (z_3 - z_1) + y_3 (z_1 - z_2) \\ B &= z_1 (x_2 - x_3) + z_2 (x_3 - x_1) + z_3 (x_1 - x_2) \\ C &= x_1 (y_2 - y_3) + x_2 (y_3 - y_1) + x_3 (y_1 - y_2) \\ D &= -x_1 (y_2 z_3 - y_3 z_2) - x_2 (y_3 z_1 - y_1 z_3) - x_3 (y_1 z_2 - y_2 z_1) \end{aligned} \tag{10–4}$$

Values for A, B, C, and D are stored in the data structure containing the coordinate and attribute information about the polygon defined in this plane.

Orientation of a plane surface in space is specified by the normal vector to the plane, as shown in Fig. 10–4. This three-dimensional normal vector has Cartesian coordinates (A, B, C).

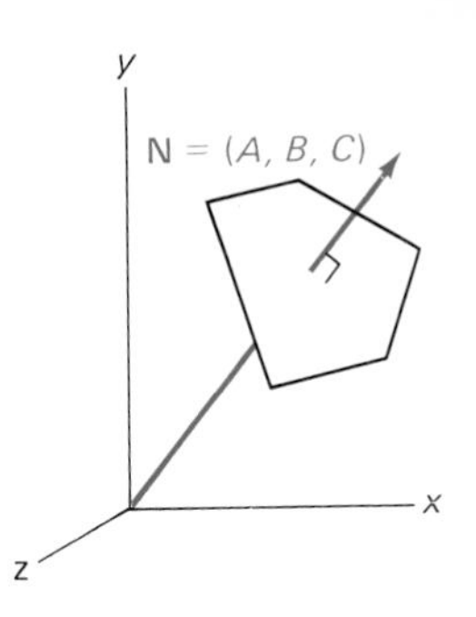

FIGURE 10–4
The vector ***N***, normal to the surface of a plane described by the equation $Ax + By + Cz + D = 0$, has Cartesian coordinates (A, B, C).

Since we are often dealing with polygon surfaces that enclose an object interior, we need to distinguish between the two sides of the surface. The side of the plane that faces the object interior is called the "inside," and the visible or outward side is the "outside." If vertices are specified in a counterclockwise direction when viewing the outer side of the plane in a right-handed coordinate system, the direction of the normal vector will be from inside to outside. This is demonstrated for one plane of a unit cube in Fig. 10–5.

To determine the components of the normal vector for the shaded surface shown in Fig. 10–5, we select three of the four vertices along the boundary of the polygon. These points are selected in a counterclockwise direction as we view from outside the cube toward the origin. The coordinates for these vertices, in the order selected, are used in Eqs. 10–4 to obtain the plane coefficients: $A = 1$, $B = 0$, $C = 0$, $D = -1$. The normal vector for this plane is in the direction of the positive x axis.

Plane equations are used also to identify inside and outside points. Any point (x, y, z) on the outside of a plane satisfies the inequality

$$Ax + By + Cz + D > 0 \tag{10–5}$$

Similarly, any points on the inside of the plane produce a negative value for the expression $Ax + By + Cz + D$. For the shaded surface in Fig. 10–5, any point outside the plane satisfies the inequality $x - 1 > 0$, while any point inside the plane has an x-coordinate value less than 1.

FIGURE 10–5
The shaded polygon surface of the unit cube has plane equation $x - 1 = 0$ and normal vector $(1, 0, 0)$.

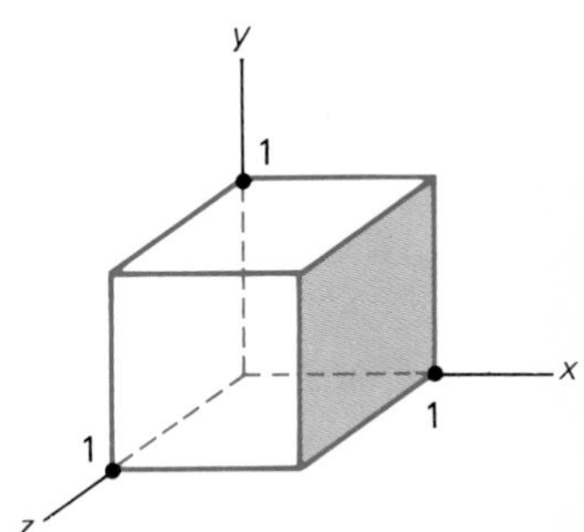

10–2 Curved Surfaces

Three-dimensional displays of curved surfaces can be generated from an input set of mathematical functions defining the surfaces or from a set of user-specified data points. When curve functions are specified, a package can use the defining equations to locate and plot pixel positions along the curve paths, much the same as with two-dimensional curves. An example of the kind of surfaces that can be generated from a functional definition is given in Fig. 10–6. From a set of input data points, a package determines the functional description of the curve that best fits the data points according to the constraints of the application. Figure 10–7 shows an object whose curved surfaces can be defined by an input set of data points.

We can represent a three-dimensional curved line in an analytical form with the pair of functions

$$y = f(x), \qquad z = g(x) \tag{10–6}$$

with coordinate x selected as the independent variable. Values for the dependent variables y and z are then determined from Eqs. 10–6 as we step through values for x from one line endpoint to the other endpoint. This representation has some disadvantages. If we want a smooth plot, we must change the independent variable whenever the first derivative (slope) of either $f(x)$ or $g(x)$ becomes greater than 1. This means that we must continually check values of the derivatives, which may become infinite at some points. Also, Eqs. 10–6 provide an awkward format for

FIGURE 10–6
Three-dimensional surfaces can be generated from a functional definition. Courtesy Los Alamos National Laboratory.

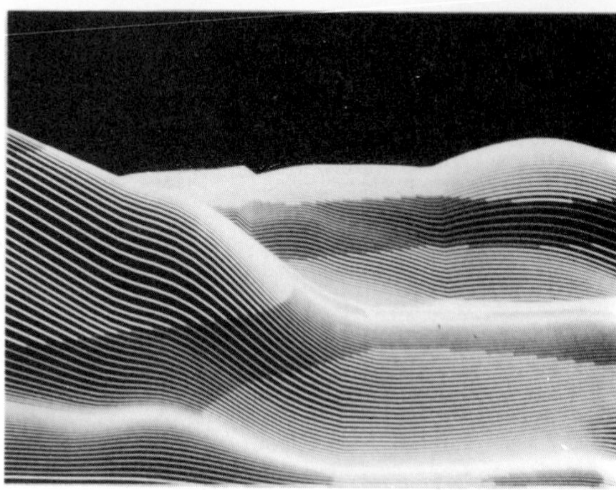

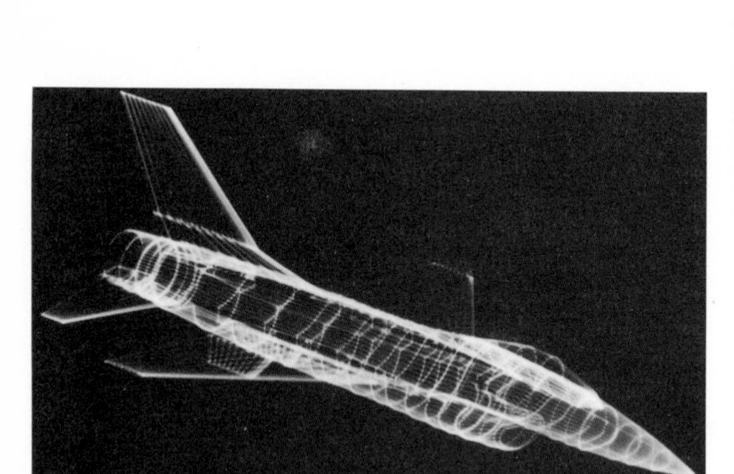

FIGURE 10–7
Line drawing of an object with curved surfaces that can be generated from a set of data points. Courtesy Evans & Sutherland.

representing multiple-valued functions. A more convenient representation of curves for graphics applications is in terms of parametric equations.

Parametric Equations

By introducing a fourth parameter, u, into the coordinate description of a curve, we can express each of the three Cartesian coordinates in parametric form. Any point on the curve can then be represented by the vector function

$$\mathbf{P}(u) = (x(u),\ y(u),\ z(u)) \tag{10–7}$$

Usually, the parametric equations are set up so that parameter u is defined in the range from 0 to 1. For example, a circle in the xy plane with center at the coordinate origin could be defined in parametric form as

$$x(u) = r\cdot\cos(2\pi u), \quad y(u) = r\cdot\sin(2\pi u), \quad z(u) = 0$$

Other parametric forms are also possible for describing circles and circular arcs.

For an arbitrary curve, it may be difficult to devise a single set of parametric equations that completely defines the shape of the curve. But any curve can be approximated by using different sets of parametric functions over different parts of the curve. Usually these approximations are formed with polynomial functions. Such a piecewise construction of a curve must be carefully implemented to ensure that there is a smooth transition from one section of the curve to the next. The smoothness of a curve can be described in terms of curve **continuity** between the sections. Zero-order continuity means simply that the curves meet. First-order continuity means that the tangent lines (first derivatives) of two adjoining curve sections are the same at the joining point. Second-order continuity means that the curvatures (second derivatives) of the two curve sections are the same at the intersection. Figure 10–8 shows examples of the three orders of continuity.

Parametric equations for surfaces are formulated with two parameters, u and v. A coordinate position on a surface is then represented by the parametric vector function

$$\mathbf{P}(u, v) = (x(u, v), y(u, v), z(u, v)) \tag{10–8}$$

The equations for coordinates x, y, and z are often arranged so that parameters u and v are defined within the range 0 to 1. A spherical surface, for example, can be described with the equations

$$\begin{aligned} x(u, v) &= r \sin(\pi u) \cos(2\pi v) \\ y(u, v) &= r \sin(\pi u) \sin(2\pi v) \\ z(u, v) &= r \cos(\pi u) \end{aligned} \tag{10–9}$$

where r is the radius of the sphere. Parameter u describes lines of constant latitude over the surface, while parameter v describes lines of constant longitude. By keeping one of these parameters fixed while varying the other over any values within the range 0 to 1, we could plot latitude and longitude lines for any spherical section (Fig. 10–9).

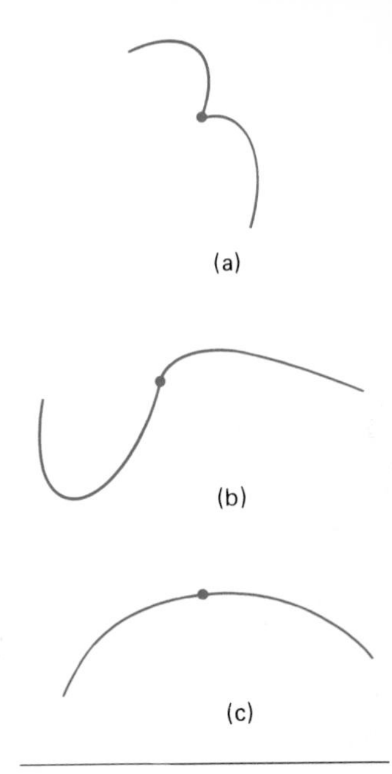

FIGURE 10–8
Piecewise specification of a curve by joining two curve segments with varying orders of continuity: (a) zero-order continuity only, (b) first-order continuity, and (c) second-order continuity.

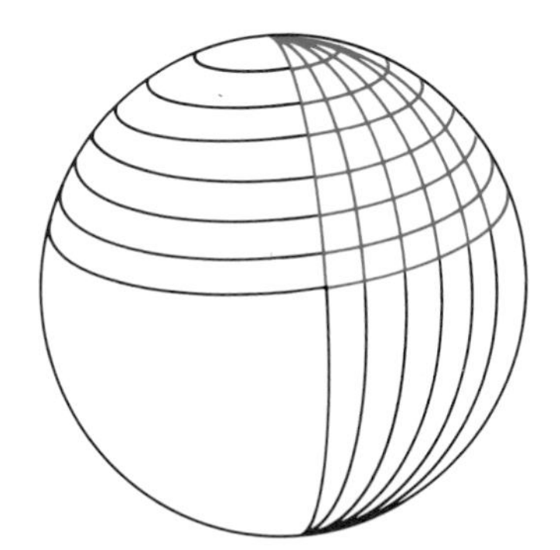
FIGURE 10–9
Section of a spherical surface described by lines of constant u and lines of constant v in Eqs. 10–9.

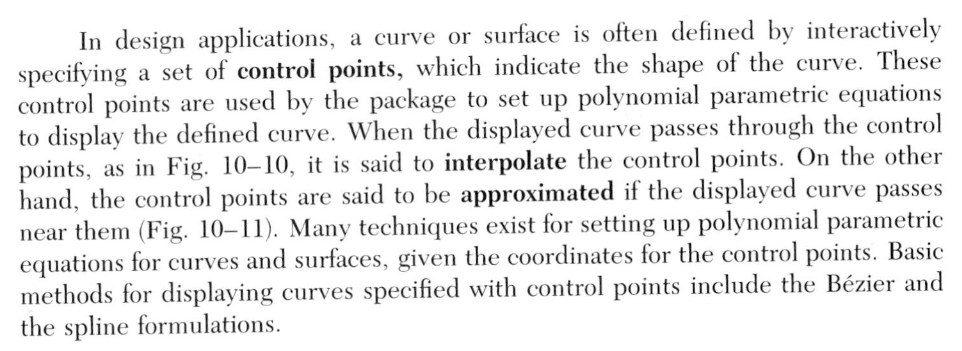
In design applications, a curve or surface is often defined by interactively specifying a set of **control points,** which indicate the shape of the curve. These control points are used by the package to set up polynomial parametric equations to display the defined curve. When the displayed curve passes through the control points, as in Fig. 10–10, it is said to **interpolate** the control points. On the other hand, the control points are said to be **approximated** if the displayed curve passes near them (Fig. 10–11). Many techniques exist for setting up polynomial parametric equations for curves and surfaces, given the coordinates for the control points. Basic methods for displaying curves specified with control points include the Bézier and the spline formulations.

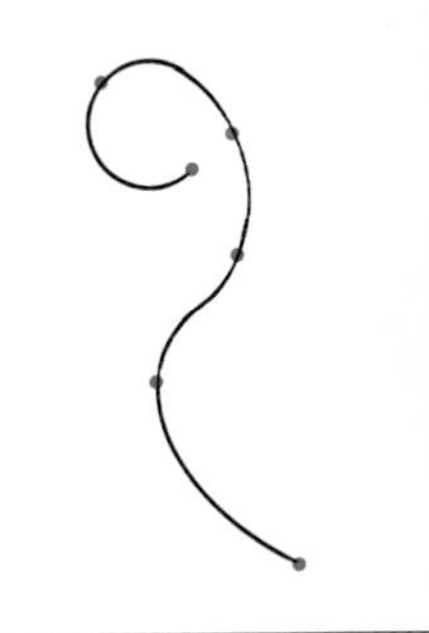
FIGURE 10–10
A set of six control points interpolated by a curve.

FIGURE 10–11
A set of four control points approximated by a curve.

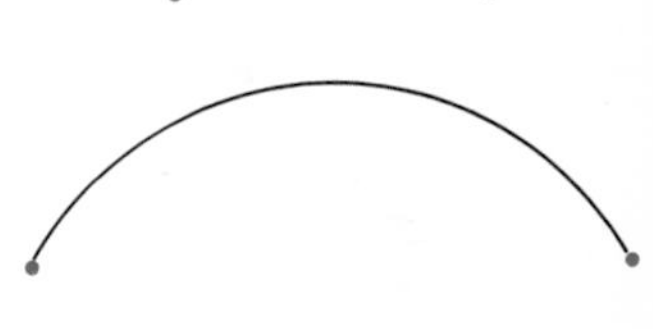

Bézier Curves

This method for constructing curves was developed by the French engineer Bézier for use in the design of Renault automobile bodies. For any input set of control points, an approximating curve is formed by adding a sequence of polynomial functions formed from the coordinates of the control points.

Suppose $n + 1$ control points are input and designated as the vectors $p_k = (x_k, y_k, z_k)$, for k varying from 0 to n. From these coordinate points, we calculate an approximating Bézier vector function $\boldsymbol{P}(u)$, which represents the three parametric equations for the curve that fits the input control points p_k. This Bézier coordinate function is calculated as

$$\boldsymbol{P}(u) = \sum_{k=0}^{n} p_k \, B_{k,n}(u) \tag{10–10}$$

Each $B_{k,n}(u)$ is a polynomial function defined as

$$B_{k,n}(u) = C(n,k)u^k (1 - u)^{n-k} \tag{10–11}$$

and the $C(n,k)$ represent the binomial coefficients

$$C(n,k) = \frac{n!}{k!\,(n - k)!} \tag{10–12}$$

Equation 10–10 can be written in explicit form as a set of parametric equations for the individual curve coordinates:

$$\begin{aligned} x(u) &= \sum_{k=0}^{n} x_k B_{k,n}(u) \\ y(u) &= \sum_{k=0}^{n} y_k B_{k,n}(u) \\ z(u) &= \sum_{k=0}^{n} z_k B_{k,n}(u) \end{aligned} \tag{10–13}$$

The polynomials $B_{k,n}(u)$ are called **blending functions** because they blend the control points to form a composite function describing the curve. This composite function is a polynomial of degree one less than the number of control points used. Three points generate a parabola, four points a cubic curve, and so forth. Figure 10–12 demonstrates the appearance of some Bézier curves for various selections of control points in the xy plane ($z = 0$). An important property of any Bézier curve is that it lies within the **convex hull** (polygon boundary) of the control points (Fig. 10–13). This ensures that the curve smoothly follows the control points without erratic oscillations.

In Fig. 10–14 we show the shapes of the four Bézier polynomials used in the construction of a curve to fit four control points. The form of each blending function determines how the control points influence the shape of the curve for values of parameter u. At $u = 0$, the only nonzero blending function is $B_{0,3}$, which has the value 1. At $u = 1$, the only nonzero function is $B_{3,3}$, with a value of 1 at that point. Thus the Bézier curve will always pass through the endpoints p_0 and p_3. The other functions, $B_{1,3}$ and $B_{2,3}$, influence the shape of the curve at intermediate values of the parameter u, so that the resulting curve tends toward the points p_1 and p_2. Blending function $B_{1,3}$ is maximum at $u = 1/3$ and $B_{2,3}$ is maximum at $u = 2/3$.

Bézier curves are useful for a variety of design applications. Closed curves can be generated by specifying the first and last control points at the same position, as in the example shown in Fig. 10–15. Specifying multiple control points at a single position gives more weight to that position. In Fig. 10–16, a single coordinate position is input as two control points, and the resulting curve is pulled nearer to this position.

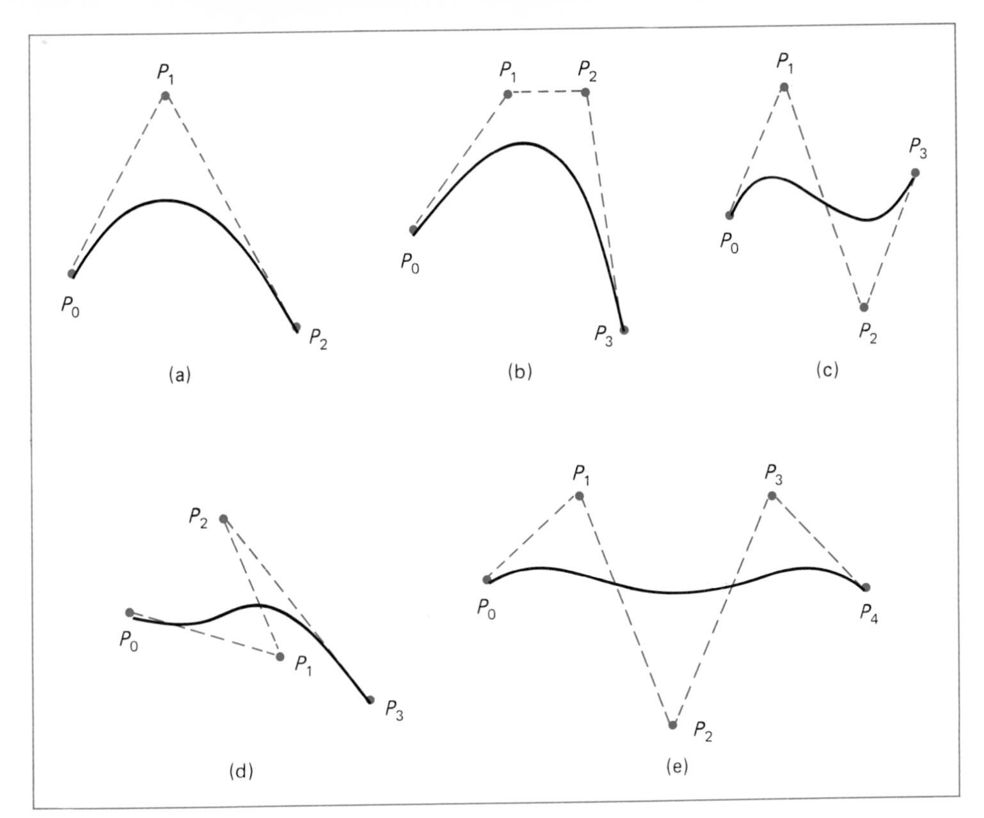

FIGURE 10–12
Examples of Bézier curves generated from three, four, and five control points in the *xy* plane. Dashed lines show the straight-line connection of the control points.

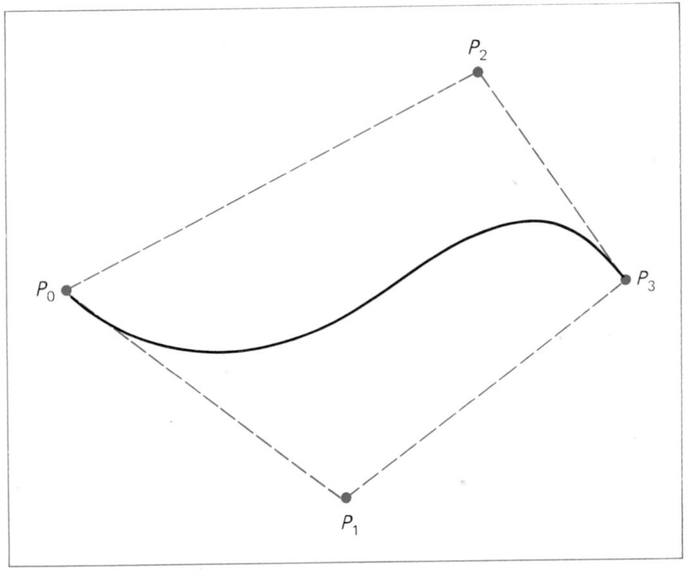

FIGURE 10–13
Convex hull (dashed line) of the control points for a Bézier curve.

In general, any number of control points could be specified for a Bézier curve, but this requires the calculation of polynomial functions of higher degree. When complicated curves are to be generated, they can be formed by piecing several Bézier sections of lower degree together. Piecing together smaller sections also gives a user better control over local variations that may be desired in a curve. Since Bézier curves pass through endpoints, it is easy to match curve sections (zero-

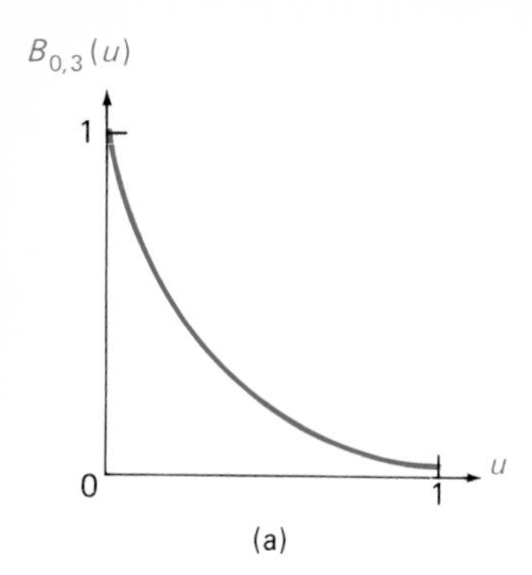

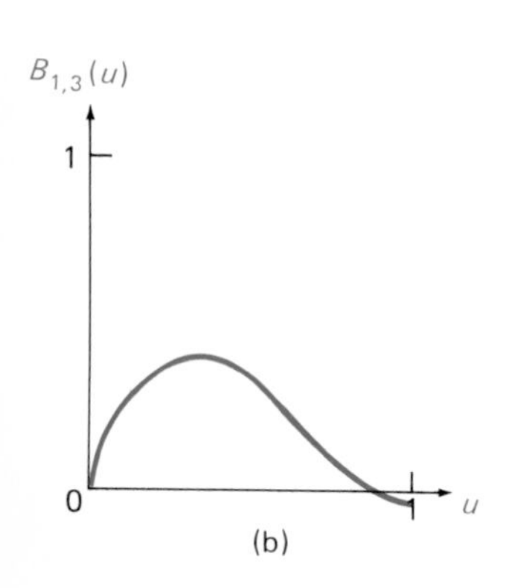

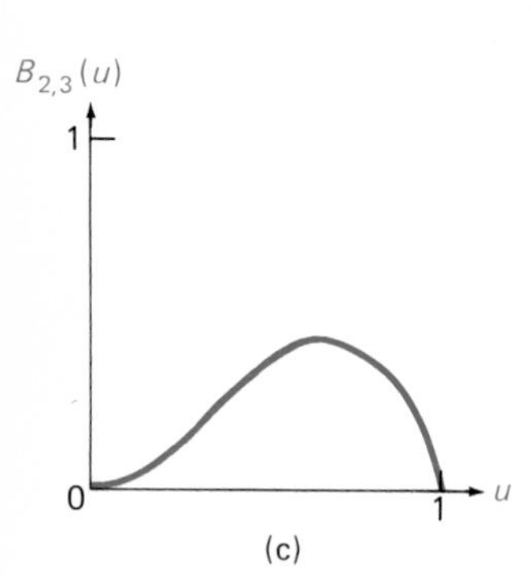

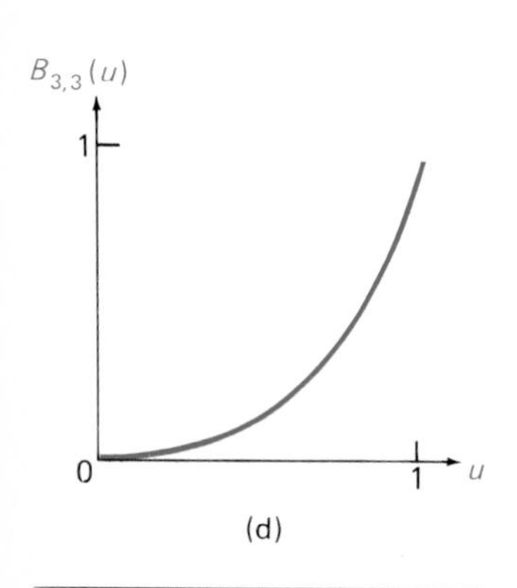

FIGURE 10–14
The four Bézier functions used in the construction of a curve with four control points.

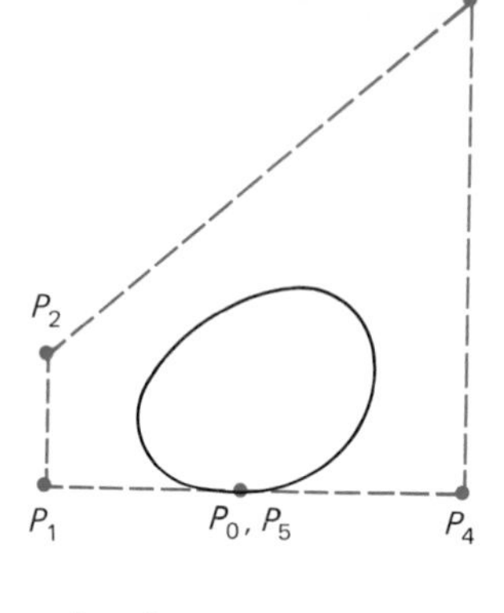

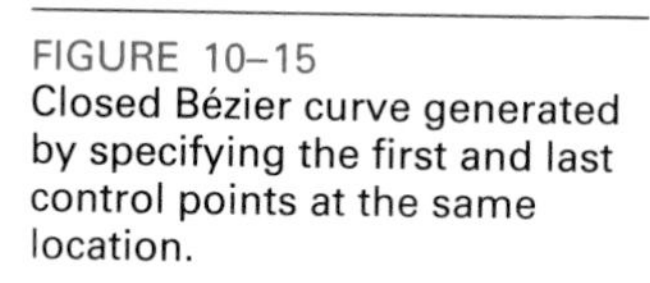

FIGURE 10–15
Closed Bézier curve generated by specifying the first and last control points at the same location.

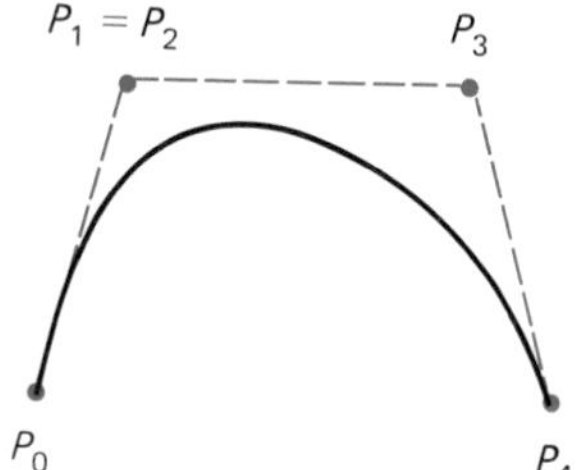

FIGURE 10–16
A Bézier curve can be made to pass closer to a given position by assigning multiple control points to that position.

order continuity). Also, Bézier curves have the important property that the tangent to the curve at an endpoint is along the line joining that endpoint to the adjacent control point. Therefore, to obtain first-order continuity between curve sections, a user can pick control points so that the positions p_{n-1} and p_n of one section are along the same straight line as positions p_0 and p_1 of the next section (Fig. 10–17). Bézier curves, in general, do not possess second-order continuity.

The following example program illustrates a method for generating Bézier curves. More efficient algorithms can be developed using recursive calculations to obtain successive points along the curves.

```
procedure bezier (n, m : integer);
  {Uses n + 1 control points. Generates curve by finding
   m + 1 points along the interval of 0 ≤ u ≤ 1}
  type
    {control points}
    control_array = array [0..max_controls, 1..3] of real;
    {n!/(k!(n-k)!)}
    coefficient_array = array [0..max_controls] of integer;
    {points to plot}
    curve_points = array [0..max_points] of real;
  var
    control : control_array;
    c : coefficient_array;
    x, y, z : curve_points;
    j : integer;

  procedure get_control_points (n : integer;
                                       var controls : control_array);
    begin
      {returns n+1 control points in controls}
```

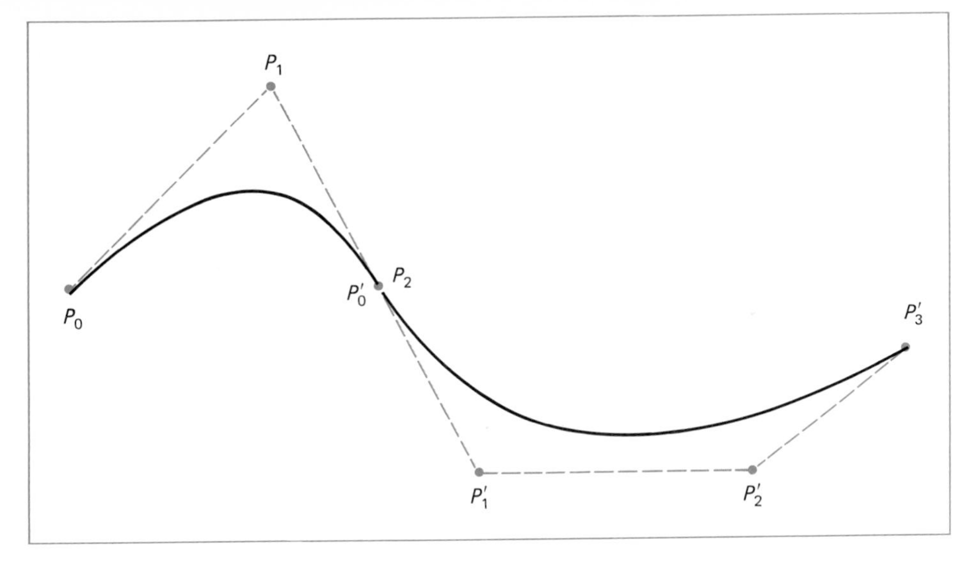

FIGURE 10–17
Curve formed with two Bézier sections. Zero-order and first-order continuity is attained between curve sections by setting $P'_0 = P_2$ and by making the points P_1, P_2, and P'_1 collinear.

```
  end;  {get_control_points}

procedure compute_coefficients (n : integer;
                               var c : coefficient_array);
  var k, m: integer;
  begin
    for k := 0 to n do begin
        {compute n!/(k!(n-k))}
        c[k] := 1;
        for m := n downto k+1 do c[k] := c[k] * m;
        for m := n-k downto 2 do c[k] := c[k] div m
      end {for k}
  end;  {compute_coefficients}

procedure compute_point (n : integer; u : real;
                         controls : control_array;
                         c : coefficient_array;
                         var x, y, z : real);
  var k : integer; b : real;
  function blending_value (n, k : integer; u : real);
    {returns blending value for this control
     point at this point in the interval}
    var bv : real; m : integer;
    begin
      {compute c[k] * (u to kth power) * ((1-u) to (n-k) power)}
      bv := c[k];
      for m := 1 to k do bv := bv * u;
      for m := 1 to n-k do bv := bv * (1 - u);
      blending_value := bv
    end;  {blending_value}

  begin  {compute_point}
    x := 0; y := 0; z := 0;
    for k := 0 to n do begin
        {add in influence of each control point}
```

```
        b := blending_value (n, k, u);
        x := x + controls[k,1] * b;
        y := y + controls[k,2] * b;
        z := z + controls[k,3] * b
      end {for k}
  end; {compute_point}

begin {bezier}
  get_control_points (n, controls);
  compute_coefficients (n, c);
  for j := 0 to m do {generate m+1 points}
      compute_point (n, j/m, controls, c, x[j+1], y[j+1], z[j+1]);
  polyline_3 (m+1, x, y, z)
end; {bezier}
```

Spline Curves

In drafting terminology, a spline is a flexible strip used to produce a smooth curve through a set of plotted control points. The term **spline curves,** or spline functions, refers to the resulting curves drawn in this manner. Such curves can be described mathematically as piecewise approximations of cubic polynomial functions with all three orders of continuity (zero-order, first-order, and second-order continuity), although many other types of approximating functions are also referred to as spline curves.

B-splines are a class of spline curves that are particularly useful in graphics applications. Given an input set of $n + 1$ control points $\boldsymbol{p}_k$, with k varying from 0 to n, we define points on the approximating B-spline curve as

$$\boldsymbol{P}(u) = \sum_{k=0}^{n} \boldsymbol{p}_k N_{k,t}(u) \tag{10–14}$$

where the B-spline blending functions $N_{k,t}$ can be defined as polynomials of degree $t - 1$. A method for setting up the polynomial form of the blending functions is to define them recursively over various subintervals of the range for parameter u. This range now depends on the number of control points n and the choice for t, so that u varies from 0 to $n - t + 2$ (instead of from 0 to 1). Setting up $n + k$ subintervals, we define the blending functions recursively as

$$\begin{aligned} N_{k,1} &= \begin{cases} 1 & \text{if } u_k \leqslant u < u_{k+1} \\ 0 & \text{otherwise} \end{cases} \\ N_{k,t}(u) &= \frac{u - u_k}{u_{k+t-1} - u_k} N_{k,t-1}(u) + \frac{u_{k+t} - u}{u_{k+t} - u_{k+1}} N_{k+1,\, t-1} \end{aligned} \tag{10–15}$$

Since the denominators in the recursive calculations can have a value of 0, this formulation assumes that any terms evaluated as 0/0 are to be assigned the value 0.

The defining positions u_j for the subintervals of u are referred to as **breakpoints,** and points on the B-spline curve corresponding to the breakpoints are called **knots.** Breakpoints can be defined in various ways. A uniform spacing of the breakpoints is achieved by setting u_j equal to j. Another method for defining uniform intervals is with breakpoints set to the values

$$u_j = \begin{cases} 0 & \text{if } j < t \\ j - t + 1 & \text{if } t \leqslant j \leqslant n \\ n - t + 2 & \text{if } j > n \end{cases} \tag{10–16}$$

for values of j ranging from 0 to $n + t$.

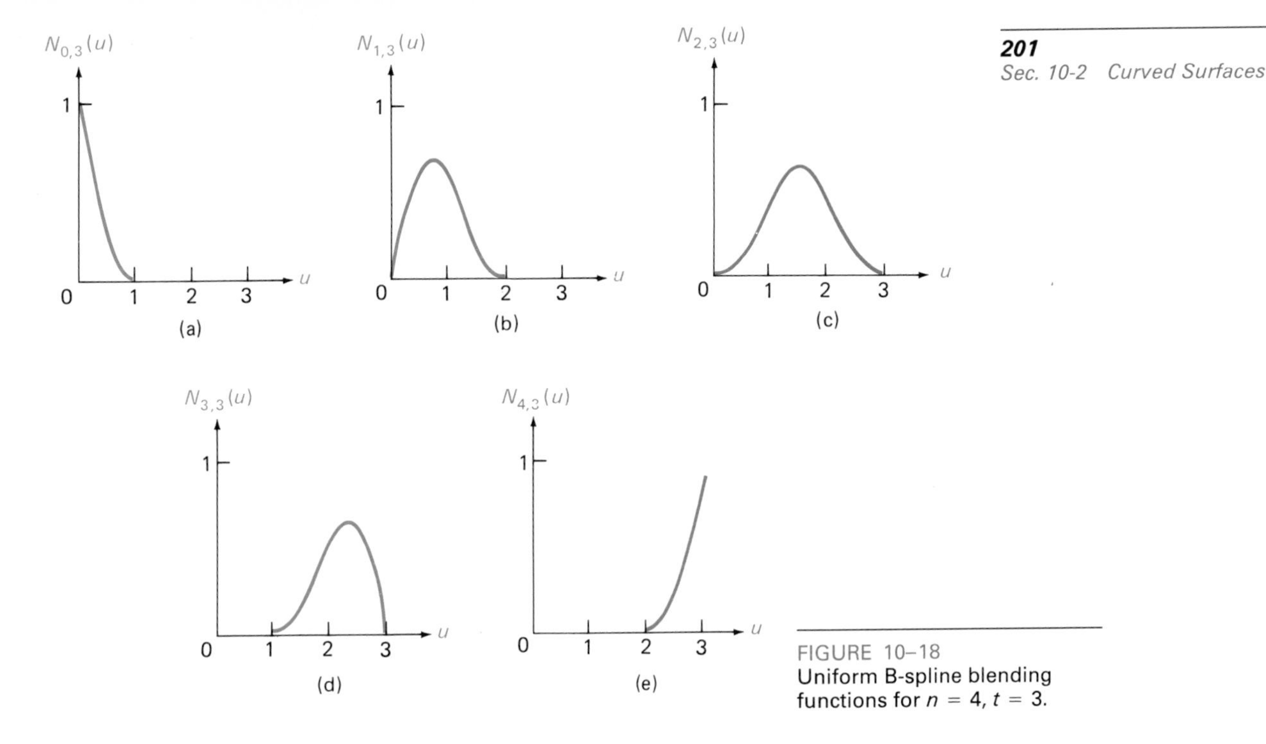

FIGURE 10–18
Uniform B-spline blending functions for $n = 4$, $t = 3$.

Figure 10–18 illustrates the shape of uniform, quadratic B-spline blending functions ($t = 3$) using five control points ($n = 4$) and subintervals as defined in Eq. 10–16. Parameter u is then varied over the interval from 0 to 3, with breakpoints u_0 to u_7 as 0, 0, 0, 1, 2, 3, 3, 3, respectively. An important property of B-spline curves, demonstrated in this example, is that the blending functions are each zero over part of the range of u. Blending function $N_{0,3}$, for instance, is nonzero only in the subinterval from 0 to 1, so the first control point influences the curve only in this interval. Thus B-spline curves allow localized changes to be made easily. If a user changes the position of the first control point, the shape of the curve near that point is changed without greatly affecting other parts of the curve. Figure 10–19 demonstrates this local property using the blending functions of Fig. 10–18.

B-spline curves have the advantage that any number of control points can be specified by a designer without increasing the degree of the curve. A cubic function ($t = 4$) could then be used for many different curve shapes, without the need to

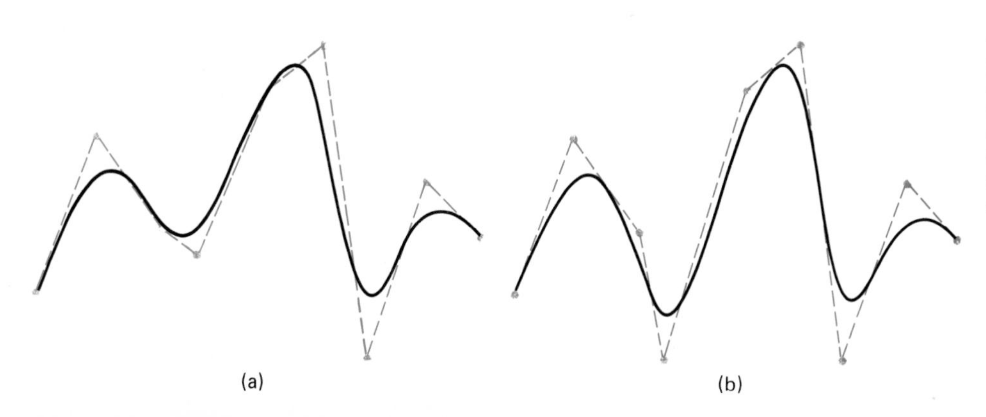

FIGURE 10–19
Local modification of a B-spline curve. Changing one of the control points in (a) produces curve (b), which is modified only in the neighborhood of the altered control point.

FIGURE 10–20
Cubic B-spline curve ($t = 4$) with multiple control points specificed at the same coordinate position to emphasize that position: (a) one control point at each position; (b) two control points at the center position; (c) three control points at the center position.

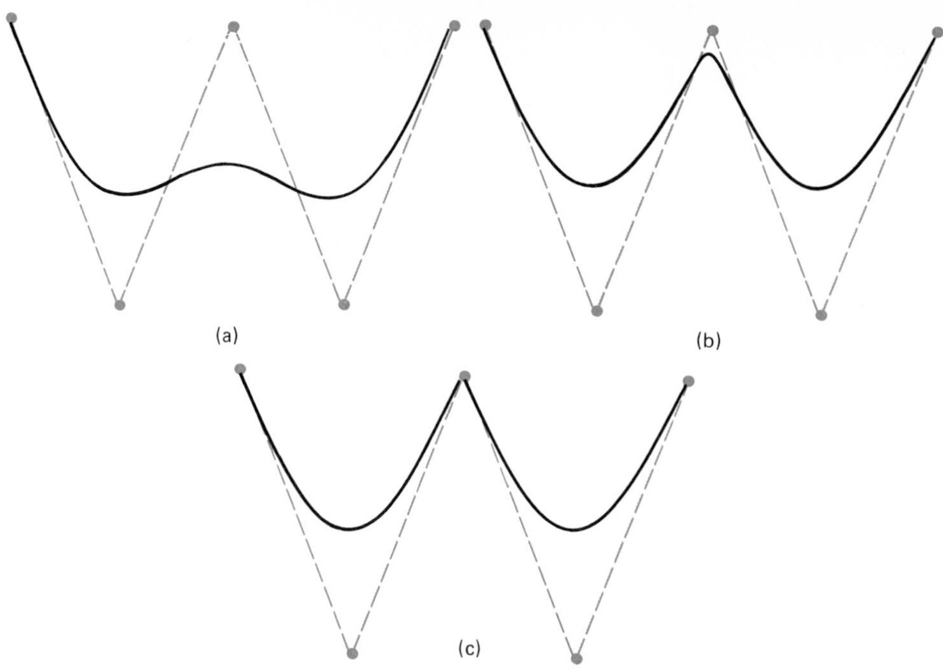

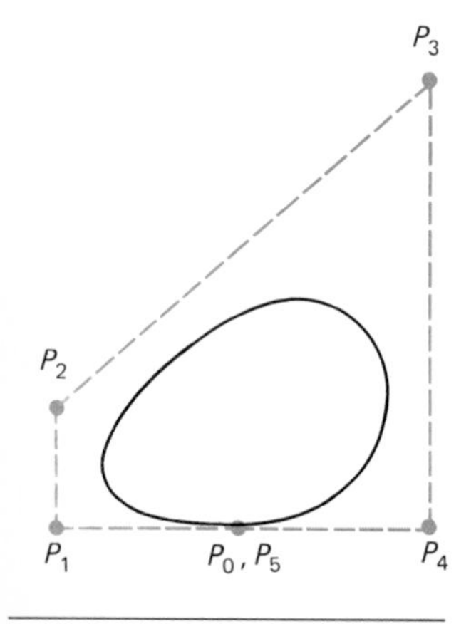

FIGURE 10–21
Closed cubic B-spline curve ($t = 4$), specified by setting the first and last control points to the same position.

piece curve segments together. Any number of control points can be added or modified to manipulate the curve shape.

As with Bézier curves, specifying multiple control points at the same coordinate position pulls the B-spline curve closer to that position (Fig. 10–20). Also, closed curves are obtained by specifying the first and last control points at the same position, as in Fig. 10–21. B-splines also lie within the convex hull of the control points. In fact, any B-spline curve with parameter t lies within the convex hull of at most $t + 1$ of its control points, so that this class of curves is tightly bound to the input positions.

Another type of spline curve is the Beta-spline. This spline method introduces two additional parameters, called beta-1 and beta-2, that give the user added capabilities for manipulating curve shapes. The two beta parameters are used to adjust the shape of the curve relative to the convex hull. Depending on the values assigned to these parameters, the curve can be made to move away from the convex hull or toward it. For certain values of beta-1 and beta-2, a Beta-spline curve reduces to a B-spline.

Bézier Surfaces

Two sets of Bézier curves can be used to represent surfaces of objects specified by input control points. The parametric vector function for the Bézier surface is formed as the Cartesian product of Bézier blending functions:

$$\mathbf{P}(u,v) = \sum_{j=0}^{m} \sum_{k=0}^{n} \mathbf{p}_{j,k} B_{j,m}(u)\, B_{k,n}(v) \tag{10–17}$$

with $\mathbf{p}_{j,k}$ specifying the location of the $(m + 1)$ by $(n + 1)$ control points.

Figure 10–22 illustrates two Bézier surface plots. The control points are connected by lighter lines, and the heavier lines show curves of constant u and con-

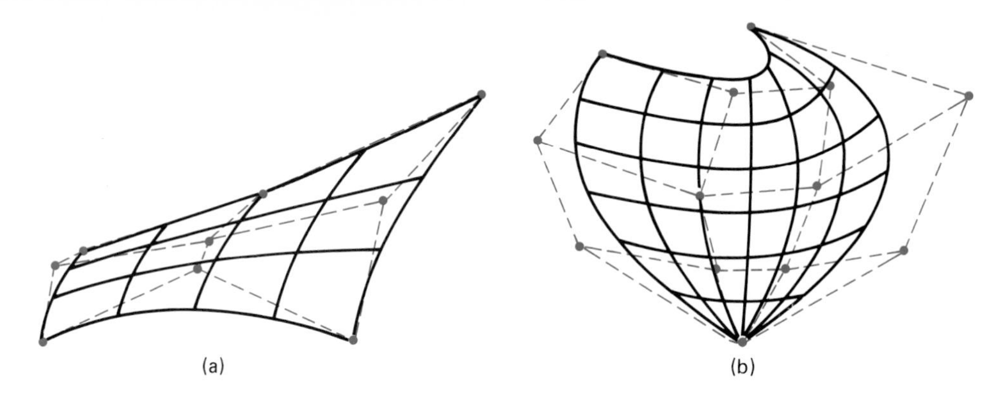

FIGURE 10–22
Bézier surfaces constructed for (a) $m = 3, n = 3$ and (b) $m = 4, n = 4$. Dotted lines connect the control points.

stant v. Each curve of constant u is plotted by varying v over the interval from 0 to 1, with u fixed at one of the values in this unit interval. Curves of constant v are plotted similarly.

Bézier surfaces have the same properties as Bézier curves, and they provide a convenient method for interactive design applications. Figure 10–23 illustrates a surface formed with two Bézier surface sections. As with curves, a smooth transition from one section to the other is assured by establishing both zero-order and second-order continuity at the boundary line. Zero-order continuity is obtained by matching control points at the boundary. First-order continuity is obtained by choosing control points along a straight line across the boundary and by maintaining a constant ratio of collinear line segments for each such line crossing the surface boundary line.

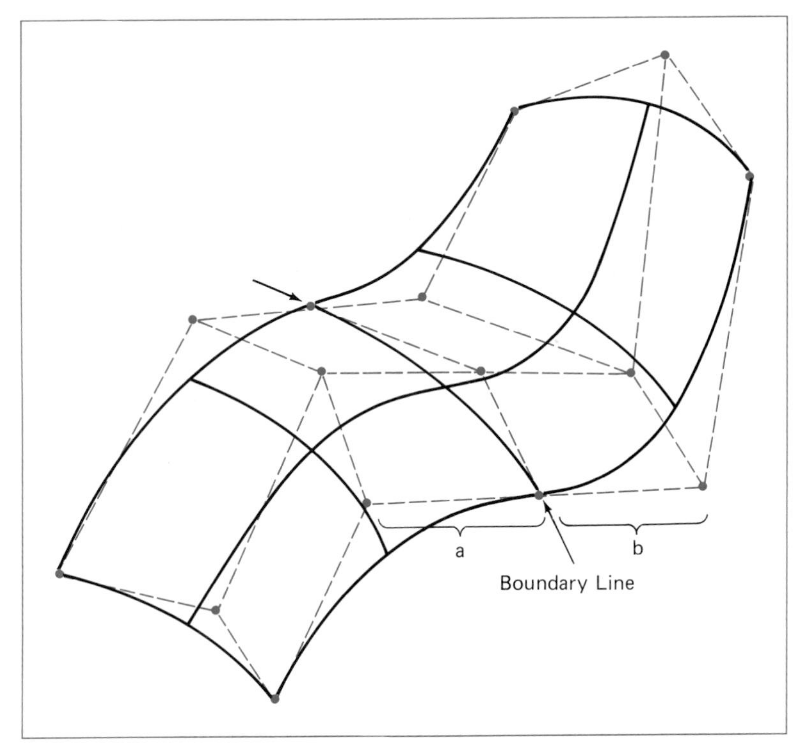

FIGURE 10–23
Bézier surface constructed with two Bézier sections, joined at the indicated boundary line. The dashed lines connect specified control points. First-order continuity is established by making the ratio of length a to length b constant for each collinear line of control points across the boundary line.

Spline Surfaces

Formulation of a B-spline surface is similar to that for a Bézier curve. The Cartesian product of B-spline blending functions defines a parametric vector function over a B-spline surface:

$$P(u, v) = \sum_{j=0}^{m} \sum_{k=0}^{n} p_{j,k} N_{j,s}(u)\, N_{k,t}(v) \tag{10–18}$$

As before, vector values for $p_{j,k}$ specify the $(m + 1)$ by $(n + 1)$ control points.

B-spline surfaces exhibit the same properties as those of their component B-spline curves. A surface is constructed from selected values for parameters s and t (which determine the polynomial degrees to be used) by evaluating the blending functions according to Eqs. 10–15.

Methods for Surface Generation

Since the equations describing Bézier and spline surfaces involve a great many calculations, it is important to devise algorithms that use efficient methods for generating successive points along the curved lines of the surfaces. Several techniques are available for reducing the calculations involved in evaluating parametric polynomial equations.

One method for reducing the number of arithmetic operations in a polynomial equation is Horner's rule, which specifies an order for factoring the terms in the polynomial. The cubic equation, $f(u) = a_0u^3 + a_1u^2 + a_2u + a_3$ for example, is evaluated in the order:

$$f(u) = [(a_0u + a_1)u + a_2]u + a_3 \tag{10–19}$$

which requires three multiplications and three additions. Although this factoring technique reduces the number of arithmetic operations, it does involve floating-point multiplications.

A more efficient method for evaluating polynomial equations is to recursively generate each succeeding value of the function by incrementing the previously calculated value:

$$f_{i+1} = f_i + \Delta f_i \tag{10–20}$$

where Δf_i is called the **forward difference.** The function f_i is evaluated at u_i, and f_{i+1} is evaluated at $u_{i+1} = u_i + \delta$, where δ is the step size for incrementing parameter u. As an example, the linear equation $f(u) = a_0u + a_1$ can be incrementally evaluated using Eq. (10–20) and the forward difference $\Delta f = a_0\delta$. In this case, the forward difference is a constant. With higher-order polynomials, the forward difference is itself a polynomial function of parameter u.

For a cubic curve, the forward difference evaluates to

$$\Delta f_i = 3a_0\delta u_i^2 + (3a\delta^2 + 2a_1\delta)u_i + a_0\delta^3 + a_1\delta^2 + a_2\delta \tag{10–21}$$

which is a quadratic function of parameter u. However, we can use the same incremental procedure to obtain successive values of Δf. That is,

$$\Delta f_{i+1} = \Delta f_i + \Delta^2 f_i \tag{10–22}$$

where the second forward difference is a linear function of u:

$$\Delta^2 f_i = 6a_0\delta^2 u_i + 6a_0\delta^3 + 2a_1\delta^2 \qquad (10\text{–}23)$$

Repeating this process once more, we can write

$$\Delta^2 f_{i+1} = \Delta^2 f_i + \Delta^3 f_i \qquad (10\text{–}24)$$

with the third forward difference as the constant

$$\Delta^3 f_i = 6a_0\delta^3 \qquad (10\text{–}25)$$

Equations (10–20), (10–22), (10–24), and (10–25) are used to incrementally obtain points along the curve from $u = 0$ to $u = 1$ with a step size δ. The initial values at $i = 0$ and $u = 0$ are

$$\begin{aligned} f_0 &= a_3 \\ \Delta f_0 &= a_0\delta^3 + a_1\delta^2 + a_2\delta \\ \Delta^2 f_0 &= 6a_0\delta^3 + 2a_1\delta^2 \end{aligned} \qquad (10\text{–}26)$$

Calculations for successive points are then carried out as a series of additions.

To apply this incremental procedure to Bézier and spline curves, three sets of calculations are needed for the coordinates $x(u)$, $y(u)$, and $z(u)$. For surfaces, incremental calculations are applied for both parameter u and parameter v.

10–3 Fractal-Geometry Methods

The techniques discussed in the preceding section generate smooth curves and surfaces. But many objects, such as mountains and clouds, have irregular or fragmented features. Such objects can be modeled using the **fractal-geometry methods** developed by Mandelbrot. Basically, fractal objects are described as geometric entities that cannot be represented with Euclidean-geometry methods. This means that a fractal curve cannot be described as one-dimensional, and a fractal surface is not two-dimensional. Fractal shapes have a fractional dimension.

Smooth curves are one-dimensional objects whose length can be precisely defined between two points. A fractal curve, on the other hand, contains an infinite variety of detail at each point along the curve, so we cannot say exactly what its length is. In fact, as we continue to zoom in on more and more detail, the length of the fractal curve grows longer and longer. An outline of a mountain, for example, shows more variation the closer we get to it (Fig. 10–24). As we near the mountain, the detail in the individual ledges and boulders becomes apparent. Moving even

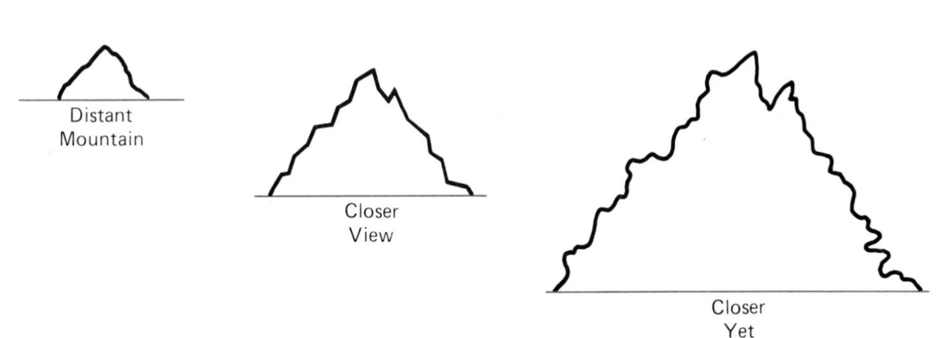

FIGURE 10–24
The ragged appearance of a mountain outline increases as we get closer.

closer, we see the outlines of rocks, then stones, and then grains of sand. At each step, the outline reveals more twists and turns, and the overall length of the curve tends to infinity as we move closer and closer. Similar types of curves describe coastlines and the edges of clouds. Such curves, represented in a two-dimensional coordinate system, can be described mathematically with fractional dimensions between 1 and 2. When a fractal curve is described in a three-dimensional space, it has a dimension between 1 and 3.

A fractal curve is generated by repeatedly applying a specified transformation function to points within a region of space. The amount of detail included in the final display of the curve depends on the number of iterations performed and the resolution of the display system. If $\boldsymbol{P}_0 = (x_0, y_0)$ is a selected initial point, each iteration of a transformation function F generates the next level of detail with the calculations

$$\boldsymbol{P}_1 = F(\boldsymbol{P}_0), \quad \boldsymbol{P}_2 = F(\boldsymbol{P}_1), \quad \boldsymbol{P}_3 = F(\boldsymbol{P}_2), \quad \ldots \tag{10–27}$$

The transformation function can be specified in various ways to generate either regular or random variations along the curve at each iteration.

An example of a fractal curve generated with a regular pattern is shown in Fig. 10–25. We start with a curve containing the two line segments shown. First the pattern for the original curve is reduced by a factor of 1/3, and this reduced pattern is used to replace the middle third of the two line segments in the original curve. The resulting curve is the second pattern shown in this figure. We then scale the reduced pattern by a factor of 1/3 again and use it to replace the middle

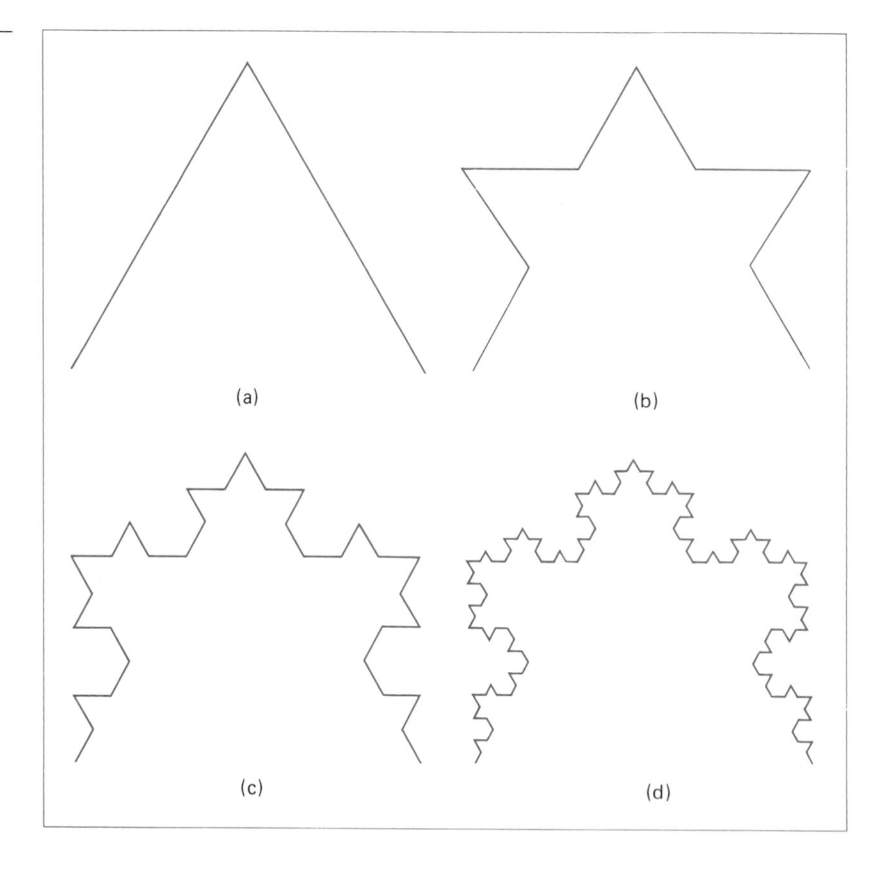

FIGURE 10–25
Producing a fractal curve by repeatedly scaling the original curve and using the reduced pattern to replace the middle third of the line segments in the curve.

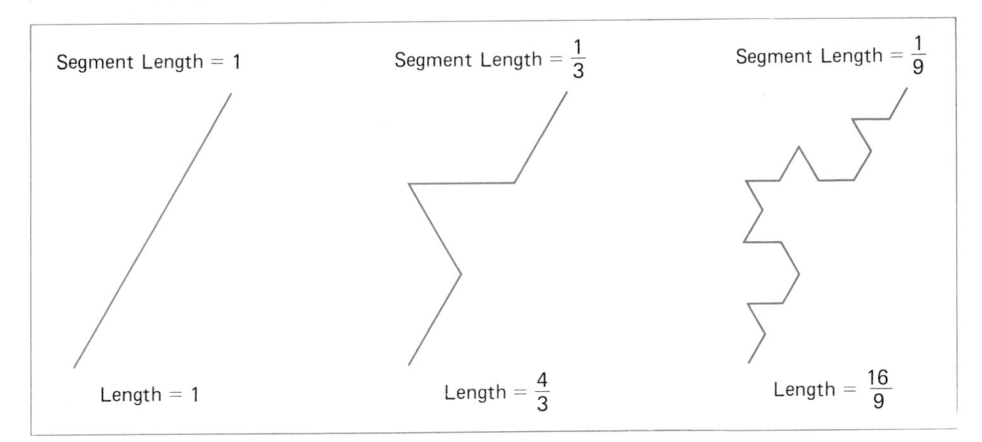

FIGURE 10–26
Length of the fractal line increases by a factor of 4/3 at each iteration, when line-segment lengths are reduced by a factor of 1/3 at each step.

third of the line segments in the transformed curve to produce the third curve shown. This process is continued by repeatedly scaling the original pattern by a factor of 1/3 and using the scaled patterns to modify line segments in the successive curves generated. At each step, the number of lines in the curve is increased by a factor of 4, and the length of each new line segment generated is one-third the length of line segments in the previous curve. This increases the total length of the fractal curve by a factor of 4/3 at each step (Fig. 10–26). Therefore, the length of the fractal curve tends to infinity as more detail is added to the curve.

One method for determining the dimension of a fractal curve such as that in Fig. 10–25 is the calculation

$$D = \frac{\ln N}{\ln (1/S)} \qquad (10\text{–}28)$$

where N is the number of subdivisions at each step and S is the scaling factor. For our example, we subdivided line segments into four parts, so $N = 4$, and we used a scaling factor of $S = 1/3$. This gives the fractal curve in Fig. 10–25 a dimension of $D = \ln 4/\ln 3 = 1.2619$. If we subdivide a line into equal parts without changing its length, $N = 1/S$ and the dimension of the line as calculated by Eq. 10–20 is $D = 1$, as described in Euclidean geometry. Similarly, when a surface is subdivided into N equal parts without changing its area, $N = 1/S^2$ and the dimension of the surface is 2. Figure 10–27 illustrates subdivision of a line and a surface without a change in size.

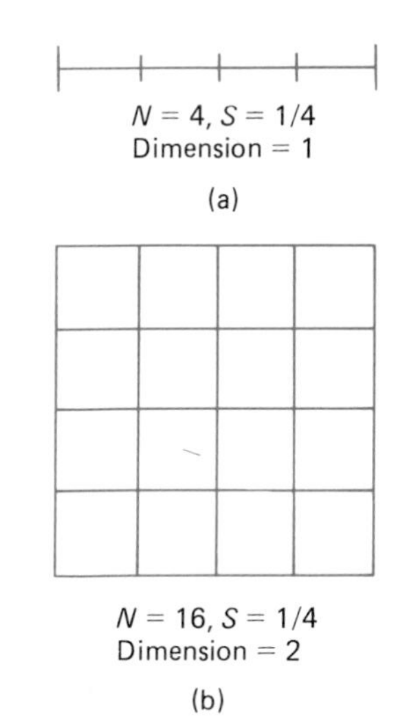

FIGURE 10–27
Subdividing a line (a) and a surface (b) into equal parts without changing the size of the objects.

Although fractal objects, by definition, contain infinite detail, we generate a fractal curve with a finite number of iterations. Therefore, the curves we display with repeated patterns actually have finite length. Our representation approaches a true fractal curve as the number of transformations is increased to produce more and more detail.

Many fractal curves are generated with functions in the complex plane. That is, each two-dimensional point (x, y) is represented as the complex expression $z = x + iy$, where x and y are real numbers and i is used to represent the square root of -1. A complex function $f(z)$ is then used to map points repeatedly from one position to another. Depending on the initial point selected, this iteration could cause points to diverge to infinity, or the points could converge to a finite limit, or the points could remain on some curve. For example, the function $f(z) = z^2$ transforms points according to their relation to the unit circle (Fig. 10–28). Any point z

FIGURE 10–28
A unit circle in the complex plane. The function $f(z) = z^2$ moves points that are inside the circle toward the origin, while points outside the circle are moved farther away from the circle.

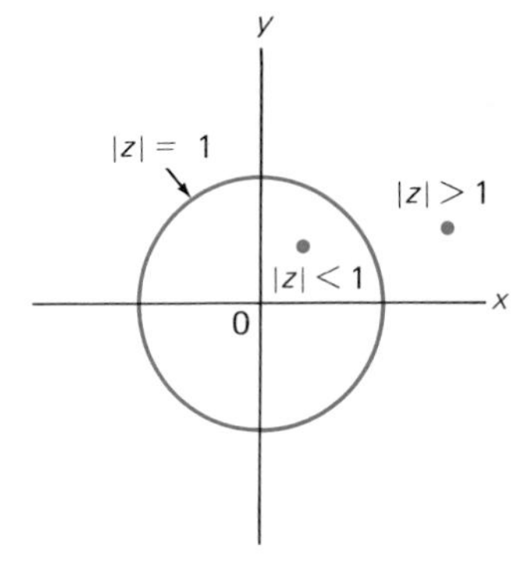

whose magnitude $|z|$ is greater than 1 is transformed through a sequence of points that tend to infinity. A point with $|z| < 1$ is transformed closer to the origin. Points on the circle, $|z| = 1$, remain on the circle. For some functions, the boundary between those points that move toward infinity and those that tend toward a finite limit is a fractal curve.

A function that is rich in fractal curves is the transformation

$$z' = f(z) = \lambda z(1 - z) \tag{10–29}$$

with λ assigned any constant complex value. Testing the behavior of transformed points in the complex plane to determine the position of the fractal curve can be a time-consuming process. A quicker method for calculating points on the fractal curve is to use the inverse of the transformation function. An initial point chosen on the inside or outside of the curve will then converge to points on the fractal curve. We can obtain the inverse function by rewriting Eq. 10–29 as

$$\lambda z^2 - \lambda z + z' = 0 \tag{10–30}$$

Solving this quadratic equation for z, we have

$$z = \frac{1}{2}[1 \pm \sqrt{1 - (4z')/\lambda}] \tag{10–31}$$

Using complex arithmetic operations, we solve this equation for the real and imaginary parts of z; that is, $x = R(z)$ and $y = I(z)$. Coordinate positions (x, y) are then plotted along the fractal curve.

The following procedure demonstrates plotting of fractal curves generated from Eq. 10–31. Since this inverse function provides two possible transformed positions, we randomly choose one or the other at each step in the iteration. Also, whenever the imaginary part of $-4z'/\lambda$ is negative, one of the possible transformed positions is in the second quadrant ($x < 0$ and $y > 0$). The other position is situated 180° away, in the fourth quadrant ($x > 0$, $y < 0$). Figure 10–29 shows two fractal curves output by this procedure.

```
{procedure fractal accepts lambda and an initial point
 z_prime, and calculates points on a fractal curve using the
 inverse of the function z_prime = lambda * z * (1 - z). Random
 is a predefined function returning values between 0 and 1}
```

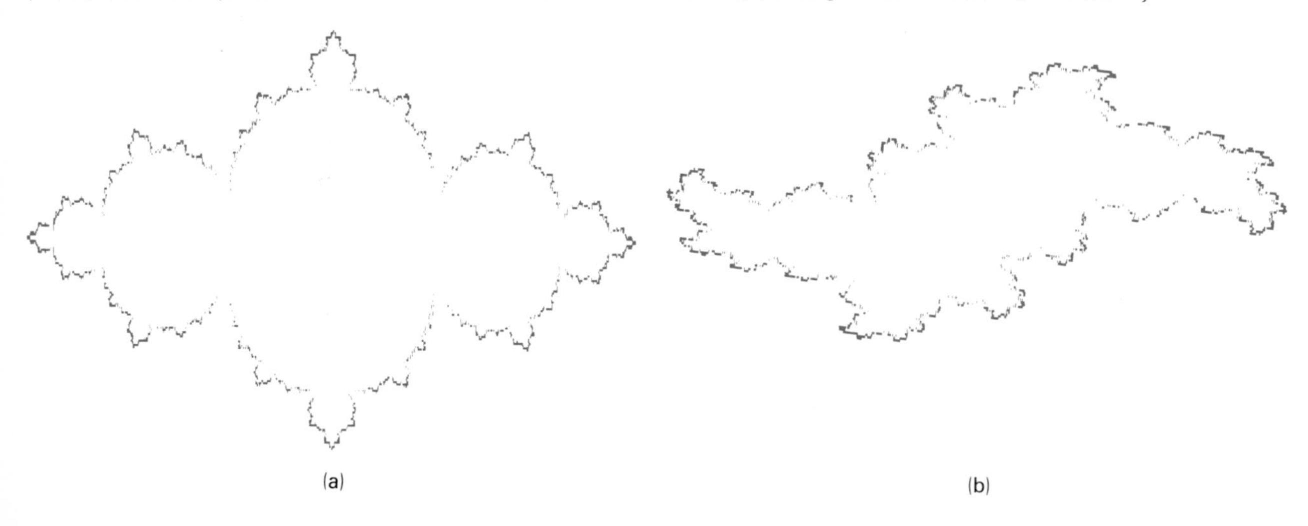

FIGURE 10–29
Two fractal curves generated with the inverse of the function $f(z) = \lambda z(1 - z)$ by PROCEDURE FRACTAL: (a) $\lambda = 3$; (b) $\lambda = 2 + i$. Each curve is plotted with 10,000 points.

```
type
  complex = record
    a : real;
    b : real
  end;  {record}
var  lambda, z_prime, z : complex;

procedure fractal (lambda, z_prime : complex; count : integer);
  var
    k : integer;
    lambda_magnitude : real;
    four_over_lambda : complex;

  {calculate_point accepts a point z_prime and returns
   a point z that satisfies inverse of z' = λz(1-z)}
  procedure calculate_point (z_prime : complex; var z : complex);
    var  save_b : real;

    procedure complex_multiply (c1, c2 : complex;
                                var result : complex);
      begin
        result.a := c1.a * c2.a - c1.b * c2.b;
        result.b := c1.a * c2.b + c1.b * c2.a
      end;  {complex_multiply}

    procedure complex_square_root (c : complex;
                                   var result : complex);
      var  c_magnitude : real;
      begin
        c_magnitude := sqrt(c.a * c.a + c.b * c.b);
        if (c_magnitude + c.a) < 0 then result.a := 0
          else result.a := sqrt((c_magnitude + c.a) / 2);
        if (c_magnitude - c.a) < 0 then result.b := 0
          else result.b := sqrt((c_magnitude - c.a) / 2)
      end;  {complex_square_root}

    begin  {calculate_point}
      {compute 4 / lambda * z_prime. Return result in z}
      complex_multiply (z_prime, four_over_lambda, z);
      {subtract from 1}
      z.a := 1 - z.a;
      {retain current y part}
      save_b := z.b;
      {take square root of z. Return result in z}
      complex_square_root (z, z);
      {determine correct sign for square root of x part}
      if save_b < 0 then z.a := -z.a;
      {choose one of the two points satisfying inverse function}
      if random < 0.5 then begin
          z.a := -z.a;
          z.b := -z.b
        end;  {if random}
      {subtract from 1}
      z.a := 1 - z.a;
      {divide by 2}
```

```
    z.a := z.a / 2;
    z.b := z.b / 2
  end; {calculate_point}

procedure plot_point (x, y : real);
  begin
    {device-dependent routine to map point to device}
  end; {plot_point}
  begin {fractal}
    {compute 4 divided by lambda}
    lambda_magnitude := lambda.a * lambda.a +
                        lambda.b * lambda.b;
    four_over_lambda.a := 4 * lambda.a / lambda_magnitude;
    four_over_lambda.b := -4 * lambda.b / lambda_magnitude;
    {calculate but don't plot the first few points.
    Gets to a point that's actually on the curve}
    for k := 1 to 10 do begin
        calculate_point (z_prime, z);
        z_prime := z            {update z_prime}
      end; {for k}
    {calculate successive points (z) on the
    curve.  For each new z, plot the real (z.a)
    and imaginary (z.b) parts as x and y}
    for k := 1 to count do begin
        calculate_point (z_prime, z);
        plot_point (z.a, z.b);
        z_prime := z            {update z_prime}
    end {for k}
end; {fractal}
```

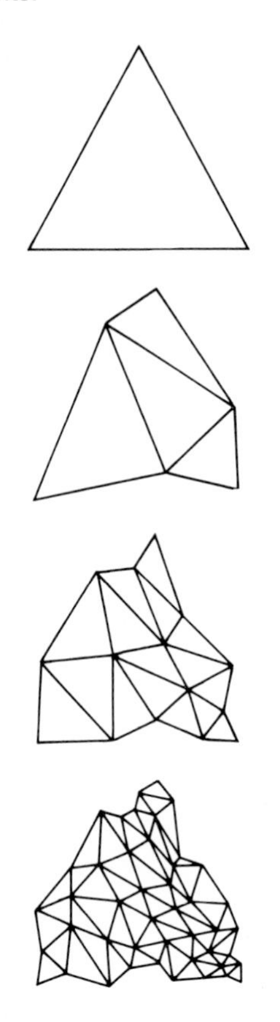

FIGURE 10–30
Creating a fractal mountain from a triangle by random displacement of boundary points.

Fractal surfaces represented in three dimensions can be described as objects with a dimension between 2 and 3. We can create fractal surfaces with procedures similar to those for generating fractal curves. Points on the surface are successively transformed at each step to produce increasingly more detail. In Fig. 10–30, a triangle is turned into a mountain by randomly displacing points on the triangle boundary. First, a random point is selected on each leg of the triangle. Then random displacement distances are applied to each of the three points, and the displaced points are joined with straight lines. As shown, in the second step, this turns the triangle into four smaller contiguous triangles. The process is repeated for each of the new triangles formed at each step. Final display of the object is a surface modeled with a mesh composed of small polygon sections. Many other types of transformations can be used to produce fractal surfaces with different features.

Complex-function transformations, such as Eq. (10–29), can also be extended to produce fractal surfaces and fractal solids. Methods for generating these objects often use quaternion representations for transforming points in three- and four-dimensional space. A quaternion has four components, one real and three imaginary, expressed as

$$q = q_0 + iq_1 + iq_2 + iq_3 \tag{10–32}$$

Using this representation and iteration methods to transform points in space, surfaces of fractal objects are generated. As successive points are generated, they are tested to determine whether they are inside or outside the surface. A basic procedure is to start with an interior point and generate successive points until an exterior point is identified. The previous interior point is then retained as a surface point. Neighbors of this surface point are then tested to determine whether they are inside or outside. Any inside point that connects to an outside point is a surface point. In this way, the procedure threads its way along the fractal boundary without generating points that are too far from the surface. When four-dimensional fractals are generated, three-dimensional slices are projected onto the two-dimensional surface of the video monitor.

Procedures for generating fractal objects require considerable computation time for evaluating the iteration function and for testing points. Each point on a surface can be represented as a small cube, giving the inner and outer limits of the surface. Output from such programs for three-dimensional fractal representations typically contain over a million vertices for the surface cubes. Display of the fractal objects is performed by applying illumination models that determine the lighting and color for each surface cube. Hidden-surface methods are then applied so that only visible surfaces of the objects are displayed.

Figure 10–31 illustrates some highly realistic scenes produced with fractal-geometry methods, using a probabilistic function to transform points in space. Frac-

(a)

(b)

(c)

FIGURE 10–31
Scenes generated by fractal-geometry methods, with a random iteration procedure. Courtesy R. V. Voss and B. B. Mandelbrot, adapted from *The Fractal Geometry of Nature* by Benoit B. Mandelbrot (New York: Freeman Press, 1982).

tal-generated displays of natural objects are often used in making movies containing animated scenes. Terrain, clouds, shorelines, and other objects are modeled with the same basic outlines. Differences in the appearances of these objects are due to the kind of lighting model used to provide object shading. Terrain reflects most of the incident light, while clouds allow light to penetrate partially and scatter, producing a diffuse effect.

The infinite detail contained in a description of a three-dimensional fractal is illustrated in the sequence of scenes in Fig. 10–32. As with two-dimensional fractals, each part of a three-dimensional fractal object has the same degree of detail as the entire object, no matter how many times we might enlarge sections of it. The sequence of scenes demonstrates this property of fractals by expanding a section of each display to produce the next display.

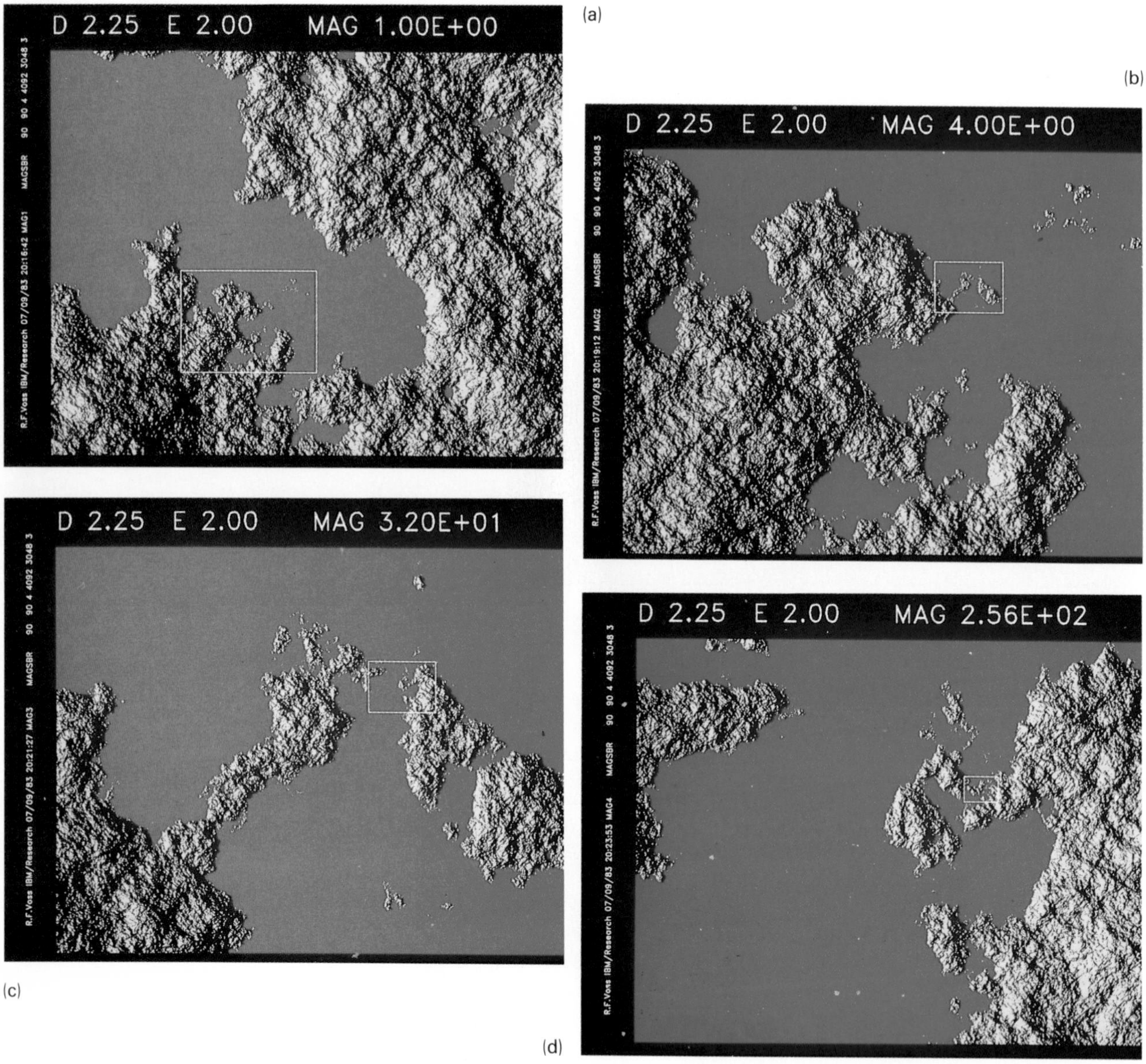

FIGURE 10–32
Fractal objects contain the same degree of detail in each part as is contained in the entire object. Each succeeding picture here is an expanded section of the preceding picture. Courtesy R. V. Voss and B. B. Mandelbrot.

(a)

(b)

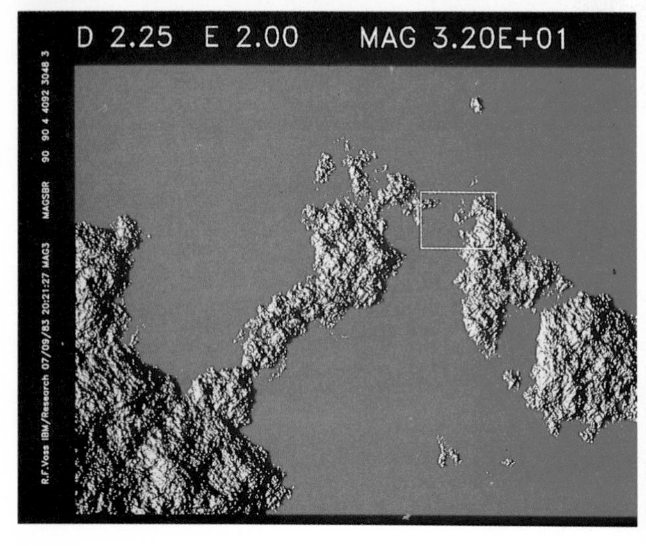

(c)

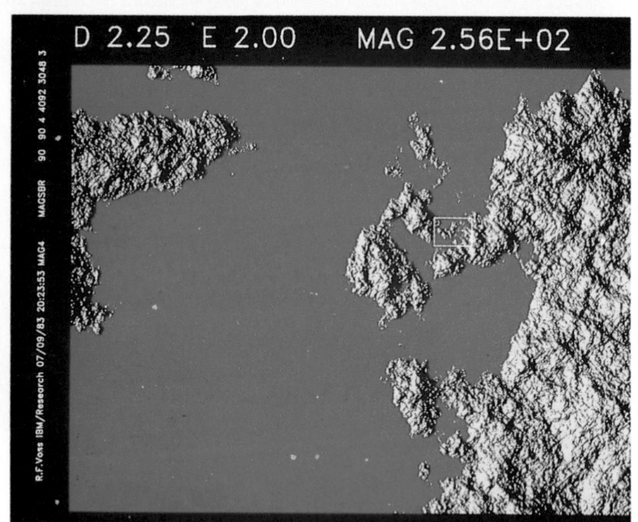

(d)

10–4 Sweep Representations

Graphics design packages can allow solid objects to be constructed from simpler objects in a number of ways. Solids with translational or rotational symmetry can be formed by sweeping a two-dimensional figure through a region of space.

Figure 10–33 shows a translational sweep. A flat object is translated over a specified distance to generate a solid. With this method, a user creates any desired two-dimensional shape and then specifies a translation direction. The volume of space over which the two-dimensional shape is swept defines the solid. Objects with translational symmetry can be defined in this way by specifying a shape that corresponds to a cross-sectional view of the object.

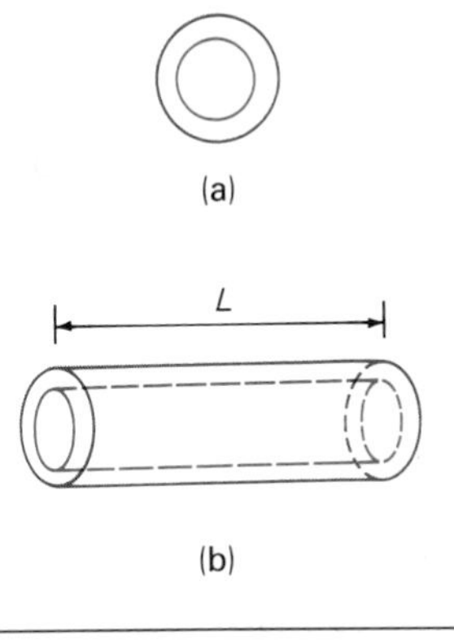

FIGURE 10–33
Representing a solid with a translational sweep. By translating the flat ring in (a) through a distance L, the hollow tube (b) is described.

A method for creating a solid with a rotational sweep is shown in Fig. 10–34. Here a hollow tube is described with a specified rotation of a defined plate. Objects with rotational symmetry can be represented with this method by specifying a flat shape corresponding to a cross section of the object parallel to the specified axis of rotation.

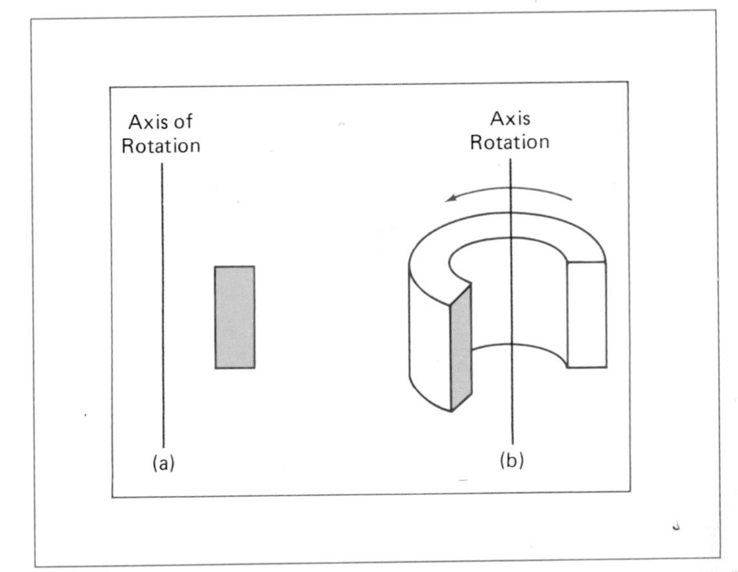

FIGURE 10–34
Representing a solid with a rotational sweep. The flat plate in (a) is rotated about a specified axis of rotation to produce the solid in (b).

Other construction methods based on symmetries can be provided by a design package. For example, the ring in Fig. 10–33 (a) is symmetric about its center. A user could describe the ring with a specification for a quarter section and the symmetry operations for completing the rest of the ring (Fig. 10–35).

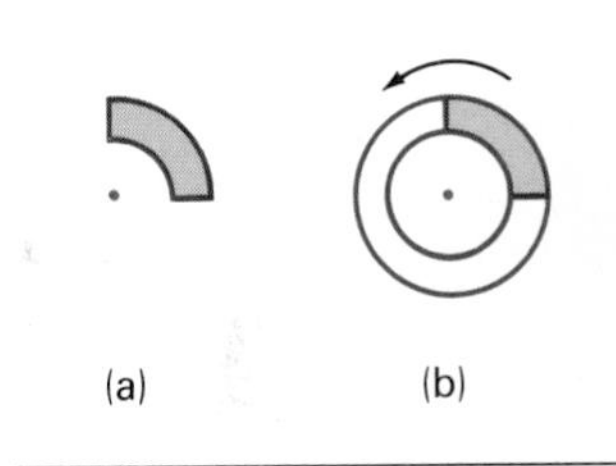

FIGURE 10–35
A description for a quarter section of a ring (a) can be used to create the shape in (b) with specified symmetry operations.

10–5 Constructive Solid-Geometry Methods

When a set of three-dimensional primitives, such as blocks, pyramids, cylinders, cones, and spheres, are provided to a user, object definitions are specified as combinations of the primitive solids. To create an object, the user specifies the construction methods to be used and the particular primitives to be combined.

Solid constructions can be accomplished with three-dimensional set operations. A user picks the solids to be combined and the set operation to be performed. Set operations include those for the union, intersection, and difference of two solids. For example, by positioning a block and pyramid as shown in Fig.

10–36 (a) and specifying the union operation, a user creates the solid in Fig. 10–36 (b). Surface definitions of the two original objects are combined to define the new, composite solid, which can then be combined with other objects until the final structure is attained.

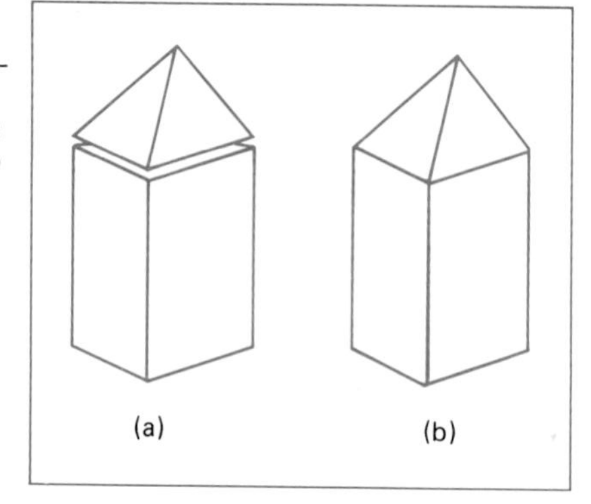

FIGURE 10–36
Combining two objects (a) with a union operation joins the two objects to produce a single solid (b).

An intersection operation describes a solid as the overlap volume of two objects, as demonstrated in Fig. 10–37. The user interactively positions the two objects so that an overlap region is created; then an intersection specification generates a new solid as the common subset of the original objects. This operation can be implemented in a graphics package using a procedure that tests the points within the two objects to determine which points are within the boundaries of both objects. Another implementation technique is based on the octree representation of solids, discussed in Section 10–6. Figure 10–38 illustrates the difference operation. In this case, the volume of the second object specified is subtracted from the volume of the first object.

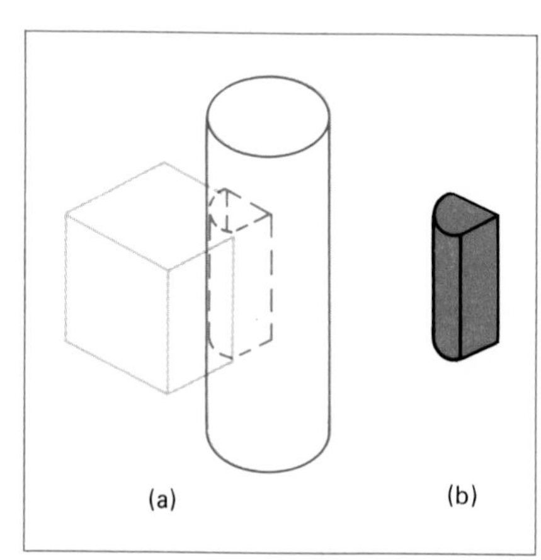

FIGURE 10–37
An intersection operation specified for the block and cylinder in (a) produces the wedge-shaped solid in (b).

FIGURE 10–38
A solid generated with a difference operation applied to the two overlapping objects in Fig. 10–37 (a).

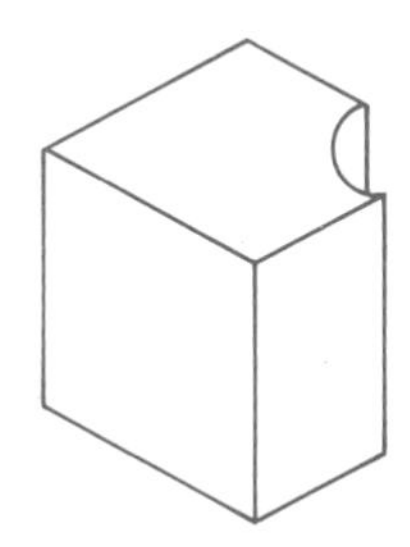

Similar construction methods can be used to generate solids from surfaces. In such packages, the set operations are used to combine various types of surface specifications. Surfaces combined with these operations can include polygons, spherical or cylindrical surfaces, or surfaces defined with polynomial functions.

10–6 Octrees

Hierarchical tree structures, called **octrees,** are used to represent solid objects in some graphics packages. The tree structure is organized so that each node corresponds to a region of three-dimensional space. This representation for solids takes advantage of spatial coherence to reduce storage requirements for three-dimensional objects. It also provides a convenient representation for displaying objects with hidden surfaces removed and for performing various object manipulations.

The octree encoding procedure for a three-dimensional space is an extension of an encoding scheme for two-dimensional space, called **quadtree** encoding. Quadtrees are generated by successively dividing a two-dimensional region into quadrants. Each node in the quadtree has four data elements, one for each of the quadrants in the region (Fig. 10–39). If all pixels within a quadrant have the same color (a homogeneous quadrant), the corresponding data element in the node stores that color. In addition, a flag is set in the data element to indicate that the quadrant is homogeneous. Suppose all pixels in quadrant 2 of Fig. 10–39 are found to be red. The color code for red is then placed in data element 2 of the node. Otherwise the quadrant is said to be heterogeneous, and that quadrant is itself divided into quadrants (Fig. 10–40). The corresponding data element in the node now flags the quadrant as heterogeneous and stores the pointer to the next node in the quadtree.

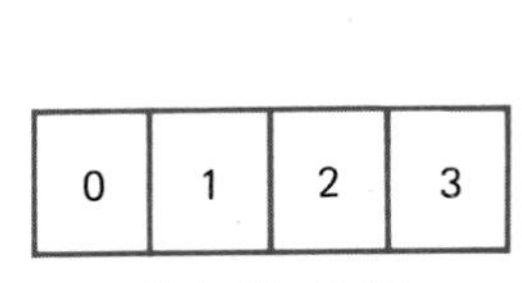

FIGURE 10–39
Region of a two-dimensional space divided into numbered quadrants and the associated quadtree node with four data elements.

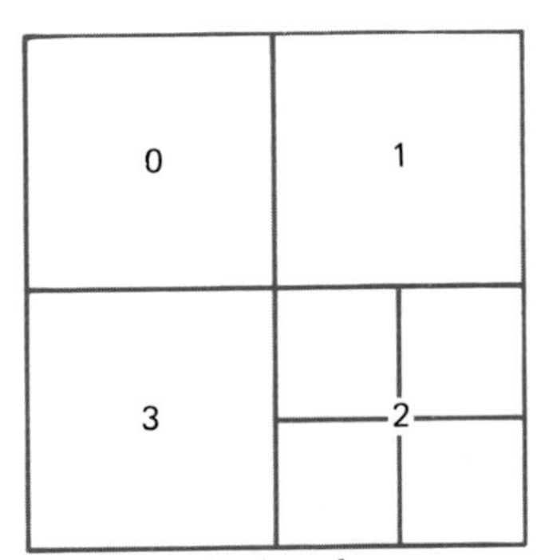

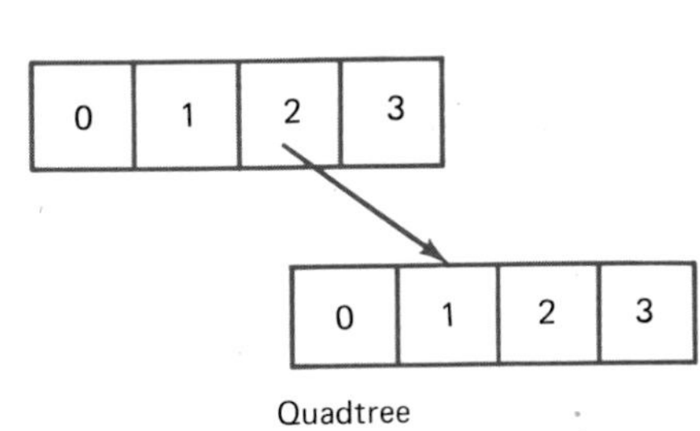

FIGURE 10–40
Region of a two-dimensional space with two levels of quadrant divisions and the associated quadtree representation.

An algorithm for generating a quadtree tests pixel intensity values and sets up the quadtree nodes accordingly. If all quadrants in the original space have a single color specification, the quadtree has only one node. For a heterogeneous region of space, the successive subdivisions into quadrants continues until all quadrants are homogeneous. Figure 10–41 shows a quadtree representation for a region containing one area with a solid color that is different from the uniform color specified for all other areas in the region.

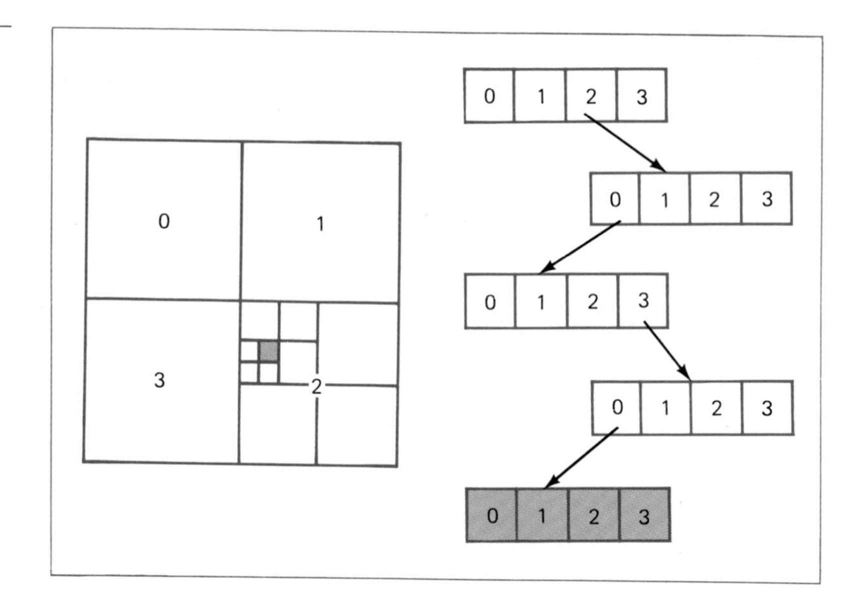

FIGURE 10–41
Quadtree representation for a region containing one foreground-color pixel on a solid background.

Quadtree encodings provide considerable savings in storage when large color areas exist in a region of space, since a single node can represent a single-color area. For an area containing 2^n by 2^n pixels, a quadtree representation contains at most n levels. Each node in the quadtree has at most four immediate descendants.

An octree encoding scheme divides regions of three-dimensional space into octants and stores eight data elements in each node of the tree (Fig. 10–42). Individual elements of a three-dimensional space are called **volume elements,** or **voxels.** When all voxels in an octant are of the same type, this type value is stored in the corresponding data element of the node. Empty regions of space are represented by voxel type "void." Any heterogeneous octant is subdivided into octants, and the corresponding data element in the node points to the next node in the octree. Procedures for generating octrees are similar to those for quadtrees: Voxels in each octant are tested, and octant subdivisions continue until the region of space contains only homogeneous octants. Each node in the octree can now have from zero to eight immediate descendants.

Algorithms for generating octrees can be structured to accept definitions of objects in any form, such as a polygon mesh, curved surface patches, or solid-geometry constructions. Using the minimum and maximum coordinate values of the object, a box (parallelepiped) can be defined around the object. This region of three-dimensional space containing the object is then tested, octant by octant, to generate the octree representation.

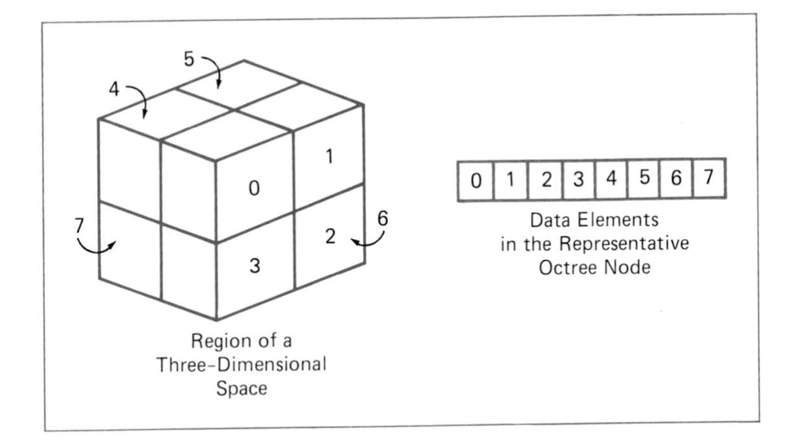

FIGURE 10–42
Region of a three-dimensional space divided into numbered octants and the associated octree node with eight data elements.

Once an octree representation has been established for a solid object, various manipulation routines can be applied to the solid. An algorithm for performing set operations accepts two octree representations as input and performs a specified operation, such as union, intersection, or difference. The octree representation for the resulting solid is output. Other algorithms perform three-dimensional transformations, hidden-surface removal, or shading on a single input octree. When an octree representation of an object is to be displayed on a video monitor, the octree can be converted into a quadtree representation. Then the quadtree structure is mapped into a frame buffer for display.

REFERENCES

Sources of information on parametric curve representations include Barsky (1983 and 1984), Bézier (1972), Burt and Adelson (1983), Farouki and Hinds (1985), Huitrie and Nahas (1985), Kochanek and Bartels (1984), and Pavlidis (1983).

Other solid-modeling references include Casale and Stanton (1985), Requicha (1980), Requicha and Voelcker (1982 and 1983), Tilove (1984), and Weiler (1985).

For further information on fractals see Mandelbrot (1977 and 1982) and Norton (1982). Related methods for modeling objects with irregular outlines are discussed in Fournier, Fussel, and Carpenter (1982) and in Reeves (1983).

Octrees and quadtrees are discussed by Carlbom, Chakravarty, and Vanderschel (1985), Doctor (1981), Frieder, Gordon, and Reynold (1985), Pavlidis (1982), and by Yamaguchi, Kunii, and Fujimura (1984).

EXERCISES

10–1. Set up geometric data tables as in Fig. 10–2 for a unit cube.

10–2. Develop alternate representations for the unit cube in Ex. 10–1 using only: (a) vertex and polygon tables, and (b) a single polygon table. Compare the three methods for representing the unit cube and estimate storage requirements for each.

10–3. Define an efficient polygon representation for a cylinder. Justify your choice of representation.

10–4. Set up a procedure for establishing polygon tables for any input set of data points defining an object.

10–5. Devise routines for checking the data tables in Fig. 10–2 for consistency and completeness.

10–6. Write a program that calculates parameters *A, B, C,* and *D* for any set of three-dimensional plane surfaces defining an object.

10–7. Given the plane parameters *A, B, C,* and *D* for all surfaces of an object, devise an algorithm to determine whether any specified point is inside or outside the object.

10–8. How would the values for the parameters *A, B, C, D* in the equation of a plane surface have to be altered if the coordinate reference is changed from a right-handed system to a left-handed system?

10–9. Using Eqs. 10–11 and 10–12, express the Bézier blending functions for four control points as explicit polynomial functions. Plot each function and label the maximum and minimum values.

10–10. Carry out the operations in Ex. 10–9 for five control points.

10–11. Revise the routines for calculating points along a Bézier curve to make use of forward differences. Limit the output to cubic curves.

10–12. Implement the routines in Ex. 10–11 to display Bézier curves for control points specified in the *xy* plane.

10–13. Using Eqs. 10–15 and 10–16, determine explicit polynomial expressions for uniform, quadratic B-spline blending functions using five control points. Plot each function and label the maximum and minimum values.

10–14. Carry out the operations in Ex. 10–13 for six control points.

10–15. Write a program using forward differences to calculate points along a cubic B-spline curve, given an input set of control points.

10–16. Implement the program in Ex. 10–15 to display cubic B-spline curves using control points defined in the *xy* plane.

10–17. Develop an algorithm for calculating the normal vector to a Bézier surface at the point $\mathbf{P}(u, v)$.

10–18. Devise an algorithm that uses forward differences to calculate points along a quadratic curve.

10–19. Derive the expressions given in Eqs. (10–21), (10–23), and (10–25) for calculating the three forward differences of a cubic curve.

10–20. Write a program for generating a fractal "snowflake" by applying the procedure shown in Fig. 10–25 to an equilateral triangle.

10–21. Using the method shown in Fig. 10–25 as a guide, develop a fractal curve using a different transformation pattern. What is the dimension of your curve?

10–22. Write a program to generate fractal curves using the equation $f(z) = z^2 + \lambda$, where λ is any selected complex constant.

10–23. Write a program to generate fractal curves using the equation $i(z^2 + 1)$, where $i = \sqrt{-1}$.

10–24. Set up procedures for generating the description of a three-dimensional object from input parameters that define the object in terms of a translational sweep.

10–25. Develop procedures for generating the description of a three-dimensional object using input parameters that define the object in terms of a rotational sweep.

10–26. Devise an algorithm for generating solid objects as combinations of three-dimensional primitive shapes, each defined as a set of surfaces, using constructive solid-geometry methods.

10–27. Develop an algorithm for performing constructive solid-geometry modeling using a primitive set of solids defined in octree structures.

10–28. Develop an algorithm for encoding a two-dimensional scene as a quadtree representation.

10–29. Set up an algorithm for loading a quadtree representation of a scene into a frame buffer for display of the scene.

10–30. Write a routine to convert the polygon definition of a three-dimensional object into an octree representation.

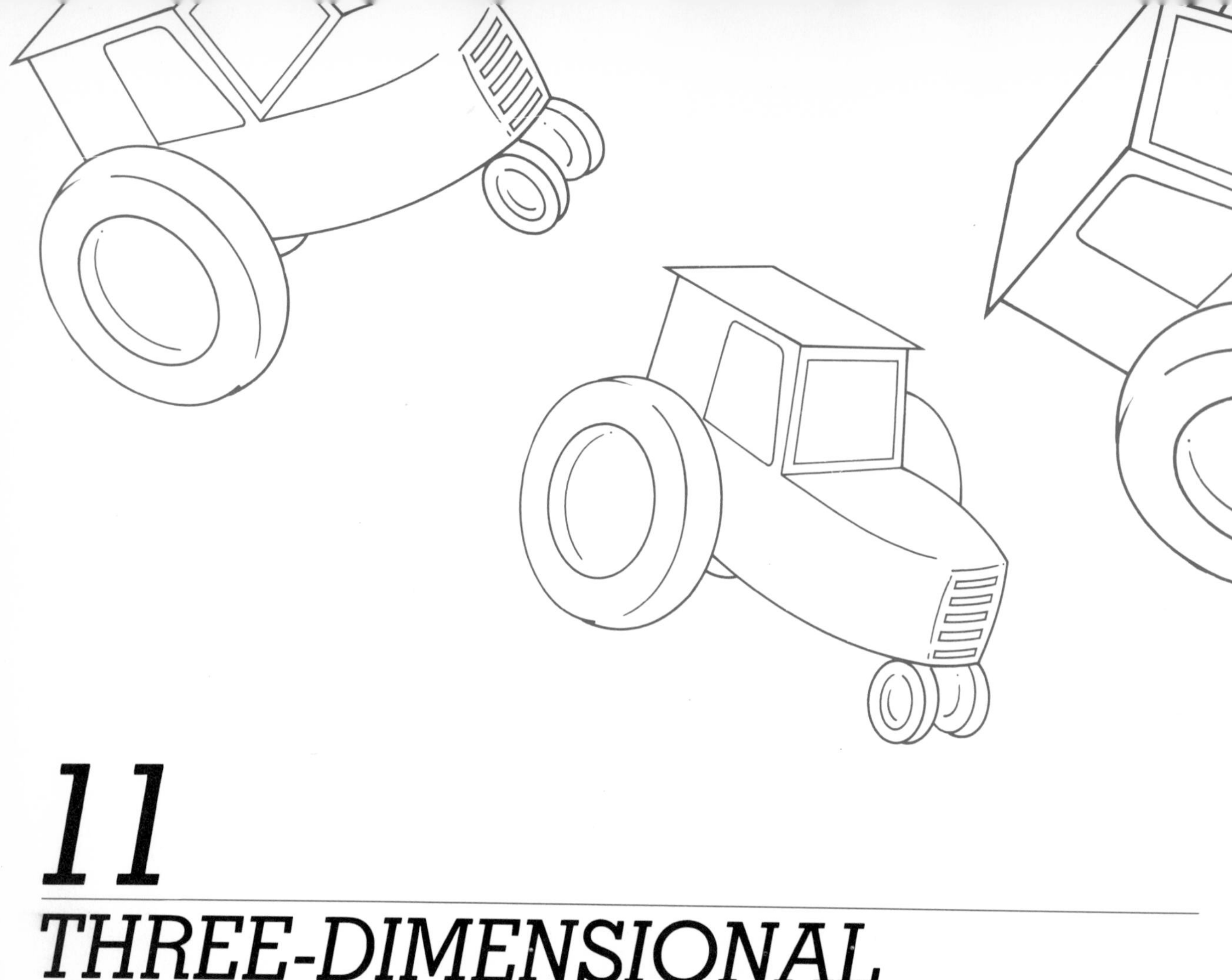

11

THREE-DIMENSIONAL TRANSFORMATIONS

Methods for translating, scaling, and rotating objects in three dimensions are extended from two-dimensional methods by including considerations for the z coordinate. A translation is now accomplished by specifying a three-dimensional translation vector, and scaling is performed by specifying three scaling factors. Extensions for three-dimensional rotations are less straightforward, since rotations can now be performed about axes with any spatial orientation. As in the two-dimensional case, geometric transformation equations can be expressed in terms of transformation matrices. Any sequence of transformations is then represented as a single matrix, formed by concatenating the matrices for the individual transformations in the sequence.

11–1 Translation

In a three-dimensional homogeneous coordinate representation, a point is translated (Fig. 11–1) from position (x, y, z) to position (x', y', z') with the matrix operation

$$[x' \; y' \; z' \; 1] = [x \; y \; z \; 1] \begin{bmatrix} 1 & 0 & 0 & 0 \\ 0 & 1 & 0 & 0 \\ 0 & 0 & 1 & 0 \\ Tx & Ty & Tz & 1 \end{bmatrix} \tag{11–1}$$

Parameters Tx, Ty, Tz, specifying translation distances for the coordinates, are assigned any real values. The matrix representation in Eq. 11–1 is equivalent to the three equations

$$x' = x + Tx, \quad y' = y + Ty, \quad z' = z + Tz \tag{11–2}$$

An object is translated in three dimensions by transforming each defining point of the object. Translation of an object represented as a set of polygon surfaces is carried out by translating the coordinate values for each vertex of each surface (Fig. 11–2). The set of translated coordinate positions for the vertices then defines the new position of the object.

We obtain the inverse of the translation matrix in Eq. 11–1 by negating the translation distances Tx, Ty, and Tz. This produces a translation in the opposite direction, and the product of a translation matrix and its inverse produces the identity matrix.

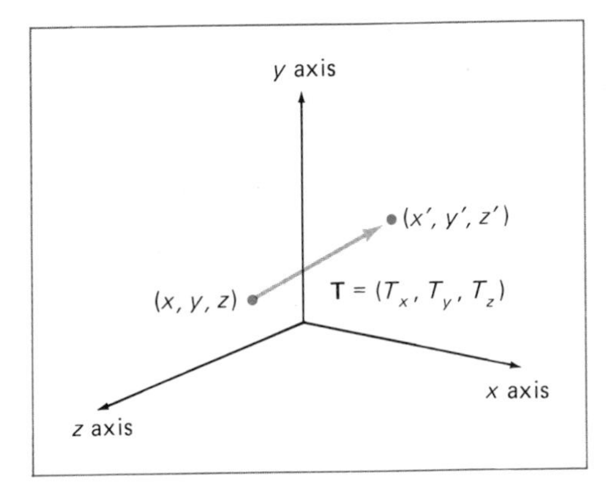

FIGURE 11–1
Translating a point with translation vector (Tx, Ty, Tz).

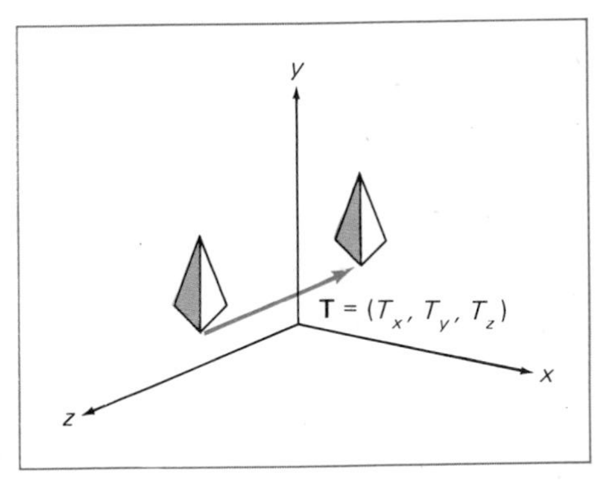

FIGURE 11–2
Translating an object with translation vector **T**.

FIGURE 11–3
Doubling the size of an object with transformation 11–3 also moves the object farther from the origin.

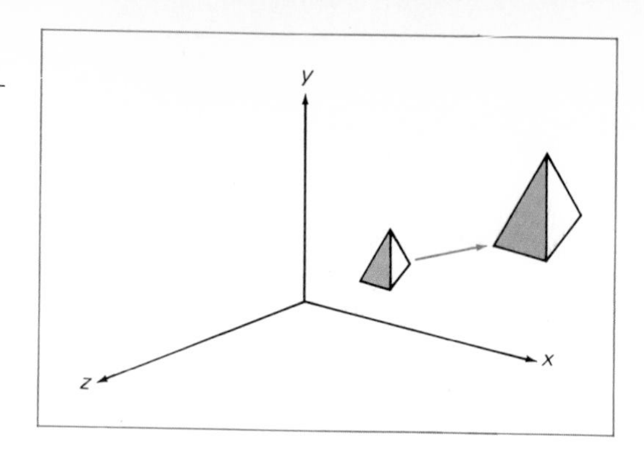

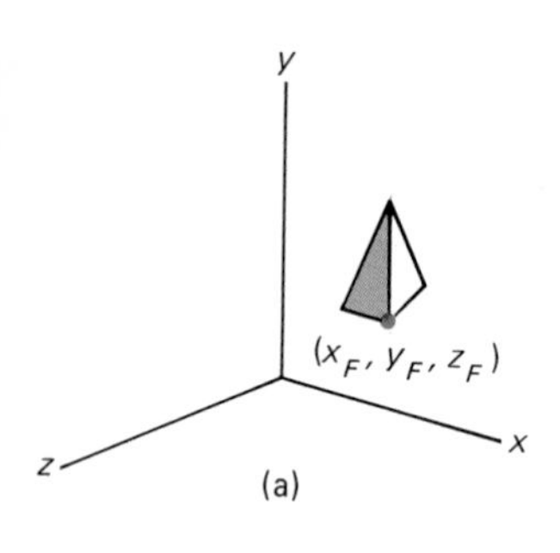

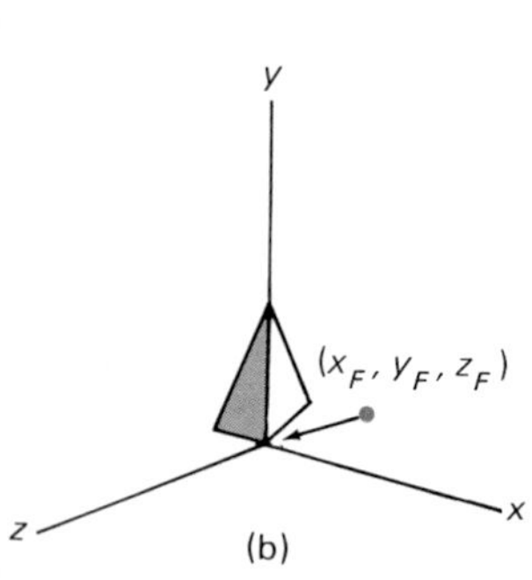

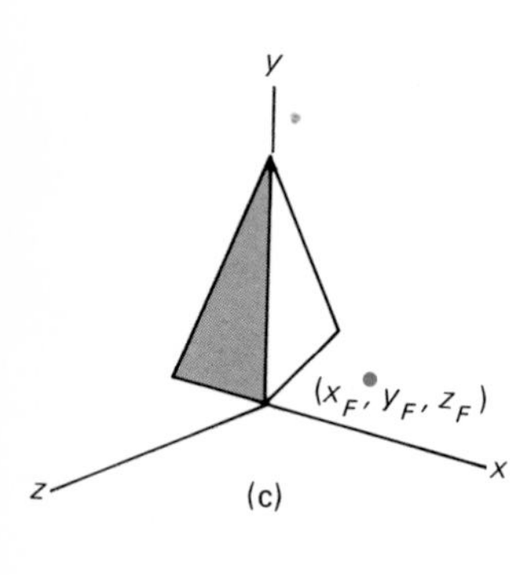

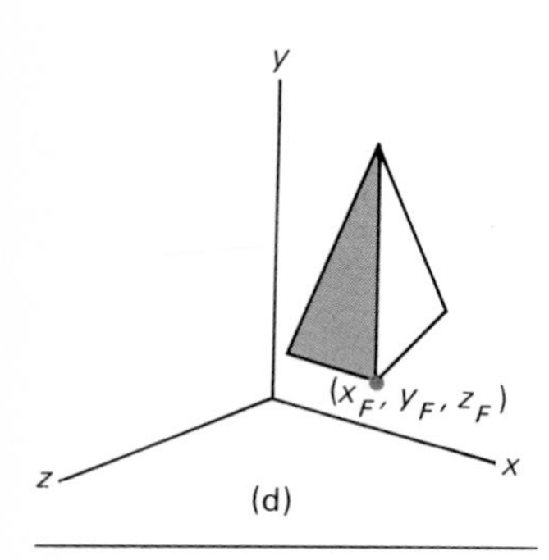

FIGURE 11–4
Scaling an object relative to a selected fixed point is equivalent to the sequence of transformations shown.

11–2 Scaling

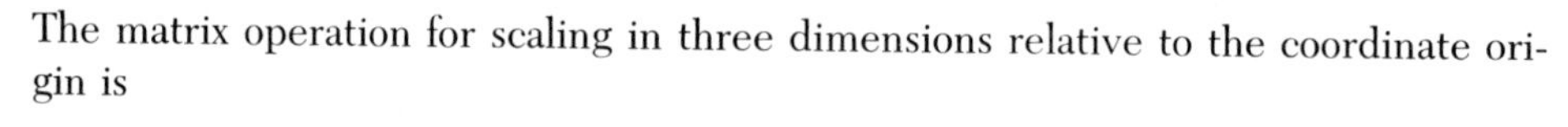

The matrix operation for scaling in three dimensions relative to the coordinate origin is

$$[x' \; y' \; z' \; 1] = [x \; y \; z \; 1] \begin{bmatrix} Sx & 0 & 0 & 0 \\ 0 & Sy & 0 & 0 \\ 0 & 0 & Sz & 0 \\ 0 & 0 & 0 & 1 \end{bmatrix} \tag{11–3}$$

where scaling parameters Sx, Sy, and Sz are assigned any positive values. This matrix operation scales a point at (x, y, z) to position (x', y', z') with the scaling equations

$$x' = x \cdot Sx, \quad y' = y \cdot Sy, \quad z' = z \cdot Sz \tag{11–4}$$

When transformation 11–3 is applied to defining points in an object, the object is scaled and moved relative to the coordinate origin. If the transformation parameters are not all equal, relative dimensions in the object are also changed. A uniform scaling ($Sx = Sy = Sz$) preserves the original shape of an object. The result of scaling an object uniformly with each scaling parameter set to 2 is shown in Fig. 11–3.

Scaling with respect to any fixed position (x_F, y_F, z_F) can be obtained by carrying out the following sequence of transformations: (1) translate the fixed point to the origin, (2) scale with Eq. 11–3, and (3) translate the fixed point back to its original position. This sequence of transformations is demonstrated in Fig. 11–4. The transformation matrix for arbitrary fixed-point scaling is formed as the concatenation of these individual transformations:

$$T(-x_F, -y_F, -z_F) \cdot S(Sx, Sy, Sz) \cdot T(x_F, y_F, z_F) = \begin{bmatrix} Sx & 0 & 0 & 0 \\ 0 & Sy & 0 & 0 \\ 0 & 0 & Sz & 0 \\ (1 - Sx)x_F & (1 - Sy)y_F & (1 - Sz)z_F & 1 \end{bmatrix} \tag{11–5}$$

where T represents the translation matrix and S the scaling matrix.

We form the inverse scaling matrix for either Eq. 11–3 or Eq. 11–5 by replacing the scaling parameters Sx, Sy, and Sz with their reciprocals. The inverse matrix generates an opposite scaling transformation, so the concatenation of any scaling matrix and its inverse produces the identity matrix.

11–3 Rotation

To specify a rotation transformation for an object, we must designate an axis of rotation (about which the object is to be rotated) and the amount of angular rotation. For two-dimensional applications, the axis of rotation is always perpendicular to the xy plane. In three dimensions, an axis of rotation can have any spatial orientation. The easiest rotation axes to handle are those that are parallel to the coordinate axes. Also, we can use rotations about the three coordinate axes to produce a rotation about any arbitrarily specified axis of rotation.

We adopt the convention that counterclockwise rotations about a coordinate axis are produced with positive rotation angles, if we are looking along the positive half of the axis toward the coordinate origin (Fig. 11–5). This convention is in accord with our earlier discussion of rotation equations in two dimensions, which specify rotations about the z axis.

The two-dimensional z-axis rotation equations are easily extended to three dimensions:

$$\begin{aligned} x' &= x\cos\theta - y\sin\theta \\ y' &= x\sin\theta + y\cos\theta \\ z' &= z \end{aligned} \qquad (11\text{–}6)$$

Parameter θ specifies the rotation angle. In homogeneous coordinate form, the three-dimensional z-axis rotation equations are expressed as

$$[x' \; y' \; z' \; 1] = [x \; y \; z \; 1] \begin{bmatrix} \cos\theta & \sin\theta & 0 & 0 \\ -\sin\theta & \cos\theta & 0 & 0 \\ 0 & 0 & 1 & 0 \\ 0 & 0 & 0 & 1 \end{bmatrix} \qquad (11\text{–}7)$$

Figure 11–6 illustrates rotation of an object about the z axis.

Transformation equations for rotations about the other two coordinate axes can be obtained with a cyclic permutation of the coordinate parameters in Eqs. 11–6. We replace x with y, y with z, and z with x. Using this permutation in Eqs. 11–6, we get the equations for an x-axis rotation:

$$\begin{aligned} y' &= y\cos\theta - z\sin\theta \\ z' &= y\sin\theta + z\cos\theta \\ x' &= x \end{aligned} \qquad (11\text{–}8)$$

which can be written in the homogeneous coordinate form

$$[x' \; y' \; z' \; 1] = [x \; y \; z \; 1] \begin{bmatrix} 1 & 0 & 0 & 0 \\ 0 & \cos\theta & \sin\theta & 0 \\ 0 & -\sin\theta & \cos\theta & 0 \\ 0 & 0 & 0 & 1 \end{bmatrix} \qquad (11\text{–}9)$$

Rotation around the x axis is demonstrated in Fig. 11–7.

Cyclically permuting coordinates in Eqs. 11–8 gives the transformation equations for a y-axis rotation:

$$\begin{aligned} z' &= z\cos\theta - x\sin\theta \\ x' &= z\sin\theta + x\cos\theta \\ y' &= y \end{aligned} \qquad (11\text{–}10)$$

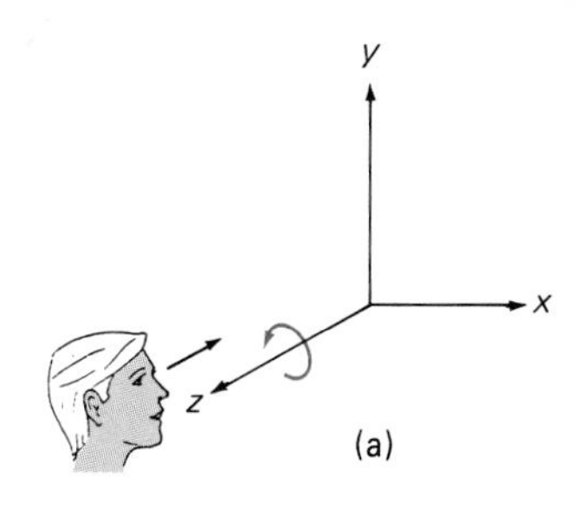

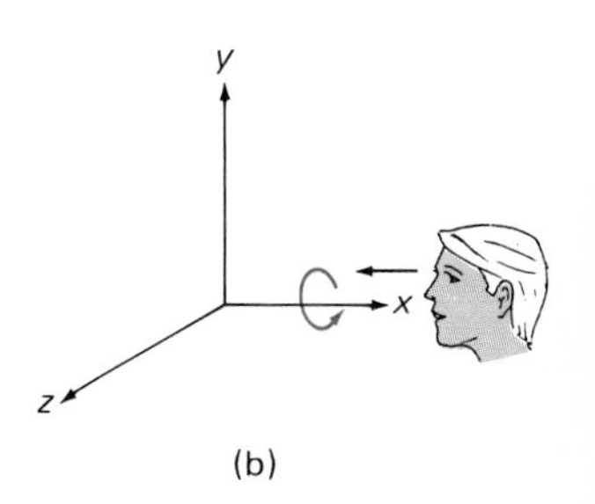

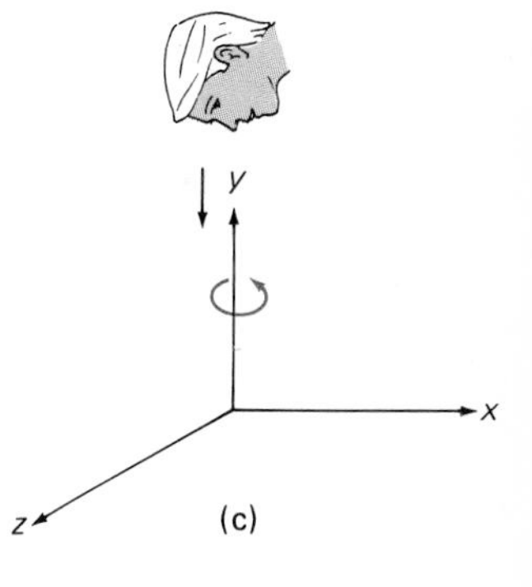

FIGURE 11–5
Positive rotation directions about the coordinate axes are counterclockwise, as viewed along the positive position of each axis toward the origin.

FIGURE 11–6
Rotation of an object about the z axis.

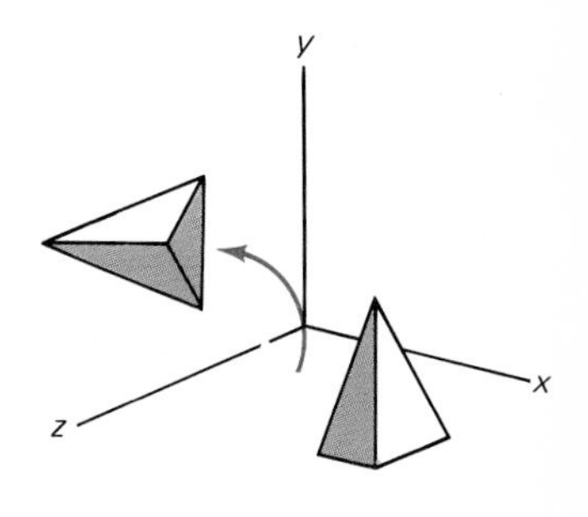

The matrix representation for y-axis rotation is

$$[x' \; y' \; z' \; 1] = [x \; y \; z \; 1] \begin{bmatrix} \cos\theta & 0 & -\sin\theta & 0 \\ 0 & 1 & 0 & 0 \\ \sin\theta & 0 & \cos\theta & 0 \\ 0 & 0 & 0 & 1 \end{bmatrix} \tag{11–11}$$

An example of y-axis rotation is shown in Fig. 11–8.

An inverse rotation matrix is formed by replacing the rotation angle θ by $-\theta$. Negative values for rotation angles generate rotations in a clockwise direction, so the identity matrix is produced when any rotation matrix is multiplied by its inverse. Since only the sine function is affected by the change in sign of the rotation angle, the inverse matrix can also be obtained by interchanging rows and columns. That is, we can calculate the inverse of any rotation matrix R by evaluating its transpose ($R^{-1} = R^T$). This method for obtaining an inverse matrix holds also for any composite rotation matrix.

FIGURE 11–7
Rotation of an object about the x axis.

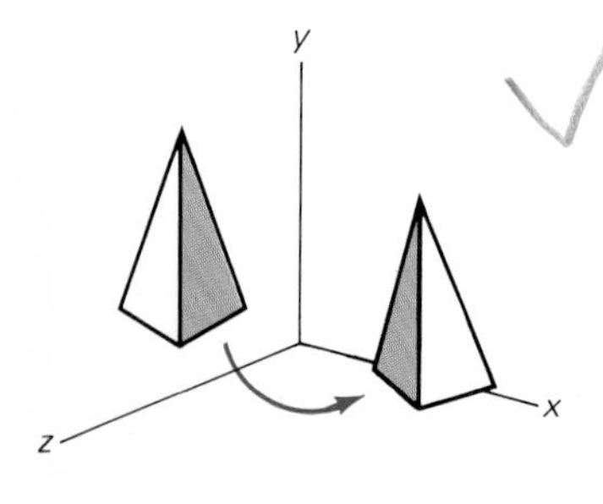

FIGURE 11–8
Rotation of an object about the y axis.

11–4 Rotation About an Arbitrary Axis

Objects can be rotated about any arbitrarily selected axis by applying a composite transformation matrix whose components perform a sequence of translations and rotations about the coordinate axes. This composite matrix is formed with combinations of the translation matrix and transformations 11–7, 11–9, and 11–11. The correct sequence of transformations can be determined by transforming coordinate positions of the object so that the selected rotation axis is moved onto one of the coordinate axes. Then the object is rotated about that coordinate axis through the

FIGURE 11–9
Sequence of transformations for rotating an object about an axis that is parallel to the x axis.

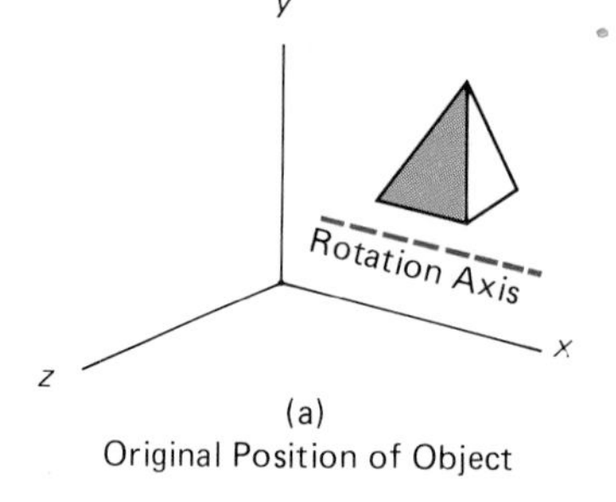

(a)
Original Position of Object

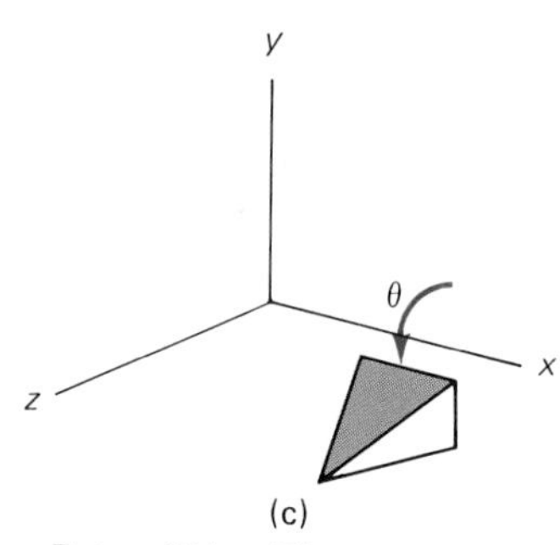

(c)
Rotate Object Through Angle θ

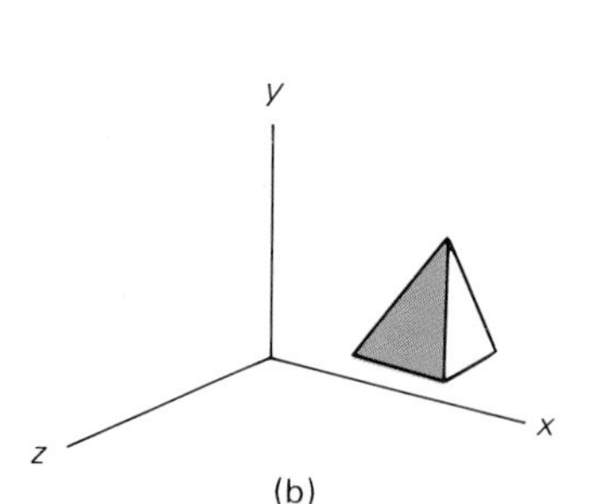

(b)
Translate Rotation Axis onto x axis

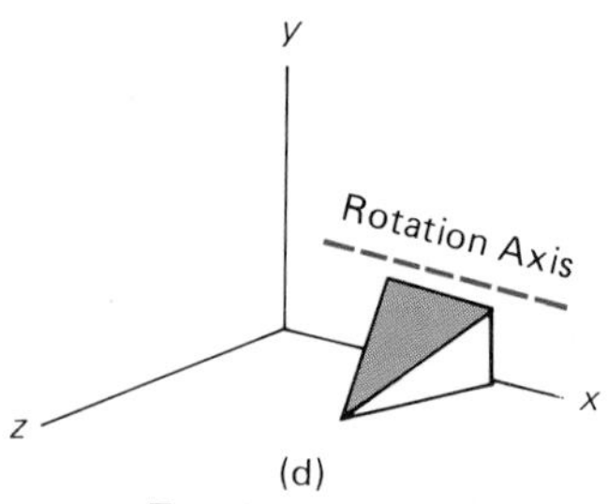

(d)
Translate Rotation Axis to Original Position

```
a := a / length ;  b := b / length;  c := c / length;
d := sqrt(b * b + c * c);

  {translate to origin}
create_translation_matrix_3 (-x1, -y1, -z1, tm);

  {rotate around x to lie in xz plane}
set_identity (m);
m[2,2] :=  c / d;  m[2,3] := b / d;
m[3,2] := -b / d;  m[3,3] := c / d;
accumulate_matrices_3 (tm, m, tm);

  {rotate around y to line up with z axis}
set_identity (m);
m[1,1] :=  d;  m[1,3] := a;
m[3,1] := -a;  m[3,3] := d;
accumulate_matrices_3 (tm, m, tm);

  {rotate around z axis using input angle}
create_z_rotation_matrix_3 (angle, m);
accumulate_matrices_3 (tm, m, tm);

  {do inverse of rotation around y}
set_identity (m);
m[1,1] := d;  m[1,3] := -a;
m[3,1] := a;  m[3,3] :=  d;
accumulate_matrices_3 (tm, m, tm);

{do inverse of rotation around x}
set_identity (m);
m[2,2] := c / d;  m[2,3] := -b / d;
m[3,2] := b / d;  m[3,3] :=  c / d;
accumulate_matrices_3 (tm, m, tm);

  {do inverse of translation}
create_translation_matrix_3 (x1, y1, z1, m);
accumulate_matrices_3 (tm, m, tm)
end;  {create_3d_rotation_matrix}
```

REFERENCES

For further discussions of homogeneous coordinates and three-dimensional transformation methods see Blinn (1977), Catmull and Smith (1980), and Pavlidis (1982).

EXERCISES

11–1. Write a program to implement the *accumulate_matrices_3* function by calculating the product of any two input transformation matrices.

11–2. Extend the program of Ex. 11–1 by allowing the input transformation matrices to be established by the transformation operations for translation, scaling, and rotation. That is, the program is to be expanded to implement the individual routines for setting these transformation matrices.

11–3. Expand the program of Ex. 11–2 to implement the *set_segment_transformation_3* operation. Given a definition of a three-dimensional object and a sequence of transformations, the program is to determine the composite matrix and apply this matrix to the object definition to obtain the transformed object.

11–4. Derive the transformation matrix for scaling an object by a scaling factor S in a direction defined by the direction cosines α, β, γ.

11–5. Develop an algorithm for scaling an object defined in an octree representation.

11–6. Write a procedure for rotating a given object about any specified rotation axis.

11–7. Develop a procedure for animating an object by incrementally rotating it about any specified axis. Use appropriate approximations to the trignometric equations to speed up the calculations, and reset the object to its initial position after each complete revolution about the axis.

11–8. Devise a procedure for rotating an object that is represented in an octree structure.

?11–9. Develop a routine to reflect an object about an arbitrarily selected plane.

11–10. Write a program to shear an object with respect to any of the three coordinate axes, using input values for the shearing parameters.

11–11. Develop a procedure for converting an object definition in one coordinate reference to any other coordinate system defined relative to the first system.

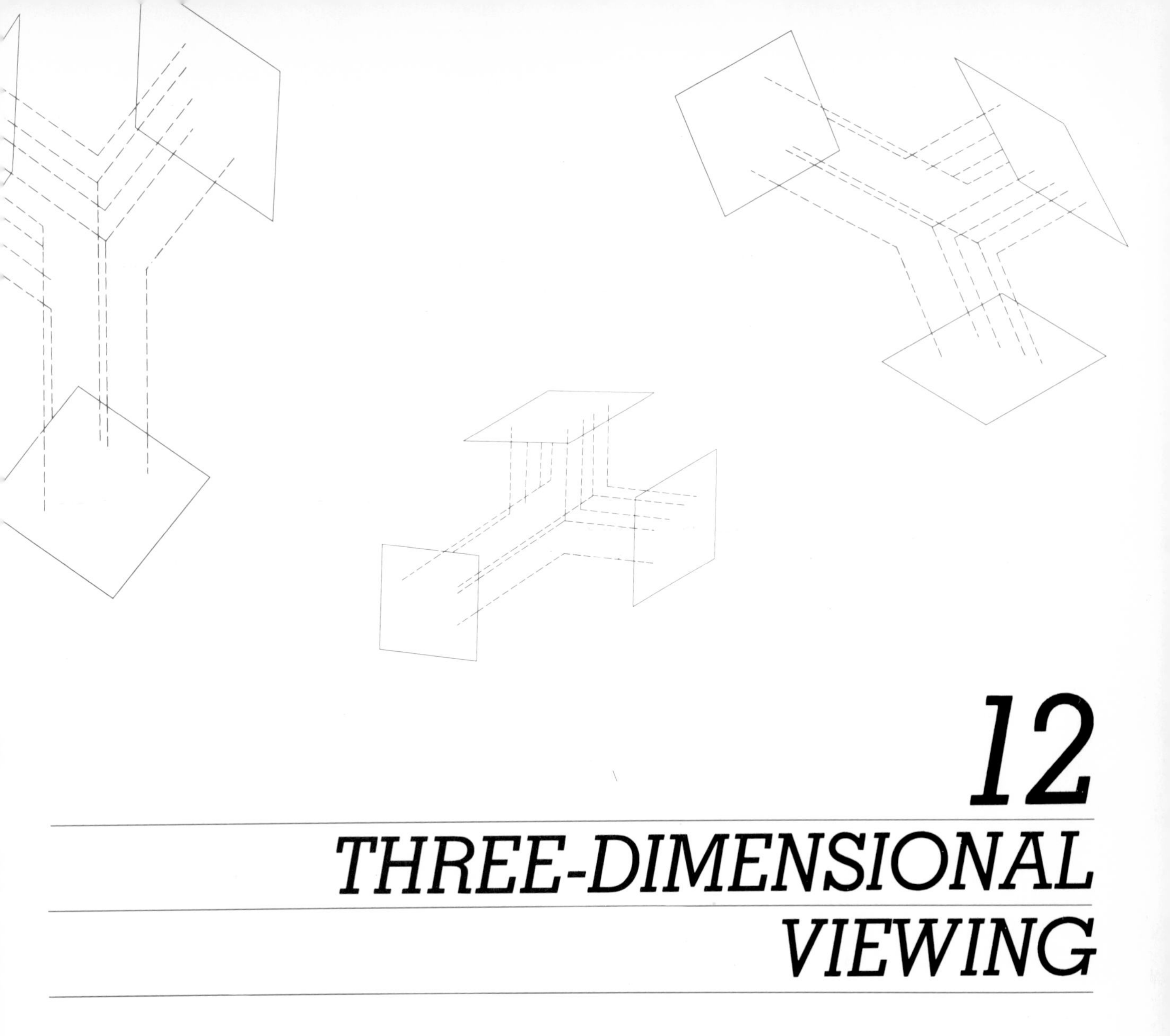

12 THREE-DIMENSIONAL VIEWING

In two dimensions, viewing operations transfer two-dimensional points in the world coordinate plane to two-dimensional points in the device coordinate plane. Object definitions, clipped against a window boundary, are mapped into a viewport. These normalized device coordinates are then converted to device coordinates, and the object is displayed on the output device. For three dimensions, the situation is a bit more complicated, since we now have some choices as to how views are to be generated. We could view a scene from the front, from above, or from the back. Or we could generate a view of what we would see if we were standing in the middle of a group of objects. Additionally, three-dimensional descriptions of objects must be projected onto the flat viewing surface of the output device. In this chapter, we first discuss the mechanics of projection. Then the operations involved in the viewing transformation are explored, and a full three-dimensional viewing pipeline is developed.

12–1 Projections

There are two basic methods for projecting three-dimensional objects onto a two-dimensional viewing surface. All points of the object can be projected to the surface along parallel lines, or the points can be projected along lines that converge to a position called the **center of projection.** The two methods, called **parallel projection** and **perspective projection,** respectively, are illustrated in Fig. 12–1. In both cases, the intersection of a projection line with the viewing surface determines the coordinates of the projected point on this projection plane. For the present, we assume that the view projection plane is the $z = 0$ plane of a left-handed coordinate system, as shown in Fig. 12–2.

A parallel projection preserves relative dimensions of objects, and this is the technique used in drafting to produce scale drawings of three-dimensional objects. This method is used to obtain accurate views of the various sides of an object, but a parallel projection does not give a realistic representation of the appearance of a three-dimensional object. A perspective projection, on the other hand, produces realistic views but does not preserve relative dimensions. Distant lines are projected as smaller than those closer to the projection plane, as seen in Fig. 12–3.

FIGURE 12–1
Two methods for projecting a line onto a viewing surface (projection plane).

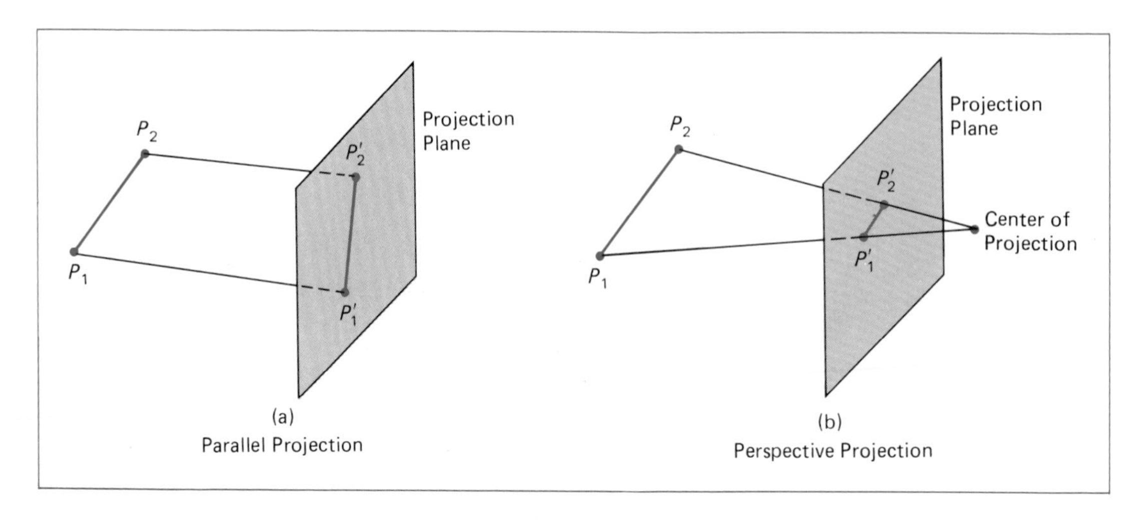

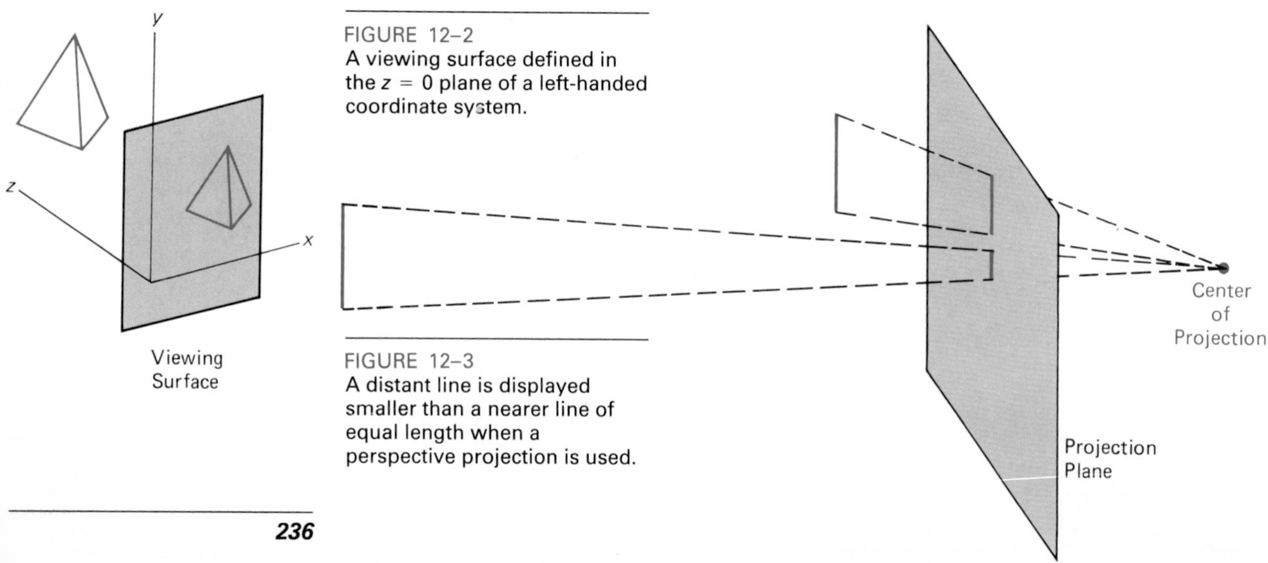

FIGURE 12–2
A viewing surface defined in the $z = 0$ plane of a left-handed coordinate system.

FIGURE 12–3
A distant line is displayed smaller than a nearer line of equal length when a perspective projection is used.

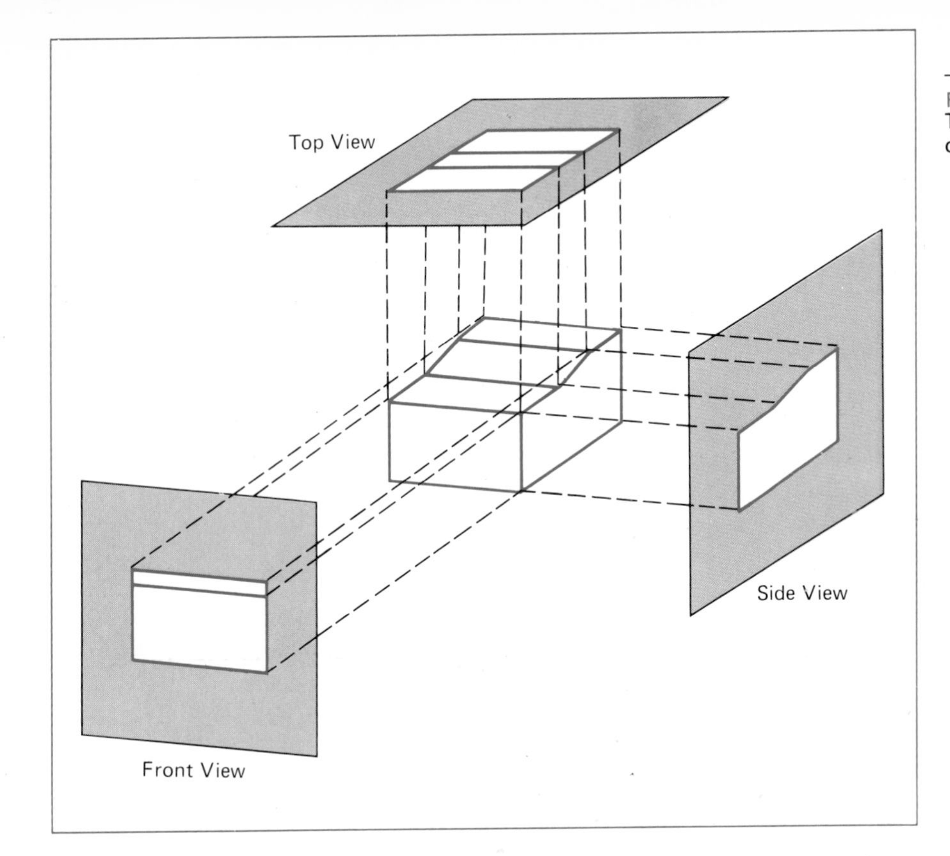

FIGURE 12–4
Three orthographic projections of an object.

Parallel Projections

Views formed with parallel projections can be characterized according to the angle that the direction of projection makes with the projection plane. When the direction of projection is perpendicular to the projection plane, we have an **orthographic projection.** A projection that is not perpendicular to the plane is called an **oblique projection.**

Orthographic projections are most often used to produce the front, side, and top views of an object, as shown in Fig. 12–4. Orthographic front, side, and rear views of an object are called "elevations", and the top views are called "plans". Engineering drawings commonly employ these orthographic projections, since lengths and angles are accurately depicted and can be measured from the drawings.

We can also form orthographic projections that show more than one face of an object. Such views are called **axonometric** orthographic projections. The most commonly used axonometric projection is the **isometric** projection. An isometric projection is obtained by aligning the projection plane so that it intersects each coordinate axis in which the object is defined (called the principal axes) at the same distance from the origin. Figure 12–5 shows an isometric projection. There are eight positions, one in each octant, for obtaining an isometric view. All three principal axes are foreshortened equally in an isometric projection so that relative proportions are maintained. This is not the case in a general axonometric projection, where scaling factors may be different for the three principal directions.

Transformation equations for performing an orthographic parallel projection are straightforward. For any point (x, y, z), the projection point (x_p, y_p, z_p) on the viewing surface is obtained as

$$x_p = x, \qquad y_p = y, \qquad z_p = 0 \tag{12-1}$$

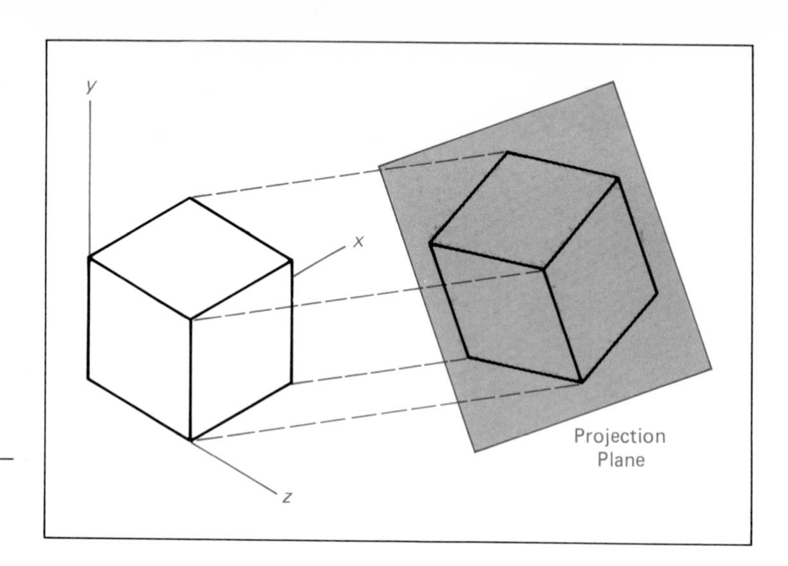

FIGURE 12–5
Isometric projection of an object onto a viewing surface.

An oblique projection is obtained by projecting points along parallel lines that are not perpendicular to the projection plane. Figure 12–6 shows an oblique projection of a point (x, y, z) along a projection line to position (x_p, y_p). Orthographic projection coordinates on the plane are (x, y). The oblique projection line makes an angle α with the line on the projection plane that joins (x_p, y_p) and (x, y). This line, of length L, is at an angle ϕ with the horizontal direction in the projection plane. We can express the projection coordinates in terms of x, y, L, and ϕ:

$$x_p = x + L\cos\phi \qquad (12\text{–}2)$$
$$y_p = y + L\sin\phi$$

A projection direction can be defined by selecting values for the angles α and ϕ. Common choices for angle ϕ are 30° and 45°, which display a combination view of the front, side, and top (or front, side, and bottom) of an object. Length L is a function of the z coordinate, and we can evaluate this parameter from the relationships

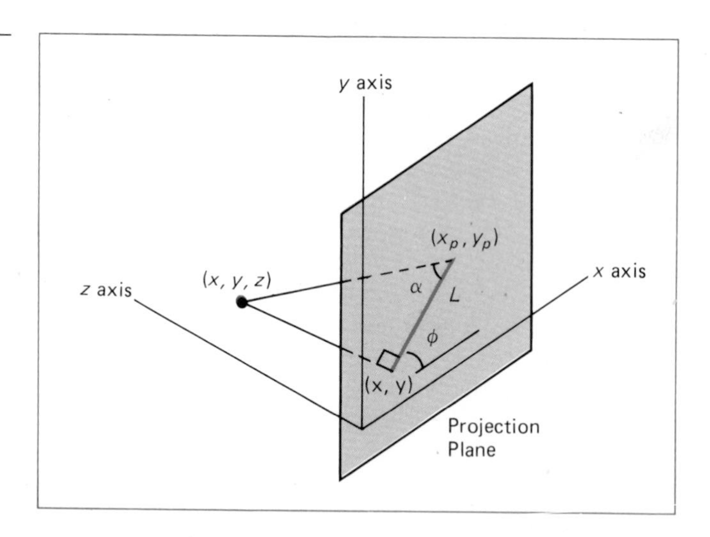

FIGURE 12–6
Oblique projection of point (x, y, z) to position (x_p, y_p) on the projection plane.

$$\tan \alpha = \frac{z}{L} = \frac{1}{L_1} \tag{12-3}$$

where L_1 is the length of the projection line from (x, y) to (x_p, y_p) when $z = 1$. From Eq. 12–3, we have

$$L = z\, L_1 \tag{12-4}$$

and the oblique projection equations 12–2 can be written as

$$\begin{aligned} x_p &= x + z(L_1 \cos\phi) \\ y_p &= y + z(L_1 \sin\phi) \end{aligned} \tag{12-5}$$

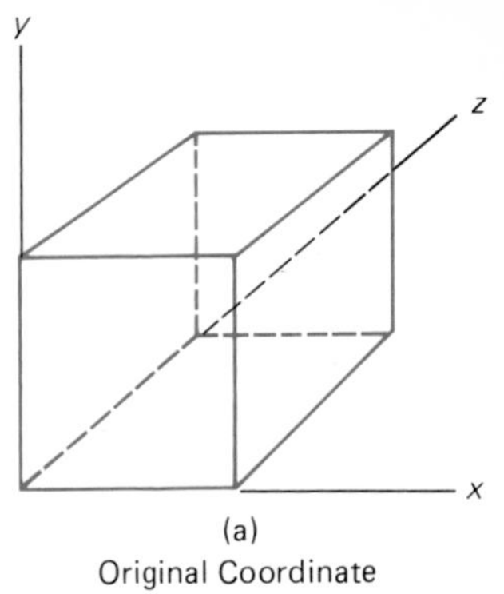

(a)
Original Coordinate Description of Object

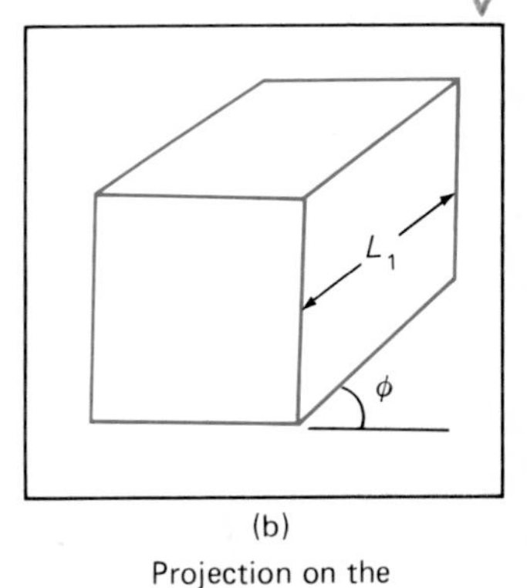

(b)
Projection on the Viewing Surface

FIGURE 12–7
Oblique projection of a box onto a viewing surface in the $z = 0$ plane.

The transformation matrix for producing any parallel projection can be written as

$$P_{\text{parallel}} = \begin{bmatrix} 1 & 0 & 0 & 0 \\ 0 & 1 & 0 & 0 \\ L_1 \cos\phi & L_1 \sin\phi & 0 & 0 \\ 0 & 0 & 0 & 1 \end{bmatrix} \tag{12-6}$$

An orthographic projection is obtained when $L_1 = 0$ (which occurs at a projection angle α of 90°). Oblique projections are generated with nonzero values for L_1. Projection matrix 12–6 has a structure similar to that of a z-axis shear matrix. In fact, the effect of this projection matrix is to shear planes of constant z and project them onto the view plane. The x- and y-coordinate values within each plane of constant z are shifted by an amount proportional to the z value of the plane so that angles, distances, and parallel lines in the plane are projected accurately. This effect is shown in Fig. 12–7, where the back plane of the box is sheared and overlapped with the front plane in the projection to the viewing surface. An edge of the box connecting the front and back planes is projected into a line of length L_1 that makes an angle ϕ with a horizontal line in the projection plane.

Two commonly used angles in oblique projections are those for which $\tan \alpha = 1$ and $\tan \alpha = 2$. For the first case, $\alpha = 45°$ and the views obtained are called **cavalier** projections. All lines perpendicular to the projection plane are projected with no change in length. Examples of cavalier projections for a cube are given in Fig. 12–8.

When the projection angle is chosen such that $\tan \alpha = 2$, the resulting view is called a **cabinet** projection. This projection angle of approximately 63.4° causes lines perpendicular to the viewing surface to be projected at one-half their length.

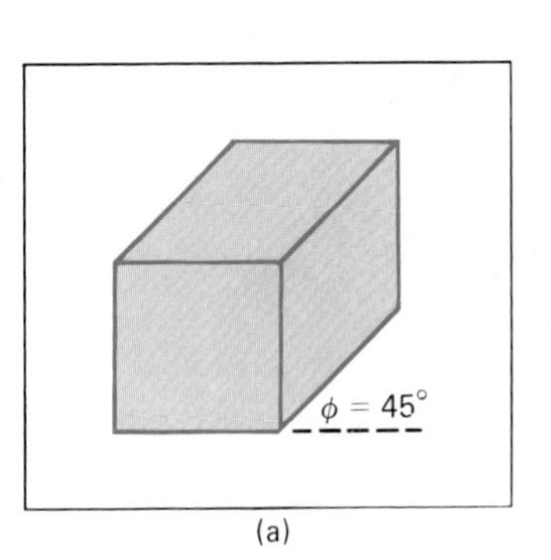

(a)

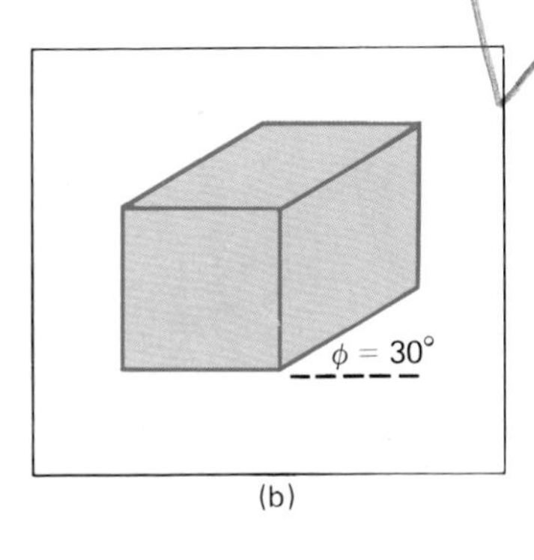

(b)

FIGURE 12–8
Cavalier projections of a cube onto a projection plane for two values of angle ϕ. Depth of the cube is projected equal to the width and height.

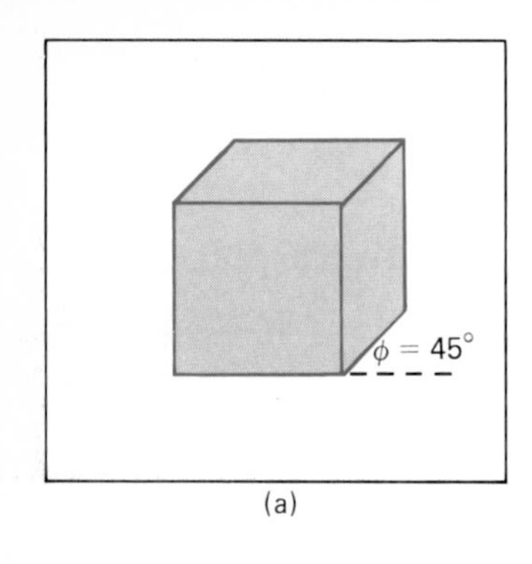

(a)

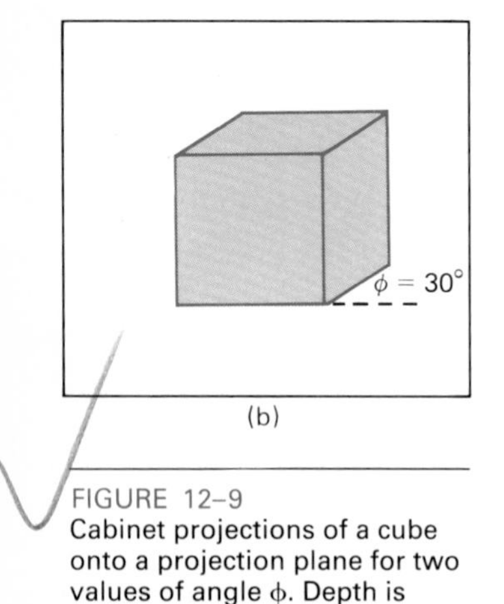

(b)

FIGURE 12–9
Cabinet projections of a cube onto a projection plane for two values of angle ϕ. Depth is projected as one-half that of the width and height.

Cabinet projections appear more realistic than cavalier projections because of this reduction in the length of perpendiculars. Figure 12–9 shows a cabinet projection for a cube.

Perspective Projections

To obtain a perspective projection of a three-dimensional object, we project points along projection lines that meet at the center of projection. In Fig. 12–10, the center of projection is on the negative z axis at a distance d behind the projection plane. Any position can be selected for the center of projection, but choosing a position along the z axis simplifies the calculations in the transformation equations.

We can obtain the transformation equations for a perspective projection from the parametric equations describing the projection line from point P to the center of projection in Fig. 12–10. The parametric form for this projection line is

$$\begin{aligned} x' &= x - xu \\ y' &= y - yu \\ z' &= z - (z + d)u \end{aligned} \tag{12–7}$$

Parameter u takes values from 0 to 1, and coordinates (x', y', z') represent any position along the projection line. When $u = 0$, Eqs. 12–7 yield point P at coordinates (x, y, z). At the other end of the line $u = 1$, and we have the coordinates for the center of projection, $(0, 0, -d)$. To obtain the coordinates on the projection plane, we set $z' = 0$ and solve for parameter u:

$$u = \frac{z}{z + d} \tag{12–8}$$

This value for parameter u produces the intersection of the projection line with the projection plane at $(x_p, y_p, 0)$. Substituting Eq. 12–8 into Eqs. 12–7, we obtain the perspective transformation equations

$$x_p = x\left(\frac{d}{z + d}\right) = x\left(\frac{1}{z/d + 1}\right)$$

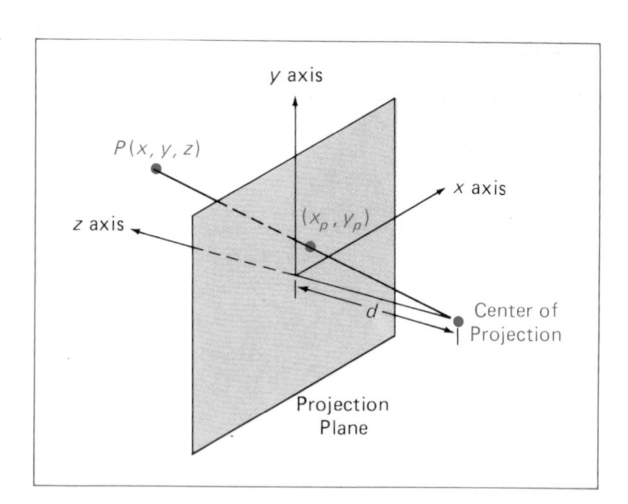

FIGURE 12–10.
Perspective projection of a point P at coordinates (x, y, z) to position $(xp, yp, 0)$ on a projection plane.

viewing coordinate xy plane that is at a specified distance from the view reference point. For our discussions, we assume that the view plane is the xy plane at the viewing coordinate origin. This allows us to project onto the $z = 0$ plane.

To generate a view from the user-specified vantage point, positions defined relative to the world coordinate origin must be redefined relative to the viewing coordinate origin. That is, we must transform the coordinates from the world coordinate system to the viewing coordinate system. This transformation is accomplished with the sequence of translations and rotations that map the viewing system axes onto the world coordinate axes. When applied to the world coordinate definition of the objects in the scene, this sequence converts them to their positions within the viewing coordinate system. The matrix representing this sequence of transformations can be obtained by concatenating the following transformation matrices:

1. Reflect relative to the xy plane, reversing the sign of each z coordinate. This changes the left-handed viewing coordinate system to a right-handed system.
2. Translate the view reference point to the origin of the world coordinate system.
3. Rotate about the world coordinate x axis to bring the viewing coordinate z axis into the xz plane of the world coordinate system.
4. Rotate about the world coordinate y axis until the z axes of both systems are aligned.
5. Rotate about the world coordinate z axis to align the viewing and world y axes.

The effect of each of these transformations is shown in Fig. 12–18. This sequence has much in common with the transformation sequence that rotates an object about

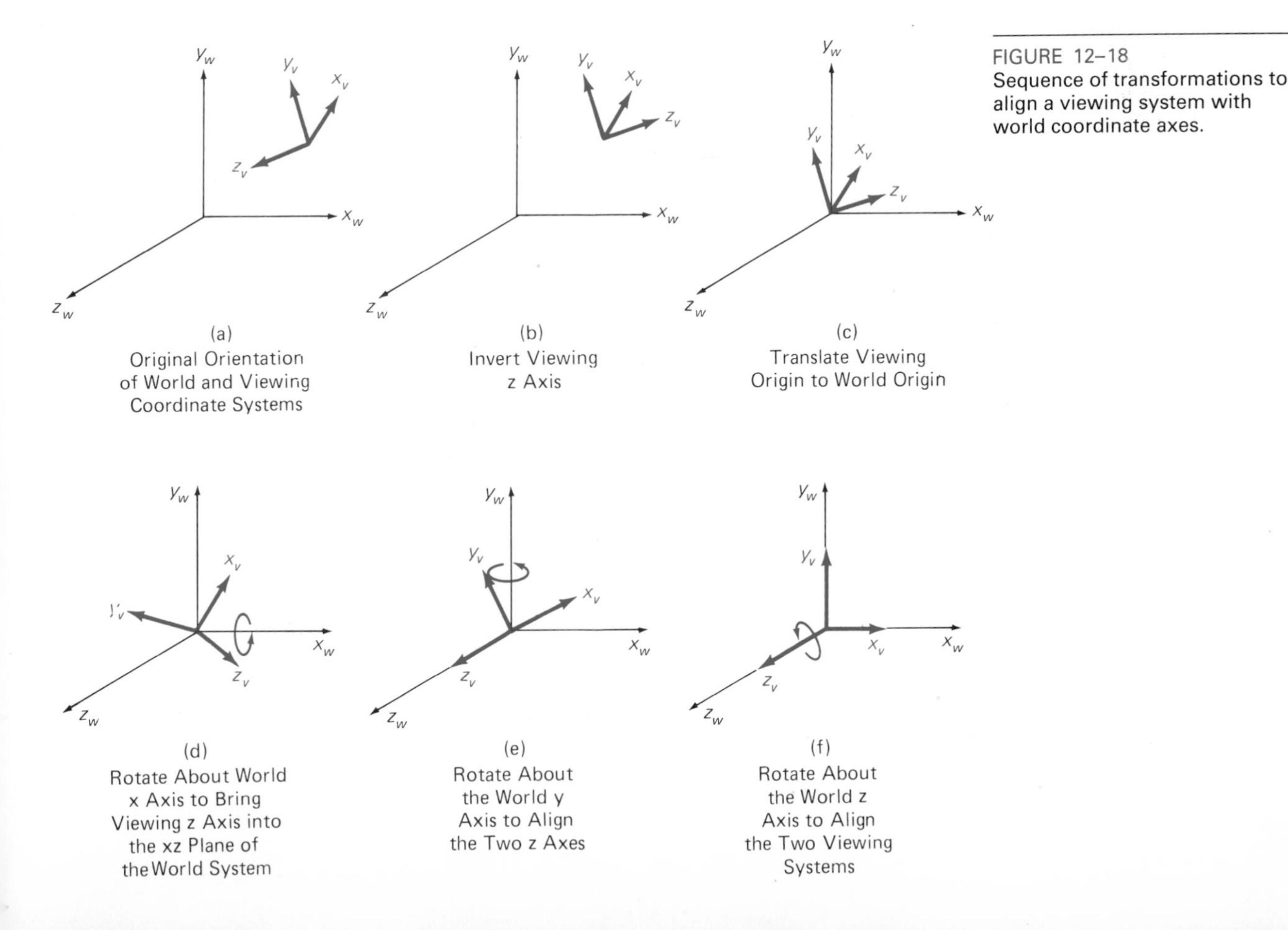

(a)
Original Orientation of World and Viewing Coordinate Systems
(b)
Invert Viewing z Axis
(c)
Translate Viewing Origin to World Origin
(d)
Rotate About World x Axis to Bring Viewing z Axis into the xz Plane of the World System
(e)
Rotate About the World y Axis to Align the Two z Axes
(f)
Rotate About the World z Axis to Align the Two Viewing Systems

FIGURE 12–18
Sequence of transformations to align a viewing system with world coordinate axes.

an arbitrary axis, and the viewing matrix components can be determined using techniques similar to those for arbitrary rotation. For packages that use a right-handed viewing coordinate system, the z coordinate inversion in step 1 is unnecessary.

View Volumes

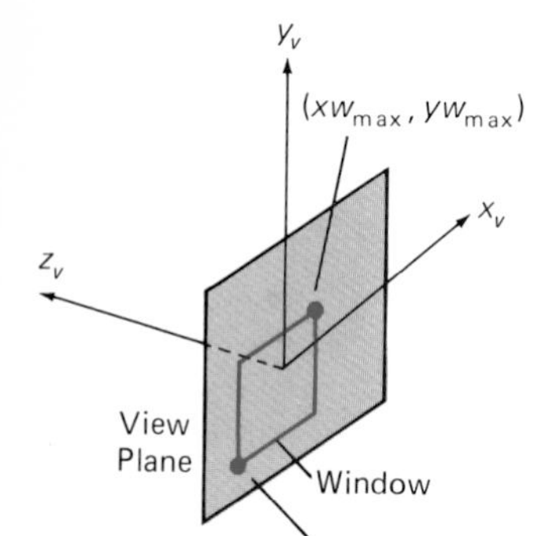

FIGURE 12–19
Window specification on the view plane, with minimum and maximum coordinates given in the viewing reference system.

In the camera analogy, the type of lens used on the camera is one factor that determines how much of the scene is caught on film. A wide-angle lens takes in more of the scene than a regular lens. In three-dimensional viewing, a **projection window** is used to the same effect. The window is defined by minimum and maximum values for x and y on the view plane, as shown in Fig. 12–19. Viewing coordinates are used to give the limits of the window, which may appear anywhere on the view plane.

The projection window is used to define a **view volume.** Only those objects within the view volume are projected and displayed on the view plane. The exact shape of the view volume depends on the type of projection requested by a user. In any case, four sides of the volume pass through the edges of the window. For a parallel projection, these four sides of the view volume form an infinite parallelepiped, as in Fig. 12–20. A truncated pyramid, with apex at the center of projection (Fig. 12–21), is used as the view volume for a perspective projection. This truncated pyramid is called a **frustum.**

Some packages restrict the coordinates of the center of projection to positions along the z axis of the viewing coordinate system. We take a more general approach and allow the center of projection to be located at any position in the viewing system. Figure 12–22 shows two orientations of the view volume pyramid relative to the viewing axes. In Fig. 12–22 (b), no points project to the view plane, since the center of projection and the objects to be viewed are on the same side of the view plane. In this case, nothing is displayed.

In parallel projections, the direction of projection defines the orientation of the view volume. By giving a position relative to the viewing coordinate origin, a user defines a vector that sets the orientation of the view volume relative to the view plane. Figure 12–23 shows the shape of view volumes for both orthographic and oblique parallel projections.

Often, one or two additional planes are used to further define the view volume. Including a **near plane** and a **far plane** produces a finite view volume bounded by six planes, as shown in Fig. 12–24. The near and far planes are always parallel to the view plane, and they are specified by distances from the view plane in viewing coordinates. Alternate names for the near and far planes are hither and yon planes and front and back planes.

With these planes, the user can eliminate parts of the scene from the viewing

FIGURE 12–20
View volume for a parallel projection.

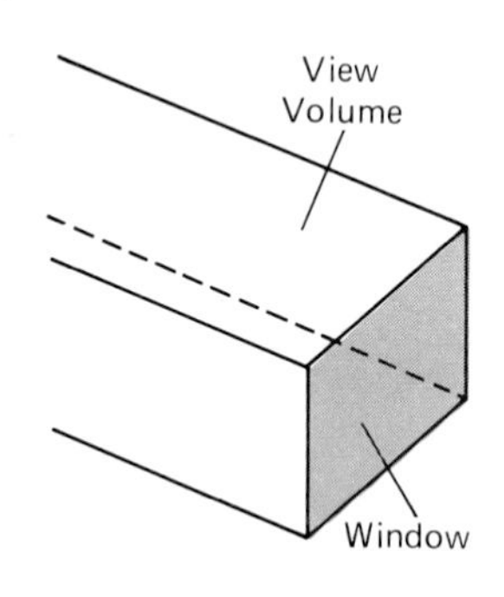

FIGURE 12–21
View volume for a perspective projection.

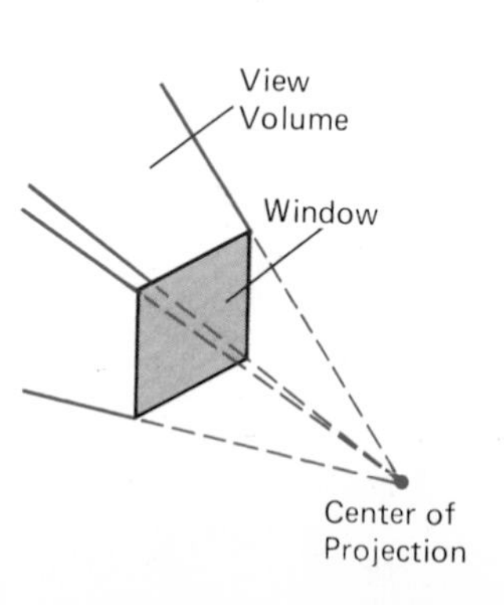

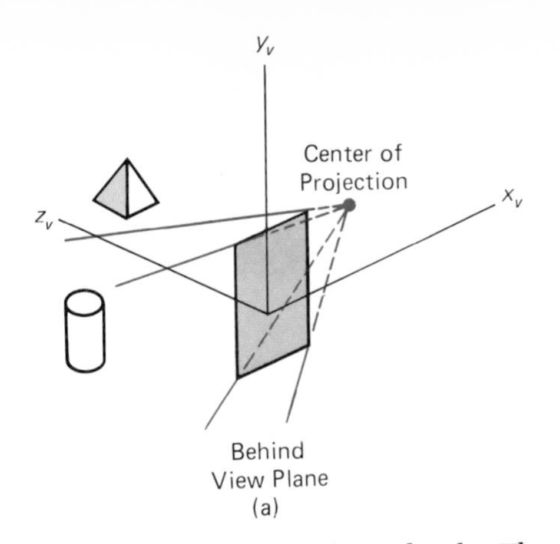

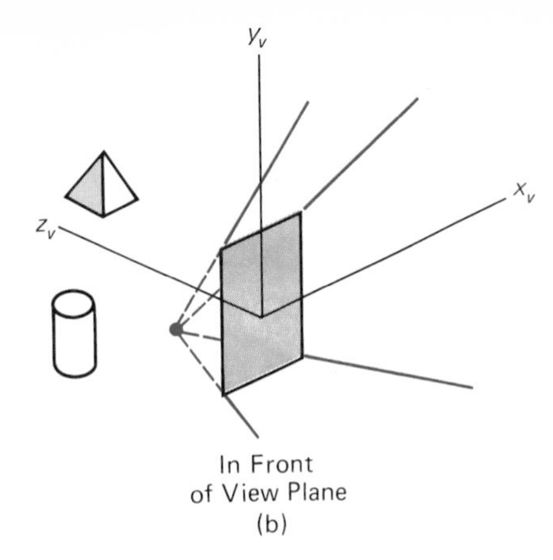

FIGURE 12–22
Perspective view volumes for two positions of the center of projection.

operations based on their depth. This is a particularly good idea when using a perspective projection. Objects that are very far from the view plane might project to only a single point. Very near objects could block out the other objects that a user wants to view. Or, when projected, near objects could be so large that they extend beyond the window boundaries and cannot be recognized.

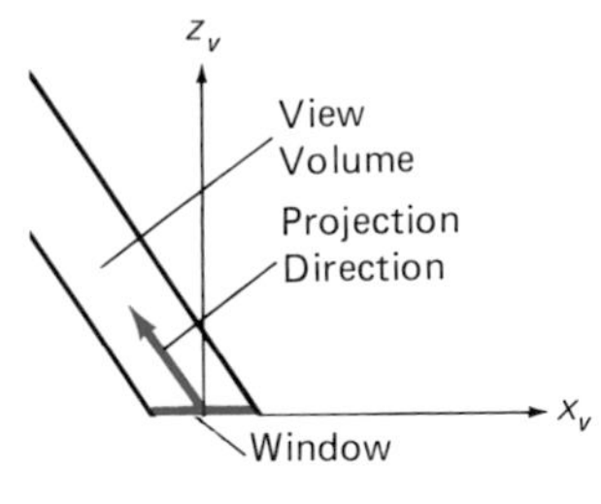

Clipping

An algorithm for three-dimensional clipping identifies and saves all line segments within the view volume for projection onto the view plane. All line segments outside the view volume are discarded. Clipping can be accomplished using an extension of two-dimensional line-clipping or polygon-clipping methods. The plane equations defining each boundary of the view volume can be used to test the relative positions of line endpoints and to locate intersection points.

By substituting the coordinates of a line endpoint into the plane equation of a boundary, we can determine whether the endpoint is inside or outside the boundary. An endpoint (x, y, z) of a line segment is outside a boundary plane if $Ax + By + Cz + D > 0$, where A, B, C, and D are the plane parameters for that boundary. Similarly, the point is inside the boundary if $Ax + By + Cz + D < 0$. Lines with both endpoints outside a boundary plane are discarded, and those with both endpoints inside all boundary planes are saved. The intersection of a line with a boundary is found using the line equations along with the plane equation. Intersection coordinates (x_I, y_I, z_I) are values that are on the line and that satisfy the plane equation $Ax_I + By_I + Cz_I + D = 0$.

Once the system has identified the objects that are interior to the view volume, they are projected to the view plane. All objects in the view volume will fall within the projection window. As in two dimensions, the contents of the window are mapped to a user-specified projection viewport. This normalizes the coordinates, which are then converted to the appropriate device coordinates for a particular display.

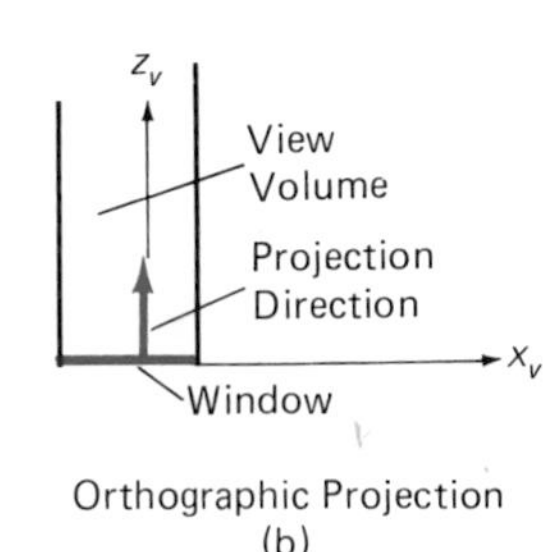

FIGURE 12–23
View volumes for oblique and orthographic parallel projections, as viewed in the *xz* plane.

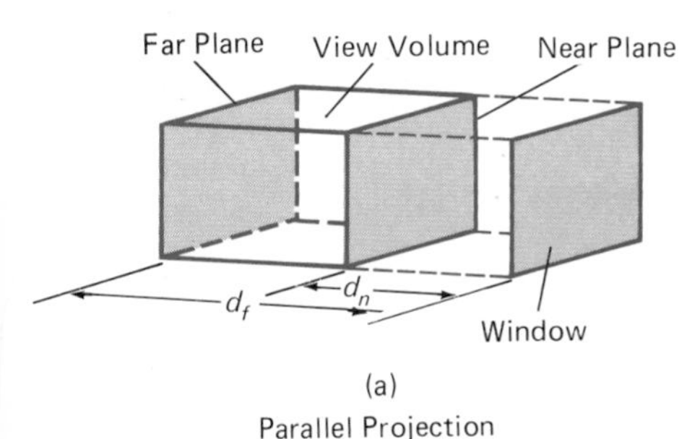

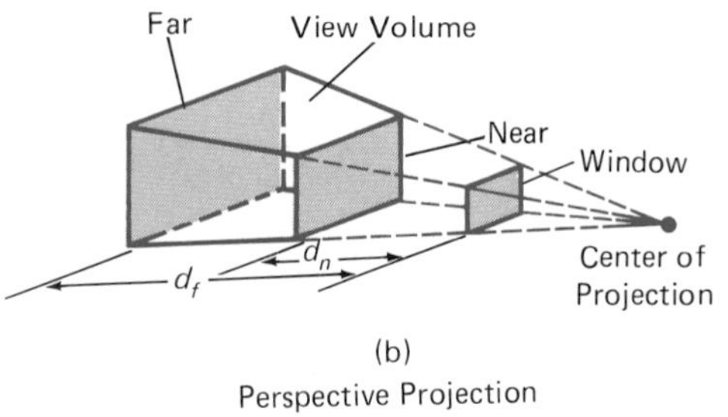

FIGURE 12–24
View volumes bounded by near and far planes and by top, bottom, and side planes. Distances to the near and far planes are specified by d_n and d_f.

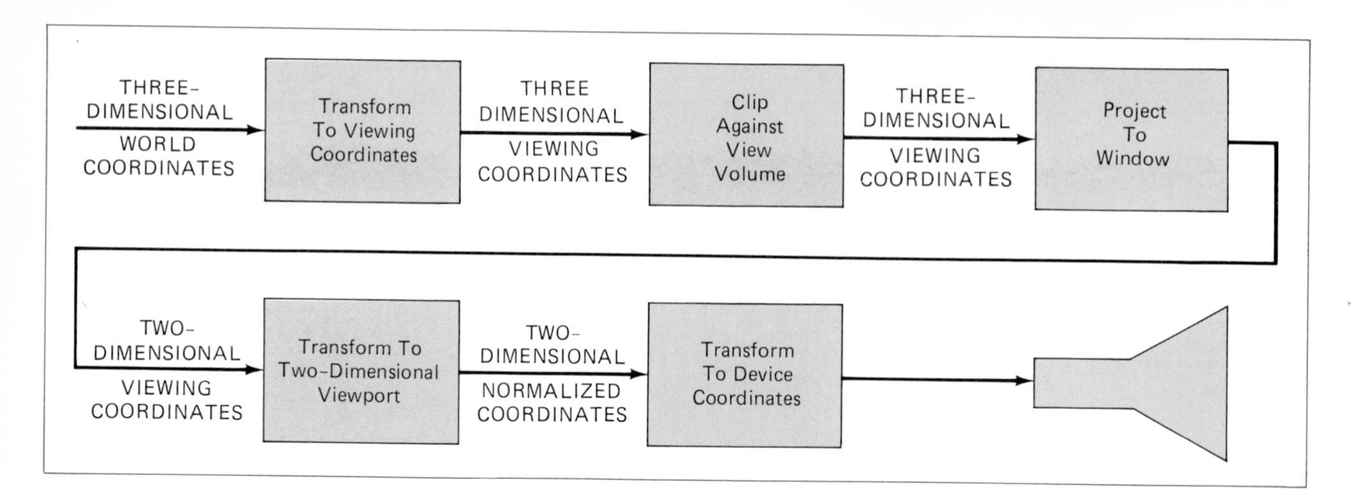

FIGURE 12–25
Logical operations in three-dimensional viewing.

12–3 Implementation of Viewing Operations

We can conceptualize a sequence of operations for performing viewing as in Fig. 12–25. First, world coordinate descriptions are converted to viewing coordinates. Next, the scene to be viewed is clipped against a view volume and projected into the window area defined on the view plane. This window is then mapped onto a viewport, designated in normalized device coordinates. The final step is to convert the normalized coordinate description into device coordinates and display the view on an output device.

The model presented in Fig. 12–25 is useful as a programmer's model or for first conceptualizing three-dimensional viewing operations. However, to provide efficiency, the actual implementation of three-dimensional viewing in a graphics package takes on a considerably different form. In this section, we look at those places where implementation concerns lead us to diverge from the basic model of three-dimensional viewing.

Normalized View Volumes

Clipping in two dimensions is generally performed against an upright rectangle; that is, the window is aligned with the x and y axes. This greatly simplifies the clipping calculations, since each window boundary is defined by one coordinate value. For example, the intersections of all lines crossing the left boundary of the window have an x coordinate equal to the left boundary.

In the three-dimensional programmer's model, clipping is performed against the view volume as defined by the projection window, the type of projection, and the near and far planes. Since the near and far planes are parallel to the view plane, each has a constant z-coordinate value. The z coordinate of the intersections of lines with these planes is simply the z coordinate of the corresponding plane. But the other four sides of the view volume can have arbitrary spatial orientations. To find the intersection of a line with one of these sides requires finding the equation for the plane containing the view volume side. This becomes unnecessary, however, if we convert the view volume before clipping to a regular parallelepiped.

FIGURE 12–26
A regular parallelepiped view volume.

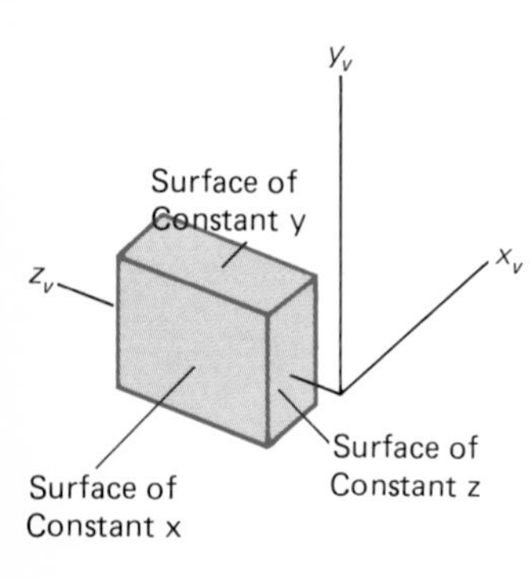

Clipping against a regular parallelepiped is simpler because each surface is now perpendicular to one of the coordinate axes. As seen in Fig. 12–26, the top

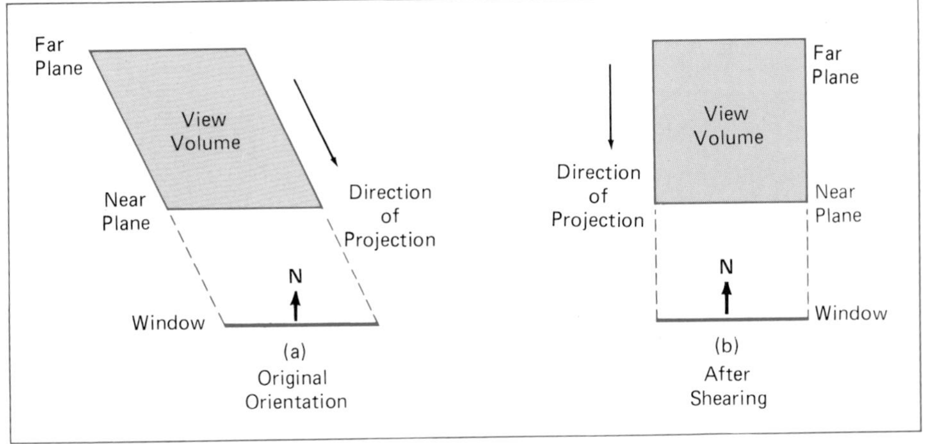

FIGURE 12–27
Shearing an oblique, parallel projection view volume into a regular parallelepiped (top view).

and bottom of such a view volume are planes of constant y, the sides are planes of constant x, and the near and far planes have a fixed z value. All lines intersecting the top plane of the parallelepiped, for example, now have the y-coordinate value of that plane. In addition to simplifying the clipping operation, converting to a regular parallelepiped reduces the projection process to a simple orthogonal projection. We first consider how to convert a view volume to a regular parallelepiped, then discuss the projection operation.

In the case of an orthographic parallel projection, the view volume is already a rectangular parallelepiped. For an oblique parallel projection, we shear the view volume to align the projection direction with the view plane normal vector, **N**. This shearing transformation brings the sides of the view volume perpendicular to the view surface, as seen in Fig. 12–27.

For the view volume in a perspective projection, shearing and scaling transformations are needed to produce the rectangular parallelepiped. We first shear in x and y directions to bring the center of projection onto the line that is normal to the center of the window (Fig. 12–28). With the apex at this point, the opposing sides of the frustum (left versus right and top versus bottom) have the same dimensions. We then apply a scaling transformation to convert the sides of the frustum into the rectangular sides of a regular parallelepiped.

Figure 12–29 shows a side view of the frustum for a parallel projection. To convert this frustum into the regular parallelepiped with height equal to that of the window, we apply a scaling transformation relative to the fixed point $(x_F, y_F, 0)$ at the window center. This transformation must scale points in the frustum that are

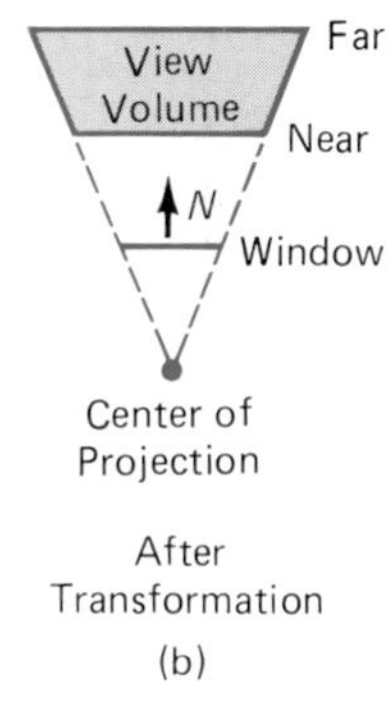

FIGURE 12–28
Shearing a perspective projection view volume to bring the center of projection onto the line perpendicular to the center of the window (top view).

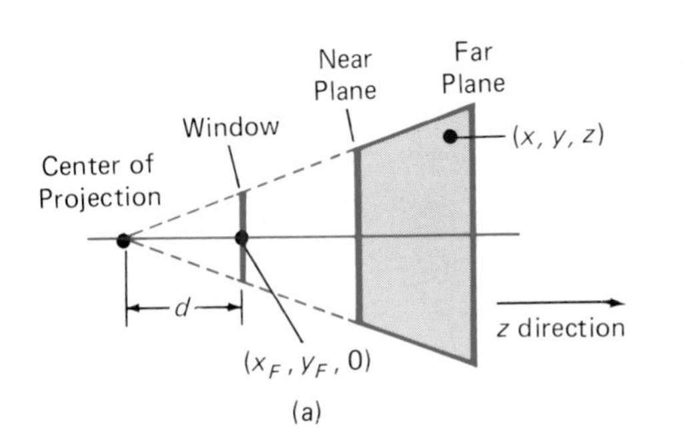

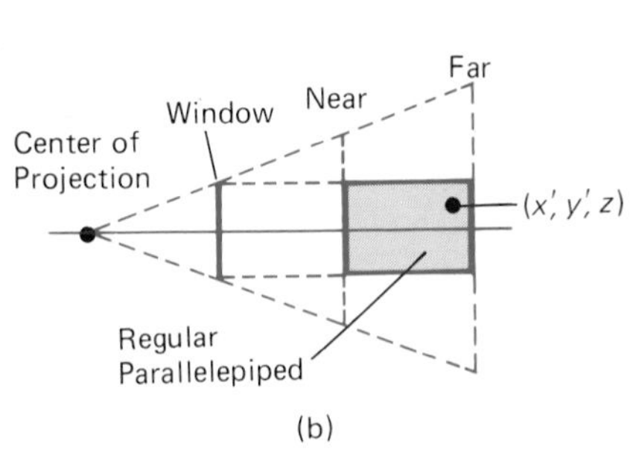

FIGURE 12–29
Side view of a frustum view volume and the resulting regular parallelepiped. A point *(x, y, z)* is scaled into the position *(x′, y′, z)* by the scaling transformation (12–13).

farther from the window more than points that are closer to the window to bring them into the parallelepiped region. In fact, the amount of scaling required is inversely proportional to the distance from the window. For a center of projection at a distance d behind the window, the required scaling factor is $d/(z + d)$, with z as the distance of a point from the window. The matrix representation for this scaling is

$$\begin{bmatrix} S & 0 & 0 & 0 \\ 0 & S & 0 & 0 \\ 0 & 0 & 1 & 0 \\ (1 - S)x_F & (1 - S)y_F & 0 & 1 \end{bmatrix} \tag{12-13}$$

where the scaling parameter is $S = d/(z + d)$. All x- and y-coordinate values in a scene are scaled by this transformation. Those points that are in the volume are mapped into points in the parallelepiped with no change in z value (Fig. 12–29).

Converting to a regular parallelepiped has another important benefit. The transformations applied accomplish a large measure of the work required to project the original points onto the view plane. For example, the transformation to convert the frustum to a regular parallelepiped essentially performs the perspective transformation. Coordinate positions are converted to the projection x and y values but retain nonzero z values. Points that fall within the z values for the near and far plane can be projected by simply removing the z coordinate. Thus by converting the view volume to a regular parallelepiped, the projection operation is reduced to a simple orthogonal projection.

We can make the viewing operations still more efficient. Converting the view volume to a regular parallelepiped (which is like doing the projection operation) occurs immediately following the mapping from world coordinates to viewing coordinates. If we concatenate the matrices to do each of these operations, each coordinate position can be transformed from its position in world coordinates to the proper position within the parallelepiped in one step.

Some graphics packages perform clipping operations using the regular parallelepiped as just described. Object parts that are within the parallelepiped are projected to the front plane and then mapped to a two-dimensional viewport.

Other packages map the regular parallelepiped onto the **unit cube** (Fig. 12–30) before clipping and projection. The unit cube is a volume defined by the following planes:

FIGURE 12–30
Transformation of a regular parallelepiped (a) to a unit cube (b).

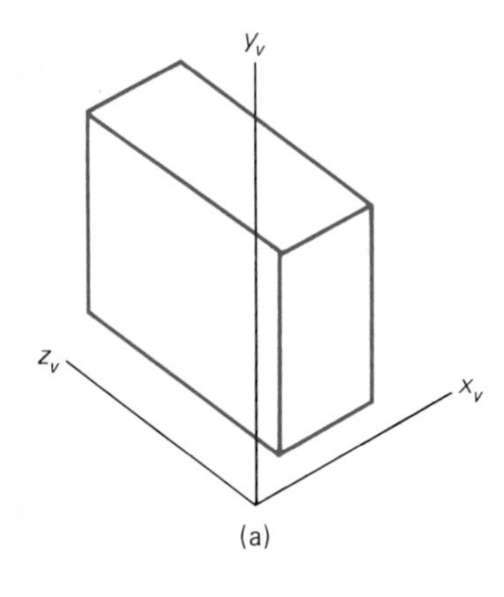

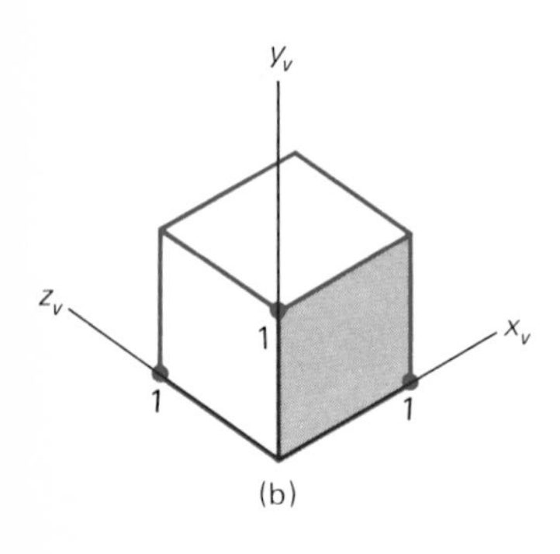

$$x = 0,\ x = 1,\ y = 0,\ y = 1,\ z = 0,\ z = 1 \tag{12-14}$$

Since the unit cube is defined by values within the range of 0 to 1, it can be referred to as a **normalized view volume.** As with a regular parallelepiped, once the volume interior has been projected to the front plane, those points are mapped to a two-dimensional viewport.

As another alternative, the regular parallelepiped, defined by the view plane window, can be mapped to a **three-dimensional viewport** before clipping. This viewport is a regular parallelepiped defined in normalized coordinates. The three-dimensional window-to-viewport mapping required is accomplished with a transformation combining scaling and translation operations similar to those for a two-dimensional window-to-viewport mapping. We can express the three-dimensional transformation matrix for these operations in the form

$$\begin{bmatrix} D_x & 0 & 0 & 0 \\ 0 & D_y & 0 & 0 \\ 0 & 0 & D_z & 0 \\ K_x & K_y & K_z & 1 \end{bmatrix} \tag{12-15}$$

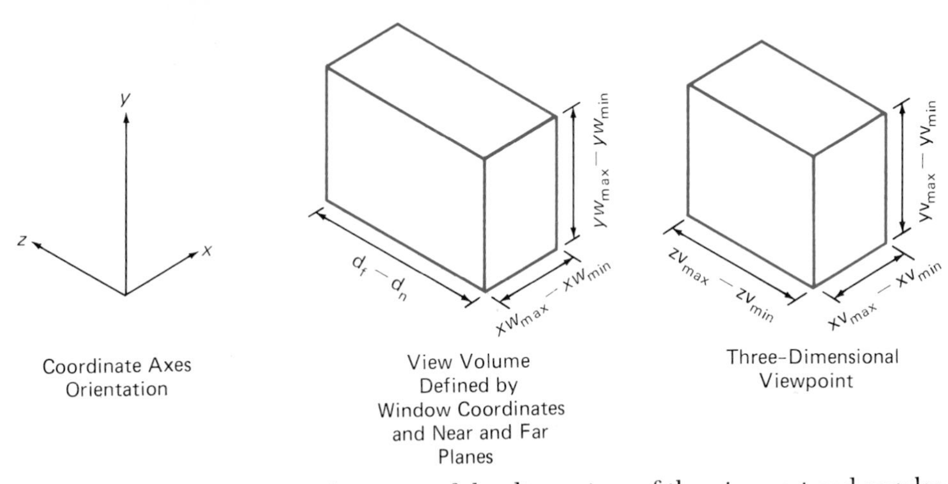

FIGURE 12–31
Dimensions of the view volume and three-dimensional viewport.

Factors D_x, D_y, and D_z are the ratios of the dimensions of the viewport and regular parallelepiped view volume in the x, y, and z directions (Fig. 12–31):

$$\begin{aligned} D_x &= \frac{xv_{\max} - xv_{\min}}{xw_{\max} - xw_{\min}} \\ D_y &= \frac{yv_{\max} - yv_{\min}}{yw_{\max} - yw_{\min}} \\ D_z &= \frac{zv_{\max} - zv_{\min}}{d_f - d_n} \end{aligned} \tag{12–16}$$

where the view volume boundaries are established by the window limits ($xw_{\min}$, $xw_{\max}$, $yw_{\min}$, $yw_{\max}$), and the positions d_n and d_f of the near and far planes. Viewport boundaries are set with the coordinate values $xv_{\min}$, $xv_{\max}$, $yv_{\min}$, $yv_{\max}$, $zv_{\min}$, and $zv_{\max}$. The additive factors K_x, K_y, and K_z in the transformation are

$$\begin{aligned} K_x &= xv_{\min} - xw_{\min} \cdot D_x \\ K_y &= yv_{\min} - yw_{\min} \cdot D_y \\ K_z &= zv_{\min} - d_n \cdot D_z \end{aligned} \tag{12–17}$$

Performing the window-to-viewport mapping before clipping orders the operations as in Fig. 12–32. The advantage of this is that the normalization transformation matrix (view volume–to–viewport mapping) can be concatenated with the matrix that converts world coordinate positions to positions within the parallelepi-

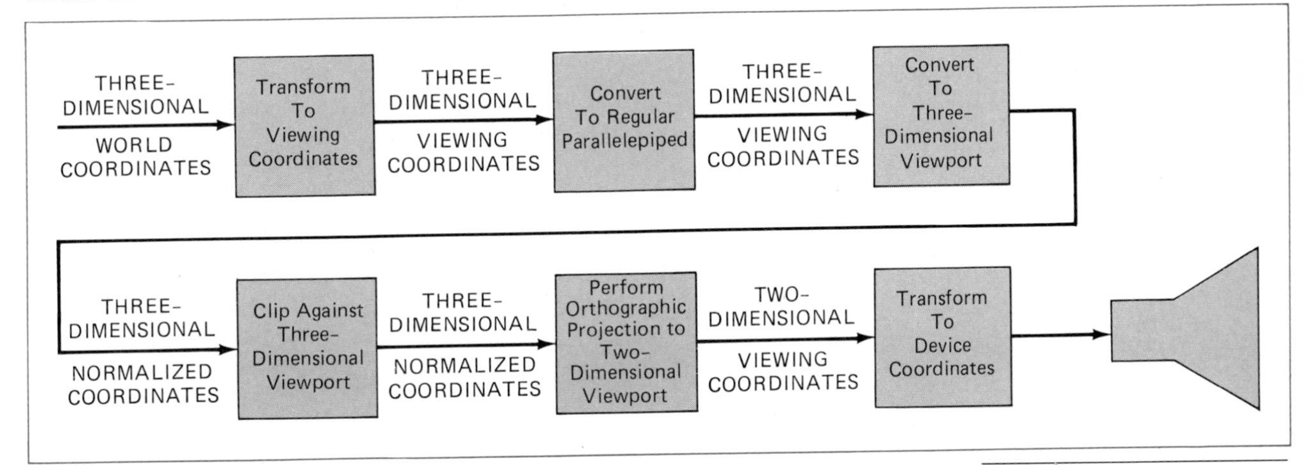

FIGURE 12–32
Implementing the viewing transformation so that operations can be combined into a single transformation matrix, applied before clipping.

ped. This resultant matrix converts positions within world coordinates to their x and y projection coordinates within the viewport. Each coordinate of the original scene needs to be transformed only once. These transformed points are clipped against the viewport. The x and y values of points inside the viewport volume are then converted to device coordinates for display (Fig. 12–33).

Clipping Against a Normalized View Volume

Surfaces can be clipped against the viewport boundaries with procedures similar to those used for two dimensions. Either line-clipping or polygon-clipping procedures could be adapted to three-dimensional viewport clipping. Curved surfaces are processed using the defining equations for the surface boundary and locating the intersection lines with the parallelepiped planes. We now consider how two-dimensional clipping procedures can be modified to account for the third dimension.

The two-dimensional concept of region codes can be extended to three dimensions by considering positions in front and in back of the three-dimensional viewport, as well as positions that are left, right, below, or above the volume. For two-dimensional clipping, we used a four-digit binary region code to identify the position of a line endpoint relative to the window boundaries. For three-dimensional points, we need to expand the region code to six bits. Each point in the description of a scene is then assigned a six-bit region code that identifies the relative position of the point with respect to the viewport. For a line endpoint at position (x, y, z), we assign the bit positions in the region code from right to left as

$$
\begin{aligned}
&\text{bit } 1 = 1 \qquad \text{if } x < xv_{\text{min}} \text{ (left)} \\
&\text{bit } 2 = 1 \qquad \text{if } x > xv_{\text{max}} \text{ (right)} \\
&\text{bit } 3 = 1 \qquad \text{if } y < yv_{\text{min}} \text{ (below)} \\
&\text{bit } 4 = 1 \qquad \text{if } y > yv_{\text{max}} \text{ (above)} \\
&\text{bit } 5 = 1 \qquad \text{if } z < zv_{\text{min}} \text{ (front)} \\
&\text{bit } 6 = 1 \qquad \text{if } z > zv_{\text{max}} \text{ (back)}
\end{aligned}
$$

For example, a region code of 101000 identifies a point as above and behind the viewport, while the region code 000000 indicates a point within the volume.

A line segment can be immediately identified as completely within the viewport if both endpoints have a region code of 000000. If either endpoint of a line segment does not have a region code of 000000, we perform the logical AND operation on the two endpoint codes. The result of this AND operation will be non-

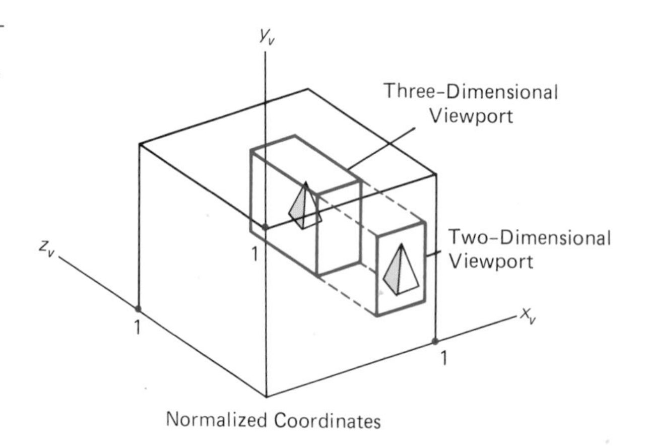

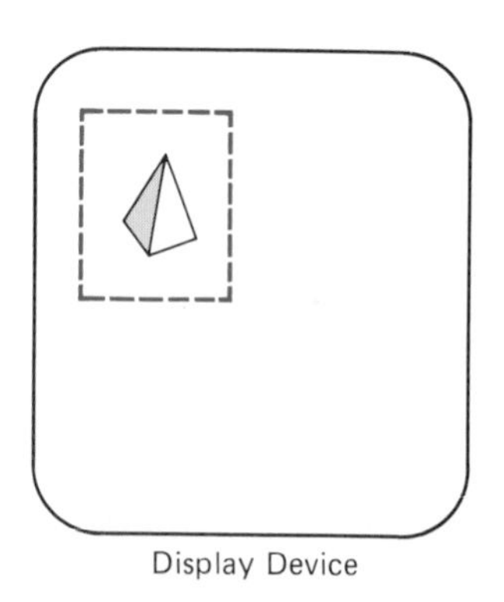

FIGURE 12–33
Mapping the interior of a three-dimensional viewport (in normalized coordinates) to device coordinates.

zero for any line segment completely outside the viewport volume. If we cannot identify a line segment as completely inside or completely outside the volume, we test for intersections with the bounding planes of the volume.

As in two-dimensional line clipping, we use the calculated intersection of a line with a viewport plane to determine how much of the line can be thrown away. The remaining part of the line is checked against the other planes, and we continue until either the line is totally discarded or a section is found inside the volume.

Determination of intersection points in line clipping, and in polygon-clipping procedures as well, should be done efficiently. Equations for three-dimensional line segments are conveniently expressed in parametric form. For a line segment with endpoints $P_1 = (x_1, y_1, z_1)$ and $P_2 = (x_2, y_2, z_2)$, we can write the parametric line equations as

$$\begin{aligned} x &= x_1 + (x_2 - x_1)u \\ y &= y_1 + (y_2 - y_1)u \\ z &= z_1 + (z_2 - z_1)u \end{aligned} \tag{12–18}$$

Coordinates (x, y, z) represent any point on the line between the two endpoints, and parameter u varies from 0 to 1. The value $u = 0$ produces the point P_1, and $u = 1$ gives point P_2.

To find the intersection of a line with a plane of the viewport, we substitute the value of the coordinate that is constant for the plane into the appropriate parametric expression of Eqs. 12–18 and solve for u. For instance, suppose we are testing a line against the front plane of the viewport. Then $z = zv_{\min}$, and

$$u = \frac{zv_{\min} - z_1}{z_2 - z_1} \tag{12–19}$$

When the value for u as calculated by Eq. 12–19 is not in the range from 0 to 1, this means that the line segment does not intersect the front plane at any point between endpoints P_1 and P_2 (line A in Fig. 12–34). If the calculated value for u is in the interval from 0 to 1, we calculate the intersection's x and y coordinates as

$$\begin{aligned} x_I &= x_1 + (x_2 - x_1)\left(\frac{zv_{\min} - z_1}{z_2 - z_1}\right) \\ y_I &= y_1 + (y_2 - y_1)\left(\frac{zv_{\min} - z_1}{z_2 - z_1}\right) \end{aligned} \tag{12–20}$$

If either x_I or y_I is not in the range of the boundaries of the viewport, then this line intersects the front plane beyond the boundaries of the volume (line B in Fig. 12–34).

The Liang-Barsky line-clipping algorithm discussed in Chapter 6 can be extended to three-dimensions by considering the effects of the near and far planes. These planes add two additional tests in the processing of the intersection parameters u_1 and u_2.

12–4 Hardware Implementations

Graphics chip sets, employing VLSI (very large scale integration) circuitry techniques, are used in some systems to perform the viewing operations. These customized chip sets are designed to transform, clip, and project objects to the output device for either three-dimensional or two-dimensional applications.

FIGURE 12–34
A side view in the yz plane of two line segments that are to be clipped against the front plane of the viewport. For line A, Eq. (12–19) produces a value of $\boldsymbol{u}$ that is outside the range from 0 to 1. For line B, Eqs. (12–20) produce intersection coordinates that are outside the range from yv_{min} to yv_{max}.

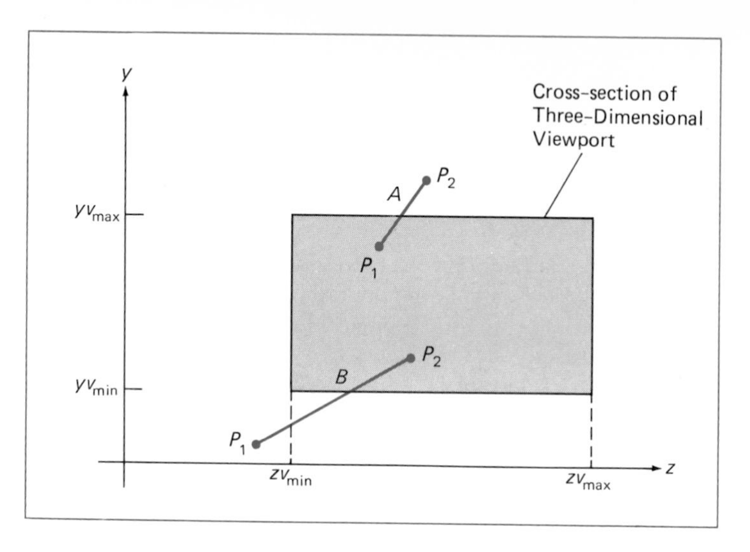

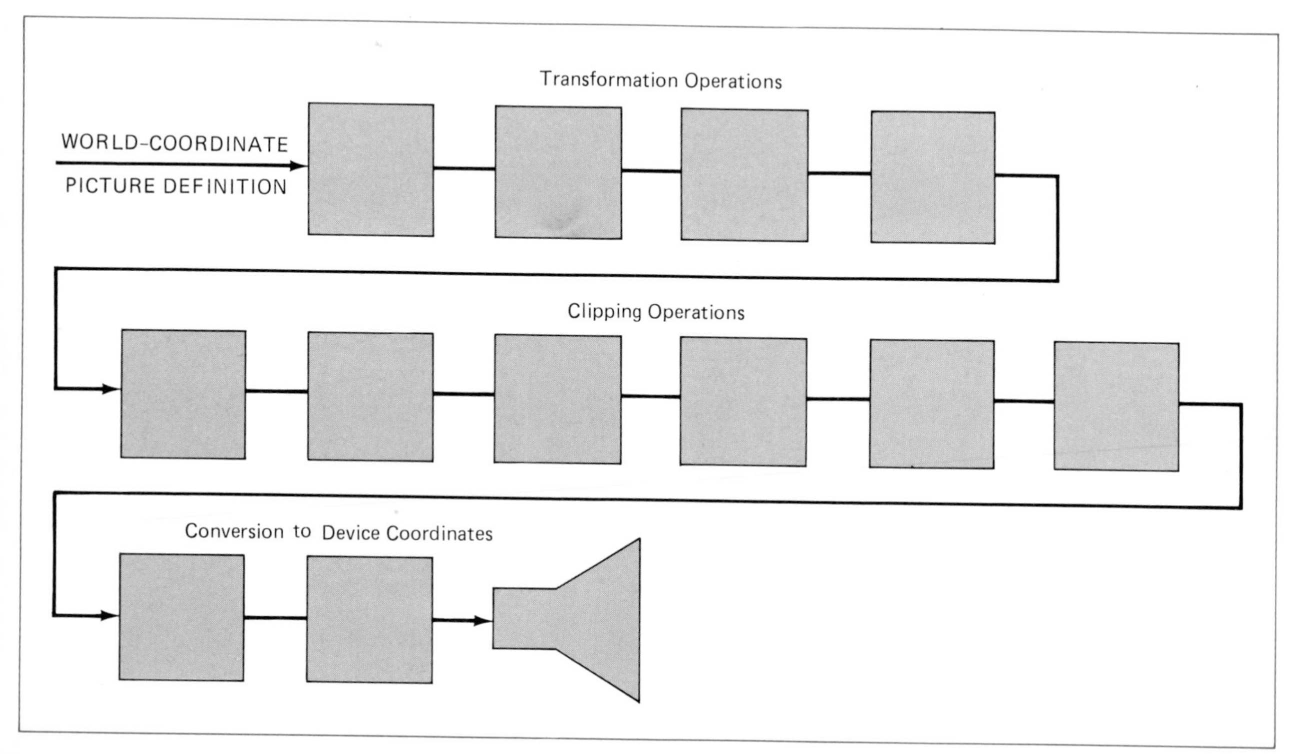

FIGURE 12–35
A graphics chip set with 12 chips to perform the various viewing operations.

Figure 12–35 shows the components in one type of graphics chip set. The chips are organized into a pipeline for accomplishing the transformation, clipping, and coordinate-conversion operations. Four initial chips are provided for matrix operations involving scaling, translation, rotation, and the transformations needed for orthographic and perspective projections. Each of the next six chips performs clipping against one of the viewport boundaries. Four of these chips are used in two-dimensional applications, and the other two are needed for clipping against the near and far planes of the three-dimensional viewport. The last two chips in the pipeline convert viewport coordinates to output device coordinates.

12–5 Programming Three-Dimensional Views

Several commands can be provided in three-dimensional graphics packages to enable an application program to set the parameters for viewing transformations. With parameters specified relative to the world coordinate origin, the matrix for transforming world coordinate descriptions to the viewing system coordinates is established using

```
create_view_matrix (xo, yo, zo, xn, yn, zn, xv, yv, zv, view_matrix)
```

Parameters *xo*, *yo*, and *zo* specify the origin (view reference point) of the viewing system. The positive *z* axis of the viewing system is established in the direction of the vector from the world coordinate origin to the point (*xn*, *yn*, *zn*). And coordinate position (*xv*, *yv*, *zv*) specifies the view up vector. The projection of this vector onto the view plane defines the positive *y* axis of the viewing coordinate system. These parameters are used to construct a *view_matrix* for transforming world coordinate positions to viewing coordinates.

To specify a second viewing coordinate system, the user can redefine some or all of the coordinate parameters and invoke *create_view_matrix* with a new matrix designation. In this way, any number of world coordinate to viewing coordinate mappings can be defined.

Once a matrix for transforming world to viewing coordinates has been defined, the projection parameters can be specified with the function

```
set_view_representation (view_index, view_matrix,
    projection_type, xp, yp, zp, xw_min, xw_max,
    yw_min, yw_max, near, far, xv_min, xv_max,
    yv_min, yv_max, zv_min, zv_max)
```

Parameter *view_index* serves as an identifying number for the viewing transformation. The transformation matrix for mapping world coordinates to viewing coordinates is specified in *view_matrix*, and *projection_type* is assigned a value of either *parallel* or *perspective*. Coordinate position (*xp*, *yp*, *zp*) establishes either the direction of projection or the center of projection, depending on the input value for parameter *projection_type*. The projection window limits are defined with coordinate values *xw_min*, *xw_max*, *yw_min*, and *yw_max*, specified relative to the viewing coordinate origin. Parameters *near* and *far* specify the location of the corresponding planes. Finally, the three-dimensional viewport boundaries are given with parameters *xv_min*, *xv_max*, *yv_min*, *yv_max*, *zv_min*, and *zv_max*, specified in normalized coordinates. An additional parameter could be included in this function to allow a user to position the view plane at any distance from the view origin. Any number of viewing transformations can be defined with this function, using different *view_index* values.

A user selects a particular viewing transformation with

```
set_view_index (vi)
```

View index number *vi* identifies the set of viewing transformation parameters that are to be applied to subsequently specified output primitives.

Clipping routines in the viewing transformation can be made optional. When a scene is to be viewed in its entirety, it is more efficient to bypass clipping since no part of the scene is to be eliminated. Or the user may know that the scene is completely on the inside of, for example, the far plane. To allow a user selectively

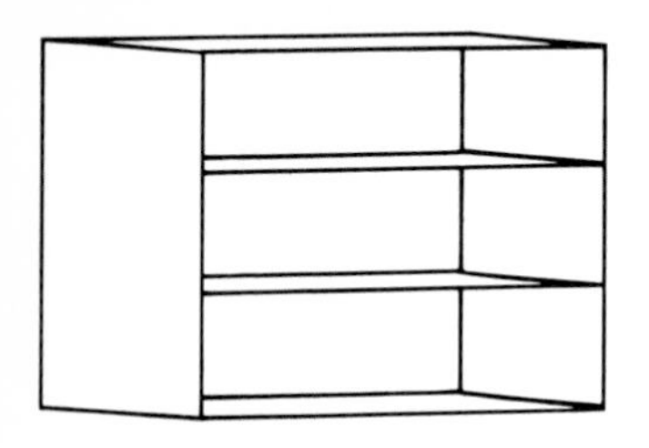

FIGURE 12–36
Object displayed by procedure drawcase using viewing index 2.

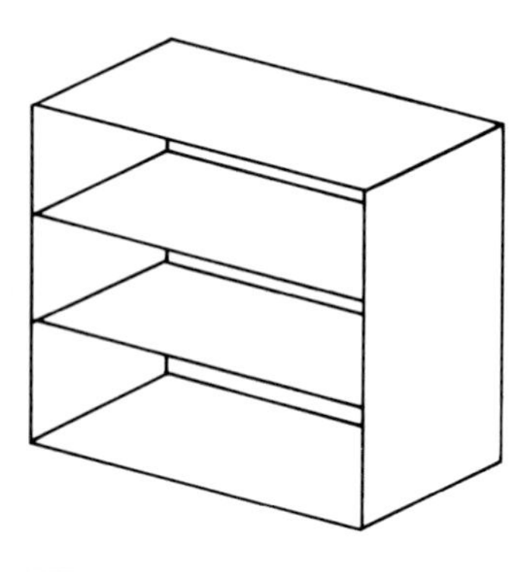

FIGURE 12–37
Object displayed by procedure drawcase using viewing index 3.

to enable and disable clipping against various planes of the view volume, additional parameters can be included in the *set_view_representation* function.

In the following program example, the use of viewing commands is illustrated by generating two perspective views of an object. The two views (shown in Figs. 12–36 and 12–37) are referenced by viewing indices 2 and 3.

```
type
  matrix = array [1..4,1..4] of real;
  projtype = (parallel, perspective);

procedure bookcase;
  begin
    { Defines bookcase with calls to                }
    { fill_area   for the back, sides, top,         }
    { bottom, and 2  shelves. Bookcase is           }
    { defined in feet, as 3'  wide, 4' high,        }
    { and 1' deep, with the back,  bottom,          }
    { left corner at (0, 0, 0).                     }
  end;  { bookcase }

procedure establish_views;
  var  viewtr1, viewtr2 : matrix;
  begin
    { first view --                                 }
    { view reference point is (-8, 3, 6)            }
    { view plane normal is (-1, 0, 1)               }
    { view up vector is (0, 0, 1)                   }
    { Store world-to-viewing transformation         }
    { matrix in viewtr1.                            }

    create_view_matrix (-8,3,6, -1,0,1, 0,0,1, viewtr1);

    { Use this world-to-viewing transformation }
    { and additional projection parameters to  }
    { fully specify view 2.                     }
    { center of projection is (-12, 3, 12)      }
    { window goes from (2,2) to (8,8)           }
    { put near plane at 10 and far at 12        }
    { viewport is (.5,.5,0) to (1,1,1)          }

    set_view_representation (2, viewtr1, perspective, -12,3,12,
        2,8,2,8, 10,12, 0.5,1, 0.5,1, 0,1);

    { second view -
    { view reference point is now (8, 10, 6)   }
    { view plane normal is now (1, 1, 1)        }
    { Store matrix in viewtr2.                  }

    create_view_matrix (8,10,6, 1,1,1, 0,0,1, viewtr2);

    { Use viewtr2 and projection para-          }
    { meters to fully specify view 3.           }
    { center of projection is now (20, 20, 20) }

    set_view_representation (3, viewtr2, perspective, 20,20,20,
        2,8,2,8, 10,12, 0.5,1, 0.5,1, 0,1)
```

```
  end;  { establish_views }

procedure drawcase;
  begin
    establish_views;
    set_view_index (2);     { generate view using transform 2 }
    bookcase;
       .
       .
       .
    set_view_index (3);     { generate view using transform 3 }
    bookcase
  end;  { drawcase }
```

12–6 Extensions to the Viewing Pipeline

To this point, our discussion has concentrated on the central part of the viewing operation, often referred to as the viewing transformation. This includes mapping to the viewing coordinate system, projecting, and clipping. We turn now to operations that may come before or after the viewing transformation and that affect the final view of an object.

Graphics packages that allow transformation matrices to be associated with segments typically apply the matrices before the viewing operations. We can think of this as rotating or repositioning the object in front of the camera. If a number of segments are to be transformed, each is transformed by the appropriate matrix, and the collection of objects is then projected by the viewing transformation to form the final view. When the transformation matrix associated with any segment is changed, the entire viewing process must be repeated.

In some situations, the user may only want to vary the appearance of the view on the output device. Maybe an engineer wants to rotate the cutaway view of some three-dimensional part that has been projected. Or perhaps an animation application needs to move an object from one area of the screen to another. The engineer could use a segment transformation or request a new view of the part using new viewing parameters. In both cases, a second trip through the entire viewing pipeline would be necessary. For such situations, graphics systems sometimes provide for **image transformations:** changes that are applied to the final two-dimensional projection. Image transformations are applied in two dimensions, allowing a user to relocate an object on the monitor but not to turn it around to see its back side. Since these changes do not require three-dimensional viewing or clipping, they are accomplished quickly.

REFERENCES

For further discussions of three-dimensional clipping and viewing transformations, see Blinn and Newell (1978), Cyrus and Beck (1978), Liang and Barsky (1984), Michener and Carlbom (1980), and Pavlidis (1982).

EXERCISES

12–1. Set up the definition of a polyhedron (solid formed with plane surfaces) in the first octant of a left-handed coordinate system (that is, all coordinate values defining the vertices of the object are to be positive). Develop a procedure for performing any specified parallel projection of the object onto the *xy* plane.

12–2. Extend the procedure of Ex. 12–1 to obtain different views of the object by first performing specified rotations of the object about rotation axes that are parallel to the projection plane, then projecting the object onto the viewing surface.

12–3. Set up the procedure in Ex. 12–1 to generate a one-point perspective projection of the object onto the projection plane, using Eq. 12–10 and any specified viewing distance d along the negative z axis.

12–4. Extend the transformation matrix in Eq. 12–10 so that the center of projection can be selected at any position $(x, y, -d)$ behind the projection plane.

12–5. Set up the procedure in Ex. 12–1 to generate a one-point perspective projection of the object onto the projection plane, using the transformation matrix of Ex. 12–4.

12–6. Assuming that a projection surface is defined as the xy plane of a left-handed coordinate system, set up the coordinate definition of a rectangular parallelepiped in this system so that it is in front of the projection plane. Using the projection matrix of Ex. 12–4, orient the block to obtain one-point and two-point perspective projections. Write a program to display these two perspective views on your system. Which of the two perspective views appears more realistic?

12–7. Extend the routine in Ex. 12–6 to obtain a three-point perspective projection. Can you detect much difference between the two-point and three-point projections?

12–8. Develop a set of routines to convert an object description in world coordinates to specified viewing coordinates. That is, implement the function for setting the *view_matrix,* given the coordinates for the view reference point, normal vector, and view up vector.

12–9. Expand the procedures in Ex. 12–8 to obtain a specified parallel projection of the object onto a defined window in the xy plane of the viewing system. Then transform the window to a viewpoint area on the screen. Assume that the object is in front of the view plane and that no clipping against a view volume is to be performed.

12–10. Expand the procedures in Ex. 12–8 to obtain a specified perspective projection of the object onto a defined window in the xy plane of the viewing system, followed by a transformation to a viewport area on the screen. Assume that the object is in front of the view plane and that no clipping against a view volume is to be performed.

12–11. Devise an algorithm to clip objects in a scene against a defined frustum. Compare the operations needed in this algorithm to those needed in an algorithm that clips the scene against a regular parallelepiped.

12–12. Write a program to convert a perspective projection frustum to a regular parallelepiped.

12–13. Modify the two-dimensional Liang-Barsky line-clipping algorithm to clip three-dimensional lines against a specified parallelepiped.

12–14. Extend the algorithm of Ex. 12–13 to clip a specified polyhedron against a parallelepiped.

12–15. For both parallel and perspective projections, discuss the conditions under which three-dimensional clipping followed by projection onto the view plane would be equivalent to first projecting then clipping against the window.

12–16. Using any clipping procedure, write a program to perform a complete viewing transformation from world coordinates to the viewport for an orthographic parallel projection of an object.

12–17. Extend the procedure of Ex. 12–16 to perform any specified parallel projection of an object onto a defined viewport.

12–18. Develop a program to implement a complete viewing operation for a perspective projection. The program is to transform the world coordinate specification of an object onto a defined two-dimensional viewport for display on a section of a video monitor.

12–19. Implement the *set_view_representation* and *set_view_index* functions to perform any specified projection operation on a defined object in world coordinates to obtain the viewport display on the screen.

12–20. Modify the routines in Ex. 12–19 to allow clipping against any view volume plane to be optional. This can be accomplished with additional parameters to set the clipping condition for each plane as either *clip* or *noclip.*

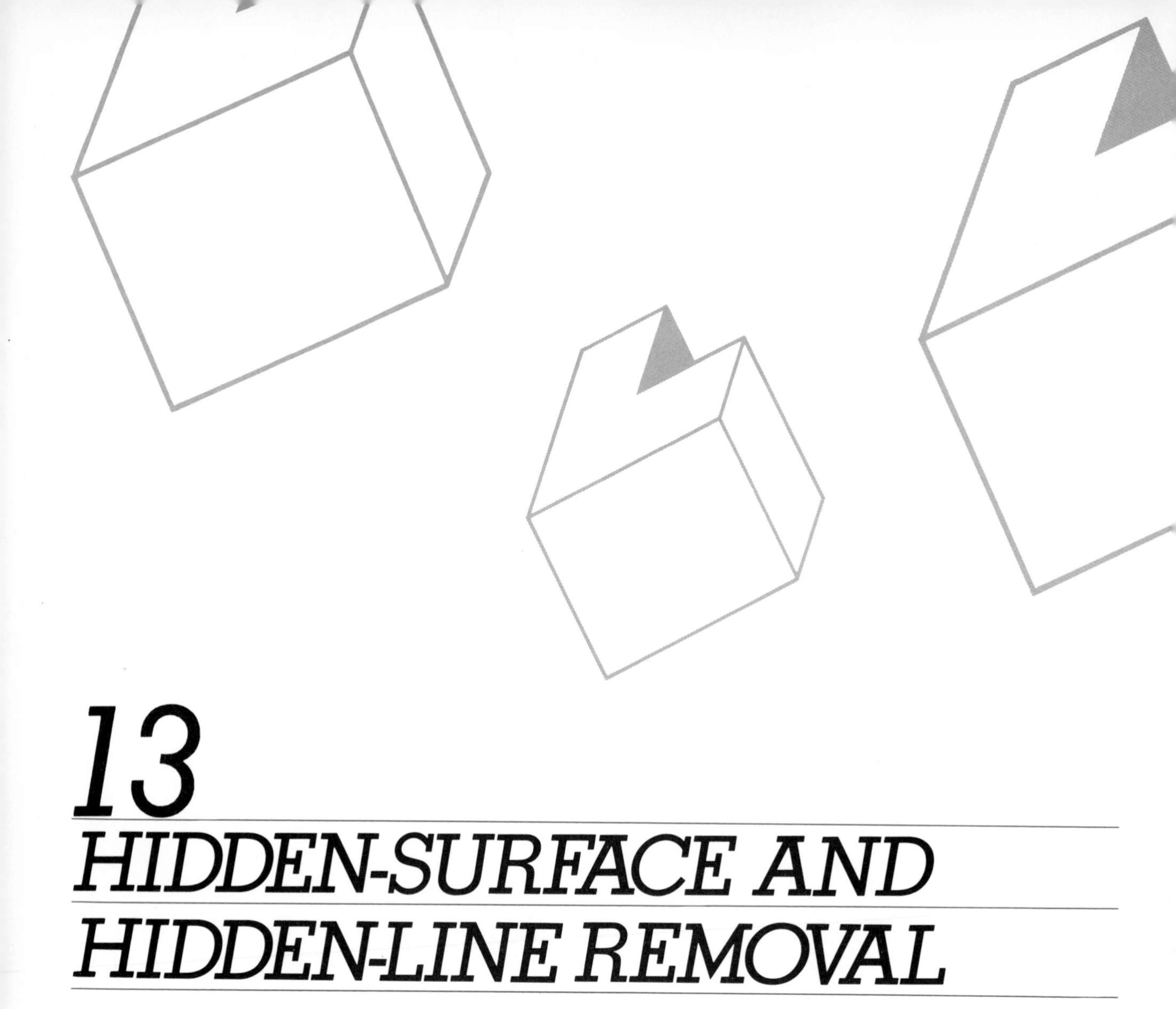

13 HIDDEN-SURFACE AND HIDDEN-LINE REMOVAL

A major consideration in the generation of realistic scenes is the identification and removal of the parts of the picture definition that are not visible from a chosen viewing position. There are many approaches we can take to solve this problem, and numerous algorithms have been devised to remove hidden parts of scenes efficiently for different types of applications. Some methods require more memory, some involve more processing time, and some apply only to special types of objects. The method chosen for a particular application depends on such factors as the complexity of the scene, type of objects to be displayed, available equipment, and whether static or animated displays are to be generated. In this chapter, we explore some of the most commonly used methods for removing hidden surfaces and lines.

13–1 Classification of Algorithms

Hidden-surface and hidden-line algorithms are often classified according to whether they deal with object definitions directly or with their projected images. These two approaches are called **object-space** methods and **image-space** methods, respectively. An object-space method compares objects and parts of objects to each other to determine which surfaces and lines, as a whole, should be labeled as invisible. In an image-space algorithm, visibility is decided point by point at each pixel position on the projection plane. Most hidden-surface algorithms use image-space methods, but object-space methods can be used effectively in some cases. Hidden-line algorithms generally employ object-space methods, although many image-space hidden-surface algorithms can be readily adapted to hidden-line removal.

Although there are major differences in the basic approach taken by the various hidden-surface and hidden-line algorithms, most use sorting and coherence methods to improve performance. Sorting is used to facilitate depth comparisons by ordering the individual lines, surfaces, and objects in a scene according to their distance from the view plane. Coherence methods are used to take advantage of regularities in a scene. An individual scan line can be expected to contain intervals (runs) of constant pixel intensities, and scan-line patterns often change little from one line to the next. Animation frames contain changes only in the vicinity of moving objects. And constant relationships can often be established between objects and surfaces in a scene.

13–2 Back-Face Removal

A simple object-space method for identifying the **back faces** of objects is based on the equation of a plane:

$$Ax + By + Cz + D = 0 \tag{13–1}$$

As we noted in Chapter 10, any point (x', y', z') specified in a right-handed coordinate system is on the "inside" of this plane if it satisfies the inequality

$$Ax' + By' + Cz' + D < 0 \tag{13–2}$$

If the point (x', y', z') is the viewing position, then any plane for which inequality 13–2 is true must be a back face. That is, it is one that we cannot see from the viewing position.

We can perform a simpler back-face test by looking at the normal vector to a plane described by equation 13–1. This normal vector has Cartesian components (A, B, C). In a right-handed viewing system with the viewing direction along the negative z_v axis (Fig. 13–1), the normal vector has component C parallel to the viewing direction. If $C < 0$, the normal vector points away from the viewing position, and the plane must be a back face.

Similar methods can be used in packages that employ a left-handed viewing system. In these packages, plane parameters A, B, C, and D can be calculated from vertex coordinates specified in a clockwise direction (instead of the counterclockwise direction used in a right-handed system). Inequality 13–2 then remains a valid test for inside points. Also, back faces have normal vectors that point away from the viewing position and are identified by $C > 0$ when the viewing direction is

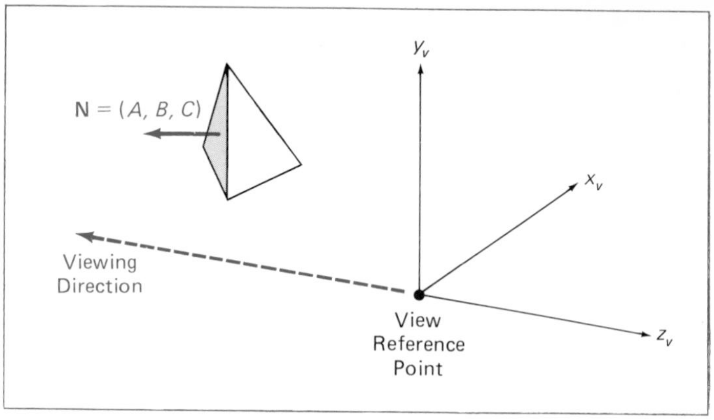

FIGURE 13–1
A plane with parameter $C < 0$ in a right-handed viewing coordinate system is identified as a back face when the viewing direction is along the negative z_v axis.

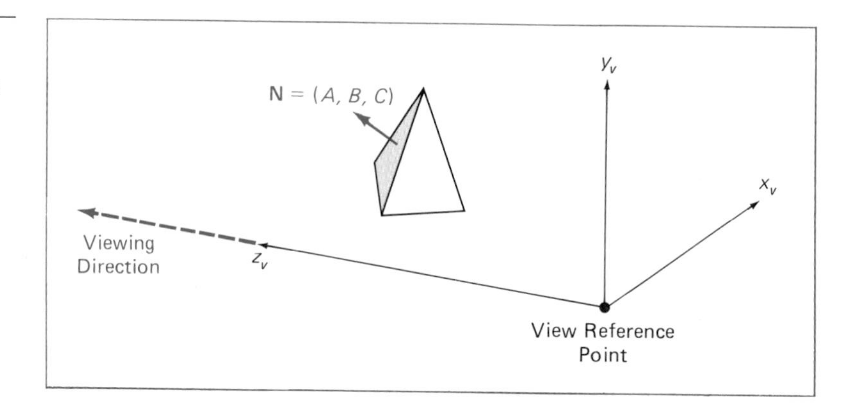

FIGURE 13–2
In a left-handed viewing system with viewing direction along the positive z_v axis, a back face is one with plane parameter $C > 0$.

along the positive z_v axis (Fig. 13–2). For all further discussions in this chapter, we assume that a left-handed viewing system is to be used.

By examining parameter C for the different planes defining an object, we can immediately identify all the back faces. For a single convex polyhedron, such as the pyramid in Fig. 13–1, this test identifies all the hidden surfaces on the object, since each surface is either completely visible or completely hidden. For other objects, more tests need to be carried out to determine whether there are additional faces that are totally or partly hidden (Fig. 13–3). Also, we may need to determine whether some objects are partially or completely obscured by other objects. In general, back-face removal can be expected to eliminate about half of the surfaces in a scene from further visibility tests.

FIGURE 13–3
View of an object with one face partially hidden.

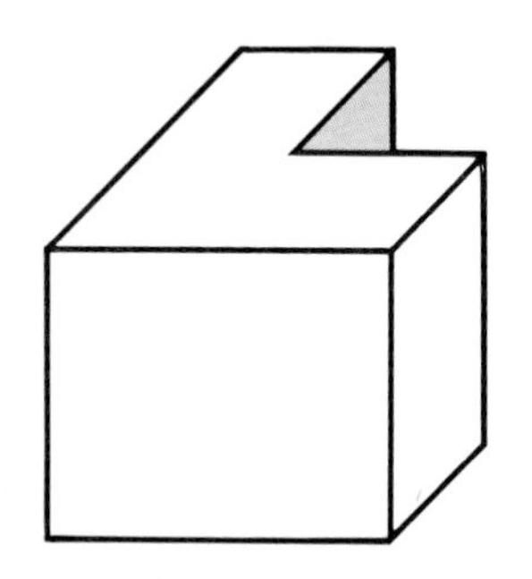

13–3 Depth-Buffer Method

A commonly used image-space approach to eliminating hidden surfaces is the **depth-buffer** method, also called the **z-buffer** method. Basically, this algorithm tests the visibility of surfaces one point at a time. For each pixel position (x, y) on the view plane, the surface with the smallest z coordinate at that position is visible. Figure 13–4 shows three surfaces at varying depths with respect to position (x, y) in a left-handed viewing system. Surface S_1 has the smallest z value at this position, so its intensity value at (x, y) is saved.

Two buffer areas are required for implementation of this method. A depth buffer is used to store z values for each (x, y) position as surfaces are compared, and the refresh buffer stores the intensity values for each position.

This method can be implemented conveniently in normalized coordinates, with depth values varying from 0 to 1. Assuming that a projection volume has been mapped into a normalized parallelepiped view volume, the mapping of each surface onto the view plane is an orthographic projection. The depth at points over the surface of a polygon is calculated from the plane equation. Initially, all positions in the depth buffer are set to 1 (maximum depth), and the refresh buffer is initialized to the background intensity. Each surface listed in the polygon tables is then processed, one scan line at a time, calculating the depth, or z value, at each (x, y) position. The calculated z value is compared to the value previously stored in the depth buffer at that position. If the calculated z value is less than the value stored in the depth buffer, the new z value is stored, and the surface intensity at that position is placed in the same location in the refresh buffer.

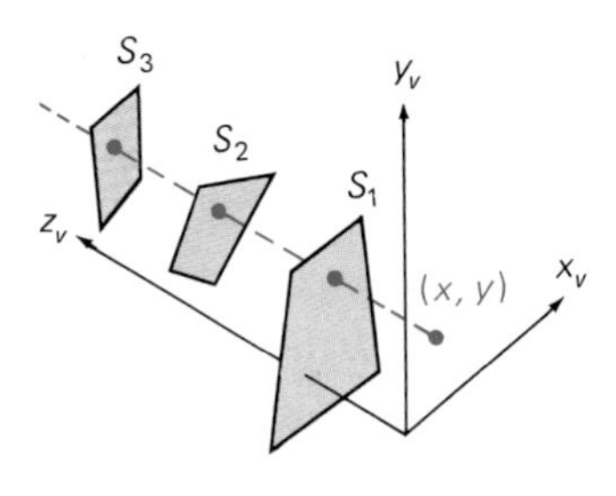

FIGURE 13–4
At position (x, y), surface S_1 has the smallest depth value and so is visible at that position.

We can summarize the steps of a depth-buffer algorithm as follows:

1. Initialize the depth buffer and refresh buffer so that for all coordinate positions (x, y), $depth(x, y) = 1$ and $refresh(x, y) = background$.
2. For each position on each surface, compare depth values to previously stored values in the depth buffer to determine visibility.
   a. Calculate the z value for each (x, y) position on the surface.
   b. If $z < depth(x, y)$, then set $depth(x, y) = z$ and $refresh(x, y) = i$, where i is the value of the intensity on the surface at position (x, y).

In the last step, if z is not less than the value of the depth buffer for that position, the point is not visible. When this process has been completed for all surfaces, the depth buffer contains z values for the visible surfaces and the refresh buffer contains only the visible intensity values.

Depth values for a position (x, y) are calculated from the plane equation for each surface:

$$z = \frac{-Ax - By - D}{C} \tag{13-3}$$

For any scan line (Fig. 13–5), x coordinates across the line differ by 1 and y values between lines differ by 1. If the depth of position (x, y) has been determined to be z, then the depth z' of the next position $(x + 1, y)$ along the scan line is obtained from Eq. 13–3 as

$$z' = \frac{-A(x + 1) - By - D}{C}$$

or

$$z' = z - \frac{A}{C} \tag{13-4}$$

The ratio A/C is constant for each surface, so succeeding depth values across a scan line are obtained from preceding values with a single subtraction.

We can obtain depth values between scan lines in a similar manner. Again, suppose that position (x, y) has depth z. Then at position $(x, y - 1)$ on the scan line immediately below, the depth value is calculated from the plane equation as

FIGURE 13–5
From position (x, y) on a scan line, the next position across the line has coordinates $(x + 1, y)$, and the position immediately below on the next line has coordinates $(x, y - 1)$.

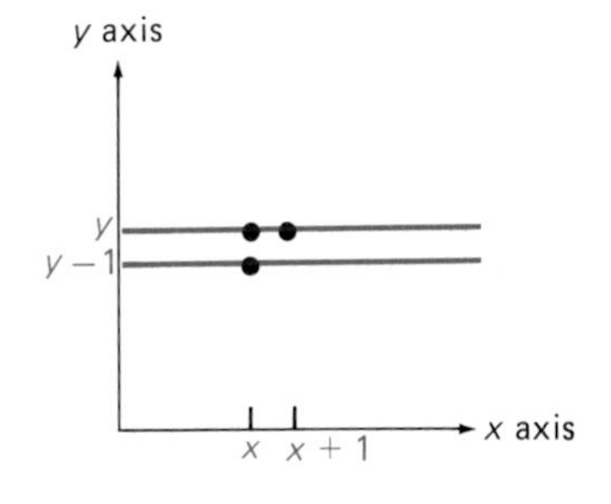

$$z'' = \frac{-Ax - B(y - 1) - D}{C}$$

or (13–5)

$$z'' = z + \frac{B}{C}$$

which requires a single addition of the constant B/C to the previous depth value z.

The depth-buffer method is easy to implement, and it requires no sorting of the surfaces in a scene. But it does require the availability of a second buffer in addition to the refresh buffer. A system with a resolution of 1024 by 1024, for example, would require over a million positions in the depth buffer, with each position containing enough bits to represent the number of z-coordinate increments needed. One way to reduce storage requirements is to process one section of the scene at a time, using a smaller depth buffer. After each view section is processed, the buffer is reused for the next section.

13–4 Scan-Line Method

This image-space method for removing hidden surfaces is an extension of the scan-line algorithm for filling polygon interiors. Instead of filling just one surface, we now deal with multiple surfaces. As each scan line is processed, all polygon surfaces intersecting that line are examined to determine which are visible. At each position along a scan line, depth calculations are made for each surface to determine which is nearest to the view plane. When the visible surface has been determined, the intensity value for that position is entered into the refresh buffer.

As we saw in Chapter 10, representation for the polygon surfaces in a three-dimensional scene can be set up to include both an edge table and a polygon table. The edge table contains coordinate endpoints for each line in the scene, the inverse slope of each line, and pointers into the polygon table to identify the surfaces bounded by each line. The polygon table contains coefficients of the plane equation for each surface, intensity information for the surfaces, and possibly pointers into the edge table.

To facilitate the search for surfaces crossing a given scan line, we can set up an active list of edges from information in the edge table. This active list will contain only edges that cross the current scan line, sorted in order of increasing x. In addition, we define a flag for each surface that is set *on* or *off* to indicate whether a position along a scan line is inside or outside of the surface. Scan lines are processed from left to right. At the leftmost boundary of a surface, the surface flag is turned on; and at the rightmost boundary, it is turned off.

Figure 13–6 illustrates the scan-line method for locating visible portions of surfaces along a scan line. The active list for scan line 1 contains information from the edge table for edges AB, BC, HE, and EF. For positions along this scan line between edges AB and BC, only the flag for surface S_1 is on. Therefore, no depth calculations are necessary, and intensity information for surface S_1 is entered from the polygon table into the refresh buffer. Similarly, between edges HE and EF, only the flag for surface S_2 is on. No other positions along scan line 1 intersect surfaces, so the intensity values in the other areas are set to the background intensity. The background intensity can be loaded throughout the buffer in an initialization routine.

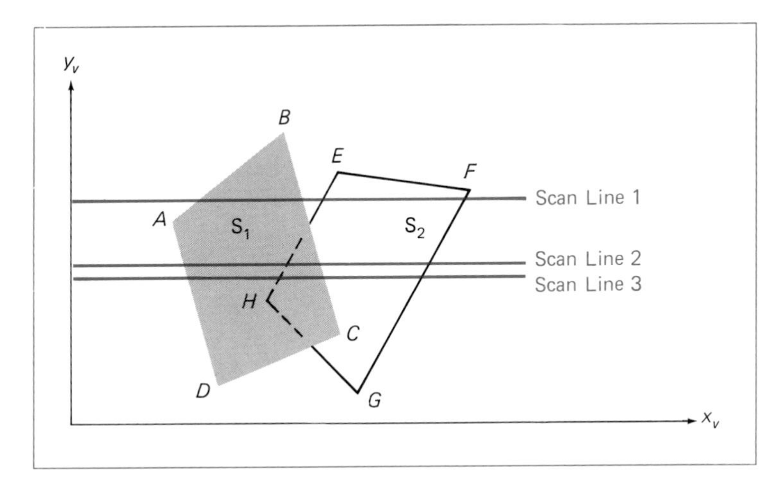

FIGURE 13–6
Scan lines crossing the projection of two surfaces, S_1 and S_2, in the view plane. Dashed lines indicate the boundaries of hidden surfaces.

The active list for scan lines 2 and 3 in Fig. 13–6 contain edges *DA*, *HE*, *BC*, and *FG*. Along scan line 2 from edge *DA* to edge *HE*, only the flag for surface S_1 is on. But between edges *HE* and *BC*, the flags for both surfaces are on. In this interval, depth calculations must be made using the plane coefficients for the two surfaces. For this example, the depth of surface S_1 is assumed to be less than that of S_2, so intensities for surface S_1 are loaded into the refresh buffer until boundary *BC* is encountered. Then the flag for surface S_1 goes off, and intensities for surface S_2 are stored until edge *FG* is passed.

We can take advantage of coherence along the scan lines as we pass from one scan line to the next. In Fig. 13–6, scan line 2 has the same active list of edges as scan line 1. Since no changes have occurred in line intersections, it is unnecessary again to make depth calculations between edges *HE* and *BC*. The two surfaces must be in the same orientation as determined on scan line 1, so the intensities for surface S_1 can be entered without further calculations.

Any number of overlapping polygon surfaces can be processed with this scan-line method. Flags for the surfaces are set to indicate whether a position is inside or outside, and depth calculations are performed when surfaces overlap. In some cases, it is possible for surfaces to obscure each other alternately (Fig. 13–7). When coherence methods are used, we need to be careful to keep track of which surface section is visible on each scan line. One way of dealing with this situation is to divide the surfaces so that conflicts are resolved. For instance, plane *ABC* in Fig. 13–8 could be divided into the three surfaces *ABED*, *DEGF*, and *CFG*. Each can be treated as a distinct surface, so that no two planes can be alternately hidden and visible.

FIGURE 13—7
Examples of surface orientations that alternately obscure one another.

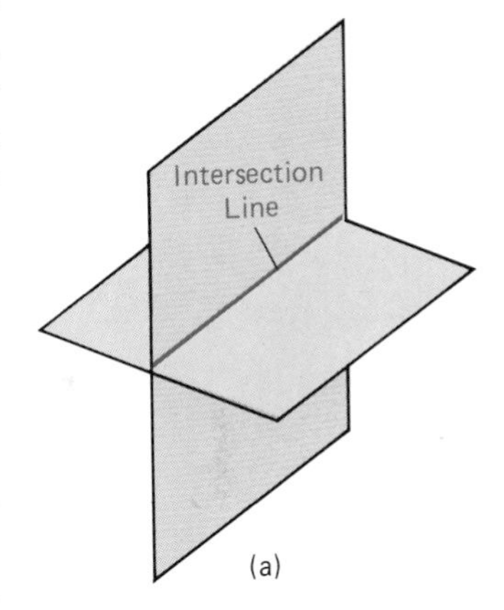

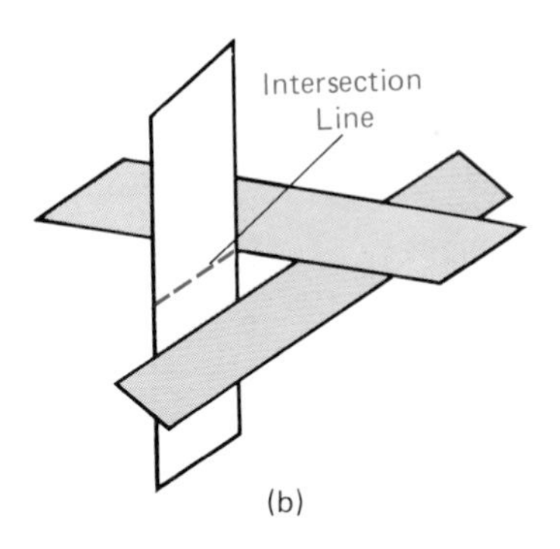

13–5 Depth-Sorting Method

It is possible to use both image-space and object-space operations in a hidden-surface algorithm. The **depth-sorting** method is a combination of these two approaches that performs the following basic functions:

1. Surfaces are sorted in order of decreasing depth.
2. Surfaces are scan-converted in order, starting with the surface of greatest depth.

FIGURE 13–8
Dividing a surface into multiple surfaces to avoid alternate visibility and invisibility problems between the two planes.

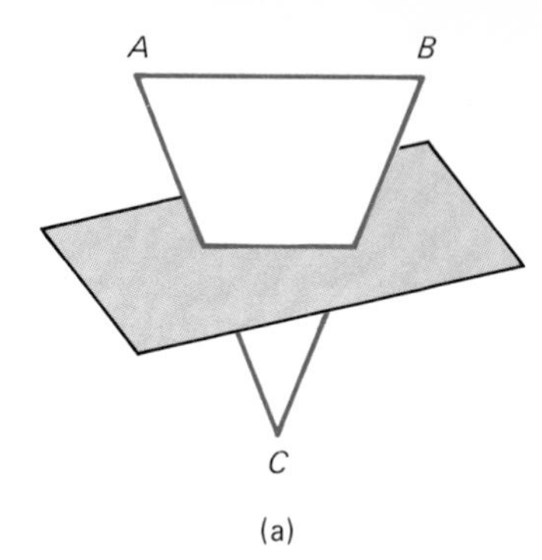

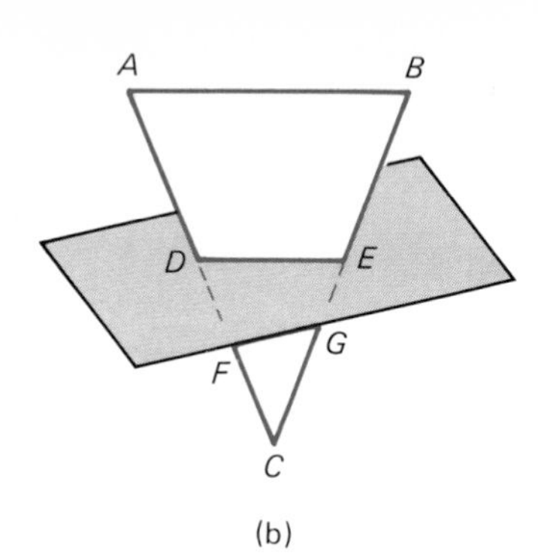

FIGURE 13–9
Two surfaces with no depth overlap.

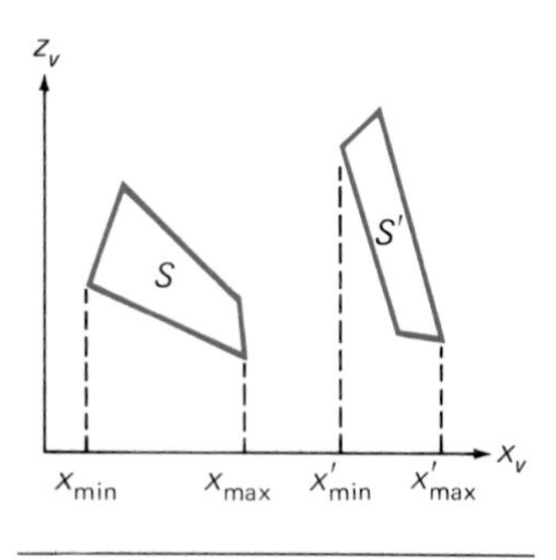

FIGURE 13–10
Two surfaces with depth overlap but no overlap in the *x* direction.

FIGURE 13–11
Surface *S* is completely "outside" the overlapping surface *S*,' relative to the view plane.

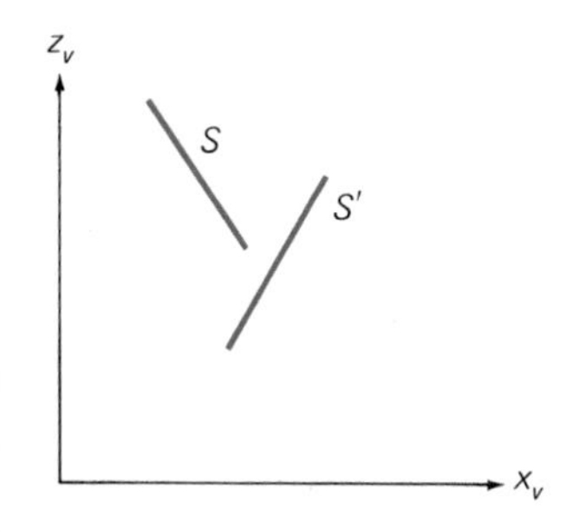

Sorting operations are carried out in object space, and the scan conversion of the polygon surfaces is performed in image space.

This method for solving the hidden-surface problem is sometimes referred to as the **painter's algorithm.** In creating an oil painting, an artist first paints the background colors. Next, the most distant objects are added. Finally, the foreground objects are painted on the canvas over the more distant objects. Each layer of paint covers up the previous layer. Using a similar technique, we first sort surfaces according to their distance from the view plane. The intensity values for the farthest surface are then entered into the refresh buffer. Taking each succeeding surface in turn (in decreasing-depth order), we "paint" the surface intensities onto the frame buffer over the intensities of the previously processed surfaces.

Painting polygon surfaces onto the frame buffer according to depth is carried out in several steps. On the first pass, surfaces are ordered according to the largest z value on each surface. The surface with the greatest depth (call it S) is then compared to the other surfaces in the list to determine whether there are any overlaps in depth. If no depth overlaps occur, S is scan converted. Figure 13–9 shows two surfaces with no depth overlap that have been projected onto the xz plane. This process is then repeated for the next surface in the list. As long as no overlaps occur, each surface is processed in depth order until all have been scan-converted. If a depth overlap is detected at any point in the list, we need to make some additional comparisons to determine whether any of the surfaces should be reordered.

For each surface that overlaps with S, we make the following tests. If any one of these tests is true, no reordering is necessary for that surface. The tests are listed in order of increasing difficulty.

1. The bounding rectangles in the xy plane for the two surfaces do not overlap.
2. Surface S is on the "outside" of the overlapping surface, relative to the view plane.
3. The overlapping surface is on the "inside" of surface S, relative to the view plane.
4. The projections of the two surfaces onto the view plane do not overlap.

As soon as one test is found to be true for an overlapping surface, we know that surface is not behind S. So we proceed to the next surface that overlaps S. If all the overlapping surfaces pass at least one of these tests, no reordering is necessary and S can be scan-converted.

Test 1 is performed in two parts. We first check for overlap in the x direction, then for overlap in the y direction. If either of these directions shows no overlap,

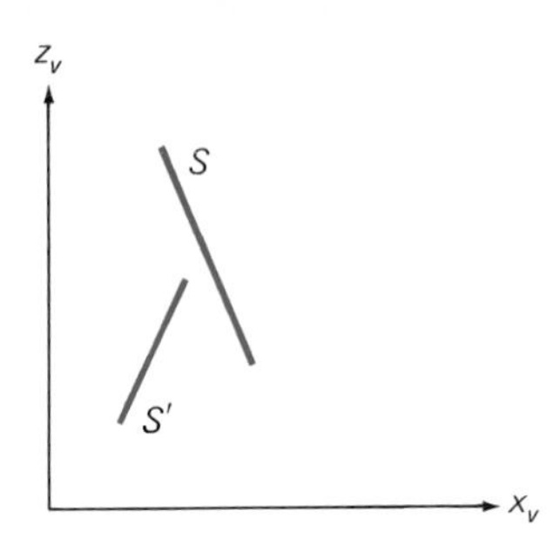

FIGURE 13–12
Overlapping surface S' is completely "inside" surface S, relative to the view plane.

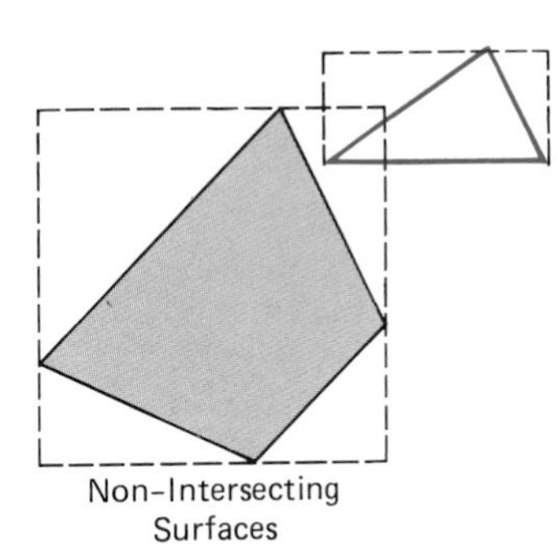

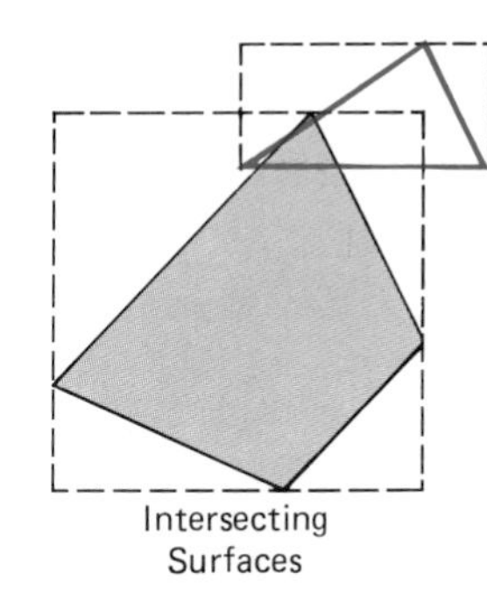

FIGURE 13–13
Two surfaces with overlapping bounding rectangles in the xy plane.

the two planes cannot obscure each other. An example of two surfaces that overlap in the z direction but not in the x direction is shown in Fig. 13–10.

We can perform test 2 by substituting the coordinates for all vertices of S into the plane equation for the overlapping surface and testing the sign of the result. Suppose that the overlapping surface has plane coefficients A', B', C', and D'. If $A'x + B'y + C'z + D' > 0$ for each vertex of S, then surface S is "outside" the overlapping surface (Fig. 13–11). As noted earlier, the coefficients A', B', C', and D' must be specified so that the normal to the overlapping surface points away from the view plane.

Inside test 3 is carried out using the plane coefficients A, B, C, and D for surface S. If the coordinates for all vertices of the overlapping surface satisfy the condition $Ax + By + Cz + D < 0$, then the overlapping surface is "inside" surface S (providing the surface normal to S points away from the view plane). Figure 13–12 shows an overlapping surface S' that satisfies this test. In this example, surface S is not "outside" S' (test 2 is not true).

If tests 1 through 3 have all failed, we try test 4 by checking for intersections between the bounding edges of the two surfaces using line equations in the xy plane. As demonstrated in Fig. 13–13, two surfaces may or may not intersect even though their bounding volumes overlap in the x, y, and z directions.

Should all four tests fail with a particular overlapping surface S', we interchange surfaces S and S' in the sorted list. An example of two surfaces that would be reordered with this procedure is given in Fig. 13–14. However, we still do not know for certain that we have found the farthest surface from the view plane. Figure 13–15 illustrates a situation in which we would first interchange S and S''. But since S'' obscures part of S', we need to interchange S'' and S' to get the three surfaces into the correct depth order. Therefore, we need to repeat the testing process for each surface that is reordered in the list.

It is possible for the algorithm just outlined to get into an infinite loop if two or more surfaces alternately obscure each other, as in Fig. 13–7. In such situations,

FIGURE 13–14
Surface S has greater depth but obscures surface S'.

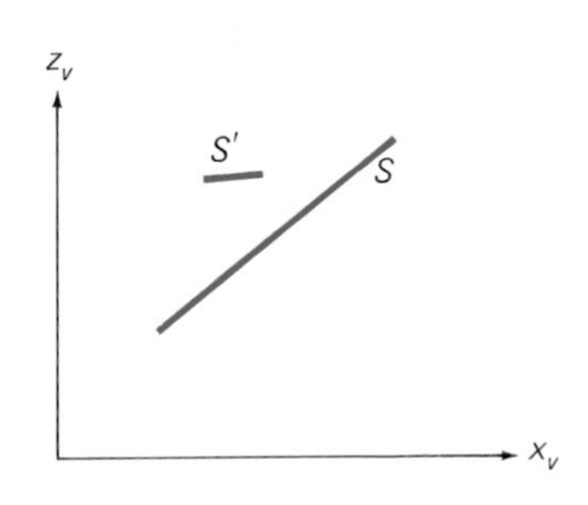

FIGURE 13–15
Three surfaces entered into the sorted surface list in the order S, S', S'' should be reordered S', S'', S.

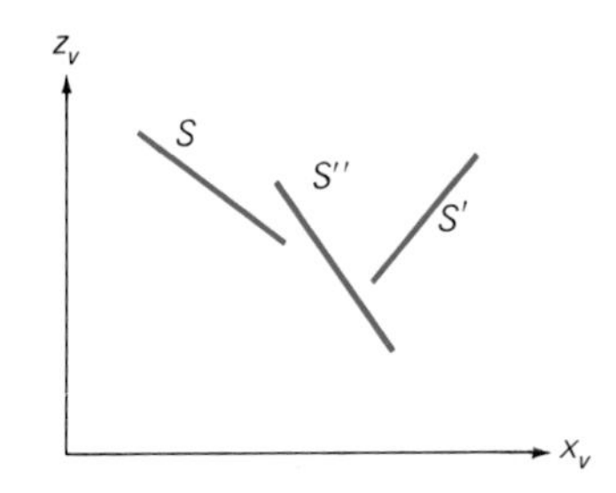

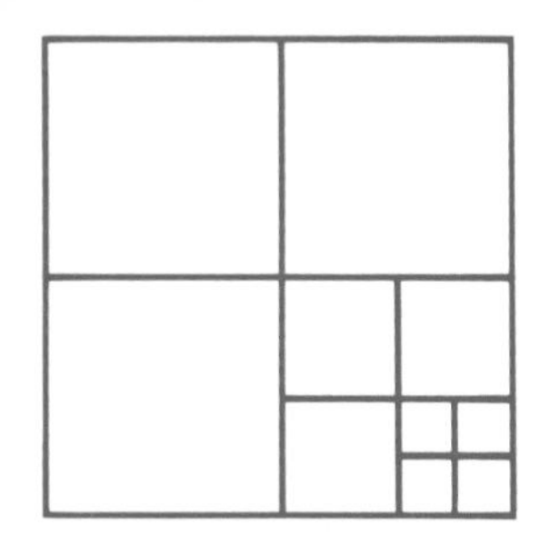

FIGURE 13–16
Area subdivisions performed with successive divisions by 2.

the algorithm would continually reshuffle the positions of the overlapping surfaces. To avoid such loops, we can flag any surface that has been reordered to a farther depth position so that it cannot be moved again. If an attempt is made to switch the surface a second time, we divide it into two parts at the intersection line of the two planes that are under comparison. The original surface is then replaced by the two new surfaces, and we continue processing as before.

13–6 Area-Subdivision Method

This technique for hidden-surface removal is essentially an image-space method, but object-space operations can be used to accomplish depth ordering of surfaces. The **area-subdivision** method takes advantage of area coherence in a scene by locating those view areas that represent part of a single surface. We apply this method by successively dividing the total viewing area into smaller and smaller rectangles until each small area is the projection of part of a single visible surface or no surface at all.

To implement this method, we need to establish tests that can quickly identify the area as part of a single surface or tell us that the area is too complex to analyze easily. Starting with the total view, we apply the tests to determine whether we should subdivide the total area into smaller rectangles. If the tests indicate that the view is sufficiently complex, we subdivide it. Next, we apply the tests to each of the smaller areas, subdividing these if the tests indicate that visibility of a single surface is still uncertain. We continue this process until the subdivisions are easily analyzed as belonging to a single surface or until they are reduced to the size of a single pixel. One way to subdivide an area is successively to divide the dimensions of the area by 2, as shown in Fig. 13–16. This approach is similar to that used in constructing a quadtree. A viewing area with a resolution of 1024 by 1024 could be subdivided ten times in this way before a subdivision is reduced to a point.

Tests to determine the visibility of a single surface within a specified area are made by comparing surfaces to the boundary of the area. There are four possible relationships that a surface can have with a specified area boundary. We can describe these relative surface characteristics in the following way (Fig. 13–17):

A **surrounding surface** is one that completely encloses the area.

An **overlapping surface** is one that is partly inside and partly outside the area.

An **inside surface** is one that is completely inside the area.

An **outside surface** is one that is completely outside the area.

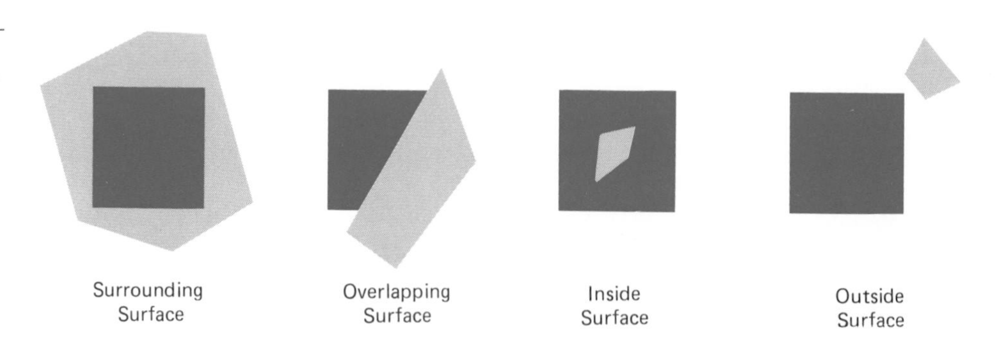

FIGURE 13–17
Possible relationships between polygon surfaces and a rectangular area.

The tests for determining surface visibility within an area can be stated in terms of these four classifications. No further subdivisions of a specified area are needed if one of the following conditions is true:

1. All surfaces are outside the area.
2. Only one inside, overlapping, or surrounding surface is in the area.
3. A surrounding surface obscures all other surfaces within the area boundaries.

Test 1 can be carried out by checking the bounding rectangles of all surfaces against the area boundaries. Test 2 can also use the bounding rectangles in the *xy* plane to identify an inside surface. For other types of surfaces, the bounding rectangles can be used as an initial check. If a single bounding rectangle intersects the area in some way, additional checks are used to determine whether the surface is surrounding, overlapping, or outside. Once a single inside, overlapping, or surrounding surface has been identified, its pixel intensities are transferred to the appropriate area within the frame buffer.

One method for implementing test 3 is to order surfaces according to their minimum depth. For each surrounding surface, we then compute the maximum z value within the area under consideration. If the maximum z value of one of these surrounding surfaces is less than the minimum z value of all other surfaces within the area, test 3 is satisfied. Figure 13–18 shows an example of the conditions for this method.

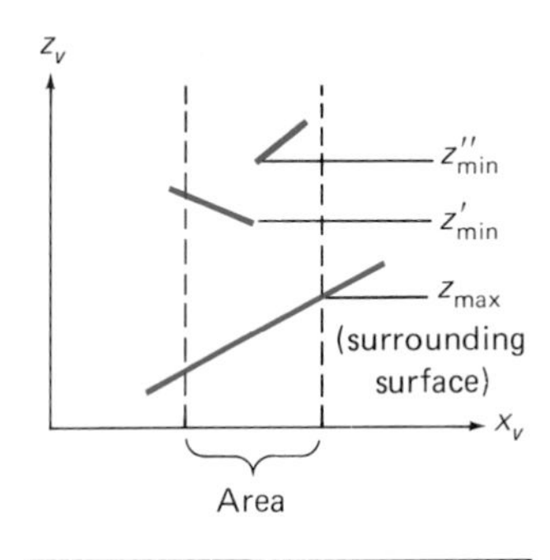

FIGURE 13–18
A surrounding surface with a maximum depth of z_{max} within an area obscures all surfaces that have a minimum depth greater than z_{max}.

Another method for carrying out test 3 that does not require depth sorting is to use plane equations to calculate z values at the four vertices of the area for all surrounding, overlapping, and inside surfaces. If the calculated z values for one of the surrounding surfaces is less than the calculated z values for all other surfaces, test 3 is true. Then the area can be filled with the intensity values of the surrounding surface.

For some situations, both methods of implementing test 3 will fail to identify correctly a surrounding surface that obscures all the other surfaces. Further testing could be carried out to identify the single surface that covers the area, but it is faster to subdivide the area than to continue with more complex testing. Once

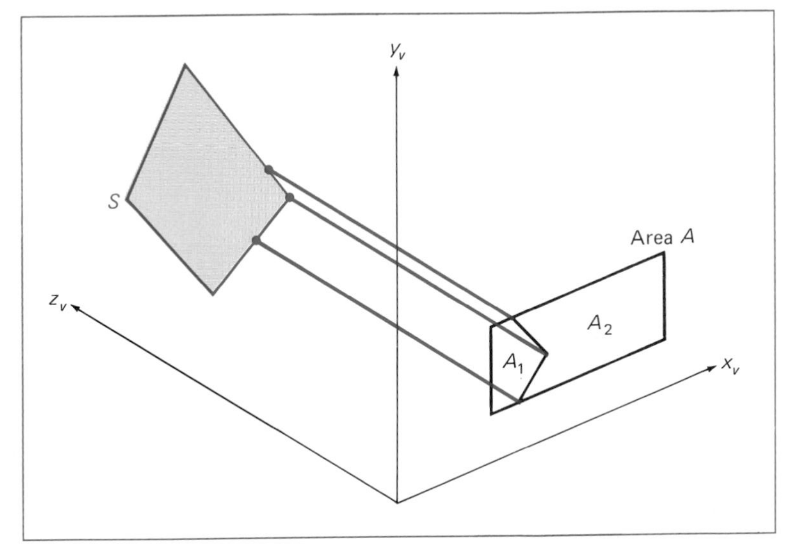

FIGURE 13–19
Area A is subdivided into A_1 and A_2 using the boundary of surface S on the view plane.

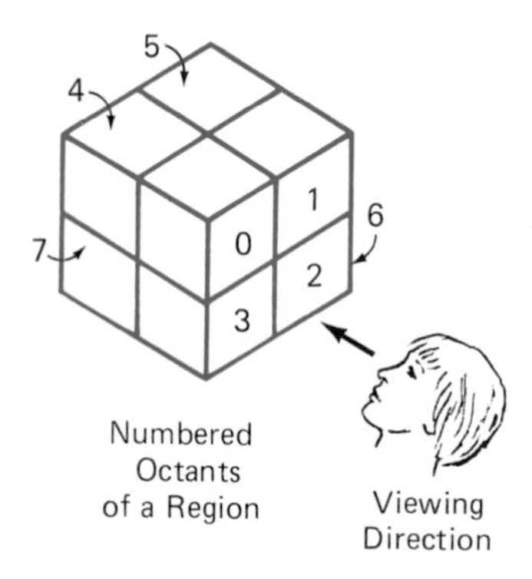

FIGURE 13–20
Objects in octants 0, 1, 2, and 3 obscure objects in the back octants (4, 5, 6, 7) when the viewing direction is as shown.

outside and surrounding surfaces have been identified for an area, they will remain outside and surrounding surfaces for all subdivisions of the area. Furthermore, some inside and overlapping surfaces can be expected to be eliminated as the subdivision process continues, so that the areas become easier to analyze. In the limiting case, when a subdivision the size of a pixel is produced, we simply calculate the depth of each relevant surface at that point and transfer the intensity of the nearest surface to the frame buffer.

As a variation on the basic subdivision process, we could subdivide areas along surface boundaries instead of dividing them in half. If the surfaces have been sorted according to minimum depth, we can use the surface of smallest z value to subdivide a given area. Figure 13–19 illustrates this method for subdividing areas. The projection of the boundary of surface S is used to partition the original area into the subdivisions A_1 and A_2. Surface S is then a surrounding surface for A_1 and visibility tests 2 and 3 can be applied to determine whether further subdividing is necessary. In general, fewer subdivisions are required using this approach, but more processing is needed to subdivide areas and to analyze the relation of surfaces to the subdivision boundaries.

13–7 Octree Methods

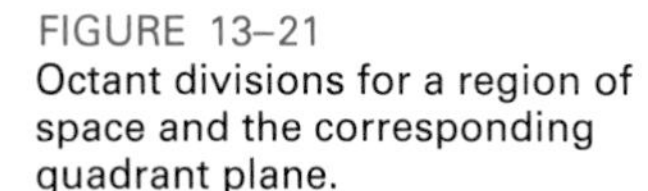
FIGURE 13–21
Octant divisions for a region of space and the corresponding quadrant plane.

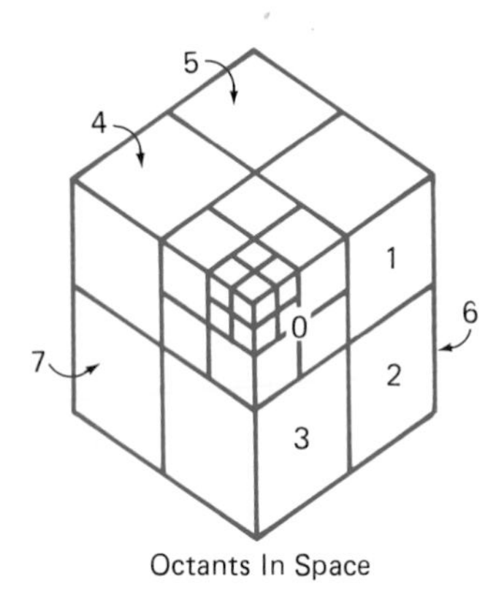

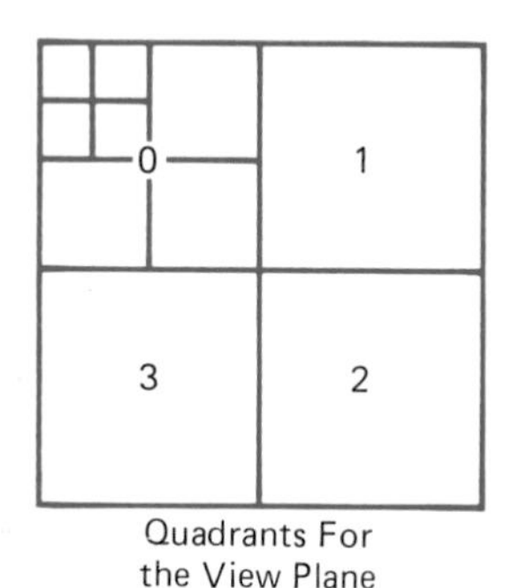

When an octree representation is used for the viewing volume, hidden-surface elimination is accomplished by projecting octree nodes onto the viewing surface in a front-to-back order. In Fig. 13–20, the front face of a region of space (the side toward the viewer) is formed with octants 0, 1, 2, and 3. Surfaces in the front of these octants are visible to the viewer. Any surfaces toward the rear of the front octants or in the back octants (4, 5, 6, and 7) may be hidden by the front surfaces.

Back surfaces are eliminated, for the viewing direction given in Fig. 13–20, by processing data elements in the octree nodes in the order 0, 1, 2, 3, 4, 5, 6, 7. This results in a depth-first traversal of the octree, so that nodes representing octants 0, 1, 2, and 3 for the entire region are visited before the nodes representing octants 4, 5, 6, and 7. Similarly, the nodes for the front four suboctants of octant 0 are visited before the nodes for the four back suboctants. The traversal of the octree continues in this order for each octant subdivision.

When a color value is encountered in an octree node, the pixel area in the frame buffer corresponding to this node is assigned that color value only if no values have previously been stored in this area. In this way, only the front colors are loaded into the buffer. Nothing is loaded if an area is void. Any node that is found to be completely obscured is eliminated from further processing, so that its subtrees are not accessed.

Different views of objects represented as octrees can be obtained by applying transformations to the octree representation that reorient the object according to the view selected. We assume that the octree representation is always set up so that octants 0, 1, 2, and 3 of a region form the front face, as in Fig. 13–20.

A method for displaying an octree is first to map the octree onto a quadtree of visible areas by traversing octree nodes from front to back in a recursive procedure. Then the quadtree representation for the visible surfaces is loaded into the frame buffer. Figure 13–21 depicts the octants in a region of space and the corresponding quadrants on the view plane. Contributions to quadrant 0 come from octants 0 and 4. Color values in quadrant 1 are obtained from surfaces in octants 1 and 5, and values in each of the other two quadrants are generated from the pair of octants aligned with each of these quadrants.

Recursive processing of octree nodes is demonstrated in procedure *convert_oct_to_quad*, which accepts an octree description and creates the quadtree representation for visible surfaces in the region. In most cases, both a front and a back octant must be considered in determining the correct color values for a quadrant. However, we can dispense with processing the back octant if the front octant is homogeneously filled with some color. For heterogeneous regions, the procedure is recursively called, passing as new arguments the child of the heterogeneous octant and a newly created quadtree node. If the front is empty, it is necessary only to process the child of the rear octant. Otherwise, two recursive calls are made, one for the rear octant and one for the front octant.

```
type
  oct_node_ptr = ↑oct_node;
  oct_entry = record
    case homogeneous : boolean of
      true : (color : integer);
      false : (child : oct_node_ptr)
    end; {record}
  oct_node = array [0..7] of oct_entry;

  quad_node_ptr = ↑quad_node;
  quad_entry = record
    case homogeneous : boolean of
      true : (color : integer);
      false : (child : quad_node_ptr)
    end; {record}
  quad_node = array [0..3] of quad_entry;

var
  newquadtree : quad_node_ptr;
  backcolor : integer;

{Assumes a frontal view of an octree (with octants 0, 1, 2,
 and 3 in front) and, as preparation for display, converts
 it to a quadtree. Accepts an octree as input, where each
 element of the octree is either a color value (homogeneous
 = true and the octant is filled with this color), the
 number -1 (homogeneous = true and the octant is empty), or
 a pointer to a child octant node (homogeneous is false).}
procedure convert_oct_to_quad (octree : oct_node;
                               var quadtree : quad_node);
  var k : integer;
  begin
    for k := 0 to 3 do begin
      quadtree[k].homogeneous := true;
      if octree[k].homogeneous then
        if (octree[k].color > -1) then              {front octant full}
          quadtree[k].color := octree[k].color
        else                                        {front octant empty}
          if octree[k+4].homogeneous then
            if (octree[k+4].color > -1) then{front empty, back full}

              quadtree[k].color := octree[k+4].color
            else                                    {front & back empty}
              quadtree[k].color := backcolor
          else begin                         {front empty, back hetero}
              quadtree[k].homogeneous := false;
```

```
                new(newquadtree);
                quadtree[k].child := newquadtree;
                convert_oct_to_quad (octree[k+4].child ↑, newquadtree ↑)
            end
        else begin                          {front hetero, back unknown}
            quadtree[k].homogeneous := false;
            new(newquadtree);
            quadtree[k].child := newquadtree;
            convert_oct_to_quad (octree[k+4].child ↑, newquadtree ↑);
            convert_oct_to_quad (octree [k  ].child ↑, newquadtree ↑)
        end
    end {for}
end;
```

13–8 Comparison of Hidden-Surface Methods

The effectiveness of a hidden-surface method depends on the characteristics of a particular application. If the surfaces in a scene are spread out in the z direction so that there is very little overlap in depth, a depth-sorting method may be best. For scenes with surfaces fairly well separated horizontally, a scan-line or area-subdivision method might be the best choice. In this way methods can be chosen that use sorting and coherence techniques to take advantage of the natural properties of a scene.

Because sorting and coherence considerations are important to the overall efficiency of a hidden-surface method, techniques for performing these operations should be chosen carefully. Whenever objects are known to be in approximate order, such as the list for an active edge table used in the scan-line method, a bubble sort can be employed effectively to perform the few exchanges required. Similarly, coherence techniques applied to scan lines, areas, or frames can be powerful tools for increasing the efficiency of hidden-surface methods.

As a general rule, the depth-sorting method is a highly effective approach for scenes with only a few surfaces. This is due to the fact that these scenes usually have few surfaces that overlap in depth. The scan-line method also performs well when a scene contains a small number of surfaces. Either the scan-line method or the depth-sorting method can be used effectively for scenes with several thousand faces. With scenes that contain more than a few thousand surfaces, the depth-buffer method or octree approach performs best. The depth-buffer method has a nearly constant processing time, independent of the number of surfaces in a scene. This is because the size of the surface areas decreases as the number of surfaces in the scene increases. Therefore, the depth-buffer method exhibits relatively low performance with simple scenes and relatively high performance with complex scenes. This approach is simple to implement, but it does require more memory than most methods. For that reason, another method, such as octrees or area subdivision, might be preferred for scenes with many surfaces.

When octree representations are used in a system, the hidden-surface elimination process is fast and simple. Only integer additions and subtractions are used in the process, and there is no need to perform sorting or intersection calculations. Another advantage of octrees is that they store more than surfaces. The entire solid region of an object is available for display, which makes the octree representation useful for obtaining cross-sectional slices of solids.

It is possible to combine and implement the various hidden-surface methods in various ways. In addition, algorithms are implemented in hardware, and special systems utilizing parallel processing are employed to increase the efficiency of these methods. Special hardware systems are usually used when processing speed is an especially important consideration, as in the generation of animated views for flight simulators.

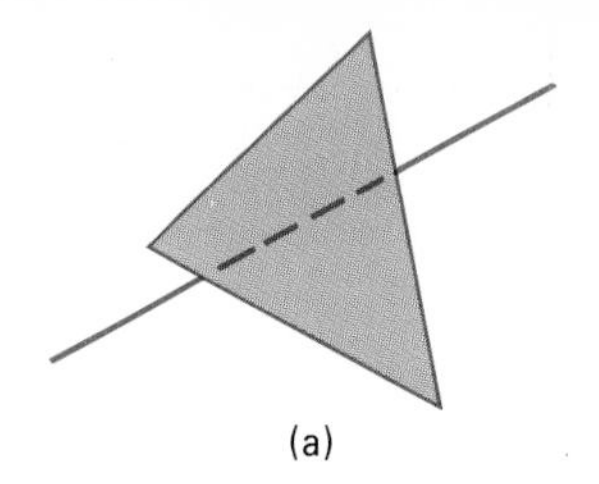

(b)

FIGURE 13–22
Hidden-line sections (dashed) for lines that (a) pass behind a surface and (b) penetrate a surface.

13–9 Hidden-Line Elimination

When only the outline of an object is to be displayed, hidden-line methods are used to remove the edges of objects that are obscured by surfaces closer to the view plane. Methods for removing hidden lines can be developed by considering object edges directly or by adapting hidden-surface methods.

A direct approach to eliminating hidden lines is to compare each line to each surface in a scene. The process involved here is similar to clipping lines against arbitrary window shapes, except that we now want to clip out the parts hidden by surfaces. For each line, depth values are compared to the surfaces to determine which line sections are not visible. We can use coherence methods to identify hidden line segments without actually testing each coordinate position. If both line intersections with the projection of a surface boundary have greater depth than the surface at those points, the line segment between the intersections is completely hidden, as in Fig. 13–22 (a). When the line has greater depth at one boundary intersection and less depth than the surface at the other boundary intersection, the line must penetrate the surface interior, as in Fig. 13–22 (b). In this case, we calculate the intersection point of the line with the surface using the plane equation and display only the visible sections.

Some hidden-surface methods are readily adapted to hidden-line removal. Using a back-face method, we could identify all the back surfaces of an object and display only the boundaries for the visible surfaces. With a depth-sorting method, surfaces can be painted into the refresh buffer so that surface interiors are in the background color, while boundaries are in the foreground color. By processing the surfaces from back to front, hidden lines are erased by the nearer surfaces. An area-subdivision method can be adapted to hidden-line removal by displaying only the boundaries of visible surfaces. Scan-line methods can be used to display visible lines by setting points along the scan line that coincide with boundaries of visible surfaces. Any hidden-surface method that uses scan conversion can be modified to a hidden-line method in a similar way.

13–10 Curved Surfaces

An effective method for handling hidden-surface elimination for objects with curved surfaces is the octree representation. Once the octrees have been established from the input definition of the objects, all hidden-surface eliminations are carried out with the same processing procedures. No special considerations need be given to curved surfaces.

Another method for dealing with hidden parts of a scene containing curved surfaces is to approximate each object as a set of plane, polygon surfaces. In the list of surfaces, we then replace each curved surface with its polygon representation and use one of the other hidden-surface methods previously discussed. This same approach can be used with Bézier and spline surfaces, where each surface section is divided into polygons. A scan-line method can then be used to load visible intensities into the refresh buffer.

We can also deal directly with the curve equations defining the surface boundaries. For most objects of interest, such as spheres, ellipsoids, cylinders, and cones, we can use quadratic equations to describe surfaces. Spherical and cylindrical surfaces are commonly used in modeling applications involving molecules, roller bearings, rings, and shafts. The quadratic equations can be expressed in parametric form, and numerical approximation techniques are often used to locate intersections of the curved surface with a scan line.

13–11 Hidden-Line and Hidden-Surface Command

A three-dimensional graphics package can accommodate several hidden-line or hidden-surface algorithms by providing a command to select the method that is to be used in a particular application. The following function is defined for this purpose:

```
set_hlhs_method_index (i)
```

Parameter i is assigned an integer code to identify the hidden-line or hidden-surface method to be used in displaying visible parts of objects. In an implementation of this function, an index table is used to define the range of allowable values for i and the type of algorithms available.

For some applications, it is useful to display hidden lines as dashed or dotted. This technique sometimes can give a better indication of the structure of complex objects than simply erasing all back parts. A system can accommodate this technique by expanding the index table to include these options.

REFERENCES

Additional sources of information on hidden-surface methods include Atherton (1983), Carpenter (1984), Crocker (1984), Frieder, Gordon, and Reynold (1985), Hamlin and Gear (1977), Sutherland, Sproull, and Schumacker (1973 and Mar. 1974), and Weiler and Atherton (1977).

EXERCISES

13–1. Develop a procedure, based on a back-face removal technique, for identifying all the front faces of a convex polyhedron with different-colored surfaces relative to a viewing plane. Assume that the object is defined in a left-handed viewing system with the *xy* plane as the viewing surface.

13–2. Implement the procedure of Ex. 13–1 in a program to orthographically project the visible faces of the object onto a window in the view plane. To simplify the procedure, assume that all parts of the object are in front of the view plane. Map the window onto a screen viewport for display.

13–3. Implement the procedure of Ex. 13–1 in a program to produce a perspective projection of the visible faces of the object in the view plane window, assuming that the object lies completely in front of the view plane. Map the window onto a screen viewport for display.

13–4. Write a program to implement the procedure of Ex. 13–1 for an animation application that incrementally rotates the object about an axis that passes through the object and is parallel to the view plane. Assume that the object lies completely in front of the view plane. Use an orthographic parallel projection to map the views successively onto the view screen.

13–5. Use the depth-buffer method to display the visible faces of any object that is defined in normalized coordinates in front of the viewport. Equations (13–4) and (13–5) are to be used to obtain depth values for all points on each surface once an initial depth has been determined. How can the storage requirements for the depth buffer be determined from the definition of the objects to be displayed?

13–6. Develop a program to implement the scan-line algorithm for displaying the visible surfaces of any object defined in front of the viewport. Use polygon and edge tables to store the definition of the object, and use coherence techniques to evaluate points along and between scan lines.

13–7. Set up a program to display the visible surfaces of a convex polyhedron using the painter's algorithm. That is, surfaces are to be sorted on depth and painted on the screen from back to front.

13–8. Extend the program of Ex. 13–7 to display any defined object with plane faces, using the depth-sorting checks to get surfaces in their proper order.

13–9. Give examples of situations where the two methods discussed for test 3 in the area-subdivision algorithm will fail to identify correctly a surrounding surface that obscures all other surfaces.

13–10. Develop an algorithm that would test a given plane surface against a rectangular area to decide whether it is a surrounding, overlapping, inside, or outside surface.

13–11. Extend the methods in Ex. 13–10 into an algorithm for generating a quadtree representation for the visible surfaces of an object by applying the area-subdivision tests to determine the values of the quadtree elements.

13–12. Set up an algorithm to load the quadtree representation of Ex. 13–11 into a raster for display.

13–13. Write a program on your system to display an octree representation for an object so that hidden surfaces are removed.

13–14. Discuss how antialiasing methods can be incorporated into the various hidden-surface elimination algorithms.

13–15. Develop an algorithm for hidden-line removal by comparing each line in a scene to each surface.

13–16. Discuss how hidden-line removal can be accomplished with the various hidden-surface methods.

13–17. Set up a procedure for displaying the hidden edges of an object with plane faces as dashed lines.

13–18. Develop an algorithm for displaying a set of three-dimensional primitives (block, sphere, and cylinder) in any relative positions so that hidden parts of objects are not displayed.

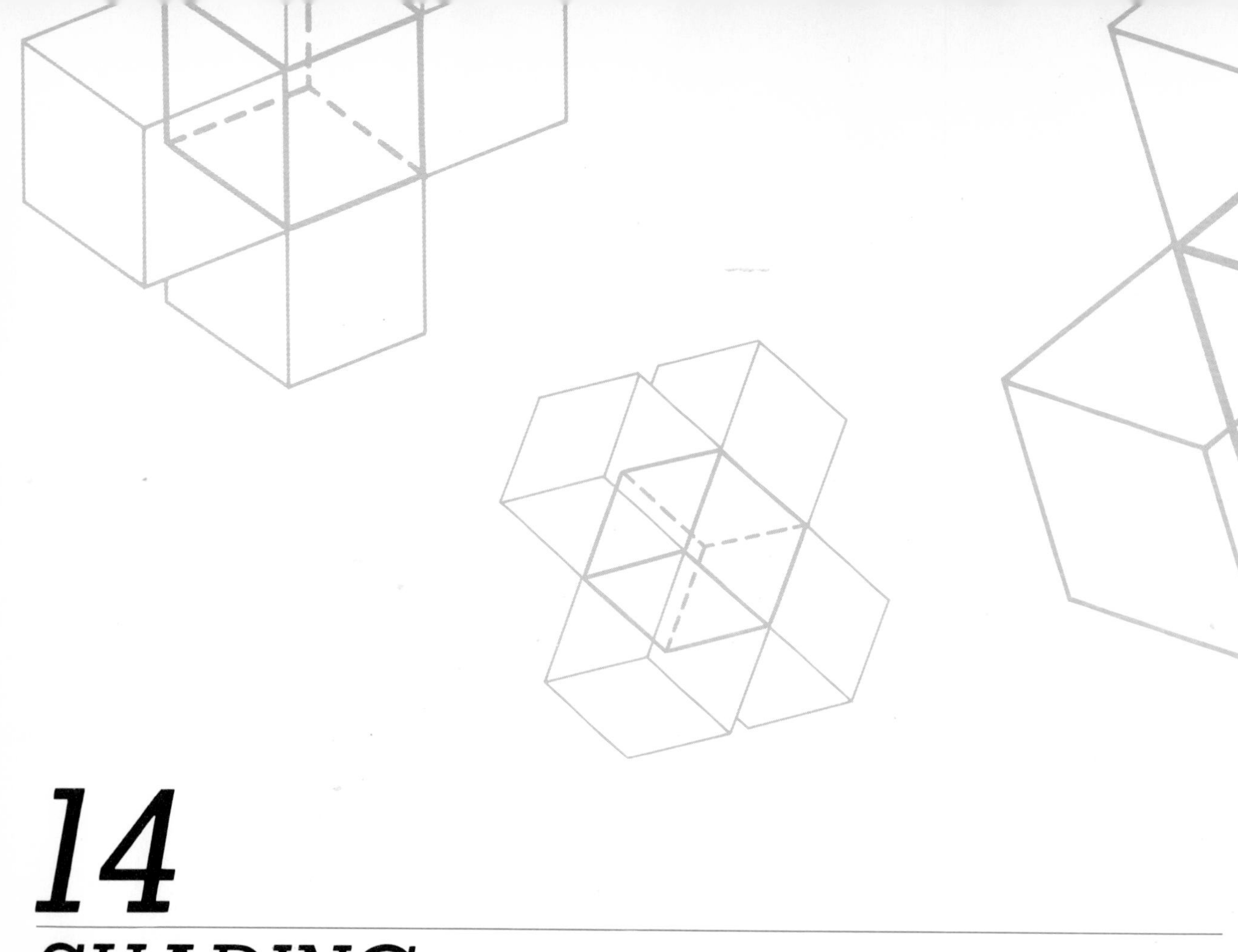

14 SHADING AND COLOR MODELS

Realistic displays of objects are obtained by generating perspective projections with hidden surfaces removed and then applying shading and color patterns to the visible surfaces. A **shading model** is used to calculate the intensity of light that we should see when we view a surface. These intensity calculations are based on the optical properties of surfaces, the relative positions of the surfaces, and their orientation with respect to light sources. We first consider how intensity calculations can be modeled from the laws of optics, then explore some of the techniques for applying calculated intensities to surfaces. Finally, we discuss the structure and applications of color models useful in graphics packages.

14–1 Modeling Light Intensities

The intensity of light seen on each surface of an object depends on the type of light sources in the vicinity and the surface characteristics of the object. Some objects have shiny surfaces, and some have dull, or matte, surfaces. In addition, some objects are constructed of opaque materials, while others are more or less transparent. A shading model to produce realistic intensities over the surfaces of an object must take these various properties into consideration.

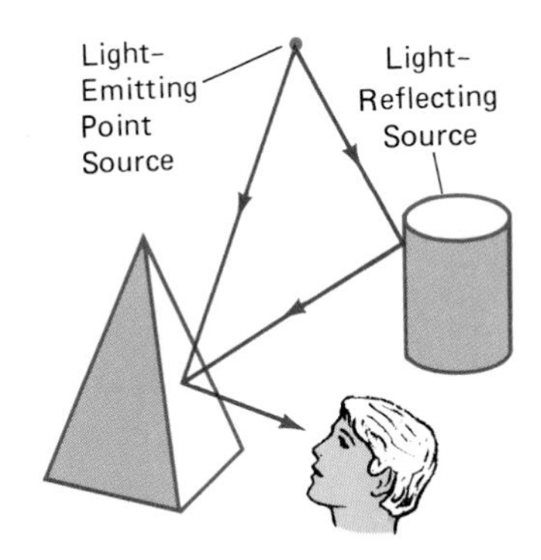

FIGURE 14–1
Surfaces of objects are illuminated by both light-emitting and light-reflecting sources.

Light Sources

When we view an object, we see the intensity of reflected light from the surfaces of the object. The light reflected from the surfaces comes from the various light sources around the object. If the object is transparent, we also see light from any sources that may be behind the object.

Light sources that illuminate an object are of two basic types, **light-emitting sources** and **light-reflecting sources** (Fig. 14–1). Light-emitting sources include light bulbs and the sun. Light-reflecting sources are illuminated surfaces of other objects, such as the walls of a room, that are near the object we are viewing. A surface that is not exposed directly to a light-emitting source will still be visible if nearby objects are illuminated. The multiple reflections of light from such nearby objects combine to produce a uniform illumination called **ambient light,** or **background light.**

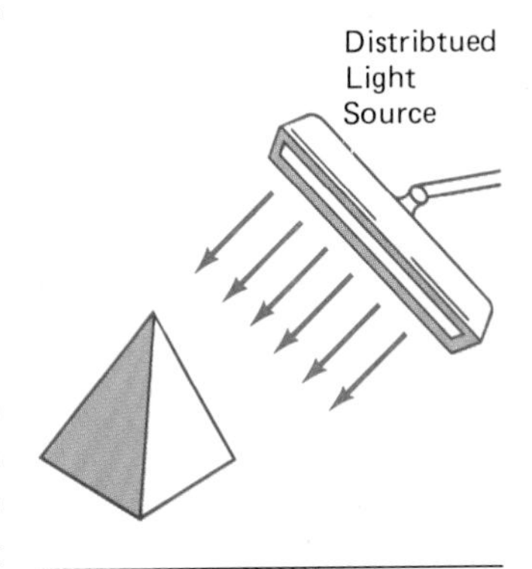

FIGURE 14–2
An object illuminated with a distributed light source.

When the dimensions of a light source are small compared to the size of an object, we can model it as a **point source,** as in Fig. 14–1. This approximation is used for most sources, such as the sun, that are sufficiently far from the object. In other cases, we have a **distributed light source.** This occurs when we have a large, nearby source, such as the long neon light in Fig. 14–2, whose area cannot be considered as infinitely small compared to the size of the illuminated object. Shading models based on the intensity laws for ambient light and point sources provide highly effective surface shading for objects with distant light sources. Only slight modifications to this model are needed to accommodate most distributed sources that may be included in a scene.

A shading model for calculating the intensity of light reflected from a surface can be established by considering contributions from the ambient light sources and point sources in the vicinity of the surface. Both of these sources produce light reflections that are scattered in all directions. This scattered light is called **diffuse reflection,** and it results from the surface roughness, or graininess. A matte surface produces primarily diffuse reflections, so that the surface appears equally bright from all viewing directions. In addition to diffuse reflection, point sources create highlights, or bright spots, called **specular reflection.** This highlighting effect is more pronounced on shiny surfaces than on dull surfaces. These two types of reflections are illustrated in Fig. 14–3.

FIGURE 14–3
Diffuse and specular reflections.

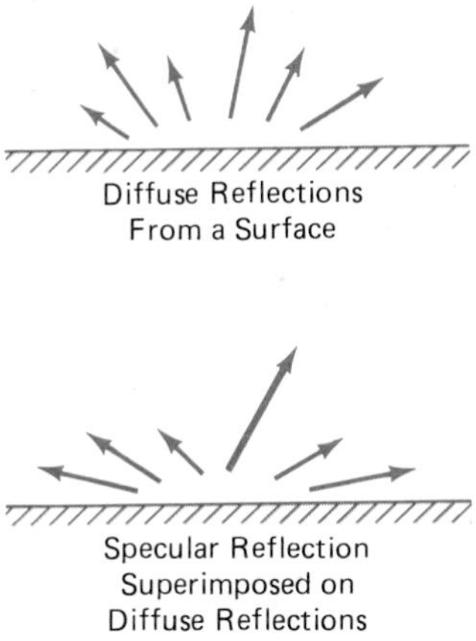

Diffuse Reflections
From a Surface

Specular Reflection
Superimposed on
Diffuse Reflections

For transparent objects, we can expand the reflection-shading model to include light transmission effects. As in reflection, both diffuse and specular transmission of light occur.

Diffuse Reflection

We first consider the effects of ambient light. Since this background light is the result of multiple reflections from nearby objects, we can consider it to be of uniform intensity I_a in all directions. When ambient light is reflected from a sur-

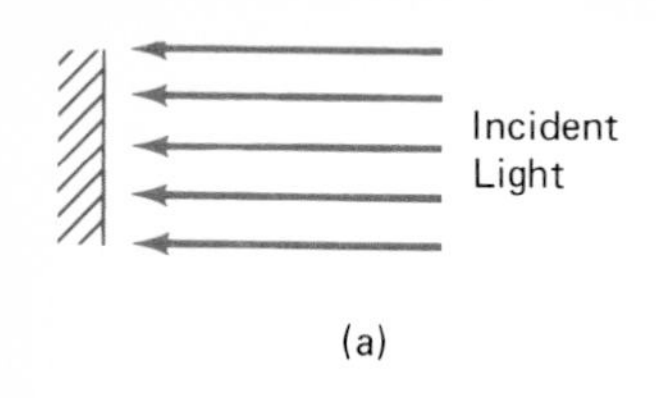

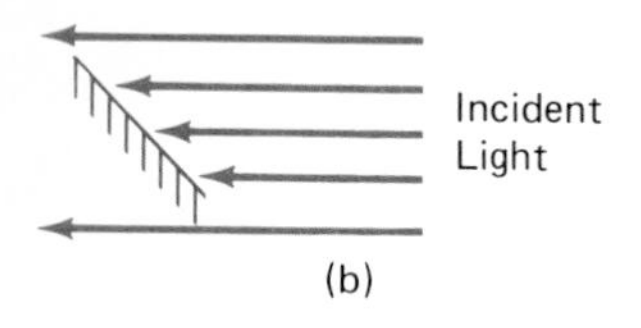

FIGURE 14–4
A surface perpendicular to the direction of the incident light (a) is more illuminated (appears brighter) than a surface at an angle (b) to the light direction.

face, it produces a uniform illumination of the surface at any viewing position from which the surface is visible. If a surface is exposed only to ambient light, we can express the intensity of the diffuse reflection at any point on the surface as

$$I = k_d I_a \tag{14–1}$$

Parameter k_d is the **coefficient of reflection,** or **reflectivity,** for the surface. It is assigned a constant value in the interval 0 to 1, according to the reflecting properties of the surface. Highly reflective surfaces have a reflectivity near 1, so that the intensity of the reflected light is nearly that of the incident light. Surfaces that absorb most of the incident light have a reflectivity near 0.

Rarely are surfaces illuminated with ambient light alone. A shading model that calculates intensities according to Eq. 14–1 would shade all visible surfaces of an object with the same intensity. More realistic shading is obtained by including the effects of point sources in the shading model.

The calculation of diffuse reflection due to a point source of light is based on Lambert's cosine law, which states that the intensity of the reflected light depends on the angle of illumination. A surface that is perpendicular to the direction of the incident light appears brighter than a surface that is at an angle to the direction of the incoming light. As the angle increases, less of the incident light falls on the surface, as shown in Fig. 14–4. This figure shows a group of parallel light rays incident on two surfaces with different spatial orientations relative to the incident light direction from a distant source. A light source that is close to the surface, however, produces incident light rays that are not parallel (Fig. 14–5). To simplify calculations in our model, we assume that each point source in a scene is sufficiently far from the surface so that the light rays from that source are parallel as they strike the surface. Later we consider extensions to the model for other types of sources.

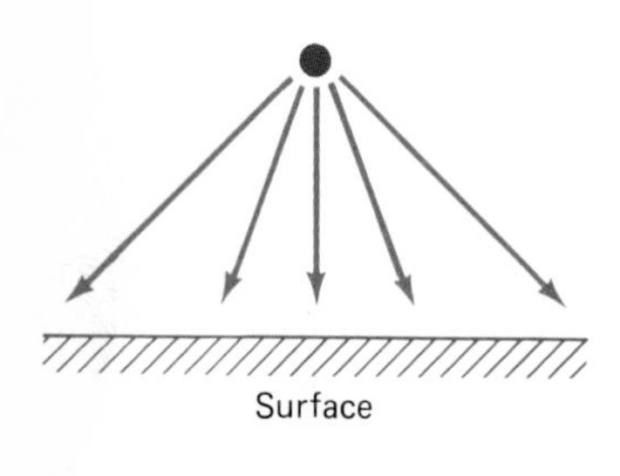

FIGURE 14–5
A nearby light source illuminates a surface with nonparallel rays of incident light.

We can describe the orientation of a surface with a unit normal vector $\boldsymbol{N}$ and the direction to the light source with a unit vector $\boldsymbol{L}$. The angle θ between these two vectors is called the **angle of incidence** (Fig. 14–6), and Lambert's cosine law states that the intensity of reflected light is proportional to $\cos\theta$. We can calculate $\cos\theta$ from the dot product of these two unit vectors:

$$\cos\theta = \boldsymbol{N} \cdot \boldsymbol{L} \tag{14–2}$$

A surface is illuminated by a point source only if the angle of incidence is between 0° and 90° ($\cos\theta$ is in the interval from 0 to 1). When $\cos\theta$ is negative, the light source is "behind" the surface.

We can expect the brightness of an illuminated surface to depend on the distance to the light source, since more distant sources are fainter than those that are nearer. If d represents the distance from a light source to a point on the surface and I_p is the intensity of the source, the intensity of the diffuse reflection at that position on the surface can be modeled as

$$I = \frac{k_d I_p}{d + d_0}(\boldsymbol{N} \cdot \boldsymbol{L}) \tag{14–3}$$

where parameter d_0 is a constant that is included to prevent the denominator from approaching zero when d is small. In an implementation of this intensity calculation, it is often convenient to assume that a point source at the viewer's position is illuminating a scene. Then d can be set to the distance from the surface position to the projection reference point, and d_0 can be adjusted until satisfactory shading patterns are obtained.

FIGURE 14–6
Angle of incidence θ between the light direction L and the surface normal N.

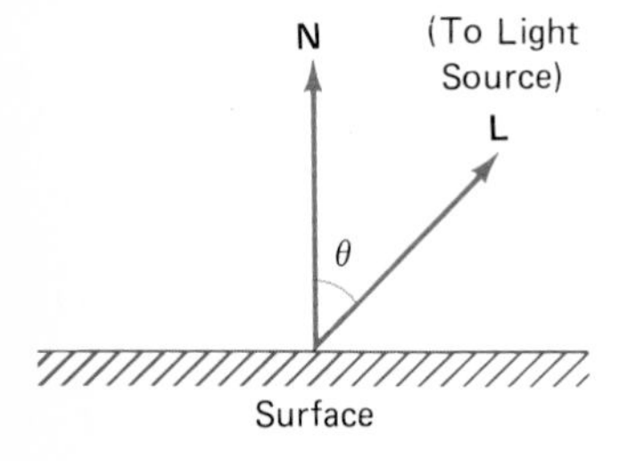

Equation 14–3 is an adaptation of Lambert's cosine law that has been found to produce realistic shading of surfaces. Theoretically, the light intensity arriving at

a surface is proportional to $1/d^2$, where d is the distance from the surface to the point source. However, most light sources are larger than points, and we can expect the intensity to decrease less rapidly. The factor $d + d_0$ in the denominator of Eq. 14–3 more accurately models the intensity reflections for surfaces at varying distances from a nearby light source.

Total diffuse reflection for a surface illuminated by ambient light and one point source is given by

$$I = k_d I_a + \frac{k_d I_p}{d + d_0}(\mathbf{N} \cdot \mathbf{L}) \tag{14–4}$$

If more than one point source is to be included in a scene, Eq. 14–4 is expanded to include terms for the additional light sources.

When color is to be included in a scene, Eq. 14–4 must be expressed in terms of the color components of the intensity. For an RGB video monitor, color components are red, green, and blue. Parameters for intensity and reflectivity then become three-element vectors, with one element for each of the color components. The vector representing the coefficient of reflection has components (k_{dr}, k_{dg}, k_{db}). A green surface, for example, has a nonzero value for the green reflectivity component, k_{dg}, while the red and blue components are set to zero ($k_{dr} = k_{db} = 0$). For any light falling on this surface, the red and blue components of the light are absorbed, and only the green component is reflected. The intensity calculation for this example reduces to the single expression

$$I_g = k_{dg} I_{ag} + \frac{k_{dg} I_{pg}}{d + d_0}(\mathbf{N} \cdot \mathbf{L}) \tag{14–5}$$

using the green components of the intensity and reflectivity vectors. In general, a surface can reflect all three color components of the incident light, and three equations would be needed to calculate the three color components of the reflected light. Calculated intensity levels for each color component are then used to adjust the corresponding electron gun in the RGB monitor.

Specular Reflection

At certain viewing angles, a shiny surface reflects all incident light, independently of the reflectivity values. This phenomenon, called specular reflection, produces a spot of reflected light that is the same color as the incident light. Normally, objects are illuminated with white light, so that the specular reflection is a bright white spot. For an ideal reflector (perfect mirror), the angle of incidence and the angle of specular reflection are the same (Fig. 14–7). We use unit vector $\boldsymbol{R}$ to represent the direction for specular reflection. Unit vector **V** points in the direction of the viewer, and unit vector $\boldsymbol{L}$ points to the light source. Specular reflection can be seen with a perfect reflector only when **V** and $\boldsymbol{R}$ coincide ($\phi = 0$).

Real objects exhibit specular reflection over a range of positions about the

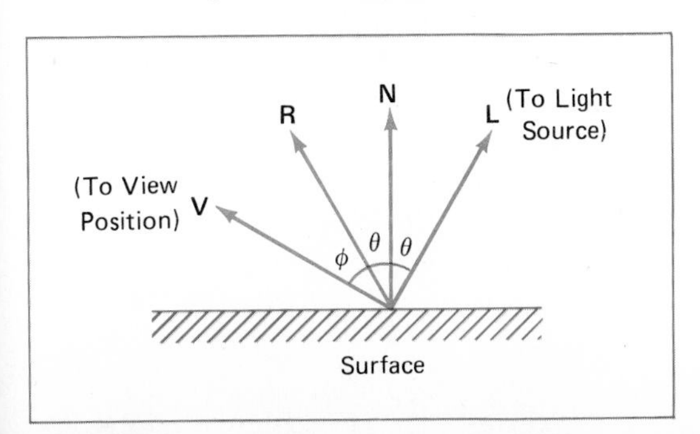

FIGURE 14–7
For a perfect reflector, angle of incidence θ is the same as the angle of reflection.

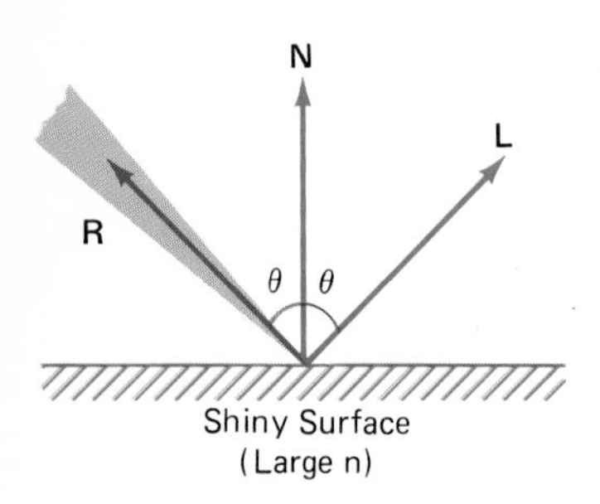

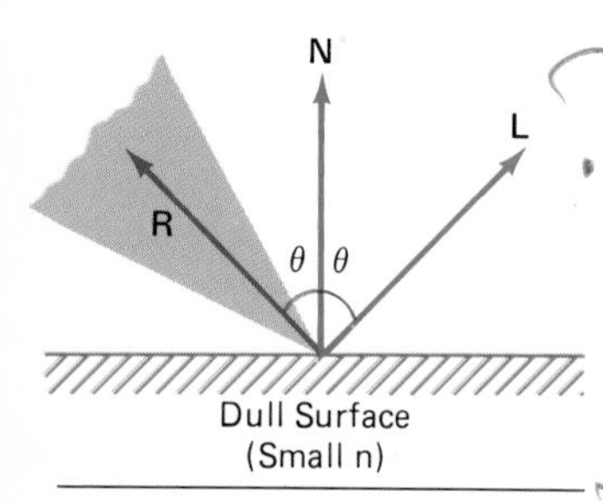

FIGURE 14–8
Modeling specular reflection (shaded area) with parameter *n*.

vector ***R***. Shiny surfaces have a narrow reflection range, and dull surfaces have a wider reflection range. Specular-reflection models must produce the highest intensity in the direction of ***R***, with the intensity decreasing rapidly as the viewing angle ϕ increases.

One method for modeling specular reflection, developed by Phong Bui Tuong and called the Phong model, sets the intensity of specular reflection proportional to $\cos^n\phi$. The value assigned to n determines the type of surface that is to be viewed. A very shiny surface is modeled with a large value for n (200 or more), and smaller values (down to 1) are used for duller surfaces. For a perfect reflector, n is infinite. A rough surface, such as cardboard, might be assigned a value near 1. The effect of n on the angular range for viewing specular reflection is shown in Fig. 14–8.

Specular reflection also depends on the angle of incidence. In general, the intensity increases as the angle of incidence increases. This effect is incorporated into the specular-reflection model by making the intensity proportional to a reflection function $W(\theta)$, so that the complete reflection model is written as

$$I = k_d I_a + \frac{I_p}{d + d_0}[k_d(\mathbf{N} \cdot \mathbf{L}) + W(\theta)\cos^n\phi] \tag{14–6}$$

The functional form for $W(\theta)$ depends on the surface material. Some materials, such as glass, exhibit very little specular reflection at smaller angles of incidence but increase the intensity of the specular reflection as θ approaches 90°. For these materials, W should vary from a value near 0 up to a value of 1 as the angle of incidence varies from 0° to 90°. Some materials have nearly constant specular reflection for all incidence angles, so that W could be assigned a constant value in the interval 0 to 1.

Since ***V*** and ***R*** are unit vectors in the viewing and reflection directions, we can set $\cos\phi = \mathbf{V} \cdot \mathbf{R}$ in Eq. 14–6. Also, for many applications, we can simplify the intensity calculations by setting $W(\theta)$ to a constant value k_s for the surface. The complete intensity model for reflection, due to ambient light and a single point source, can then be written as

$$I = k_d I_a + \frac{I_p}{d + d_o}[k_d(\mathbf{N} \cdot \mathbf{L}) + k_s(\mathbf{V} \cdot \mathbf{R})^n] \tag{14–7}$$

In this model, constant values are assigned to parameters k_d, k_s, and d_o for each illuminated surface. Intensity values for the ambient light and the point sources are set, and values for the unit vectors are established. For each point on an illuminated surface, we calculate the relevant dot products and determine the intensity of the reflected light.

If the point source is far from a plane surface, the dot product $\mathbf{N} \cdot \mathbf{L}$ is approximately constant over the surface. Similarly, if the view reference point is sufficiently far from the surface, the product $\mathbf{V} \cdot \mathbf{R}$ is constant. When these simplifications can be made, the number of calculations is significantly reduced. If one or both of these simplifications cannot be made, dot products must be evaluated at each point. Since we can expect the values of these products to change only slightly from one point to the next, coherence methods can be used to calculate the dot products across a scan line.

FIGURE 14–9
Vector ***B*** is along the bisector of the angle between vectors ***V*** and ***L***.

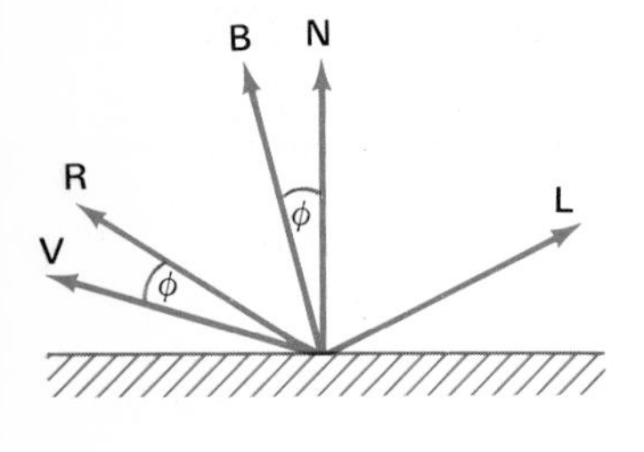

Vectors ***L*** and ***N*** can be used to determine vector ***R*** for the calculation of $\cos\phi$. Alternatively, we can evaluate $\cos\phi$ directly in terms of vectors ***L***, ***N***, and ***V*** as follows. The direction of the normal vector ***N*** is along the bisector of the angle between ***R*** and ***L***, and we can define a vector ***B*** along the bisector of the angle between ***V*** and ***L*** (Fig. 14–9) as

$$B = \frac{\mathbf{V} + \mathbf{L}}{|\mathbf{V} + \mathbf{L}|} \tag{14-8}$$

Then, vector $\mathbf{B}$ can then be used to calculate cos ϕ, since

$$\mathbf{V} \cdot \mathbf{R} = \mathbf{N} \cdot \mathbf{B} \tag{14-9}$$

Other methods for modeling light intensities have been developed. One technique, developed by Torrance and Sparrow and adapted to graphics applications by Blinn, divides each surface in a scene into a set of tiny planes. Each of the small planes is assumed to be an ideal reflector, and the planes are oriented randomly over the total surface. A Gaussian distribution function is used to set the orientation of each plane. The specular reflection for the surface is calculated as the total contribution from the small planes as a function of the intensity I_p from a distant point source and the vectors **N**, **V**, and **L**.

Refracted Light

When a transparent object is to be modeled, the intensity equations must be modified to include contributions from light sources in back of the object. In most cases, these sources are light-reflecting surfaces of other objects, as in Fig. 14–10. Reflected light from these surfaces passes through the transparent object and modifies the object intensity, as calculated by Eq. 14–6 or Eq. 14–7. Light passing through a surface is called **transmitted light** or **refracted light.**

Both **diffuse refraction** and **specular refraction** can take place at the surfaces of an object. Diffuse effects are important when a partially transparent surface, such as frosted glass, is to be modeled. Light passing through such materials is scattered so that a blurred image of background objects is obtained. Diffuse refractions can be generated by decreasing the intensity of the refracted light and spreading intensity contributions at each point on the refracting surface onto a finite area. These manipulations are time-consuming, and most shading models employ only specular effects.

Realistic models of transparent materials, such as clear glass, can be developed by adding specular-refraction contributions to the reflected intensity calculations. When light is incident upon a transparent surface, part of it is reflected and

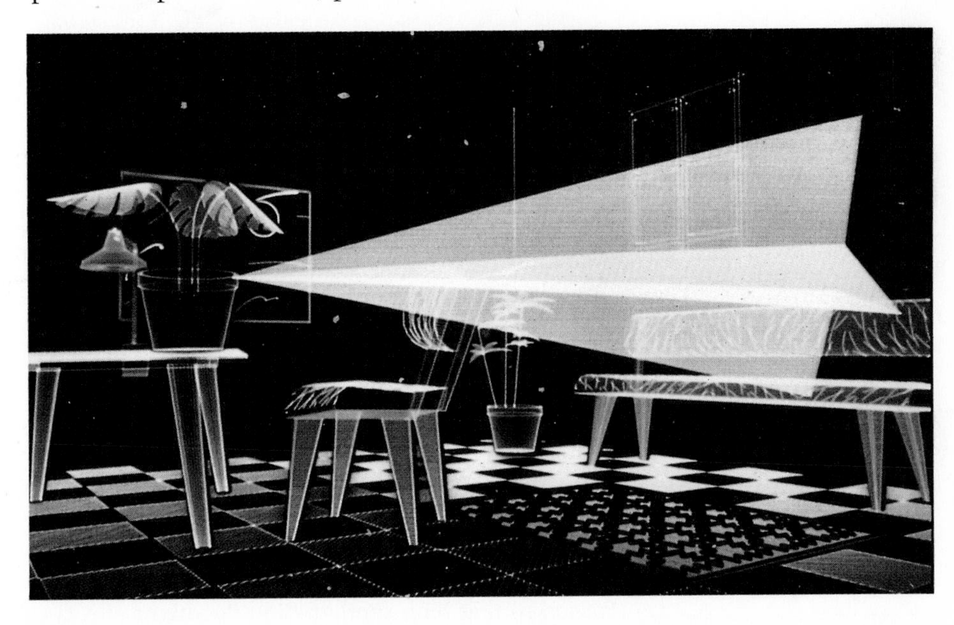

FIGURE 14–10
A transparent object, modeled here as a thin sheet of paper, transmits some light from surfaces behind the object. From Panasonic "Glider," courtesy Robert Abel & Associates.

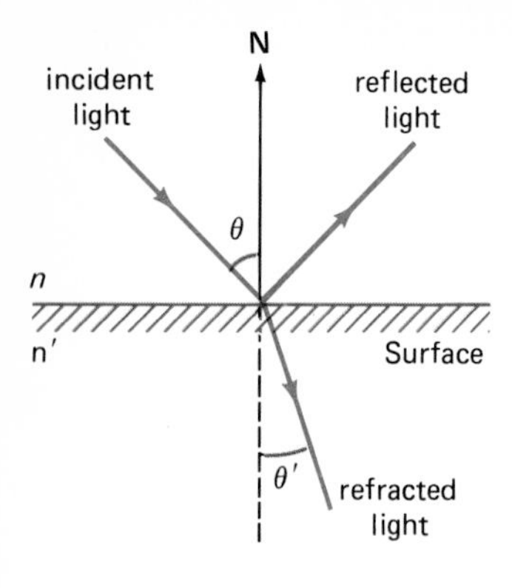

FIGURE 14–11
A ray of light incident upon a surface is partially reflected and partially refracted.

part is refracted (Fig. 14–11). Because the speed of light is different in different materials, the path of the refracted light is different from that of the incident light. The direction of the refracted light, specified by the **angle of refraction,** is a function of the **index of refraction** of a material. Specifically, the angle of refraction θ' is calculated from the angle of incidence θ, the index of refraction n of the material outside the surface (usually air), and the index of refraction n' of the surface material according to the law of refraction:

$$n \sin\theta = n' \sin\theta' \tag{14–10}$$

Actually, the index of refraction of a material is a function of the wavelength of the incident light, so that the different color components of a light ray will be refracted at different angles. For most applications, we can use an average index of refraction for the different materials that are modeled in a scene. The index of refraction of air is approximately 1, and that of crown glass is about 1.5. Using these values in Eq. 14–10 with an angle of incidence of 30° yields an angle of refraction of about 19°. Figure 14–12 illustrates the changes in the path direction for a light ray refracted through a glass object. The overall effect of the refraction is to shift the incident light to a parallel path. Since the calculations of the trigonometric functions in Eq. 14–10 are time-consuming, refraction effects could be modeled by simply shifting the path of the incident light a small amount.

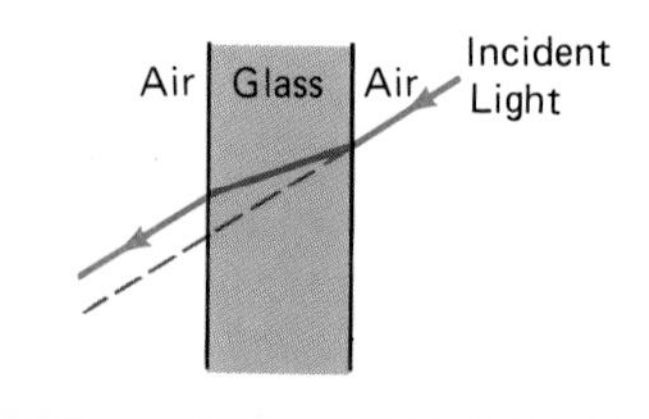

FIGURE 14–12
Refraction of light through a glass object. The emerging refracted ray travels along a path that is parallel to the incident light path (dashed line).

A simpler procedure for modeling transparent objects is to ignore the path shifts altogether. In effect, this approach assumes that there is no change in the index of refraction from one material to another, so that the angle of refraction is always the same as the angle of incidence. This method speeds up the calculation of intensities and produces realistic displays for thin objects, such as champagne glasses.

We can modify the reflection-shading model to include refraction effects by projecting the intensity of background objects to the front surface of the transparent object, as shown in Fig. 14–13. The intensity I_b of a background object is added to the intensity I_t of the transparent object. A more realistic shading pattern is obtained by using a "refraction coefficient" r to weight the reflected and refracted intensity contributions to produce a total intensity I, calculated as

$$I = rI_t + (1 - r)I_b \tag{14–11}$$

For highly transparent objects, we assign r a value near 0. Opaque objects transmit no light from background objects, and we can set $r = 1$ for these materials. It is also possible to allow r to be a function of position over the surface, so that different parts of an object can transmit more or less background intensity according to the values assigned to r.

FIGURE 14–13
The intensity of a background object at point P can be added to the intensity of a transparent object along a projection line (dashed).

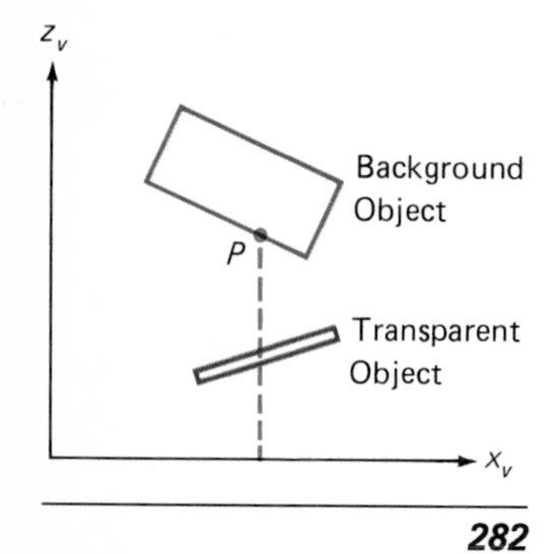

Background objects relative to a transparent object can be identified with a hidden-surface method that sorts surfaces according to depth. Once the surfaces have been sorted, those that are visible through the transparent object are identified. Then the intensity of corresponding points along the viewing direction is combined according to Eq. 14–11 to produce the total shading pattern for the transparent object.

An example of the application of an intensity model to the surfaces of an automobile is shown in Fig. 14–14. The surfaces of the car display specular reflections (bright spots and lines), diffuse reflections, and refracted light through the windows. An intensity model employing multiple light sources and "light controls" was used to reproduce the lighting effects typically available in a photography studio. The light controls allowed the light direction and the light concentration to be varied so that both spotlight and floodlight effects could be simulated.

FIGURE 14–14
Illumination patterns produced with an intensity model, displaying diffuse reflections, specular reflections, and refracted light. Courtesy David Warn, General Motors Research Lab.

Texture and Surface Patterns

The shading model we have discussed for calculating light intensities provides a smooth shading for every surface in a scene. However, many objects do not have smooth surfaces. Surface texture is needed to model accurately such objects as brick walls, gravel roads, and shag carpets. In addition, some surfaces contain patterns that must be included in the shading model. The surface of a vase could contain a painted design; a water glass might have the family crest engraved into the surface; a tennis court contains markings for the alleys, service areas, and base line; and a four-lane highway has dividing lines and other markings, such as oil spills and tire skids. Figure 14–15 illustrates the display of surfaces modeled with texture and patterns.

Intensity values furnished by a shading model can be adjusted to accommodate surface texture by altering the surface normal so that it is a function of position over the surface. If we allow the surface normal to vary randomly, we can obtain an irregularly textured surface such as that of a raisin. A repeating function can be used to model a more regular surface, such as a sculptured carpet that contains a repeated texture pattern. An irregular surface can also be modeled by dividing the surface into a collection of small, randomly oriented surfaces. In addition, we could allow the coefficient of reflection to vary with position so as to obtain greater variations in intensity.

FIGURE 14–15
Combining surface patterns and textures with an intensity model. Courtesy Robert Abel & Associates.

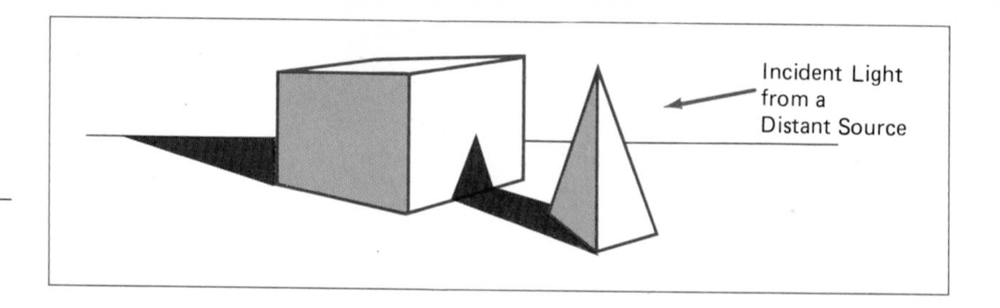

FIGURE 14–16
Objects modeled with shadows.

Texture-mapping methods are also used to model objects. Patterns can be applied to surfaces of three-dimensional objects with methods similar to those for applying patterns to two-dimensional objects. A shading pattern is defined in an array, and the array is mapped onto an object at a designated position. The array pattern can be treated as a plane surface, and the position and orientation of the pattern specified relative to the object to contain the texture pattern. This is accomplished with commands in a graphics package to set the pattern reference point and two vectors defining the orientation of the array plane, such as the x-axis and y-axis directions of the array. The difference between two-dimensional and three-dimensional objects is that now the pattern must be wrapped around the three-dimensional object. Pattern planes can be stored and linked to the appropriate object surfaces so that they can be processed by a hidden-surface method along with the object surfaces. Intensity values stored in pattern arrays are used to modify or replace intensity values calculated in the shading model.

Shadows

Hidden-surface methods can be used to locate areas where light sources produce shadows. By applying a hidden-surface method with a light source at the view position, we can determine which surface sections cannot be "seen" from the light source. These are the shadow areas. Once we have determined the shadow areas for all light sources, the shadows could be treated as surface patterns and stored in pattern arrays. Figure 14–16 illustrates the generation of shading patterns for two objects on a table and a distant light source. All shadow areas in this figure are surfaces that are not visible from the position of the light source. An example of a scene modeled with shadows (and surface patterns) is given in Fig. 14–17.

The shadow patterns generated by a hidden-surface method are valid for any selected viewing position, so long as the light source positions are not changed. Surfaces that are visible from the view position are shaded according to the intensity model, with surface patterns and shadow patterns added. Visible surfaces that are not illuminated by a point source have only ambient light intensity applied. Visible surfaces that are illuminated by a point source are shaded by combining the intensity model and the pattern arrays. Projected shadow areas are shaded with the ambient light intensity only.

14–2 Displaying Light Intensities

Values of intensity calculated by a shading model for each surface in a scene must be converted to one of the allowable intensity levels for the particular graphics system in use. Some systems are capable of displaying several intensity levels di-

FIGURE 14–17
A scene modeled with various intensities, surface patterns, and shadows. Courtesy Evans & Sutherland and Rediffusion Simulation.

rectly, while others are capable of only two levels for each pixel (on or off). In the first case, we convert intensities from the shading model into one of the available levels for storage in the refresh buffer. For bilevel systems, we can represent shading intensities with a **halftoning** method that converts the intensity of each point on a surface into a rectangular pixel grid that can display a number of intensity levels. The number of intensity levels that can be displayed with this method depends on how many pixels we include in the grid.

Assigning Intensity Levels

Allowable pixel intensity levels in a system are usually distributed over the range from 0 to 1. The value 0 indicates that a pixel position is off, and the value 1 signifies maximum intensity. All remaining values are spaced over the range 0 to 1 so that the ratio of successive intensities is constant. This method of assigning intensity levels is chosen so that intensity changes on a video screen correspond to the way our eyes perceive changes in intensity. If n intensity levels are to be assigned in this way, with the lowest level (above 0) called I_1 and the highest I_n, we require that

$$\frac{I_2}{I_1} = \frac{I_3}{I_2} = \cdots = \frac{I_n}{I_{n-1}} = r \tag{14–12}$$

The constant ratio r is determined from the values for I_1 and n by expressing I_n in terms of I_1:

$$I_n = r^{n-1} I_1 \tag{14–13}$$

But $I_n = 1$, so

$$r = \left(\frac{1}{I_1}\right)^{1/(n-1)} \tag{14–14}$$

with I_1 chosen as any convenient value near 0. For example, we could set $I_1 = 1/8$ for a system with four intensity levels (above 0). Since $n - 1 = 3$, we have $r = 2$, and the five possible intensity levels for a pixel are assigned the values 0, 1/8, 1/4, 1/2, and 1.

(a)

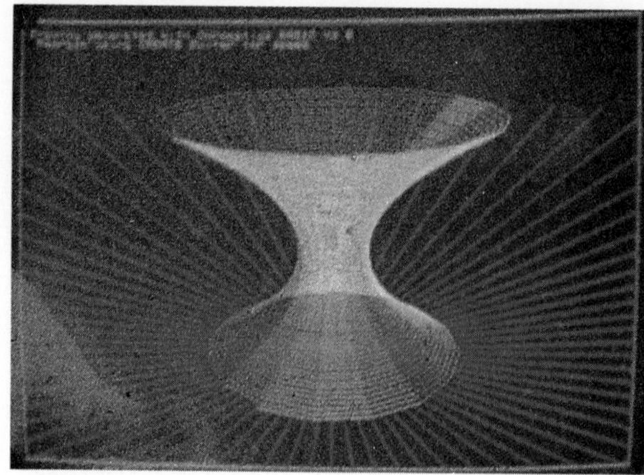

(b)

FIGURE 14–18
A photograph reproduced with a halftoning method. The marked area in the picture at left is magnified on the right, showing how the original tones in the picture are represented with varying size dots. Courtesy Chromatics, Inc.

The number of intensity levels above zero is usually chosen to be a multiple of 2. A four-level system provides minimum shading capability, and high-quality shading patterns are generated on systems that are capable of from 32 to 256 different intensity levels per pixel.

An intensity value calculated from a shading model is converted to the nearest allowable level, I_k, for the system. If there are very many allowable intensity levels, a binary search could be used to locate the nearest level. Since an intensity value must be converted to a control grid voltage for the video monitor in use, there are several alternatives for storing intensity levels in the refresh buffer. We could store the intensity value I_k, or we could simply store the level number k. Another possibility is to store a value that is directly proportional to the control grid voltage. If a voltage value is not stored in the raster, a lookup table is used to convert the raster entry into a voltage value.

Halftoning

When a graphics output device is capable of displaying only two intensity levels per pixel (on and off), a halftoning method can be used to furnish the intensity variations in a scene. Each intensity position in the original scene is replaced with a rectangular pixel grid. This method is used in printing to reproduce photographs for publication in magazines, newspapers, or books. Figure 14–18 illustrates a photograph reproduced with a halftone technique.

A graphics package employing a halftoning technique displays a scene by replacing each position in the original scene with an n by n grid of pixels. The number of pixels that are turned on in each grid is determined by the intensity level of the corresponding position in the scene. Using this technique, we can turn a two-level system into one with the five possible intensitities shown in Fig. 14–19. These

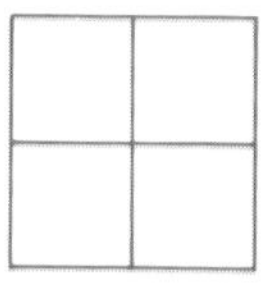

0

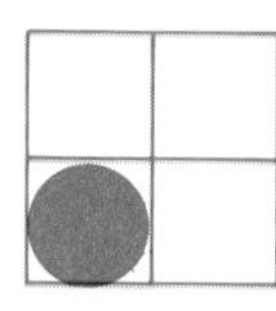

1

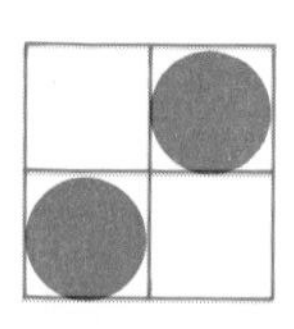

2

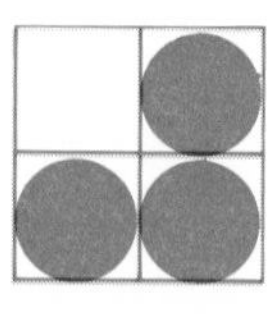

3

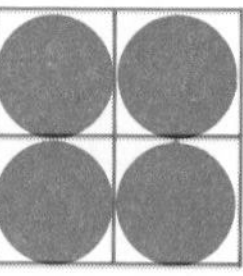

4

FIGURE 14–19
A 2 by 2 pixel grid can be used to produce five intensity levels for a halftone representation of a scene.

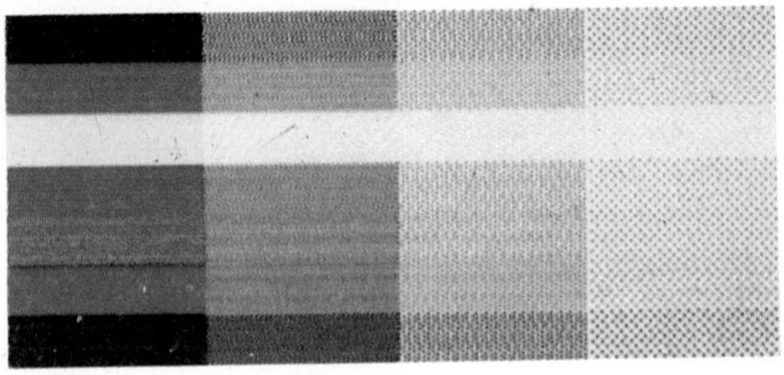

FIGURE 14–20
A gray scale (top) and color scales produced with halftoning. Courtesy Diablo Systems Inc., A Xerox Company.

levels are labeled 0 through 4. For any n by n grid, we obtain n^2 intensity levels above zero with this technique. Figure 14–20 illustrates a range of gray shades generated with a halftoning method. Also shown are some color variations obtained by halftoning, where the number of color pixels turned on in the grid controls the color shade to be displayed.

As a general rule, symmetrical pixel arrangements in a grid are to be avoided whenever possible. Otherwise, patterns can be introduced into the halftone representation that were not in the original picture. This is especially important for larger pixel grids. For a 3 by 3 pixel grid, the third intensity level above zero would be better represented by the pattern in Fig. 14–21 (a) than by any of the symmetrical arrangements in Fig. 14–21 (b). The symmetrical patterns in this figure would produce either vertical, horizontal, or diagonal streaks in a halftone representation. One method for avoiding the introduction of such patterns is randomly to select different pixel arrangements for the representation of an intensity level. The four possibilities in Fig. 14–22 can be alternately selected to represent the second intensity level above zero instead of using only a single arrangement for this level. Another method that has been used to minimize extraneous patterns is to form successive grid patterns with the same pixels turned on. That is, if a pixel is on for one grid level, it is on for all higher levels. This scheme was used to set pixel

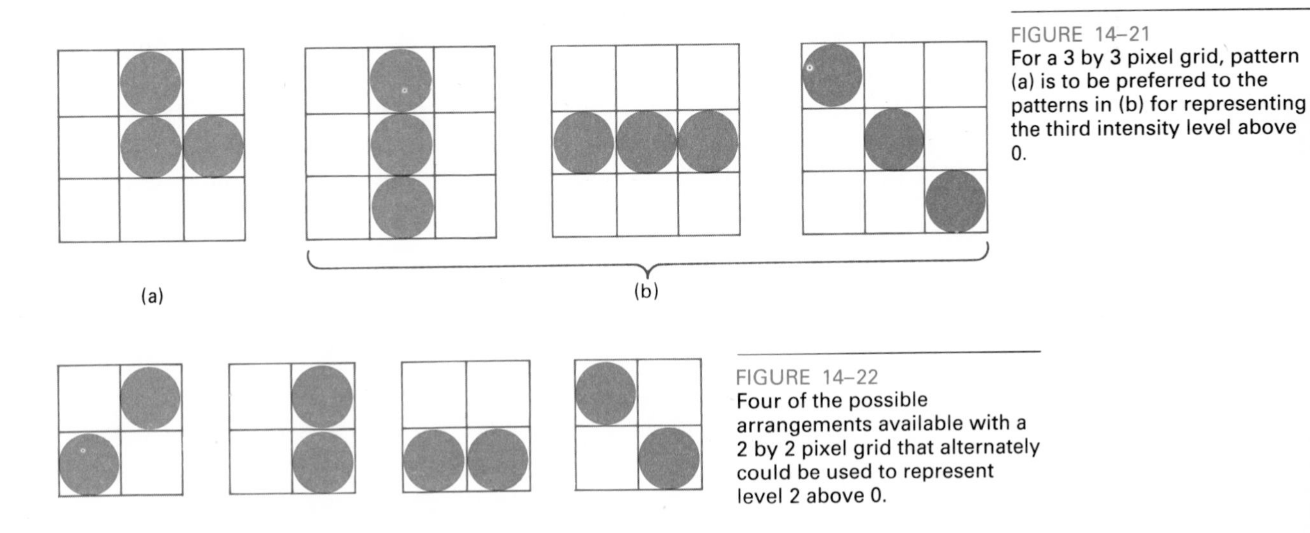

FIGURE 14–21
For a 3 by 3 pixel grid, pattern (a) is to be preferred to the patterns in (b) for representing the third intensity level above 0.

FIGURE 14–22
Four of the possible arrangements available with a 2 by 2 pixel grid that alternately could be used to represent level 2 above 0.

patterns in Fig. 14–19, where the pixel in the lower left corner is on for level 2 and for all subsequent levels.

Halftoning techniques can also be used to increase the number of intensity options on systems that are capable of displaying more than two intensities per pixel. For example, on a system that can display four intensity levels per pixel, we can extend the available intensity levels from four to 13 by employing a halftoning method with 2 by 2 pixel grids. In Fig. 14–19, the four grids above zero now represent several levels each, since each pixel can take three intensity values above zero. Figure 14–23 shows one way to assign the pixel intensities to obtain the 13 distinct levels. Intensity levels for individual pixels are labeled 0 through 3, and the overall levels for the system are labeled 0 through 13.

Normally 2 by 2 or 3 by 3 grids are used in halftoning representations. Although larger grids provide more intensity levels, they tend to introduce new patterns into a scene, and the overall resolution of the output is reduced. A video monitor with a resolution of 1024 by 1024 would be reduced to 512 by 512 when 2 by 2 grids are used, while 4 by 4 grids reduce the resolution to 256 by 256.

Since the use of pixel grids reduces resolution, halftoning methods are more easily applied when the resolution of the original scene is less than that of the output device. When the resolution of the scene is equal to or greater than that of the output device, intensity areas of the scene can be mapped into the pixel grid. A simple method for accomplishing this is to divide the scene into enough rectangular areas to correspond to the pixel grid resolution. The average intensity in each area is then mapped into one of the allowable intensity levels.

Dithering techniques are used with halftoning methods to smooth the edges of displayed objects. We can accomplish this by adding a **dither intensity,** or **dither noise,** to the calculated intensity of points. The amount of dither noise added to the intensity can be determined with a random process or with a calculation based on the coordinate position of the point. Adding a dither value to calculated intensities tends to break up the contours of objects, and the overall appearance of a scene is improved.

Another way to apply a dithering method is to compare the intensity of a point in a scene to a dither value. The corresponding pixel on the output device is turned on if the intensity is greater than the dither value. As with dither noise, these dither values can be generated using a random process, or they can depend on the coordinates of the point. As an example of this approach, suppose 2 by 2 pixel grids are to be used in the halftone process. We can define a dither matrix as

$$D = \begin{bmatrix} 3 & 1 \\ 0 & 2 \end{bmatrix} \tag{14–15}$$

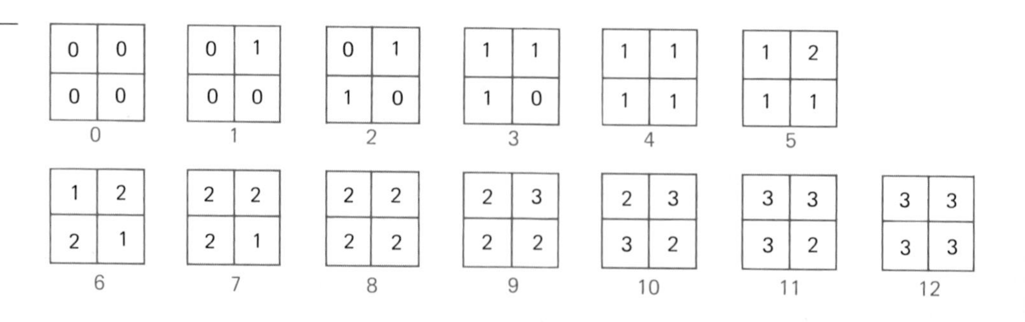

FIGURE 14–23
Intensity levels 0 through 12 obtained with halftoning techniques using 2 by 2 pixel grids on a four-level system.

The sequence of integers 0 through 3 has been assigned to matrix positions in D in the same order that pixels were added to the 2 by 2 grid in Fig. 14–19. Suppose (x, y) is the coordinate position of a point in a scene with intensity I. We calculate a position (i, j) in the dither matrix as

$$i = x \bmod 2, \qquad j = y \bmod 2 \tag{14–16}$$

If $I > D(i, j)$, we turn on the pixel at position (x, y). Otherwise it is not turned on. Higher-order dither matrices can be used to display greater intensity ranges.

14–3 Surface-Shading Methods

An intensity model can be applied to surface shading in various ways, depending on the type of surface and the requirements of a particular application. Objects with plane surfaces can sometimes be shaded realistically with a single intensity for each surface. For a curved surface, we could apply the intensity model to each point on the surface. This can produce a highly realistic display, but it is a time-consuming process. If we wish to speed up the intensity calculations, we could represent a curved surface as a set of polygon planes and apply the shading model to each plane. Each of these polygon surfaces could then be shaded with constant intensity, or we could vary the shading with interpolated intensity values.

Constant Intensity

Under certain conditions, an object with plane surfaces can be shaded realistically using constant surface intensities. In the case where a surface is exposed only to ambient light and no surface designs, textures, or shadows are to be applied, constant shading generates an accurate representation of the surface. For a surface illuminated by a point source, the shading model produces a constant surface intensity, provided that the point source and the view reference point are sufficiently far from the surface. When the point source is far from the surface, there is no change in the direction to the source ($\mathbf{N} \cdot \mathbf{L}$ is constant). Similarly, the direction to a distant viewing point will not change over a surface, so $\mathbf{V} \cdot \mathbf{R}$ is constant.

A curved surface that is represented as a set of plane surfaces can be shaded with constant surface intensities if the planes subdividing the surface are made small enough. This can generate a reasonably good display in many cases, especially when surface curvature changes gradually and light sources and view position are far from the surface. Figure 14–24 shows an object modeled with constant shading. With this method, the intensity is calculated at an interior point of each plane, and the entire surface is shaded with the calculated intensity. When the orientation between adjacent planes changes abruptly, the difference in surface intensities can produce a harsh and unrealistic effect. We can smooth the intensity discontinuities in these cases by varying the intensity over each surface according to some interpolation scheme.

Gouraud Shading

This **intensity interpolation** scheme, developed by Gouraud, removes intensity discontinuities between adjacent planes of a surface representation by linearly varying the intensity over each plane so that intensity values match at the plane boundaries. In this method, intensity values along each scan line passing through a

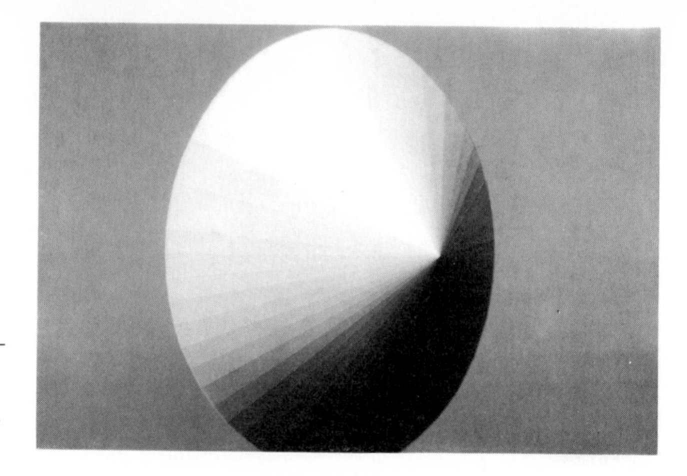

FIGURE 14–24
Solid modeling using constant shading over polygon surfaces. Courtesy Megatek Corp.

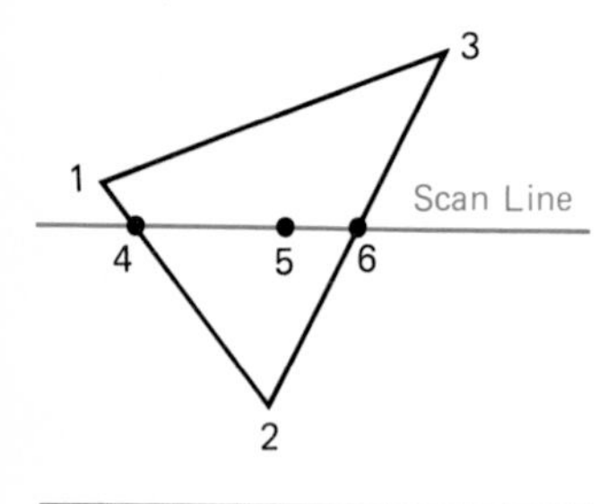

FIGURE 14–25
For interpolated shading, the intensity value at point 4 is determined from intensity values at points 1 and 2, intensity at point 6 is determined from values at points 2 and 3, and intensities at other points (such as 5) along the scan line are interpolated between values at points 4 and 6.

surface are interpolated from the intensities at the intersection points with the surface.

Figure 14–25 demonstrates this interpolation scheme. Intensity values at the vertices of each polygon are determined from a shading model. All other intensities for the surface are then calculated from these values. In Fig. 14–25, the intensity at the left intersection of the scan line with the polygon (point 4) is interpolated from the intensities at vertices 1 and 2 with the calculation

$$I_4 = I_1 \frac{y_4 - y_2}{y_1 - y_2} + I_2 \frac{y_1 - y_4}{y_1 - y_2} \tag{14–17}$$

For this example, the intensity at point 4 has a value that is closer to the intensity of point 1 than that of point 2. Similarly, intensity at the right intersection of the scan line (point 6) is interpolated from intensity values at vertices 2 and 3. Once these bounding intensities are established for a scan line, an interior point (such as point 5) is interpolated from the bounding intensities at points 4 and 6 with the calculation

$$I_5 = I_4 \frac{x_6 - x_5}{x_6 - x_4} + I_6 \frac{x_5 - x_4}{x_6 - x_4} \tag{14–18}$$

This process is repeated for each scan line passing through the polygon.

With this interpolation method, surface normals must first be approximated at each vertex of a polygon. This is accomplished by averaging the surface normals for each polygon containing the vertex point, as shown in Fig. 14–26. These vertex normal vectors are then used in the shading model to generate the vertex intensity values. When color is used to shade a surface, the intensity of each color component is calculated at the vertices. The method can be combined with a hidden-surface algorithm to fill in the visible polygons along each scan line. An example of an object shaded with the Gouraud method appears in Fig. 14–27.

Gouraud shading removes the intensity discontinuities associated with the constant shading model, but it has some other deficiencies. Highlights on the surface are sometimes displayed with anomalous shapes, and the linear intensity interpolation can cause bright or dark intensity streaks, called Mach bands, to appear on the surface. These effects can be reduced by dividing the surface into a greater

FIGURE 14–26
The normal vector at the common vertex point P is calculated as the average of the surface normals for each plane: $\mathbf{N}_p = (\mathbf{N}_1 + \mathbf{N}_2 + \mathbf{N}_3) / 3$.

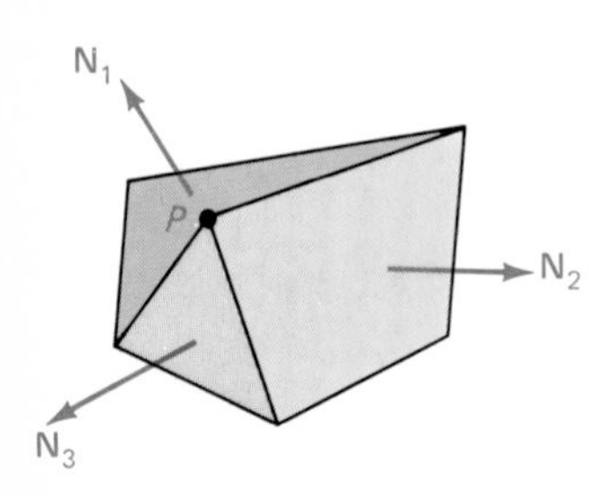

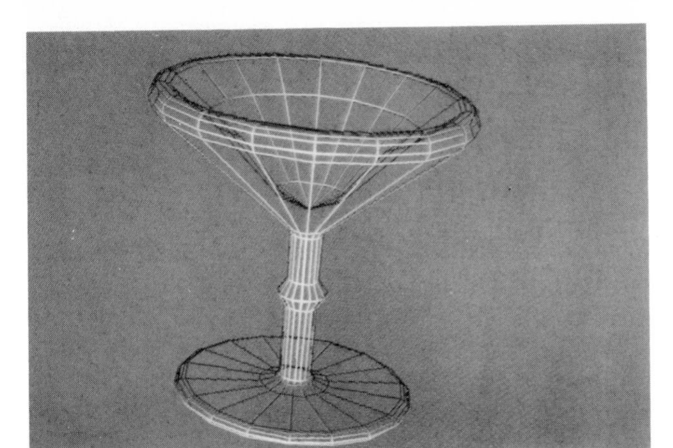

(a)

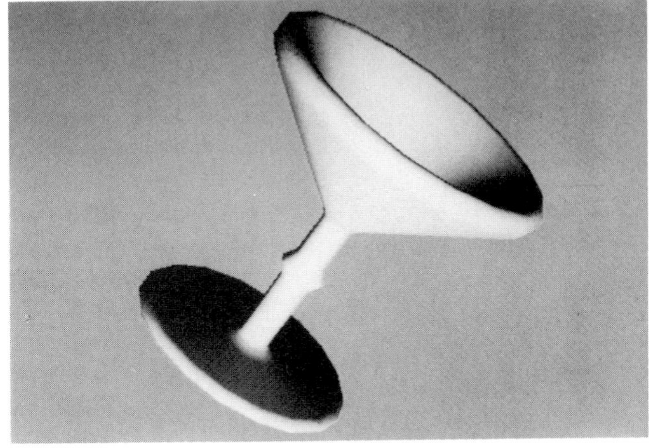

(b)

FIGURE 14–27
An example of Gouraud shading. The object is divided into the planes shown in (a) and shaded with the intensity interpolation method (b). Courtesy Megatek Corp.

number of polygon faces or by using other methods, such as Phong shading, that require more calculations.

Phong Shading

Improvements in the Gouraud shading patterns can be accomplished by approximating the surface normal at each point along a scan line, then calculating the intensity using the approximated normal vector at that point. This method, developed by Phong Bui Tuong, is also referred to as a **normal-vector interpolation** scheme. It displays more realistic highlights on a surface and greatly reduces the Mach-band effect.

Phong shading first interpolates the normal vectors at the bounding points along a scan line. For the polygon shown in Fig. 14–25, the normal vector at point 4 is interpolated from the calculated normals at points 1 and 2. The normal at point 6 is calculated similarly from the normals at points 2 and 3. All interior points along the scan line are then assigned normal vectors that are interpolated from those at points 4 and 6. Intensity calculations using an approximated normal vector at each point along the scan line produce more accurate results than the direct interpolation of intensities from the vertex intensity values. However, the trade-off is that Phong shading requires considerably more calculations. The automobile shown in Fig. 14–14 is an example of the application of Phong shading.

Ray-Tracing Algorithms

Since an infinite number of intensity points could be generated over the various surfaces in a scene, an effective method for determining the specular intensities at visible surface positions is to trace rays backward from the viewing position

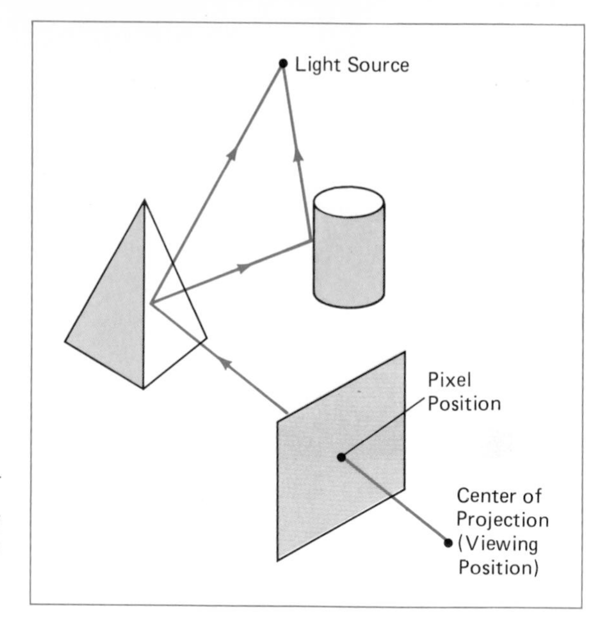

FIGURE 14–28
Tracing a ray from the center of projection to a light source with multiple reflections from the objects in a scene.

to the light source. This technique, referred to as **ray tracing,** is illustrated in Fig. 14–28.

Starting from the viewing position (or center of projection), the ray passing through each pixel in the view plane is traced back to a surface in the three-dimensional scene. The ray is then reflected backward from this surface to determine whether it came from another surface or from a point source. This backward tracing continues until the ray ends at a light source or passes out of the scene. Each pixel is processed in this way, and the number of rays that must be traced is equal to the number of pixels to be displayed.

When transparent objects are encountered in the ray-tracing process, the intensity contributions from both specular reflection and specular refraction are taken into account. At a transparent surface, the ray is divided into the two components shown in Fig. 14–29. Each ray is then traced individually to its source.

After a ray has been processed to determine all specular intensity contributions, the intensity of the corresponding pixel is set. Ray tracing requires

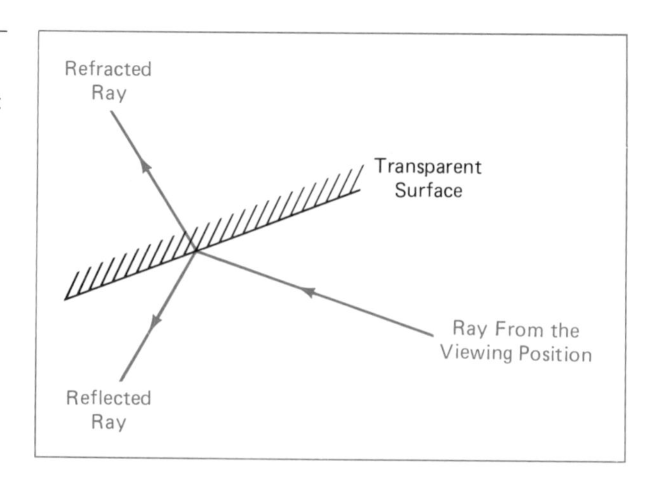

FIGURE 14–29
Ray tracing follows both reflected and refracted paths at the surface of a transparent object.

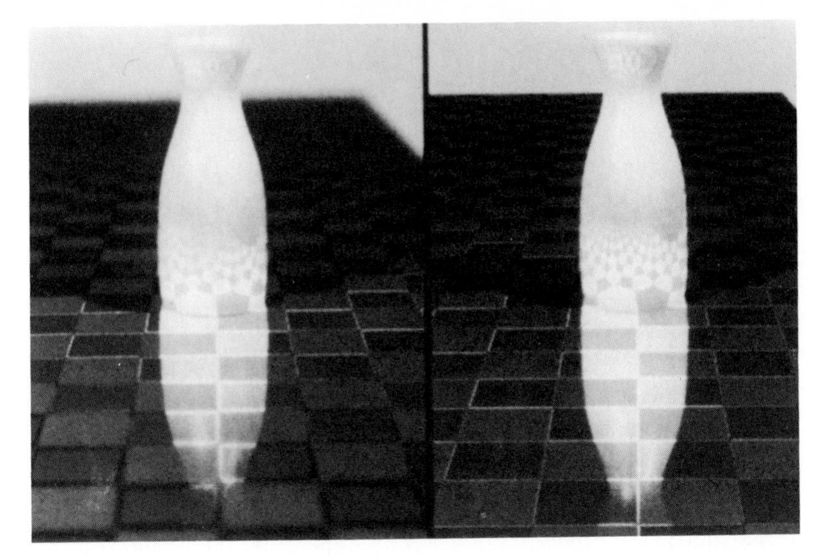

FIGURE 14–30
Scenes generated with ray-tracing methods, using (a) a pinhole camera at the viewing position and (b) a finite-aperture camera. Courtesy Gould Inc., Imaging & Graphics Division and Michael Potmesil, Rensselaer Polytechnic Institute.

considerable computation, but it produces highly realistic results. Figure 14–30 shows two views of a scene generated with ray-tracing techniques. The view in Fig. 14–30 (a) was generated with rays emanating from the center of projection (a single point), while Fig. 14–30 (b) simulates a camera with a lens of finite area. That is, instead of projecting rays back from a single viewing point (a pinhole camera), we suppose that a camera with a finite lens aperture is at the viewing position. By including lens parameters in the viewing model, the projected view can be focused at any plane in the scene. This provides a more realistic projection, as in Fig. 14–30 (b), where the background and near foreground are not focused as sharply as the surface of the vase.

Octree Methods

The application of an intensity model to scenes represented in octrees requires some special considerations. Each volume element, or voxel, in a scene contains information about the type of material at that position, but no information about surface orientations is explicitly stored in the octree. However, we have seen that intensity calculations depend on the surface normals. To include surface shading, we therefore need to modify the procedures that map an octree onto a quadtree for display.

Surface orientation information can be extracted from the octree by examining the regions around each voxel as it is processed for display. If some of the neighboring regions are void, the voxel is part of a surface. Also, the surface normal must have some component in the direction of the void region. An algorithm for testing regions around a voxel can start with the front of the octree (as discussed previously) and examine the four neighboring regions (left, right, top, and bottom) shown in Fig. 14–31. The data stored in the octree for the voxel and its four neighbors are passed to an intensity illumination procedure that first checks to see whether the voxel is part of a surface. If it is, surface normals are determined, and the voxel illumination is calculated from an intensity model.

Assuming that all octree data elements are homogeneous, the intensity for a visible, nonvoid voxel is stored in the quadtree for display. This intensity is calculated by the intensity illumination function using information about the regions

FIGURE 14–31
A voxel and four neighboring regions in an octree.

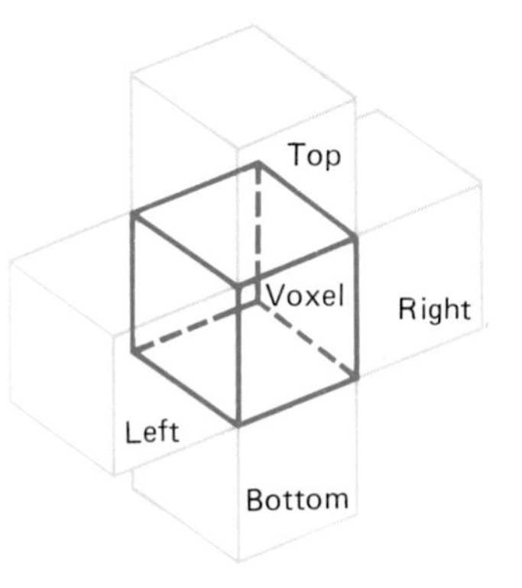

around the voxel. Transparent materials are handled by superimposing intensities for transparent voxels and background voxels in the same quadtree position. A refraction coefficient, r, with a value between 0 and 1, is used in Eq. 14–11 to combine the intensity values of the transparent and background voxels.

Fractal Surfaces

As we have seen, fractal surfaces are generated by nonrandom or random iteration processes that are used to model objects with irregular features, such as clouds and terrain. To determine intensity settings for the various points over the surface of a fractal object, we need some method for determining surface normals. One method for accomplishing this is to represent a fractal surface as a number of small planes, with a surface normal set for each plane.

In general, however, fractal representations contain infinite detail at each surface point of the object. This means that we cannot precisely define a surface normal at any point. For these representations, approximate surface normals can be calculated using the coordinate differences between neighboring points on the surface. Figure 14–32 shows the relationship between a point P at position (x, y, z) and a neighboring point P' at (x', y', z') for a cross-sectional view in the yz plane. We assume that the component $\mathbf{N}_{yz}$ of a surface normal at P makes an angle α with the surface at that point. The tangent of this angle can be calculated as the ratio of the z-coordinate and y-coordinate differences of the two points:

$$\tan\alpha = \frac{\Delta z}{\Delta y} \tag{14–19}$$

where $\Delta z = z - z'$ and $\Delta y = y - y'$. A similar calculation determines the other component $\mathbf{N}_{xz}$ of the normal vector. This determines a normal vector at P relative to point P'. We can then repeat this process for the other points around P and average the various normal-vector directions to obtain the final approximation for the normal at P. Once the approximate surface normal has been established for the points over the fractal surface, the intensity model is applied.

Antialiasing Surface Boundaries

We have seen that lines and polygon edges can be smoothed with antialiasing techniques that either adjust pixel positions (pixel phasing) or set the pixel intensities according to the percent of pixel-area coverage at each point. Similar antialiasing methods can be applied to smooth out the bondaries in a scene containing a collection of surfaces.

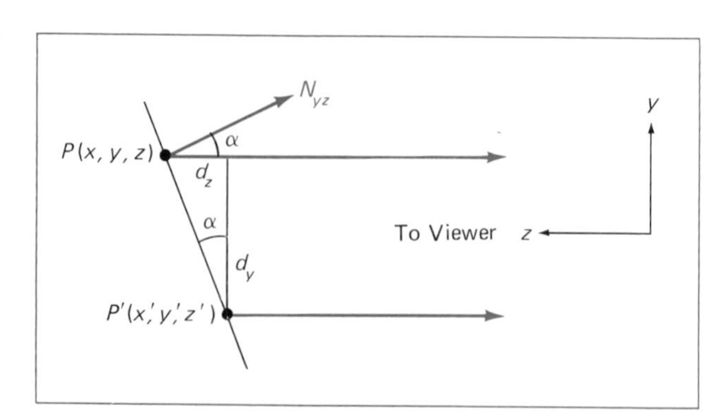

FIGURE 14–32
Approximating the fractal surface normal at position P by examining the position of neighboring surface points.

When multiple objects are to be displayed in a scene, we can expect that some pixel areas will be overlapped by two or more surface boundaries, as in Fig. 14–33. The intensity to be assigned to such pixels is determined by sampling the intensities of all such overlapping surfaces. Contributions from the individual overlapping surfaces can be combined according to the percent of the pixel area that each surface covers. Any of the methods discussed in Chapter 4 for estimating pixel coverage could be used to determine the percent of intensity contribution from each overlapping surface.

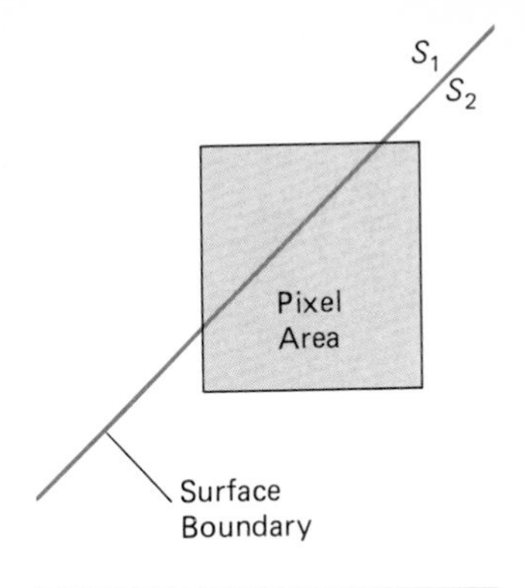

FIGURE 14–33
A boundary line between two surfaces passing through a pixel area. Surface S_1 covers about one-third of the pixel area, and surface S_2 covers the other two-thirds.

A method that provides better estimates of interior pixel areas is to treat the rectangular area of a pixel as a small window. Polygons are then clipped against each of these windows, and the area of a polygon inside the pixel window is calculated. This antialiasing technique could also be applied to surfaces with curved boundaries.

Some other approaches to antialiasing have been used to take advantage of the capabilities of a particular intensity method, such as ray tracing or the shading of fractal surfaces. In ray tracing, instead of projecting each ray through the center of the corresponding pixel to be displayed, rays can be projected through the corners of the pixel (Fig. 14–34). The average of the four intensities obtained in this way is assigned to the pixel. This averaging method antialiases the pixels along surface boundaries. When a pixel area projects onto a part of the scene containing a great deal of detail, the area can be further subdivided so that more projection rays are used in the averaging process.

With fractal objects, each cube used to represent a three-dimensional surface can be treated as slightly transparent. This allows cubes behind illuminated cubes to receive some incident light. Intensities over the fractal surface can then have contributions from more than one cube, and the stairstep appearance of boundaries can be reduced, especially when the incident direction of light is parallel to one surface of the cubes.

14–4 Color Models

Our discussions of color up to this point have concentrated on the mechanisms for generating color displays with combinations of red, green, and blue light. This model is helpful in understanding how color is represented on a video monitor, but several other color models are useful as well in graphics applications. Some models are used to describe color output on printers and plotters, and other models provide a more intuitive color-parameter interface for the user.

A **color model** is a method for explaining the properties or behavior of color within some particular context. No single color model has yet been devised that explains all aspects of color, so we make use of different models to help describe the different perceived characteristics of color.

FIGURE 14–34
Rays projected through the corners of a pixel area provide four intensities that are averaged to a value for that pixel.

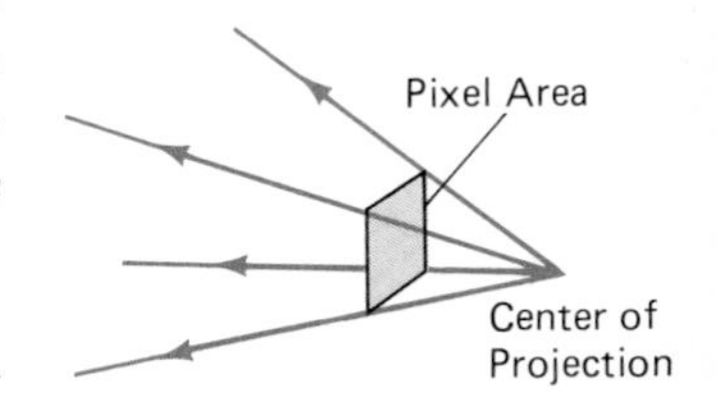

Properties of Light

What we perceive as "light," or different colors, is a narrow frequency band within the electromagnetic spectrum. A few of the other frequency bands within this spectrum are called radio waves, microwaves, infrared waves, and X-rays. Figure 14–35 shows the approximate frequency ranges for some of the electromagnetic bands.

Each frequency value within the visible band corresponds to a distinct color. At the low-frequency end is a red color (4.3×10^{14} hertz), and the highest fre-

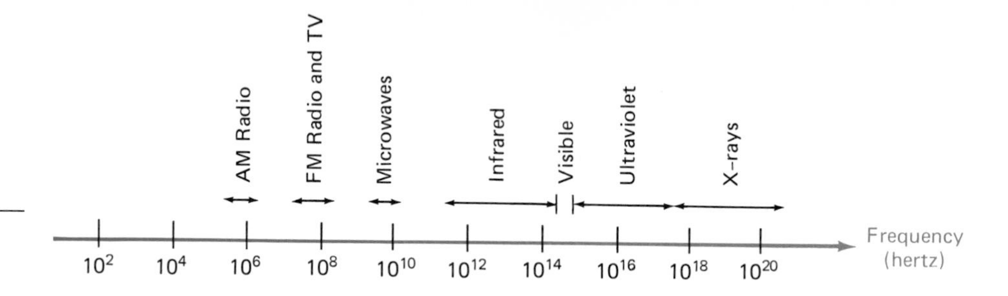

FIGURE 14–35
Electromagnetic spectrum.

quency we can see is a violet color (7.5×10^{14} hertz). Between these frequency limits, the human eye can distinguish nearly 400,000 different colors. These colors range from the reds through orange and yellow at the low frequency end to greens, blues, and violet at the high end.

Since light is an electromagnetic wave, we can describe the various colors in terms of either the frequency or the wavelength of the wave. The two quantities are inversely proportional to each other, with the proportionality constant as the speed of light in the material in which the light is traveling. Light at the red end of the spectrum has a wavelength of approximately 700 nanometers (nm), and the wavelength of the violet light at the other end of the spectrum is about 400 nm.

A light source such as the sun or a light bulb emits all frequencies within the visible range to produce white light. When white light is incident upon an object, some frequencies are reflected and some are absorbed by the object. The combination of frequencies present in the reflected light determines what we perceive as the color of the object. If low frequencies are predominant in the reflected light, the object is described as red. In this case, we say the perceived light has a **dominant frequency** (or **dominant wavelength**) at the red end of the spectrum. The dominant frequency is also called the **color,** or **hue,** of the light.

Other properties besides frequency are useful for describing the characteristics of light. When we view a source of light, our eyes respond to the color (or dominant frequency) and two other basic sensations. One of these we call the **luminance,** or **brightness** of the light. Brightness is related to the intensity of the light: The higher the intensity (or energy), the brighter the source appears. The other characteristic is the **purity,** or **saturation,** of the light. Purity describes how washed out or how "pure" the color of the light appears. Pastels and pale colors are described as less pure. These three characteristics—dominant frequency, brightness, and purity—are commonly used to describe the different properties we perceive in a source of light. The term **chromaticity** is used to refer collectively to the two properties describing the color characteristics, purity and dominant frequency.

The energy emitted by a white-light source has a distribution over the visible frequencies as shown in Fig. 14–36. Each frequency component within the range from red to violet contributes more or less equally to the total energy, and the color of the source is described as white. When a dominant frequency is present, the energy distribution for the source takes a form such as that in Fig. 14–37. We would now describe the light as having the color corresponding to the dominant frequency. The energy of the dominant light component is labeled as E_D in this figure, and the contributions from the other frequencies produce white light of intensity E_W. The brightness of the source can be calculated as the area under the curve, giving the total energy emitted. Purity depends on the difference between

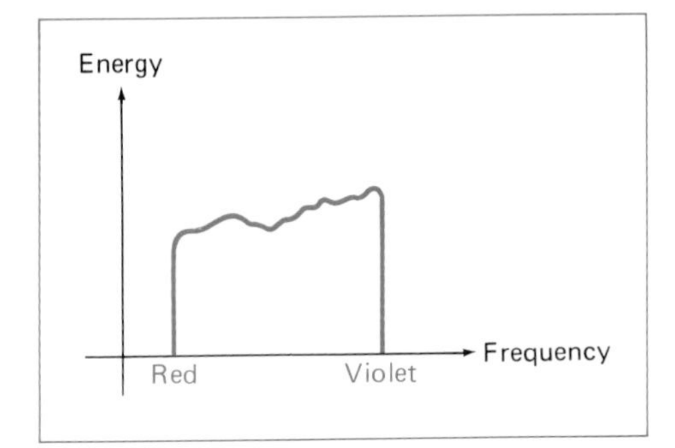

FIGURE 14–36
Energy distribution of a white-light source.

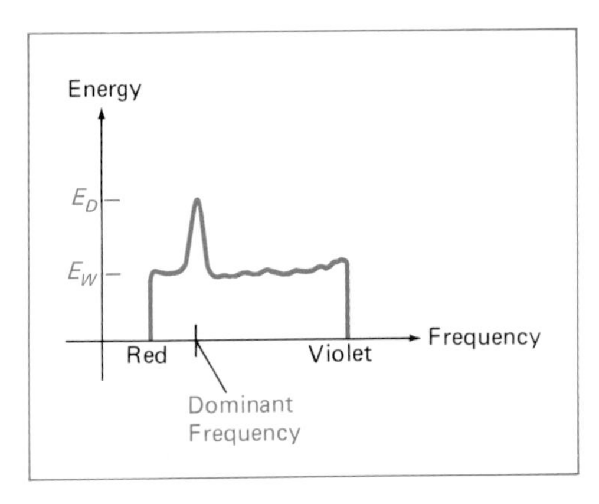

FIGURE 14–37
Energy distribution of a light source with a dominant frequency.

E_D and E_W. The larger the energy E_D of the dominant frequency compared to the white light component E_W, the more pure the light. We have a purity of 100 percent when $E_W = 0$ and a purity of 0 percent when $E_W = E_D$.

When we view light that has been formed by a combination of two or more sources, we see a resultant light with characteristics determined by the original sources. Two different-color light sources with suitably chosen intensities can be used to produce a range of other colors. If the two color sources combine to produce white light, they are referred to as **complementary colors.** Examples of complementary color pairs are red and cyan, green and magenta, and blue and yellow. With a judicious choice of two or more starting colors, we can form a wide range of other colors. Typically, color models that are used to describe combinations of light in terms of dominant frequency (the hue or color) use three colors to obtain a reasonably wide range of colors, called the **color gamut** for that model. The two or three colors used to describe other colors in such a color model are referred to as **primary colors.**

No unique set of primary colors exists that will describe all possible frequencies within the visible spectrum. More colors could be added to the list of primaries to widen the color gamut, but it would still require an infinite number of primaries to include all possible colors. However, three primaries are sufficient for most purposes, and colors not in the color gamut for a specified set of primaries can still be described by extended methods. For example, if a certain color cannot be obtained from three given primaries, we can mix one or two of the primaries with that color to obtain a match with the remaining primaries. In this extended sense, a set of primary colors can be considered to describe all colors.

Standard Primaries and the Chromaticity Diagram

An international standard for primary colors was established in 1931 by the International Commission on Illumination (Commission Internationale de l'Éclairage, also called the CIE). The purpose of this standard is to allow all other colors to be defined as the weighted sum of the three primaries. Since no three colors in the visible spectrum are able to accomplish this, the standard primary "colors" established by the CIE do not correspond to any real colors. The fact that these primaries represent imaginary colors is not really important, since any real color is only defined by the amounts of each primary needed mathematically to produce that color. Each primary in this international standard is defined by its energy distribution curve.

If we let A, B, and C represent the amounts of the standard primaries needed to match a given color within the visible spectrum, we can express the components of the color with the calculations

$$x = \frac{A}{A + B + C}, \quad y = \frac{B}{A + B + C}, \quad z = \frac{C}{A + B + C} \tag{14-20}$$

Since $x + y + z = 1$, any two of these three quantities are sufficient to define a color. This makes it possible for us to represent all colors on a two-dimensional diagram. When we plot x and y values for colors in the visible spectrum, we obtain the tongue-shaped curve shown in Fig. 14–38. This curve is called the CIE **chromaticity diagram.** Color points are labeled along the curve according to wavelength in nanometers, from the red end to the violet end of the spectrum. Point C in the diagram corresponds to the white-light position. Actually, this point is plotted for a white-light source known as **illuminant C,** which is used as a standard approximation for "average daylight."

The curve in Fig. 14–38 is called the chromaticity diagram because it provides a means for quantitatively defining purity and dominant wavelength. For any color point, such as C_1 in Fig. 14–39, we define the purity of the color as the relative distance of the color point from the white-light point C along the straight line joining C to the curve (representing colors in the visible spectrum). Color C_1 in this figure is about 25 percent pure, since it is situated at about one-fourth the

FIGURE 14–39
Defining purity and dominant wavelength on the chromaticity diagram.

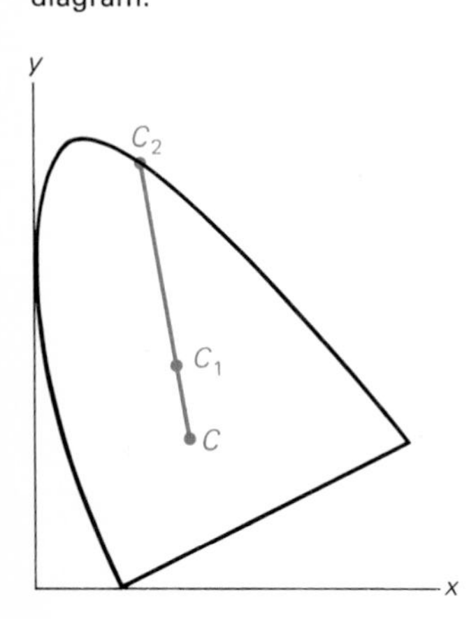

FIGURE 14–38
The CIE chromaticity diagram, with color positions within the visible spectrum labeled in wavelength units (nm).

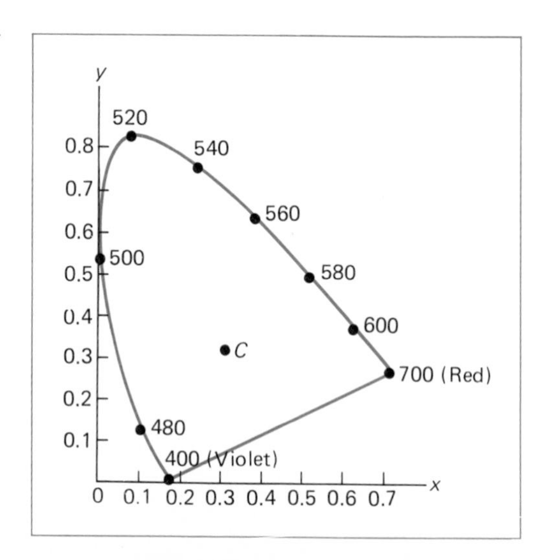

total distance from C to C_2. Dominant wavelength of any color is defined as the wavelength on the spectral curve that intersects the line joining C and that color point. For color point C_1, the dominant wavelength is at C_2, since the line joining C and C_1 intersects the spectral curve at point C_2.

Complementary colors are represented by the two endpoints of a line passing through point C, as in Fig. 14–40. When the two colors C_1 and C_2 are combined in proper proportions, white light results.

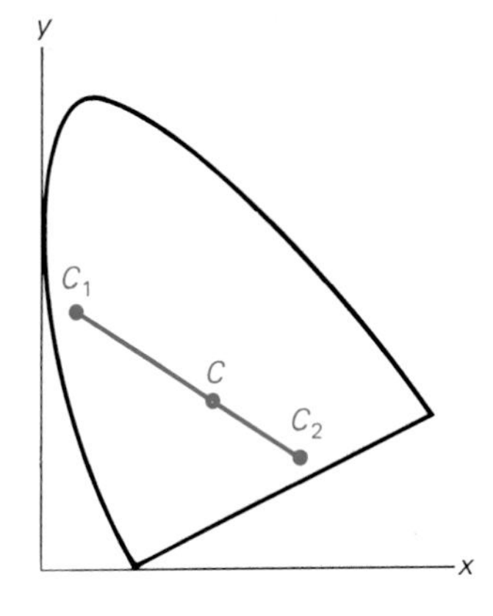

FIGURE 14–40
Representing complementary colors on the chromaticity diagram.

Color gamuts are represented on the chromaticity diagram with lines joining the color points defining the gamut. All colors along the line joining points C_1 and C_2 in Fig. 14–41 can be obtained by mixing amounts of the colors represented by the endpoints. If a greater proportion of C_1 is used, the resultant color is closer to C_1 than to C_2. The color gamut for the three points C_3, C_4, and C_5 is the shaded triangle formed with vertices for the three colors. These three colors can be used to generate any color within the triangle, but no points outside the triangle could be generated with these colors. Thus the chromaticity diagram helps us understand why no set of three primaries can be used to generate all colors, since no triangle within the diagram can encompass all colors. Such triangular color gamuts are typically defined on chromaticity diagrams for video monitors and hard-copy devices. This provides a convenient means for comparing the color gamuts for the various devices.

Intuitive Color Concepts

An artist creates a color painting by mixing color pigments with white and black pigments to form the various shades, tints, and tones in the scene. Starting with the pigment for a "pure color" (or "pure hue"), a black pigment is added to produce different **shades** of that color. The more black pigment the artist adds, the darker the shade produced. Similarly, different **tints** of the color are obtained by adding a white pigment to the original color, making it lighter as more white is added. **Tones** of the color are produced by adding both black and white pigments.

To many, these color concepts are more intuitive than describing a color as a set of three numbers that give the relative proportions of the primary colors. It is generally much easier to think of making a color lighter by adding white and making a darker color by adding black. Therefore, graphics packages providing color palettes to a user often employ two or more color models. One model provides an intuitive color interface for the user, and the others describe the color components for the output devices.

FIGURE 14–41
Color gamuts defined on the chromaticity diagram for a two-color and a three-color system of primaries.

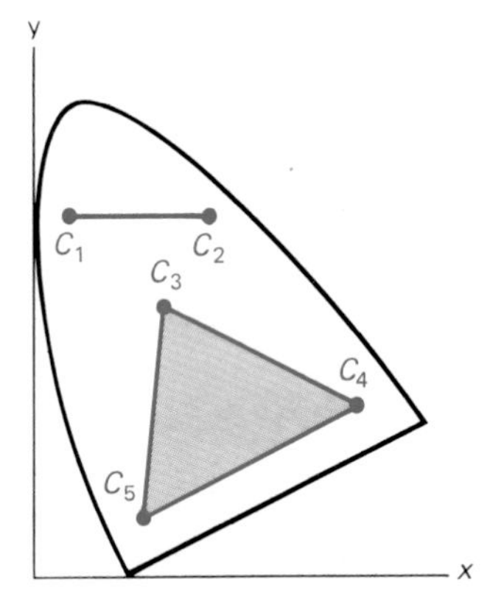

RGB Color Model

Our eyes perceive color through the stimulation of three visual pigments in the cones of the retina. These visual pigments have a peak sensitivity at wavelengths of about 630 nm (red), 530 nm (green), and 450 nm (blue). By comparing intensities in a light source, we perceive the color of the light. This tri-stimulus theory of vision is the basis for displaying color output on a video monitor using the three color primaries red, green, and blue, referred to as the RGB color model.

We can represent this model with the unit cube defined on R, G, and B axes, as shown in Fig. 14–42. The origin represents black, and the vertex with coordinates (1, 1, 1) is white. Vertices of the cube on the axes represent the primary colors, and the remaining vertices represent the complementary color for each of the primary colors.

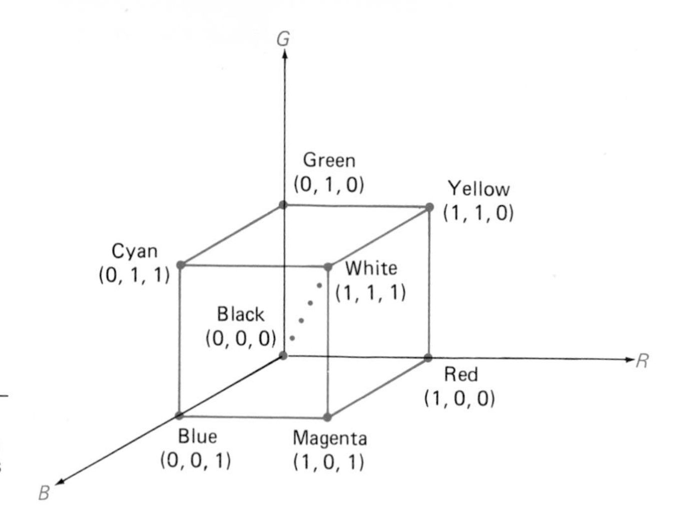

FIGURE 14–42
The RGB color model, defining colors with an additive process within a unit cube.

This color scheme is an additive model: Intensities of the primary colors are added to produce other colors. Each color point within the bounds of the cube can be represented as the triple (R, G, B), where values for R, G, and B are assigned in the range from 0 to 1. The magenta vertex is obtained by adding red and blue to produce the triple (1, 0, 1), and white at (1, 1, 1) is the sum of the red, green, and blue vertices. Shades of gray are represented along the main diagonal of the cube from the origin (black) to the white vertex. Each point along this diagonal has an equal contribution from each primary color, so that a gray shade halfway between black and white is represented as (0.5, 0.5, 0.5). The color graduations along the front and top planes of the RGB cube are illustrated in Fig. 14–43.

When only a few color choices are available on an RGB system, the color options can be expanded by using halftoning methods. This approach with colors is similar to that for developing gray patterns, but now we can set the individual RGB color dots within each pixel as well. For example, suppose we have a bilevel system, so that the intensity of each color dot within a pixel is either on or off. Without halftoning, we have eight color combinations. But if we expand each position in a

FIGURE 14–43
Three planes of the RGB color cube defined in Fig. 14–42. Shown are the front two planes (red-magenta-white-yellow and blue-cyan-white-magenta) and the top plane (green-yellow-white-cyan). Courtesy Diablo Systems Inc., A Xerox Company.

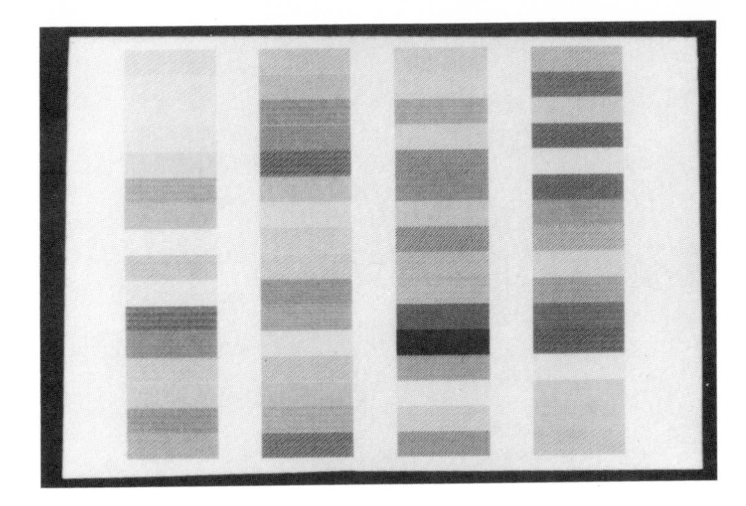

FIGURE 14–44
Color patterns obtained with a halftoning technique on an RGB system with two intensity settings for each RGB color dot. Courtesy Diablo Systems Inc., A Xerox Company.

scene to a 2 by 2 pixel grid, we now have five possible settings for the red color dots within the 2 by 2 grid, five possible settings for the green dots, and five for the blue dots. This gives us a total of 125 different color combinations. An example of the color patterns possible with this method is shown in Fig. 14–44. This color technique is commonly referred to as dithering, although the term is also applied to the halftoning methods previously discussed for extending the intensity levels of a scene with resolution equal to that of the output device.

CMY Color Model

A color model defined with the primary colors cyan, magenta, and yellow (CMY) is useful for describing color output to hard-copy devices. Unlike video monitors, which produce a color pattern by combining light from the screen phosphors, hard-copy devices such as plotters produce a color picture by coating a paper with color pigments. We see the colors by reflected light, a subtractive process.

As we have noted, cyan can be formed by adding green and blue light. Therefore, when white light is reflected from cyan-colored ink, the reflected light must have no red component. That is, red light is absorbed, or subtracted, by the ink. Similarly, magenta ink subtracts the green component from incident light, and yellow subtracts the blue component. A unit cube representation for the CMY model is illustrated in Fig. 14–45.

In the CMY model, the point (1, 1, 1) represents black, because all components of the incident light are subtracted. The origin represents white light. Equal amounts of each of the primary colors produce grays, along the main diagonal of the cube. A combination of cyan and magenta ink produces blue light, because the red and green components of the incident light are absorbed. Other color combinations are obtained by a similar subtractive process.

The printing process often used with the CMY model generates a color point with a collection of four ink dots, somewhat as an RGB monitor uses a collection of three phosphor dots. One dot is used for each of the primary colors (cyan, magenta, and yellow), and one dot is black. A black dot is included because the combination of cyan, magenta, and yellow inks typically produce dark gray instead of black. Some plotters produce different color combinations by spraying the ink for the three primary colors over each other and allowing them to mix before they dry.

FIGURE 14–45
The CMY color model, defining colors with a subtractive process inside a unit cube.

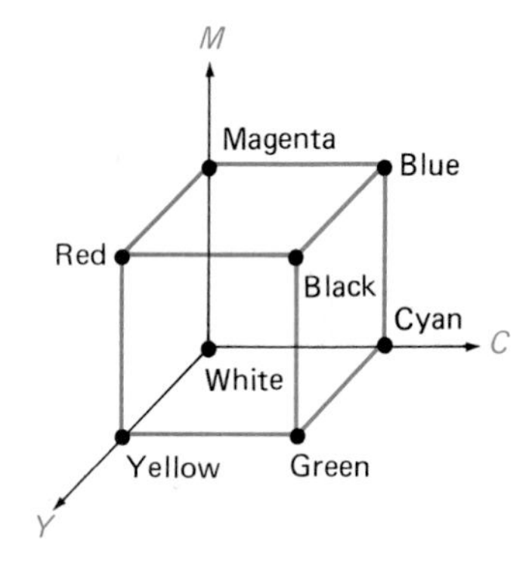

Conversion Between RGB and CMY Models

When an RGB representation is to be converted to the CMY model, each RGB primary value can be subtracted from the RGB representation (1, 1, 1) for white light to give the corresponding CMY components. We write this in a vector form as

$$\begin{bmatrix} C \\ M \\ Y \end{bmatrix} = \begin{bmatrix} W \\ W \\ W \end{bmatrix} - \begin{bmatrix} R \\ G \\ B \end{bmatrix} \tag{14–21}$$

For example, $C = (1, 1, 1) - R$, and the other primaries are converted similarly.

Conversion from CMY to RGB can be performed by subtracting CMY components from the CMY representation (1, 1, 1) for black. In vector form, this is expressed as

$$\begin{bmatrix} R \\ G \\ B \end{bmatrix} = \begin{bmatrix} B \\ B \\ B \end{bmatrix} - \begin{bmatrix} C \\ M \\ Y \end{bmatrix} \tag{14–22}$$

For the red component, we have $R = (1, 1, 1) - C$, with similar calculations for the other primaries.

FIGURE 14–46
When the RGB color cube (a) is viewed along the diagonal from white to black, a hexagon outline (b) is seen.

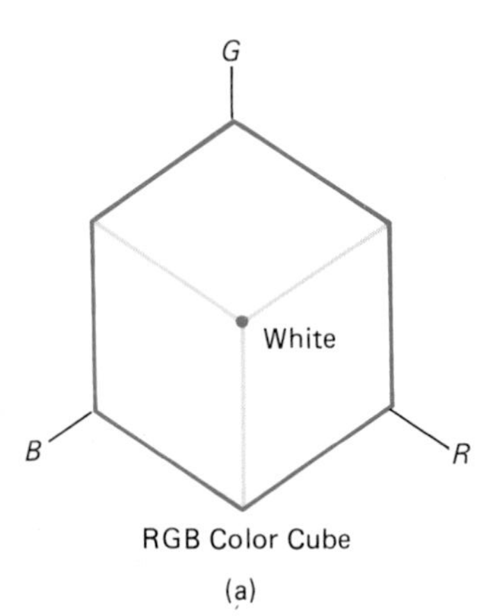

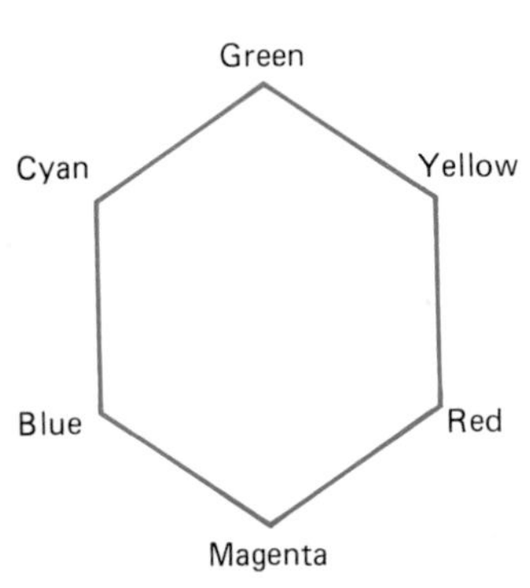

HSV Color Model

This model employs color descriptions that have a more intuitive appeal to a user. Instead of choosing colors according to their RGB components, users can specify a hue (or color) and the amount of white and black to add to the color to obtain different shades, tints, and tones. The three color parameters in this model that are presented to a user are called *hue*, *saturation*, and *value* (HSV).

The three-dimensional representation of the HSV model is derived from the RGB cube. If we imagine viewing the cube along the diagonal from the white vertex to the origin (black), we see an outline of the cube that has the hexagon shape shown in Fig. 14–46. The boundary of the hexagon represents the various hues, and it is used as the top of the HSV hexcone (Fig. 14–47). In the hexcone, saturation is measured along a horizontal axis, and value is along a vertical axis through the center of the hexcone.

Hue is represented as an angle about the vertical axis, ranging from 0° at red through 360°. Vertices of the hexagon are separated by 60° intervals. Yellow is at 60°, green at 120°, and cyan opposite red at $H = 180°$. Complementary colors are 180° apart.

Saturation S varies from 0 to 1. It is represented in this model as the ratio of the purity of a selected hue to its maximum purity at $S = 1$. A selected hue is said to be one-quarter pure at the value $S = 0.25$. At $S = 0$, we have the gray scale.

Value V varies from 0 at the apex of the hexcone to 1 at the top. The apex represents black. At the top of the hexcone, colors have their maximum intensity. When $V = 1$ and $S = 1$, we have the "pure" hues. White is the point at $V = 1$ and $S = 0$.

This is a more intuitive model for most users. Starting with a selection for a pure hue, which specifies the hue angle H and sets $V = S = 1$, the user describes the color desired in terms of adding either white or black to the pure hue. Adding black decreases the setting for V while S is held constant. To get a dark blue, V could be set to 0.4 with $S = 1$ and $H = 240°$. Similarly, when white is to be added

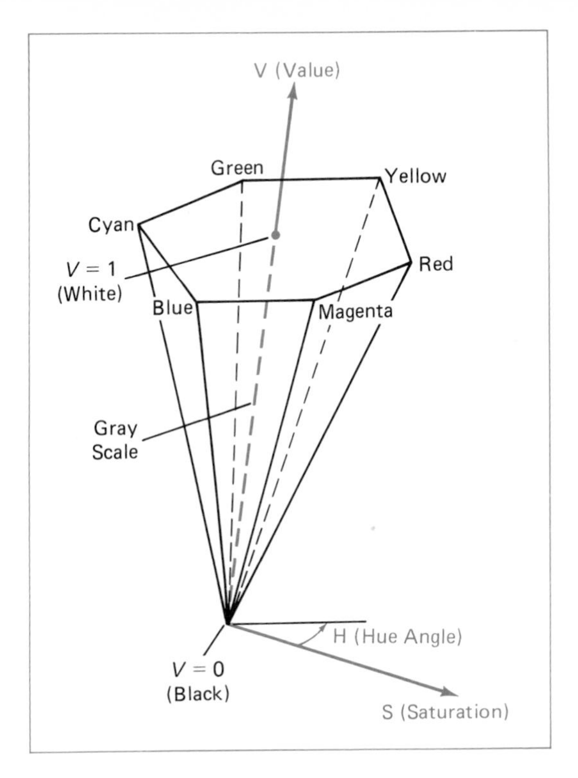

FIGURE 14–47
The HSV hexcone.

to the hue selected, parameter S is decreased while keeping V constant. A light blue could be designated with $S = 0.3$ while $V = 1$ and $H = 240°$. If a user wants some black and some white added, both V and S are decreased. An interface for this model typically presents the HSV parameter choices in a color palette.

Color concepts associated with the terms *shades, tints,* and *tones* are represented in a cross-sectional plane of the HSV hexcone (Fig. 14–48). Adding black to a pure hue decreases V down the side of the hexcone. Thus various shades are represented with values $S = 1$ and $0 \leq V \leq 1$. Adding white to a pure tone produces different tints across the top plane of the hexcone, where parameter values are $V = 1$ and $0 \leq S \leq 1$. Various tones are specified by adding both black and white, producing color points within the triangular cross-sectional area of the hexcone.

The human eye can distinguish about 128 different hues and about 130 different tints (saturation levels). For each of these, a number of shades (value settings) can be detected, depending on the hue selected. About 23 shades are discernible with yellow colors, and about 16 different shades can be seen at the blue end of the spectrum. This means that we can distinguish about $128 \times 130 \times 23 = 382{,}720$ different colors. For most graphics applications, 128 hues, 8 saturation levels, and 15 value settings are sufficient. With this range of parameters in the HSV color model, 16,384 colors would be available to a user, and the system would need 14 bits of color storage per pixel. Color lookup tables could be used to reduce the storage requirements per pixel and to increase the number of available colors.

FIGURE 14–48
Cross section of the HSV hexcone, showing the regions representing shades, tints, and tones.

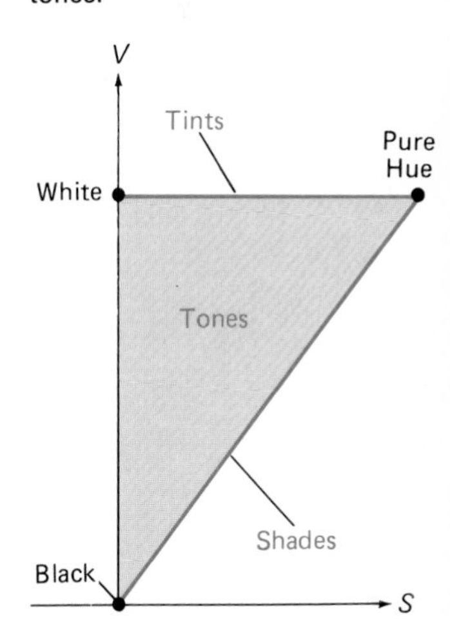

Conversion Between HSV and RGB Models

If HSV color parameters are made available to a user of a graphics package, these parameters are transformed to the RGB settings needed for the color monitor. To determine the operations needed in this transformation, we first consider how the HSV hexcone can be derived from the RGB cube. The diagonal of this cube from black (the origin) to white corresponds to the V axis of the hexcone. Also, each subcube of the RGB cube corresponds to a hexagonal cross-sectional area of the hexcone. At any cross section, all sides of the hexagon and all radial lines from the V axis to any vertex have the value V. For any set of RGB values, V is equal to the maximum value in this set. The HSV point corresponding to the set of RGB values lies on the hexagonal cross section at value V. Parameter S is then determined as the relative distance of this point from the V axis. Parameter H is determined by calculating the relative position of the point within each sextant of the hexagon. An algorithm for mapping any set of RGB values into the corresponding HSV values is given in the following procedure.

```
{Converts a color specification given in rgb parameters to the
 equivalent specification in hsv parameters. As input, r, g, and
 b are in the range from 0 to 1}
procedure convert_rgb_to_hsv (r, g, b : real; var h, s, v : real);
  var  m, rl, gl, bl : real;
  begin
    v := max(r, g, b);                     {set v}
    m := min(r, g, b);
    if v <> 0 then s := (v - m) / v
        else s = 0;                        {set s}
    if s <> 0 then begin
        rl := (v - r) / (v - m);           {distance of color from red}
        gl := (v - g) / (v - m);           {distance of color from green}
        bl := (v - b) / (v - m);           {distance of color from blue}
        if v = r then if m = g then h := 5 + bl
                      else h := 1 - gl;
        if v = g then if m = b then h := 1 + rl
                      else h := 3 - bl
        else if m = r then h := 3 + gl
                 else h := 5 - rl;
        h := h * 60                        {convert to degrees}
      end  {if s <> 0}
    else
      {return h as undefined}
  end; {convert_rgb_to_hsv}
```

We obtain the transformation from HSV parameters to RGB parameters by determining the inverse of the equations in the above procedure. These inverse operations are carried out for each sextant of the hexcone. The resulting transformation equations are summarized in the following algorithm.

```
{Converts a color specification given in hsv parameters to
 the equivalent specification in rgb parameters. As input,
 h is in the range 0 to 360, s and v are in the range 0 to 1.}
procedure convert_hsv_to_rgb (h, s, v : real; var r, g, b : real);
  var  p1, p2, p3 : real;
  begin
    if h = 360 then h := 0;
    h := h / 60;           {convert h to be in [0,6)}
    i := floor(h);         {i = greatest integer <= h}
    f := h - i;            {f = fractional part of h}
    p1 := v * (1 - s);
    p2 := v * (1 - (s * f));
    p3 := v * (1 - (s * (1 - f)));
    case i of
```

```
        0 : begin
               r := v;   g := p3; b := p1 end;
        1 : begin
               r := p2; g := v;   b := p1 end;
        2 : begin
               r := p1; g := v;   b := p3 end;
        3 : begin
               r := p1; g := p2; b := v  end;
        4 : begin
               r := p3; g := p1; b := v  end;
        5 : begin
               r := v;   g := p1; b := p2 end
    end  {case}
end; {convert}
```

HLS Color Model

Another model based on intuitive color parameters is the HLS system used by Tektronix. This model has the double-cone representation shown in Fig. 14–49. The three color parameters in this mode are called *hue*, *lightness*, and *saturation* (HLS).

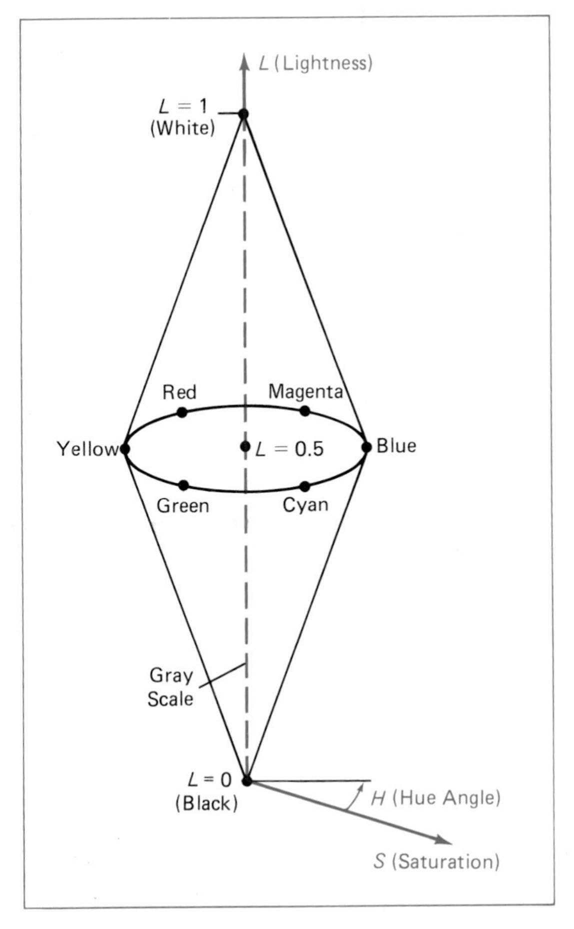

FIGURE 14–49
The HLS double cone.

Hue has the same meaning as in the HSV model. It specifies an angle about the vertical axis that locates a chosen hue. In this model, $H = 0°$ corresponds to blue. The remaining colors are specified around the perimeter of the cone in the same order as in the HSV model. Magenta is at 60°, red is at 120°, and cyan is located at $H = 180°$. Again, complementary colors are 180° apart on the double cone.

The vertical axis in this model is called lightness, L. At $L = 0$, we have black, and white is at $L = 1$. Gray scale is along the L axis, and the "pure hues" lie on the $L = 0.5$ plane.

Saturation parameter S again specifies relative purity of a color. This parameter varies from 0 to 1, and pure hues are those for which $S = 1$ and $L = 0.5$. As S decreases, the hues are said to be less pure. At $S = 0$, we have the gray scale.

This model, like the HSV model, allows a user to think in terms of making a selected hue darker or lighter. A hue is selected with hue angle H, and the desired shade, tint, or tone is obtained by adjusting L and S. Colors are made lighter by increasing L and made darker by decreasing L. When S is decreased, the colors move toward gray.

Color Selection

A graphics package can provide color capabilities to a user in a way that aids the user in making color selections. The color model in the package can allow users freely to select any color combinations, but the system can be designed to help the user to select harmonizing colors. In addition, the designer of a package can follow some basic color rules when designing the color displays that are to be presented to the user.

One method for obtaining a set of coordinating colors is to generate the set from some subspace of a color model. If colors are selected at regular intervals along any straight line within the RGB or CMY cube, for example, we can expect to obtain a set of well-matched colors. Randomly selected hues can be expected to produce harsh and clashing color combinations. Another consideration in color displays is the fact that we perceive colors at different depths. This occurs because our eyes focus on colors according to their frequency. Blues, in particular, tend to recede. Displaying a blue pattern next to a red pattern can cause eye fatigue, since we continually need to refocus when our attention is switched from one area to the other. This problem can be reduced by separating these colors or by using colors from one-half or less of the color hexagon in the HSV model. With this technique, a display contains either blues and greens or reds and yellows.

As a general rule, the use of a smaller number of colors produces a more pleasing display than a large number of colors, and tints and shades blend better than pure hues. For a background, gray or the complement of one of the foreground colors is usually best.

REFERENCES

Additional information on intensity models is to be found in Blinn (1977, 1978, and 1982), Blinn and Newell (1976), Phong (1975), Cook (1984), Cook and Torrance (1982), Potmesil and Chakravarty (1983), Torrance and Sparrow (1967), and Warn (1983). Methods for modeling shadows are discussed in Brotman and Badler (1984) and in Crow (1977). For further information on texturing see Blinn (1976), Crow (1984), Haruyama and Barsky, and Schweitzer (1983).

Ray-tracing techniques are discussed in Cook, Porter, and Carpenter (1984), Glassner (1984), Hanrahan (1983), Kajiya (1983), Potmesil and Chakravarty (1982), and in Weghorst, Hooper, and Greenberg (1984).

Halftoning algorithms are presented in Jarvis and Roberts (1976), Jarvis, Judice, and Ninke (1976), and Judice, Jarvis, and Ninke (1974).

Sources of information on color models include Baldwin (1984), Joblove and Greenberg (1978), Murch (1984), and Smith (1978 and 1979).

EXERCISES

14–1. Write a routine to implement the shading model of Eq. (14–7) assuming that the light source and view position are "infinitely" far from the object to be shaded.

14–2. Develop an algorithm for shading a surface using the calculations in Eq. 14–7 when the light source and view position cannot be assumed to be infinitely far from the surface.

14–3. Devise a scan-line algorithm that takes advantage of coherence to evaluate the dot products in Eq. 14–7 along the scan lines.

14–4. Devise an algorithm for identifying surfaces that are visible through a transparent object.

14–5. Outline a scan-line method for obtaining the intensity values over the surface of a transparent object that is in front of one or more background objects.

14–6. Discuss how the different hidden-surface removal methods can be combined with an intensity model for displaying objects with opaque surfaces.

14–7. Discuss how the various hidden-surface removal methods can be modified to process transparent objects. Are there any hidden-surface methods that cannot handle transparent surfaces?

14–8. Modify the routine of Ex. 14–1 to include a texture function. Apply both a random and a repeating texture function to display two different types of surfaces.

14–9. Develop an algorithm for mapping a specified texture-intensity pattern onto a defined surface.

14–10. Set up an algorithm, based on one of the hidden-surface removal methods, that will display shadow areas around an object illuminated by a distant point source.

14–11. For a system with n intensity levels above zero, show that any calculated intensity value I, between 0 and 1, can be mapped onto one of the n discrete levels by calculating the integer level number:

$$k = \text{round}\,(\log_r (I/I_1)) + 1$$

14–12. How many intensity levels can be displayed with a halftoning method using *n* by *n* pixel grids where each pixel can be displayed with *m* different intensities?

14–13. Given a surface of intensity values, write a routine to display the surface using a 3 by 3 halftoning method with two intensity levels per pixel.

14–14. Develop an algorithm that uses 3 by 3 dither matrices to set intensity values over the surface of an object.

14–15. Write a routine to implement a constant shading method for the surfaces of a defined polyhedron.

14–16. Write a procedure for implementing Gouraud shading over the surfaces of a specified polyhedron.

14–17. Implement the Phong shading method for an object with plane faces.

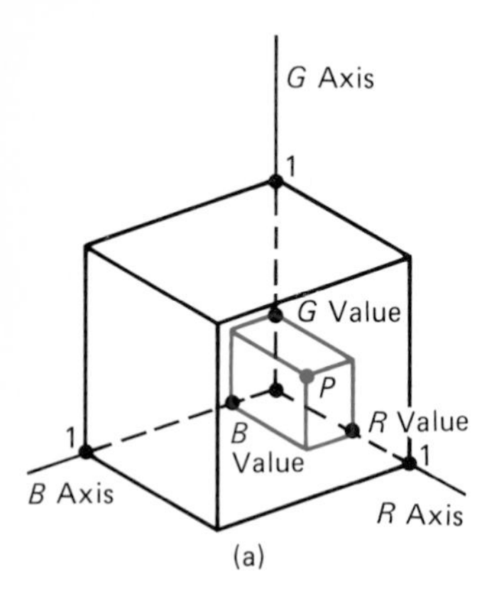

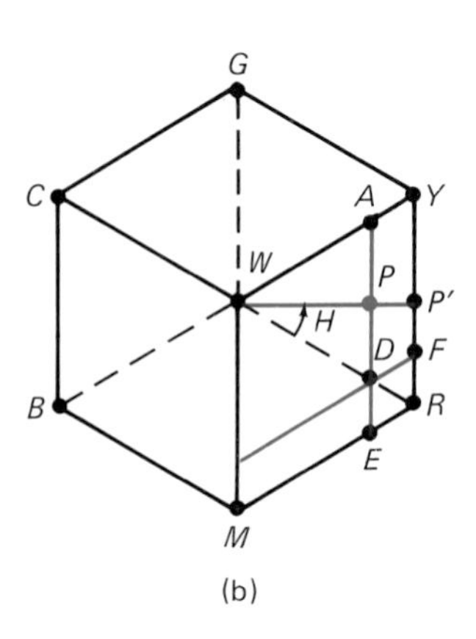

FIGURE 14–50
Mapping point P in the RGB color space (a) into the corresponding position (b) on a hexagon cross section of the HSV hexcone.

14–18. Set up an algorithm that determines surface intensities using a ray-tracing method for a scene containing a single object illuminated by a single point source.

14–19. Devise an algorithm to process an intensity model and an octree representation of an object, storing visible surface intensity values in a quadtree.

14–20. How many different color combinations can be generated using halftoning on a two-level RGB system with a 2 by 2 pixel grid? How many combinations can be obtained with a 3 by 3 pixel grid?

14–21. Derive the expressions for converting RGB color parameters to HSV values using the diagrams in Fig. 14–50. Position P in this diagram marks a color point specified by assigned values for R, G, and B. In this example, R is assumed to have the maximum value and B the minimum value, so that color point P is mapped onto the first sextant of an HSV hexagon cross section as shown in Fig. 14–50 (b). The length of each side of this hexagon is equal to the value of parameter V, which is equal to the maximum of the RGB values. This value is also equal to each of the distances from the hexagon center W (the white point) to the color vertices. Parameter S is calculated as the relative distance of point P from the center point W of the hexagon: $S = |WP|/|WP'| = |WA|/|WY| = (|WY| - |AY|)/|WY| = (R - B)/R = [\max(R,G,B) - \min(R,G,B)]/\max(R,G,B)$. Similarly, parameter H can be calculated as a number between 0 and 1 by determining the relative distance of P from the line WR. That is, $H = |DP|/|DA| = (|EP| - |ED|)/(|EA| - |ED|) = (G - B)/(R - B) = [G - \min(R,G,B)]/[\max(R,G,B) - \min(R,G,B)]$. Perform a similar analysis for each of the other sextants of the hexagon to obtain the complete transformation calculations for any input set of RGB values.

14–22. Derive the inverse operations of the procedures in Ex. 14–21 to obtain the expressions for converting HSV color values to RGB values. As an example of this approach, consider a color point in the first sextant of the HSV hexagon (Fig. 14–50). For this case, parameter S can be written as $(V - B)/V$. The inverse of this expression yields: $B = V(1 - S)$. Using this expression in $H = (G - B)/(V - B)$, we obtain $G = V[1 - S(1 - H)]$. Similar methods can be applied in each of the other sextants.

14–23. Write a procedure that would allow a user to select HSV color parameters from a displayed menu, then convert the color choices to RGB values for storage in a raster.

14–24. Revise the procedures outlined in Ex. 14–21 to convert a set of RGB color values to HLS color parameters.

14–25. Set up an algorithm for converting the HLS color parameters to RGB color parameters. The proper conversion procedures can be determined by obtaining the inverse operations of the procedures in Ex. 14–24.

14–26. Implement the algorithm of Ex. 14–25 in a program that allows a user to select HLS values from a color menu then converts these values to corresponding RGB values.

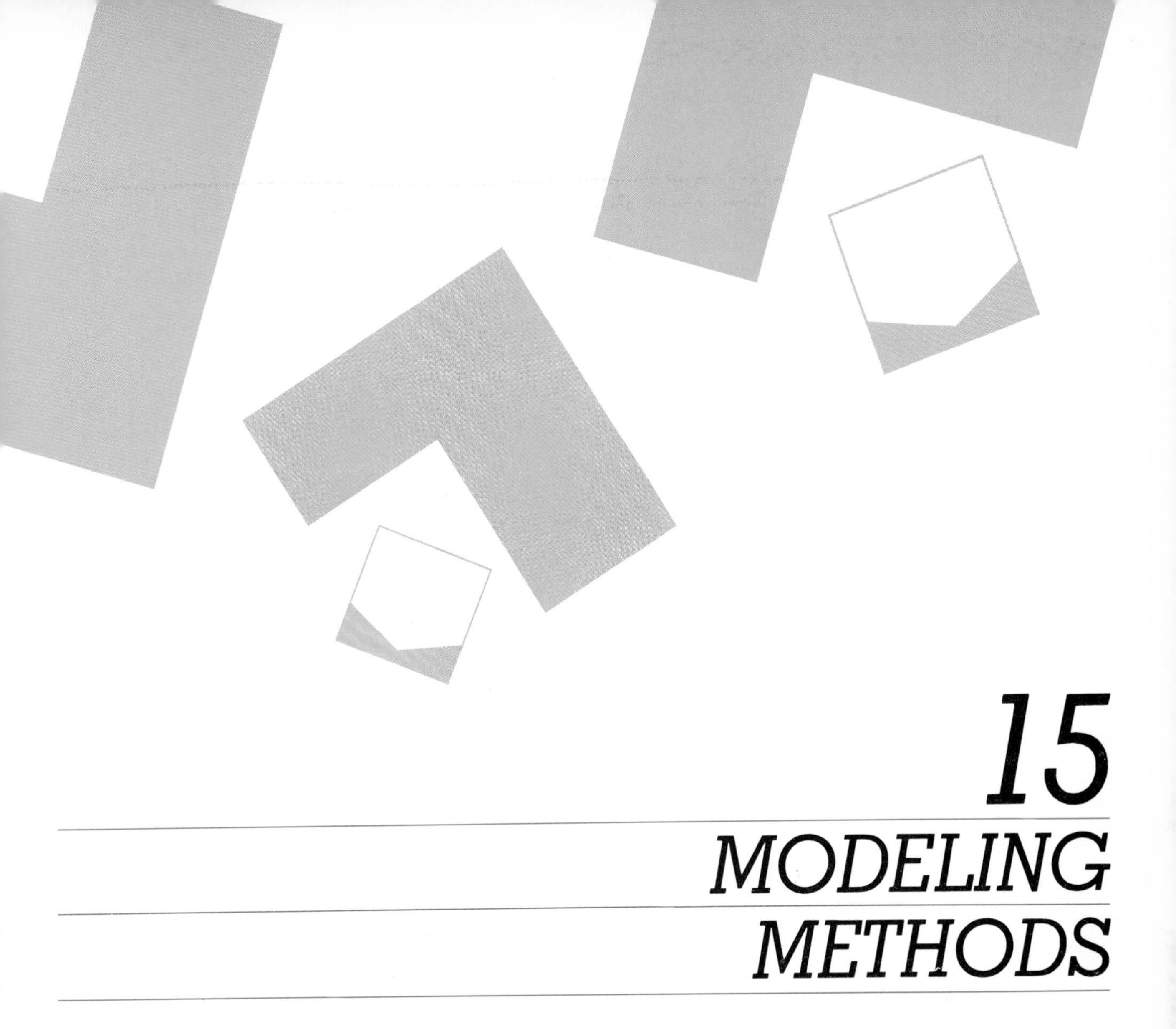

15 MODELING METHODS

An important use of graphics is in the design and representation of different types of systems. Architectural and engineering systems, such as building layouts and electronic circuit schematics, are commonly put together using computer-aided design methods. Graphical methods are used also for representing economic, financial, organizational, scientific, social, and environmental systems. Representations for these systems are often constructed to simulate the behavior of a system under various conditions. The outcome of the simulation can serve as an instructional tool or as a basis for making decisions about the system. To be effective in these various applications, a graphics package must possess efficient methods for constructing and manipulating the graphical system representations.

15–1 Basic Modeling Concepts

The creation and manipulation of a system representation is termed **modeling.** Any single representation is called a **model** of the system. Models for a system can be defined graphically, or they can be purely descriptive, such as a set of equations that define the relationships between system parameters. Graphical models are often referred to as **geometric models,** because the component parts of a system are represented with geometric entities such as lines, polygons, or circles. We are concerned here only with graphics applications, so we will use the term *model* to mean a computer-generated, geometric representation of a system.

Model Representations

Figure 15–1 shows a representation for a logic circuit, illustrating the features common to many system models. Component parts of the system are displayed as geometric structures, called **symbols,** and relationships between the symbols are represented in this example with a network of connecting lines. Three standard symbols are used to represent logic gates for the Boolean operations: AND, OR, and NOT. The connecting lines define relationships in terms of input and output flow (from left to right) through the system parts. One symbol, the AND gate, is displayed at two different positions within the logic circuit. Repeated positioning of a few basic symbols is a common method for building complex models. Each such occurrence of a symbol within a model is called an **instance** of that symbol. We have one instance for the OR and NOT symbols in Fig. 15–1, and two instances of the AND symbol.

In many cases, the particular symbols chosen to represent the parts of a system are dictated by the system description. For circuit models, standard electrical or logic symbols are used. With models representing abstract concepts, such as political, financial, or economic systems, symbols may be any convenient geometric pattern.

Information describing a model is usually provided as a combination of geometric and nongeometric data. Geometric information includes coordinate positions for locating the component parts, output primitives and attribute functions to define the structure of the parts, and data for constructing connections between the parts. Nongeometric information includes text labels, algorithms describing the operating characteristics of the model, and rules for determining the relationships or connections between component parts, if these are not specified as geometric data.

The information needed to construct and manipulate a model is often stored in some type of data structure, such as a table or linked list. This information could also be specified in procedures. In general, a model specification will contain both

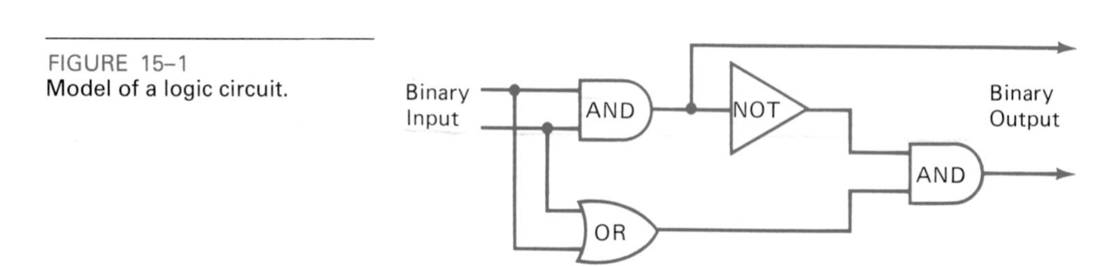

FIGURE 15–1
Model of a logic circuit.

SYMBOL CODE	GEOMETRIC DESCRIPTION	IDENTIFYING LABEL
Gate 1	(Coordinates and other Parameters)	AND
Gate 2	⋮	OR
Gate 3	⋮	NOT
Gate 4	⋮	AND

FIGURE 15–2
A data table defining the structure and position of each gate in the circuit of Fig. 15–1.

data structures and procedures, although some models are defined completely with data structures and others use only procedure specifications. An application to perform solid modeling of objects might use mostly information taken from some data structure to define coordinate positions, with very few procedures. A weather model, on the other hand, may need mostly procedures to calculate plots of temperature and pressure variations.

As an example of how combinations of data structures and procedures can be used, we consider some alternative model specifications for the logic circuit of Fig. 15–1. One method is to define the logic components in a data table (Fig. 15–2), with processing procedures used to specify how the network connections are to be made and how the circuit operates. Geometric data in this table include coordinates and parameters necessary for drawing and positioning the gates. These symbols could all be drawn as polygon shapes, or they could be formed as combinations of straight line segments and elliptical arcs. Labels for each of the component parts also have been included in the table, although the labels could be omitted if the symbols are displayed as commonly recognized shapes. Procedures would then be used to display the gates and construct the connecting lines, based on the coordinate positions of the gates and a specified order for connecting them. An additional procedure is used to produce the circuit output (binary values) for any given input. This procedure could be set up to display only the final output, or it could be designed to display intermediate output values to illustrate the internal functioning of the circuit.

Another method for specifying the circuit model is to define as much of the system as possible in data structures. The connecting lines, as well as the gates, could be defined in a data table, which explicitly lists endpoints for each of the lines in the circuit. A single procedure might then display the circuit and calculate the output. Going to the other extreme, the model could be defined completely in procedures using no external data structures.

Symbol Hierarchies

Many models can be organized as a hierarchy of symbols. The basic "building blocks" for the model are defined as simple geometric shapes appropriate to the type of model under consideration. These basic symbols can be used to form composite objects, called **modules,** which themselves can be grouped to form higher-level modules, and so on, for the various components of the model. In the simplest case, we can describe a model by a one-level hierarchy of component parts, as in Fig. 15–3. For this circuit example, we assume that the gates are positioned and connected to each other with straight lines according to connection rules that are

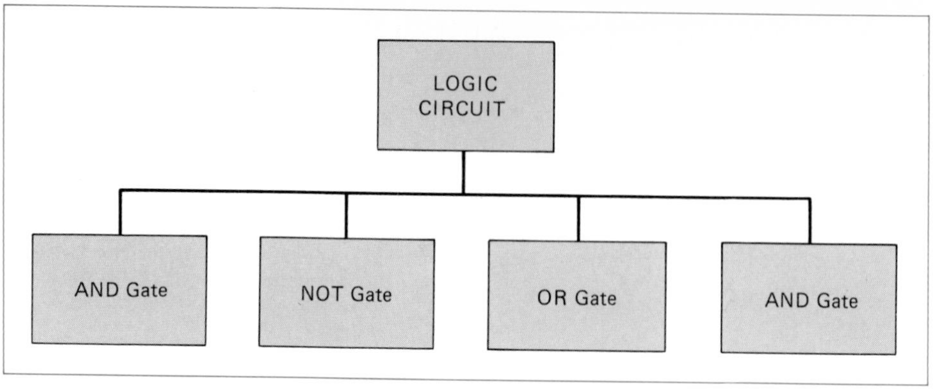

LOGIC CIRCUIT

AND Gate
NOT Gate
OR Gate
AND Gate

FIGURE 15–3
A one-level hierarchical description of a circuit formed with logic gates.

specified with each gate description. The basic symbols in this hierarchical description are the logic gates. Although the gates themselves could be described as hierarchies—formed from straight lines, elliptical arcs, and text—that sort of description would not be a convenient one for constructing logic circuits, in which the simplest building blocks are gates. For an application in which we were interested in designing different geometric shapes, the basic symbols could be defined as straight line segments and arcs.

An example of a two-level symbol hierarchy appears in Fig. 15–4. Here a facility layout is planned as an arrangement of work areas. Each work area is outfitted with a collection of furniture. The basic symbols are the furniture items: worktable, chair, shelves, file cabinet, and so forth. Higher-order objects are the work areas, which are put together with different furniture organizations. An instance of a basic symbol is defined by specifying its size, position, and orientation within each work area. For a facility layout package with fixed sizes for objects, only position and orientation need be specified by a user. Positions are given as coordinate locations in the work areas, and orientations are specified as rotations that determine which way the symbols are facing. At the second level up the hierarchy, each work area is defined by specifying its size, position, and orientation within the facility layout. The boundary for each work area might be fitted with a divider, which encloses the work area and provides aisles within the facility.

More complex symbol hierarchies are formed by repeated grouping of symbol clusters at each higher level. The facility layout of Fig. 15–4 could be extended to

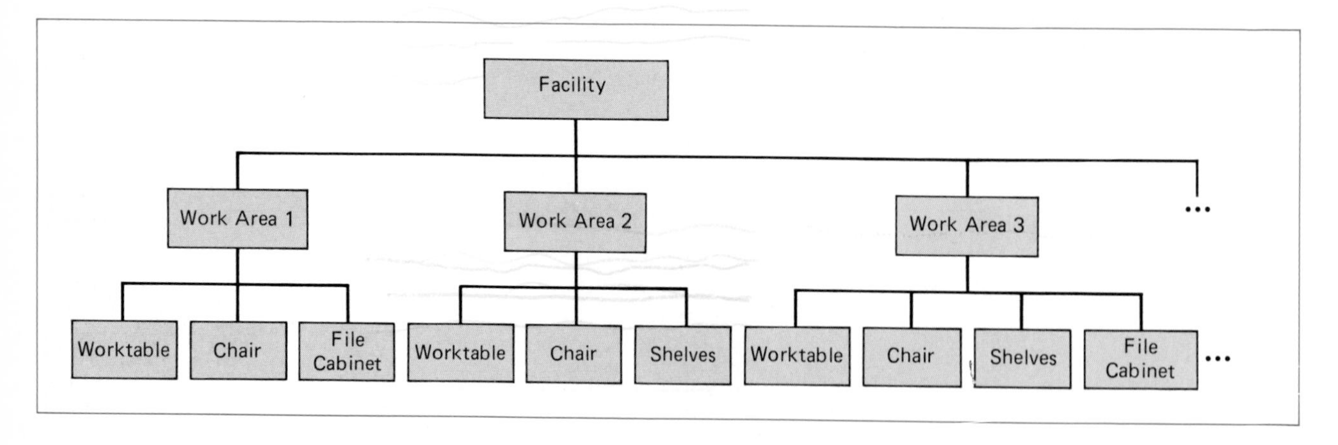

FIGURE 15–4
A two-level hierarchical description of a facility layout.

include symbol clusters that form different rooms, different floors of a building, different buildings within a complex, and different complexes at widely separated physical locations.

Modeling Packages

Standard general-purpose graphics systems are usually not designed to accommodate extensive modeling applications. Routines necessary to handle modeling procedures and data structures are often set up as separate **modeling packages,** and standard graphics packages then can be adapted to interface with the modeling package. The purpose of graphics routines is to provide methods for generating and manipulating final output displays. Modeling routines, by contrast, provide a means for defining and rearranging model representations in terms of symbol hierarchies, which are then processed by the graphics routines for display. Systems designed specifically for modeling may integrate the modeling and graphics functions into one package.

Symbols available in a modeling package are defined and structured according to the type of application the package has been designed to handle. Modeling packages can be designed for either two-dimensional or three-dimensional displays. Figure 15–5 illustrates a two-dimensional layout used in circuit design. A three-dimensional molecular modeling display is shown in Fig. 15–6. Three-dimensional facility

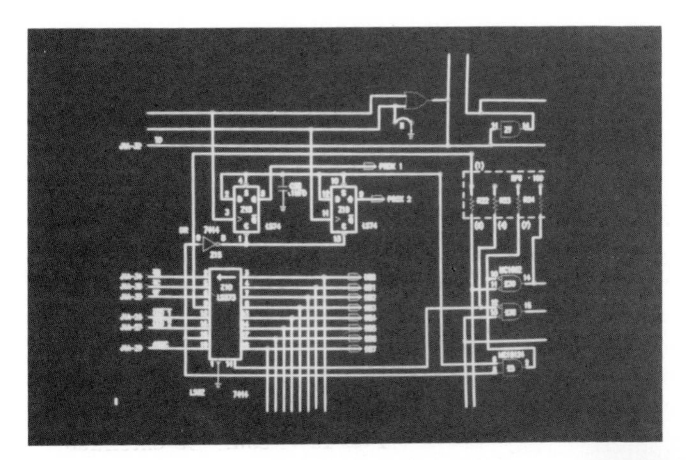

FIGURE 15–5
Two-dimensional modeling layout used in circuit design. Courtesy Summagraphics.

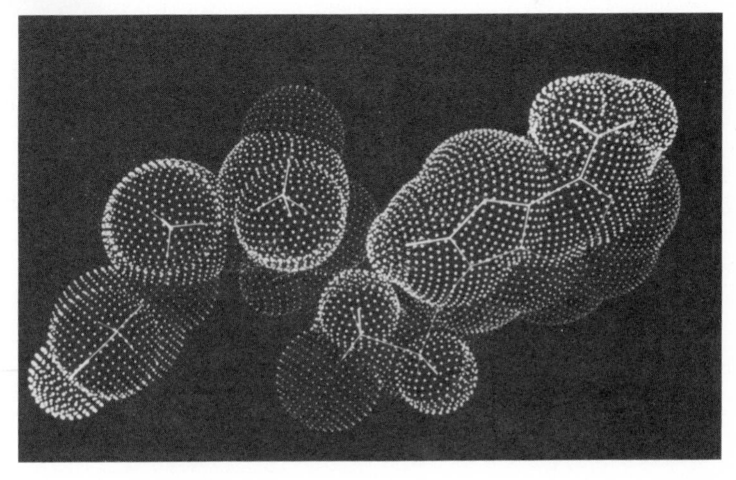

FIGURE 15–6
MIDAS molecular modeling system display of a color-coded, three-dimensional representation of a component of adenosine triphosphate as it is rotated about the glycosyl bond. Courtesy Evans & Sutherland and Computer Graphics Laboratory, University of California at San Francisco.

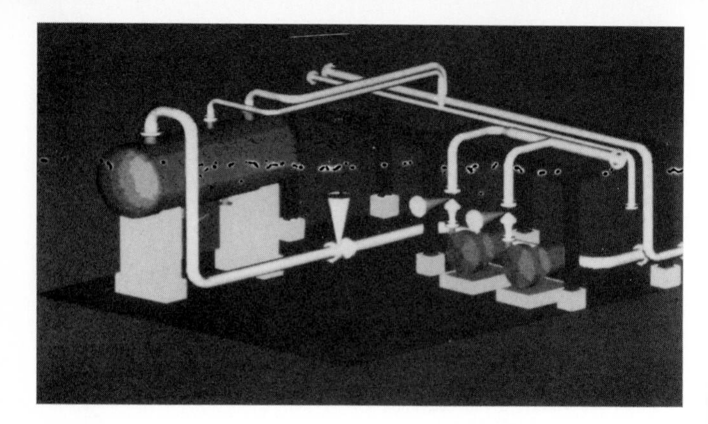

FIGURE 15–7
Three-dimensional, color-coded layout used in plant design. Courtesy Intergraph Corp.

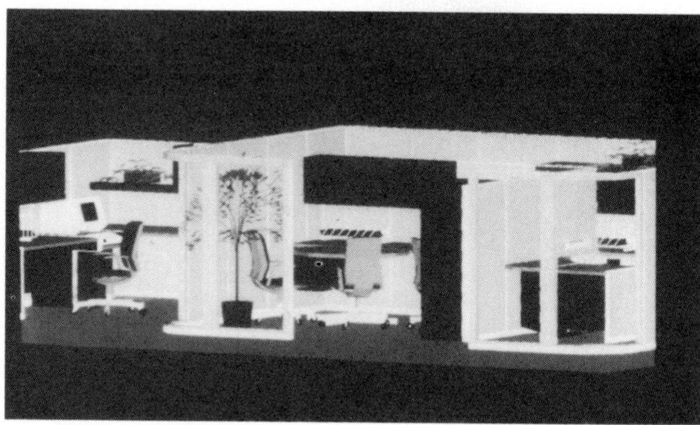

FIGURE 15–8
Office facility design using three-dimensional displays. Courtesy Intergraph Corp.

layouts are demonstrated in Figs. 15–7 and 15–8, which illustrate models for plant design and office design. Such three-dimensional displays can give a designer a better appreciation of the appearance of a layout. In the following sections, we explore the characteristic features of modeling packages and the methods for interfacing or integrating modeling functions with graphics routines.

15–2 Master Coordinates and Modeling Transformations

In many design applications, models are constructed with instances (transformed copies) of the geometric shapes that are defined in a basic symbol set. Instances are created by positioning the basic symbols within the world coordinate reference of the model. Since we do not know in advance where symbol instances are to be placed in the world coordinate system, the definition of a basic symbol must be stated in an independent coordinate reference, which is then transformed to world coordinates. This independent reference is called the **master coordinate** system for each symbol. Figure 15–9 illustrates master coordinate definitions for two symbols that could be used in a two-dimensional facility-layout application. Arrays *worktable* and *chair* are used to store the master coordinate definitions. An example of instance placement within a facility world coordinate reference is shown in Fig. 15–10, where the worktable and chair of Fig. 15–9 have been rotated and translated.

Any transformations applied to the master coordinate definition of a symbol are referred to as **modeling transformations.** A particular kind of modeling transformation, the creation of an instance in world coordinates from a master coordinate symbol definition, is termed an **instance transformation.** In general, modeling packages allow applications programs to perform both instance transformations (master to world coordinates) and transformations within a symbol master system (master to master coordinates).

Arrays for Chair Coordinates

x_chair	y_chair
−3	−3
−3	3
3	3
3	−3
1	3
2	0
1	−3

5
0
−5
−5 0 5

Chair

(a)

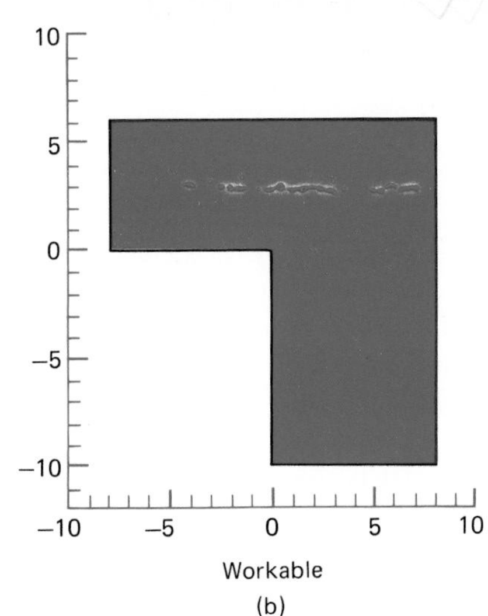

Arrays for Workable Coordinates

x_worktable	y_worktable
0	0
−8	0
−8	6
8	6
8	−10
0	−10

FIGURE 15–9
Master coordinate definitions of two symbols for use in a two-dimensional facility design package.

Modeling Transformations

Master coordinate symbol definitions are typically transformed by a modeling package into world coordinates using matrix transformation methods. These packages are designed to concatenate individual transformations into a homogeneous coordinate **modeling transformation matrix,** MT. An instance transformation is carried out by applying MT to points in the master coordinate definition of a symbol to produce the corresponding world coordinate position of the object:

$$(x_{\text{world}}, y_{\text{world}}, z_{\text{world}}, 1) = (x_{\text{master}}, y_{\text{master}}, z_{\text{master}}, 1) \cdot MT \tag{15–1}$$

A matrix can be designated as the modeling transformation matrix in a modeling package with

```
set_modeling_transformation (mt)
```

This matrix can be used to produce instance transformations, as in Eq. 15–1, or to generate a master-to-master transformation, if the original size or orientation of a symbol is to be altered.

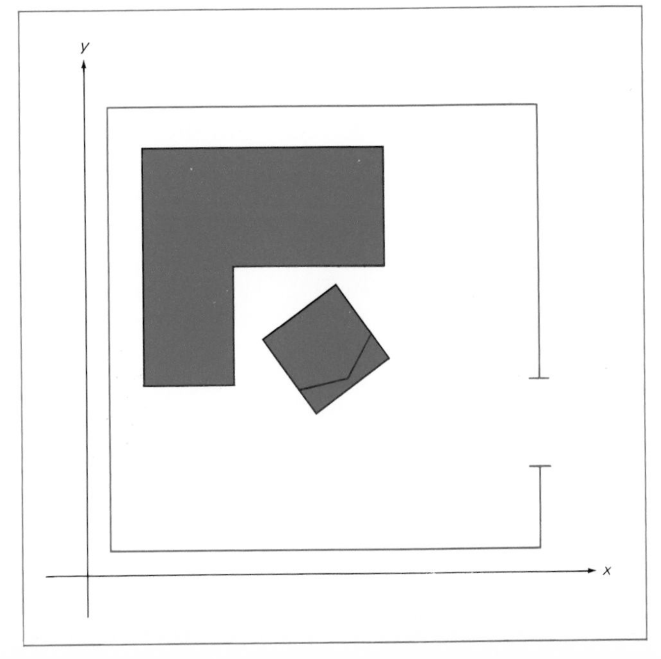

FIGURE 15–10
Instances of furniture symbols (Fig. 15–9) in the world coordinates of a facility layout.

A set of routines to perform individual transformations can be designated in a modeling package as

```
set_modeling_translation (tx, ty, tz)
set_modeling_scale_factors (sx, sy, sz)
set_modeling_rotation (ax, ay, az)
```

Translation distances *tx*, *ty*, and *tz* are applied to a master coordinate definition to move an object into the world coordinate description of a module. If the object is to be scaled or rotated, these transformations are applied relative to the master coordinate origin. Scaling parameters are designated as *sx*, *sy*, and *sz*, and rotation angles about the master coordinate axes are denoted as *ax*, *ay*, and *az*.

Transformation routines are used to update the current modeling transformation matrix. An update is performed by concatenating the matrix representation for the specified individual transformation with the modeling matrix. This composition of matrices is carried out in a modeling package with the individual transformation matrix on the left. For example, a translation specification would produce an updated modeling matrix MT through the operation

$$MT = T \cdot MT$$

As we shall see, multiplying matrices in this order facilitates processing of multilevel hierarchies. However, this arrangement requires that we reverse the logical order of the calls to the transformation routines. Suppose an object is to be rotated and then translated to a world coordinate system. This means that the order of the transformation calls should be translate first, then rotate. If we denote the transformation matrices for translation and rotation as T and R, this sequence of calls would update the transformation matrix in the order

$$\begin{aligned} MT &= T \cdot MT \\ MT &= R \cdot MT \\ MT &= R \cdot T \cdot MT \end{aligned}$$

so that the object symbol is first rotated, then translated.

To demonstrate the use of modeling transformations, we consider a method for creating instances of basic symbols for a solid-modeling application. Basic symbols available in this package are the three-dimensional primitives, such as a block, cylinder, sphere, cone, and other solids. These symbols can be defined in procedures that contain the output primitives for constructing the object in master coordinates. We assume that a user can move any number of symbols into a world coordinate system. Once the transformation of symbols from master coordinates to world coordinates is completed, solid-construction operations are to be performed on the instances to form new objects as the union, intersection, or difference of those symbols that have been transformed to world coordinates.

When a user selects a symbol and the associated transformation parameters, this information is stored in an array of records, which we shall call *instances*. Figure 15–11 shows the structure of records in this array. The record entries include a code field, identifying the symbol to be instanced, and the instance transformation parameters for converting the symbol master coordinates into world coordinates. An interactive process could be used to create this array. A pick device might select a symbol from a symbol menu. Then scaling and rotations parameters could be selected from transformation-parameter menus. Finally, the desired position of the instance (scaled and rotated symbol) within a displayed world coordinate representation would be selected. The parameters for the symbol code, scaling factors, rotation angles, and translation distances are then entered into the *instances* array.

FIGURE 15–11
Record fields for the array instances, used by procedure display_instances.

Symbol Code (1,2,3,· · ·)
Translation Parameters (T_x, T_y, T_z)
Scaling Parameters (S_x, S_y, S_z)
Rotation Angles (A_x, A_y, A_z)

Once the *instances* array is complete, the modeling package can transform the master coordinate definition of each symbol and pass the new instance to the graphics package for display. The following procedures illustrate one way in which this might be done.

```
type
  instance = record
    symbol                                 : integer;
    tx, ty, tz, sx, sy, sz, ax, ay, az : real
  end;  {instance}
var
  instances : array [1..max_instances] of instance;

procedure display_instance;
  var k : integer;
  begin
    for k := 1 to max_instances do begin
        create_segment (k);
        with instances [k] do begin
            set_modeling_transformation (identity);
            set_modeling_translation (tx, ty, tz);
            set_modeling_rotation (ax, ay, az);
            set_modeling_scale_factors (sx, sy, sz);
            case symbol of
                1 : cylinder;
                2 : block;
                    .
                    .
                    .
              end  {case}
          end;  {with instances}
        close_segment
      end  {for k}
  end;  {display_instance}

procedure cylinder;
  begin  {definition of cylinder}  end;

procedure block;
  begin  {definition of block}  end;
```

In this example, each instance (in world coordinates) is stored in a separate segment. When segments are passed to the graphics package, they are further modified by the viewing transformation. The object is transformed to viewing coordinates, then clipped and mapped into device coordinates for display. In some systems, these viewing transformations are combined with the modeling transformations to form a single transformation operation.

Our example procedure for creating instance segments first scales and rotates a given symbol within its master coordinate reference. Then the object is translated to world coordinates. This is accomplished by specifying the transformation operations in the reverse order: first translate, then rotate, then scale. This ordering of the transformation routines combines the matrix representations for these operations in the order

$$MT = S \cdot R \cdot T \cdot MT$$

so that the final matrix is equivalent to scaling followed by rotation then translation.

Modeling Symbol Hierarchies

As we have seen, modeling applications may require the composition of basic symbols into groups, called modules; these modules may be combined into higher-level modules; and so on. Figure 15–12 illustrates the tree structure for a three-level symbol hierarchy. Such symbol hierarchies can be handled by applying the methods of our previous solid-modeling example to each succeeding level in the tree. We first define a module as a list of symbol instances and their transformation parameters. At the next level, we define each higher-level module as a list of the lower-module instances and their transformation parameters. This process is continued up to the root of the tree, which represents the total picture in world coordinates.

To illustrate a technique for setting up the definition of a symbol hierarchy, we extend the solid-modeling application of our previous example to a two-level hierarchy. Transformations must now be applied at two levels. Instances are positioned within a module, and modules are positioned within the overall picture. We assume that the modules represent subassemblies that are to be fitted together to form the composite structure (piece of equipment). The subassemblies (modules) are created by combining the basic symbols to obtain the shapes needed for the component parts of the total assembly.

Transformation matrices must now be constructed for each module, as well as for each instance within each module. To transform a symbol into its proper position in the total assembly, it is first instanced into the subassembly, then moved into final position using the module-transformation parameters specified for that subassembly. This process can be performed by concatenating the module transformation with the instance transformation necessary to place a symbol in the module, so that the modeling transformation at each step is calculated as

$$MT = MT_{\text{master,module}} \cdot MT_{\text{module,world}} \tag{15–2}$$

Applying this modeling transformation matrix MT to a symbol is then equivalent to first performing the instance transformation (master to module coordinates), then moving the symbol into the final assembly (module to world coordinates).

The module transformation matrix $MT_{\text{module,world}}$ must be applied to all instances making up the subassembly. To be able repeatedly to apply this module transformation matrix to its component instances, the module matrix must be saved

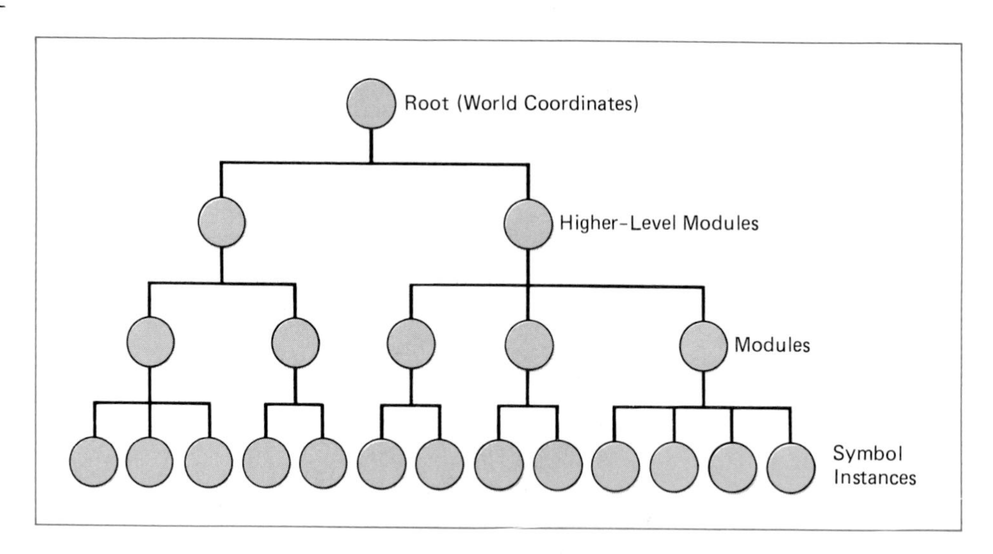

FIGURE 15–12
Tree-structure representation for a three-symbol hierarchy.

so that the modeling matrix *MT* can be restored to its original form after each instance has been moved into final position. For this purpose, two additional routines are used in the modeling package:

```
save_modeling_transformation (mt, m_stack)
restore_modeling_transformation (mt, m_stack)
```

The first routine saves the current modeling transformation matrix for later use. When the modeling matrix is to be restored to its original form, the second routine is called to retrieve the stored matrix from the designated stack area. These two operations allow the modeling package to combine symbols into modules and then transform the modules into the final world coordinate representation.

An example of how a two-level hierarchy can be implemented for a solid-modeling application is given in the following procedures. The record array *modules* lists the modeling transformation parameters for positioning each module within the facility, using a pointer *(instance_start)* to instances comprising the module. *Instances* is a list of the symbols and transformation parameters needed to form the instances within the module.

```
type
  instance = record
      symbol_type                      : integer;
      tx, ty, tz, sx, sy, sz, ax, ay, az : real
    end;  {instance}
  module  = record
      tx, ty, tz, sx, sy, sz, ax, ay, az : real;
      instance_count                   : integer;
      instance_start                   : integer

    end;  {module}
var
  instances : array [1..max_instances] of instance;
  modules   : array [1..max_modules] of module;

procedure display_picture;
  var j : integer;
  begin
    for j := 1 to max_modules do begin
        create_segment (j);  {put each module in a segment}
        set_modeling_transformation (identity);
        with modules [j] do begin
            set_modeling_translation (tx, ty, tz);
            set_modeling_rotation (ax, ay, az);
            set_modeling_scale_factor (sx, sy, sz);
            display_module (instance_start,
                            instance_start+instance_count-1)
          end;  {with modules}
        close_segment
      end  {for j}
   end;  {display_picture}

procedure display_module (start, finish : integer);
  var  k : integer;
  begin
    for k := start to finish do begin
        save_modeling_transformation (mt, m_stack);
        with instances [k] do begin
```

```
            set_modeling_translation (tx, ty, tz);
            set_modeling_rotation (ax, ay, az);
            set_modeling_scale_factors (sx, sy, sz);
            case symbol of
                1 : cylinder;
                2 : block;
                    .
                    .
                    .
              end {case}
          end; {with instances [k]}
        restore_modeling_transformation (mt, mt_stack)
      end {for k}
    end; {display_module}
```

The first procedure in this example opens a new segment and creates a module transformation matrix. Scaling, rotation, and translation parameters for this transformation are obtained, as in the previous example, from the array of module records. Each module record also points to the record array for the symbol instances forming that module. The second procedure concatenates the module transformation matrix with the transformation matrices for each symbol instance, with the instance transformation matrix on the left. This produces the composite matrix needed to transform symbols into a module and then to world coordinates. Since the same module-to-world transformation must be used with each symbol that is moved into the module reference system, the module-to-world transformation is saved so that it can be restored for application to each succeeding symbol.

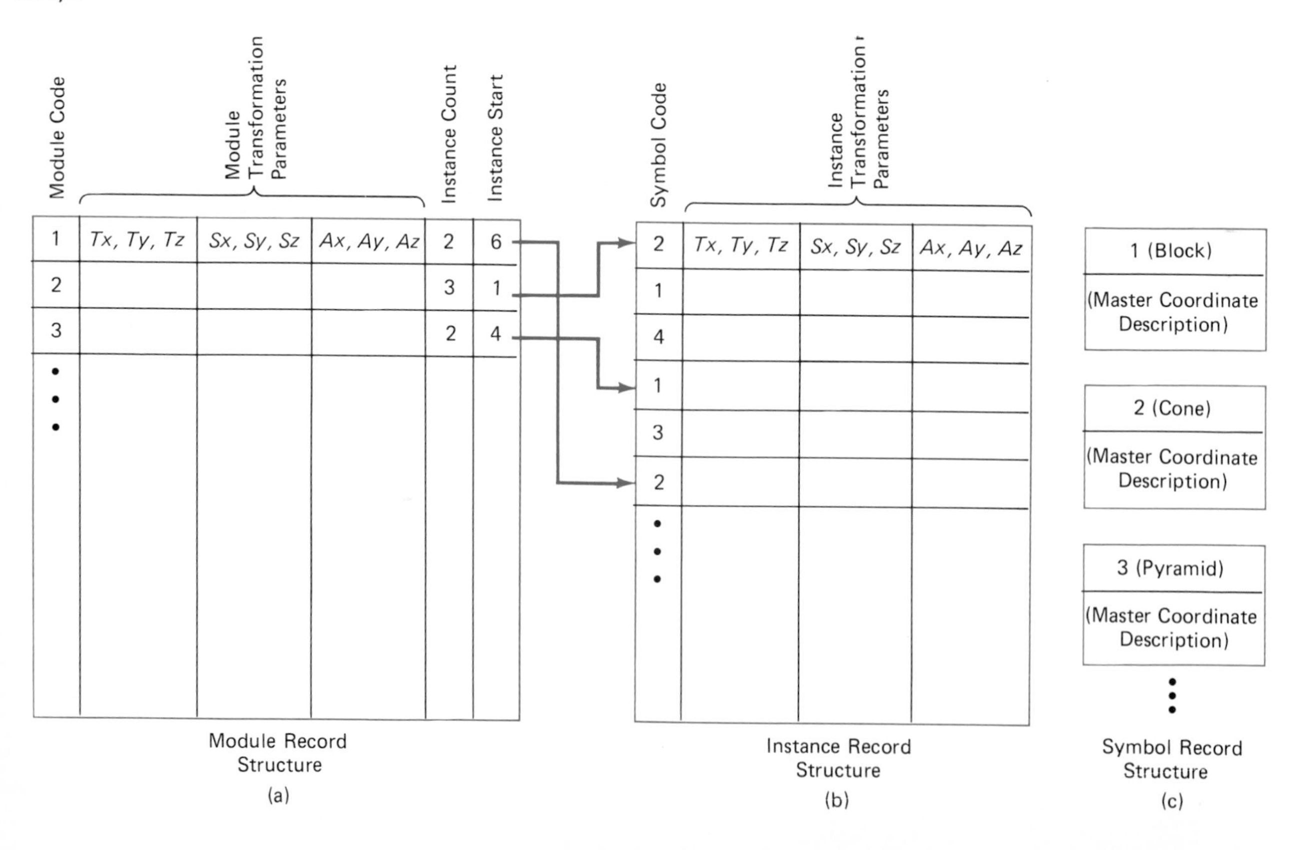

FIGURE 15–13
Record layouts for (a) modules, (b) instances, and (c) symbol arrays.

In addition to the two procedures listed, the modeling package would also need the individual symbol-definition procedures for the three-dimensional primitives (block, cone, etc.). Figure 15–13 illustrates how the data structures that supply data to the display procedures could be set up for this two-level symbol hierarchy. One array is used for listing the modules and one for listing symbol instances, and the several symbol arrays provide master coordinate definitions. The modules could be predefined, or they could be user-defined at the time the picture is created.

An alternative arrangement for storing the module and instance data is to set up the data records in a linked list. The records could be linked so as to represent the tree structure of the symbol hierarchy. This would provide greater flexibility for reorganizing the modules within a picture. New symbol or module instances could be inserted more easily, and existing instances are deleted by changing pointer fields.

We can generalize this two-level hierarchy to include additional levels by setting up procedures for handling the further transformations needed to move instances to the final world coordinate representation. Starting at the top of the tree, appropriate transformation matrices at each level are concatenated to the left of the current transformation matrix to produce a new modeling matrix. This matrix is stored on the stack for subsequent use, and modules at the next lower level are processed. At the lowest level, the matrix applied to the output primitives will be a composite matrix formed from all the higher levels. When a branch of the tree has been completed, the tree is ascended back through the various levels until the start of an unprocessed branch is reached. As processing moves back up the tree, previously stored transformation matrices are restored (popped from the stack).

Display Procedures

The steps required to create a symbol instance within a module can be summarized as

1. Save the current modeling transformation matrix.
2. Combine the instance transformations with the current modeling transformation matrix.
3. Call the symbol procedure.
4. Restore the original modeling transformation matrix.

Since these operations are standard for each symbol instance to be created, a modeling package could provide a single routine to accomplish the same results. This routine is referred to as a **display procedure,** and the call to this routine is termed a **display procedure call.**

A display procedure call specifies the symbol name and the instance transformation parameters in some convenient format, such as

```
display (symbol_name, sx, sy, sz, ax, ay, az, tx, ty, tz)
```

In this form, the name of the symbol becomes a parameter that is passed to the procedure *display*. Other formats could be set up for this operation.

Display procedure calls provide a convenient shorthand for creating symbol instances. However, some flexibility in specifying the order of transformations could be sacrificed. For the format stated, the display procedure might carry out the instance transformation operations in the fixed order: scaling first, followed by rotation, then translation. A particular implementation could allow the order of the transformations to be changed to fit different application needs.

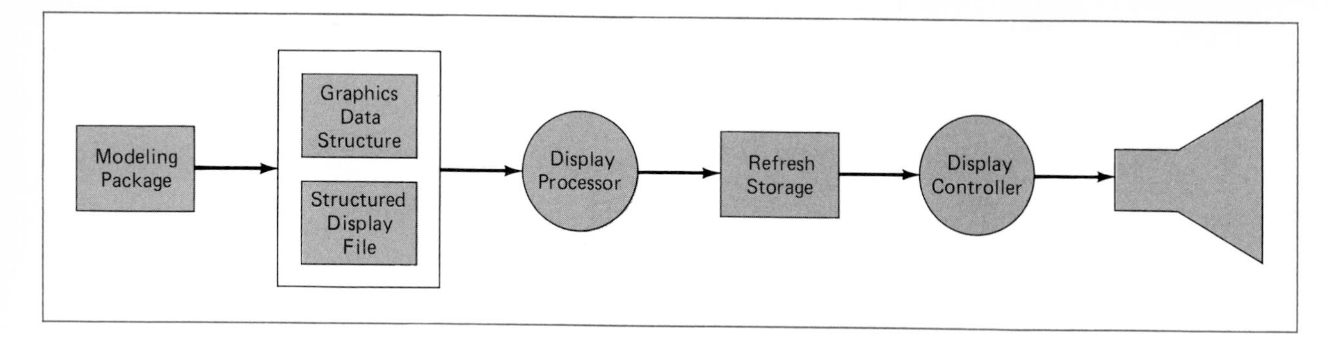

FIGURE 15–14
A structured display file formed from the graphical data structure by a modeling package.

15–3 Structured Display Files

When models formed with symbol hierarchies are to be displayed on a refresh monitor, a modeling package can build a display file that reflects the symbol and module relationships employed in the model. This **structured display file** is accessed by the display processor, which uses the file to create and update the display information in the refresh storage area (Fig. 15–14). Refresh information can be organized either as a linear arrangement of segments or as a frame buffer, depending on whether a vector or a raster monitor is in use. A display controller then refreshes the screen using the contents of the refresh storage area.

The structured display file is organized so that the display processor can make rapid changes to the refresh area without processing through all the nongeometric data that may be in the graphical data structure representation of the model. Segments in this display file contain only master coordinate geometric data and the transformation parameters needed to position instances. These segments are organized into a hierarchical relationship that represents the tree structure of the model. The display processor accesses the structured display file, performing the instance transformations, viewing operations, and clipping necessary for displaying the model on the video monitor.

15–4 Symbol Operations

We have seen how hierarchical models can be constructed using standard procedures to set up symbol and module specifications. Another method that can be used in a modeling package is to allow symbols and modules to be defined with operations similar to those used with segments. For example, the following operations could be used to set up the definition of a symbol in exactly the same way that segments are created:

```
create_symbol (id);
        .
        .
        .
close_symbol;
```

Once a symbol has been defined with *create_symbol*, it can be included in a segment using an *insert_symbol* operation in the same way that we previously called a symbol procedure. This method is illustrated in the following segment definition:

```
create_segment (12);
        .
        .
        .
save_modeling_transformation (mt);
        .
    {perform instance transformations}
        .
insert_symbol (5);
restore_modeling_transformation (mt);
        .
        .
        .
close_segment;
```

Here symbol 5 is inserted into a larger structure defined within segment 12.

In a similar way, commands can be provided to form groups (or modules) of symbols. The following statements could be used for this purpose:

```
create_group (id);
    .
    .
    .
close_group;
```

Groups could be formed with symbols or other groups, using *insert* routines and transformation operations in much the same way that we previously used procedures. In the following example, group 8 is defined as a combination of group 7 and symbol 10:

```
create_group (8);
      .
      .
      .
insert_group (7);
      .
      .
      .
insert_symbol (10);
      .
      .
      .
close_group;
```

Instance transformations would be used to position group 7 and symbol 10 within group 8, and the *insert* operations set pointers to the corresponding symbol and group definitions.

Creating symbols and groups with the preceding modeling routines provides a convenient method for setting up hierarchies and has some advantages over procedure definitions. Permanent symbol libraries can be created and stored with this approach. A user could define an individual symbol library, and a general facility symbol library also could be available to all users at any time. Attributes and other options can be specified for defined symbols. For example, the *insert* operation could include specification of the transformation parameters for the symbol or module referenced, as is done with display procedure calls.

In designing a modeling package, symbol management considerations must be taken into account. As with segments, we need efficient methods for assigning

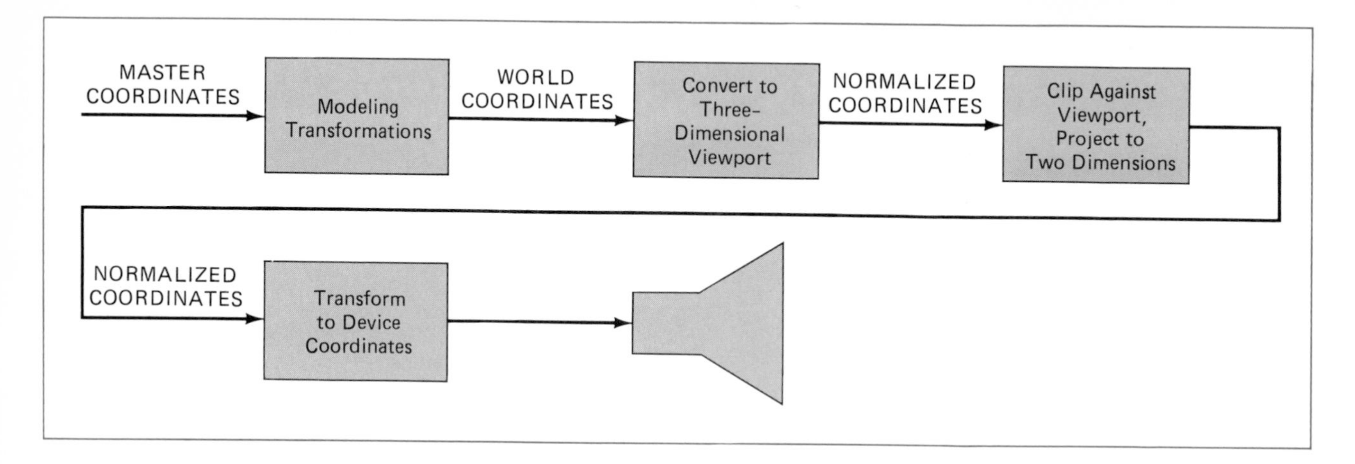

FIGURE 15–15
Transformation of master coordinates into device coordinates.

storage and for conducting symbol searches. Routines to accomplish these functions can be designed as part of the modeling package. If a host system is to be used, these tasks could be performed either by the operating system resource management routines or by a relational database system.

15–5 Combining Modeling and Viewing Transformations

Modeling transformations turn master coordinate definitions into world coordinates, which then are passed to the graphics package for the final transformation into device coordinates (clipping and window-to-viewport mapping). Figure 15–15 shows a sequence for transforming a master coordinate definition into device coordinates when modeling routines are interfaced to a graphics package. It is possible to perform the various operations shown in a different order, depending on the structure of the packages and whether the system is designed for two-dimensional or three-dimensional applications.

Master Coordinate Clipping

When modeling routines are built into a graphics package, the instance transformations can be combined with the viewing transformations. In many cases, it is then possible to carry out the clipping operations before any transformations are applied (Fig. 15–16). This can be an efficient approach, since unnecessary processing of clipped primitives by the transformation routines is avoided. For two-dimen-

FIGURE 15–16
Combined modeling and viewing transformations with clipping performed first.

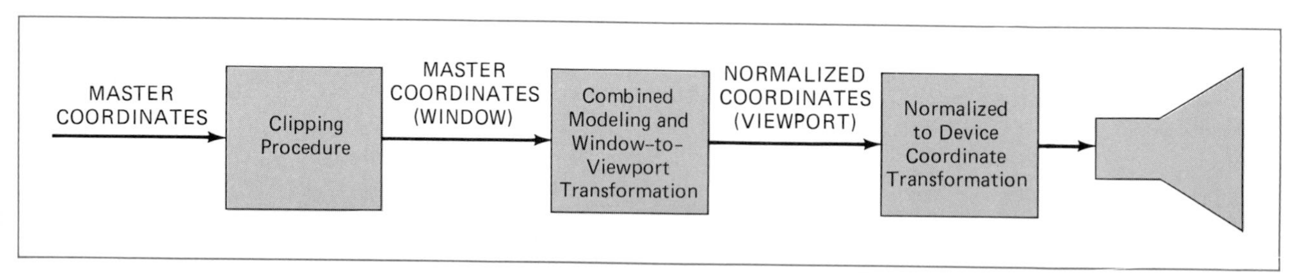

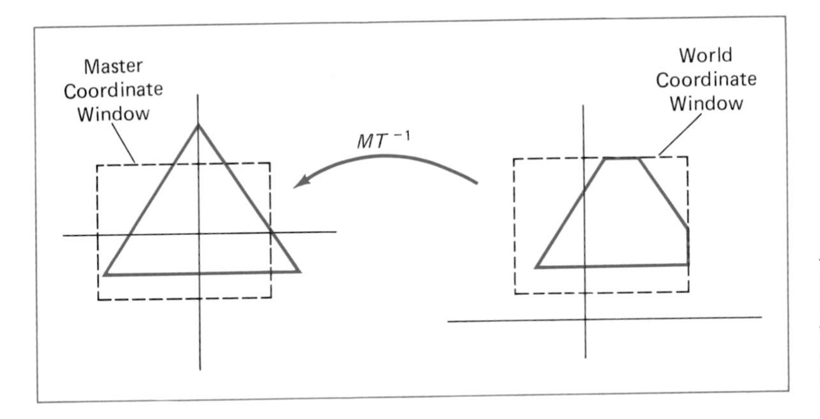

FIGURE 15–17
Mapping a world coordinate window onto the master coordinate reference of a symbol.

sional applications with no rotations and standard window and viewport rectangles, clipping can easily be done in master coordinates. The package calculates the inverse of the instance transformation (MT^{-1}) and maps the window boundary onto the master coordinate definition of each symbol to be displayed. This process is illustrated in Fig. 15–17. Each instance in the model is clipped in master coordinates; then the modeling and windowing transformations are applied as one transformation matrix operation.

When rotations are involved, it is more efficient to transform first, then clip. A two-dimensional graphics package that allows rotated windows could make checks to determine whether a rotated window or any rotation transformations are to be used. If either a rotated window or a rotation transformation has been specified, clipping is performed after the transformations. Otherwise, clipping can be performed first, as in Fig. 15–16.

Bounding Rectangles for Symbols

To speed up the clipping process, symbols (and symbol groups) can be processed through a first clipping pass using bounding rectangles, as shown in Fig. 15–18. This provides a fast identification of the symbols that are completely inside or outside the window. The same process can be used for groups of symbols by determining the outer boundaries of the group.

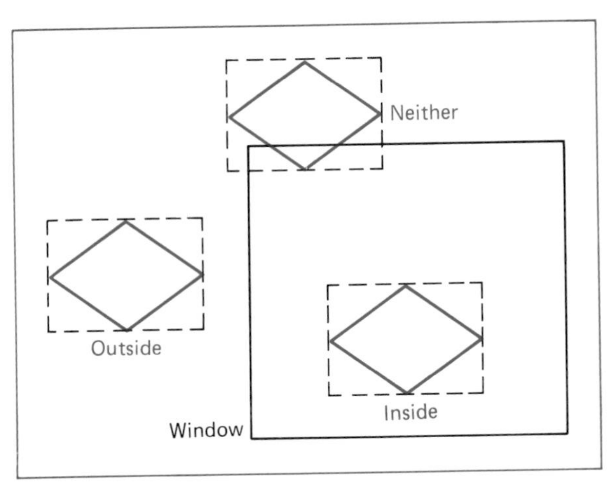

FIGURE 15–18
Bounding rectangles used to determine whether a symbol is completely inside or outside a window.

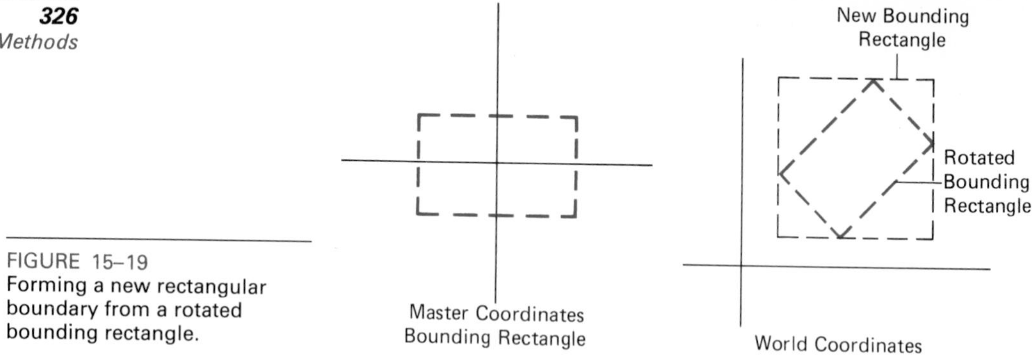

FIGURE 15–19
Forming a new rectangular boundary from a rotated bounding rectangle.

As each symbol and module is created, the bounding rectangle can be determined and stored with the symbol or module definition. When clipping is applied, these bounding rectangles are referenced by the clipping routines. The bounding rectangles can be used also for fast pick identification of symbols or groups.

When rotations are involved in the instance transformations, new boundary rectangles can be calculated after the transformation. Figure 15–19 illustrates how a new rectangle is formed from a rotated one.

REFERENCES

Modeling methods and applications are discussed in the following papers: Clark (1976) and Renfrow (1977).

EXERCISES

15–1. Discuss the model representations that would be appropriate for several distinctly different types of systems. Also discuss how each system might be implemented in a graphics modeling package.

15–2. For a logic-circuit modeling application, such as that in Fig. 15–1, give a detailed description of the data structures needed for specifying all modeling information. Then describe how the same information could be given in procedures and compare the advantages and disadvantages of each approach.

15–3. Develop a modeling package for electrical design that will allow a user to position electrical symbols within a circuit network. Only translations need be applied to place an instance of one of the electrical menu shapes into the network. Once a component has been placed in the network, it is to be connected to other specified components with straight line segments.

15–4. Write a two-dimensional facility layout package. A menu of furniture shapes is to be provided to the user, who can place the objects in any location within a single room. A one-level hierarchy is to be used, and instance transformations are limited to rotations and translations.

15–5. Expand the package of Ex. 15–4 to generate a two-level hierarchy. The furniture objects are to be placed into work areas, and the work areas are to be arranged within a larger area.

15–6. Expand the package of Ex. 15–4 to generate a three-level hierarchy. Furniture objects are arranged into work areas, which are then combined into department areas, and the various departments are fitted into a larger area.

15–7. Develop a complete algorithm for implementing the procedures for constructive solid modeling by combining three-dimensional primitives to generate subassemblies, then combining the subassemblies into the final assembly. As the last step, the created assemblies are to be mapped onto a designated area of a video monitor.

15–8. Write routines to implement a display procedure, given a set of input transformation parameters and a symbol identification.

15–9. Develop a procedure for maintaining and updating a structured display file.

15–10. Write a routine for converting the contents of a structured display file into appropriate intensity values in a frame buffer.

15–11. Develop implementation routines for the commands to create symbol definitions.

15–12. Devise an implementation for the commands to create groups as combinations of symbols and other groups.

15–13. Set up a clipping procedure to clip objects in master coordinates by mapping a given window boundary into the master coordinate system, given the transformation parameters.

15–14. Develop an algorithm for the steps in a modeling package that first transforms objects, then clips.

15–15. Devise a clipping procedure using bounding rectangles around the symbols.

15–16. Develop a two-dimensional modeling package that combines modeling and viewing transformations.

15–17. Extend the package of Ex. 15–16 to three dimensions.

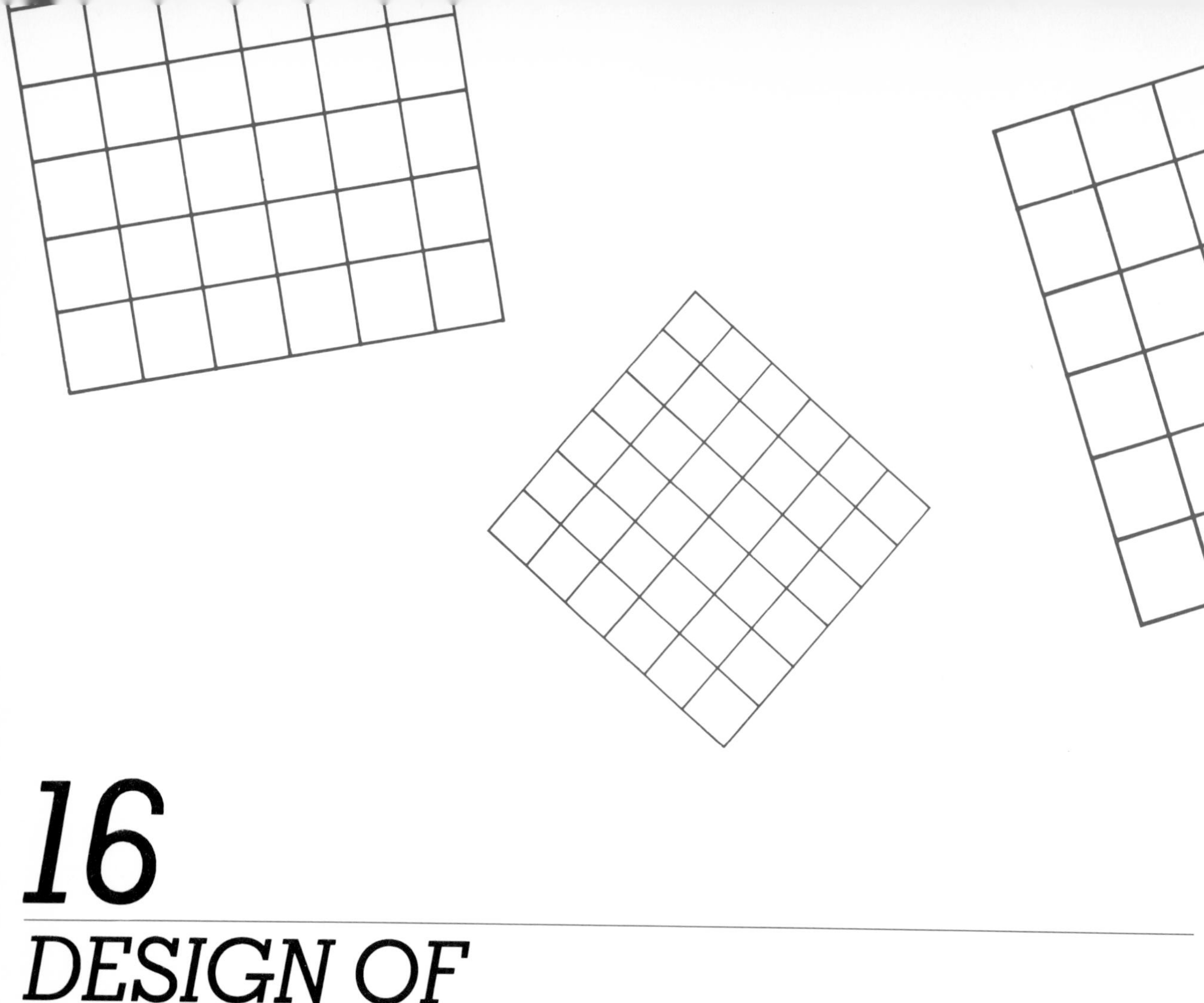

16 DESIGN OF THE USER INTERFACE

In designing a graphics package, we need to consider not only the graphics operations to be performed but also how these operations are to be made available to a user. This interface should be designed so as to provide a convenient and efficient means for the user to access basic graphics functions, such as displaying objects, setting attributes, or performing transformations. The graphics package might be set up to produce engineering designs, architectural plans, drafting layouts, or business graphs; or it could be designed as an artist's paintbrush program. Whatever the type of application intended, we need to decide on the interactive dialogue that best suits the intended user, the kind of manipulation routines that are to be used, and the output devices that are appropriate for the type of applications involved. A poorly designed interface increases the chances for user error and can significantly increase the time it takes a user to complete a task.

16–1 Components of a User Interface

There are many factors to be considered in the design of a user interface. In addition to the specific operations that are to be made available to the user, we must consider how menus are to be organized, how the graphics package is to respond to input and errors, how the output display is to be organized, and how the package is to be documented and explained to the user. To aid us in exploring these factors, we consider the design of a user interface in terms of the following components:

User model
Command language
Menu formats
Feedback methods
Output formats

The **user model** provides the definition of the concepts involved in the graphics package. This model helps the user to understand how the package operates in terms of application concepts. It explains to the user what type of objects can be displayed and how they can be manipulated.

Operations available to a user are defined in the **command language,** which specifies the object manipulation functions and file operations. Typical object manipulation functions are those for rearranging and transforming objects in a scene. File operations can provide for the creation, renaming, and copying of segments. How the user commands are to be structured will depend on the type of input and output devices chosen for the graphics system.

Processing options can be presented to a user in a **menu format.** The menus could be used to list both the available operations and the objects to be manipulated.

An important consideration in the design of an interface is how the system is to respond, or give **feedback,** to user input. Feedback assists a user in operating the system by acknowledging the receipt of commands, by sending various messages to the user, and by signaling when menu selections have been received. Some types of feedback are an integral part of the structure of the command language, while other forms of feedback are provided to help a user understand the operation of the system.

How information is to be presented to the user is determined by the **output formats.** An output picture should be organized so as to provide information to the user in the most effective manner possible. Factors to be considered in the design of the output formats include the choice of geometric patterns that are used to represent objects and the overall arrangement of the output on a display device.

16–2 User Model

Design of a user interface starts with the user model. This determines the conceptual framework that is to be presented to a user. The model describes what the system is designed to accomplish and what graphics operations are available. For example, if the graphics system is to be used as a tool for architectural design, the model describes how the package can be used to construct and display views of buildings. Once the user model has been established, the other components of the interface can be developed. The final step in the design of the model is to prepare the user manual, which explains the system and provides help in using it.

Basically, the user model defines the graphics system in terms of objects and the operations that can be performed on objects. For a facility layout system, objects could be defined as a set of furniture items (tables, chairs, etc.), and the available operations would include those for positioning and removing different pieces of furniture within the facility layout. Similarly, a circuit design program might use electrical or logic elements for objects, with positioning operations available for adding or deleting elements within the overall circuit design.

Such objects as furniture items and circuit elements are referred to as **application objects.** In addition to these objects, the user model could contain other objects that are used to control graphics operations. These are called **control objects.** Examples of control objects are cursors for selecting screen positions, menu selection symbols, and positioning grids (Fig. 16–1). Operations available in a package allow for manipulation of both the application objects and the control objects. Users can be provided with the capability to move the control objects around, to delete these objects, or to change their size.

Only familiar concepts should be employed in the user model. If we are setting up a graphics package as an aid to architectural design, the operation of the package should be described in terms of the positioning of walls, doors, windows, and other building components to form the architectural structures. The model should not contain references to concepts that may be unfamiliar to a user. There is no reason to assume, for example, that an architect would be familiar with data structure terms. Introducing detailed information concerning the operation of the system in terms of tree structures, linked lists, and segments would place an unnecessary burden on the user. All information in the user model should be presented in the language of the application.

Overall, the user model should be as simple and consistent as possible. A complicated model is difficult for a user to understand and to work with in an effective way. The number of objects and graphics operations in the model should be minimized to only those that are necessary to the application. This makes it easier for the user to learn the system. On the other hand, if the package is oversimplified, it can be easy to learn but difficult to apply. The designer of the user model should also strive for consistency. Objects and operations should not be defined in different ways when used in different contexts. A single symbol, for example, should not serve both as an application object and as a control object, depending on the interaction mode. This can make it difficult for a user to keep track of symbol meanings. It is much easier for a user to apply the package if objects and operations are defined and used in a consistent manner.

FIGURE 16–1
Examples of control objects.

+
Screen Cursor

⟶
Menu Selection Symbol

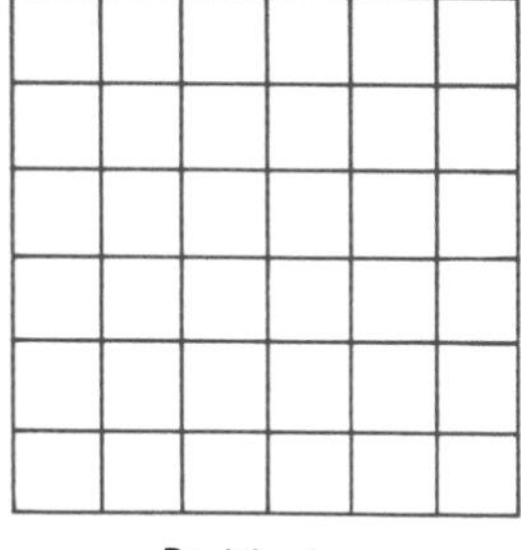
Positioning Grid

As a starting point for developing a user model, a task analysis is often performed. The task analysis is a study of the user's environment and needs. This study is typically carried out by interviewing prospective users and observing how they perform their tasks. Conclusions from the task analysis can form the basis for deciding on the type of objects and operations needed and the way that the graphics package should be presented to the user.

16–3 Command Language

The interactive language chosen for a graphics package should be as natural as possible for the user to learn, with all operations specified in terms relating to the application area. A circuit design package specifies operations in terms of manipulating circuit elements, while the operations in a drafting package are those for drawing geometric figures. The commands should be designed so that the user does not have to learn new concepts, as well as the new language.

Minimizing Memorization

Each operation in the command language should be structured so that it is easy for a user to understand and remember the purpose of the operation. Obscure, complicated, inconsistent, and abbreviated command formats should be avoided. They only confuse a user and reduce the effectiveness of the graphics package. A command such as *select_object* is much easier for a user to remember than the abbreviated format *so*. One key, or button, that deletes any command is easier to use than a number of delete keys for different operations. In addition, the command language should be structured so that the user is not required constantly to shift attention from one input device to another.

For a less experienced user, a command language with a few easily understood operations is usually more effective than a large, comprehensive language set. A simplified command set is easy to learn and remember, and the user can concentrate on the application instead of on the language. However, an experienced user might find some applications hard to handle with a small command set designed for the novice. To accommodate a range of users, command languages can be designed in several levels. Beginners can use the lowest level, which contains the minimum command set, and experts can use the larger command sets at the higher levels. As experience is gained, a beginner can move up the levels, expanding the command language a little at a time.

User Help Facilities

It is very important that help facilities be included in the command language. Different levels of help allow beginners to obtain detailed instructions, while more experienced users can get short prompts that do not interrupt their concentration.

Help facilities can include a tutorial session that provides instruction in how to use the system. A beginner can sit down with the package and get an overview of what the package is designed to do and how the basic command set operates. Several example applications can be provided so that the user can see how the operations really work in typical applications.

If different help levels are available, a beginner can select the lowest level and receive detailed prompts and explanations at each step during the application of the package. The prompts can tell the beginner exactly what to do next. An experienced user could select less prompting or turn off the prompting altogether. Prompts can be given in the form of menus or displayed messages. For the expert user, more subtle prompts can be given. A blinking cursor could indicate when a coordinate input is required, and a scale could be displayed when a scalar value is to be selected.

Backup and Error Handling

During any sequence of operations, some easy mechanism for backing up, or aborting, should be available. Often an operation can be canceled before execution is completed, with the system restored to the state it was in before the operation was started. With the ability to back up at any point, a user can confidently explore the capabilities of the system, knowing that the effects of any mistake can be erased.

Backup can be provided in many forms. Positions of windows and viewports can be tested, and objects can be dragged into different positions before deciding on the final placement. Sometimes a system can be backed up through several operations, allowing the user to reset the system to some specified point. In a system with extensive backup capabilities, all inputs could be saved so that a user can back up and "rerun" any part of a session.

In the absence of a backup facility, other methods can be used to help users overcome the effects of errors. A user could be asked to verify some commands before executing the instructions. It would be too time-consuming to ask the user *Are you sure you want to do this?* after every input, but it would be appropriate for actions that either cannot be undone or would take too great an effort to restore.

Good diagnostics and error messages should be incorporated into the command language to allow users to avoid errors and to understand what went wrong when an error has been made. Obscure error messages cannot help a user correct an error. The error message should provide a clear statement of what is wrong and what needs to be done to correct the situation. Typically, the recovery system will reject an incorrect input and inform the user of the error in a way that helps the user determine the proper input at that point.

Response Time

The time that it takes the system to respond to a user input depends on the complexity of the task requested. For many routine input requests, the system is able to respond immediately. When a user inputs a complicated processing request, some delay can be expected, and this delay time could be used to plan the next phase of the application.

Regardless of the complexity of the input request, users expect systems to make some type of immediate response; otherwise, they cannot be sure that the input was received and that the system is processing. A graphics package should be designed to make this "instantaneous" response to user input in about one-tenth of a second. A response time longer than this can disrupt a user's train of thought, since the user begins wondering what the system is doing instead of thinking about the next step in the application. If processing time is to be longer than one-tenth of a second, the immediate response simply lets the user know that the input has been received. For the beginning user, this response could be a message stating that the input is being processed. With experienced users, a blinking cursor or a change in color or intensity can serve the same purpose.

Some systems are designed so that the variability in response times between different types of processing is not too great. A system that alternates between instantaneous responses and delays of several seconds (or minutes) could be more disconcerting to a user than a slower-responding system with less variability in response times.

Command Language Styles

There are several possible styles for the command language, and the choice of format for the input commands depends on a number of factors. These factors include the design goals of the package, the type of input devices to be used, and the type of anticipated user.

Overall, the command language can be set up so that the sequence of input actions is directed either by the graphics package or by the user. When the package directs the input, the user is told what type of action is expected at each step. This is a particularly effective approach for beginners, where prompts and menus are used to explain what is required and how to enter the input. In some cases, users may be restricted to a limited number of responses, such as replying with *yes*, *no*, or a numeric value. With other systems, a user may be directed to select an action from a list of alternatives. This selection could be made from a displayed menu, using a light pen or cursor control from a keyboard, or the selection could be made using function keys. Once the action has been chosen, such as selecting an object, the user is directed to enter appropriate parameters (coordinates, transformation values, etc.).

For user-initiated languages, the system actions are directed by inputs from the user. Little or no prompting is given to the user, who independently selects the type of action to be performed at each step. This approach provides the greatest flexibility and is best suited to the experienced user. Some menus and prompts could be used to remind the user of options available with any selected operation, but in general the user is free to explore the capabilities of the system without following a preset action sequence.

When a package is designed only to carry on a dialogue, no input command set is made available to the user. Input is achieved by selecting options from displayed menus and by making simple responses to prompts. This method for obtaining input is suitable for beginners but is inefficient for those with more experience, who can better utilize the capabilities of a package with an input command set. The commands can be designed to be input with function keys or by typing them on a standard keyboard.

To minimize user learning time, the syntax of input commands should be simple and straightforward. For example, to remove an object from a display, a user could enter the command *delete_object*, followed by the parameter values that identify the object to be deleted. Parameter designations could give an object number or the object name. Alternatively, the user might point to the object with a light pen or move the screen cursor to the location of the object. Commands that specify the action first, as in the example just given, are called prefix commands. Postfix commands specify parameters first and the action second. When postfix commands are used, the parameter list can be corrected in case of an error before the action is given.

In some applications, user input can be reduced by selecting a command syntax that allows for multiple parameter specifications. For example, instead of a separate delete command for each object, a user could enter a single delete command that allows several objects to be deleted:

```
delete
  object_name1
  object_name2
  object_name3
      .
      .
      .
```

The action of the delete command would be terminated when the next operation is entered.

Another possibility for reducing user input is to eliminate the parameter list within commands. Instead of the command *delete_object* followed by the object

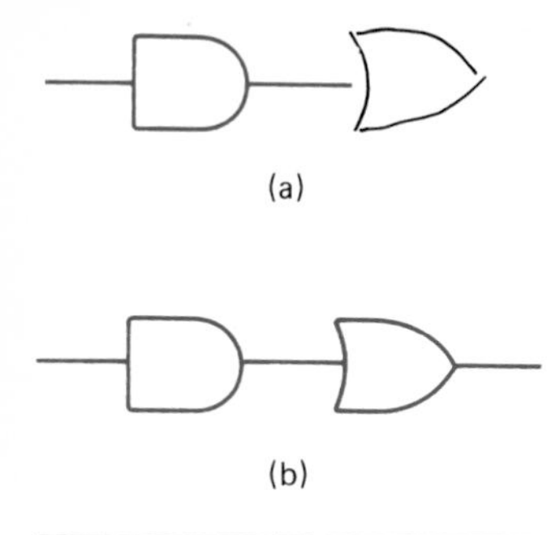

FIGURE 16–2
An input symbol (a) sketched by a designer is identified by the pattern recognition program and replaced with the corresponding circuit element shape (b).

designation, the name of the object could be incorporated into the command. This requires a separate command for each object, but the commands could be implemented as function keys, so that only one input is required to enter the command. As an example, a facility layout package could allow each furniture item to be deleted with a separate command, such as *delete_worktable*, *delete_shelves*, and so forth.

A sketch pattern is another form of input that is useful in many design applications. Here the user draws a pattern with a stroke device, such as a graphics tablet; the pattern is displayed on a video monitor and processed by the design program. The input pattern could be used with a pattern recognition program to select a shape, or this type of input could be the basis for a painting package that allows a user to create a design or picture.

Pattern recognition programs take an input pattern and match it against a library of predefined shapes. A logic circuit designer could use such a program to sketch circuit layouts. Each circuit element is sketched, as in Fig. 16–2 (a), and the program replaces the sketched element with the corresponding library shape, as in Fig. 16–2 (b). This technique can be faster than selecting a shape from a menu and positioning it within the circuit if the library of stored shapes is not too large (about 20 times or less) and no rotations are involved.

A pattern recognition program is used with some systems to allow a parts designer to determine whether a needed part already exists or could be produced by modifying a similar part. Figure 16–3 illustrates a designer's sketch of a plate with five holes and a matching part found by the program. In this example, the designer might decide that the matching part is close enough so that it could be modified by adding the four smaller holes. To establish the size of input patterns, the user could be required to sketch objects within the bounds of a displayed grid.

FIGURE 16–3
An input pattern (a) sketched by a parts designer is matched against the library of existing parts to locate a similar shape (b).

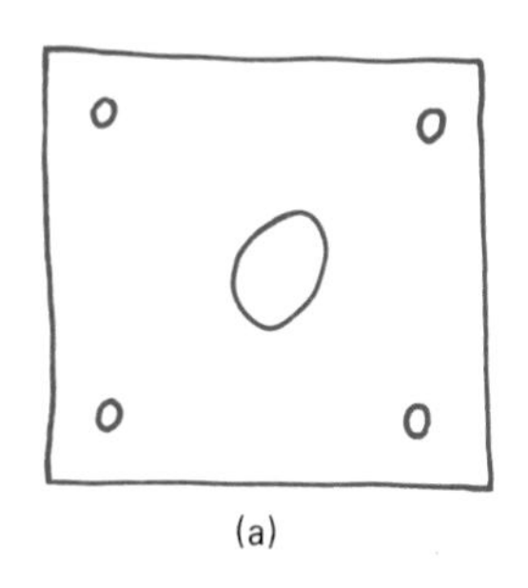

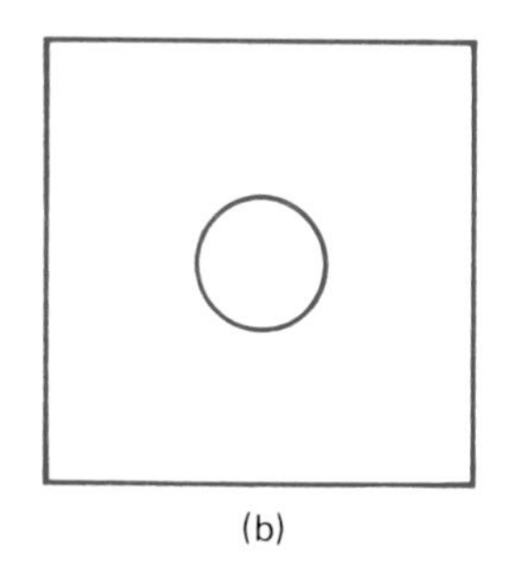

Painting programs can be useful for displaying some types of networks (piping and wiring layouts) and for creative art applications. In network drawing, the user sketches the connecting lines, and the program adjusts the sketch into a layout of horizontal and vertical connections (Fig. 16–4). For art applications, the user can be provided with two menus and an area in which to create the picture, as shown in Fig. 16–5. One menu presents a set of brushes allowing the creation of different line sizes and line textures (sharp lines, airbrush, etc.). The other menu lists a palette of available colors and shades. By selecting different brushes and colors, an artist can produce pictures like the creative art examples in Chapter 1.

16–4 Menu Design

Most graphics packages make use of menus. In some cases, all user input is specified with menu selections. When menus are employed in a program, the user is relieved of the burden of remembering input options. Not only does this cut down on the amount of memorization required of the user by listing the range of available options, but it also prevents the user from electing options that are not valid at that point. In addition, menus can easily be changed to accommodate different applications, whereas function keys or buttons must be reprogrammed and relabeled if they are to be changed. Menus can be used as the input mechanism for both operations and parameter values.

Interactive menu selection can be accomplished with most types of input devices. The light pen and touch panel are used to make very rapid selections,

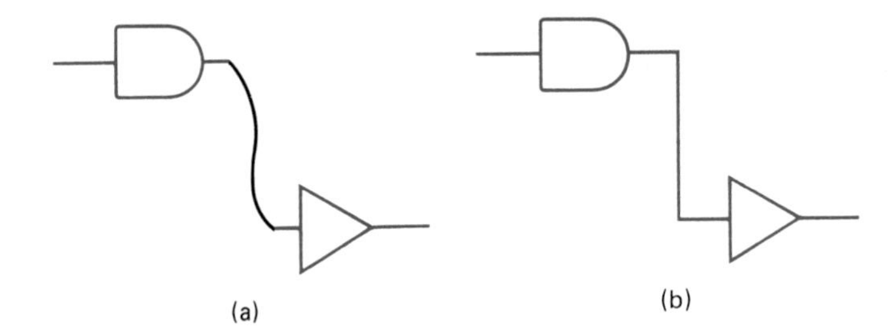

FIGURE 16–4
In a network drawing program, a user's sketch of a connecting line (a) is adjusted to a set of horizontal and vertical lines (b).

since the user simply touches the screen area containing the menu option to be selected. Keyboards, joysticks, and other devices can be used to make selections by positioning a cursor or other symbol at a menu position. A keyboard can also be used to type in the identifying name or number of a menu item. For a small number of menu items, a voice entry system can provide an efficient method for selecting options.

In general, menus with fewer options are more effective, since they reduce the amount of searching time needed to find a particular option, and they take up less screen space. Usually, menus are placed to one side of the screen so that they do not interfere with the displayed picture. When an extensive menu with long descriptions for each option is to be presented, it may have to occupy the entire screen so that the picture and menu are alternately displayed. This can have a very disruptive effect on the user's train of thought, since visual continuity with the picture is lost each time a menu selection is to be made. Such full-screen menus should be avoided by shortening the descriptions of the menu options and breaking the selections into two or more submenus.

One method for organizing a menu into smaller submenus is to arrange the menu options into a multilevel structure. A selection from the first menu brings up a menu at the second level, and so on. This is an effective method for dealing with operations and parameters. Once an action is selected from the operation menu, a parameter menu (scales or dials) can be presented. More than two or three levels should be avoided; otherwise, the menu structure could confuse the user.

Items listed in a menu can be presented either as character strings or as graphical icons (geometric shapes). Such symbols are useful in presenting menus for circuit design and facility layout planning. Icons can also be used to describe the meanings of some actions, as illustrated in Fig. 16–6. A straight arrow is used in this example to indicate a translation direction, and a curved arrow indicates rotation direction. Advantages of icons are that they generally take up less space and are more quickly recognized than corresponding text descriptions. A beginning user may find icons more difficult to use initially, but once the icon set is learned, inputs can be made faster and with fewer errors.

For most applications, menus are always placed in one position, such as at the side or bottom of the screen. Items are always placed in the same position, so that the user gets used to making each selection in a fixed location. Another possibility is to use a "movable" (or "pop-up") menu that is placed near the current position of the screen cursor (Fig. 16–7). Since the user is assumed to be working with the screen cursor, the movable menus allow selections to be made with a minimum of eye and hand movement. The movable menu is placed over part of the picture and so should be displayed only when needed. Movable menus can be implemented on a raster system by replacing a rectangular part of the picture with

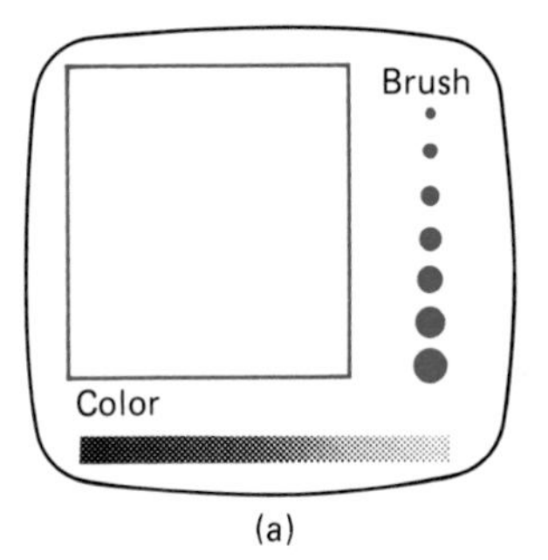

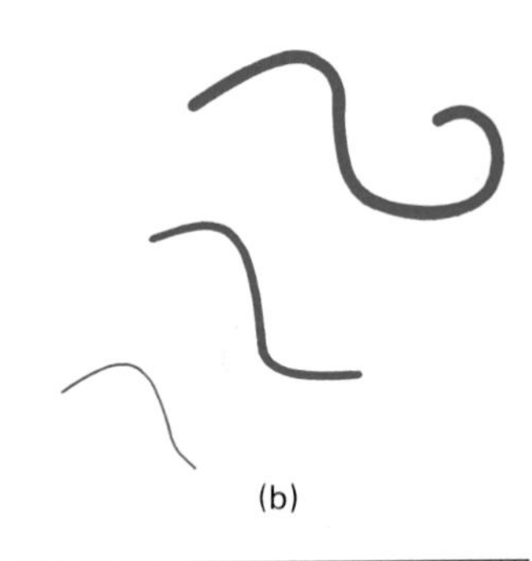

FIGURE 16–5
(a) Menus used in a painting program to select colors and brush textures and sizes. (b) Lines drawn with different-sized brushes.

FIGURE 16–6
Icons that could be used to select directions for translation and rotation.

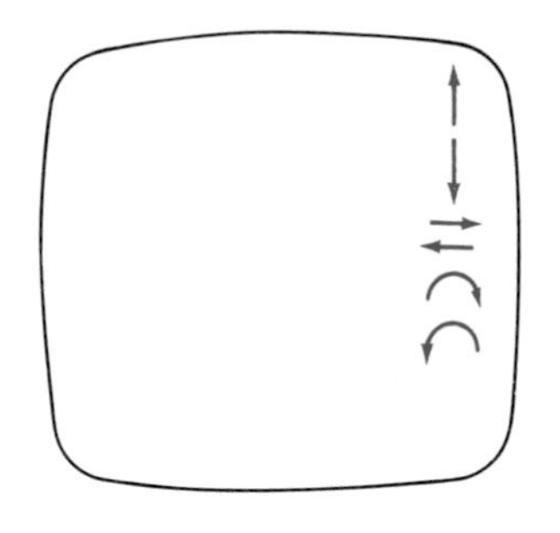

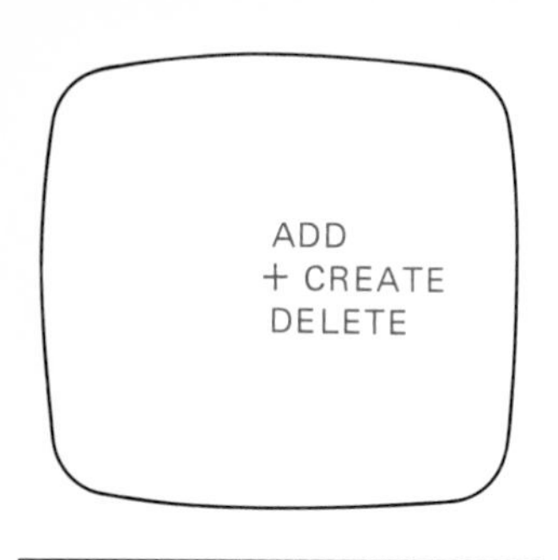

FIGURE 16–7
A movable menu placed near the current position of the screen cursor.

the menu information. With a vector system, the menu and picture may overlap and make it difficult for the user to identify menu items.

16–5 Feedback

An important part of any graphics system is the amount of feedback supplied to a user. The system needs to carry on a continual interactive dialogue and inform the user what the system is doing at each step. This is particularly important when the response time is high. Without feedback, a user may begin to wonder what the system is doing and whether the input should be given again.

As each input is received from the user, a response should immediately appear on the screen. The message should be brief and clearly indicate the type of processing in progress. This not only informs the user that the input has been received, but it also tells the user what the system is working on so that any input errors can be corrected. If processing cannot be completed within a few seconds, several feedback messages might be displayed to keep the user informed of the progress of the system. In some cases this could be a flashing message to tell the user that the system is still working on the input request. It may also be possible for the system to display partial results as they are completed, so that the final display is built up a piece at a time. The system might also allow the user to input other commands or data while one instruction is being processed.

Feedback messages should be given clearly enough so that they have little chance of being overlooked. On the other hand, they should not be so overpowering that the user's concentration is interrupted. With function keys, feedback can be given as an audible click or by lighting up the key that has been pressed. Audio feedback has the advantage that it does not use up screen space, and the user does not have to take attention from the work area to receive the message. When messages are displayed on the screen, a fixed message area can be used so that the user always knows where to look for messages. In some cases, it may be advantageous to place feedback messages in the user work area near the cursor. Different colors can be used to distinguish the feedback from other objects in the display.

To speed system response, feedback techniques can be chosen to take advantage of the operating characteristics of the type of devices in use. With a raster display, pixel intensities are easily inverted (Fig. 16–8), so this method can be used to provide rapid feedback for menu selections. Other methods that could be used for menu selection feedback include highlighting, changing the color, and blinking the item selected.

FIGURE 16–8
Inverting pixel intensities as a means for providing feedback to confirm a menu selection.

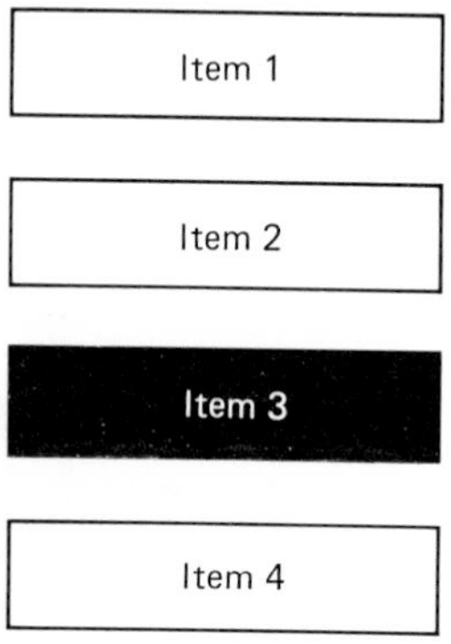

Item 1

Item 2

Item 3

Item 4

Special symbols can be designed for different types of feedback. For example, a cross or a "thumbs down" symbol could be used to indicate an error, and a blinking "at work" sign could be used to tell the user that the system is processing the input. This type of feedback can be very effective with a more experienced user, but the beginner may need more detailed feedback that not only clearly indicates what the system is doing but also what the user should input next.

With some types of input, **echo** feedback is desirable. Typed characters can be displayed on the screen as they are input so that a user can detect and correct errors immediately. Button and dial input can be echoed in the same way. Scalar values that are selected with dials or from displayed scales are usually echoed on the screen to let the user check input values for accuracy. Position selections could be echoed with a cursor or other symbol that appears at the selected position. If positions are selected with the screen cursor, coordinate values of the cursor location can be displayed to indicate precisely which position has been selected.

16–6 Output Formats

Information presented to the user of a graphics package includes a combination of output pictures, menus, messages, and other forms of dialogue generated by the system. There are many possibilities for arranging and presenting this output information to the user, and the designer of a graphics package must consider how best to design the output formats to achieve the greatest visual effectiveness. Considerations in the design of output formats include menu and message structures, icon and symbol shapes, and the overall screen layouts. As previously discussed, the menu and message structures depend on a number of factors. Besides the experience level of the user and the type of application, the structure of menus and messages will be influenced by the layout chosen for screen output.

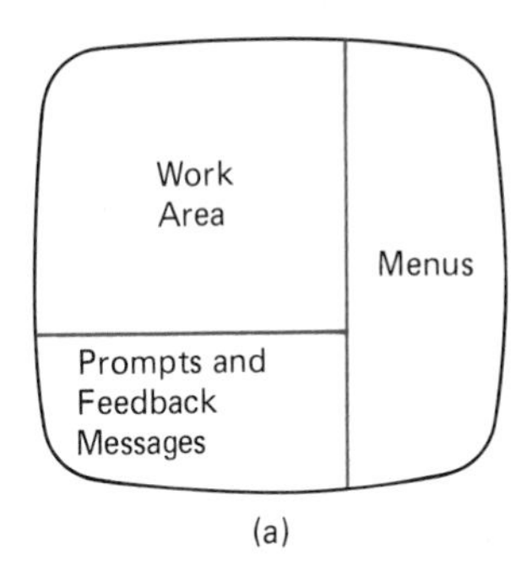

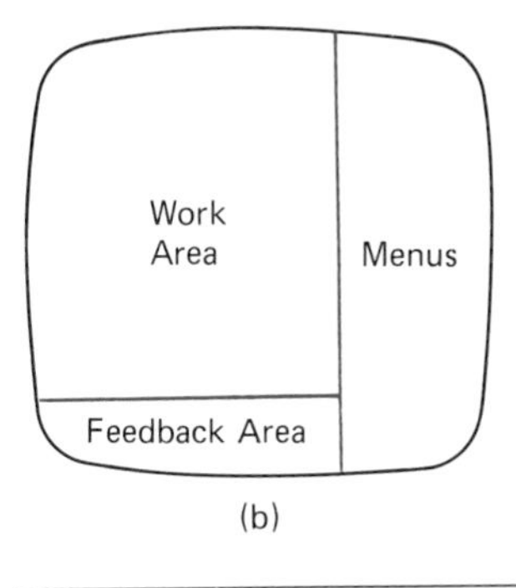

FIGURE 16–9
Screen layout possibilities: (a) large message area; (b) reduced message area.

Icon and Symbol Shapes

The structure of many symbols used in a graphics package depends on the type of application for which the package is intended, such as electrical design, architectural planning, or facility layouts. Shapes of the symbols are chosen so that they provide a simple yet clear picture of the object or operation they are meant to represent. Other symbols, such as cursors or menu pointers, should be designed so that they are clearly different from the other icons. In some cases, a package may allow a user to specify the shape of some symbols.

Screen Layout

Three basic components of the screen layout are the user work area, the menu area, and the area for displaying prompts and feedback messages. Fixed sections of the screen can be designated for each of these three areas, as shown in Fig. 16–9 (a). In this layout, menus are always presented at the right, and messages are displayed at the bottom of the screen. To make the work area as large as possible, the menu and message areas should be minimized. If a very large work area is desirable, the menu and message areas could be deleted when not needed so that the work area can expand to fill the screen. One way to allow maximum use of the screen is to give the user some control over the size of the menu and message areas. An experienced user, for example, could reduce the size of the message area by eliminating unnecessary prompting, as in Fig. 16–9 (b).

FIGURE 16–10
Screen layout with overlapping windows.

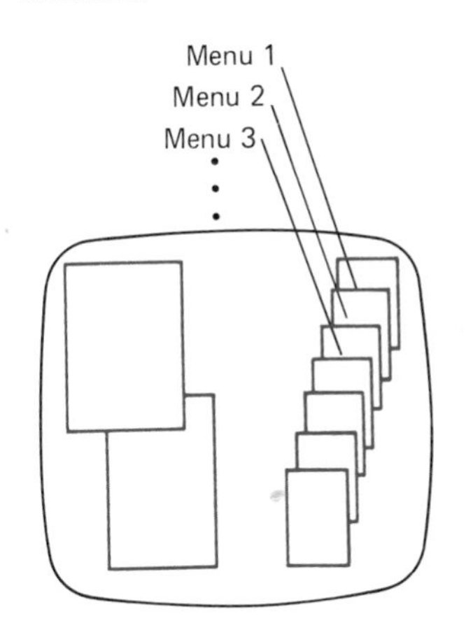

Greater flexibility in organizing screen layouts can be provided to the user by allowing any number of overlapping window areas to be set up. In this scheme, the user specifies the area of the screen in which to display a menu or picture. As each new window is positioned on the screen, it may overlap and obscure previously created windows. Figure 16–10 shows a layout with several overlapping windows. A set of operations must be provided to the user for creating, deleting, and positioning the window areas.

With some applications, it is convenient to have a zoom capability available. This allows a user to expand selected parts of a picture or to show a wider view of a scene, as if the user were moving away from the screen. The zoom feature can be used to blow up a small region of a larger picture so that it fills the work area. Enlargement can be carried out to display additional detail that cannot be seen in the total picture. Sometimes the user would like to view both the entire picture and the expanded section at the same time. One way to do this is to superimpose the total picture as a small window inset on the screen, as shown in Fig. 16–11. Another method is to include two video monitors in the graphics system. A DVST can be used to display the total picture, and a refresh display can be used to show enlargements.

In addition to a zoom capability, it is convenient in some modeling applications to be able to simplify a display by removing some of the detail to provide a better presentation of the essential structure. For example, a layout of a city could be shown with all streets or with only the major streets. By displaying only larger streets, the major thoroughfares in the city can be emphasized (Fig. 16–12).

Overall, the screen layouts chosen should avoid a cluttered appearance. Menus and other areas should be kept simple and easy to understand, and the user should be presented with familiar patterns in a consistent way. As an aid to making the screen layout easy to understand, different colors and line styles can help to distinguish the different menus, prompts, feedback messages, and other display items.

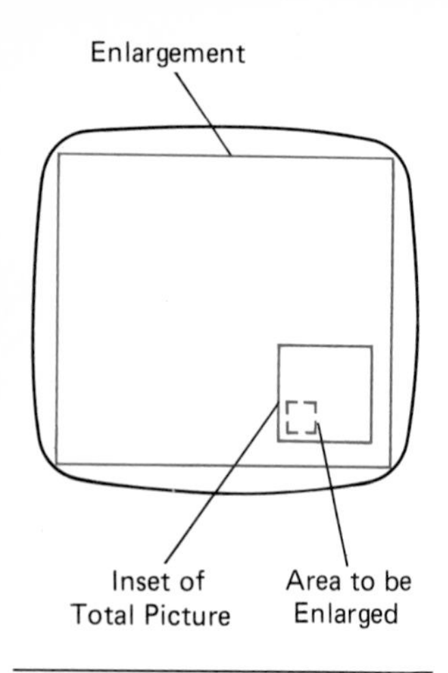

FIGURE 16–11
Enlarged picture section containing an inset showing the total picture.

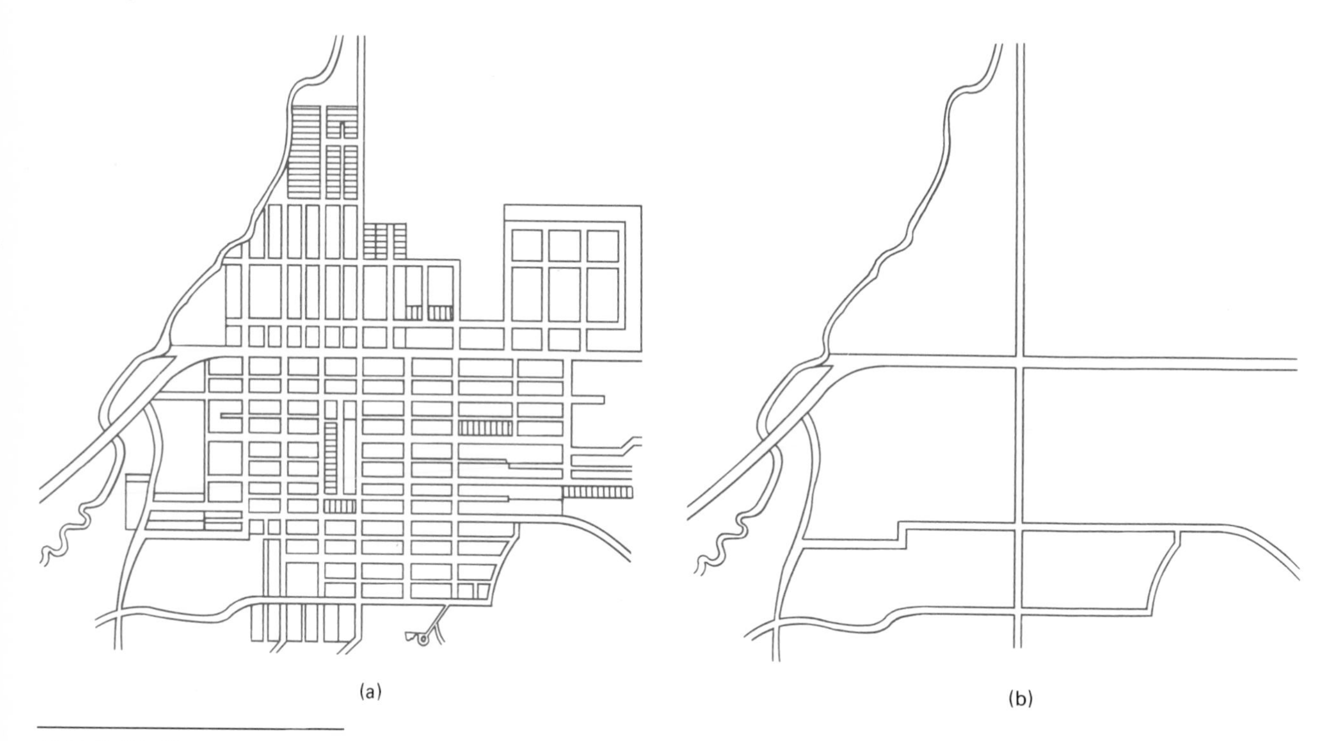

FIGURE 16–12
Display of the street layout for a city: (a) all streets included; (b) only major thoroughfares displayed for emphasis.

REFERENCES

Techniques for user-interface design are presented in Carroll and Carrithers (1984), Foley, Wallace, and Chan (1984), Good, Whiteside, Wixon, and Jones (1984), Goodman and Spence (1978), Lodding (1983), Phillips (1977), and in Swezey and Davis (1983).

EXERCISES

16–1. Select some graphics application with which you are familiar and set up a user model that will serve as the basis for the design of a user interface for graphics applications in that area.

16–2. Summarize the possible formats for the command language and discuss the relative advantages and disadvantages of each for three different applications areas.

16–3. List the possible help facilities that can be provided in a user interface and explain the type of facilities that would be appropriate for different levels of users.

16–4. Summarize the possible ways of handling backup and errors. State which approaches are more suitable for the beginner and which are better suited to the experienced user.

16–5. List the possible formats for presenting menus to a user and explain under what circumstances each might be appropriate.

16–6. Discuss alternatives for feedback in terms of the various levels of users.

16–7. For several applications areas, devise a screen layout that would be suitable for each application area.

16–8. List the various considerations in the choice of icon and symbol shapes.

16–9. Design a user interface for a paintbrush program.

16–10. Design a user interface for a two-level hierarchical modeling package.

16–11. For any area with which you are familiar, design a complete user interface to the graphics package providing capabilities to any users in that area.

BIBLIOGRAPHY

AMANTIDES, J. (July 1984). "Ray Tracing with Cones," SIGGRAPH '84 proceedings, *Computer Graphics,* 18(3), 129–135.

ANDERSON, D. P. (July 1983). "Techniques for Reducing Pen Plotting Time," *ACM Transactions on Graphics,* 2(3), 197–212.

ANGEL, I. O. (1981). *A Practical Introduction to Computer Graphics.* New York: Halstead Press.

APPEL, A. (1968). "Some Techniques for Shading Machine Renderings of Solids," *AFIPS Conference Proceedings,* 32, 37–45.

ARTWICK, B. A. (1984). *Applied Concepts in Microcomputer Graphics.* Englewood Cliffs, N.J.: Prentice-Hall.

ATHERTON, P., K. WEILER, AND D. GREENBERG (Aug. 1978). "Polygon Shadow Generation," SIGGRAPH '78 proceedings, *Computer Graphics,* 12(3), 275–281.

ATHERTON, P. R. (July 1983). "A Scan-Line Hidden Surface Removal Procedure for Constructive Solid Geometry," SIGGRAPH '83 proceedings, *Computer Graphics,* 17(3), 73–82.

BADLER, N., J. O'ROURKE, AND B. KAUFMAN (July 1980). "Special Problems in Human Movement Simulation," SIGGRAPH '80 proceedings, *Computer Graphics,* 14(3), 189–197.

BALDWIN, L. (Sept. 1984). "Color Considerations," *BYTE,* 9(10), 227–246.

BARNHILL, R. E., AND R. F. RIESENFELD, eds. (1974). *Computer-Aided Geometric Design.* New York: Academic Press.

BARNHILL, R. E., J. H. BROWN, AND I. M. KLUCEWICZ (Aug. 1978). "A New Twist in Computer-Aided Geometric Design," *Computer Graphics and Image Processing,* 8(1), 78–91.

BARRETT, R. C., AND B. W. JORDAN (March 1974). "Scan-Conversion Algorithms for a Cell-Organized Raster Display," *Communications of the ACM,* 17(3), 157–163.

BARSKY, B. A., AND J. C. BEATTY (April 1983). "Local Control of Bias and Tension in Beta-Splines," *ACM Transactions on Graphics,* 2(2), 109–134.

BARSKY, B. A. (Jan. 1984). "A Description and Evaluation of Various 3-D Models," *IEEE Computer Graphics and Applications,* 4(1), 38–52.

BEATTY, J. C., J. S. CHIN, AND H. F. MOLL (Aug. 1979). "An Interactive Documentation System," SIGGRAPH '79 proceedings, *Computer Graphics,* 13(2), 71–82.

BÉZIER, P. (1972). *Numerical Control: Mathematics and Applications.* A. R. Forrest, trans. London: Wiley.

BLINN, J. F., AND M. E. NEWELL (Oct. 1976). "Texture and Reflection in Computer-Generated Images," *Communications of the ACM,* 19(10), 542–547.

BLINN, J. F. (Summer 1977). "A Homogeneous Formulation for Lines in 3-Space," SIGGRAPH '77 proceedings, *Computer Graphics,* 11(2), 237–241.

——— (Summer 1977). "Models of Light Reflection for Computer-Synthesized Pictures," SIGGRAPH '77 proceedings, *Computer Graphics,* 11(2), 192–198.

BLINN, J. F., AND M. E. NEWELL, (Aug. 1978). "Clipping Using Homogeneous Coordinates," SIGGRAPH '78 proceedings, *Computer Graphics* 12(3), 245–251.

BLINN J. F. (Aug. 1978). "Simulation of Wrinkled Surfaces," SIGGRAPH '78 proceedings, *Computer Graphics,* 12(3), 286–292.

——— (July 1982). "A Generalization of Algebraic Surface Drawing," *ACM Transactions on Graphics,* 1(3), 235–256.

——— (July 1982). "Light Reflection Functions for Simulation of Clouds and Dusty Surfaces," SIGGRAPH '82 proceedings, *Computer Graphics,* 16(3), 21–29.

BONO, P. R., ET AL. (July 1982). "GKS: The First Graphics Standard," *IEEE Computer Graphics and Applications,* 2(5), 9–23.

BORUFKA, H. G., H. W. KUHLMANN, AND P. J. W. TEN HAGEN (July 1982). "Dialog Cells: A Method for Defining Interactions," *IEEE Computer Graphics and Applications,* 2(5), 25–33.

BOUKNIGHT, W. J. (Sept. 1970). "A Procedure for Generation of Three-Dimensional Halftoned Computer Graphics Representations," *Communications of the ACM,* 13(9), 537–545.

BOUQUET, D. L. (Aug. 1978). "An Interactive Graphics Application to Advanced Aircraft Design," SIGGRAPH '78 proceedings, *Computer Graphics,* 12(3), 330–335.

BOYSE, J. W., AND J. E. GILCHRIST (March 1982). "GMSolid: Interactive Modeling for Design and Analysis of Solids," *IEEE Computer Graphics and Applications,* 2(2), 27–40.

BRESENHAM, J. E. (1965). "Algorithm for Computer Control of Digital Plotter," *IBM Systems Journal,* 4(1), 25–30.

——— (Feb. 1977). "A Linear Algorithm for Incremental Digital Display of Circular Arcs," *Communications of the ACM,* 20(2), 100–106.

BREWER, J. A., AND D. C. ANDERSON (Summer 1977). "Visual Interaction with Overhauser Curves and Surfaces," SIGGRAPH '77 proceedings, *Computer Graphics,* 11(2), 132–137.

BRODLIE, K. W., ed. (1980). *Mathematical Methods in Computer Graphics and Design.* London: Academic Press.

BROTMAN, L. S., AND N. I. BADLER (Oct. 1984). "Generating Soft Shadows with a Depth Buffer Algorithm," *IEEE Computer Graphics and Applications,* 4(10), 5–12.

BROWN, M. H. (July 1984). "A System for Algorithm Animation," SIGGRAPH '84 proceedings, *Computer Graphics,* 18(3), 177–186.

BURT, P. J., AND E. H. ADELSON (Oct. 1983). "A Multiresolution Spline with Application to Image Mosaics," *ACM Transactions on Graphics,* 2(4), 217–236.

CARLBOM, I., I. CHAKRAVARTY, AND D. VANDERSCHEL, (April 1985). "A Hierarchical Data Structure for Representing the Spatial Decomposition of 3-D Objects," *IEEE Computer Graphics and Applications,* 5(4), 24–31.

CARPENTER, L. (July 1984). "The A-Buffer: An Antialiased Hidden-Surface Method," SIGGRAPH '84 proceedings, *Computer Graphics,* 18(3), 103–108.

CARROLL, J. M., AND C. CARRITHERS (Aug. 1984). "Training Wheels in a User Interface," *Communications of the ACM,* 27(8), 800–806.

CARSON, G. S. (Sept. 1983). "The Specification of Computer Graphics Systems," *IEEE Computer Graphics and Applications,* 3(6), 27–41.

CARUTHERS, L. C., J. VAN DEN BOS, AND A. VAN DAM (Summer 1977). "GPGS: A Device-Independent General-Purpose Graphic System for Stand-Alone and Satellite Graphics," SIGGRAPH '77 proceedings, *Computer Graphics,* 11(2), 112–119.

CASALE, M. S., AND E. L. STANTON (Feb. 1985). "An Overview of Analytic Solid Modeling," *IEEE Computer Graphics and Applications,* 5(2), 45-56.

CATMULL, E. (May 1975). "Computer Display of Curved Surfaces," *Proceedings of the IEEE Conference on Computer Graphics, Pattern Recognition, and Data Structure.*

——— (Aug. 1978). "A Hidden-Surface Algorithm with Antialiasing," SIGGRAPH '78 proceedings, *Computer Graphics,* 12(3), 6–11.

——— (Aug. 1978). "The Problems of Computer-Assisted Animation," SIGGRAPH '78 proceedings, *Computer Graphics,* 12(3), 348–353.

——— AND A. R. SMITH (July 1980). "3-D Transformations of Images in Scanline Order," SIGGRAPH '80 proceedings, *Computer Graphics,* 14(3), 279–285.

——— (July 1984). "An Analytic Visible Surface Algorithm for Independent Pixel Processing," SIGGRAPH '84 proceedings, *Computer Graphics,* 18(3), 109–115.

CHASEN, S. H. (1978). *Geometric Principles and Procedures for Computer Graphic Applications.* Englewood Cliffs, N.J.: Prentice-Hall.

CLARK, J. H. (Oct. 1976). "Hierarchical Geometric Models for Visible Surface Algorithms," *Communications of the ACM,* 19(10), 547–554.

——— (July 1982). "The Geometry Engine: A VLSI Geometry System for Graphics," SIGGRAPH '82 proceedings, *Computer Graphics,* 10(3), 127–133.

COOK, R. L., AND K. E. TORRANCE (Jan. 1982). "A Reflectance Model for Computer Graphics," *ACM Transactions on Graphics,* 1(1), 7–24.

COOK, R. L., T. PORTER, AND L. CARPENTER (July 1984). "Distributed Ray Tracing," SIGGRAPH '84 proceedings, *Computer Graphics,* 18(3), 137–145.

COOK R. L. (July 1984). "Shade Trees," SIGGRAPH '84 proceedings, *Computer Graphics,* 18(3), 223–231.

COONS, S. A. (June 1967). *Surfaces for Computer-Aided Design of Space Forms,* Project MAC, M.I.T., TR–41.

CROCKER, G. A. (July 1984). "Invisible Coherence for Faster Scan-Line Hidden-Surface Algorithms," SIGGRAPH '84 proceedings, *Computer Graphics,* 18(3), 95–102.

CROW, F. C. (Summer 1977). "Shadow Algorithms for Computer Graphics," SIGGRAPH '77 proceedings, *Computer Graphics,* 11(2), 242–248.

——— (Nov. 1977). "The Aliasing Problem in Computer-Synthesized Shaded Images," *Communications of the ACM,* 20(11), 799–805.

——— (Aug. 1978). "The Use of Grayscale for Improved Raster Display of Vectors and Characters," SIGGRAPH '78 proceedings, *Computer Graphics,* 12(3), 1–5.

——— (Jan. 1981). "A Comparison of Antialiasing Techniques," *IEEE Computer Graphics and Applications,* 1(1), 40–49.

——— (July 1984). "Summed-Area Tables for Texture Mapping," SIGGRAPH '84 proceedings, *Computer Graphics,* 18(3), 207–212.

CYRUS, M., AND J. BECK (1978). "Generalized Two- and Three-Dimensional Clipping," *Computers and Graphics,* 3(1), 23–28.

DOCTOR, L. J., AND J. G. TORBERG (July 1981). "Display Techniques for Octree-Encoded Objects," *IEEE Computer Graphics and Applications,* 1(3), 29–38.

DUNGAN, W., A. STENGER, AND G. SUTTY (Aug. 1978). "Texture Tile Considerations for Raster Graphics," SIGGRAPH '78 proceedings, *Computer Graphics,* 12(3), 130–134.

DUNLAVEY, M. R. (Oct. 1983). "Efficient Polygon-Filling Algorithms for Raster Displays," *ACM Transactions on Graphics,* 2(4), 264–274.

EASTMAN, C., AND M. HENRION (Summer 1977). "GLIDE: A Language for Design Information Systems," SIGGRAPH '77 proceedings, *Computer Graphics,* 11(2), 24–33.

ENCARNAÇAO, J. (July 1980). "The Workstation Concept of GKS and the Resulting Conceptual Differences to the GSPC CORE System," SIGGRAPH '80 proceedings, *Computer Graphics,* 14(3), 226–230.

ENDERLE, G., K. KANSY, AND G. PFAFF, (1984). *Computer Graphics Programming: GKS—The Graphics Standard.* Berlin: Springer-Verlag.

FAROUKI, R. T., AND J. K. HINDS, (May 1985). "A Hierarchy of Geometric Forms," *IEEE Computer Graphics and Applications,* 5(5), 51–78.

FISHKIN, K. P., AND B. A. BARSKY (July 1984). "A Family of New Algorithms for Soft Filling," SIGGRAPH '84 proceedings, *Computer Graphics,* 18(3), 235–244.

FOLEY, J. D., AND A. VAN DAM (1982). *Fundamentals of Interactive Computer Graphics,* Reading, Mass.: Addison-Wesley.

FOLEY, J. D., V. L. WALLACE, AND P. CHAN (Nov. 1984). "The Human Factors of Computer Graphics Interaction Techniques," *IEEE Computer Graphics and Applications,* 4(11), 13–48.

FORREST, A. R. (Dec. 1972). "On Coons and Other Methods for the Representation of Curved Surfaces," *Computer Graphics and Image Processing,* 1(4), 341–354.

FOURNIER, A., D. FUSSEL, AND L. CARPENTER (June 1982). "Computer Rendering of Stochastic Models," *Communications of the ACM,* 25(6), 371–384.

FRIEDER, G., D. GORDON, AND R. A. REYNOLD (Jan. 1985). "Back-to-Front Display of Voxel-Based Objects," *IEEE Computer Graphics and Applications,* 5(1), 52–60.

FU, K. S., AND A. ROSENFELD (Oct. 1984). "Pattern Recognition and Computer Vision," 17(10), 274–282.

FUCHS, H., G. D. ABRAM, AND E. D. GRANT (July 1983). "Near Real-Time Shaded Display of Rigid Objects," SIGGRAPH '83 proceedings, *Computer Graphics,* 17(3), 65–72.

FUJIMOTO, A., AND K. IWATA (Dec. 1983). "Jag-Free Images on Raster Displays," *IEEE Computer Graphics and Applications,* 3(9), 26–34.

FUJIMOTO, A., C. G. PERROT, AND K. IWATA (June 1984). "A 3-D Graphics Display System with Depth Buffer and Pipeline Processor," *IEEE Computer Graphics and Applications,* 4(6), 11–23.

GARDNER, T. N., AND H. R. NELSON (March-April 1983). "Interactive Graphics Developments in Energy Exploration," *IEEE Computer Graphics and Applications,* 3(2), 33–44.

GILOI, W. K. (1978). *Interactive Computer Graphics: Data Structures, Algorithms, Languages.* Englewood Cliffs, N.J.: Prentice-Hall.

GLASSNER, A. S. (Oct. 1984). "Space Subdivision for Fast Ray Tracing," *IEEE Computer Graphics and Applications,* 4(10), 15–22.

GLINERT, E. P., AND S. L. TANIMOTO (Nov. 1984). "Pict: An Interactive Graphical Programming Environment," *IEEE Computer Graphics and Applications,* 17(11), 7–25.

GOOD, M. D., ET AL. (Oct. 1984). "Building a User-Derived Interface," *Communications of the ACM,* 27(10), 1032–1043.

GOODMAN, T., AND R. SPENCE (Aug. 1978). "The Effect of System Response Time on Interactive Computer-Aided Problem Solving," SIGGRAPH '78 proceedings, *Computer Graphics,* 12(3), 100–104.

GRAPHICS STANDARDS PLANNING COMMITTEE (1977). "Status Report of the Graphics Standards Planning Committee," *Computer Graphics,* 11.

——— (Aug. 1979). "Status Report of the Graphics Standards Committee," *Computer Graphics,* 13(3).

GROTCH, S. L. (Nov. 1983). "Three-Dimensional and Stereoscopic Graphics for Scientific Data Display and Analysis," *IEEE Computer Graphics and Applications,* 3(8), 31–43.

GUIBAS, L. J., AND J. STOLFI (July 1982). "A Language for Bitmap Manipulation," *ACM Transactions on Graphics,* 1(3), 191–214.

HABER, R. N., AND L. WILKINSON (May 1982). "Perceptual Components of Computer Displays," *IEEE Computer Graphics and Applications,* 2(3), 23–35.

HACKATHORN, R. J. (Summer 1977). "Anima II: A 3-D Color Animation System," SIGGRAPH '77 proceedings, *Computer Graphics,* 11(2), 54–64.

HALL, R. A., AND D. P. GREENBERG (Nov. 1983). "A Testbed for Realistic Image Synthesis," *IEEE Computer Graphics and Applications,* 3(8), 10–20.

Hamlin, G., and C. W. Gear (Summer 1977). "Raster-Scan Hidden-Surface Algorithm Techniques," SIGGRAPH '77 proceedings, *Computer Graphics,* 11(2), 206–213.

Hanna, S. L., J. F. Abel, and D. P. Greenberg (Oct. 1983). "Intersection of Parametric Surfaces by Means of Lookup Tables," *IEEE Computer Graphics and Applications,* 3(7), 39–48.

Hanrahan, P. (July 1983). "Ray Tracing Algebraic Surfaces," SIGGRAPH '83 proceedings, *Computer Graphics,* 17(3), 83–90.

Harrington, S. (1983). *Computer Graphics: A Programming Approach.* New York: McGraw-Hill.

Haruyama, S., and B. A. Barsky (March 1984). "Using Stochastic Modeling for Texture Generation," *IEEE Computer Graphics and Applications,* 4(3), 7–19.

Hatfield, L., and B. Herzog (Jan. 1982). "Graphics Software: From Techniques to Principles," *IEEE Computer Graphics and Applications,* 2(1), 59–79.

Hawrylyshyn, P. A., R. R. Tasker, and L. W. Organ (Summer 1977). "CASS: Computer-Assisted Stereotaxic Surgery," SIGGRAPH '77 proceedings, *Computer Graphics,* 11(2), 13–17.

Heckbert, P. (July 1982). "Color Image Quantization for Frame Buffer Display," SIGGRAPH '82 proceedings, *Computer Graphics,* 16(3), 297–307.

——— and P. Hanrahan (July 1984). "Beam Tracing Polygonal Objects," SIGGRAPH '84 proceedings, *Computer Graphics,* 18(3), 119–127.

Herbison-Evans, D. (Nov. 1982). "Real-Time Animation of Human Figure Drawings with Hidden Lines Omitted," *IEEE Computer Graphics and Applications,* 2(9), 27–33.

Hopgood, F. R. A., et al. (1983). *Introduction to the Graphical Kernel System (GKS).* London: Academic Press.

Hubschman, H., and S. W. Zucker (April 1982). "Frame-to-Frame Coherence and the Hidden Surface Computation: Constraints for a Convex World," *ACM Transactions on Graphics,* 1(2), 129–162.

Huitric, H., and M. Nahas (March 1985). "B-Spline Surfaces: A Tool for Computer Painting," *IEEE Computer Graphics and Applications,* 5(3), 39–47.

Ikedo, T. (May 1984). "High-Speed Techniques for a 3-D Color Graphics Terminal," *IEEE Computer Graphics and Applications,* 4(5), 46–58.

Jarvis, J. F., C. N. Judice, and W. H. Ninke (March 1976). "A Survey of Techniques for the Image Display of Continuous Tone Pictures on Bilevel Displays," *Computer Graphics and Image Processing,* 5(1), 13–40.

Jarvis, J. F., and C. S. Roberts (Aug. 1976). "A New Technique for Displaying Continuous Tone Images on a Bilevel Display," *IEEE Transactions,* COM-24(8), 891–898.

Joblove, G. H., and D. Greenberg (Aug. 1978). "Color Spaces for Computer Graphics," SIGGRAPH '78 proceedings, *Computer Graphics,* 12(3), 20–25.

Jordan, B. W., and R. C. Barrett (Nov. 1973). "A Scan Conversion Algorithm with Reduced Storage Requirements," *Communications of the ACM,* 16(11), 676–679.

Judice, J. N., J. F. Jarvis, and W. Ninke (Oct.–Dec. 1974). "Using Ordered Dither to Display Continuous Tone Pictures on an AC Plasma Panel," *Proc. SID,* 161–169.

Kajiya, J. T. (July 1983). "New Techniques for Ray Tracing Procedurally Defined Objects," *ACM Transactions on Graphics,* 2(3), 161–181.

Kay, D. S., and D. Greenberg (Aug. 1979). "Transparency for Computer-Synthesized Images," SIGGRAPH '79 proceedings, *Computer Graphics,* 13(2), 158–164.

Kochanek, D. H. U., and R. H. Bartels (July 1984). "Interpolating Splines with Local Tension, Continuity, and Bias Control," SIGGRAPH '84 proceedings, *Computer Graphics,* 18(3), 33–41.

Korein, J., and N. I. Badler (July 1983). "Temporal Antialiasing in Computer-Generated Animation," SIGGRAPH '83 proceedings, *Computer Graphics,* 17(3), 377–388.

Lantz, K., and W. I. Nowicki (Jan. 1984). "Structured Graphics for Distributed Systems," *ACM Transactions on Graphics,* 3(1), 23–51.

Levoy, M. (Summer 1977). "A Color Animation System Based on the Multiplane Technique," SIGGRAPH '77 proceedings, *Computer Graphics,* 11(2), 65–71.

Levy, H. M. (Jan. 1984). "VAXstation: A General-Purpose Raster Graphics Architecture," *ACM Transactions on Graphics,* 3(1), 70–83.

Liang, Y. -D., and B. A. Barsky (Nov. 1983). "An Analysis and Algorithm for Polygon Clipping," *Communications of the ACM,* 26(11), 868–877.

——— (Jan. 1984). "A New Concept and Method for Line Clipping," *ACM Transactions on Graphics,* 3(1), 1–22.

Lieberman, H. (Aug. 1978). "How to Color in a Coloring Book," SIGGRAPH '78 proceedings, *Computer Graphics,* 12(3), 111–116.

Lipkie, D. E., et al. (July 1982). "Star Graphics: An Object-Oriented Implementation," SIGGRAPH '82 proceedings, *Computer Graphics,* 16(3), 115–124.

Lippman, A. (July 1980). "Movie-Maps: An Application of the Optical Videodisc to Computer Graphics," SIGGRAPH '80 proceedings, *Computer Graphics.* 14(3), 32–42.

LODDING, K. N. (March-April 1983). "Iconic Interfacing," *IEEE Computer Graphics and Applications,* 3(2), 11–20.

LOOMIS, J., ET AL. (July 1983). "Computer Graphic Modeling of American Sign Language," SIGGRAPH '83 proceedings, *Computer Graphics,* 17(3), 105–114.

MAGNENAT-THALMANN, N., AND D. THALMANN (Dec. 1983). "The Use of High-Level 3-D Graphical Types in the Mira Animation System," *IEEE Computer Graphics and Applications,* 3(9), 9–16.

MANDELBROT, B. B. (1977). *Fractals: Form, Chance, and Dimension.* San Francisco: Freeman Press.

——— (1982). *The Fractal Geometry of Nature.* New York: Freeman Press.

MARCUS, A. (July 1983). "Graphic Design for Computer Graphics," *IEEE Computer Graphics and Applications,* 3(4), 63–70.

MARGOLIN, B. (Feb. 1985) "New Flat Panel Displays," *Computers and Electronics,* 23(2), 66–108.

MATHERAT, P. (Aug. 1978). "A Chip for Low-Cost Raster-Scan Graphic Display," SIGGRAPH '78 proceedings, *Computer Graphics,* 12(3), 181–186.

MCILROY, M. D. (Oct. 1983). "Best Approximate Circles on Integer Grids," *ACM Transactions on Graphics,* 2(4), 237–263.

MCKEOWN, K., AND N. I. BADLER (July 1980). "Creating Polyhedral Stellations," SIGGRAPH '80 proceedings, *Computer Graphics,* 14(3), 19–24.

MEHL, M. E., AND S. J. NOLL (Aug. 1984). "A VLSI Support for GKS," *IEEE Computer Graphics and Applications,* 4(8), 52–55.

MICHENER, J. C., I. B. CARLBOM (July 1980). "Natural and Efficient Viewing Parameters," SIGGRAPH '80 proceedings, *Computer Graphics,* 14(3), 238–245.

MINSKY, A. R. (July 1984). "Manipulating Simulated Objects with Real-World Gestures Using a Force and Position Sensitive Screen," SIGGRAPH '84 proceedings, *Computer Graphics,* 18(3), 195–203.

MITROO, J. B., N. HERMAN, AND N. I. BADLER (Aug. 1979). "Movies from Music: Visualizing Musical Compositions," SIGGRAPH '79 proceedings, *Computer Graphics,* 13(2), 218–225.

MURCH, G. M. (Nov. 1984). "Physiological Principles for the Effective Use of Color," *IEEE Computer Graphics and Applications,* 4(11), 49–54.

MYERS, W. (July 1984). "Staking Out the Graphics Display Pipeline," *IEEE Computer Graphics and Applications,* 4(7), 60–65.

NEWMAN, W. M. (1968). *A System for Interactive Graphical Input.* SJCC, Washington, D.C.: Thompson Books, 47–54.

——— (Oct. 1971). "Display Procedures," *Communications of the ACM,* 14(10), 651–660.

——— AND R. F. SPROULL (1974). "An Approach to Graphics System Design," *Proceedings of the IEEE,* IEEE.

——— (1979). *Principles of Interactive Computer Graphics.* New York: McGraw-Hill.

NG, N., AND T. MARSLAND (1978). "Introducing Graphics Capabilities to Several High-Level Languages," *Software Practice and Experience,* 8, 629–639.

NIIMI, H., ET AL. (July 1984). A Parallel Processor System for Three-Dimensional Color Graphics," SIGGRAPH '84 proceedings, *Computer Graphics,* 18(3), 67–76.

NORTON, A. (July 1982). "Generation and Display of Geometric Fractals in 3-D," SIGGRAPH '82 proceedings, *Computer Graphics,* 16(3), 61–67.

PAO, Y. C. (1984). *Elements of Computer-Aided Design.* New York: Wiley.

PARKE, F. I. (July 1980). "Adaptation of Scan and Slit-Scan Techniques to Computer Animation," SIGGRAPH '80 proceedings, *Computer Graphics,* 14(3), 178–181.

——— (Nov. 1982). "Parameterized Models for Facial Animation," *IEEE Computer Graphics and Applications,* 2(9), 61–68.

PAVLIDIS, T. (Aug. 1978). "Filling Algorithms for Raster Graphics," SIGGRAPH '78 proceedings, *Computer Graphics,* 12(3), 161–166.

——— (Aug. 1981). "Contour Filling in Raster Graphics," SIGGRAPH '81 proceedings, *Computer Graphics,* 15(3), 29–36.

——— (1982). *Algorithms for Graphics and Image Processing.* Rockville, Md.: Computer Science Press.

——— (Jan. 1983). "Curve Fitting with Conic Splines," *ACM Transactions on Graphics,* 2(1), 1–31.

PHILLIPS, R. L. (Summer 1977). "A Query Language for a Network Data Base with Graphical Entities," SIGGRAPH '77 proceedings, *Computer Graphics,* 11(2), 179–185.

PHONG B. T. (June 1975). "Illumination for Computer-Generated Images," *Communications of the ACM,* 18(6), 311–317.

PIKE, R. (April 1983). "Graphics in Overlapping Bitmap Layers," *ACM Transactions on Graphics,* 2(2), 135–160.

PITTEWAY, M. L. V., AND D. J. WATKINSON, (Nov. 1980). "Bresenham's Algorithm with Gray Scale," *Communications of the ACM,* 23(11), 625–626.

PITTMAN, J. H., AND D. P. GREENBERG (July 1982). "An Interactive Graphics Environment for Architectural Energy Simulation," SIGGRAPH '82 proceedings, *Computer Graphics,* 16(3), 233–241.

Potmesil, M., and I. Chakravarty (April 1982). "Synthetic Image Generation with a Lens and Aperture Camera Model," *ACM Transactions on Graphics,* 1(2), 85–108.

——— (July 1983). "Modeling Motion Blurs in Computer-Generated Images," SIGGRAPH '83 proceedings, *Computer Graphics,* 17(3), 389–399.

Preiss, R. B. (Nov. 1978). "Storage CRT Display Terminals: Evolution and Trends," *Computer,* 11(11), 20–28.

Preston, K., et al. (Oct. 1984). "Computing in Medicine," *Computer,* 17(10), 294–313.

Prosser, C. J., and A. C. Kilgour (July 1983). "An Integral Method for the Graphical Output of Conic Sections," *ACM Transactions on Graphics,* 2(3), 182–191.

Reeves, W. T. (April 1983). "Particle Systems: A Technique for Modeling a Class of Fuzzy Objects," *ACM Transactions on Graphics,* 2(2), 91–108.

Reisenfeld, R. F. (1973). *Applications of B-Spline Approximation to Geometric Problems of Computer-Aided Design,* Ph.D. dissertation, Syracuse University.

Renfrow, N. V. (Summer 1977). "Computer Graphics for Facilities Management," SIGGRAPH '77 proceedings, *Computer Graphics,* 11(2), 42–47.

Requicha, A. A. G. (Dec. 1980). "Representations of Rigid Solids: Theory, Methods, and Systems," *ACM Computing Surveys,* 12(4), 437–464.

——— and H. B. Voelcker (March 1982). "Solid Modeling: A Historical Summary and Contemporary Assessment," *IEEE Computer Graphics and Applications,* 2(2), 9–24.

——— (Oct. 1983). "Solid Modeling: Current Status and Research Directions," *IEEE Computer Graphics and Applications,* 3(7), 25–37.

Reynolds, C. W. (July 1982). "Computer Animation with Scripts and Actors," SIGGRAPH '82 proceedings, *Computer Graphics,* 16(3), 289–296.

Rhodes, M. L., et al. (Aug. 1983). "Computer Graphics and an Interactive Stereotactic System for CT-Aided Neurosurgery," *IEEE Computer Graphics and Applications,* 3(5), 31–37.

Roese, J. A., and L. E. McCleary (Aug. 1979). "Stereoscopic Computer Graphics for Simulation and Modeling," SIGGRAPH '79 proceedings, *Computer Graphics,* 13(2), 41–47.

Rogers, D. F., and J. A. Adams (1976). *Mathematical Elements for Computer Graphics.* New York: McGraw-Hill.

Rogers, D. F. (1985). *Procedural Elements for Computer Graphics.* New York: McGraw-Hill.

Rosenthal, D. S. H., et al. (July 1982). "The Detailed Semantics of Graphics Input Devices," SIGGRAPH '82 proceedings, *Computer Graphics,* 16(3), 33–38.

Rubin, S. M., and T. Whitted (July 1980). "A 3-Dimensional Representation for Fast Rendering of Complex Scenes," SIGGRAPH '80 proceedings, *Computer Graphics,* 14(3), 110–116.

Sakurai, H., and D. C. Gossard (July 1983). "Solid Model Input Through Orthographic Views," SIGGRAPH '83 proceedings, *Computer Graphics,* 17(3), 243–252.

Schacter, B. J., ed. (1983). *Computer Image Generation.* New York: Wiley.

Schweitzer, D. (July 1983). "Artificial Texturing: An Aid to Surface Visualization," SIGGRAPH '83 proceedings, *Computer Graphics,* 17(3), 23–29.

Scott, J. E. (1982). *Introduction to Interactive Computer Graphics.* New York: Wiley.

Sechrest, S., and D. P. Greenberg (Jan. 1982). "A Visible Polygon Reconstruction Algorithm," *ACM Transactions on Graphics,* 1(1), 25–42.

Sherr, S. (1979). *Electronics Displays.* New York: Wiley.

Singh, B., et al. (July 1983). "A Graphics Editor for Benesh Movement Notation," SIGGRAPH '83 proceedings, *Computer Graphics,* 17(3), 51–62.

Slottow, H. G. (July 1976). "Plasma Displays," *IEEE Transactions on Electron Devices,* ED–23(7).

Smith, A. R. (Aug. 1978). "Color Gamut Transform Pairs," SIGGRAPH '78 proceedings, *Computer Graphics,* 12(3), 12–19.

——— (Aug. 1979). "Tint Fill," SIGGRAPH '79 proceedings, *Computer Graphics,* 13(2), 276–283.

——— (July 1984). "Plants, Fractals, and Formal Languages," SIGGRAPH '84 proceedings, *Computer Graphics,* 18(3), 1–10.

Sorensen, P. R. (Sept. 1984). "Fractals," *BYTE,* 157–172.

Sproull, R. F., and I. E. Sutherland (1968). *A Clipping Divider.* AFIPS Fall Joint Computer Conference.

Sproull, R. F., et al. (Jan 1983). "The 8 by 8 Display," *ACM Transactions on Graphics,* 2(1), 32–56.

Sutherland, I. E. (1963). "Sketchpad: A Man-Machine Graphical Communication System," *AFIPS Spring Joint Computer Conference,* 23, pp. 329–346.

———, R. F. Sproull, and R. A. Schumacker (1973). "Sorting and the Hidden-Surface Problem," *National Computer Conference Proceedings.*

SUTHERLAND, I. E., AND G. W. HODGMAN (Jan. 1974). "Reentrant Polygon Clipping," *Communications of the ACM,* 17(1), 32–42.

SUTHERLAND, I. E., R. SPROULL, AND R. SCHUMACKER (March 1974). "A Characterization of Ten Hidden Surface Algorithms," *ACM Computing Surveys,* 6(1), 1–55.

SWEZEY, R. W., AND E. G. DAVIS (Nov. 1983). "A Case Study of Human Factors Guidelines in Computer Graphics," *IEEE Computer Graphics and Applications,* 3(8), 21–30.

TANNAS, L. (July 1978). "Flat Panel Displays: A Critique," *IEEE Spectrum,* 15(7).

TILLER, W. (Sept. 1983). "Rational B-Splines for Curve and Surface Representation," *IEEE Computer Graphics and Applications,* 3(6), 61–69.

TILOVE, R. (July 1984). "A Null-Object Algorithm for Constructive Solid Geometry," *Communications of the ACM,* 27(7), 684–694.

TORRANCE, K. E., AND E. M. SPARROW (Sept. 1967). "Theory for Off-Specular Reflection from Roughened Surfaces," *J. Opt. Soc. Am.,* 57(9), 1105–1114.

TURKOWSKI, K. (July 1982). "Antialiasing Through the Use of Coordinate Transformations," *ACM Transactions on Graphics,* 1(3), 215–234.

VICKERS, D. (1970). "Head-Mounted Display Terminal," *Proceedings of the 1970 IEEE International Computer Group Conference,* IEEE.

VOELCKER, H., ET AL. (Aug. 1978). "The PADL-1.0/2 System for Defining and Displaying Solid Objects," SIGGRAPH '78 proceedings, *Computer Graphics,* 12(3), 257–263.

WALLACE, B. (Aug. 1981). "Merging and Transformation of Raster Images for Cartoon Animation," SIGGRAPH '81 proceedings, *Computer Graphics,* 15(3), 253–262.

WALLACE, V. L. (Spring 1976). "The Semantics of Graphics Input Devices," *Computer Graphics,* 10(1).

WARN, D. (July 1983). "Lighting Controls for Synthetic Images," SIGGRAPH '83 proceedings, *Computer Graphics,* 17(3), 13–21.

WARNER, J. R. (Oct. 1981). "Principles of Device-Independent Computer Graphics Software," *IEEE Computer Graphics and Applications,* 1(4), 85–100.

WEGHORST, H., G. HOOPER, AND D. P. GREENBERG (Jan. 1984). "Improved Computational Methods for Ray Tracing," *ACM Transactions on Graphics,* 3(1), 52–69.

WEILER, K., AND P. ATHERTON (Summer 1977). "Hidden Surface Removal Using Polygon Area Sorting," SIGGRAPH '77 proceedings, *Computer Graphics,* 11(2), 214–222.

WEILER, K. (July 1980). "Polygon Comparison Using a Graph Representation," SIGGRAPH '80 proceedings, *Computer Graphics,* 14(3), 10–18.

——— (Jan. 1985). "Edge-Based Data Structures for Solid Modeling in Curved-Surface Modeling Environments," *IEEE Computer Graphics and Applications,* 5(1), 21–40.

WEINBERG, R. (Aug. 1978). "Computer Graphics in Support of Space Shuttle Simulation," SIGGRAPH '78 proceedings, *Computer Graphics,* 12(3), 82–86.

WHITTED, T. (June 1980). "An Improved Illumination Model for Shaded Display," *Communications of the ACM,* 23(6), 343–349.

WOLFRAM, S. (Sept. 1984). "Computer Software in Science and Mathematics," *Scientific American,* 251(3), 188–203.

WOODSFORD, P. A. (July 1976). "The HRD-1 Laser Display System," SIGGRAPH '76 proceedings, *Computer Graphics,* 10(2), 68–73.

YAMAGUCHI, K., T. L. KUNII, AND K. FUJIMURA (Jan. 1984). "Octree-Related Data Structures and Algorithms," *IEEE Computer Graphics and Applications,* 4(1), 53–59.

YESSIOS, C. I. (Aug. 1979). "Computer Drafting of Stones, Wood, Plant, and Ground Materials," SIGGRAPH '79 proceedings, *Computer Graphics,* 13(2), 190–198.

SUBJECT INDEX

A

B

C

I

J

K

L

M

N

O

P

Q

R

S

T

U

V

W

Z

FUNCTION INDEX